The cover painting suggests the molecular unity within the organismal diversity of life. A ribbon of DNA, symbol of the "universal" genetic code that directs formation of all living things on Earth, spirals among five representative organisms, one from each of the "five kingdoms"—bacteria, protists, fungi, plants, and animals. Also shown are models of some of life's basic ingredients: a water molecule, an α helix—fundamental component of proteins, the primary structural and catalytic constituents of life—and a molecule of ATP, the energy currency that drives biochemical events. The ubiquitous viral form (an information molecule in a protein coat), which exists on the border between living and non-living systems, is represented by a model of the adenovirus.

DNA

Water molecule

α helix

ATP

Adenovirus

ANIMALIA
Cnemidophorus inornatus heptagrammus
seven-striped whiptail

PLANTAE
Passiflora × alatocaerulea
passionflower

FUNGI
Schizophyllum commune

PROTISTA
Emiliania huxleyi

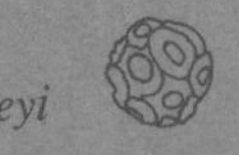

MONERA
Escherichia coli

PRINCIPLES OF BIOCHEMISTRY

Grey is all theory. Green grows
the golden tree of life.

Goethe

PRINCIPLES OF BIOCHEMISTRY

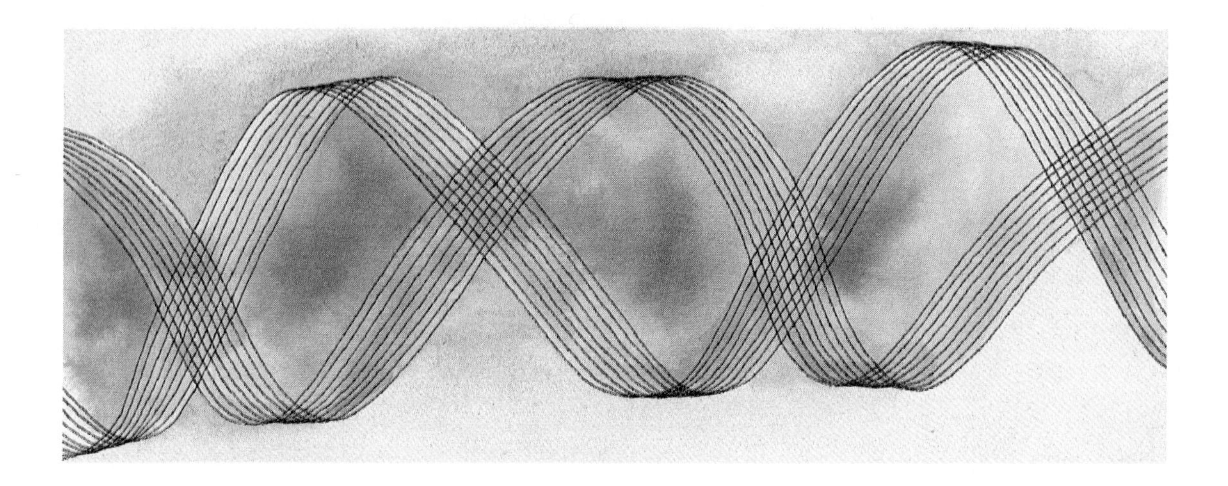

H. Robert Horton
North Carolina State University

Laurence A. Moran
University of Toronto

Raymond S. Ochs
Case Western Reserve University

J. David Rawn
Towson State University

K. Gray Scrimgeour
University of Toronto

NEIL PATTERSON PUBLISHERS
PRENTICE HALL
Englewood Cliffs NJ 07632

Principles of Biochemistry

ISBN 0–13–042409–9
ISBN 0–13–065665–8 (international edition)

Printed in the United States of America: December, 1992

Library of Congress Cataloging-in-Publication Data
Principles of biochemistry / H. Robert Horton . . . [et al.].
p. cm.
Includes bibliographical references and index.
ISBN 0–13–042409–9.—ISBN 0–13–065665–8 (pbk)
1. Biochemistry. I. Horton, H. Robert, 1935– .
QP514.2.P753 1993 574.19′2—dc20 92-39438
CIP

Color separations and printing and binding are by Arcata Graphics Kingsport. The text is set in Times Roman; the figure legends are set in Frutiger.

NEIL PATTERSON PUBLISHERS
PRENTICE HALL
Englewood Cliffs, NJ 07632

Prentice-Hall International (UK) Limited, *London*
Prentice-Hall of Australia Pty, Limited, *Sydney*
Prentice-Hall Canada Inc., *Toronto*
Prentice-Hall Hispanoamerica, S.A., *Mexico*
Prentice-Hall of India Private Limited, *New Delhi*
Prentice-Hall of Japan, Inc., *Tokyo*
Simon & Schuster Asia Pte. Ltd., *Singapore*
Editora Prentice-Hall do Brasil, Ltda., *Rio de Janeiro*
Prentice-Hall, Inc., *Englewood Cliffs, New Jersey*

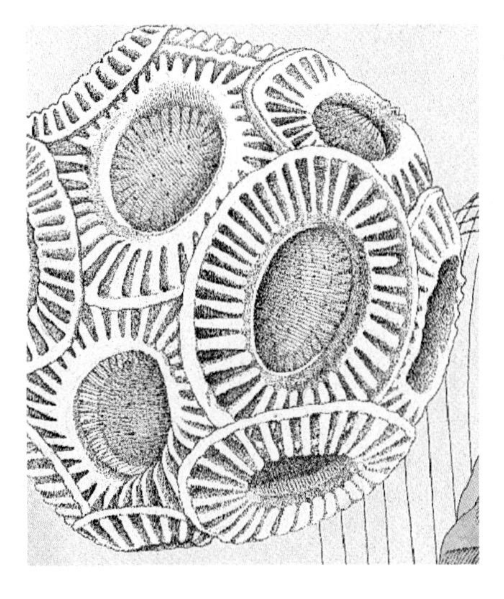

The Authors

H. Robert Horton

Dr. Horton, who received his Ph.D. from the University of Missouri in 1962, is William Neal Reynolds Professor and Undergraduate Coordinator in the Department of Biochemistry at North Carolina State University, where he has served on the faculty for nearly 30 years. He was appointed Alumni Distinguished Professor in 1979. Professor Horton's research includes protein conformation and enzyme mechanisms.

Laurence A. Moran

After earning his Ph.D. from Princeton University in 1974, Professor Moran spent four years at the Université de Genève in Switzerland. He has been a member of the faculty at the University of Toronto since 1978, specializing in molecular biology. His research findings on heat-shock genes have been published in many scholarly journals.

Raymond S. Ochs

Professor Ochs, who earned his Ph.D. from Indiana University, is Assistant Professor in the Department of Nutrition at Case Western Reserve University. He is an expert on metabolic regulation and has edited a monograph on this topic—Ochs, R. S, Hanson, R. W., and Hall, J., eds. *Metabolic Regulation* (1985, Elsevier) and authored and co-authored numerous research papers and reports.

J. David Rawn

Professor Rawn, who earned his Ph.D. from Ohio State University in 1971, has taught and done research in the Department of Chemistry at Towson State University for the past 20 years. He did not write chapters for *Principles of Biochemistry,* but his textbook *Biochemistry* (1989, Neil Patterson) served as a source of information and ideas concerning content and organization and the use of stereo images to illuminate molecular structures and conformational change.

K. Gray Scrimgeour

Professor Scrimgeour received his doctorate from the University of Washington in 1961 and has been a faculty member at the University of Toronto since 1967. He is the author of *The Chemistry and Control of Enzymatic Reactions* (1977, Academic Press), and his work on enzymatic systems has been published in more than 50 professional journal articles during the past 30 years. From 1984–1992, he was editor of the journal *Biochemistry and Cell Biology.*

The chapter on photosynthesis (16) was written by Dr. Willy Kalt. She wrote the first version while a postdoctoral fellow at Duke University and the present version at Agriculture Canada in Kentville, Nova Scotia, where she is now a research scientist.

Credits

Publisher:	Neil Patterson
Editorial Director:	Sherri Foster
Senior Editor:	Morgan Ryan
Editor:	Katherine Hodgin
Editorial Assistant:	Donna Curasi
Production Manager:	Donna Young
Production Assistant:	Kate Dunlap
Principal Artists:	Lisa Shoemaker (illustrations rendered by hand) George Sauer (illustrations rendered electronically)
Cover Artist:	Ippy Patterson
Designer:	Neil Patterson*

Contributing Editors: Matthew Barnett-Lawrence, John Challice, Lori Harmon, Terri O'Quin, Charlotte Pratt. *Contributing Artists:* Susan Averill, Mike Dulude, Tom Edgerton, Brian Eller, Sarah McQueen, Linda Murray, Andrea Weaver, Mike Webb. *Business Manager:* Christina Faircloth. *Research and Administrative Assistance:* Tom Flynn, Eve Jones, Cris Sexton, Kay White.

Reviewers and Advisors: Please see pages x and xi.

Special thanks to Ben Banks and Kim Gernert for the computer-generated stereo images developed specifically for this text at

Molecular Graphics and Modelling Shared Resource
Duke Comprehensive Cancer Center
Duke University

*gratefully acknowledging the powerful influence of Malcolm Grear Designers, a studio that revolutionized college textbook design in the 1960s, using form in ways that enhance function and setting an enduring standard for the rest of us.

Seek simplicity and distrust it.

Alfred North Whitehead

Preface

The ballooning of biochemistry texts in the past couple of decades inspired us to follow Whitehead's imperative. This is a short book—less than 700 text pages—in territory dominated by giant tomes.

Probably most teachers, certainly most students, would prefer a shorter text, but the ideal edition would have to be tailored to fit the array of topics and emphases each teacher prefers. So, perversely, the wish for brevity engenders increasingly fatter volumes as publishers encourage authors to write encyclopedic texts in hopes that teachers will carve from this bulk a diversity of shapes, each one suited to a particular course.

The "one-size-fits-all" strategy seems to be adaptive—megatexts abound. But there is a cost to those who must use them. Students can get lost in thickets of detail and fail to capture essential principles. It is this dilemma that motivated our effort to "seek simplicity" by focusing on principles while remaining ever mindful of the second part of Whitehead's advice.

This book is intended to serve teachers and students engaged in full-year courses as well as short ones. We treat basic processes with a depth comparable to that achieved by much longer texts. Our coverage of biomolecules, enzymes, energetics, and metabolism is substantial enough to suit standard introductory courses, whatever their length. Only elaborations and tangential excursions have been set aside. Our story of biological information flow is told in three chapters, enough molecular biology for the beginning biochemistry student since this elegant subject receives detailed treatment in subsequent courses.

We count on students having behind them introductory courses in general and organic chemistry, and we build a solid chemical foundation by elucidating reaction mechanisms where these will improve understanding without confusing or discouraging our readers. The physiologically important forms of ionizable groups are shown consistently, and stereochemically correct structures are emphasized where they are important. Computer-generated stereo views (many developed especially for this text by Jane and David Richardson's group at Duke University) clarify the shape and function of biomolecules. A stereoviewer comes with each copy of the text.

The narrative is up-to-date. There are a number of new descriptions of three-dimensional protein structures based on recent X-ray crystallographic work. We also include, to name a few examples, new information on the role of chaperones in protein folding, new molecular data revealing evolutionary relationships among organisms, new work on the role of transition-state stabilization in enzyme catalysis and on transition-state analogs and the generation of catalytic antibodies, newly identified prosthetic groups, the role of histidine in the activity of triose phosphate isomerase, new images of DNA visualized by scanning tunnelling electron microscopy, new insights into the energetics and regulation of metabolism, new work on the Q cycle in oxidative phosphorylation and photophosphorylation, new research on the formation of spliceosome complexes and their role in RNA processing, and new information on the structure and function of ribosomes.

The text also offers significant innovations in the teaching of broad principles, revealing unity and pattern within the details. The chapter on enzyme mechanisms and the following one on coenzymes categorize the major types of reaction mechanisms students will encounter in their study of metabolism later in the book. These mechanisms, when they appear within metabolic pathways, can then be welcomed as familiar roadmarks rather than endured as unfamiliar detours.

Coenzymes are treated as a central feature of pathways rather than simply as side-products of reactions. The distinction between mobile cosubstrates and bound prosthetic groups is clearly drawn in the chapter on coenzymes and reinforced throughout the metabolism chapters. In particular, the role of FAD is clarified. Many books show reduced FAD as a coproduct of reactions, suggesting a parallelism between it and mobile cosubstrates such as NAD and ubiquinone (Q). Our text emphasizes that FAD, as an enzyme-bound prosthetic group, is not released to solvent but must be converted back to its oxidized form with each catalytic cycle by donating its electrons (and protons) to a mobile cosubstrate such as Q. We identify the mobile carrier, rather than FAD, as the product in such reactions. Q, often introduced only within the context of the respiratory electron-transport chain, is introduced in this text in the chapter on coenzymes. It appears as a mobile cosubstrate of the succinate dehydrogenase complex in the citric acid cycle and is a familiar player by the time students learn of its role in oxidative phosphorylation.

The chapter on bioenergetics clearly delineates the difference between standard and actual changes in free energy and explains why standard values cannot be used as predictors of the direction of metabolic flux in vivo. We explain that the actual change in free energy in the metabolic steady state indicates whether a reaction is near equilibrium or metabolically irreversible and that only metabolically irreversible reactions are subject to regulation. We emphasize biochemical events in vivo and contrast them with experimental observations in vitro that have led to misconceptions about certain physiological processes. In this chapter, we also make clear the connection between reduction potential and free-energy change, thus giving students an overview of the primary energy transactions prior to discussing details of metabolic pathways. Many texts explain reduction potential only within the chapter on oxidative phosphorylation, thereby delaying (if not preventing) an understanding of energy flux between catabolic and biosynthetic processes.

Our chapter on oxidative phosphorylation is unique among current texts in developing an understanding of chemiosmosis *before* describing the details of electron transport coupled to the synthesis of ATP. An insightful comparison of protonmotive force to electromotive force gives the student an immediate understanding of the energy that drives oxidative phosphorylation.

Principles of Biochemistry has four parts. Part One contains two chapters that introduce the subject. In Chapter 1, we review fundamental aspects of cell structure and function and give an overview of biomolecular structures, touching upon historical landmarks in the development of biochemistry and molecular biology. In Chapter 2, we consider the properties of water, introducing its role in solution chemistry, its ionization, and its contribution to the stability and reactivity of biomolecules and macromolecular assemblies.

Part Two contains eight chapters on the structures of biomolecules and the relationship of molecular architecture and chemical reactivity to biological function. We first examine amino acids, their structures, ionic properties, reactivities, and linkages within proteins (Chapter 3); this discussion serves as background for the next topic—the three-dimensional structures of proteins in relation to their physiological functions (Chapter 4). In the next three chapters, we examine the fundamental principles of enzyme catalysis: the general properties and classification of enzymes, enzyme kinetics, and allosteric transitions (Chapter 5); mechanisms of enzymatic catalysis and the roles of amino acid residues within active sites, as both chemical reactants and binding sites (Chapter 6); and the involvement of coenzymes as cosubstrates and prosthetic groups within the active sites of enzymes (Chapter 7). We continue Part Two by examining the structure and function of three other major classes of biomolecules: carbohydrates (Chapter 8); lipids and biological membranes (Chapter 9), which includes a discussion of membrane transport; and nucleic acids (Chapter 10).

Part Three contains nine chapters dealing with bioenergetics and transformations of biomolecules. In Chapter 11, we give an overview of metabolism and guide the student through thermodynamic principles that underlie bioenergetics, emphasizing the roles of energy-rich metabolites, such as ATP, and reduced coenzymes, such as NADH. We next describe glycolysis (Chapter 12) and introduce hormone action and second messengers; then we cover the citric acid cycle, describing its central role in aerobic metabolism (Chapter 13), and follow with an explanation of oxidative phosphorylation (Chapter 14). Glycogen metabolism, gluconeogenesis, and the pentose phosphate pathway come next (Chapter 15), followed by a modern view of photosynthesis (Chapter 16), metabolism of fatty acids (Chapter 17), metabolism of amino acids (Chapter 18), and metabolism of purine and pyrimidine nucleotides, the biochemical precursors of nucleic acids (Chapter 19).

Part Four concludes the book with three chapters on the flow of biological information; in these chapters, we sustain our emphasis on central principles, covering recent discoveries and showing the chemical mechanisms underlying the molecular biology of gene expression and its regulation. In Chapter 20, we describe replication and repair of DNA, followed by transcription and the processing of RNA (Chapter 21), and protein synthesis (Chapter 22).

Science is more and more a collaborative effort, several workers joining forces in one and sometimes two or more laboratories. Science textbook publishing is likewise an increasingly collaborative enterprise. There still are estimable texts by single authors, but the trend is toward sharing the work. Each author of this text wrote chapters in his particular area of expertise but also collaborated with the other authors and the publishing group to ensure a unified presentation. A brief biographical account of the authors is given on page v; the identity of each player on the publishing side—editors, artists, and the like—is shown on page vi; and the invaluable reviewers and advisors who helped guide this project are listed on pages x and xi.

Whoever said "I wrote a long letter because I didn't have time to write a short one" nicely caught the irony of our experience in producing *Principles of Biochemistry*. It took us more than a sensible amount of time and labor to draw from this burgeoning science the core of ideas and processes that a first course in biochemistry must offer in these days of ferment and discovery.

Neil Patterson

Chapel Hill

December 1992

Reviewers and Advisors*

Deborah A. Adams, Agouron Institute
George C. Allen, Stanford University Medical Center
G. Harvey Anderson, University of Toronto
Laurens Anderson, University of Wisconsin
Rashid A. Anwar, University of Toronto
Dipali V. Apte, University of Illinois at Urbana-Champaign
Roy R. Baker, University of Toronto
James R. Bamburg, Colorado State University
Robert W. Baughman, Harvard Medical School
Bruce P. Bean, Harvard Medical School
G. Vann Bennett, Duke University School of Medicine
Ludwig Brand, Johns Hopkins University
Greg G. Brown, McGill University
John M. Buchanan, Massachusetts Institute of Technology
Kent O. Burkey, U.S. Department of Agriculture
P. Jonathan G. Butler, Medical Research Council, Cambridge, England
Judith L. Campbell, California Institute of Technology
Nicholas C. Carpita, Purdue University
Frank C. Church, University of North Carolina School of Medicine
Steven G. Clarke, University of California, Los Angeles
James E. Darnell, Jr., The Rockefeller University
Christian de Duve, Rockefeller University and International Institute of Cellular and Molecular Pathology, Belgium
Daniel V. DerVartanian, University of Georgia
Richard E. Dickerson, University of California, Los Angeles
Walter J. Dobrogosz, North Carolina State University
David H. Dressler, Harvard Medical School
Paul T. Englund, Johns Hopkins University School of Medicine
Shelagh M. Ferguson-Miller, Michigan State University
Robert J. Fogelsong, Duke University
Irwin Fridovich, Duke University School of Medicine
Naba K. Gupta, University of Nebraska, Lincoln
James H. Hageman, New Mexico State University
Nancy V. Hamlett, Swarthmore College
Franklin M. Harold, National Jewish Center for Immunology and Respiratory Medicine
Edward D. Harris, Texas A & M University
Robert A. Harris, Indiana University School of Medicine
Ari H. Helenius, Yale University School of Medicine
Kenneth Hellman, Smith College
Henri G. Hers, de l'Universite Catholique de Louvain, Belgium
Robert L. Hill, Duke University School of Medicine
Peter C. Hinkle, Cornell University
George E. Hoch, University of Rochester
Jan B. Hoek, Thomas Jefferson University
Johns W. Hopkins, III, Washington Univeristy
Ching-hsien Huang, University of Virginia School of Medicine
Evan E. Jones, North Carolina State University
Kenneth M. Jones, University of Leicester, England
Mary Ellen Jones, University of North Carolina at Chapel Hill
Manfred L. Karnovsky, Harvard Medical School
Robert A. B. Keates, University of Guelph

*The affiliations shown are those that obtained when these reviewers and advisors made their contributions.

Judith A. Kelly, The University of Connecticut
Debra A. Kendall, The University of Connecticut
Sue C. Kinnamon, Colorado State University
Aaron Klug, Medical Research Council, Cambridge, England
Ronald Kluger, University of Toronto
Arthur Kornberg, Stanford University Medical School
Nicholas A. Kredich, Duke University School of Medicine
Monty Krieger, Massachusetts Institute of Technology
LeRoy R. Kuehl, University of Utah
James A. Lake, University of California, Los Angeles
M. Daniel Lane, Johns Hopkins University School of Medicine
Robert N. Lindquist, San Francisco State University
Rose G. Mage, National Institutes of Health
John H. Mangum, Brigham Young University
Lynn A. Margulis, University of Massachusetts
James B. Meade, University of North Carolina School of Medicine
William C. Merrick, Case Western Reserve University School of Medicine
Richard Ogden, Agouron Institute
Philip Oliver, University of Cambridge, England
Marvin R. Paule, Colorado State University
Henry P. Paulus, Boston Biomedical Research Institute
Allen T. Phillips, Pennsylvania State University
Simon J. Pilkis, State University of New York at Stony Brook
Huntington Potter, Harvard Medical School
Jane H. Potter, University of Maryland
Gary L. Powell, Clemson University
Charlotte W. Pratt, University of North Carolina School of Medicine
Margaret Rand, University of Toronto
Charles C. Richardson, Harvard Medical School
Phillips W. Robbins, Massachusetts Institute of Technology
Nadia A. Rosenthal, Children's Hospital/Harvard Medical School
Milton H. Saier, Jr., University of California, San Diego
Marvin L. Salin, Mississippi State University
Vern G. Schirch, Medical College of Virginia
B. Trevor Sewell, University of Cape Town, South Africa
Gordon C. Shore, McGill University
James N. Siedow, Duke University
Lewis M. Siegel, Duke University School of Medicine
Gerald R. Smith, Fred Hutchinson Cancer Research Center
Jaro Sodek, University of Toronto
Ronald L. Somerville, Purdue University
Franklin W. Stahl, University of Oregon
Deborah A. Steege, Duke University School of Medicine
Clarence H. Suelter, Michigan State University
Mary C. Sugden, Queen Mary and Westfield College, England
Judith A. Swan, Duke University
George Taborsky, University of California at Santa Barbara
Keith E. Taylor, University of Windsor
Dean R. Tolan, Boston University
George Tomlinson, University of Winnipeg
Thomas W. Traut, University of North Carolina at Chapel Hill
Peter H. von Hippel, University of Oregon
Robert W. Wheat, Duke University
George B. Witman, Worcester Foundation for Experimental Biology
Owen N. Witte, University of California, Los Angeles
Charles Yanofsky, Stanford University
Douglas C. Youvan, Massachusetts Institute of Technology
James K. Zimmerman, Clemson University

List of Chapters

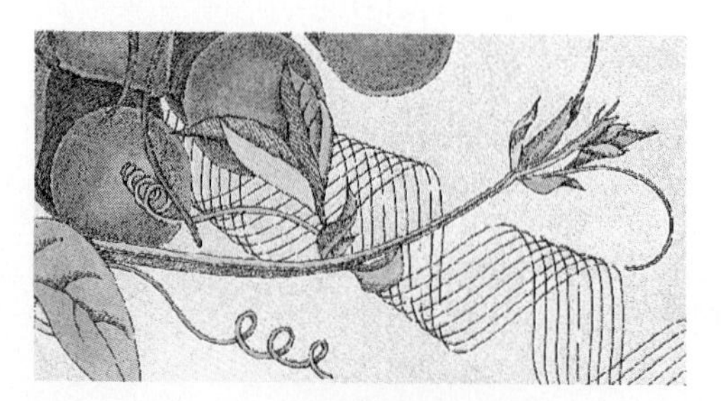

Part Three
Metabolism and Bioenergetics

Part Four
Biological Information Flow

Contents

Part One
Introduction

Part Two
Structures and Functions of Biomolecules

Chapter 8 Carbohydrates 8·1

Chapter 9 Lipids and Biological Membranes 9·1

Chapter 10 Nucleic Acids 10·1

Part Three
Metabolism and Bioenergetics

Chapter 11
Bioenergetics: ATP and Other Energy-Rich Metabolites 11·1

Chapter 12
Glycolysis 12·1

Chapter 13 The Citric Acid Cycle 13·1

Chapter 16 Photosynthesis 16·1

Chapter 17 Lipid Metabolism 17·1

Chapter 18
Amino Acid Metabolism

Chapter 19 Nucleotide Metabolism 19·1

Part Four Biological Information Flow

Chapter 20 DNA Replication and Repair 20·1

Chapter 22
Protein Synthesis 22·1

Part One

Introduction

1

Introduction to Biochemistry

Biochemistry is the study of the molecules and chemical reactions of life. Biochemists use the methods of all scientists: observations are made and hypotheses are constructed and tested. The scientific approach to the study of life has led to an understanding of the basic biochemistry that is common to all living organisms. Although scientists usually concentrate their research efforts on particular organisms, the results of these studies often apply to other species as well. For example, the central metabolic processes that function in the cyanobacterium *Scytonema* (Figure 1·1) are very much like those found in distantly related organisms such as ourselves.

The basic principles of biochemistry form the subject of this book. We shall describe the most important and fundamental concepts governing the chemistry of life, those that are common to most species, including bacteria such as *Scytonema* and mammals such as ourselves. Where appropriate, we shall point out important differences that distinguish some groups of organisms from others. For example, the mechanism of gene expression in a bacterium such as *Scytonema* differs in some ways from that in plants and animals. Furthermore, *Scytonema* is capable of both photosynthesis and fixation of nitrogen, biochemical processes that are absent in most other organisms.

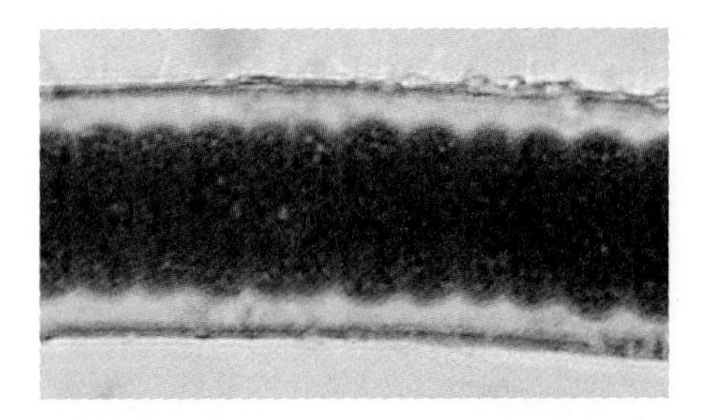

Figure 1·1
Scytonema, a filamentous cyanobacterium. The basic biochemical processes of *Scytonema* are similar to those of distantly related organisms, even though *Scytonema* also has capabilities—photosynthesis and fixation of nitrogen—that are absent in most other organisms. (Courtesy of John M. Kingsbury.)

To understand the chemistry of life, it is first necessary to acquire a knowledge of fundamental chemical principles. Advances in this area took place rapidly from the 1500s through the 1800s, when experiments led to workable theories of reaction kinetics and thermodynamics and to the discovery of the atomic compositions of some molecules. By the end of this period, many chemicals produced in living organisms had been identified; the stage was set for the birth of biochemistry.

1·1 Biochemistry Is a Relatively New Science

Figure 1·2
Friedrich Wöhler (1800–1882). By synthesizing urea, Wöhler showed that a chemical of life could be made in the laboratory. (Courtesy of The Bettmann Archive.)

By the beginning of the nineteenth century, the world was broadly divided into the "organic," living world and the "inorganic," nonliving world. The difference between the living and the inanimate was attributed to the action of a vital force, which led to the belief that the organic and inorganic realms obeyed quite different chemical laws, a belief known as *vitalism*. In 1828, however, Friedrich Wöhler (Figure 1·2) made the momentous discovery that heating the inorganic compound ammonium cyanate produced the organic compound urea. Vitalism held that urea could be obtained only from urine produced by living organisms. The synthesis of urea and other organic compounds dealt a severe blow to vitalism, but the concept lived on in the idea that only living organisms could carry out the complicated reactions of life. According to this assumption, the mechanisms of biochemistry could only be explained on the basis of principles that did not apply to the nonliving world.

Belief in *spontaneous generation* was also widespread before the middle of the last century. The appearance of maggots in rotting meat or microorganisms in other decaying organic material was attributed to the de novo formation of new life. In 1862, Louis Pasteur (Figure 1·3) showed that microorganisms did not appear in sterilized solutions of organic materials unless they were exposed to air (and thus to living microorganisms). From this, Pasteur concluded that living organisms only arise from other living organisms.

Pasteur was a vitalist in the sense that he believed that life was necessary for biochemical reactions to occur. This belief was put to rest in 1897 by the experiments of Eduard and Hans Buchner, who showed that nonliving extracts of cells could catalyze biochemical reactions. They used the word "enzyme" to describe the molecules from cells that are responsible for this catalysis. (The term had been coined twenty years earlier by Wilhelm Kühne.)

Figure 1·3
Louis Pasteur (1822–1895). Pasteur disproved the concept of the spontaneous generation of life. He, and many other prominent scientists of his era, also showed that living cells carry out complex biochemical reactions. (Courtesy of The Bettmann Archive.)

Emil Fischer (Figure 1·4) is also one of the founders of biochemistry. During the latter part of the past century, he studied the structures of biological compounds and made important contributions to our understanding of the function of enzymes. In particular, he suggested that a substrate (the reactant in an enzyme-catalyzed reaction) is complementary in shape to its enzyme, much as a key fits into a lock. Many other nineteenth-century scientists made major contributions to the emergence of biochemistry by showing that chemical reactions in living cells followed the same principles that applied to nonbiological molecules.

By the beginning of the twentieth century, the first departments of biochemistry had been founded in some universities in Europe and North America. Researchers in the new discipline of biochemistry devoted most of their studies to the isolation of biomolecules and characterization of their structures and functions and to *metabolism,* the network of enzyme-catalyzed reactions in living cells. Many metabolic pathways were discovered in the first half of this century.

James B. Sumner was the first to demonstrate, in 1926, that an enzyme could be crystallized. He showed that urease is a protein. After several more enzymes were identified as proteins, it was concluded that all enzymes are proteins. The first detailed molecular structures of proteins were not elucidated until the 1950s, by Max Perutz and John C. Kendrew, although many workers, including Linus Pauling, made important earlier contributions.

The demonstration in 1944 that *genes* (until then purely theoretical units of biological inheritance) are composed of *deoxyribonucleic acid* (DNA) stimulated interest in nucleic acids. These studies led to the solving of the three-dimensional structure of DNA by James D. Watson and Francis H. C. Crick in 1953. The recognition that biological function is inextricably linked to information encoded in the genes led to the emergence of molecular biology within the past fifty years.

Figure 1·4
Emil Fischer (1852–1919). Fischer made many contributions to our understanding of the structures of biological molecules. (Courtesy of The Bettmann Archive.)

1·2 Biochemistry Is Empirical and Reductionist

Biochemistry is reductionist in the sense that its goal is to understand the basic chemical and physical principles that apply to living organisms. The approach often involves the study of individual reactions in a metabolic pathway or the structure of the components of complex assemblies of molecules in cells. The ultimate goal, however, is to reassemble these individual components into the large structures found inside cells in order to understand how reactions are coordinated and controlled. In recent years, biochemists have made remarkable advances by using this strategy.

To many scientists, the universe is remarkable not only for its beauty but also for its consistency, its reasonableness, and its accessibility to rational analysis. As you study biochemistry, patterns will emerge that are logical and self-consistent. Biochemistry is still more an empirical than a theoretical science, but as it matures, a sound theoretical basis is forming. In spite of the incredible diversity of organisms, it is possible to make useful generalizations that appear to apply to all of life.

1. *Life requires energy.* Living organisms are constantly transforming energy into useful work. They grow and multiply. Much of this energy is ultimately supplied by the sun. Sunlight is "captured" by plants, algae, and photosynthetic bacteria and used for the synthesis of biological compounds. When these organisms die, the compounds are ingested and broken down by protozoa, fungi, bacteria, and animals that are incapable of photosynthesis (Figure 1·5).

2. *Biochemical reactions require catalysts.* Nearly all biochemical reactions are catalyzed by specific enzymes. Reactions that would proceed extremely slowly in the absence of such catalysts can occur rapidly in a living organism. A large portion of biochemical research is devoted to understanding the structures and functions of enzymes.

3. *Life depends on information encoded in genes.* The structure of specific proteins and control of their synthesis is passed from one generation to the next. This information is contained in the genome, the sum total of genetic information in a cell. The flow of information is from DNA (genes) to ribonucleic acid (RNA) to proteins, a sequence known as the Central Dogma (Figure 1·6).

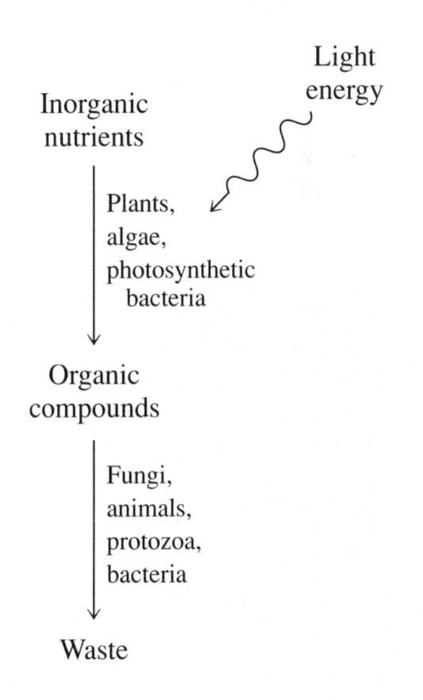

Figure 1·5
Energy flow. Photosynthetic organisms capture the energy of sunlight to synthesize organic compounds. The energy contained in these compounds is in turn utilized by other organisms that ingest them as food.

$$\text{DNA} \longrightarrow \text{RNA} \longrightarrow \text{Protein}$$

Figure 1·6
The Central Dogma. Information flows from DNA to RNA to protein.

The specific examples given for each generality are not comprehensive; some catalysts, for example, are composed of RNA, not protein, and there are other sources of energy in addition to sunlight. Nevertheless, these generalities have been assembled from a massive amount of information collected by scientists over the years, and they form the basis of our current understanding of the principles of biochemical processes.

Biochemistry is an empirical science, and new experiments are reported daily in the scientific literature. The results of some of these experiments affect our understanding of how life works. Biochemistry is an active and rapidly evolving discipline and much remains to be learned.

1·3 Living Organisms Obey the Laws of Physics and Chemistry

One of the main themes of biochemistry is that processes in living organisms obey the same laws of physics and chemistry that are observed elsewhere in the universe. This theme has developed over the past several hundred years as experiments revealed the absence of any peculiar "vital force" that might govern biological activity.

Biological compounds are made from the same elements found in other molecules, but the prevalence of certain elements in cells differs substantially from their abundance on Earth. The elemental composition of three organisms is shown in Table 1·1. Of 92 naturally occurring elements, only about two dozen are present and essential in living cells and of those the six shown in Table 1·1—carbon, hydrogen, nitrogen, oxygen, phosphorus, and sulfur—are the most common. Water is a major component of cells and accounts for the high percentage (by weight) of oxygen. Carbon is much more abundant in living organisms than it is in the rest of the universe, and some elements, such as silicon, aluminum, and iron are present in only trace amounts in cells although they are abundant on Earth. One of the characteristics of life is that cells concentrate certain elements and molecules and exclude others. Even the composition of extracellular bodily fluids, such as blood plasma, differs from that of nonbiological liquids, such as sea water.

The chemical reactions that occur inside cells are the same kinds of reactions that occur with molecules in nonenzymatic chemical reactions. Chemical bonds are made and broken according to mechanisms common to all chemistry. However, the reactions that occur in living cells are catalyzed and thus proceed at very fast rates.

Table 1·1 Elemental composition of three organisms

Element	Human	Alfalfa	Bacterium
Carbon	19.37%	11.34%	12.14%
Hydrogen	9.31	8.72	9.94
Nitrogen	5.14	0.83	3.04
Oxygen	62.81	77.90	73.68
Phosphorus	0.63	0.71	0.60
Sulfur	0.64	0.10	0.32
Total	97.90%	99.60%	99.72%

[Adapted from Curtis, H., and Barnes, S. (1989). Biology, 5th ed. (New York: Worth Publishers), p. 35.]

Biochemical reactions involve specific chemical bonds or parts of a molecule. These sites of reactivity, or *functional groups,* can be classified into a few common types, as shown in Figure 1·7. The mechanisms of the reactions between functional groups in biomolecules have been studied by a variety of techniques, including comparison to well studied chemical reactions.

(a) *Functional groups*

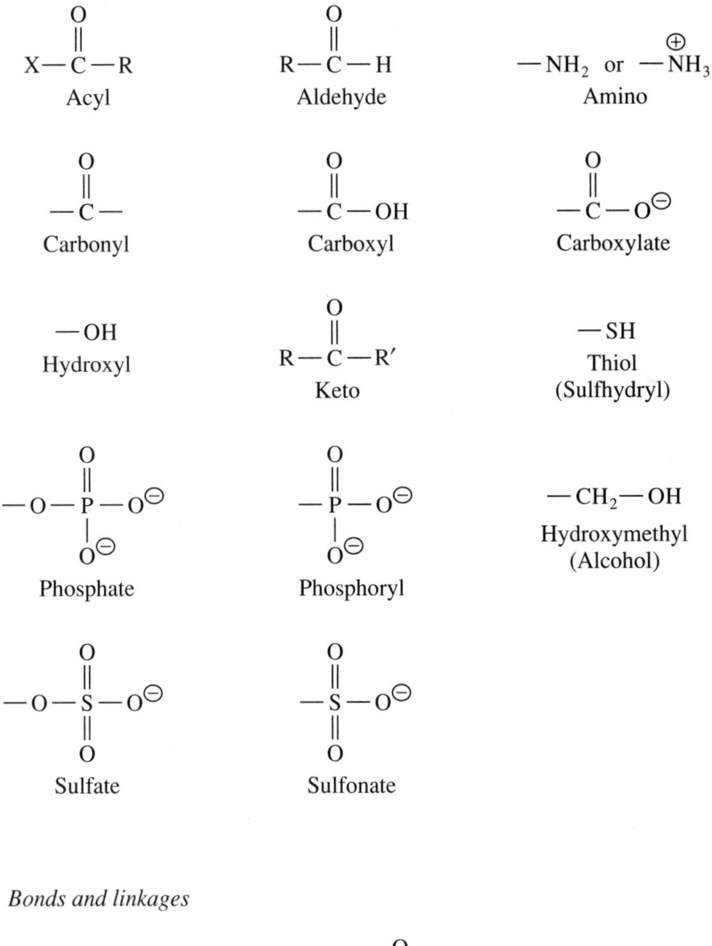

(b) *Bonds and linkages*

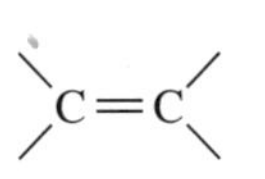

Double

Ester

Ether

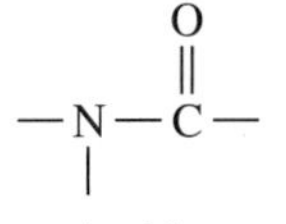

Amide

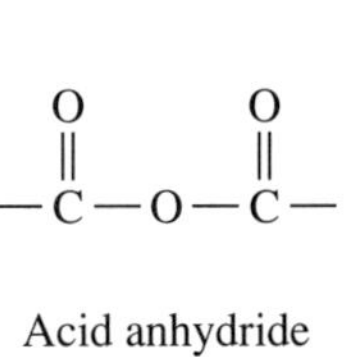

Acid anhydride

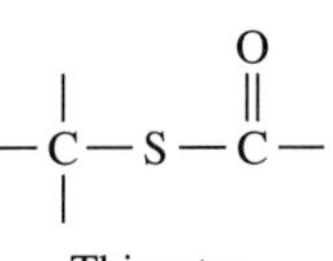

Thioester

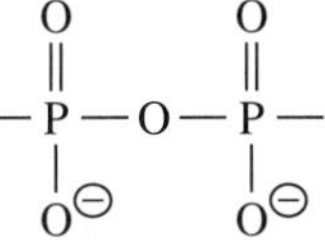

Phosphoanhydride

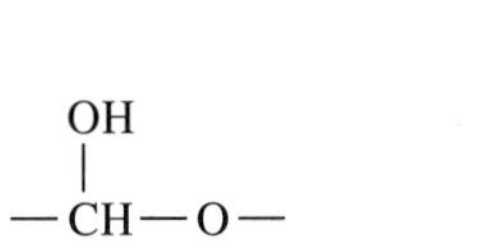

Hemiacetal (from an aldehyde plus an alcohol)

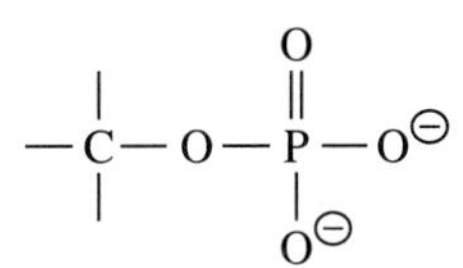

Phosphomonoester

Figure 1·7
Important functional groups **(a)** and bonds and linkages **(b)** in biochemistry. R represents an alkyl group, a group that contains carbon atoms plus hydrogen. X represents any group other than an alkyl group or hydrogen.

The study of energy in molecules and of changes in energy during biochemical reactions is called *bioenergetics*. Bioenergetics is part of the field of thermodynamics, a branch of physical science that deals with energy changes. The basic thermodynamic principles that apply to energy flow in nonliving systems also apply to the chemistry of life. An appreciation of these principles is one of the keys to understanding biochemistry.

Knowledge of the energy changes associated with a reaction or process allows us to predict the equilibrium position or direction of the reaction. One of the thermodynamic properties that influences direction is *entropy,* the degree of disorder. Reactions or processes are more likely to occur spontaneously if they are associated with an increase in entropy.

However, thermodynamic considerations cannot reveal how quickly a reaction will occur. The kinetics of a reaction, or its specific rate, depends on the energy content of individual molecules. Potential reactants must have sufficient energy to overcome an energy barrier that hinders their reaction.

1·4 Structure and Function Are Inseparable

Living cells contain biopolymers created by joining together, or condensing, many smaller organic molecules or monomers. Each *condensation* proceeds by the removal of the elements of a water molecule to create a chain of *residues*. In some cases, such as proteins and nucleic acids, the order of the different residues is very important; in other cases, such as some carbohydrates, a single residue is repeated many times.

Biopolymers have distinctive properties that are very different from those of their constituent building blocks. For example, starch is not soluble in water and does not taste sweet although it is a polymer of the sugar glucose. Observations such as this have led to the general principle of hierarchical organization of life, which states that each new level of organization in biology generates structures with emergent properties that cannot be predicted from the properties of their constituents. The levels of complexity can be represented as atoms, molecules, biopolymers, organelles, cells, and on to organs and organisms in more complex species.

It is obvious that cells can carry out some functions that are beyond the capability of their isolated molecular components. As stated earlier, one of the goals of biochemistry is to study the structure of individual biomolecules and then use the resulting information to piece together the function of larger structures. The principle of organizational hierarchy should not be construed to mean that cells do not obey the laws of physics and chemistry but merely that the function of biomolecules is closely related to structure.

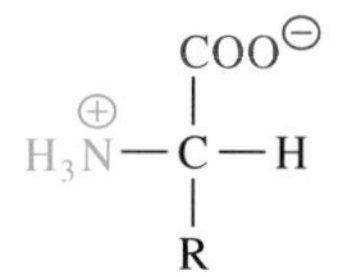

Figure 1·8
General structure of an amino acid. All amino acids contain at least two functional groups—an amino group (blue) and a carboxylate group (red). During protein synthesis, the amino group of one amino acid reacts with the carboxylate group of another to form a peptide bond. Different amino acids contain different side chains (designated —R).

A. Proteins Are Biopolymers of Amino Acids

There are 20 common amino acids in all cells. These are synthesized from simpler precursors or absorbed as nutrients. All amino acids have at least two functional groups—an amino group and a carboxylate group (Figure 1·8). These groups condense during protein synthesis to form a peptide bond, resulting in the end-to-end joining of amino acids to form a polypeptide or protein.

Proteins are large biopolymers, many of which function as enzymes. Others serve as major structural components of cells and organisms. Their structure is determined, in large part, by the order of the amino acids that are joined together. This order, in turn, is encoded by the genes.

The function of a protein depends upon a proper three-dimensional shape. The three-dimensional structures of many proteins have been determined, and several principles governing the relationship between structure and function have become

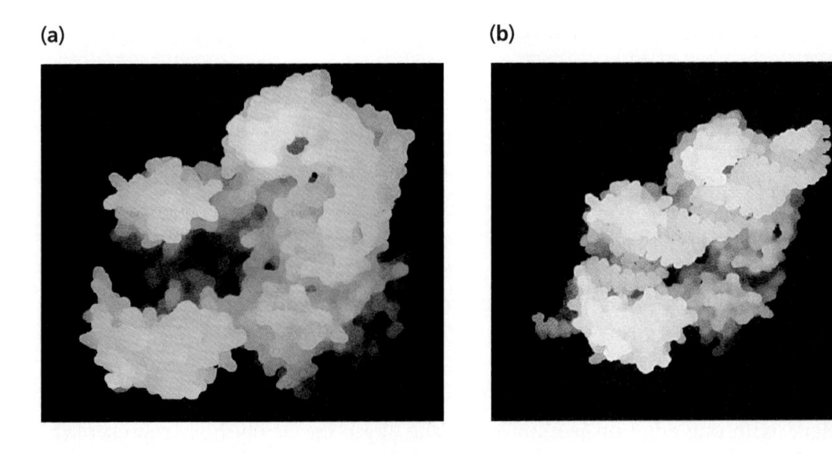

Figure 1·9
Computer-generated images of HIV reverse transcriptase. **(a)** Structure of the enzyme. Note the large groove where the substrate binds. **(b)** A model of the enzyme with bound substrate—a DNA/RNA hybrid molecule. (Courtesy of Thomas A. Steitz.)

clear. For example, most enzymes contain a cleft or groove that binds the substrates of a reaction. Figure 1·9 shows the structure of the enzyme reverse transcriptase, which is encoded by the HIV (human immunodeficiency virus) genome. During infection, reverse transcriptase catalyzes synthesis of DNA by promoting the copying of the viral RNA molecule. The resulting DNA/RNA hybrid lies in a large groove in the middle of the enzyme. Knowledge of the structure of this enzyme has assisted in development of specific drugs that can inhibit its function and possibly prevent the spread of acquired immune deficiency syndrome (AIDS).

Proteins can associate to form larger structures. This next level of organization produces complexes that can carry out a series of reactions. We shall consider such complexes in our descriptions of metabolism and DNA replication.

B. Carbohydrates Are Simple Sugars, Their Polymers, or Other Sugar Derivatives

Carbohydrates are an abundant and diverse group of biomolecules that share the common property of being composed primarily of carbon, oxygen, and hydrogen. Carbohydrates include simple sugars, their polymers, and other sugar derivatives, which often contain several hydroxyl groups. Cellulose, for example, is a linear polymer of glucose residues (Figure 1·10). One of the major components of plant cell walls, cellulose is the most abundant biological macromolecule. Other carbohydrate polymers also play a role in the structure of cells and tissues, and some, such as glycogen and starch, serve as storage molecules.

The structure of cellulose reveals that both covalent and noncovalent bonds are important in biological structures. Each glucose residue is joined covalently to the next. The molecule is further stabilized by intrachain hydrogen bonding between a hydroxyl group of one unit and an oxygen atom of another. The importance of such weak interactions will be emphasized throughout this text. Figure 1·10 also illustrates how important it is to understand the three-dimensional structures of biochemical molecules. The structure of glucose can be represented on a two-dimensional page in several ways; some of these representations fail to reveal crucial aspects of the actual structure (Box 1·1).

Figure 1·10
Cellulose. Cellulose is a linear polymer of glucose residues. Each glucose unit is joined covalently to the next. In addition, the structure is stabilized by intrachain hydrogen bonds between adjacent residues.

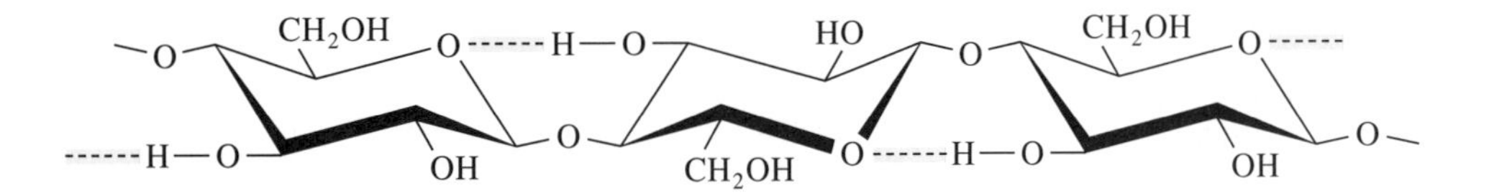

Box 1·1 Representations of biochemical structures

The structures of biochemicals can be represented on a page in several ways. For example, one form of glucose can be shown as a linear molecule with a carbon chain substituted by hydroxyl groups and an aldehyde group, as shown in Figure 1a. This linear representation is called a *Fischer projection* (after Emil Fischer, mentioned earlier). In solution, however, most glucose molecules are in a ring form as a result of a covalent bond between the carbon of the carbonyl group and the oxygen of the C-5 hydroxyl group. The ring form can be drawn as a Fischer projection (Figure 1b) but is more commonly shown as a *Haworth projection* (Figure 1c). The Haworth projection is rotated in a clockwise direction from the Fischer projection. It portrays the ring part of the molecule as a plane with one edge projecting out of the page; the heavy lines represent the parts of the molecule extending toward the viewer. Hydroxyl groups pointing upward lie above the plane of the ring, those pointing downward lie below the plane. Even the standard Haworth projection, commonly drawn in this text, does not convey a complete understanding of the three-dimensional structure of glucose, however. The actual molecule can adopt several different conformations, including the *chair conformation* shown in Figure 1d. The chair conformation is found in cellulose. It is important to realize that by drawing a molecular structure on a flat sheet of paper, we often lose information that may be important in understanding the function of the molecule.

The image in Stereo S1 conveys a better sense of three-dimensional structure than the two-dimensional drawings. We shall present many such stereo images throughout the text in order to illustrate more accurately the structures and shapes of biomolecules.

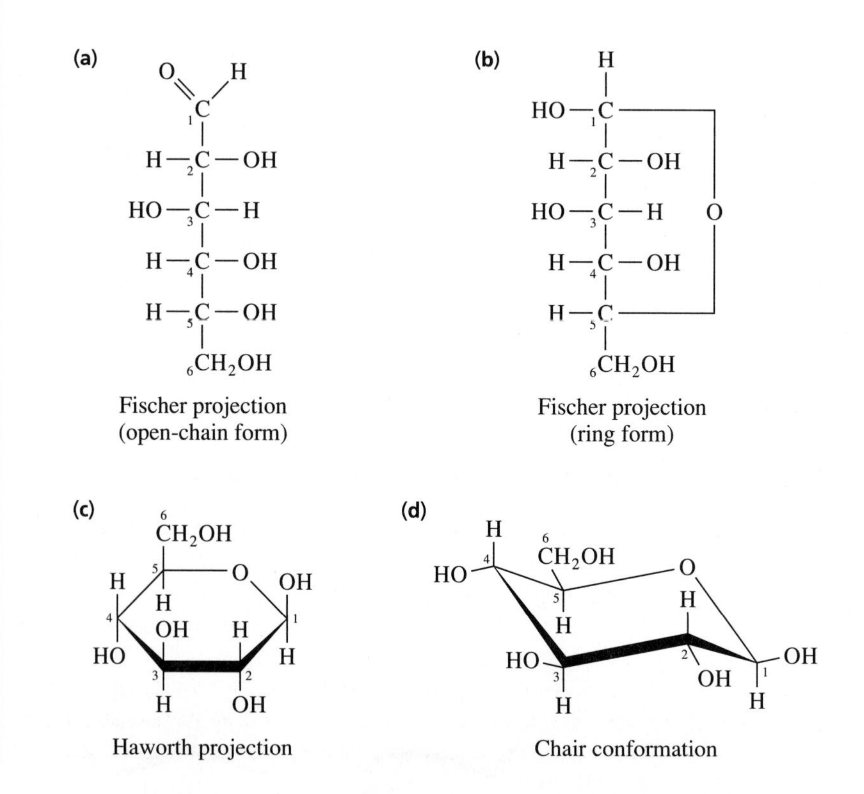

Box 1·1
Figure 1
Representations of the structure of glucose. **(a)** In the Fischer projection, glucose is drawn as a linear molecule in order to show all of the functional groups. **(b)** In solution, most glucose forms a ring structure, shown here as a Fischer projection. **(c)** In a Haworth projection, the ring is almost perpendicular to the page as indicated by the thick lines, which represent bonds closer to the viewer. **(d)** The ring can exist in several different conformations, including the chair conformation shown. Each of these representations emphasizes different aspects of the structure of glucose. Sugars will usually be drawn as Haworth projections in this text.

Box 1·1
Stereo S1
Chair conformation of one of the forms of glucose. (Stereo image by Ben N. Banks and Kim M. Gernert.)

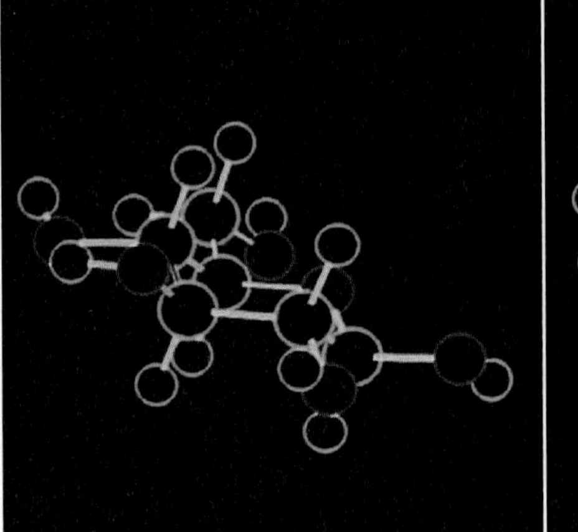
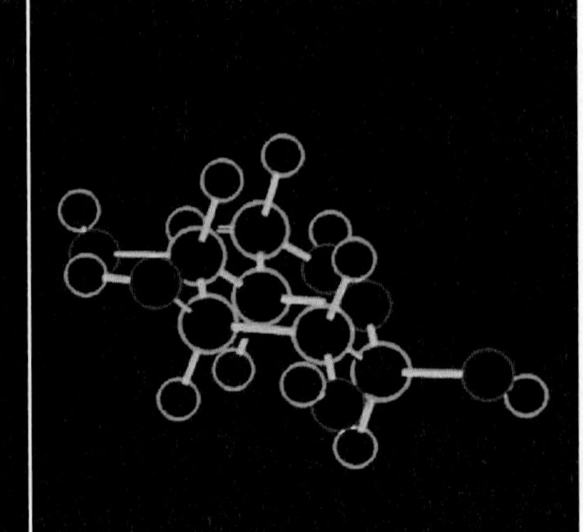

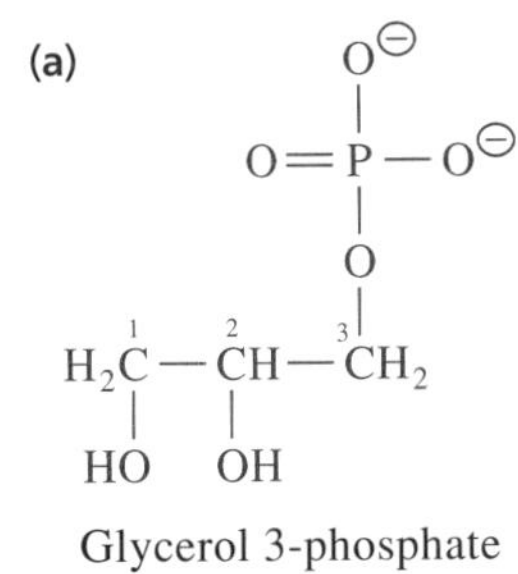

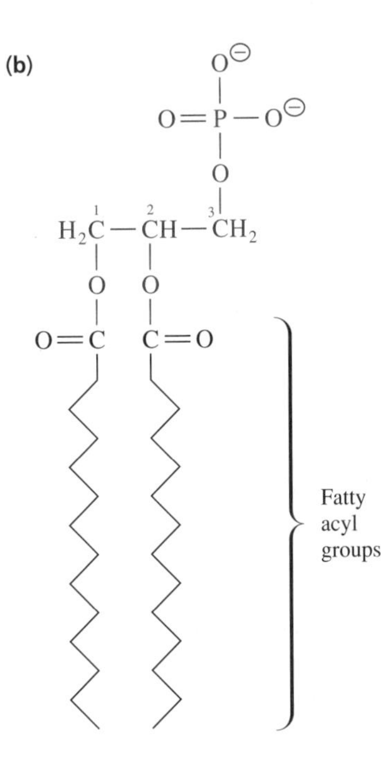

Figure 1·11
Glycerophospholipid. Glycerol 3-phosphate is shown in **(a)**. Two long-chain fatty acids are bound to glycerol 3-phosphate through ester linkages to form the glycerophospholipid shown in **(b)**. The glycerol 3-phosphate portion of the glycerophospholipid is polar, and the fatty acyl groups are nonpolar.

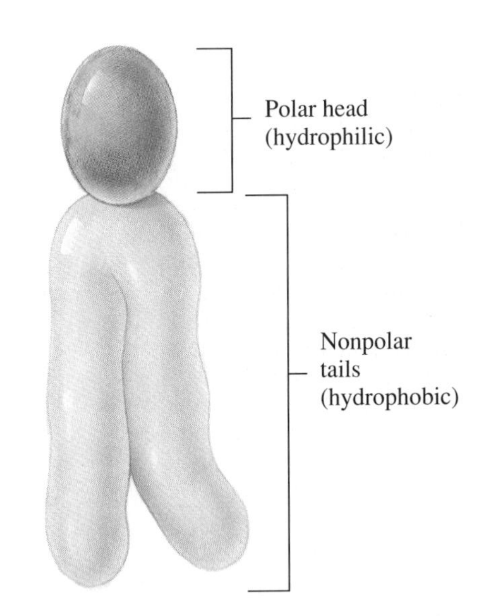

Figure 1·12
General structure of a membrane lipid. The molecule consists of a polar head (blue), which can interact with an aqueous environment, and a nonpolar tail (yellow). (Illustrator: Lisa Shoemaker.)

C. Lipids Are Important Components of Biological Membranes

We shall encounter many different types of lipids. The simplest lipids are the *fatty acids*. These are long-chain hydrocarbons with a carboxyl group at one end. Fatty acids are most commonly found as part of larger molecules called *glycerophospholipids,* which consist of glycerol 3-phosphate and two fatty acyl groups (Figure 1·11). Glycerophospholipids are major components of biological membranes.

Membrane lipids usually have a polar, hydrophilic (water-loving) head that can interact with an aqueous environment and a nonpolar, hydrophobic (water-fearing) tail (Figure 1·12). The hydrophobic tails of such lipids associate, producing sheets

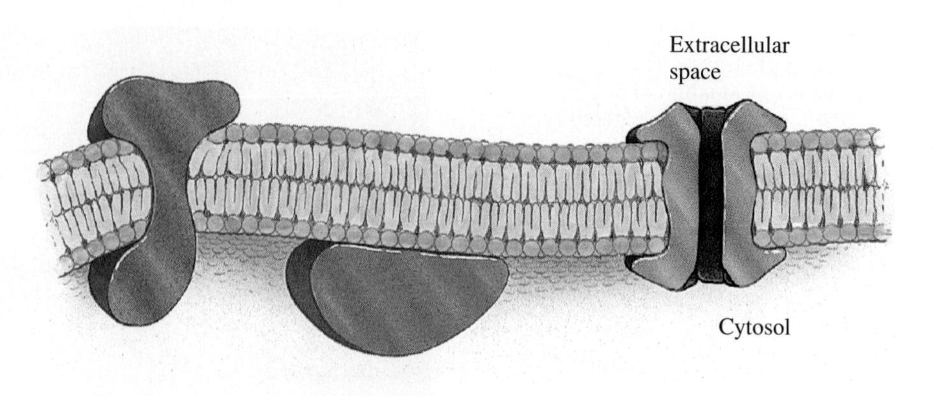

Figure 1·13
General structure of a plasma membrane. The plasma membrane consists of a lipid bilayer with associated proteins. The hydrophobic tails of individual lipid molecules associate to form the core of the membrane. Some proteins span the bilayer, whereas others are attached to the surface in various ways. (Illustrator: Lisa Shoemaker.)

of lipid bilayers, which form membranes. These membranes serve to separate a cell or a compartment within a cell from its environment. The membrane surrounding the cell, called the *plasma membrane,* has proteins embedded within the bilayer. Some membrane proteins serve as channels for the entry of nutrients and exit of waste, whereas others catalyze reactions that occur specifically at the cell surface (Figure 1·13). Intracellular membranes also contain proteins that serve various biological functions. Membranes are among the largest and most complex cellular structures and are the sites of many important biochemical reactions.

D. Nucleic Acids Are Biopolymers of Nucleotides

Nucleic acids are polymers composed of monomers called *nucleotides*. Nucleotides contain a sugar, a heterocyclic nitrogenous base, and at least one phosphoryl group. The sugar in ribonucleotides is the five-carbon carbohydrate ribose; that in deoxyribonucleotides is its closely related derivative deoxyribose. The nitrogenous bases of nucleotides belong to two families known as *purines* and *pyrimidines:* the major purines are adenine and guanine; the major pyrimidines are cytosine, thymine, and uracil.

The structure of the nucleotide adenosine triphosphate (ATP) is shown in Figure 1·14. In addition to serving as a building block of nucleic acids, ATP is the central purveyor of energy in living cells. Cleavage of the bonds between the phosphoryl groups of ATP can be coupled to other reactions so that the energy released can be used elsewhere. We shall encounter many biochemical reactions that involve the breakdown and formation of ATP.

To form nucleic acids, nucleotides are joined through a covalent bond formed between a phosphoryl group of one nucleotide and one of the sugar hydroxyl groups of another. This produces a backbone consisting of alternating sugars and phosphates. In DNA, the bases of two different strands interact to form a helical structure (Figure 1·15). It is the specific sequence of base pairs in DNA that carries genetic information.

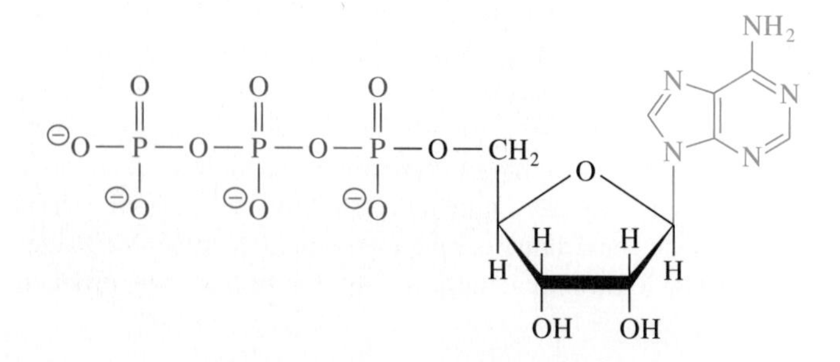

Figure 1·14
Structure of adenosine triphosphate (ATP). The nitrogenous base adenine is shown in blue. It is condensed with the sugar ribose (black). Three phosphoryl groups (red) are also bound to the ribose moiety.

Nucleic acids can be huge. Some human chromosomes, for example, consist of DNA molecules with over one hundred million individual nucleotides. Bacteria, which have fewer genes, usually contain a single, smaller DNA molecule. In most bacteria, the single chromosome is circular, that is, the ends of the molecule are joined.

Ribonucleic acid (RNA) is the other polynucleotide commonly found in living organisms. There are many different kinds of RNA molecules, including the familiar *messenger RNA,* which is involved directly in the transfer of information from DNA to protein. *Transfer RNAs* (tRNAs) are smaller molecules required for protein synthesis, and *ribosomal RNAs* are major components of ribosomes, the RNA:protein complexes on which proteins are synthesized.

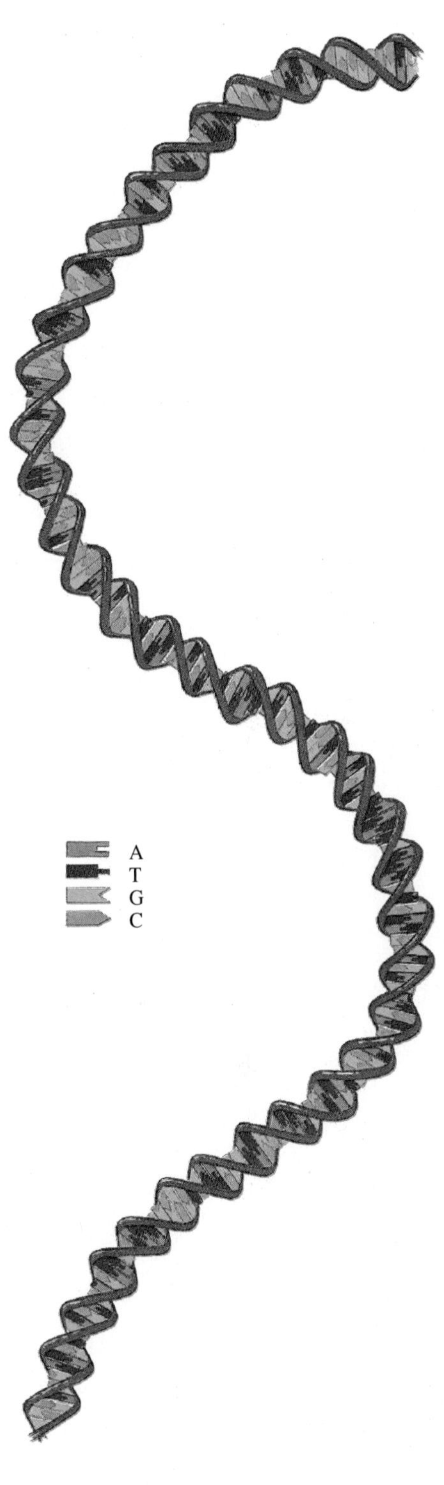

Figure 1·15
Deoxyribonucleic acid (DNA). Two different polymers of nucleotides associate to form a double helix. Complementary bases from each strand interact in the middle of the structure. The specific sequence of base pairs in DNA carries genetic information. (Illustrator: Lisa Shoemaker.)

1·5 Cells Are the Basic Units of All Life

Most biochemical reactions take place within membrane-bounded compartments called cells. This makes sense from a biochemical perspective because it allows the concentrations of biomolecules within cells to be much higher than could be achieved in the absence of membrane-bounded compartments. Every organism is either a single cell or is composed of many cells.

The realization that cells are the basic unit of life had to await the invention of the microscope. Almost 300 years ago, Robert Hooke used the microscope he had built to examine a thin slice of cork. He saw that the plant tissue was divided into small compartments, which he named cells (Latin, *cella,* small room) (Figure 1·16).

Over 150 years elapsed between Hooke's observation and Mathias Schleiden's proposal that the structures of all plant tissues are based upon an organization of cells. Shortly thereafter, Theodor Schwann proposed that all animal tissues are also organizations of cells. At about the same time, bacteria and algae were recognized as unicellular organisms.

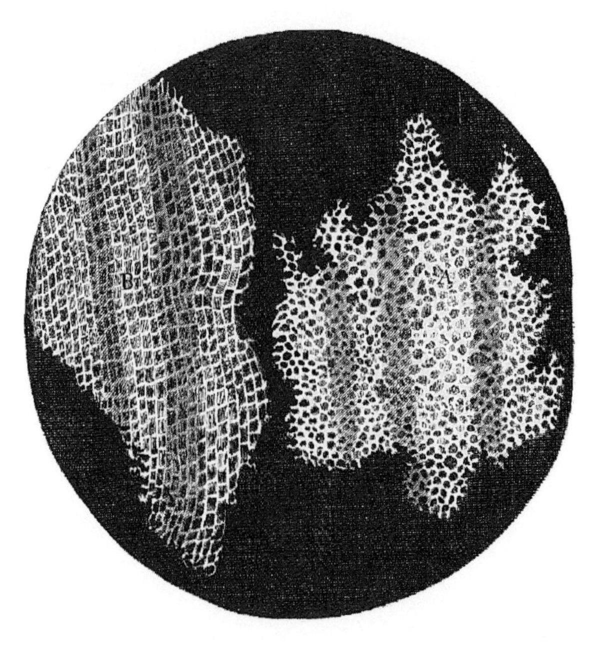

Figure 1·16
Drawing by Robert Hooke of the microscopic structure of a slice of cork. This drawing was published in 1665 in his book *Micrographia.* (Courtesy of Rare Books and Manuscripts Division, The New York Public Library; Astor, Lenox, and Tilden Foundations.)

Cells exist in a remarkable variety of sizes and shapes. Despite such diversity, all cells can be broadly categorized as either prokaryotic or eukaryotic. *Prokaryotes* (Greek, *pro,* before; *karyon,* kernel) are usually single-celled organisms and lack a membrane-bounded nucleus. In contrast, *eukaryotic cells* (Greek, *eu,* proper) are usually much larger and typically have membrane-bounded nuclei. They also usually contain internal membranes that divide the cell into organelles such as mitochondria and chloroplasts. These organelles have specific functions necessary for the life of the cell.

A. *E. coli* Is a Representative Prokaryote

There are many different species of prokaryotes. They can be found in almost every conceivable environment from hot sulfur springs to the insides of larger cells. Bacteria have been detected in the atmosphere, in the deepest parts of the oceans, and even several kilometers underground. A few examples of the many different sizes and shapes of bacteria are shown in Figure 1·17.

The most extensively studied organism is the prokaryote *Escherichia coli (E. coli),* a bacterium that resides in the intestinal tracts of humans and other mammals. This species evolved fairly recently and, like most bacteria, it exhibits some specialized adaptations to its environment. Nevertheless, *E. coli* has many of the features common to all prokaryotes.

The *E. coli* cell is about 0.5 μm in diameter and 1.5 μm long. It has numerous proteinaceous fibers that extend outward from the surface. *Flagella* (Latin, *flagellum,* whip) provide cell motility; they can propel *E. coli* at a velocity of 30 $\mu m\,s^{-1}$, or about 20 body lengths per second. Pili (Latin, *pilus,* hair) are shorter than flagella. *Pili* aid in sexual conjugation and appear to enable cells to adhere to surfaces (Figure 1·18).

Cell walls are common in prokaryotes and confer the different shapes that characterize individual bacteria. The *E. coli* cell wall consists of a network of covalently linked carbohydrate and peptide chains. In some prokaryotes, including *E. coli,* this *peptidoglycan layer* has large molecules called lipopolysaccharides deposited over it, forming an outer membrane.

The bacterial cell wall provides the cell with mechanical strength and protection from osmotic shock. The concentrations of metabolites in most bacterial cells greatly exceed the concentrations of solutes in the media surrounding them, and this

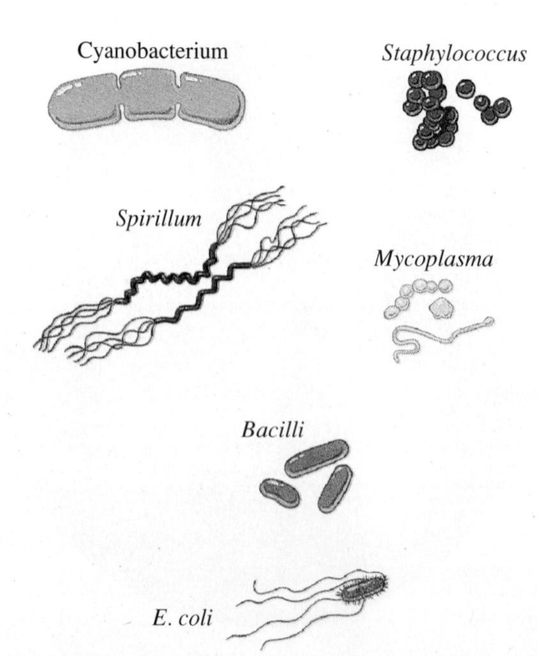

Figure 1·17
Representative types of bacteria. (Illustrator: Lisa Shoemaker.)

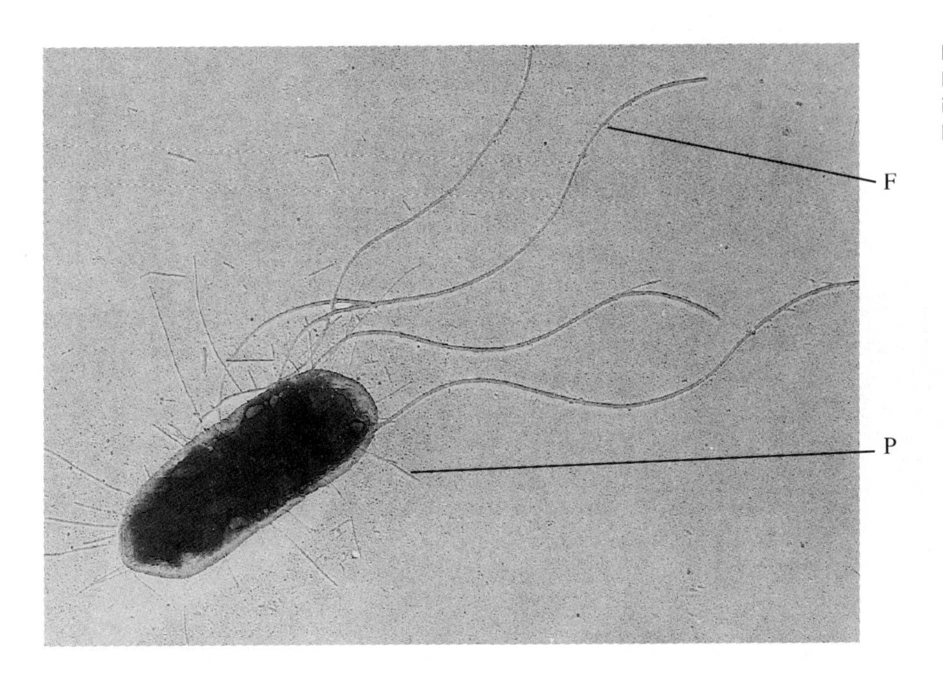

Figure 1·18
Electron micrograph of an *E. coli* cell, showing flagella (F) and pili (P). (Courtesy of Howard Berg.)

imbalance can result in high internal osmotic pressure. Without the cell wall, a bacterial cell subjected to such osmotic imbalance would swell and burst.

Underneath the cell wall, bacterial cells, like all cells, are surrounded by a *plasma membrane* composed of a lipid bilayer with embedded proteins. This membrane is selectively permeable, and some of the proteins contained within it mediate the passage of certain molecules and ions into and out of cells. The plasma membrane of many bacteria also contains electron-transport proteins that are involved in energy production. In certain bacteria, the plasma membrane is highly convoluted and extends into the interior of the cell; in photosynthetic bacteria, this convoluted membrane contains the pigments that absorb light energy during photosynthesis.

The part of the *E. coli* cell enclosed by the plasma membrane, exclusive of the genetic material in the particular region of the cell termed the nucleoid region, is called the *cytoplasm*. The *cytosol*, or cell sap, is the fluid portion of the cytoplasm. It contains about 20% protein by weight, in addition to many small molecules and ions. A sense of the diversity of smaller biomolecules (MW of 1000 or less) contained in an *E. coli* cell can be obtained from Table 1·2. Among these smaller biomolecules are amino acids, nucleotides, and fatty acids—the building blocks of proteins, nucleic acids, and lipids, respectively.

Table 1·2 Number of types of small molecules present in an *E. coli* cell whose sole source of carbon is glucose

Molecular category	Number of types
Amino acids and their precursors and derivatives	120
Nucleotides and their precursors and derivatives	100
Fatty acids and their precursors	50
Carbohydrates	250
Vitamins, coenzymes, isoprenoids, porphyrins, and their precursors	300

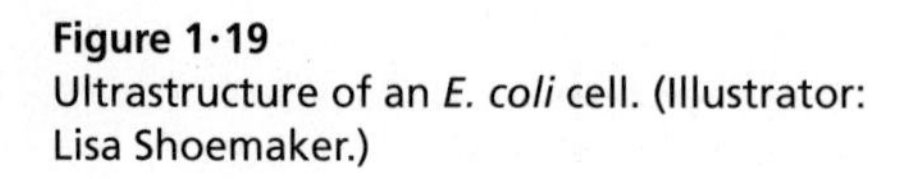

Figure 1·19
Ultrastructure of an *E. coli* cell. (Illustrator: Lisa Shoemaker.)

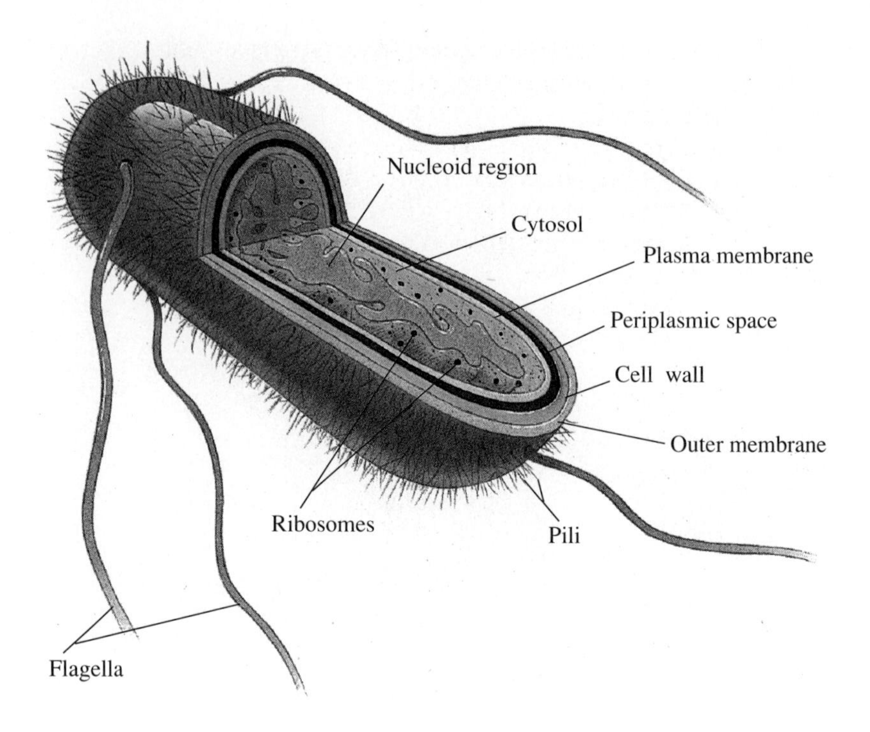

In addition to these smaller molecules, an *E. coli* cell contains some 15 000 ribosomes and several hundred thousand enzyme molecules. Many of the biochemical reactions catalyzed by these enzymes are common to all living cells. The cytosol also contains mRNA and tRNA molecules that, along with ribosomes, are essential in protein synthesis.

The principal genetic material in *E. coli* is a large, circular DNA molecule that contains about 2000 genes. It is localized in the nucleoid region (Figure 1·19).

B. Eukaryotic Cells Contain Membrane-Bounded Organelles

Eukaryotic cells are usually much larger than prokaryotic cells. One of the chief characteristics of eukaryotic cells is the presence of *organelles,* membrane-bounded compartments. For example, the *nucleus,* which consists of the genetic material, is surrounded by a double-membrane structure called the *nuclear envelope.* Most eukaryotic cells also contain internal membrane complexes called the *endoplasmic reticulum* (ER). This complex of internal tubules and sheets is the site of many specific biochemical reactions. The ER also plays an important role in sorting and secreting proteins. Near the nucleus, the ER merges with the outer membrane of the nuclear envelope. The aqueous region enclosed by the membranes of the ER is known as the *lumen.*

Most, but not all, eukaryotes contain *mitochondria,* the sites of most of the energy production in aerobic eukaryotic cells (Figure 1·20). A mitochondrion, like a nucleus, has a double membrane. The outer mitochondrial membrane surrounds the organelle, and the inner mitochondrial membrane is folded, increasing the surface area. The aqueous phase enclosed by the inner membrane is called the *matrix.* It contains many of the enzymes involved in aerobic energy metabolism. As carbohydrates, fatty acids, and amino acids are oxidized to carbon dioxide and water in the mitochondria, the released energy is conserved by phosphorylation of ADP to produce the energy-rich molecule ATP. In turn, ATP can be used by the cell for such *endergonic* (energy-requiring) processes as biosynthesis, active transport of certain molecules and ions, and muscle contraction. The number of mitochondria found in cells varies considerably. Whereas some eukaryotic cells may contain only a few, others may have thousands.

Mitochondria are derived from bacteria that entered into a symbiotic relationship with primitive eukaryotic cells over one billion years ago. In modern organisms, mitochondria cannot exist outside the eukaryotic cell, but they still contain vestiges of their bacterial origin. For example, the mitochondrial matrix contains DNA and ribosomes, which function in the synthesis of some of the proteins required by the organelle. A few species of eukaryotes lack mitochondria and must incorporate bacteria into their cytoplasm in order to carry out efficient metabolism.

(a)

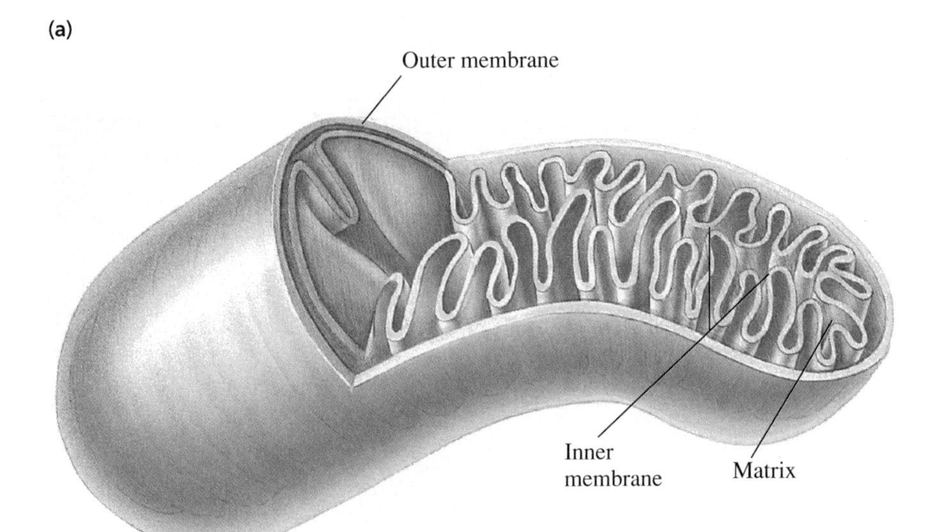

(b)

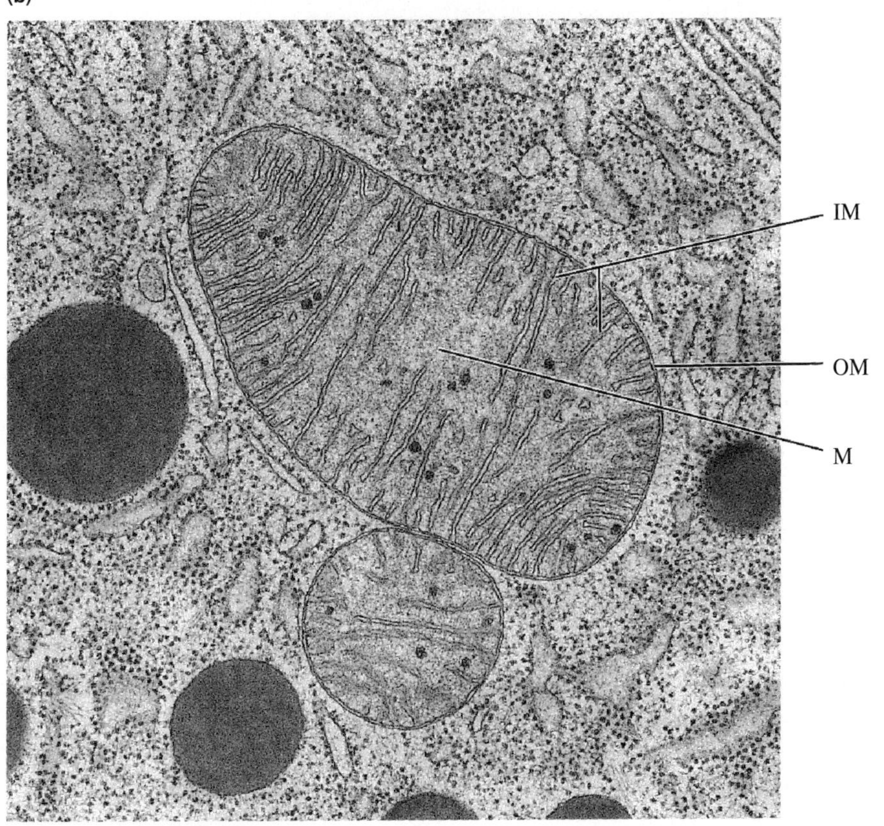

Figure 1·20
Mitochondrion. Mitochondria are the principal sites of energy production in aerobic eukaryotic cells. Carbohydrates, fatty acids, and amino acids are oxidized in this organelle. **(a)** Illustration. (Illustrator: Lisa Shoemaker.) **(b)** Electron micrograph of mitochondria from a bat pancreas cell. Visible are the outer membrane (OM), which surrounds the mitochondrion, and the highly folded inner membrane (IM), which protrudes into the matrix (M). (Courtesy of Keith R. Porter.)

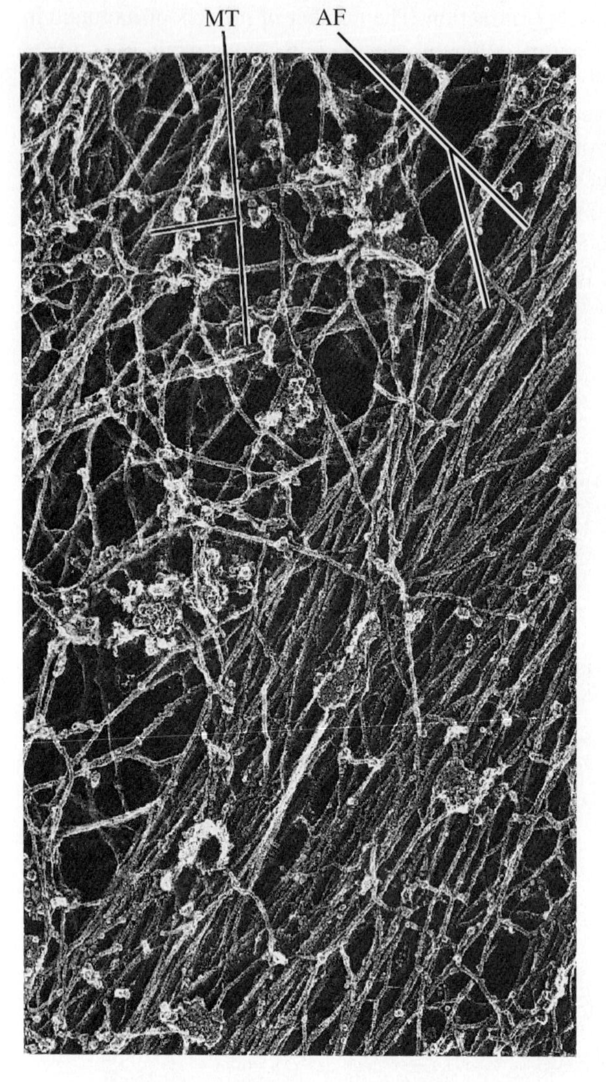

Figure 1·21
Electron micrograph of the cytoskeleton of a fibroblast cell. This protein scaffold consists of a network of microtubules, actin filaments, and intermediate filaments. Note the thicker microtubules (MT) and bundles of actin filaments (AF). (Courtesy of John E. Heuser and M. Kirschner.)

In many species, the ER is closely associated with the *Golgi apparatus,* a membrane system composed of flattened, fluid-filled sacs. The cell also contains many membrane vesicles that bud off from the the ER and eventually fuse with the Golgi. Additional vesicles are given off from the Golgi, and these fuse with other internal membrane-bounded compartments such as lysosomes or peroxisomes.

Lysosomes (Greek, *lysis,* dissolution) have acidic (low pH) interiors containing a variety of enzymes that catalyze the breakdown of cellular molecules, such as proteins and nucleic acids, and that digest particles, such as bacteria, ingested by the cell. *Peroxisomes* carry out oxidation reactions that produce toxic hydrogen peroxide (H_2O_2). Peroxisomes contain the enzyme catalase, which catalyzes the breakdown of H_2O_2 to water and O_2, thereby destroying toxic H_2O_2 as it is generated. The sequestering of peroxide-generating metabolism in specialized organelles protects the rest of the cell from toxic by-products of oxidation, an excellent example of the value of intracellular compartmentalization.

The shape of a eukaryotic cell is maintained by the *cytoskeleton,* a protein scaffold that consists of a network of microtubules, actin filaments, and intermediate filaments (Figure 1·21). In addition to providing structure, cytoskeletal components are involved in changes in the shape of a cell, the movement of organelles within a cell, and mitosis and meiosis. A drawing of a typical animal cell is shown in Figure 1·22.

(a)

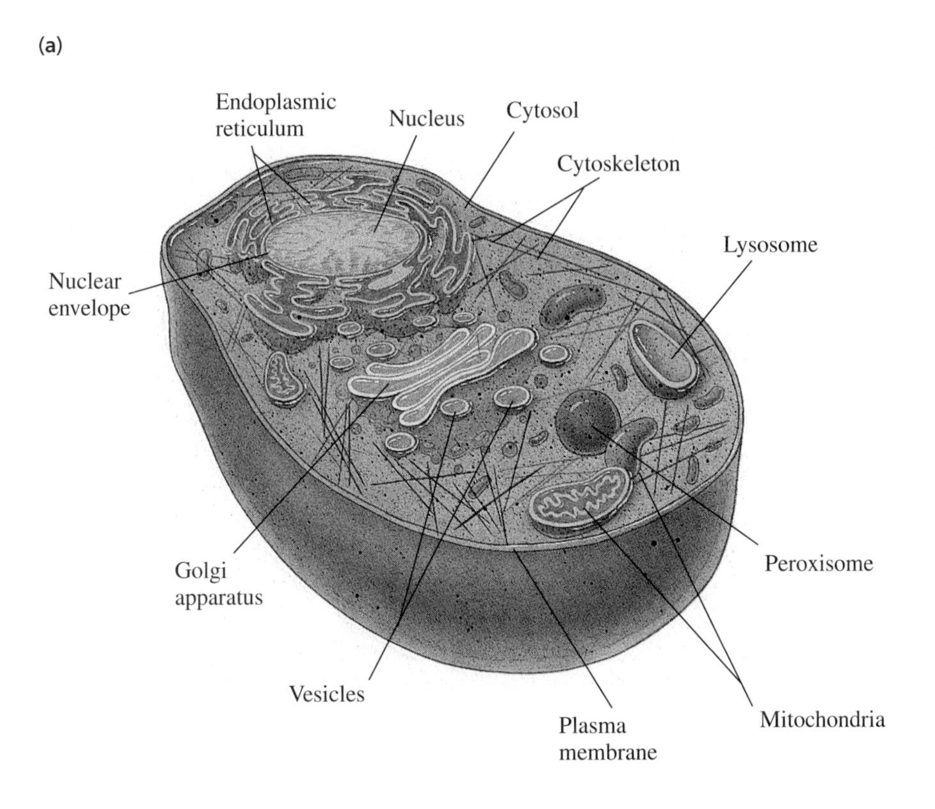

(b)

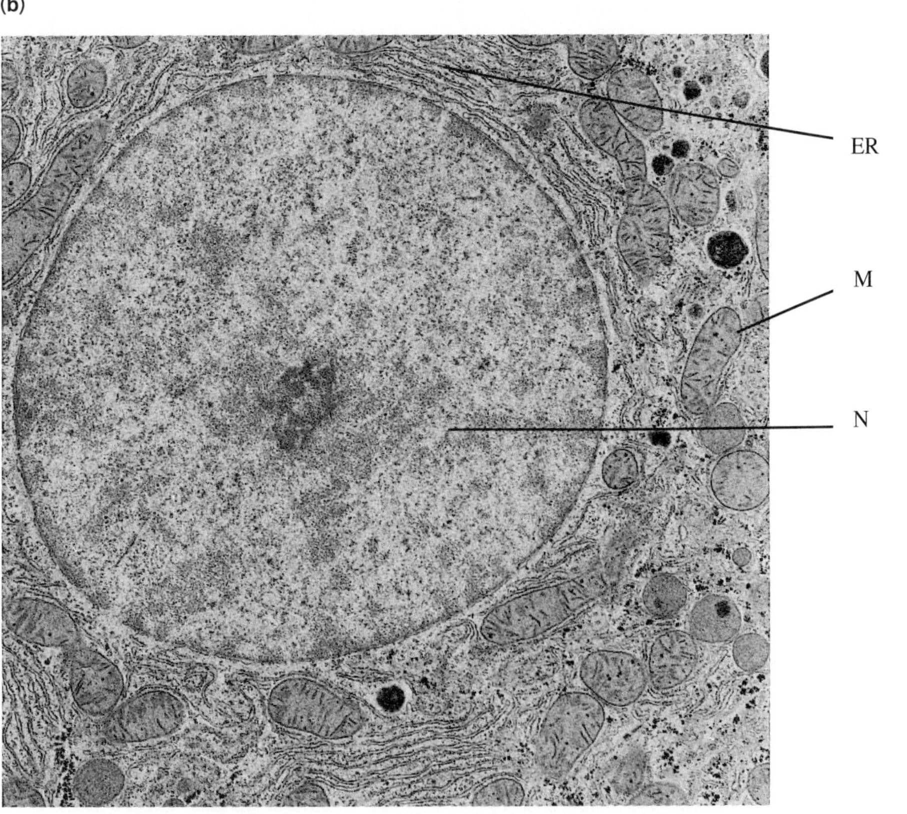

Figure 1·22
Eukaryotic cell (animal). (**a**) Illustration. Eukaryotic cells are much larger and much more complex than prokaryotic cells. (Illustrator: Lisa Shoemaker.) (**b**) Electron micrograph of a portion of a rat liver cell. The major ultrastructural features shown are: nucleus (N), mitochondria (M), and endoplasmic reticulum (ER). (Courtesy of Keith R. Porter.)

Plant cells are similar to bacteria in that they usually have a rigid cell wall. In flowering plants, this wall is composed primarily of cellulose. The chambered structure of cork observed by Robert Hooke was composed of cell walls with the cellular interiors gone.

Plant cells may also contain *chloroplasts,* the sites of photosynthesis. Chloroplasts are surrounded by a double membrane that encloses an aqueous region called the *stroma.* The chloroplast also contains a network of internal membranes, called *thylakoid membranes,* which are highly folded into a network of flattened, saclike vesicles. Stacks of these vesicles, known as *grana,* are present in the chloroplasts of many plants (Figure 1·23).

Chloroplasts have evolved from photosynthetic bacteria that invaded primitive eukaryotic cells. This event probably occurred more recently than the capture of mitochondria; as a result, chloroplasts often contain large amounts of genetic material that encodes many proteins. The chloroplasts of modern plants and algae carry out photosynthetic reactions similar to those in cyanobacteria, such as *Scytonema.* The products of these reactions are closely integrated with the metabolism of the rest of the cell.

(a)

Outer membrane

Inner membrane

Stroma

Thylakoid membrane

Grana

(b)

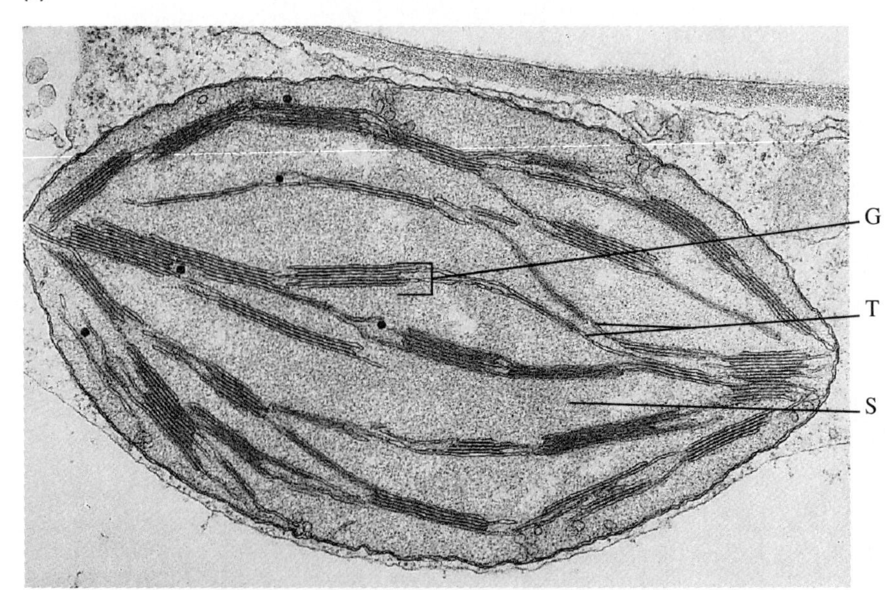

Figure 1·23
Chloroplast. Chloroplasts are the sites of photosynthesis in green plants. Light energy is captured by pigments embedded in the thylakoid membrane and is utilized to reduce carbon dioxide to simple carbohydrates. **(a)** Illustration. (Illustrator: Lisa Shoemaker.) **(b)** Electron micrograph of a chloroplast from a spinach leaf cell. The major features shown are: grana (G), thylakoid membrane (T), and stroma (S). (Courtesy of A. D. Greenwood.)

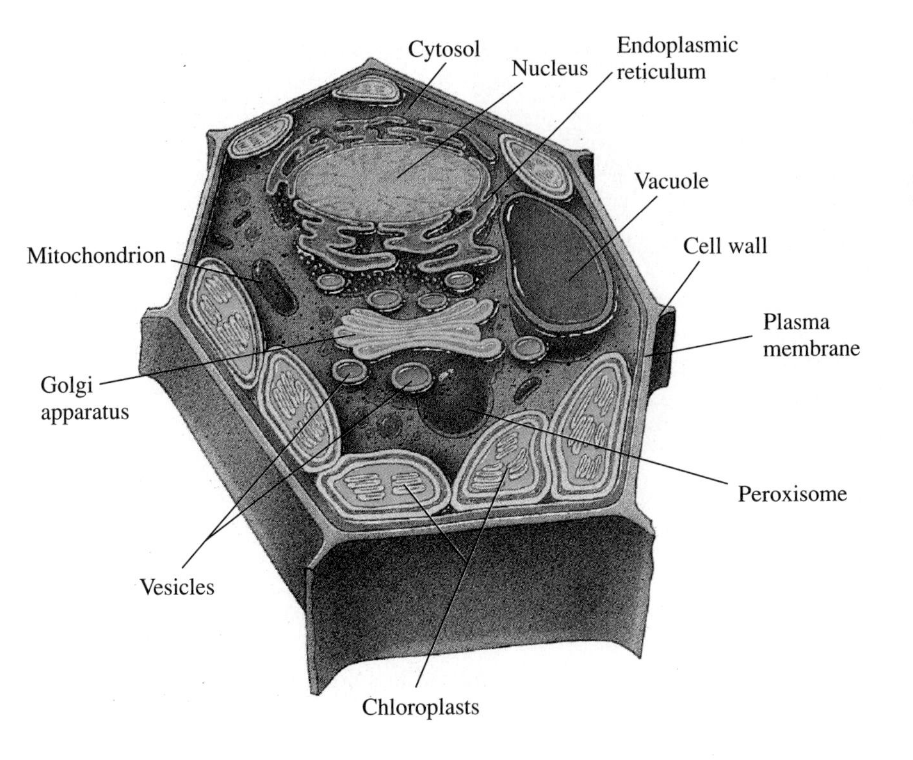

Figure 1·24
Plant cell. Unique features of plant cells include chloroplasts, the sites of photosynthesis in green plants; rigid cell walls composed of cellulose; and vacuoles, large, fluid-filled spaces containing solutes and cellular wastes. (Illustrator: Lisa Shoemaker.)

Electron micrographs of mature plant cells often reveal large, nonstaining areas called *vacuoles,* which are surrounded by a single lipid bilayer. These fluid-filled organelles, which are related to lysosomes, appear to be storage sites for water-soluble plant pigments, salts, carbohydrates, proteins, and metabolic waste products. Enlargement of these fluid-filled organelles is directly related to enlargement of the cell itself. A drawing of a typical cell from a flowering plant is shown in Figure 1·24.

Box 1·2 Infections by Viruses Are Models for Certain Biochemical Processes in Living Organisms

In its simplest form, a virus consists of a nucleic acid molecule surrounded by a protein coat. Viruses lack cellular structure and cannot carry out independent metabolic reactions, but they are able to use the metabolic and genetic machinery of living cells for their own propagation. Thus, viruses may be regarded as genetic parasites; they are not free living organisms. Although viruses are usually not among the organisms included in biological classification systems, they serve as models for the study of biochemical processes in living organisms.

Viruses contain a relatively small number of genes. The size of the viral genome varies with the type of virus, but it typically contains between 3 and 100 genes. Unlike the genomes of living organisms, which are composed exclusively of DNA, some viral genomes are composed of RNA.

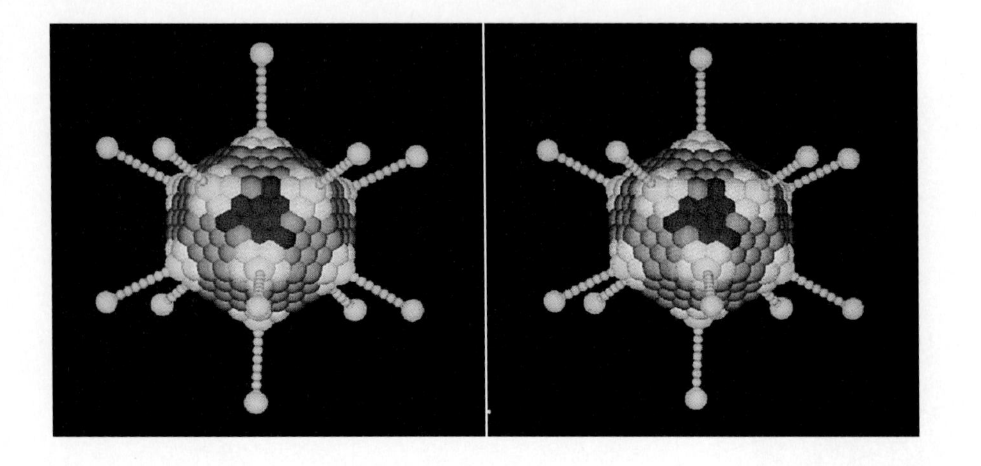

Box 1·2
Stereo S1
Adenovirus. Adenoviruses cause acute respiratory disease, pharyngitis, and conjunctivitis. Note the protein spikes that project from the capsid. (Stereo image by Richard J. Feldmann.)

The viral nucleic acid is wrapped in a protein coat, or capsid, which is composed of multiple copies of one or more types of proteins. In addition, the capsid of many viruses contains protein spikes that project from its surface (Stereo S1). These spikes are important in the recognition of host cells and in the transfer of the viral genome into host cells. The capsids of some more complex viruses are surrounded by a membrane envelope composed of membrane lipids and virus-specific membrane-embedded proteins (Figure 1). This organization is more characteristic of viruses that infect eukaryotic cells; viruses that infect prokaryotic cells rarely have membrane envelopes.

The structure, assembly, and genetics of viruses are subjects of intense study, in part because they serve as models that are used to understand living organisms. Studies of the genetics of bacteriophages, or simply phages (Greek, *phagos,* one that eats), which are viruses that attack bacteria, have provided valuable insights into gene expression and gene regulation in bacteria. Similarly, viruses that infect eukaryotes have been useful as experimental models for eukaryotic genetic mechanisms. The structures of viruses such as that shown in Stereo S1 provide us with useful examples of the assembly and organization of complex biological structures.

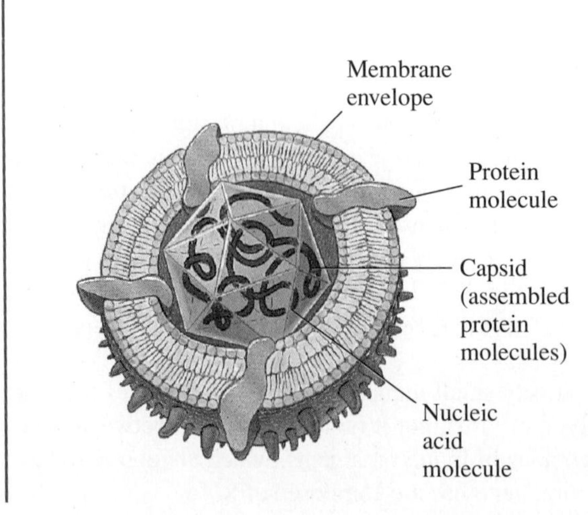

Box 1·2
Figure 1
Example of a virus with a membrane envelope. (Illustrator: Lisa Shoemaker.)

As causative agents of disease in plants and animals, viruses are of immense agricultural and medical importance. Smallpox, chicken pox, and certain tumors are caused by specific DNA-containing viruses, and poliomyelitis, certain leukemias and myelomas, a variety of plant diseases, and acquired immune deficiency syndrome (AIDS) are caused by specific RNA-containing viruses. The name virus (Latin for poison) signifies the action of these genetic parasites only too well.

C. Molecules, Organelles, and Cells Vary Greatly in Size

An understanding of the relative sizes of biomolecules and cells is important in the study of biochemistry. Table 1·3 lists units of length that are used to express cellular and macromolecular dimensions, and Figure 1·25 (next two pages) contrasts the sizes of an *E. coli* cell, a bacterial ribosome, and various constituents of a eukaryotic cell. A typical eukaryotic cell has a diameter of 25 μm (or 25 000 nm). Completely surrounding the cell is the plasma membrane, which is only approximately 6.0 nm thick; yet this seemingly fragile structure separates the living cell from its environment. Water (the most abundant biomolecule) has a molecular weight of 18 and a diameter of about 0.4 nm. Amino acids, with an average molecular weight of about 130, range in size from about 0.6 to 1.2 nm. Hemoglobin, a protein in red blood cells that transports oxygen from the lungs to the rest of the body, consists of 574 amino acid residues; it has a molecular weight of about 64 000 and a diameter of 6.4 nm. A double-stranded DNA molecule is approximately 2.4 nm in diameter, whereas ribosomes, composed of some 70 different proteins and several molecules of RNA, have a diameter of about 25 nm. The much larger mitochondrion has a diameter of 500 nm and is similar in size to an *E. coli* cell. A chloroplast is some three to five times longer than a mitochondrion. The nucleus can range in size from about the size of a chloroplast to many times larger.

Table 1·3 Units of length commonly used in biochemistry

SI Units* (Système International d'Unités)

Name	Symbol	Submultiple
millimeter	mm	10^{-3} m
micrometer	μm	10^{-6} m
nanometer	nm	10^{-9} m

*Basic unit of length is the meter (m).

Older units still in use

Name	Symbol	SI equivalent
micron	μ	micrometer (μm)
millimicron	mμ	nanometer (nm)
angstrom	Å	10^{-1} nanometer (0.1 nm)

Figure 1·25
Scale drawings of *E. coli,* a bacterial ribosome, and various constituents of a eukaryotic cell. The scale of the images on the left page is 100 times that on the right page. A ribosome has been included on both pages to provide a frame of reference. Note the width of the plasma membrane relative to the width of the ribosome on the left page. Then note the width of the ribosome on the right page and imagine the much thinner plasma membrane in relation to structures such as the mitochondrion and chloroplast. (Illustrator: Lisa Shoemaker.)

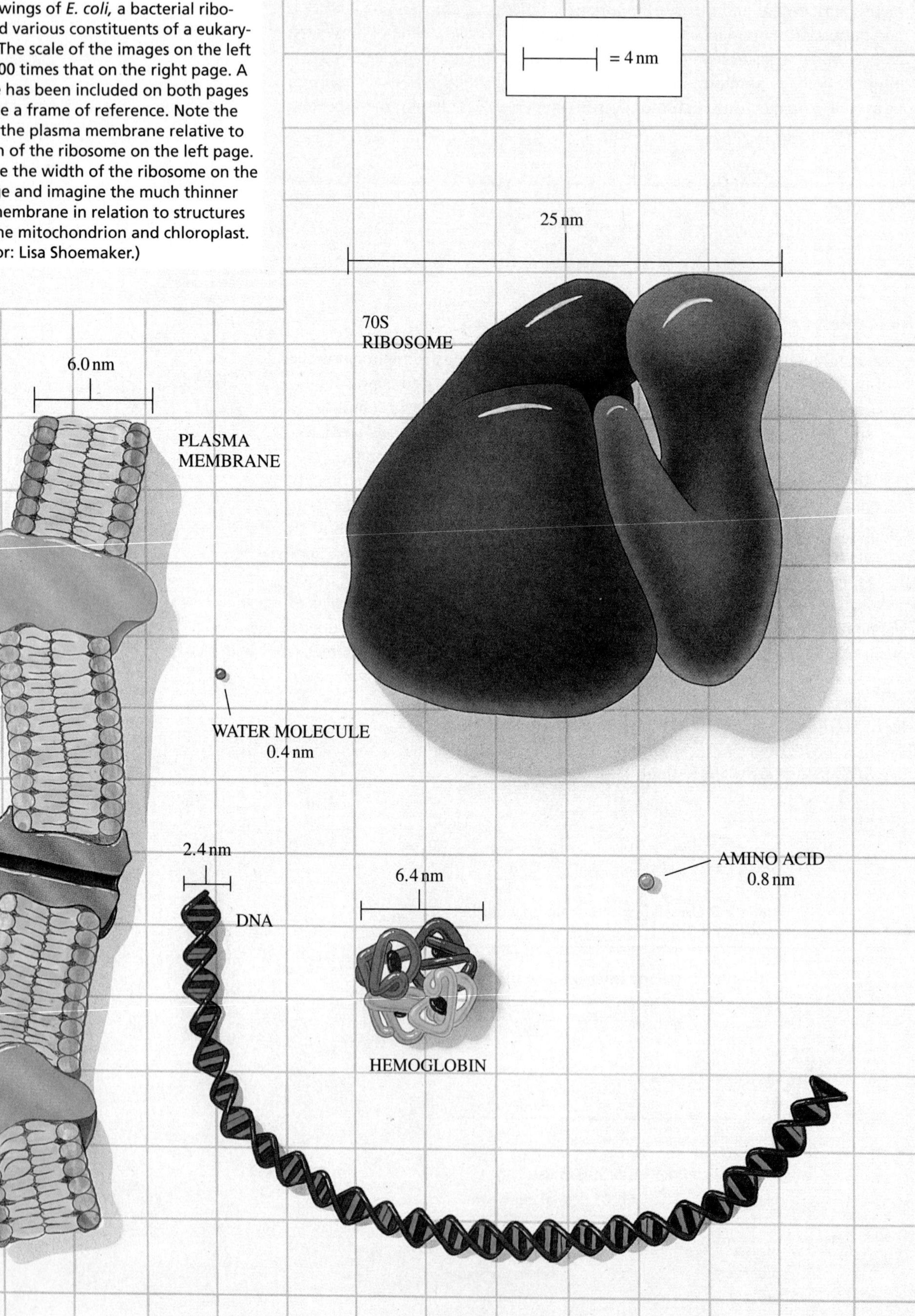

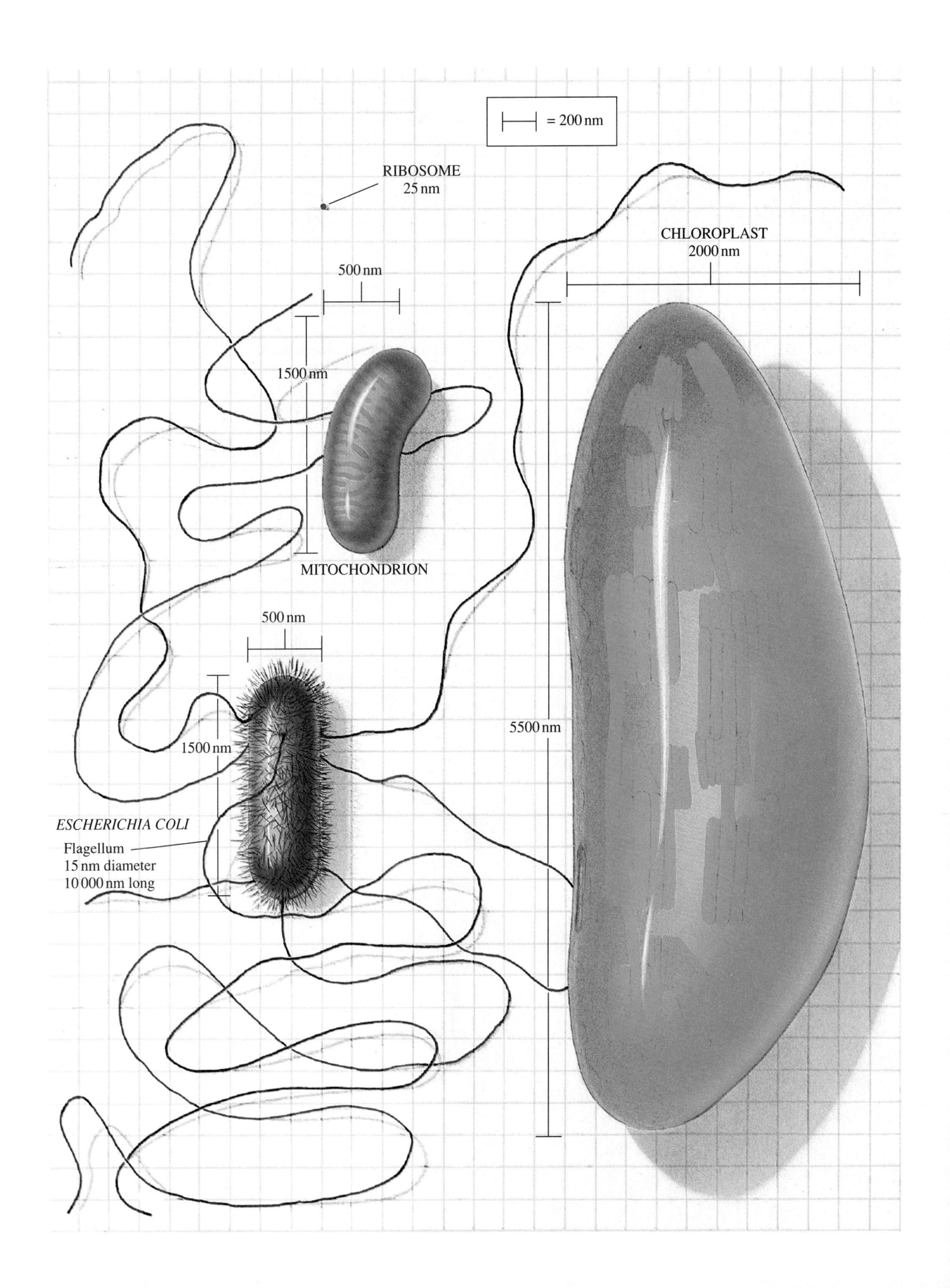

= 200 nm
RIBOSOME
25 nm
CHLOROPLAST
2000 nm
500 nm
1500 nm
MITOCHONDRION
500 nm
5500 nm
1500 nm
ESCHERICHIA COLI
Flagellum
15 nm diameter
10 000 nm long

1·6 All Organisms Have Evolved from an Ancient Ancestor

The idea that information is passed on from one generation to the next implies that progeny inherit exact copies of the genes of their parents. This may be true in the short term but not on the scale of millions of years. Genetic information can be altered by mutation, and new mutations can become fixed in a population through natural selection and genetic drift.

Gradually, populations of similar organisms diverge from each other and new species evolve. Biological evolution can be traced through the fossil record or by directly comparing the sequences of genes and proteins. These observations suggest that all of the millions of species that exist today have descended from a single ancestor that lived several billion years ago. This ancient ancestral cell was undoubtedly capable of glycolysis (the breakdown of glucose) and many of the other fundamental biochemical processes that are common to all cells. It could synthesize amino acids and lipids and almost certainly used ATP as the fundamental unit of energy. It used the same genetic code that we find in its modern descendants. How the ancestral cell evolved from simpler organisms is an unsolved problem. The origin of life itself, an event that occurred more than three billion years ago, is the subject of much speculation.

The study of living things is a huge task—the diversity of nature is overwhelming. Yet even a child recognizes the tiger in the house cat, the wolf in the dog, the eagle in the sparrow. Biological classification represents a search, in the midst of nature's diversity, for its underlying unity. There is general agreement on the placement of organisms into groups based on shared characteristics. One of the largest groups is the phylum; organisms within a particular phylum are descended from an ancestor that lived hundreds of millions of years ago. We humans are placed in the phylum Chordata, which includes all other vertebrates, as well as additional species that have a notochord, a stiff supportive rod. Evolution and divergence gave rise to further groups within phyla. These are referred to as classes, orders, families, and genera, with each subdivision representing an increasing number of shared characteristics. Organisms that diverged relatively recently are classified within the same genus but are divided into different species. A complete classification of our species is shown in Table 1·4.

Table 1·4 Classification of humans in a common taxonomic system

Category	Taxon	Characteristics
Kingdom	Animalia	Multicellular organisms requiring organic plant and animal substances for food
Phylum	Chordata	Animals with notochord; dorsal hollow nerve cord; gill slits in pharynx at some stage of life history
Subphylum	Vertebrata	Spinal cord enclosed in a vertebral column; body basically segmented; brain enclosed by skull
Superclass	Tetrapoda	Land vertebrates; four limbs
Class	Mammalia	Young nourished by milk glands; breathing by lungs; skin with hair or fur; body cavity divided by diaphragm; red blood cells without nuclei; constant body temperature
Order	Primates	Tree dwellers or their descendants; usually with fingers and flat nails; reduced sense of smell
Family	Hominidae	Flat face, eyes forward; color vision; upright, bipedal locomotion; hands and feet with different specialization
Genus	*Homo*	Large brain; speech; long childhood
Species	*Homo sapiens*	Prominent chin; high forehead; sparse body hair

Groups of related phyla are termed kingdoms, which have been presumed to represent the most fundamental and ancient distinctions between organisms. The popular five-kingdom classification system described in most introductory biology books places prokaryotes in a single kingdom called Monera. Eukaryotes are divided into four kingdoms: Fungi, Plantae, Animalia, and Protista. (Protists are a diverse group of eukaryotic organisms, mostly unicellular, that do not fall within the other three eukaryotic kingdoms.)

Recently, molecular data clarified the relationships of many groups of bacteria that were difficult to classify on the basis of morphology alone. This information has also led to increased understanding of the relationships between eukaryotes and prokaryotes and has led to a revised classification. A classification scheme promoted by Carl Woese and his co-workers divides organisms at a level above kingdoms into three phylogenetic domains that are the most evolutionarily divergent. Within the three domains, organisms are divided into kingdoms, which do not completely correspond to those of the current five-kingdom system. On a molecular level, all eukaryotes appear to be quite similar; therefore, the eukaryotic kingdoms are placed within a single domain, with the suggested name Eucarya. The biochemistry of prokaryotes indicates that they are a much more diverse group of organisms than eukaryotes. They can be divided into two groups, with a domain proposed for each: Eubacteria, consisting of the common bacteria, including those species that gave rise to mitochondria and chloroplasts within eukaryotic cells; and Archaea, prokaryotes that live in what appear to be primeval environments. The proposed domain Archaea includes the methanogens (methane producers), extreme thermophiles (Greek, *therme,* heat; *philos,* lover), and extreme halophiles (Greek, *hals,* salt). The name "archaea" is proposed as a more precise term than the current name for such organisms, "archaebacteria," to avoid the perception that these organisms are closely related to common bacteria. In fact, archaea appear to be more closely related, on a molecular level, to eukaryotes than to common bacteria. A comparison of the highest-level taxa of the two classification schemes is shown in Figure 1·26 (next page).

The three-domain system uses molecular data as the ordering principle for a scheme that accurately expresses evolutionary relationships, whereas the five-kingdom system is based upon homologies and differences at a considerably higher level of organization. The ability to determine evolutionary relationships from the interpretation of molecular data is one of the many exciting advances in biochemistry in the past decade, and the discussion of a revised system of biological classification is emblematic of the broad implications of biochemical research.

1·7 Biochemistry Is Also Holistic

One of the goals of biochemists is to integrate a large body of knowledge into a molecular explanation of life. This has been and continues to be a challenging task. Nevertheless, biochemists have made a great deal of progress toward defining the basic reactions that are common to all cells and how these reactions are interrelated. We shall describe a number of biochemical pathways in later chapters. In many cases, it will become obvious that the pathway, not the individual reactions, is the functional unit.

One of the characteristics of life is that its biochemistry consists of a highly regulated and controlled set of reactions. The activities of enzymes are modified according to conditions within the cell. The rate of a biochemical pathway can be affected by the concentration of the product, the substrate, or even molecules from a competing pathway.

2·1 The Water Molecule Is Polar

The structure of a molecule of water, H_2O, is not linear but V-shaped (Figure 2·1a). The angle between the two covalent O—H bonds is 104.5°. The angled shape of the water molecule allows it to form the intermolecular bonds that give water its unusual properties. Two orbitals of the oxygen atom participate in covalent bonds with the two hydrogen atoms, whereas each of the remaining two orbitals has an unshared electron pair (Figure 2·1b). Bonds from water to neighboring interacting atoms will be at two of the corners of a tetrahedron that surrounds the central atom of oxygen.

Oxygen attracts electrons within a covalent bond more strongly than does hydrogen; that is, it is more *electronegative* than hydrogen. As a result, an uneven distribution of charge occurs within each O—H bond of the water molecule, with oxygen bearing a partial negative charge and hydrogen bearing a partial positive charge. This uneven distribution of charge within a bond is known as a *dipole*, and the bond is said to be *polar*.

The polarity of a molecule depends both on the polarity of its covalent bonds and the geometry of the molecule. The nonlinear arrangement of the polar O—H bonds of water creates a permanent dipole for the molecule as a whole (Figure 2·2). Thus, even though it is electrically neutral, the water molecule is polar. Carbon dioxide also has polar covalent bonds, but the linear arrangement of the molecule makes CO_2 symmetric and causes the polarities to cancel. As a result, carbon dioxide is not polar.

2·2 The Water Molecule Forms Hydrogen Bonds

An important consequence of the polarity of the water molecule is the attraction of water molecules for one another. The attraction between one of the slightly positive hydrogen atoms of one water molecule and the slightly negative oxygen atom of

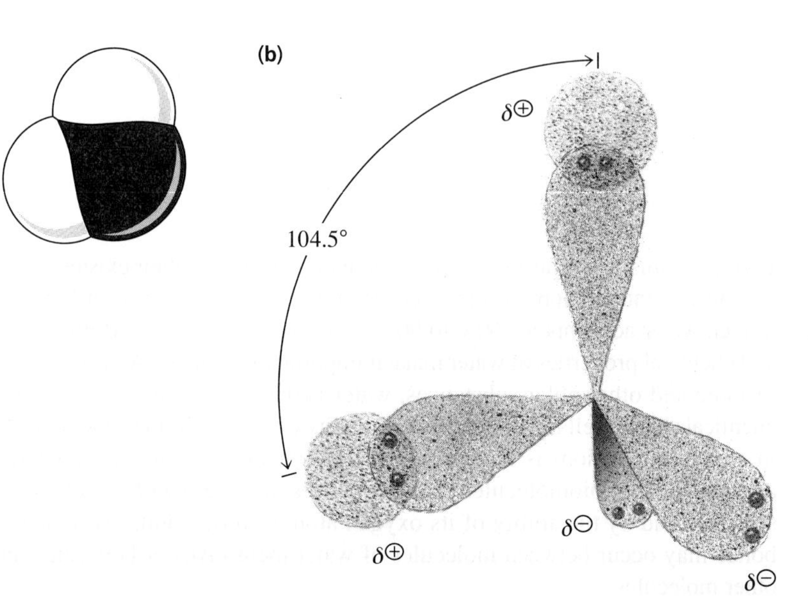

Figure 2·1
(a) Space-filling structure of a water molecule. **(b)** Angle between the covalent bonds of a water molecule. (Illustrator: Lisa Shoemaker.)

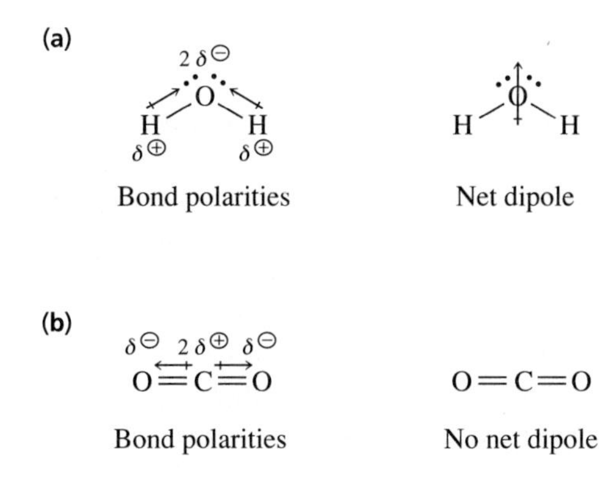

Figure 2·2
Polarity. **(a)** The geometry of a water molecule is such that the polar covalent bonds create a permanent dipole for the molecule as a whole, with the oxygen bearing a partial negative charge (symbolized by 2 δ⊖) and each hydrogen bearing a partial positive charge (symbolized by δ⊕). **(b)** Although a molecule of carbon dioxide also has two polar covalent bonds, its symmetric arrangement causes their polarities to cancel. As a result, CO_2 is not polar. (Arrows depicting dipoles point toward the negative charge, with the cross at the positive end.)

another produces a <u>*hydrogen bond*</u> (Figure 2·3). In a hydrogen bond between two water molecules, the hydrogen atom remains covalently bonded to its oxygen atom and lies almost twice as far from the oxygen atom that is the hydrogen acceptor atom.

Hydrogen bonds are much weaker than typical covalent bonds. For example, the energy required to break a hydrogen bond is about 3–6 kcal mol^{-1}, whereas the energy required to break a covalent O—H bond is about 110 kcal mol^{-1} and that required to break a covalent C—H bond is about 100 kcal mol^{-1}.

Orientation is important in hydrogen bonding. A hydrogen bond is most stable when the hydrogen atom and the two highly electronegative atoms associated with it (in the case of hydrogen bonding between two water molecules, the two oxygen atoms) are aligned or close to being in line.

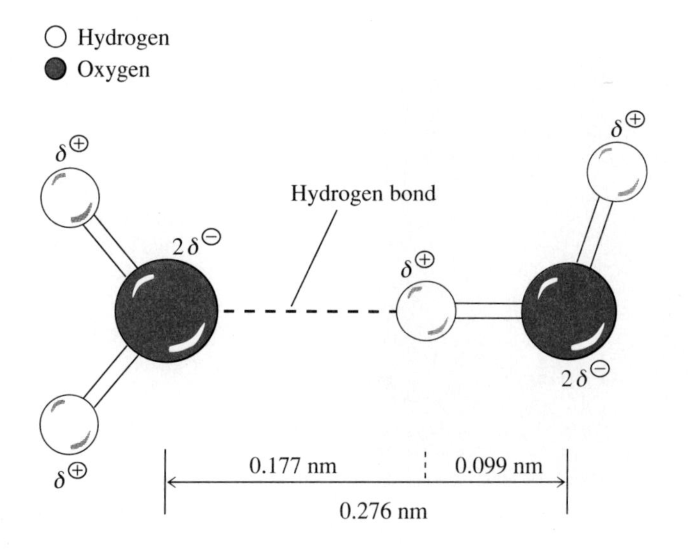

Figure 2·3
Hydrogen bonding between two water molecules. A partially positive (δ⊕) hydrogen atom of one water molecule attracts the partially negative (2 δ⊖) oxygen atom of a second water molecule, forming a hydrogen bond.

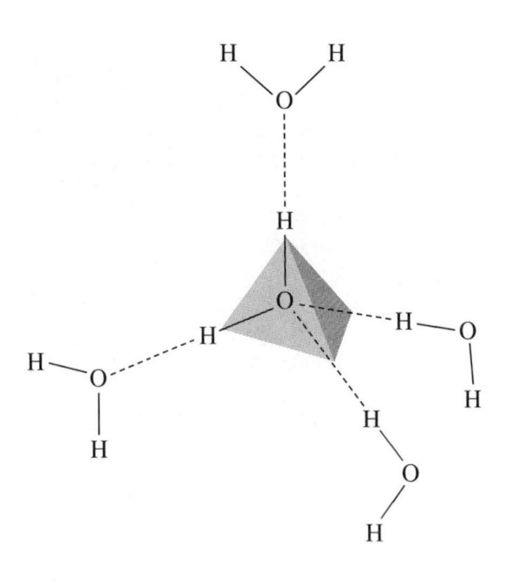

Figure 2·4
Hydrogen bonds of a water molecule. A water molecule can form up to four hydrogen bonds. The oxygen atom of a water molecule accepts two hydrogens, and each O—H group serves as a hydrogen donor. Hydrogen bonds are indicated by dashed lines.

The three-dimensional interactions of liquid water have been difficult to study, but much has been learned by examining the structure of ice crystals (Stereo S2·1). In the common form of ice, each molecule of water participates in four hydrogen bonds. Each water molecule is a hydrogen donor in two of these bonds and a hydrogen acceptor in the other two. The hydrogen bonds in ice are arranged tetrahedrally around the oxygen atom of each water molecule. The average energy required to break each hydrogen bond in ice has been estimated to be 5.5 kcal mol^{-1}.

Because of the many intermolecular interactions in ice, the melting point of ice is much higher than expected for a molecule of its molecular weight (MW = 18). In other words, a large amount of energy, in the form of heat, is required to disrupt the hydrogen-bonded lattice of ice. As ice melts, the majority of its hydrogen bonds are retained in liquid water, so it is presumed that a molecule of liquid water can also form up to four hydrogen bonds with its neighbors (Figure 2·4). The tetrahedral hydrogen-bonding pattern of ice is also retained in liquid water, but there is substantial disorder and rapid fluctuation of the structure. This accounts for the fluidity of liquid water compared to the rigidity of ice.

An important property of water derived from its hydrogen-bonding characteristics is its specific heat, the amount of heat needed to raise the temperature of 1 g of water by 1°C. A relatively large amount of heat is required to raise the temperature of water because each water molecule participates in multiple hydrogen bonds that must be broken in order for the kinetic energy of the water molecules to increase. Since most of the mass of large multicellular organisms is water, and smaller organisms live in water, temperature fluctuations within cells are thus minimized. This feature is of critical biological importance since the rates of most biochemical reactions are sensitive to temperature.

Another property of water derived from its structure is the heat of vaporization, which, like the specific heat of water, is much higher than that of many other liquids. As is the case with melting, a large amount of heat is required to evaporate water because hydrogen bonds must be broken to permit water molecules to dissociate from one another and enter the gas phase. Because the evaporation of water absorbs so much heat, perspiration is an effective mechanism for decreasing body temperature. These unusual thermal properties of water make it an excellent environment for living organisms, as well as an excellent medium for the chemical processes of life.

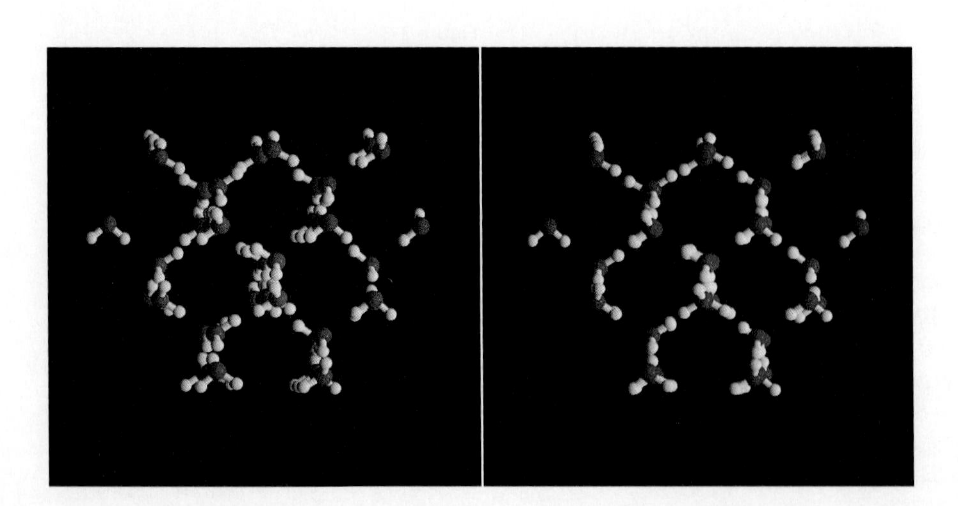

Stereo S2·1
Structure of ice. In ice, water molecules form an open, hexagonal lattice in which every water molecule is hydrogen bonded to four others. Color key: hydrogen, blue; oxygen, red. (Courtesy of National Institutes of Health.)

The density of most substances increases upon freezing. As molecular motion slows, tightly packed crystals form. The density of water also increases as it cools—until it reaches a maximum at 4°C (277 K). However, below 4°C, water expands as it cools. This expansion is caused by the formation of the more open hydrogen-bonded ice crystal in which each water molecule is hydrogen bonded rigidly to four others (Stereo S2·1). As a result, ice with its open lattice is less dense than liquid water, whose molecules can move enough to pack more closely. Because ice is less dense than liquid water, ice floats, and water freezes from the top down. A layer of ice on a pond thus serves as an insulator that protects the creatures below from extreme cold.

(a)

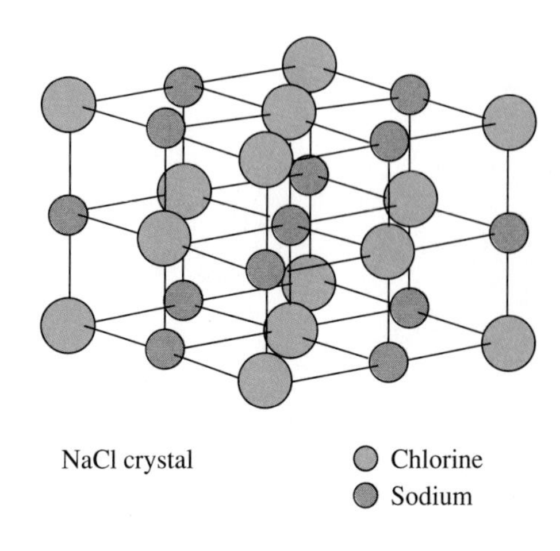

(b)

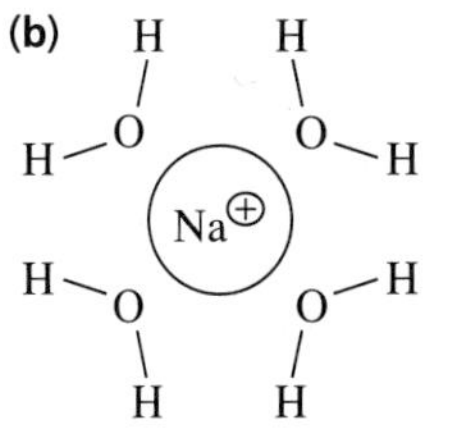

Figure 2·5
Dissolution of sodium chloride (NaCl) in water. **(a)** In the solid state, sodium chloride is a crystalline array held together by electrostatic forces. **(b)** When NaCl is placed in water, the forces between the positive and negative ions that dissociate from the crystal are weakened by water, and these ions are not attracted back to the surface of the crystal. As a result, the crystal dissolves. Each $Na^{\oplus}$ in aqueous solution is surrounded by a shell of solvent molecules called a solvation sphere. Actually, there are several layers of water molecules in the solvation sphere.

2·3 Ionic and Polar Substances Readily Dissolve in Water

Because of its polarity, water readily interacts with ionic species and with other polar substances. This interaction diminishes electrostatic interactions between the ions or molecules of the substance. For example, when sodium chloride (NaCl) is dissolved in water, the lattice structure of the crystal (Figure 2·5a) is disrupted. Because the many polar water molecules compete with the relatively few sodium and chloride ions, the opposite electric charges of $Na^{\oplus}$ and $Cl^{\ominus}$ are very much weaker in water than they are in the intact crystal. As a result, the sodium and chloride ions dissociate from one another. Each $Na^{\oplus}$ attracts the negative ends of several water molecules (Figure 2·5b), whereas each $Cl^{\ominus}$ may attract the positive ends of several water molecules. *Cations* (positive ions), usually being smaller, attract more water molecules than do *anions* (negative ions). The shell of water molecules that surrounds each ion is called a solvation sphere. A molecule or ion surrounded by solvent molecules is said to be *solvated*. When the solvent is water, such molecules or ions are said to be *hydrated*.

The solubility of molecules is determined chiefly by their polarity and their ability to form hydrogen bonds with water. Nonionic organic molecules with polar functional groups, such as short-chain alcohols and carbohydrates, are very soluble in water. Such molecules disperse among the water molecules, with which their polar functional groups form intermolecular hydrogen bonds. Substances that readily dissolve in water, including ionic substances such as sodium chloride and polar molecules such as glucose and table sugar (sucrose), are said to be *hydrophilic* (Greek, *hydōr,* water; *philos,* loving).

2·4 Nonpolar Substances Are Insoluble in Water

In contrast to hydrophilic substances, hydrocarbons, such as octane (from gasoline), and other nonpolar substances have very low solubility in water because molecules of water tend to interact with other molecules of water rather than with nonpolar substances. As a result, water molecules tend to exclude nonpolar substances, forcing nonpolar molecules together. For example, oil tends to form a single droplet in water, thereby minimizing the area of the interface between the two substances. This phenomenon of exclusion of nonpolar substances by water is called the hydrophobic effect, and nonpolar molecules are said to be *hydrophobic* (Greek, *phobos,* fear). The hydrophobic effect is critical in the folding of proteins and the self-assembly of biological membranes.

Soaps, such as sodium palmitate (Figure 2·6a), are alkali metal salts of long-chain fatty acids. Soaps represent a class of substances that are both hydrophilic (the carboxylate anion of sodium palmitate) and hydrophobic (the hydrocarbon tail). Such molecules are said to be *amphipathic* (Greek, *amphi,* on both sides; *pathos,* passion). Amphipathic cleansing compounds, such as soaps, are generally termed *detergents.* One of the most commonly encountered synthetic detergents in biochemistry is sodium dodecyl sulfate (SDS) (Figure 2·6b). When a detergent is gently spread on the surface of water, an insoluble monolayer forms in which the hydrophobic, nonpolar tails of the detergent molecules extend into the air, whereas the hydrophilic, ionic heads are hydrated, extending into the water (Figure 2·7). When detergent is dispersed in water at a sufficiently high concentration rather than layered on the surface, groups of detergent molecules aggregate into *micelles.* In one common structure of micelles, the nonpolar tails of the detergent molecules associate with one another in the center of the structure. Such aggregation results in minimal contact with water molecules, so that the core of a micelle is a liquid hydrocarbon. The ionic heads project into the aqueous solution and are therefore hydrated. Small, compact micelles contain about 80–100 molecules of detergent. The cleansing action of soap and other detergents is accomplished by the entrapment of water-insoluble grease and oils within the hydrophobic interiors of micelles. Suspension of nonpolar compounds in water in this fashion is termed *solubilization.* As we shall discover in Chapter 4, the formation of micelles has a parallel in the folding of proteins, many of which have hydrophobic interiors and hydrophilic surfaces.

2·5 Noncovalent Interactions Are Important in Cells

Two types of noncovalent interactions, hydrogen bonds and hydrophobic interactions, have been introduced in this chapter. Such weak interactions play extremely important roles in the structures of biopolymers, in the recognition of one polymer by another, and in the binding of reactants to enzymes. The four major noncovalent bonds or forces are electrostatic interactions, hydrogen bonds, van der Waals and related forces, and hydrophobic interactions.

Electrostatic interactions between two charged particles are potentially the strongest noncovalent forces, but their strengths vary widely. The attractions between ions of opposite charge also can extend over greater distances than other attractions. The stabilization of NaCl crystals by ionic interactions is an example of electrostatic forces. Weaker electrostatic attractions are those between dipoles and other charges. Electrostatic forces are important in the recognition of a substrate by an enzyme. Most enzymes have either anionic or cationic sites to attract oppositely charged reactants. Charge-charge attractions between internal ionic regions of proteins are sometimes called salt linkages or salt bridges. However, these attractions are not as strong as those in salt crystals. The most accurate term for such attractions is *ion pairs.* The strength of an electrostatic interaction depends on the nature of the solvent. Water, as noted in Section 2·3, greatly weakens these interactions. As a result, strong electrostatic forces do not play a major role in the stability of biological polymers.

Hydrogen bonds, which are a type of electrostatic interaction, are among the strongest of common noncovalent bonds in biological systems. The energy required to break a hydrogen bond is usually about 3–6 kcal mol^{-1}, as noted in Section 2·2. The strength of hydrogen bonds, intermediate between covalent bonds and ionic bonds, allows them to be quite stable but break readily. In general, a hydrogen bond can form when a hydrogen atom that is covalently bonded to a strongly electronegative atom, such as nitrogen or oxygen, lies approximately

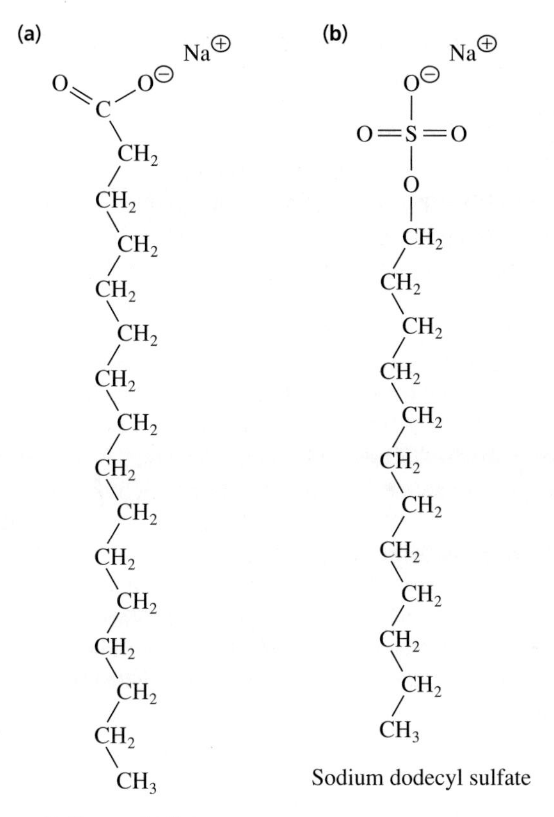

Figure 2·6
(**a**) Sodium palmitate, a soap. (**b**) Sodium dodecyl sulfate (SDS), a synthetic detergent.

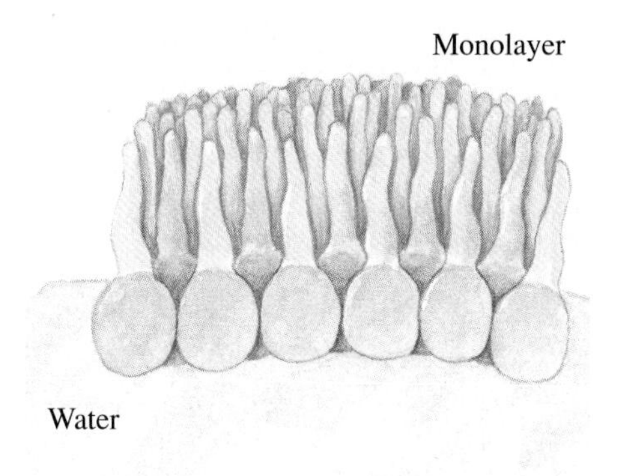

Figure 2·7
Cross-sectional views of structures formed by soaps and other detergents in water. Detergents can form monolayers at the air-water interface. They can also form micelles, which are aggregates of the detergents in which the hydrocarbon tails associate away from water. (Illustrator: Michael R. Dulude.)

0.2 nm from another strongly electronegative atom that has an unshared electron pair (Figure 2·8). In contrast, a C—H bond is expected to be insufficiently polar for the hydrogen atom to form a hydrogen bond with an electronegative atom. However, a few weak C—H------O and C—H------N hydrogen bonds have been found in proteins. As we shall see, hydrogen bonds are common within (*intra*molecular) and between (*inter*molecular) biological macromolecules such as proteins and nucleic acids (Chapters 4 and 10, respectively). For example, complementary base pairing within DNA molecules depends on precise hydrogen bonding (Figure 2·9).

Weaker electrostatic forces also exist. These include the interactions between the dipoles of two uncharged polarized bonds and the interactions between a dipole and a transient dipole induced in a neighboring molecule. These forces are of short range and low magnitude, about 0.3 kcal mol^{-1} and 0.2 kcal mol^{-1}, respectively.

Weak intermolecular forces are produced between all neutral atoms by transient electrostatic interactions. These *van der Waals forces*, named after the Dutch physicist Johannes Diderick van der Waals who discovered them, occur only when atoms are very close together. They originate from the infinitesimal dipole generated in atoms by the movement of negatively charged electrons about the positively charged nucleus. Thus, van der Waals forces are electrostatic attractions of the nucleus of one atom or molecule for the electrons of the other. The attractive interactions between these transient dipoles are called London dispersion forces. The

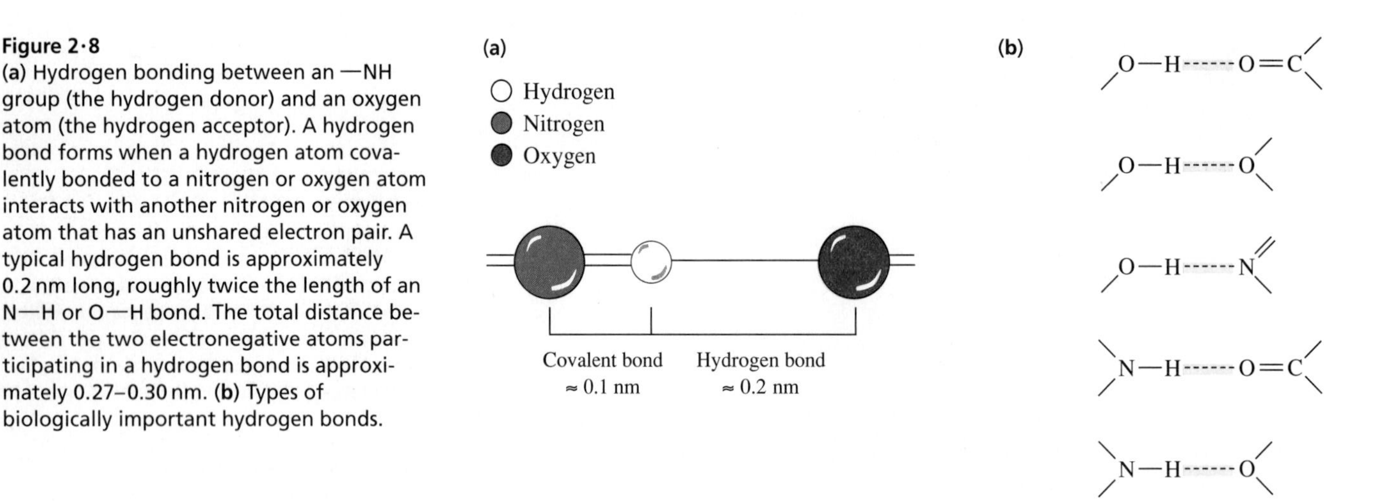

Figure 2·8
(a) Hydrogen bonding between an —NH group (the hydrogen donor) and an oxygen atom (the hydrogen acceptor). A hydrogen bond forms when a hydrogen atom covalently bonded to a nitrogen or oxygen atom interacts with another nitrogen or oxygen atom that has an unshared electron pair. A typical hydrogen bond is approximately 0.2 nm long, roughly twice the length of an N—H or O—H bond. The total distance between the two electronegative atoms participating in a hydrogen bond is approximately 0.27–0.30 nm. **(b)** Types of biologically important hydrogen bonds.

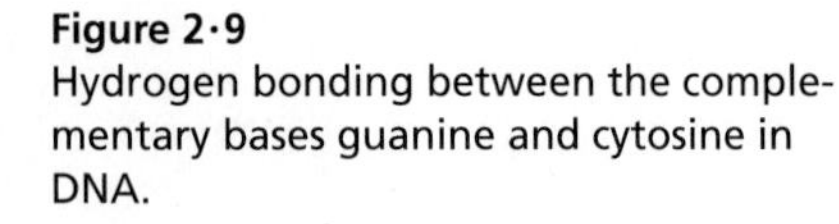
Figure 2·9
Hydrogen bonding between the complementary bases guanine and cytosine in DNA.

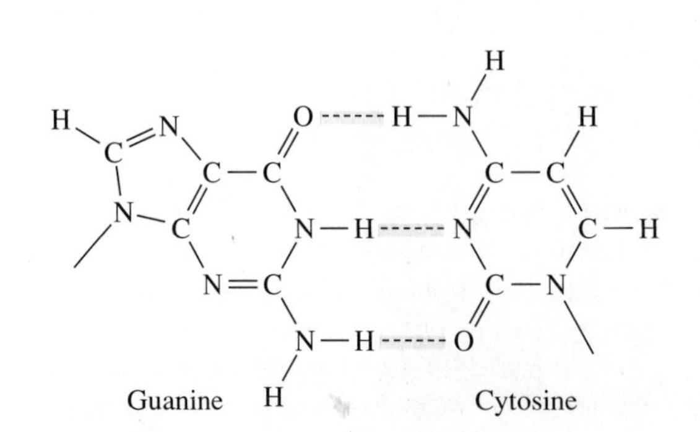

magnitude of an induced dipole–induced dipole interaction between nonpolar molecules, such as methane, is about 0.1 kcal mol^{-1} at an internuclear separation of 0.3 nm. If atoms involved in van der Waals forces approach one another too closely, there is strong repulsion. Conversely, there is hardly any attraction if the atoms are farther apart than the optimal packing distance, called the *van der Waals radius* (Table 2·1). When two atoms are separated by the sum of their van der Waals radii, they are said to be in van der Waals contact, and the attractive force between them is maximal (Figure 2·10).

Hydrophobic interactions are also very weak forces, having a magnitude of about 1 kcal mol^{-1} per interaction. Although they sometimes are called hydrophobic *bonds,* this description is incorrect. Nonpolar molecules or groups aggregate not because of mutual attraction but because of the preference of water molecules trapped around the nonpolar compounds for each other. Water molecules that surround a nonpolar molecule in solution are relatively immobile, whereas water molecules in the bulk solvent phase are much more mobile, or disordered. In thermodynamic terms, there is a net gain in the combined entropy of the solvent and the nonpolar solute when water is freed from around the nonpolar groups and these groups aggregate.

Although van der Waals forces and hydrophobic interactions are weak, the clustering of nonpolar groups within a protein, nucleic acid molecule, or biological membrane permits many of these weak interactions to form. In these cases, such forces play important roles in maintaining the structures of the molecules.

Often several weak forces combine to maintain the shapes of biological polymers. For example, the heterocyclic bases of nucleic acids, even though polar, are relatively insoluble in water and interact more readily with each other than with water. In DNA, bases are paired horizontally by hydrogen bonds (Figure 2·9). Such pairs of bases are also stacked one above another by forces known collectively as stacking interactions. This stacking arrangement is stabilized by a variety of noncovalent bonds that include van der Waals forces and hydrophobic interactions (Section 10·2C).

Table 2·1 Van der Waals radii of several atoms

Atom	Radius (nm)
Hydrogen	0.12
Oxygen	0.14
Nitrogen	0.15
Carbon	0.17
Sulfur	0.185
Phosphorus	0.19

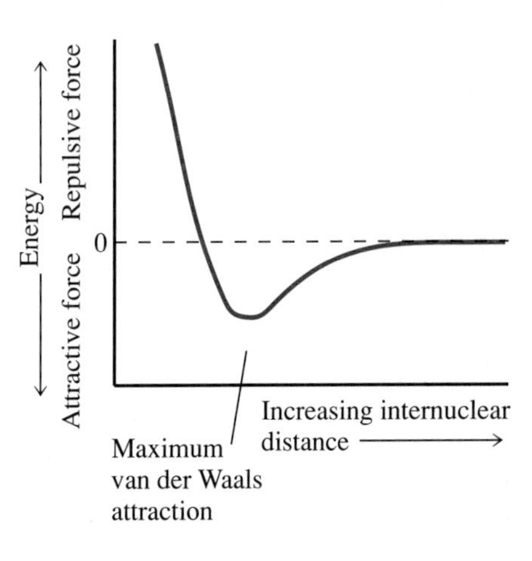

Figure 2·10
Effect of internuclear separation on van der Waals forces. Van der Waals forces are strongly repulsive at short internuclear distances and very weak at long internuclear distances. When two atoms are separated by the sum of their van der Waals radii, the van der Waals attraction is maximal.

2·6 Water Is Nucleophilic

Electron-rich chemicals are called *nucleophiles* because they seek positively charged or electron-deficient species, called *electrophiles,* such as nuclei. Nucleophiles are either negatively charged or have unshared pairs of electrons. They attack positive centers in addition or displacement reactions. Compounds containing oxygen, nitrogen, or sulfur are the most common biological nucleophiles. Water has two unshared pairs of electrons and therefore is nucleophilic. Although water is a relatively weak nucleophile, its cellular concentration is so high that many biological compounds, such as polymers, could be degraded by nucleophilic attack by water. An example is the degradation of a protein by water, catalyzed by a proteolytic enzyme. In such a reaction, known as *hydrolysis,* the protein is eventually degraded to its monomeric units, amino acids. This type of reaction involves the transfer of an acyl group from the polymer to water. Hydrolysis is essentially an irreversible reaction; that is, the equilibrium for a hydrolytic reaction lies far in the direction of degradation.

Several questions must then be asked. If there is so much water in cells, why are biopolymers not rapidly degraded to their components? Similarly, if the equilibrium lies toward breakdown, how does biosynthesis occur in aqueous cells? There are several simple ways in which cells prevent unwanted hydrolytic reactions or overcome unfavorable equilibria. First, the groups that link the monomeric units of biopolymers are unusually stable in neutral solution. It has been estimated that, for survival, no more than one in every million ester linkages in DNA should hydrolyze during the lifetime of an organism. The amide bonds of proteins are much less susceptible to spontaneous hydrolysis. Thus, although the hydrolysis of polymers is favored, polymers are quite stable in the absence of the enzymes that catalyze their hydrolysis. These enzymes, from the group of enzymes known as hydrolases, are stored in inactive forms or enclosed in organelles and thus are not always available to catalyze the degradation of the biopolymers.

Biosynthesis of polymers occurs by transfer of an acyl or carbonyl group to some nucleophile other than water. Because these reactions must proceed against a very unfavorable equilibrium, they usually require two chemical steps and involve the use of the chemical potential energy of adenosine triphosphate (ATP). A simple but useful example is the biosynthesis of an amide bond. In the reaction catalyzed by the enzyme glutamine synthetase, the amino acid glutamate is amidated to form the related amino acid glutamine by addition of ammonia:

$$\text{Glutamate} + \text{ATP} + \text{NH}_3 \xrightarrow{\text{Glutamine synthetase}} \text{Glutamine} + \text{ADP} + \text{P}_i \qquad \textbf{(2·1)}$$

where P_i represents inorganic phosphate. As you may recall from organic chemistry, carboxylic acids can be activated to acyl chlorides to form amides in a similar reaction (Figure 2·11). In the enzymatic condensation of glutamate with ammonia to form glutamine, the chemical potential energy of ATP is used to form a mixed phosphate-carboxylate anhydride called γ-glutamyl phosphate (Figure 2·12). If γ-glutamyl phosphate were in aqueous solution, it would be rapidly hydrolyzed to glutamate and phosphate. However, it remains bound to the enzyme in a cavity called the active site, where it is shielded from water molecules. Because

Figure 2·11
Chemical synthesis of an amide from a carboxylic acid. This reaction requires activation of the acid. The activated intermediate, an acyl chloride, undergoes nucleophilic attack by ammonia to produce the amide.

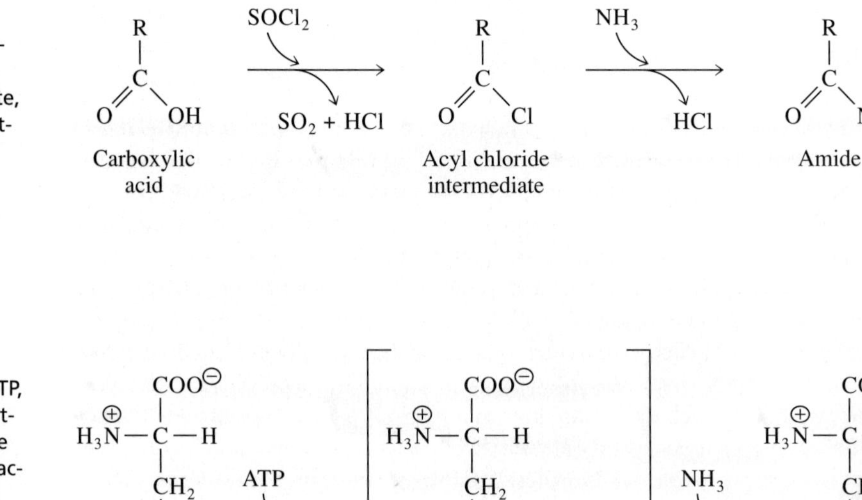

Figure 2·12
Synthesis of glutamine from glutamate, ATP, and ammonia. This two-step reaction is catalyzed by glutamine synthetase. The active site of glutamine synthetase protects the activated intermediate from hydrolysis. This intermediate is extremely unstable in aqueous solution.

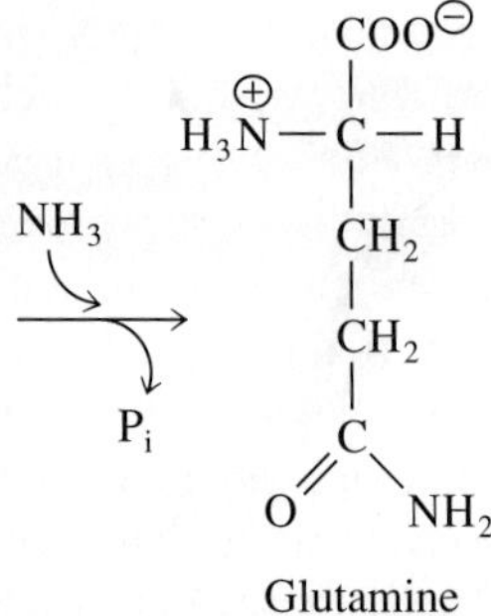

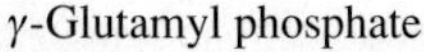

of its extreme reactivity with nucleophiles, γ-glutamyl phosphate readily forms the amide (glutamine) by reacting with ammonia, which is also bound within the active site of the enzyme. We shall see that both the activation of a reactant by transfer of a phosphoryl group from ATP and the exclusion of water within active sites of enzymes are characteristic of the syntheses of virtually all biopolymers.

2·7 Water Undergoes Ionization

Among the properties of water that are important to the life of a cell is the slight tendency of water to ionize. Pure water contains not only H_2O but also a low concentration of *hydronium ions* ($H_3O^{\oplus}$) and an equal concentration of hydroxide ions ($OH^{\ominus}$).

$$H_2O + H_2O \rightleftharpoons H_3O^{\oplus} + OH^{\ominus} \quad (2\cdot2)$$

Although a hydronium ion is often written as $H_3O^{\oplus}$, it actually has several more molecules of water associated with it. For convenience, hydronium ions will hereafter be referred to simply as protons and represented as $H^{\oplus}$.

According to the Brønsted-Lowry concept of acids and bases, an *acid* is defined as a substance that can donate protons, and a *base* as a substance that can accept protons. As Equation 2·2 indicates, water can function as either a proton donor (acid) or a proton acceptor (base).

The ionization of water can be analyzed quantitatively. The *equilibrium constant* (K_{eq}) for the ionization of water is given by

$$K_{eq} = \frac{[H^{\oplus}][OH^{\ominus}]}{[H_2O]} \quad (2\cdot3)$$

One liter of water has a mass of 1000 g, and the molecular weight of water is 18.015. Therefore, water has a concentration of approximately 55.5 M. Equation 2·3 thus leads to

$$55.5\, K_{eq} = [H^{\oplus}][OH^{\ominus}] \quad (2\cdot4)$$

Electrical conductivity measurements have established that the equilibrium constant for the ionization of water is 1.8×10^{-16} M. Because this equilibrium constant is so small, the concentration of water remains virtually unchanged by ionization. Substituting the above equilibrium constant in Equation 2·4 gives

$$1.0 \times 10^{-14} = [H^{\oplus}][OH^{\ominus}] \quad (2\cdot5)$$

The value 1.0×10^{-14} (M^2) in Equation 2·5 is called the ion product of water and is designated K_w.

$$K_w = [H^{\oplus}][OH^{\ominus}] = 1.0 \times 10^{-14} \quad (2\cdot6)$$

Since pure water is electrically neutral, its ionization produces an equal number of protons and hydroxide ions; thus, $[H^{\oplus}] = [OH^{\ominus}]$. Equation 2·6 can therefore be rewritten as

$$K_w = [H^{\oplus}]^2 = 1.0 \times 10^{-14} \quad (2\cdot7)$$

Taking the square root of Equation 2·7 leads to

$$[H^{\oplus}] = 1.0 \times 10^{-7}\,M \tag{2·8}$$

Since $[H^{\oplus}] = [OH^{\ominus}]$, the ionization of pure water produces 10^{-7} M $H^{\oplus}$ and 10^{-7} M $OH^{\ominus}$. When any aqueous solution contains equal concentrations of $H^{\oplus}$ and $OH^{\ominus}$, it is said to be neutral. If there is more $H^{\oplus}$ than $OH^{\ominus}$, the solution is acidic, and if $OH^{\ominus}$ predominates, the solution is basic, or alkaline.

2·8 The pH Scale Provides a Measure of Acidity and Basicity

Many biochemical processes depend upon the concentration of $H^{\oplus}$, even though $[H^{\oplus}]$ may not appear explicitly in the description of a reaction. The transport of oxygen in the blood, the catalysis of reactions by enzymes, and the generation of metabolic energy during respiration or photosynthesis are some of the many biological phenomena that depend upon the concentration of $H^{\oplus}$. Because the range of $[H^{\oplus}]$ in aqueous solutions is enormous, it is convenient to use a logarithmic scale called *pH* to measure the concentration of $H^{\oplus}$. pH is defined as the negative logarithm of the concentration of $H^{\oplus}$.

$$pH = -\log [H^{\oplus}] \tag{2·9}$$

Thus, a neutral solution has a pH value of 7.0 since $-\log(10^{-7}) = 7.0$. Because pH is the negative logarithm of $[H^{\oplus}]$, acidic solutions have pH values less than 7.0, and basic solutions have pH values greater than 7.0. The lower the pH below 7.0, the more acidic the solution; the higher the pH above 7.0, the more basic the solution. Since the pH scale is logarithmic, a change in pH of one unit correlates with a tenfold change in the concentration of $H^{\oplus}$. Figure 2·13 shows the pH values of various fluids.

Many dyes, such as litmus, phenol red, and phenolphthalein, change color at characteristic pH values and can be used to determine the approximate pH of aqueous solutions. More accurate measurements of pH are routinely made using pH meters. A pH meter is an instrument that incorporates a selective glass electrode, which is sensitive to the concentration of $H^{\oplus}$. Measurement of pH is sometimes important in the diagnosis of disease. The normal pH of human blood is 7.4, which is frequently referred to as physiological pH. The blood of patients suffering from certain diseases, such as diabetes, can have a lower pH, a condition called acidosis. The condition in which the pH of the blood is higher than 7.4, called alkalosis, can result from persistent, prolonged vomiting (loss of hydrochloric acid from the stomach) or during hyperventilation (excessive loss of carbonic acid as carbon dioxide).

2·9 The Acid Dissociation Constants of Weak Acids Are Determined by Titration

Acids and bases that dissociate completely in water, such as hydrochloric acid and sodium hydroxide, are called strong acids and strong bases. Many acids and bases, such as the amino acids from which proteins are made and the purines and pyrimidines from which DNA and RNA are made, do not dissociate completely in water. These substances are known as *weak acids* and *weak bases*.

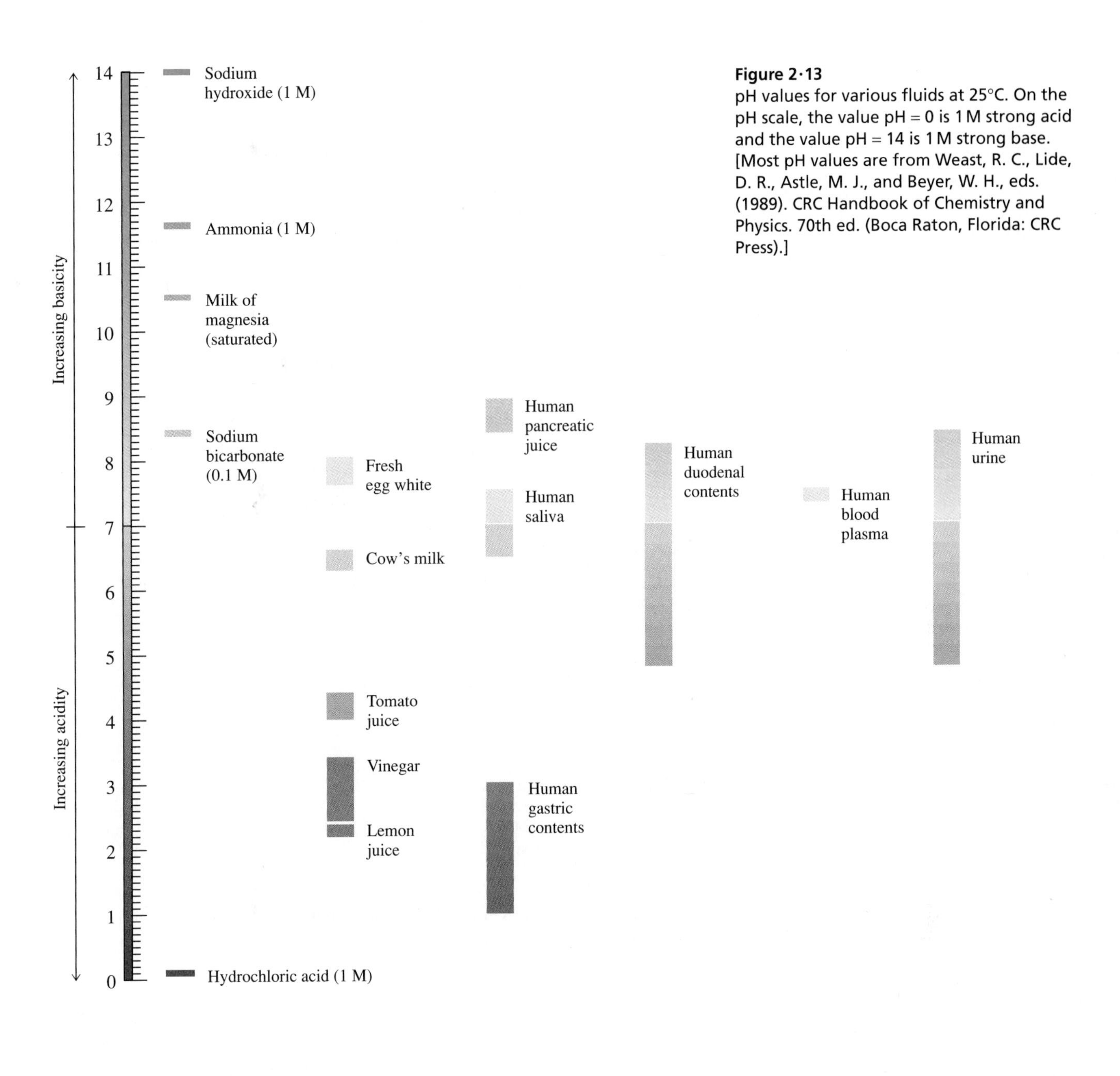

Figure 2·13
pH values for various fluids at 25°C. On the pH scale, the value pH = 0 is 1 M strong acid and the value pH = 14 is 1 M strong base. [Most pH values are from Weast, R. C., Lide, D. R., Astle, M. J., and Beyer, W. H., eds. (1989). CRC Handbook of Chemistry and Physics. 70th ed. (Boca Raton, Florida: CRC Press).]

Let us consider the ionization of acetic acid, the weak acid in vinegar.

$$\underset{\text{(weak acid)}}{\underset{\text{Acetic acid}}{CH_3COOH}} \underset{}{\overset{K_a}{\rightleftharpoons}} H^{\oplus} + \underset{\text{(conjugate base, or salt)}}{\underset{\text{Acetate anion}}{CH_3COO^{\ominus}}} \qquad \textbf{(2·10)}$$

As this equation illustrates, an acid can be converted to its *conjugate base* by loss of a proton, and in the reverse reaction, a base can be converted to its *conjugate acid* by addition of a proton.

The equilibrium constant for the dissociation of a proton from an acid is called the *acid dissociation constant*, designated K_a. To simplify calculations and make easy comparisons, bases are considered in their protonated forms. These conjugate acids are very weak acids. For example, the K_a of the base ammonia (NH_3) is the

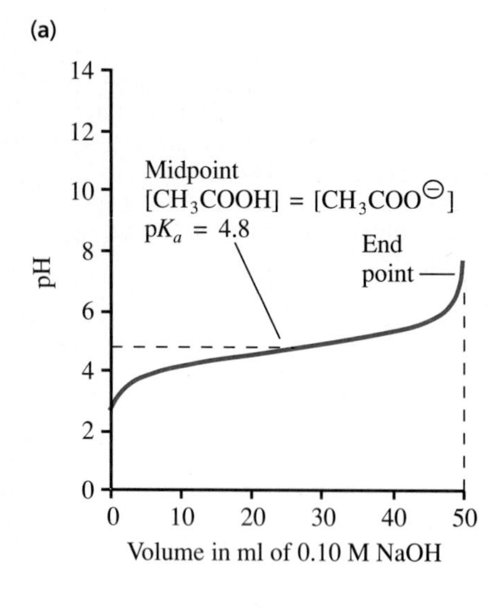

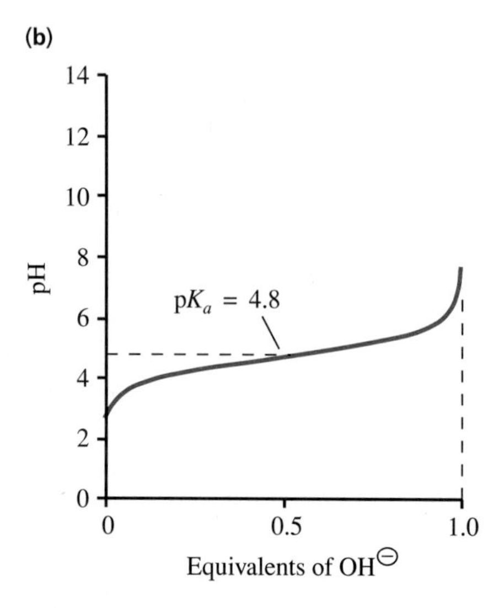

Figure 2·14
Titration curves for acetic acid. **(a)** Titration of 50 ml of 0.10 M acetic acid (CH_3COOH) with 0.10 M aqueous sodium hydroxide (NaOH) at 25°C. There is an inflection point (a point of minimum slope) at the midpoint of titration (25 ml of NaOH added), at which $[CH_3COOH] = [CH_3COO^{\ominus}]$ and pH = pK_a. At the end point (50 ml of NaOH added), all the molecules of acetic acid have been titrated to sodium acetate. **(b)** Generalized titration curve for acetic acid. Here, the same data are expressed in terms of the number of molar equivalents of strong base ($OH^{\ominus}$) added rather than the volume of a solution of specified concentration. Again, an inflection point occurs when one-half of the acetic acid has been titrated to its conjugate base, acetate, by the addition of 0.5 equivalent of strong base to the solution of acetic acid. This is the point at which $[CH_3COOH] = [CH_3COO^{\ominus}]$ and pH = pK_a. The pK_a of acetic acid is thus 4.8.

measure of the acid strength of its conjugate acid, the ammonium ion ($NH_4^{\oplus}$). For acetic acid,

$$K_a = \frac{[H^{\oplus}][CH_3COO^{\ominus}]}{[CH_3COOH]} \tag{2·11}$$

The value of K_a for acetic acid at 25°C is 1.76×10^{-5} M. Table 2·2 lists the K_a values for several common acids. Because these values are numerically small and clumsy in calculations, it is useful to place them on a logarithmic scale. Thus, a new parameter, pK_a, is defined by analogy with pH. A pH value is a measure of acidity, and a pK_a value is a measure of acid strength.

$$pK_a = -\log K_a \tag{2·12}$$

From Equation 2·11, we see that K_a is related to the concentration of $H^{\oplus}$ and to the ratio of the concentrations of the acetate ion and undissociated acetic acid. Taking the logarithm of Equation 2·11 gives

$$\log K_a = \log \frac{[H^{\oplus}][A^{\ominus}]}{[HA]} \tag{2·13}$$

where HA represents the undissociated acid (in this case, acetic acid) and $A^{\ominus}$ represents its conjugate base (in this case, the acetate anion). Since log (xy) = log x + log y, Equation 2·13 can be rewritten as

$$\log K_a = \log [H^{\oplus}] + \log \frac{[A^{\ominus}]}{[HA]} \tag{2·14}$$

Rearranging Equation 2·14 gives

$$-\log [H^{\oplus}] = -\log K_a + \log \frac{[A^{\ominus}]}{[HA]} \tag{2·15}$$

The negative logarithms in Equation 2·15 have previously been defined as pH and pK_a (Equations 2·9 and 2·12, respectively). Thus,

$$pH = pK_a + \log \frac{[A^{\ominus}]}{[HA]} \tag{2·16}$$

or

$$pH = pK_a + \log \frac{[\text{conjugate base}]}{[\text{weak acid}]} \tag{2·17}$$

Equation 2·17 is called the *Henderson-Hasselbalch equation*. It defines the pH of a solution in terms of the pK_a of the weak acid and the logarithm of the ratio of concentrations of the dissociated species (conjugate base) to the protonated species (weak acid). It is useful to note that the greater the concentration of the conjugate base, the higher the pH, whereas the greater the concentration of the weak acid, the lower the pH. When the concentrations of a weak acid and its conjugate base are exactly the same, the pH of the solution is equal to the pK_a of the acid (since the ratio of concentrations equals 1.0 and the logarithm of 1.0 equals zero).

The pK_a values of weak acids are determined by *titration*. A typical titration curve is presented in Figure 2·14a. In this example, a solution of acetic acid is titrated by measuring the pH while small aliquots of a strong base of known concentration (0.10 M NaOH) are added. The pH is plotted versus the volume of NaOH added. A more generalized presentation of the same data is shown in Figure 2·14b, in which pH is plotted as a function of the number of equivalents of strong base added during the titration. Note that since acetic acid has only one ionizable group (its carboxyl group), only one equivalent of a strong base is needed to titrate acetic acid to its conjugate base, the acetate anion. When the acid has been titrated

Table 2·2 Dissociation constants and pK_a values of weak acids in aqueous solutions at 25°C

Acid	K_a (M)	pK_a
HCOOH (Formic acid)	1.77×10^{-4}	3.8
CH_3COOH (Acetic acid)	1.76×10^{-5}	4.8
$CH_3CHOHCOOH$ (Lactic acid)	1.37×10^{-4}	3.9
H_3PO_4 (Phosphoric acid)	7.52×10^{-3}	2.1
$H_2PO_4^{\ominus}$ (Dihydrogen phosphate ion)	6.23×10^{-8}	7.2
HPO_4^{2-} (Monohydrogen phosphate ion)	2.20×10^{-13}	12.7
H_2CO_3 (Carbonic acid)	4.30×10^{-7}	6.4
$HCO_3^{\ominus}$ (Bicarbonate ion)	5.61×10^{-11}	10.2
$NH_4^{\oplus}$ (Ammonium ion)	5.62×10^{-10}	9.2
$CH_3NH_3^{\oplus}$ (Methylammonium ion)	2.70×10^{-11}	10.7

with one-half an equivalent of base, the concentration of undissociated acetic acid exactly equals the concentration of the acetate anion, and the resulting pH, 4.8, is thus the experimentally determined pK_a for acetic acid.

Similar titration curves can be obtained for each of the monoprotic acids (acids having only one ionizable group) listed in Table 2·2. All would exhibit the same general shape as Figure 2·14b, but the inflection point representing the midpoint of titration (one-half an equivalent titrated) would fall lower on the pH scale for a stronger acid (such as formic acid or lactic acid) and higher for a weaker acid (such as ammonium ion or methylammonium ion).

Many biologically important acids and bases have two or more ionizable groups. The number of pK_a values for such substances is equal to the number of ionizable groups, which can be experimentally determined by titration. For example, phosphoric acid requires three equivalents of strong base for complete titration, and three pK_a values are evident from its titration curve (Figure 2·15). The

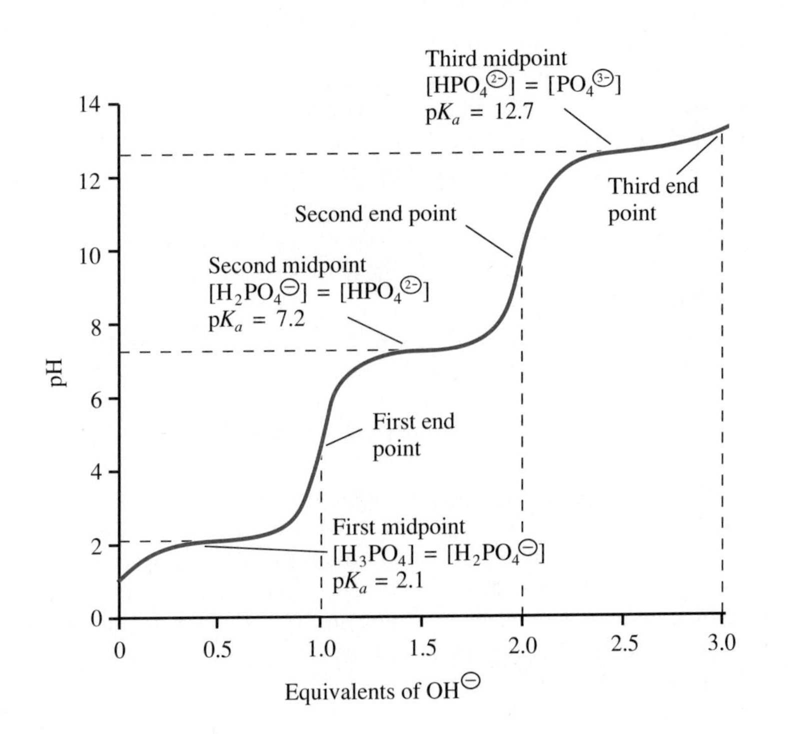

Figure 2·15
Titration curve for phosphoric acid (H_3PO_4). There are three inflection points (at 0.5, 1.5, and 2.5 equivalents of strong base titrated), corresponding to the three pK_a values for phosphoric acid (2.1, 7.2, and 12.7). The three titrations can be distinguished easily, although H_3PO_4 and HPO_4^{2-} are not ideal weak acids.

three pK_a values reflect the three equilibrium constants and thus the existence of four possible species (conjugate acids and bases) of inorganic phosphate.

$$\mathrm{HO{-}P(=O)(OH){-}OH} \underset{2.1}{\overset{pK_1}{\rightleftharpoons}} \mathrm{HO{-}P(=O)(OH){-}O^{\ominus}} + \mathrm{H^{\oplus}} \underset{7.2}{\overset{pK_2}{\rightleftharpoons}} \mathrm{{}^{\ominus}O{-}P(=O)(OH){-}O^{\ominus}} + \mathrm{H^{\oplus}} \underset{12.7}{\overset{pK_3}{\rightleftharpoons}} \mathrm{{}^{\ominus}O{-}P(=O)(O^{\ominus}){-}O^{\ominus}} + \mathrm{H^{\oplus}}$$

(2·18)

At a physiological pH of 7.4, the predominant species of inorganic phosphate (often symbolized by P_i) are HPO_4^{2-} and $H_2PO_4^{\ominus}$. At pH 7.2, these two species exist in equal concentrations. The concentrations of H_3PO_4 and PO_4^{3-} are so low at pH 7.4 that they can be ignored. This is generally the case for a minor species if the pH is more than two units away from its pK_a.

2·10 Buffered Solutions Resist Changes in pH

If the pH of a solution remains nearly constant when small amounts of strong acid or strong base are added, the solution is said to be *buffered*. The ability of a solution to resist changes in pH is known as its buffer capacity. Inspection of the titration curves of acetic acid (Figure 2·14) and phosphoric acid (Figure 2·15) reveals that the most effective buffering, indicated by the region of minimum slope on the curve, occurs when the concentrations of a weak acid and its conjugate base are equal—in other words, when the pH equals the pK_a. The effective range of buffering by a mixture of a weak acid and its conjugate base is usually considered to be from one pH unit below to one pH unit above the pK_a.

An excellent example of buffer capacity is found in blood plasma of mammals, which has a remarkably constant pH of 7.4. Consider the results of an experiment that compares the addition of an aliquot of strong acid to a volume of blood plasma with a similar addition of strong acid to an equal volume of either physiological saline (0.154 M NaCl) or water. When 1.0 ml of 10 M HCl (hydrochloric acid) is added to 1000 ml of physiological saline or water that is initially at pH 7.0, the pH is lowered to 2.0 (in other words, $[H^{\oplus}]$ is diluted to 10^{-2} M). However, when 1.0 ml of 10 M HCl is added to 1000 ml of blood plasma at pH 7.4, the pH is again lowered, but only to 7.2—impressive evidence for the effectiveness of physiological buffering.

The pH of the blood is primarily regulated by the carbon dioxide–carbonic acid–bicarbonate buffer system. A plot of the percentages of carbonic acid (H_2CO_3) and each of its conjugate bases as a function of pH is shown in Figure 2·16. Note that the major forms of carbonic acid at pH 7.4 are carbonic acid and the bicarbonate anion ($HCO_3^{\ominus}$).

The buffer capacity of blood depends upon equilibria between gaseous carbon dioxide (CO_2(gaseous)), which is present in the air spaces of the lungs, aqueous carbon dioxide (CO_2(aqueous)), which is produced by respiring tissues and dissolved in blood, carbonic acid, and bicarbonate. As can be seen in Figure 2·16, the equilibrium between bicarbonate and carbonate (CO_3^{2-}) does not contribute substantially to the buffer capacity of blood because the pK_a of bicarbonate is 10.2, which is too high to have an effect on physiological buffering.

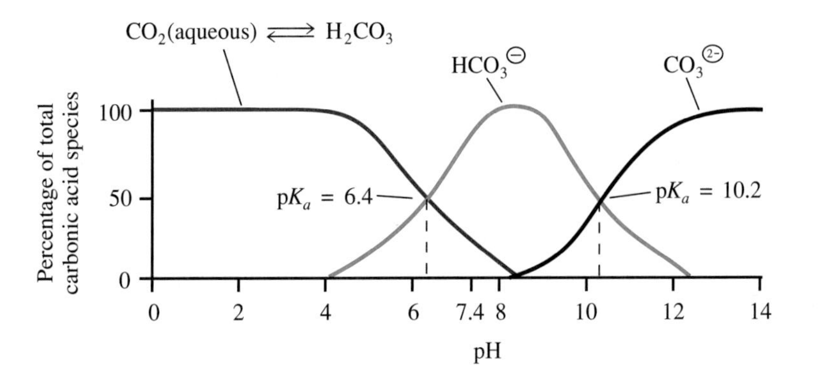

Figure 2·16
Percentages of carbonic acid and its conjugate bases as a function of pH. At pH 7.4 (the pH of blood), the concentrations of carbonic acid (H_2CO_3) and bicarbonate ($HCO_3^{\ominus}$) are substantial, but the concentration of carbonate ($CO_3^{2\ominus}$) is negligible.

The first of the three relevant equilibria of the carbon dioxide–carbonic acid–bicarbonate buffer system is the dissociation of carbonic acid to bicarbonate and $H^{\oplus}$ with an effective pK_a of 6.4.

$$H_2CO_3 \rightleftharpoons H^{\oplus} + HCO_3^{\ominus} \qquad \textbf{(2·19)}$$

This equilibrium is affected by a second equilibrium in which CO_2(aqueous) is in equilibrium with its hydrated form, carbonic acid.

$$CO_2\text{(aqueous)} + H_2O \rightleftharpoons H_2CO_3 \qquad \textbf{(2·20)}$$

Finally, CO_2(gaseous) is in equilibrium with CO_2(aqueous).

$$CO_2\text{(gaseous)} \rightleftharpoons CO_2\text{(aqueous)} \qquad \textbf{(2·21)}$$

The regulation of the pH of blood afforded by these three equilibria is pictured schematically in Figure 2·17. If the pH of blood falls due to a metabolic process that produces excess $H^{\oplus}$, the concentration of H_2CO_3 increases momentarily, but H_2CO_3 rapidly loses water to form dissolved CO_2(aqueous), which enters the gaseous phase in the lungs and is expired as CO_2(gaseous). An increase in the partial pressure of CO_2 (pCO_2) in the air expired from the lungs thus compensates for the increased hydrogen ions. Conversely, if the pH of the blood rises, the concentration of $HCO_3^{\ominus}$ increases transiently, but the pH is rapidly restored as the breathing rate changes and the reservoir of CO_2(gaseous) in the lungs is converted to CO_2(aqueous) and then to H_2CO_3 in the capillaries of the lungs. Again, the equilibrium of the blood buffer is rapidly restored by changing the partial pressure of CO_2 in the lungs.

Within cells, both proteins and inorganic phosphate contribute to intracellular buffering. Hemoglobin is the strongest buffer in blood other than the carbon dioxide–carbonic acid–bicarbonate buffer. As mentioned at the end of Section 2·9, the major species of inorganic phosphate present at a physiological pH of 7.4 are $HPO_4^{2\ominus}$ and $H_2PO_4^{\ominus}$, reflecting the second pK_a value for phosphoric acid, 7.2.

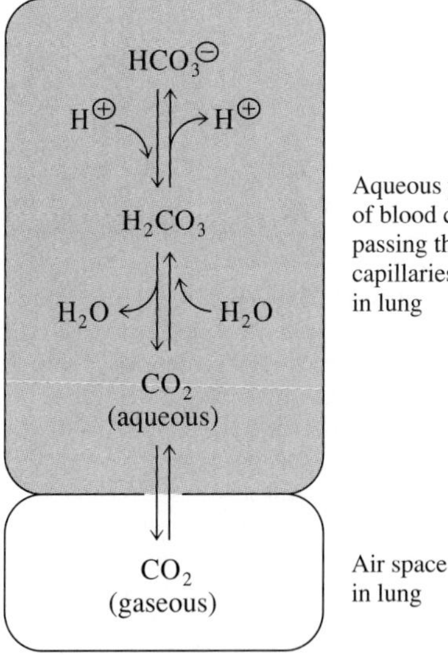

Figure 2·17
Regulation of the pH of blood in mammals. The pH of blood is closely controlled by the ratio of [$HCO_3^{\ominus}$] and pCO_2 in the air spaces of the lungs. If the pH of blood decreases due to the presence of excess $H^{\oplus}$, pCO_2 increases in the lungs, restoring the equilibrium. If, on the other hand, the concentration of $HCO_3^{\ominus}$ rises because the pH of blood increases, CO_2(gaseous) dissolves in the blood, again restoring the equilibrium.

2·11 Aqueous Solutions May Develop Osmotic Pressures

Another important property of solutions is osmotic pressure. *Osmosis* is the spontaneous transport of solvent by diffusion through a semipermeable membrane from a more dilute to a more concentrated solution. A semipermeable membrane allows passage of solvent but only some solutes. The pressure that must be exerted to prevent this flow of solvent is called *osmotic pressure*. Consider two solutions, one of higher solute concentration and one of lower solute concentration, separated by a semipermeable membrane (Figure 2·18). The solution of higher solute concentration expands as solvent crosses the membrane from the solution of lower solute concentration. This expansion continues until the system reaches equilibrium. If the liquid outside the tube in Figure 2·18 were pure water, the pressure that would have to be applied to the top of the liquid in the vertical tube to prevent such expansion would be equal to the osmotic pressure of the solution within the tube. If the liquid outside the tube were a solution of low solute concentration rather than pure water, the pressure needed to prevent expansion of the liquid within the tube would be equal to the difference in the osmotic pressures of the two solutions.

Osmotic pressure depends on the total concentration of solute particles (molecules or ions) rather than the chemical nature of the solute. Solutions that have equal osmotic pressures are called *iso-osmotic*. However, two solutions that would otherwise be iso-osmotic may not be so when they occur on opposite sides of a cell membrane because cell membranes have selective permeability. A solution in which cells are osmotically balanced is called *isotonic*. Cells must be in an isotonic medium to prevent swelling or shrinking. For example, physiological saline, which contains 0.154 M NaCl, is isotonic with human blood cells. Blood plasma must obviously be isotonic with blood cells. Solutions of glucose and/or salts administered intravenously to a patient must also be isotonic with human blood.

When two solutions are not isotonic, the solution with the lower osmotic pressure is called *hypotonic*, and that with the higher osmotic pressure is called *hypertonic*. If living cells are placed in a hypotonic solution, a net influx of water into the cells causes the cells to swell and possibly burst. In the case of red blood cells, such bursting is known as hemolysis. If red blood cells are placed in a hypertonic solution, water leaves the cells, and the cells shrink.

The effects of osmotic pressure and the use of semipermeable membranes will be encountered in several other situations in biochemistry. Animal and bacterial cells control their volumes by pumping $Na^{\oplus}$ out of the cell to maintain the correct osmotic pressure. Plant cells have rigid cell walls that are strong enough to overcome osmotic pressure and force water back out when it enters. Some cells or organelles can be broken and their contents released for analysis by treatment with a hypotonic solution that breaks their membranes. Dialysis of a solution of proteins to remove unwanted salts makes use of semipermeable membranes that allow water, buffers, and salts, but not proteins, to pass through.

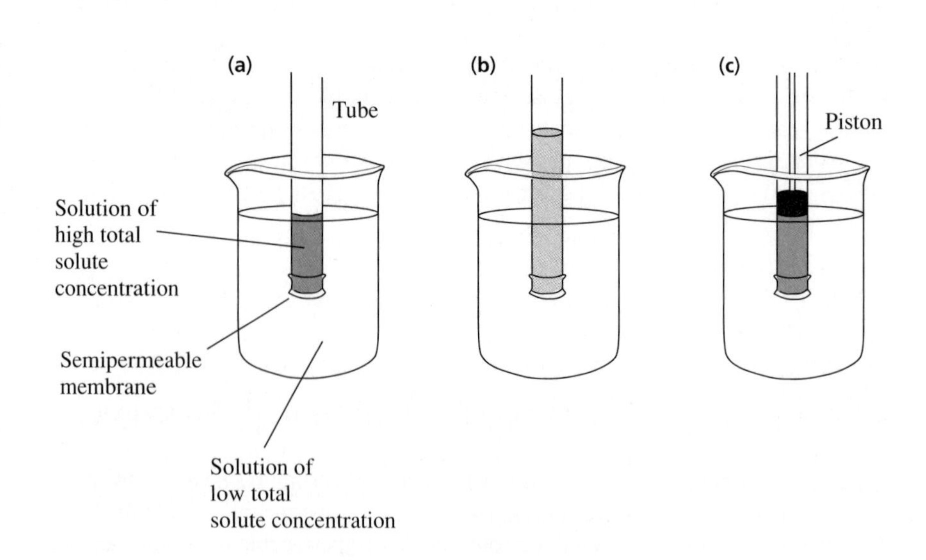

Figure 2·18
Osmotic pressure. **(a)** When two solutions of unequal solute concentrations are separated by a semipermeable membrane, water will flow across the membrane from the solution of lower total solute concentration to the solution of higher total solute concentration. **(b)** Water will continue to flow across the membrane until the system reaches equilibrium. **(c)** If the solution outside the tube were pure water, the pressure that would have to be applied to the top of the liquid within the tube to prevent osmotic flow (illustrated here by a piston) would be equal to the osmotic pressure of the solution within the tube. If the liquid outside the tube were a solution of low solute concentration rather than pure water, the pressure needed to prevent osmotic flow would be equal to the difference in the osmotic pressures of the two solutions.

Part Two

Structures and Functions of Biomolecules

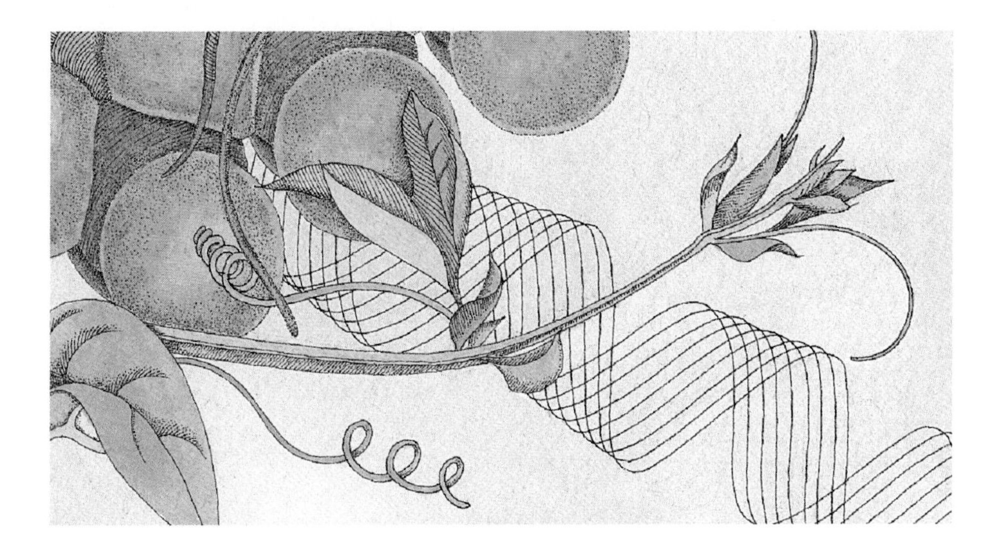

3

Amino Acids and the Primary Structures of Proteins

In the early 1800s, the Dutch chemist Gerardus Johannes Mulder was investigating the properties of albumins, substances from milk and eggs that coagulate upon heating. Mulder found that albumins contain carbon, oxygen, hydrogen, and nitrogen. In 1838, the Swedish scientist Jöns Jacob Berzelius suggested to Mulder that these substances be called *proteins*, derived from the Greek *proteios* meaning first or primary, because he suspected that these compounds might be the most important of biological substances. The name was prophetic. The following list, although not exhaustive, gives a glimpse of the wide range of biological functions in which proteins are now known to be involved.

1. Many proteins function as biochemical catalysts known as enzymes. Enzymes catalyze nearly all reactions that occur in living organisms.

2. Proteins can bind other molecules for the purposes of storage and transport. For example, myoglobin binds and stores oxygen in skeletal and cardiac muscle cells, and hemoglobin binds and transports O_2 and CO_2 in red blood cells.

3. Structural proteins provide mechanical support and shape to cells and hence to tissues and organisms.

4. Assemblies of proteins can do mechanical work, such as movement of flagella, separation of chromosomes at mitosis, and contraction of muscles.

5. Many proteins play a role in decoding information in the cell. Some, such as the ribosomal proteins, are required for translation, whereas others play a role in regulating gene expression by binding to nucleic acids.

6. Some proteins are hormones, which regulate biochemical activities in target cells or tissues; other proteins serve as receptors for hormones.

7. Some proteins serve other specialized functions. For example, immunoglobulins, one of the classes of proteins within the immune system, defend against bacterial and viral infections in vertebrates.

3·1 Proteins Are Made from 20 Different Amino Acids

Proteins are macromolecules consisting of linear polymers, or chains, of *amino acids*. All organisms use the same 20 amino acids as building blocks for the assembly of protein molecules. These 20 amino acids are therefore often cited as the common or standard amino acids. Despite the limited number of amino acid types, the variations in the order in which they are connected and in the numbers of amino acids per protein allow an almost limitless variety of proteins.

The *primary structure* of a protein is the sequence in which amino acids are covalently connected to form a polypeptide chain. We shall consider higher levels of protein structure in Chapter 4, after we have examined the nature of amino acids and the formation of polypeptides.

The 20 common amino acids are termed α-amino acids because they have an amino group and an acidic carboxyl group attached to C-2, which is also known as the α-carbon.

$$\overset{\oplus}{H_3N}-\underset{2}{\overset{\overset{R}{|}}{CH}}-\underset{1}{COOH} \qquad \overset{\oplus}{H_3N}-\underset{\alpha}{\overset{\overset{R}{|}}{CH}}-COOH \tag{3·1}$$

In addition, a hydrogen atom and a side chain are also attached to the α-carbon. The side chain, commonly represented as —R, is distinctive for each amino acid. Figure 3·1a shows the general structure of an amino acid in perspective; Figure 3·1b shows a representative amino acid, serine, which has $-CH_2OH$ as its side chain. The carbons of a side chain can be lettered sequentially as β, γ, δ, and ε, which refer to carbons 3, 4, 5, and 6, respectively.

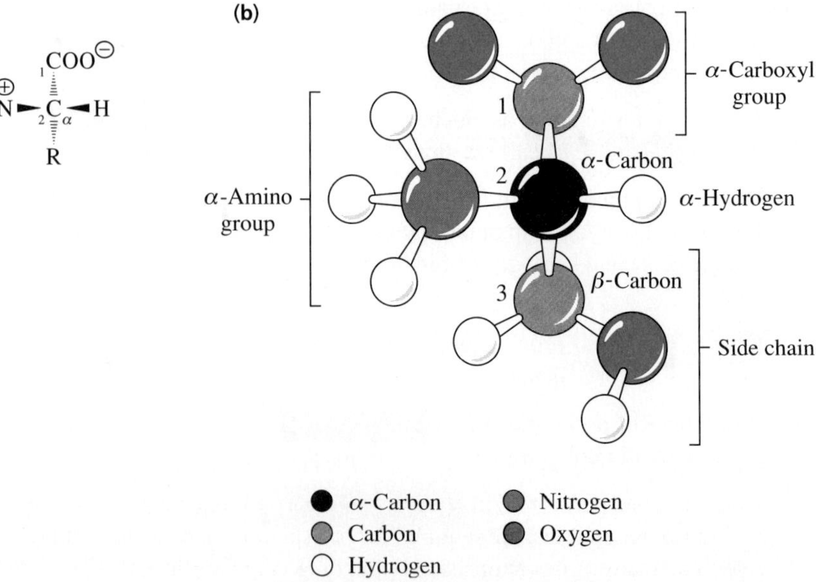

Figure 3·1
Two representations of an amino acid at neutral pH. **(a)** General structure. An amino acid has a carboxyl group, an amino group, a hydrogen atom, and a side chain, designated —R, all attached to C-2, the α-carbon. The numbering or lettering of the carbon skeleton of an amino acid is based upon the reference carbon of the carboxylic acid (C-1, or the carbon to which the α-carbon is attached). Solid wedges indicate bonds above the plane of the paper; dashed wedges, bonds below the plane of the paper. The blunt ends of wedges are nearer the viewer than are the pointed ends. **(b)** Three-dimensional ball-and-stick drawing of serine (in which $-R = -CH_2OH$). Note the alternative numbering and lettering systems for designating carbons.

At neutral pH, the amino group is protonated (—$NH_3^{\oplus}$), whereas the carboxyl group is ionized (—$COO^{\ominus}$). The pK_a values of α-carboxyl groups range from 1.8 to 2.5, and the pK_a values of α-amino groups range from 8.7 to 10.7. Thus, at a physiological pH of 7.4, amino acids are *zwitterions* (German, *zwitter*, hybrid), or dipolar ions, even though their net charge may be zero. We shall see in the next section that some side chains can also ionize.

In 19 of the 20 amino acids used for biosynthesis of proteins, the α-carbon atom is *chiral* (Greek, *cheir*, handed), or asymmetric, since four different groups are bonded to it. The exception is glycine, for which —R is simply a hydrogen atom (thus, the α-carbon is bonded to two hydrogen atoms). Accordingly, amino acids can exist as *stereoisomers*, compounds with the same molecular formula that differ in the arrangement, or *configuration*, of their atoms in space. A change in configuration requires the breaking of a bond or bonds. Two stereoisomers that are nonsuperimposable mirror images can exist for each chiral amino acid. Such stereoisomers are called *enantiomers*. By convention, the mirror-image pairs of amino acids are designated D (for dextro from Latin, *dexter*, right) and L (for levo, Latin, *laevus*, left). The configuration of the amino acid structure in Figure 3·1a is L; its mirror image would be D. To assign the stereochemical designation properly, one must draw the amino acid vertically with its α-carboxyl group at the top. It is then compared to the stereochemical reference compound, glyceraldehyde, which will be further described in Section 8·1. In this orientation, the α-amino group of the L isomer is on the left of the α-carbon and that of the D isomer is on the right. Figure 3·2a shows the general structure of L- and D-amino acids, and Figure 3·2b

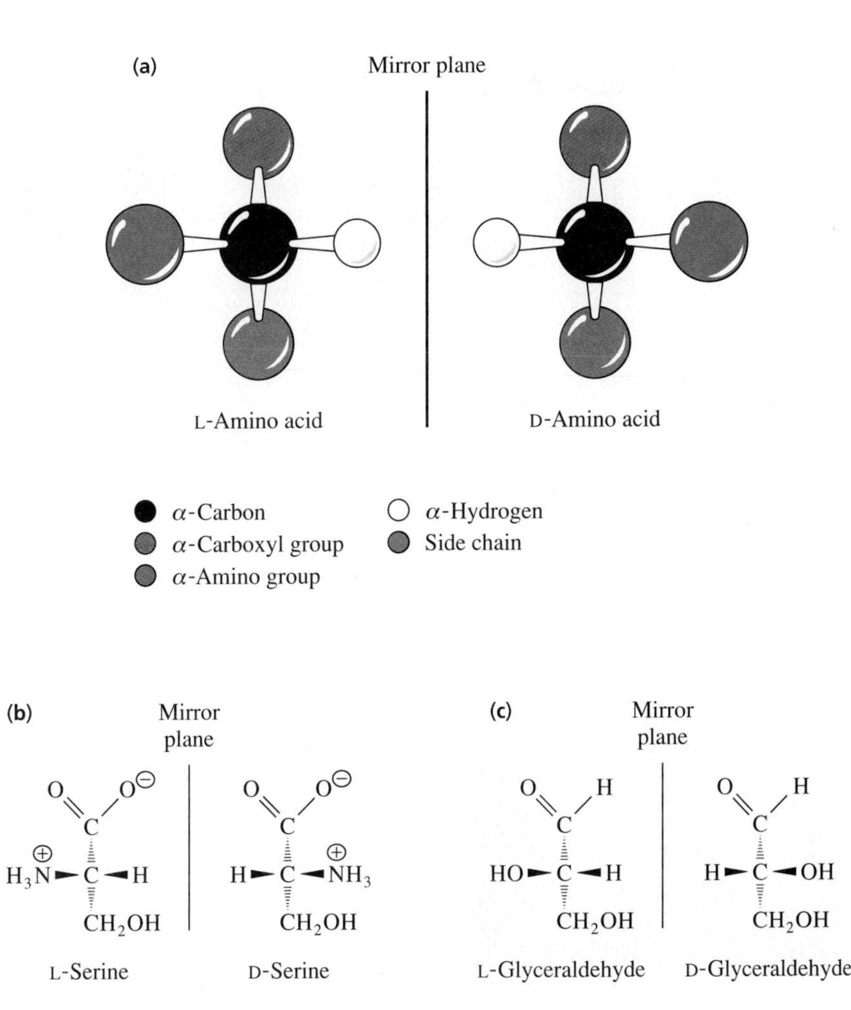

Figure 3·2
Mirror-image pairs of compounds containing a single chiral carbon atom. **(a)** General structures of L- and D-amino acids. **(b)** L-Serine and D-serine. **(c)** The historical reference compounds L-glyceraldehyde and D-glyceraldehyde.

shows a particular example, L- and D-serine. These can be compared to the carbohydrates L- and D-glyceraldehyde shown in Figure 3·2c. Stereo S3·1 shows the L and D configurations of serine.

Although a few D-amino acids occur in nature, the 19 chiral amino acids used in the assembly of proteins are all of the L configuration. Therefore, by convention, amino acids are assumed to be in the L configuration unless specifically designated D. The D and L designations are more useful in the case of amino acids than the *R* and *S* designations that most students learn in their study of organic chemistry because different side chains result in different priorities in the *R* and *S* system of designating absolute configuration. For example, L-alanine is the *S* enantiomer, but L-cysteine is the *R* enantiomer. (Box 13·1 provides a discussion of *RS* nomenclature.)

Often it is convenient to draw the structures of L-amino acids in a form that is stereochemically uncommitted. For example, alanine—the amino acid with a methyl side chain—could be drawn simply as

$$\overset{\oplus}{H_3N}-\underset{}{\overset{\overset{\displaystyle CH_3}{|}}{CH}}-COO^{\ominus} \qquad (3\cdot2)$$

This format is particularly useful for short peptides. Uncommitted structures for chemical compounds are commonly used when a correct representation of stereochemistry is not critical to a given discussion.

3·2 The 20 Amino Acids Have Different Side Chains

The structures of the 20 amino acids commonly found in proteins are shown in Figure 3·3 as *Fischer projections*. In Fischer projections, horizontal bonds at a chiral center extend toward the viewer, and vertical bonds extend away. Examination of the structures in Figure 3·3 reveals considerable variation in the side chains of the 20 amino acids. Some side chains are polar or ionized at neutral pH and are therefore hydrophilic, whereas others are nonpolar and thus hydrophobic. Since individual amino acids are zwitterions, they are water-soluble. However, it is essential to realize that when amino acids are combined by polymerization into peptides and proteins (Section 3·4), the ionic charges on the α-amino and α-carboxyl groups that form the links are lost. As a result, the ionic, polar, or nonpolar characteristics

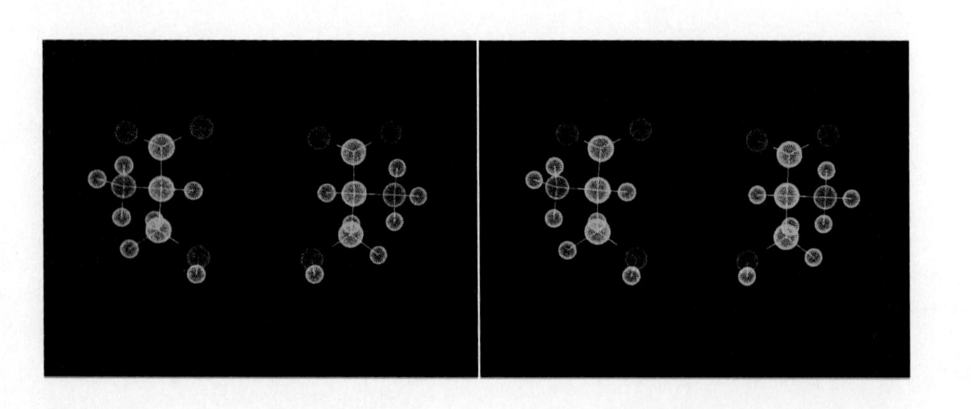

Stereo S3·1
L-Serine (left) and D-serine. Color key: carbon, green; hydrogen, light blue; nitrogen, dark blue; oxygen, red. (Stereo image by Ben N. Banks and Kim M. Gernert.)

Figure 3·3
Structures of the 20 common amino acids at pH 7, their names, and one- and three-letter abbreviations. Amino acids are classified by the polarity of their side chains. Polar side chains are further classified as neutral, basic, or acidic.

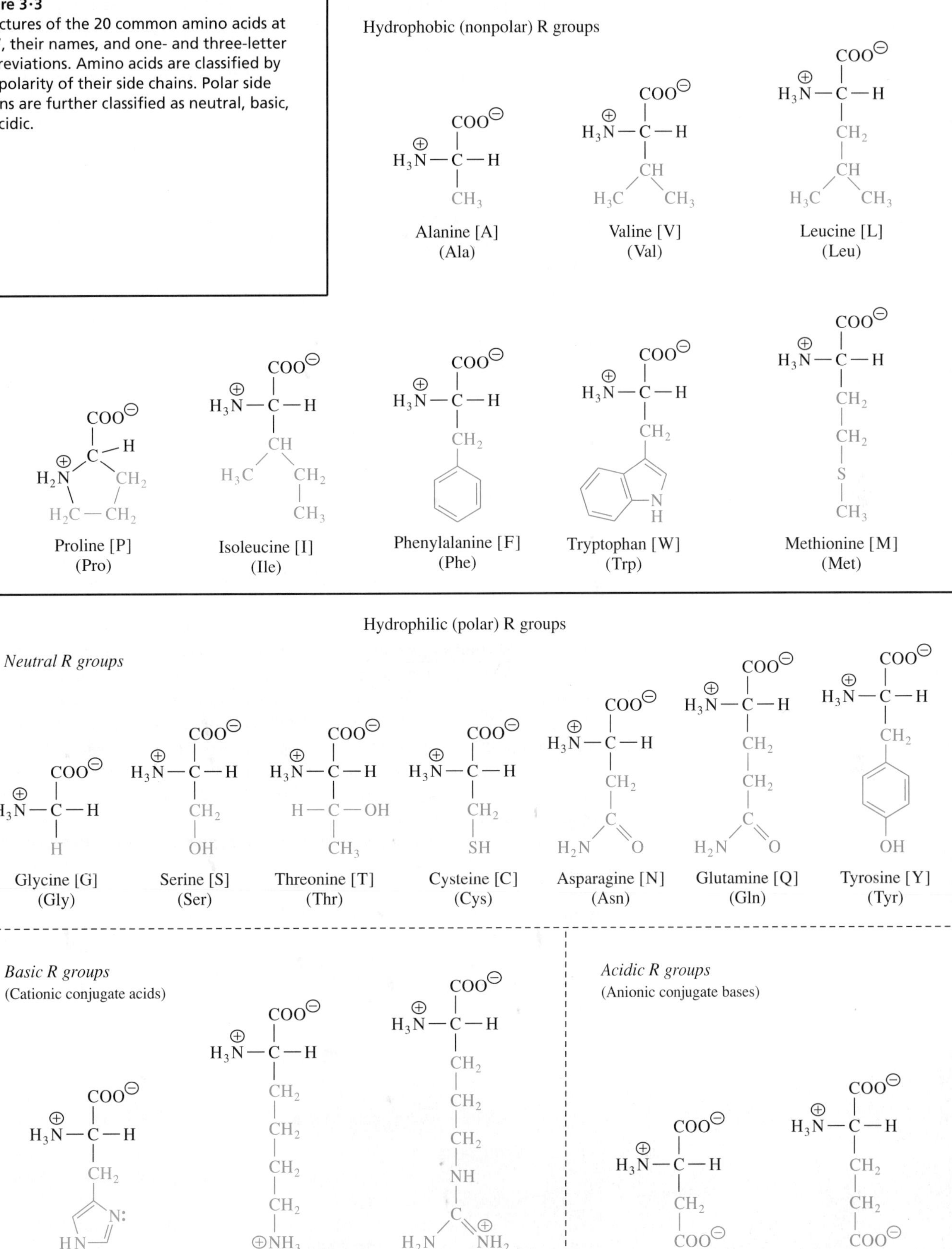

of the side chains greatly influence the overall three-dimensional shape, or conformation, of a protein. For example, water-soluble globular proteins typically contain hundreds of amino acid residues in a compact conformation. Most of the hydrophilic side chains of such a protein are on the surface of the macromolecule and are thus exposed to water, whereas many of the hydrophobic side chains are in the interior of the protein and are shielded from water.

To understand the functions of individual amino acids in proteins, it is important to learn their structures and to be able to relate these structures to the one- and three-letter abbreviations commonly used in referring to them. Four amino acids, *alanine* (Ala, A), *valine* (Val, V), *leucine* (Leu, L), and the structural isomer of leucine, *isoleucine* (Ile, I), have aliphatic (in this case, saturated hydrocarbon) side chains. The side chain of alanine is simply a methyl group, whereas valine, leucine, and isoleucine each contain a branched aliphatic side chain. Note that both the α- and β-carbon atoms of isoleucine are asymmetric. Thus, isoleucine has two chiral centers and four possible stereoisomers: L-isoleucine, its enantiomer D-isoleucine, L-*allo*isoleucine, and D-*allo*isoleucine. (The prefix *allo-* simply indicates an isomeric form or variety.) L-Isoleucine is the only one of the four isomers normally found in proteins. Although the side chains of alanine, valine, leucine, and isoleucine have no reactive functional groups, these amino acids play an important role in establishing and maintaining the three-dimensional structures of proteins because of their tendency to cluster away from water.

Proline (Pro, P) differs markedly from the other 19 amino acids in that its cyclic side chain, a saturated hydrocarbon, is bonded to the nitrogen of its α-amino group as well as to the α-carbon. Thus, strictly speaking, proline is an *α-imino acid* since it contains a secondary amino group rather than a primary amino group. Nonetheless, the proline residues found in proteins are all of the L configuration and have the same chirality about their α-carbons as the other L-amino acids. The heterocyclic pyrrolidine ring of proline restricts the geometry of polypeptides, sometimes introducing abrupt changes in the direction of the peptide chain.

Phenylalanine (Phe, F), *tyrosine* (Tyr, Y), and *tryptophan* (Trp, W) have aromatic side chains that absorb ultraviolet (UV) light. At neutral pH, both tyrosine and tryptophan absorb UV light at 280 nm, whereas phenylalanine is almost transparent at 280 nm but absorbs weakly at 260 nm. Since most proteins contain tyrosine and/or tryptophan, absorbance of solutions at 280 nm is routinely used to estimate protein concentration (Section 3·5). Note that tyrosine is structurally similar to phenylalanine; the *para*-hydrogen of phenylalanine is replaced in tyrosine by a hydroxyl group (—OH), making tyrosine a phenol (a very weak acid; pK_a of the side chain = 10.5). In alkaline solutions, loss of the phenolic proton leads to a shift in the spectrum of tyrosine, increasing its absorption maximum from 280 to 296 nm. The bicyclic structure in the side chain of tryptophan is indole; thus, tryptophan can be regarded as indolealanine.

Methionine (Met, M) and *cysteine* (Cys, C) are the two sulfur-containing amino acids. Methionine contains a nonpolar methyl thioether group in its side chain. In cysteine, a sulfur atom takes the place of the hydroxyl oxygen of serine; thus, cysteine can be regarded as thioserine. Alternatively, since compounds containing sulfhydryl groups (—SH) are known as mercaptans, cysteine can also be regarded as mercaptoalanine. Because the sulfur atom is polarizable, the sulfhydryl group of cysteine is chemically reactive and can form weak hydrogen bonds with oxygen and nitrogen. Moreover, the sulfhydryl group of cysteine is a weak acid ($pK_a = 8.4$). Thus, the sulfhydryl group can lose its proton to become a negatively charged thiolate ion. Furthermore, when some proteins are hydrolyzed, a compound called cystine can be isolated. Cystine is an oxidized form consisting of two cysteine molecules linked by a *disulfide bond* (Figure 3·4). Disulfide bonds play an important role in stabilizing the three-dimensional structures of some proteins by serving as covalent cross-links between oxidized cysteine residues at different positions in peptide chains.

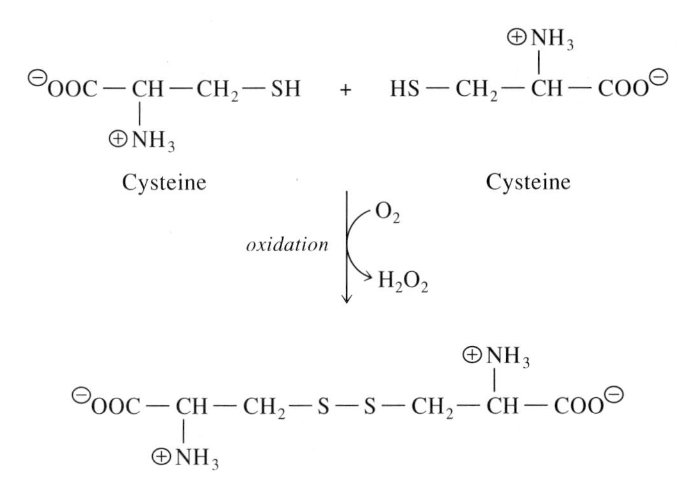

Figure 3·4
Oxidation of the sulfhydryl groups of two cysteine molecules. The oxidation of cysteine proceeds most easily at alkaline pH values at which the thiol group is ionized. When oxidation links the sulfhydryl groups of two cysteine molecules, the resulting compound is a disulfide called cystine.

Glycine (Gly, G) has the least complex structure of the amino acids, since —R is simply a hydrogen atom. As a result, the α-carbon of glycine is not chiral and does not have the L configuration. Glycine has a unique role within the conformations of many proteins because its side chain is small enough to fit into niches that accommodate no other amino acid.

Aspartate (Asp, D) and *glutamate* (Glu, E) are dicarboxylic amino acids. In addition to their α-carboxyl groups, aspartate possesses a β-carboxyl group, and glutamate possesses a γ-carboxyl group. Because the side chains of aspartate and glutamate are ionized at pH 7, they confer negative charges on proteins and are usually found on the surfaces of protein molecules. Aspartate and glutamate are often referred to as *aspartic acid* and *glutamic acid*. However, under most physiological conditions, they are anionic, not in the form of their conjugate acids, and, like all salts, have the "-ate" suffix. Glutamate is probably best known as its monosodium salt, monosodium glutamate (MSG), which is used as a flavor enhancer.

Asparagine (Asn, N) and *glutamine* (Gln, Q) are the amides of aspartic acid and glutamic acid, respectively. Although the side chains of asparagine and glutamine are uncharged, these amino acids are highly polar and are often found on the surfaces of proteins, where they can interact with water molecules. In addition, the polar amide groups of asparagine and glutamine can form hydrogen bonds with atoms in the side chains of other polar amino acids.

Serine (Ser, S) and *threonine* (Thr, T) have uncharged polar side chains containing β-hydroxyl groups. Unlike the phenolic side chain of tyrosine, the hydroxyl groups of serine and threonine have the properties of primary and secondary aliphatic alcohols, respectively. Although the hydroxymethyl group of serine ($-CH_2OH$) does not have a measurable pK_a value in aqueous solutions, this alcohol can react in the active sites of a number of enzymes as though it were ionized. Note that threonine, like isoleucine, has two chiral centers, the α- and β-carbon atoms. Only one of the four stereoisomers, L-threonine, commonly occurs in proteins.

The remaining three amino acids, *lysine* (Lys, K), *arginine* (Arg, R), and *histidine* (His, H), have hydrophilic side chains that are nitrogenous bases. Lysine is a diamino acid, having both α- and ε-amino groups. The ε-amino group exists as an alkylammonium ion ($-CH_2-NH_3^{\oplus}$) at neutral pH and thus confers a positive charge on proteins. Arginine is the most basic of the 20 amino acids; that is, its side-chain guanidinium ion has the highest pK_a value (pK_a = 12.5). Thus, arginine is also positively charged in proteins. Finally, histidine has an imidazole ring in its side chain and thus can be regarded as imidazolealanine. The imidazole group (Im) is ionizable ($pK_a = 6.0$), and the protonated form of this ring is called an imidazolium ion ($ImH^{\oplus}$).

3·3 The Ionic States of Amino Acids Depend on the Ambient pH

The physical properties of amino acids are influenced by the ionic states of the α-carboxyl and α-amino groups and any ionizable groups in the side chains. Seven of the common amino acids have ionizable side chains with measurable pK_a values. Each amino acid has either two or three pK_a values, and these values differ among the amino acids. Consequently, at a given pH, amino acids frequently have different net charges. Even slight differences in net charges on amino acids and proteins can be exploited in order to separate and purify them, as we shall see in Section 3·5. The ionic states of amino acid side chains strongly influence the three-dimensional structures and biochemical functions of proteins. In particular, a number of ionizable amino acid residues are involved in catalysis by enzymes; thus, an understanding of the ionic properties of amino acids is necessary in order to understand enzyme mechanisms.

The α-COOH group of an amino acid is a weak acid. Accordingly, we can use the Henderson-Hasselbalch equation (Section 2·9) to calculate the fractions of these groups that are ionized at any given pH.

$$\mathrm{pH} = \mathrm{p}K_a + \log \frac{[\text{conjugate base}]}{[\text{weak acid}]} \tag{3·3}$$

As can be seen in Table 3·1, the pK_a values of the α-carboxyl groups of free amino acids range from 1.8 to 2.5. These values are lower than those of typical carboxylic acids, such as acetic acid ($pK_a = 4.8$), due to the inductive influence of the neighboring $—NH_3^{\oplus}$ group. For a typical amino acid whose α-COOH group has a pK_a of 2.0, the ratio of conjugate base (carboxylate anion, salt, or the ionized species) to weak acid (carboxylic acid, or the protonated species) at pH 7.0 can be calculated:

$$7.0 = 2.0 + \log \frac{[\mathrm{RCOO}^{\ominus}]}{[\mathrm{RCOOH}]} \tag{3·4}$$

Therefore, the ratio of carboxylate anion to carboxylic acid is 100 000:1. The predominant species of the α-carboxyl group of amino acids at neutral pH is thus the carboxylate anion.

Next, we consider the ionization of the α-amino groups of free amino acids, whose pK_a values range from 8.7 to 10.7. For an amino acid whose α-amino group might have a pK_a of 10.0, the ratio of conjugate base, or free amine ($—NH_2$), to conjugate acid, or protonated amine ($—NH_3^{\oplus}$), at pH 7.0 would be 1:1000. Such calculations verify our earlier statement that free amino acids exist predominantly as zwitterions at neutral pH. Thus, it is inappropriate to draw the structure of an amino acid with both —COOH and $—NH_2$ groups, since there is no pH at which the carboxyl group will be predominantly protonated while the amino group is in its unprotonated, basic form. Note that the imino group of proline ($pK_a = 10.6$) is also protonated at neutral pH, so that proline, despite the bonding of the side chain to the α-amino group, is also zwitterionic at pH 7, as shown in Figure 3·3.

Ionization of the side chains of those amino acids having three readily ionizable groups—aspartate, glutamate, tyrosine, cysteine, lysine, arginine, and histidine—obeys the same principles as the ionizations of the α-carboxyl and α-amino

Table 3·1 pK_a values of acidic and basic constituents of free amino acids at 25°C*

Amino acid	pK_a values		
	α-Carboxyl group	α-Amino group	Side chain
Glycine	2.4	9.8	
Alanine	2.4	9.9	
Valine	2.3	9.7	
Leucine	2.3	9.7	
Isoleucine	2.3	9.8	
Methionine	2.1	9.3	
Proline	2.0	10.6	
Phenylalanine	2.2	9.3	
Tryptophan	2.5	9.4	
Serine	2.2	9.2	
Threonine	2.1	9.1	
Cysteine	1.9	10.7	8.4
Tyrosine	2.2	9.2	10.5
Asparagine	2.1	8.7	
Glutamine	2.2	9.1	
Aspartic acid	2.0	9.9	3.9
Glutamic acid	2.1	9.5	4.1
Lysine	2.2	9.1	10.5
Arginine	1.8	9.0	12.5
Histidine	1.8	9.3	6.0

*Values have been rounded off.
[Values from Dawson, R. M. C., Elliott, D. C., Elliott, W. H., and Jones, K. M. (1986). *Data for Biochemical Research*, 3rd ed. (Oxford: Clarendon Press).]

groups. Thus, the Henderson-Hasselbalch equation can be applied to each. Figure 3·5 depicts the ionization of the γ-carboxyl group of glutamate (pK_a = 4.1). Note that the negative charge of the carboxylate anion is delocalized. Note also that, being further removed from the influence of the α-ammonium ion, the γ-carboxyl group is a weak acid, similar in strength to acetic acid (pK_a = 4.8), whereas the α-carboxyl group is a stronger acid (pK_a = 2.1).

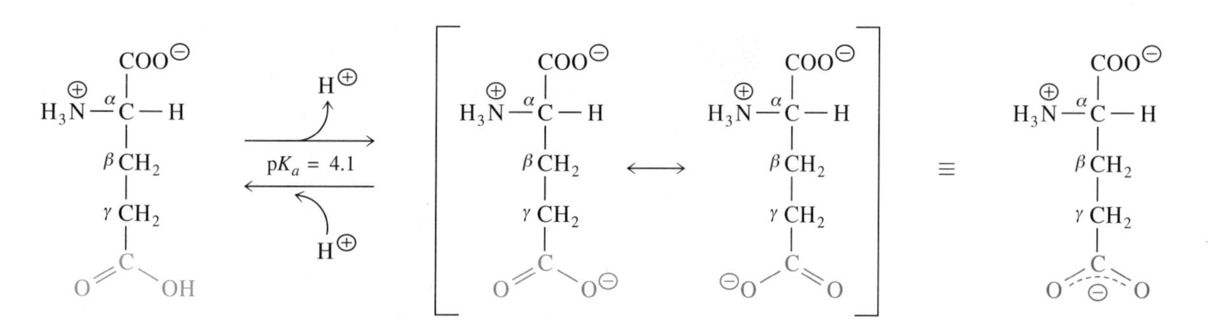

Figure 3·5
Ionization of the protonated γ-carboxyl group of glutamate. The measured pK_a values of 2.1 and 4.1, obtained from titration of fully protonated glutamic acid, reflect the first and second ionizations. Actually, at a pH midway between 2.1 and 4.1, 93% of the α-COOH group is deprotonated to —COO⊖ and 7% of the γ-COOH is deprotonated to —COO⊖. Such overlapping pK_a values are thus not strictly assignable to individual groups.

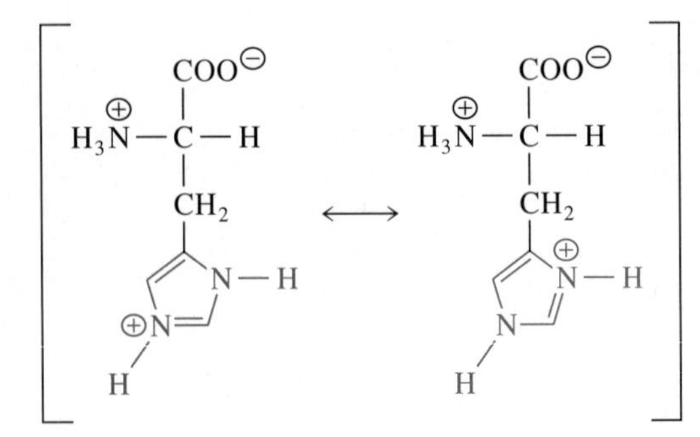

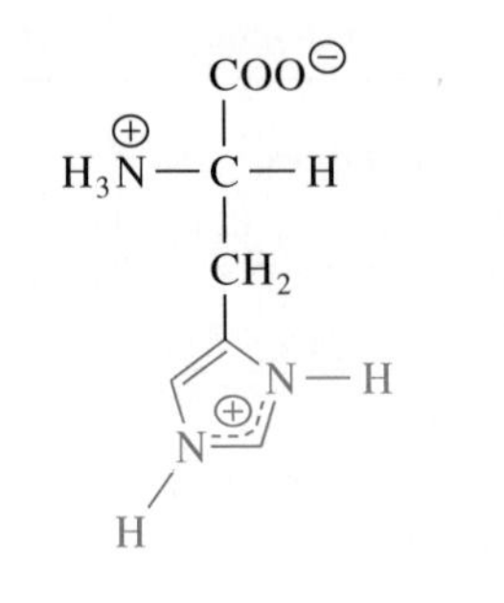

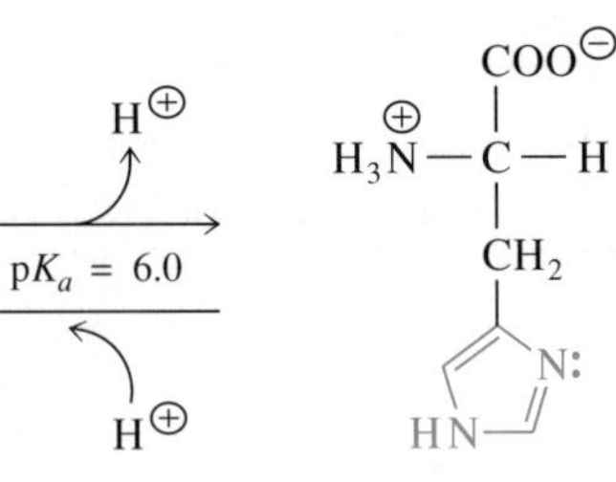

Imidazolium ion (protonated form) of histidine side chain

Imidazole (deprotonated form) of histidine side chain

Figure 3·6
Dissociation of a proton from the imidazolium ring of the side chain of histidine. Charge delocalization occurs in the imidazolium ion.

Figure 3·6 shows the deprotonation of the imidazolium ion of the side chain of histidine and depicts charge delocalization in the imidazolium ion, which predominates below pH 6.0. At pH 7.0, the ratio of imidazole, Im (conjugate base), to imidazolium ion, ImH⊕ (conjugate acid), is 10:1. Thus, the protonated and neutral forms of the side chain of histidine are both present in significant concentrations near physiological pH. A given histidine side chain in a protein may be either protonated or unprotonated, depending on its immediate environment within the protein. This property makes the side chain of histidine ideal for the transfer of protons within the catalytic sites of enzymes.

Figure 3·7 shows the deprotonation of the guanidinium group of the side chain of arginine in strong alkali. Charge delocalization in the guanidinium ion contributes to its very high pK_a of 12.5.

Titration curves such as those given for weak acids in Section 2·9 provide data for determining the pK_a values of amino acids. Two examples are shown in Figure 3·8. Alanine, which has two ionizable groups, exhibits two pK_a values, 2.4 and 9.9, each of which is at the center of a buffering region. Histidine, which has three ionizable groups, exhibits pK_a values of 1.8, 6.0, and 9.3, each of which is associated with a buffering zone.

Figure 3·7
Deprotonation of the side chain of arginine in strong alkali. The equilibrium between the guanidinium ion with its delocalized charge and the unprotonated guanidine group of arginine lies overwhelmingly in the direction of the protonated form at pH 7.

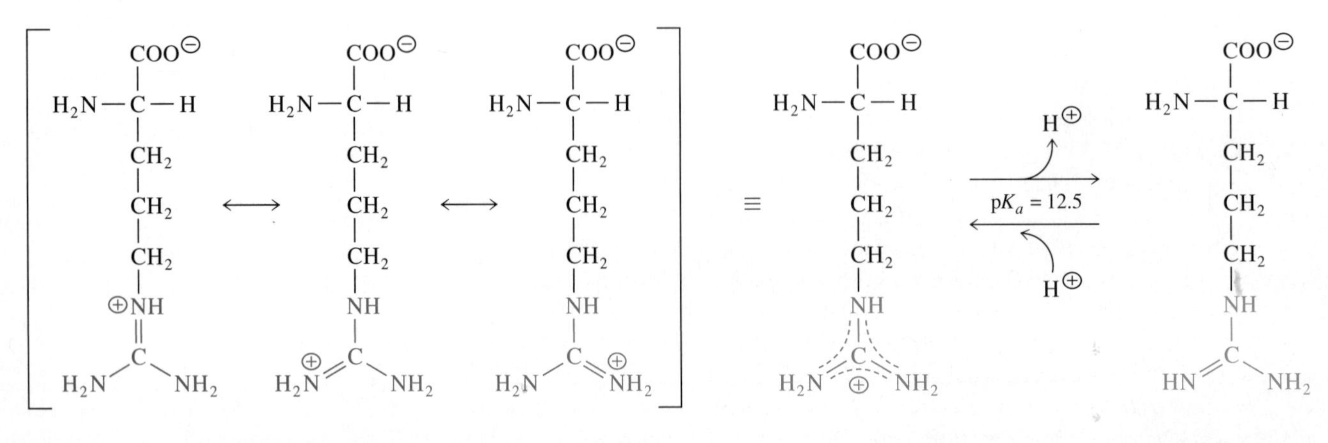

Guanidinium ion (protonated form) of arginine side chain

Guanidine group (deprotonated form) of arginine side chain

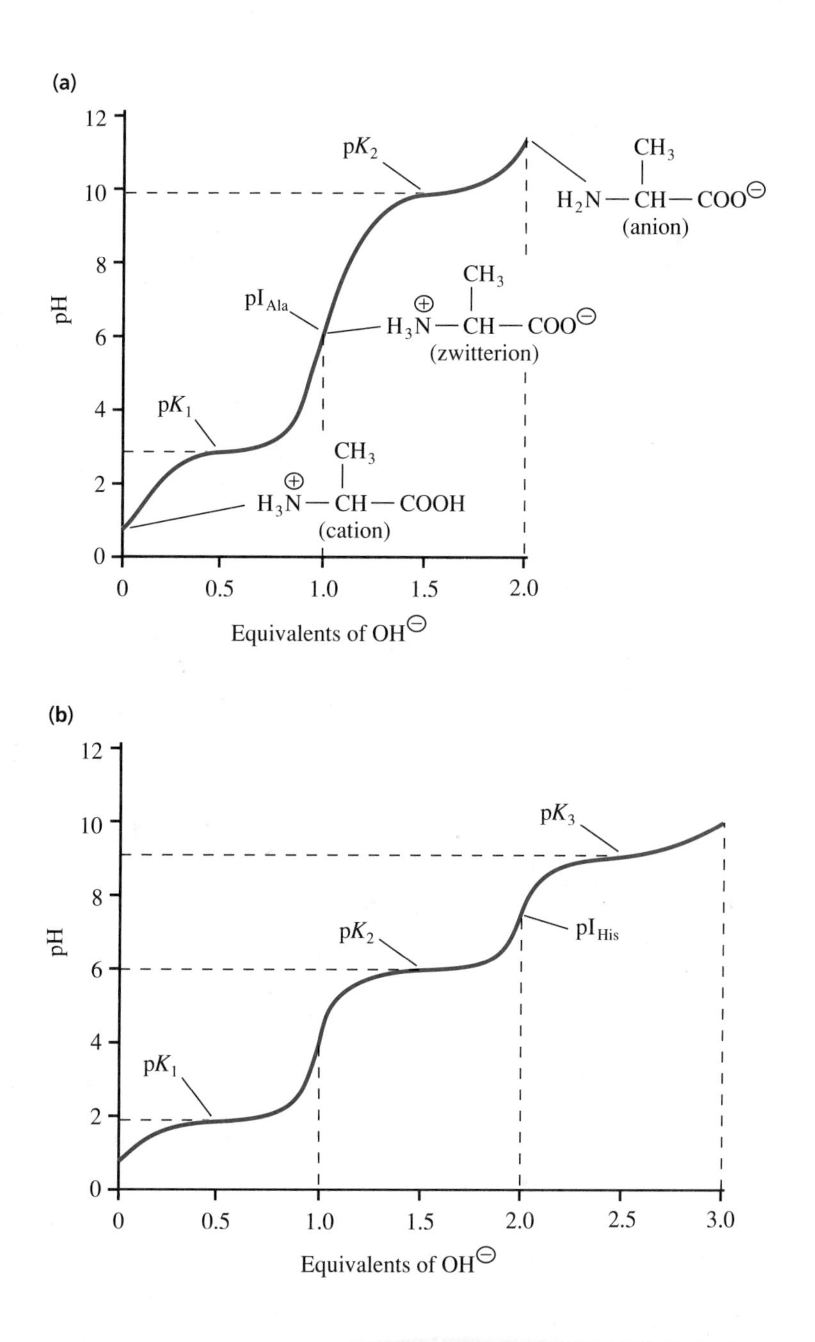

Figure 3·8
Titration of amino acids. **(a)** Titration curve for alanine. The first pK_a value is 2.4; the second is 9.9. **(b)** Titration curve for histidine. The three pK_a values are 1.8, 6.0, and 9.3. pI_{Ala} and pI_{His} represent the isoelectric points of the two amino acids.

Recall from Section 2·9 that a pK_a value equals the pH at the midpoint in the titration of an ionizable group. In other words, it is the point at which the concentration of a weak acid exactly equals the concentration of its conjugate base. One can deduce that the net charge on alanine molecules at pH 2.4 averages +0.5 and that the net charge at pH 9.9 averages −0.5. Midway between pH 2.4 and pH 9.9, at pH 6.15, the average net charge on alanine molecules in solution is zero. For this reason, pH 6.15 is referred to as the *isoelectric point* (pI), or isoelectric pH, of alanine. If alanine were placed in an electric field, such as that generated for electrophoresis (Section 3·5), at a pH below its pI, it would carry a net positive charge (in other words, its cationic form would predominate), and it would therefore migrate toward the *cathode* (the negative electrode of the electrophoresis apparatus). At a pH higher than its pI, alanine would carry a net negative charge and would migrate toward the *anode* (the positive electrode). At its isoelectric point (pH = 6.15), alanine would not migrate in either direction. The isoelectric point of an amino acid that, like alanine, contains only two ionizable groups (the α-amino and the α-carboxyl groups) is the arithmetic mean of its two pK_a values. However, for an amino acid such as histidine, which contains three ionizable groups, one must assess the average net charge at each of the three pK_a values in order to identify the isoelectric point correctly. For histidine at pH 1.8, the net charge averages +1.5; at pH 6.0, +0.5; and at pH 9.3, −0.5. Thus, the isoelectric point for histidine is midway between 6.0 and 9.3, or 7.65.

In proteins, the pK_a values of ionizable side chains can vary from those of the free amino acids. Two factors cause this perturbation of ionization constants. First, because α-amino and α-carboxyl groups lose their charges once they are linked by peptide bonds in proteins, they no longer exert strong inductive effects on their neighboring side chains. Second, the position of an ionizable side chain within the three-dimensional structure of a protein can affect its pK_a. As an example, the enzyme ribonuclease A has four histidines, but the side chain of each has a slightly different pK_a value as a result of differences in their microenvironments.

3·4 Amino Acids in Proteins Are Linked by Peptide Bonds

Amino acids are linked together in a chain by condensation of the α-carboxyl group of one amino acid with the α-amino group of another. The linkage thus formed between the amino acids is a secondary amide bond called a *peptide bond* (Figure 3·9). Note that a water molecule is lost from the condensing amino acids in the reaction. Unlike the carboxyl and amino groups of free amino acids in solution, the groups involved in peptide bonds carry no ionic charges.

The linked amino acid moieties are called amino acid *residues*. The names of residues in a polypeptide chain are formed by replacing the ending "-ine" or "-ate" with "-yl." Thus, a glycine residue in a polypeptide is called glycyl; a glutamate residue is called glutamyl. In the cases of asparagine, glutamine, and cysteine, "-yl" is simply added to the end to form asparaginyl, glutaminyl, and cysteinyl, respectively. The "-yl" ending indicates that the residue is an acyl unit (a structure that lacks the hydroxyl of the carboxyl group).

The free amino group and free carboxyl group at opposite ends of a peptide chain are called the *N-terminus* (amino terminus) and the *C-terminus* (carboxyl terminus), respectively. At neutral pH, each terminus carries an ionic charge. By convention, amino acid residues in a peptide chain are numbered from the N-terminus to the C-terminus and are usually written from left to right.

The primary structure of a protein is the linear sequence of amino acid residues linked by peptide bonds. Since both the standard three-letter abbreviations for the amino acids (for example, Gly–Arg–Phe–Ala–Lys) and the one-letter abbreviations (for example, GRFAK) are used to denote the primary structures of peptide chains, it is important to know both abbreviation systems. The terms *dipeptide*, *tripeptide*, *oligopeptide*, and *polypeptide* refer to chains of two, three, several (up

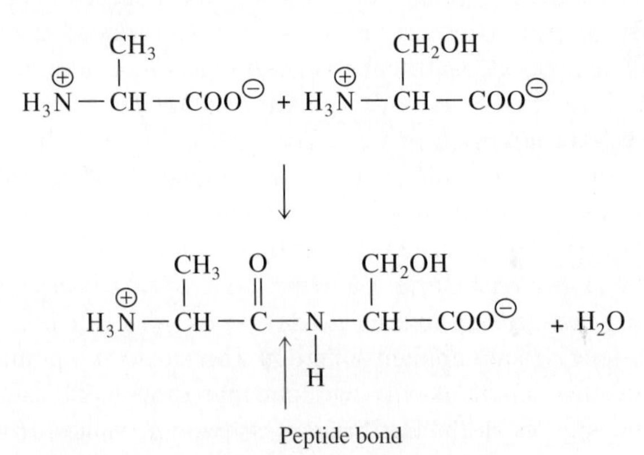

Figure 3·9
Peptide bond between two amino acids. The α-carboxyl group of one amino acid condenses with the α-amino group of another, with loss of a water molecule. The result is a dipeptide in which the amino acids are linked by a peptide bond. Here, alanine is condensed with serine to form alanylserine. (The chemical potential energy of ATP drives the reactions of peptide-bond synthesis.)

to about 20), or many (usually more than 20) amino acid residues, respectively. Note that a dipeptide (two residues) contains one peptide bond, a tripeptide contains two peptide bonds, a pentapeptide contains four peptide bonds, and so on. As a general rule, each peptide chain, whatever its length, possesses one free α-amino group and one free α-carboxyl group. (Exceptions include covalently modified terminal residues and circular peptide chains.) Most of the ionic charges associated with a protein molecule are contributed by the side chains of the constituent amino acids, and thus the ionic properties of a protein, as well as its solubility, depend on its amino acid composition. Furthermore, as we shall see in Chapter 4, interactions between side chains contribute to the stabilization of the three-dimensional structure of a protein molecule.

Although research on peptides declined for quite a few years, several discoveries have made peptide research popular again. The biochemistry of neuropeptides, such as endorphins, the natural pain killers, is a major field of research. Very simple peptides are being used as useful food additives. For example, the sweetening agent aspartame, the methyl ester of aspartylphenylalanine, is now used widely in diet drinks. Many other dipeptides are being tested for their taste, so perhaps a salty-tasting peptide will someday replace sodium chloride for those persons who require a low-sodium diet.

3·5 Proteins Can Be Purified by a Variety of Biochemical Techniques

In order to study a particular protein in the laboratory, one must often separate that protein from all other cell components, including other, often very similar, proteins. The steps of purification vary for different proteins but usually involve similar biochemical techniques. These techniques exploit minor differences in the solubilities, net charges, sizes, and binding specificities of proteins. In this section, we shall consider some of the common methods of protein purification. For most proteins, these methods must be applied at low temperatures, in the range from 0 to 4°C. Low temperatures minimize protein degradation during purification by impeding the activity of *proteases* (enzymes that cleave peptide bonds) and by reducing the likelihood that proteins will denature (many proteins are extremely heat-sensitive).

The first step in protein purification is to obtain a solution of proteins. The source of a protein is often whole cells or tissues in which the target protein may account for less than 0.1% of the total dry weight. Isolation of an intracellular protein requires that separated cells or chopped tissue be suspended in buffer and homogenized, or disrupted into cell fragments. Disruption of cells can be accomplished mechanically, chemically, or enzymatically. Large cell debris may then be removed, often by filtration through cheesecloth, and subcellular components may be separated by centrifugation, which separates principally on the basis of mass and density. To obtain proteins from subcellular organelles, the organelles must be disrupted, usually by treatment with a solution containing buffer and detergent. Most proteins will be dissolved in the buffer solution in which the cells were suspended for homogenization. For the following discussion, let us assume that the desired protein is one of a mixture of many proteins in this buffered solution.

Usually, the next step in protein purification is a relatively crude separation, or *fractionation*, procedure that makes use of the different solubilities proteins have in solutions of salts. Ammonium sulfate, a protein-stabilizing salt, is used most often in fractionation. Enough ammonium sulfate is mixed with the solution of proteins to precipitate the less soluble impurities. The target protein and other, more soluble, proteins are recovered by centrifugation and remain in the fluid, termed the supernatant fraction. Then, more ammonium sulfate is added to the supernatant fraction, until the desired protein is precipitated. The mixture is centrifuged, the fluid removed, and the precipitate dissolved in a minimal volume of buffer. Typically, fractionation using ammonium sulfate gives a two- to threefold purification (that is, one-half to two-thirds of the impurities have been removed from the resulting enriched protein fraction). Although the extent of purification is not large, this procedure concentrates the target protein and provides a solution that has a smaller volume, which is more suitable for column chromatography. At this point, the solvent is usually exchanged by dialysis for a buffer solution that has no residual ammonium sulfate and has the composition needed for chromatography. In dialysis, the protein solution is placed in a sealed sack composed of a semipermeable membrane, and the sack is suspended in a very large volume of buffer. The sack is actually a cylinder of cellophane tubing, held shut by knots. The proteins remain inside the sack because of their high molecular weights, but their previous medium is gradually exchanged for that in which the sack is suspended.

Column chromatography is used next to fractionate the mixture of proteins in solution. A cylindrical column is filled with an insoluble matrix, often consisting of substituted cellulose fibers or beads of synthetic resin. A protein mixture is applied to the column and washed through the matrix by the addition of solvent. As solvent flows through the column, the exiting liquid, or eluate, is collected in many fractions, a few of which are represented in Figure 3·10a. The rate at which proteins travel through the matrix depends on interactions between matrix and protein.

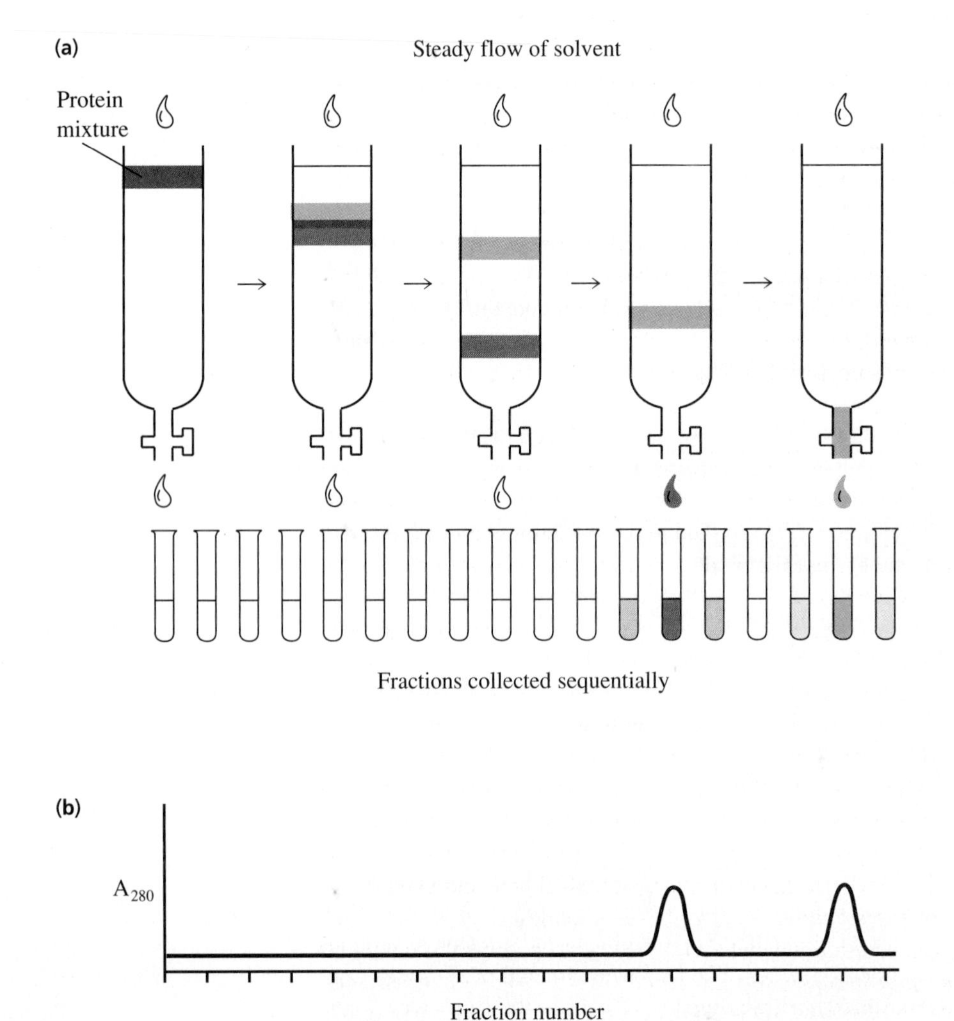

Figure 3·10
Column chromatography. **(a)** Schematic view of fractionation using a chromatographic column. Initially, a mixture of proteins is added to the column, which houses an insoluble, solid matrix. Solvent then flows steadily into the column from a reservoir. Washed by solvent, different proteins (represented by red and blue bands) travel through the column at different rates, depending on their interactions with the matrix. Eluate is collected in a series of fractions. **(b)** The protein concentration of each fraction is determined by measuring spectrophotometric absorbance at 280 nm. The peaks correspond to the elution of the protein bands shown in (a). The fractions are then tested for the activity of the target protein.

For a given column, different proteins will be eluted at different rates. The concentration of protein in each fraction can be determined by measuring the spectrophotometric absorbance of the eluate at 280 nm (Figure 3·10b). (Recall from Section 3·2 that at neutral pH, tyrosine and tryptophan absorb UV light at 280 nm.) The fractions containing protein must then be assayed for biological activity to locate the target protein.

Several types of column chromatography are classified according to the type of matrix. In *ion-exchange chromatography*, the matrix consists of beads or fibers carrying positive charges (anion-exchange resins) or negative charges (cation-exchange resins). Anion-exchange supports bind negatively charged proteins, retaining them in the matrix for subsequent elution. Conversely, cation-exchange materials bind positively charged proteins. The bound proteins can be serially eluted by gradually increasing the salt concentration in the solvent, since like-charged salt ions and proteins bind to the matrix competitively.

Gel-filtration chromatography, which separates proteins on the basis of molecular size, utilizes a gel resin consisting of porous beads. Proteins that are smaller than the average pore size penetrate much of the internal volume of the beads and are therefore retarded by the matrix. The smaller the protein, the later it is eluted from the column. Fewer of the pores are accessible to larger protein molecules. Consequently, the largest proteins flow past the beads and are eluted first. Because of this separation of proteins on the basis of molecular size, gel-filtration chromatography is also known as molecular exclusion chromatography. The choice of pore size of a gel-filtration matrix depends on the molecular weight of the protein to be purified.

Affinity chromatography, the most selective type of column chromatography, relies on binding interactions between a protein and a ligand. A *ligand* is a molecule, group, ion, or atom that binds usually noncovalently to another molecule or atom. In affinity chromatography, the ligand is covalently attached to the matrix. The ligand may be a reactant or product to which an enzyme binds in vivo, or it may be an antibody that recognizes the protein of interest. As a mixture of proteins passes through the column, only the target protein specifically binds to the matrix. The column is then washed several times with the buffer in which the proteins were dissolved to rid the column of nonspecifically bound proteins. Finally, the target protein is eluted by passing through the column a solvent containing a high concentration of the free ligand, or perhaps a concentrated solution of salt. Affinity chromatography alone can sometimes purify a protein 1000- to 10 000-fold.

The resolution of conventional chromatographic methods is greatly increased by the use of *high-pressure liquid chromatography* (HPLC). The matrix in HPLC consists of beads that are smaller, more uniform in size, and more tightly packed than the beads used in conventional chromatographic columns. High pressure (hundreds of pounds per square inch) is applied to pump a protein mixture and solvent through a column, which is usually made of glass. Although speed, resolution, and sensitivity are greatly increased using HPLC, the separations are still based on the same principles that govern ion-exchange, gel-filtration, and affinity chromatography.

Electrophoresis separates proteins based on their migration in an electric field. In *polyacrylamide gel electrophoresis* (PAGE), protein samples are placed on a highly cross-linked gel matrix, and an electric field is applied. The matrix is buffered to a mildly alkaline pH so that proteins are anionic and migrate toward the anode. Typically, several samples are run at once, together with a reference sample. The gel matrix retards the migration of large molecules as they move in the electric field. Hence, proteins are fractionated on the basis of both charge and mass.

A modification of the standard electrophoresis technique uses the negatively charged detergent sodium dodecyl sulfate (SDS) to overwhelm the native charge on proteins so that they are separated only on the basis of mass. *SDS-polyacrylamide gel electrophoresis* (SDS-PAGE) is used to assess the purity and to

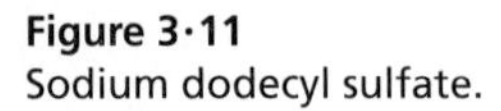

Figure 3·11
Sodium dodecyl sulfate.

estimate the molecular weight of a protein. In SDS-PAGE, the detergent is added to the polyacrylamide gel as well as to the protein samples. Before loading the protein samples on the gel, 2-mercaptoethanol is added to the sample, which is heated. The combination of SDS, 2-mercaptoethanol, and heat breaks any disulfide bonds and denatures (unfolds) the proteins, thus allowing for electrophoretic separation of protein subunits. The dodecyl sulfate anion, which has a long hydrophobic tail (Figure 3·11), binds to hydrophobic side chains of amino acid residues in the polypeptide chain. It binds at a ratio of approximately one molecule of SDS for every two residues of a typical protein. Since larger proteins bind proportionately more SDS, the charge-to-mass ratios of all treated proteins are approximately the same, and all of the SDS-protein complexes are highly negatively charged.

After the protein samples are loaded onto the gel and an electric field is applied, all SDS-protein complexes move toward the anode, as diagrammed in Figure 3·12a. However, their rate of migration through the gel is inversely proportional to the logarithm of their molecular weight—larger proteins encounter more resistance and therefore migrate more slowly than smaller proteins. This sieving effect is similar to gel-filtration chromatography with one exception. In gel filtration, larger molecules are excluded from the pores of the gel and thus travel faster. In SDS-PAGE, all molecules must penetrate the pores of the continuous gel; thus, the largest proteins travel most slowly. The protein bands that result from this differential migration, as pictured in Figure 3·12b, can be visualized by staining. Molecular weights of unknown proteins can be estimated by comparing their migrations to the migrations of reference proteins electrophoresed on the same gel.

Isoelectric focusing, a modified form of electrophoresis, employs buffers to create a pH gradient within a polyacrylamide gel. When protein samples are electrophoresed in such a gel, each protein will migrate to the point in the pH gradient at which it is no longer charged. In other words, each protein will band at its isoelectric point (pI). The separated proteins can then be visualized as in SDS-PAGE.

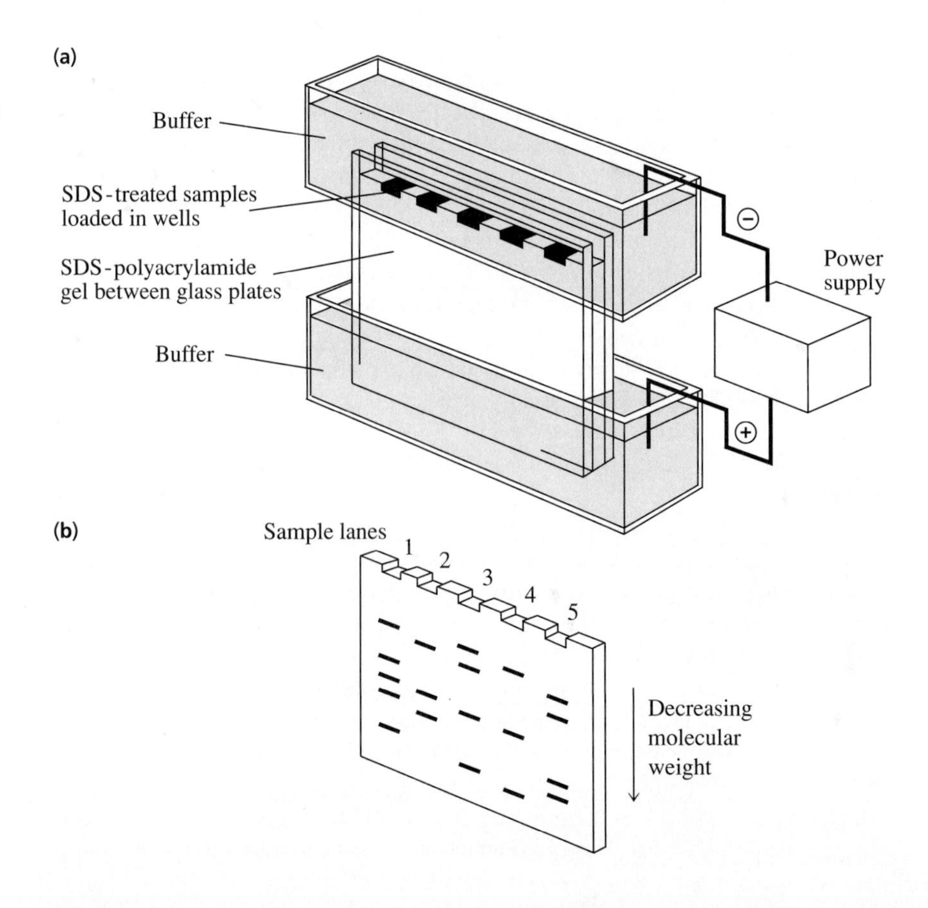

Figure 3·12
SDS-PAGE. **(a)** An electrophoresis apparatus includes an SDS-polyacrylamide gel between two glass plates and buffer in upper and lower reservoirs, as shown. Samples treated with SDS are loaded into the wells of the gel, and voltage is applied. Since proteins complexed with SDS are negatively charged, they migrate toward the anode. **(b)** The banding pattern of the proteins after electrophoresis can be visualized by staining. Each of the five samples shown contains three or more protein components. Since the smallest proteins migrate fastest, the proteins of lowest molecular weight are at the bottom of the gel.

3·6 The Amino Acid Composition of Proteins Can Be Determined Quantitatively

Once a protein has been isolated in a purified state, its amino acid composition can be determined. First, the peptide bonds of the protein are cleaved by acid hydrolysis, typically using 6 M HCl at 110°C in vacuo for 16 to 72 hours (Figure 3·13). Next, the hydrolyzed mixture, or hydrolysate, is subjected to a chromatographic procedure during which each of the amino acids is separated and quantitated, a process called *amino acid analysis*. One method of amino acid analysis involves treatment of the protein hydrolysate with phenylisothiocyanate (PITC) at pH 9.0 to generate phenylthiocarbamoyl (PTC)-amino acid derivatives (Figure 3·14). The PTC-amino acid mixture is then subjected to HPLC in a column of finely divided silica to which short hydrocarbon chains have been attached. The amino acids are separated by the hydrophobic properties of their side chains. As each PTC-amino acid derivative is eluted, it is detected, and its concentration is determined by measuring the absorbance of the eluate at 254 nm (the peak absorbance of the PTC moiety). A plot showing the absorbance of the eluate from an HPLC column as a

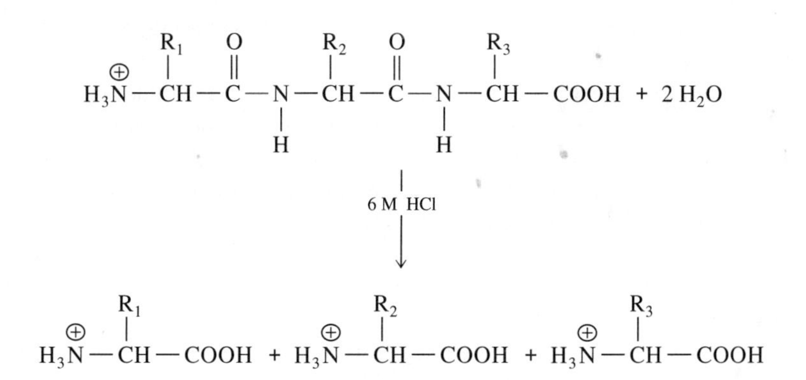

Figure 3·13
Acid-catalyzed hydrolysis of a peptide. Incubation with 6 M HCl at 110°C for 16 to 72 hours releases the constituent amino acids of a peptide.

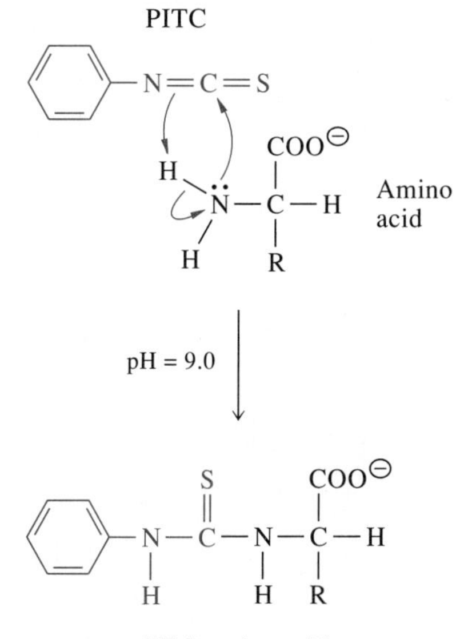

Figure 3·14
Amino acid treated with phenylisothiocyanate (PITC). The α-amino group of an amino acid reacts with phenylisothiocyanate to give a phenylthiocarbamoyl-amino acid (PTC-amino acid).

function of time is given in Figure 3·15. The peaks, which correspond to amino acids, are identified by standard one-letter abbreviations. Since different PTC-amino acid derivatives are eluted at different rates, the timing of the peaks identifies the amino acids. The amount of each amino acid present in the aliquot of the hydrolysate subjected to HPLC is proportional to the area under its peak. With this method, amino acid analysis can be performed on samples as small as 1 picomole (10^{-12} mol) of a protein that contains approximately 200 residues.

Despite its usefulness, acid hydrolysis under one set of hydrolytic conditions cannot yield a complete amino acid analysis. Since the side chains of asparagine and glutamine contain amide bonds, the acid used to cleave the peptide bonds of the protein also converts asparagine to aspartic acid plus ammonium ion (with a chloride counterion if HCl is used) and glutamine to glutamic acid plus ammonium ion. When acid hydrolysis is used, the combined totals of [glutamate + glutamine] are designated with the abbreviations Glx or Z, and the combined totals of [aspartate + asparagine] are designated Asx or B, as in the chromatogram shown in Figure 3·15.

At elevated temperatures, the indole side chain of tryptophan is especially sensitive to oxidation by air. During acid hydrolysis of proteins, which typically employs temperatures of about 110°C, the side chain of tryptophan is almost totally destroyed, even in evacuated sealed tubes. Thus, the tryptophan content of a protein is often estimated on the basis of its ultraviolet spectrum. Alternatively, it may be analyzed following alkaline hydrolysis or, more often, by including an antioxidant in acid hydrolysis.

Small losses of serine (averaging 10% per 20 hours) and threonine and tyrosine (5% per 20 hours) are also experienced during conventional acid hydrolysis. Conversely, the peptide bonds of valine and isoleucine, which are sterically shielded by hydrocarbon branching at their β-carbons, are slower to hydrolyze than other peptide bonds. For these reasons, several samples of a purified protein are subjected to acid hydrolysis for periods ranging from 16 to 72 hours, and the results of amino acid analyses are usually interpreted by extrapolation of the data obtained.

Cysteine cannot be determined accurately as a component of the acid hydrolysate. Therefore, cysteine residues are oxidized or carboxymethylated (Section 3·7) before hydrolysis of the protein, thereby forming derivatives that can be measured quantitatively after acid hydrolysis.

The amino acid compositions of many proteins have been determined, and dramatic differences in these compositions have been found, illustrating the tremendous potential for diversity based on different combinations of the 20 amino acids. The amino acid compositions of five relatively small proteins are given in Table 3·2. The data given are based on the complete primary structures that have been obtained for these proteins. Note that proteins need not contain all 20 of the common amino acids.

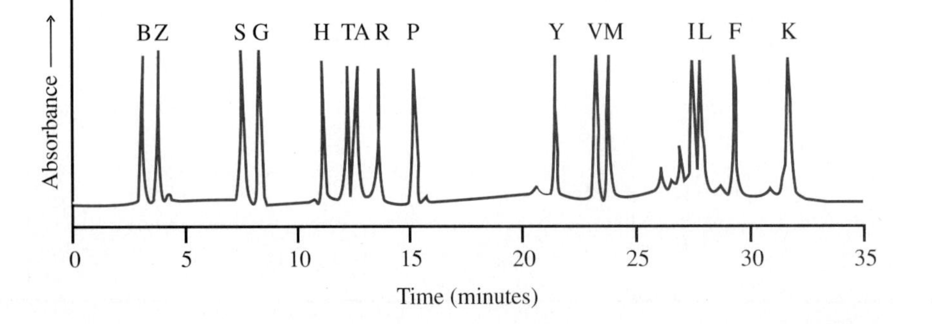

Figure 3·15
Chromatogram obtained from HPLC separation of PTC-amino acids. PTC-amino acids in the column eluate are detected by their absorbance of ultraviolet light at a wavelength of 254 nm. Peaks are labelled with one-letter abbreviations. The symbols B and Z indicate totals of [asparagine + aspartate] and [glutamine + glutamate], respectively. [Adapted from Hunkapiller, M. W., Strickler, J. E., and Wilson, K. J. (1984). Contemporary methodology for protein structure determination. *Science* 226:304–311.]

A surprising discovery was that a 21st amino acid, *selenocysteine* (which contains selenium in place of the sulfur of cysteine), is incorporated into a few proteins. In addition, some of the amino acid residues incorporated into certain proteins are biochemically altered after they have been assembled into polypeptide chains. One example, discussed earlier in this chapter, is the formation of cystine residues from two cysteine residues. Thus, although only 20 amino acids are used in the biosynthesis of most proteins, more than 20 species of amino acid residues are found in a number of proteins. Also, many other amino acids occur in cells as free compounds or as components of molecules other than proteins.

3·7 The Edman Degradation Procedure Is Used to Determine the Sequence of Amino Acid Residues

Amino acid analysis provides information on the composition of a protein but not on the primary structure (sequence of residues). Pehr Edman developed a technique that permits sequential removal and identification of one residue at a time from the N-terminus of a protein. This technique revolutionized protein-sequence analysis. The *Edman degradation procedure* involves treating a protein or polypeptide with phenylisothiocyanate (PITC), also known as the Edman reagent, at pH 9.0. PITC reacts under these conditions with the free N-terminus of the chain to form a phenylthiocarbamoyl derivative, or PTC-peptide (Figure 3·16). (Recall that PITC is also used in the measurement of free amino acids, as shown in Figure 3·14.)

Table 3·2 Amino acid composition of several proteins

Amino acid	Number of residues per molecule of protein				
	Lysozyme (hen egg white)	Cytochrome *c* (human)	Ferredoxin (spinach)	Insulin (bovine)	Hemoglobin, α subunit (human)
Nonpolar					
Ala	12	6	9	3	21
Val	6	3	7	5	13
Leu	8	6	8	6	18
Ile	6	8	4	1	0
Pro	2	4	4	1	7
Met	2	3	0	0	2
Phe	3	3	2	3	7
Trp	6	1	1	0	1
Polar, neutral					
Cys	8	2	5	6	1
Gly	12	13	6	4	7
Ser	10	2	7	3	11
Thr	7	7	8	1	9
Tyr	3	5	4	4	3
Asn	13	5	2	3	4
Gln	3	2	4	3	1
Polar, acidic					
Asp	8	3	11	0	8
Glu	2	8	9	4	4
Polar, basic					
Lys	6	18	4	1	11
Arg	11	2	1	1	3
His	1	3	1	2	10
Total residues	129	104	97	51	141

Figure 3·16
Edman degradation procedure. At pH 9.0, the N-terminal residue of a polypeptide chain reacts with phenylisothiocyanate to give a phenylthiocarbamoyl peptide. Treating this derivative with trifluoroacetic acid (F_3CCOOH) releases an anilinothiazolinone derivative of the N-terminal amino acid residue without cleaving the other peptide bonds of the polypeptide chain. The anilinothiazolinone is extracted and treated with aqueous acid, which rearranges the derivative so that it forms a stable phenylthiohydantoin derivative. This derivative can then be identified chromatographically, usually by HPLC. The remainder of the polypeptide chain is returned to alkaline conditions, and the amino acid residue formerly in the second position (now the new N-terminal residue) is subjected to the next cycle of Edman degradation.

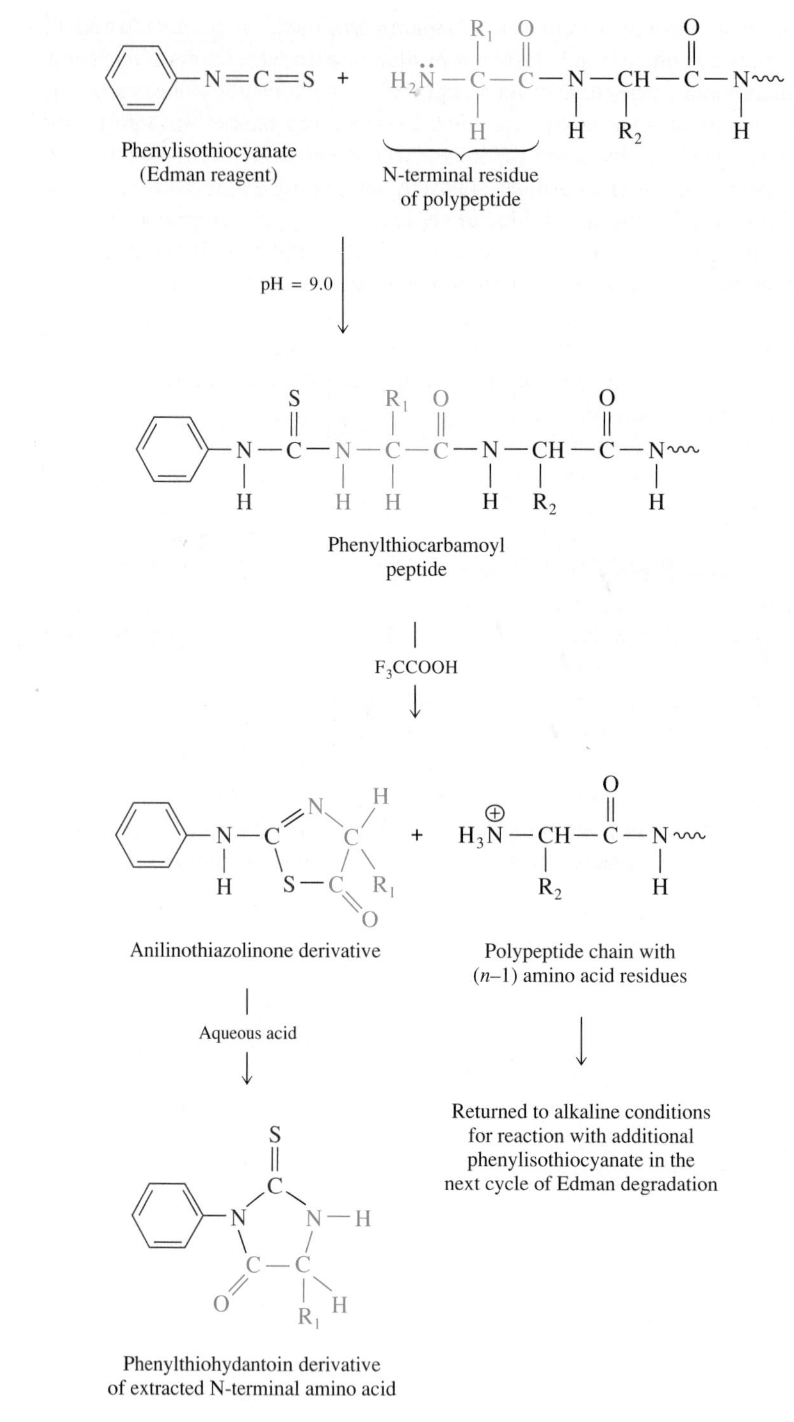

When the PTC-peptide is treated with an anhydrous acid, such as trifluoroacetic acid, the peptide bond of the N-terminal residue is selectively cleaved, releasing an anilinothiazolinone derivative of the residue. This derivative can be extracted with an organic solvent, such as butyl chloride, leaving the remaining peptide in the aqueous phase. The unstable anilinothiazolinone derivative is treated with aqueous acid, which converts it to a stable phenylthiohydantoin derivative of the amino acid that had been the N-terminal residue (PTH-amino acid). The polypeptide chain in the aqueous phase, now one residue shorter (residue 2 of the original protein is now the N-terminus), can be adjusted back to pH 9.0 and treated again with PITC. The entire procedure (coupling, cleavage, extraction, and conversion) can be repeated serially using an automated instrument known as a sequenator. Each cycle yields a PTH-amino acid that can be identified chromatographically, usually by HPLC.

If a protein contains one or more cystine residues (pairs of cysteine residues cross-linked by disulfide bonds), the disulfide bonds must be cleaved to permit release of the half-cystine residues as PTH-amino acids during the appropriate cycles of Edman degradation. One method of cleavage involves treating the protein with performic acid, which oxidizes cystine to two cysteic acid residues (Figure 3·17). Performic acid oxidation also oxidizes cysteine residues to cysteic acid and methionine residues to methionine sulfone, each of which is stable and can either be quantitated after acid hydrolysis or converted to a PTH-amino acid by Edman degradation. However, performic acid destroys tryptophan and is therefore not favored for most protein-sequencing tasks. (Of historical interest, tryptophan is not present in either of the first two proteins successfully sequenced, bovine insulin and bovine ribonuclease A. Performic acid oxidation was thus a suitable choice for cleavage of the disulfide bonds in those proteins.) A preferred method for cleaving disulfide bonds without destroying tryptophan involves treating the protein with an excess of a thiol compound, such as 2-mercaptoethanol, which reduces cystine

Figure 3·17
Cleaving disulfide bonds with performic acid. Performic acid cleaves disulfide bonds of proteins and prevents the bonds from re-forming by oxidizing a cystine residue to two cysteic acid residues.

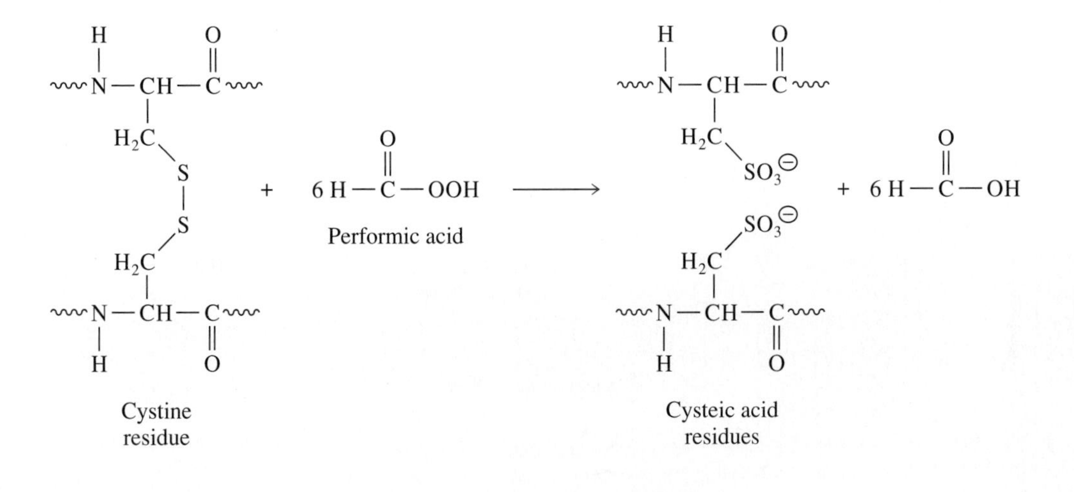

residues to pairs of cysteine residues (Figure 3·18a). The reactive sulfhydryl groups of the cysteine residues are then blocked by treatment with an alkylating agent, such as iodoacetate, which converts oxidizable cysteine residues to stable *S*-carboxymethylcysteine residues, thereby preventing the reformation of disulfide bonds in the presence of oxygen (Figure 3·18b).

The yield of the Edman degradation procedure under carefully controlled conditions approaches 100%, and a few picomoles of sample protein can yield sequences of 30 residues or more (higher for larger samples of a given protein) before further measurement is obscured by the increasing concentration of unrecovered sample from previous cycles of the procedure. However, despite the power of the Edman degradation procedure, it must be supplemented by fragmentation procedures to obtain the primary structure of large proteins.

3·8 Proteins Can Be Selectively Cleaved to Form Shorter Peptides

Most proteins contain too many residues to be sequenced in their entirety by Edman degradation proceeding from the N-terminus. Therefore, proteases or certain chemical reagents can be used to selectively cleave some of the peptide bonds of a protein. The smaller peptides formed can then be isolated and subjected to sequencing by the Edman degradation procedure.

The chemical reagent cyanogen bromide (BrCN), for example, cleaves by reacting specifically with methionine residues to produce peptides with C-terminal homoserine lactone residues and new N-terminal residues (Figure 3·19). Since most proteins contain relatively few methionine residues, treatment with BrCN usually produces only a few peptide fragments. For example, reaction of BrCN with a polypeptide chain containing three internal methionine residues and a

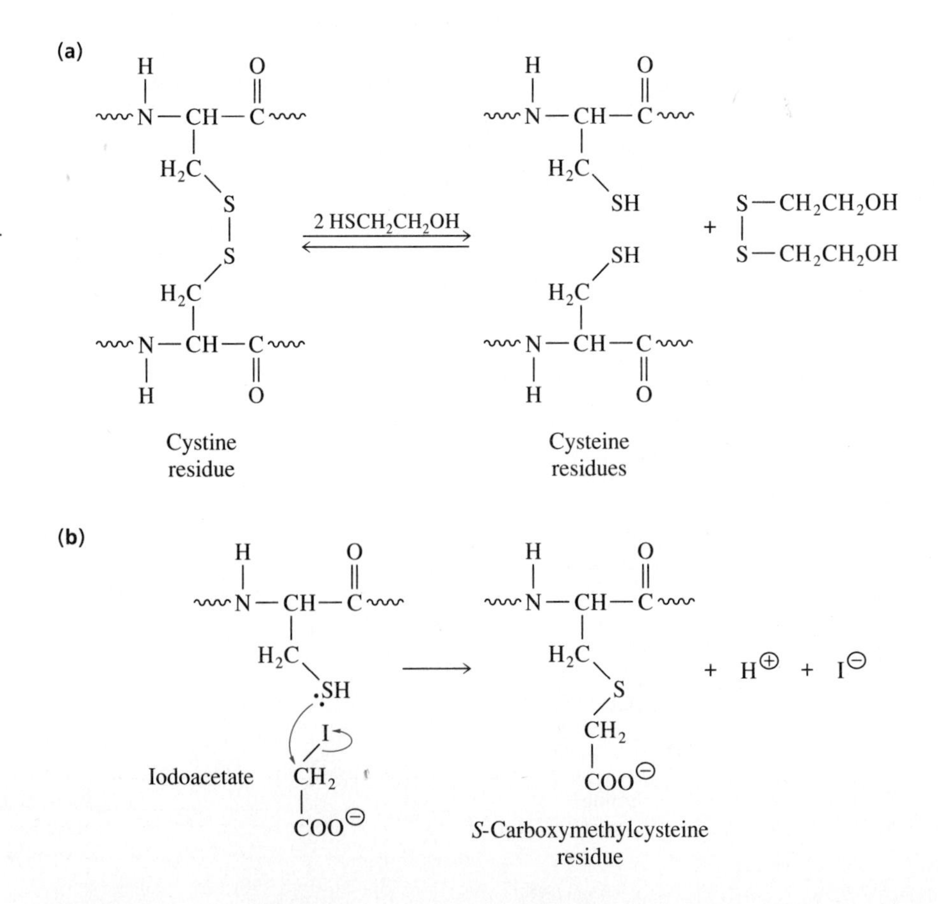

Figure 3·18
Cleaving disulfide bonds with 2-mercaptoethanol and blocking them with iodoacetate. **(a)** When a protein is treated with excess 2-mercaptoethanol ($HSCH_2CH_2OH$), a disulfide-exchange reaction occurs in which each cystine residue is reduced to two cysteine residues and 2-mercaptoethanol is oxidized to a disulfide. **(b)** Treating the reduced protein with the alkylating agent iodoacetate converts all free cysteine residues to stable *S*-carboxymethylcysteine residues, thus preventing the reformation of disulfide bonds in the presence of oxygen.

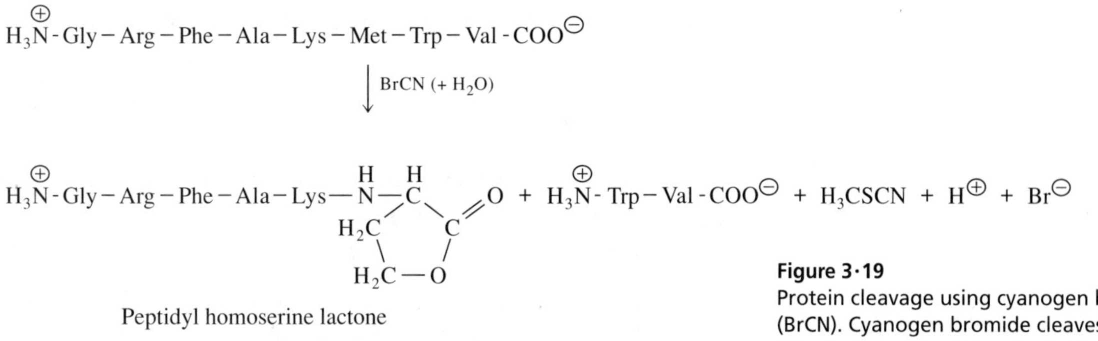

$$\overset{\oplus}{H_3N}\text{-Gly-Arg-Phe-Ala-Lys-Met-Trp-Val-COO}^{\ominus}$$

$$\downarrow \text{BrCN } (+ H_2O)$$

$$\overset{\oplus}{H_3N}\text{-Gly-Arg-Phe-Ala-Lys-}\text{(Peptidyl homoserine lactone)} + \overset{\oplus}{H_3N}\text{-Trp-Val-COO}^{\ominus} + H_3CSCN + H^{\oplus} + Br^{\ominus}$$

Peptidyl homoserine lactone

Figure 3·19
Protein cleavage using cyanogen bromide (BrCN). Cyanogen bromide cleaves polypeptide chains on the C-terminal side of methionine residues. The reaction produces a peptidyl homoserine lactone and generates a new amino terminus.

C-terminal alanine residue should generate four peptide fragments, three that possess a C-terminal homoserine lactone residue and one that possesses a C-terminal alanine residue (the original C-terminus of the protein).

Many proteases are available, and trypsin, *Staphylococcus aureus* V8 protease, and chymotrypsin are often used in protein sequencing. Trypsin specifically catalyzes the hydrolysis of peptide bonds on the carbonyl side of lysine residues (lysyl bonds) and arginine residues (arginyl bonds), both of which bear positively charged side chains (Figure 3·20a). *S. aureus* V8 protease catalyzes cleavage of peptide bonds on the carbonyl side of negatively charged residues (glutamyl and aspartyl bonds) and, under appropriate conditions (50 mM ammonium bicarbonate), selectively cleaves only glutamyl bonds. Chymotrypsin, the least specific of the three proteases, preferentially catalyzes the hydrolysis of peptide bonds on the carbonyl side of uncharged residues with aromatic or bulky hydrophobic side chains, for example, phenylalanyl, tyrosyl, and tryptophanyl bonds (Figure 3·20b).

By judicious application of cyanogen bromide, trypsin, *S. aureus* V8 protease, and chymotrypsin to individual samples of a large protein whose disulfide bonds have been reduced and alkylated, one can generate many peptide fragments of various sizes, which can then be isolated and sequenced by Edman degradation. In the final stage of sequence determination, the amino acid sequence of a large polypeptide chain can be deduced by lining up matching sequences of overlapping peptide fragments, as illustrated with a short peptide in Figure 3·20c.

For proteins that contain disulfide bonds, the complete covalent structure is not fully resolved until the positions of the disulfide bonds have been established.

Figure 3·20
Cleavage of an oligopeptide and sequencing of its amino acids. **(a)** Trypsin catalyzes cleavage of peptides on the carbonyl side of the basic residues arginine and lysine. **(b)** Chymotrypsin catalyzes cleavage of peptides on the carbonyl side of aromatic residues, including phenylalanine, tyrosine, and tryptophan, and some other residues with bulky side chains. **(c)** By using the Edman degradation procedure to determine the sequence of each fragment (highlighted in boxes) and then lining up the matching sequences of overlapping fragments, one can find the order of the fragments and thus deduce the sequence of the entire oligopeptide chain, seen at the top of both (a) and (b).

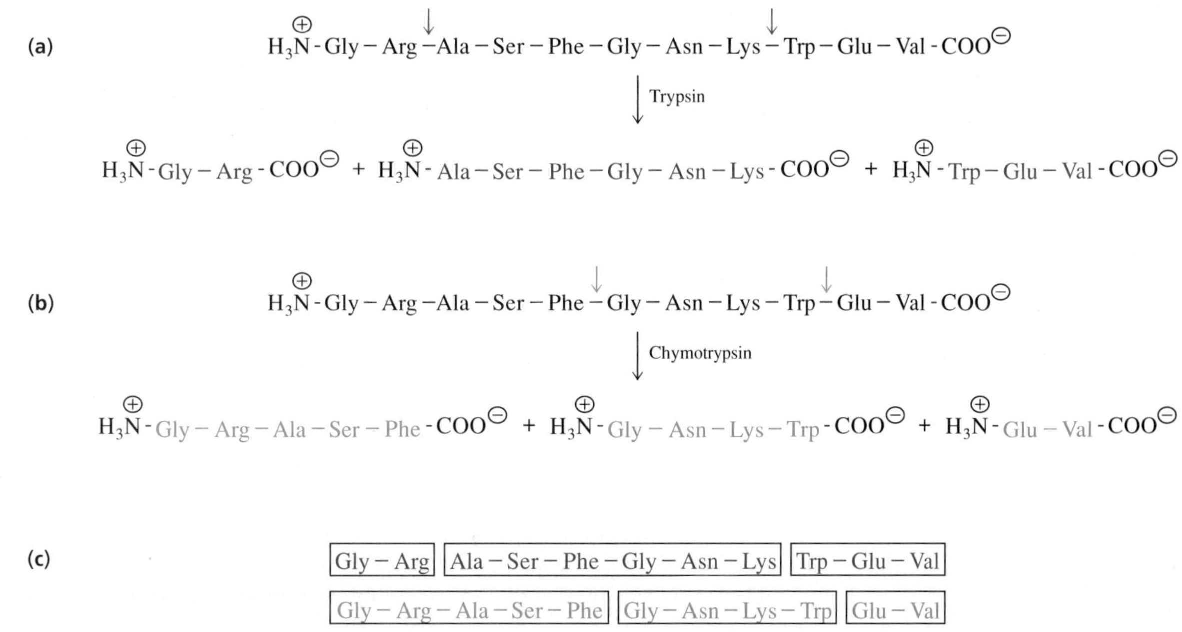

(a)

$$\overset{\oplus}{H_3N}\text{-Gly-Arg}\downarrow\text{-Ala-Ser-Phe-Gly-Asn-Lys}\downarrow\text{-Trp-Glu-Val-COO}^{\ominus}$$

$$\downarrow \text{Trypsin}$$

$$\overset{\oplus}{H_3N}\text{-Gly-Arg-COO}^{\ominus} + \overset{\oplus}{H_3N}\text{-Ala-Ser-Phe-Gly-Asn-Lys-COO}^{\ominus} + \overset{\oplus}{H_3N}\text{-Trp-Glu-Val-COO}^{\ominus}$$

(b)

$$\overset{\oplus}{H_3N}\text{-Gly-Arg-Ala-Ser-Phe}\downarrow\text{-Gly-Asn-Lys-Trp}\downarrow\text{-Glu-Val-COO}^{\ominus}$$

$$\downarrow \text{Chymotrypsin}$$

$$\overset{\oplus}{H_3N}\text{-Gly-Arg-Ala-Ser-Phe-COO}^{\ominus} + \overset{\oplus}{H_3N}\text{-Gly-Asn-Lys-Trp-COO}^{\ominus} + \overset{\oplus}{H_3N}\text{-Glu-Val-COO}^{\ominus}$$

(c)

Gly – Arg	Ala – Ser – Phe – Gly – Asn – Lys	Trp – Glu – Val

Gly – Arg – Ala – Ser – Phe	Gly – Asn – Lys – Trp	Glu – Val

Determining the positions of the disulfide cross-links requires a multistep procedure.

1. Free sulfhydryl groups of the native protein are blocked by treatment with iodoacetate.

2. The modified protein is cleaved using a specific reagent or enzyme, such as cyanogen bromide or trypsin.

3. Each peptide fragment is isolated, and its disulfide bonds are oxidized with performic acid, which splits the cross-linked peptide fragments into two chains containing cysteic acid residues.

4. The sequences of the individual fragments obtained by Edman degradation are compared with the established sequence of the complete polypeptide chain. Positions occupied by cysteic acid residues were originally involved in disulfide bonds, and positions that contain *S*-carboxymethylcysteine originally were free sulfhydryl groups.

The process of generating and sequencing peptide fragments has one other major application. The N-terminal α-amino groups of many enzymes, especially those of mammalian tissues, are acetylated. These substituted amines do not react at all when subjected to the Edman degradation procedure. Selective cleavage produces peptide fragments with unblocked N-termini that can be separated and sequenced, so at least internal sequences can be obtained.

In recent years, it has become relatively easy to deduce the amino acid sequence of a protein by determining the sequence of nucleotides in the gene that encodes the protein. (DNA sequencing is described in Box 20·1.) However, direct protein sequencing retains its importance because DNA sequences do not reveal when and where disulfide bonds occur or whether amino acid residues are modified after synthesis of the protein.

In 1953, Frederick Sanger was the first scientist to determine the complete sequence of a protein, the hormone insulin, which has 51 residues. In 1955, he was awarded a Nobel prize for this work. Years later, Sanger also won a Nobel prize for pioneering the sequencing of nucleic acids. Results from both types of sequencing are now being obtained rapidly and at accelerating rates. Table 3·3 reflects the progress made in sequencing proteins, RNA molecules, and DNA molecules from 1935 to 1984. Millions of residues of proteins and nucleic acids have now been sequenced, and work is underway to sequence the entire human genome—some three billion base pairs.

3·9 Comparisons of the Primary Structures of Proteins Can Reveal Evolutionary Relationships

The amino acid sequence of a protein is determined by the gene that encodes it. Therefore, differences among primary structures of proteins reflect evolutionary change.

The sequences of amino acids in proteins from closely related species are likely to be quite similar. The number of differences within the amino acid sequences of these proteins gives an idea of how far various species have diverged in the course of evolution. Distantly related species will contain proteins that show many differences.

The protein cytochrome *c*, which consists of a single polypeptide chain of from 104 to 111 amino acid residues, provides an excellent opportunity for evolutionary comparisons at the molecular level because it is found in all aerobic organisms. Figure 3·21 illustrates the similarities between cytochrome *c* sequences in different species by depicting them as a tree whose branches are proportional in

Table 3·3 Progress in the sequencing of proteins, RNA molecules, and DNA molecules

Year	Protein	RNA	DNA	Number of residues sequenced
1935	Insulin			1
1945	Insulin			2
1947	Gramicidin S			5
1949	Insulin			9
1953	Insulin			51
1960	Ribonuclease			124
1965		$tRNA_{Ala}$		75
1967		5S RNA		120
1968			Bacteriophage λ	12
1978			Bacteriophage ϕX 174	5 386
1981			Mitochondria	16 569
1982			Bacteriophage λ	48 502
1984			Epstein-Barr virus	172 282

[Adapted from Sanger, F. (1988). Sequences, sequences, and sequences. *Annu. Rev. Biochem.* 57:1–28.]

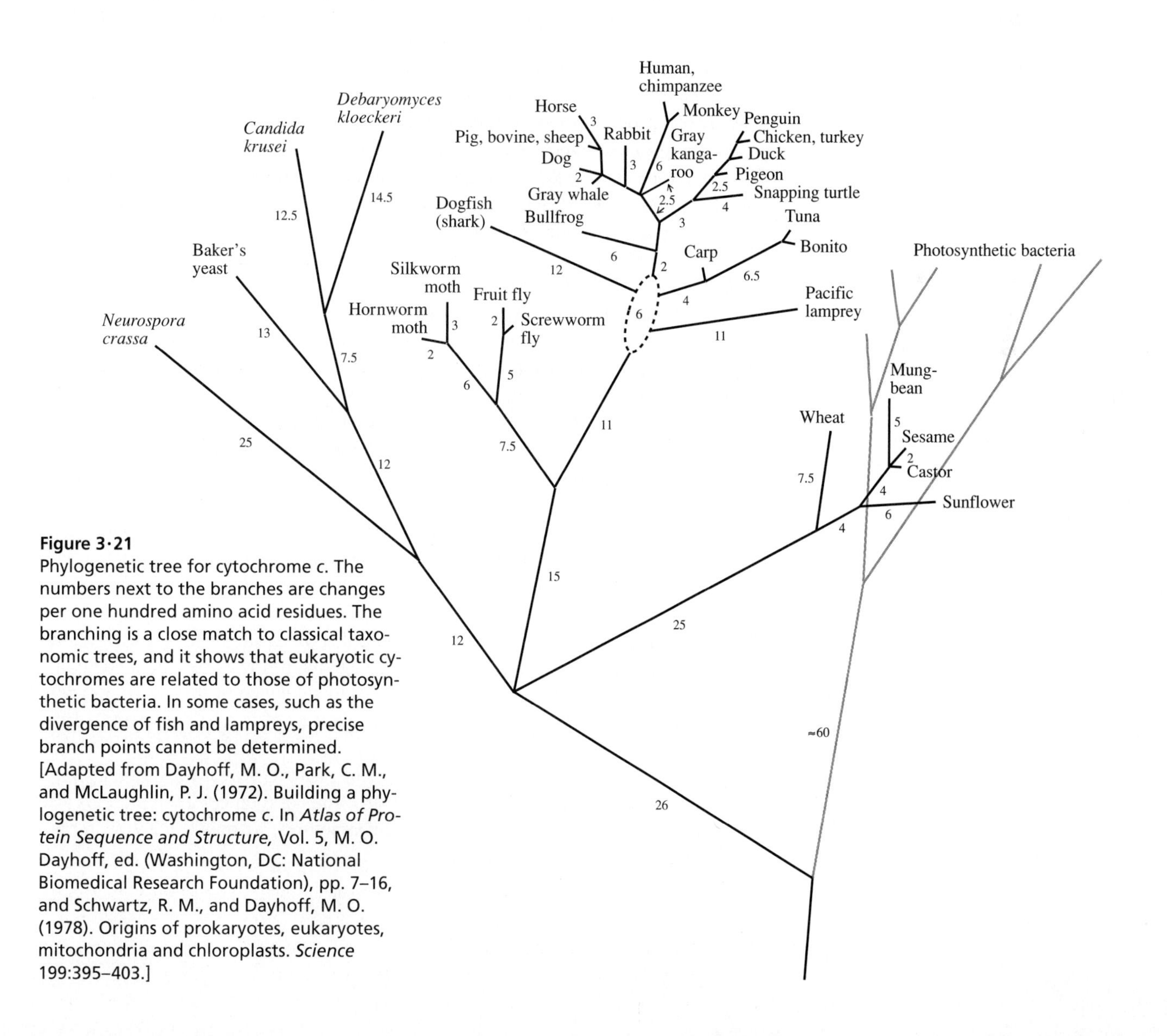

Figure 3·21
Phylogenetic tree for cytochrome *c*. The numbers next to the branches are changes per one hundred amino acid residues. The branching is a close match to classical taxonomic trees, and it shows that eukaryotic cytochromes are related to those of photosynthetic bacteria. In some cases, such as the divergence of fish and lampreys, precise branch points cannot be determined. [Adapted from Dayhoff, M. O., Park, C. M., and McLaughlin, P. J. (1972). Building a phylogenetic tree: cytochrome *c*. In *Atlas of Protein Sequence and Structure,* Vol. 5, M. O. Dayhoff, ed. (Washington, DC: National Biomedical Research Foundation), pp. 7–16, and Schwartz, R. M., and Dayhoff, M. O. (1978). Origins of prokaryotes, eukaryotes, mitochondria and chloroplasts. *Science* 199:395–403.]

length to the number of differences in the amino acid sequences of cytochrome *c*. Species that are closely related have relatively few differences in the primary structures of their cytochrome *c* molecules. For example, the penguin and chicken proteins differ at fewer than six positions. At great evolutionary distances, the number of differences in the same protein may be fairly large; nevertheless, many amino acids in cytochrome *c* are identical in every species. These *conserved* positions are essential for the proper function of the protein.

The phylogenetic trees constructed from a comparison of protein sequences closely resemble those constructed by evolutionary biologists using morphological data.

Summary

Proteins are made from 20 amino acids, some of which may be modified after synthesis of the protein. Differences in the properties of amino acids reflect differences among their side chains, which are often designated —R. Except for glycine, which has no chiral carbon, all amino acids in proteins are of the L configuration. The side chains of amino acids can be classified according to their interactions with water: hydrophobic side chains are derived from compounds that are sparingly soluble in water, and hydrophilic side chains, from compounds that are soluble in water. Hydrophilic amino acid residues are further subdivided into basic, acidic, or polar nonionizable side chains. The properties of the side chains of amino acids are important factors in stabilizing the conformations and determining the functions of proteins.

The ionic state of the acidic and basic groups of amino acids and polypeptides depends on the pH. At pH 7, the α-carboxyl group is anionic ($—COO^{\ominus}$) and the α-amino group is cationic ($—NH_3^{\oplus}$). The charges of ionizable side chains depend on both the pH and their pK_a values. Differences in charges can be used to separate amino acids and proteins.

Amino acid residues in proteins are linked by peptide bonds. The sequence of residues is called the primary structure. The amino acid composition of a protein can be determined quantitatively by hydrolyzing the peptide bonds and analyzing the hydrolysate chromatographically.

The sequence of a polypeptide chain can be determined by the Edman degradation procedure. In each cycle of this procedure, the N-terminal residue of the protein reacts with phenylisothiocyanate to form a phenylthiocarbamoyl peptide. The modified N-terminal residue is then cleaved and treated with an aqueous acid to form a phenylthiohydantoin derivative, which can be identified chromatographically. The polypeptide chain, now one amino acid shorter, is again treated with phenylisothiocyanate. Often, sequences of 30 or more residues can be determined by using the Edman degradation procedure.

Sequences of larger polypeptides can be determined by selective cleavage of peptide bonds using proteases or chemical reagents, followed by Edman degradation of the resulting fragments. The amino acid sequence of the entire polypeptide is then deduced by lining up matching sequences of overlapping peptide fragments.

Comparisons of the primary structures of proteins reveal evolutionary relationships. As species diverge, the primary structures of their common proteins also diverge. By comparing differences in protein structures, we may reach a clearer understanding of evolutionary history.

Selected Readings

General References

Creighton, T. E. (1983). *Proteins: Structures and Molecular Principles* (New York: W. H. Freeman and Company), pp. 1–60. This section of Creighton's monograph presents an excellent description of the chemistry of polypeptides.

Dickerson, R. E., and Geis, I. (1969). *The Structure and Action of Proteins* (Menlo Park, California: Benjamin/Cummings Publishing Company). An excellent review on the early characterization of protein molecules.

Haschemeyer, R. H., and Haschemeyer, A. E. V. (1973). *Proteins. A Guide to Study by Physical and Chemical Methods* (New York: John Wiley & Sons). Still an excellent source of general information about proteins.

Amino Acid Composition of Proteins

Hill, R. L. (1965). Hydrolysis of proteins. *Adv. Protein Chem*. 20:37–107.

Ozols, J. (1990). Amino acid analysis. *Methods Enzymol*. 182:587–601.

Determination of Amino Acid Sequences

Hunkapiller, M. W., Strickler, J. E., and Wilson, K. J. (1984). Contemporary methodology for protein structure determination. *Science* 226:304–311.

Sanger, F. (1988). Sequences, sequences, and sequences. *Annu. Rev. Biochem*. 57:1–28. The story of the sequencing of protein, RNA, and DNA.

Walsh, K. A., Ericsson, L. H., Parmelee, D. C., and Titani, K. (1981). Advances in protein sequencing. *Annu. Rev. Biochem*. 50:261–284.

Molecular Evolution

Doolittle, R. F. (1981). Similar amino acid sequences: chance or common ancestry? *Science* 214:149–159. This paper describes the ways in which evolutionary information can be gleaned from comparisons of the primary structures of proteins and the pitfalls that are to be avoided in making such comparisons.

4

Proteins: Three-Dimensional Structure and Function

On the basis of both their physical characteristics and functions, proteins can be divided into two major classes, fibrous and globular. These two classes differ markedly in their behavior: one is usually static, and the other dynamic. *Fibrous proteins* are water insoluble and are usually physically tough. They provide mechanical support to individual cells and to entire organisms. Typically, fibrous proteins are built upon a single, repetitive structure, which assembles into cables or threads. Examples of fibrous proteins are α-keratin, the major component of hair and nails, and collagen, the major protein component of tendons, skin, bones, and teeth.

Globular proteins are usually compact, roughly spherical macromolecules whose polypeptide chains are tightly folded. Most soluble globular proteins are located in the cytosol, in the aqueous regions of organelles, or in extracellular fluids. Other globular proteins are bound to or part of biological membranes. Globular proteins characteristically have a hydrophobic interior and a hydrophilic surface. They possess indentations or clefts, which specifically interact with and transiently bind, or recognize, other compounds. By selectively binding other molecules, globular proteins serve as dynamic agents of biological action. Globular proteins include *enzymes,* the biochemical catalysts of cells (Chapters 5 and 6), and a large number of proteins that serve noncatalytic roles.

In this chapter, we shall consider the molecular architecture of both fibrous and globular proteins, from the simple linkage of amino acids by peptide bonds to the three-dimensional shape of fully formed protein molecules. We shall learn that simple shapes found in fibrous proteins—the α helix and the β sheet—often occur as elements of structure in the more complex globular proteins. We shall see that the binding clefts of globular proteins are complementary in shape and chemical properties to the molecules with which they specifically interact. We shall examine how the biological functions of proteins are related to and dependent on their structures. Above all, we shall learn that proteins have properties beyond those of free amino acids.

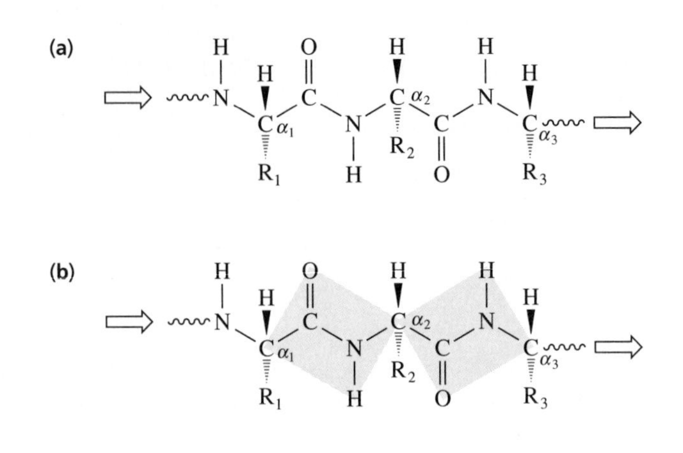

Figure 4·1
Generalized structure of a polypeptide chain. Arrows point in the N- to C-terminal direction. **(a)** Repeating N—C_α—C units connected by peptide bonds form the backbone of the polypeptide chain. **(b)** The peptide group consists of the N—H and C=O groups involved in formation of the peptide bond as well as the α-carbons on each side of the peptide bond. Two peptide groups are highlighted in the diagram.

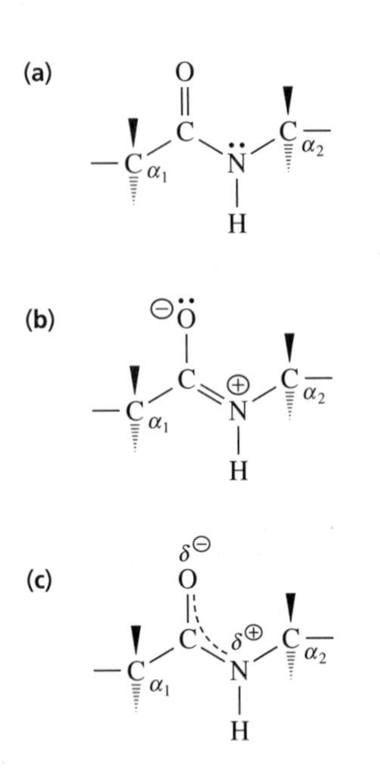

Figure 4·2
Structure of the peptide bond. **(a)** This structure shows the peptide bond as a single C—N bond; the bond between the carbonyl carbon and oxygen is a double bond, and the amide nitrogen has an unshared electron pair. **(b)** In this structure, the peptide bond is a double bond, and the bond between the carbonyl carbon and the oxygen is a single bond, with the carbonyl oxygen having an unshared electron pair. **(c)** This resonance hybrid provides a truer representation of the peptide bond. Electrons are shared by the carbonyl oxygen, the carbonyl carbon, and the amide nitrogen. All six atoms of the peptide group are in a plane because of the partial double-bond character of the peptide bond. The peptide group is polar since the carbonyl oxygen has a partial negative charge and the amide nitrogen has a partial positive charge.

4·1 The Peptide Group Is Polar and Planar

As described in Chapter 3, amino acid residues are linked by peptide bonds to form linear polypeptide chains. The *backbone* of a polypeptide chain consists of repeating N—C_α—C units connected by peptide bonds (Figure 4·1a). Attached to the backbone are the amide hydrogens, carbonyl oxygens, and various side chains connected to the α-carbons. It is often helpful to discuss not only the two atoms involved in the peptide bond but also their four substituents: the carbonyl oxygen atom, the amide hydrogen atom, and the two adjacent α-carbon atoms. These six atoms constitute the *peptide group* (Figure 4·1b).

Although it is customary to draw the carbonyl group of a peptide or amide with a double bond between the carbon and oxygen, as shown in Figure 4·2a, the actual nature of the atoms involved in the peptide bond and their substituents lies between this structure and the one shown in Figure 4·2b. A truer representation is depicted in the resonance hybrid shown in Figure 4·2c. Measurements reveal that the carbonyl carbon–nitrogen bond of a peptide group is shorter than typical carbon-nitrogen single bonds but longer than typical carbon-nitrogen double bonds. The bond appears to have about 40% double-bond character. Since oxygen is more electronegative than nitrogen, the delocalized electrons of the peptide bond are shifted toward oxygen. For this reason, the peptide group is polar. The carbonyl oxygen has a partial negative charge and can serve as a hydrogen acceptor in hydrogen bonds. The nitrogen has a partial positive charge, so that the weakly acidic —NH group can serve as a hydrogen donor in hydrogen bonds.

The partial double-bond character of peptide bonds is sufficient to prevent free rotation around the C—N bond. As a result, the peptide group is planar. However, rotation can occur about each N—C_α bond and each C_α—C bond in proteins.

The peptide group can have one of two possible configurations, either the *trans* or *cis* geometric isomer (Figure 4·3). In the *trans* configuration, the two α-carbons of adjacent amino acid residues are on opposite sides of the peptide bond and at opposite corners of the rectangle formed by the planar peptide group. In the *cis* configuration, the two α-carbons are on the same side of the peptide bond and closer together. Steric interference between the side chains attached to the two α-carbons makes the *cis* configuration less favorable than the extended *trans* configuration. Consequently, nearly all peptide groups in proteins are *trans*. Rare exceptions occur, however, usually at bonds involving the amide nitrogen of proline, for which the *cis* configuration creates only slightly more steric interference than the *trans* configuration (Figure 4·4). About 10% of proline residues in proteins analyzed by X-ray crystallography have been found to be in the *cis* configuration.

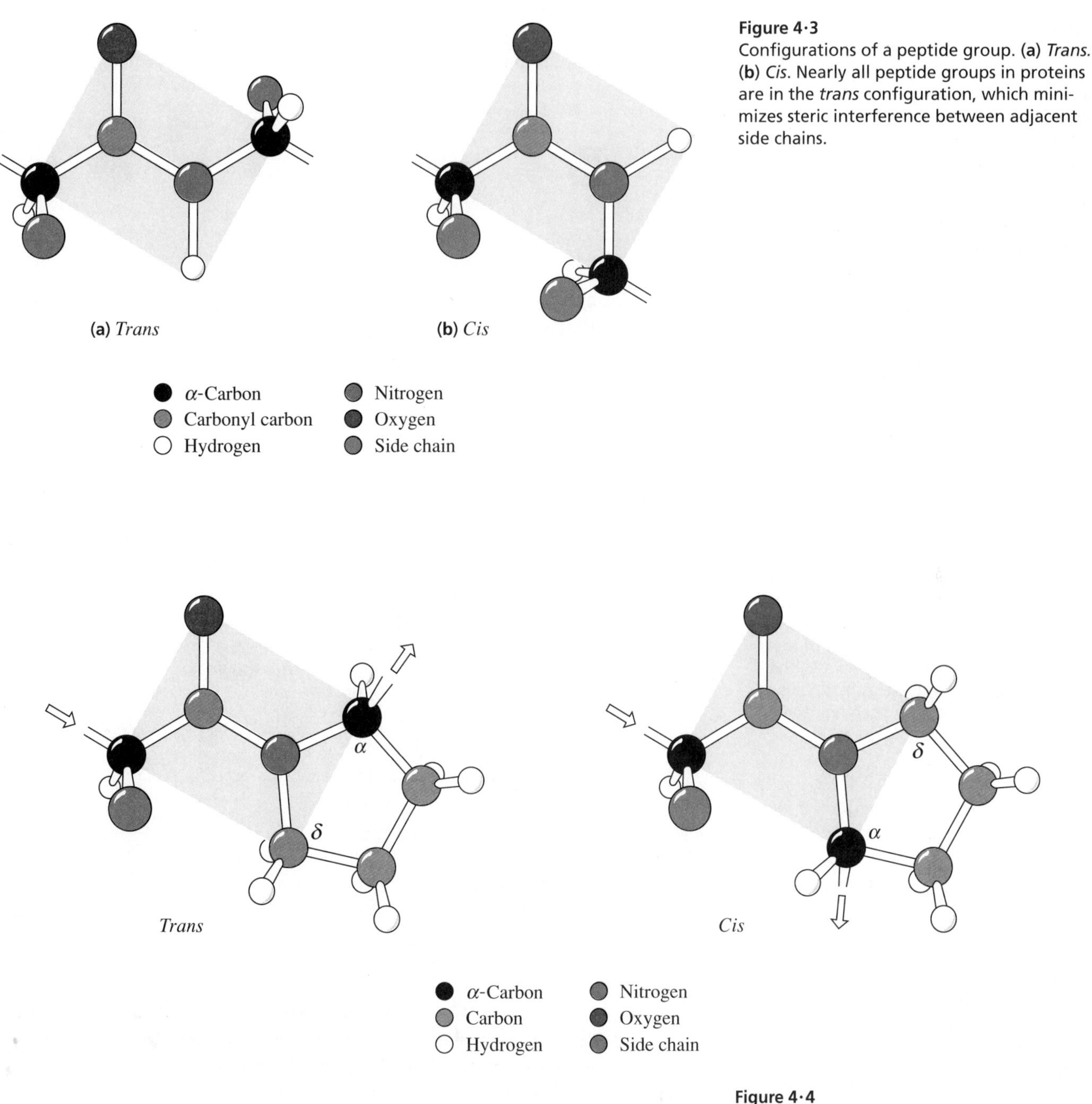

Figure 4·3
Configurations of a peptide group. **(a)** *Trans.* **(b)** *Cis.* Nearly all peptide groups in proteins are in the *trans* configuration, which minimizes steric interference between adjacent side chains.

Figure 4·4
Trans and *cis* configurations of peptide groups involving proline. In the *trans* configuration, steric interference between the δ position of the pyrrolidine ring and the side chain of the adjacent amino acid residue is only slightly less than that existing in the *cis* configuration between the α-carbon and the side chain of the adjacent amino acid residue. Most *cis* peptide groups in proteins involve proline residues. The arrows indicate the direction from the N- to the C-terminus.

4·2 Peptide Chains Assume a Restricted Number of Conformations

The biological activity of a protein is dependent on its overall conformation, or shape after folding. A *conformation* is a spatial arrangement that depends on rotation of a bond or bonds. Thus, unlike the *configuration* of a molecule, the conformation of a protein can change without the breaking of bonds. Considering the possible rotations of each bond, the number of potential conformations for a protein

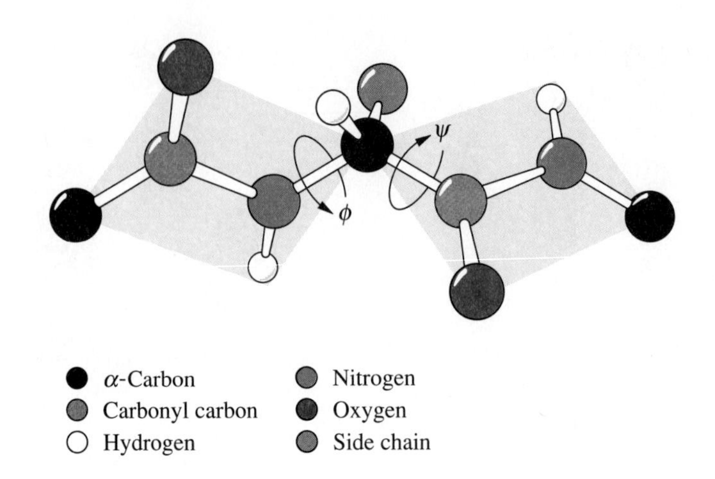

Figure 4·5
Rotation around the N—C_α and the C_α—C bonds, which link peptide groups in a polypeptide chain. Each α-carbon of an internal residue is a pivot that links two adjacent planar peptide groups. The rotation angle about the N—C_α bond is called ϕ and that about the C_α—C bond is called ψ. The planes of the peptide groups are shaded and the substituents of the outer α-carbons have been omitted for clarity.

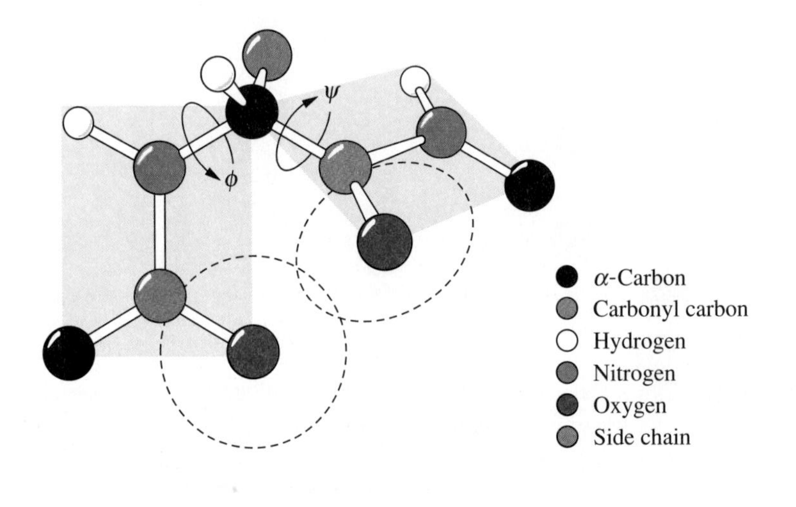

Figure 4·6
An unstable conformation caused by steric interference between carbonyl oxygens of adjacent residues. In this example, rotation around the N—C_α and the C_α—C bonds is restricted. The van der Waals radii of the carbonyl-oxygen atoms are shown by the dotted lines. For clarity, the substituents of the outer α-carbons have been omitted.

molecule is astronomical. However, under physiological conditions, the protein assumes a single stable shape, known as the *native conformation*. Conformational flexibility in proteins is provided by rotation around the N—C_α and the C_α—C bonds, which link the rigid peptide groups in a polypeptide chain (Figure 4·5). This rotation itself is limited by steric interference between main-chain and side-chain atoms of adjacent residues (Figure 4·6). In the case of proline, the N—C_α bond is fixed by its inclusion in the pyrrolidine ring of the side chain. The rotation angle around the N—C_α bond of a peptide group is designated ϕ (phi), and that around the C_α—C bond is designated ψ (psi).

4·3 There Are Four Levels of Protein Structure

Depending upon its complexity, a protein may be described as having up to four levels of structure (Figure 4·7). As we noted in Chapter 3, *primary structure* is the

(a) Primary structure

–Ala–Glu–Val–Thr–Asp–Pro–Gly–

(b) Secondary structure

α Helix

β Sheet

(c) Tertiary structure

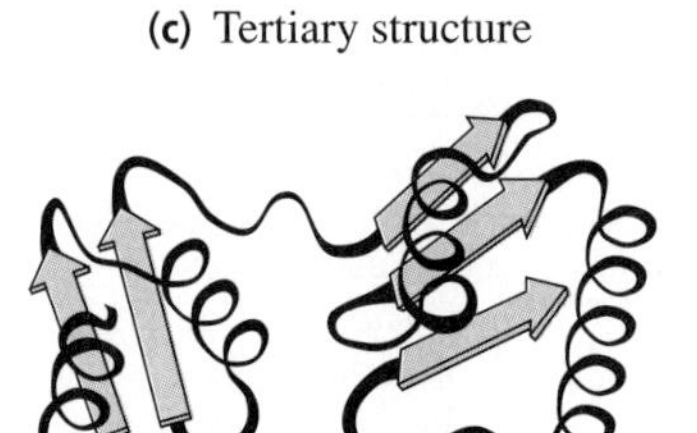

(d) Quaternary structure

Figure 4·7
Levels of protein structure. **(a)** The linear sequence of amino acid residues defines the primary structure. **(b)** Secondary structure consists of regularly repeating conformations of the peptide chain, such as α helices and β sheets. **(c)** Tertiary structure refers to the native conformation of an entire polypeptide chain. **(d)** Quaternary structure refers to the arrangement of two or more separate polypeptide chains into a multisubunit molecule.

sequence of covalently linked amino acid residues; it describes the linear or one-dimensional structure of a protein. The other levels, which describe the three-dimensional structure of a protein, include *secondary structure,* regularities in local conformations; *tertiary structure,* the biologically active, three-dimensional structure of an entire polypeptide; and for some proteins, *quaternary structure,* organization of two or more polypeptide chains into a multisubunit protein. These four structural levels are most clearly evident in globular proteins.

4·4 The α Helix and β Sheet Are the Common Secondary Structures

In the early 1950s, using data from X-ray crystallographic studies of simple compounds such as amino acids and di- and tripeptides, Linus Pauling (Nobel Prize in Chemistry, 1954) and Robert Corey proposed several types of secondary structures. Their proposals took into account possible steric constraints and opportunities for stabilization by formation of hydrogen bonds. It is now known that two of these proposed structures are the major secondary structures of many fibrous and globular proteins. The existence of these two secondary structures explains the earlier classification by William Astbury of fibrous proteins into two categories, the α class and the β class. The elastic proteins in the α class have a repeat unit of 0.5–0.55 nm. The inelastic or extended β class have a repeat unit of 0.7 nm. For the α class, Pauling and Corey proposed a structure called the α helix; for the β class, they proposed a structure called the β sheet.

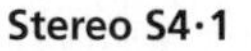

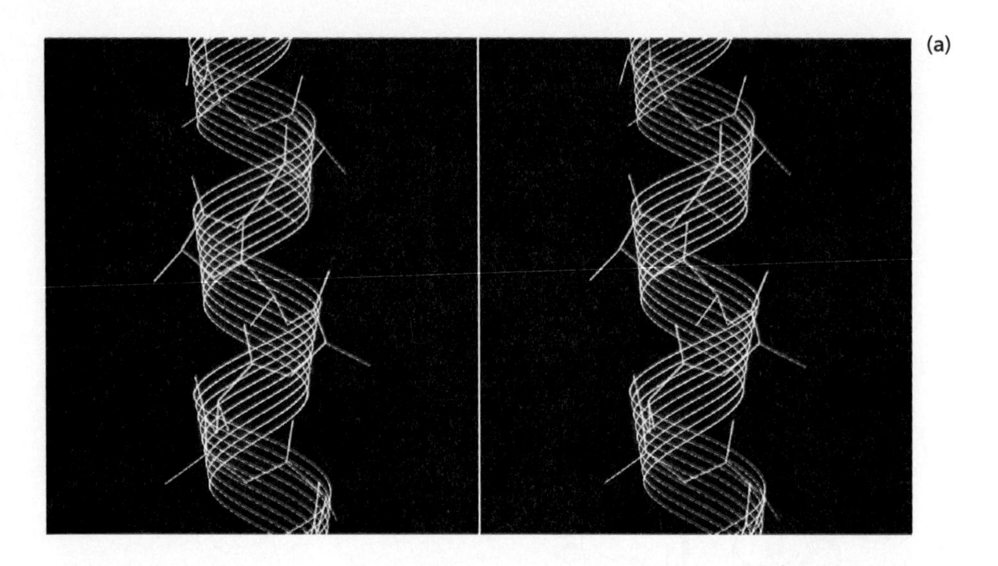

(a)

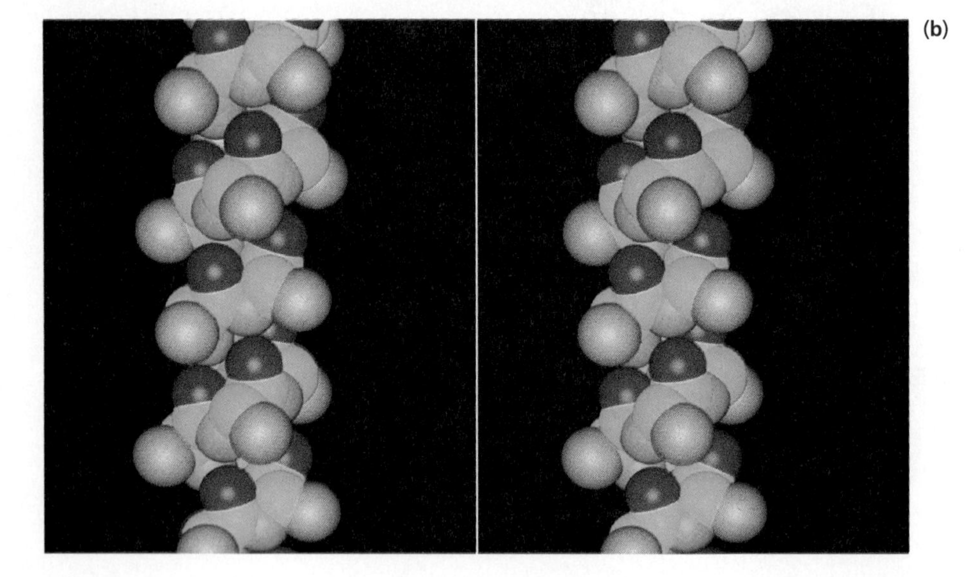

(b)

Stereo S4·1
Right-handed α helix.
(a) The ribbon highlights the shape of the helix All side chains (orange) have been clipped at the β-carbon to make the backbone and carbonyl oxygens (both green) more clearly visible. **(b)** Space-filling model. Notice how tightly packed the atoms are in the α helical conformation. Color key: carbon, green; β-carbons, orange; nitrogen, blue; oxygen, red. (Based upon coordinates provided by B. Shaanan; stereo images by Ben N. Banks and Kim M. Gernert.)

A. The α Helix Is a Right-Handed Helix Stabilized by Hydrogen Bonds

A right-handed *α helix* is depicted in Stereo S4·1 and Figure 4·8. Theoretically, an α helix might be right-handed or left-handed, but, for L amino acid residues, the left-handed conformation is destabilized by steric interference between carbonyl oxygens and side chains. Hence, the α helices found in protein structures are nearly always right-handed. Some residues, usually glycines, have been found in left-handed α-helical conformations, but only in stretches not longer than four residues.

Within an α helix, each carbonyl oxygen of residue n is hydrogen bonded to the α-amino nitrogen of residue $n + 4$ in the conventional N- to C-terminal direction (Figure 4·8). The chain of atoms closed by the hydrogen bond can be regarded as a 13-atom ring structure: the carbonyl oxygen, 11 backbone atoms, and the amide hydrogen. Note that the intrahelical hydrogen bonds are nearly parallel to the long axis of the helix, with the carbonyl groups all pointing toward the C-terminal end.

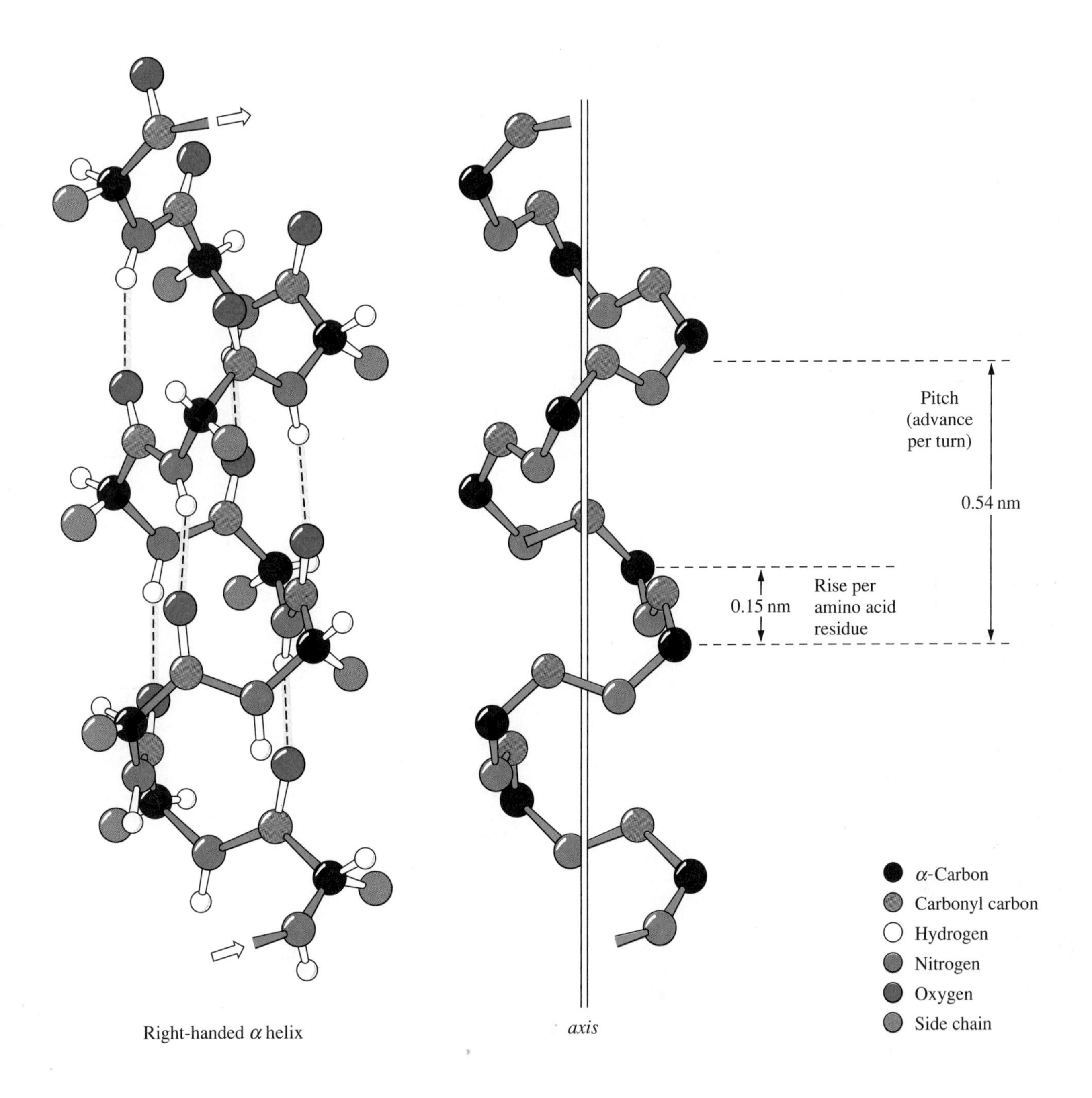

Figure 4·8
Right-handed α helix. Each carbonyl oxygen forms a hydrogen bond with the amide hydrogen of the fourth residue further toward the C-terminus of the polypeptide chain. The N, O, and H atoms of the hydrogen bonds are in line, forming bonds that are approximately parallel to the long axis of the helix. Note that all the —C=O groups point toward the C-terminus. In an ideal α helix, equivalent positions recur every 0.54 nm (this distance is called the pitch of the helix), each amino acid residue advances the helix by 0.15 nm along the long axis of the helix (a dimension called the rise), and there are 3.6 amino acid residues per turn. The arrows at the ends of the helix indicate the direction from the N- to the C-terminus.

Although a single intrahelical hydrogen bond would not provide appreciable structural stability, the cumulative effect of many hydrogen bonds within an α helix stabilizes this conformation, especially in hydrophobic regions within the interior of a protein where water molecules do not compete for hydrogen bonds. In fact, given an appropriate amino acid sequence (discussed below), the α helix can be the most stable secondary structure for a polypeptide chain.

In an ideal α helix, the *rise,* or the distance each residue advances a helix along its axis, is 0.15 nm, and the number of amino acid residues required for one complete turn is 3.6. In actual protein structures, α helices may be slightly distorted, but they still have 3.5 to 3.7 residues per turn. Equivalent positions recur approximately every 0.54 nm, a distance called the *pitch* of the helix. Since the orientation of the polar peptide groups is preserved throughout the regularly repeating conformation of the helix, the entire helix may be considered a dipole with a positive N-terminus and a negative C-terminus.

Table 4·1 Positional preferences of amino acids in helices

Amino acid	N-terminal end	Middle	C-terminal end
Pro	0.8	<u>0.3</u>	0.7
Gly	**1.8**	<u>0.5</u>	**3.9**
Ser	**2.3**	0.6	0.8
Thr	1.6	1.0	0.3
Asn	**3.5**	0.9	1.6
Gln	0.4	1.3	0.9
Asp	**2.1**	1.0	0.7
Glu	0.4	0.8	0.3
Lys	0.7	1.1	1.3
Arg	0.4	1.3	0.9
His	1.1	1.0	1.3
Ala	0.5	**1.8**	0.8
Leu	<u>0.2</u>	1.2	0.7
Val	<u>0.1</u>	1.2	<u>0.2</u>
Ile	0.2	1.2	0.7
Phe	0.2	1.3	0.5
Tyr	0.8	0.8	0.8
Met	0.8	1.5	0.8
Trp	0.3	1.5	0
Cys	0.6	0.7	0.4

Data were obtained from 215 helices in crystal structures of 45 proteins. The values are reported as relative preferences, that is, the ratio of observed occurrences to the number of occurrences expected based upon the percentage composition of the proteins examined. Boldfaced values are statistically higher, and underlined values statistically lower, than expected. The middle group tabulates the positions of residues more than five residues from either end of the helix. The end residues are the residues at the N- and C-terminal positions of the helices. [Adapted from Richardson, J. S., and Richardson, D. C. (1989). Principles and patterns of protein conformation. In *Prediction of Protein Structure and the Principles of Protein Conformation,* G. D. Fasman, ed. (New York: Plenum Press), pp. 15–16.]

As shown in Stereo S4·1 and Figure 4·8, all side chains of the amino acids point outward from the cylinder of the helix, thereby minimizing steric interference. Nevertheless, the stability of an α-helical structure is affected by the identity of the side chains, and some amino acid residues are found in α-helical conformations more often than others. Moreover, amino acid residues may have preferred positions within a helix (Table 4·1). For example, alanine, which has a small, uncharged side chain, fits well into the α-helical conformation and is prevalent in the helices of globular proteins. In fact, the synthetic polypeptide poly-L-alanine spontaneously forms an α helix in aqueous solution. In contrast, glycine, whose side chain consists of a single hydrogen atom, destabilizes α-helical structures by allowing greater freedom of rotation about its α-carbon than is possible for other residues. For this reason, many α helices begin or end with glycine. Proline is the least common residue in an α helix because its ring side chain disrupts or bends the right-handed helical conformation by occupying space that a neighboring residue of the helix would otherwise occupy. In addition, the lack of a hydrogen atom on its amide nitrogen keeps proline from fully participating in the hydrogen-bond network of an α helix. Thus, proline is found more often at the ends of α helices than in the interior.

Soon after Pauling and Corey proposed the α-helical conformation, Max Perutz verified its existence when he observed in the fibrous protein α-keratin a minor repeating unit (0.15 nm) corresponding to the rise of the α helix. α-Keratin, a major component of wool, hair, skin, and fingernails, is a fibrous protein composed almost entirely of α helices. The basic unit of α-keratin is the protofibril, consisting of four right-handed α helices interrupted by intervening nonhelical sections. These helices are paired into two left-handed cables, or supercoils, that intertwine to form a larger left-handed supercoil (Figure 4·9). The four individual helices and the cables are twisted in opposite directions. This structural pattern, also used in the manufacture of rope, resists unwinding and gives the fiber unusual strength. Protofibrils are in turn arrayed in a larger structure called a microfibril, which appears to consist of nine protofibrils surrounding a core of two other protofibrils.

Protofibrils and microfibrils are cross-linked by disulfide bonds that increase the stability of the overall structure. Keratins with many disulfide bonds, such as those of nails, are hard and rather inflexible. In contrast, keratins with relatively few disulfide bonds, such as those of wool, are flexible and stretch easily. Hair may be curled or uncurled by treating it with a solution that reduces disulfide bonds, setting it, and then treating it with a solution that oxidizes cysteine residues so that they form new disulfide bonds.

Globular proteins vary in their α-helical content. Some, such as the oxygen-binding proteins myoglobin and hemoglobin, contain over 75% of their residues in α helices (Stereos S4·14 and S4·16); others, such as the enzyme chymotrypsin, contain very little α-helical structure. The average content of α helix in the globular proteins that have been examined is about 25%.

In addition to the α helix, some globular proteins contain short regions of *3_{10} helix*. Like the α helix, the 3_{10} helix is right-handed. However, it has a tighter ring structure than the α helix—10 atoms rather than 13—and has fewer residues per turn (3.0) and a longer pitch (0.60 nm). The 3_{10} helix is also slightly less stable than

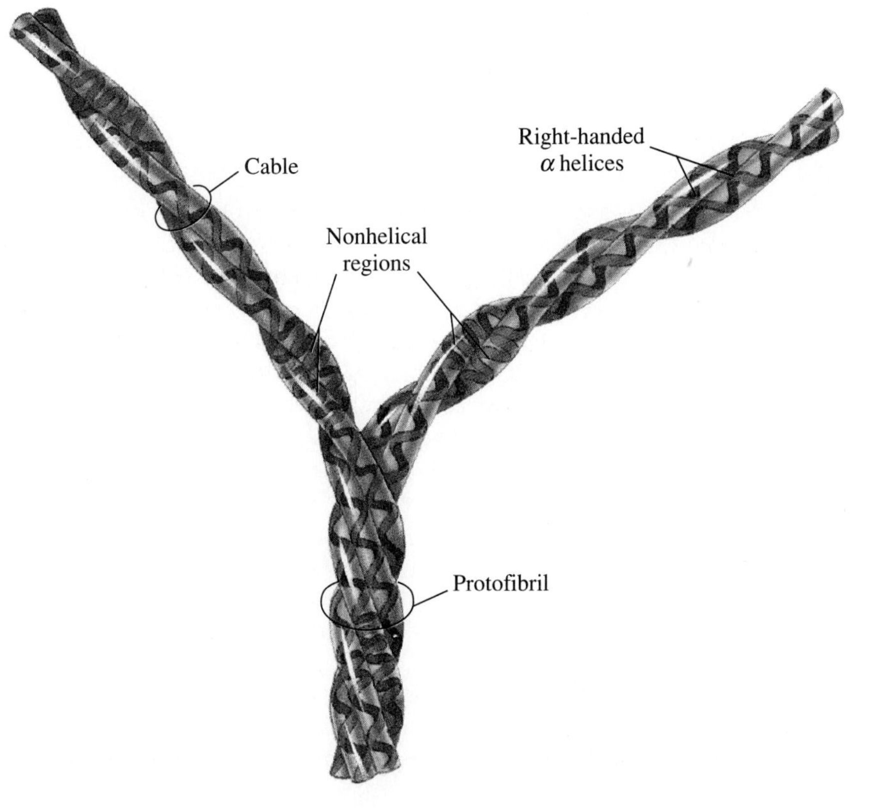

Figure 4·9
Protofibril of α-keratin. Two left-handed cables are intertwined to form a left-handed protofibril. Each cable consists of two chains that are mostly right-handed α helices. (Illustrator: Lisa Shoemaker.)

the α helix because of steric hindrances and the awkward geometry of its hydrogen bonding, which is between the carbonyl oxygen of residue n and the α-amino group of residue $n + 3$ (versus $n + 4$ in the α helix). Although the 3_{10} helix is not as common as the α helix, it is not rare. When it occurs, it is usually only a few residues in length and is often at the C-terminal end of an α helix.

B. β Sheets Are Composed of Extended Polypeptide Chains

Another secondary structure proposed by Pauling and Corey, the *β sheet,* consists of extended polypeptide chains (called *β strands*) stabilized by hydrogen bonds between carbonyl oxygens and amide hydrogens. These bonds may link two or more adjacent polypeptide chains or different segments of the same chain. As shown in Figure 4·10, such hydrogen bonds are nearly perpendicular to the extended polypeptide chains, which may be either parallel (running in the same N- to C-terminal direction, as shown in Figure 4·10a) or antiparallel (running in the opposite N- to C-terminal direction, as shown in Figure 4·10b). Although extended, the β sheet is not absolutely planar but slightly pleated due to the bond angles between peptide groups (Stereo S4·2). The side chains point alternately above and below the plane of the sheet.

The fibrous protein silk fibroin, produced by the silkworm *Bombyx mori,* contains polypeptide chains arrayed in antiparallel β sheets (Stereo S4·3). The primary structures of most silk fibroins contain long stretches of the repeating sequence –Gly–Ser–Gly–Ala–Gly–Ala–. Since the side chains of amino acid residues in β

Figure 4·10
β sheets. The arrows point in the N-terminal to C-terminal direction. **(a)** Structure of a parallel β sheet. The hydrogen bonds are evenly spaced but slanted. **(b)** Structure of an antiparallel β sheet. The hydrogen bonds are essentially perpendicular to the strands, and the space between hydrogen-bonded pairs is alternately wide and narrow.

(a)

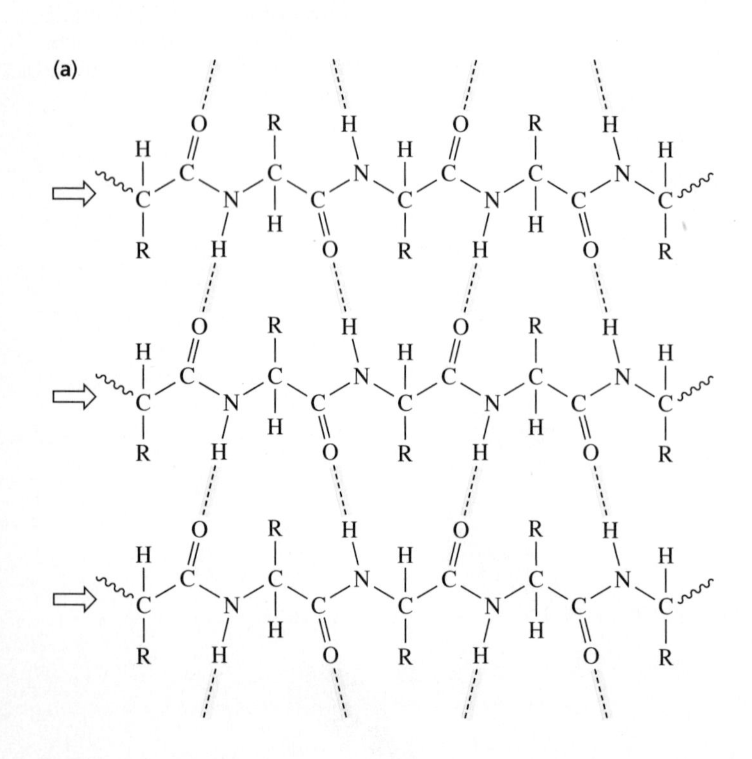

(b)

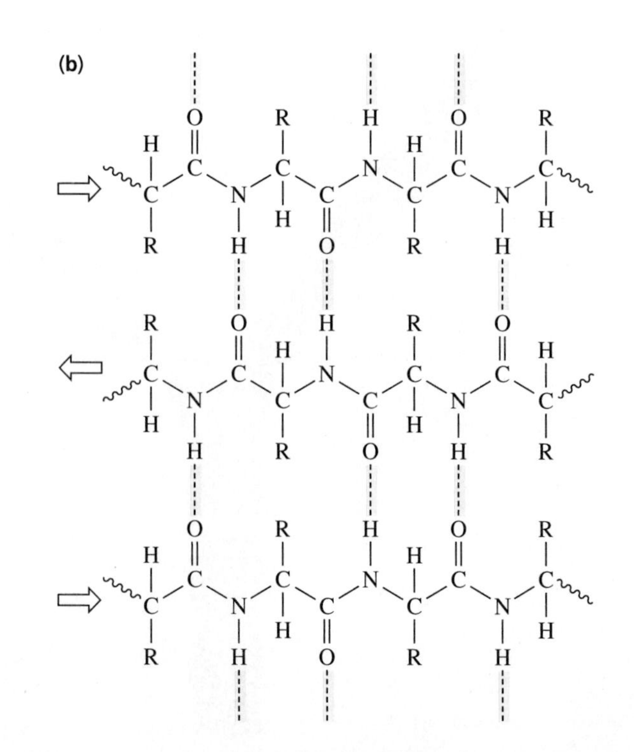

Stereo S4·2
Side view of a parallel β sheet. In this side view, the pleated nature of the parallel β sheet can be seen. Note that the side chains (here clipped at the β-carbon) point alternately above and below the plane of the sheet. Color key: carbon, green; hydrogen, light blue; nitrogen, dark blue; oxygen, red. (Based upon coordinates provided by Karl D. Hardman; stereo image by Richard J. Feldmann.)

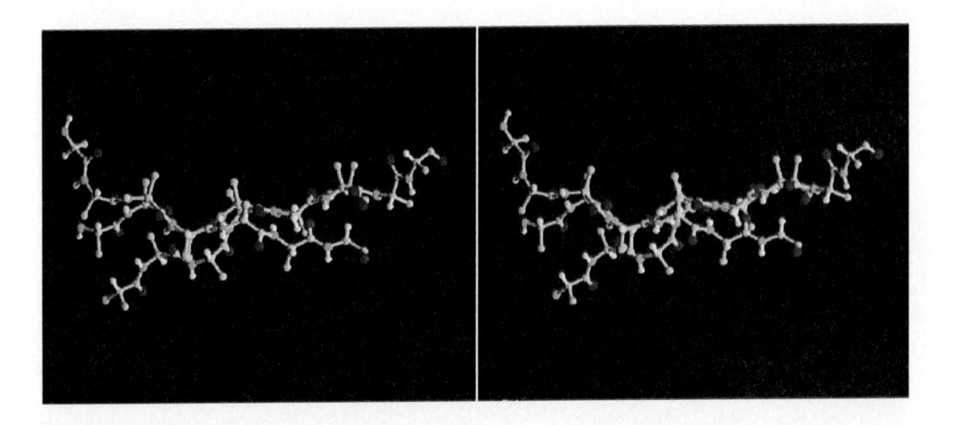

Stereo S4·3
Model of silk fibroin. Silk fibroin consists of an array of antiparallel β sheets. β sheets are stabilized by hydrogen bonding between carbonyl oxygens in one chain and amide hydrogens in an adjacent chain. Each chain of silk fibroin contains repeats of the amino acid sequence –Gly–Ser–Gly–Ala–Gly–Ala–. Side-chain hydrogen atoms of glycine residues lie on one side of each sheet and methyl and hydroxymethyl side chains of the alanine and serine residues, respectively, lie on the other side. Sheets are stacked such that side chains of glycine residues of neighboring sheets face one another, as do the side chains of alanine and serine. Noncovalent interactions between side chains stabilize the stacked sheets. Color key: carbon, green; hydrogen, light blue; nitrogen, dark blue; oxygen, red. (Stereo image by Ben N. Banks and Kim M. Gernert.)

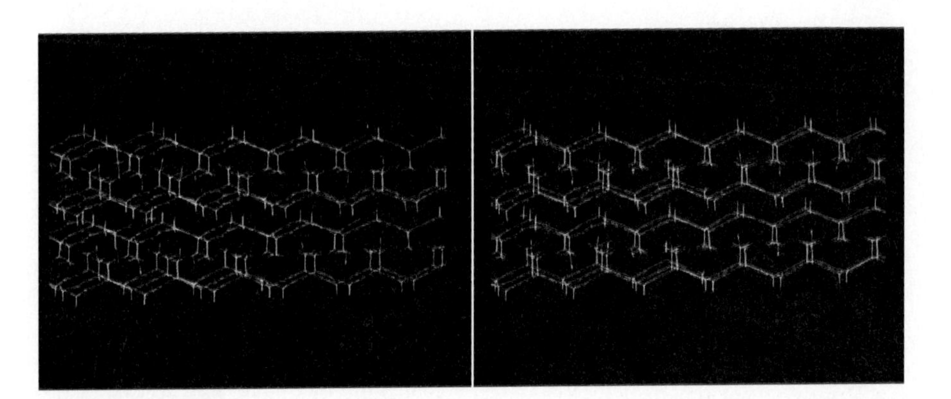

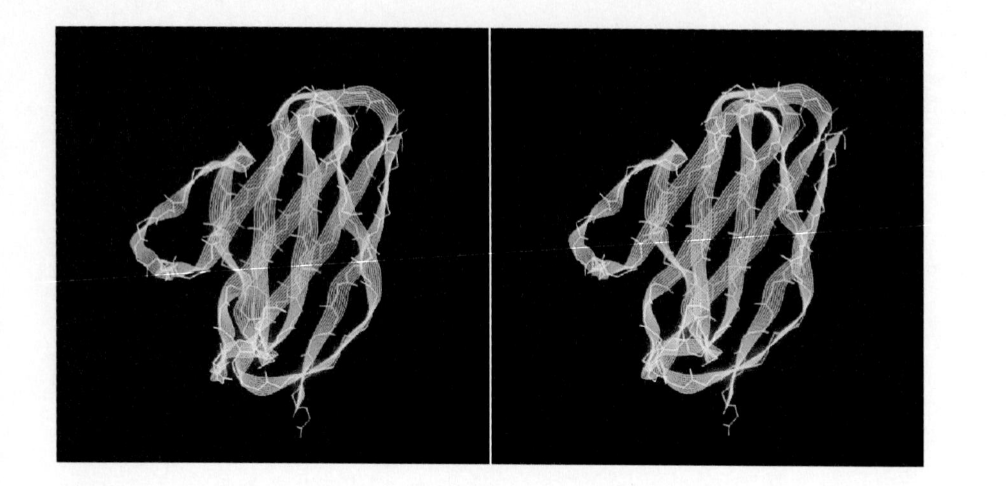

Stereo S4·4
Conformation of one-half of a dimeric Bence-Jones protein. Excreted in the urine of persons suffering from myeloma, a type of cancer, Bence-Jones proteins are polypeptide chains derived from the immunoglobulin called the myeloma protein. This half of the dimer consists largely of β-sheet structure. Two antiparallel β sheets (backbone in blue) are linked by intervening nonrepetitive sections (backbone in yellow). (Based upon coordinates provided by O. Epp and Reuben E. Huber; stereo image by Ben N. Banks and Kim M. Gernert.)

sheets extend alternately above and below the plane of the sheet, the side chain hydrogens of glycine residues lie on one side and the methyl and hydroxymethyl side chains of the alanine and serine residues lie on the other side, allowing a close stacking of sheets. Silk fibroin is flexible because the stacked sheets are held together only by noncovalent interactions between the side chains, which are just the right size to nestle together.

The secondary structures of many globular proteins contain regions of β-sheet conformation. For example, one-half of the dimeric Bence-Jones protein (derived from a myeloma immunoglobulin) consists largely of antiparallel β sheets (Stereo S4·4). The right-handed twist that occurs in β-sheet regions of the Bence-Jones protein is typical of the β-sheet regions of globular proteins.

4·5 A Different Helical Structure Is Found in Collagen

The fibrous protein collagen contains a different helical structure. Collagen is the major protein component of most connective tissue and the most abundant vertebrate protein, constituting about 25% to 35% of the total protein in mammals. There are many different types of collagen proteins with remarkably diverse functions and forms. The hard substance of bone contains collagen and a calcium-phosphate polymer. Collagen in tendons forms stiff, ropelike fibers of tremendous tensile strength. In skin, collagen takes the form of loosely woven fibers, permitting expansion in all directions, and in blood vessels, collagen fibers are arranged in elastic networks. Familiar products derived from collagen include gelatin and glue.

The native aggregate of collagen is a molecule consisting of three chains having left-handed helices coiled around each other to form a right-handed supercoil

(Stereo S4·5). This supercoil is stabilized by interchain hydrogen bonding and by the opposing twist of the helices and the supercoil. A typical collagen molecule is a rod 300 nm long and 1.5 nm in diameter. Within each collagen chain, the left-handed helix has 3.0 amino acid residues per turn and a pitch of 0.94 nm, giving a distance along the axis of 0.31 nm per residue. The left-handedness of the helix and much of the rigidity of collagen arises from steric constraints imposed by many proline residues (Stereo S4·6). The ends of collagen molecules contain nonhelical sections, which appear to be necessary for proper alignment and cross-linking within collagen fibrils.

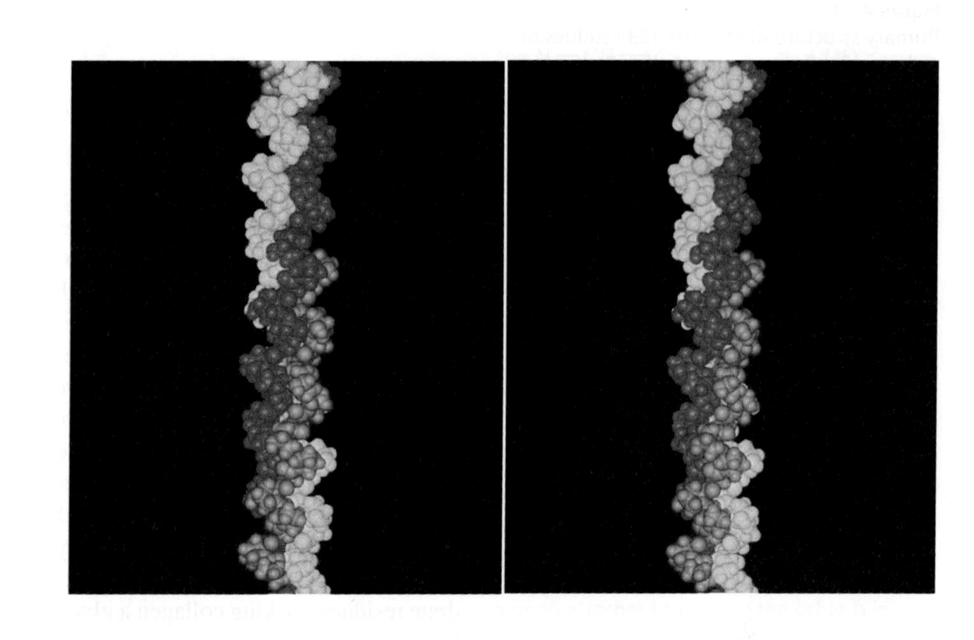

Stereo S4·5
Model of collagen triple helix. Three left-handed collagen helices are coiled around one another in a right-handed supercoil. (Based upon coordinates provided by Barbara Brodsky and Cynthia G. Long; stereo image by Ben N. Banks and Kim M. Gernert.)

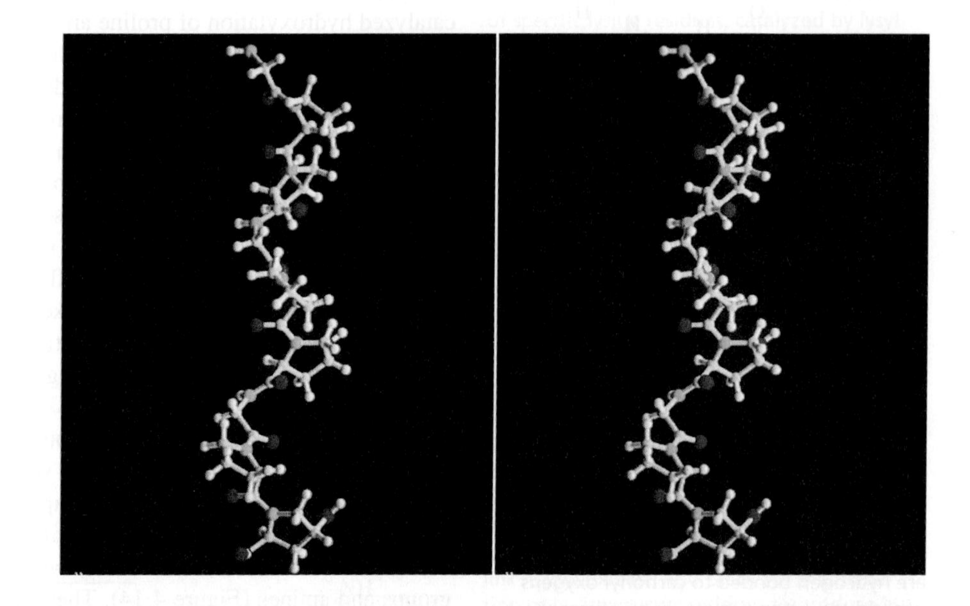

Stereo S4·6
Collagen helix. The collagen helix is an extended left-handed helix in which the pyrrolidine rings of proline project away from the long axis of the chain. There are no intrachain hydrogen bonds in the collagen helix. Color key: carbon, green; hydrogen, light blue; nitrogen, dark blue; oxygen, red. (Stereo image by Richard J. Feldmann.)

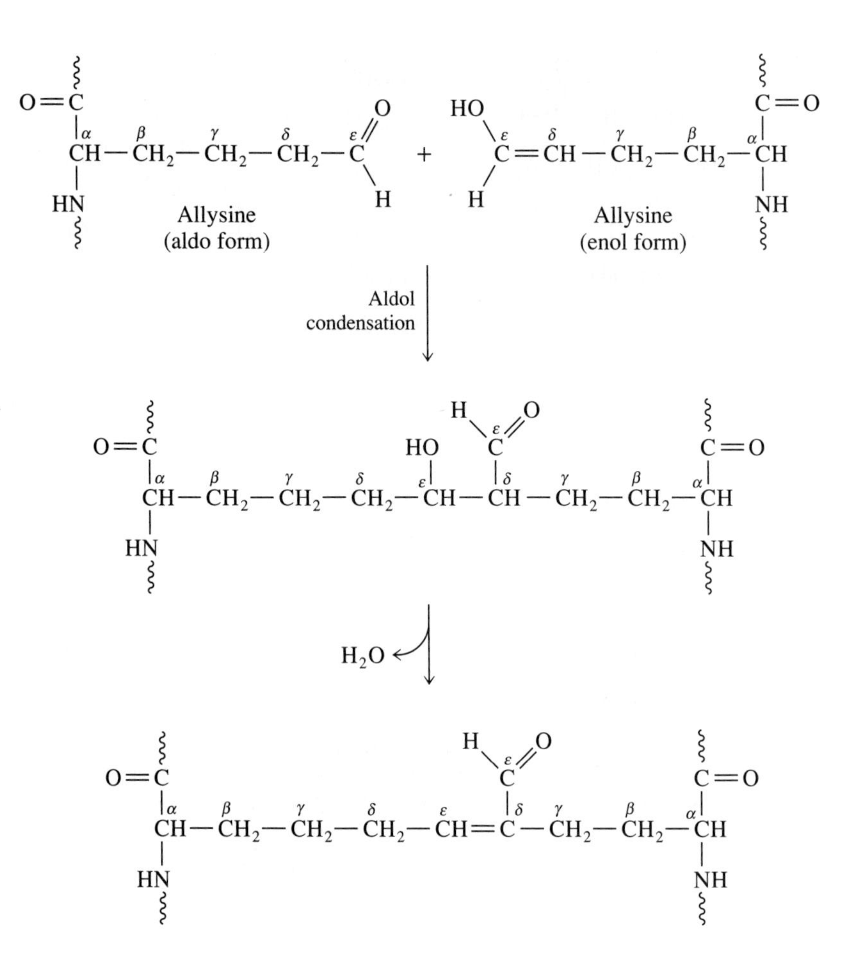

Figure 4·15
Cross-linking in collagen via aldol condensation between two allysine residues. Both the hydrated and dehydrated forms shown probably occur in vivo.

condensation to form cross-links, usually within collagen molecules (Figure 4·15). Both types of cross-links are converted to more stable bonds, but the chemistry of these conversions is unknown.

4·6 Nonrepetitive Regions Are Essential in Globular Proteins

As we have noted, globular proteins contain elements of secondary structure, that is, regions of consecutive residues with repeating conformation. In addition to having regions of secondary structure, globular proteins contain stretches of nonrepetitive three-dimensional structure. These nonrepetitive regions connect secondary structures and provide directional changes necessary for a protein to fold into its globular shape. These changes occur at *loops,* compact connecting regions with nonrepetitive structure ranging from 2 to 16 residues in length. Loops often contain many hydrophilic residues and are found on the surfaces of globular proteins, where they are exposed to solvent and form hydrogen bonds with water. The term *turn* is applied to loops having only a few residues. The two most common types of turns, Type I and Type II (or glycine turn), also called β bends, are loops that span four amino acid residues and are stabilized by hydrogen bonding between the carbonyl oxygen of the first residue and the amide nitrogen of the fourth residue (Stereo S4·7). In the Type II turn, the third residue is glycine; in both types of turns, proline is often the second residue. The tight turns connecting two adjacent antiparallel β strands are often called *hairpin loops*.

A number of globular proteins have many residues in nonrepeating conformations. For example, in cytochrome *c*, a heme-containing globular protein that participates in electron transfer, nearly 50% of the residues of the protein are in regions of nonrepetitive structure (Stereo S4·8).

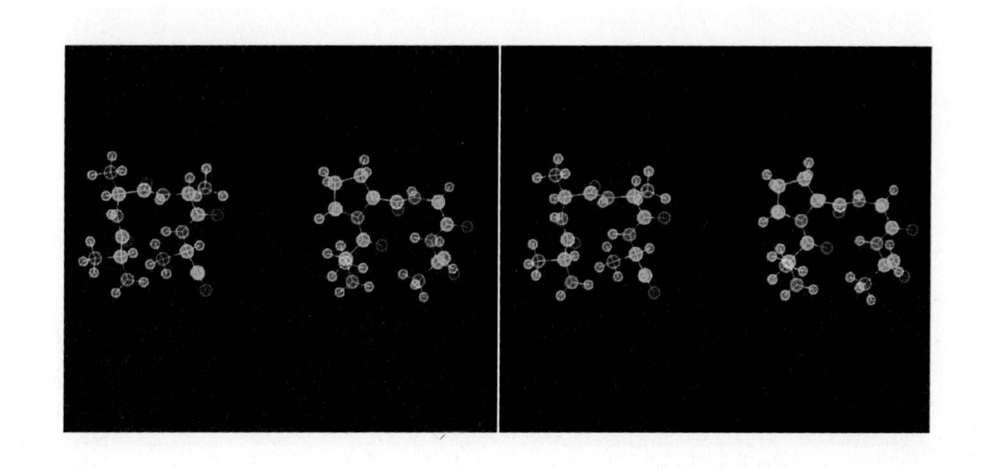

Stereo S4·7
Structures of two turns. Type I (on the left) is the most common turn. Type II (or glycine turn, on the right) has glycine as the third residue and frequently proline as the second residue. Both turns are stabilized by hydrogen bonding between the carbonyl oxygen of the first residue and the amide hydrogen of the fourth residue. Color key: carbon, green; β-carbon, orange; hydrogen, light blue; nitrogen, dark blue; oxygen, red. (Type I is based upon coordinates provided by Brian W. Matthews and M. A. Holmes; type II is based upon coordinates provided by M. N. G. James and A. R. Sielecki; stereo image by Ben N. Banks and Kim M. Gernert.)

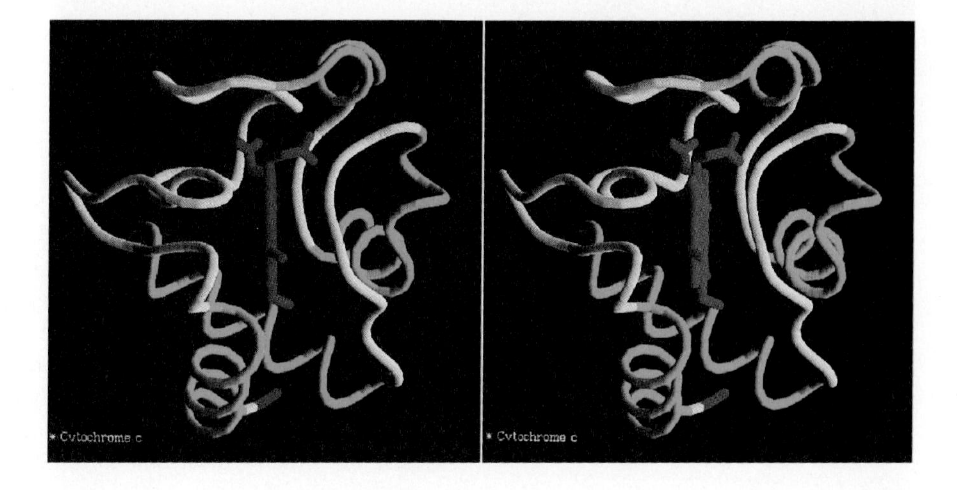

Stereo S4·8
Conformation of tuna heart cytochrome *c*, a heme-containing protein that participates in electron transfer. The heme group (in red) is surrounded by a number of nonrepetitive regions (in yellow) and several α-helical regions (in green). (Based upon coordinates provided by Richard E. Dickerson; stereo image by Richard J. Feldmann.)

4·7 Ramachandran Plots Indicate That Amino Acid Residues Can Assume Few Conformations

Because the peptide bond has some double-bond character and is therefore rigid, the local conformation of a polypeptide backbone can be described by ϕ and ψ. Recall that these are the angles of the bonds connecting peptide groups, that is, the N—C_α and the C_α—C bonds, respectively. Each of these angles is defined by the relative positions of four atoms of the backbone. Stereo S4·9 and its corresponding caption describe how ϕ can be measured, and Stereo S4·10, how ψ can be measured. Clockwise angles are positive, and counterclockwise angles are negative, with each having a 180° sweep.

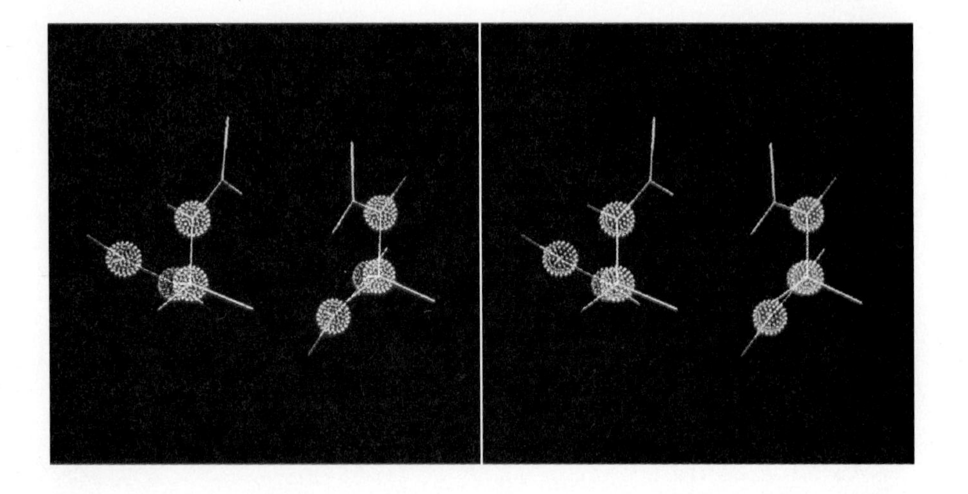

Stereo S4·9
Examples showing how ϕ is measured in peptides. To measure ϕ, the peptide is viewed down the length of the C_α—N bond in the C- to N-terminal direction and the angle between the C—C_α bond and the next peptide bond toward the N-terminus is measured. The peptide on the left shows ϕ typical of a residue in an α helix; the peptide on the right shows ϕ typical of a residue in a parallel β sheet. Counterclockwise angles, such as these, are negative. Color key: backbone bonds, pink; carbon, green; nitrogen, blue; oxygen, red. (Stereo image by Ben N. Banks and Kim M. Gernert.)

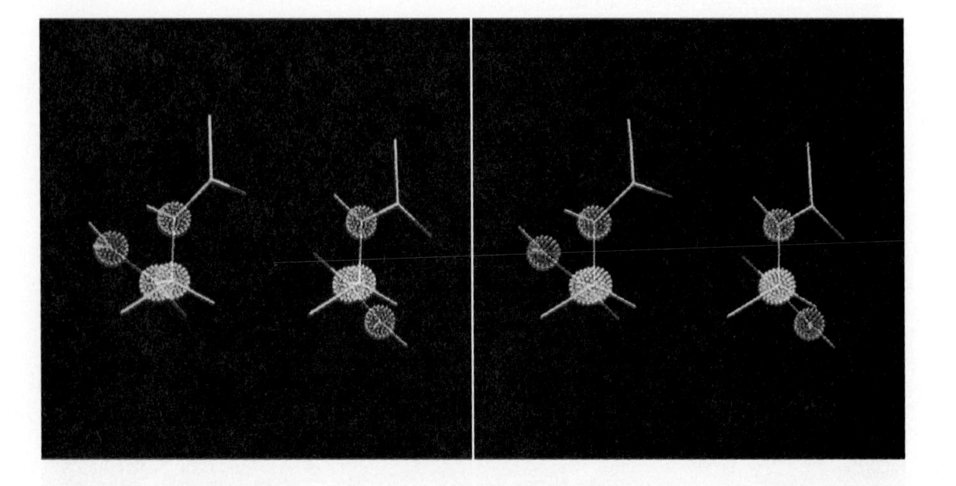

Stereo S4·10
Examples showing how ψ is measured in peptides. To measure ψ, the peptide is viewed down the length of the C_α—C bond in the N- to C-terminal direction and the angle between the N—C_α bond and the next peptide bond toward the C-terminus is measured. The peptide on the left shows ψ typical of a residue in an α helix; the peptide on the right shows ψ typical of a residue in a parallel β sheet. The counterclockwise angle of the peptide on the left is negative, and the clockwise angle of the peptide on the right is positive. Color key: backbone bonds, pink; carbon, green; nitrogen, blue; oxygen, red. (Stereo image by Ben N. Banks and Kim M. Gernert.)

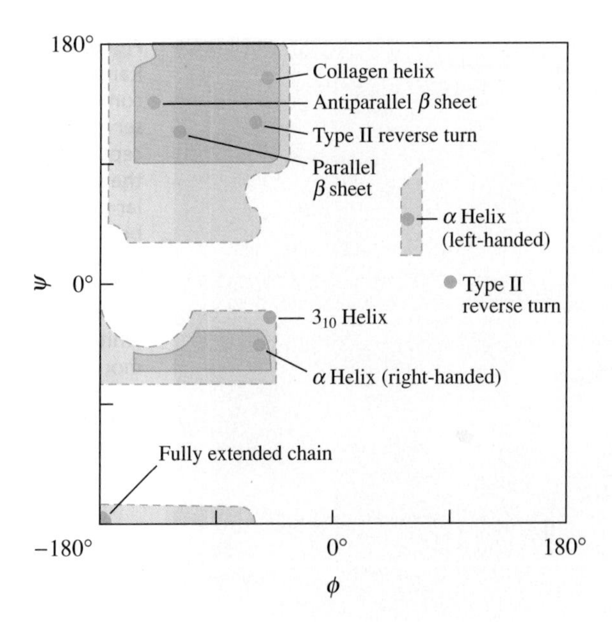

Figure 4·16
Ramachandran plot showing a variety of conformations. Solid lines indicate the range of commonly observed ϕ and ψ values. Dotted lines give the outer limits for an alanine residue. Large blue dots correspond to values of ϕ and ψ that produce recognizable conformations such as the α helix and the β sheet. The white portions of the plot correspond to values of ϕ and ψ that rarely or never occur.

Plots of ψ versus ϕ for the residues in a polypeptide chain are called *Ramachandran plots,* after the biochemist G. N. Ramachandran. Ramachandran and his colleagues used space-filling models of peptides and made calculations to determine which values of ϕ and ψ are sterically permitted in the polypeptide chain. Figure 4·16 is a Ramachandran plot showing ϕ and ψ values associated with several recognizable conformations. Table 4·2 lists these ϕ,ψ values.

Secondary structures result only when a consecutive stretch of amino acid residues have similar ϕ,ψ values. In native protein conformations, slight distortions in these values often occur. Substantial variation from these values, however, usually disrupts the secondary structure. Table 4·2 gives ideal values for the recognizable conformations listed. Vacant areas on the Ramachandran plot represent conformations that are impossible or rare because atoms or groups would be too close together. Most amino acid residues fall within the shaded, or permitted, areas shown on the plot, which are based on alanine as a typical amino acid. Some bulky amino acids have smaller permitted areas. Proline is restricted to a ϕ value of about −60° to −77° because its N—C_α bond is constrained by inclusion in the pyrrolidine ring of the side chain. In contrast, glycine residues are exempt from many steric restrictions because they lack β-carbons; thus, they are very flexible and have ϕ,ψ values that often fall outside the shaded regions of the plot.

Table 4·2 ϕ and ψ values for some recognizable conformations

Conformation	ϕ	ψ
α helix (right-handed)	−57°	−47°
α helix (left-handed)	+57°	+47°
3_{10} helix (right-handed)	−49°	−26°
Antiparallel β sheet	−139°	+135°
Parallel β sheet	−119°	+113°
Collagen helix	−51°	+153°
Type II reverse turn (second residue)	−60°	+120°
Type II reverse turn (third residue)	+90°	0°
Fully extended chain	−180°	−180°

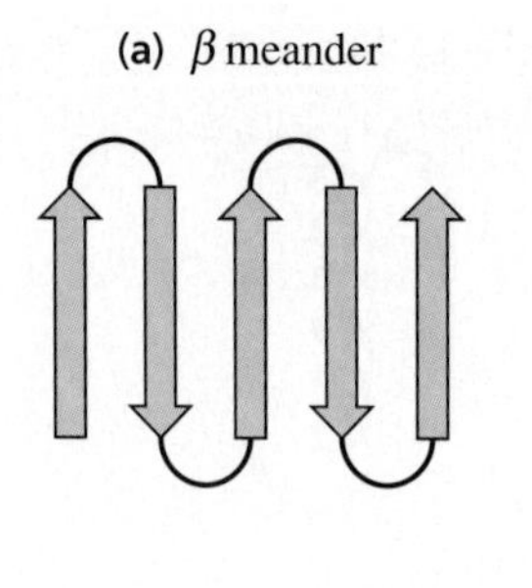

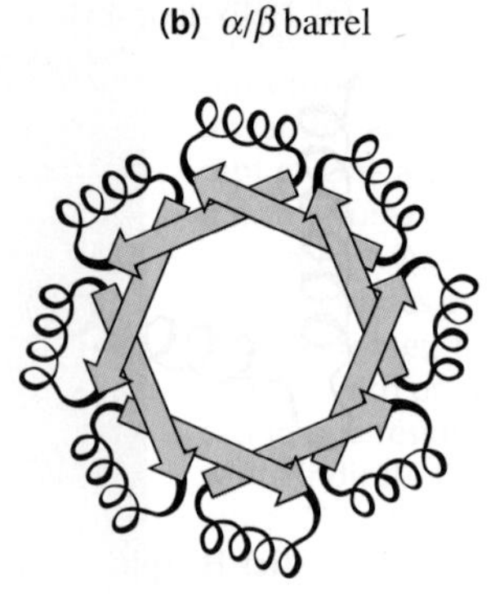

Figure 4·19
Two examples of domain structures. **(a)** The β meander is composed of adjacent antiparallel β strands connected in a simple, direct pattern. It can be formed into a barrel shape. **(b)** The α/β barrel shown here forms a structure with eight parallel β strands connected by eight α helices. Arrows indicate the N- to C-terminal direction of the peptide chain.

or up-and-down sheets. It contains antiparallel β strands connected by hairpin loops (Figure 4·19a). This structure sometimes forms a barrel. Such a structure in which β strands form the staves of a barrel is called a *β barrel*. The most common recognizable domain structure is an *α/β barrel*, which is made up of repeating $\alpha\beta$ units (Figure 4·19b). Close packing and suitable amino acid composition of the parallel β strands in an α/β barrel often make the core of the barrel a hydrophobic binding or reaction site.

Some proteins have one domain; others have two or more. An example of a protein having only one domain is the enzyme triose phosphate isomerase, which has an α/β barrel structure (Stereo S4·11). The extent of contact between domains varies from protein to protein. At one extreme, domains may exist as separate subunits of a protein that has quaternary structure. In multilobed proteins, domains may be connected by a single loop sometimes called a *hinge*. At the other extreme, domains may be joined by extensive and close contact. For example, the enzyme papain, which catalyzes selective hydrolytic cleavage of peptide bonds, consists of two interlocking domains (Stereo S4·12). The region or regions between domains often form crevices, or clefts, that may serve as binding sites for other molecules. For instance, the cleft between the two domains of papain, visible in Stereo S4·12, functions as a binding site for certain polypeptide sequences. Folding of a polypeptide chain within one domain is presumed to occur more or less independently of folding in other domains.

Just as the proposals by Pauling and Corey established secondary structure, elucidation of the structure of the oxygen-binding protein myoglobin (discussed below) laid the foundation for our present knowledge of tertiary structure of proteins. A number of observations made about the structure of myoglobin are typical of water-soluble globular proteins. Generally, water molecules are excluded from the interior of globular proteins. The interior is made up almost exclusively of hydrophobic amino acids, particularly those that are highly hydrophobic (valine, leucine, isoleucine, phenylalanine, and methionine). The surface contains both hydrophilic and hydrophobic amino acids. Most of the charged polar residues are located on the surface. However, several ionizable residues may be located internally in a cleft accessible to molecules that the protein specifically binds; these exceptional residues are important functional units of the protein. For example, as we shall see, internal ionizable residues play an important role in enzyme mechanisms (Chapter 6).

Note that whereas ribbon models of globular proteins such as that shown in Stereo S4·11a permit tracing of the three-dimensional winding of the polypeptide backbone, space-filling models such as that shown in Stereo S4·11b reveal that the structure of a globular protein is actually very compact, with side chains nestled together in such a way that the interior of the protein is nearly impenetrable, even by a small molecule such as water.

(a)

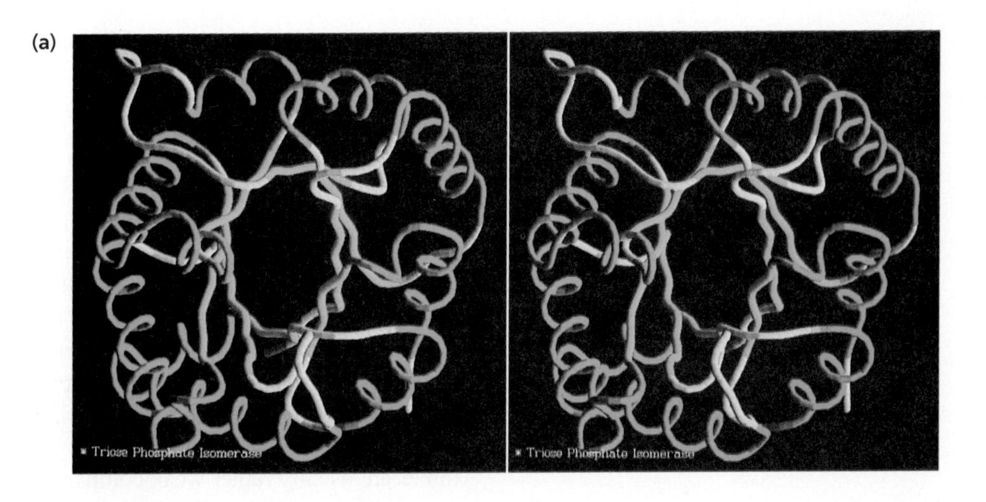

(b)

Stereo S4·11
Conformation of triose phosphate isomerase, an enzyme involved in the metabolism of sugars. (**a**) A cylinder of β strands (in blue) is surrounded by a wreath of α helices (in green), to make an α/β barrel. Regions of nonrepetitive structure are shown in yellow. (Based upon coordinates provided by Ian Wilson; stereo image by Richard J. Feldmann.) (**b**) Space-filling model. Color key: α helices, green; β barrel, blue; nonrepetitive structure, yellow. (Based upon coordinates provided by T. Alber, Gregory A. Petsko, and E. Lolis; stereo image by Ben N. Banks and Kim M. Gernert.)

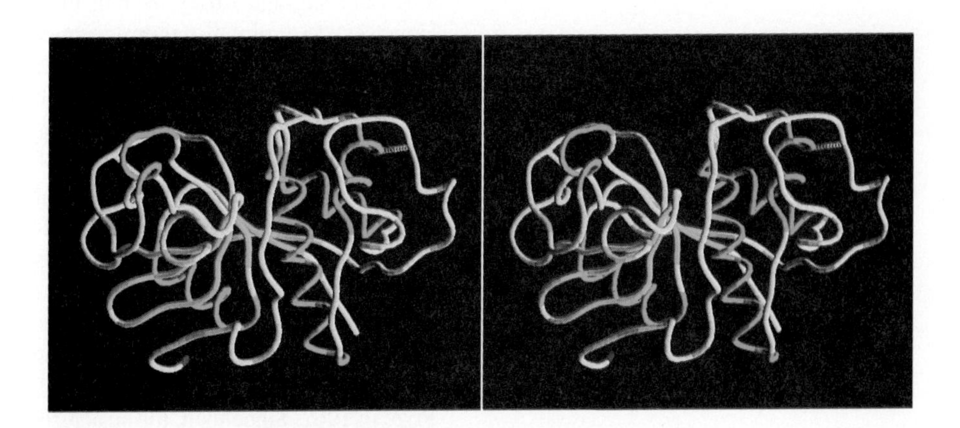

Stereo S4·12
Tertiary structure of papain. Two domains interlock in the native conformation of the protein. One domain contains three α helices and the other a single α helix and an antiparallel β-sheet. The β-sheet regions are shown in blue, the α-helical regions in green, and the nonrepetitive regions in yellow. (Based upon coordinates provided by Jan Drenth; stereo image by Richard J. Feldmann.)

B. Proteins with Quaternary Structure Are Aggregates of Globular Subunits

Some globular proteins have a further level of organization called quaternary structure. Quaternary structure is limited to proteins with multiple subunits and refers to the organization of the subunits. Each subunit is a separate polypeptide and may be called either a monomer or simply a chain. A multisubunit protein is referred to as an oligomer. The arrangement of the subunits within an oligomeric protein always has a defined stoichiometry and usually displays symmetry. The monomers of the protein may be identical or different. The first example we shall encounter of a protein having quaternary structure is hemoglobin (Section 4·16 and Stereo S4·16), which has four polypeptide chains of two slightly different types. Hydrophobic interactions are the principal forces holding subunits together, although electrostatic forces may contribute to the proper alignment of the subunits. Because intersubunit forces are usually rather weak compared to the forces stabilizing tertiary structure, the subunits of an oligomeric protein can often be isolated in a laboratory. However, in vivo the subunits are fairly tightly associated, as are many proteins that form complexes.

Determination of the subunit composition of an oligomeric protein is an essential step in the physical description of a protein. Typically, the molecular weight of the native oligomer is estimated by gel-filtration chromatography, and then the molecular weight of each chain is determined by SDS-polyacrylamide gel electrophoresis (Section 3·5). For a protein having only one type of chain, the ratio of the two values provides the number of chains per oligomer.

4·10 X-Ray Crystallography and Nuclear Magnetic Resonance Spectroscopy Are Used to Determine the Three-Dimensional Structures of Proteins

Whereas chemical methods such as Edman degradation are useful for determining the primary structures of proteins and the locations of covalent cross-links, X-ray crystallography is the most powerful tool available for determining the secondary, tertiary, and quaternary structures of biological macromolecules (Figure 4·20). Using information obtained from X-ray diffraction patterns of suitable crystals, it is possible to construct an electron density map that shows the positions of the atoms in a protein. From this map and from a knowledge of the sequence of the protein, it is possible to determine the three-dimensional structure of the protein.

X-ray diffraction studies of fibrous proteins were first attempted in the 1930s. Measurements of the simple repeating units of fibrous proteins from these studies aided Pauling and Corey in their proposal of the α helix and β sheet structures. However, determining the three-dimensional structure of a globular protein molecule containing thousands of atoms presented formidable technical difficulties. Chief among these was the difficulty of calculating atomic positions from the positions and intensities of diffracted X-ray beams. Therefore, the development of X-ray crystallogaphy of macromolecules closely followed the development of computers. In 1959, scientists in the laboratory of John C. Kendrew were able to obtain crystals of a globular protein suitable for X-ray crystallographic studies and to analyze the results obtained. Using crystals of sperm-whale myoglobin, they ultimately elucidated the structure of myoglobin at a resolution of 0.2 nm. Elucidation of the structure of myoglobin by Kendrew, and of hemoglobin a few years later by Kendrew's colleague Max Perutz, provided the first insights into the nature of the tertiary structure of globular proteins. Their efforts earned these scientists the Nobel Prize in 1962. Since then, the structures of hundreds of proteins have been revealed by X-ray crystallography. Although analysis of X-ray diffraction patterns has been facilitated by major technical advances, including the use of high-speed computers,

(a)

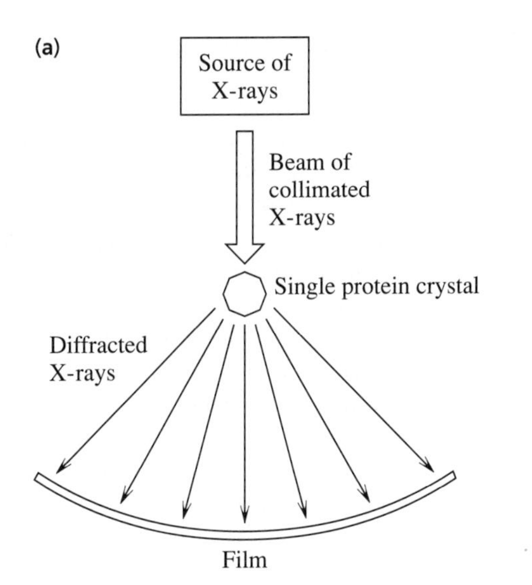

(b)

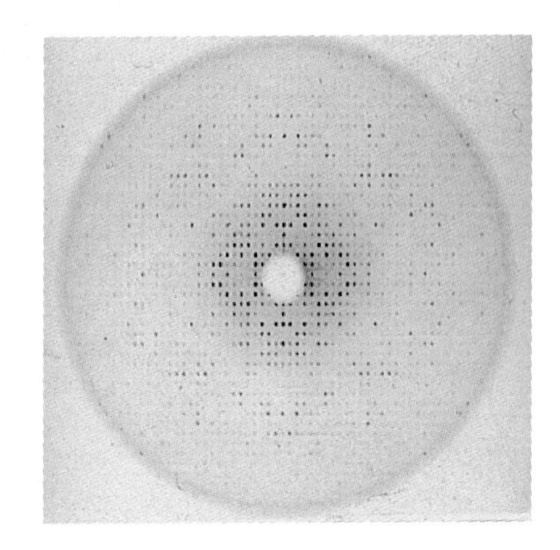

Figure 4·20
X-ray crystallography. **(a)** A beam of collimated, or parallel, X rays impinges on a protein crystal, which diffracts the rays onto cylindrical film. The atomic structure is then deduced by mathematical analysis of the diffraction pattern. **(b)** X-ray diffraction pattern of a crystal of adult human deoxyhemoglobin. (Courtesy of Eduardo Padlan.)

determination of protein structures is limited by the preparation of crystals of a quality suited to X-ray diffraction.

Unlike X-ray crystallography, *nuclear magnetic resonance* (NMR) spectroscopy permits the study of proteins in solution. NMR spectroscopy is a technique that uses absorption of electromagnetic radiation by molecules in magnetic fields of varying frequencies to find the spin states of certain atomic nuclei. In the study of protein structure, NMR is usually used to measure the spin states of ^{1}H atoms. Specialized NMR techniques record interactions between hydrogen atoms that are close together. Combining these results with a knowledge of the protein sequence allows determination of conformations. The complexity of NMR spectra has thus far precluded determination by NMR spectroscopy of the structures of proteins substantially larger than MW 15 000.

Years before NMR was used for structural determinations, it was used for titration of ionizable groups of enzymes. Since the observed frequency of energy absorption for a nucleus depends on its chemical environment, identification of spin states could be used in combination with NMR spectra of amino acids and peptides to determine pK_a values of individual ionic residues. In combination with other data, these pK_a values may be used to indicate the location of a residue, for example, whether a residue is located in the hydrophobic interior or the solvated exterior of the protein. Values of pK_a also provide evidence for the roles of ionizable residues at the active site (Section 6·3).

4·11 Folding and Stabilization of Globular Proteins Depend on a Variety of Interactions

Protein folding and stabilization of the biologically functional conformation of proteins depend on a number of factors, including the hydrophobic effect, hydrogen bonding, van der Waals interactions, and ionic interactions. It is thought that, as a protein folds, the first few interactions initiate subsequent interactions by assisting in alignment of groups. This process is known as *cooperativity of folding,* the phenomenon whereby the formation of one part of a structure leads to the formation of the remaining parts of the structure. Although noncovalent interactions are individually weak, the sum of these interactions stabilizes the native shapes of proteins. Furthermore, the weakness of each noncovalent interaction gives proteins the resilience and flexibility to undergo small conformational changes needed for the functioning of the protein.

A. The Hydrophobic Effect Is the Principal Driving Force in Protein Folding

Nonpolar (hydrophobic) side chains are more stable when they are aggregated than when they are solvated by water. The tendency of hydrophobic groups to associate with one another by virtue of their exclusion from water is called the hydrophobic effect (Chapter 2). This effect is the major driving force in protein folding. Because water molecules interact more strongly with each other than with nonpolar groups, the nonpolar side chains aggregate, causing the protein to fold. Nonpolar side chains are driven into the interior of the protein, and most polar side chains remain in contact with water on the surface of the protein. The sections of the polar backbone that are forced into the interior of a protein neutralize their polarity by hydrogen bonding to each other and forming secondary structures. Thus, the hydrophobic nature of the interior not only accounts for the association of hydrophobic residues but also the stabilization of helices and sheets.

The hydrophobic effect can be explained in thermodynamic terms. The water molecules that surround nonpolar groups are relatively well ordered, forming an enclosing structure, or cage. The water molecules are hydrogen bonded to each other but only weakly attracted to the group within the cage. When nonpolar groups are removed from contact with water, the disruption of the cage structure is accompanied by an increase in the entropy of the water molecules as they leave the ordered cages to become part of the bulk solvent. The increase in solvent entropy more than offsets the decreased entropy of the folded protein and provides the most significant driving force for protein folding.

B. Hydrogen Bonds and van der Waals Forces Stabilize Globular Proteins

Hydrogen bonds contribute to the cooperativity of folding and help stabilize the native conformations of globular proteins. As noted previously, the carbonyl and amide groups of the polypeptide backbone, especially those in the interior of a globular protein, often form hydrogen bonds with each other to produce α helices and β sheets. In addition, hydrogen bonds can form between the polypeptide backbone and water, between the polypeptide backbone and polar side chains, between two polar side chains, and between polar side chains and water. Table 4·3 lists some

Table 4·3 Examples of hydrogen bonds formed by polar amino acid residues

Type of hydrogen bond		Distance between donor and acceptor atom (nm)
Hydroxyl-hydroxyl	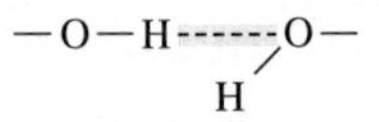 —O—H------O(H)—	0.28
Hydroxyl-carbonyl	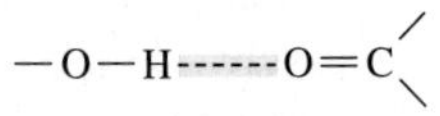 —O—H------O=C<	0.28
Amide-carbonyl	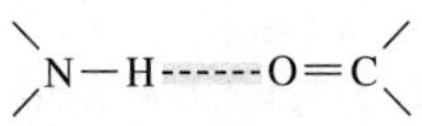 >N—H------O=C<	0.29
Amide-hydroxyl	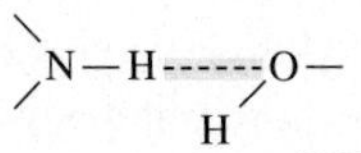 >N—H------O(H)—	0.30
Amide-imidazole nitrogen	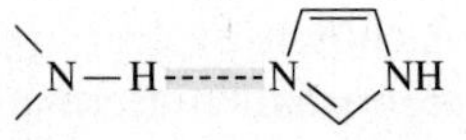 >N—H------N(imidazole)NH	0.31

of the many types of hydrogen bonds found in proteins. Efficient packing that maximizes van der Waals contacts between nonpolar residues also appears to contribute to the stability of globular proteins.

C. Covalent Cross-Links and Ionic Interactions Sometimes Help Stabilize Globular Proteins

In addition to hydrogen bonds, covalent cross-links, such as disulfide bonds, help stabilize the native conformations of some globular proteins. Although disulfide bonds are not usually found in intracellular proteins, they are sometimes found in proteins that are secreted from cells. When these proteins leave the intracellular environment, the presence of disulfide bonds makes the proteins less susceptible to unfolding and subsequent degradation.

To a small extent, ionic interactions between oppositely charged side chains may help stabilize globular proteins. Ionic side chains typically occur on the surface and are thus solvated and contribute minimally to the overall stabilization of the protein. However, occasionally two oppositely charged ions will form an ion pair within the interior of a protein. These ion pairs are typically associated with a special function in the molecular architecture of the protein. For example, in chymotrypsin, an ion pair forms between an α-amino group (Ile-16) and a β-carboxyl group (Asp-194) during activation of the enzyme precursor chymotrypsinogen (Section 5·10).

D. Protein Folding Is a Sequential, Cooperative Process

Protein-folding experiments using simple proteins, such as ribonuclease A, which we shall discuss in the next section, have allowed us to make some general observations regarding the folding of polypeptides into biologically active proteins. First, proper folding of a polypeptide chain does not occur by a random search. Rather, protein folding appears to be a cooperative, sequential process, in which formation of the first few structural elements assists in the alignment of subsequent structural features. Second, the folding pattern of a protein is dependent upon its primary structure. For some proteins, the information encoded in the sequence is sufficient to direct proper folding. For other proteins, particularly some larger proteins, proper sequence information alone seems insufficient for correct folding. In these cases, proper folding requires enzymes, such as those that reshuffle disulfide bonds via thiol-disulfide interchanges, or *chaperones,* binding proteins that assist correct folding by binding newly synthesized polypeptides before they fold, thereby preventing incorrect folding. Chaperones also speed up some stages of folding.

Although there have been many attempts to predict the secondary or tertiary structures of proteins from their primary structures, these attempts have met with limited success, primarily because so many variables determine the conformation of the polypeptide chain, including interactions between residues that are widely separated in the primary structure of the protein. Protein folding therefore continues to be a major research problem in protein chemistry.

4·12 Denaturing Agents Cause Proteins to Unfold

Environmental changes or chemical treatments may cause a disruption in the native conformation of a protein, with concomitant loss of biological activity. Such a disruption is called *denaturation*. Because the native conformation is only marginally stable, the energy needed to cause denaturation is often small, perhaps equivalent to the disruption of three or four hydrogen bonds.

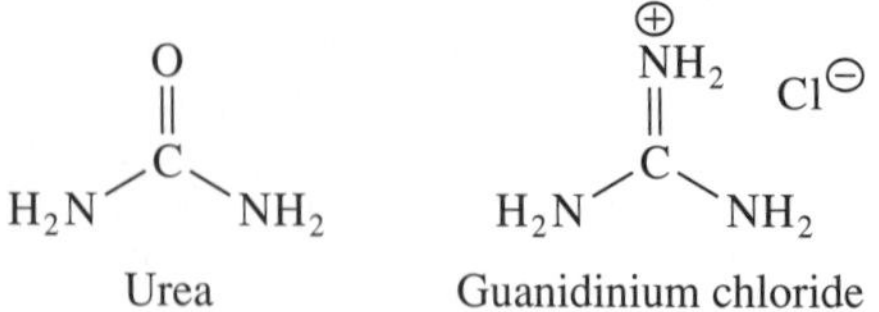

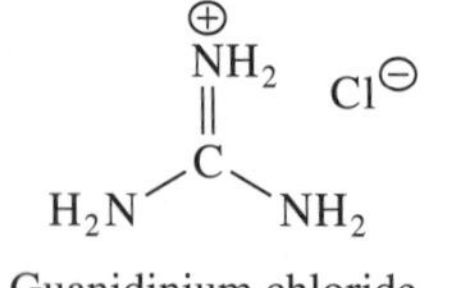

Figure 4·21
Urea and guanidinium chloride.

Several methods can be used to denature a protein. Raising or lowering the pH can change the ionic state of ionizable side chains in a protein, thereby breaking hydrogen bonds, creating regions of charge repulsion, and disrupting ion pairs, all of which contribute to denaturation. Heating a protein solution causes an increase in vibrational and rotational energy that can upset the delicate balance of weak interactions stabilizing the functional, folded conformation. High temperatures or treatment with strong acid or alkali can cause irreversible inactivation through covalent changes, such as deamidation of asparagine or glutamine residues and destruction of disulfide bonds.

Two types of chemicals, chaotropic ("chaos-promoting") agents and detergents, cause denaturation of proteins under less harsh conditions. Since these chemicals do not cleave covalent bonds, they disrupt only secondary, tertiary, and quaternary structures, not primary structure. Consequently, the effects of these chemicals can sometimes be reversed and studies using these denaturing agents can provide insight into protein folding. High concentrations of chaotropic agents such as guanidinium salts and urea (Figure 4·21) allow water molecules to penetrate into the interior of proteins, thereby disrupting the hydrophobic interactions that normally stabilize the native conformation. Detergents such as sodium dodecyl sulfate (SDS, Figure 2·6b) denature proteins at lower concentrations than chaotropic agents. The hydrophobic tails of such molecules penetrate the hydrophobic interior of a protein, disrupting the hydrophobic interactions and thereby denaturing the protein.

Dithiothreitol (DTT)

Oxidized dithiothreitol (a cyclic disulfide)

Figure 4·22
Dithiothreitol.

Complete denaturation of proteins that contain disulfide bonds requires the cleavage of these bonds in addition to disruption of hydrophobic interactions and hydrogen bonds. 2-Mercaptoethanol and other thiol reagents, such as dithiothreitol (DTT, Figure 4·22), can be added to a denaturing medium (8 M urea, 5 M guanidinium chloride, or 1% SDS) in order to reduce any disulfide bonds to sulfhydryl groups. The reduction of the disulfide bonds of the protein is accompanied by oxidation of the thiol reagent.

One early study of protein folding described the regeneration of the native conformation of a protein from its denatured state. This study, conducted by Christian B. Anfinsen and his co-workers, demonstrated the importance of primary structure in directing folding of a simple protein into its native conformation (Figure 4·23). Anfinsen used the protein ribonuclease A, a pancreatic enzyme that catalyzes digestion of ribonucleic acids. Ribonuclease A consists of a single chain of 124 amino acid residues, cross-linked by four disulfide bonds (Stereo S4·13).

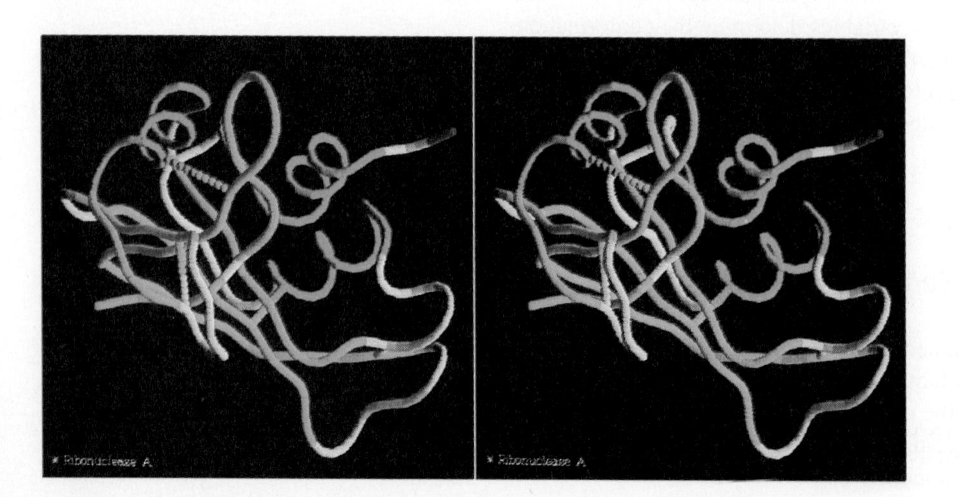

Stereo S4·13
Ribonuclease A. There are four disulfide bonds (shown as rods) in ribonuclease A. If the polypeptide is reductively denatured, the cysteine residues must be paired correctly during renaturation for the enzyme to become active. Color key: α-helical regions, green; β-sheet regions, blue; nonrepetitive regions, yellow. (Based upon coordinates provided by Frederic M. Richards; stereo image by Richard J. Feldmann.)

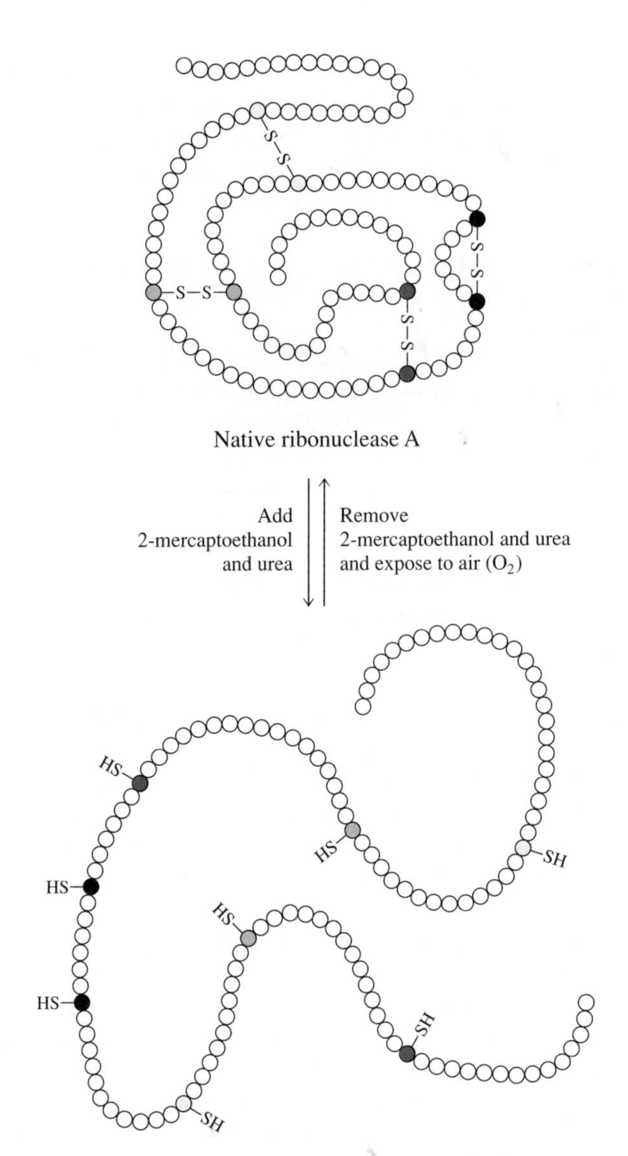

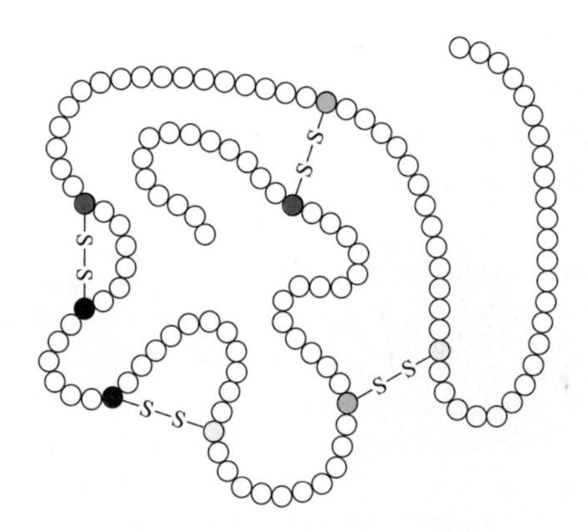

Figure 4·23
Denaturation and renaturation of ribonuclease. Treatment of native ribonuclease A (top) with 8 M urea containing 2-mercaptoethanol unfolds the protein and disrupts disulfide bonds to produce reduced, reversibly denatured ribonuclease (middle). If 2-mercaptoethanol alone is removed, ribonuclease A reoxidizes in the presence of air (O_2), but the disulfide bonds form randomly to produce a scrambled, inactive protein (such as the form shown at the bottom). If urea and 2-mercaptoethanol are removed simultaneously under appropriate conditions, the protein returns to its native conformation and in the presence of air the correct disulfide bonds are formed (top). Such renaturation suggests the importance of primary structure in directing the folding of this simple protein.

Denaturation of ribonuclease A with 8 M urea containing 2-mercaptoethanol results in complete loss of tertiary structure and enzymatic activity and yields polypeptide chains containing eight sulfhydryl groups. If the eight sulfhydryl groups paired randomly, 105 possible disulfide-bonded structures could be produced (7 possible pairings for the first bond, 5 for the second, 3 for the third, and 1 for the fourth; $7 \times 5 \times 3 \times 1 = 105$), with only 1 out of 105 molecules being active. In fact, when reductant is removed and oxidation is allowed to occur in the presence of denaturant (8 M urea), disulfide bonds form between incorrect partners in about 99% of the protein population, generating a solution of protein that has about 1% of its original enzymatic activity. However, when urea and the reductant are removed simultaneously and dilute solutions of the reduced protein are then exposed to air at physiological pH, ribonuclease A spontaneously regains its native conformation, its correct set of disulfide bonds, and its full enzymatic activity.

4·13 The Structures of Globular Proteins Allow Them to Selectively and Transiently Bind Other Molecules

The shapes of globular proteins, with their indentations, interdomain interfaces, or other crevices, allow them to fulfill dynamic functions by selectively and transiently binding other molecules. This property is best exemplified by the interaction between specific reactants (substrates) and enzymes at substrate-binding sites or active sites, but it is also true of other globular proteins. Because binding sites are hydrophobic and positioned toward the interior of the protein, they are relatively free of water. In some cases, binding at one site can change the conformation of another site and therefore affect binding at the distant site. The diversity of folding and of side-chain structures from protein to protein allows for diversity in binding-site conformation, binding-site specificity, and binding-site interactions.

As part of their binding sites, some globular proteins contain *cofactors,* relatively small, typically nonprotein molecules that are required for the protein to function (Chapter 7). Some cofactors dissociate and reassociate during the course of a physiological reaction; those that do not dissociate are called *prosthetic groups.* An active protein possessing all of its cofactors is often referred to as a *holoprotein;* a protein whose cofactors are absent is called an *apoprotein.*

In the course of this book, we shall explore in detail the structure and function of a number of globular proteins. We shall begin here with myoglobin and hemoglobin, the first two proteins whose structures were revealed in atomic detail.

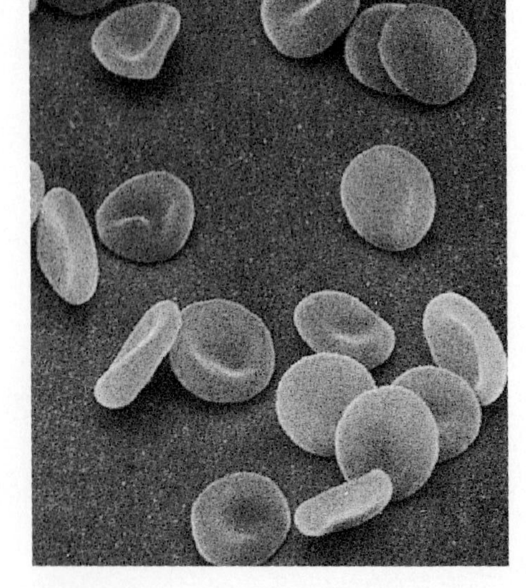

Figure 4·24
Scanning electron micrograph of red blood cells. Mature red blood cells are biconcave disks. Each red blood cell contains approximately 300 million hemoglobin molecules. (Courtesy of Keith R. Porter.)

4·14 Myoglobin and Hemoglobin Are Oxygen-Binding Proteins

Like most globular proteins, myoglobin (Mb) and hemoglobin (Hb) carry out their biological functions by selectively binding other molecules. By binding and then releasing molecular oxygen (O_2), myoglobin stores oxygen and facilitates its diffusion within muscle, and hemoglobin transports oxygen in the blood of vertebrates. In mammals and birds, hemoglobin itself is transported by red blood cells, or erythrocytes (Greek, *erythros,* red; *kytos,* hollow vessel). Viewed under a microscope, a mature mammalian erythrocyte is a biconcave disk that lacks a nucleus or other internal membrane-bounded compartments (Figure 4·24). A human erythrocyte is filled with approximately 3×10^8 molecules of hemoglobin.

The red color associated with the oxygenated forms of myoglobin and hemoglobin (for instance, the red color of skeletal and cardiac muscle and oxygenated blood) is due to the prosthetic group, heme (Figure 4·25). Although the term *prosthetic group* usually refers to a molecule required for an enzyme to function, it is also applied to the heme of myoglobin and hemoglobin (which are not enzymes) since heme is essential for the binding and release of oxygen by these proteins. Heme consists of a tetrapyrrole ring system called protoporphyrin IX complexed with ionic iron. The four pyrrole rings of this system are linked by methene (—CH═) bridges, so that the entire porphyrin structure is unsaturated, highly conjugated, and planar. Four methyl groups, two vinyl groups, and two carboxyethyl (or propionate) substituents are attached to the tetrapyrrole ring system. The iron, which is in the reduced or ferrous (Fe^{2+}) oxidation state, replaces two protons of protoporphyrin IX. The Fe-protoporphyrin IX complex is a resonance hybrid in which the iron is bound equally to the four nitrogens of protoporphyrin IX.

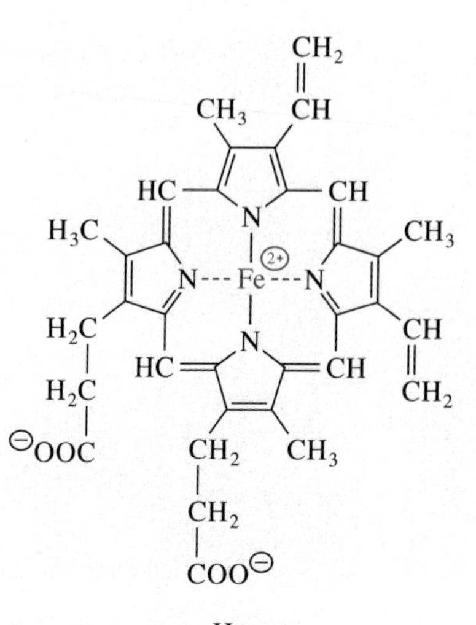

Figure 4·25
Structure of heme (Fe-protoporphyrin IX). The iron atom is in the ferrous oxidation state. Although the figure shows iron covalently bonded to only two nitrogen atoms, actually iron is equally bound to all four nitrogen atoms.

4·15 Myoglobin Is a Monomeric Heme Protein

Myoglobin accounts for about 8% of the total protein in the muscles of diving mammals, such as seals and whales, that store large amounts of oxygen. A relatively small protein ($4.4 \times 3.5 \times 2.5$ nm), myoglobin consists of a single polypeptide chain in addition to the heme prosthetic group. The polypeptide component, or apoprotein, is called globin and consists of 153 amino acid residues. Figure 4·26 depicts the tertiary structure of sperm-whale myoglobin, with its eight α helices designated A through H in the N- to C-terminal direction. Intervening bends and nonrepetitive regions are designated AB, BC, etc. Although amino acid residues may be numbered sequentially in the N- to C-terminal direction (1–153), to provide greater clarity for comparisons of similar globins from various sources, they are usually referred to by their positions relative to the eight helices. For example, residue E7 is the seventh residue of helix E; residue CD2 is the second residue of the

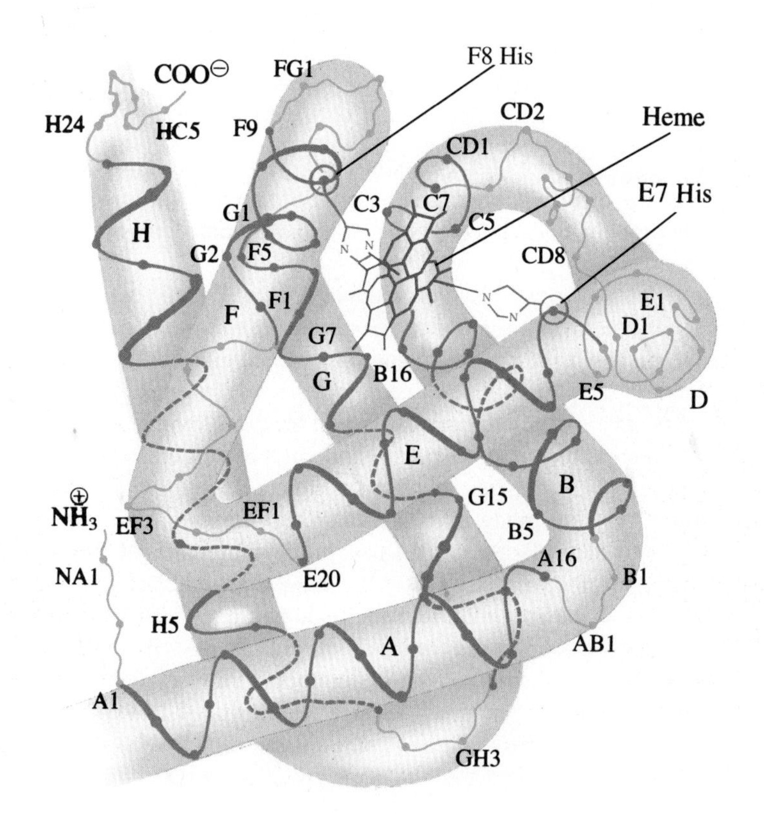

Figure 4·26
Diagram of sperm-whale myoglobin. Myoglobin consists of eight α helices (in blue) connected by short nonrepetitive segments (in green). The helices are labelled A through H, starting from the N-terminus. Nonhelical segments are labelled with the letters of the α helices they connect (AB, CD, etc.). The heme group (in red) binds oxygen. It is wedged between the E and F helices. His-E7 forms a hydrogen bond with oxygen, and His-F8 is complexed to the iron atom within the heme. (Illustrator: Tom Edgerton.)

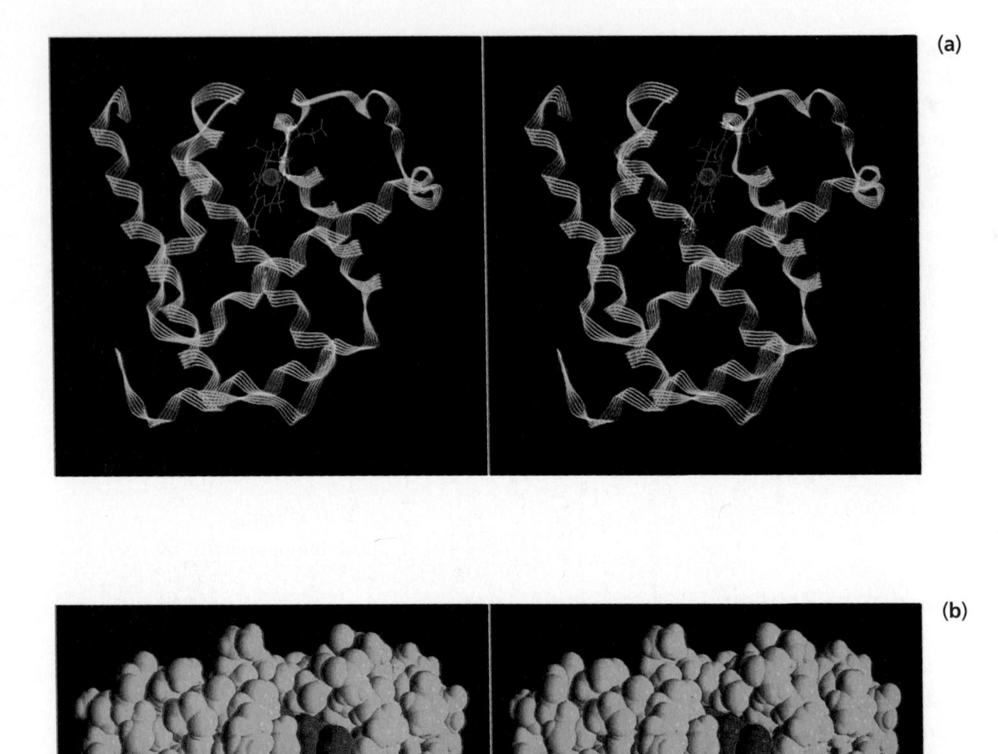

Stereo S4·14
Myoglobin. The heme group, shown in red, is almost completely buried. **(a)** Ribbon model. (Based upon coordinates provided by H. C. Watson and John C. Kendrew; stereo image by Ben N. Banks and Kim M. Gernert.) **(b)** Space-filling model. (Based upon coordinates provided by John C. Kendrew; stereo image by Richard J. Feldmann.)

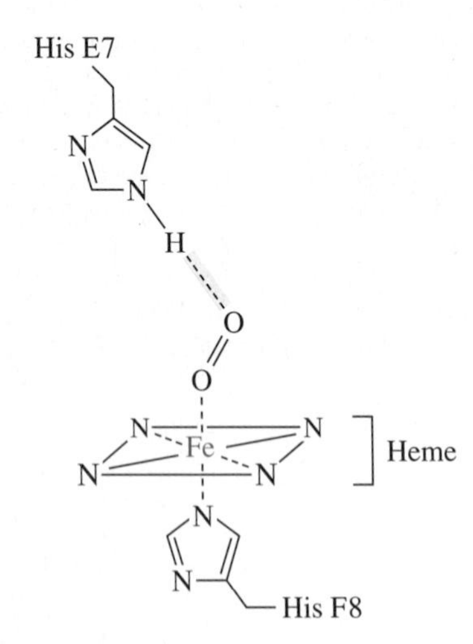

Figure 4·27
The oxygen-binding site of oxymyoglobin. The heme prosthetic group is represented by a parallelogram with a nitrogen atom at each corner.

nonrepetitive segment between helix C and helix D; and residue NA1 is the N-terminal residue of the molecule, preceding helix A. Stereo S4·14a shows the helical structure of myoglobin, and Stereo S4·14b reveals the compactness of the molecule.

With the exception of two histidine residues, E7 and F8, all polar residues are located on the surface, and the interior of myoglobin is composed almost entirely of nonpolar residues. The heme prosthetic group is wedged into a hydrophobic, cage-like cleft formed by the protein. The two anionic carboxyl groups of the propionate substituents of heme are exposed on the surface of the myoglobin molecule, and are thus hydrated, whereas the nonpolar methyl and vinyl groups are buried in the hydrophobic interior. In oxygenated myoglobin, called oxymyoglobin, the ferrous iron is coordinated by six ligands, making the complex octahedral (Figure 4·27 and Stereo S4·15). Four of the ligands of the iron in myoglobin are the four nitrogen atoms of the tetrapyrrole ring system; the fifth ligand is an imidazole nitrogen from His-F8 (referred to as the proximal, or near, histidine); and the sixth is molecular oxygen bound between the iron and the imidazole side chain of His-E7 (referred to as the distal, or distant, histidine, since it is slightly too far from the iron to be part of its coordination sphere). The nonpolar side chains of Val-E11 and Phe-CD1, shown in Stereo S4·15, contribute to the steric constraint and hydrophobicity of the oxygen-binding pocket and help hold the heme group in place.

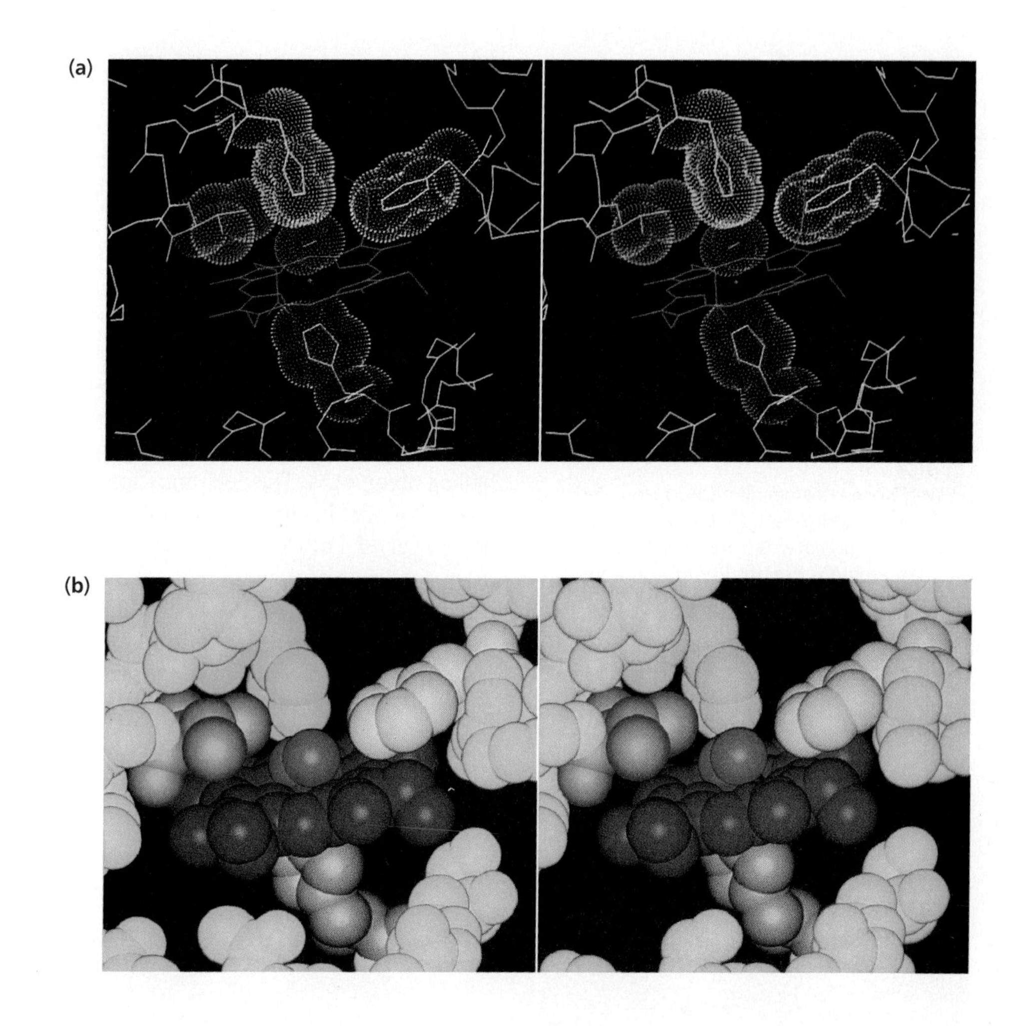

Stereo S4·15
The oxygen-binding site of oxymyoglobin. Fe^{2+} (red), lying almost in the heme plane (also red), is bound to oxygen (pink) and is flanked by His-E7 (green) and His-F8 (orange). Val-E11 (dark blue) and Phe-CD1 (yellow) contribute to the hydrophobic environment of the oxygen-binding site. The backbones of neighboring residues are shown in light blue. **(a)** Stick model. **(b)** Space-filling model. (Based upon coordinates provided by T. Takano; stereo images by Ben N. Banks and Kim M. Gernert.)

The hydrophobic crevice of the globin holds the key to the ability of myoglobin (and hemoglobin) to suitably bind and release oxygen. In aqueous solution, heme does not reversibly bind oxygen; instead, the Fe^{2+} of the heme is almost instantly oxidized to Fe^{3+}. The structure of the globin of either myoglobin or hemoglobin prevents formal transfer of an electron and precludes irreversible oxidation, thereby assuring *oxygenation*, the reversible binding of molecular oxygen for transport.

4·16 Hemoglobin Is a Tetramer Whose Polypeptide Chains Are Similar to Those of Myoglobin

In 1865, long before any protein had been isolated as a pure substance, Wilhelm Kühne proposed that the oxygen-binding proteins of muscle and blood (now known to be myoglobin and hemoglobin) were related, if not identical. A century later, Max Perutz determined the structure of horse hemoglobin by X-ray crystallography, a task that required over two decades of effort. At last, the similarities and differences between myoglobin and hemoglobin were revealed.

Hemoglobin is a tetramer composed of two each of two types of globin chains, α and β, both similar to, but slightly shorter than, the single chain of myoglobin

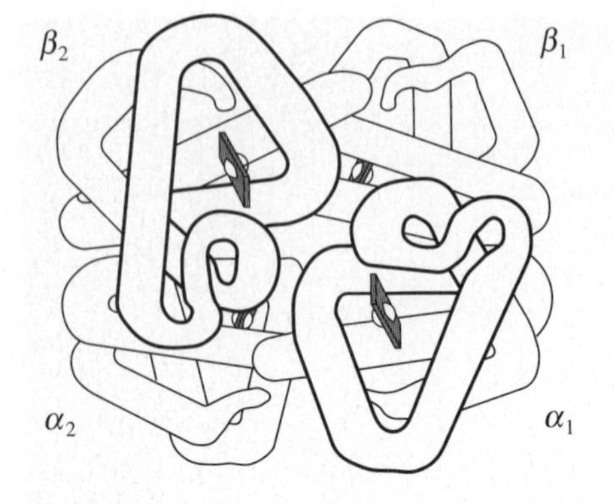

Figure 4·28
Hemoglobin tetramer. [Adapted from Dickerson, R. E., and Geis, I. (1969). *The Structure and Action of Proteins* (Menlo Park, California: Benjamin/Cummings Publishing Company).]

(Figure 4·28 and Stereo S4·16). Each of the two α chains consists of 141 amino acid residues; each of the two β chains consists of 146 residues. The tertiary structure of each of the four chains is almost identical to that of myoglobin. For this reason, the structural features of a subunit of hemoglobin are labelled in the same way as myoglobin. The striking similarity in structure between a myoglobin molecule and a hemoglobin subunit can be seen by comparing Stereos S4·14a and S4·17. In fact, Perutz described hemoglobin as "just four myoglobin molecules put together." Hemoglobin, however, is not simply a tetramer of myoglobin molecules. Because an α chain interacts with a β chain much more strongly than α interacts with α or β with β, hemoglobin is actually a dimer of $\alpha\beta$ subunits.

The structures of globins in many species are very similar. For example, leghemoglobin (Lb), an oxygen-binding monomeric protein found in leguminous plants (Stereo S4·18), has a three-dimensional structure much the same as mammalian myoglobin. The determination of the amino acid sequences of globins from dozens of species has revealed the evolutionary relationships of this family of globin genes. For example, the mammalian α and β globins of hemoglobins are more similar to each other than either one is to the globin of myoglobin. This observation shows that there was a duplication of the globin gene in early ancestors of mammals; one gene gave rise to modern myoglobin and the other to α and β globins. A more recent duplication gave rise to the separate α-globin and β-globin genes. Some invariant residues in the primary structures of globin molecules can be

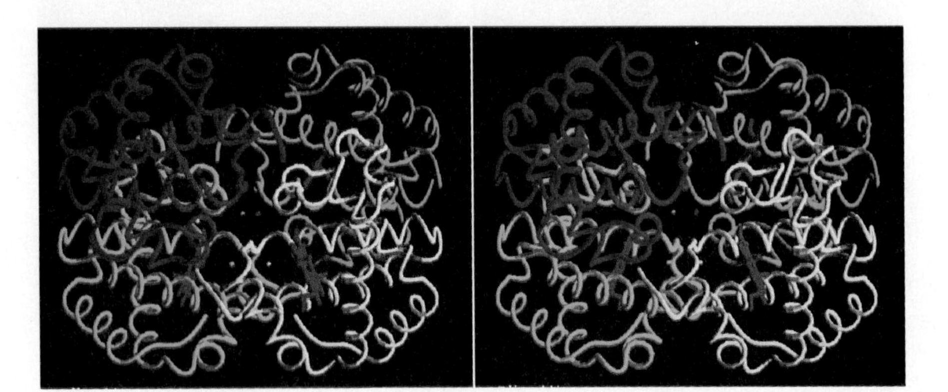

Stereo S4·16
Hemoglobin tetramer. The α and β chains face each other across a central cavity. Color key: α_1, yellow; α_2, light blue; β_1, green; β_2, dark blue; heme group, red. (Based upon coordinates provided by Max Perutz; stereo image by Richard J. Feldmann.)

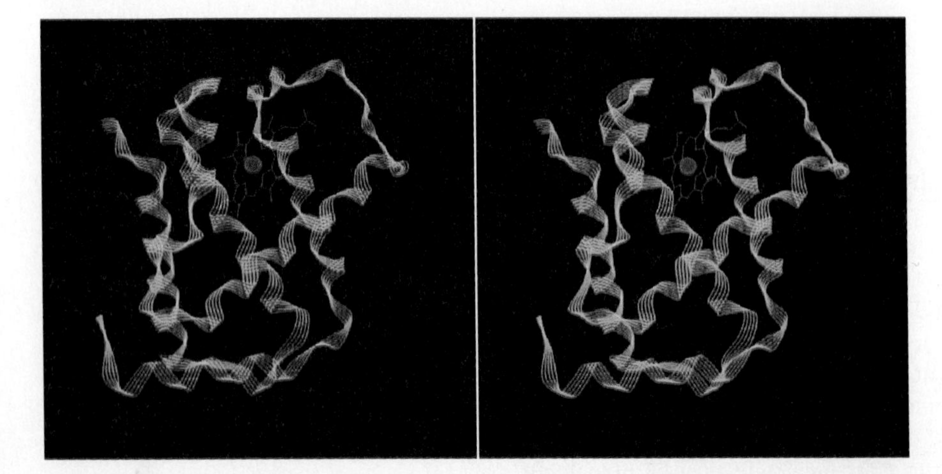

Stereo S4·17
α Subunit of horse hemoglobin. The heme group is shown in red. (Based upon coordinates provided by B. Shaanan; stereo image by Ben N. Banks and Kim M. Gernert.)

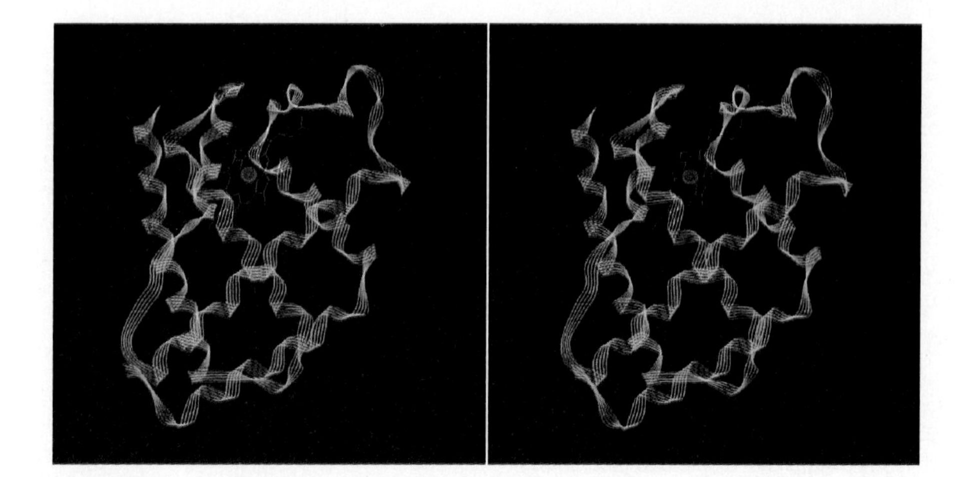

Stereo S4·18
Leghemoglobin. The heme group is shown in red. (Based upon coordinates provided by B. K. Vainshtein and E. H. Harutyunyan; stereo image by Ben N. Banks and Kim M. Gernert.)

related to their function. For example, all globins contain histidine residues at positions E7 (the oxygen-binding site) and F8 (the fifth ligand to iron). Similarly, Phe-CD1, which contributes to the hydrophobic cage surrounding the heme, is invariant in all known vertebrate myoglobins.

An evolutionary relationship based on identities in amino acid sequence inadequately conveys the similarity of tertiary structures because the substitution of one residue by a similar residue may have little effect on the conformation and function of a polypeptide chain. A substitution that does not significantly affect conformation or function is said to be *conservative*. For example, substitution of the valine at position E11 (present in most globins) by isoleucine (in the α chains of kangaroo hemoglobin) or by leucine (in soybean leghemoglobin) is conservative since it does not appreciably affect the protein. In contrast, substitution of a hydrophilic residue (such as glutamate) by a hydrophobic one (such as valine) is *nonconservative* since the substitution may have a profound effect on the conformation or function of a protein, as is the case with persons suffering from sickle cell anemia (Section 4·19).

In addition to the genes for the α and β globins of adult hemoglobin, some mammals have genes that encode an ε-globin subunit and a γ-globin subunit, which appear during embryonic or fetal development (Section 4·18).

4·17 Myoglobin and Hemoglobin Have Different Oxygen-Binding Curves

The physiological functions of myoglobin and hemoglobin depend on reversible binding of oxygen to the heme prosthetic groups, a process that can be depicted by oxygen-binding curves.

The equilibrium constant (K_{eq}) for the binding of molecular oxygen (O_2) by myoglobin (Mb)

$$\text{Mb} + \text{O}_2 \rightleftharpoons \text{MbO}_2 \qquad (4\cdot1)$$

is given by

$$K_{eq} = \frac{[\text{MbO}_2]}{[\text{Mb}][\text{O}_2]} \qquad (4\cdot2)$$

The fractional saturation (Y) is the proportion of total myoglobin that is oxygenated:

$$\text{Y} = \frac{[\text{MbO}_2]}{[\text{Mb}] + [\text{MbO}_2]} \qquad (4\cdot3)$$

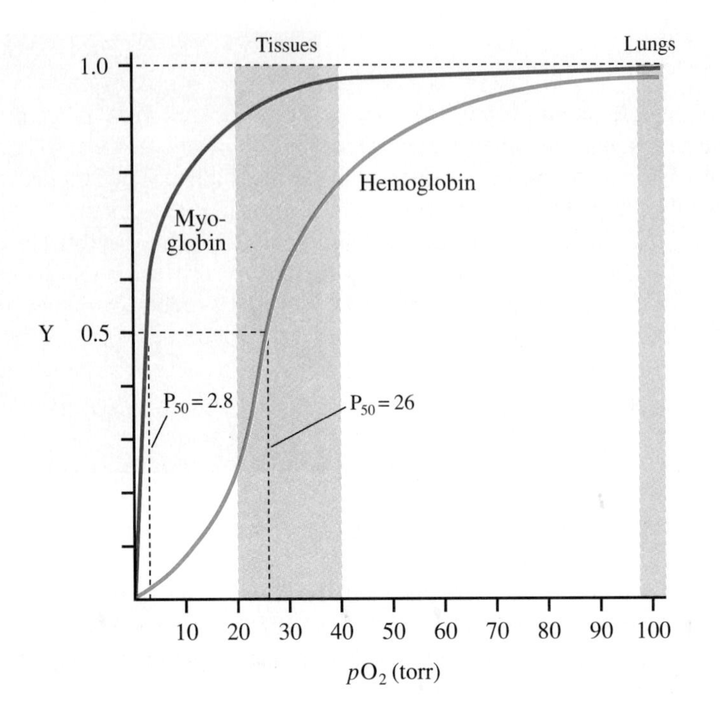

Figure 4·29
Oxygen-binding curves of myoglobin and hemoglobin. The curves show a plot of the fractional saturation, Y, of each protein versus the concentration of oxygen measured as the partial pressure, pO_2 (torr). When Y = 0.5, the protein is half saturated with oxygen. The oxygen-binding curve of myoglobin is hyperbolic, with half-saturation at an oxygen pressure of 2.8 torr. The oxygen-binding curve of hemoglobin is sigmoidal, with half-saturation at an oxygen pressure of 26 torr. The sigmoidal shape of the oxygen-binding curve indicates that there is positive cooperativity in the binding of oxygen by hemoglobin.

A comparison of the two plots shows that myoglobin has a greater affinity for oxygen than does hemoglobin at all oxygen pressures. Nevertheless, in the lungs, where the oxygen pressure is high, hemoglobin is nearly saturated with oxygen. In tissues, where the partial pressure of oxygen is low, oxygen is released from oxygenated hemoglobin and transferred to myoglobin.

When the fractional saturation of a solution containing myoglobin is plotted versus the concentration of oxygen (measured as the partial pressure of gaseous oxygen, pO_2), a hyperbolic curve is obtained (Figure 4·29). The myoglobin molecules are half saturated (Y = 0.5) at a pO_2 of 2.8 torr. This constant is usually referred to as the half-saturation pressure, P_{50}. One can generalize that when there is a single equilibrium constant for the binding of a ligand to a protein, a plot of fractional saturation versus concentration of ligand will be hyperbolic.

Whereas only one molecule of oxygen binds to each molecule of myoglobin, up to four molecules of oxygen bind to hemoglobin, one per heme group of the tetrameric protein. The equation for binding of O_2 to hemoglobin is more complicated than that for myoglobin. As shown in Figure 4·29, a plot of the fractional saturation of hemoglobin versus pO_2 results in a sigmoidal (S-shaped) curve rather than a hyperbolic curve. The sigmoidal curve indicates that the oxygen-binding sites of hemoglobin interact such that binding of one molecule of oxygen to one heme group facilitates binding of additional molecules of oxygen to the other three heme groups. This interactive binding phenomenon is termed *positive cooperativity of binding*.

The physiological roles of myoglobin and hemoglobin are directly related to their relative affinities for oxygen at low oxygen pressures. As Figure 4·29 shows, at the high pO_2 found in the lungs (about 100 torr), both myoglobin and hemoglobin would have a high affinity for oxygen and would be nearly saturated. However, at all pO_2 values below 50 torr, myoglobin has a significantly greater affinity for oxygen than hemoglobin. Within the capillaries of tissues such as muscles, where pO_2 is low (ranging from 20 to 40 torr), a significant fraction of the oxygen carried by hemoglobin in erythrocytes is released because of the lower affinity of hemoglobin for oxygen. However, at this concentration of oxygen, myoglobin in the muscle tissue binds the oxygen released by hemoglobin. The differential affinities of myoglobin and hemoglobin for oxygen thus lead to an efficient system for oxygen delivery from the lungs to muscle.

The same heme prosthetic group is present in myoglobin and hemoglobin, but its affinity for oxygen differs because the environment provided by the protein cages of myoglobin and hemoglobin are slightly different. Protein environments can change reactions qualitatively as well as quantitatively. As we shall see in our discussion of cytochromes in Chapter 14, under the influence of different apoproteins, heme undergoes oxidation and reduction instead of oxygenation.

Deoxygenated hemoglobin, called deoxyhemoglobin, has a rather low affinity for oxygen because ion pairs cross-linking the subunits of hemoglobin stabilize the structures of the individual chains such that the protein resists binding oxygen. As oxygen binds to hemoglobin, a significant conformational change occurs that disrupts the cross-linking between subunits and thus increases the affinity of the protein for oxygen. When oxygen binds to a heme site within each $\alpha\beta$ dimer, one $\alpha\beta$ subunit of hemoglobin rotates about 15° with respect to the other $\alpha\beta$ pair, as illustrated in Figure 4·30. In deoxyhemoglobin, the iron is somewhat out of the plane of the porphyrin ring (about 0.06 nm toward the proximal histidine F8), and the six electrons of the *d* orbital of iron are arranged as four unpaired electrons and one electron pair, thus accommodating binding to five ligands (the four porphyrin nitrogen atoms and an imidazole nitrogen of His-F8). When oxygen binds, the electronic structure of the iron changes with the result that the bond between the iron and His-F8 becomes shorter, the four bonds to the porphyrin nitrogen atoms are strengthened, and iron moves about 0.04 nm closer to the porphyrin plane (Figure 4·31). Consequently, the entire F helix of the β chain moves toward the H helix, disrupting ion pairs that cross-link the subunits in deoxyhemoglobin. Once oxygen binds to a heme within each $\alpha\beta$ dimer, the disruption of ion pairs leads to a change in quaternary structure (the 15° rotation of one dimer) that increases the oxygen-binding affinity of the remaining unoxygenated heme sites. The positive cooperativity of hemoglobin thus promotes its full oxygenation in the high pO_2 of the lungs and the efficient unloading or dissociation of oxygen in the low pO_2 of the capillary beds of other tissues.

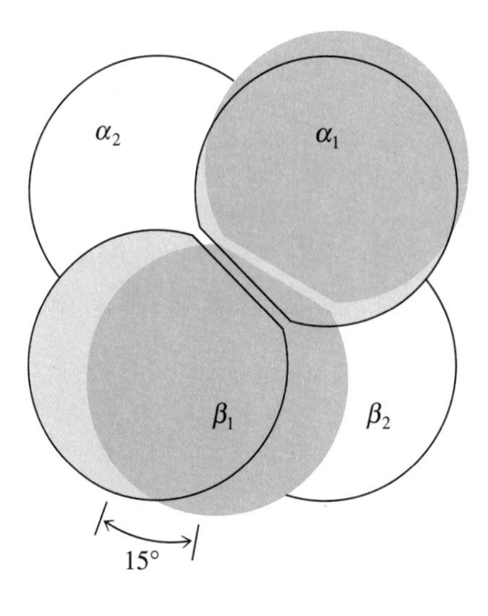

Figure 4·30
Oxygen-induced conformational change in hemoglobin. The oxygenation of hemoglobin causes a conformational change in which one $\alpha\beta$ dimer ($\alpha_1\beta_1$) rotates 15° relative to the other dimer ($\alpha_2\beta_2$). Deoxygenated $\alpha_1\beta_1$, light blue; oxygenated $\alpha_1\beta_1$, dark blue.

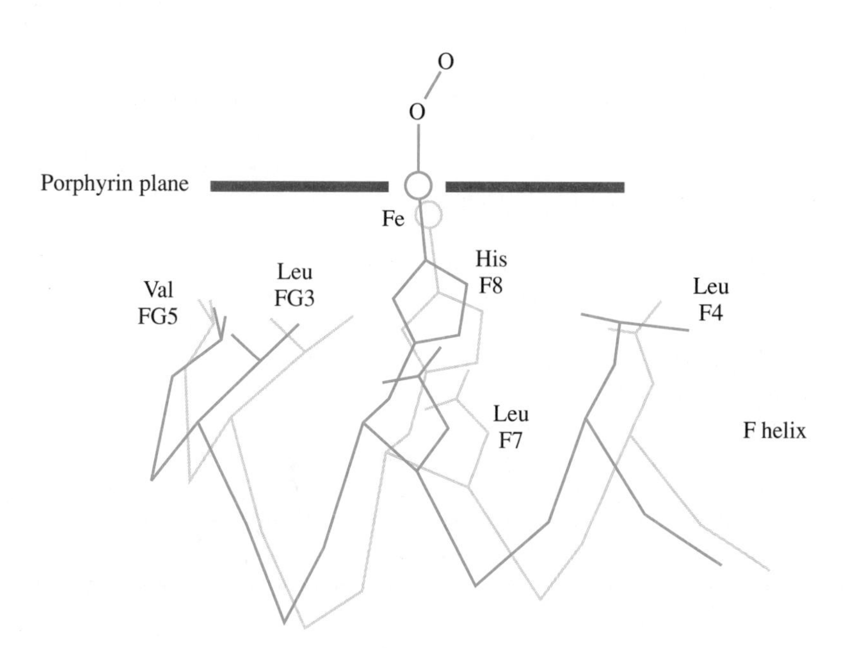

Figure 4·31
Conformational changes induced by oxygenation. When the heme iron of hemoglobin is oxygenated, the proximal histidine (His-F8) is pulled toward the porphyrin ring. The remainder of the F helix also shifts position, disrupting ion pairs that cross-link the subunits of deoxyhemoglobin. When oxygen binds to a chain of each dimer, the disruption of ion pairs leads to a change in quaternary structure (the 15° rotation of one dimer) that increases the oxygen-binding affinity of the remaining unoxygenated heme sites. Deoxyhemoglobin is shown in light blue; oxyhemoglobin in dark blue. [Adapted from Baldwin, J., and Chothia, C. (1979). Haemoglobin: the structural changes related to ligand binding and its allosteric mechanism. *J. Mol. Biol.* 129:175–220.]

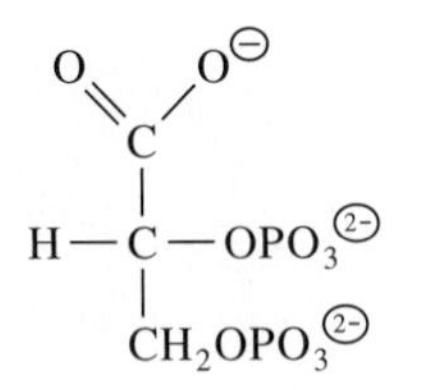

Figure 4·32
2,3-*Bis*phospho-D-glycerate (2,3BPG).

4·18 Hemoglobin Is an Allosteric Protein

In addition to positive cooperativity of binding, *allosteric interactions* (Greek, *allos*, other; *stereos*, solid) are important in regulation of the binding and release of oxygen by hemoglobin. Allosteric interactions occur when a specific small molecule, called an *allosteric effector* or modulator, binds to a protein (usually an enzyme) and modulates its activity. The allosteric effector binds reversibly at a site separate from the functional binding site of the protein. Its binding transmits information about effector concentration to the functional portion of the protein. An effector may be further classified as an allosteric activator or an allosteric inhibitor, depending on its effect on the allosteric protein. A protein whose activity is modulated by allosteric effectors is called an *allosteric protein*. Because allosteric effectors bind reversibly to allosteric protein molecules, the activity of the total population of the protein solution responds to changes in the concentrations of the effectors.

Allosteric regulation is accomplished by small but significant changes in the native conformations of allosteric proteins. The binding of an allosteric *inhibitor* causes a protein to rapidly change from its active shape (called the R or relaxed state) to its inactive shape (called the T or taut state). The binding of an allosteric *activator* causes the reverse change. The reversible change from R to T or from T to R is termed the allosteric transition. The change in conformation of an allosteric protein caused by binding or release of an effector extends from the allosteric site to the functional binding site. The activity level of an allosteric protein depends on the relative proportions of its R and T forms, and these in turn depend on the relative concentrations of the ligands (either substrates or effectors) that bind to each form. In hemoglobin, the deoxy conformation, which resists oxygen binding, is considered the inactive (T) state and the oxy conformation, which facilitates oxygen binding, is considered the active (R) state.

The molecule 2,3-*bis*phospho-D-glycerate (2,3BPG) is an allosteric effector of hemoglobin in erythrocytes (Figure 4·32). The presence of 2,3BPG raises the P_{50} for binding of oxygen to adult hemoglobin in whole blood to about 26 torr—much higher than the P_{50} for oxygen binding to pure hemoglobin in aqueous solution, about 12 torr. In other words, 2,3BPG in erythrocytes substantially lowers the affinity of deoxyhemoglobin for oxygen. The concentrations of 2,3BPG and hemoglobin within erythrocytes are nearly equal (about 4.7 mM).

A molecule of 2,3BPG can bind in the central cavity of the deoxy conformation of hemoglobin, which is lined with the positively charged side chains of a lysine (Lys-EF6) and two histidine residues (His-NA2 and His-H21) and the N-terminal α-amino group of each β chain (Figure 4·33). The 2,3BPG molecule forms ionic bonds with these residues, whose positively charged side chains are complementary in position to the negative charges of 2,3BPG. When 2,3BPG is bound, the deoxy conformation is stabilized. Conversely, oxygenation precludes binding of 2,3BPG, since the central cavity of oxyhemoglobin is too small to accommodate 2,3BPG. Thus, oxygen and 2,3BPG have opposite effects on the $R \rightleftharpoons T$ equilibrium. Binding of oxygen increases the proportion of hemoglobin molecules in the oxy (R) conformation, and binding of 2,3BPG increases the proportion of hemoglobin molecules in the deoxy (T) conformation. Because oxygen and 2,3BPG have different binding sites, 2,3BPG is an allosteric effector.

2,3BPG has an important physiological role. In the absence of 2,3BPG, hemoglobin is nearly saturated at an oxygen pressure of about 20 torr. Thus, at the low partial pressure of oxygen that prevails in the tissues (20–40 torr), hemoglobin without 2,3BPG would not unload its oxygen to myoglobin. In the presence of equimolar 2,3BPG, however, hemoglobin is only about one-third saturated at 20 torr. Thus, the allosteric effect of 2,3BPG allows hemoglobin to transfer oxygen to myoglobin at the low partial pressures of oxygen in the tissues.

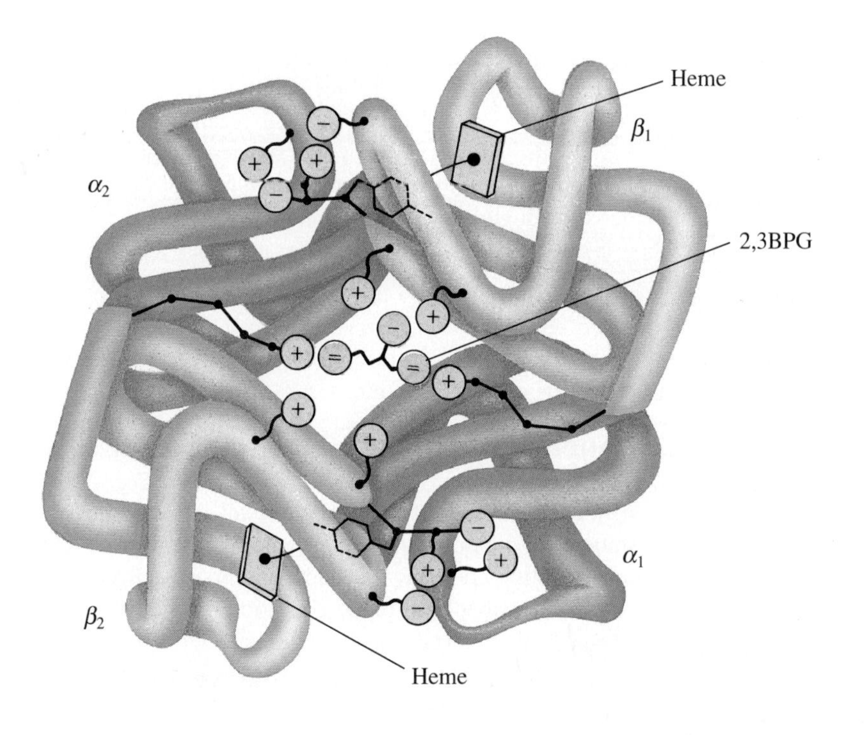

Figure 4·33
Binding of 2,3BPG to deoxyhemoglobin. The central cavity of deoxyhemoglobin is lined with positively charged groups that are complementary to the carboxylate and phosphate groups of 2,3BPG. Both 2,3BPG and the ion pairs shown help stabilize the deoxy conformation. Only hemes of the β subunits are shown. (Illustrator: Tom Edgerton.) [Adapted from Dickerson, R. E. (1972). X-ray studies of protein mechanisms. *Annu. Rev. Biochem.* 41:815–842.]

In mammals, the regulation of hemoglobin oxygenation by 2,3BPG has a special role in the delivery of oxygen to the fetus from the placenta. A unique hemoglobin, designated hemoglobin F (Hb F), is produced in fetuses. Like adult hemoglobin, Hb F is tetrameric and contains two α chains. However, the other two chains are γ globins, not the β globins found in adult hemoglobin. In γ chains, the H21 position is serine rather than histidine. Consequently, Hb F has two fewer positive charges in its central cavity and thus a lower affinity for 2,3BPG and a greater affinity for oxygen (lower P_{50}) than maternal hemoglobin. As a result, oxygen released by maternal hemoglobin in the placenta can be bound by fetal hemoglobin and transported to fetal tissues. Shortly before birth, the synthesis of β globin sharply increases and that of γ globin sharply decreases. Normally, by the end of the first year, γ globin synthesis has completely ceased.

Additional regulation of the binding of oxygen to hemoglobin involves carbon dioxide and protons, both of which are products of aerobic metabolism. Early in this century, Christian Bohr (father of the celebrated physicist Niels Bohr) observed that carbon dioxide decreases the affinity of hemoglobin for oxygen. Carbon dioxide does this indirectly by lowering the pH inside the blood cells. Enzyme-catalyzed hydration of carbon dioxide within red blood cells produces carbonic acid (H_2CO_3), a weak acid that dissociates to form bicarbonate and a proton, thereby lowering pH. This lowering of pH increases protonation of several groups in hemoglobin and consequently increases the ion pairing that helps stabilize the deoxy conformation. Thus, an increase in carbon dioxide and the concomitant lowering of pH increases the P_{50} of hemoglobin (Figure 4·34). This phenomenon, called the *Bohr effect*, increases the efficiency of the oxygen-delivery system: in the inhaling lung, where the carbon dioxide level is low, oxygen is readily picked up by hemoglobin; in metabolizing tissues, where the carbon dioxide level is high, oxygen is readily unloaded by hemoglobin.

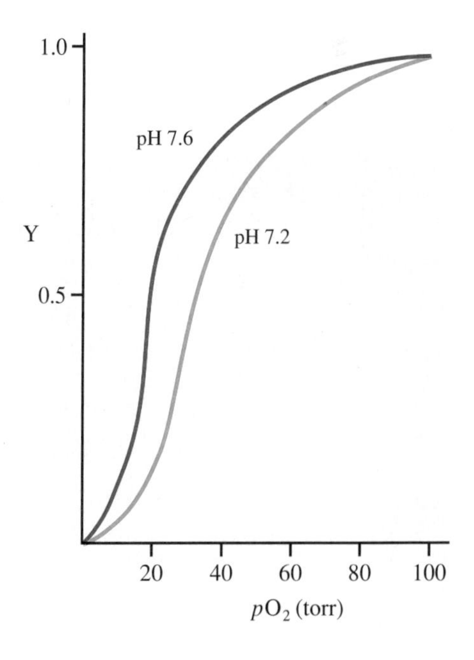

Figure 4·34
The Bohr effect. Lowering the pH decreases the affinity of hemoglobin for oxygen.

Carbon dioxide is transported from the tissues to the lungs in two ways. Much of the carbon dioxide produced by metabolism is transported as dissolved bicarbonate ions to the lungs, where it reassociates with protons released from hemoglobin and is exhaled as carbon dioxide. Some carbon dioxide, however, is carried via hemoglobin, itself. Since the N-terminal α-amino groups of deoxyhemoglobin

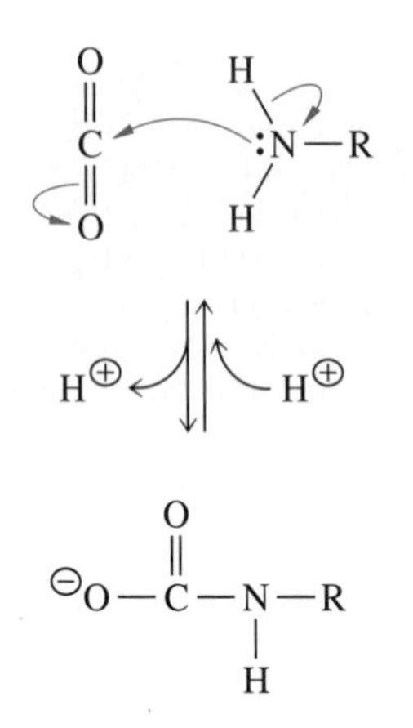

Figure 4·35
Carbamate adduct. Carbon dioxide produced by metabolizing tissues can react reversibly with the N-terminal residues of the globin chains of hemoglobin, converting them to carbamate adducts.

have a pK_a of 8.0, less than one-sixth are unprotonated at pH 7.2, the pH of blood cells. Some of these neutral amino groups react reversibly with carbon dioxide to form carbamate adducts (Figure 4·35). When hemoglobin reaches the lungs, where the partial pressure of carbon dioxide is low and that of oxygen is high, hemoglobin is converted to its oxygenated state. Since oxyhemoglobin has a lower affinity for carbon dioxide than deoxyhemoglobin has, the bound carbon dioxide is released. Thus, whereas carbon dioxide in the peripheral tissues increases the efficiency of oxygen release from hemoglobin, oxygen in the lungs increases the efficiency of carbon dioxide release from the blood.

4·19 Sickle-Cell Anemia Is a Molecular Disease

In 1904, a severely anemic student visited a Chicago physician named James Herrick. Examining the student's red blood cells under a microscope, Herrick discovered that among the normal cells were many that had a highly unusual sickle shape (Figure 4·36). He described the abnormal blood condition for the first time and called it *sickle-cell anemia*. Unlike normal erythrocytes, sickled cells cannot pass efficiently through the capillaries. Consequently, circulation is impaired and serious tissue damage can occur. Moreover, sickled cells tend to rupture, resulting in fewer red blood cells.

It took over four decades to establish that sickle-cell anemia is the result of a molecular alteration in hemoglobin transmitted by a mutated recessive gene on an autosomal (nonsex) chromosome. Persons who inherit two mutated genes are *homozygous* and develop sickle-cell anemia, whereas those who inherit one mutated gene and one normal gene are *heterozygous* and are said to have *sickle-cell trait*. Individuals with sickle-cell trait do not generally suffer from the symptoms of sickle-cell anemia.

The first clue to the nature of the molecular alteration of sickle-cell hemoglobin (Hb S) was obtained by Linus Pauling and his co-workers, who used electrophoresis to compare Hb S with normal adult hemoglobin, Hb A. The results indicated that Hb S had a greater net positive charge than Hb A. Sequence analysis by Vernon Ingram revealed that the β chains of Hb S molecules contain a nonpolar valine residue in position 6 in place of the negatively charged glutamate residue found in position 6 of the β chains of Hb A. This substitution, caused by a single nucleotide substitution in the gene encoding the β chain, accounts for the difference in ionic charge and is responsible for the disease.

In Hb A, the anionic glutamate $\beta 6$ residue lies on the surface of hemoglobin away from either the oxygen- or 2,3BPG-binding sites. Why should replacement of this glutamate residue by a valine residue have such profound consequences? In the deoxy conformation of Hb S, the hydrophobic valine residue substituted at this site on each of the two β chains forms a hydrophobic contact with a pocket in a neighboring Hb S molecule (in the oxy conformation, the pocket is inaccessible). This interaction leads to polymerization of deoxy Hb S at low oxygen pressures typical of capillary beds in metabolically active tissues. The resulting double-stranded polymers aggregate into long helical fibers containing 14 to 16 strands (Figure 4·37). These insoluble aggregates deform the red blood cells into the sickle shape. Sickle-cell crisis results when the deformed erythrocytes block small blood vessels, causing oxygen deprivation that can lead to permanent tissue damage and even death. Heterozygotes do not develop the disease because the rate of fiber formation in heterozygotes is about 1000 times slower than that in homozygotes.

Sickle-cell trait and sickle-cell anemia are most prevalent in populations who live in or whose origins are the tropics, where malaria is prevalent. Individuals who carry the sickle-cell gene have increased resistance to malaria, and, as a consequence, constitute a larger portion of the population. In regions of Africa where the

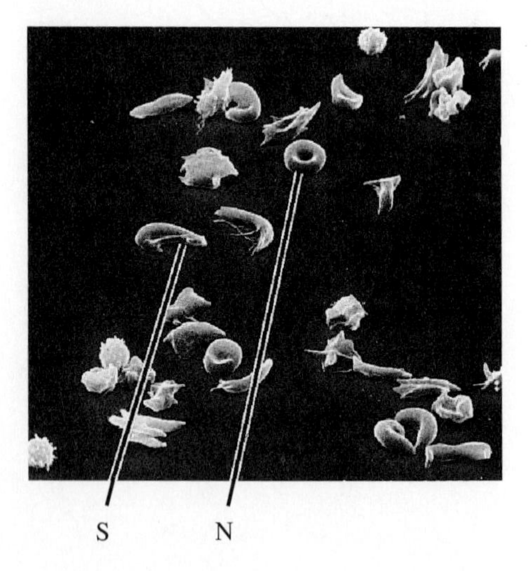

Figure 4·36
Scanning electron micrograph of normal (N) and sickled (S) red blood cells. (Courtesy of Susan Shyne.)

Figure 4·37
Structure of deoxyhemoglobin S fiber. Each circle represents a hemoglobin molecule. In sickled fibers, strands of hemoglobin tetramers are twisted into helical fibers of 14 to 16 strands. A cutaway view shows an inner core of four strands surrounded by ten outer strands. (Illustrator: Lisa Shoemaker.)

incidence of malaria is particularly high, 20% or more of the population carry the sickle-cell trait. In contrast, among African Americans, who are no longer exposed to malaria, only about 9% carry the sickle-cell gene, and only about 1% suffer from sickle-cell anemia. Individuals with the sickle-cell gene are resistant to malaria because the malarial parasite spends part of its life cycle in the red blood cell, and the alteration in red blood cells in persons with the sickle-cell gene prevents the malarial parasite from thriving. Heterozygotes with one sickle-cell gene thus have an advantage over homozygotes with the sickle-cell gene and homozygotes with two Hb A β genes since they do not suffer severely from either sickle-cell anemia or malaria.

Summary

Proteins can be classified as fibrous or globular. Fibrous proteins are generally water insoluble, physically tough, and built of repetitive structures. They usually have static functions. Globular proteins, most of which are soluble in aqueous solutions, are compact, roughly spherical macromolecules whose polypeptide chains are tightly folded. They act dynamically by transiently binding specific molecules.

The primary structural link of all proteins is the peptide bond, which connects adjacent amino acid residues. Peptide bonds are polar and planar and have some double-bond character. Because of steric restraints, peptide groups are usually in the *trans* configuration. Rotation around the N—C_α and C_α—C bonds gives polypeptide chains conformational flexibility within limits imposed by steric restrictions.

Proteins may have up to four levels of structure: primary (sequence of amino acid residues), secondary (local conformation, stabilized by hydrogen bonds), tertiary (biologically active, three-dimensional structure of an entire polypeptide chain), and quaternary (noncovalent aggregation of two or more polypeptide chains into a multisubunit protein). The highly complex three-dimensional structures of proteins, some of which have been determined by X-ray crystallography, must be preserved to maintain biological activity.

The right-handed α helix is a common secondary structure found in some fibrous and globular proteins. It contains 3.6 amino acid residues per turn and has a pitch of 0.54 nm. The other major type of secondary structure is the β sheet, which can be either parallel or antiparallel. A less common secondary structure is the 3_{10} helix.

A different helical structure is found in collagen, a fibrous protein of the connective tissue. A collagen molecule consists of three left-handed polypeptide helices intertwined to form a right-handed supercoil. Covalent cross-linking through modification of lysine residues stabilizes the protein as does interchain hydrogen bonding.

Most globular proteins have considerable stretches of residues in nonrepeating conformations. These regions include turns and loops needed to connect α helices and β strands. The secondary structural elements are often connected into recognizable combinations called supersecondary structures or motifs. Larger globular units called domains or lobes are usually associated with a particular function.

Folding of a globular protein into its biologically active state is a sequential, cooperative process involving the hydrophobic effect and, to a lesser extent, hydrogen bonding, van der Waals interactions, and ion pairing. In some cases, enzymes and chaperones assist folding. The native conformation of a protein can be disrupted by the addition of denaturing agents.

The compact, folded structures of globular proteins allow them to selectively bind other molecules. For example, the globular structures of the heme-containing proteins myoglobin and hemoglobin allow these proteins to bind and release oxygen in a manner that facilitates delivery of oxygen to respiring tissues.

A monomeric protein, myoglobin contains a single polypeptide chain of 153 residues that is folded into a compact globular structure composed of eight α helices. Its heme prosthetic group, which binds oxygen, is shielded from water in a cleft, or hydrophobic cage, formed by the protein. Hemoglobin consists of four chains (two α and two β chains in adult hemoglobin), each similar to the globin chain of myoglobin. The deoxy (T) and oxy (R) conformations of hemoglobin differ in their affinity for oxygen. Due to structural interactions associated with its tertiary and quaternary structure, hemoglobin displays positive cooperativity in its binding of oxygen and is subject to allosteric regulation.

Slight differences in the primary structures of hemoglobin molecules can result in significant functional differences: substitution of a serine residue in fetal hemoglobin (Hb F) for a histidine residue found in normal adult hemoglobin (Hb A) increases its affinity for oxygen, thereby increasing the efficiency of oxygen delivery from maternal blood to the fetus. A mutation leading to replacement of a glutamate residue at position 6 in the β chains of normal adult hemoglobin with a valine residue results in Hb S, the hemoglobin responsible for sickle-cell anemia.

Selected Readings

General References

Creighton, T. E. (1984). *Proteins, Structures and Molecular Principles* (New York: W. H. Freeman and Company). Chapters 4 to 7.

Doolittle, R. F. (1985). Proteins. *Sci. Am.* 253(4):88–99. A brief but clear review of proteins and their binding of ligands.

The α Helix

Pauling, L., Corey, R. B., and Branson, H. R. (1951). The structure of proteins: two hydrogen-bonded helical configurations of the polypeptide chains. *Proc. Natl. Acad. Sci. USA* 37:205–211. A classic paper in which the structure of the α helix was defined for the first time.

The 3_{10} Helix

Toniolo, C., and Benedetti, E. (1991). The polypeptide 3_{10}-helix. *Trends Biochem. Sci.* 16:350–353.

Collagen

Eyre, D. R. (1980). Collagen: molecular diversity in the body's protein scaffold. *Science* 207:1315–1322. A review of the structures and functions of various classes of collagen.

Eyre, D. R., Paz, M. A., and Gallop, P. M. (1984). Cross-linking in collagen and elastin. *Annu. Rev. Biochem.* 53:717–748.

Miller, A. (1982). Molecular packing in collagen fibrils. *Trends Biochem. Sci.* 7:13–18. A short review of collagen structure.

Piez, K. A., and Reddi, A. H., eds. (1984). *Extracellular Matrix Biochemistry* (New York: Elsevier Science Publishing Company).

Structure of Globular Proteins

Branden, C., and Tooze, J. (1991). *Introduction to Protein Structure* (New York: Garland Publishing). A comprehensive analysis of three-dimensional structures of proteins, based upon X-ray crystallographic analyses.

Burley, S. K., and Petsko, G. A. (1988). Weakly polar interactions in proteins. *Adv. Protein Chem.* 39:125–189.

Chothia, C., and Finkelstein, A. V. (1990). The classification and origins of protein folding patterns. *Annu. Rev. Biochem.* 59:1007–1039. How α helices and β sheets pack together in supersecondary structures.

Leszczynski, J. F., and Rose, G. D. (1986). Loops in globular proteins: a novel category of secondary structure. *Science* 234:849–855.

Richardson, J. S., and Richardson, D. C. (1989). Principles and patterns of protein conformation. In *Prediction of Protein Structure and the Principles of Protein Conformation,* Fasman, G. D., ed. (New York: Plenum Publishing Corporation), pp. 1–98. An excellent review of secondary and tertiary structures of globular proteins.

Rose, G. D., Geselowitz, A. R., Lesser, G. J., Lee, R. H., and Zehfus, M. H. (1985). Hydrophobicity of amino acid residues in globular proteins. *Science* 229:834–838.

Denaturation

Ahern, T. J., and Klibanov, A. M. (1985). The mechanism of irreversible enzyme inactivation at 100°C. *Science* 228:1280–1284.

Protein Folding

Dill, K. A. (1990). Dominant forces in protein folding. *Biochemistry* 29:7133–7155.

Ellis, R. J., and von der Vies, S. M. (1991). Molecular chaperones. *Annu. Rev. Biochem.* 60: 321–347.

Gething, M., and Sambrook, J. (1992). Protein folding in the cell. *Nature* 355:33–45.

Kim, P. S., and Baldwin, R. L. (1990). Intermediates in the folding reactions of small proteins. *Annu. Rev. Biochem.* 59:631–660.

Nilsson, B., and Anderson, S. (1991). Proper and improper folding of proteins in the cellular environment. *Annu. Rev. Microbiol.* 45:607–35.

Richards, F. M. (1991). The protein folding problem. *Sci. Am.* 264(1):54–63.

Rossmann, M. G., and Argos, P. (1981). Protein folding. *Annu. Rev. Biochem.* 50:497–532.

Myoglobin

Phillips, S. E. V. (1980). Structure and refinement of oxymyoglobin at 1.6 Å resolution. *J. Mol. Biol.* 142:531–554.

Hemoglobin

Ackers, G. K., Doyle, M. L., Myers, D., Daugherty, M. A. (1992). Molecular code for cooperativity in hemoglobin. *Science* 255:54–63.

Baldwin, J., and Chothia, C. (1979). Haemoglobin: the structural changes related to ligand binding and its allosteric mechanism. *J. Mol. Biol.* 129:175–179.

Dickerson, R. E., and Geis, I. (1983). *Hemoglobin: Structure, Function, Evolution, and Pathology* (Menlo Park, CA: Benjamin/ Cummings Publishing Company).

Perutz, M. F. (1978). Hemoglobin structure and respiratory transport. *Sci. Am.* 239(6):92–125.

Sickle-Cell Hemoglobin

Eaton, W. A., and Hofrichter, J. (1988). Sickle cell hemoglobin polymerization. *Adv. Protein Chem.* 40:63–279.

Ingram, V. M. (1957). Gene mutations in human hemoglobin: the chemical difference between normal and sickle cell hemoglobin. *Nature* 180:326–328.

Pauling, L., Itano, H. A., Singer, S. J., and Wells, I. C. (1949). Sickle cell anemia, a molecular disease. *Science* 110:543–548.

5

Properties of Enzymes

Enzymes are catalysts of extraordinary efficiency and specificity. You may recall from earlier studies that a catalyst accelerates the approach of a reaction toward equilibrium without changing the position of that equilibrium. Enzymes are so effective as biological catalysts that most of the reactions they catalyze would not proceed in their absence in a reasonable time without extremes of temperature, pressure, or pH. Enzyme-catalyzed or *enzymatic* reactions are 10^3 to 10^{17} times faster than the corresponding uncatalyzed reactions. Enzymes also typically catalyze reactions orders of magnitude faster than nonenzymatic catalysts, such as those used in chemical synthesis.

Enzymes are highly specific for the reactants, or *substrates,* they act upon. The degree of substrate specificity varies. One enzyme might react with a group of substrates of closely related structures, whereas another might act only upon a single molecular species. Many enzymes exhibit *stereospecificity,* meaning they act upon only a single stereoisomer of the substrate. Perhaps the most important general aspect of enzyme specificity is *reaction specificity,* that is, the lack of formation of wasteful by-products. Reaction specificity is reflected in exceptional product yields, which are essentially 100%. The efficiency of enzymes not only saves energy for the living cell, but also precludes buildup of potentially toxic by-products of *metabolism,* the reactions of the cell.

Some enzymes function as control points in metabolism. As we shall see, metabolism is regulated in a variety of ways, including alterations in the concentrations of enzymes, substrates, and enzyme inhibitors, and modulation of the activity levels of certain enzymes.

Even the simplest living organisms contain multiple copies of a thousand different enzymes. In multicellular organisms, it is the particular complement of enzymes present that differentiates one cell type from another. Most of the enzymes discussed in this book are among the several hundred enzymes common to virtually all cells. These enzymes catalyze the reactions of the central metabolic pathways necessary for cellular maintenance.

Since James B. Sumner first crystallized the enzyme urease in 1926 and established that urease is a protein, enzymes have consistently been shown to be proteins or proteins plus cofactors. We now know, however, that certain RNA molecules also

Table 5·1 Illustrative examples of the six major enzyme classes

Major class	Enzyme	Reaction description	Example of reaction catalyzed	Coenzyme involved
1. Oxidoreductase	Lactate dehydrogenase	Oxidation of the secondary alcohol L-lactate to pyruvate, a ketone	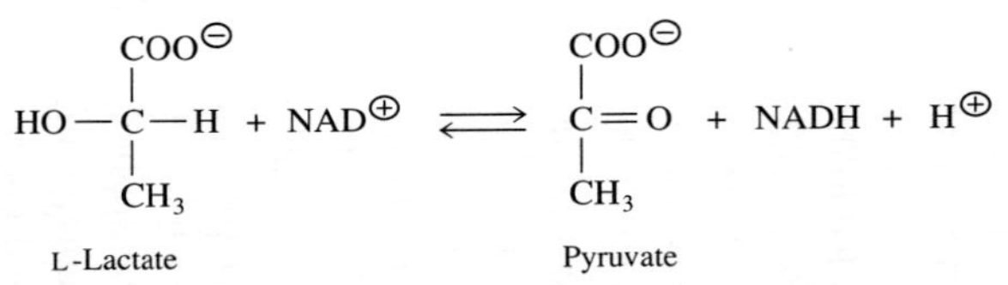 L-Lactate ($HO{-}CH(COO^{\ominus}){-}CH_3$) + $NAD^{\oplus}$ $\rightleftharpoons$ Pyruvate ($CH_3{-}C(=O){-}COO^{\ominus}$) + NADH + $H^{\oplus}$	$NAD^{\oplus}$ (Nicotinamide adenine dinucleotide)
2. Transferase	Alanine transaminase (Alanine aminotransferase)	Transfer of an amino group	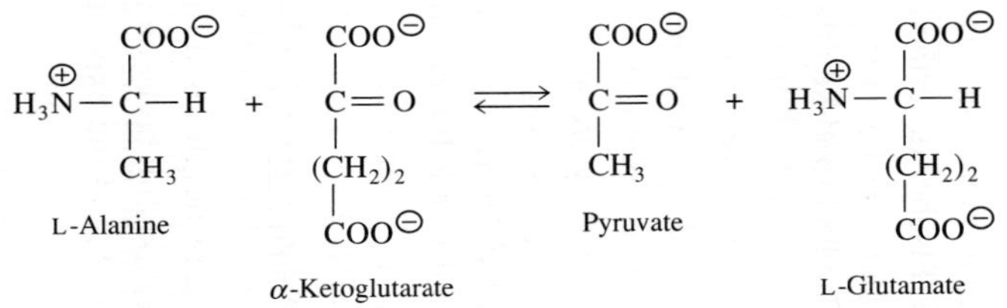 L-Alanine ($H_3N^{\oplus}{-}CH(COO^{\ominus}){-}CH_3$) + α-Ketoglutarate ($^{\ominus}OOC{-}C(=O){-}(CH_2)_2{-}COO^{\ominus}$) $\rightleftharpoons$ Pyruvate ($CH_3{-}C(=O){-}COO^{\ominus}$) + L-Glutamate ($H_3N^{\oplus}{-}CH(COO^{\ominus}){-}(CH_2)_2{-}COO^{\ominus}$)	Pyridoxal phosphate
3. Hydrolase	Trypsin	Hydrolysis of Lys–Y (or Arg–Y) peptide bonds, where Y ≠ Pro	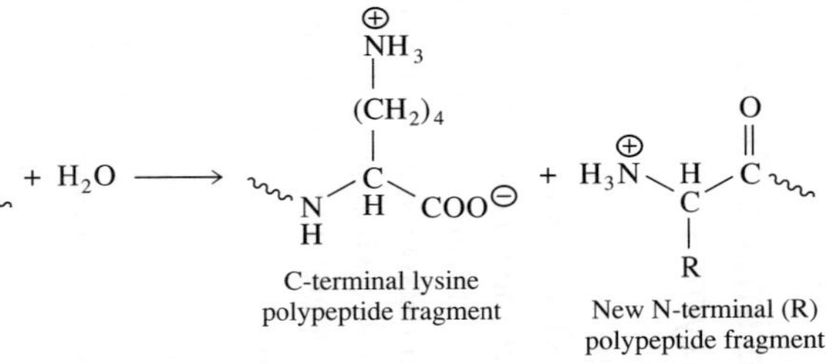 Lysine residue within polypeptide chain ($\sim NH{-}CH((CH_2)_4{-}NH_3^{\oplus}){-}C(=O){-}NH{-}CHR{-}C(=O)\sim$) + H_2O $\longrightarrow$ C-terminal lysine polypeptide fragment ($\sim NH{-}CH((CH_2)_4{-}NH_3^{\oplus}){-}COO^{\ominus}$) + New N-terminal (R) polypeptide fragment ($H_3N^{\oplus}{-}CHR{-}C(=O)\sim$)	None
4. Lyase	Pyruvate decarboxylase	Decarboxylation of pyruvate	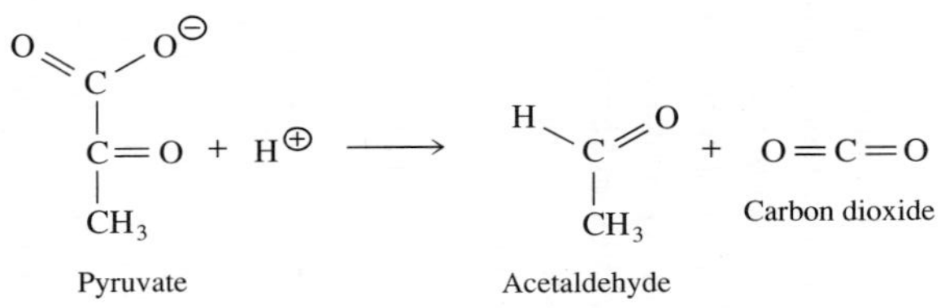 Pyruvate ($CH_3{-}C(=O){-}COO^{\ominus}$) + $H^{\oplus}$ $\longrightarrow$ Acetaldehyde ($CH_3{-}CHO$) + Carbon dioxide ($O{=}C{=}O$)	Thiamine pyrophosphate

Table 5·1 (Continued) Illustrative examples of the six major enzyme classes

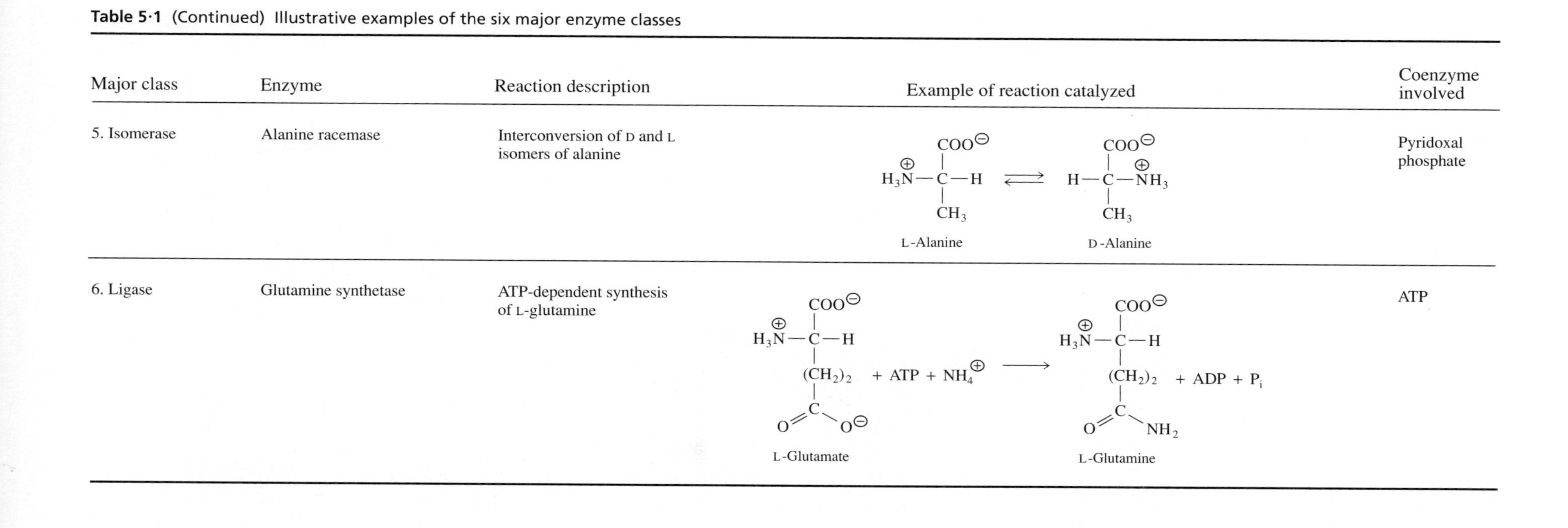

Major class	Enzyme	Reaction description	Example of reaction catalyzed	Coenzyme involved
5. Isomerase	Alanine racemase	Interconversion of D and L isomers of alanine	L-Alanine ($H_3\overset{\oplus}{N}-CH(COO^{\ominus})-CH_3$) $\rightleftharpoons$ D-Alanine ($H-C(COO^{\ominus})(\overset{\oplus}{N}H_3)-CH_3$)	Pyridoxal phosphate
6. Ligase	Glutamine synthetase	ATP-dependent synthesis of L-glutamine	L-Glutamate ($H_3\overset{\oplus}{N}-CH(COO^{\ominus})-(CH_2)_2-C(=O)O^{\ominus}$) + ATP + $NH_4^{\oplus}$ $\longrightarrow$ L-Glutamine ($H_3\overset{\oplus}{N}-CH(COO^{\ominus})-(CH_2)_2-C(=O)NH_2$) + ADP + P_i	ATP

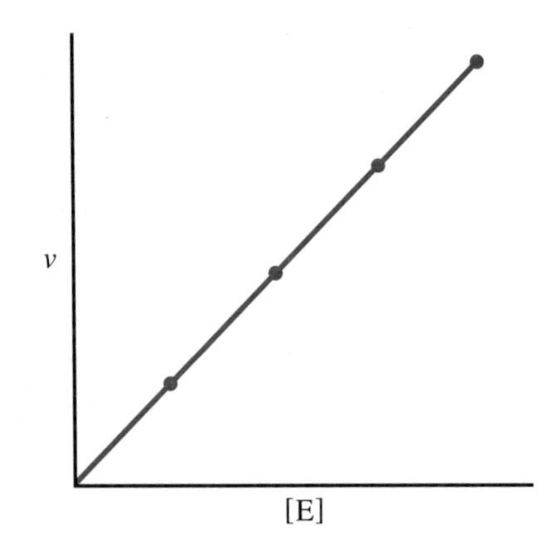

Figure 5·1
Effect of enzyme concentration [E] on the velocity, *v*, of an enzyme-catalyzed reaction at a fixed saturating [S]. Because the reaction rate is affected by the concentration of enzyme but is not increased by increasing the concentration of the other reactant, S, this bimolecular reaction is pseudo first-order.

If the concentration of one reactant is so high that it remains essentially constant during the reaction, then the reaction is said to be *zero-order* with respect to that reactant, and the term for the reactant can be eliminated from the rate equation. The reaction then becomes an artificial or *pseudo first-order* reaction:

$$v = k[S_1]^1[S_2]^0 = k'[S_1] \quad (5·3)$$

An example of a pseudo first-order reaction is the nonenzymatic hydrolysis of table sugar, sucrose, in aqueous acid solution. Sucrose is composed of a residue each of the sugars glucose and fructose.

$$\underset{(C_{12}H_{22}O_{11})}{\text{Sucrose}} + \underset{(H_2O)}{\text{Water}} \longrightarrow \underset{(C_6H_{12}O_6)}{\text{Glucose}} + \underset{(C_6H_{12}O_6)}{\text{Fructose}} \quad (5·4)$$

The concentration of water, which is both the solvent and a substrate, is so high that it remains effectively constant during the reaction. Under these reaction conditions, the reaction rate depends only on the concentration of sucrose. The concentration of water can be made a limiting factor by replacing much of the water with a nonreacting solvent, reinforcing the idea that the original reaction was artificially first-order.

Pseudo first-order conditions are used in analyses called *enzyme assays* that determine concentrations of enzymes. In enzyme assays, all reactants except the enzyme are present in excess to ensure that reaction conditions are pseudo first-order. The assay is performed at a constant temperature that is high enough to produce high activity but not so high that the rate of denaturation of the enzyme is appreciable. The straight line in Figure 5·1 illustrates the effect of changes in concentration of a single reactant, the enzyme, in a pseudo first-order reaction. Under these conditions, there are sufficient substrate molecules to convert all molecules of E to ES, a condition called *saturation* of E with S. Enzymes are normally present in assays at very low concentrations compared to the concentrations of substrates. Enzymes are not, however, necessarily saturated inside cells. The concentration of enzyme in a test sample can be easily determined by comparing its activity to a reference curve similar to the model curve in Figure 5·1.

Let us now consider the kinetic variables for a simple enzymatic reaction, the conversion of a substrate S to a product P, catalyzed by an enzyme E. Although most enzymatic reactions have two or more substrates, the principles of enzyme kinetics can be elucidated by assuming the simple case of one substrate and one product. The initial moments of this bimolecular reaction can be written as:

$$E + S \underset{k_{-1}}{\overset{k_1}{\rightleftharpoons}} ES \xrightarrow{k_{cat}} E + P \quad (5·5)$$

Notice that the conversion of the ES complex into free enzyme and product is shown by a one-way arrow. During the initial period when measurements are made, little product has been formed, so the reverse reaction is negligible. The velocity measured during this short period is called the *initial velocity,* v_0. The use of v_0 measurements simplifies the interpretation of kinetic data and avoids complications associated with product inhibition and slow denaturation of enzyme. The rate constants k_1 and k_{-1} in Equation 5·5 govern the rates of association of S with E and dissociation of S from ES, respectively. The formation and dissociation of ES complexes are usually very rapid reactions because only noncovalent bonds are formed and broken. The substrate is chemically altered in the second step. For the simplest case, the rate constant for this step is k_{cat}, the *catalytic constant* or *turnover number,* the number of catalytic events per second per active site or enzyme molecule.

Initial velocities are obtained from *progress curves,* graphs of the concentration of product against time (Figure 5·2). The initial velocity is the tangent or slope $\Delta[P]/\Delta t$ at the origin of a progress curve.

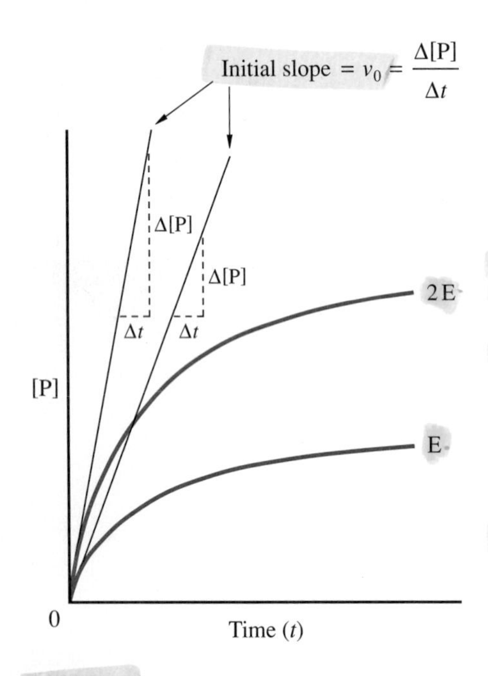

Figure 5·2
Progress curve for an enzyme-catalyzed reaction in which a substrate is converted to product P. [P], the concentration of product, increases as the reaction proceeds. The initial velocity of the reaction (v_0) is the slope of the initial linear portion of the curve. Note that the rate of the reaction doubles when twice as much enzyme (2 E, the upper curve) is added to an otherwise identical reaction mixture.

5·3 The Michaelis-Menten Equation Is a Rate Equation for Enzymatic Catalysis

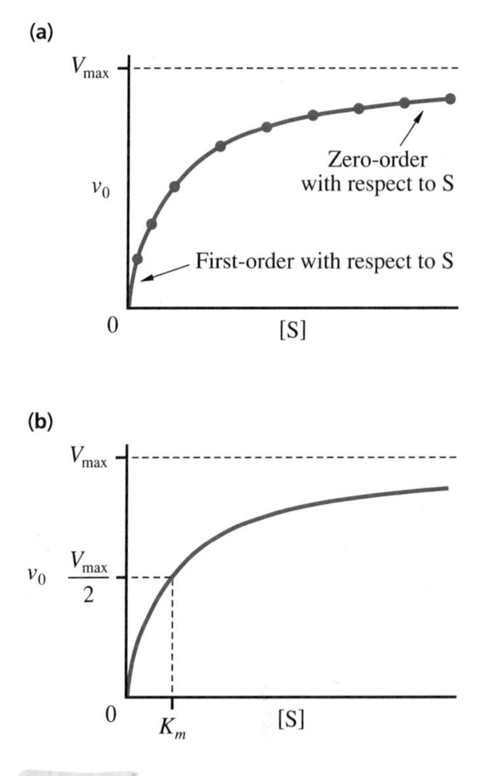

Figure 5·3
(a) Plot of initial velocity, v_0, versus substrate concentration, [S], for an enzyme-catalyzed reaction. Each experimental point was obtained from a separate progress curve. The shape of the curve is hyperbolic. At low substrate concentrations, the curve approximates a straight line that rises steeply. In this region of the curve, the reaction is first-order with respect to substrate. At high concentrations of substrate, the enzyme is saturated and the reaction is zero-order with respect to substrate. **(b)** The concentration of substrate that corresponds to half-maximum velocity is called the Michaelis constant, K_m. The enzyme is half-saturated when [S] = K_m.

In the early 1900s, workers in several laboratories were examining the effects of variations in substrate concentrations in order to derive rate equations for enzymatic processes. Most were using extracts of yeast to catalyze the hydrolysis of sucrose to a molecule each of glucose and fructose (Equation 5·4). Two major observations were made. First, at high concentrations of substrate, E is saturated by S, and the reaction rate is independent of the concentration of substrate. The value of v_0 for a solution of E that is saturated with S is called the *maximum velocity,* V_{max}. Second, at low concentrations of substrate, the reaction is first-order with respect to substrate (making the reaction second-order overall: first-order with respect to S and first-order with respect to E). At intermediate substrate concentrations, the order with respect to S is fractional, decreasing from first-order toward zero-order as [S] increases. The shape of the v_0 versus [S] curve from low to high [S] is a rectangular hyperbola (Figure 5·3a). Thus, the rate equation for Equation 5·5 is the equation for a hyperbola of this type.

The equation for a rectangular hyperbola is:

$$y = \frac{ax}{b + x} \qquad (5\cdot6)$$

where a is the asymptote of the curve (the value of y at an infinite value of x) and b is the point on the x axis corresponding to a value of $y_{infinity}/2$. We can obtain the rate equation for the simple bimolecular reaction $E + S \rightleftharpoons ES \longrightarrow E + P$ by substituting four terms from enzyme kinetics into the general equation for a rectangular hyperbola. Three of the terms have already been introduced: $y = v_0$, $x = [S]$, and $a = V_{max}$. The fourth term, b in the general equation, is the *Michaelis constant,* K_m, defined as the concentration of substrate when v_0 is equal to one-half V_{max} (Figure 5·3b). The complete rate equation is written:

$$v_0 = \frac{V_{max}[S]}{K_m + [S]} \qquad (5\cdot7)$$

This is called the *Michaelis-Menten equation*, named after Leonor Michaelis and Maud Menten who provided strong evidence for this enzymatic relationship. Before deriving the Michaelis-Menten equation by a kinetic approach, let us consider the meaning of the constants V_{max} and K_m.

A. The Maximum Velocity, V_{max}, Is Achieved when E Is Saturated with S

In Figure 5·3a, at the highest concentrations of substrate S, the velocity approaches a constant value as [S] becomes very large. In this region, the enzyme is nearing saturation with S and the reaction is zero-order with respect to S (adding more S has virtually no effect). Only the addition of more enzyme can increase the velocity when [S] is very large. The constant velocity that is asymptotically approached is the maximum velocity, V_{max}. V_{max} is defined for any given amount of enzyme as the initial velocity v_0 at saturating concentrations of substrate. The rate equation for this pseudo first-order region of the curve can be written in these equivalent forms:

$$v_0 \text{ (at saturation)} = V_{max} = k[E][S]^0 = k[E]_{total} = k_{cat}[ES] \qquad (5\cdot8)$$

The rate constant k is equal to the catalytic constant k_{cat} (the rate constant for conversion of ES to free enzyme and product) because at saturation essentially all molecules of E are present as ES. Another simple relationship shown in Equation 5·8 is that $V_{max} = k_{cat}\,[E]_{total}$. By rearranging, we obtain the definition for k_{cat}:

$$k_{cat} = \frac{V_{max}}{[E]_{total}} \tag{5·9}$$

The catalytic constant is the maximum velocity, V_{max}, divided by the original concentration of enzyme present, $[E]_{total}$, or the number of moles of substrate converted to product under saturating conditions per second per mole of enzyme (or per mole of active site for an oligomeric enzyme). The unit for k_{cat} is s^{-1}. The reciprocal of k_{cat} is the time required for one catalytic event. Therefore, the catalytic constant is a measure of how quickly an enzyme can catalyze a reaction.

B. The Michaelis Constant, K_m, Is [S] at Half-Saturation of E

The Michaelis constant has a number of meanings. As we saw above, K_m is the initial concentration of substrate at half-maximum velocity, or at half-saturation of E with S. This can be verified by substituting $[S] = K_m$ into Equation 5·7. Solving, v_0 is found to be $V_{max}/2$. For some enzymes, K_m can be further defined in terms of the constants for the formation and breakdown of the ES complex, as we shall see in the following two sections.

C. The Michaelis-Menten Equation Can Be Derived by Assuming Steady-State Reaction Conditions

The Michaelis-Menten equation has been derived in several ways, but the usual derivation is that provided by George E. Briggs and J. B. S. Haldane, termed the steady-state derivation. Soon after a small amount of enzyme E is mixed with substrate S, there is a period called the *steady state* when [ES] is constant because the rates of decomposition of ES to either E + S or E + P are equal to the rate of formation of the ES complex from E + S. The rate of the latter reaction depends on the concentration of free enzyme (enzyme molecules not in the form of ES), which is $([E]_{total} - [ES])$. Expressing these statements about the steady state algebraically provides this equivalence:

$$(k_{-1} + k_{cat})[ES] = k_1([E]_{total} - [ES])[S] \tag{5·10}$$

Equation 5·10 is rearranged to collect the rate constants, and the ratio of constants obtained defines the Michaelis constant, K_m:

$$\frac{k_{-1} + k_{cat}}{k_1} = K_m = \frac{([E]_{total} - [ES])[S]}{[ES]} \tag{5·11}$$

Next, this equation is solved for [ES] in several steps:

$$([E]_{total} - [ES])[S] = [ES]K_m \tag{5·12}$$

Expanding,

$$[ES]K_m = ([E]_{total}[S]) - ([ES][S]) \tag{5·13}$$

Collecting [ES] terms,

$$[ES](K_m + [S]) = [E]_{total}[S] \tag{5·14}$$

and

$$[ES] = \frac{[E]_{total}[S]}{K_m + [S]} \tag{5·15}$$

Since the velocity of an enzyme-catalyzed reaction depends on the rate of breakdown of ES to E + P (Equation 5·8),

$$v = k_{cat}[ES] \quad (5\cdot16)$$

The velocity, and hence the rate equation, can be obtained by substituting the value of [ES] from Equation 5·15 into Equation 5·16, with $v = v_0$:

$$v_0 = \frac{k_{cat}[E]_{total}[S]}{K_m + [S]} \quad (5\cdot17)$$

Also, because $k_{cat}\,[E]_{total} = V_{max}$ (Equation 5·8), Equation 5·17 can be rewritten in the most familiar form of the Michaelis-Menten equation:

$$v_0 = \frac{V_{max}[S]}{K_m + [S]} \quad (5\cdot18)$$

D. K_m Can Be the Dissociation Constant for the ES Complex

From the steady-state derivation, we see that K_m is the ratio of the constants for breakdown of ES divided by the constant for its formation (Equation 5·11). If the rate of the chemical step, k_{cat}, is slower than either k_1 or k_{-1}, as is often the case, then k_{cat} can be neglected, and K_m becomes k_{-1}/k_1, the equilibrium constant for the dissociation of the ES complex to E + S. Similarly, in their restrictive derivation, Michaelis and Menten assumed that ES is in rapid equilibrium with E + S and that K_m is the dissociation constant of ES. K_m is thus a measure of the affinity of E for S. The lower the value of K_m, the more tightly the substrate is bound. K_m values are sometimes used to differentiate enzymes having the same function. For example, there are several different forms of lactate dehydrogenase in mammals, each with distinctive K_m values for its substrates.

In some instances, the K_m of an enzyme for a substrate can be used to obtain a rough estimate of the concentration of the substrate in a cell because the values are frequently similar. The K_m value falls near the middle of the steeply rising portion of the rate curve for the enzyme, and when [S] approximates K_m, the enzyme is most able to both respond proportionally to changes in [S] and use most of its catalytic potential. Physiologically, this can help prevent the accumulation of substrate under changing conditions.

E. The Ratio k_{cat}/K_m Describes Rates and Substrate Specificity

In the region of the hyperbolic curve where the concentration of S is very low and the curve still approximates a straight line, reactions are second-order: first-order with respect to S and first-order with respect to E. The rate equation for this region is:

$$v_0 = k[E][S] \quad (5\cdot19)$$

Michaelis and Menten first wrote the full rate equation in the form that included $k_{cat}[E]$ (Equation 5·17) rather than V_{max}. If we consider only the region of the Michaelis-Menten curve at very low [S], Equation 5·17 can be simplified by neglecting the term for [S] in the denominator, a very small value:

$$v_0 = \frac{k_{cat}}{K_m}[E][S] \quad (5\cdot20)$$

Therefore, the rate constant for Equation 5·19 is k_{cat}/K_m. This rate constant is a measure of two features of enzymatic catalysis. First, it is a measure of the specificity or preference of an enzyme for different substrates, which can be assessed by comparing the k_{cat}/K_m values of the enzyme/substrate pairs. Suppose that two different substrates, A and B, are competing for catalysis by the same enzyme. When the concentrations of A and B are equal, the ratio of the rates of their conversion to product by a single enzyme is equal to the ratio of their k_{cat}/K_m values. For this reason, the ratio k_{cat}/K_m is often called the *specificity constant*. Second, it is the rate constant for the reaction of dilute concentrations of substrate—an indication of the overall rate of the enzymatic reaction $E + S \longrightarrow E + P$ at very low substrate concentrations. The full significance and application of this meaning for k_{cat}/K_m will be discussed in the next section.

5·4 Rate Constants Indicate the Catalytic Efficiency of Enzymes

The kinetic constants of enzymatic reactions, K_m and k_{cat}, can be used to gauge the catalytic efficiency of enzymes. K_m is a measure of the stability of the ES complex. It is equal to the ratio of $[E]_{free}[S]$ divided by [ES] under steady-state reaction conditions (Equation 5·11). k_{cat} is the first-order rate constant for the conversion of ES to E + P. It is a measure of the catalytic activity of an enzyme, telling how many reactions per second a molecule of enzyme can catalyze. Values for k_{cat} of about $10^3\ s^{-1}$ are typical. The ratio k_{cat}/K_m is an apparent second-order rate constant for the formation of E + P from E + S when the overall reaction is limited by the encounter of S with E. This constant approaches 10^8 to $10^9\ M^{-1}\ s^{-1}$, the fastest rate at which two uncharged solutes can approach each other by diffusion. Enzymes that can catalyze reactions at this extremely rapid rate are discussed in Section 6·7. Both k_{cat}/K_m and k_{cat} will be used in Chapter 6 in discussing the rates of enzymatic reactions. The meanings of the terms are summarized in Figure 5·4. Comparisons of enzymatic reactions with nonenzymatic reactions can be made with these constants. The absolute values of these constants tell how powerful each enzyme is as a catalyst.

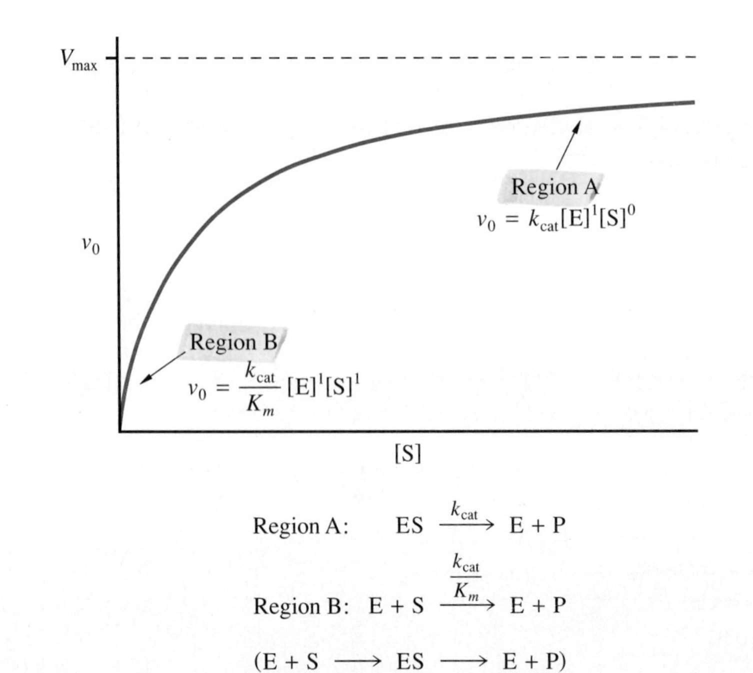

Figure 5·4
Meanings of the two rate constants, k_{cat} and k_{cat}/K_m. The catalytic constant, k_{cat}, is the first-order rate constant for the conversion of the ES complex to E + P. It is measured most easily when the enzyme is saturated with substrate (region A on the Michaelis-Menten curve shown). The ratio k_{cat}/K_m is the second-order rate constant at very low concentrations of substrate for conversion of E + S to E + P (region B). The reactions measured by these rate constants are summarized below the graph.

Table 5·2 Rate accelerations of some enzymes

	Nonenzymatic rate (k_n in s^{-1})	Enzymatic rate (k_{cat} in s^{-1})	Rate acceleration (k_{cat}/k_n)
Chymotrypsin	4×10^{-9}	4×10^{-2}	10^7
Lysozyme	3×10^{-9}	5×10^{-1}	2×10^8
Triose phosphate isomerase	6×10^{-7}	2×10^3	3×10^9
Fumarase	2×10^{-8}	2×10^3	10^{11}
β-Amylase	3×10^{-9}	1×10^3	3×10^{11}
Urease	3×10^{-10}	3×10^4	10^{14}
Adenosine deaminase	10^{-12}	10^2	10^{14}
Alkaline phosphatase	10^{-15}	10^2	10^{17}

Enzymes are extremely efficient catalysts. Their efficiency is measured by the *rate acceleration* that they provide. This value is the ratio of the rate of a reaction in the presence of the enzyme (k_{cat}) divided by the rate of the same reaction in the absence of enzyme (k_n). Surprisingly few values of rate acceleration are known because most cellular reactions occur extremely slowly in the absence of enzymes, and therefore their nonenzymatic rates are difficult to measure.

Several examples of rate accelerations are provided in Table 5·2. Typical values are in the range of 10^8 to 10^{12}, but some are quite a bit higher (up to 10^{17}). The difficulty in obtaining rate constants for nonenzymatic reactions is exemplified by the half-time for deamination of adenosine at 20°C and pH 7, about 20000 years! Because the nonenzymatic rate of this reaction is ultraslow, adenosine deaminase has an extraordinarily high rate acceleration despite a moderate k_{cat} of $10^2\,s^{-1}$.

The catalytic constant k_{cat}, the number of catalytic events that an enzyme molecule can catalyze per second, also called the turnover number, is a measure of the absolute, not relative, activity of an enzyme. Table 5·3 lists representative values of k_{cat}. Most enzymes are potent catalysts having k_{cat} values of 10^2 to $10^3\,s^{-1}$. Some enzymes, however, are much faster. For example, the last three examples in Table 5·3 have k_{cat} values of $10^6\,s^{-1}$ or greater. Extremely rapid catalysis is essential for the physiological function of certain enzymes. Carbonic anhydrase must act very rapidly in order to maintain equilibrium between aqueous CO_2 and bicarbonate (Section 2·10). As discussed in Section 6·7, both superoxide dismutase and catalase are responsible for rapid removal of the toxic oxygen metabolites superoxide anion and hydrogen peroxide.

Table 5·3 Examples of turnover numbers

Enzyme	Turnover number (k_{cat} in s^{-1})
Papain	10
Ribonuclease	10^2
Carboxypeptidase	10^2
Trypsin	10^2 (to 10^3)
Acetylcholinesterase	10^3
Kinases	10^3
Dehydrogenases	10^3
Transaminases	10^3
Carbonic anhydrase	10^6
Superoxide dismutase	10^6
Catalase	10^7

The values of k_{cat} are the first-order rate constants for the reaction ES $\longrightarrow$ E + P. Most values were obtained from Eigen, M. and Hammes, G. G. (1963). Elementary steps in enzyme reactions (as studied by relaxation spectrometry). *Adv. Enzymol.* 25:1–38. They are given only as orders of magnitude.

5·5 K_m and V_{max} Are Easily Measured

K_m and V_{max} can be measured in several ways. Both values are obtained by analysis of initial velocities at a series of substrate concentrations and one concentration of enzyme. In order to obtain reliable values for the kinetic constants, the [S] points must be spread out, both below and above K_m, to produce a hyperbola. Using a suitable computer program, values for K_m and V_{max} are arrived at by fitting the experimental results to the equation for the hyperbola.

Before computers were available, values were obtained by using graphic transformations of the Michaelis-Menten equation that give straight lines. From these lines, one can extrapolate or obtain a slope to estimate the value of the asymptote of the v_0 versus [S] plot, V_{max}. Linear transformations, common before computers, are

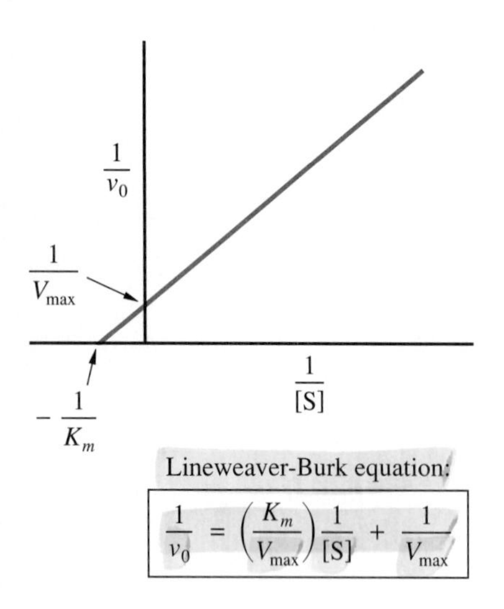

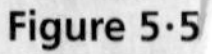
Figure 5·5
The Lineweaver-Burk (double-reciprocal) plot is derived from a linear transformation of the Michaelis-Menten equation. Values of $1/v_0$ are plotted as a function of 1/[S] values.

still encountered. The most often used transformation of the Michaelis-Menten equation is the Lineweaver-Burk or double-reciprocal plot (Figure 5·5). The values of $1/v_0$ are plotted against 1/[S]. The absolute value of $1/K_m$ is obtained from the intercept of the line at the x axis, and the value of $1/V_{max}$ is obtained from the y intercept. Although values obtained from double-reciprocal plots are less precise than computer-generated values, double-reciprocal plots are easily understood and provide recognizable patterns for the study of enzyme inhibition, an extremely important aspect of enzymology that we shall examine in the next section.

Values of k_{cat} can be obtained from measurements of V_{max} only when the absolute concentration of the enzyme is known. Values of K_m can be determined even when enzymes have not been purified, as long as only one enzyme in the impure preparation can catalyze the observed reaction.

5·6 Reversible Inhibitors Are Bound to Enzymes by Noncovalent Forces

An inhibitor (I) is a compound that binds to an enzyme and interferes with its activity by preventing either the formation of the ES complex or its breakdown to E + P. Inhibitors are used experimentally to investigate enzyme mechanisms and to decipher metabolic pathways. Natural inhibitors serve as regulators of metabolism, and many medicinal drugs are enzyme inhibitors.

Inhibitors can be either reversible or irreversible. Reversible inhibitors are bound to enzymes by the same noncovalent forces that bind substrates and products. Because they bind noncovalently, reversible inhibitors are easily removed from solutions of enzymes by dialysis or gel filtration (Section 3·5). The constant

Table 5·4 Effects of reversible inhibitors on kinetic constants

Type of inhibitor	Effect	Rate law
Competitive (I binds to E only)	Raises K_m; V_{max} remains unchanged	$v_0 = \frac{V_{max}[S]}{K_m\left(1 + \frac{[I]}{K_i}\right) + [S]}$
Uncompetitive (I binds to ES only)	Lowers V_{max} and K_m; ratio of V_{max}/K_m remains unchanged	$v_0 = \frac{V_{max}[S]}{K_m + [S]\left(1 + \frac{[I]}{K_i}\right)}$
Noncompetitive (I binds to E and ES)		$v_0 = \frac{V_{max}[S]}{K_m\left(1 + \frac{[I]}{K_i}\right) + [S]\left(1 + \frac{[I]}{K_i'}\right)}$
Pure noncompetitive (I binds to E and ES equally)	Lowers V_{max}; K_m remains unchanged	$E + I \overset{K_i}{\rightleftharpoons} EI$
Mixed noncompetitive (I binds to E and ES unequally)	Lowers V_{max}; raises or lowers K_m	$ES + I \overset{K_i'}{\rightleftharpoons} ESI$

for the dissociation of I from the EI complex, called the *inhibition constant*, K_i, is described by the equation:

$$K_i = \frac{[E][I]}{[EI]} \tag{5·21}$$

There are several types of reversible inhibition, which can be distinguished experimentally by their effects on the kinetic behavior of enzymes. These effects are summarized in Table 5·4 and discussed below. The rate laws for the reactions in the presence of reversible inhibitors are slightly expanded forms of the Michaelis-Menten equation with K_i and [I] incorporated.

A. Competitive Inhibitors Bind Only to Free E

Competitive inhibitors are commonly encountered in biochemistry and medicine and therefore deserve the greatest study. In competitive inhibition, I can bind only to unliganded molecules of enzyme, E. Competitive inhibition can be shown with the kinetic scheme in Figure 5·6a, an expansion of Equation 5·5 that includes formation of the EI complex. In this scheme, only ES can lead to the formation of product.

When a competitive inhibitor is bound to a molecule of enzyme, it prevents the binding of substrate to that enzyme molecule. Conversely, binding of substrate prevents binding of the inhibitor; that is, S and I compete for binding to the enzyme. Most commonly, S and I bind at the same site, the active site. However, some inhibitors bind at a different site on the enzyme, but the inhibition exhibits competitive characteristics. When both I and S are present in a solution, the proportion of the enzyme that is able to form ES complexes depends on the relative concentrations of S and I and their relative affinities for the enzyme.

The formation of EI can be reversed by increasing the concentration of S. At sufficiently high concentrations of S, saturation of E with S can still be achieved. Therefore, the maximum velocity has the same value as that in the absence of inhibitor. The more competitive inhibitor present, the more substrate is needed for half-saturation. We have seen that the concentration of substrate at half-saturation is K_m. Thus, in the presence of increasing concentrations of a competitive inhibitor, K_m increases. The new value is usually referred to as the observed or *apparent* K_m, K_m^{app}. On a double-reciprocal plot, this change is seen as a decrease in the absolute value of the intercept at the x axis ($1/K_m$), whereas the y intercept ($1/V_{max}$) remains the same at any concentration of I (Figure 5·6b).

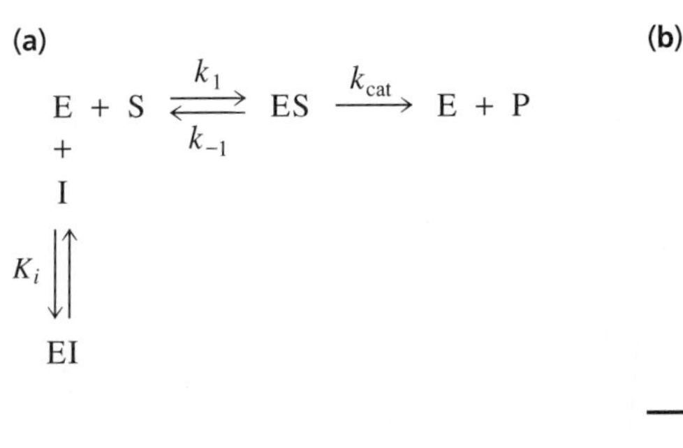

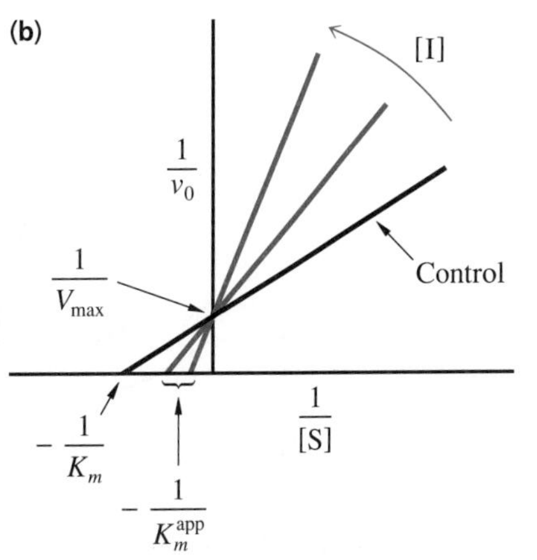

Figure 5·6
Competitive inhibition. **(a)** Kinetic scheme illustrating binding of I to E in competitive inhibition. **(b)** Double-reciprocal plot for competitive inhibition. V_{max} remains unchanged, and K_m increases. The black line labelled "Control" is the result in the absence of inhibitor. The red lines are results in the presence of inhibitor, with the arrow showing the direction of increase of [I].

Figure 5·7
Structures of benzamidine and indole. Benzamidine is an analog of an arginine residue in a peptide, and indole is an analog of a tryptophan residue.

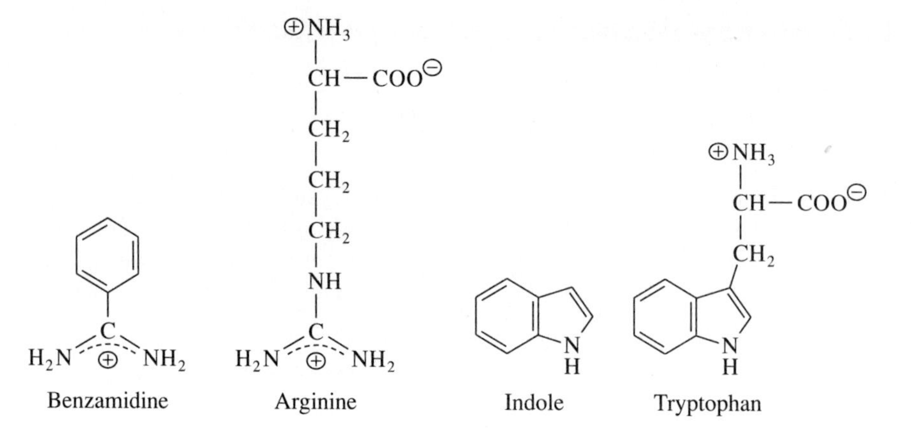

Substrate analogs are compounds that can bind to enzymes because of their similarity to substrates but that react very slowly or not at all. These compounds are usually competitive inhibitors. Two examples of nonmetabolizable substrate analogs are benzamidine, a competitive inhibitor of trypsin, and indole, a competitive inhibitor of chymotrypsin (Figure 5·7). Trypsin catalyzes hydrolysis of peptide bonds on the carboxyl side of arginine and lysine, and benzamidine is an analog of arginine that resembles the alkylguanidyl side chain of the amino acid. Chymotrypsin is specific for bulky aromatic and hydrophobic amino acids including tryptophan. Recall that tryptophan can be regarded as indolealanine. Benzamidine and indole act as competitive inhibitors by competing with their analogous substrates for the binding pocket of an active site. By systematically varying the structures of substrate analogs and measuring their potencies as inhibitors, one can sometimes draw conclusions about the structure of the active site of the enzyme.

B. Uncompetitive Inhibitors Bind Only to ES

Uncompetitive inhibitors bind only to ES, not to free enzyme (Figure 5·8a). In uncompetitive inhibition, V_{max} is decreased ($1/V_{max}$ is increased) by the conversion of some molecules of E to the inactive form ESI. Since it is the ES complex that binds I, the decrease in V_{max} is not reversed by the addition of more substrate. Uncompetitive inhibitors also decrease the K_m (seen as an increase in the absolute value of $1/K_m$ on a double-reciprocal plot) because the equilibria for the formation of both ES and ESI are shifted toward the complexes by the binding of I. Experimentally, the lines on a reciprocal plot representing varying concentrations of an uncompetitive inhibitor all have the same slope, indicating proportionally decreased values for K_m and V_{max} (Figure 5·8b). This type of inhibition usually occurs only with multisubstrate reactions.

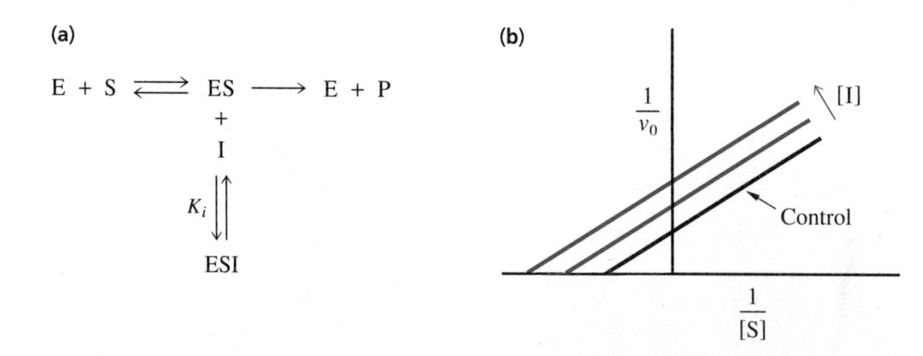

Figure 5·8
Uncompetitive inhibition. **(a)** Kinetic scheme illustrating binding of I to ES in uncompetitive inhibition. **(b)** Double-reciprocal plot for uncompetitive inhibition. V_{max} and K_m both decrease (i.e., the absolute values of $1/V_{max}$ and $1/K_m$, obtained from the y and x intercepts, respectively, both increase). The ratio of V_{max}/K_m remains unchanged.

C. Noncompetitive Inhibitors Bind to Both E and ES

Noncompetitive inhibitors bind to both E and ES, so that EI and ESI complexes—both inactive—can be formed (Figure 5·9a). When these inhibitors, which do not structurally resemble substrates, bind to E and ES with equal affinity ($K_i = K_i'$, the inhibition constant for ES $\rightleftharpoons$ ESI, as shown in Figure 5·9a), the result is termed *pure noncompetitive inhibition*. This type of inhibition is characterized by a decrease in V_{max} (increase in $1/V_{max}$) with no change in K_m. On a double-reciprocal plot, the lines for pure noncompetitive inhibition intersect at a point on the 1/[S] axis (Figure 5·9b). The effect of pure noncompetitive inhibition is to reversibly titrate E and ES with I, in essence removing active enzyme molecules from solution. This inhibition cannot be overcome by the addition of S. Pure noncompetitive inhibition is rare, but examples are known among allosteric enzymes.

When an inhibitor binds to E and ES with unequal affinity, the result is termed *mixed noncompetitive inhibition*. In this case, V_{max} decreases and K_m may either increase or decrease, depending on whether K_i in Figure 5·9a is lower or higher, respectively, than K_i'. On a double-reciprocal plot, the lines for mixed noncompetitive inhibition intersect to the left of the $1/v_0$ axis, anywhere but on the 1/[S] axis (Figure 5·9c). Mixed noncompetitive inhibitors will be encountered in the next section.

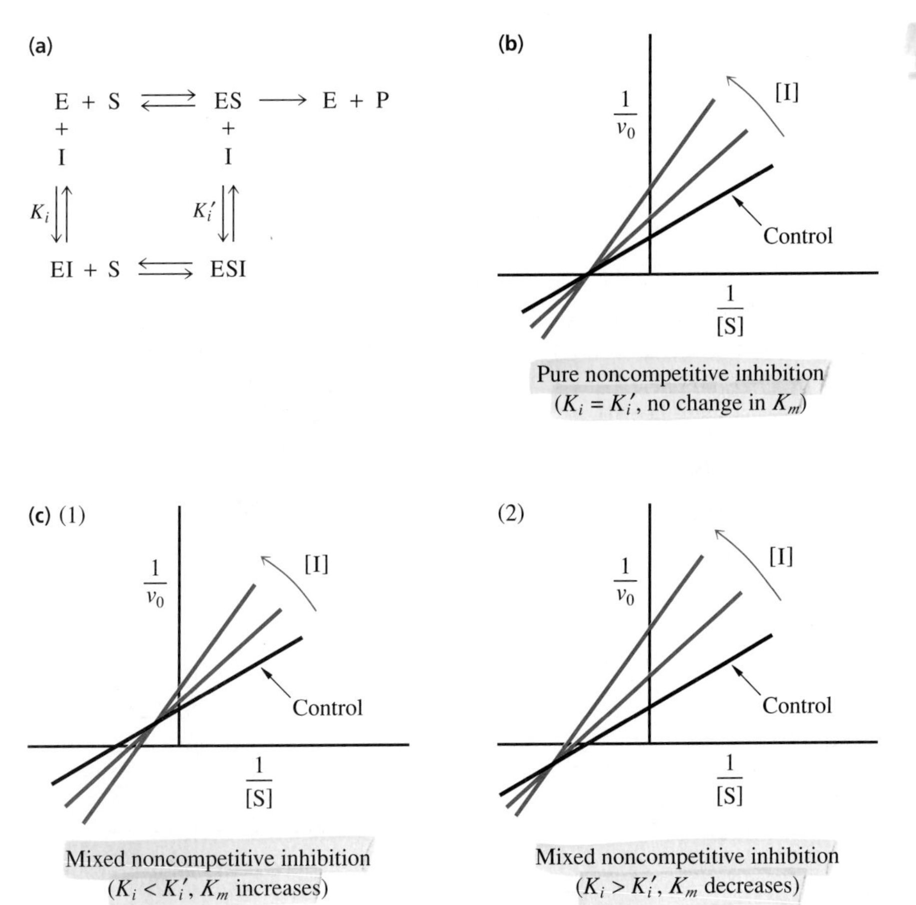

Figure 5·9
Noncompetitive inhibition. **(a)** Kinetic scheme illustrating binding of I to E or ES in noncompetitive inhibition. **(b)** Double-reciprocal plot for pure noncompetitive inhibition, in which I binds to E and ES with equal affinity. V_{max} decreases, K_m remains the same. **(c)** Double-reciprocal plots for mixed noncompetitive inhibition, in which I binds to E and ES with different affinity. (1) If I has greater affinity for E than ES, K_m increases; (2) If I has greater affinity for ES than E, K_m decreases.

5·7 Kinetic Measurements Can Determine the Order of Binding of Substrates and Release of Products in Multisubstrate Reactions

Kinetic measurements of multisubstrate reactions are a little more complicated than simple one-substrate enzyme kinetics. However, for many purposes, such as designing an enzyme assay, it is sufficient simply to determine the K_m of each substrate in the presence of saturating amounts of each of the other substrates. Furthermore, the simple enzyme kinetics discussed in this chapter can be extended to distinguish among several mechanistic possibilities for group-transfer reactions. This is done by measuring the effect of variations in the concentration of one substrate on the kinetic results obtained for the other. Additional information is obtained by determining the type of reversible inhibition caused by each product of the reaction.

There are several different kinetic schemes by which a multisubstrate reaction could occur. W. W. Cleland introduced a notation that abbreviates a kinetic scheme to a horizontal line or group of lines (Figure 5·10). The sequence of steps proceeds from left to right. The addition of substrate molecules (A, B, C...) to the enzyme and the release of products (P, Q, R...) from the enzyme are indicated by vertical arrows. The various forms of the enzyme, free E or ES complexes, are written under the line. The ES complexes that undergo chemical transformation when the active site is filled are written in parentheses. *Sequential reactions* (Figure 5·10a) are those that require all of the substrates to be present before any product is released. Sequential reactions can be either *ordered,* in which there is an obligatory order for addition of substrates and release of products, or *random,* in which there is no obligatory order of binding or release. In *ping-pong* reactions (Figure 5·10b), a product is released before all of the substrates are bound. For a bisubstrate ping-pong reaction, the first substrate is bound, the enzyme is altered by substitution, and the first

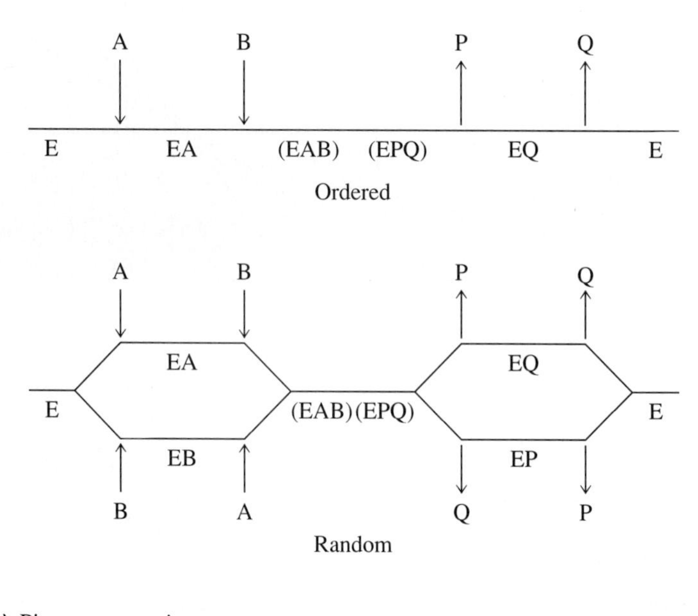

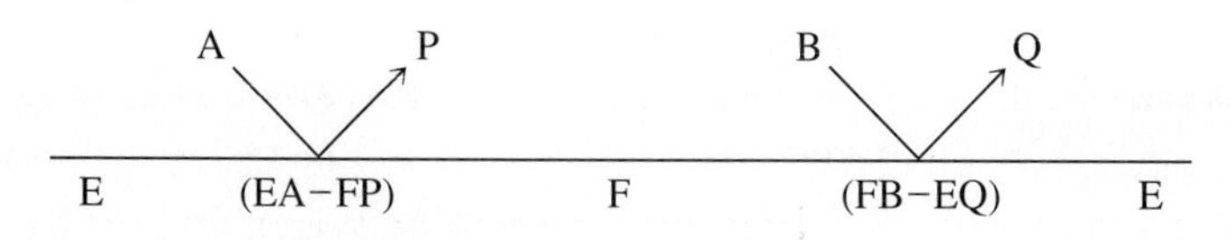

Figure 5·10
Notation devised by W. W. Cleland for bisubstrate reactions. **(a)** In sequential reactions, all substrates are bound before a product is released. The binding of substrates may be either ordered or random. **(b)** In ping-pong reactions, one substrate is bound and a product is released, leaving a substituted enzyme that binds a second substrate and releases a second product, restoring the enzyme to its original form.

product is released, after which the second substrate is bound, the altered enzyme is restored to its original form, and the second product is released. The binding and release of ligands in a ping-pong mechanism are usually indicated by slanted lines. The two forms of the enzyme are indicated by E (unsubstituted) and F (substituted).

Sequential and ping-pong reactions can be distinguished by initial-velocity experiments. The concentration of either substrate is varied at several levels of the other substrate, and double-reciprocal plots are drawn, with each line representing a different level of the second substrate. The lines intersect for a sequential mechanism and are parallel for a ping-pong mechanism (Figure 5·11).

To complete the kinetic mechanism (that is, to distinguish ordered from random reactions and to find the order of ligand binding and release), the effect of products as reversible inhibitors with respect to each substrate is measured. This straightforward method was first suggested by Robert Alberty in 1953. Because products are the substrates of the reverse reaction, they act as inhibitors of the forward reaction by binding to the enzyme and preventing the reaction in the forward direction. Each product is tested as a reversible inhibitor at several concentrations, with variable concentrations of one substrate and fixed concentrations of the other. The results can be analyzed as shown in Table 5·5. For example, if a product affects only K_m^{app} of the variable substrate (competitive inhibition), then the product and the substrate are competing for the same form of the enzyme. Q in the ordered reaction in Figure 5·10 can only bind to free E and therefore affects only the K_m^{app} of the leading substrate, A. Product P in the same reaction can only bind to EQ and thus is a mixed noncompetitive inhibitor of both A and B. Fitting the observations of inhibition studies to Table 5·5 reveals the order of ligand binding and release by identifying the ligands as A or B, P or Q, and distinguishes the type of kinetic mechanism.

Kinetic mechanism studies like these can be performed with very little enzyme. They supplement information found by other experiments and can assist in the elucidation of the chemical mechanisms of enzymes.

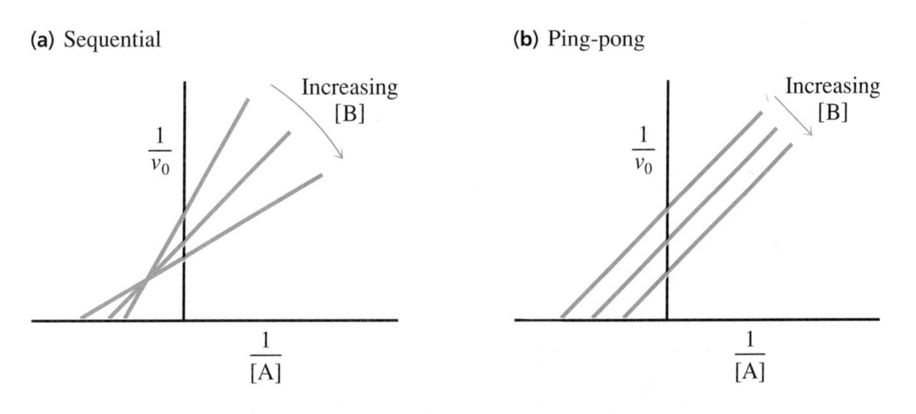

Figure 5·11
Double-reciprocal plots for sequential and ping-pong reactions. In the experiment, the concentration of one substrate (A in the figure) is varied at several different concentrations of the second substrate (B), and the initial velocity of each reaction is measured. **(a)** Intersecting lines indicate a sequential kinetic mechanism. **(b)** Parallel lines indicate a ping-pong kinetic mechanism.

Table 5·5 Determination of kinetic mechanism from types of product inhibition for the general reaction $A + B \rightleftharpoons P + Q$

Product tested as reversible inhibitor:	P	Q	P	Q	
Variable substrate:	A	A	B	B	**Kinetic mechanism**
Observations	mixed noncompetitive	competitive	mixed noncompetitive	mixed noncompetitive	ORDERED
	competitive	competitive	competitive	competitive	RANDOM
	mixed noncompetitive	competitive	competitive	mixed noncompetitive	PING-PONG

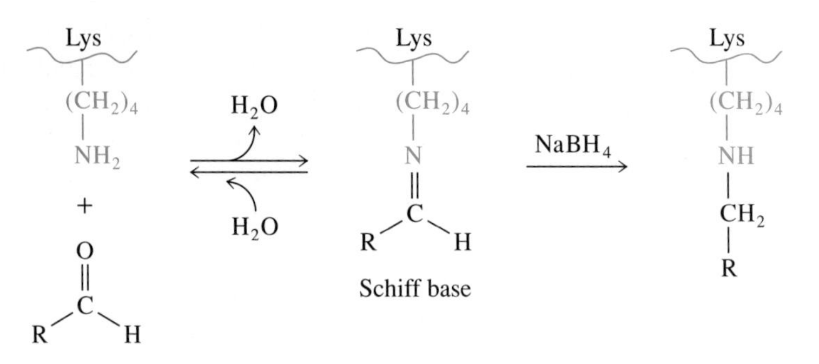

Figure 5·12
Reaction of the ε-amino group of a lysine residue with an aldehyde. Reduction of the Schiff base with sodium borohydride ($NaBH_4$) forms a stable substituted enzyme.

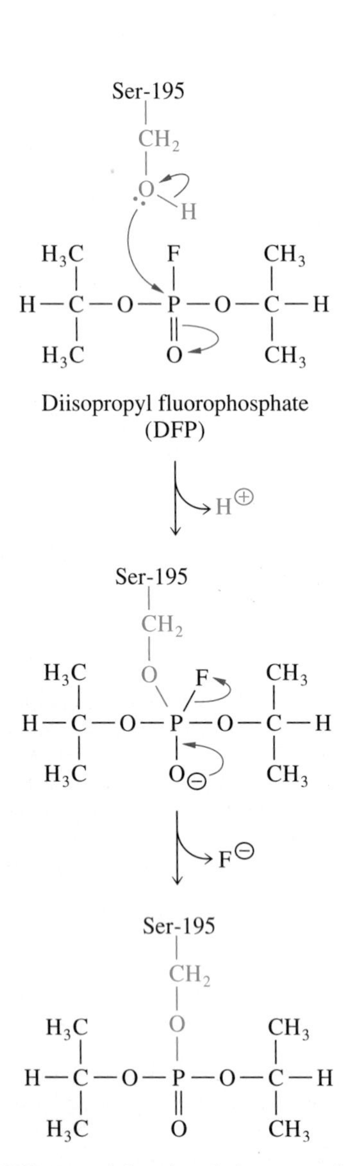

Figure 5·13
Reaction of diisopropyl fluorophosphate (DFP) with a single, highly nucleophilic serine residue (Ser-195) at the active site of chymotrypsin, producing inactive diisopropylphosphoryl-chymotrypsin. DFP similarly inactivates other serine proteases and serine esterases.

5·8 Irreversible Inhibitors Covalently Modify Enzymes

An irreversible inhibitor forms a stable covalent bond with an enzyme molecule. An important use of irreversible inhibitors is to identify active-site amino acid residues by specific substitution of their reactive side chains. A covalent inhibitor that reacts with only one type of amino acid is added to a solution of enzyme, which is tested for loss of activity after incubation. Ionizable side chains in their nucleophilic forms are modified by acylation or alkylation reactions. For example, free amino groups such as the ε-amino group of lysine react with an aldehyde to form a Schiff base that can be stabilized by reduction with sodium borohydride, $NaBH_4$ (Figure 5·12).

The nerve gas diisopropyl fluorophosphate (DFP) was used to identify a serine residue as an active-site component of chymotrypsin (Figure 5·13). Only one serine residue out of the 27 in the enzyme is sufficiently nucleophilic to attack the fluorophosphate of DFP and form a stable phosphoester. When DFP labelled with radioactive (^{32}P) phosphate was used in this reaction, one mole of the phosphate compound was found to be incorporated per mole of chymotrypsin, and the enzyme was inactivated. After the resulting substituted chymotrypsin was partially hydrolyzed in acid, a radioactive peptide with the sequence Asp–^{32}P-Ser–Gly was isolated. When the enzyme was completely sequenced, the substituted serine was identified as Ser-195. Inactivation by DFP and similar compounds is used as a test for active-site serine hydroxyl groups in hydrolytic enzymes.

More useful than general substituting reagents are irreversible inhibitors with structures that allow them to bind specifically to an active site. These inhibitors are referred to as *active-site directed reagents* or *affinity labels*.

An affinity label was synthesized to study the active-site residues of chymotrypsin. Chymotrypsin catalyzes the hydrolysis of peptide bonds whose carbonyl group is contributed by a bulky, hydrophobic residue, although it can also catalyze the hydrolysis of esters or simple amides. Tosylphenylalanylglycine (Figure 5·14a) is a substrate for which chymotrypsin has shown specificity. Replacement of the nitrogen-containing leaving group of this substrate with a hydrocarbon group results in a ketone, 1-tosylamido-2-phenylethyl methyl ketone (Figure 5·14b). This compound is a competitive inhibitor of chymotrypsin because it binds to the hydrophobic pocket of the enzyme but is not subject to nucleophilic attack by the catalytic residues of the enzyme. The ketone can be further modified by substituting a very electronegative group for a hydrogen of its methyl group, making it susceptible to attack by a nucleophilic residue of chymotrypsin but not subject to subsequent hydrolysis. Hence, the chloromethyl ketone (1-tosylamido-2-phenylethyl chloromethyl ketone, TPCK), synthesized by Guenther Schoellman and Elliott Shaw in the 1960s and pictured in Figure 5·14c, reacts irreversibly with chymotrypsin, forming a covalent bond with a nucleophilic residue in the active site. When Schoellman and Shaw analyzed the covalently modified enzyme, they found that the chymotrypsin was inactive and that His-57 was modified, suggesting its involvement in catalysis.

(a)

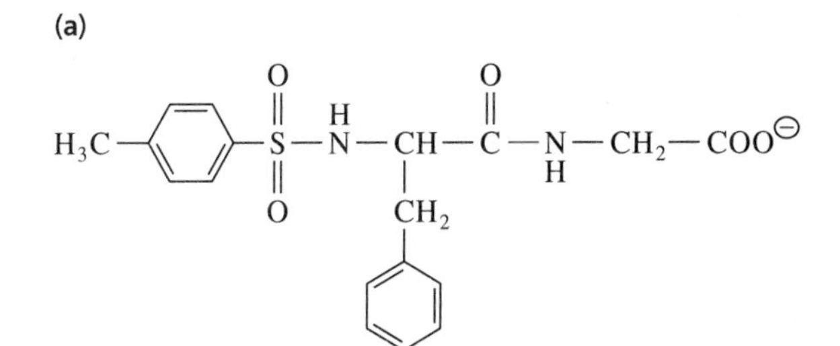

Tosylphenylalanylglycine

(b)

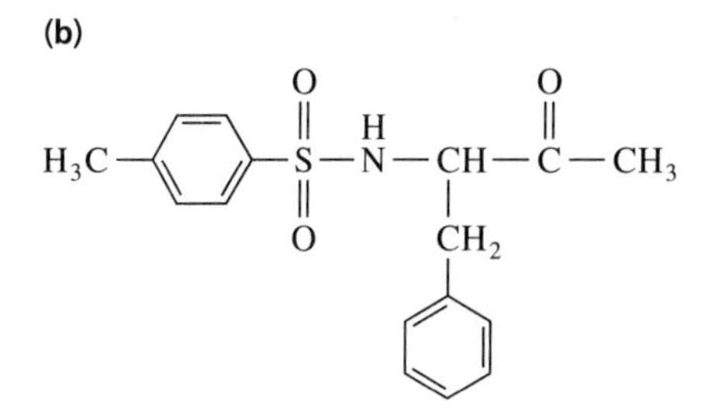

Tosylamidophenylethyl methyl ketone

(c)

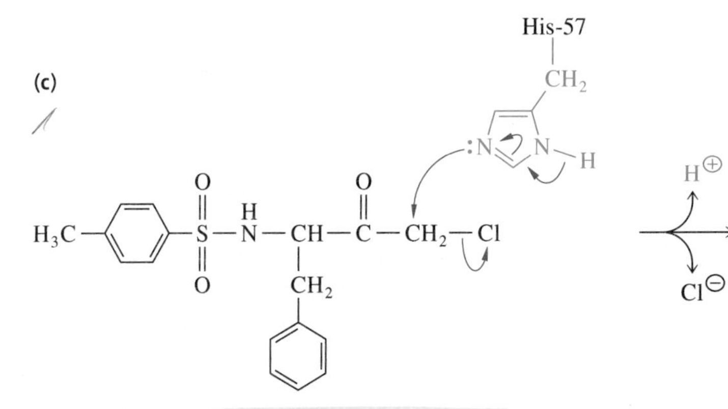

Tosylamidophenylethyl chloromethyl ketone (TPCK)

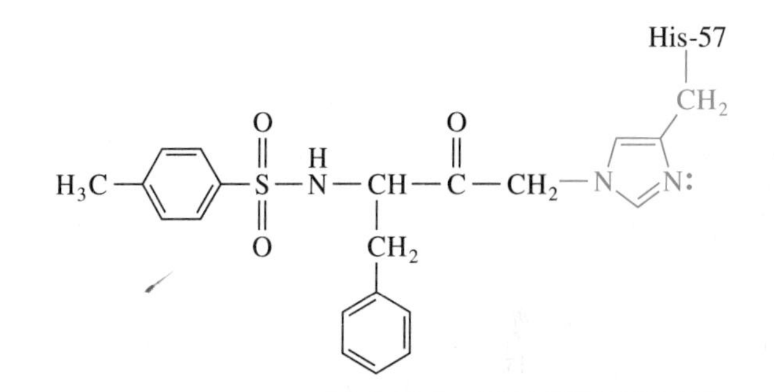

Alkylated derivative of His-57 at active site of chymotrypsin (inactive)

Figure 5·14
Structural comparison of a substrate, a reversible competitive inhibitor, and an irreversible inactivator of chymotrypsin. **(a)** Structure of tosylphenylalanylglycine, an artificial substrate of chymotrypsin. The peptide bond cleaved by the action of chymotrypsin is indicated in red. **(b)** Structure of *N*-tosylamidophenylethyl methyl ketone. This substrate analog acts as a competitive inhibitor of chymotrypsin. **(c)** Structure and reactivity of *N*-tosylamidophenylethyl chloromethyl ketone, TPCK, an irreversible inactivator of chymotrypsin. The phenylalanyl side chain of TPCK gives it affinity for the substrate-binding site of chymotrypsin, while its chloromethyl ketone group is susceptible to attack by nucleophiles such as the imidazole side chain of a histidine residue or the sulfhydryl group of a cysteine residue. In the case of chymotrypsin, enzyme activity is lost as one of its two histidine residues, His-57, becomes alkylated by reaction with TPCK.

5·9 Site-Directed Mutagenesis Is Used to Produce Modified Enzymes

The latest technique for individually testing the functions of the amino acid side chains of an enzyme is *site-directed mutagenesis*. Mutagenesis is more specific than irreversible inhibition and also allows the testing of amino acids for which there are no specific chemical inactivators. In this procedure, one amino acid is specifically replaced by another through the biosynthesis of a modified enzyme. The gene that encodes an enzyme is isolated and sequenced. The enzyme can be synthesized in bacterial cells by inserting the gene into a suitable *vector*, such as a virus, that can be used to transform bacteria. After verifying that the unmodified or *wild-type* enzyme is synthesized by the bacteria, DNA is prepared with the base sequence altered to encode a modified protein. The bacteria are infected with the new DNA, just as they were infected in the control experiment. The mutant protein is synthesized, isolated, and tested. If possible, the enzyme is purified and its structure checked by X-ray crystallography to make sure that the mutation has not altered its conformation.

A successful example of site-directed mutagenesis was the alteration of the bacterial peptidase subtilisin to make it more resistant to chemical oxidation. Subtilisin has been added to detergent powders to help remove protein stains such as chocolate and blood. Resistance to oxidation, by bleach for instance, increases the suitability of subtilisin as a detergent additive. Subtilisin has a methionine residue at position 222 in the active-site cleft that readily oxidizes, leading to inactivation of the enzyme. In a series of mutagenic experiments, Met-222 was systematically replaced by each of the other common amino acids. All 19 possible mutant subtilisins were isolated and tested. Most had greatly diminished activity. The Cys-222 mutant had high activity but was also subject to oxidation. The Ala-222 and Ser-222 mutants, with side chains that are nonoxidizable, were not inactivated by oxidation and had relatively high activity. They were the only active and oxygen-stable mutant subtilisins. Several studies of active sites by site-directed mutagenesis will be discussed in later sections.

Having examined the general properties of enzymes as revealed by kinetic methods, we now turn to the specific example of serine proteases to explore the relationship between protein structure and catalytic function.

5·10 Chymotrypsin, Like Many Proteases, Is Synthesized as an Inactive Precursor

The protease chymotrypsin is a member of a family of *serine proteases*. These proteases, which include the digestive enzymes chymotrypsin, trypsin, and elastase, all have a functional serine residue in their active sites and a common catalytic mechanism. In mammals, serine proteases act outside the cells in which they are synthesized. This poses a problem that was raised earlier (Section 2·6): how are the proteins of the cells in which proteases are assembled protected from hydrolytic destruction catalyzed by these enzymes? Cells have solved this problem by a simple type of regulation, the synthesis of the proteases (and certain other potentially destructive enzymes) as *zymogens,* inactive precursors that must be covalently modified to become active. The pancreatic zymogens trypsinogen, chymotrypsinogen, proelastase, and procarboxypeptidase are activated extracellularly under appropriate physiological conditions by *selective proteolysis*—enzymatic cleavage of one or a few specific peptide bonds. This activation process effectively controls the activity of these proteases.

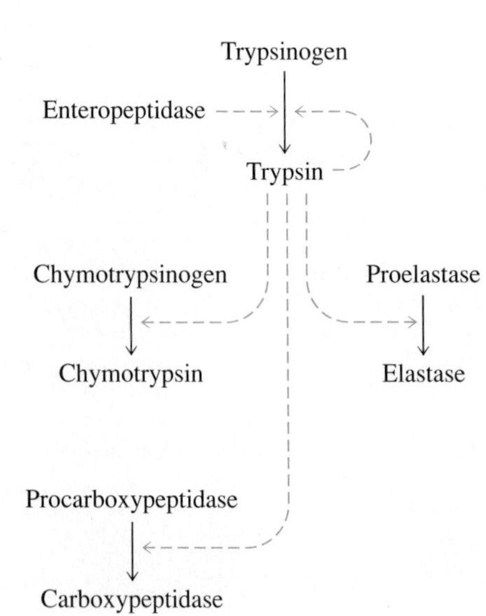

Figure 5·15
Activation of pancreatic zymogens in the duodenum. Enteropeptidase initiates this process by catalyzing the cleavage of a bond in trypsinogen to form trypsin. Trypsin is the common activator of the pancreatic zymogens. Trypsin can activate trypsinogen as well as the pancreatic zymogens chymotrypsinogen, proelastase, and procarboxypeptidase.

The pancreatic zymogens become hydrolases that catalyze the digestion of dietary proteins. They are produced by the acinar cells of the pancreas and delivered to the duodenum (upper portion of the small intestine) as inactive precursors in the pancreatic juice. All of the peptide bonds of food proteins are made accessible to intestinal proteolysis by denaturation. After food has been mechanically disrupted by chewing and moistened with saliva, it is swallowed and the proteins are denatured by mixing with hydrochloric acid in the stomach. The food proteins are then subjected to hydrolysis catalyzed by pepsin, a protease with high activity in the acidic medium. The partially hydrolyzed mixture passes into the intestine. In the duodenum, the intestinal protease enteropeptidase triggers the activation of the pancreatic zymogens by catalyzing the cleavage of a particular peptide bond in trypsinogen, thereby producing a fully activated trypsin molecule. Enteropeptidase, like trypsin, is a protease that catalyzes cleavage of peptide bonds on the carboxyl side of lysine and arginine residues. Once activated, trypsin proteolytically activates the other pancreatic zymogens as well as activating additional trypsinogen molecules autocatalytically (Figure 5·15). Thus, the formation of active trypsin by the action of enteropeptidase can be viewed as the master activation step.

There is an additional control to prevent premature activation of the pancreatic zymogens. The pancreas produces small amounts of a molecule called *trypsin inhibitor* to protect against the rapid destruction of pancreatic cells that might occur if even a single molecule of trypsinogen were prematurely activated. The potency of trypsin stems not only from its own proteolytic activity but also from the proteolytic activity of the zymogens it activates. Trypsin inhibitor, a small protein (MW 6000), binds noncovalently but extremely tightly to the active site of trypsin, forming an EI complex with a dissociation constant of 10^{-13} M. The inhibitor protein, an analog of the peptide substrates of trypsin, is bound by one of its lysine residues to the active site of trypsin. The molecular structure of the inhibitor matches the shape of the active site of the enzyme extremely well—so well that no water can enter the cleft to cause hydrolysis of the inhibitor (Stereo S5·1).

As an example of the covalent changes that take place upon activation of a zymogen, let us examine the conversion of chymotrypsinogen by sequential cleavages to form active α-chymotrypsin, whose primary structure is shown in Figure 5·16 (next page). For convenience, the numbering of the chymotrypsinogen peptide sequence is retained here for chymotrypsin and is also used for reference to corresponding amino acid residues in homologous proteins. Trypsin-catalyzed cleavage of a single peptide bond in chymotrypsinogen between Arg-15 and Ile-16 results in a fully active but rather unstable enzyme known as π-chymotrypsin, in which the 15-residue N-terminal peptide remains bound by a disulfide bond between Cys-1 and Cys-122. Three additional peptide bonds of π-chymotrypsin, between residues 13 and 14, 146 and 147, and 148 and 149, are susceptible to cleavage by active chymotrypsin molecules. Examination of the three-dimensional structure of chymotrypsinogen showed that these three bonds are available for hydrolysis because they are on the surface of π-chymotrypsin. Their cleavage results in the excision of

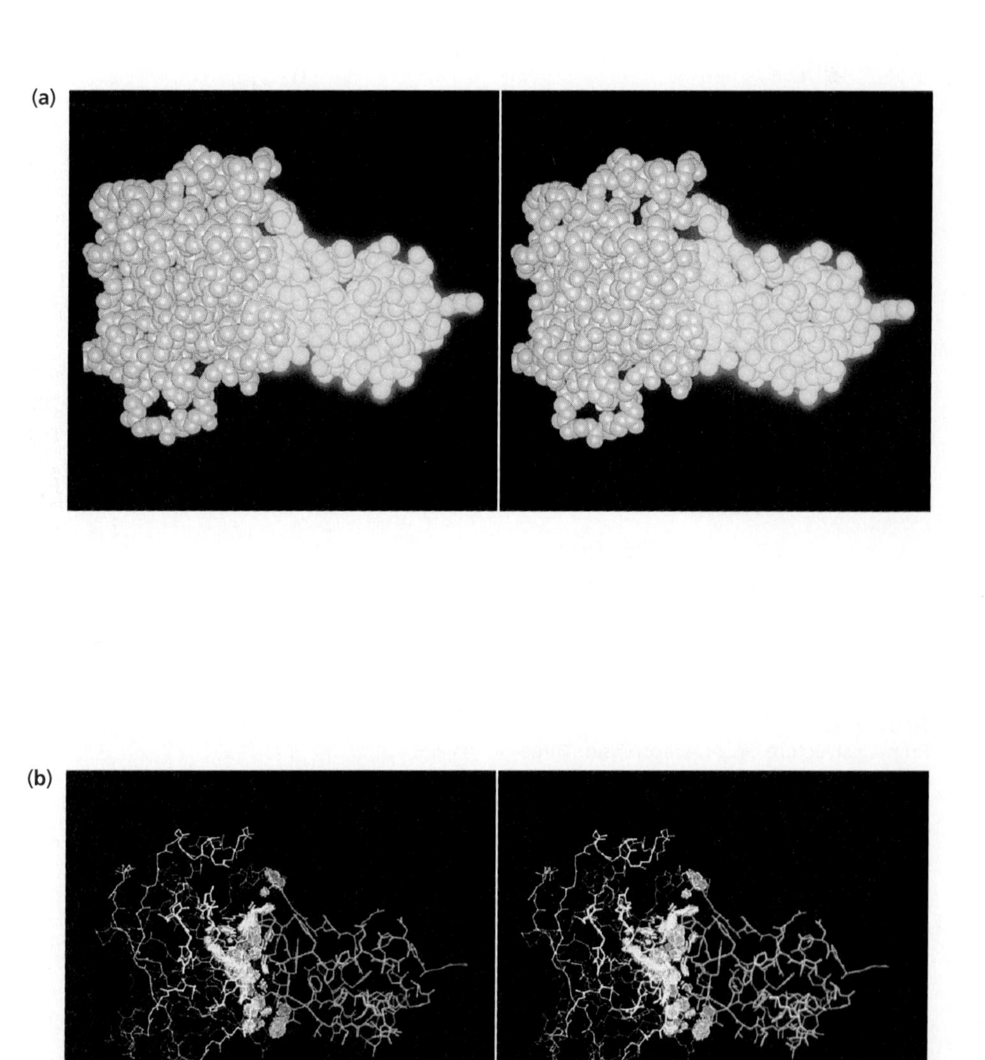

Stereo S5·1
Trypsin–trypsin inhibitor complex. (**a**) Space-filling model showing trypsin in gold and the inhibitor in blue. (**b**) Backbone model of trypsin (gold) and complete structure of trypsin inhibitor (blue), with dot surfaces showing van der Waals contacts between residues comprising the enzyme-inhibitor interface (carbon, green; nitrogen, blue; oxygen, red). Lys-15 of the inhibitor binds to trypsin at the same site that a normal substrate would bind. The very close match between the two proteins makes the binding of the pancreatic trypsin inhibitor to trypsin one of the strongest known noncovalent interactions between protein molecules. (Based on coordinates provided by R. Huber and J. Deisenhofer; stereo image (a) by Richard J. Feldmann, stereo image (b) by Ben N. Banks and Kim M. Gernert.)

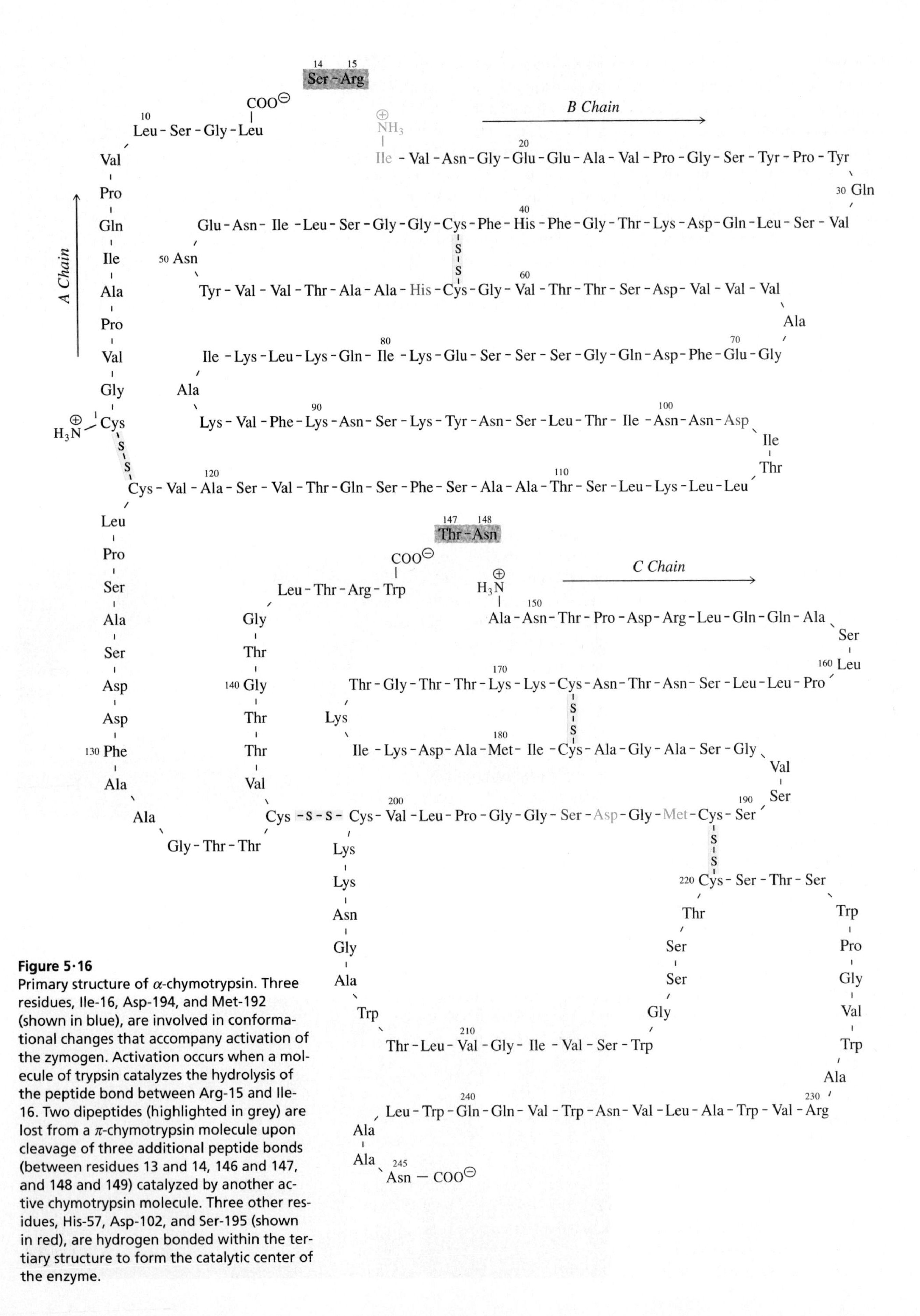

Figure 5·16
Primary structure of α-chymotrypsin. Three residues, Ile-16, Asp-194, and Met-192 (shown in blue), are involved in conformational changes that accompany activation of the zymogen. Activation occurs when a molecule of trypsin catalyzes the hydrolysis of the peptide bond between Arg-15 and Ile-16. Two dipeptides (highlighted in grey) are lost from a π-chymotrypsin molecule upon cleavage of three additional peptide bonds (between residues 13 and 14, 146 and 147, and 148 and 149) catalyzed by another active chymotrypsin molecule. Three other residues, His-57, Asp-102, and Ser-195 (shown in red), are hydrogen bonded within the tertiary structure to form the catalytic center of the enzyme.

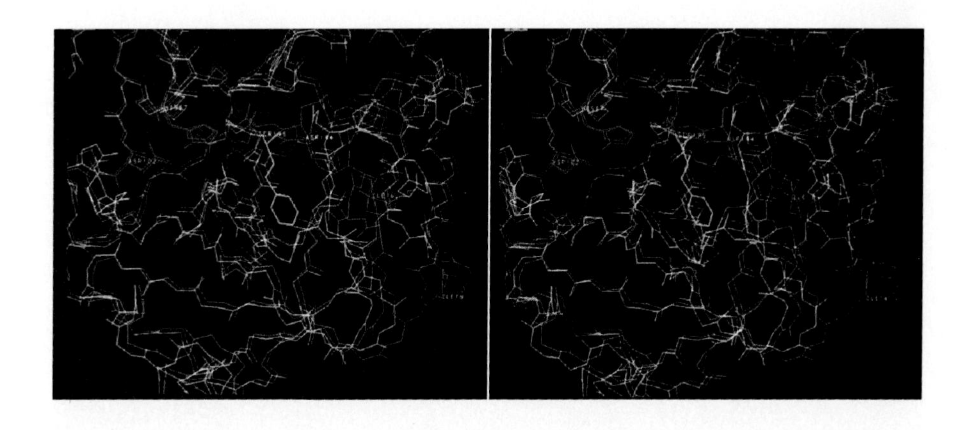

Stereo S5·2
Overlay of backbone for chymotrypsinogen (purple) and α-chymotrypsin (blue). Ile-16, Asp-194, and the catalytic-site residues (Asp-102, His-57, and Ser-195) in both the zymogen and the active enzyme are shown in red. A portion of the substrate is shown in green. (Coordinates for chymotrypsin provided by A. Tulinsky and R. Blevins, those for chymotrypsinogen provided by D. Wang, W. Bode, and R. Huber; stereo image by Ben N. Banks and Kim M. Gernert.)

dipeptides Ser-14–Arg-15 and Thr-147–Asn-148. Hence, both in the small intestine and in vitro, the most stable form of the enzyme, known as α-chymotrypsin or simply chymotrypsin, is a three-chained protein containing 241 amino acid residues. The tertiary structure of chymotrypsin is stabilized by its five disulfide bonds joining residues 1 and 122, 42 and 58, 136 and 201, 168 and 182, and 191 and 220.

X-ray crystallography has revealed one major difference between the conformation of chymotrypsinogen and chymotrypsin: the lack of a hydrophobic substrate-binding pocket in the zymogen. The differences in structure are shown in Stereo S5·2. Upon zymogen activation, the newly generated α-amino group of Ile-16 turns inward and interacts with the β-carboxyl group of Asp-194 to form an ion pair. This local conformational change pulls the side chain of Met-192 away from its hydrophobic contacts in the zymogen conformation, thereby generating a relatively hydrophobic substrate-binding pocket. Near this substrate-binding site are three ionizable side chains (Asp-102, His-57, and Ser-195) that participate in catalysis.

A number of zymogens have roles in *enzyme cascades*. Activation of a small concentration of initial zymogen results in the sequential activation of several enzymes, ultimately leading to a large concentration, and thus high activity, of an active primary enzyme. Enzyme cascades result in rapid signal amplification, as illustrated in Figure 5·17.

In mammals, an elaborate enzyme cascade is responsible for blood clotting, an important protective mechanism against excessive loss of blood. Clotting involves at least a dozen proteins known as *coagulation factors*. A number of these factors are proteases whose sole function is to cleave one or two specific peptide bonds of a single zymogen substrate, thereby activating another member of the cascade system. Enzyme cascades in which activation occurs by mechanisms other than proteolytic activations of zymogens have important regulatory roles in metabolism.

	Number of active enzyme molecules
Initiating enzyme	1
$A_i \longrightarrow A_a$	10
$B_i \longrightarrow B_a$	10^2
$C_i \longrightarrow C_a$	10^3
$D_i \longrightarrow D_a$	10^4

Figure 5·17
Principle of enzyme cascades. A cascade of enzyme activity results when several enzymes are sequentially activated. A very small concentration of the initiating enzyme (enteropeptidase in the case of pancreatic zymogen activation) converts inactive enzyme A_i to active A_a, which in turn converts inactive B_i to active B_a, and so forth. The number of activated enzymes increases exponentially (arbitrarily shown here as tenfold), so that the effect of the original activating enzyme is enormously amplified. Such cascades are observed in the activation of pancreatic zymogens, the activation of blood coagulation factors, and in metabolic regulatory systems.

5·11 X-ray Crystallographic Analysis Revealed the Basis for the Substrate Specificity of Serine Proteases

Many of the serine proteases share homologies in primary, secondary, and tertiary structure, as revealed by sequence analysis and X-ray crystallography. Similarities in the backbone conformations and active-site residues of chymotrypsin, trypsin, and elastase may be seen by comparing Stereos S5·3a, b, and c. Each is a bilobed structure with the active site located in a cleft between the two lobes. In addition to the active-site serine and histidine residues found by chemical substitution reactions (Section 5·8), an aspartate residue, Asp-102, was found by X-ray crystallographic experiments. The side chains of these three amino acid residues are hydrogen bonded. The alignment of Ser-195, His-57, and Asp-102 is highlighted in Stereos S5·3a, b, and c.

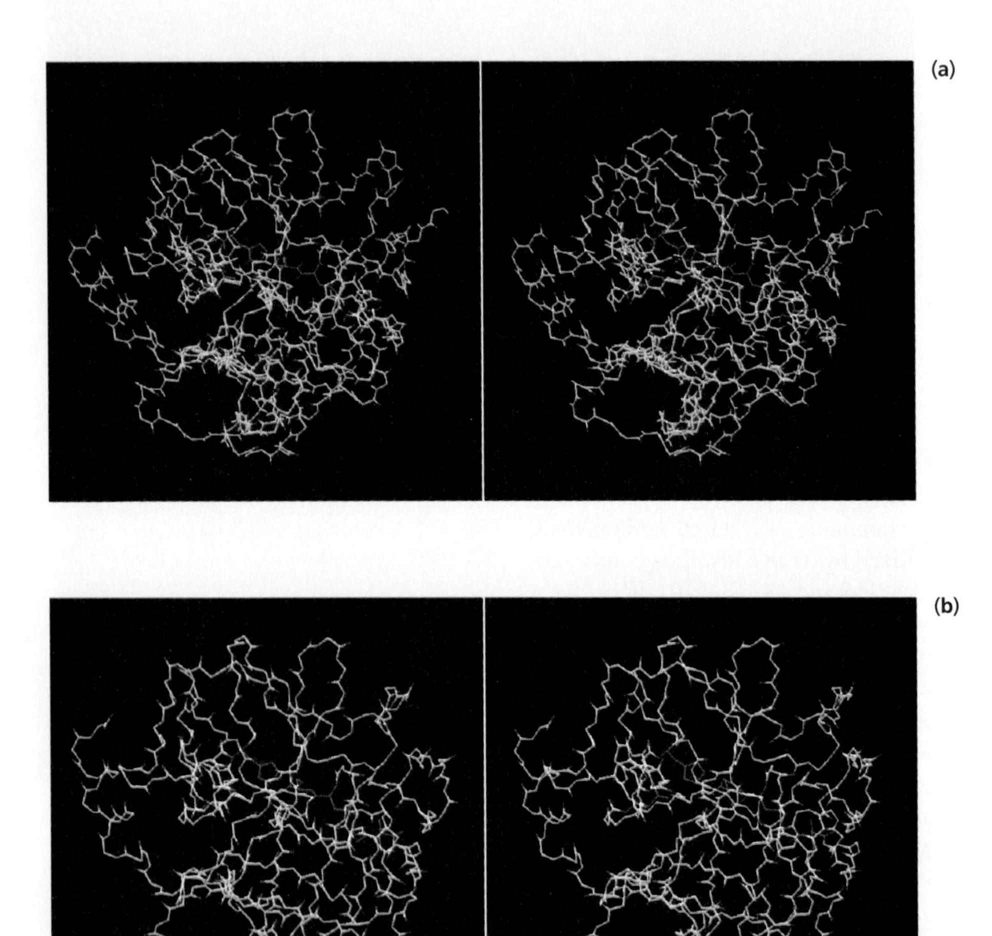

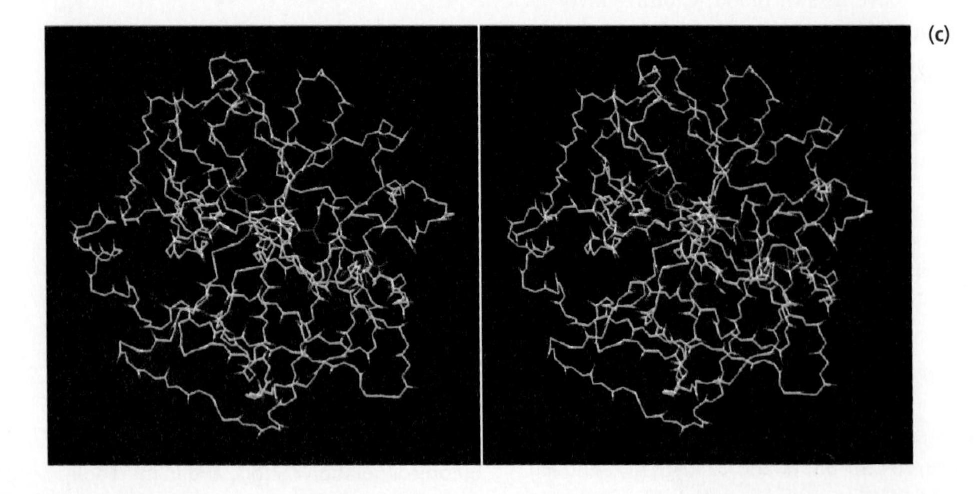

Stereo S5·3
Backbone models of chymotrypsin, trypsin, and elastase with residues at the catalytic center shown in red. (Stereo images by Ben N. Banks and Kim M. Gernert.) **(a)** Chymotrypsin. (Based on coordinates provided by A. Tulinsky and R. Blevins.) **(b)** Trypsin. (Based on coordinates provided by J. Walter, R. Huber, and W. Bode.) **(c)** Elastase. (Based on coordinates provided by T. Prange and I. Li De La Sierra.)

The substrate specificities of chymotrypsin, trypsin, and elastase can be accounted for by relatively small structural differences in the enzymes. Recall that trypsin catalyzes hydrolysis of peptide bonds whose carbonyl groups are contributed by arginine or lysine. Like chymotrypsin, trypsin contains a binding pocket that correctly positions its substrates for nucleophilic attack by the active-site residue Ser-195. Each protease has a similar extended region into which polypeptides fit, but the so-called specificity pocket near the active-site serine is markedly different for each enzyme. The key difference in structure between trypsin and chymotrypsin is that an uncharged amino acid residue at the base of the hydrophobic binding pocket of chymotrypsin (Figure 5·18a) is replaced in trypsin by an aspartate residue (Figure 5·18b). This negatively charged aspartate residue is responsible for the substrate specificity of trypsin. In the ES complex, it forms an ion pair with the positively charged side chains of arginine and lysine residues of substrates.

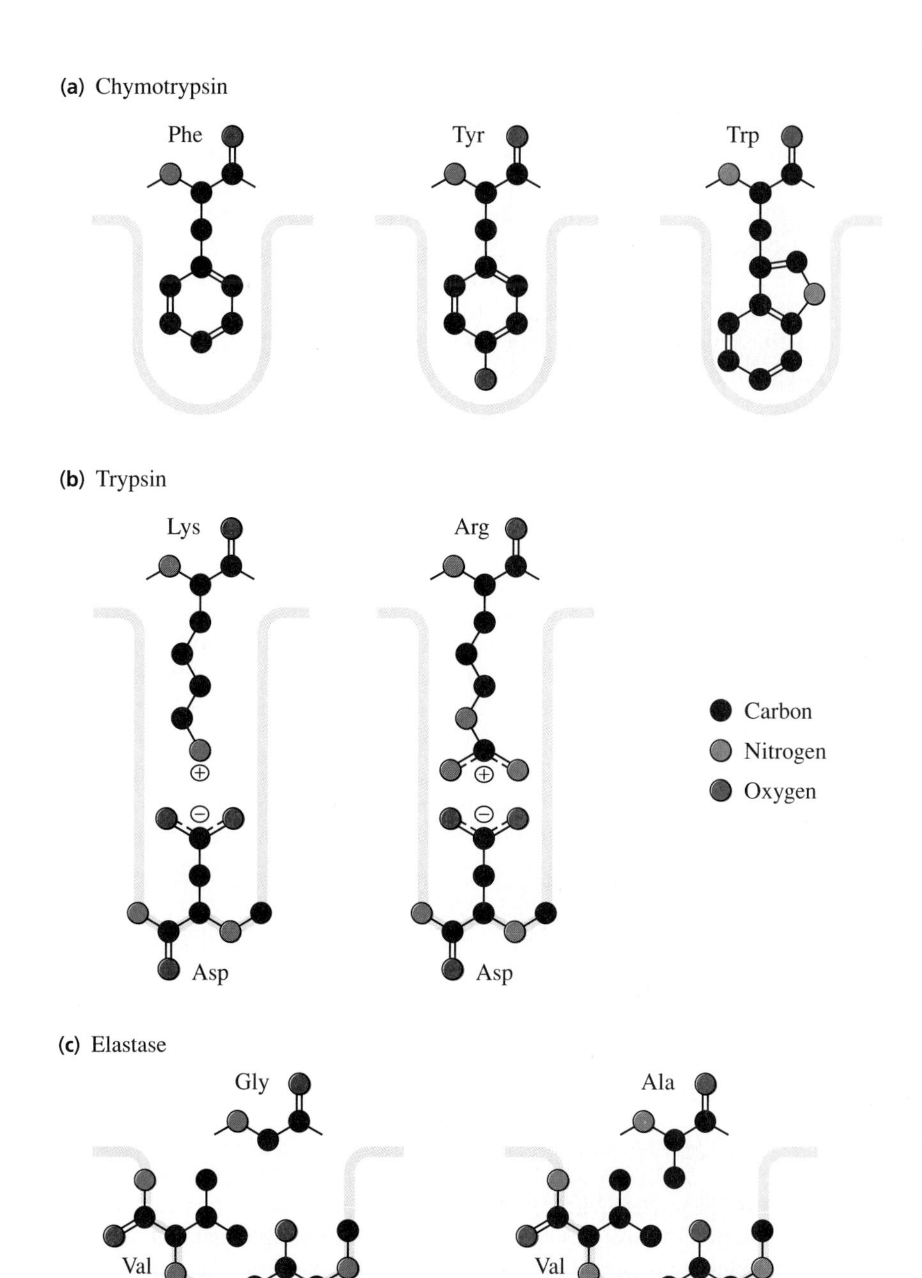

Figure 5·18
Binding sites of chymotrypsin, trypsin, and elastase. The differences in the substrate specificities of chymotrypsin, trypsin, and elastase result from differences in their binding sites. **(a)** Chymotrypsin has a hydrophobic pocket that binds the side chains of aromatic amino acid residues. **(b)** A negatively charged aspartate residue sits at the bottom of the binding pocket of trypsin. As a result, trypsin binds the positively charged side chains of lysine and arginine residues. Because the side chain of lysine is one atom shorter than that of arginine, its ion pair is mediated through a molecule of water (not shown). **(c)** The binding pocket of elastase is more shallow due to the presence of valine and threonine residues. Thus, elastase binds only amino acids with small side chains, especially glycine and alanine. Residues with larger side chains cannot be positioned properly for catalysis.

Elastase, so named because it catalyzes the degradation of elastin, a fibrous protein that is rich in glycine and alanine, specifically cleaves peptide bonds in which the carbonyl group is from a small residue with an uncharged side chain. Its tertiary structure is similar to that of chymotrypsin except that the binding pocket of elastase is much shallower. The binding pocket is smaller because two glycine residues found at the entrance of the binding site of chymotrypsin and trypsin are replaced in elastase by much larger valine and threonine residues (Figure 5·18c). These residues keep potential substrates with large side chains away from the catalytic center.

5·12 Regulatory Enzymes Are Usually Oligomers

Certain enzymes, critically placed in metabolic pathways, have an important property in addition to their catalytic efficiency and specificity: their activity can be regulated reversibly. Generally, allosteric phenomena (Section 4·18) are responsible for control of the activity of these enzymes. Regulatory enzymes are often located at the first step that is unique to a metabolic pathway, called the first committed step of the metabolic sequence. This location allows the complete pathway to be regulated by the activity of the first enzyme. Inhibition of the first enzyme prevents the accumulation of intermediates and the ultimate end product.

In the linear pathway below, which produces P from A, the ultimate product P is an allosteric inhibitor of the regulatory enzyme E_1, which catalyzes the first committed step:

$$A \xrightarrow[(-)]{E_1} B \xrightarrow{E_2} C \xrightarrow{E_3} D \xrightarrow{E_4} \longrightarrow \longrightarrow \longrightarrow P \qquad (5\cdot22)$$

This type of regulation is called *feedback inhibition*. When pathways are branched, regulation may be slightly more complicated. For example, there may be several enzymes catalyzing the first committed step, each one regulated by a different modulator, or complete inhibition of a regulatory enzyme may require binding of several modulators.

We have seen how the conformation of hemoglobin and its affinity for oxygen change when 2,3-*bis*phosphoglycerate is bound (Section 4·18). Regulatory enzymes also undergo allosteric transitions between active R (relaxed) states and inactive T (taut) states. A regulatory enzyme has a second ligand-binding site away from its catalytic center. This second site is called the *regulatory site*. The conformational change associated with the binding of an allosteric inhibitor to its regulatory site is transmitted to the active site of the enzyme, which changes shape sufficiently to alter its activity. The regulatory and catalytic sites are physically distinct regions of the protein, often on separate domains or even on separate subunits.

Examination of regulatory enzymes has shown that they have a number of general features.

1. The activities of regulatory enzymes are sensitive to metabolic inhibitors and activators. These *modulators* or *effectors* seldom resemble the substrates or products of the controlled enzyme. In fact, consideration of the structural differences between substrates and metabolic inhibitors suggested that regulatory compounds are bound to regulatory sites separate from the catalytic sites.

2. Allosteric modulators bind noncovalently to the enzymes they regulate. Some modulators alter the K_m of the enzyme for a substrate, others the V_{max} of the

enzyme. Modulators are not altered chemically by the enzyme. Note, however, that certain regulatory enzymes are controlled by covalent modification of the enzyme itself (Section 5·15).

3. Often, a regulatory enzyme can be partially denatured in vitro with loss or impairment of its regulatory properties but no loss of catalytic activity. This process, called *desensitization,* reinforces the conclusion that catalytic and regulatory events occur at different sites on the enzyme.

4. A regulatory enzyme usually has a v_0 versus [S] curve for at least one substrate that is sigmoidal rather than hyperbolic. A sigmoidal curve is caused by cooperative binding of substrate (Section 4·17), which in turn suggests the presence of multiple binding sites in such an enzyme.

   Figure 5·19 illustrates the regulatory role that cooperative binding can play. Addition of an activator can shift the sigmoidal curve toward a hyperbolic shape, lowering the apparent K_m, the concentration of substrate required for half-saturation, and raising the activity at a given [S]. Addition of an inhibitor can raise the apparent K_m of the enzyme and lower its activity at any particular concentration of substrate.

5. With few exceptions, regulatory enzymes possess quaternary structure. Not all enzymes composed of subunits are regulatory enzymes, however. The individual polypeptide chains of a regulatory enzyme may be identical or different. For those with identical subunits, each polypeptide chain contains both the catalytic and regulatory sites, and the oligomer is a simple symmetric complex—most often a dimer or a tetramer. The oligomers of regulatory enzymes that contain nonidentical subunits have a variety of more complex but still symmetric aggregations.

The allosteric R $\rightleftharpoons$ T transition between the active and the inactive conformations of a regulatory enzyme is rapid. In the simplest cases, substrate and activator molecules bind only to the R state, and inhibitor molecules bind only to the T state. The position of the R-T equilibrium is controlled by the relative affinities and concentrations of the various ligands. Some allosteric inhibitors are competitive inhibitors even though they are not substrate analogs and do not bind at the active site. For example, in the presence of the inhibitor in Figure 5·19, the enzyme has a higher apparent K_m for its substrate but an unaltered V_{max}. Therefore, this allosteric effector is a competitive inhibitor (Section 5·6A).

Some regulatory enzymes exhibit noncompetitive inhibition patterns. Binding of a modulator at the allosteric site does not prevent substrate from binding, but it appears to distort the conformation of the active site sufficiently to decrease the activity of the enzyme.

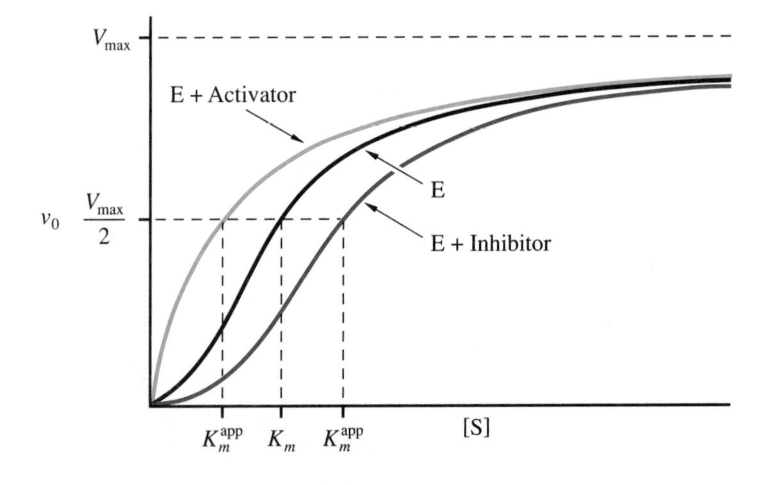

Figure 5·19
The role of cooperativity of binding in regulation. The activity of an allosteric enzyme that has a sigmoidal binding curve can be altered markedly when either an activator or an inhibitor is bound to the enzyme. Addition of an activator can lower the apparent K_m, raising the activity at a given [S]. Conversely, addition of inhibitor can raise the apparent K_m, producing much less activity at any given [S].

5·13 Two Models Have Been Proposed to Describe Allosteric Regulation

Two models that account for the cooperativity of binding of ligands to oligomeric proteins have gained general recognition. Both the *concerted theory* and the *sequential theory* describe the cooperative transitions in simple quantitative terms. The latter is a more general theory; the former is adequate to explain many allosteric enzymes.

The concerted theory, or symmetry-driven theory, was devised to explain the cooperative binding of identical ligands, such as substrates. It supposes that there is one binding site per subunit for each ligand, that the conformation of each subunit is constrained by its association with other subunits, and that when the protein changes conformation, it retains its molecular symmetry (Figure 5·20a). Thus, there are two conformations in equilibrium, R (with a high affinity for substrate) and T (a low-affinity state). The binding of substrate shifts the equilibrium. When the conformation of the protein changes, the affinity of its ligand-binding sites also changes. The theory was extended to include the binding of allosteric modulators, and it can be simplified quantitatively by assuming that S binds only to the R state and I binds only to the T state. The concerted theory is based on the observed structural symmetry of regulatory enzymes. It suggests that all subunits of a given protein molecule have the same conformation; that is, they retain their symmetry. Experimental data obtained with a number of enzymes can be explained by this simple theory.

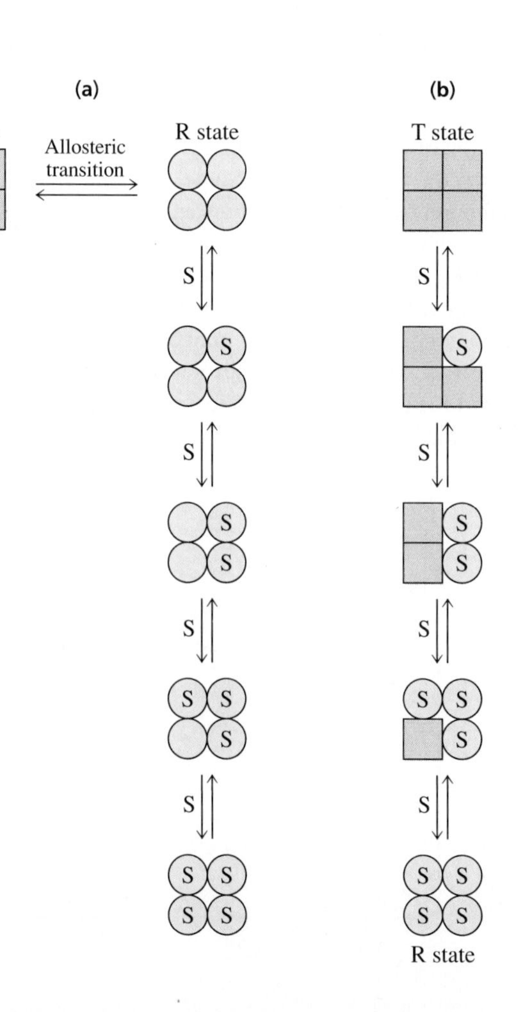

Figure 5·20
Two models for cooperativity of binding of substrate (S) to a tetrameric protein. **(a)** In the concerted model, subunits are all either R or T, and S binds only to the R state. **(b)** In the sequential model, binding of S to a subunit converts only that subunit to the R conformation. Neighboring subunits might remain in the T state or might assume conformations between T and R.

The sequential theory, or ligand-induced theory, is a later and more general proposal. It is based on the idea that binding of a ligand may induce a change in the tertiary structure of the subunit to which it binds. This subunit-ligand complex may change the conformations of neighboring subunits to varying extents. This theory too assumes that only one shape has high affinity for the ligand, but it differs from the concerted theory in allowing the existence of both high- and low-affinity subunits in an oligomeric molecule that has fractional saturation (Figure 5·20b). The sequential theory can account for negative cooperativity, a decrease in affinity as subsequent molecules of ligand are bound to an oligomer. Negative cooperativity occurs with a relatively small number of enzymes. The sequential theory treats the concerted theory, with its symmetry requirement, as a limiting and simple case.

5·14 Aspartate Transcarbamoylase Was the First Allosteric Enzyme to Be Thoroughly Characterized

Aspartate transcarbamoylase (ATCase) from the bacterium *Escherichia coli* is the most thoroughly studied regulatory enzyme. In *E. coli,* ATCase catalyzes the first committed step of the biosynthesis of pyrimidine nucleotides (Figure 5·21). (Nucleotide biosynthesis is discussed in detail in Chapter 19.) This reaction forms carbamoyl aspartate from carbamoyl phosphate and aspartate. The nucleophilic amino group of aspartate attacks the carbonyl group of carbamoyl phosphate, releasing the phosphate and forming a compound that can cyclize to form the six-membered pyrimidine ring. Arthur Pardee and his co-workers showed that the most potent metabolic inhibitor of ATCase is the end product of the pathway, cytidine triphosphate (CTP). The same workers also found that ATP is an activator of ATCase. ATP

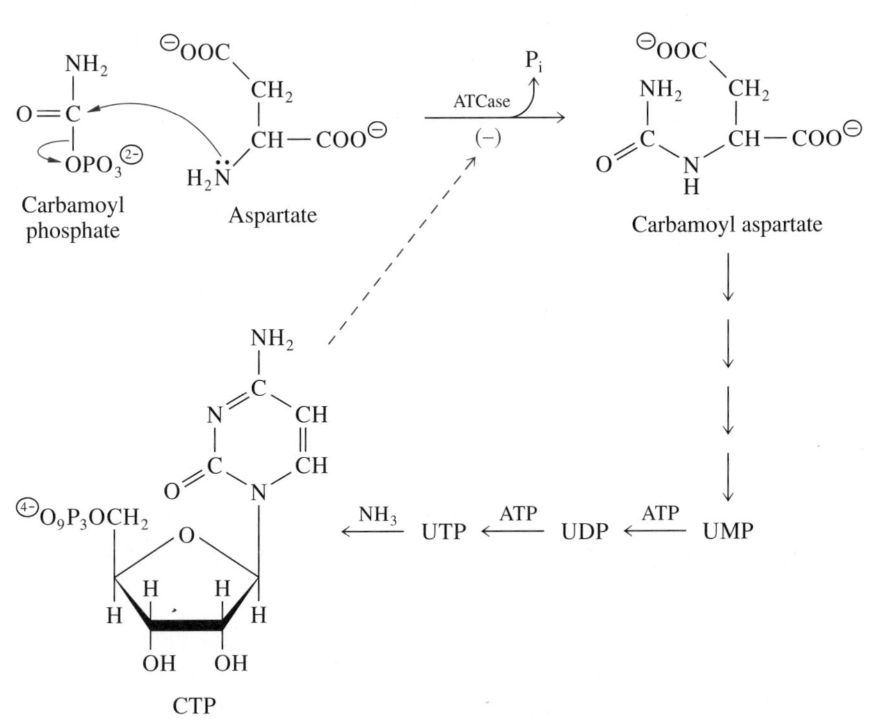

Figure 5·21
Abbreviated pathway for biosynthesis of pyrimidine nucleotides. Aspartate transcarbamoylase (ATCase) catalyzes the formation of carbamoyl aspartate from carbamoyl phosphate and aspartate. Carbamoyl aspartate is converted to the nucleotides UMP, UDP, and UTP, and the ultimate product, CTP. Aspartate and carbamoyl aspartate have been drawn in a configuration that corresponds to the position of the atoms in the pyrimidine ring that is formed.

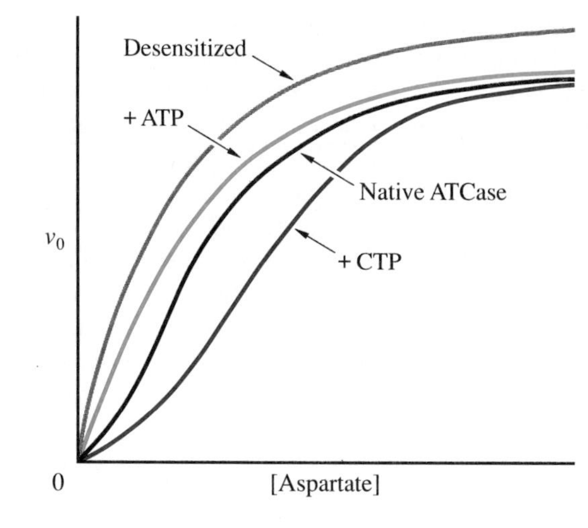

Figure 5·22
Allosteric regulation of ATCase from *E. coli.* ATP is an activator and CTP is an inhibitor of the native enzyme. The desensitized ATCase is unaffected by addition of ATP or CTP.

is an end product of the pathway by which purine nucleotides are synthesized. Activation of ATCase by ATP presumably enhances the production of pyrimidine nucleotides to balance the supply of both types of nucleotides needed for synthesis of nucleic acids. Both CTP and ATP affect the binding of the substrate aspartate. The velocity versus [aspartate] plot for the ATCase reaction is sigmoidal (Figure 5·22). CTP raises the apparent K_m for aspartate without changing the V_{max} of the enzyme; thus, CTP is a competitive inhibitor, one that binds away from the active site. CTP makes the curve more sigmoidal, indicating greater cooperativity in the binding of aspartate to ATCase. The presence of ATP shifts the curve toward a hyperbolic shape, decreasing the cooperativity of binding to the enzyme.

Several methods of mild denaturation or chemical modification desensitize ATCase. The selective destruction of the regulatory activity is accompanied by conversion of the sigmoidal aspartate-saturation curve for ATCase to a normal hyperbola (the upper curve in Figure 5·22). In addition, there is a slight increase in the activity of ATCase, as though some physical restraint were removed.

Studies of purified ATCase have revealed both the quaternary structure and the path of desensitization of the enzyme (Figure 5·23a). The native form of ATCase has a molecular weight of 310 000. ATCase is desensitized by treatment with *p*-hydroxymercuribenzoate, a compound that reacts specifically with thiol groups (Figure 5·23b). In ATCase, each of six zinc ions is complexed to four cysteine residues. The Zn^{2+} ions stabilize a conformation necessary for proper subunit interaction. Treatment with *p*-hydroxymercuribenzoate releases the Zn^{2+} and results in the formation of subunits of two sizes, having molecular weights of about 100 000 and 34 000. The larger subunit possesses all of the catalytic activity of ATCase and therefore is called the catalytic subunit. It is not inhibited by and does not bind CTP. Only the smaller subunit, the regulatory subunit, can bind CTP or ATP. When analyzed by SDS-polyacrylamide gel electrophoresis (Section 3·5), the components of the catalytic subunit migrate as one chain with a molecular weight of 33 000. The regulatory subunit, when similarly tested, is denatured to a single polypeptide chain of molecular weight 17 000. By considering the molecular weights of the chains and the amounts of protein in each subunit fraction, it was deduced that the native form of ATCase contains 12 polypeptide chains: six catalytic (C) chains and six regulatory (R) chains, arranged as two trimers of C chains and three dimers of R chains. The quaternary structure of ATCase is summarized in Table 5·6 and illustrated in Figure 5·24. Each chain of a catalytic trimer is connected to a chain of the second catalytic trimer through a regulatory dimer.

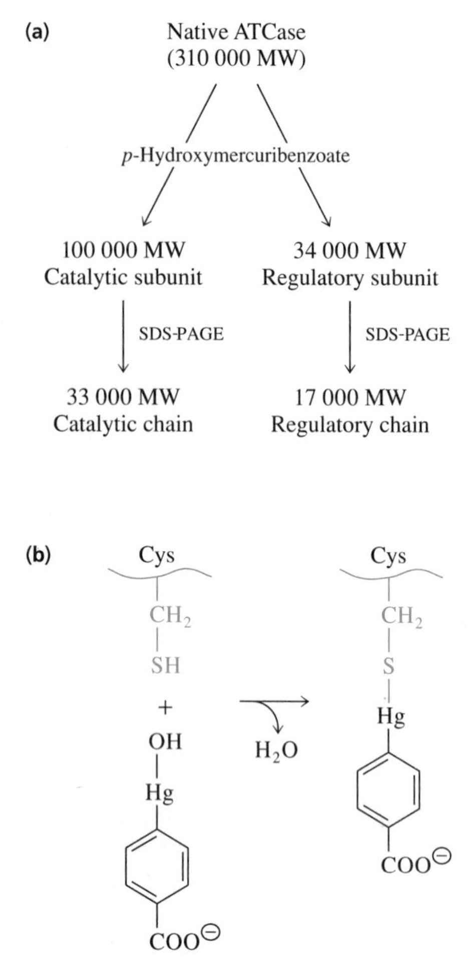

Figure 5·23
Quaternary structure of ATCase. **(a)** The stoichiometry of the quaternary structure of ATCase was arrived at by determining the molecular weights of the protein fragments formed after desensitization and after complete denaturation. **(b)** ATCase is desensitized by *p*-hydroxymercuribenzoate, which reacts with the thiol groups of cysteine residues.

The most important finding from studies of ATCase is that the active site is physically distinct from the regulatory site but coupled to it. In this case, the regulatory and catalytic sites are even located on different polypeptide chains. Recent

Table 5·6 Quaternary structure of aspartate transcarbamoylase from *E. coli*

	Molecular weight	Number of chains	Composition
Catalytic chain	33 000	1	C
Catalytic subunit	100 000	3	C_3
Regulatory chain	17 000	1	R
Regulatory subunit	34 000	2	R_2
Native enzyme	310 000	12	$(C_3)_2(R_2)_3$

X-ray crystallographic studies by William Lipscomb and his colleagues on ATCase and several enzyme-ligand complexes have shown that the active site is composed of amino acids from two adjacent C chains of the catalytic subunit; that is, the active sites are at the interfaces between C chains. They have also shown that the T ⟶ R transition involves movement of one trimer away from the other, with the CRRC interchain interactions being maintained during this conformational change. They suggest that binding of the first molecule of aspartate, in the presence of the second substrate, carbamoyl phosphate, causes all six of the catalytic chains to move into a conformation having high affinity for substrate and increased catalytic activity. Thus, ATCase is converted from the T to R state in a concerted manner. The two rigid domains of each catalytic chain move closer together so that the active site in the R form is partially closed, and thus more suitable for reaction, compared to its shape in the T form.

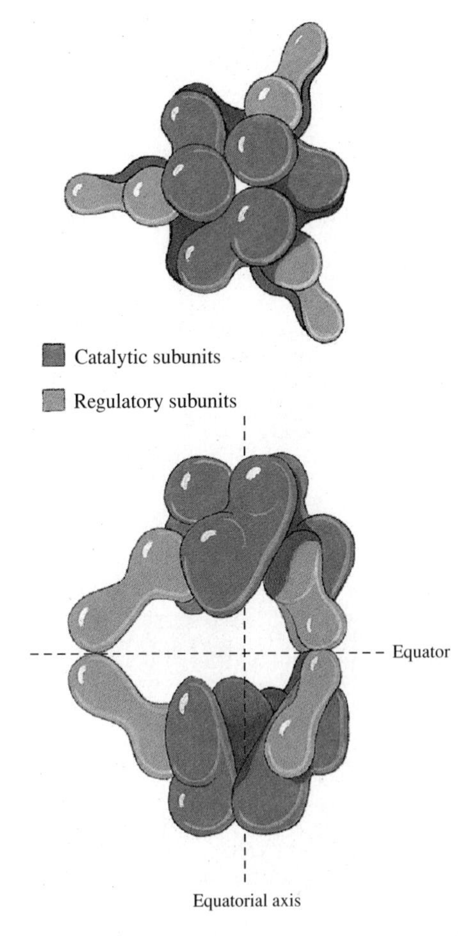

Figure 5·24
Structure of ATCase from *E. coli*. ATCase contains both catalytic and regulatory subunits. The two catalytic subunits are stacked and nearly eclipsed trimers that are held together by three regulatory dimers. Only two regulatory dimers are shown in the lower view. Each regulatory dimer is bound to two catalytic chains. Note that each chain, both catalytic and regulatory, is composed of two domains. (Illustrator: Lisa Shoemaker.) [Adapted from Krause, K. L., Volz, K. W., and Lipscomb, W. N. (1985). Structure at 2.9-Å resolution of aspartate carbamoyltransferase complexed with the bisubstrate analogue *N*-(phosphonacetyl)-L-aspartate. *Proc. Natl. Acad. Sci. U.S.A.* 82:1643–1647.]

5·15 Some Regulatory Enzymes Undergo Covalent Modification

In addition to simple allosteric interactions, there are two other ways that enzyme activity can be regulated. The amount of an enzyme can be controlled by regulating the rate of its synthesis or degradation. This type of control is much slower than the allosteric R ⇌ T transition. For some enzymes, an intermediate rate of control is provided by covalent modification of the enzyme. Regulation by covalent modification usually requires one enzyme for activation and another for inactivation. The activating/deactivating enzymes may be the end components of longer regulatory sequences.

Enzymes controlled by covalent modification, termed *interconvertible enzymes*, are believed to generally undergo R ⇌ T transitions. They are frozen in one conformation or the other by a covalent substitution. The substitution reaction is catalyzed by an accessory enzyme called a *converter enzyme*, and the release of the substituent is catalyzed by another converter enzyme (Figure 5·25a, next page). Converter enzymes are usually controlled by allosteric effectors, although in some cases, converter enzymes within a regulatory sequence are themselves subject to covalent modification, with the entire sequence under allosteric control.

The most common type of covalent modification is phosphorylation of a specific serine residue of the interconvertible enzyme. An enzyme called a protein kinase catalyzes the transfer of the terminal phosphoryl group from ATP to the appropriate serine residue of the regulated enzyme. The activity of the protein kinase is itself regulated. The phosphoserine of the interconvertible enzyme is hydrolyzed by the activity of a protein phosphatase, releasing phosphate and returning the interconvertible enzyme to its dephosphorylated state. Individual enzymes differ as to

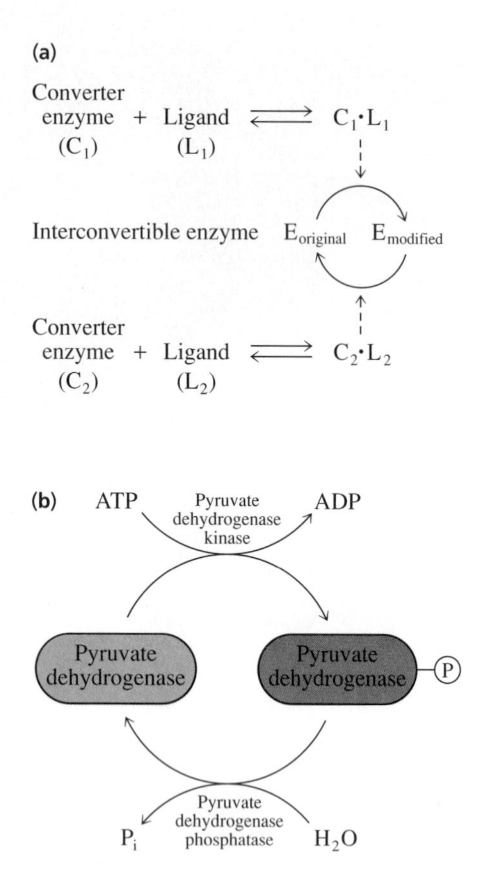

Figure 5·25
Pathway of covalent modification of interconvertible enzymes. **(a)** The general scheme for covalent modification utilizes two allosterically controlled converter enzymes, C_1 and C_2, which catalyze modifications of the interconvertible enzyme, E, the substrate of the covalent substitution. The interconvertible enzyme exists in two forms, original and modified. C_1 and C_2 are activated when they bind their respective modulator ligands, L_1 and L_2. **(b)** Control of the activity of pyruvate dehydrogenase is an example of regulation by covalent modification. Phosphorylation of pyruvate dehydrogenase, the interconvertible enzyme, is catalyzed by pyruvate dehydrogenase kinase. Pyruvate dehydrogenase is inactivated by phosphorylation. It is reactivated by hydrolysis of its phosphoserine, catalyzed by an allosteric hydrolase called pyruvate dehydrogenase phosphatase.

whether it is their phospho- or dephospho- forms that are active. The reactions involved in the regulation of pyruvate dehydrogenase by covalent modification are shown in Figure 5·25b. Pyruvate dehydrogenase catalyzes a key step in the breakdown of glucose, the decarboxylation of pyruvate leading to the formation of acetyl coenzyme A and CO_2. This reaction connects the pathway of glycolysis to the citric acid cycle. Phosphorylation of pyruvate dehydrogenase, catalyzed by the allosteric enzyme pyruvate dehydrogenase kinase, inactivates the dehydrogenase. The kinase can be activated by any of several metabolites. Phosphorylated pyruvate dehydrogenase is reactivated under different metabolic conditions by hydrolysis of its phosphoserine residue, catalyzed by pyruvate dehydrogenase phosphatase. This phosphatase is activated by Ca^{2+}.

The three-dimensional structures of the phospho- and dephospho- forms of glycogen phosphorylase and isocitrate dehydrogenase, determined by X-ray crystallography, demonstrate two different ways in which covalent modification can alter catalytic activity. A specific kinase catalyzes the phosphorylation of Ser-14 of the inactive glycogen phosphorylase *b* to form active glycogen phosphorylase *a*. Phosphorylation causes a previously disordered segment of the protein to become helical, a shape stabilized by ion pairs formed between the phosphate and two arginines and by the formation of several new hydrogen bonds. The conformational changes that occur after the covalent modification of glycogen phosphorylase alter the sites where the substrates and modulators are bound. Phosphorylation of this enzyme produces, as expected, a T $\longrightarrow$ R transition.

An alternative mechanism for regulation by phosphorylation has been found. When isocitrate dehydrogenase from *E. coli* is inactivated by phosphorylation, the binding of its substrate, isocitrate, is prevented. The one covalently modified residue of the enzyme is a serine that, in the unsubstituted enzyme, forms a hydrogen bond with a carboxylate group of isocitrate. When the enzyme is covalently modified, the hydrogen bond cannot be formed, and there is electrostatic repulsion between the carboxylate group and phosphate. X-ray crystallography has shown that the shape of the enzyme does not change appreciably upon covalent modification. Thus, kinase-catalyzed covalent modification is known at both regulatory and catalytic sites of interconvertible enzymes.

5·16 Multienzyme Complexes and Multifunctional Enzymes Have Increased Efficiency

In a number of cases, enzymes that catalyze sequential reactions in the same metabolic pathway have been found to be physically associated. This phenomenon offers several potential advantages. Enzyme complexes allow *channelling* of reactants between active sites—the product of one reaction can be transferred to the next active site without entering the bulk solvent. Channelling can speed up the reactions of the pathway and protect unstable intermediates. There are several ways in which enzymes are associated. For some, several active sites are found on a single, multifunctional polypeptide chain. The fusion of related enzyme activities on one protein allows organisms to regulate a series of reactions by coordinating the expression of a single gene. In some pathways, several individual enzymes are noncovalently but fairly tightly associated. Several examples of these *multienzyme complexes* will be encountered later in this text. Enzymes of some other pathways are associated by weaker interactions, mainly hydrophobic. Because such complexes dissociate easily, it has been difficult to demonstrate their association and only a few have been incontrovertibly established. Attachment to membranes is another way that enzymes may be associated. Aggregations of enzyme proteins will be encountered a number of times in later chapters. The search for enzyme complexes and the evaluation of their catalytic and regulatory roles is an extremely active area of research in enzyme chemistry.

Summary

Enzymes, the catalysts of living organisms, are remarkable for their catalytic efficiency and their substrate and reaction specificity. With the exception of certain catalytic RNA molecules, enzymes are proteins, or proteins plus auxiliary compounds or ions called cofactors. Enzymes can be grouped into six major classes according to the nature of the reactions they catalyze: oxidoreductases (dehydrogenases), transferases, hydrolases, lyases, isomerases, and ligases (synthetases).

Chemical kinetic experiments involve the systematic variation of reaction conditions, especially the concentration of substrate, and measurement of the alteration in the rate of formation of the product. Enzyme kinetic measurements are usually made by measuring initial velocities, which are obtained from progress curves that show the amount of product formed over time. The first step in an enzyme-catalyzed reaction is the formation of a noncovalent enzyme-substrate complex. As a result, enzymatic reactions are characteristically first-order with respect to enzyme concentration and typically show hyperbolic dependence on substrate concentration. Maximum velocity (V_{max}) is reached when the substrate concentrations are saturating. The Michaelis-Menten equation describes such kinetic behavior. The Michaelis constant (K_m) is equal to the concentration of substrate that gives half-maximum reaction velocity—that is, half saturation of E with S. Values for the kinetic constants K_m and V_{max} can be determined by computer analysis of kinetic data or they can be estimated graphically. The double-reciprocal or Lineweaver-Burk plot of $1/v_0$ versus 1/[S] represents a linear transformation of the hyperbolic Michaelis-Menten equation.

The catalytic constant, k_{cat}, or turnover number for an enzyme, is the maximum number of molecules of substrate that can be transformed into product per molecule of enzyme (or per active site) per second and thus is equal to V_{max} divided by enzyme concentration. The unit for k_{cat} is s^{-1}. The ratio k_{cat}/K_m is an apparent second-order rate constant that governs the reaction of an enzyme with a substrate in dilute, nonsaturating solutions. Its value can provide a measure of the catalytic efficiency and substrate specificity of an enzyme.

The rates of enzyme-catalyzed reactions are also affected by the presence of inhibitors. Enzyme inhibitors may be classified as reversible or irreversible. Reversible inhibitors are bound noncovalently by the enzyme. They can be competitive (those that increase the apparent value of K_m, with no change in V_{max}), uncompetitive (those that decrease K_m and V_{max} proportionally), or noncompetitive (those that decrease V_{max}, with either no change in K_m—pure noncompetitive—or an increase or decrease in K_m—mixed noncompetitive). Inhibition studies can be used to determine the kinetic scheme of multisubstrate reactions (whether they are ordered, random, or ping-pong) and to determine the order of binding of substrates to an enzyme and the order of release of products.

Irreversible inhibitors form covalent bonds with the enzyme. By treating an enzyme with an irreversible inhibitor and then sequencing a segment of the protein, it is often possible to determine the identity of reactive amino acid residues, which can contribute to deciphering catalytic mechanisms. Representing another approach to investigating enzymes, including their catalytic mechanisms, site-directed mutagenesis can be used to change the identity of a single amino acid residue within a protein, generating a new protein whose properties may be revealing with respect to the role of the original amino acid residue.

Many proteases are synthesized as inactive zymogens that are activated extracellularly under appropriate conditions by selective proteolysis, the specific enzymatic cleavage of one or a few peptide bonds. Enteropeptidase catalyzes cleavage of trypsinogen to form the active enzyme, trypsin. Trypsin, in turn, activates other pancreatic zymogens, such as chymotrypsinogen. Blood clotting depends on a cascade of zymogen activation involving over a dozen coagulation factors. Such cascades provide enormous signal amplification.

Regulation of metabolism is often provided by allosteric modulators that act on certain key oligomeric enzymes. Allosteric modulators bind to a site other than the active site. They alter either the apparent K_m or V_{max} values of regulatory enzymes through control of the R $\rightleftharpoons$ T equilibrium. Aspartate transcarbamoylase catalyzes the first committed step in the biosynthesis of pyrimidine nucleotides by *E. coli*. Its control properties have been elucidated at the molecular level. Covalent modification of certain regulatory enzymes provides a control mechanism that is slower than allosteric interactions but more rapid than control achieved by changes in the concentration of enzyme, which is regulated at the level of gene expression.

Selected Readings

Enzyme Catalysis

Boyer, P. D., ed. (1970–1990). *The Enzymes.* Vols. 1, 2, 3, and 19, 3rd ed. (New York: Academic Press). These are the volumes containing general topics, from a comprehensive series of reviews.

Fersht, A. (1985). *Enzyme Structure and Mechanism.* 2nd ed. (New York: W. H. Freeman and Company). A description of the function of enzymes, with many original proposals.

Lipscomb, W. N. (1983). Structure and catalysis of enzymes. *Annu. Rev. Biochem.* 52:17–34.

Scrimgeour, K. G. (1977). *Chemistry and Control of Enzyme Reactions.* (London: Academic Press). An outline of the principles of modern enzymology.

Enzyme Kinetics

Cornish-Bowden, A., and Wharton, C. W. (1988). *Enzyme Kinetics.* (Oxford: IRL Press). A brief outline of enzyme kinetics.

Piszkiewicz, D. (1977). *Kinetics of Chemical and Enzyme-Catalyzed Reactions.* (New York: Oxford University Press). An introductory text on kinetics.

Segal, H. L. (1959). The development of enzyme kinetics. In *The Enzymes,* Vol. 1, 2nd ed., P. D. Boyer, H. Lardy, and K. Myrbäck, eds. (New York: Academic Press). pp. 1–48. A history of the development of enzyme kinetics.

Segel, I. H. (1975). *Enzyme Kinetics: Behavior and Analysis of Rapid Equilibrium and Steady State Enzyme Systems.* (New York: Wiley Interscience). A comprehensive description of kinetic behavior and analysis.

Wong, J. T.-F. (1975). *Kinetics of Enzyme Mechanisms.* (London: Academic Press). An advanced mathematical examination of kinetic mechanisms.

Site-Directed Mutagenesis

Johnson, K. A., and Benkovic, S. J. (1990). Analysis of protein function by mutagenesis. In *The Enzymes,* Vol. 19, 3rd ed., P. D. Boyer, ed. (New York: Academic Press). pp. 159–211.

Serine Proteases

Light, A., and Janska, H. (1989). Enterokinase (enteropeptidase): comparative aspects. *Trends Biochem. Sci.* 14:110–112.

Navia, M. A., McKeever, B. M., Springer, J. P., Lin, T.-Y., Williams, H. R., Fluder, E. M., Dorn, C. P., and Hoogsteen, K. (1989). Structure of human neutrophil elastase in complex with a peptide chloromethyl ketone inhibitor at 1.84-Å resolution. *Proc. Natl. Acad. Sci. USA* 86:7–11.

Neurath, H. (1984). Evolution of proteolytic enzymes. *Science* 224:350–357.

Steitz, T. A., and Shulman, R. G. (1982). Crystallographic and NMR studies of the serine proteases. *Annu. Rev. Biophys. Bioeng.* 11:419–444.

Zymogen Activation

Huber, R., and Bode, W. (1978). Structural basis of the activation and action of trypsin. *Acc. Chem. Res.* 11:114–122.

Sharma, S. K., and Hopkins, T. R. (1981). Recent developments in the activation process of bovine chymotrypsinogen A. *Bioorg. Chem.* 10:357–374.

Regulatory Enzymes

Barford, D. (1991). Molecular mechanisms for the control of enzymic activity by protein phosphorylation. *Biochim. Biophys. Acta* 1133:55–62.

Edelman, A. M., Blumenthal, D. K., and Krebs, E. G. (1987). Protein serine/threonine kinases. *Annu. Rev. Biochem.* 56:567–613. Properties of the kinases that catalyze phosphorylation of interconvertible enzymes.

Hurley, J. H., Dean, A. M., Sohl, J. L., Koshland, D. E., Jr., and Stroud, R. M. (1990). Regulation of an enzyme by phosphorylation at the active site. *Science* 249:1012–1016. Control of the activity of isocitrate dehydrogenase by covalent modification of a binding residue.

Kantrowitz, E. R., and Lipscomb, W. N. (1988). *Escherichia coli* aspartate transcarbamylase: the relation between structure and function. *Science* 241:669–674.

6

Mechanisms of Enzymes

In this chapter, we examine how enzymes catalyze reactions. This involves studying their *mechanisms,* the atomic or molecular events that occur during reactions. Individual enzyme mechanisms have been deduced from kinetic experiments and more recently from protein structural studies, especially X-ray crystallography. The general processes of catalysis by enzymes have been elucidated by integrating mechanistic information derived from individual enzymes with studies of suitable nonenzymatic model reactions. A complete explanation of catalysis by enzymes requires quantitative methods. Mechanisms are evaluated by interpretation of the kinetic constants k_{cat} and K_m.

6·1 Enzymatic Mechanisms Are Described with the Terminology of Chemistry

The same symbolism used in organic chemistry to represent chemical-bond transformation is employed in representing enzymatic mechanisms. With the short review of terms given in this section and the specific points mentioned with individual mechanisms, you will be able to understand all of the enzyme-catalyzed reactions that are presented throughout this book.

Since covalent chemical bonds consist of pairs of electrons shared between two atoms, bonds can be cleaved in two ways. In most reactions, both electrons stay with one atom, so that an ionic intermediate and a *leaving group* are formed. In ionic reactions, two species are involved: one is electron-rich, or *nucleophilic,* and the other electron-poor, or *electrophilic* (Section 2·6). A nucleophile, which has a negative charge or unshared electron pair, attacks the electrophilic center of the other reactant. The convention in mechanistic chemistry is to show a curved arrow pointing from a nucleophile to the electron-deficient position of an electrophile. Hence, transfer of an acyl group can be written as this general mechanism:

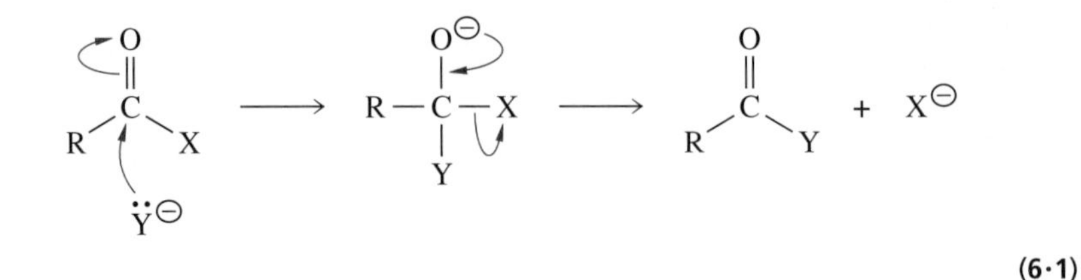

(6·1)

The nucleophile $Y^{\ominus}$ attacks the carbonyl carbon (adds to the C=O bond) to form a tetrahedral addition intermediate from which the group $X^{\ominus}$ is eliminated. $X^{\ominus}$ is the leaving group. In hydrolysis of an amide bond, the nucleophile is $OH^{\ominus}$ from water and $X^{\ominus}$ is an amine. Several examples of this type of mechanism are described in this chapter.

Ionization of carbons involved in covalent bonds occurs in two ways. If a carbon atom retains both electrons, a *carbanion* is produced:

$$R_3C{-}X \rightleftharpoons R_3C{:}^{\ominus} + X^{\oplus} \qquad (6\cdot2)$$

If the bond broken in Equation 6·2 is a C—H bond, then a proton is released, and the carbanion is the conjugate base of the organic compound. Formation of a carbanion is the most common pathway for cleavage of carbon bonds in biology.

If the carbon atom loses both electrons, a cationic *carbonium ion,* sometimes called a carbocation, is formed:

$$R_3C{-}Y \rightleftharpoons R_3C^{\oplus} + {:}Y^{\ominus} \qquad (6\cdot3)$$

If Y in Equation 6·3 is a hydrogen atom, then $Y^{\ominus}$ is a hydride ion, $H^{\ominus}$ (a proton accompanied by two electrons). Many oxidation reactions proceed by this mechanism. Substitution reactions can occur this way, with $Y^{\ominus}$ (the leaving group) being replaced by another anion. In this substitution mechanism, the ionization step is slow compared to addition of the other anion. Therefore, the overall rate of substitution is first-order because it depends on the concentration of only *one* compound, the original reactant.

Another type of substitution, the direct displacement mechanism, involves *two* molecules reacting in the initial slower step. The rate of this type of reaction depends on the concentrations of both reactants. The attacking group or molecule adds to the face of the carbon atom opposite from the leaving group to form a transition state that has five groups attached to the central carbon atom. The *transition state,* shown below in square brackets, is an unstable high-energy state. It has a structure between that of the reactant and the product:

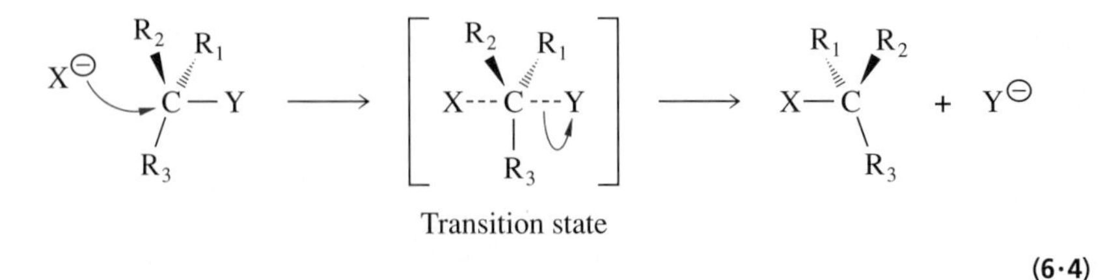

(6·4)

In the less common type of bond cleavage, one electron remains with each product to form two free radicals that are usually very unstable:

$$RO{-}OR_1 \longrightarrow RO\cdot + \cdot OR_1 \qquad (6\cdot5)$$

Oxidation-reduction reactions are central to the supply of biological energy. Oxidations can take several forms: addition of oxygen, removal of hydrogen, or removal of electrons. In the oxidation of formaldehyde to formic acid, an atom of oxygen is added:

$$\mathrm{H{-}C({=}O){-}H} + \tfrac{1}{2}\,O_2 \longrightarrow \mathrm{H{-}C({=}O){-}OH} \qquad \textbf{(6·6)}$$

In the oxidation of lactate to pyruvate (Table 5·1), two hydrogen atoms are removed from the alcohol group of lactate. Mechanistically, they are removed as a proton and a hydride ion. Dehydrogenations of this type are the most common biological oxidation-reduction reactions. In the third type of oxidation, the valence of metal ions is increased by removal of electrons. For example, the hemoprotein cytochrome *c* undergoes cyclic oxidation and reduction between its reduced ferrous (Fe^{2+}) form and its oxidized ferric (Fe^{3+}) form.

6·2 Catalysts Stabilize Transition States

The rate of a chemical reaction depends on how effectively reactants collide to form a transition state. For the collisions to lead to product, the colliding substances must be in the correct orientation and must possess sufficient energy to approach the physical configuration of the atoms and bonds of the product. The transition state is an unstable, energized arrangement of atoms in which chemical bonds are being formed or broken. The energy required to reach the transition state from the *ground state* of the reactants is called the *energy of activation* of the reaction and is often referred to as the activation barrier. The progress of a reaction is usually represented by a graph called a *reaction diagram*. An example showing conversion of a substrate to a product in a single step is given in Figure 6·1. The *y* axis shows the energies of the reacting species. The *x* axis, called the reaction coordinate, measures the progress of the reaction. The transition state occurs at the peak of the activation barrier.

Transition states are not yet detectable experimentally, but their structures, which are somewhat similar to those of unstable intermediates, can be predicted.

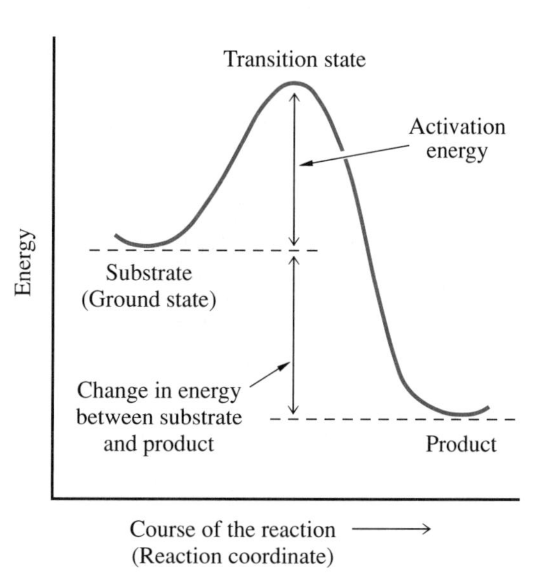

Figure 6·1
Reaction diagram for a single-step reaction. This curve shows the lowest energy path between the substrate and the product. The upper arrow shows the energy of activation for the forward reaction. Molecules of substrate that have more energy than the activation energy pass over the activation barrier and become molecules of product.

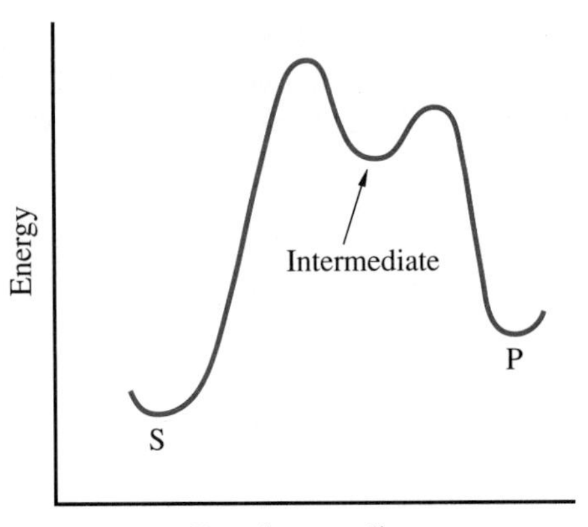

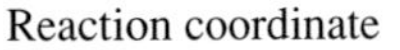

Figure 6·2
Reaction diagram showing a reaction with an intermediate. The metastable intermediate occurs in the trough between the two transition states. The rate-determining step in the forward direction is formation of the first transition state, the step that has the higher activation energy.

Like unstable intermediates, the structures of transition states are often drawn inside square brackets. Often the structures drawn for transition states indicate bonds that are in the process of forming or breaking.

Intermediates differ from transition states. They are compounds that are sufficiently stable to be detected or isolated. If there is an intermediate in a reaction, the energy diagram will have a trough (Figure 6·2). There would then be two transition states in the reaction, one preceding formation of the intermediate and one preceding its conversion to product. The slowest or *rate-determining step* in the formation of the product P from the substrate S is the step with the highest activation energy. In Figure 6·2, it is the formation of the intermediate. Relatively little energy is required for the intermediate to either continue to product or revert to the original reactant.

Catalysts provide reaction pathways that have lower activation energies than uncatalyzed reactions. An enzyme lowers the overall activation energy by providing a multistep pathway, each step of which has a lower energy of activation than corresponding stages in the nonenzymatic reaction.

The first step in an enzymatic reaction is the formation of a noncovalent enzyme-substrate complex, ES. When reactants bind to enzymes, the reactants lose

Figure 6·3
Enzymatic catalysis of the reaction A + B ⟶ A-B. **(a)** Reaction diagram for an uncatalyzed reaction. **(b)** By suitably binding A and B in its active site to form the EAB complex, the enzyme has brought the two reactants closer to the transition state, making formation of the transition state more frequent. As shown by comparison of the activation arrows in (a) and (b), the collecting of reactants lowers the energy of activation compared to that for the uncatalyzed reaction of A with B. **(c)** In addition to the effect in (b), an enzyme binds the transition state more tightly than it binds substrates, further lowering the energy of activation. Thus, the enzymatic reaction has a much lower activation energy than the uncatalyzed reaction.

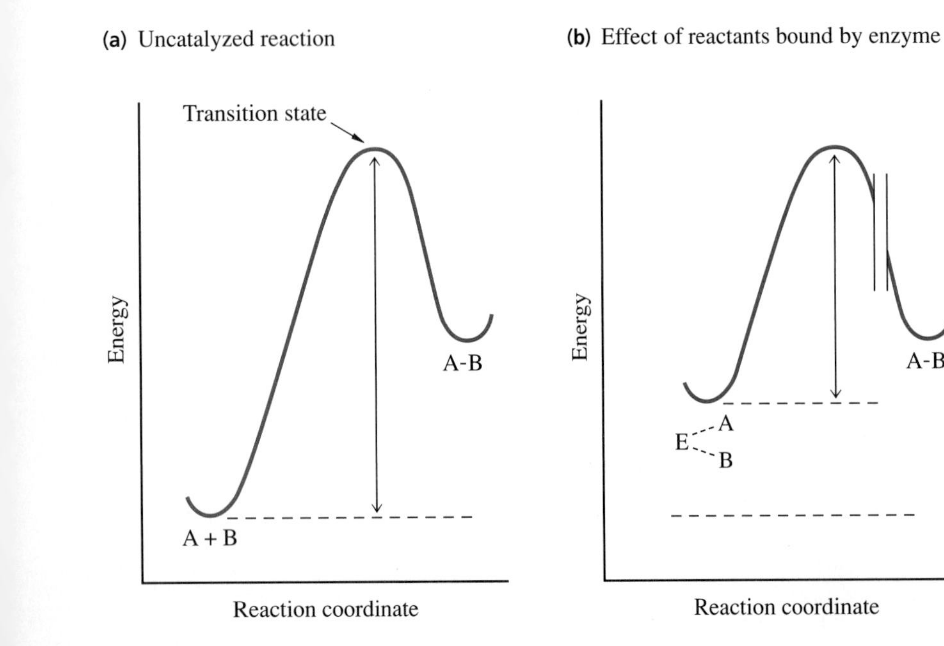

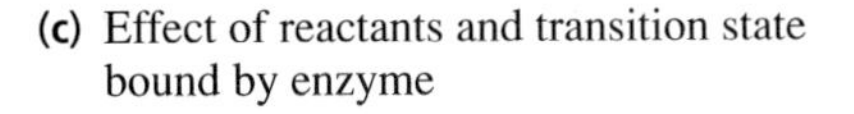

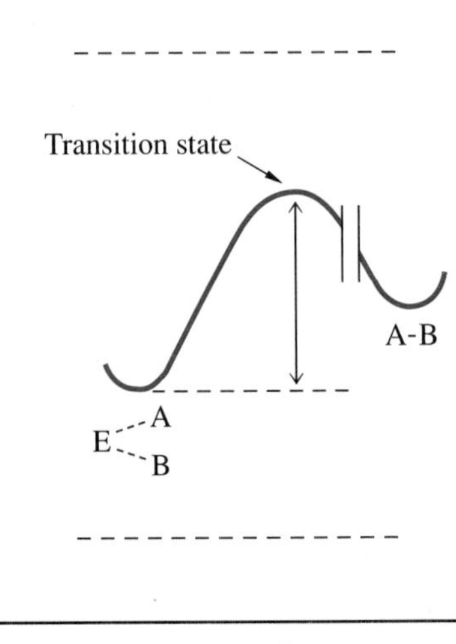

a great deal of entropy compared to their entropy in free solution. In a reaction between A and B, formation of the EAB complex makes the probability of reaction much higher for the enzyme-catalyzed than for the uncatalyzed reaction (Figure 6·3a and b). Bringing the reactants together accounts for a large part of the catalytic power of enzymes. In addition, the active sites of enzymes bind not only substrates and products but also transition states. In fact, transition states bind to active sites much more tightly than do substrates. The extra binding interactions stabilize the transition state, further lowering the energy of activation (Figure 6·3c). As discussed in Sections 6·8 to 6·11, the binding by enzymes first of substrates and then of transition states provides the greatest effect in enzyme catalysis.

6·3 Polar Amino Acids Form the Catalytic Centers of Enzymes

After formation of the ES complex, active-site amino acid residues act as chemical catalysts. We shall now examine the chemical processes that contribute to the function of enzymes. Structural studies have verified that, although the active-site cavity of an enzyme—as part of the interior of the protein—is lined with hydrophobic amino acids, a few polar, ionizable amino acids (and a few molecules of water) are usually also found in the cleft. The polar amino acids that undergo chemical changes during enzymatic catalysis are called the *catalytic center* of the enzyme.

Table 6·1 lists the ionizable and reactive residues found in active sites and some of their roles. Histidine, which has a pK_a value of about 6 to 7 in proteins, is often an acceptor or donor of protons. Aspartate, glutamate, and occasionally lysine also can participate in transfer of protons. Certain amino acids such as serine and cysteine have reactivity suitable for covalently transferring groups from one substrate to a second substrate. At neutral pH, aspartate and glutamate usually have negative charges, and lysine and arginine have positive charges. These anions and cations can serve as sites for electrostatic binding of oppositely charged groups on substrates.

As discussed in Section 3·3, the pK_a values of the ionizable groups of amino acid residues in proteins may differ—usually slightly—from the values that the same groups have in free amino acids. Table 6·2 (next page) lists the typical pK_a values of ionizable residues in proteins. Compare these ranges to the exact values given for free amino acids in Table 3·1. Because of differences in their microenvironments, individual groups can differ from similar groups elsewhere in a protein. Occasionally, the side chain of a catalytic amino acid will exhibit a pK_a value quite

Table 6·1 Catalytic functions of ionizable amino acids

Amino acid	Reactive group	Net charge at pH 7	Principal functions
Aspartate	$-COO^{\ominus}$	–1	Cation binding; proton transfer
Glutamate	$-COO^{\ominus}$	–1	Cation binding; proton transfer
Histidine	Imidazole	Near 0	Proton transfer
Cysteine	$-S^{\ominus}$	Near 0	Covalent binding of acyl groups
Tyrosine	$-OH$	0	Hydrogen bonding to ligands
Lysine	$-NH_3^{\oplus}$	+1	Anion binding
Arginine	Guanidinium	+1	Anion binding
Serine	$-CH_2OH$	0	Covalent binding of acyl groups

Table 6·2 Typical pK_a values of ionizable groups of amino acids in proteins

Group	Expected pK_a
Terminal α-carboxyl	3–4
Side-chain carboxyl	4–5
Imidazole	6–7
Terminal α-amino	7.5–9
Thiol	8–9.5
Phenol	9.5–10
ε-Amino	~10
Guanidine	~12
Hydroxymethyl	~16

different from that shown in Table 6·2. Bearing in mind that pK_a values may be perturbed, one method of indirectly testing for involvement of individual amino acids in a reaction is to examine the effect of pH on the rate of the reaction. If the change in rate correlates with the pK_a value of a certain ionic amino acid (Section 6·6), a residue of that amino acid may be involved in the catalytic mechanism.

6·4 Almost All Enzyme Mechanisms Include Acid-Base Catalysis

In active sites, ionic side chains participate in two kinds of chemical catalysis: acid-base catalysis and covalent catalysis. *Acid-base catalysis* is the acceleration of a reaction achieved by transfer of a proton. Many nonenzymatic chemical reactions proceed only when they are performed in strong acid because a proton must be added to the reactant, or they proceed only in strong alkali because a proton must be removed. Indeed, the most common forms of nonenzymatic catalysis involve transfer of protons—proton addition by an acid, proton removal by a base. Acid-base catalysis is also common in enzymatic reactions. Enzymes utilize the side chains of amino acids that can donate and accept protons under the nearly neutral pH conditions of cells. In this manner, the active site of an enzyme can provide the biological equivalent of a solution of strong acid or strong base.

It is convenient to use B: to represent a base, or proton acceptor, and BH⊕ to represent its conjugate acid, a proton donor. A proton acceptor can assist reactions in two ways. It can cleave a C—H bond by removing the proton to produce a carbanion:

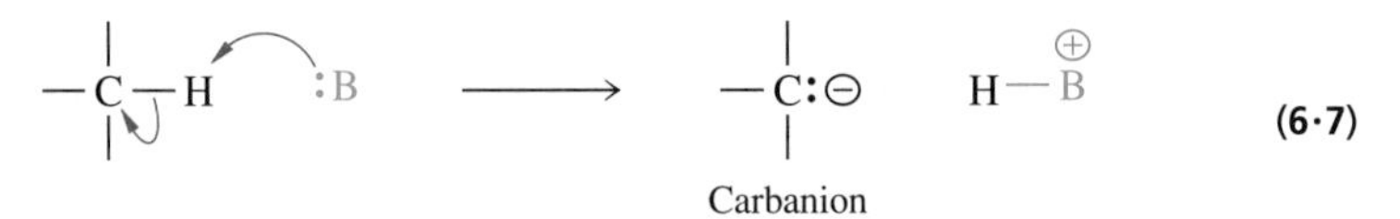

(6·7)

It can also participate in the cleavage of other bonds to carbon, such as a C—N bond, by generating the equivalent of OH⊖ in neutral solution by removal of a proton from a molecule of water:

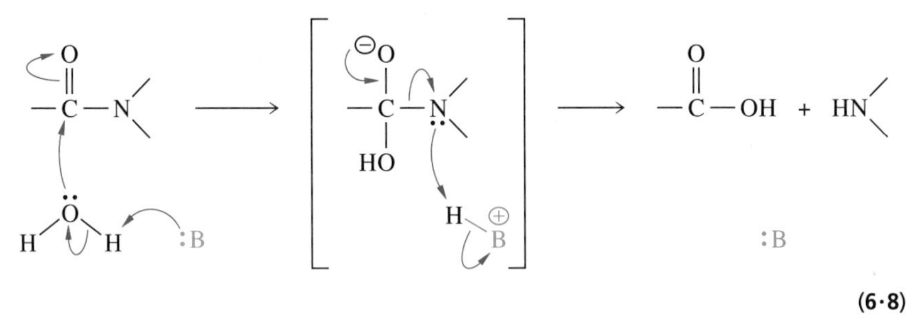

(6·8)

Conversely, BH⊕ can donate a proton when the reaction proceeds in the opposite direction. Because the imidazole/imidazolium of the side chain of histidine has a pK_a of about 6 to 7 in most proteins, it is an ideal group for proton transfer at neutral pH values.

6·5 Many Enzymatic Group-Transfer Reactions Proceed by Covalent Catalysis

Covalent catalysis refers to reactions in which the substrate or part of it forms a covalent bond with the catalyst. In enzymatic covalent catalysis, a portion of one substrate is transferred first to the enzyme and then to a second substrate. For example, a group X can be transferred from molecule A-X to molecule B in the following two steps via the covalent ES complex X-E:

$$\text{A-X} + \text{E} \rightleftharpoons \text{A} + \text{X-E} \tag{6·9}$$

and

$$\text{X-E} + \text{B} \rightleftharpoons \text{B-X} + \text{E} \tag{6·10}$$

The reaction catalyzed by bacterial sucrose phosphorylase (Equation 6·11) is an example of group transfer by covalent catalysis.

$$\text{Sucrose} + \text{P}_i \rightleftharpoons \text{Glucose 1-phosphate} + \text{Fructose} \tag{6·11}$$

This type of reaction is called *phosphorolysis* because the glucosyl group from sucrose is transferred to phosphate, rather than to water as in the hydrolysis of sucrose (Equation 5·4). The first chemical step in the sucrose phosphorylase reaction is formation of a covalent glucosyl-enzyme intermediate:

$$\text{Sucrose} + \text{Enzyme} \rightleftharpoons \text{Glucosyl-Enzyme} + \text{Fructose} \tag{6·12}$$

This covalent ES intermediate can either donate the glucose unit to another molecule of fructose, in the reverse of Equation 6·12, or to phosphate:

$$\text{Glucosyl-Enzyme} + \text{P}_i \rightleftharpoons \text{Glucose 1-phosphate} + \text{Enzyme} \tag{6·13}$$

It appears that about 20% of enzymes use this mode of catalysis. Proof of covalent catalysis in an enzyme mechanism relies on the isolation or demonstration of the reactive intermediate. The observation of ping-pong kinetics (Section 5·7) is another indication of covalent catalysis.

6·6 The Rates of Enzymatic Reactions Are Affected by pH

The effect of pH on the reaction rate of an enzyme can suggest which ionic amino acid residues are in its active site. The sensitivity to pH usually reflects an alteration in the ionization state of one or more residues involved in catalysis and occasionally in substrate binding. Of course, extremes of pH can cause denaturation of enzymes. However, under pH conditions that do not denature the enzyme, when the initial velocity of the reaction (or in some cases the maximum velocity) is plotted against pH, a bell-shaped curve is often obtained—a curve with information that can help delineate a mechanism. The pH at the point of maximum activity is called the *pH optimum* of the enzyme. Routine assays are usually performed at this pH, which is maintained using appropriate buffers. The pH-rate profile can often be explained by assuming that the ascending sigmoidal curve is caused by the deprotonation of an active-site amino acid (B) and that the sigmoidal curve descending from the pH optimum is caused by the deprotonation of a second active-site amino acid residue (A). Thus, a simple bell-shaped curve is the result of two overlapping titrations. The side chain of A must be protonated for activity and the side chain of B must be unprotonated.

$$\underset{\text{Inactive}}{{}^{\oplus}\text{HA}\cdots\text{BH}^{\oplus}} \rightleftharpoons \underset{\text{Active}}{{}^{\oplus}\text{HA}\cdots\text{B}} \rightleftharpoons \underset{\text{Inactive}}{\text{A}\cdots\text{B}} \tag{6·14}$$

The inflection points of the two curves approximate the pK_a values of the two ionizable residues. At the pH optimum, midway between the two pK_a values, the greatest number of enzyme molecules are in the active ${}^{\oplus}$HA···B form. Ribonuclease (Section 10·6A), which has two active-site histidine residues, is an example of an enzyme that exhibits a bell-shaped pH-rate profile; the two histidines of ribonuclease have pK_a values of 5.8 (B) and 6.2 (${}^{\oplus}$HA), and the pH optimum is 6.0. The pH-rate profile is not bell-shaped if only one or if more than two ionizable amino acids participate in the catalytic mechanism.

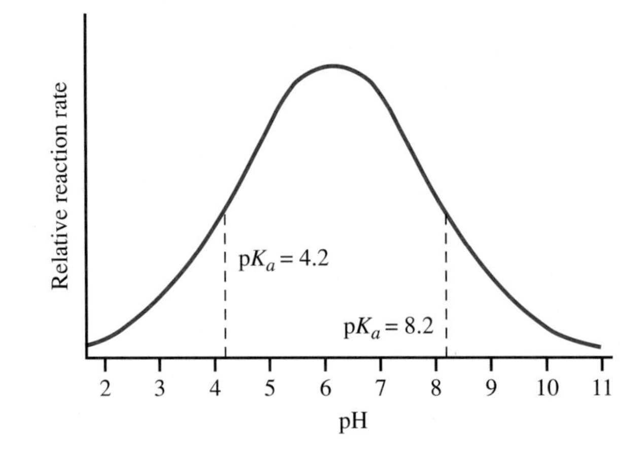

Figure 6·4
pH-rate profile for papain. The pH-rate profile for papain is bell-shaped. The left and right segments of the pH-rate curve correspond to titration curves for the side chains of two active-site amino acids. The inflection point at pH 4.2 corresponds to the considerably perturbed pK_a of Cys-25, and the inflection point at pH 8.2 to the pK_a of His-159. The enzyme is active only when these ionic groups are present as the thiolate-imidazolium ion pair [—S⊖....⊕HIm—].

Papain is a protease found in papaya fruit and used in meat tenderizers. Its mechanism exhibits both acid-base catalysis and covalent catalysis. It has a typical pH-rate profile with inflection points at pH 4.2 and pH 8.2 (Figure 6·4). This profile suggests that the activity of papain depends on two active-site amino acid residues that have pK_a values of about 4 and 8.

Many proteases catalyze the hydrolysis of peptides by the general mechanism shown in Figure 6·5, which is an expansion of Equation 6·1. After the noncovalent ES complex is formed by binding of the substrate in the active site, an active-site nucleophilic group (—X:, the covalent catalyst) attacks the carbonyl carbon of the peptide bond to form an unstable tetrahedral intermediate. Cleavage of the peptide bond occurs when a proton is donated by another active-site amino acid (⊕HB), the acid-base catalyst, to the nitrogen atom of the tetrahedral intermediate to produce the amine leaving group. The amine diffuses out of the active site and is replaced by

Figure 6·5
General mechanism for enzymatic hydrolysis of a peptide. In this mechanism, the enzyme reacts with the acyl group of the substrate to form a covalent enzyme-substrate intermediate, the acyl-nucleophile adduct. The nucleophile (—X:) is the side chain of an ionizable amino acid residue in the active site. A second active-site residue (—B) acts as a proton donor and acceptor.

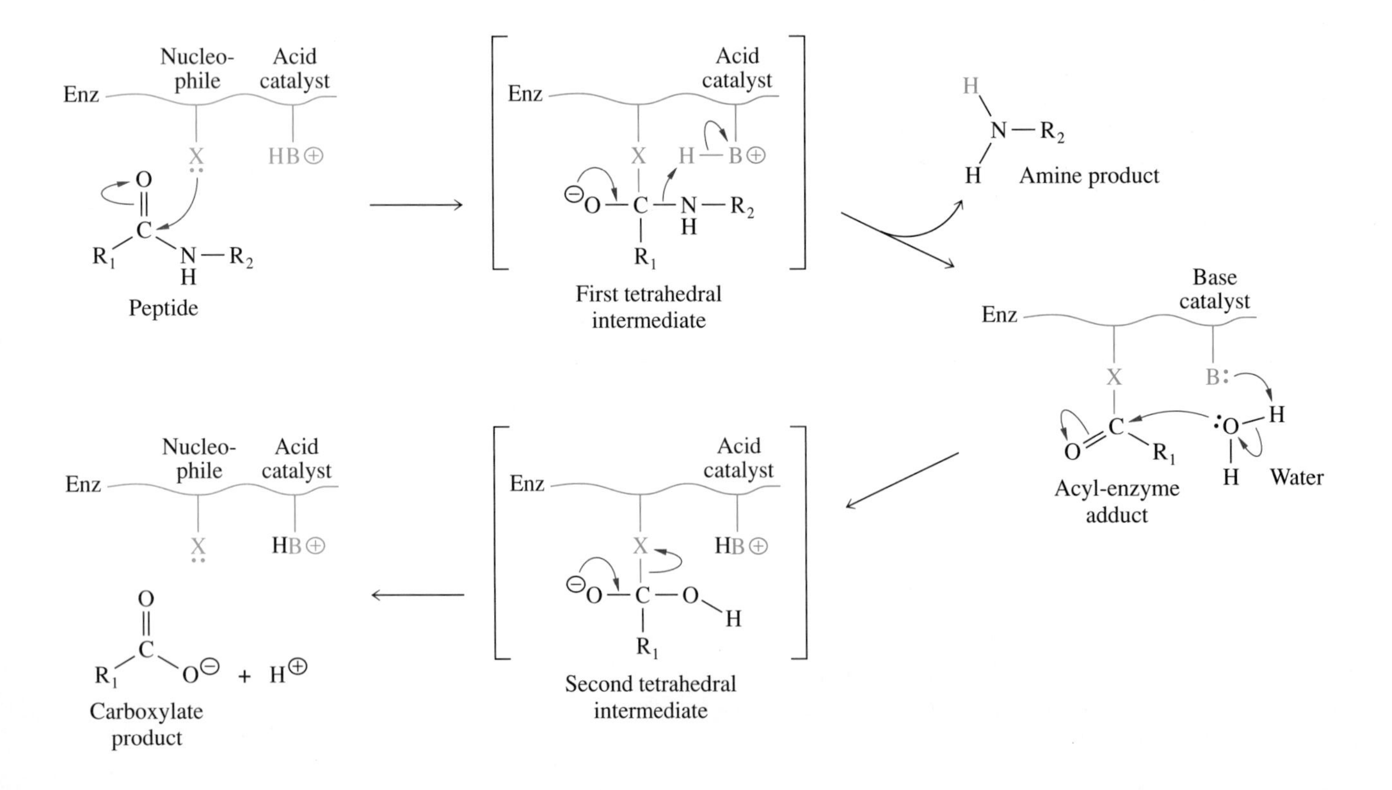

a molecule of water. The other portion of the substrate, the acyl group, has substituted the nucleophilic group of the enzyme, producing the covalent acyl-enzyme adduct. As depicted in the lower half of Figure 6·5, the carboxylic acid from the peptide substrate is released from the substituted enzyme by nucleophilic attack on the carbonyl group by $OH^{\ominus}$ from water. The tetrahedral intermediate formed as a result of this addition of $OH^{\ominus}$ decomposes to form the carboxylate, a proton, and regenerated free enzyme.

Papain follows a ping-pong kinetic mechanism (Equation 6·15) and the chemical mechanism of Figure 6·5. The acyl group of its peptide substrate becomes bound to the enzyme and is then transferred to a molecule of water.

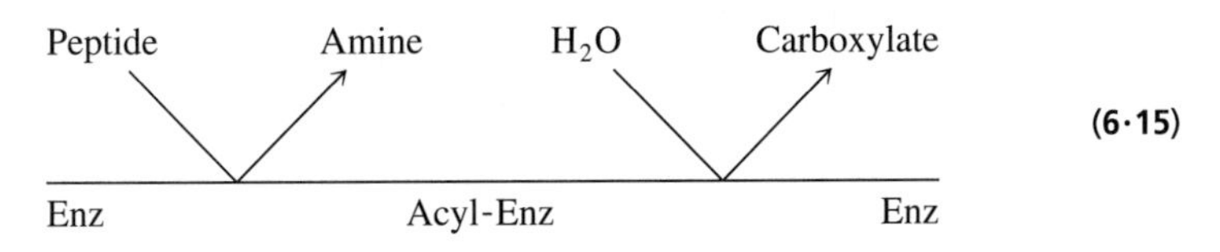

(6·15)

In papain, the two active-site ionizable residues are a nucleophilic cysteine and a proton-donating imidazolium group of histidine. Although the side chain of cysteine normally has a pK_a of 8 to 9.5, in the cleft of papain it forms a thiolate-imidazolium ion pair with histidine, causing perturbation of its pK_a to a value of about 4 and raising that of the histidine residue to about 8. The three ionic forms of the catalytic center of papain are shown in Figure 6·6. Only the upper tautomer of the intermediate form is active. It is only in this form that papain has *both* the nucleophilic cysteine and the protonated histidine required for catalyzing the hydrolysis reaction.

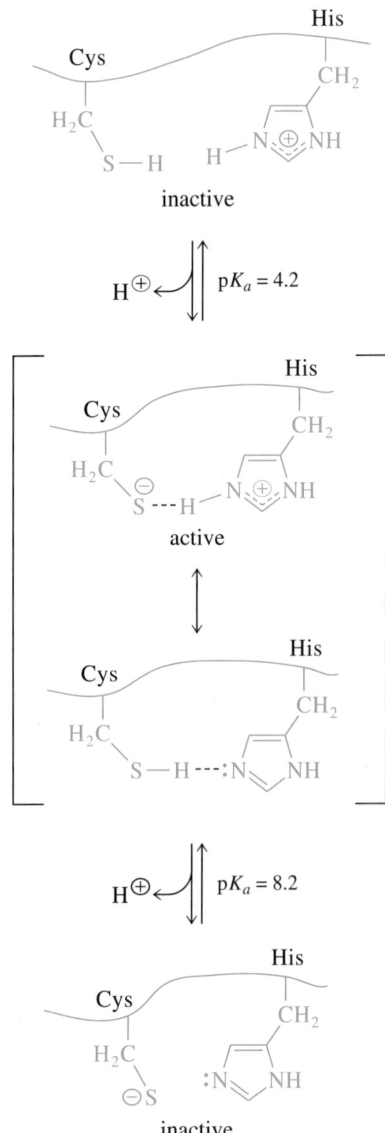

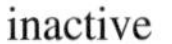

Figure 6·6
Three ionic forms of active-site residues of papain. Only the upper tautomer of the intermediate form is active.

6·7 The Upper Limit for Catalysis Is the Rate of Diffusion

A few enzymes catalyze reactions at rates approaching the upper physical limit of reactions in solution, the rate of diffusion of reactants toward each other. A reaction that occurs with every collision between reactant molecules is termed a *diffusion-controlled reaction*. The frequency of encounter between a typical solute and an enzyme has been calculated to be about 10^8 to $10^9\ M^{-1}\ s^{-1}$. (For reference, the upper limit for a molecular vibration—the fastest unimolecular reaction—is about 10^{12} to $10^{13}\ s^{-1}$.) The frequency of encounter can be higher if there is electrostatic attraction between the reactants. The apparent second-order rate constants for six very fast enzymes are listed in order of increasing rate in Table 6·3. As you can see, only one

Table 6·3 Enzymes with second-order rate constants near the upper limit

Enzyme	Substrate	k_{cat}/K_m ($M^{-1}\ s^{-1}$)
Catalase	H_2O_2	4×10^7
Carbonic anhydrase	CO_2	8.3×10^7
Acetylcholinesterase	Acetylcholine	1.6×10^8
Fumarase	Fumarate	1.6×10^8
Triose phosphate isomerase	Glyceraldehyde 3-phosphate	4×10^8
Superoxide dismutase	O_2	2×10^9

The apparent second-order rate constants for the enzyme-catalyzed reaction E + S ⟶ E + P were obtained from the ratio k_{cat}/K_m. For these enzymes, the formation of the ES complex is the rate-determining step.

of these constants—for superoxide dismutase—greatly exceeds a rate of $10^8\ M^{-1}\ s^{-1}$. The explanation for faster catalysis by superoxide dismutase is discussed below.

The binding of substrate to an enzyme is a very fast second-order reaction. If this step is also the rate-determining step, and if the rest of the reaction is simple and fast, the enzyme may approach the upper limit for catalysis. Only a limited number of types of chemical reactions can proceed this quickly. These include association reactions, some proton transfers, electron transfers, and conformational changes in proteins. All of the reactions described in Table 6·3 are so simple that their rate-determining steps are roughly as fast as the binding of their substrate molecules by the enzymes.

The reaction catalyzed by triose phosphate isomerase is listed as the second fastest in Table 6·3. This enzyme, found in the pathway of glycolysis (Chapter 12), catalyzes the rapid interconversion of dihydroxyacetone phosphate (DHAP) and glyceraldehyde 3-phosphate (G3P).

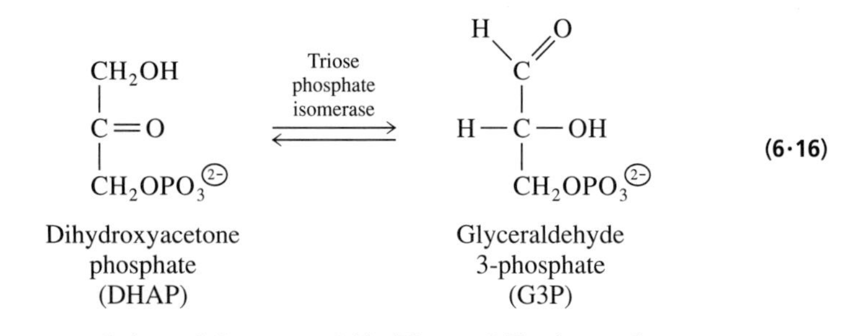

(6·16)

The reaction proceeds by a 1,2-proton shift (Figure 6·7). An analogous nonenzymatic interconversion in alkaline solution is a two-step process. The enzymatic interconversion involves two extra steps, the formation of the noncovalent E-DHAP complex and the dissociation of the E-G3P complex. Triose phosphate isomerase has two ionizable active-site residues, glutamate and histidine. Upon binding of substrate, the carbonyl oxygen of DHAP forms a hydrogen bond with the neutral imidazole group of the histidine residue. The carboxylate group of the glutamate residue removes a proton from C-1 of DHAP to form an enediolate intermediate. The oxygen at C-2 is protonated by the histidine residue, which then abstracts a proton from the oxygen at C-1 to form another unstable enediolate intermediate. In this proton transfer, the conjugate acid form of histidine appears to be the neutral species, and the conjugate base appears to be the imidazolate:

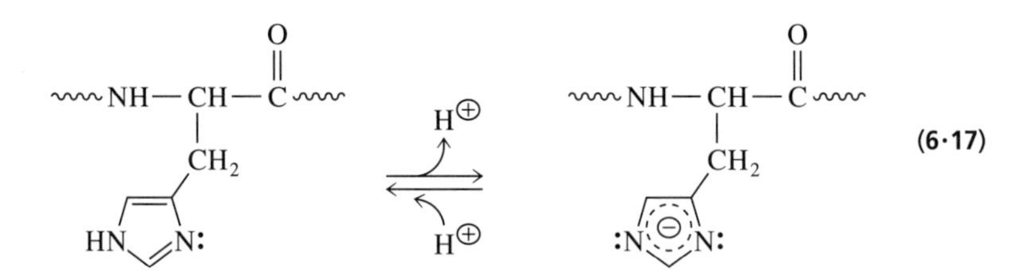

(6·17)

This ionization of a histidine residue is unusual and has not been implicated in the catalytic mechanism of any other enzyme.

Elaborate kinetic measurements performed in the laboratory of Jeremy Knowles have determined the rate constants of all four kinetically measurable enzymatic steps in both directions:

$$E + DHAP \rightleftharpoons E\text{-}DHAP \rightleftharpoons E\text{-}Intermediate \rightleftharpoons E\text{-}G3P \rightleftharpoons E + G3P$$

(6·18)

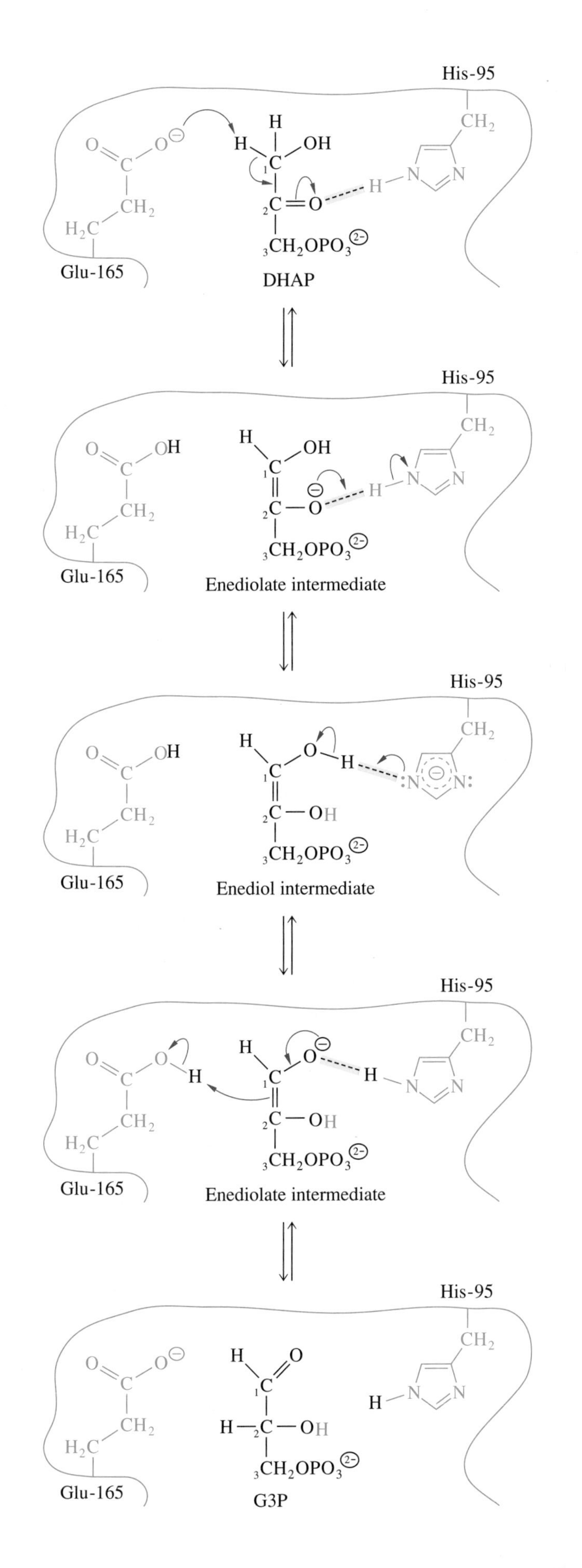

Figure 6·7
Mechanism of the reaction catalyzed by triose phosphate isomerase. The active-site glutamate residue of the enzyme removes a proton from C-1 of the enzyme-bound dihydroxyacetone phosphate (DHAP) and donates a proton to C-2 to form enzyme-bound glyceraldehyde 3-phosphate (G3P). The neutral histidine residue in the active site transfers a proton between the oxygen atoms on C-1 and C-2.

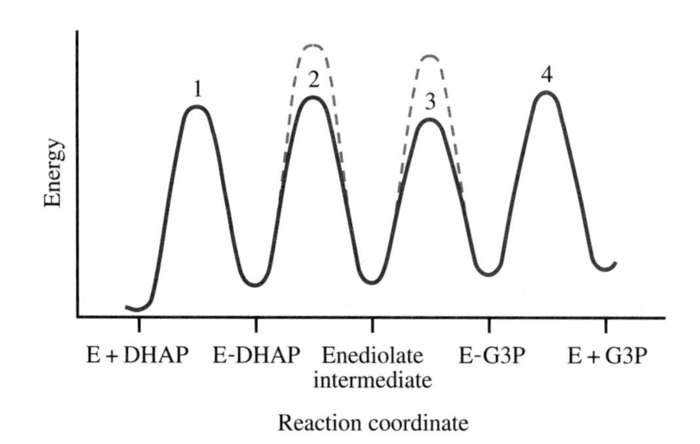

Figure 6·8
Reaction diagram for the reaction catalyzed by triose phosphate isomerase. The solid line represents the profile for the wild-type (naturally occurring) enzyme. The dotted line shows the profile for a mutant enzyme in which the active-site glutamate residue has been replaced by an aspartate residue. With the mutant enzyme, the activation energies for the proton-transfer reactions (Steps 2 and 3) are significantly higher. The wild-type enzyme catalyzes the reaction about one thousand times faster than the mutant enzyme. [Adapted from Raines, R. T., Sutton, E. L., Strauss, D. R., Gilbert, W., and Knowles, J. R. (1986). Reaction energetics of a mutant triosephosphate isomerase in which the active-site glutamate has been changed to aspartate. *Biochemistry* 25:7142–7154.]

The reaction diagram constructed from these rate constants is shown in Figure 6·8, along with the profile of a mutant triose phosphate isomerase (dashed lines) to be discussed below. Notice that all of the barriers for the nonmutant reaction are approximately the same height; that is, the steps are balanced. This is an ideal strategy for enzymatic catalysis. The value of the second-order rate constant k_{cat}/K_m for the conversion of G3P to DHAP is $4 \times 10^8\ M^{-1}\ s^{-1}$, near the rate of a diffusion-controlled reaction. The physical step of S binding to E is rapid but not much faster than any of the chemical steps in the reaction sequence. However, improvement of any of the chemical steps would not greatly increase the efficiency of the enzyme. It appears that through evolution this isomerase has virtually reached its maximum possible efficiency as a catalyst.

Knowles and his colleagues have converted the active-site glutamate residue of triose phosphate isomerase to an aspartate residue by site-directed mutagenesis (Section 5·9). In the modified isomerase, the same reactive group ($—COO^{\ominus}$) is present as in the wild-type enzyme, but it is slightly farther from the substrate. The wild-type and mutant enzymes have fairly similar K_m values. However, the values for k_{cat}, measured for both the forward and reverse reactions catalyzed by the mutant enzyme, are greatly decreased. This experiment verified that the glutamate residue in the active site is essential for full activity. It also changed the enzyme from diffusion-controlled to a more typical enzyme that exhibits rapid binding of reactants and slower chemical catalysis steps.

Superoxide dismutase is an even faster catalyst than triose phosphate isomerase, apparently because its negatively charged substrate is attracted by a positively charged electric field near the active site. Superoxide dismutase catalyzes the very rapid removal of the toxic superoxide radical anion, $\cdot O_2^{\ominus}$, a by-product of oxidative metabolism, by catalyzing its conversion to molecular oxygen and hydrogen peroxide, which is rapidly removed by the subsequent action of enzymes such as catalase.

$$4\,\cdot O_2^{\ominus} \xrightarrow[\text{Superoxide dismutase}]{4\,H^{\oplus} \;\; 2\,O_2} 2\,H_2O_2 \xrightarrow[\text{Catalase}]{} 2\,H_2O + O_2 \qquad (6\cdot19)$$

The reaction catalyzed by superoxide dismutase proceeds in two steps, in which an atom of copper bound to the enzyme is oxidized and then reduced:

$$\text{E-Cu}^{2+} + \cdot O_2^{\ominus} \longrightarrow \text{E-Cu}^{\oplus} + O_2 \qquad (6\cdot20)$$

and

$$\text{E-Cu}^{\oplus} + \cdot O_2^{\ominus} + 2\,H^{\oplus} \longrightarrow \text{E-Cu}^{2+} + H_2O_2 \qquad (6\cdot21)$$

The overall reaction includes binding of the anionic substrate molecules, electron and proton transfers, and release of the uncharged products—all very rapid reactions. The value of k_{cat}/K_m for superoxide dismutase at 25°C is near $2 \times 10^9\,M^{-1}\,s^{-1}$, *faster* than expected for association of the substrate with the enzyme based on typical diffusion rates. The active-site copper atom is at the bottom of a deep channel in the protein. Positively charged amino acids in and near the active site seem to direct the $\cdot O_2^{\ominus}$ to the active-site channel by electrostatic attraction. The electric field around superoxide dismutase enhances the rate of association about 30-fold. With this enzyme, and perhaps with others, more than simple collision is involved in formation of the ES complex, making catalysis extremely fast.

6·8 Binding of Reactants Plays the Most Important Role in Enzymatic Catalysis

Hypothetical explanations for enzymatic catalysis have been in circulation for many years, but sufficient scientific evidence to support these ideas has been obtained only recently. We have already seen two *chemical* modes by which enzymes catalyze reactions, acid-base catalysis (Section 6·4) and covalent catalysis (Section 6·5). Based upon the effects of acid-base catalysts on nonenzymatic reactions, it is estimated that acid-base catalytic residues probably accelerate enzymatic reactions by about 10 to 100 times. From similar considerations, covalent catalysis provides about the same rate acceleration as does proton transfer.

As important as these chemical modes are, they leave a large unexplained gap in rate accelerations. The ability of globular proteins to specifically bind and orient ligands can completely account for the remainder. The binding of reactants in the active sites of enzymes provides not only substrate and reaction specificity but also much of the catalytic power of enzymes. Two interrelated catalytic modes operate, both based on the interaction of reactant molecules with a matching enzyme. For multisubstrate reactions, the collecting of substrate molecules in the active site raises their effective concentrations from their concentrations in free solution. In the same way, binding of substrates near catalytic active-site residues increases the effective concentrations of these two reactants. High effective concentrations favor the more frequent formation of transition states. This phenomenon has had many names, the most widely accepted being the *proximity effect*. The proximity effect requires weak binding of reactants to enzymes, since extremely tight binding would defeat catalysis. The second major catalytic mode arising from the ligand-enzyme interaction is the increased binding of transition states to enzymes compared to the binding of substrates or products. This catalytic mode has been referred to as strain or distortion but is more precisely called *transition-state stabilization*. There is an equilibrium (*not* the reaction equilibrium) between ES and the enzymatic transition state, $ES^{\ddagger}$. Interaction between the enzyme and its ligands in the transition state shifts this equilibrium toward $ES^{\ddagger}$ and lowers the energy of activation.

The effects of proximity and transition-state stabilization were outlined in Figure 6·3. Experiments, some of which we shall describe, have suggested that proximity contributes to enzymatic catalysis an enhancing effect of about 10^4 to 10^5 and transition-state stabilization at least that much. When these effects are multiplied together with acid-base catalysis, we can see clearly how enzymes achieve their extraordinary rate accelerations.

The binding forces responsible for formation of ES complexes and for stabilization of $ES^{\ddagger}$ are familiar from Chapters 2 and 4: electrostatic forces, hydrogen bonds, hydrophobic interactions, and van der Waals forces. Electrostatic interactions are stronger in nonpolar environments than in water. Because active sites are largely nonpolar, electrostatic forces in the clefts of enzymes can be quite strong. Hydrogen bonds, next in bond strength, are often formed between substrates and

Table 6·4 Potential interactions between substrates and amino acid residues of proteins. The binding characteristics shown arise from the side chains. In addition, all amino acid residues can participate in hydrogen bonding through the peptide backbone.

Amino acid	Electrostatic	Hydrogen bonding	Hydrophobic
Ionizable:			
Aspartate	+	+	
Glutamate	+	+	
Lysine	+	+	
Arginine	+	+	
Histidine	+	+	
Cysteine		+	
Polar, nonionizable:			
Asparagine		+	
Glutamine		+	
Serine		+	
Threonine		+	
Polypeptide backbone		+	
Some hydrophobic character:			
Tyrosine		+	+
Tryptophan		+	+
Methionine		+	+
Glycine (little binding, other than through the backbone)			
Hydrophobic:			
Alanine			+
Valine			+
Leucine			+
Isoleucine			+
Phenylalanine			+
Proline			+

The + sign indicates that the side chain of this amino acid can participate in the type of interaction in that column. [Adapted from Baker, B. R. (1967). *Design of Active-Site-Directed Irreversible Enzyme Inhibitors* (New York: John Wiley & Sons), p. 25.]

enzymes. Although hydrogen bonds are fairly stable, they are weak enough that bound reactants can dissociate readily. There are hydrophobic interactions between nonpolar groups of substrates and hydrophobic regions in the clefts of enzymes. A large number of weak van der Waals interactions also helps to bind substrates. Keep in mind that both the chemical properties of the amino acid components and the shape of the active site of a globular protein also determine the substrate specificity of the enzyme. Table 6·4 provides a list of the interactions possible between individual amino acids and ligands such as substrates.

6·9 The Proximity Effect Is an Antientropy Phenomenon

Two molecules coming together from dilute solution to form a transition state is an improbable process, accompanied by the loss of a great deal of entropy. William Jencks and his colleagues made a conceptual advance by considering the reaction of two molecules at the active site as equivalent to converting a nonenzymatic bimolecular reaction into an intramolecular or unimolecular reaction. Correct positioning of two reactants produces a large loss of entropy, sufficient to account for a large rate acceleration. The acceleration is expressed in terms of the enhanced relative concentration, called the *effective molarity,* of the reacting groups in the unimolecular reaction. Experiments have been performed to compare the reaction rates of compounds having the correct preassembly of reactants into a single molecule

(k_1) with the rates of a bimolecular counterpart (k_2). The effective molarity can be obtained from this ratio:

$$\text{Effective molarity} = \frac{k_1 \text{ (in s}^{-1}\text{)}}{k_2 \text{ (in M}^{-1}\text{s}^{-1}\text{)}} \qquad \textbf{(6·22)}$$

All of the units in this equation except M cancel, so the ratio is expressed in molar units. Because the effective molarities of suitable chemical model reactions often greatly exceed concentrations that can be attained in solution, the large rate increases characteristic of enzymatic catalysis must include both high local concentration and very favorable geometry of the reacting groups.

The proximity effect in a chemical model system can be illustrated by experiments performed by Thomas Bruice and Upendra K. Pandit. In these experiments, the reactivities of a series of intramolecular reactants were compared to that of a bimolecular reaction (Figure 6·9). The unimolecular substrates had progressively greater restriction of rotation around a bridging arm and progressively higher

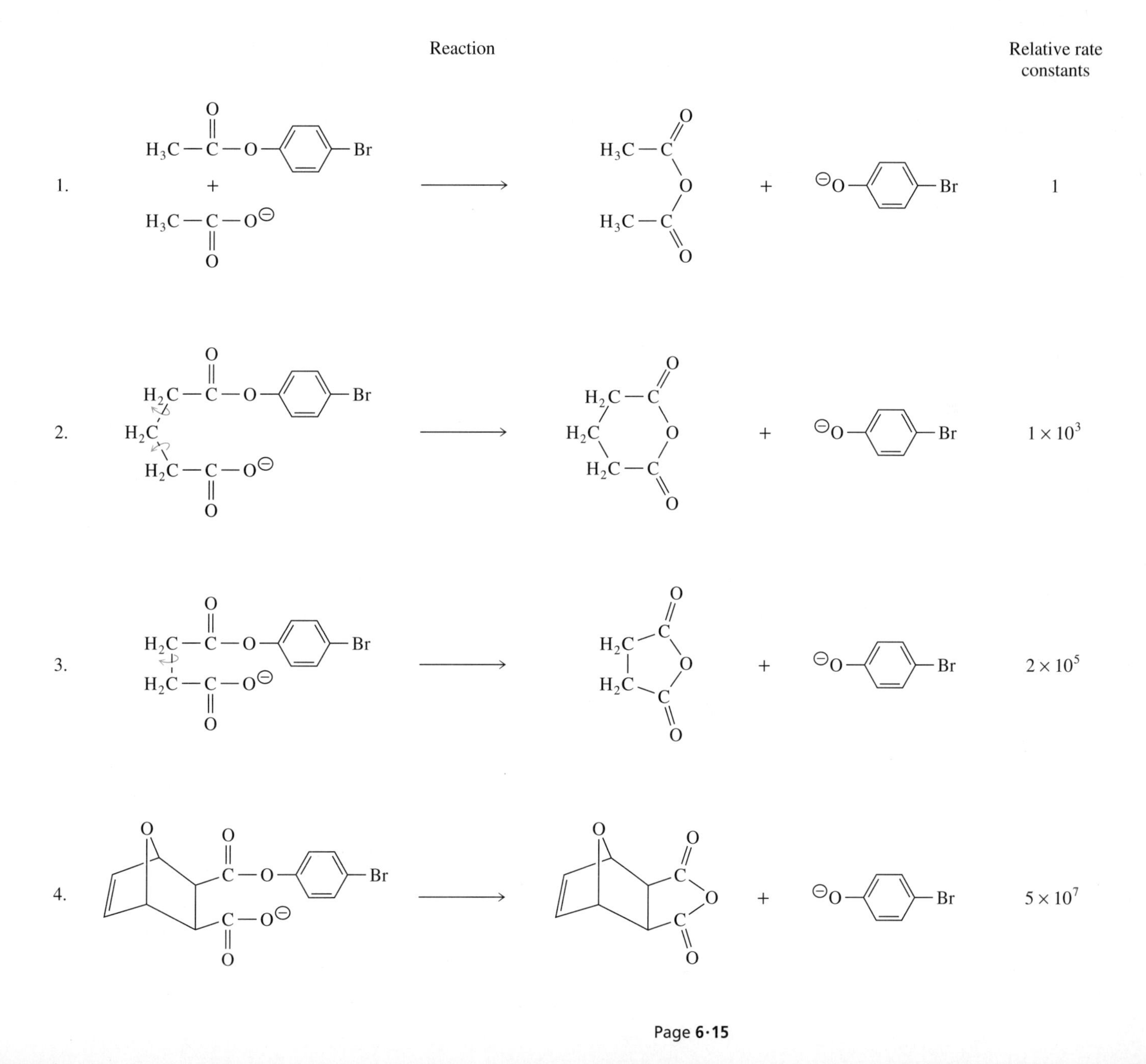

Figure 6·9
Reactions of a series of carboxylates with substituted phenyl esters. The proximity effect is illustrated by the increase in rate observed when the reactants are held more rigidly in proximity. The rate constant for reaction 4 is 50 million times that of Reaction 1, the bimolecular reaction.

effective molarities. The bimolecular reaction was the two-step hydrolysis of *p*-bromophenyl acetate, catalyzed by acetate and proceeding via the formation of acetic anhydride. (The second step, hydrolysis of acetic anhydride, is not shown in Figure 6·9.) The rate of the first step was determined by measuring the release of a colored phenolate. Each of the unimolecular reactions involves formation of either a five- or six-membered ring in the rate-limiting step, covalent catalysis by formation of an acid anhydride. With each restriction placed upon the substrate molecule, the relative rate constant—measured by k_1/k_2—increased markedly. Note that the glutarate ester, compound 2, has two bonds that allow rotational freedom, whereas the succinate ester, compound 3, has only one. The most restricted compound, the rigid bicyclic compound 4, has no rotational freedom. Its rigid structure positions the carboxylate close to the ester. With an extremely high probability of reaction, this compound showed an effective molarity of the carboxylate group of 5×10^7 M. Theoretical considerations suggest that the greatest rate acceleration that can be expected from the proximity effect is about 10^8. All of this rate acceleration can be attributed to loss of entropy.

6·10 Extremely Tight Binding of Substrate to an Enzyme Would Interfere with Catalysis

The insight gained from the previous section about unimolecular nonenzymatic reactions can be applied to enzymes. Reactions of ES complexes can be seen as analogous to unimolecular reactions. The prepositioning of substrates in an enzyme active site produces a large rate acceleration. However, the maximum 10^8-fold acceleration that can be generated by the proximity effect cannot be exerted by enzymes. Typically, the change in entropy upon binding of substrate provides an acceleration of roughly 10^4. In ES complexes, the reactants are brought toward but not extremely close to the transition state. This conclusion is based upon both mechanistic reasoning and measurements of the tightness of binding of substrates to enzymes, usually employing K_m values. The following discussion explains why the binding of substrates to enzymes cannot be extremely tight; that is, why K_m values cannot be extremely low.

The reaction diagram in Figure 6·10, that for a simple unimolecular reaction, compares the energy path of a nonenzymatic reaction with the multistep path followed in enzymatic catalysis. The transition state $ES^{\ddagger}$ in the enzymatic reaction is stabilized by the amount shown by the upper arrow (1). If the ES complex were stabilized by a similar amount (lower arrow, (3)), the enzyme would not be a catalyst

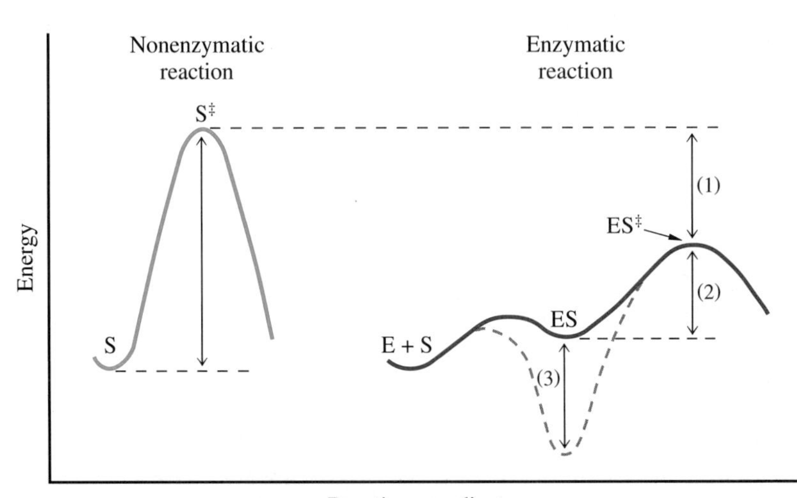

Figure 6·10
Energy of substrate binding. In this hypothetical reaction, the enzyme accelerates the rate of the reaction by stabilizing the transition state. The amount of stabilization is indicated by the upper arrow (1). The activation barrier for the formation of the transition state $ES^{\ddagger}$ from ES (2) is low, so this enzyme is a good catalyst. If the enzyme were to bind the substrate tightly, lower dashed line and arrow (3), the activation barrier from the stabilized ES complex would be close to that of the nonenzymatic reaction. Tight binding of this type is termed a thermodynamic pit.

because it would take just as much energy to reach ES‡ from ES as is required for the nonenzymatic reaction. If substrate were bound extremely tightly, there would be little or no catalysis. Because the energy difference between ES and ES‡ is significantly less than that between S and S‡ (the transition state in the nonenzymatic reaction), k_{cat} is greater than k_n (the rate constant for the nonenzymatic reaction). Therefore, enzymes must position reactants fairly rigidly before the transition state is reached but not so rigidly that they have lost all of their entropy by forming an ES complex that is too stable. The situation of excessive ES stability is termed a thermodynamic pit.

Consideration of the K_m values of a wide variety of enzymes for their substrates shows that enzymes avoid the thermodynamic pit. Most K_m values are fairly high, on the order of 0.1 mM. They would be lower if all parts of the substrates were bound to the enzymes. For example, the K_m values observed for the binding of ATP to most ATP-requiring enzymes are about 0.1 mM or higher, but there is a protein (not an enzyme) that binds ATP with a dissociation constant of 10^{-10} mM for the ATP-protein complex. The billionfold difference indicates that in an ES complex, in which the ground state of a substrate is bound to the enzyme, not all parts of the substrate are bound. This is very important, because it leads to the next major force that drives enzymatic catalysis, increased binding of reactants in the transition state, ES‡.

Enzymes that are nonspecific exhibit relatively high K_m values for their substrates. Enzymes specific for small substrate molecules such as urea, carbon dioxide, and superoxide anion exhibit high K_m values for these compounds because the substrates are so small that they can form few noncovalent bonds with enzymes. In contrast, larger substrate molecules and enzyme-bound intermediates are bound quite tightly to enzymes. Typically, enzymes have lower K_m values for coenzymes—molecules larger than many substrates. Enzymes requiring prosthetic groups (coenzymes that remain bound to the enzyme as a permanent component of the active site) exhibit very low K_m values. For example, the K_m of a typical transaminase for pyridoxal phosphate is 4 μM, reflecting binding that is at least 25-fold tighter than that of ATP with most of its enzymes.

Because enzymes have relatively high K_m values for substrates, the concentrations of substrates inside cells often are well below the K_m values of their corresponding enzymes. This means the equilibrium of the reaction $E + S \rightleftharpoons ES$ favors E + S. In other words, the formation of the ES complex is slightly uphill energetically (Figures 6·3 and 6·10), and the ES complex is closer to the energy of the transition state than is the ground state. Kinetically, this weak binding of substrates accelerates reactions. K_m values are optimized by evolution for effective catalysis—low enough that proximity is achieved, but high enough that the ES complex is not too stable.

6·11 Transition States Bind Tightly to Enzymes

Since the 1920s, enzyme chemists have suggested that enzymes bind reactants in the transition state with great affinity. Evidence in favor of this theory came in the early 1970s from experiments by Richard Wolfenden and Gustav Lienhard. They showed that chemical analogs of transition states are potent inhibitors of enzymes. Other experimental support is now available to support transition-state stabilization as a major catalytic mode.

Recall Emil Fischer's lock-and-key theory of enzyme specificity (Section 5·2). Fischer proposed that enzymes were rigid templates that accepted only certain keys—substrates. The lock-and-key theory is still a generally accepted explanation for enzymatic specificity, and also for enzymatic catalysis, with one stipulation—that it is the transition state (or sometimes an unstable intermediate) that is the key, not the substrate molecule. When a substrate binds to an enzyme, the enzyme produces a distortion of the substrate that forces it toward the transition state. There can be maximal interaction with the substrate molecule only in $ES^{\ddagger}$. A portion of this binding in $ES^{\ddagger}$ can be between the enzyme and *nonreacting* portions of the substrate. The effects of binding nonreacting portions of substrates can be seen by comparing chymotrypsin-catalyzed hydrolysis of the amide bond of phenylalanine in *N*-acetyl-Phe-Gly-NH_2 with that in *N*-acetyl-Phe-Ala-NH_2. These synthetic substrates have similar K_m values, but the latter substrate, with a methyl side chain replacing a hydrogen atom, has a k_{cat} that is 20 times larger.

For catalysis to occur, the transition state must be stabilized. The enzyme must be complementary in shape and chemical character to the transition state. Figure 6·10 shows that the energy of activation is lowered by the tight binding of the transition state to an enzyme. At the same time, the binding of S to E is weak. The task, then, has been to find examples of differences in the interactions in ES and $ES^{\ddagger}$.

There are several ways in which the comparative stabilization of $ES^{\ddagger}$ could occur:

1. An enzyme could have an active site with a shape that is a closer match to the transition state than to the substrate. An undistorted substrate molecule would not be fully bound.

2. The enzyme could have sites that bind the partial charges present only in the transition state. In a related manner, a nonpolar active site could stabilize a transition state in which there is a decrease in localized charge compared to the substrate.

Examples of these phenomena will be discussed in the next three sections.

6·12 Transition-State Analogs Are Potent Inhibitors

Transition-state analogs are compounds that resemble the transition state. The transition state itself, with a half-life of about 10^{-13} second, is too short-lived to isolate in the laboratory. However, chemicals can be synthesized that resemble these activated species. If the modified lock-and-key theory is correct, a transition-state analog should bind extremely tightly to the appropriate enzyme and thus be a potent inhibitor. One of the first examples of a successfully designed transition-state analog was 2-phosphoglycolate (Figure 6·11), a compound proposed to be an analog of a transition state in the reaction catalyzed by triose phosphate isomerase (Section 6·7). This analog binds to the isomerase at least 100 times more tightly than does either of the substrates of the enzyme. Some of the additional binding is through a partially negative oxygen atom of the carboxylate group of 2-phosphoglycolate, a feature shared with the transition state but not the substrate.

Experiments with adenosine deaminase have demonstrated a transition-state analog that binds to the enzyme with amazing affinity because it resembles the transition state very closely. Adenosine deaminase catalyzes the hydrolytic conversion of the purine nucleoside adenosine to inosine. The first step of this reaction, before

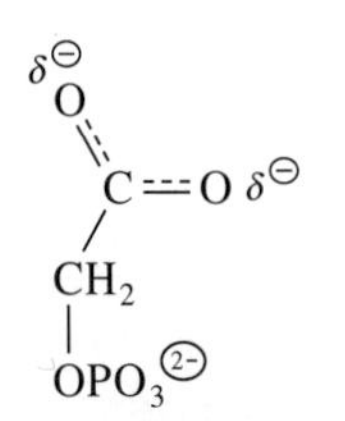

2-Phosphoglycolate (transition-state analog)

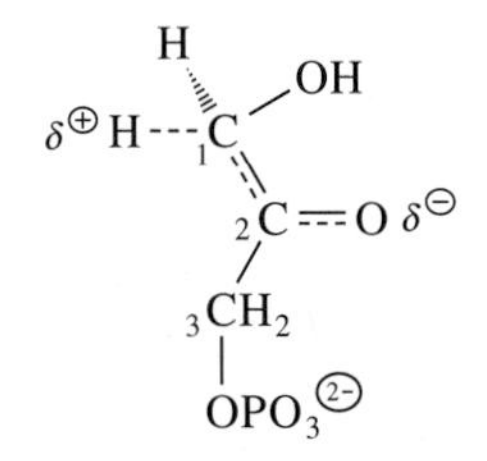

Transition state

Dihydroxyacetone phosphate (substrate)

Figure 6·11
2-Phosphoglycolate, a transition-state analog for the enzyme triose phosphate isomerase. 2-Phosphoglycolate is presumed to be an analog of C-2 and C-3 of the transition state (center) between dihydroxyacetone phosphate (right) and the initial enediolate intermediate in the reaction.

the loss of the amino group as ammonia, is the addition of a molecule of water (Figure 6·12a). The complex with water, called a covalent hydrate, does not form in aqueous solution but does form as soon as adenosine is bound to the enzyme. The hydrate is a tetrahedral intermediate that quickly decomposes to products. Adenosine deaminase has fairly broad substrate specificity and consequently catalyzes the hydrolytic removal of various groups from the 6-position of purine nucleosides. However, the inhibitor purine ribonucleoside (Figure 6·12b) has no 6-substituent and only undergoes the first enzymatic step of hydrolysis, the addition of the water molecule. The covalent hydrate that is formed is a transition-state analog, a competitive inhibitor having a K_i of 3×10^{-13} M. The binding of this analog exceeds that of either the substrate or the product by a factor of more than 10^8. A very similar reduced inhibitor, 1,6-dihydropurine ribonucleoside (Figure 6·12c), cannot undergo a hydration reaction and thus lacks the hydroxyl group at C-6; it has a K_i of only 5×10^{-6} M. Therefore, one can conclude that the enzyme must specifically and avidly bind the transition-state analog—and also the transition state—through interaction with the hydroxyl group at C-6.

Recently, transition-state analogs bound to proteins have been used as antigens to induce the formation of antibodies that have catalytic activity. Antibodies are the defense molecules of vertebrates. Their synthesis is initiated when foreign macromolecules (antigens) are present in the body. Each functions by binding its antigen, which leads to removal of the antigen from the body. Normally, antibodies do not exhibit catalytic activity. It was proposed that antibodies raised against transition-state analogs would bind the transition states well and should therefore be catalysts. Subsequent experiments have led to the production of some antibodies with catalytic activity. They show the predicted substrate specificity and relatively modest rate enhancements, up to about 10^5. The results support the theory that transition-state stabilization is a major factor in catalysis.

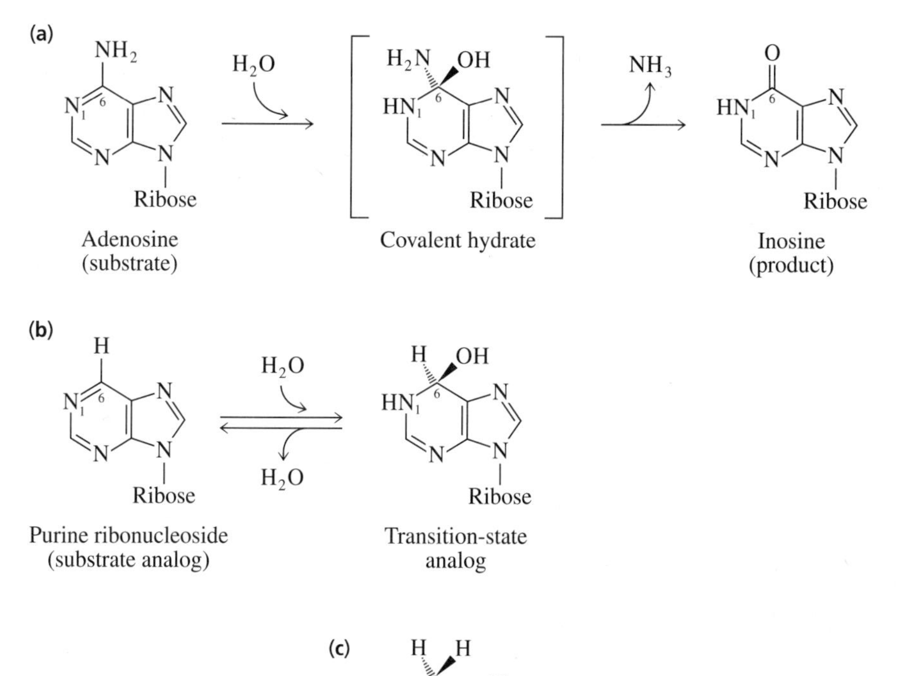

Figure 6·12
Inhibition of adenosine deaminase by a transition-state analog. **(a)** In the deamination of adenosine, a proton adds to N-1 and a hydroxide ion to C-6 to form an unstable covalent hydrate, which decomposes to produce inosine and ammonia. **(b)** When the inhibitor purine ribonucleoside is added to a solution of adenosine deaminase, it also rapidly forms a covalent hydrate, 6-hydroxy-1,6-dihydropurine ribonucleoside. This covalent hydrate is a transition-state analog that binds over a million times more avidly than another competitive inhibitor, 1,6-dihydropurine ribonucleoside **(c)**, which differs from the transition state analog only by lacking the 6-hydroxyl group.

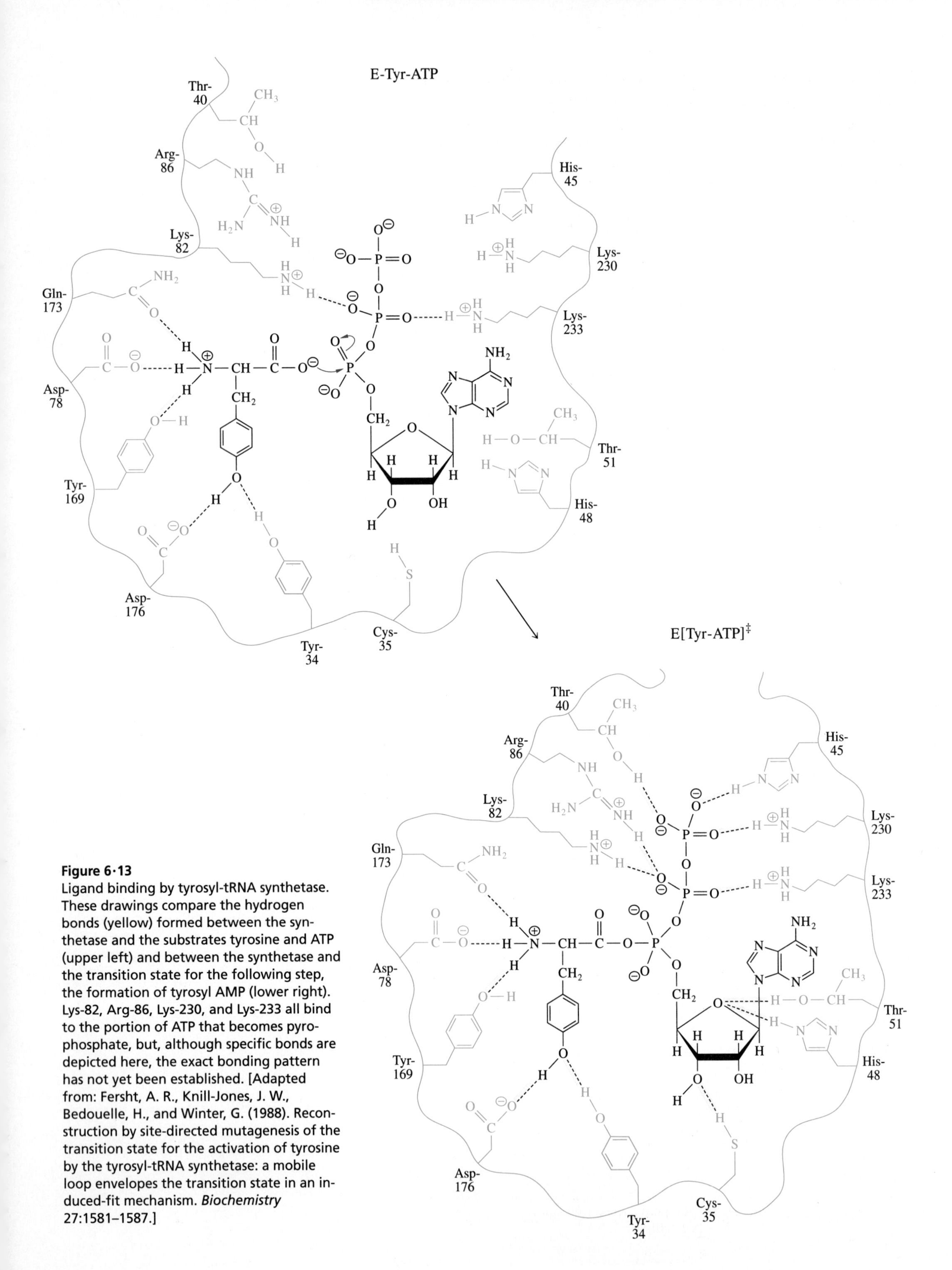

Figure 6·13
Ligand binding by tyrosyl-tRNA synthetase. These drawings compare the hydrogen bonds (yellow) formed between the synthetase and the substrates tyrosine and ATP (upper left) and between the synthetase and the transition state for the following step, the formation of tyrosyl AMP (lower right). Lys-82, Arg-86, Lys-230, and Lys-233 all bind to the portion of ATP that becomes pyrophosphate, but, although specific bonds are depicted here, the exact bonding pattern has not yet been established. [Adapted from: Fersht, A. R., Knill-Jones, J. W., Bedouelle, H., and Winter, G. (1988). Reconstruction by site-directed mutagenesis of the transition state for the activation of tyrosine by the tyrosyl-tRNA synthetase: a mobile loop envelopes the transition state in an induced-fit mechanism. *Biochemistry* 27:1581–1587.]

6·13 Experiments with a tRNA Synthetase Have Shown the Role of Hydrogen Bonding in Catalysis

Aminoacyl-tRNA synthetases catalyze the covalent attachment of amino acids to specific transfer RNA (tRNA) molecules, preceding incorporation of the amino acid into a protein (Section 22·3). Tyrosyl-tRNA synthetase catalyzes the activation of the amino acid tyrosine. The enzymatic charging of this tRNA proceeds in two steps, the formation of the energy-rich enzyme-bound intermediate tyrosyl adenylate (Tyr-AMP) and then the transfer of tyrosine to its tRNA:

$$\text{E} + \text{Tyr} + \text{ATP} \rightleftharpoons \text{E-Tyr-AMP} + \text{PP}_i \quad \textbf{(6·23)}$$

and

$$\text{E-Tyr-AMP} + \text{tRNA} \rightleftharpoons \text{E} + \text{Tyr-tRNA} + \text{AMP} \quad \textbf{(6·24)}$$

The mechanism of this enzyme has been studied in the laboratory of Alan Fersht by systematic site-directed mutagenesis accompanied by kinetic measurements, assisted by knowledge of the three-dimensional structure of the synthetase (with and without tyrosine) and the enzyme-bound Tyr-AMP complex. Each side chain suspected of forming a hydrogen bond with the substrate, the intermediate, or the transition state was replaced, one at a time or in various combinations, by a smaller side chain, usually one that does not form hydrogen bonds. The binding contribution of each residue was estimated from the kinetic effect of each alteration, measured by determinations of k_{cat} and K_m values.

Because no ionizable amino acid residues that could serve as chemical catalysts have been found to be properly located in the active site of the tyrosyl-tRNA synthetase, it has been proposed that this enzyme uses transition-state stabilization as its principal mechanism of catalysis. Mutagenic alteration of a side chain that binds to a substrate only in the transition state should lower k_{cat} but leave K_m unchanged. This result was obtained when Thr-40 was altered to Ala-40, and when His-45 was changed to Gly-45. It was concluded that these side chains must bind the γ-phosphate of ATP in the transition state but not in the E-Tyr-ATP enzyme-substrate complex. Similarly, Cys-35, His-48, and Thr-51 interact with ribose in the transition state but not in the ES complex. After ATP has bound to the enzyme, it attracts Arg-86 and Lys-230. These amino acids are part of mobile protein loops that open to allow substrate to enter or product to leave but in $ES^{\ddagger}$ completely enclose the reactants to contribute to transition-state stabilization. Figure 6·13 illustrates the differences in binding between the E-Tyr-ATP complex and the $E[\text{Tyr-ATP}]^{\ddagger}$ transition state. There are at least seven additional hydrogen bonds in the transition state.

6·14 The Mechanisms of Serine Proteases Illustrate Both Chemical and Binding Modes of Catalysis

Chymotrypsin and the other serine proteases are probably the most thoroughly studied enzymes. Because many of the mechanistic properties of serine proteases apply to enzymes in general, they are suitable examples of how the molecular structures of proteins are responsible for their enzymatic functions.

Covalent modification experiments have shown that both serine and histidine occur in the active site of chymotrypsin (Section 5·8), and X-ray crystallographic studies have revealed that an aspartate residue is also present in the active site (Stereo 5·2 and Section 5·10). The only other charged amino acid side chains in the interior of the protein are the α-amino group of Ile-16 and the β-carboxylate group

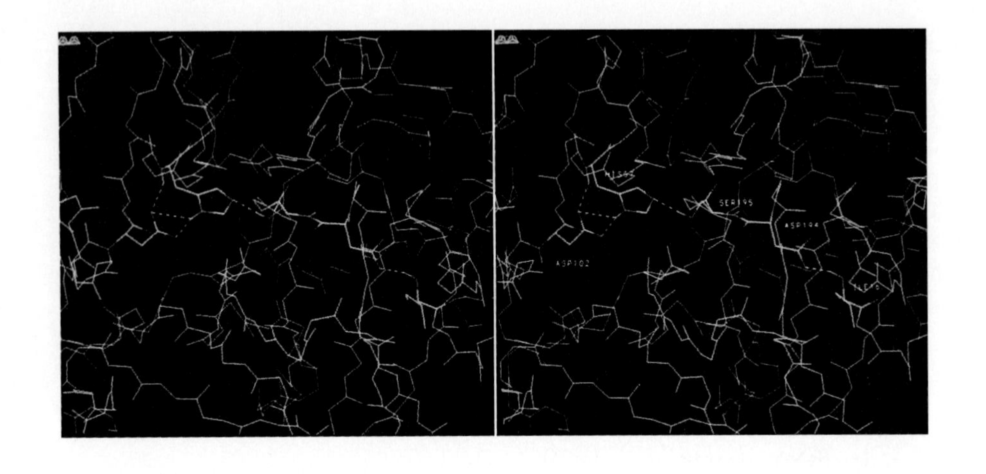

Stereo S6·1
Structure of the catalytic site of chymotrypsin. Active-site residues Asp-102, His-57, and Ser-195 are arrayed in a hydrogen-bonded network. The conformation of these three residues is stabilized by a hydrogen bond between the carbonyl oxygen of the carboxylate side chain of Asp-102 and the peptide-bond nitrogen of His-57 and by formation of an ion pair between the β-carboxyl group of Asp-194 and the α-amino group of Ile-16. (Based on coordinates provided by A. Tulinsky and R. Blevins; stereo image by Ben N. Banks and Kim M. Gernert.)

of Asp-194, which form an ion pair involved in the activation of chymotrypsinogen (Section 5·10). The discovery that Asp-102, buried in a rather hydrophobic environment, is hydrogen bonded to His-57, which in turn is hydrogen bonded to Ser-195 (Stereo S6·1), was particularly exciting. In this grouping, called the *catalytic triad,* His-57—stabilized by Asp-102—abstracts a proton from Ser-195 (Figure 6·14). A common feature of all serine proteases is, in fact, the presence of a Ser-His-Asp catalytic triad.

The discovery that Ser-195 is a catalytic residue of chymotrypsin was surprising because the side chain of serine is usually not a strong enough nucleophile to attack an amide bond. The pK_a of the hydroxymethyl group of a serine residue is usually ~16, similar in reactivity to the hydroxyl group of ethanol. You may recall from organic chemistry that although ethanol can form ethoxides such as sodium ethoxide, this ionization requires the presence of an extremely strong base, such as metallic sodium.

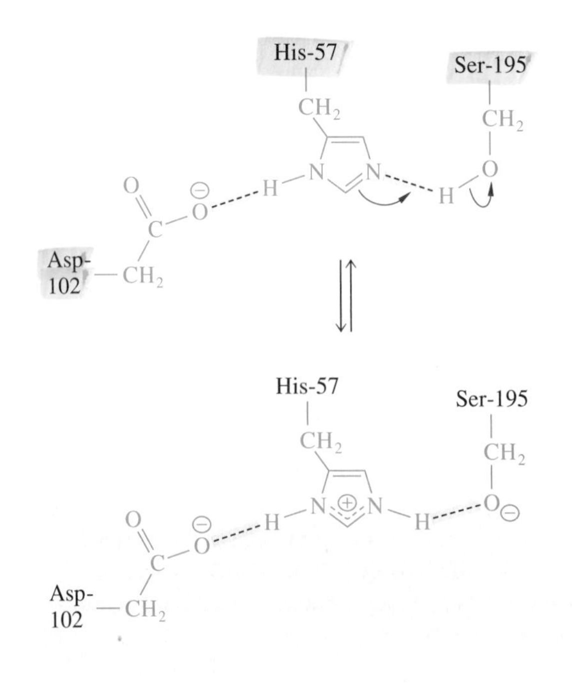

Figure 6·14
The catalytic triad of chymotrypsin. The imidazole ring of His-57 removes the proton from the hydroxymethyl side chain of Ser-195 (to which it is hydrogen bonded), thereby making Ser-195 a powerful nucleophile. This interaction is facilitated by interaction of the imidazolium ion with its other hydrogen-bonded partner, the buried β-carboxyl group of Asp-102.

Augmented by later experiments that showed that Asp-102 does not become protonated during catalysis, a fairly detailed mechanism for chymotrypsin and related serine proteases has been proposed. The six steps of the proposed mechanism are illustrated in Figure 6·15 (next page):

1. The enzyme-substrate complex is formed, orienting the substrate for reaction. Interactions holding the substrate in place include binding of the R_1 group in the specificity pocket. The binding interactions position the carbonyl carbon of the *scissile* peptide bond, or bond susceptible to cleavage, next to the oxygen of Ser-195.

2. A proton is removed from the hydroxyl group of Ser-195 by the basic His-57, and the nucleophilic oxygen attacks the carbonyl carbon of the peptide bond to produce a tetrahedral intermediate (E-TI_1), which is believed to be similar to the transition state for this step of the reaction. When the tetrahedral intermediate is formed, the C—O bond changes from a double bond to a longer single bond. This allows the negatively charged oxygen (the *oxyanion*) of the tetrahedral intermediate to move to a previously vacant position, called the *oxyanion hole,* where it can form hydrogen bonds with the peptide-chain —NH groups of Gly-193 and Ser-195. These hydrogen bonds stabilize the transition state, the oxyanion form of the substrate, by binding it more tightly to the enzyme than the substrate was bound.

3. Next, His-57, as an imidazolium ion, acts as an acid catalyst, donating a proton to the nitrogen of the scissile peptide bond, thus cleaving it. The first product (P_1), the amine, is released. The carboxyl group from the peptide forms a covalent bond with the enzyme, producing an acyl-enzyme intermediate. Steps 1 to 3 are called the *acylation* steps.

4. As the peptide P_1 with the new amino group leaves the active site, water enters. Hydrolysis (*deacylation*) of the acyl-enzyme intermediate starts when water donates a proton to His-57 (a basic imidazole in this step) and provides an $OH^{\ominus}$ group to attack the carbonyl group. A second tetrahedral intermediate (E-TI_2) is formed and stabilized by the oxyanion hole.

5. His-57, once again an imidazolium ion, donates a proton leading to collapse of the second tetrahedral intermediate and formation of the second product—a polypeptide (P_2) with a new carboxylate group.

6. The carboxylate product is released from the active site, and free chymotrypsin is regenerated.

It was once suggested that Asp-102 becomes protonated when Ser-195 is ionized. However, physical measurements have shown that the proton stays on His-57, and Asp-102 remains negative throughout the catalytic process. Nonetheless, Asp-102 is essential for optimal catalysis by serine proteases. It probably stabilizes and orients His-57 so that it can properly accept the proton from Ser-195.

All serine proteases, not only the pancreatic enzymes but also the blood-clotting enzyme thrombin, the blood-clot–dissolving enzyme plasmin, the spermatozoan enzyme acrosin, and the insect enzyme cocoonase, share the same basic mechanism. Part of the evidence for the similarity in mechanism among the serine proteases is the observation that they are all inactivated by treatment with DFP (Section 5·8). Many of the serine proteases share homologies in primary, secondary, and tertiary structure, as revealed by sequence analysis and X-ray crystallography. Presumably, they have evolved from a common ancestral protein.

Subtilisin is one of a family of bacterial proteases that can be inactivated by exposure to DFP but whose primary structures show no homology with the vertebrate family of serine proteases. X-ray crystallography has revealed that even though the tertiary structure of subtilisin bears no resemblance to that of chymotrypsin, the reactive serine in position 221 of subtilisin is hydrogen bonded to a histidine (residue

Figure 6·15
Mechanism of chymotrypsin-catalyzed cleavage of a peptide bond. In Step 1, the initial, noncovalent enzyme-substrate complex forms. The specificity pocket is shaded. In Step 2, the oxygen of the side chain of Ser-195 attacks the carbonyl carbon of the peptide bond of the substrate to form a tetrahedral intermediate. In Step 3, the negatively charged oxygen is stabilized in the oxyanion hole, and the peptide bond of the substrate breaks, forming an amine (P_1) and an acyl-enzyme intermediate. In Step 4, a water molecule that has replaced the amine at the active site attacks the acyl-enzyme intermediate to give a second tetrahedral intermediate. In Step 5, the tetrahedral intermediate collapses, forming the carboxylate product (P_2) and regenerating active chymotrypsin. Finally, in Step 6, P_2 leaves the active site.

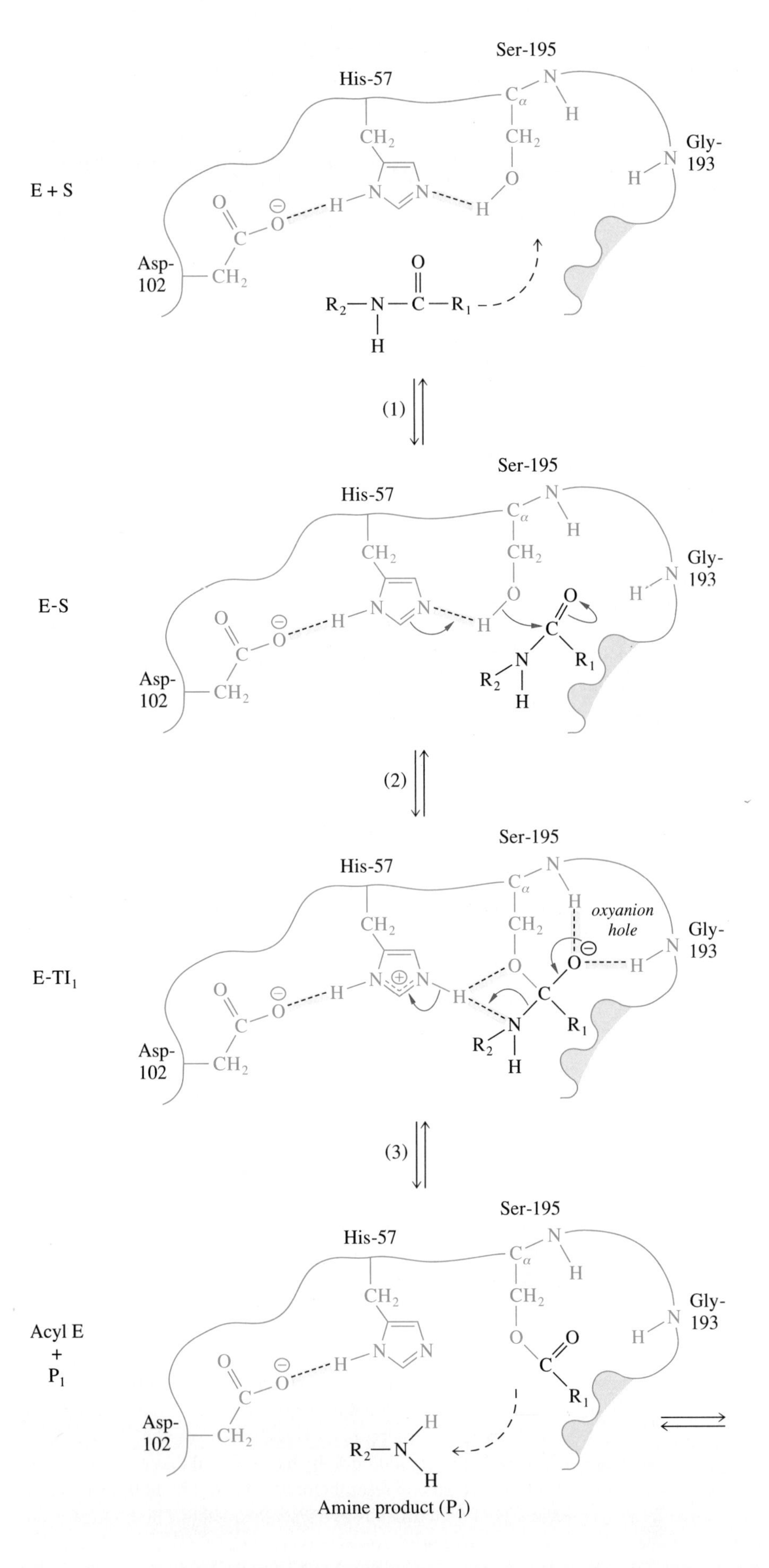

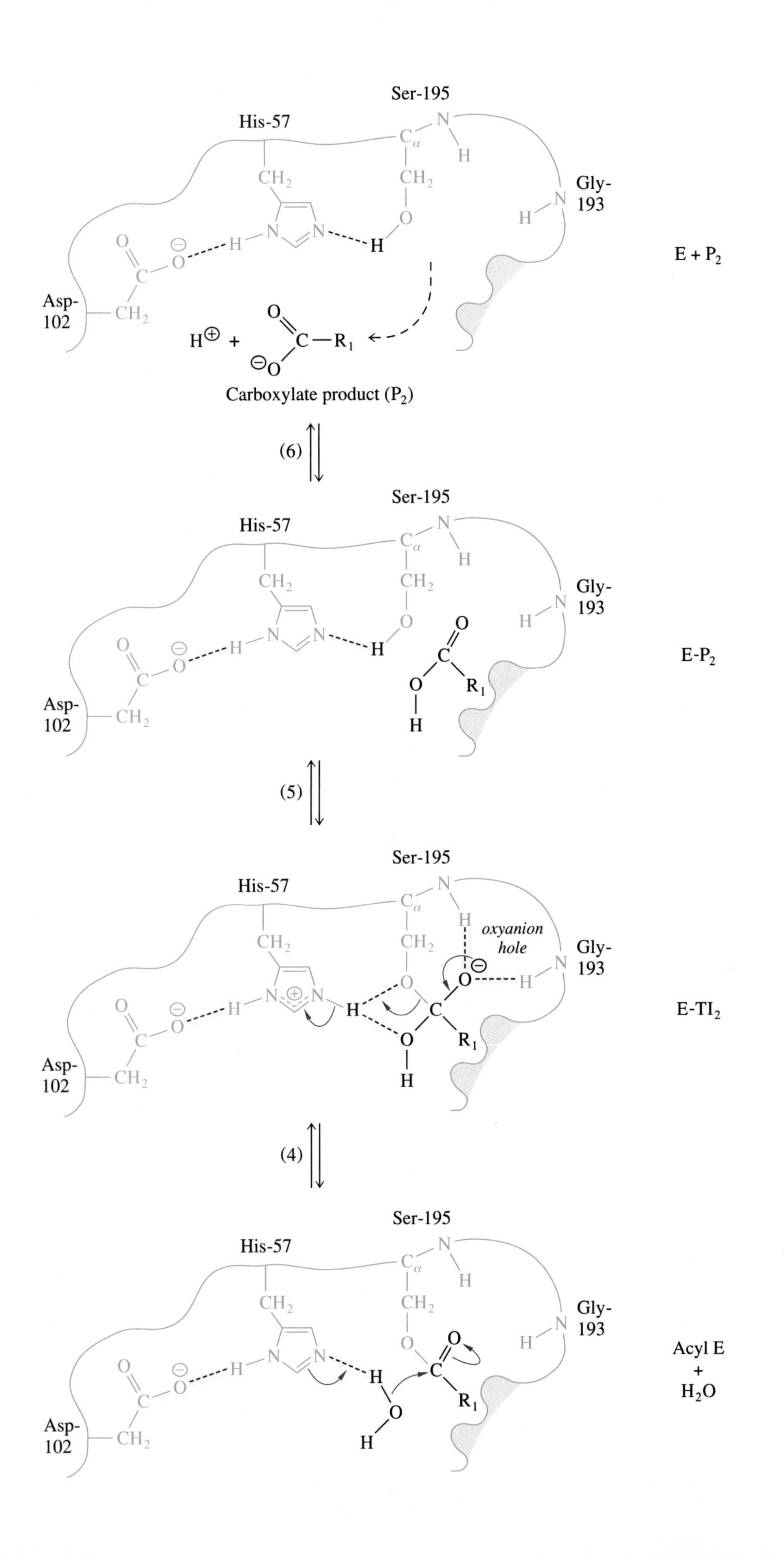

$E + P_2$

$E\text{-}P_2$

$E\text{-}TI_2$

Acyl E
+
H_2O

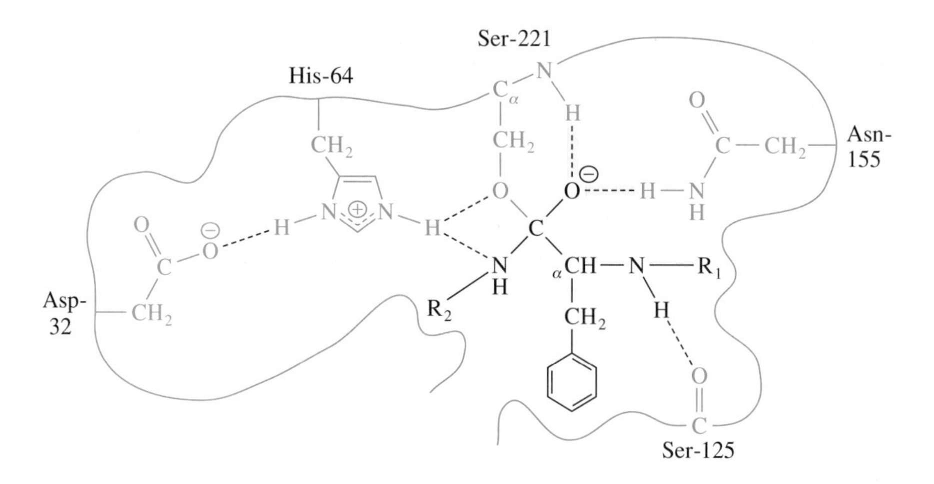

Figure 6·16
Binding of the tetrahedral intermediate by subtilisin. The oxyanion of the tetrahedral intermediate is bound by two hydrogen bonds to subtilisin, one to the backbone amide of Ser-221 and the other to the side chain of Asn-155. These bonds do not exist in the enzyme-substrate complex. This structure is an example of transition-state stabilization.

64) which, in turn, is hydrogen bonded to a buried aspartate (residue 32) (Figure 6·16). Thus, the mechanism outlined for chymotrypsin also applies to the subtilisin family of enzymes. The similarities between subtilisin and chymotrypsin in catalytic mechanism and in the structure of the catalytic triad are an example of *convergent evolution*, a phenomenon in which proteins take different evolutionary paths to a common mechanism and function.

Site-directed mutagenesis has been used to assess the catalytic importance of the amino acids in the catalytic triad. The active-site aspartate-102 of trypsin was changed to asparagine and the kinetic properties of the mutated and unmutated enzymes were compared. At neutral pH, the wild-type trypsin had values of k_{cat} and k_{cat}/K_m that were about 10^4 greater than that of the mutant enzyme. The rate of inactivation of the mutant enzyme by the inhibitor DFP decreased to the same extent, showing that conversion of the aspartate to asparagine decreased the nucleophilic properties of the active-site serine.

Subtilisin exhibits a rate acceleration of approximately 10^{10}. The catalytic triad of subtilisin was completely dissected by mutagenesis experiments carried out by Paul Carter and James Wells. Each of the three amino acid components was changed, in turn, to an alanine residue. Alanine was chosen as the replacement residue to avoid any unfavorable steric, charge, or hydrogen-bond effects. Replacement of either the covalent catalyst, serine, or the acid-base catalyst, histidine, had the largest effect, a 10^6-fold loss in activity in either case. Replacement of aspartate decreased activity 10^4-fold. The triple mutant, in which all three active-site amino acids were converted to alanine residues, had activity similar to either the serine or histidine single mutants. The combined action of the three amino acids of the triad accelerates the hydrolysis of the substrate by a factor of just over 10^6. The least active mutant forms of subtilisin had a residual activity approximately 3000 times the nonenzymatic rate. This appreciable residual catalysis is ascribed to binding effects such as transition-state stabilization.

The oxyanion hole, the site that binds the oxyanion of the tetrahedral intermediate, cannot be modified by mutagenesis of either trypsin or chymotrypsin because the hydrogen-bond donors in both these enzymes are amide groups of the protein backbones and so cannot be changed by mutation. With subtilisin, the amide side chain of an asparagine residue participates in the stabilization of the tetrahedral intermediate (Figure 6·16). The catalytic importance of the extra hydrogen bonds has been verified by mutagenesis experiments in which Asn-155 was replaced by amino acids that cannot form hydrogen bonds. The removal of one hydrogen-bonding side chain in subtilisin lowered the activity of the enzyme by a factor of about 10^3, with no change in the K_m for its substrates. This experiment verifies that transition-state stabilization by the oxyanion hole provides considerable enhancement to the rate of hydrolysis catalyzed by subtilisin.

All of the catalytic modes described in this chapter are used in the mechanisms of serine proteases. In the reaction scheme shown in Figure 6·15, Steps 1 and 4 in the forward direction use proximity to gather the reactants. For example, when a water molecule replaces the amine P_1, it is held by histidine in Step 4, providing a proximity effect. Acid-base catalysis by histidine lowers the barriers for Steps 2 and 4. Covalent catalysis using the $—CH_2OH$ of serine occurs in Steps 2 through 5. The extremely unstable tetrahedral intermediates at Steps 2 and 4 are stabilized by the oxyanion hole. The chemical catalysis and binding effects—both proximity and transition-state stabilization—all make major contributions to the enzymatic activity of serine proteases.

Having gained insight into the general mechanisms of action of enzymes, we can now examine reactions that involve coenzymes. These reactions require reactive groups that cannot be supplied by the side chains of amino acids. Chapter 7 shows how coenzymes provide additional reactive groups to the active sites of enzymes.

Summary

Reaction mechanisms describe in atomic detail how a reaction proceeds. Kinetic experiments—measurements of reaction rates under varying conditions—offer an indirect approach to examining reactions that can contribute to the elucidation of mechanisms. Studies of the structures of enzymes supply additional mechanistic information.

The rate of a reaction depends on the rate of effective collisions between reactants. For each step in a reaction, there is a transition state, or energized configuration, that the reactants must pass through. The amount of energy needed to form the transition state, called the activation energy, affects the rate of the reaction. Catalysis provides a faster reaction pathway by lowering the energy of activation.

Ionizable and reactive amino acid residues in the active site of an enzyme form its catalytic center. Two major chemical modes of enzymatic catalysis are acid-base catalysis and covalent catalysis. In acid-base catalysis, proton transfer contributes to the acceleration of the reaction, with protons either donated by a weak acid or accepted by a base in the active site. The effect of pH on the rate of an enzymatic reaction can suggest the identities of active-site components. In covalent catalysis, the substrate or a portion of it is attached to the enzyme covalently to form a reactive intermediate.

The rates of catalysis for a few enzymes are so high that they approach the upper limit set by the rate at which reactants approach each other by diffusion. For enzymatic catalysis to be so rapid, each step of the reaction must be rapid, and the activation energies for the various steps must be balanced.

Although acid-base catalysis and covalent catalysis are important, the greatest part of the rate acceleration achieved by an enzyme generally arises from the binding of reacting ligands to the enzyme. The initial formation of a noncovalent enzyme-substrate complex (ES) collects and orients reactants, which alone produces an acceleration of the reaction, a phenomenon termed the proximity effect. The binding of reactants must be relatively weak, yet strong enough that the entropy of the reactants is considerably decreased. The energy of activation is further lowered by the binding of transition states with greater affinity than the binding of substrates. Support for transition-state stabilization as a major catalytic mode comes from the potent inhibitory activity of transition-state analogs, synthetic compounds that resemble transition states structurally. Enzymes, therefore, catalyze reactions by assisting first in the formation, and then in the stabilization, of transition states.

The role of binding in catalysis has been demonstrated with tyrosyl-tRNA synthetase using structural information, kinetic experiments, and site-directed mutagenesis. The results of these experiments suggest that the transition state, $ES^{\ddagger}$, is stabilized by at least seven hydrogen bonds that form only with the transition state, not with the reactant in its ground state.

The serine proteases, exemplified by chymotrypsin, utilize both chemical and binding modes of catalysis. All serine proteases possess a hydrogen-bonded Ser–His–Asp catalytic triad in their active sites. The serine residue serves as a covalent catalyst, and the histidine residue serves as an acid-base catalyst. The aspartate residue, which is essential for maximum activity, aligns the histidine residue and stabilizes its protonated form. Anionic tetrahedral intermediates form additional hydrogen bonds with the enzyme, contributing to catalysis by stabilizing the transition state.

Selected Readings

General References on Enzymatic Mechanisms

Fersht, A. (1985). *Enzyme Structure and Mechanism,* 2nd ed. (New York: W. H. Freeman and Company).

Jencks, W. P. (1969). *Catalysis in Chemistry and Enzymology* (New York: McGraw-Hill). This monograph explains the basis of the chemistry of catalysis in great detail.

Page, M. I. (1987). Theories of enzyme catalysis. In *Enzyme Mechanisms,* M. I. Page and A. Williams, eds. (London: Royal Society of Chemistry), pp. 1–13. A short and up-to-date review.

Scrimgeour, K. G. (1977). *Chemistry and Control of Enzyme Reactions* (London: Academic Press). Chapters 4, 6, 7, and 10.

Walsh, C. (1979). *Enzymatic Reaction Mechanisms* (San Francisco: W. H. Freeman and Company).

Diffusion-Controlled Enzymes

Knowles, J. R., and Albery, W. J. (1977). Perfection in enzyme catalysis: the energetics of triosephosphate isomerase. *Acc. Chem. Res.* 10:105–111.

Lodi, P. J., and Knowles, J. R. (1991). Neutral imidazole is the electrophile in the reaction catalyzed by triosephosphate isomerase: structural origins and catalytic implications. *Biochemistry* 30:6948–6956.

Sharps, K., Fine, R., and Honig, B. (1987). Computer simulations of the diffusion of a substrate to an active site of an enzyme. *Science* 236:1460–1463. This article describes the attraction of the substrate to the active site.

Binding and Catalysis

Jencks, W. P. (1975). Binding energy, specificity, and enzymatic catalysis: the Circe effect. *Adv. Enzymol.* 43:219–410. A thorough view of binding and catalysis.

Jencks, W. P. (1987). Economics of enzyme catalysis. In *Cold Spring Harbor Symposia on Quantitative Biology* 52:65–73. A summary of recent experiments on enzymatic catalysis.

Hackney, D. D. (1990). Binding energy and catalysis. In *The Enzymes,* Vol. 19, 3rd ed., P. D. Boyer, ed. (New York: Academic Press), pp. 1–36. Section II of this article explains the interrelation of binding and catalysis very clearly.

Transition-State Stabilization

Benkovic, S. J. (1992). Catalytic antibodies. *Annu. Rev. Biochem.* 61:29–54.

Fersht, A. R., Knill-Jones, J. W., Bedouelle, H., and Winter, G. (1988). Reconstruction by site-directed mutagenesis of the transition state for the activation of tyrosine by the tyrosyl-tRNA synthetase: a mobile loop envelops the transition state in an induced-fit mechanism. *Biochemistry* 27:1581–1587.

Kraut, J. (1988). How do enzymes work? *Science* 242:533–540. A review with strong emphasis on transition-state stabilization.

Wolfenden, R. (1972). Analog approaches to the structure of the transition state in enzyme reactions. *Acc. Chem. Res.* 5:10–18.

Serine Proteases

Carter, P., and Wells, J. A. (1988). Dissecting the catalytic triad of a serine protease. *Nature* 332:564–568. Use of site-directed mutagenesis to study the individual active-site residues of subtilisin.

Kossiakoff, A. A., and Spencer, S. A. (1981). Direct determination of the protonation states of aspartic acid 102 and histidine 57 in the tetrahedral intermediate of the serine proteases: neutron structure of trypsin. *Biochemistry* 20:6462–6467.

7

Coenzymes

The catalytic activities of many enzymes depend on the presence of components called *cofactors*. Cofactors are chemicals required by inactive *apoenzymes* (proteins only) to convert them to active *holoenzymes* (Greek, *holos*, complete or whole). There are two types of cofactors, essential ions and organic compounds known as *coenzymes* (Figure 7·1). Both organic and inorganic cofactors are essential portions of the active sites to which they specifically bind. Some essential ions, called activator ions, are reversibly bound and often participate in the binding of substrates, whereas tightly bound metal ions frequently participate directly in catalytic reactions.

Coenzymes, which are larger than most metabolites, act as group-transfer reagents. They are specific for the chemical groups, called mobile metabolic groups, that they accept and donate. For some coenzymes, the mobile metabolic group is hydrogen or an electron; other coenzymes carry larger, covalently attached chemical groups. Mobile metabolic groups are attached at the *reactive center* of the coenzyme (colored red in structures presented in this chapter). Focussing on the chemical reactivity of the reactive center simplifies the study of these rather complex molecules.

Coenzymes are separated into two types based upon how they interact with the apoenzyme. Coenzymes of one type—often called *cosubstrates*—are actually substrates in enzyme-catalyzed reactions; a cosubstrate is altered in the course of the

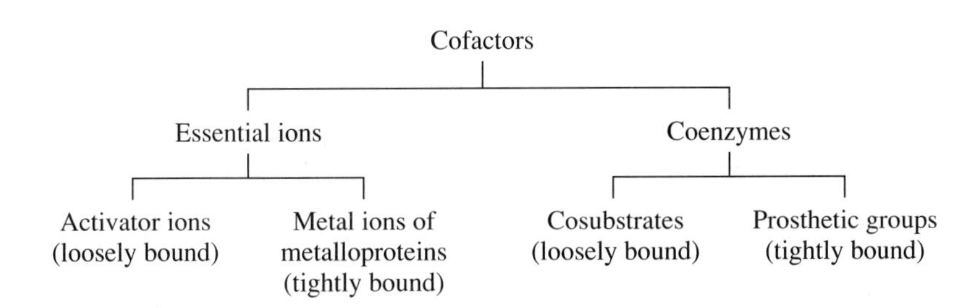

Figure 7·1
Types of cofactors. Cofactors can be divided into two types—essential ions and coenzymes—which can be further distinguished by their affinities to their apoenzymes. An essential ion may be an activator ion, which binds loosely to an enzyme, or a metal ion that is bound tightly within a metalloprotein. Similarly, a coenzyme may be a loosely bound cosubstrate or a tightly bound prosthetic group.

reaction and dissociates from the active site. The original structure of the cosubstrate is regenerated in a subsequent enzyme-catalyzed reaction so that the cosubstrate is recycled repeatedly within the cell, unlike the product derived from a simple substrate, which typically undergoes further transformation. Cosubstrates are extremely important in cellular metabolism, serving to shuttle metabolic groups among different enzyme-catalyzed reactions.

The distinction between a cosubstrate and a regular substrate can be illustrated by comparing the fates of the coenzyme adenosine triphosphate (ATP) and the substrate glucose following the reaction catalyzed by the enzyme hexokinase. This reaction is the first step in the pathway of glycolysis (Chapter 12), which converts glucose to pyruvate.

$$\text{ATP} + \text{Glucose} \longrightarrow \text{ADP} + \text{Glucose 6-phosphate} \qquad \textbf{(7·1)}$$

Although both reactants are converted to products by this phosphoryl-group transfer reaction, ADP is usually reconverted to ATP by a subsequent enzymatic reaction, whereas glucose 6-phosphate is metabolized to other substances (for instance, it may be degraded through a series of enzymatic reactions to the three-carbon acids pyruvate and lactate, or it may be utilized by certain cells for the biosynthesis of sucrose or the carbohydrate polymers starch or glycogen).

The second type of coenzyme is referred to as a *prosthetic group*. Whereas a cosubstrate associates reversibly with the coenzyme-binding site of an enzyme, a prosthetic group remains bound to the protein. Like the ionic amino acid residues that comprise an active site, a prosthetic group must be regenerated to its original form during each full catalytic event; otherwise, the holoenzyme cannot remain catalytically active. Both coenzyme types supply active sites with reactive groups that are not present on the side chains of amino acid residues.

7·1 Some Coenzymes Are Derived from Common Metabolites

Some coenzymes are synthesized from common metabolites and are thus referred to as *metabolite coenzymes*. ATP is the most familiar and most abundant of the metabolite coenzymes. Once formed, ATP can donate phosphoryl, adenosyl, or adenylyl (AMP) groups. Another metabolite coenzyme, *S*-adenosylmethionine (Figure 7·2a), is synthesized by the reaction of methionine with ATP.

$$\text{Methionine} + \text{ATP} \longrightarrow S\text{-Adenosylmethionine} + \text{PP}P_i \qquad \textbf{(7·2)}$$

In this reaction, ATP donates an adenosyl group to the sulfur of methionine to form a sulfonium compound. The inorganic tripolyphosphate (PPP_i) that is formed is

Figure 7·2
Structures of two metabolite coenzymes. **(a)** *S*-Adenosylmethionine. **(b)** The substituted pyrimidine nucleotide uridine diphosphate glucose (UDP-glucose). Recall from Chapter 1 that nucleotides are biochemical building blocks containing a nitrogenous base, a five-carbon sugar (often ribose), and one or more phosphate groups. Nucleotides are structural components of several coenzymes in this chapter.

(a)

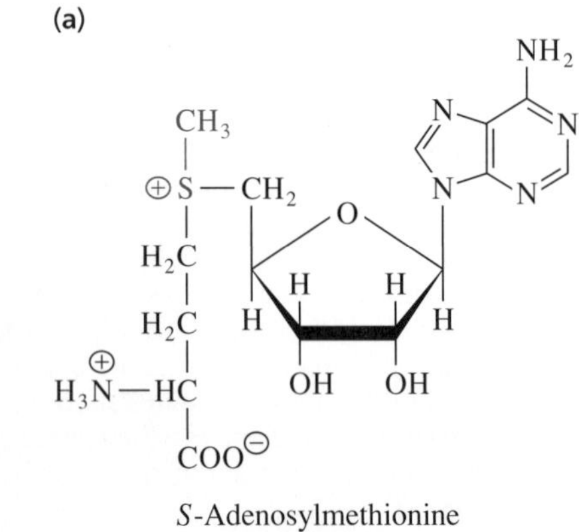

(b)

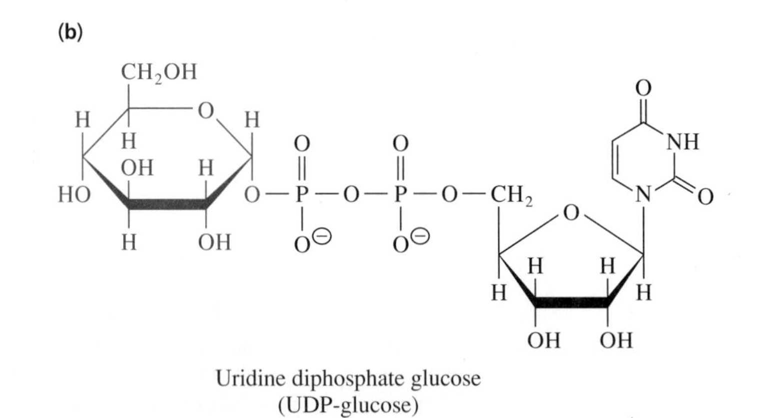

rapidly hydrolyzed to pyrophosphate (PP_i) and phosphate (P_i) by the action of the same enzyme that catalyzes the formation of *S*-adenosylmethionine. The thiomethyl group of methionine is unreactive, but the positively charged sulfonium of *S*-adenosylmethionine is a potent donor of methyl groups. Nucleophilic acceptors readily react with *S*-adenosylmethionine, which is the donor of most methyl groups used in biosynthesis.

Also among the metabolite coenzymes are the substituted pyrimidine nucleotides. Uridine nucleotides are involved in the transfer of sugar groups, and cytidine nucleotides transfer alkyl phosphates. For example, uridine diphosphate glucose (UDP-glucose, Figure 7·2b) is the source of glucose for the biosynthesis of the storage carbohydrate glycogen and the structural polysaccharide of plants, cellulose.

7·2 In Animals, Many Coenzymes Are Derived from B Vitamins

Vitamins are defined as organic substances that must be obtained by an animal as nutrients, usually in small amounts (mg or μg quantities per day). It is believed that many organisms that have vitamin requirements conserve chemical energy by relying on other organisms to supply these micronutrients, thereby avoiding the cost of biosynthesizing the numerous enzymes required to produce the vitamins metabolically. *Vitamin-derived coenzymes* are those formed from precursors that must be obtained as nutrients. In animal cells, many coenzymes are synthesized from dietary precursors known as B vitamins. The vitamin B complex comprises the water-soluble vitamins recovered from a natural source via homogenization and extraction with an aqueous solution. Although the ultimate sources of B vitamins are usually plants and microorganisms, carnivorous animals can obtain B vitamins from meat.

When a vitamin is lacking from the diet of an animal, the result is a nutritional deficiency disease, such as scurvy, beriberi, or pellagra, which can be prevented or cured by eating the appropriate vitamin. The concept of vitamins developed early in this century, yet the link between scurvy and nutrition was recognized four centuries ago. British navy physicians discovered that citrus juice was a remedy for scurvy in sailors whose diet lacked fresh fruits and vegetables. It was not until 1930, however, that ascorbic acid (vitamin C), shown in Figure 7·3, was isolated and proven to be the essential dietary component supplied by citrus juices. The carbohydrate ascorbic acid is a lactone, an internal ester in which the C-1 carboxylate group is condensed with the C-4 hydroxyl group. We now know that one of the functions of ascorbic acid is to act as a reducing agent that provides for the proper enzymatic hydroxylation of collagen (Section 4·5). Although most animals can

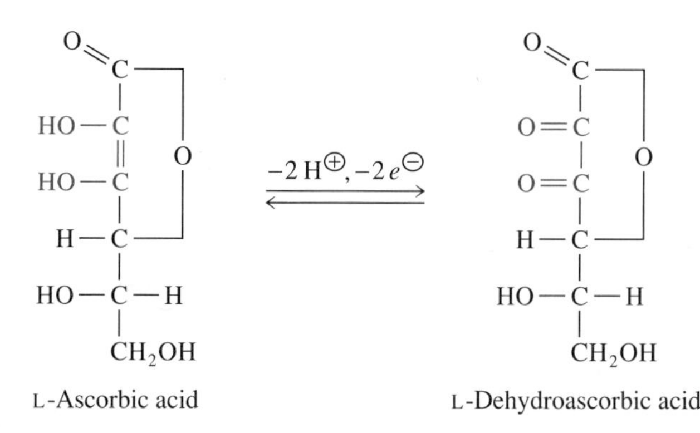

Figure 7·3
Structure of ascorbic acid (vitamin C) and its dehydro or oxidized form.

synthesize ascorbic acid in the course of carbohydrate metabolism, guinea pigs and primates (including humans) lack this ability and must therefore rely on dietary sources.

The term *vitamin* (originally spelled "vitamine") was coined by Casimir Funk in 1912 to describe a "vital amine" from rice husks that cured beriberi, a nutritional deficiency disease that results in neural degeneration (polyneuritis). Beriberi was first described in fowl, then in humans whose diets consisted largely of polished rice. When more of these organic micronutrients were discovered, the name vitamin was applied to a large group of organic compounds. Two broad classes were recognized: water-soluble vitamins and lipid- or fat-soluble vitamins. The anti-beriberi vitamin (thiamine) became known as vitamin B_1, and other components of the vitamin B complex were referred to as vitamin B_2, vitamin B_6, and so on. Even though many of these substances proved not to be amines, the term vitamin has been retained.

Since water-soluble vitamins are readily excreted in the urine, they are required daily in small amounts. Conversely, fat-soluble vitamins, which include vitamins A, D, E, and K, are stored by animals, and excessive intakes should be avoided since they can result in toxic conditions known as hypervitaminoses.

Table 7·1 lists the major vitamins of the B complex and the more common coenzymes derived from them. The table also briefly describes the metabolic role of each coenzyme. Lipoic acid is included because it is often grouped with the B vitamins. However, with the exception of some bacteria, most organisms can synthesize lipoic acid, and it is therefore not a required nutrient.

Table 7·1 Major B vitamins and related coenzymes

B vitamin	Principal coenzymes	Major metabolic roles	Mechanistic role
Niacin	Nicotinamide adenine dinucleotide ($NAD^{\oplus}$), nicotinamide adenine dinucleotide phosphate ($NADP^{\oplus}$), and their reduced forms	Oxidation-reduction reactions involving two-electron transfers	Cosubstrate
Riboflavin (B_2)	Flavin mononucleotide (FMN), flavin adenine dinucleotide (FAD), and their reduced forms	Oxidation-reduction reactions involving one- and two-electron transfers	Prosthetic group
Thiamine (B_1)	Thiamine pyrophosphate (TPP)	Aldehyde-transfer reactions	Prosthetic group
Pyridoxine (B_6)	Pyridoxal phosphate (PLP)	Reactions involving transfer of groups to and from amino acids	Prosthetic group
Biotin	Biocytin (biotin bound to ε-amino group in a biotinylated enzyme)	Reactions involving ATP-dependent carboxylation of substrates or carboxyl-group transfer between substrates	Prosthetic group
Folic acid	Tetrahydrofolate	Reactions involving transfers of one-carbon substituents, especially formyl and hydroxymethyl groups; provides the methyl group for thymine in DNA	Cosubstrate
Pantothenic acid	Coenzyme A (CoA)	Acyl-group transfer reactions	Cosubstrate
Vitamin B_{12} (cobalamin)	Adenosylcobalamin and methylcobalamin	Intramolecular rearrangements and methyl-group transfer reactions	Prosthetic group
Lipoic acid (not a vitamin)	Lipoamide residue (lipoyl group bound to ε-amino group in a protein)	Reactions involving oxidation of a hydroxyalkyl group from TPP and subsequent transfer as an acyl group	Prosthetic group

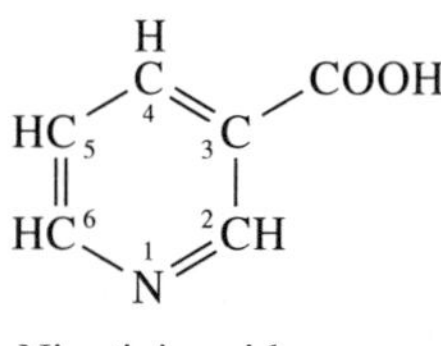

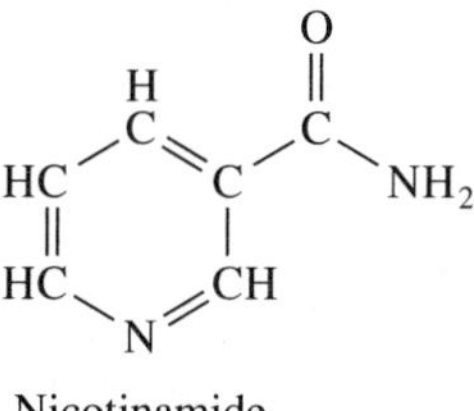

Figure 7·4
Structures of nicotinic acid (niacin) and nicotinamide.

7·3 NAD⊕ and NADP⊕ Are Nucleotides Derived from Niacin

The nicotinamide coenzymes were the first coenzymes to be recognized. By the mid-1930s, *nicotinamide adenine dinucleotide* (NAD⊕) and the closely related *nicotinamide adenine dinucleotide phosphate* (NADP⊕) had been isolated and structurally characterized in the laboratories of Hans von Euler-Chelpin and Otto Warburg. Both coenzymes contain nicotinamide, the amide of nicotinic acid (Figure 7·4). At about the same time, Conrad Elvehjem and D. Wayne Woolley, who were studying a deficiency disease of dogs called "blacktongue" and its counterpart in humans, pellagra, succeeded in identifying nicotinic acid (often referred to as niacin) as the antipellagra factor. It was established that nicotinic acid or nicotinamide are essential in the diets of animals and that they serve as precursors of the coenzymes NAD⊕ and NADP⊕. (In many species, metabolism of tryptophan can also lead to NAD⊕. Thus, dietary tryptophan can spare some of the requirement for niacin or nicotinamide.)

Structurally, nicotinic acid is the 3-carboxyl derivative of pyridine. Accordingly, nucleotide coenzymes containing nicotinamide are often referred to as pyridine nucleotide coenzymes. The structures of NAD⊕ and NADP⊕ and their reduced forms, NADH and NADPH, are shown in Figure 7·5. Note that both coenzymes contain a phosphoanhydride linkage that joins two 5′-nucleotides, adenosine monophosphate and the ribonucleotide of nicotinamide, called nicotinamide

Figure 7·5
Oxidized and reduced forms of NAD (and NADP). NAD⊕ is the major biological oxidizing agent in catabolic processes. C-4 of the nicotinamide group is reduced when NAD⊕ is converted to NADH. Oxidation of NADH in mitochondria provides energy that is recovered as ATP. In NADP⊕, the 2′-hydroxyl of the sugar ring of adenosine is substituted by a phosphoryl group. NADP⊕ is reduced to NADPH, largely by reactions of the pentose phosphate pathway (Chapter 15) or by photosynthesis (Chapter 16), and is the reducing agent for many biosynthetic processes.

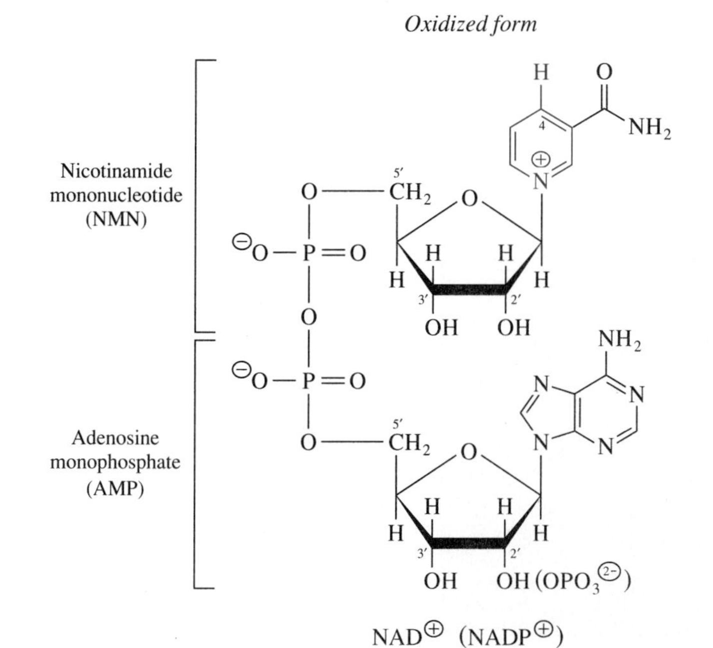

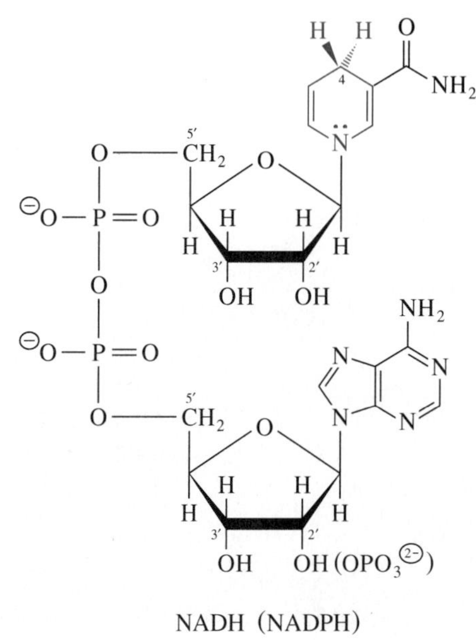

mononucleotide (NMN). In the case of NADP⊕ and NADPH, a phosphoryl group is present on the 2′-oxygen of the adenylate moiety. Pyridine nucleotide–dependent dehydrogenases catalyze the oxidation of their substrates by transferring two electrons and a proton in the form of a hydride ion (H⊖) to C-4 of the nicotinamide group of NAD⊕ or NADP⊕, generating the reduced forms, NADH or NADPH. Thus, reduction and oxidation reactions involving the pyridine nucleotides always occur two electrons at a time. NADH and NADPH are said to possess "reducing power." NADPH can supply energy and hydrogen for reduction reactions, and NADH can be oxidized in mitochondria, leading to the production of chemical potential energy in the form of ATP.

NADH and NADPH exhibit an absorbance peak at 340 nm contributed by the dihydropyridine ring, whereas NAD⊕ and NADP⊕ do not absorb light at this wavelength. The appearance and disappearance of the absorbance band at 340 nm is useful for measuring the rates of enzymatic pyridine nucleotide–linked oxidations and reductions.

Lactate dehydrogenase is an NAD⊕-dependent enzyme that catalyzes the reversible oxidation of lactate, the end product of glucose degradation under anaerobic (oxygen-free) conditions. Note that a proton is formed when either NAD⊕ or NADP⊕ is reduced.

$$\underset{\text{Lactate}}{H_3C-\overset{\overset{\displaystyle OH}{|}}{CH}-COO^{\ominus}} + NAD^{\oplus} \rightleftharpoons \underset{\text{Pyruvate}}{H_3C-\overset{\overset{\displaystyle O}{\|}}{C}-COO^{\ominus}} + NADH + H^{\oplus} \qquad (7\cdot3)$$

As can be seen in Stereo S7·1, a histidine residue, His-195, is located within the active site of the enzyme. Figure 7·6 depicts the mechanism by which this histidine residue, serving as a base catalyst, participates in the catalytic action of lactate dehydrogenase. It abstracts a proton from the C-2 hydroxyl group of lactate, facilitating transfer of the hydride ion from C-2 of the substrate to C-4 of the bound NAD⊕. In this mechanism, we see that both the enzyme *and* the coenzyme are involved in catalyzing the oxidation of lactate to pyruvate. Like most dehydrogenases, lactate dehydrogenase must bind the pyridine nucleotide cosubstrate to form the holoenzyme before it binds its simple substrate (Figure 7·7).

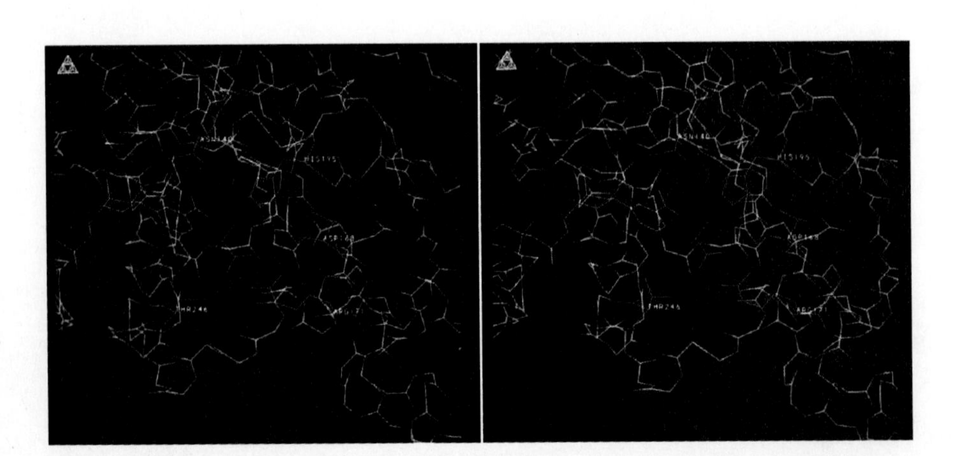

Stereo S7·1
Structure of the active-site region of lactate dehydrogenase with NAD⊕ and lactate bound. For most of the protein molecule, only the backbone is shown. The side chains of the active-site residues are shown in red and labelled. A portion of NAD⊕ is shown in orange, lactate in purple. The imidazole ring of His-195 is near C-2 of lactate. The positively charged side chain of Arg-171 binds the carboxylate group of lactate electrostatically. (Based on coordinates provided by U. Grau and M. Rossman; stereo image by Ben N. Banks and Kim M. Gernert.)

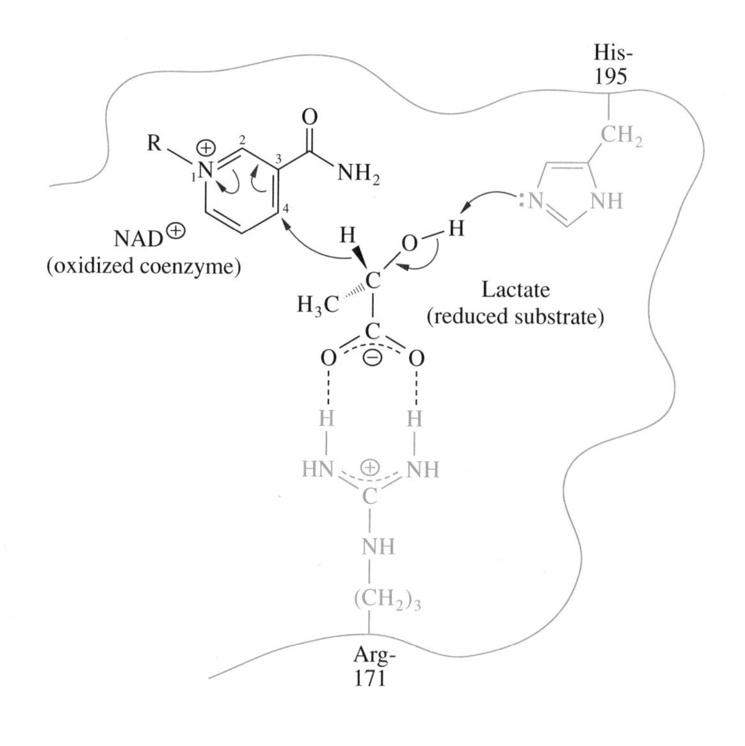

Figure 7·6
Mechanism of lactate dehydrogenase leading to the formation of pyruvate from lactate. The reversible reaction catalyzed by lactate dehydrogenase involves transfer of a hydride ion (H⊖) from the reduced substrate, lactate, to NAD⊕, as shown here, or from the reduced coenzyme, NADH, to the oxidized substrate, pyruvate.

With all NAD⊕- or NADP⊕-dependent dehydrogenases, reduction of the coenzyme is strictly stereospecific. Any given dehydrogenase transfers a hydride ion exclusively to one side (face) of the pyridine ring and not to the other. In the reverse reaction catalyzed by the same enzyme (in the case of lactate dehydrogenase, for example, reduction of pyruvate to lactate), the same hydrogen is transferred (as a hydride ion) from the reduced coenzyme to the substrate.

Although NAD⊕-dependent dehydrogenases differ in some structural aspects, many have several common features. A number of the dehydrogenases that have been studied have two domains per subunit. One domain binds the pyridine-nucleotide cosubstrate, and the other—the catalytic domain—binds the substrate. The active-site cleft lies between the two domains. In lactate dehydrogenase, the NAD⊕-binding domain includes a six-stranded β sheet and four α helices of the enzyme. The adenine moiety of NAD⊕ is buried in a hydrophobic pocket, whereas the pyrophosphate bridge is bound electrostatically to an arginine and a lysine residue. The nicotinamide ring is in a cleft in the interior of the protein. The carbonyl group of the 3-carboxamide of nicotinamide forms a specific hydrogen bond with the apoenzyme, enforcing hydride addition to only one face of the pyridine ring. In contrast to the similarities of the nucleotide binding domains, the structures of the catalytic domains of NAD⊕-dependent dehydrogenases vary widely.

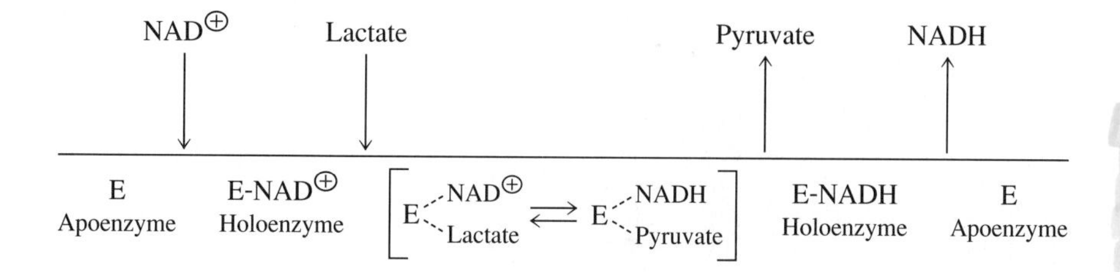

Figure 7·7
Ordered kinetic mechanism for lactate dehydrogenase. Notice that the coenzyme NAD⊕ is bound first and that its reduced form, NADH, is released last.

7·4 FAD and FMN Are Nucleotides Containing Riboflavin

During their studies of the nicotinamide coenzymes, Warburg and W. Christian observed that NADPH could be reoxidized by molecular oxygen, provided that a protein from yeast was present. Purified from brewer's yeast, this protein was yellow and consequently became known as the "old yellow enzyme." Several research groups succeeded in separating the yellow prosthetic group from the colorless apoenzyme and showed that it is the phosphate ester of riboflavin. Subsequent recombination of the two fractions restored catalytic activity. This was the first demonstration that an apoenzyme and its prosthetic group could be reversibly separated.

Riboflavin (Figure 7·8a), or vitamin B_2, consists of the five-carbon alcohol ribitol (a reduced form of ribose) linked to a nitrogen atom of isoalloxazine, the characteristic feature of flavins (Latin, *flavus*, yellow). The yellow color of the prosthetic group is imparted by the conjugated double-bond system of isoalloxazine. Riboflavin is abundant in milk, whole grains, and liver.

The two nucleotide-coenzyme forms of riboflavin are *flavin mononucleotide* (FMN, riboflavin 5′-phosphate) and *flavin adenine dinucleotide* (FAD), whose structures are shown in Figure 7·8b. Like NAD⊕ and NADP⊕, FAD contains adenine and a pyrophosphate linkage; however, unlike the pyridine nucleotides, FAD bears no positive charge.

Many enzymes, referred to as *flavoenzymes* or *flavoproteins*, require FAD or FMN as a prosthetic group that participates in the catalysis of oxidation-reduction reactions. Characteristically, the nucleotide is noncovalently but very tightly bound, though in a few instances, FAD is actually covalently bound to the protein. A number of oxidoreductases require, in addition to their flavin prosthetic group, one or more metal ions (for instance, iron or molybdenum); such enzymes are known as *metalloflavoproteins*.

Whereas the oxidized forms of riboflavin, FAD, and FMN all absorb visible light at 445–450 nm and appear yellow, upon reduction the conjugated double-bond system of the isoalloxazine is lost, so that $FADH_2$ and $FMNH_2$ are colorless. In Section 14·5, we shall find that FMN is reduced to $FMNH_2$ in mitochondria by transfer of a hydride ion from NADH. However, unlike NADH and NADPH, which

Figure 7·8
(a) Structure of riboflavin. Flavins are compounds that contain the isoalloxazine ring system. **(b)** Structures of flavin mononucleotide (FMN, black) and flavin adenine dinucleotide (FAD, black and blue). The reactive center is shown in red.

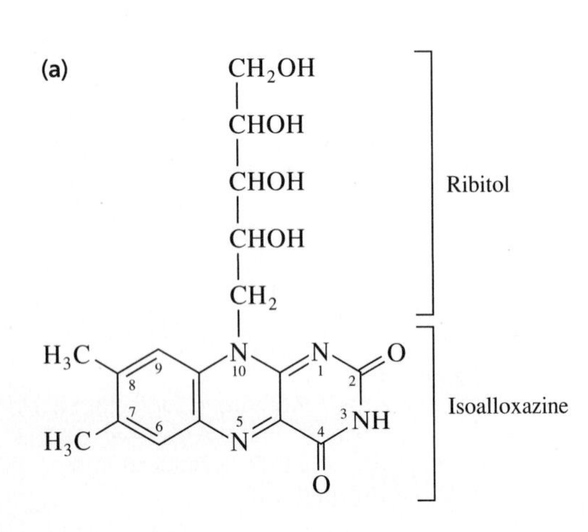

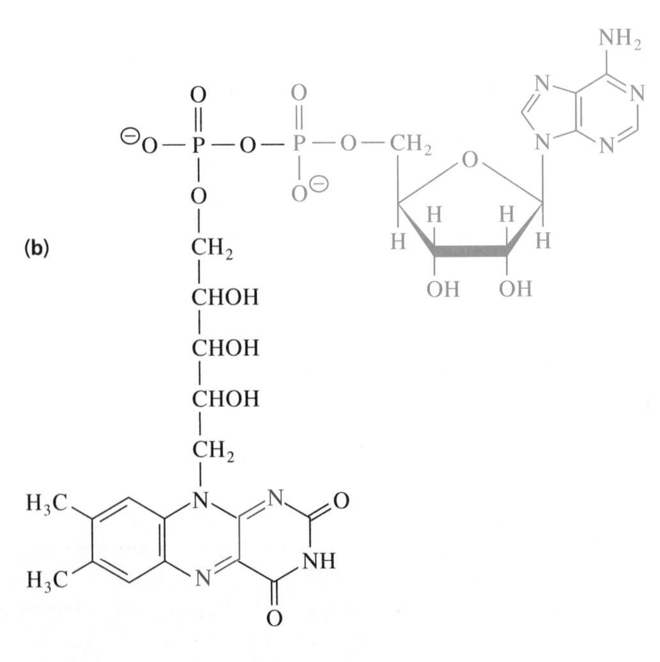

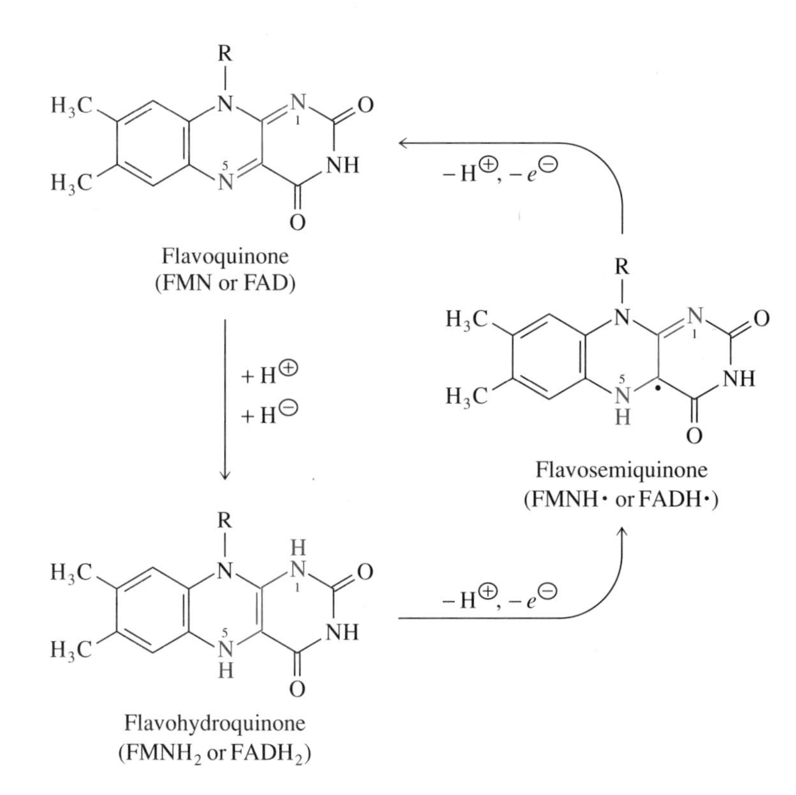

Figure 7·9
Reduction and reoxidation of FMN or FAD. The conjugated double bonds between N-1 and N-5 are reduced by addition of a hydride ion from NADH and a proton from solvent to form $FMNH_2$ or $FADH_2$, the hydroquinone forms of the coenzymes. Oxidation occurs with an intermediate step. A single electron is removed by a one-electron oxidizing agent such as Fe^{3+}, with loss of a proton, to form a fairly stable free-radical intermediate. This flavosemiquinone is then oxidized by removal of a proton and one electron to form the fully oxidized FMN or FAD.

can only participate in two-electron transfers, $FMNH_2$ and $FADH_2$ can donate electrons either one or two at a time. A partially oxidized compound, FADH· or FMNH·, is formed when the one-electron pathway is followed (Figure 7·9). These intermediates are relatively stable free radicals called *flavosemiquinones*. In mitochondria, the protein that is reduced by $FMNH_2$ has Fe^{3+} as its electron acceptor. Because Fe^{3+} can only accept one electron, forming Fe^{2+}, the reduced flavin must be oxidized in two one-electron steps via the semiquinone intermediate. The coupling through FMN of two-electron transfers with one-electron transfers is important in cellular respiration because further steps in the multistep mitochondrial process by which NADH is oxidized involve one-electron mechanisms. Thus, the redox versatility of the flavins is metabolically significant.

7·5 Thiamine Pyrophosphate Is a Derivative of Vitamin B_1

The structure of thiamine, the antiberiberi vitamin, was elucidated in 1935 by Robert R. Williams. Its structure includes a pyrimidine ring and a positively charged thiazolium ring, as shown in Figure 7·10. Thiamine (vitamin B_1) is abundant in the husks of rice, in cereals, and in liver. Beriberi is most often found in parts of the world where polished rice is a staple of the diet, since the rice husk, when removed, takes with it most of the essential thiamine.

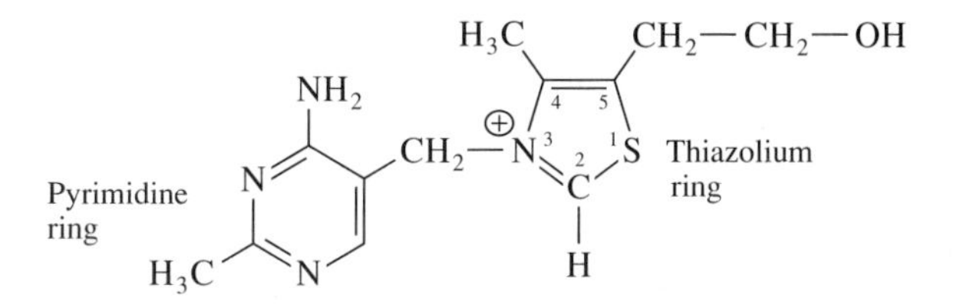

Figure 7·10
Structure of thiamine (vitamin B_1).

Figure 7·11
Formation of thiamine pyrophosphate. Thiamine pyrophosphate synthetase catalyzes pyrophosphoryl-group transfer from ATP to thiamine (vitamin B_1), converting it to the coenzyme thiamine pyrophosphate (TPP). The thiazolium ring contains the reactive portion (red) of the coenzyme.

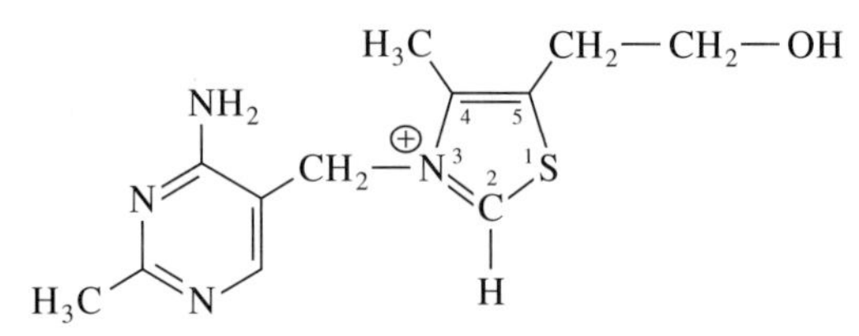

Thiamine (Vitamin B_1)

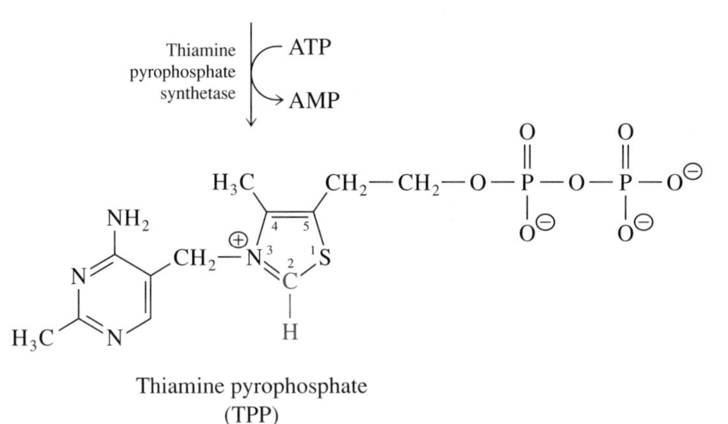

Thiamine pyrophosphate (TPP)

The coenzyme form of the vitamin is *thiamine pyrophosphate* (TPP). Within animal cells, the coenzyme is synthesized from dietary thiamine by an enzymatic transfer of pyrophosphate from ATP (Figure 7·11).The thiazolium ring contains the reactive portion of the coenzyme.

About half a dozen known decarboxylases (carboxy-lyases) require TPP as a coenzyme, and it was once referred to as "cocarboxylase." The first successful isolation of thiamine pyrophosphate was from yeast, where it was shown to be the prosthetic group of pyruvate decarboxylase. The activity of pyruvate decarboxylase allows yeast to convert pyruvate to acetaldehyde, which is subsequently reduced to ethanol. The mechanistic role of TPP can be illustrated by examining the reaction catalyzed by pyruvate decarboxylase (Figure 7·12). C-2 of TPP is acidic, but with an extremely high pK_a of 18. The positive charge of the thiazolium ring attracts electrons, weakening the bond between C-2 and hydrogen, thus allowing its ionization. A proton dissociates from C-2 of the thiazolium ring, presumably removed by a basic residue of the enzyme, generating a resonance-stabilized dipolar carbanion known as an *ylid*. Ylids are molecules that have opposite charges on adjacent atoms. The negatively charged C-2 attacks the electron-deficient carbonyl carbon of the substrate, pyruvate. The first product, CO_2, is released, shortening the carbon skeleton of the pyruvate attached to the thiazole ring to two carbons and resulting in the formation of a resonance-stabilized carbanion. In the following step, protonation of this carbanion produces the intermediate hydroxyethylthiamine pyrophosphate (HETPP), often called "active acetaldehyde." HETPP is cleaved, releasing acetaldehyde (the second product) and regenerating the ylid form of the enzyme-TPP complex. TPP is then reformed as the ylid is protonated by the enzyme. TPP will be encountered as a coenzyme required in the oxidation of α-keto acids (Sections 13·2 and 13·7). Oxidative decarboxylation of pyruvate and of α-ketoglutarate proceed by mechanisms analogous to that in Figure 7·12. In addition to its role as a coenzyme for α-keto acid decarboxylations, TPP serves as a prosthetic group for enzymes known as transketolases (Section 15·8), which catalyze transfer of two-carbon keto groups between sugar molecules.

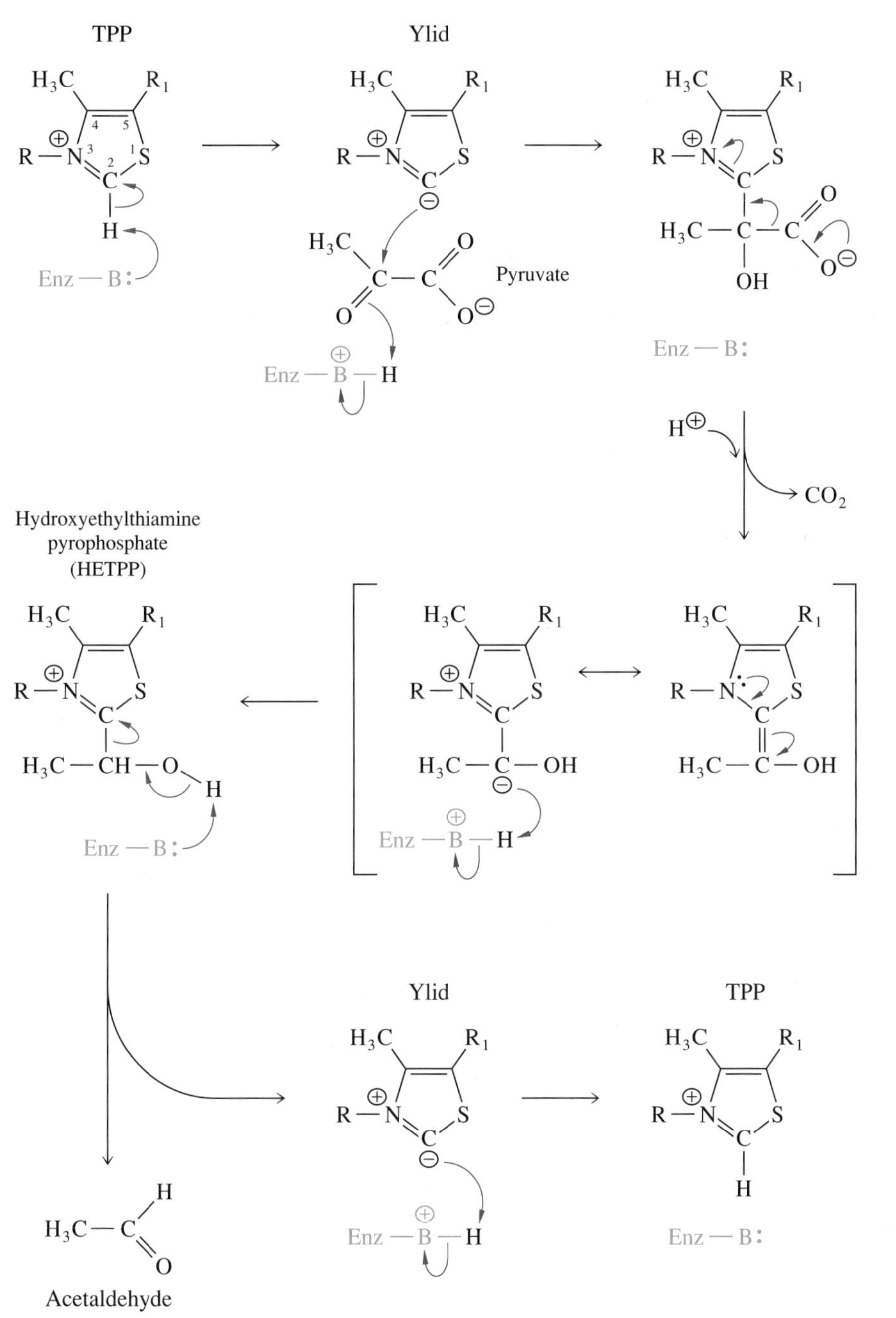

Figure 7·12
Mechanism of pyruvate decarboxylase showing the participation of thiamine pyrophosphate (TPP). Deprotonation of the thiazolium ring of TPP generates a dipolar carbanion known as an ylid. The negatively charged C-2 atom of the ylid attacks the carbonyl carbon of pyruvate, releasing CO_2 and resulting in the formation of a resonance-stabilized carbanion. Protonation of the latter carbanion produces HETPP, which is cleaved, releasing acetaldehyde. Protonation of the regenerated ylid form of TPP completes the catalytic cycle.

7·6 Pyridoxal Phosphate Is Derived from Vitamin B_6

The B_6 family of water-soluble vitamins consists of three closely related molecules. The first to be characterized was *pyridoxine*. However, most naturally occurring vitamin B_6 occurs as either *pyridoxal* or *pyridoxamine*—usually in a phosphorylated form. These three compounds differ only in the state of oxidation or amination of the carbon bound to position 4 of the pyridine ring (Figure 7·13). Vitamin B_6 is widely available from plant and animal sources. Induced deficiencies in rats result in dermatitis and a number of other disorders related to protein metabolism. Once

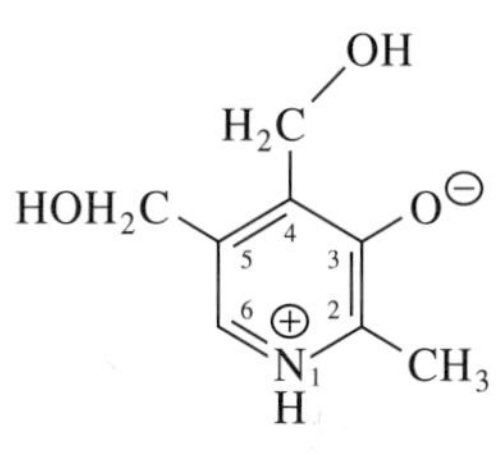
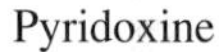

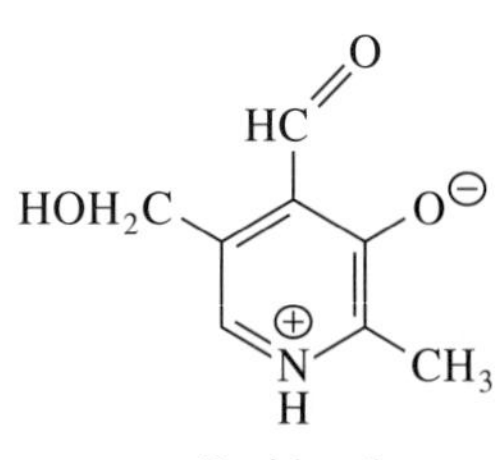
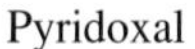

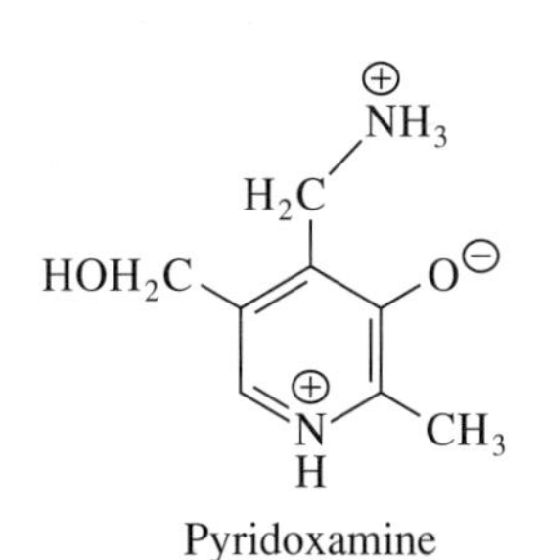

Figure 7·13
Structures of the vitamins of the B_6 family: pyridoxine, pyridoxal, and pyridoxamine.

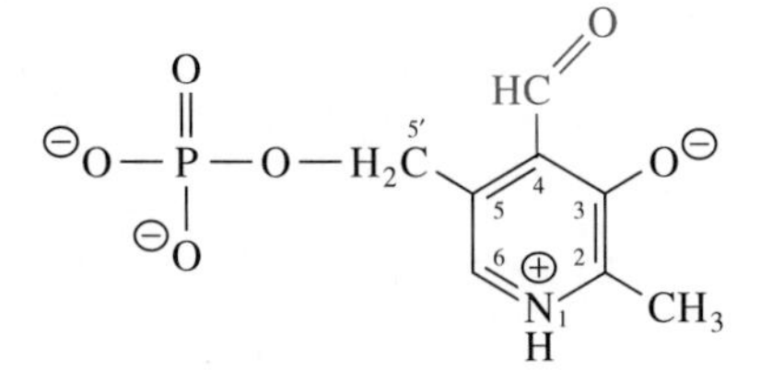

Figure 7·14
Structure of the coenzyme pyridoxal 5′-phosphate (PLP). The reactive center of PLP is the aldehyde group.

inside a cell, dietary pyridoxine is converted to pyridoxine 5′-phosphate by enzymatic transfer of the γ-phosphoryl group of ATP and then oxidized to form the coenzyme *pyridoxal 5′-phosphate* (PLP), shown in Figure 7·14.

Pyridoxal phosphate is the prosthetic group for a large number of enzymes that catalyze a variety of reactions involving amino acids, including isomerization, decarboxylation, and side-chain elimination or replacement reactions.

In PLP-requiring enzymes, the prosthetic group is bound as a Schiff base to the ε-amino group of a lysine residue at the active site (Figure 7·15). A Schiff base is a complex formed by condensation of a primary amine with an aldehyde or a ketone, producing an imine linkage; these Schiff bases are sometimes referred to as aldimines or ketimines, respectively. The enzyme-coenzyme Schiff base shown on the right in Figure 7·15 is sometimes referred to as an internal aldimine; during reactions, the coenzyme forms an external aldimine with the substrate. PLP is bound at all times by many weak noncovalent interactions; the covalent but reversible linkage of the internal aldimine gives added strength to the binding of the scarce, vitamin-derived coenzyme to the apoenzyme when the enzyme is not functioning.

The initial step in all PLP-dependent enzymatic reactions involving amino acids is the formation of an external aldimine linking PLP to the α-amino group of the amino acid. When an amino acid substrate binds to a PLP-enzyme that is in the internal aldimine form, a *transimination reaction* takes place (Figure 7·16). This transfer reaction proceeds via a geminal-diamine intermediate rather than via formation of the free aldehyde form of PLP. Note that the Schiff bases contain a system of conjugated double bonds leading to a positive charge on N-1. During subsequent steps in the mechanism of a PLP-enzyme–catalyzed reaction, the prosthetic

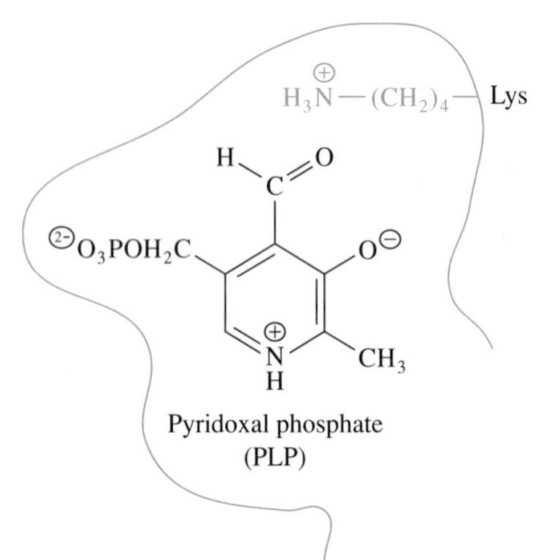

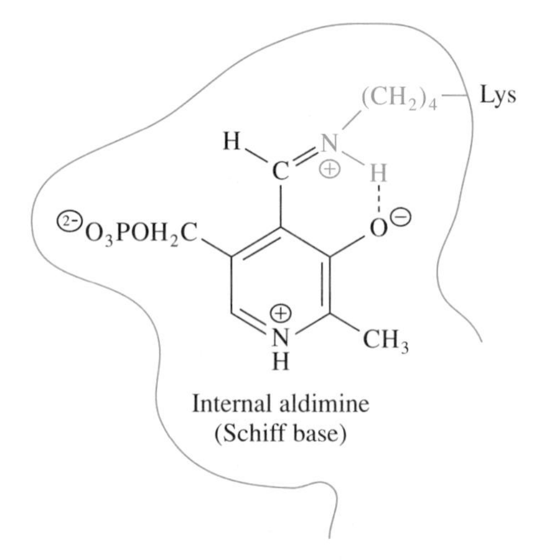

Figure 7·15
Binding of PLP to a PLP-dependent enzyme. Pyridoxal phosphate is bound to the apoenzyme by numerous noncovalent interactions and by formation of a Schiff base involving the aldehyde group of the coenzyme and the ε-amino group of a lysine residue in the active site.

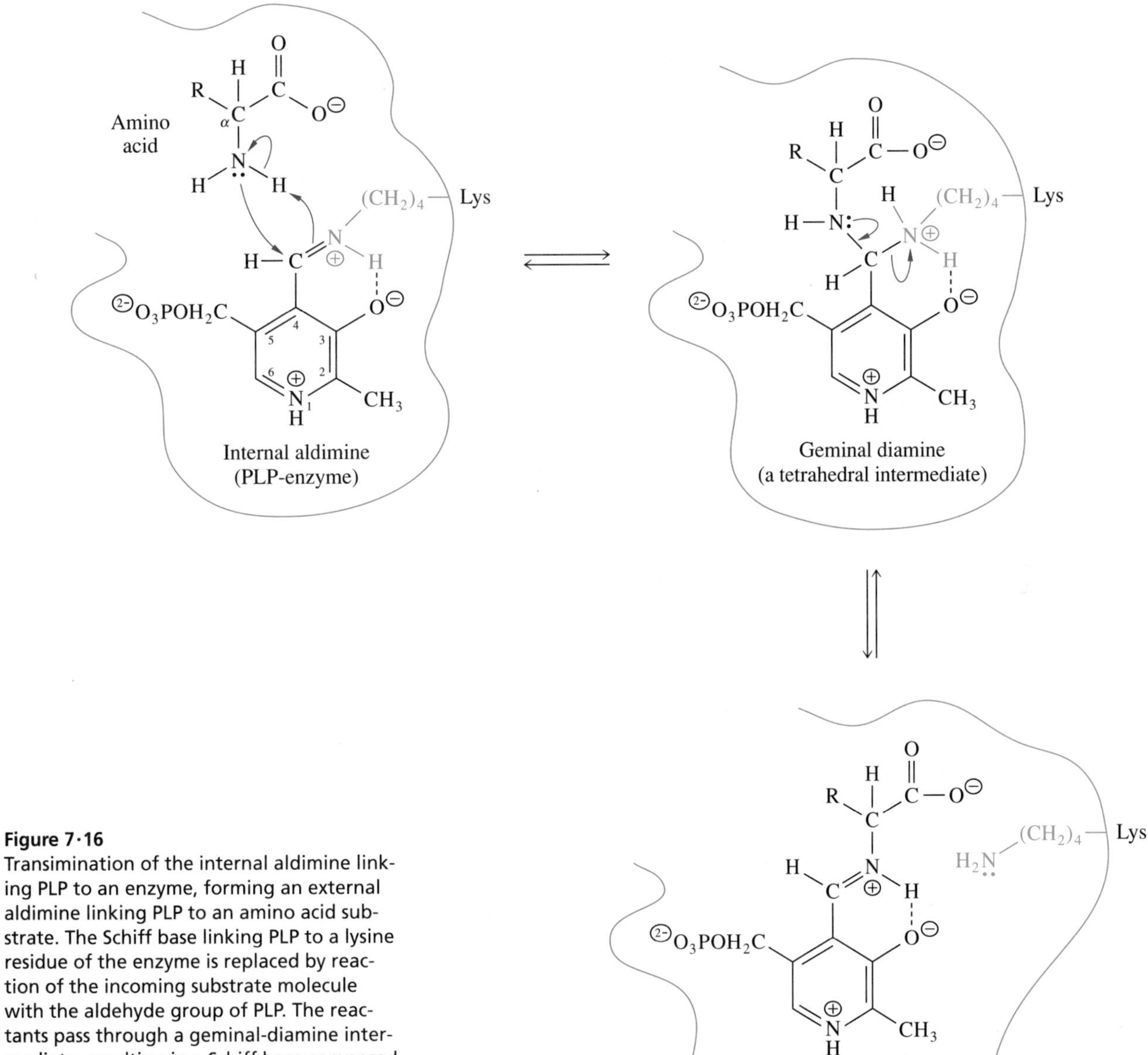

Figure 7·16
Transimination of the internal aldimine linking PLP to an enzyme, forming an external aldimine linking PLP to an amino acid substrate. The Schiff base linking PLP to a lysine residue of the enzyme is replaced by reaction of the incoming substrate molecule with the aldehyde group of PLP. The reactants pass through a geminal-diamine intermediate, resulting in a Schiff base composed of PLP and the substrate. The positively charged nitrogen atom at N-1 of PLP attracts electrons in the reaction of PLP with an amino acid.

group serves as an "electron sink." Once an α-amino acid forms a Schiff base with PLP, electron withdrawal toward N-1 weakens the three bonds to the α-carbon. In other words, the Schiff base with PLP provides a means for stabilizing a carbanion that can be formed by enzyme-directed loss of one of the three groups attached to the α-carbon of the amino acid. Of the three bonds, the one that breaks depends on the nature and location of amino acid residues contained within the active site of the enzyme. A simple illustration of the role of PLP is the racemization of an amino acid (its stereoisomerization at the α-carbon). The cell walls of many bacteria contain D-alanine. These bacteria synthesize the less common D isomer from L-alanine

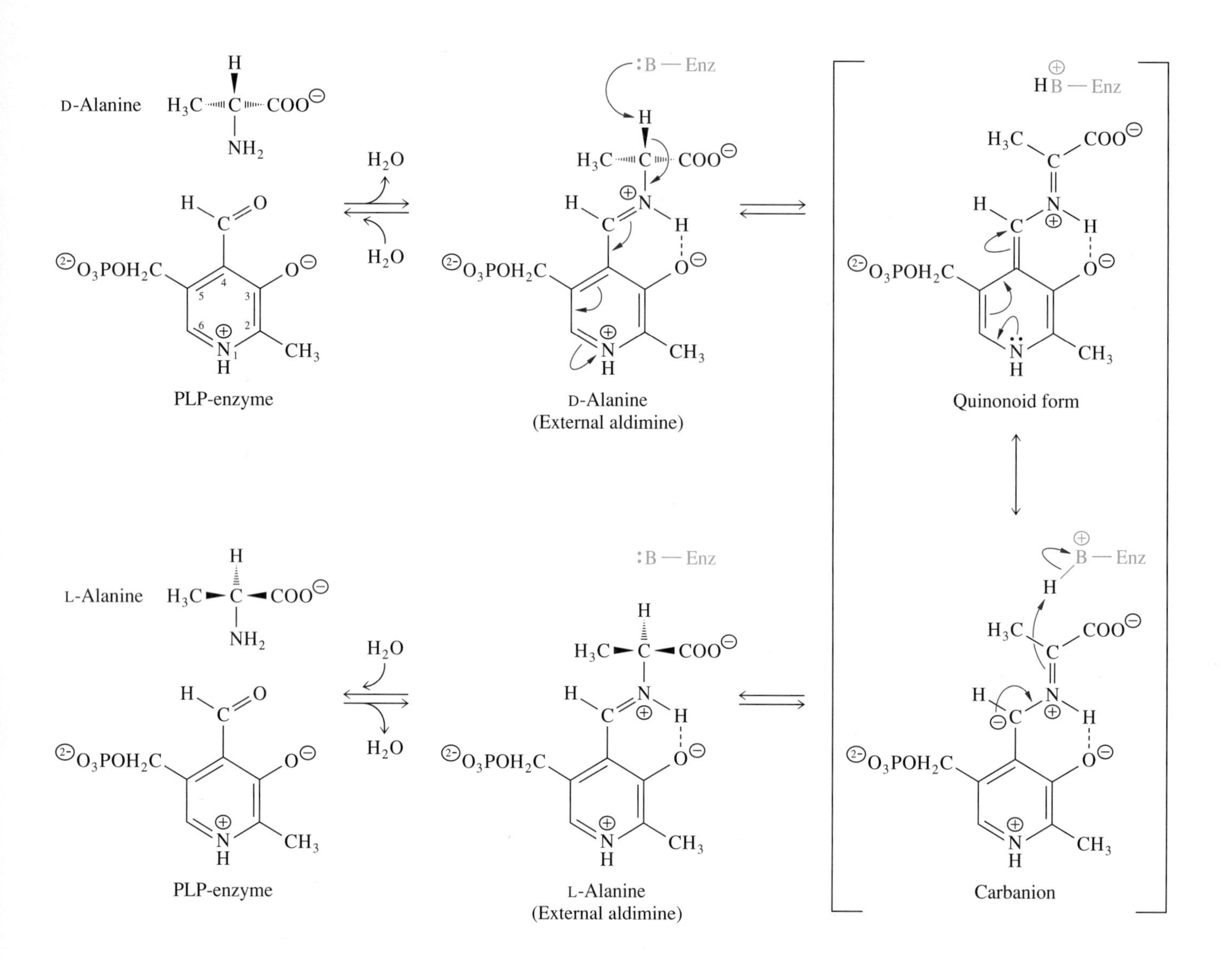

Figure 7·17
Mechanism of racemization of alanine catalyzed by a PLP-dependent racemase. The bonds about the α-carbon atom of the alanine molecule are weakened by formation of the Schiff base. A proton can then be removed from the α-carbon by a basic group of the enzyme. Two resonance forms of the carbanion intermediate are shown. The proton is returned to a different face of the α-carbon, producing the opposite stereoisomer of alanine.

in a reaction catalyzed by alanine racemase. Figure 7·17 illustrates the reversible racemization of D-alanine. The α-hydrogen of the amino acid in the Schiff base is removed, generating a resonance-stabilized carbanion intermediate. The carbanion is then protonated on a different face of the α-carbon. The new Schiff base is hydrolyzed (or cleaved by transimination with the active-site lysine) to generate the opposite stereoisomer. PLP will be encountered frequently in Chapter 18, during the discussion of the metabolism of amino acids.

7·7 Biotin Serves as a Prosthetic Group for Some Carboxylases

Biotin (once referred to as vitamin H) was initially identified as a necessary factor for the growth of yeast. In 1936, it was successfully isolated from dried egg yolks (1 mg obtained from 500 lb) and in 1940, from liver concentrates. The structure of biotin (Figure 7·18) was proposed in 1942 and confirmed by synthesis a year later. Because biotin is synthesized by intestinal bacteria and is required only in very small (μg) amounts each day, biotin deficiency is rare in humans or animals fed normal diets. A biotin deficiency can be induced, however, by ingestion of raw egg whites. Contained in egg whites is the protein *avidin* (MW 70 000), a tetramer of identical subunits, each of which binds biotin avidly (the dissociation constant is about 10^{-15} M), making it unavailable for absorption from the intestinal tract. Biotin deficiency (egg-white toxicity) has been found in people who eat large amounts of raw eggs. However, when eggs are cooked, avidin is denatured, abolishing its affinity for biotin and eliminating its toxicity.

Biotin serves as a coenzyme for enzymes that catalyze carboxyl-transfer reactions and ATP-dependent carboxylation reactions. Biotin is covalently linked at the active site of its host enzymes by an amide bond to the ε-amino group of a lysine residue (Figure 7·19). The biotinyl-lysine residue is sometimes referred to as *biocytin.*

Biotin acts as a carrier of carbon dioxide, as illustrated by its role in the pyruvate carboxylase reaction. Pyruvate carboxylase catalyzes the carboxylation of the three-carbon acid pyruvate by bicarbonate to form the four-carbon acid oxaloacetate, the substrate of several metabolic pathways. The reaction occurs in two steps: formation of carboxybiotin and transfer of CO_2 from carboxybiotin to pyruvate. As

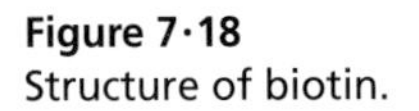

Figure 7·18
Structure of biotin.

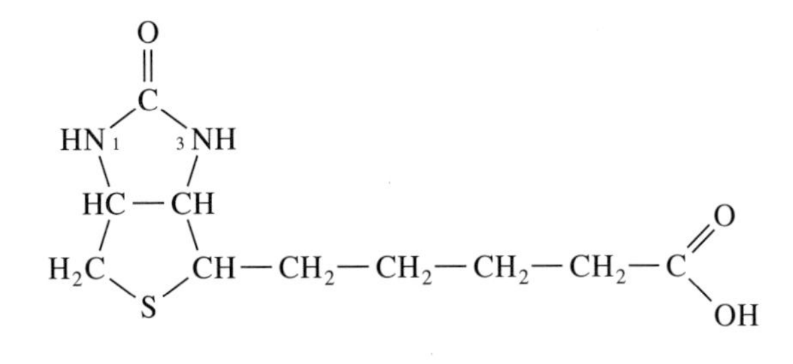

Figure 7·19
Biocytin. Biotin is covalently bound to enzymes by an amide linkage between the carboxylate group of biotin and the ε-amino group of a lysine residue (blue) at the active site of the enzyme. The biotinyl-lysyl adduct is sometimes called biocytin. The reactive center of biocytin is N-1, shown in red.

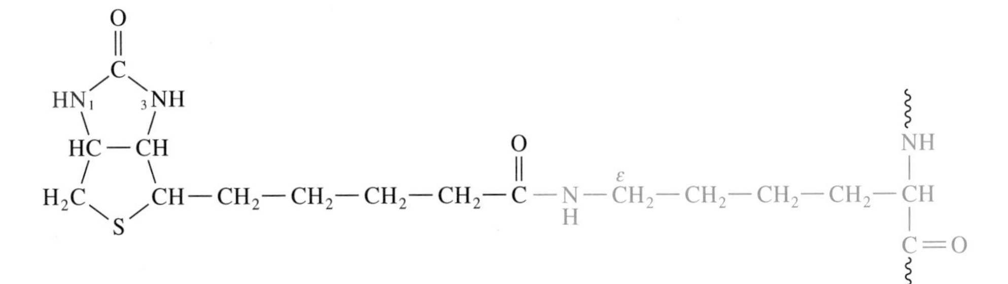

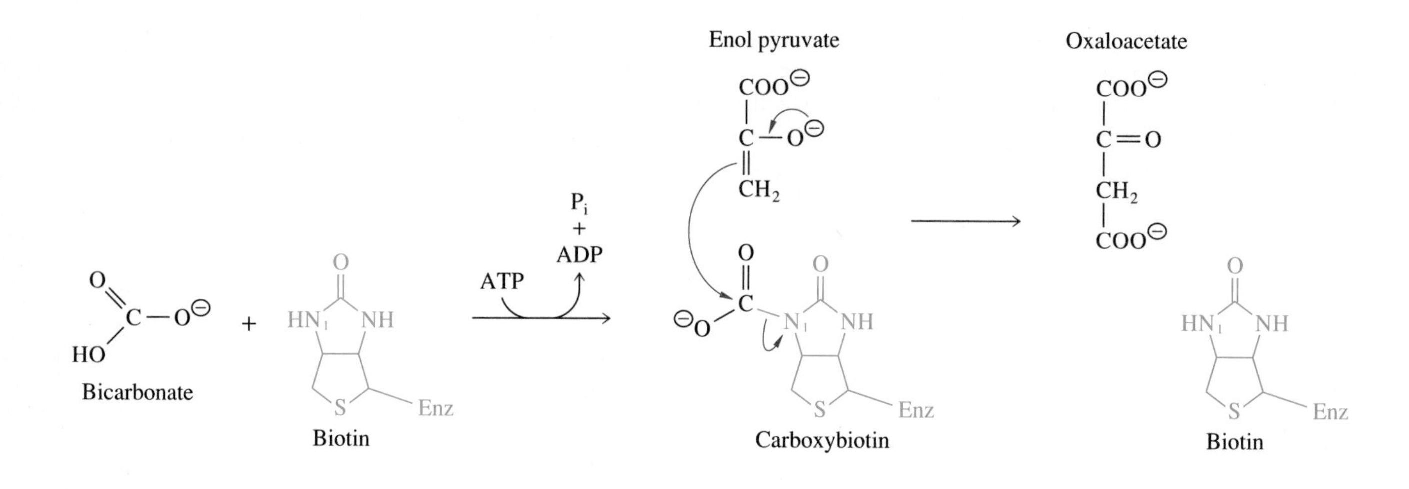

Figure 7·20
Involvement of biotin in the reaction catalyzed by pyruvate carboxylase. In the first step, carboxybiotin is formed from biotin, bicarbonate, and ATP. Carboxyl-group transfer from carboxybiotin to the enol form of pyruvate yields oxaloacetate and regenerates biotin.

shown in Figure 7·20, the enzyme-bound biotin serves as the intermediate carrier of the carboxyl mobile metabolic group. The mechanism by which biotin is carboxylated is not yet known. It has been suggested that carbonylphosphate is formed from bicarbonate and ATP, and that this reactive compound (or CO_2 formed from it) carboxylates N-1 of biotin.

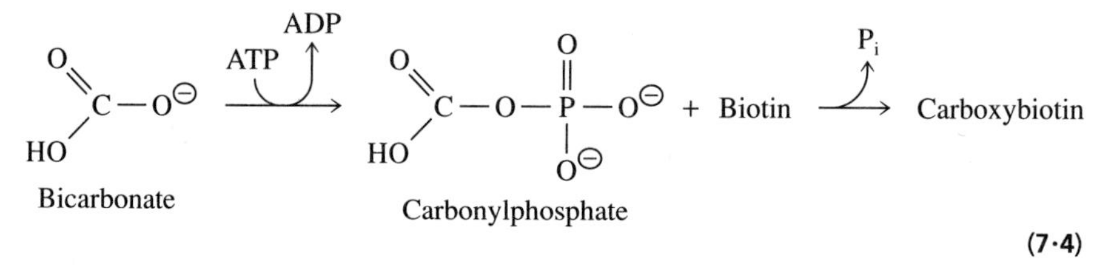

(7·4)

The N-1 carboxybiotinyl-enzyme provides a stable activated form of CO_2 that can be transferred to a second substrate, pyruvate. The enolate form of pyruvate attacks the carboxyl group of carboxybiotin to form oxaloacetate.

7·8 Tetrahydrofolate and Tetrahydrobiopterin Are Both Pterins

The vitamin *folate* (Latin, *folium*, leaf)—also known as *pteroylglutamate*—was isolated in the early 1940s from green leaves, liver, and yeast. Analysis of the isolated vitamin showed that it is composed of three main components: *pterin* (that is, 2-amino-4-oxo-substituted pteridine), a *p*-aminobenzoic acid moiety, and a glutamate residue (Figure 7·21). However, it was discovered that the coenzyme form of

Figure 7·21
Structures of (a) the pteridine ring system, (b) pterin, and (c) folate, a conjugated pterin containing *p*-aminobenzoate (shown in red) and glutamate (shown in blue).

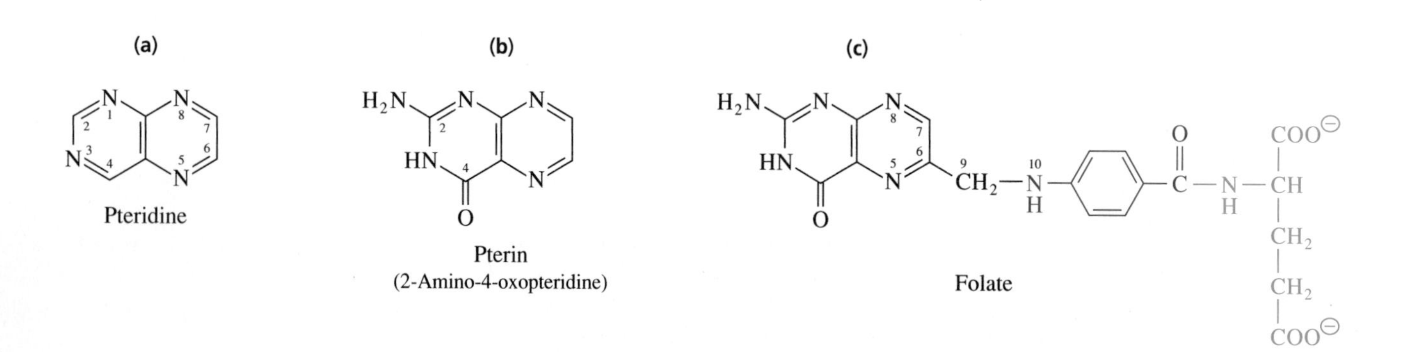

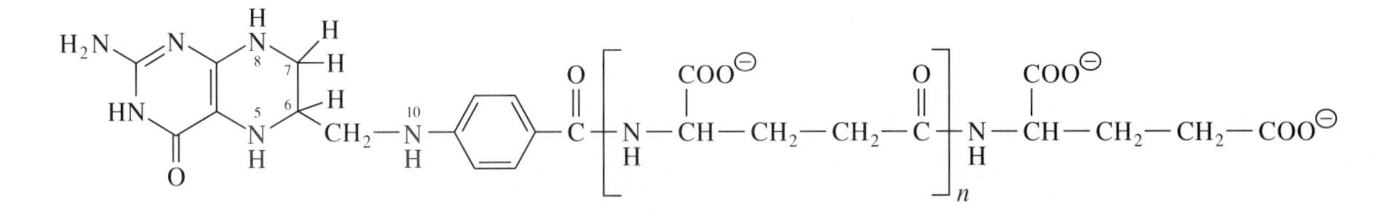

Figure 7·22
Structure of the poly-γ-glutamyl form of tetrahydrofolate. Tetrahydrofolate usually contains five or six glutamate residues. The reactive centers of the coenzyme, N-5 and N-10, are shown in red.

folate was partially degraded during its isolation. The coenzyme differs from the isolated vitamin in two respects: it is reduced to form *tetrahydrofolate,* and it is modified by the addition of glutamate residues bound to one another through γ-glutamyl amide linkages (Figure 7·22). The anionic polyglutamyl moiety, usually five to six residues long, participates in the binding of tetrahydrofolate to enzymes and assists with its retention inside cells, since nonpolar lipid membranes are impermeable to charged molecules.

Tetrahydrofolate is formed from folate by the addition of hydrogens to positions 5, 6, 7, and 8 of the pterin ring system. Folate is reduced in two steps, both involving NADPH, in a reaction catalyzed by dihydrofolate reductase:

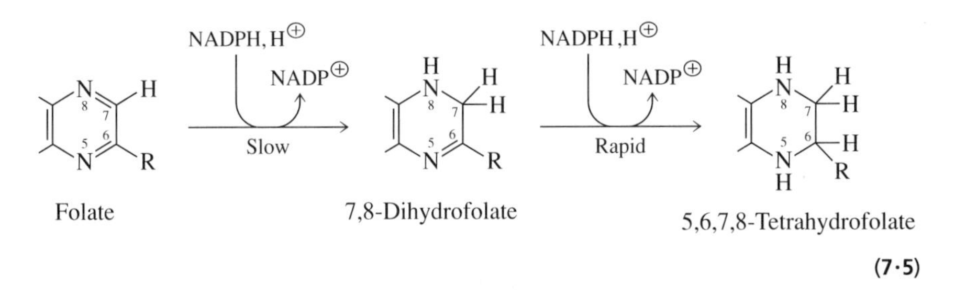

(7·5)

The primary metabolic function of dihydrofolate reductase is the reduction of dihydrofolate formed during the biosynthesis of the methyl group of deoxythymidine monophosphate (dTMP, Section 19·10). This reaction, which utilizes a derivative of tetrahydrofolate, is an essential step in the biosynthesis of DNA and hence is required for cell division. Because of its role in cell division, dihydrofolate reductase has been extensively studied as a target for chemotherapy in the treatment of cancer.

The mechanism of reduction of 7,8-dihydyrofolate by NADPH in the dihydrofolate reductase reaction exhibits a remarkable role for water in the active site. The reduction occurs by addition of a proton to N-5, giving C-6 a partial positive charge, thereby assisting the nucleophilic addition of a hydride ion to C-6.

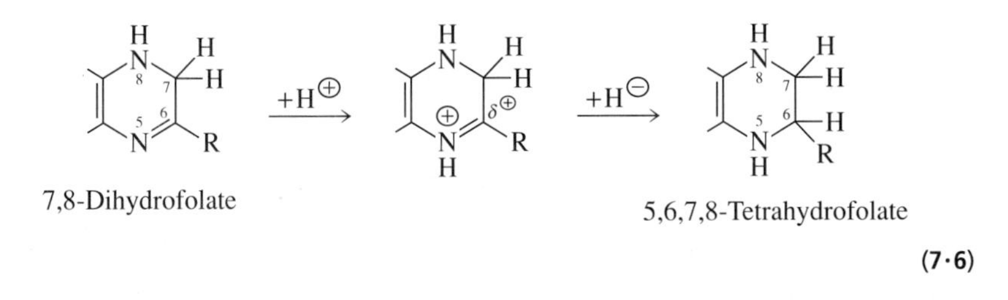

(7·6)

X-ray crystallographic, kinetic, and mutagenic experiments have demonstrated that Asp-27 in the active site of dihydrofolate reductase is an acid catalyst. Examination of the three-dimensional structure of the enzyme led to the unusual conclusion that since the aspartate residue is not close enough to N-5 to transfer the proton directly,

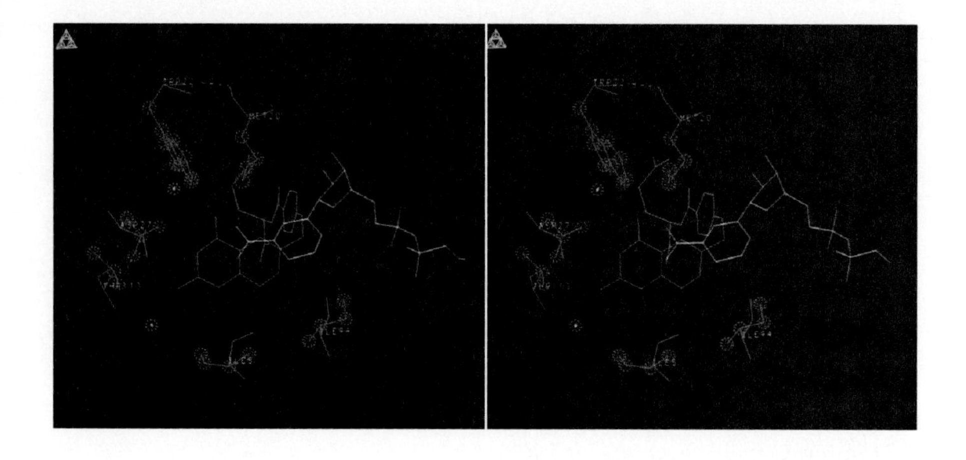

Stereo S7·2
Three-dimensional view of the active site of dihydrofolate reductase containing the oxidized substrate folate (purple) and the oxidized coenzyme NADP⊕ (green). The side chains of the active-site residues (red) are labelled. Note how C-4 of the nicotinamide ring of NADP⊕ is positioned close to the ring of folate that must be reduced. Rather than having a histidine residue as a proton-transfer group, this reductase has an aspartate residue (Asp-27), which transfers a proton to the pterin substrate via intervening water molecules in the active site. One water molecule thought to be involved in this proton transfer is shown as a blue sphere above Asp-27. (Based on coordinates provided by Joseph Kraut; stereo image by Ben N. Banks and Kim M. Gernert.)

proton transfer must involve the participation of water molecules present in the active site. The architecture of the active site, with folate and NADP⊕ bound, is shown in Stereo S7·2.

5,6,7,8-Tetrahydrofolate is required by enzymes that catalyze biochemical transfers of one-carbon units at the oxidation levels of methanol (CH_3OH), formaldehyde (HCHO), and formic acid (HCOOH). Thus, the fundamental groups bound to tetrahydrofolate are methyl, methylene, or formyl groups, respectively. The structures of several one-carbon derivatives of tetrahydrofolate and the enzymatic interconversions that occur among the various substituted forms are shown in Figure 7·23. The one-carbon metabolic groups are covalently bound to the secondary amines N-5 or N-10 of tetrahydrofolate, or to both in a ring form. 5,10-Methylenetetrahydrofolate has been referred to as "active formaldehyde," and 10-formyltetrahydrofolate as "active formate." We shall see in later chapters that tetrahydrofolate-dependent delivery of one-carbon units is important in a number of reactions in nucleic acid metabolism and in protein synthesis.

Vertebrates depend on dietary folate as the precursor of tetrahydrofolate because they have lost the ability to join a pterin to *p*-aminobenzoate. Bacteria retain this ability. Sulfanilamide and other sulfa drugs are antibacterial agents because they inhibit the joining of *p*-aminobenzoate to pterin in certain pathogenic bacteria.

$$H_2N-C_6H_4-SO_2(NH_2) \qquad (7\cdot7)$$

Sulfanilamide

Since the target of the drug is a reaction catalyzed by bacteria but not their hosts, sulfa drugs are selectively toxic, attacking only the pathogens.

5,6,7,8-Tetrahydrobiopterin is another pterin-derived coenzyme. It has a small side chain in place of the aminobenzoyl oligoglutamyl moiety of tetrahydrofolate and acts by a different mechanism (Figure 7·24, page 7·20). It is synthesized by animals as well as other organisms. The final step in the biosynthesis of 5,6,7,8-tetrahydrobiopterin, its formation from 7,8-dihydrobiopterin and NADPH, is catalyzed by dihydrofolate reductase. Tetrahydrobiopterin functions as the cofactor for

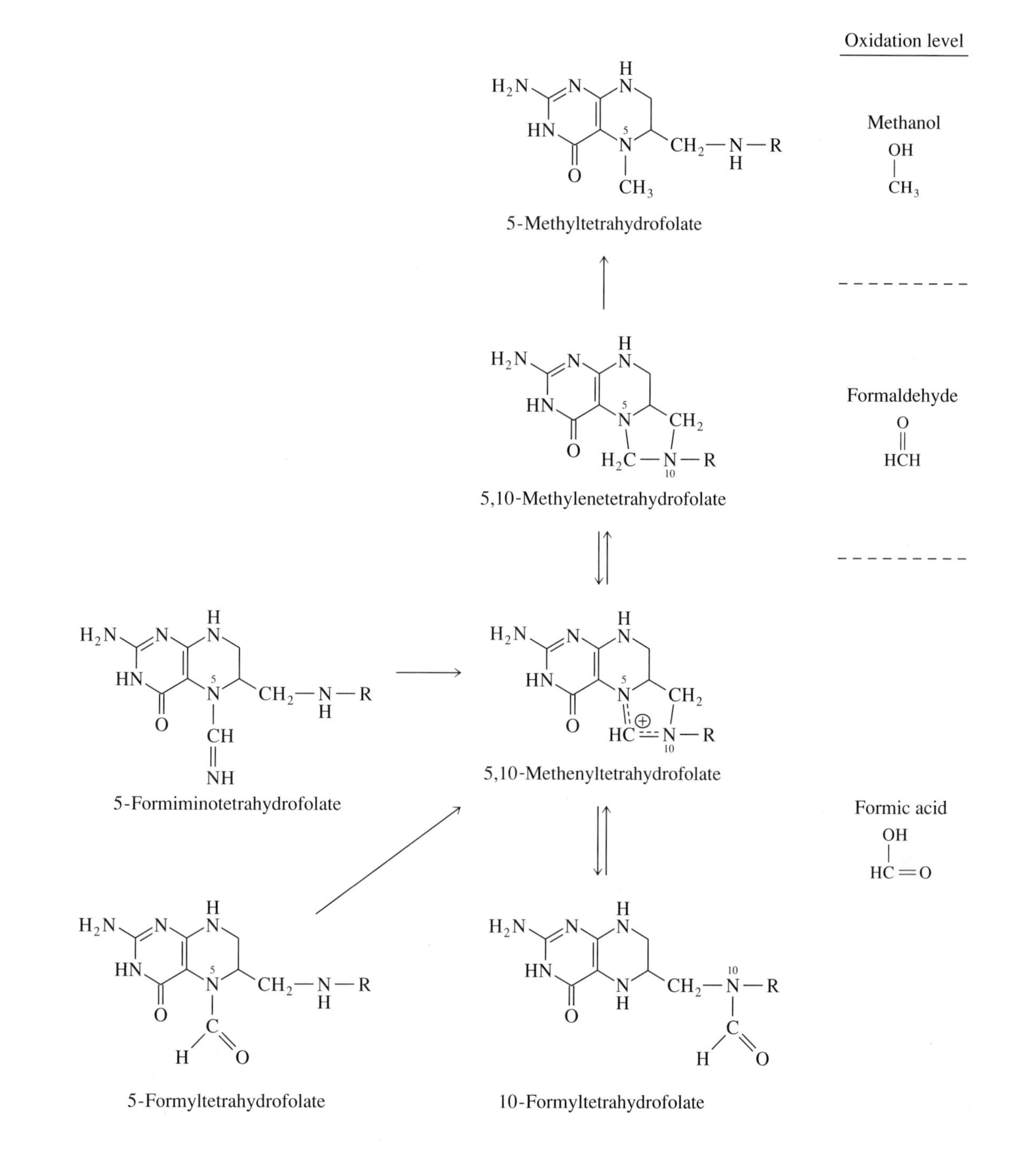

Figure 7·23
Structures of one-carbon derivatives of tetrahydrofolate. The derivatives can be interconverted enzymatically by the routes shown. (R denotes the benzoyl oligoglutamate portion of tetrahydrofolate.)

several enzymes known as *hydroxylases*. In the reactions catalyzed by these enzymes, the coenzyme is oxidized, resulting in a quinonoid dihydro form that has lost hydrogen atoms from N-3 and N-5. The quinonoid form is restored to active tetrahydrobiopterin in an NADH-dependent reaction catalyzed by dihydropteridine reductase. The reaction catalyzed by phenylalanine hydroxylase, which converts phenylalanine to tyrosine, provides an example of the role of tetrahydrobiopterin. Tetrahydrobiopterin supplies electrons to reduce O_2, which is a substrate for this reaction. One oxygen atom is incorporated into phenylalanine as an —OH group, and the other oxygen atom is reduced to H_2O.

The tetrahydropterin moiety also occurs as part of a cofactor known as *molybdopterin*, which is bound tightly to enzymes such as xanthine oxidase (Section 19·12) that contain molybdenum.

Figure 7·24
Oxidation and regeneration of tetrahydrobiopterin. The active coenzyme 5,6,7,8-tetrahydrobiopterin is oxidized in the phenylalanine hydroxylase reaction, resulting in the formation of quinonoid dihydrobiopterin. The dihydro form is restored to the active tetrahydro form in an NADH-dependent reaction catalyzed by dihydropteridine reductase.

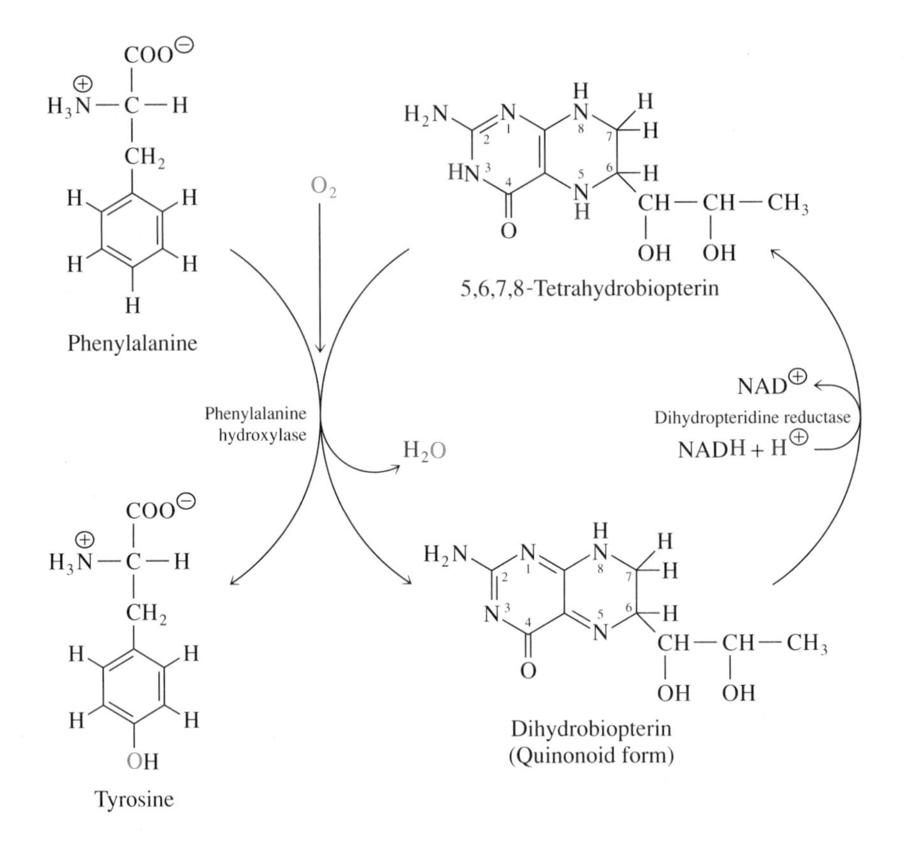

7·9 Coenzyme A Is Derived from Pantothenic Acid

Coenzyme A was first identified in the mid-1940s as a cofactor required for biological acetylations. It is now recognized as the coenzyme most prominently involved in *acyl-group transfer reactions* in which simple carboxylic acids and fatty acids are the mobile metabolic groups. Indeed, the name coenzyme A (often abbreviated CoA or CoASH) was derived from its role as the *acetylation* (now, more generally, acylation) coenzyme. Its structure, determined in the late 1940s by Fritz Lipmann, has three major portions: a 2-mercaptoethylamine unit that bears a free —SH group (the site for acylation and deacylation of the coenzyme), a pantothenic acid unit

Figure 7·25
Structure of coenzyme A. The molecule consists of 2-mercaptoethylamine bound to the vitamin pantothenic acid, which is bound in turn by a phosphoester linkage to an ADP group that has an additional 3′-phosphate. The reactive center of coenzyme A is the thiol group.

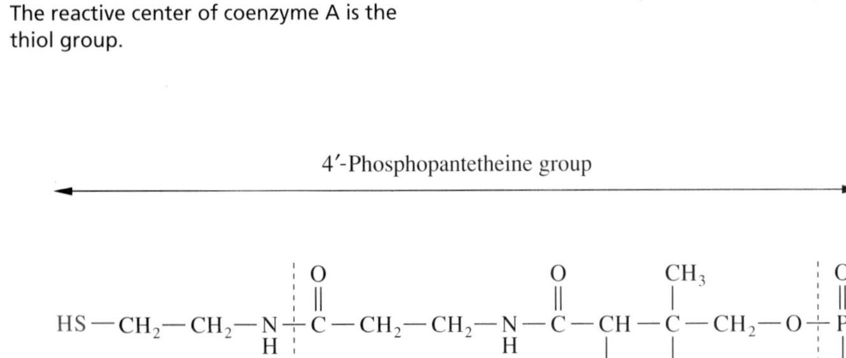

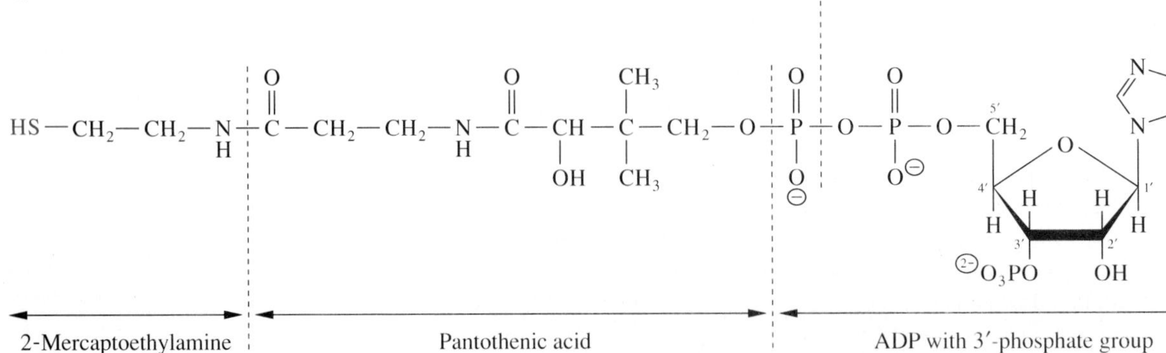

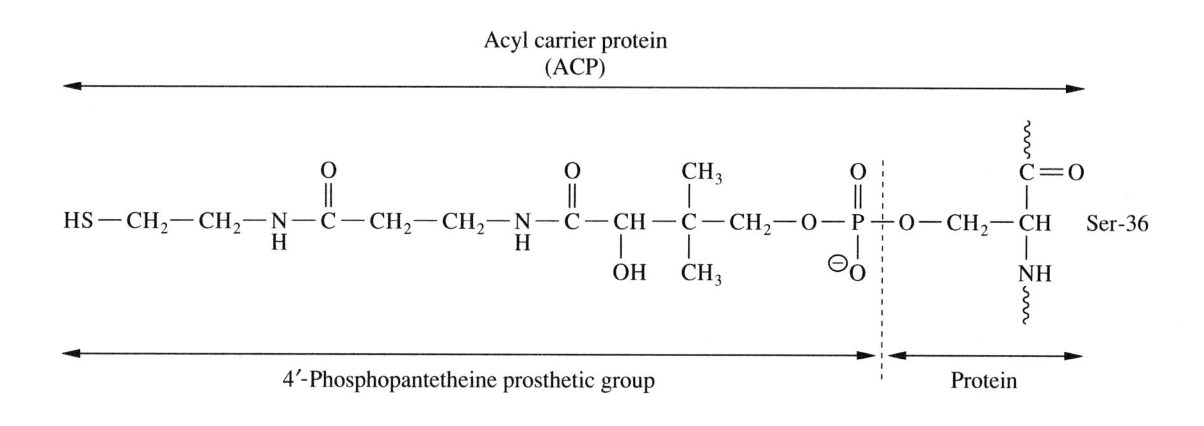

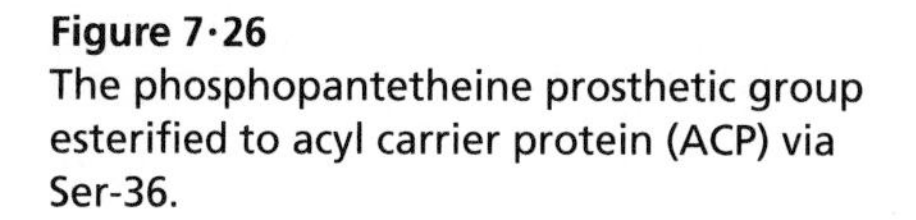
Figure 7·26
The phosphopantetheine prosthetic group esterified to acyl carrier protein (ACP) via Ser-36.

(which is an amide of β-alanine and pantoic acid), and an ADP moiety whose 3′-hydroxyl group is esterified with a third phosphate (Figure 7·25). *Pantothenate* (Greek, *pantothen,* everywhere), a B vitamin that occurs widely in nature, was extensively purified in the 1930s by Roger J. Williams (brother of Robert R. Williams, mentioned earlier as the discoverer of thiamine), who isolated it as an essential growth factor (vitamin B_3) for yeast.

Acetyl CoA has been called "active acetate." The energy of the thioester formed between acetate and the —SH group of CoA is similar to the energy of each of the phosphoanhydrides of ATP. This reflects the fact that thioesters resemble oxygen-acid anhydrides and thus are energy-rich metabolites, unlike ordinary oxygen esters of carboxylic acids. Despite the energy associated with acetyl CoA, it is quite resistant to nonenzymatic hydrolysis at neutral pH values.

If the 3′-phospho-AMP moiety is removed from the CoA structure shown in Figure 7·25 by splitting the molecule between the two phosphoryl groups, the remaining portion, a phosphate ester containing the 2-mercaptoethylamine and pantothenate residues, is known as *phosphopantetheine*. Phosphopantetheine serves as the prosthetic group of a small protein of 77 amino acid residues known as the *acyl carrier protein* (ACP), first isolated from *E. coli* by Roy Vagelos and colleagues. The prosthetic group is esterified to ACP via the side-chain oxygen of Ser-36 (Figure 7·26). The —SH of the prosthetic group of ACP serves as the site of acylation for intermediates in the biosynthesis of fatty acids (Section 17·12). Thus, in both CoA and ACP, it is the phosphopantetheine group that provides the structure needed for biological transfers of acyl groups activated as thioesters.

7·10 Vitamin B_{12} and Its Coenzyme Forms Contain Cobalt

In 1926, George Minot and William Murphy discovered that pernicious anemia—previously incurable and usually fatal—was improved in patients fed a diet that included large amounts of liver. The factor responsible for relieving the symptoms is vitamin B_{12}, the largest B vitamin and the last to be isolated.

Victims of pernicious anemia lack normal secretion by the stomach mucosa of a glycoprotein called intrinsic factor. This protein specifically binds vitamin B_{12}, and it is the B_{12}–intrinsic factor complex that is absorbed by cells of the small intestine. Currently, impaired absorption is treated by injections of vitamin B_{12} at regular intervals.

In 1948, vitamin B_{12}, also called *cobalamin,* was obtained from liver in the form of a red crystalline cyanide derivative. The cyanide ion is not part of the vitamin but comes from charcoal used in the purification of vitamin B_{12}. The complex

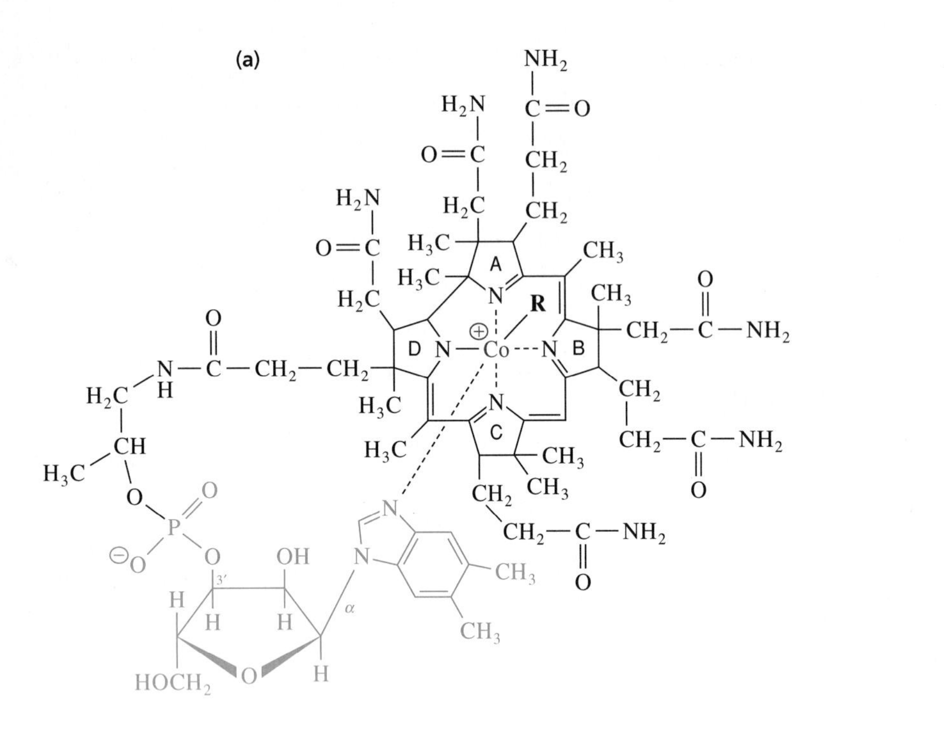

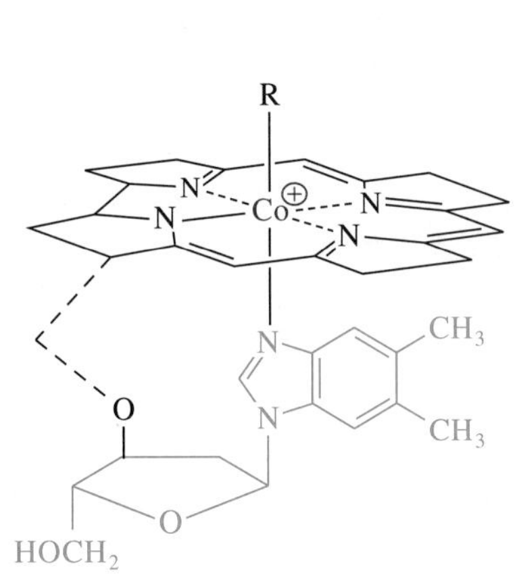

Figure 7·27
Structure of the vitamin B_{12} compounds.
(a) Vitamin B_{12} has two characteristic components: the corrin ring system (black) and the 5,6-dimethylbenzimidazole ribonucleotide (blue). The corrin ring system contains four pyrrole rings and resembles the porphyrin of heme. The major difference is that pyrrole rings A and D are joined directly rather than by a methene bridge. Furthermore, the metal coordinated by the four pyrrole nitrogens of corrin is cobalt (shown in red) rather than iron. The second component of vitamin B_{12} is the 5,6-dimethylbenzimidazole ribonucleotide group. A nitrogen atom of the benzimidazole ring is coordinated with the cobalt of the corrin ring. The benzimidazole ribonucleotide is also bound via an ester linkage to a side chain of the corrin ring system. **(b)** This abbreviated structure shows the flat corrin ring of vitamin B_{12} with an R group above the ring and the benzimidazole below the ring. In the vitamin form, the R group is either —CN or —OH. In the primary coenzymatic form of the vitamin, the R group is 5′-deoxyadenosine; in another coenzyme form, it is a methyl group.

structure of *cyanocobalamin,* the cyanide form of vitamin B_{12}, was determined by Dorothy C. Hodgkin in 1956 by means of X-ray crystallographic and chemical studies. The detailed structure of the vitamin B_{12} compounds is shown in Figure 7·27a. Note the resemblance of the *corrin* ring system to the *porphyrin* ring system (such as occurs in heme, Figure 4·25). Differences include the lack of a methene (—CH═) bridge between rings A and D and the presence of trivalent cobalt rather than the divalent iron found in heme. The abbreviated structure shown in

Table 7·2 Forms of vitamin B_{12}. The various forms of vitamin B_{12} are distinguished by the R group that is bound to the cobalt ion.

—R	Name
—CN	Cyanocobalamin (Vitamin B_{12})
—OH	Hydroxocobalamin (Vitamin B_{12a})
$—CH_3$	Methylcobalamin (Methyl B_{12})
OH OH; H, H, H, H; O; $5'\,CH_2$; N; NH_2	5′-Deoxyadenosylcobalamin (Adenosyl B_{12})

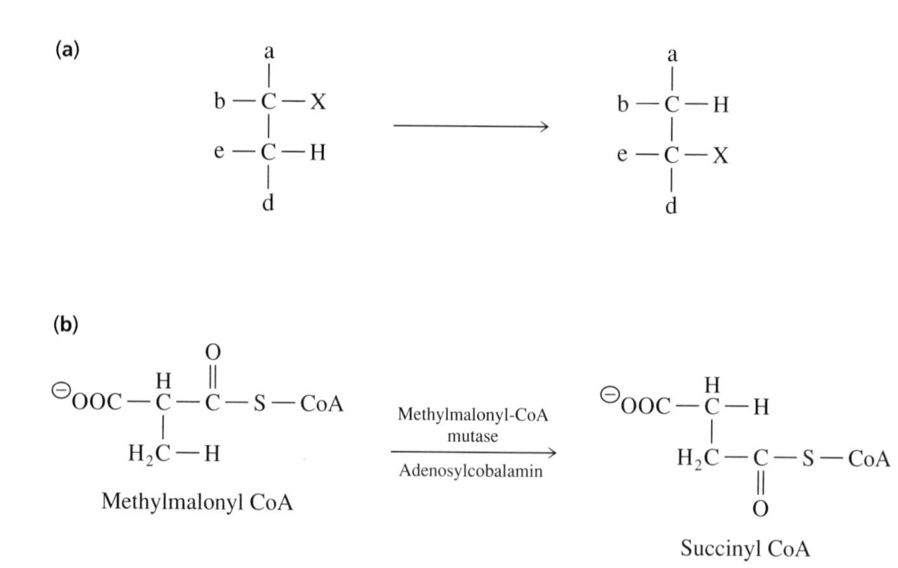

Figure 7·28
Intramolecular rearrangement catalyzed by adenosylcobalamin-dependent enzymes.
(a) Adenosylcobalamin participates in intramolecular rearrangements in which a hydrogen atom and a substituent on an adjacent carbon atom exchange places.
(b) Rearrangement of methylmalonyl CoA to succinyl CoA is catalyzed by the B_{12}-dependent methylmalonyl-CoA mutase.

Figure 7·27b emphasizes the positions of two axial ligands bound to the cobalt, a benzimidazole ribonucleotide below the cobalt atom and an R group above it. Table 7·2 illustrates the various R groups that have been isolated with cobalamin.

Vitamin B_{12} is not synthesized in plant or animal cells but is required by all animals and by some bacteria and algae. It functions in the two unstable coenzyme forms *adenosylcobalamin* and *methylcobalamin,* so called because a 5′-deoxyadenosyl or methyl group, respectively, is bound to the cobalt atom (Table 7·2). The vitamin is synthesized by only a few microorganisms. Ruminants have intestinal microorganisms in their forestomach that synthesize vitamin B_{12}, which is then absorbed in the lower gut. Carnivores acquire B_{12} from their diet. After absorption, most of the vitamin is converted via enzymatic reduction and reaction with ATP to the adenosyl-coenzyme derivative. During isolation, the 5′-deoxyadenosyl moiety of adenosylcobalamin, linked to the rest of the molecule by the unusual 5′-carbon-to-cobalt bond, is readily lost, replaced by $—OH^{\ominus}$ or $—CN^{\ominus}$.

The role of adenosylcobalamin is related to the lability of its unique C—Co bond. The coenzyme participates in several enzyme-catalyzed intramolecular rearrangements in which a hydrogen atom and a second group, bound to adjacent carbon atoms within a substrate, exchange positions (Figure 7·28a). An example is the methylmalonyl-CoA mutase reaction (Figure 7·28b), which is important in the metabolism of odd-chain fatty acids and leads to the formation of succinyl CoA, an intermediate of the citric acid cycle (Chapter 13). In partnership with tetrahydrofolate, methylcobalamin is involved in the transfer of methyl groups, for instance, in the regeneration of methionine from homocysteine in mammals (Figure 7·29). In this reaction, the methyl group of 5-methyltetrahydrofolate is passed to a reactive reduced form of vitamin B_{12} to form methylcobalamin, which can transfer the methyl group to the thiol side chain of homocysteine.

The exact molecular causes of pernicious anemia are not known. In humans, vitamin B_{12} deficiency affects the production of both red and white blood cells. In addition, prolonged deficiency leads to neurological disorders, possibly related to incorporation of methylmalonyl CoA in place of malonyl CoA into long-chain fatty acids.

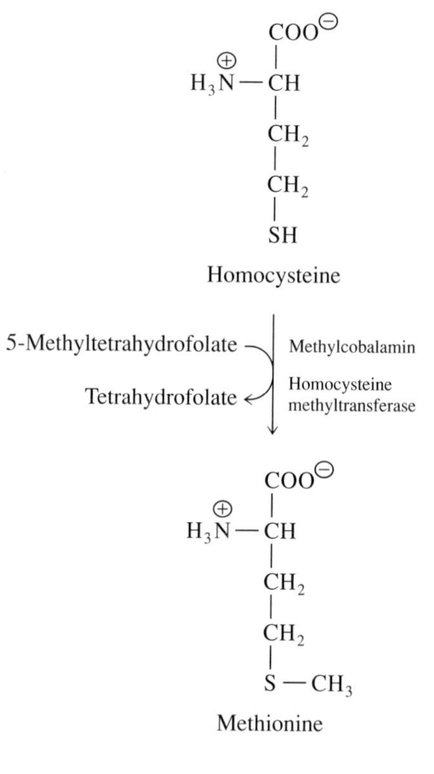

Figure 7·29
Biosynthesis of methionine from homocysteine.

Selected Readings

Blakley, R. L., and Benkovic, S. J., eds. (1985). *Folates and Pterins,* Vol. 1 and Vol. 2. (New York: John Wiley & Sons). The most recent comprehensive coverage of pterin coenzymes.

Dolphin, D., Poulson, R., and Avramović, O., eds. (1987). *Pyridine Nucleotide Coenzymes.* (New York: John Wiley & Sons). Part A contains chapters on history, nomenclature, chemical and physical properties, and spectroscopic techniques (15 chapters). Part B emphasizes biochemical, nutritional, and medical aspects (17 chapters).

Knowles, J. R. (1989). The mechanism of biotin-dependent enzymes. *Annu. Rev. Biochem.* 58:195–221.

McIntire, W. S., Wemmer, D. E., Chistoserdov, A., and Lidstrom, M. E. (1991). A new cofactor in a prokaryotic enzyme: tryptophan tryptophylquinone as the redox prosthetic group in methylamine dehydrogenase. *Science* 252:817–824. This article describes a new prosthetic group derived from two residues of tryptophan and also reviews other amino acid–derived prosthetic groups.

Scrimgeour, K. G. (1977). *Chemistry and Control of Enzyme Reactions.* (London: Academic Press). Chapters 8 and 9.

Wagner, A. F., and Folkers, K. (1964). *Vitamins and Coenzymes.* (New York: Interscience Publishers). An excellent treatment of the discovery and characterization of the vitamins.

Walsh, C. (1979). *Enzymatic Reaction Mechanisms.* (San Francisco: W. H. Freeman and Company).

8

Carbohydrates

The *carbohydrates* include simple sugars, their polymers, and other sugar derivatives. Widely distributed in nature, carbohydrates represent, on the basis of mass, the most abundant class of organic biomolecules on Earth. (Water, although more abundant, is not organic.) Most of this carbohydrate accumulates as a result of photosynthesis, the process by which certain organisms convert solar energy into chemical energy and incorporate atmospheric carbon dioxide into carbohydrate.

Carbohydrates play several crucial roles in living organisms. They are partially reduced molecules, which, upon oxidation, yield energy needed to drive metabolic processes; thus, carbohydrates can act as energy-storage molecules. Polymeric carbohydrates serve a variety of functions. For example, they are found in the cell walls and protective coatings of many organisms. Carbohydrates attached to cell membranes play a role in cellular recognition and in cell-to-cell communication. Finally, derivatives of sugars are found in a number of biological molecules, including antibiotics, coenzymes (Chapter 7), and the nucleic acids RNA and DNA (Chapter 10).

Carbohydrates, also known as saccharides, can be classified according to the number of monomeric units they contain. *Monosaccharides* are the basic monomeric units of carbohydrate structure. All monosaccharides share the empirical formula $(CH_2O)_n$, where n is three or greater. The term *carbohydrate* itself is derived from $(C \cdot H_2O)_n$, or "hydrate of carbon." *Oligosaccharides* are polymers of between 2 and about 20 monosaccharide residues. The most common oligosaccharides are the *disaccharides*, which consist of two linked monosaccharides. *Polysaccharides* are polymers that contain many (usually more than 20) monosaccharide residues. Oligosaccharides and polysaccharides do not have the empirical formula $(CH_2O)_n$ because water is eliminated during polymer formation. There are also a number of important monosaccharide derivatives that have more complex empirical formulas.

(a)

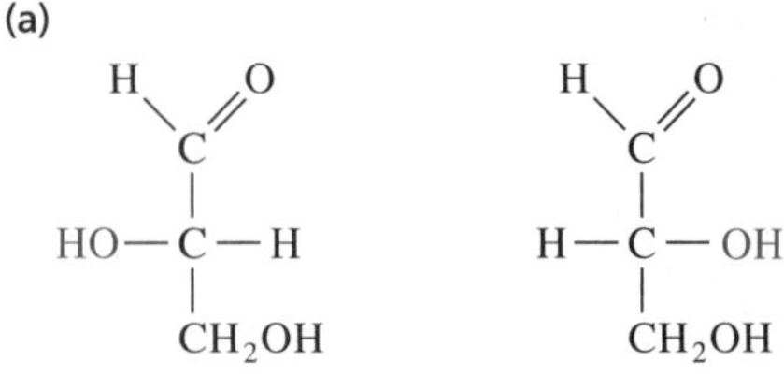

(b)

CH_2OH
|
C=O
|
CH_2OH

Dihydroxyacetone

Figure 8·1
Structures of glyceraldehyde and dihydroxyacetone. **(a)** The designations L and D for glyceraldehyde refer to the configuration at the chiral carbon (C-2), as shown. **(b)** Dihydroxyacetone is achiral.

In general, pure monosaccharides and disaccharides are water-soluble, white, crystalline solids and have a sweet taste. Examples include glucose and fructose (monosaccharides) and sucrose and lactose (disaccharides). As a rule, the suffix "-ose" is used in naming carbohydrates, although there are a number of exceptions.

8·1 Monosaccharides Can Be Classified as Either Aldoses or Ketoses

As indicated above, all monosaccharides contain at least three carbon atoms. One of these is a carbonyl carbon (that is, the monosaccharide is either an aldehyde or a ketone), and each of the remaining carbon atoms bears a hydroxyl (—OH) group. The two general classes of monosaccharides are *aldoses* and *ketoses*. In aldoses, the most oxidized carbon, designated C-1, is aldehydic. In ketoses, the most oxidized carbon, usually C-2, is ketonic.

The simplest monosaccharides are *trioses*, or three-carbon sugars. The aldehydic triose (or aldotriose) *glyceraldehyde* (Figure 8·1a) is chiral—it has an asymmetric center at its central carbon, C-2 (Recall Figure 3·2.). The ketonic triose (or ketotriose) *dihydroxyacetone* (Figure 8·1b) is achiral—it has no center of asymmetry. All other monosaccharides can be viewed as longer-chain versions of those two sugars, and, as we shall soon see, all other monosaccharides are chiral.

The enantiomers D- and L-glyceraldehyde are shown as stick models in Stereo S8·1. Chiral molecules are optically active; that is, they rotate the plane of polarized light. The convention for designating D and L isomers was originally based on the optical properties of glyceraldehyde. The form of glyceraldehyde that caused rotation to the right (dextrorotatory) was designated D; the form of glyceraldehyde that caused rotation to the left (levorotatory) was designated L. Structural knowledge was limited when this convention was established in the late nineteenth century, and configurations for the enantiomers of glyceraldehyde were assigned arbitrarily, with a 50% probability of error. It was not until the mid-twentieth century that technology became available that proved the original structural assignments were indeed correct.

Longer aldoses and ketoses may be regarded as extensions of glyceraldehyde and dihydroxyacetone, respectively, with chiral H—C—OH groups inserted between the carbonyl carbon and the primary alcohol, or "tail," carbon. Figure 8·2 shows the structures of the four-, five-, and six-carbon aldoses related to D-glyceraldehyde. Note that the numbering of the carbon atoms proceeds from the aldehydic carbon, which is assigned the number 1. By convention, sugars are said to have the D configuration when the configuration of the chiral carbon with the highest

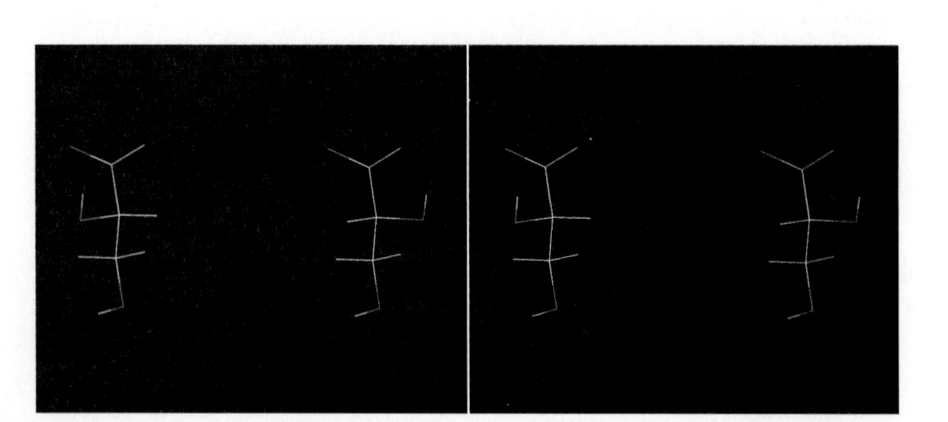

Stereo S8·1
L-Glyceraldehyde (left) and D-glyceraldehyde (right). (Stereo image by Ben N. Banks and Kim M. Gernert.)

Figure 8·2
Structures of D-aldoses, shown as Fischer projections. Those shown in blue will be of greatest importance in our study of biochemistry. Several D-aldoses are commonly referred to by their three-letter abbreviations, as indicated in parentheses.

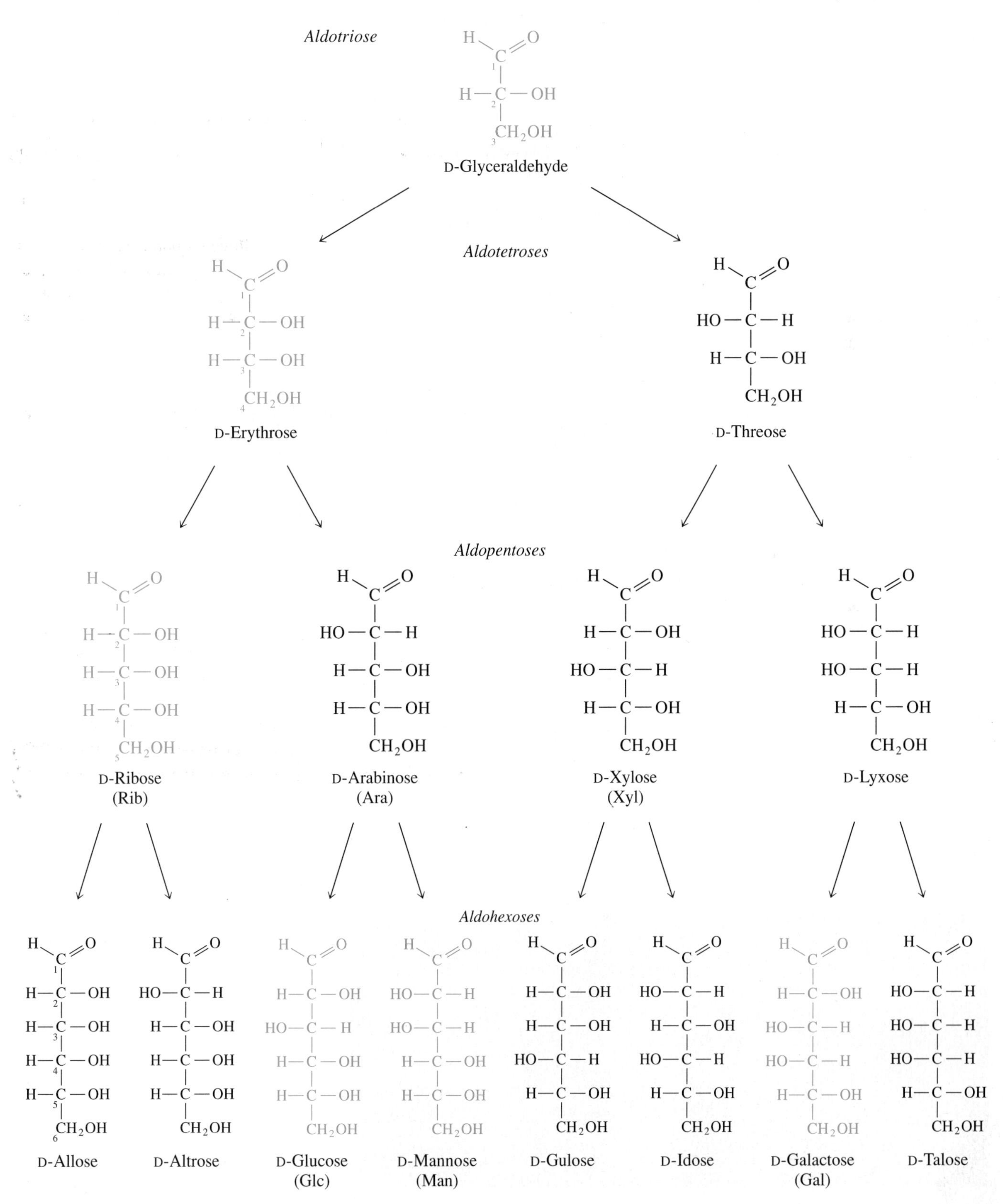

number—that is, the chiral carbon most distant from the carbonyl carbon—is D. Thus, there is no predictable association between the D and L nomenclature of a sugar and whether it is dextrorotatory or levorotatory. The arrangement of asymmetric carbon atoms is unique for each monosaccharide, giving the sugar its distinctive properties.

Not included in Figure 8·2 are the L enantiomers of the 15 aldoses shown. Note that pairs of enantiomers are mirror images at every chiral carbon; in other words, the configuration at every chiral carbon is opposite. Thus, while the hydroxyl groups bound to carbon atoms 2, 3, 4, and 5 of D-glucose point right, left, right, and right, respectively, in the Fischer projection, those of L-glucose point left, right, left, and left. The D enantiomers of sugars predominate in nature.

Whereas there are two stereoisomers for the one aldotriose (D- and L-glyceraldehyde), which possesses a single chiral carbon atom, there are four stereoisomers for *aldotetroses* (D- and L-erythrose and D- and L-threose) because erythrose and threose each possess two chiral carbon atoms. In general terms, there are 2^n possible stereoisomers for a compound with *n* chiral carbons. Thus, aldohexoses, which possess four chiral carbons, have a total of 2^4 or 16 stereoisomers (the eight D aldohexoses in Figure 8·2 and their L enantiomers). Nonenantiomeric stereoisomers are referred to as *diastereomers*.

Diastereomers that differ in configuration at *only one* of several chiral centers are called *epimers*. For example, D-mannose and D-galactose are epimers of D-glucose (at C-2 and C-4, respectively), although they are not epimers of each other (Figure 8·2 and Stereo S8·2).

Longer-chain ketoses are related to dihydroxyacetone in the same way that longer-chain aldoses are related to glyceraldehyde (Figure 8·3). Notice that a ketose has one fewer chiral carbon atom than the aldose of the same empirical formula. For example, there are only two stereoisomers of a ketotetrose (D- and L-erythrulose) and four stereoisomers of a ketopentose (D- and L-xylulose and D- and L-ribulose). Ketotetroses and ketopentoses are named by inserting "-ul-" in the name of the corresponding aldose. For example, the ketose xylulose is analogous to the aldose xylose. However, this naming scheme does not apply to the ketohexoses (tagatose, sorbose, psicose, and fructose), which bear trivial names unrelated to the names of the corresponding aldohexoses.

One final note about monosaccharide nomenclature: early experiments on the hydrolysis of sucrose (table sugar, obtained from plants such as beets and sugarcane) led to the discovery that its monosaccharide components, D-glucose and D-fructose, were dextrorotatory and levorotatory, respectively. As a result, glucose and fructose were initially called *dextrose* and *levulose*. Those names are still encountered occasionally.

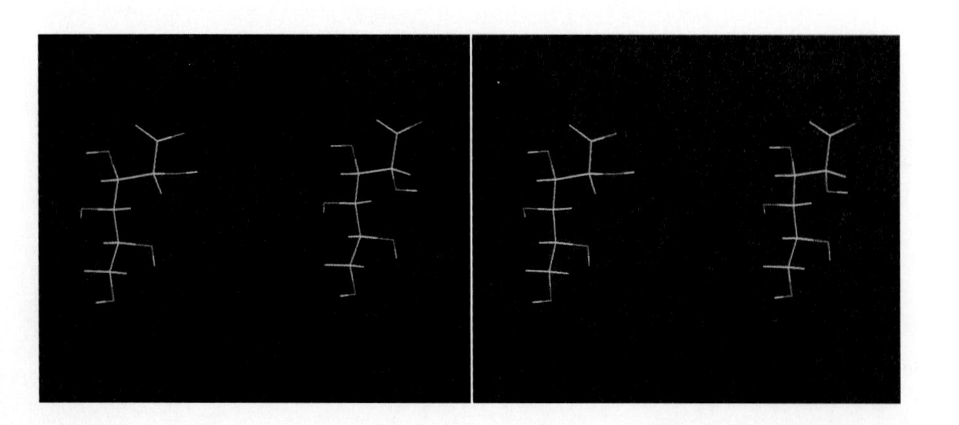

Stereo S8·2
D-Glucose (left) and D-mannose (right). D-Glucose and D-mannose are epimers—non-mirror-image stereoisomers that differ in configuration about only one carbon—in this case, C-2. As such, they represent a special type of diastereomers. (Stereo image by Ben N. Banks and Kim M. Gernert.)

Figure 8·3
Structures of D-ketoses. Those shown in blue are the ketoses of greatest importance in our study of biochemistry.

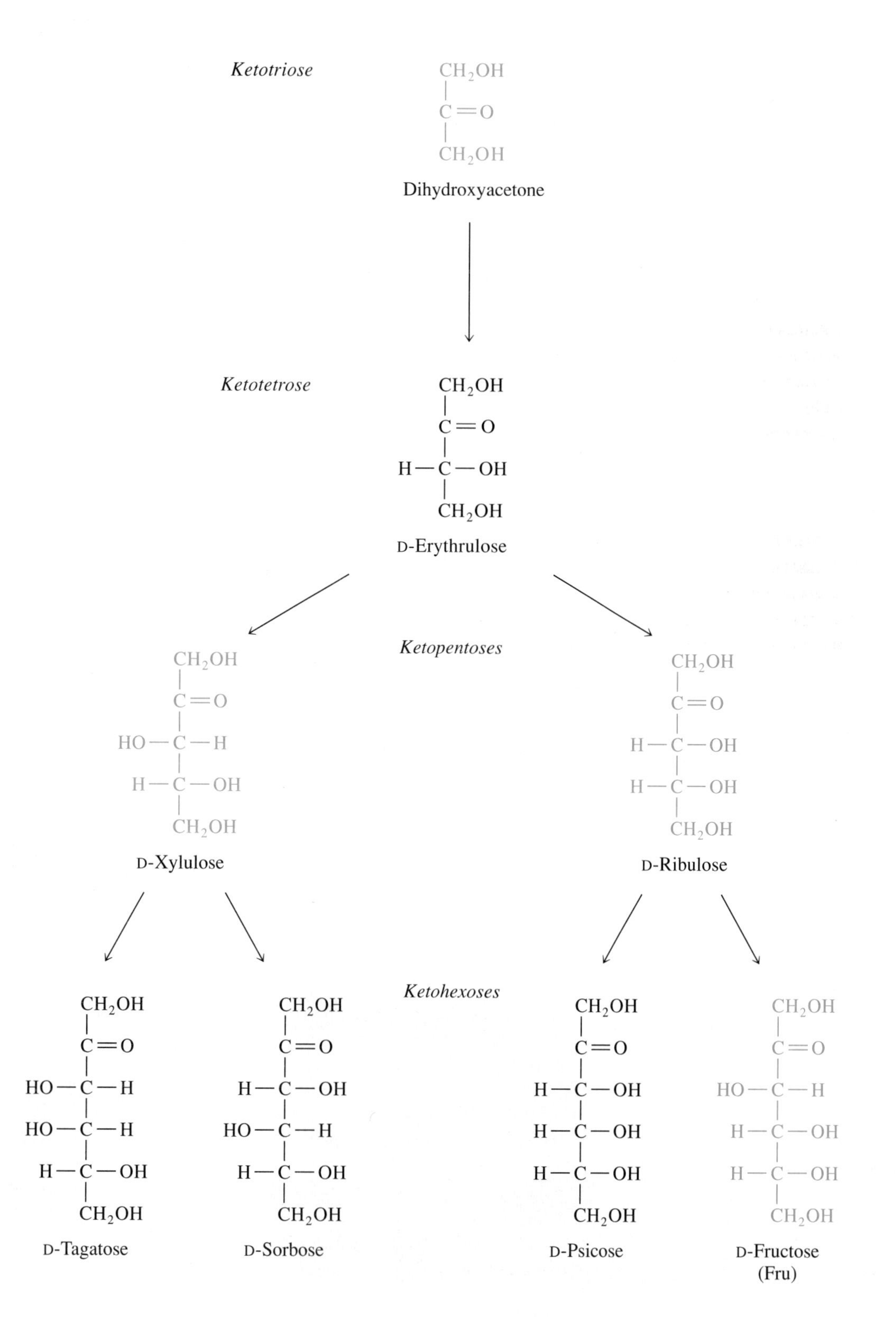

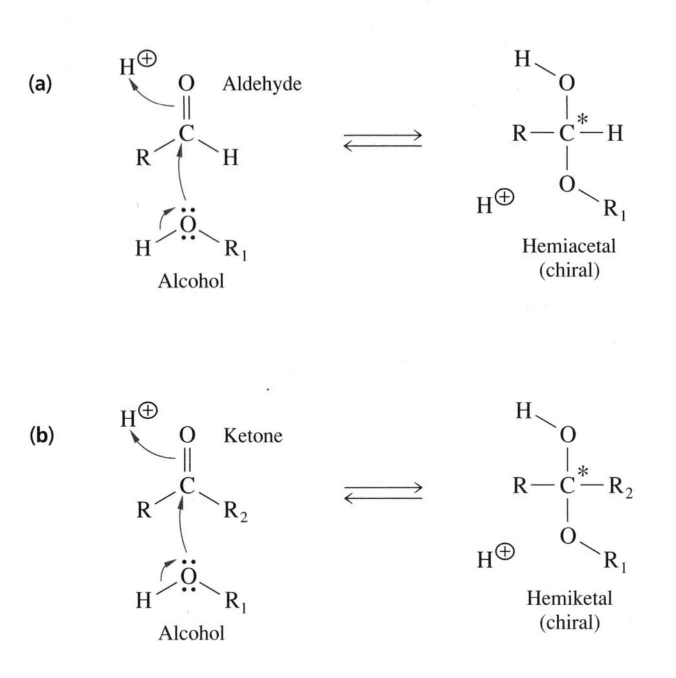

Figure 8·4
(a) Reaction of an alcohol with an aldehyde to form a hemiacetal. **(b)** Reaction of an alcohol with a ketone to form a hemiketal. In carbohydrate chemistry, both products are called hemiacetals, based on the recommendations of the IUPAC and IUB nomenclature committees. The asterisks indicate the position of the newly formed chiral center.

8·2 Aldoses and Ketoses Can Form Cyclic Hemiacetals

Decades ago, it was discovered that aldopentoses and aldohexoses behave as though they have one more chiral carbon atom than is evident from the structures shown in Figure 8·2. It was discovered, for example, that D-glucose exists in two forms that contain *five* (not four) asymmetric carbons. The source of this additional asymmetry is an intramolecular cyclization reaction that results in a new chiral center.

When an alcohol reacts with an aldehyde (to form a *hemiacetal*) or with a ketone (to form a *hemiketal*), a chiral sp^3-hybridized carbon atom is formed from the achiral sp^2-hybridized carbon atom of the carbonyl group. Figure 8·4 shows the formation of chiral hemiacetal and hemiketal structures. The nomenclature committees of the International Union of Pure and Applied Chemistry (IUPAC) and the International Union of Biochemistry (IUB) have recommended abandoning the term *hemiketal;* in carbohydrate chemistry, both products are called *hemiacetals*.

In monosaccharides, the carbonyl carbon of an aldose containing five or more carbon atoms and of a ketose containing six or more carbon atoms can react with an intramolecular hydroxyl group to form a cyclic hemiacetal in solution. Cyclic hemiacetals exist as either five- or six-membered ring structures in which one of the substituents of the ring is the oxygen atom from the hydroxyl group that reacted to form the hemiacetal. (The hemiacetal is thus heterocyclic.) Because it resembles the six-membered ring *pyran* (Figure 8·5a), the six-membered heterocyclic hemiacetal of a monosaccharide is called a *pyranose*. Similarly, because the five-membered heterocyclic hemiacetal of a monosaccharide resembles furan (Figure 8·5b), it is called a *furanose*. Note, however, that, unlike pyran and furan, the rings of carbohydrates do not contain double bonds.

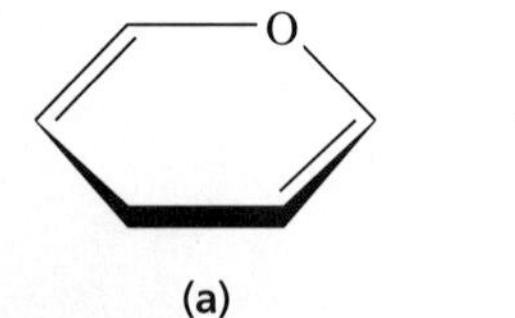

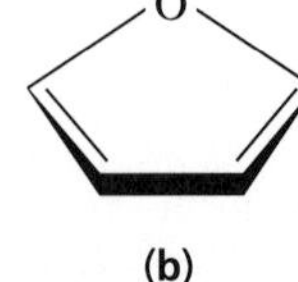

Figure 8·5
(a) Pyran. **(b)** Furan.

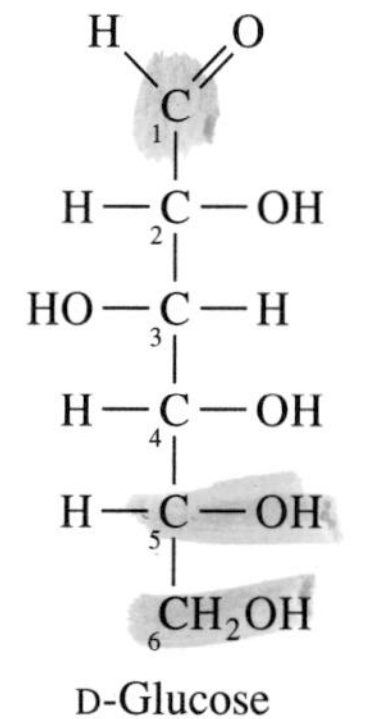

D-Glucose
(Fischer projection)

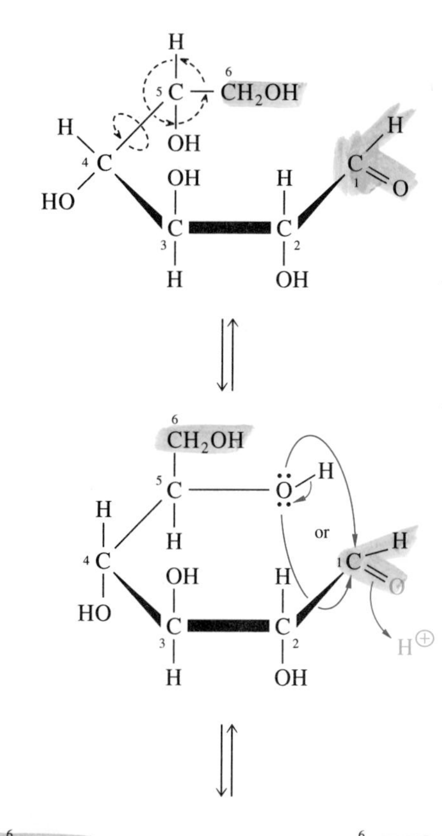

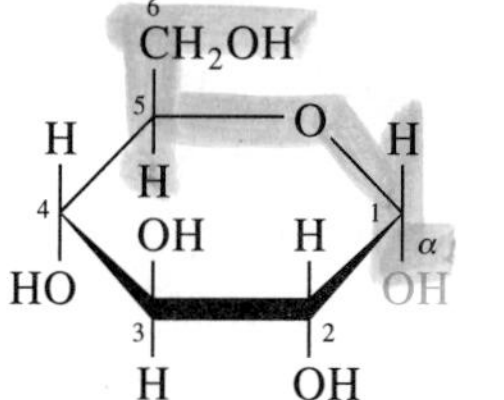

α-D-Glucopyranose
(Haworth projection)

or

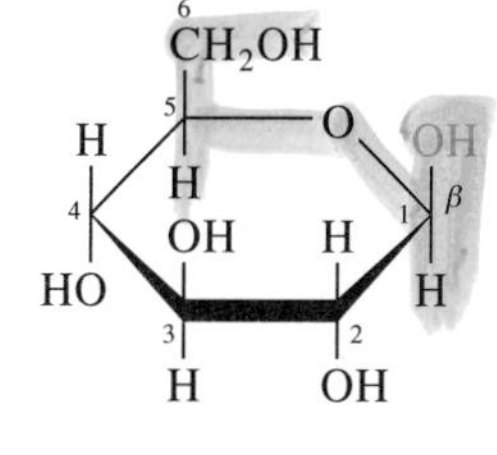

β-D-Glucopyranose
(Haworth projection)

Figure 8·6
Cyclization of D-glucose to form glucopyranose. The Fischer projection (top left) is rearranged into a three-dimensional representation (top right). Rotation of the bond between C-4 and C-5 brings the C-5 hydroxyl group in proximity to the C-1 aldehyde group. Reaction of the hydroxyl group at C-5 with one side of C-1 gives α-D-glucopyranose; reaction of the hydroxyl group with the other side gives β-D-glucopyranose. The α- and β-D-glucopyranose products as shown are known as Haworth projections. In Haworth projections, the lower edges of the ring (heavy lines) project in front of the plane of the paper and the upper edges project behind the plane of the paper. In Haworth projections of D sugars, in which the anomeric carbon is depicted on the right and numbered carbons increase in the clockwise direction, the $—CH_2OH$ group is always oriented up. Hydroxyl groups to the *right* of the carbon skeleton in the Fischer projection are *down* in the Haworth projection formula; hydroxyl groups to the *left* (Fischer) are *up* (Haworth). In the α-D anomer of glucose, the hydroxyl group at C-1 is down; in the β-D anomer, the hydroxyl group at C-1 is up.

The most oxidized carbon of a cyclized monosaccharide (which, like the carbonyl carbon of an aldehyde or a ketone, shares a total of four electrons with oxygen) is referred to as the *anomeric carbon.* In ring structures, the anomeric carbon is chiral. Thus, the cyclized aldose or ketose can adopt either one of two *anomeric configurations* (designated α or β), as illustrated for D-glucose in Figure 8·6.

In solution, aldoses and ketoses capable of forming ring structures (that is, aldoses containing five or more carbon atoms and ketoses containing six or more carbon atoms) equilibrate among their various cyclic and open-chain forms. At equilibrium and 31°C, for example, D-glucose exists as a mixture of approximately 64% β-D-glucopyranose and 36% α-D-glucopyranose, with only a tiny fraction in either the furanose or open-chain forms. Similarly, at equilibrium and 31°C, D-ribose exists as a mixture of 58.5% β-D-ribopyranose, 21.5% α-D-ribopyranose,

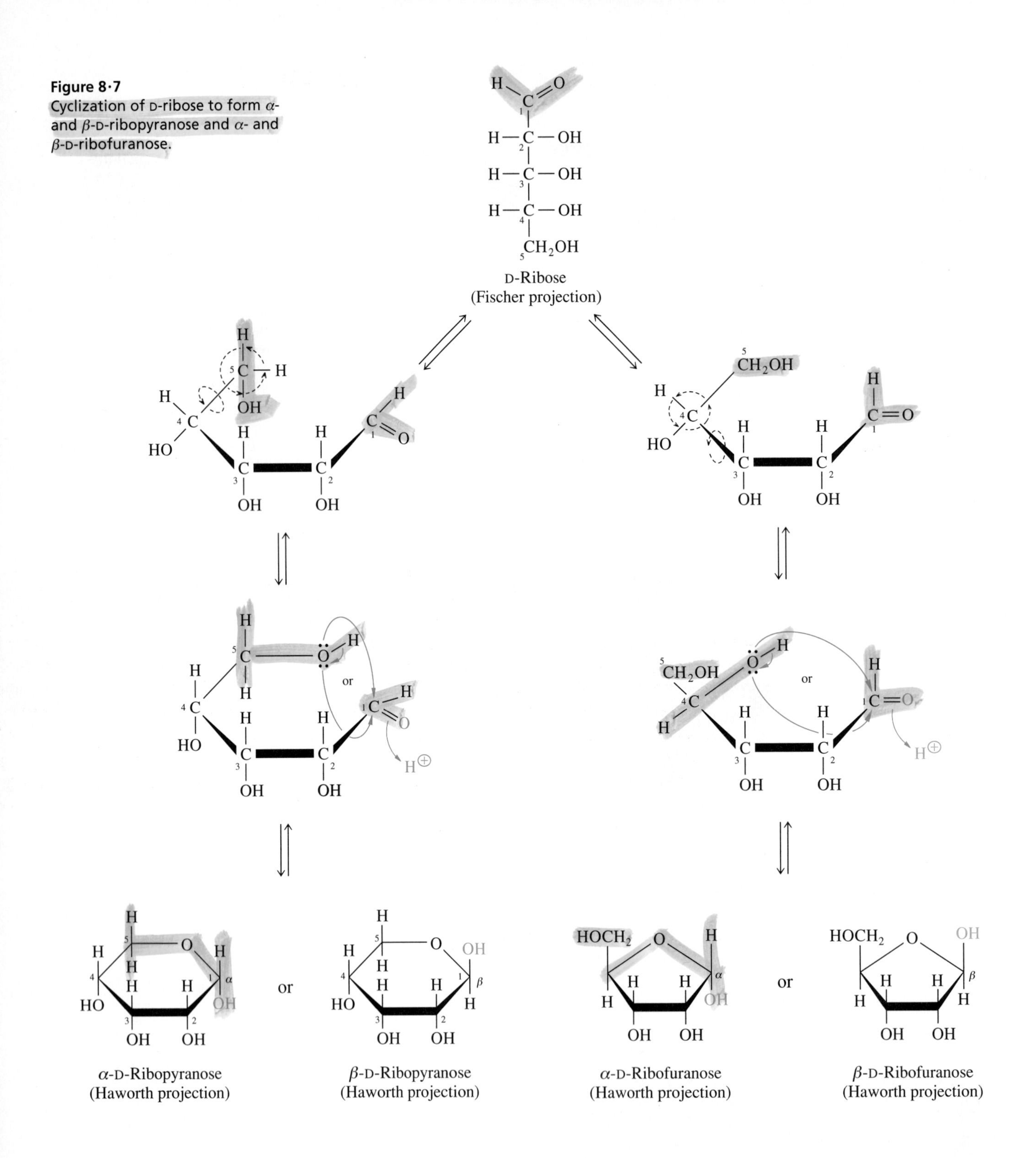

Figure 8·7
Cyclization of D-ribose to form α- and β-D-ribopyranose and α- and β-D-ribofuranose.

13.5% β-D-ribofuranose, and 6.5% α-D-ribofuranose, with a tiny fraction in the open-chain form (Figure 8·7). The relative abundance of the various forms of monosaccharides at equilibrium reflects the relative stabilities of each form. Although unsubstituted D-ribose is most stable as the β-pyranose, its structure in nucleotides (Chapter 10) is the β-furanoside. Box 8·1 discusses the use in this book of ring structures and the nomenclature of carbohydrates.

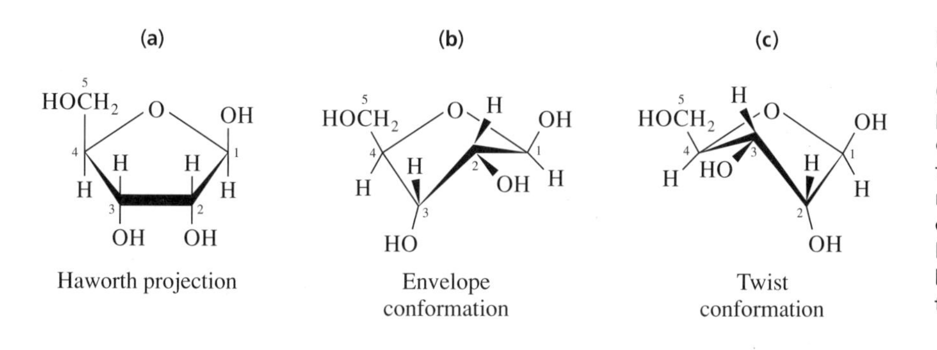

Figure 8·8
Conformations of β-D-ribofuranose. **(a)** Haworth projection. **(b)** One of ten possible envelope conformers, in which one ring atom (in this case, C-2) lies above the plane defined by the four remaining ring atoms. **(c)** One of ten possible twist conformers, in which one ring atom (C-3) lies above and one ring atom (C-2) lies below the plane defined by the remaining three ring atoms.

The ring drawings shown in Figures 8·6 and 8·7 are known as *Haworth projections,* or Haworth perspectives. For most purposes, Haworth projections are adequate representations of stereochemistry. They have the advantage that they can be related easily to Fischer projections. Imagine a cyclic monosaccharide drawn such that the carbons are numbered in ascending order in the clockwise direction. Hydroxyl groups pointing *down* in the Haworth projection point to the *right* of the carbon skeleton in the Fischer projection, whereas hydroxyl groups pointing *up* in the Haworth projection point to the *left* in the Fischer projection. The configuration of the anomeric carbon atom in a Haworth projection of a D sugar is designated α if it has the D-like configuration (hydroxyl group pointing down) and β if it has the L-like configuration (hydroxyl group pointing up). Note, however, that enantiomers are mirror images of each other, and therefore the configuration of functional groups at every carbon is opposite in a pair of enantiomers. Accordingly, the configuration of the anomeric carbon atom of an L sugar by convention is designated α if it has the L-like configuration (hydroxyl group pointing up) and β if it has the D-like configuration (hydroxyl group pointing down).

Because of their general clarity and the ease with which they are drawn, Haworth projections are commonly used in biochemistry (often abbreviated by omitting the carbon-bound hydrogen atoms). However, ring structures containing sp^3-hybridized (tetrahedral) carbons are not actually planar. More realistic views of ring structures are presented in Figures 8·8 and 8·9.

Furanose rings can adopt either an *envelope conformation,* in which one of the five ring atoms is out-of-plane with the remaining four approximately coplanar, or a *twist conformation,* in which two of the five ring atoms are out-of-plane, one on

Figure 8·9
Conformations of β-D-glucopyranose ring structures. **(a)** Haworth projection. **(b)** The chair conformation, in which all hydroxyl groups, including the one at C-1, are equatorial. **(c)** The boat conformation, in which the ring structure is destabilized due to steric repulsion between the hydroxyl group on C-1 and the hydrogen attached to C-4. There is also repulsion between the hydroxyl group on C-2 and the hydrogen atoms on C-3 and C-5.

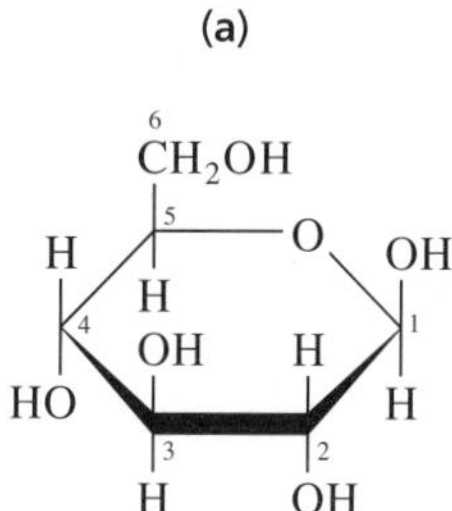

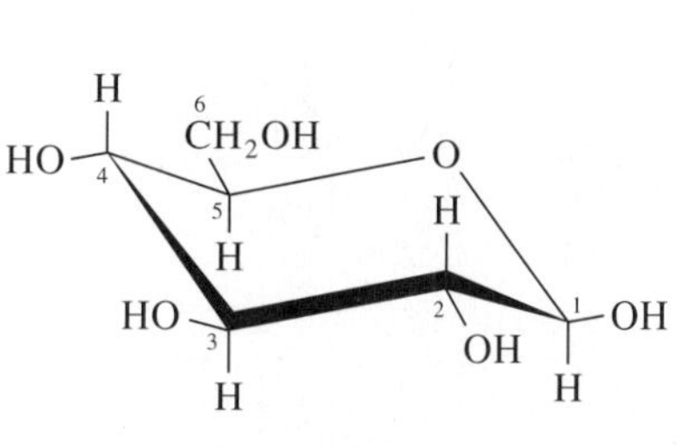

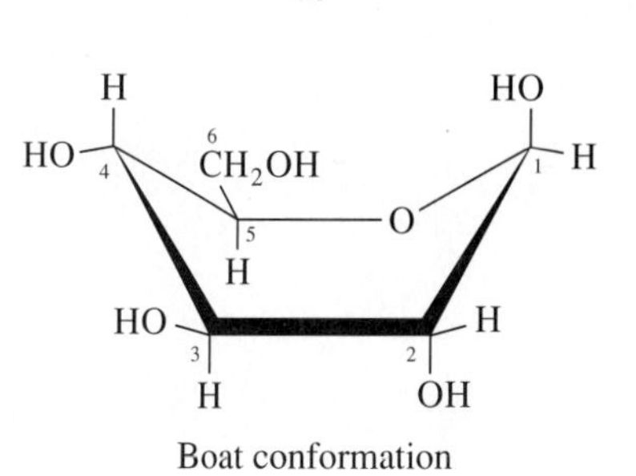

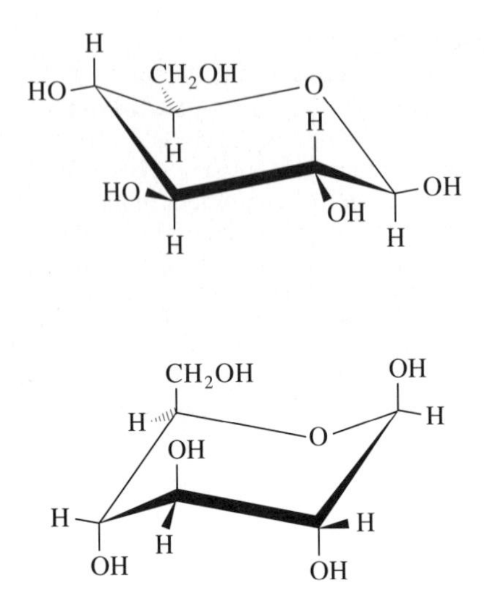

Figure 8·10
The two possible chair conformers of β-D-glucopyranose. The top conformer is more stable.

either side of the plane formed by the other three atoms (Figure 8·8 and Stereo S8·3). For each furanose, there are 10 possible envelope conformers and 10 possible twist conformers. All of these forms rapidly interconvert.

Pyranose rings tend to assume one of two distinct conformations, the *chair conformation* or the *boat conformation* (Figure 8·9 and Stereo S8·4). For each pyranose, there are six distinct boat conformers and two distinct chair conformers. Since steric repulsion among the ring substituents is minimized in the chair conformation, chair conformations are generally more stable than boat conformations. Note that there are two different positions occupied by the substituents of a pyranose ring in the chair conformation. *Axial substituents* are those that extend perpendicular to the plane of the ring, whereas *equatorial substituents* are those that extend along the plane of the ring. In the case of cyclohexane derivatives, six ring substituents are always axial (three above and three below the plane of the ring) and six are always equatorial. In the case of pyranoses, five substituents are axial and five are equatorial. However, whether a group is axial or equatorial depends on which carbon atom (C-1 or C-4) extends above the plane of the ring when the ring is in the chair conformation. Figure 8·10 and Stereo S8·5 show the two different chair conformers of β-D-glucopyranose. The conformation that is most stable is the one in which the bulkiest ring substituents are equatorial.

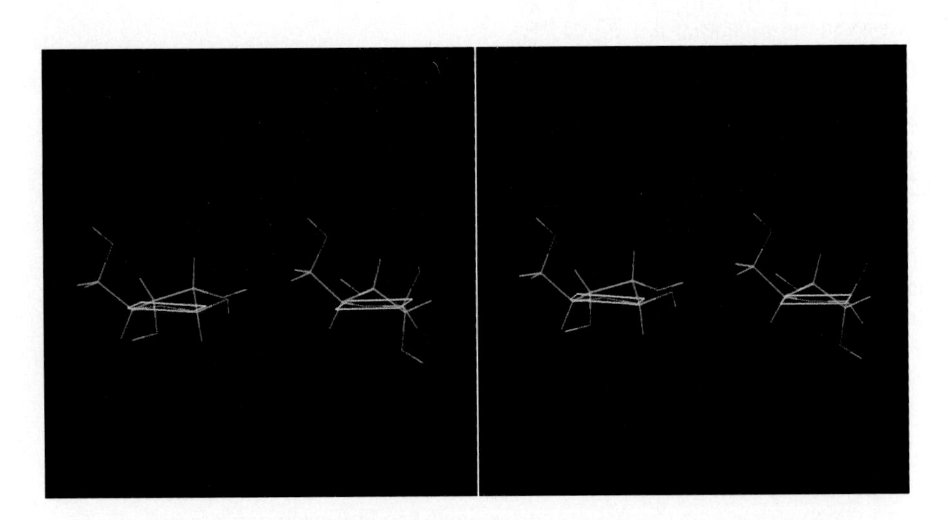

Stereo S8·3
The envelope (left) and twist (right) conformations of β-D-ribofuranose. In the envelope conformer shown, C-2 lies above the plane defined by C-1, C-3, C-4, and the ring oxygen. In the twist conformer shown, C-3 lies above and C-2 lies below the plane defined by C-1, C-4, and the ring oxygen. (Stereo image by Ben N. Banks and Kim M. Gernert.)

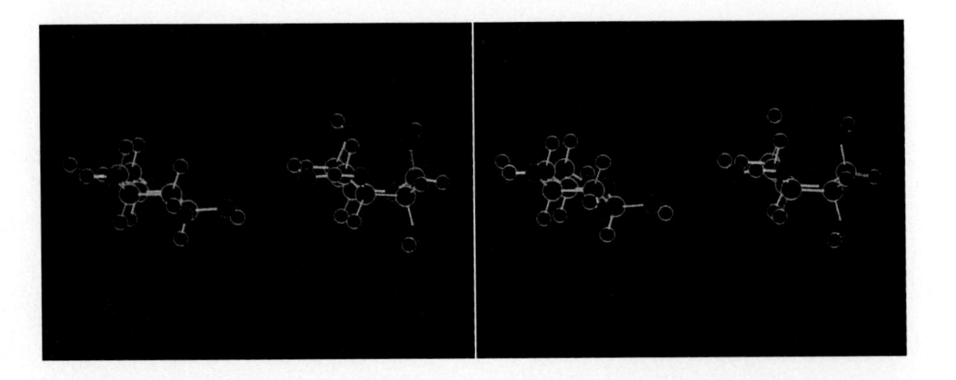

Stereo S8·4
Chair (left) and boat (right) conformations of β-D-glucopyranose. (Stereo image by Ben N. Banks and Kim M. Gernert.)

Stereo S8·5
The two chair conformers of β-D-glucopyranose. The conformation on the left is more stable. (Stereo image by Ben N. Banks and Kim M. Gernert.)

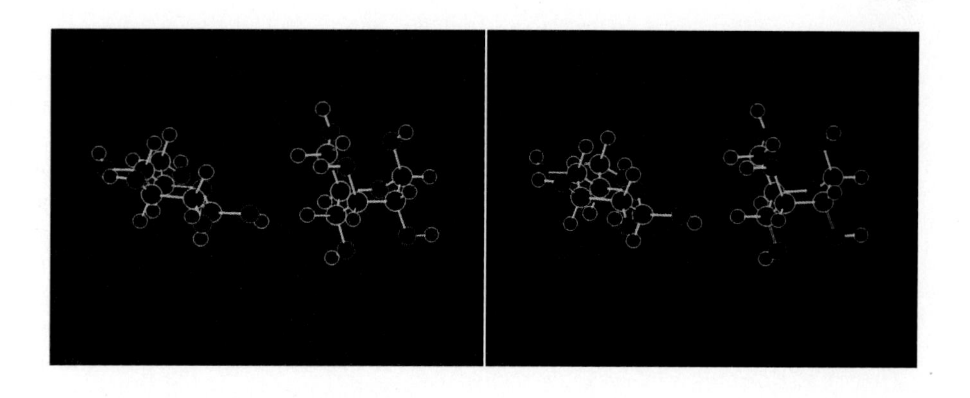

Box 8·1 A Word About Carbohydrate Nomenclature in this Book

We often draw monosaccharides in the β-D-pyranose form. However, you should remember that the anomeric forms of five- and six-carbon sugars exist in equilibrium. Thus, although we might illustrate a point using a β-D-anomer, the same point might apply to the other anomeric forms of a carbohydrate.

In order to distinguish between those times when the anomeric form of a saccharide is important and those times when it is not, we have adopted the following guidelines. Throughout this chapter and the rest of the book, we will refer to sugars in a nonspecific way when we are referring to an equilibrium mixture of the various anomeric forms. Thus, when we say "glucose," we mean both α- and β-D-glucopyranose as well as open-chain glucose and furanose forms. When we are referring to a specific form of a saccharide, however, we will refer to it precisely. Thus, if we mean only β-D-glucopyranose, we will use that name. Finally, since the D enantiomers of carbohydrates predominate in nature, we will always assume a carbohydrate to be of the D configuration unless specified otherwise.

8·3 There Are a Variety of Biologically Important Derivatives of Monosaccharides

In addition to the monosaccharides, which have the characteristic formula $(CH_2O)_n$ with $n \geq 3$, there are many monosaccharide derivatives in biological systems. These include polymerized monosaccharides, such as oligosaccharides and polysaccharides, as well as several classes of nonpolymerized derivatives. We shall consider oligosaccharides and polysaccharides in Sections 8·4 and 8·6. Here, we shall briefly introduce a few of the other monosaccharide derivatives important to living systems, including sugar phosphates, the deoxy and amino sugars, the sugar alcohols, the sugar acids, and vitamin C.

When monosaccharides are used as fuel, they are metabolized as phosphate esters. Figure 8·11 shows the structures of several sugar phosphates we shall encounter in our study of carbohydrate metabolism.

The structures of two deoxy sugars and an amino sugar are shown in Figure 8·12. In each of these, a hydrogen atom or an amino group substitutes for one of the hydroxyl groups in the parent monosaccharide. The deoxy sugars shown are 2-deoxy-D-ribose, an important building block for DNA, and α-L-fucose (6-deoxy-L-galactose), a common component of polysaccharides and one example of a naturally occurring L enantiomer. The amino sugar shown is *N*-acetylneuraminic acid, an important constituent of a family of lipids called gangliosides (Section 9·5).

In addition to deoxy sugars and amino sugars, two other classes of naturally occurring monosaccharide derivatives are the sugar alcohols and sugar acids. In sugar alcohols, the carbonyl oxygen has been reduced, resulting in a carbon-bound hydroxyl group. Figure 8·13 provides examples of sugar alcohols including glycerol and *myo*-inositol (both important components of lipids), and ribitol (a component of flavin adenine dinucleotide, or FAD). Sugar acids, such as aldonic acids and alduronic acids, are carboxylic acids derived from aldoses either by oxidation of C-1 (the aldehydic carbon) to yield an aldonic acid or by oxidation of the highest-numbered carbon (the carbon bearing the primary alcohol) to yield an alduronic acid. The structures of gluconic acid and glucuronic acid are shown in Figure 8·14. Sugar acids are important components of many polysaccharides.

Figure 8·11
Structures of several metabolically important sugar phosphates.

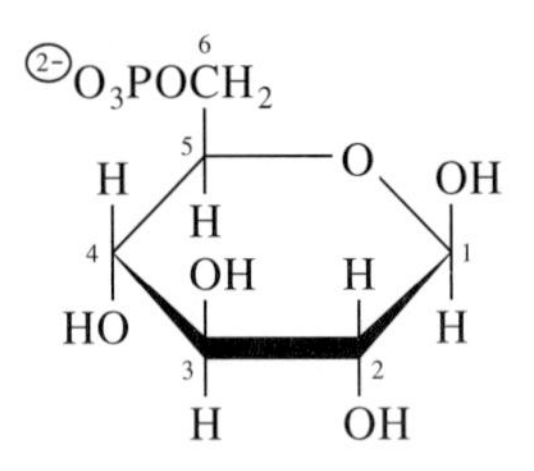

β-D-Glucose 6-phosphate

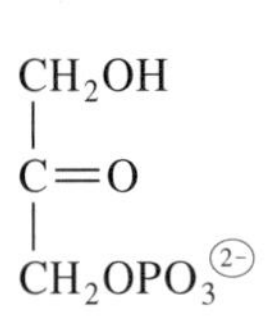

Dihydroxyacetone phosphate

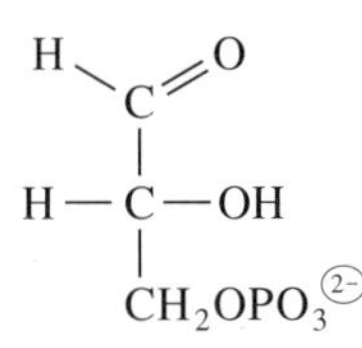

D-Glyceraldehyde 3-phosphate

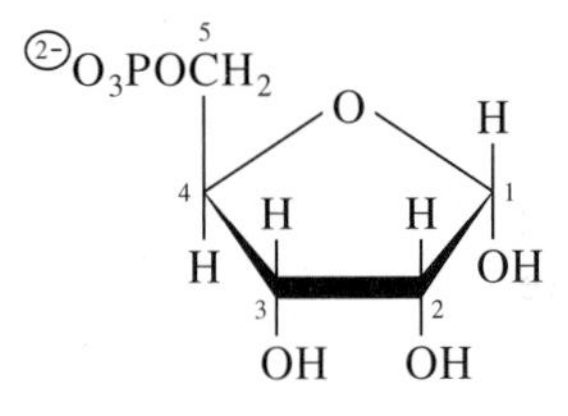

α-D-Ribose 5-phosphate

Figure 8·12
Structures of the deoxy sugars 2-deoxy-D-ribose and L-fucose and the amino sugar *N*-acetylneuraminic acid.

2-Deoxy-D-ribose
(β-D-furanose anomer)

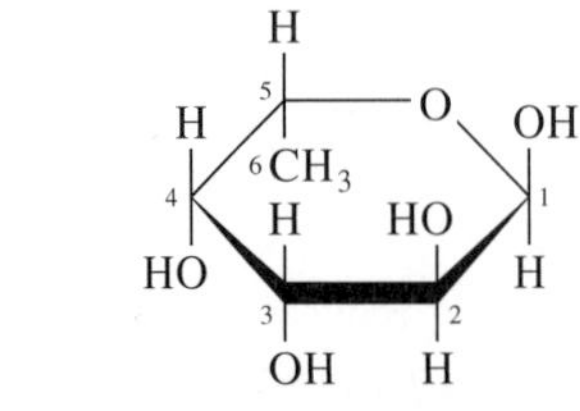

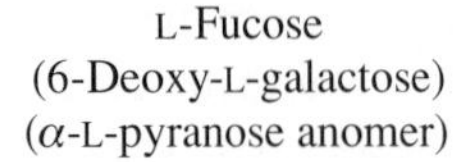

L-Fucose
(6-Deoxy-L-galactose)
(α-L-pyranose anomer)

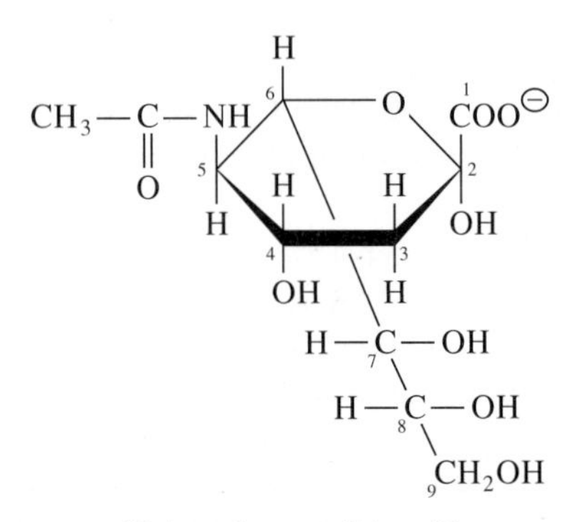

N-Acetylneuraminic acid
(α-D-pyranose anomer)

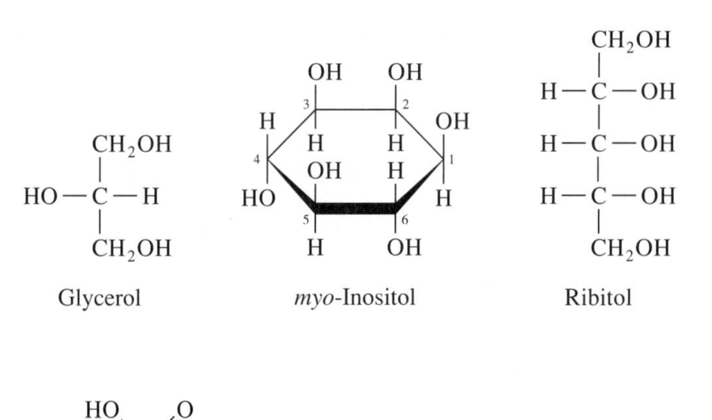

Ribitol

Figure 8·13
Structures of several sugar alcohols. Glycerol (a reduced form of glyceraldehyde) and *myo*-inositol (metabolically derived from glucose) are important constituents of many lipids (Chapter 9). Ribitol (a reduced form of ribose) is a constituent of the vitamin riboflavin and its coenzymes (Chapter 7).

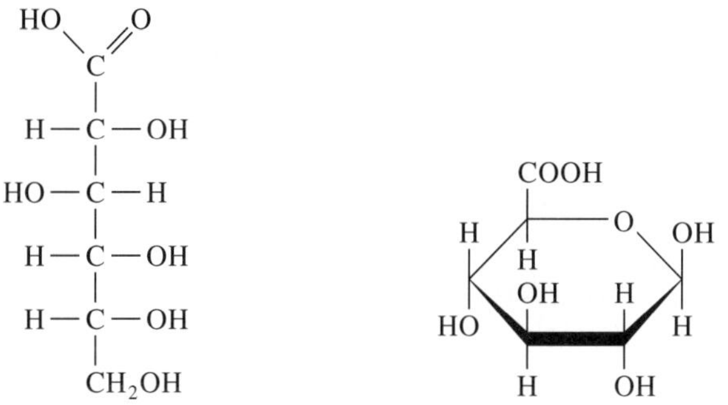

D-Gluconic acid

D-Glucuronic acid
(β pyranose anomer)

Figure 8·14
Structures of sugar acids derived from D-glucose.

Finally, L-ascorbic acid, or vitamin C, is an enediol of a lactone derived from an aldonic acid (Figure 8·15). The C-3 hydroxyl group of this sugar derivative has a relatively low pK_a of 4.2 due to resonance stabilization of the conjugate base. We noted earlier that ascorbate is an essential cofactor in the catalytic hydroxylation reactions involving proline and lysine residues during collagen synthesis (Sections 4·5 and 7·2).

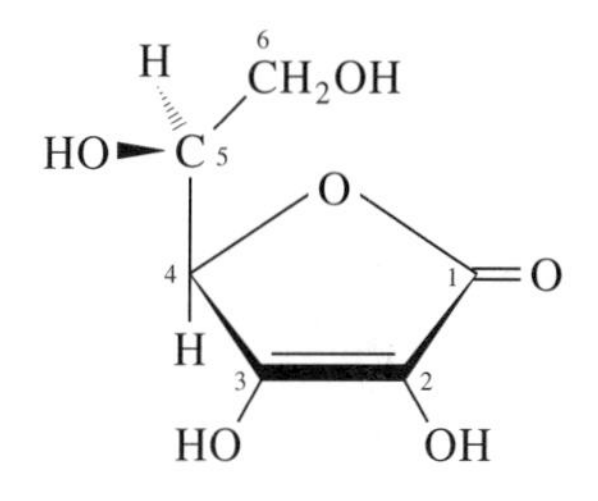

Figure 8·15
L-Ascorbic acid (Haworth projection).

8·4 Disaccharides Consist of Two Monosaccharide Residues Linked by a Glycosidic Bond

The primary structural linkage in all polymers of monosaccharides is known as the *glycosidic bond*. A glycosidic bond is formed by condensation of two monosaccharides with the elimination of water—much as a peptide bond is formed by the condensation of two amino acids. In terms of chemical structure, glycosidic bonds are *acetal linkages* involving condensation of the anomeric carbon of one monosaccharide with a hydroxyl group of another. (Recall that peptide bonds are *amide linkages* involving condensation of the α-amino group of one amino acid with the α-carboxyl group of another.) In contrast to peptide-bond formation between α-amino and α-carboxyl groups in defining the primary structure of a protein, the glycosidic bonds linking monosaccharide residues can involve any of several hydroxyl groups in addition to the hydroxyl group on the anomeric carbon atom. Thus, in considering the structure of an oligosaccharide, it is important to note not only which monosaccharide residues are involved, but also which of their atoms are linked by glycosidic bonds.

Furthermore, because there are a number of ways in which the anomeric carbon of one monosaccharide can be linked by a glycosidic bond to another monosaccharide, precise nomenclature is needed to specify the exact bonding in each case. In a disaccharide, for example, the linking atoms, the configuration of the glycosidic bond, and the name of each monosaccharide residue (including its designation as a pyranosyl or furanosyl structure) must be specified to denote a precise structure. Figure 8·16 illustrates structures and nomenclature for four common disaccharides.

Maltose (Figure 8·16a) is released during the hydrolysis of amylose (a type of starch), which is a polymer of glucose residues. Maltose is composed of two D-glucose residues joined by an α-glycosidic bond. The glycosidic bond links C-1 of one residue (left in Figure 8·16a) to C-4 of the second residue (right). Maltose is therefore α-D-glucopyranosyl-(1→4)-D-glucose. It is important to note that the anomeric carbon linked by the glycosidic bond is *not free to isomerize* but is fixed in the α configuration, whereas the glucose residue on the right (the *reducing end*, as explained in Box 8·2) freely equilibrates among the α, β, and open-chain structures, the latter present only in very small amounts. The structure shown in Figure 8·16a is the β pyranose anomer of maltose (the anomer whose reducing end is in the β configuration, the predominant anomeric form).

Cellobiose (β-D-glucopyranosyl-(1→4)-D-glucose) is another dimer of glucose (Figure 8·16b). Cellobiose is the degradation product of the structural polysaccharide cellulose. The only difference between cellobiose and maltose is that the glycosidic linkage in cellobiose is β (compared to the α linkage in maltose). The glucose residue on the right in Figure 8·16b, like the residue on the right in 8·16a, equilibrates among the α, β, and open-chain structures.

Figure 8·16
Structures of **(a)** maltose, **(b)** cellobiose, **(c)** lactose, and **(d)** sucrose. The oxygen atom of each glycosidic bond is shown in red.

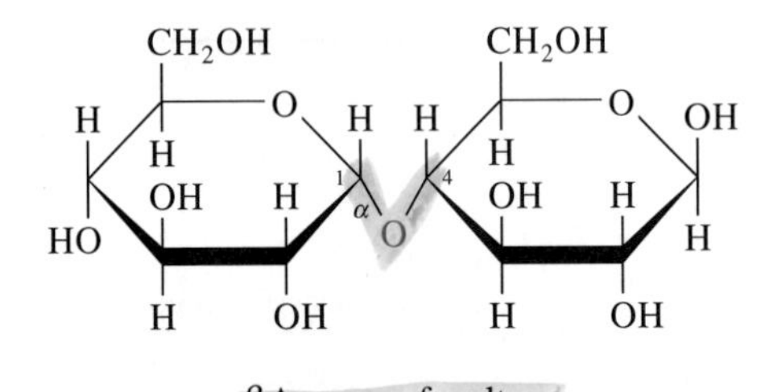

β Anomer of maltose
(α-D-Glucopyranosyl-(1→4)-β-D-glucopyranose)

(b)

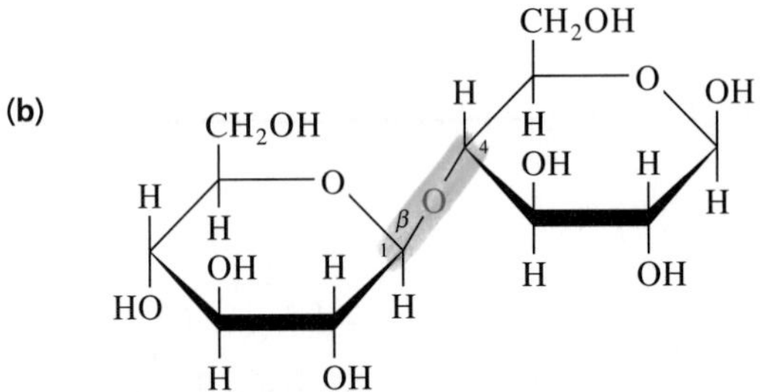

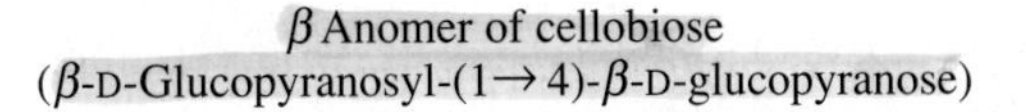

β Anomer of cellobiose
(β-D-Glucopyranosyl-(1→4)-β-D-glucopyranose)

(c)

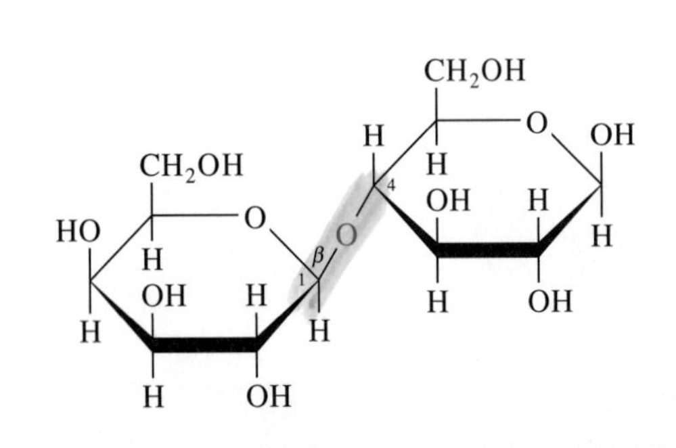

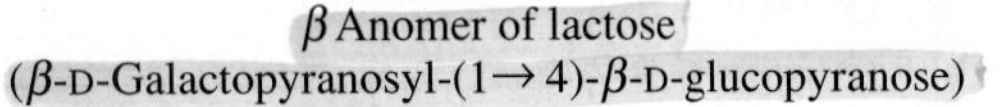

β Anomer of lactose
(β-D-Galactopyranosyl-(1→4)-β-D-glucopyranose)

(d)

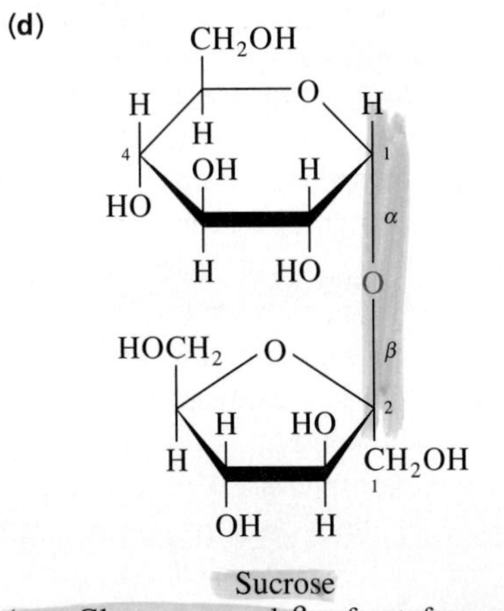

Sucrose
(α-D-Glucopyranosyl β-D-fructofuranoside)

Lactose (β-D-galactopyranosyl-($1 \rightarrow 4$)-D-glucose), a major carbohydrate found in milk, is a disaccharide of mammalian origin that is synthesized only in lactating mammary glands (Figure 8·16c). Note from its structure that lactose is an epimer of cellobiose.

Sucrose (α-D-glucopyranosyl β-D-fructofuranoside), or table sugar, is the most abundant disaccharide found in nature (Figure 8·16d). Whereas lactose is unique to mammals, sucrose is synthesized only in plants. Sucrose is distinguished from the other three disaccharides in Figure 8·16 in that its glycosidic bond links the anomeric (oxidized) carbon atoms of two monosaccharide residues. Sucrose can be regarded as either α-D-glucopyranosyl β-D-fructofuranoside or β-D-fructofuranosyl α-D-glucopyranose (equivalent terms). Note that the atoms involved in this glycosidic bond are implied by name and do not have to be specified as ($1 \rightarrow 2$) or ($2 \rightarrow 1$). The configurations of both the glucopyranose and fructofuranose residues are fixed in sucrose, and neither residue is free to equilibrate between α and β anomers.

Box 8·2 Monosaccharides and Most Disaccharides Are Reducing Sugars

Because monosaccharides and most disaccharides contain a reactive carbonyl group, they are readily oxidized to diverse products upon treatment with alkaline solutions of reducible metal ions such as Cu^{2+} or Ag^{+}. Such carbohydrates, which include glucose, maltose, cellobiose, and lactose, are sometimes classified as *reducing sugars*. Carbohydrates such as sucrose, which are not readily oxidized because both anomeric carbon atoms are fixed in a glycosidic linkage, do not reduce metal ions and are therefore classified as *nonreducing sugars*.

In Benedict's test, carbohydrate is added to an alkaline solution of copper sulfate. Reducing sugars are able to reduce the Cu^{2+}, which is blue, to Cu^{+}, which precipitates from the alkaline solution as red-orange Cu_2O. Reducing sugars can also reduce ammoniacal silver nitrate (Ag^{+}) to metallic silver (Ag^0), producing a mirror on a glass surface, such as on the inside of a test tube. Historically, such reactions were widely used to detect the presence and concentration of glucose in the urine of persons afflicted with diabetes mellitus (Latin, *mellitus,* honey-sweet). Today, most laboratories use newer and more accurate enzymological methods to conduct such tests.

8·5 Saccharides Can Form Glycosidic Linkages with Aglycons

The anomeric carbons of sugars can form glycosidic linkages with a variety of nonsugar molecules. Organic nonsugar molecules linked to sugars are referred to as *aglycons,* and the compounds formed are called *glycosides.* Examples of aglycons include hydroxybenzyl alcohol, which combines with D-glucose to form salicin; cyanidin, which combines with two residues of D-glucose to form cyanin; and digoxigenin, which combines with a β-linked trisaccharide of digitoxose to form digoxin (Figure 8·17). Salicin, a compound derived from willow and poplar bark, has been used to treat headaches. Cyanin is the pigment that gives roses and certain other flowers their red color. Digoxin, a compound extracted from the leaves of the foxglove plant, is an important drug in the treatment of heart disease.

The β-galactosides constitute an abundant class of glycosides. In these compounds, a variety of nonsugar molecules are joined in a β linkage to galactose. For example, derivatives of β-D-galactosyl 1-glycerol (Figure 8·17d) are common in eukaryotic cell membranes and can be hydrolyzed readily by enzymes called β-galactosidases.

8·6 Polysaccharides Are Long Polymers of Monosaccharide Residues

Because the cell walls of plants are made of polysaccharides, on the basis of mass, polysaccharides are the most abundant organic biomolecules on Earth. They are frequently divided into two broad classes: *homopolysaccharides,* or homoglycans, which are polymers containing only one type of monosaccharide residue, and *heteropolysaccharides,* or heteroglycans, which are polymers containing more than one type of monosaccharide residue. Unlike proteins, whose primary structures are encoded by the genome and thus have specified lengths, polysaccharides are created without a template by the addition of particular monosaccharide and oligosaccharide residues. As a result, the lengths and compositions of particular polysaccharide molecules may vary within a population of similar molecules. Such a population is said to be polydisperse.

Most polysaccharides can also be classified according to their biological roles as *storage polysaccharides,* such as starch and glycogen, or *structural polysaccharides,* such as cellulose or chitin. There are many different polysaccharides in nature. We shall consider the structures of several representative examples in the following two subsections.

A. Starch and Glycogen Are Storage Homopolysaccharides of Glucose

D-Glucose, the chief source of metabolic energy for many organisms, is stored intracellularly in polymeric form, thereby avoiding the excessive osmotic pressures that would result from large accumulations of the free monosaccharide. The most common homopolysaccharide of glucose in plants and fungi is called *starch,* and that in animals is called *glycogen.* Both types of polysaccharides occur in bacteria.

In plant cells, starch is present as a mixture of amylose and amylopectin and is stored in granules whose diameters range from 3 to 100 μm. Amylose is an unbranched polymer of some 100 to 1000 D-glucose residues connected by α-(1$\rightarrow$4) glycosidic linkages, specifically termed α-(1$\rightarrow$4) glucosidic bonds because the anomeric carbon of a glucose residue is involved in the bond. These same linkages

Figure 8·17
Structures of four glycosides. The aglycons are shown in blue. **(a)** Salicin, a component of willow and poplar bark that has analgesic properties. **(b)** Cyanin, the red pigment of various flowers, including roses. **(c)** Digoxin, a compound extracted from the foxglove plant that is used to treat heart disease. **(d)** β-D-Galactosyl 1-glycerol, derivatives of which are common in eukaryotic cell membranes.

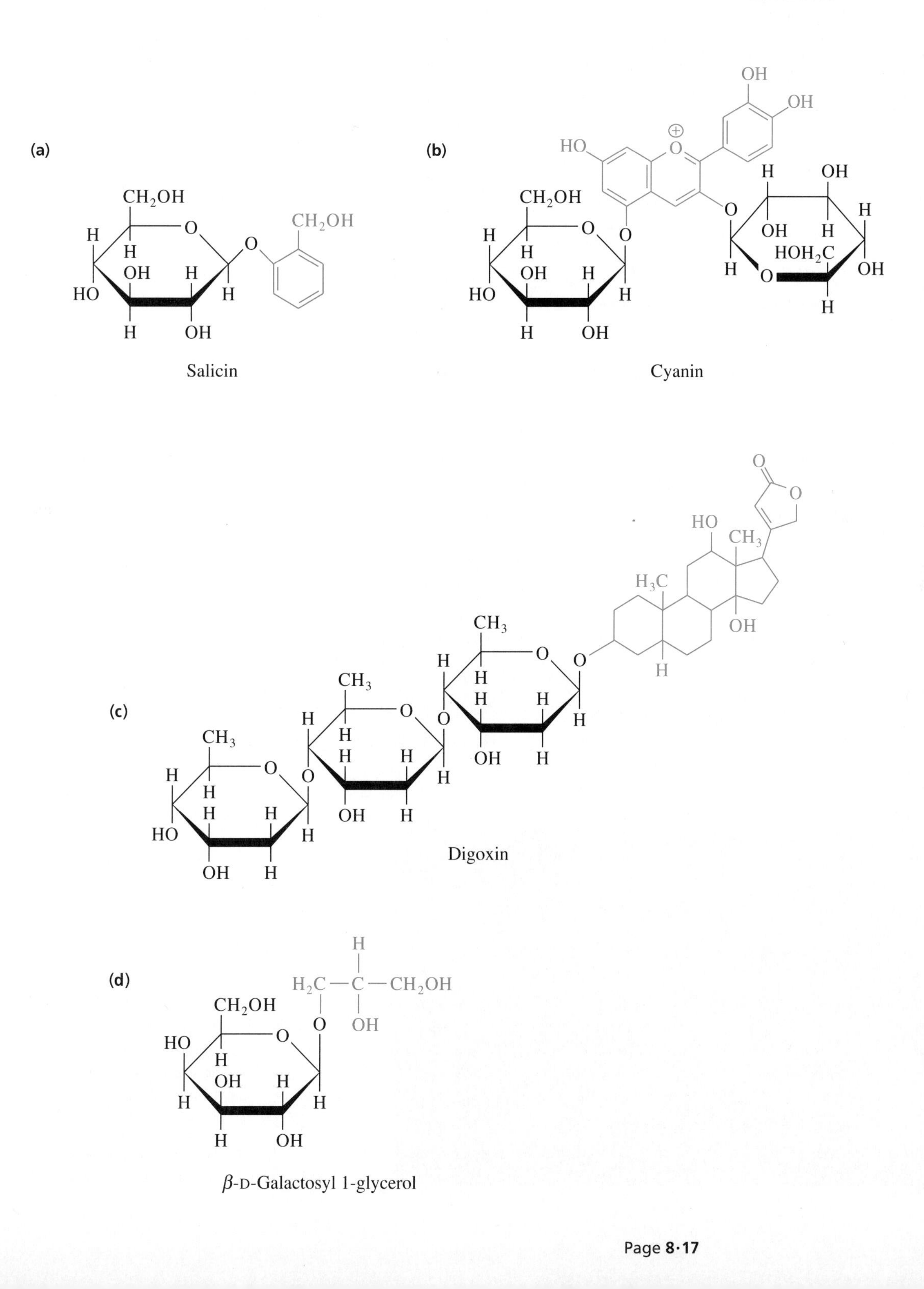

connect glucose monomers in the disaccharide maltose (Figure 8·18a and Stereo S8·6). Although it is not truly soluble in water, amylose forms hydrated aggregates in aqueous solution and can form a helical structure under some conditions.

Amylopectin is essentially a branched version of amylose (Figure 8·18b). In addition to α-(1→4) linkages, amylopectin contains branch points at which α-(1→6) bonds occur. Branching occurs, on average, once every 25 residues, and the branches, or side chains, contain some 15 to 25 glucose residues. Some side chains themselves contain branches. When isolated from living cells, amylopectin molecules range in size from 300 to 6000 glucose residues.

Glycogen, a glucose-storage polysaccharide found in animals and bacteria, is also a branched polymer of glucose residues. Glycogen contains the same α-(1→6) branching as amylopectin, but the branches occur with greater frequency, and the

Figure 8·18
Structures of amylose and amylopectin. **(a)** Amylose, one form of starch, is a linear polymer of glucose residues linked by (1→4)-α-D-glucosidic bonds. **(b)** Amylopectin, a second form of starch, is a branched polymer. The linear glucose residues of the main chain and the side chains of amylopectin are linked by (1→4)-α-D-glucosidic bonds, and the side chains are linked to the main chain by (1→6)-α-D-glucosidic bonds.

(a)

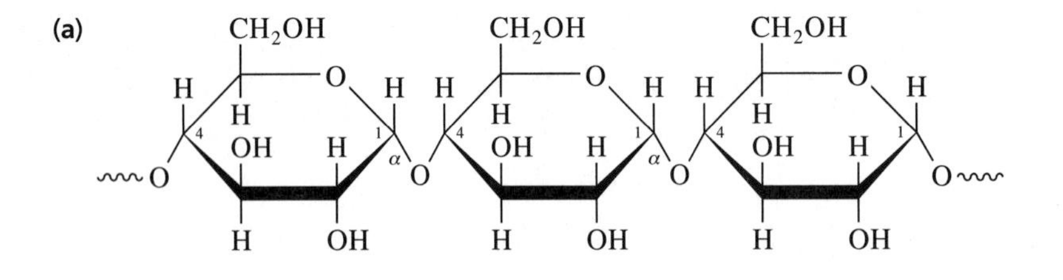

(b)

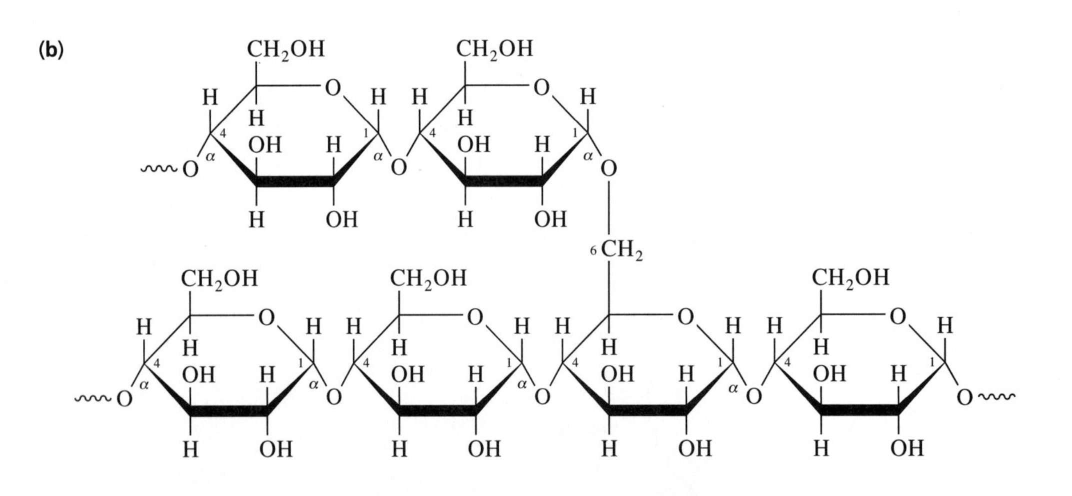

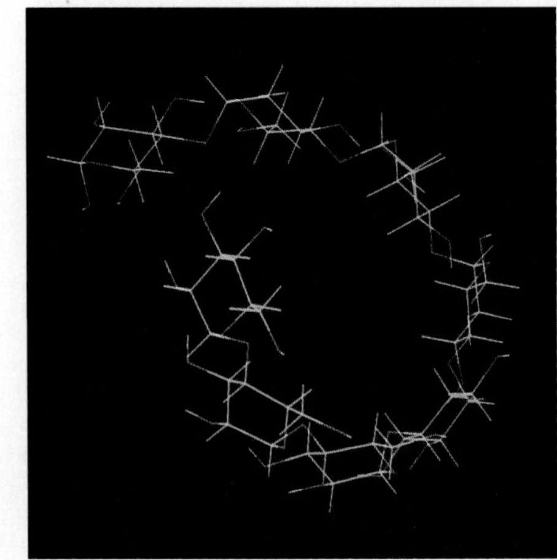

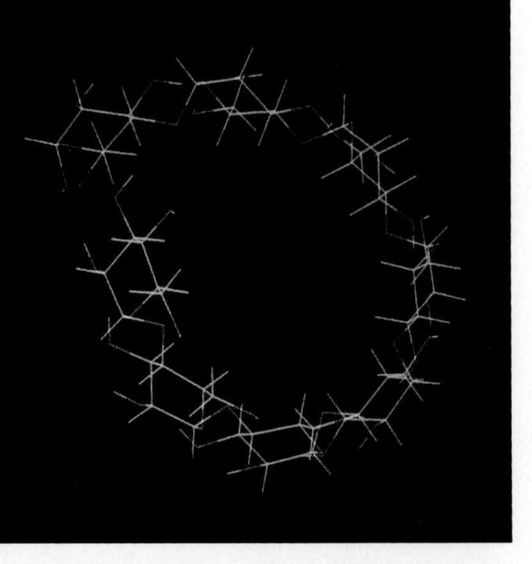

Stereo S8·6
Amylose. Amylose can assume a left-handed helical conformation, which is hydrated on the inside as well as on the outer surface. (Stereo image by Ben N. Banks and Kim M. Gernert.)

side chains of glycogen contain fewer glucose residues. In general, glycogen molecules tend to be larger than starch molecules, containing up to several hundred thousand glucose residues. In mammals, depending on the nutritional state, glycogen can account for up to 10% of the mass of the liver and 1% of the mass of muscle.

An adult human requires about 500 g of carbohydrate per day, most of which is supplied by starch and/or glycogen. Dietary starch and glycogen are degraded in the gastrointestinal tract by the actions of α-amylase and a debranching enzyme, each of which catalyzes hydrolysis of certain α-D-glycosidic bonds. Another hydrolase, known as β-amylase, exists in plants. (In the case of α- and β-amylase, the "α" and "β" designations refer to types of amylases, and *not to the configurations* of the glycosidic bonds of the substrate; both types of amylases act only on α-(1→4)-D-glycosidic bonds.)

Figure 8·19 depicts the actions of those two hydrolases on glycogen and amylopectin. α-Amylase is an *endoglycosidase* that catalyzes random hydrolysis of α-(1→4)-D-glucosidic bonds of amylose, amylopectin, and glycogen. β-Amylase is an *exoglycosidase* that catalyzes sequential hydrolysis of maltose from the free, nonreducing ends of amylopectin in plants, shortening them one maltose residue at a time. The α-(1→6) linkages at branch points are not substrates for either α-amylase or β-amylase. After amylase-catalyzed hydrolysis of glycogen or amylopectin, highly branched cores resistant to further hydrolysis, called *limit dextrins,* remain. Limit dextrins can be further degraded only after debranching enzymes have catalyzed hydrolysis of the α-(1→6) linkages at branch points.

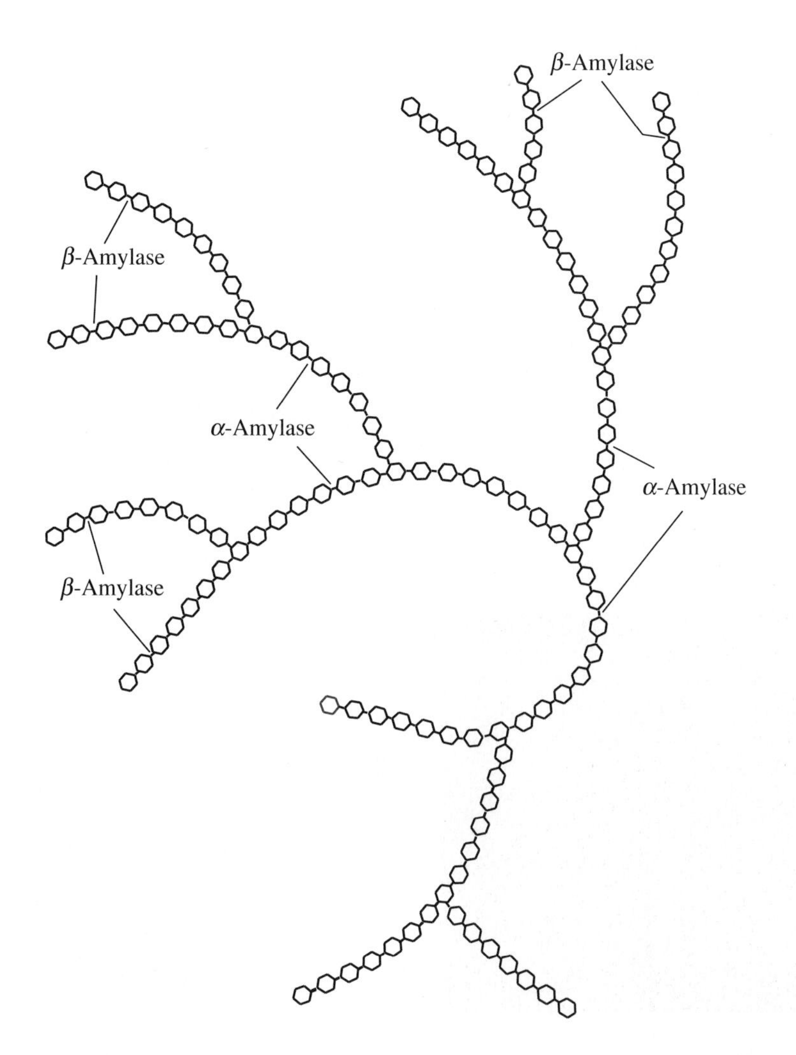

Figure 8·19
Points of hydrolysis of glycogen and amylopectin by animal and plant amylases. This schematic diagram shows the points at which glycogen and amylopectin can be hydrolyzed by incubations with α-amylase and β-amylase. Each hexagon represents a glucose residue; the red hexagon represents the single reducing end of the branched polymer. α-Amylase, a plant and animal enzyme, is an endoglycosidase; it catalyzes the random hydrolysis of α-(1→4) linkages within glycogen chains during digestion. β-Amylase, a plant enzyme, is an exoglycosidase; it catalyzes the sequential removal of maltose residues from the ends of the outer branches of amylopectin. Neither α- nor β-amylase is able to catalyze hydrolysis of α-(1→6) linkages. Thus, after the action of α- and β-amylase, a limit dextrin still remains. Limit dextrins can only be degraded by the action of debranching enzymes, which catalyze the hydrolysis of α-(1→6) linkages. (An actual glycogen or amylopectin molecule contains many more glucose residues than are represented here.)

We shall consider other enzymes that catalyze the intracellular synthesis and breakdown of glycogen in Chapter 15. At this point, it is sufficient to note that the branched structures of amylopectin and glycogen molecules possess only one reducing end but many nonreducing ends. It is at those nonreducing ends that the vast majority of enzymatic lengthening and degradation reactions occur.

B. Cellulose and Chitin Are Structural Homopolysaccharides

Plant cell walls contain a high percentage of the structural homopolysaccharide *cellulose,* which accounts for over 50% of the organic matter in the biosphere. Unlike storage polysaccharides, cellulose and other structural polysaccharides are extracellular molecules extruded by the cells in which they are synthesized. Like amylose, cellulose is a linear homopolysaccharide of glucose residues, but in cellulose the glucose residues are joined by β-(1→4) linkages rather than α-(1→4) linkages (Box 8·3). The glucose monomers of cellulose are therefore joined by the same linkages that join the two glucose monomers of the disaccharide cellobiose. The β linkages of cellulose result in a rather rigid, extended conformation in which each successive glucose residue is rotated 180° relative to its neighbors (Figure 8·20). The equatorial hydroxyl groups are extensively hydrogen bonded, forming bundles of polymeric chains, or fibrils (Stereo S8·7). The fibrils are insoluble in

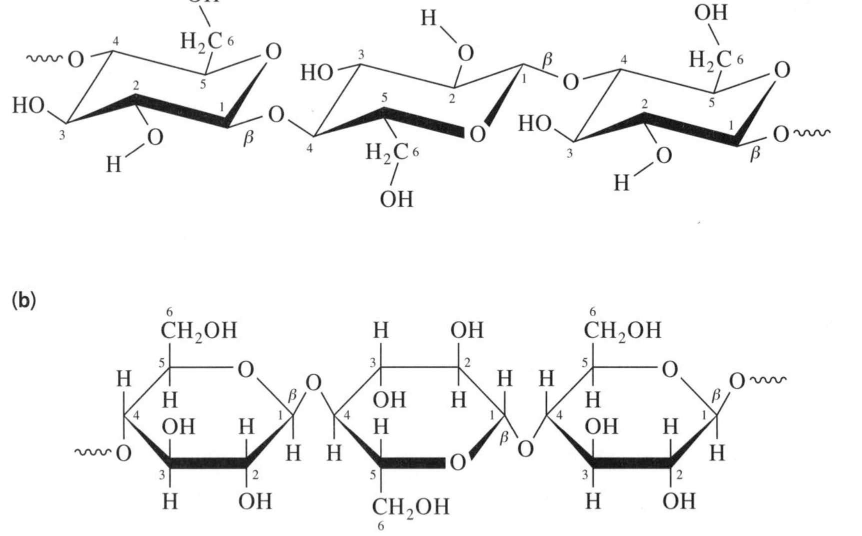

Figure 8·20
(a) Chair conformation of β-(1→4)-linked D-glucose residues in cellulose. (b) Modified Haworth projection of β-(1→4)-linked D-glucose residues in cellulose, emphasizing the alternating orientation of successive glucose residues within the overall conformation of cellulose chains.

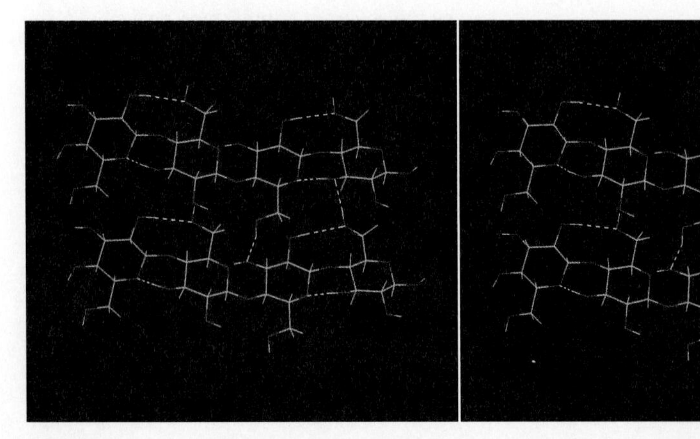

Stereo S8·7
Cellulose fibers. Cellulose is a homopolysaccharide made up of repeating cellobiose units. Hydrogen bonding among individual linear chains gives cellulose its strength and rigidity. (Stereo image by Ben N. Banks and Kim M. Gernert.)

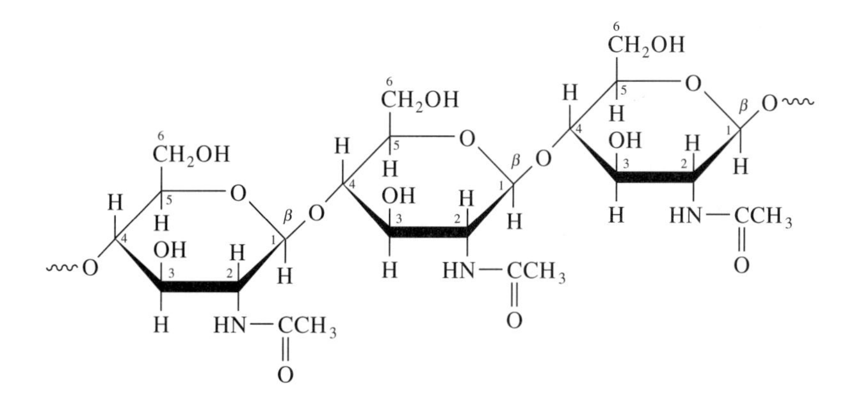

Figure 8·21
Chitin. Chitin is a homopolysaccharide consisting of repeating units of β-(1→4)-linked *N*-acetylglucosamine residues.

water and confer strength and rigidity. Cotton fibers are almost entirely cellulose, and wood averages about 50% cellulose content.

Cellulose molecules vary greatly in size, ranging from about 300 to over 15 000 glucose residues. Because of the strength it imparts, cellulose is used in a variety of commercial applications and is a component of a number of synthetic materials, including cellophane and the fabric rayon.

Chitin is a homopolysaccharide found in the exoskeletons of insects and crustaceans and also in the cell walls of most fungi and many algae. Chitin is a linear polymer, similar to cellulose, but consists of β-(1→4)-linked *N*-acetylglucosamine (GlcNAc) residues rather than glucose residues (Figure 8·21). Since the GlcNAc residues are negatively charged, chitin fibers are polyanionic and can form complexes with various cations. The binding of cations to the chitin matrix confers rigidity and strength.

Box 8·3 Stereospecific Hydrolysis of Glycosidic Bonds

In Chapter 5, we commented on the stereospecificity of many enzymes. Enzymes that catalyze the hydrolysis of α-D-glucosidic bonds (α-glucosidases, such as maltase and amylase) do not catalyze the hydrolysis of β-D-glucosidic bonds. Conversely, β-glucosidases (such as cellulase) do not catalyze hydrolysis of α-D-glucosidic bonds.

Humans and other monogastric mammals that can metabolize starch, glycogen, lactose, and sucrose as energy sources cannot metabolize cellulose because they lack enzymes capable of catalyzing the hydrolysis of β-glucosidic linkages. Ruminants such as cows and sheep, on the contrary, have microorganisms in their rumens that produce such enzymes, allowing them to digest cellulose and cellobiose to glucose. Thus, ruminants can obtain energy from eating plants such as grass that are rich in cellulose. In monogastric mammals, cellulose remains undigested and constitutes "bulk" or "fiber."

8·7 Heteropolysaccharides Are Widely Distributed in Nature

There are a wide variety of heteropolysaccharides in nature, a few examples of which we shall consider here.

One example of a heteropolysaccharide is *hyaluronic acid,* a glycosaminoglycan that is a primary component of *ground substance,* the highly viscous gel-like intercellular matrix in vertebrate connective tissues. Hyaluronic acid is also found in the vitreous humor of the eye and the synovial fluid (lubricant) of joints. Hyaluronic acid is a linear polymer consisting of alternating residues of D-glucuronic acid (GlcA) and *N*-acetyl-D-glucosamine (GlcNAc) linked as follows: β-GlcA-(1→3)-β-GlcNAc-(1→4)-β-GlcA-(1→3)-β-GlcNAc-, and so on (Figure 8·22). At physiological pH, glycosaminoglycans are polyanionic and are extensively hydrated. In fact, glycosaminoglycans function as viscous and elastic constituents partly as a result of their ability to immobilize large amounts of water.

Certain pathogenic bacteria secrete *hyaluronidases,* enzymes that catalyze the cleavage of the β-(1→3) or β-(1→4) linkages of hyaluronic acid, rendering the tissues under attack susceptible to bacterial invasion. Hyaluronidases are also found in both snake and insect venoms, where they facilitate the spread of venom toxins.

Ground substance contains another heteropolysaccharide, keratan sulfate. Keratan sulfate is composed chiefly of alternating galactose and *N*-acetylglucosamine-6-sulfate residues linked by β-(1→4) and β-(1→3) glycosidic bonds (Figure 8·23). In addition, keratan sulfate contains small amounts of fucose, mannose, sialic acid, and *N*-acetylglucosamine residues, making it perhaps the most heterogeneous of all glycosaminoglycans.

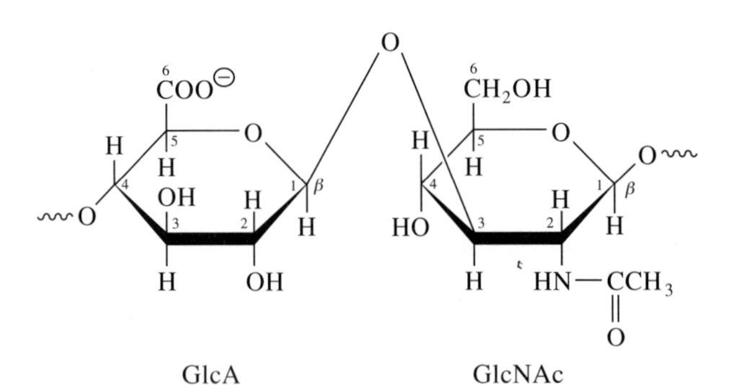

Figure 8·22
Hyaluronic acid. A heteropolysaccharide component of several highly viscous bodily fluids in vertebrates, hyaluronic acid is a linear polymer of alternating D-glucuronic acid (GlcA) and *N*-acetylglucosamine residues (GlcNAc). Each GlcA residue is linked to one GlcNAc residue through a β-(1→3) linkage, and that GlcNAc residue is in turn linked to the next GlcA residue through a β-(1→4) linkage.

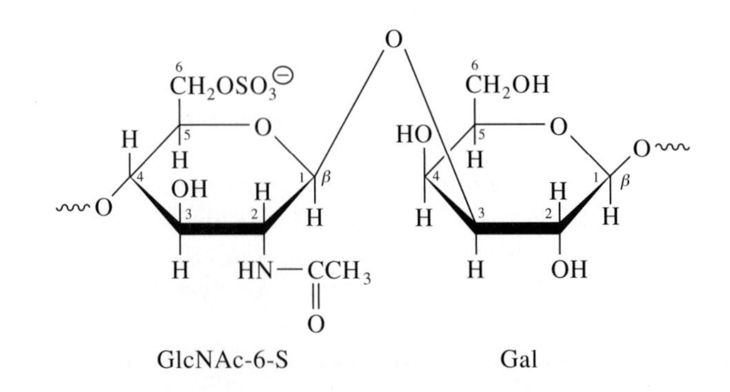

Figure 8·23
Keratan sulfate. Keratan sulfate is a heteropolysaccharide composed chiefly of a linear polymer of alternating galactose and *N*-acetylglucosamine-6-sulfate residues. Each *N*-acetylglucosamine-6-sulfate residue is linked to one galactose residue through a β-(1→3) linkage, and that galactose is in turn linked to the next *N*-acetylglucosamine residue through a β-(1→4) linkage.

8·8 Glycoproteins and Proteoglycans Are "Hybrid" Macromolecules

In Chapter 4, we noted that collagen is a *glycoprotein* since some of its hydroxylysine residues are covalently bonded to carbohydrate residues. Glycoproteins are proteins that contain carbohydrate residues covalently attached through their anomeric carbon atoms. The carbohydrate portion may account for as little as 1% to well over 50% of the molecular weight of a glycoprotein. In many cases, the oligosaccharide chains of glycoproteins are branched.

There are two types of carbohydrate-protein linkages; each type is named based on the atom of the protein to which the carbohydrate is linked. *O-Linked* oligosaccharide chains are attached to proteins by covalent bonding to the hydroxyl oxygen of amino acid side chains. Figure 8·24 shows a typical *O*-linked oligosaccharide found in collagen: *O*-α-D-glucopyranosyl-(1→2)-*O*-β-D-galactopyranosyl-(1→5)-hydroxylysine. These sugar residues are added to collagen by the successive actions of a galactosyltransferase and a glucosyltransferase. In many *O*-linked glycoproteins, the carbohydrate chains are attached to specific serine or threonine side chains. N-*Linked* oligosaccharide chains are attached to proteins by covalent bonding to the amide nitrogen of asparagine residues. The same core sugar residues are present in all known *N*-linked oligosaccharides (Figure 8·25). The core residues consist of two *N*-acetylglucosamine residues and three mannose residues. Additional sugar residues are added to generate a large number of different structures.

Many proteins—including extracellular and membrane-bound enzymes, structural proteins, receptors, transport proteins, and certain hormones—are glycoproteins. Like many of the polysaccharides, the oligosaccharide portions of glycoproteins are polydisperse. Examples of glycoproteins include those attached to the outer surface of the plasma membrane of eukaryotic cells. These glycoproteins function as receptors of extracellular signals important in cell recognition, growth control, and membrane transport. An example of a major surface glycoprotein in human erythrocytes (red blood cells) is glycophorin A, a single polypeptide chain of 131 amino acid residues to which are bound some 15 *O*-linked and one *N*-linked

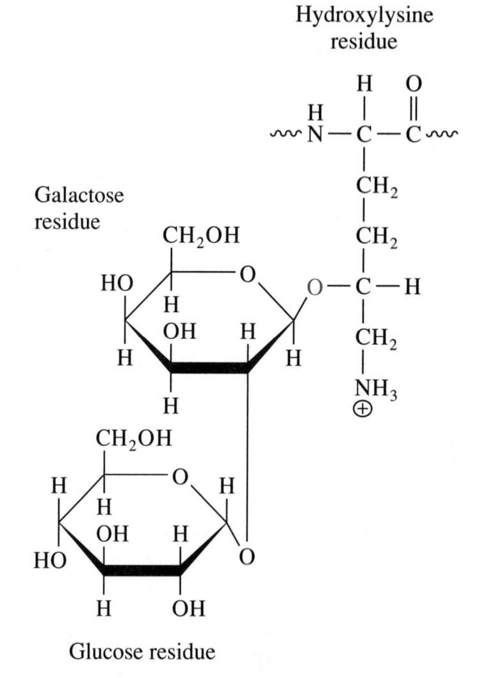

Figure 8·24
Structure of an *O*-linked oligosaccharide attached to collagen. A disaccharide of glucose and galactose residues is often attached to the hydroxyl group of a hydroxylysine residue of collagen.

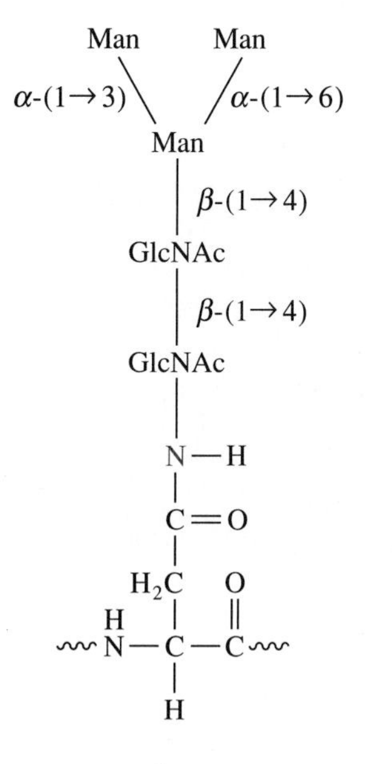

Figure 8·25
Core residues common to *N*-linked oligosaccharides. *N*-Linked oligosaccharides are characteristically heterogeneous, branched structures containing numerous carbohydrate residues linked to a $(Man)_3(GlcNAc)_2$ core. This core is in turn bound to an asparagine residue of the protein.

oligosaccharide chains. (The term *glycophorin,* from Greek, means "sugar bearer.") Altogether, glycophorin A contains about 100 monosaccharide residues, which account for about 60% of its molecular weight. The sugar moieties, all located on the outer surface of the erythrocyte membrane, form a bushlike covering and may play an important role in cell-cell and cell-matrix interactions.

The members of one subclass of carbohydrate-protein complexes, known as *proteoglycans,* contain more carbohydrate than protein and are extensively hydrated. The proteoglycan complex of cartilage consists of a long hyaluronic acid molecule that serves as a central strand to which highly glycosylated core proteins are noncovalently bound through interactions with globular link proteins. Electron micrographs of the proteoglycan substance of cartilage reveal bottlebrush-like structures in which core proteins and linking proteins are covered by "bristles" of polyanionic oligosaccharides bearing sulfate groups (esterified to sugar hydroxyls) and carboxylate groups (oxidized C-6 substituents), which are capable of binding massive amounts of water (Figure 8·26).

Carbohydrate residues are also found in *glycolipids,* structures that will be considered in the next chapter.

(a)

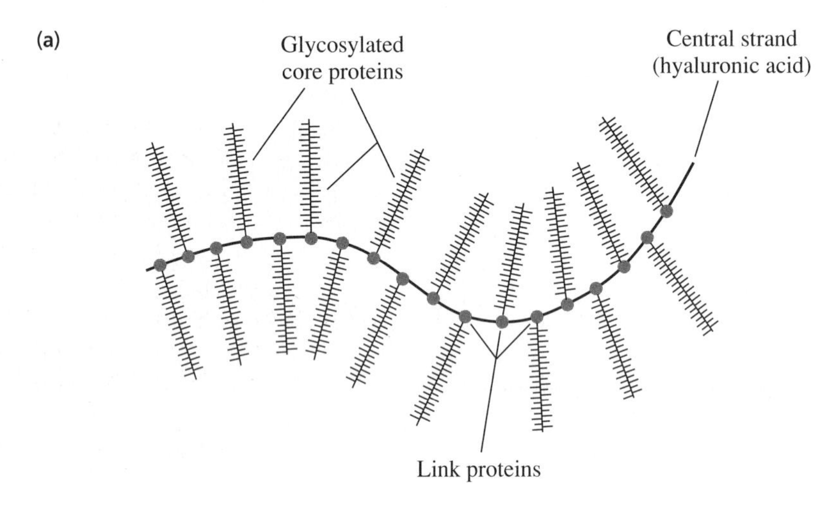

(b)

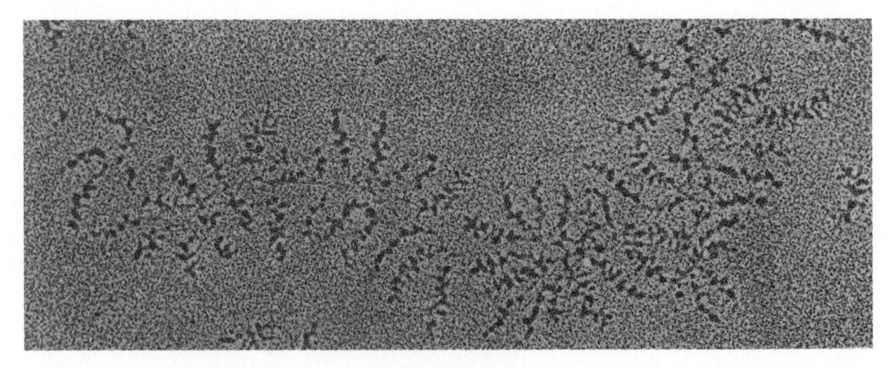

Figure 8·26
Proteoglycans. Proteoglycans are a subclass of carbohydrate-protein complexes in which the carbohydrate component exceeds the protein component by weight. **(a)** The proteoglycan substance of cartilage consists of a central strand of hyaluronic acid to which link proteins are bound. To each of these link proteins is bound in turn a long, glycosylated core protein. **(b)** Under the electron microscope, the whole structure has the appearance of a bottlebrush. (Courtesy of Arnold I. Caplan.)

Summary

Carbohydrates consist of hydroxyaldehydes (aldoses) and hydroxyketones (ketoses) and their derivatives. They include monosaccharides (simple sugars), oligosaccharides, and polysaccharides. Except for the simplest ketose, dihydroxyacetone, carbohydrates are chiral and therefore exhibit optical activity. For a given monosaccharide, there are 2^n possible stereoisomers, where n is the number of chiral carbon atoms. Among these stereoisomers, any two that are nonsuperimposable mirror images of each other are referred to as enantiomers. Two that differ in configuration at only one of several chiral centers are known as epimers, whereas stereoisomers that differ in configuration at more than one carbon atom but are not mirror images are called diastereomers. A monosaccharide is designated D or L, depending on the configuration of the chiral carbon farthest from the aldehydic (C-1) or ketonic (usually C-2) carbon atom.

Aldopentoses, aldohexoses, and ketohexoses exist principally as cyclic hemiacetals known as furanoses and pyranoses. In sugar hemiacetals, the anomeric carbon (the carbonyl carbon in the open-chain form) is sp^3-hybridized, giving these structures an additional asymmetric center. The chirality of the anomeric carbon is designated either α (having the D-like configuration for D sugars or L-like configuration for L sugars) or β (L-like for D sugars or D-like for L sugars). Two optical isomers that differ in configuration only at their anomeric carbon atoms are referred to as anomers. Although pyranoses exist primarily in chair conformations, they are often depicted in Haworth projections, which reveal the chirality of each asymmetric carbon atom including the anomeric carbon.

Among the important derivatives of monosaccharides are sugar phosphates, deoxy sugars, amino sugars, and sugar acids.

Monosaccharide residues can be linked via glycosidic bonds to form disaccharides, oligosaccharides, and polysaccharides. Four important disaccharides are maltose (α-D-glucopyranosyl-($1 \rightarrow 4$)-D-glucose), cellobiose (β-D-glucopyranosyl-($1 \rightarrow 4$)-D-glucose), lactose (β-D-galactopyranosyl-($1 \rightarrow 4$)-D-glucose), and sucrose (α-D-glucopyranosyl β-D-fructofuranoside). Lactose, an epimer of cellobiose, is the major carbohydrate in milk. Sucrose is synthesized in many plants and is the most abundant disaccharide found in nature. The anomeric carbons of both monosaccharide residues in sucrose are involved in the glycosidic linkage of the disaccharide.

Glucose is the repeating monomeric unit of the storage polysaccharides amylose, amylopectin, and glycogen, and of the structural polysaccharide cellulose, which is the most abundant organic substance in the biosphere. Chitin is another example of a homopolysaccharide. Its monomeric unit is β-($1 \rightarrow 4$)-linked *N*-acetylglucosamine. Hyaluronic acid is a polymer of alternating glucuronic acid and *N*-acetylglucosamine residues and thus is a heteropolysaccharide. The carbohydrate moieties of glycoproteins are usually oligosaccharides that are linked to the side-chain oxygen of specific serine or threonine residues (or hydroxylysine residues in collagen), or to the amide nitrogen of specific asparagine residues. Proteoglycans are complex aggregates containing more carbohydrate than protein; an example is the proteoglycan substance of cartilage.

Selected Readings

Candy, D. S. (1980). *Biological functions of carbohydrates* (New York: Halsted Press).

Collins, P. M., ed. (1987). *Carbohydrates* (London and New York: Chapman and Hall). A source book that provides names, structural and empirical formulas, physical and chemical properties, and references to the literature for several thousand carbohydrates.

El Khadem, H. S. (1988). *Carbohydrate chemistry: monosaccharides and their derivatives* (Orlando, Florida: Academic Press).

Rademacher, T. W., Parekh, R. B., and Dwek, R. A. (1988). Glycobiology. *Annu. Rev. Biochem. 57,* 785–838. A review of oligosaccharides linked to proteins and lipids, including changes in glycosylation in several diseases.

Sharon, N. (1980). Carbohydrates. *Sci. Amer. 243(5),* 90–116.

9

Lipids and Biological Membranes

Having studied proteins and carbohydrates, we are now ready to consider a third major class of biomolecules, the *lipids*. Like proteins and carbohydrates, lipids (Greek, *lipos,* fat) are found in all known living organisms and play an essential role in the maintenance of life. Unlike proteins and carbohydrates, however, lipids are highly polymorphic and difficult to define structurally. Rather, they are often defined operationally, as water-insoluble (or only sparingly soluble) organic compounds found in biological systems. Lipids are either hydrophobic (nonpolar) or amphipathic (having both nonpolar and polar substituents).

9·1 Lipids Exhibit Great Diversity in Structure and Function

Although lipid structures are often complex, there are some common architectural themes. The simplest lipids are the *fatty acids,* monocarboxylic acids of the general formula R—COOH, where R represents a hydrocarbon tail. Fatty acids are also components of many more complex types of lipids, including *triacylglycerols* (fats and oils), *glycerophospholipids* (also called phosphoglycerides), and *sphingolipids,* as well as *waxes* and *eicosanoids. Steroids* and *lipid vitamins* are structurally distinct lipids derived from a five-carbon molecule called isoprene; steroids

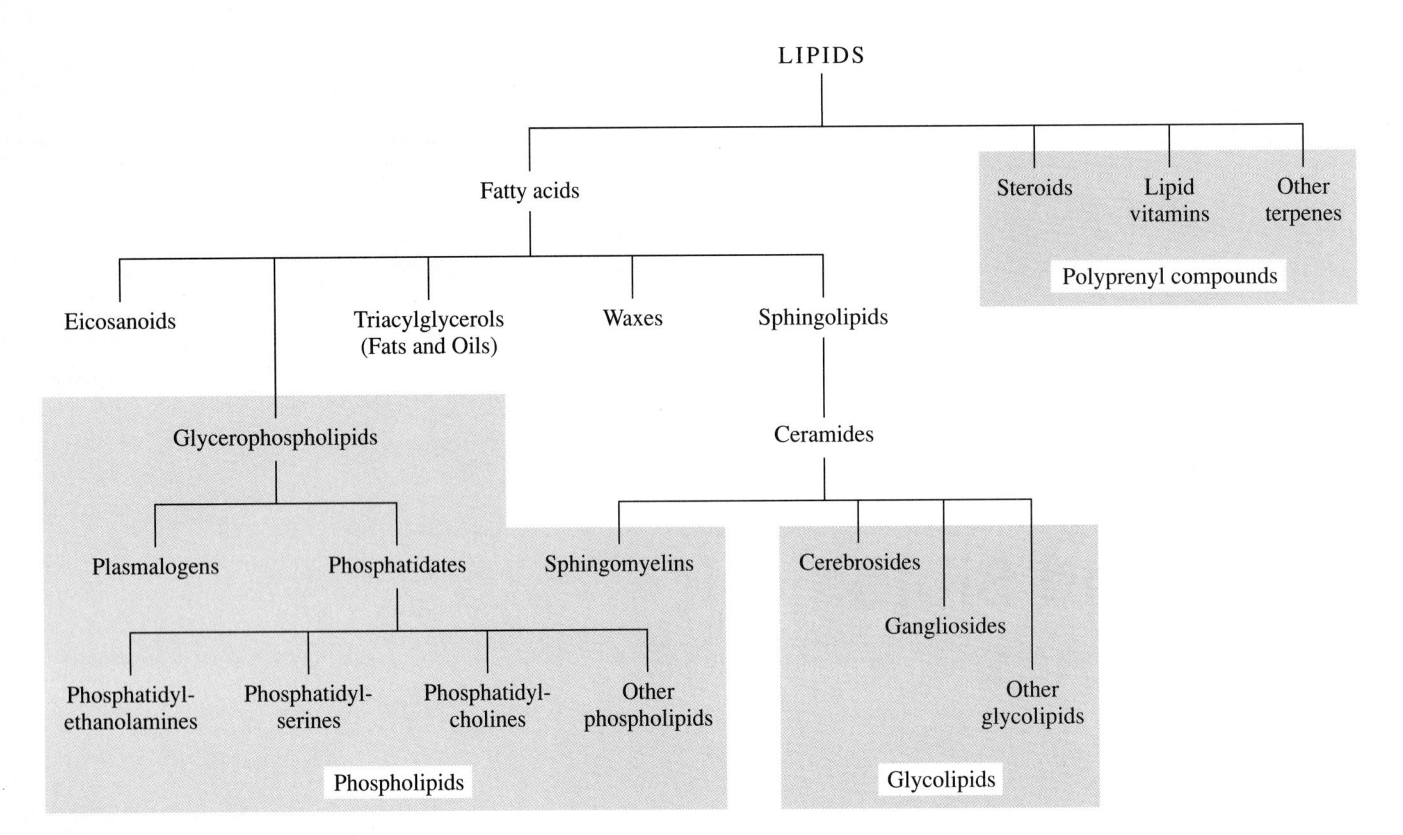

Figure 9·1
Organization of the major types of lipids based on structural relationships. Fatty acids are the simplest lipids in terms of structure. A number of other types of lipids either contain or are derived from fatty acids. These include the triacylglycerols, the glycerophospholipids, and the sphingolipids, as well as the eicosanoids and the waxes. The glycerophospholipids and the sphingomyelins all contain phosphate and are thus classified as phospholipids. Cerebrosides and gangliosides contain monosaccharide derivatives and are thus classified as glycolipids. Steroids and lipid vitamins are derived from the five-carbon molecule isoprene and are classified as polyprenyl compounds or isoprenoids.

are based on a fused, four-membered cyclic ring structure, and lipid vitamins are composed primarily of long hydrocarbon chains or fused rings. Figure 9·1 shows the major types of lipids and their relationships to one another. Note that lipids containing phosphate moieties are grouped into the general category of *phospholipids,* lipids containing carbohydrate moieties are called *glycolipids,* and lipids derived from isoprene are called *polyprenyl compounds* or *isoprenoids.*

In addition to diverse structures, lipids have diverse biological functions. A variety of amphipathic lipids, including glycerophospholipids and sphingolipids, serve as important structural components of all biological membranes. In some organisms, fats and oils (triacylglycerols) function as intracellular storage depots for metabolic energy. Fats also provide animals with thermal insulation and padding. Waxes provide protective surface barriers for living organisms as components of their cell walls, exoskeletons, and skins. A variety of specialized functions are associated with specific lipids; in mammals, for example, the steroid hormones regulate and integrate a host of metabolic activities and the eicosanoids serve in the regulation of blood pressure, body temperature, and smooth-muscle contraction.

In some cases, lipids perform their biological functions as individual molecules; in other cases, they interact with other biomolecules to function as part of a complex or aggregate. Lipid complexes include lipoproteins (particles composed of lipid and protein), and lipid aggregates include biological membranes (thin bilayers composed of lipid, protein, and sometimes carbohydrate).

9·2 Fatty Acids Are Components of Many Lipids

More than 100 different fatty acids have been identified in the lipids of microorganisms, plants, and animals. These fatty acids differ from one another by the lengths of their hydrocarbon tails, their degree of unsaturation (the number of carbon-carbon double bonds in the hydrocarbon tail), and the positions of those carbon-carbon double bonds in their fatty acid chains.

Some of the fatty acids commonly found in mammals are shown in Table 9·1. Most fatty acids have a pK_a of about 4.5 and are therefore ionized at physiological pH (anionic forms are shown in Table 9·1). Fatty acids can be referred to by either IUPAC (International Union of Pure and Applied Chemistry) or common names. The common names of the frequently encountered fatty acids are most often used in lipid nomenclature.

The number of carbon atoms in fatty acids usually ranges from 12 to 22 and is almost always even, since fatty acids are synthesized by the sequential addition of two-carbon units. (Fatty acid biosynthesis is discussed in Chapter 17.) In IUPAC nomenclature, the carboxyl carbon is labelled C-1 and the remaining carbons are numbered sequentially. In common nomenclature, Greek letters are used to identify the carbon atoms. The carbon adjacent to the carboxyl carbon (C-2 in IUPAC nomenclature) is designated α, and the other carbons are lettered in alphabetical order—β, γ, δ, ε, and so on. The Greek letter ω is used to specify the carbon atom farthest from the carboxyl group, whatever the length of the hydrocarbon tail (Figure 9·2).

Fatty acids without a carbon-carbon double bond are classified as _saturated,_ whereas those with at least one carbon-carbon double bond are classified as _unsaturated._ Unsaturated fatty acids with only one double bond are called _monounsaturated,_ and those with two or more are called _polyunsaturated._ The configuration of the double bonds in unsaturated fatty acids is generally *cis*. In IUPAC nomenclature, the positions of double bonds are indicated by the symbol Δ^N, where the superscript N indicates the lowest-numbered carbon atom of each double-bonded pair (Table 9·1). Note that the carbon atoms involved in the double bonds of most polyunsaturated fatty acids are separated by a methylene group, and the double bonds are thus *not* conjugated.

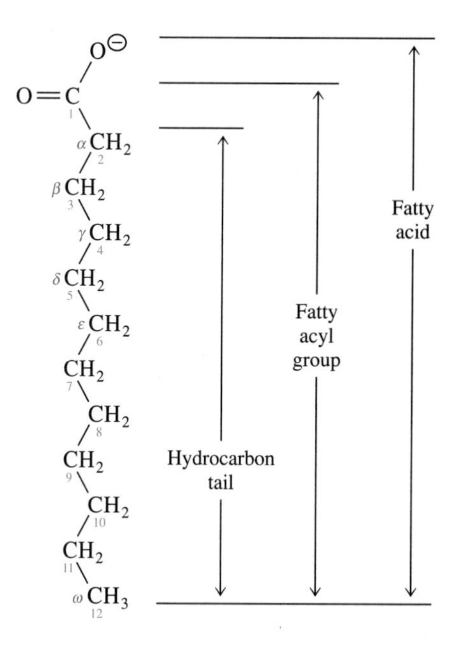

Figure 9·2
Basic structure and nomenclature of fatty acids. Fatty acids consist of a long hydrocarbon tail bound to a carboxyl group. Since the pK_a of the carboxyl group is approximately 4.5, fatty acids are anionic at physiological pH. Both the IUPAC and common systems for labelling carbons are illustrated. The fatty acid shown (laurate, or dodecanoate) has 12 carbons and is saturated since it contains no carbon-carbon double bonds.

Table 9·1 Some common fatty acids (anionic forms) incorporated in membrane lipids

Number of carbons	Number of double bonds	Common name	IUPAC name	Melting point, °C	Molecular formula
12	0	Laurate	Dodecanoate	44	$CH_3(CH_2)_{10}COO^{\ominus}$
14	0	Myristate	Tetradecanoate	52	$CH_3(CH_2)_{12}COO^{\ominus}$
16	0	Palmitate	Hexadecanoate	63	$CH_3(CH_2)_{14}COO^{\ominus}$
18	0	Stearate	Octadecanoate	70	$CH_3(CH_2)_{16}COO^{\ominus}$
20	0	Arachidate	Eicosanoate	75	$CH_3(CH_2)_{18}COO^{\ominus}$
22	0	Behenate	Docosanoate	81	$CH_3(CH_2)_{20}COO^{\ominus}$
24	0	Lignocerate	Tetracosanoate	84	$CH_3(CH_2)_{22}COO^{\ominus}$
16	1	Palmitoleate	*cis*-Δ^9-Hexadecenoate	− 0.5	$CH_3(CH_2)_5CH{=}CH(CH_2)_7COO^{\ominus}$
18	1	Oleate	*cis*-Δ^9-Octadecenoate	13	$CH_3(CH_2)_7CH{=}CH(CH_2)_7COO^{\ominus}$
18	2	Linoleate	*cis, cis*-$\Delta^{9,12}$-Octadecadienoate	− 9	$CH_3(CH_2)_4(CH{=}CHCH_2)_2(CH_2)_6COO^{\ominus}$
18	3	Linolenate	all *cis*-$\Delta^{9,12,15}$-Octadecatrienoate	−17	$CH_3CH_2(CH{=}CHCH_2)_3(CH_2)_6COO^{\ominus}$
20	4	Arachidonate	all *cis*-$\Delta^{5,8,11,14}$-Eicosatetraenoate	−49	$CH_3(CH_2)_4(CH{=}CHCH_2)_4(CH_2)_2COO^{\ominus}$

[Values from Dawson, R. M. C., Elliott, D. C., Elliott, W. H., and Jones, K. M. (1986). *Data for Biochemical Research,* 3rd ed. (Oxford: Clarendon Press).]

A shorthand notation often used to identify fatty acids is two numbers separated by a colon; the first number refers to the number of carbon atoms in the fatty acid, and the second refers to the number of carbon-carbon double bonds. Using this shorthand, palmitate can be written as 16:0, oleate as 18:1, and arachidonate as 20:4. The numbers of the lowest-numbered carbon atoms involved in double-bonded pairs are put in parentheses to distinguish two different fatty acids that have the same number of carbon atoms and the same number of carbon-carbon double bonds. For example, linolenate is written as 18:3 (9, 12, 15) and its isomer, γ-linolenate as 18:3 (6, 9, 12).

The length of the hydrocarbon chain of a fatty acid and its degree of unsaturation influence the melting point of the acid. Compare the melting points listed in Table 9·1 for the saturated fatty acids laurate (12:0), myristate (14:0), and palmitate (16:0): as the length of the hydrocarbon tail increases, the melting point of the saturated fatty acid also increases. This is because van der Waals interactions among neighboring hydrocarbon tails increase as the tails get longer and more energy is required to disrupt the increased van der Waals interactions during melting.

Compare also the structures of stearate (18:0), oleate (18:1), and linolenate (18:3) in Figure 9·3 and Stereo S9·1. Although represented here in an extended conformation, the saturated hydrocarbon tail of stearate is flexible, since every carbon-carbon bond is free to rotate. The presence of *cis* double bonds in oleate and linolenate hinders rotation about these bonds and produces a pronounced bend

Figure 9·3
Structures of three C_{18} fatty acids. **(a)** Stearate (octadecanoate), a saturated fatty acid. **(b)** Oleate (*cis*-Δ^9-octadecenoate), a monounsaturated fatty acid. **(c)** Linolenate (all *cis*-$\Delta^{9,12,15}$-octadecatrienoate), a polyunsaturated fatty acid. The *cis* double bonds put kinks in the tails of the unsaturated fatty acids.

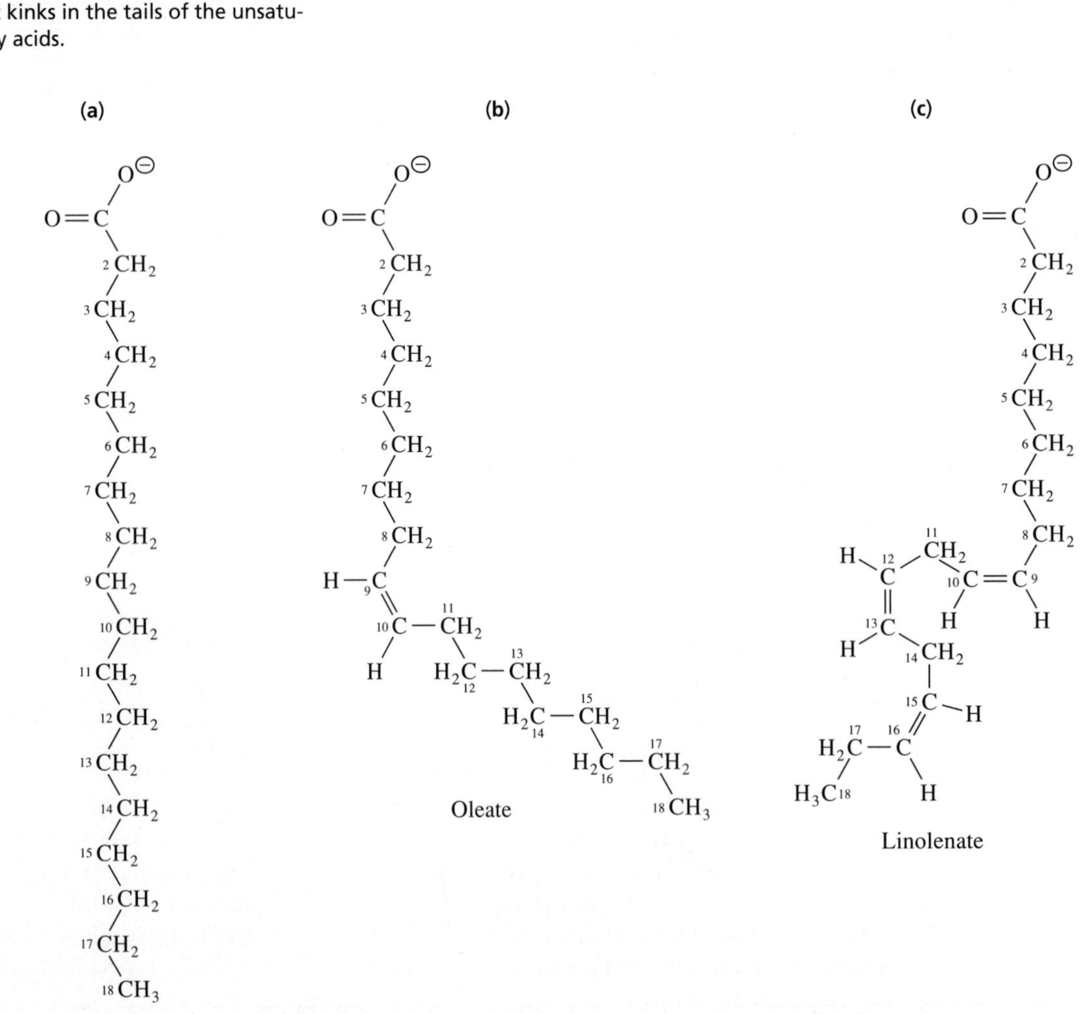

within the hydrocarbon chain. This bend prevents the formation of closely packed, well-ordered crystals and hence decreases van der Waals interactions among the hydrocarbon chains. Consequently, *cis* unsaturated fatty acids have a lower melting point than do saturated fatty acids. Notice that stearate (melting point 70°C) is a solid at body temperature, whereas oleate (melting point 13°C) and linolenate (melting point –17°C) are both liquids.

Although fatty acids are important components of many lipids, free fatty acids occur only in trace amounts in living cells. Most fatty acids are esterified to more complex lipid molecules. In esters and other derivatives of carboxylic acids, the RC=O moiety contributed by the acid is called the *acyl group*. In common lipid nomenclature, complex lipids that contain specific fatty acyl groups are named according to the length of the hydrocarbon chain and the number of double bonds of the acyl groups they contain. For example, esters based on the 12-carbon saturated fatty acid laurate are called *lauroyl esters,* and those based on the 18-carbon unsaturated fatty acid with two double bonds are called *linoleoyl esters*. (In old lipid nomenclature, the *"o"* before the *"yl"* in the names of fatty acyl groups was omitted; thus you may see lauroyl given as "lauryl," linoleoyl as "linoleyl," and so on.)

The relative abundance of particular fatty acids varies with type of organism, type of organ (in multicellular organisms), and food source. The most abundant fatty acids in animals are usually oleate (18:1), palmitate (16:0), and stearate (18:0), although polyunsaturated fatty acids are also prevalent. Mammals can synthesize saturated and monounsaturated fatty acids de novo, but they require certain polyunsaturated fatty acids in their diets. In particular, linoleate (18:2), which is abundant in plant oils, and linolenate (18:3), which is abundant in fish oils, are considered *essential fatty acids* for mammals because they cannot synthesize them. Mammals are able to synthesize other polyunsaturated fatty acids from an adequate supply of linoleate and linolenate.

A number of fatty acids are found in nature in addition to those listed in Table 9·1. For example, several branched-chain fatty acids and fatty acids containing cyclopropane rings have been found in certain bacteria. Many of these long-chain fatty acids are extremely limited in their distribution and have highly specialized functions.

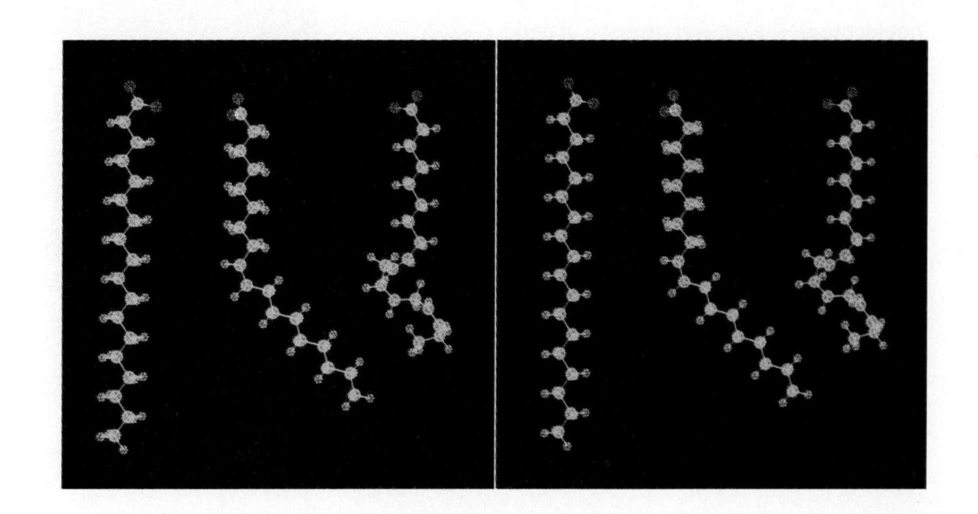

Stereo S9·1
Stearate (left), oleate (middle), and linolenate (right). (Stereo image by Ben N. Banks and Kim M. Gernert.)

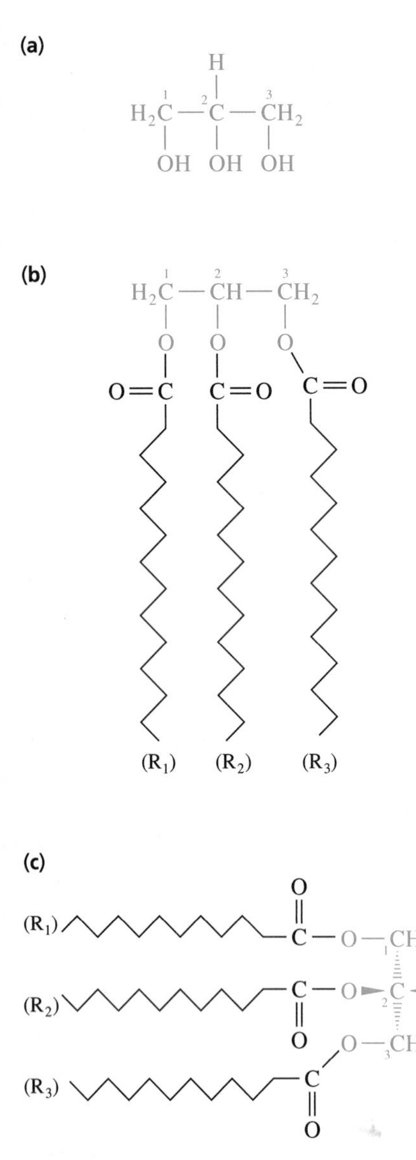

Figure 9·4
Triacylglycerol structure. Glycerol **(a)** forms the backbone to which three fatty acyl residues are esterified **(b)**. Although glycerol is nonchiral, C-2 of a triacylglycerol is chiral if the acyl groups bound to C-1 and C-3 are not identical. The general structure of a triacylglycerol is shown in **(c)** with perspective bonds around the chiral carbon; the molecule is oriented for ease of comparison with the structure of L-glyceraldehyde shown in Figure 3·2c. This orientation allows stereospecific numbering of glycerol derivatives, designated *sn*, with C-1 at the top and C-3 at the bottom.

9·3 Triacylglycerols Are Neutral, Nonpolar Lipids

Fatty acids can serve as important fuel molecules during metabolism. Oxidation of fatty acids yields more energy (~9 kcal/g) than does the oxidation of either protein or carbohydrate (~4 kcal/g each). Fatty acids are generally stored as complex lipids called *triacylglycerols*. As their name implies, triacylglycerols (historically referred to as triglycerides) are composed of three fatty acyl residues esterified to glycerol, a three-carbon alcohol (Figure 9·4). Triacylglycerols are neutral (nonionic), nonpolar (and hence hydrophobic) lipids. Being hydrophobic (unlike carbohydrates), triacylglycerols can be stored in an anhydrous environment and are not solvated by water, which would take up space and add mass, reducing the efficiency of energy storage.

Fats and oils are mixtures of triacylglycerols. Whether a triacylglycerol mixture is a solid (fat) or a liquid (oil) depends on its fatty acid composition and on the temperature. Triacylglycerols that contain only saturated, long-chain fatty acyl groups tend to be solids at body temperature, whereas those that contain unsaturated or short-chain fatty acyl groups tend to be liquids.

Triacylglycerols, because of their hydrophobicity, coalesce as fat droplets in cells. Fat droplets are sometimes seen near mitochondria in cells that rely on fatty acids to supply energy for cellular activities (Figure 9·5). In mammals, most fat is stored in adipose tissue, which is composed of specialized cells known as adipocytes. Each adipocyte contains a large fat droplet that accounts for nearly the entire volume of the cell. Although widely distributed throughout the bodies of mammals, most adipose tissue occurs immediately under the skin and in the abdominal cavity. This subcutaneous fat serves both as a storage depot for energy and as thermal padding and therefore tends to be more extensive in warm-blooded aquatic mammals, for whom such padding is especially important.

Fatty acids are released from triacylglycerols through the catalytic activity of *lipases*, a family of hydrolases. In the gastrointestinal tract, lipases release fatty acids during the digestion of triacylglycerols. In humans, lipids ingested in food are broken down in the small intestine. Because lipids are not water-soluble but the

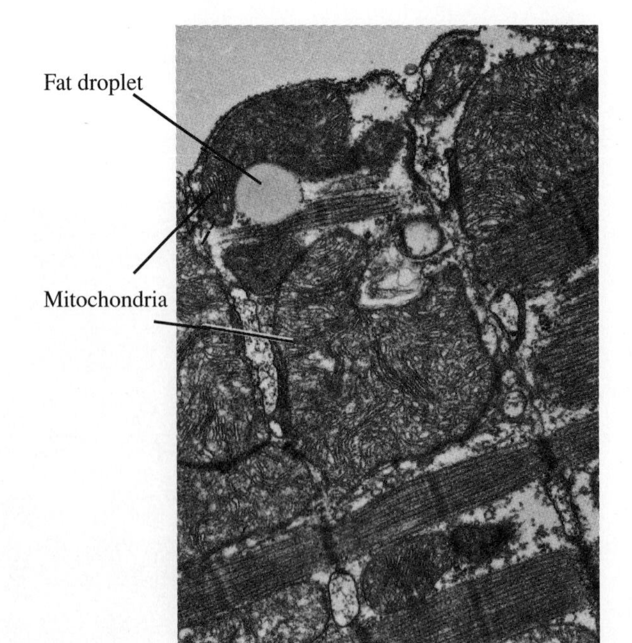

Figure 9·5
Fat droplets in cardiac muscle cells. Fat droplets are often found close to mitochondria in cells that rely on fat to supply energy for cellular activities. Hydrolysis of triacylglycerols in the fat droplets of cardiac muscle cells releases fatty acids. The oxidation of these fatty acids in the mitochondria produces ATP, which is required for muscle contraction. (Courtesy of Mary C. Reedy.)

enzymes that digest them are, lipid digestion requires the presence of strong detergents called *bile salts*. Bile salts are amphipathic derivatives of cholesterol (Section 9·6) that aid digestion by emulsifying lipids in the intestine. Digestive lipases, such as pancreatic lipase and the phospholipases, which are synthesized as zymogens in the pancreas and then secreted into the small intestine, require bile salts for lipolytic activity. Pancreatic lipase specifically catalyzes hydrolysis of the primary esters (at C-1 and C-3) of triacylglycerols, releasing free fatty acids and generating monoacylglycerols. Unlike the nonpolar triacylglycerols, monoacylglycerols have appreciable polarity (due to their two hydroxyl groups) and can form stable micelles, which aid in further lipid digestion and absorption from the small intestine.

9·4 Glycerophospholipids Are Major Components of Biological Membranes

Although triacylglycerols are the most abundant type of lipid in mammals on the basis of weight, they are not amphipathic and therefore do not form lipid bilayers, which are structural components of biological membranes (Section 9·9). The most abundant lipids in membranes are the *glycerophospholipids* (also called phosphoglycerides), which are constructed upon a backbone of glycerol 3-phosphate (Figure 9·6). The simplest type of glycerophospholipid, known as a *phosphatidate,* consists of two fatty acyl groups esterified to C-1 and C-2 of glycerol 3-phosphate. Phosphatidates usually occur only as metabolic intermediates in the biosynthesis of more complex glycerophospholipids.

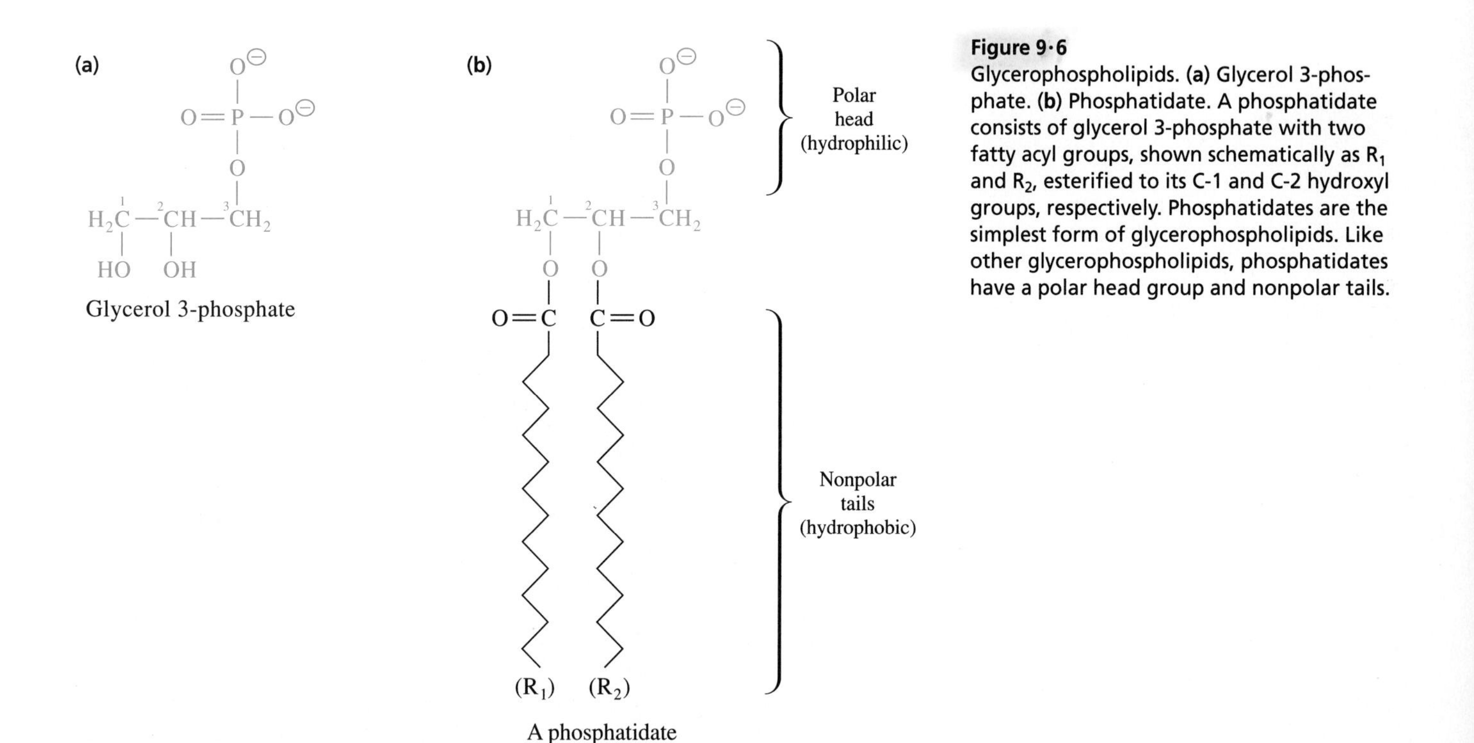

Figure 9·6
Glycerophospholipids. **(a)** Glycerol 3-phosphate. **(b)** Phosphatidate. A phosphatidate consists of glycerol 3-phosphate with two fatty acyl groups, shown schematically as R_1 and R_2, esterified to its C-1 and C-2 hydroxyl groups, respectively. Phosphatidates are the simplest form of glycerophospholipids. Like other glycerophospholipids, phosphatidates have a polar head group and nonpolar tails.

In more complex glycerophospholipids, the phosphate is esterified to both glycerol and another alcohol. Table 9·2 identifies some of the families of glycerophospholipids that can be formed by the esterification of a phosphatidate and an alcohol. The structures of three of the most common families of membrane glycerophospholipids, *phosphatidylethanolamine* (also known as cephalin), *phosphatidylserine,* and *phosphatidylcholine* (also known as lecithin), are shown in Figure 9·7.

Table 9·2 Some common substituents attached to the phosphate group of glycerophospholipids

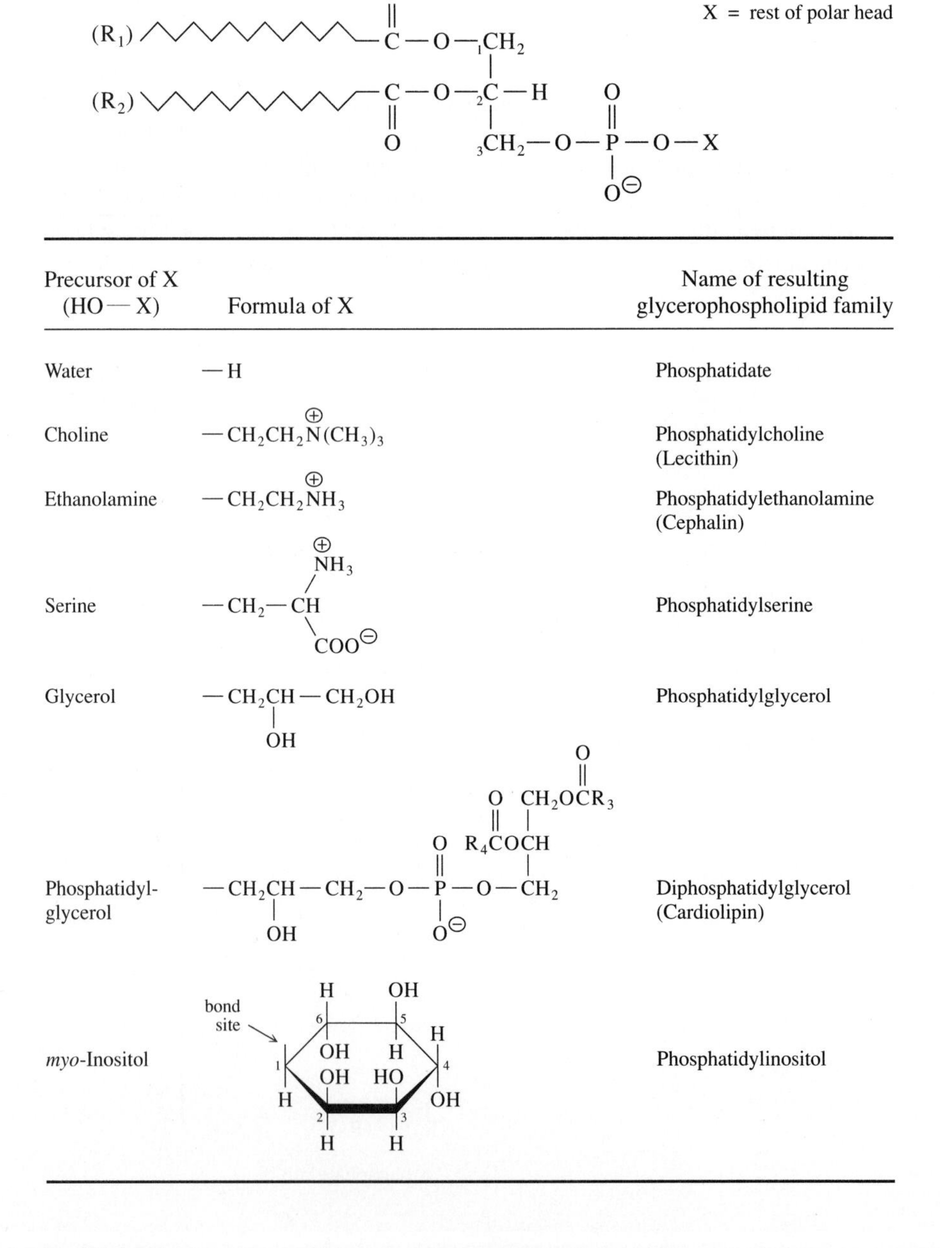

Precursor of X (HO—X)	Formula of X	Name of resulting glycerophospholipid family
Water	$-H$	Phosphatidate
Choline	$-CH_2CH_2\overset{\oplus}{N}(CH_3)_3$	Phosphatidylcholine (Lecithin)
Ethanolamine	$-CH_2CH_2\overset{\oplus}{N}H_3$	Phosphatidylethanolamine (Cephalin)
Serine	$-CH_2-CH(\overset{\oplus}{N}H_3)(COO^{\ominus})$	Phosphatidylserine
Glycerol	$-CH_2CH(OH)-CH_2OH$	Phosphatidylglycerol
Phosphatidyl-glycerol	$-CH_2CH(OH)-CH_2-O-P(=O)(O^{\ominus})-O-CH_2-CH(OCOR_4)-CH_2OCOR_3$	Diphosphatidylglycerol (Cardiolipin)
myo-Inositol		Phosphatidylinositol

The glycerophospholipids listed in Table 9·2 are not single compounds but families of compounds that have the same polar head group but different fatty acyl chains. For example, human red blood cell membranes contain at least 21 different species of phosphatidylcholine that differ from each other in the fatty acyl chains esterified at C-1 and C-2 of glycerol 3-phosphate. In general, the type of fatty acid at each position is not random—saturated fatty acids are usually esterified to C-1 and unsaturated fatty acids to C-2 in glycerophospholipids.

Notice that glycerophospholipids are amphipathic molecules, having a polar head (with an anionic phosphate and often one or two other charged groups) and long, nonpolar tails. Such amphipathic lipids can be depicted generally as long

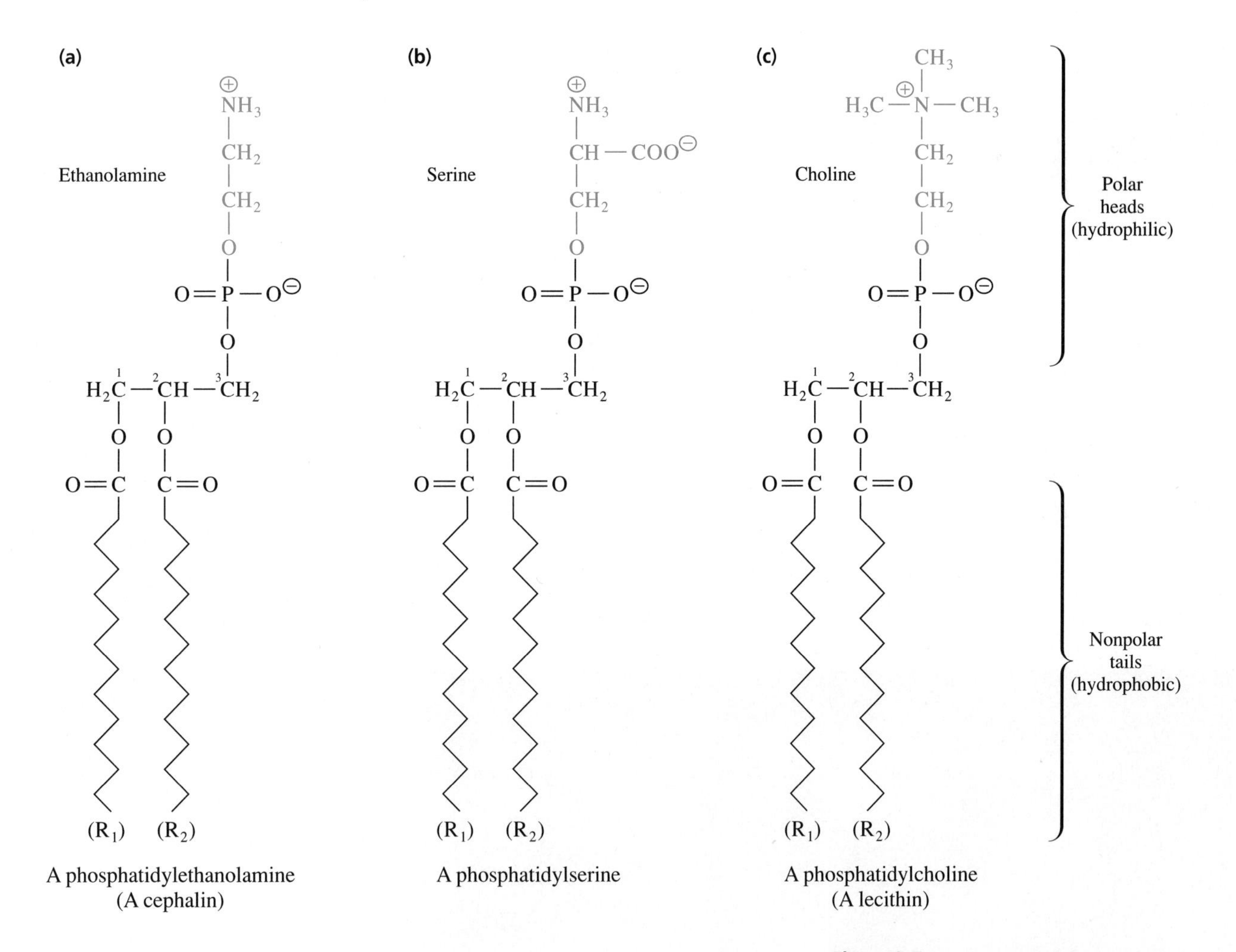

Figure 9·7
General structures of **(a)** phosphatidylethanolamine, **(b)** phosphatidylserine, and **(c)** phosphatidylcholine. Functional groups derived from the esterified alcohol are shown in blue. Since each of these lipids can contain many combinations of fatty acyl groups (shown here schematically as R_1 and R_2), the general name refers to a family of compounds, not to a single molecule. Note that all three of these molecules have a polar head group and nonpolar tails.

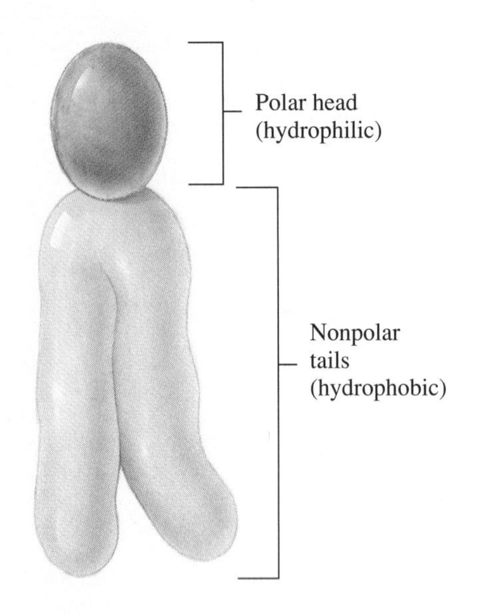

Figure 9·8
A general way to depict amphipathic lipid molecules such as the glycerophospholipids. The lipid illustrated here has two nonpolar tails. This icon is used extensively throughout this book. (Illustrator: Lisa Shoemaker.)

molecules with a polar head group and either one or two nonpolar tails (Figure 9·8). The phosphatidylcholine shown in Stereo S9·2 provides a typical example of a glycerophospholipid. This molecule, dipalmitoylphosphatidylcholine, constitutes over 50% of a secretion known as lung surfactant. Lung surfactant reduces the surface tension at the air-water interface of the alveoli, which enables oxygen to pass efficiently from the air passageways into the cells.

Although most glycerophospholipids consist of fatty acids linked to glycerol 3-phosphate by esterification, one family, called *plasmalogens,* has a hydrocarbon chain linked to C-1 of glycerol 3-phosphate by a vinyl ether bond (Figure 9·9). The most common alcohols esterified to the phosphate group of plasmalogens are ethanolamine and choline. Plasmalogens account for about 23% of the glycerophospholipids in the human central nervous system and are also found in the membranes of cells obtained from peripheral nerve and muscle tissue.

9·5 Sphingolipids Constitute a Second Class of Molecules in Biological Membranes

Although glycerophospholipids account for the bulk of the lipids in most membrane systems, *sphingolipids,* a second class of amphipathic lipids, are also present in plant and animal membranes. In mammals, they are particularly abundant in tissues of the central nervous system.

The structural backbone of sphingolipids is sphingosine (*trans*-4-sphingenine), an unbranched C_{18} alcohol with a *trans* double bond between C-4 and C-5,

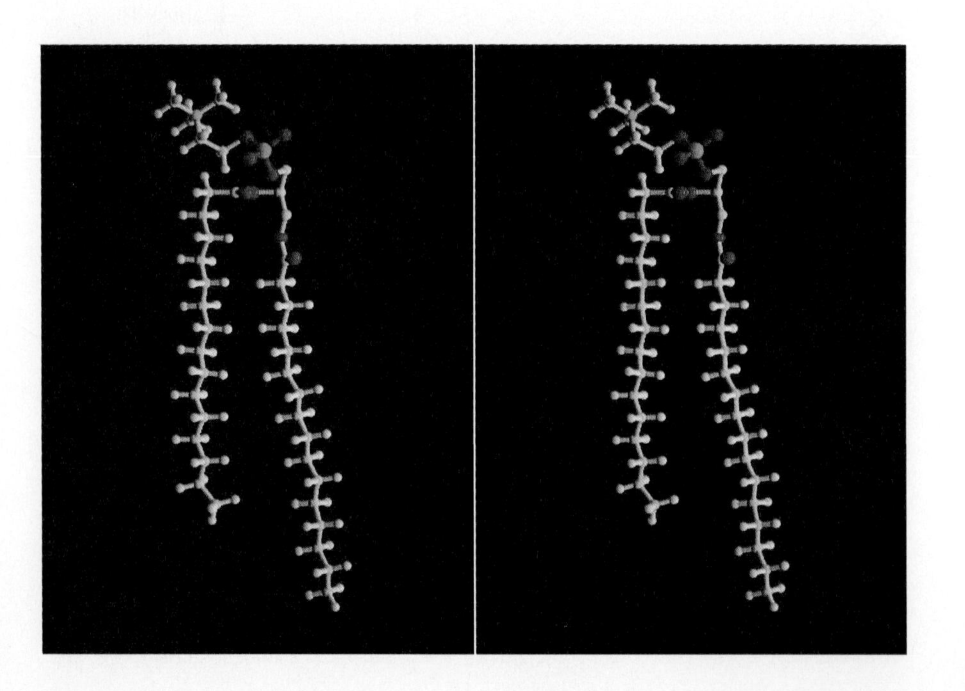

Stereo S9·2
Dipalmitoylphosphatidylcholine. (Stereo image by Richard J. Feldmann.)

an amino group at C-2, and hydroxyl groups at C-1 and C-3 (Figure 9·10a). A *ceramide* consists of a fatty acid linked to the C-2 amino group of sphingosine by an amide bond (Figure 9·10b). Ceramides are the metabolic precursors of all sphingolipids. The three major families of sphingolipids are the *sphingomyelins,* the *cerebrosides,* and the *gangliosides.* Of these, only the sphingomyelins contain phosphate and are therefore classified as phospholipids; cerebrosides and gangliosides contain carbohydrate residues and are therefore classified as glycolipids.

Sphingomyelin consists of phosphocholine attached to the C-1 hydroxyl group of a ceramide (Figure 9·10c). Note the resemblance of the amphipathic lipid molecule sphingomyelin to a phosphatidylcholine molecule (Figure 9·7c); both are zwitterions containing choline, phosphate, and two long, hydrophobic tails. Sphingomyelins are present in the plasma membranes of most mammalian cells and are a major component of the myelin sheaths that surround certain nerve cells.

Cerebrosides are glycolipids that contain one monosaccharide residue attached via a β-glycosidic linkage to C-1 of a ceramide. Galactocerebrosides, also

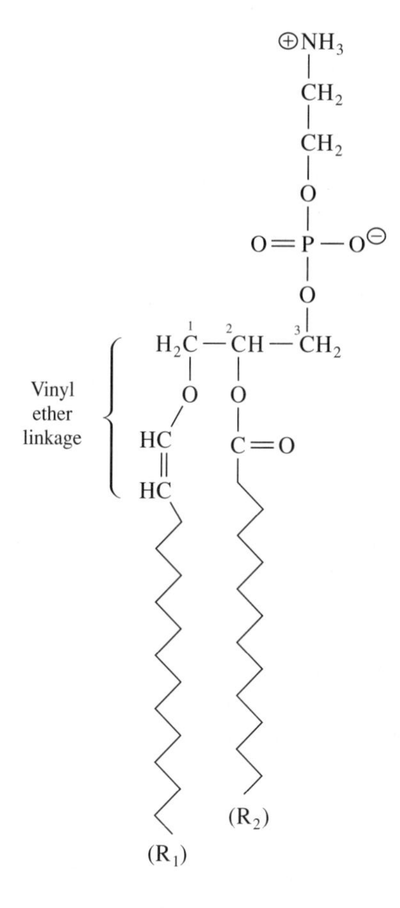

Figure 9·9
Structure of an ethanolamine plasmalogen. A hydrocarbon tail is linked to C-1 of glycerol 3-phosphate as a vinyl ether.

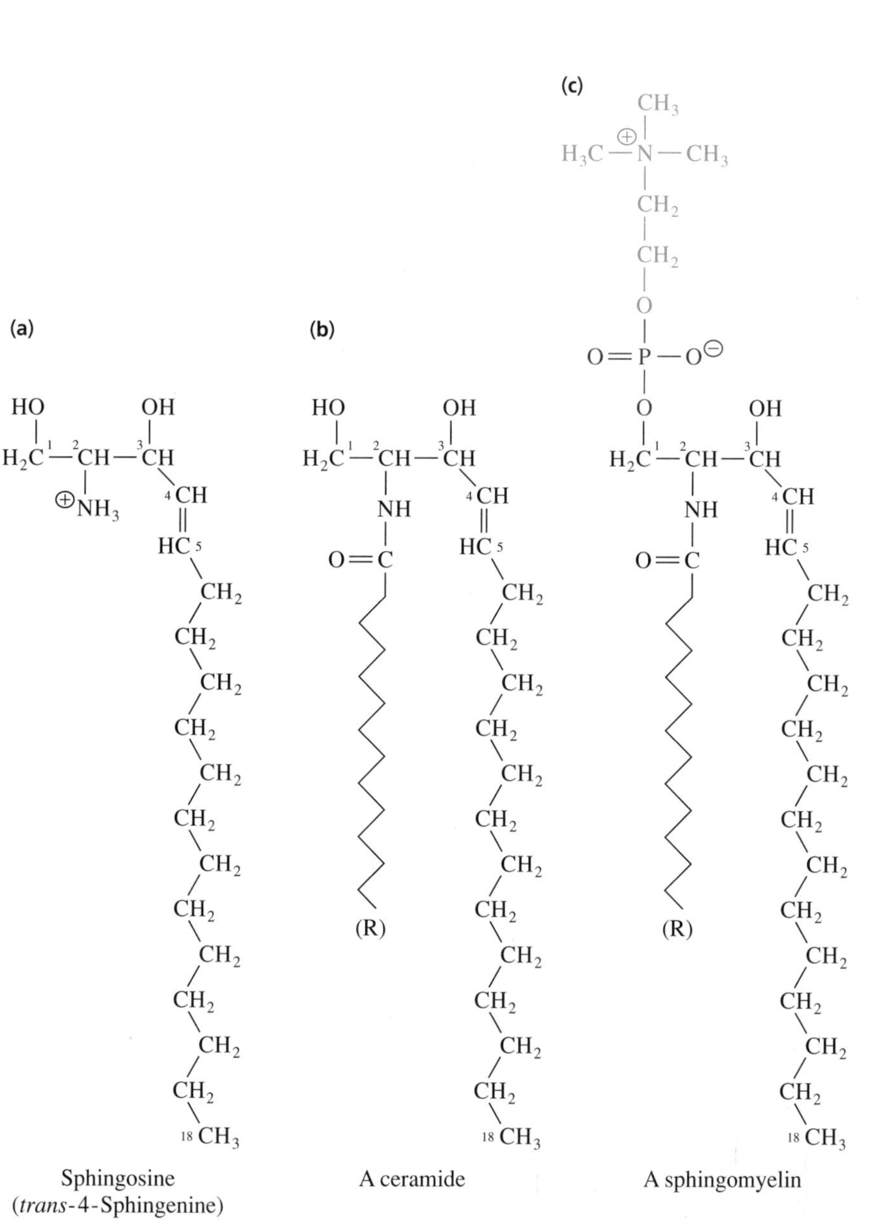

Figure 9·10
Structures of sphingosine, ceramide, and sphingomyelin. (a) Sphingosine is a long-chain alcohol with an amine at C-2. Sphingosine serves as the backbone for sphingolipids, a second class of molecules found in biological membranes. (b) Ceramides have a long-chain fatty acyl group (R) attached to the amine group of sphingosine. (c) Sphingomyelins have a phosphate group (red) attached to the C-1 hydroxyl group of a ceramide and a choline (blue) attached to the phosphate.

known as galactosylceramides, are cerebrosides whose polar head groups are single β-galactosyl residues (Figure 9·11). Galactocerebrosides are abundant in nerve tissue and account for about 15% of the lipids of myelin sheaths. Many other mammalian tissues contain glucocerebrosides, in which a β-glucosyl residue, rather than a β-galactosyl residue, is bound to C-1 of a ceramide. In some related glycolipids, a linear chain of up to four additional monosaccharide residues may be attached to the galactosyl residue of a galactocerebroside. Because they contain more than one monosaccharide residue, such glycolipids are not called cerebrosides, but have a variety of names.

Gangliosides are more complex glycolipids in which oligosaccharide chains containing *N*-acetylneuraminic acid (NANA) are attached to a ceramide. NANA is the acetyl derivative of a complex nine-carbon amino-sugar, a member of the sialic acid family of compounds (Figure 9·12a and Stereo S9·3). Gangliosides are anionic due to the presence of the carboxyl group of NANA.

There is considerable diversity in the structures of gangliosides, and well over 100 varieties have been characterized. Gangliosides are present on cell surfaces, with the two hydrocarbon chains of the ceramide moiety embedded in the hydrophobic part of the plasma membrane and the complex oligosaccharide structures forming an outside "coat." Gangliosides provide cells with distinguishing surface markers that may serve in cellular recognition and cell-to-cell communication (for example, the blood group antigens that determine whether your blood type is A, B, AB, or O are gangliosides).

Although highly varied in their molecular structures, gangliosides can be classified according to the identity of "core" oligosaccharide structures that are bound to the ceramide. In each of the four recognized core structures, the ceramide is linked through C-1 to a β-glucopyranosyl residue, which is in turn bound to a

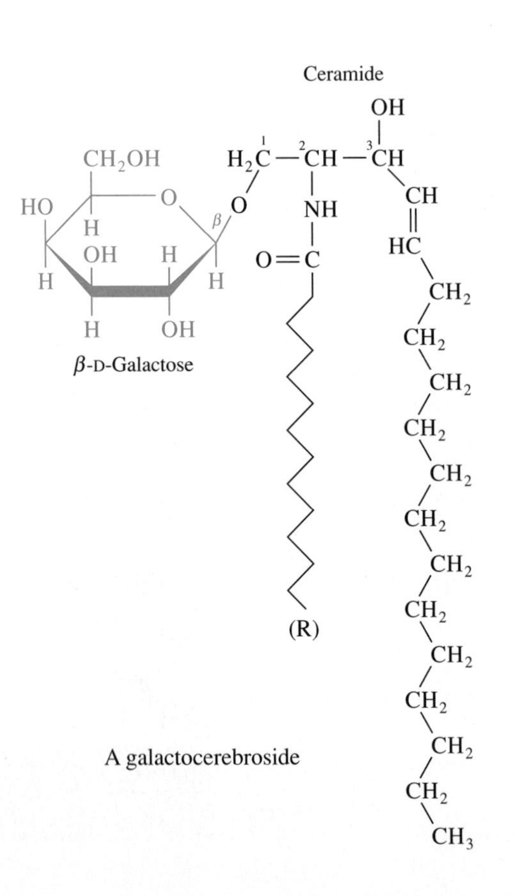

Figure 9·11
General structure of a galactocerebroside. β-D-galactose (blue) is attached to the C-1 hydroxyl group of a ceramide (black).

(a)

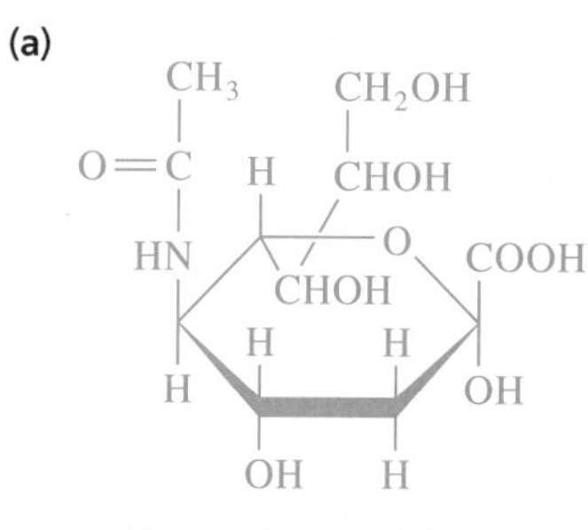

N-Acetylneuraminic acid
(NANA)

(b)

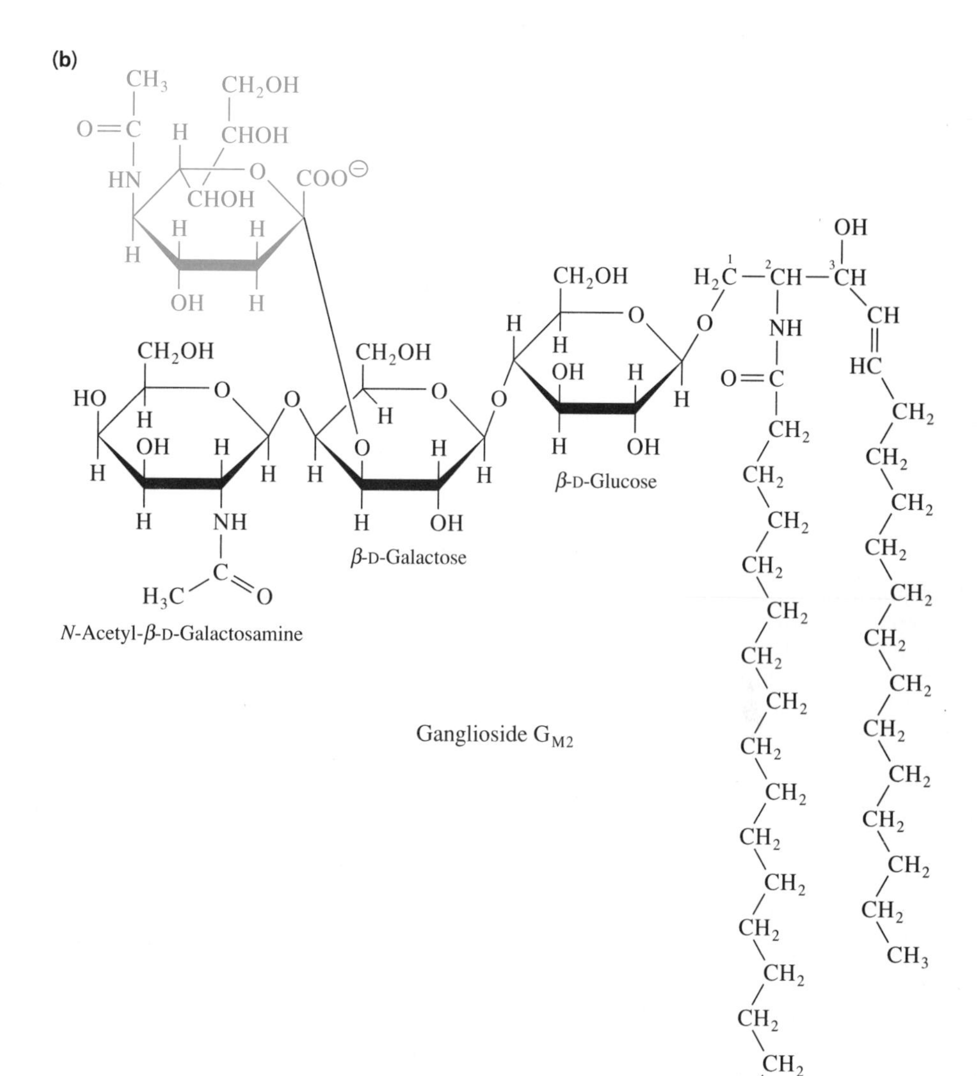

Figure 9·12
Ganglioside structure. (**a**) *N*-Acetylneuraminic acid (NANA) is a component of ganglioside structure. (**b**) G_{M2}, a representative ganglioside.

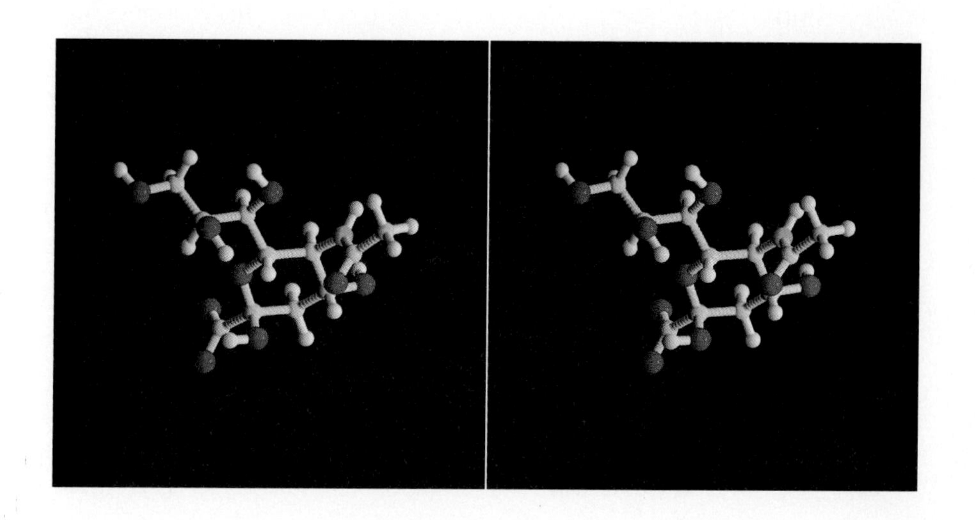

Stereo S9·3
N-Acetylneuraminic acid. (Stereo image by Richard J. Feldmann.)

β-galactosyl residue. The core structures differ in size and in the number and position of their NANA residues. One of the four core structures is exemplified by ganglioside G_{M2}, which is depicted in Figure 9·12b. The "M" in G_{M2} stands for "monosialo," indicating that the lipid has only one NANA residue in its core structure. (G_{M2} was the second monosialo ganglioside characterized, thus the subscript 2.)

Ganglioside metabolism is an active area of medical research. Recent investigations have revealed that the composition of membrane glycolipids can change dramatically during the development of malignant tumors. It is also known that genetically inherited defects in ganglioside metabolism are responsible for a number of debilitating and often lethal diseases, such as Tay-Sachs disease and generalized gangliosidosis.

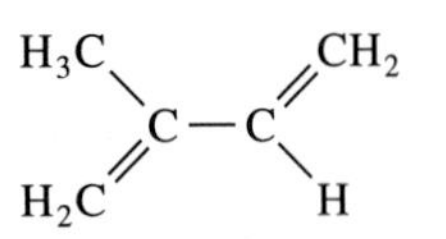

Figure 9·13
Isoprene. Isoprene forms the basic structural unit of all polyprenyl compounds, which include the steroids and lipid vitamins.

9·6 Steroids Constitute a Third Class of Membrane Lipids

The steroids constitute a third class of lipids commonly found in biological membranes. Steroids—and the lipid vitamins, which we shall discuss in Section 9·7—are more broadly classified as *polyprenyl compounds,* compounds synthesized from the five-carbon molecule isoprene (Figure 9·13). Steroids have a characteristic cyclic nucleus, which consists of four fused rings designated A, B, C, and D, as shown in Figure 9·14. *Cholesterol* is one example of a steroid; it is a common component of the plasma membranes of mammals but is only rarely found in plants and never in prokaryotes. Other steroids include the sterols of plants, yeasts, and fungi (which contain at least 27 carbon atoms and, like cholesterol, have a hydroxyl group at C-3); the mammalian steroid hormones (such as the C_{18} estrogens, the C_{19} androgens, and the C_{21} progestins and adrenal corticosteroids); and the C_{24} bile salts (the emulsifying agents noted in Section 9·3). The structures of these steroids

Figure 9·14
Structures of several steroids. The fused ring system of steroids consists of four rings (lettered A, B, C, and D). **(a)** Cholesterol. **(b)** Stigmasterol, a common sterol component of plant membranes. **(c)** Testosterone, a steroid hormone involved in male development in animals. **(d)** Sodium cholate, a bile salt. Cholate (the conjugate base of cholic acid) is important in aiding digestion of lipids.

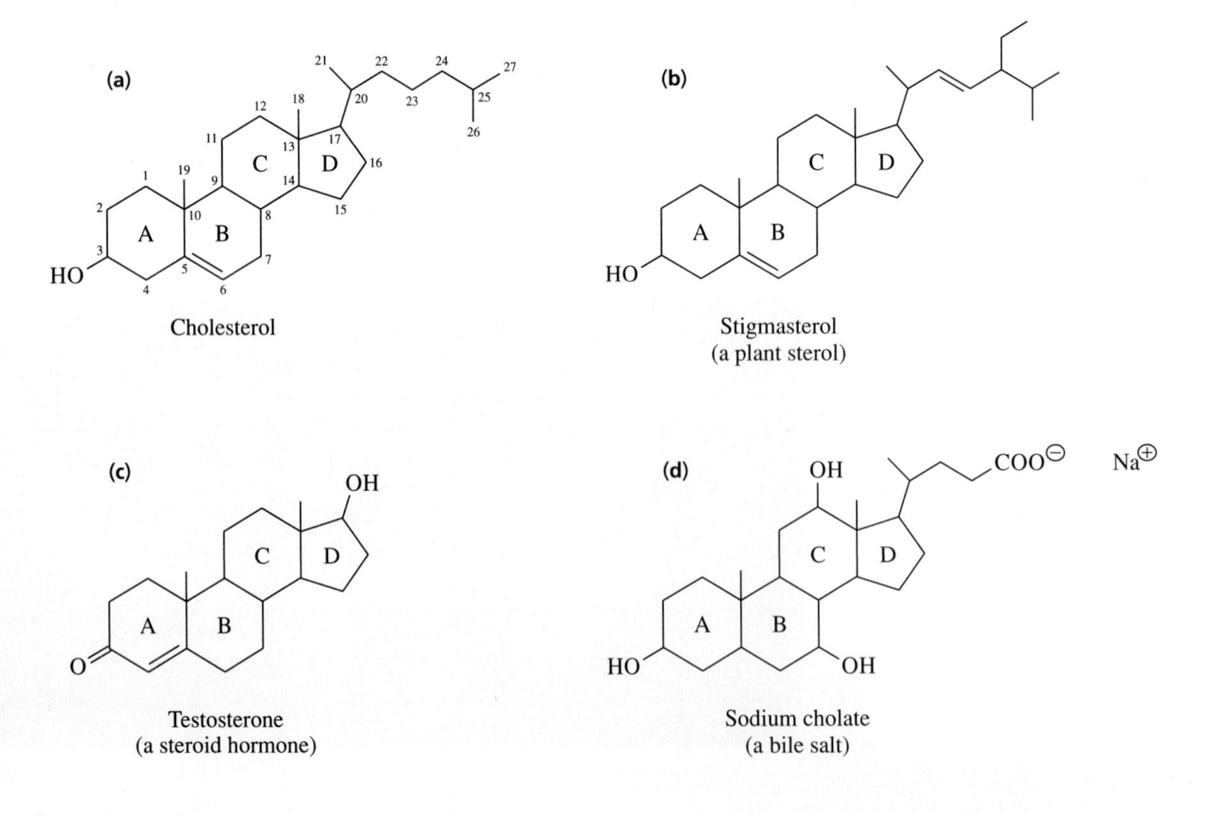

Stereo S9·4
Cholesterol, side view. The polar hydroxyl group is on the left. Note the nearly planar nature of cholesterol due to its fused ring system. (Stereo image by Richard J. Feldmann.)

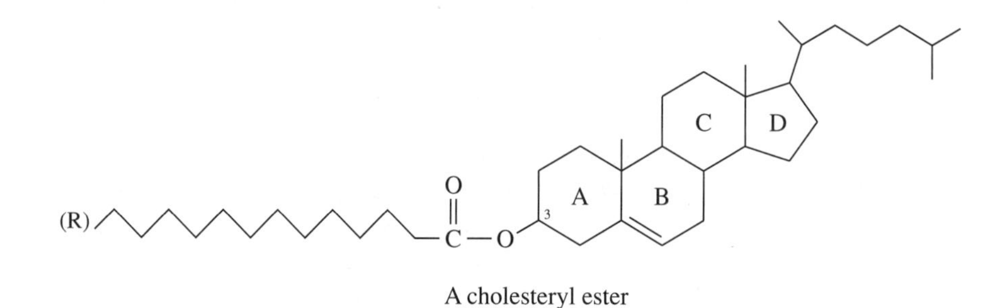

A cholesteryl ester

Figure 9·15
Cholesteryl ester. The presence of a fatty acyl group at C-3 makes cholesteryl esters even more hydrophobic than cholesterol.

differ in the length of the side chain attached to C-17 of the fused ring system and in the number and placement of methyl groups, double bonds, hydroxyl groups, and, in some cases, keto groups.

Despite its bad press, cholesterol serves an essential role in mammalian biochemistry. Cholesterol is not only a component of certain membranes, but it is also a precursor in the synthesis of the steroid hormones and bile salts. It is evident from the structure shown in Figure 9·14a and Stereo S9·4 that cholesterol is far more hydrophobic than the glycerophospholipids and sphingolipids we have previously encountered as membrane lipids. In fact, with the hydroxyl group at C-3 as its only polar constituent, cholesterol has a maximal free concentration in water of 10^{-8} M. The presence of cholesterol, with its rigid fused ring system, serves to modulate the fluidity of mammalian cell membranes (Section 9·10).

The esterification of fatty acids to the C-3 hydroxyl group of cholesterol forms *cholesteryl esters,* which are even more hydrophobic than cholesterol itself (Figure 9·15). Cholesteryl esters are found in lipoproteins in the blood and are formed when cholesterol is to be stored within cells. Lipoproteins are discussed in Section 9·8.

9·7 There Are Many Other Biologically Important Lipids

In addition to the lipids already discussed, many other lipids are important in biological systems. These include the fatty acid–derived waxes and eicosanoids and the lipid vitamins and other polyprenyl compounds.

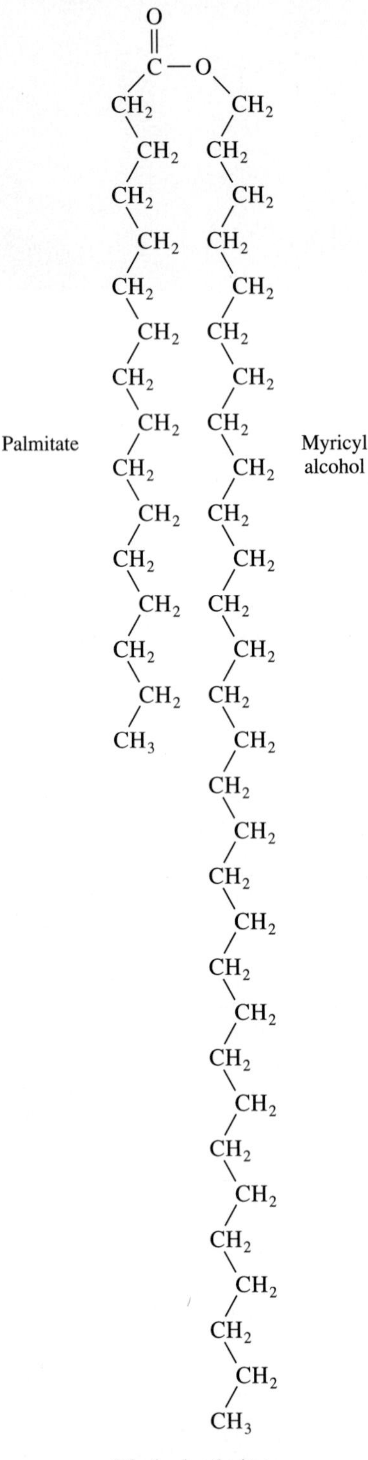

Figure 9·16
Myricyl palmitate. Myricyl palmitate is a wax, a fatty acid ester of a long-chain monohydroxylic alcohol.

Waxes are nonpolar esters of long-chain monohydroxylic alcohols and long-chain fatty acids. One example of a wax is myricyl palmitate (Figure 9·16). Myricyl palmitate, the ester of palmitate (16:0) and the 30-carbon myricyl alcohol, is a major component of beeswax. The hydrophobicity of myricyl palmitate makes beeswax highly water-insoluble, and its high melting point (as a consequence of its long, saturated hydrocarbon chains) makes beeswax hard and solid at typical outdoor temperatures. Waxes are widely distributed throughout nature; they form protective, waterproof coatings on the leaves and fruits of certain plants and on animal skins, furs, feathers, and exoskeletons.

In mammalian cells, certain C_{20} fatty acids are the precursors of prostaglandins, thromboxanes, and leukotrienes, physiologically potent regulatory molecules called *eicosanoids* (Greek, *eikosi,* twenty). Prostaglandins and thromboxanes are derived from the oxidation and cyclization of arachidonate, 20:4. One example of an eicosanoid is prostaglandin E_2 (PGE_2), whose structure is shown in Figure 9·17. Eicosanoids like PGE_2 are involved in such processes as the regulation of blood pressure, the induction of labor, and the induction of blood clotting.

The fat-soluble or lipid vitamins are vitamins A, D, E, and K. Like steroids, the lipid vitamins are polyprenyl compounds, composed primarily of long hydrocarbon chains or fused rings (Figure 9·18). Even the three lipid vitamins that contain polar groups (vitamins A, D, and E) are relatively hydrophobic and are more soluble in a lipid environment than in an aqueous one. All serve important roles in human metabolism.

A variety of other polyprenyl compounds also occur in biological systems (Figure 9·19). Limonene, for example, is the lipid component of lemons chiefly responsible for their distinctive smell; bactoprenol is a bacterial lipid that plays a role during synthesis of cell walls, and juvenile hormone I is a compound found in insects that regulates larval development.

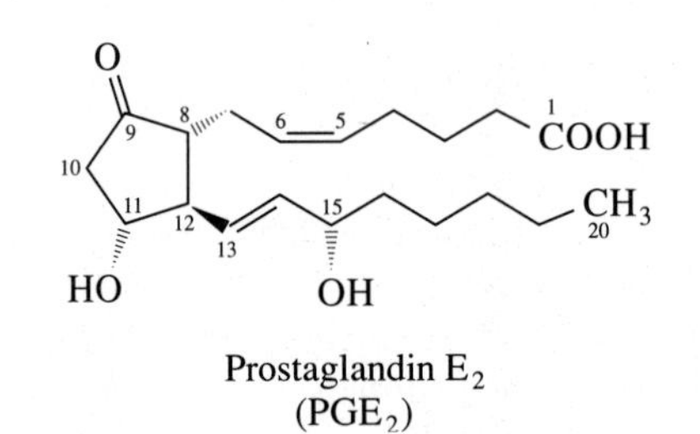

Figure 9·17
Prostaglandin E_2 (PGE_2). Prostaglandins are C_{20} acids that contain a cyclopentane ring. PGE_2 is derived metabolically from arachidonate (20:4).

9·8 Some Lipids Are Transported in Lipoproteins

Because they are essentially insoluble in water, cholesterol and its esters, as well as the nonpolar triacylglycerols, cannot be transported in blood as free molecules. Instead, these lipids are complexed with phospholipids and amphipathic proteins called *apoproteins* to form particles known as *lipoproteins.* A variety of lipoproteins are found in human plasma, and several apoproteins have been identified as constituents of those lipoprotein complexes. Lipoproteins are macromolecular assemblies with hydrophobic cores and hydrophilic surfaces. The core contains triacylglycerol and cholesteryl esters; the surface consists of amphipathic molecules: cholesterol, phospholipids, and the apoproteins (Figure 9·20).

Lipoproteins are classified according to their relative densities. Lipids are less dense than proteins; therefore the greater the lipid content of a lipoprotein, the

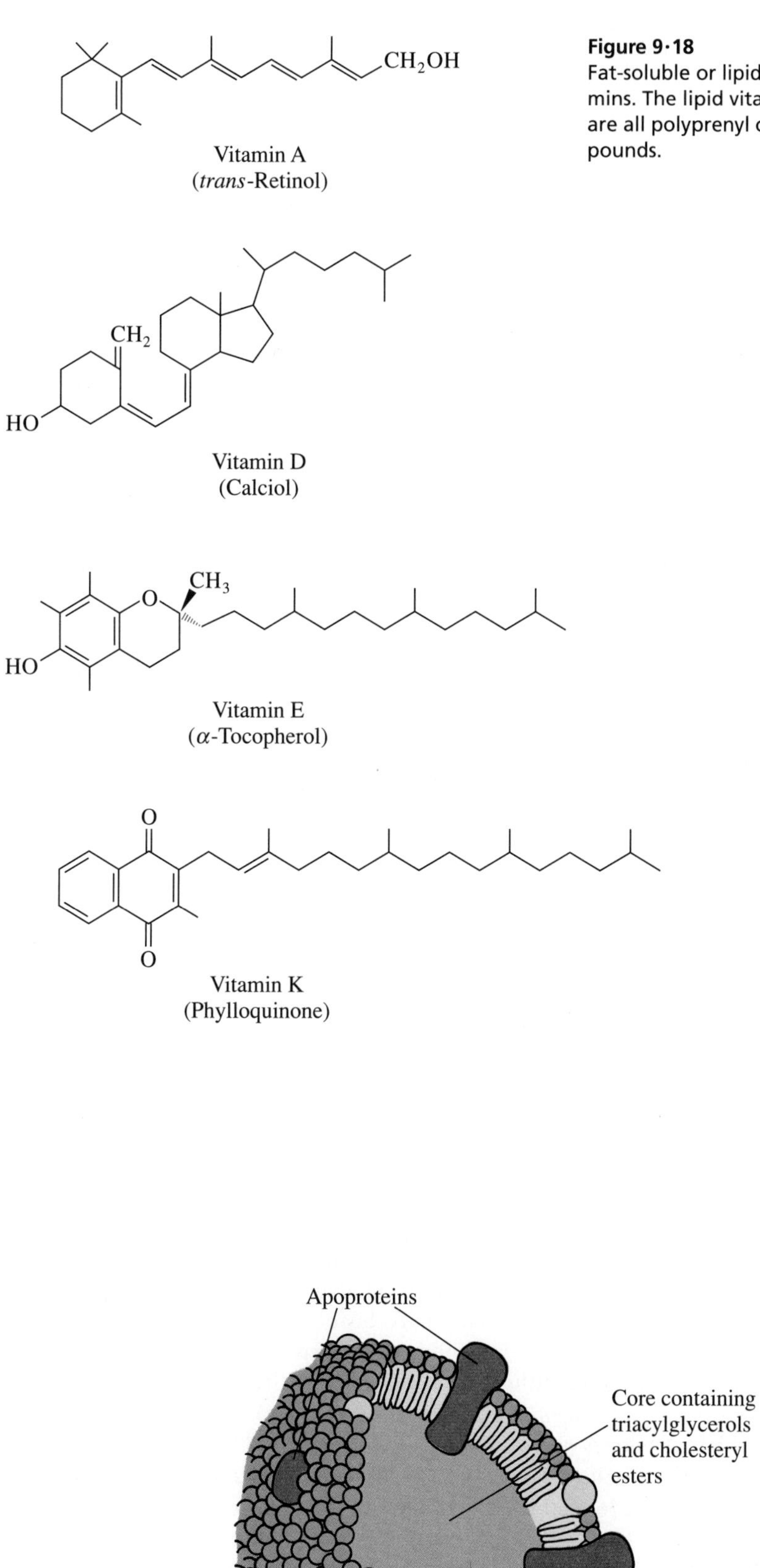

Figure 9·18
Fat-soluble or lipid vitamins. The lipid vitamins are all polyprenyl compounds.

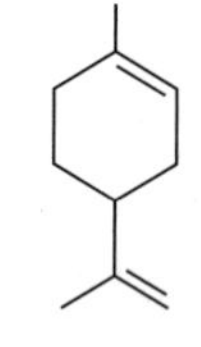

CH2OH
9
Bactoprenol
(Undecaprenyl alcohol)

O
CH3
O
O
Juvenile hormone I

Figure 9·19
Some biologically important polyprenyl compounds. Limonene is the lipid component of lemons that gives them their smell. Bactoprenol is a bacterial lipid that plays a role during the synthesis of cell walls. Juvenile hormone I regulates larval development in insects.

Figure 9·20
Structure of a lipoprotein. A lipoprotein contains a core of neutral lipids, which include triacylglycerols and cholesteryl esters. The core is coated with a monolayer of phospholipids in which apoproteins and cholesterol are embedded. (Illustrator: Lisa Shoemaker.)

Table 9·3 Characteristics of lipoproteins in human plasma

	Chylomicrons	VLDL	LDL	HDL
Molecular weight × 10^{-6}	>400	5–6	2.3	0.18–0.36
Density (g cm^{-3})	<1.006	0.95–1.006	1.006–1.063	1.063–1.210
Chemical composition (%)				
Triacylglycerol	85	50	10	4
Free cholesterol	1	7	8	2
Cholesteryl ester	3	12	37	15
Phospholipid	9	18	20	24
Protein	2	10	23	55

[Adapted from Kritchevsky, D. (1986). Atherosclerosis and nutrition. Nutrition International 2:290–297.]

lower its density (Table 9·3). The major classes of human lipoproteins are: the *chylomicrons,* which carry triacylglycerols and cholesterol from the small intestine to the tissues; the *very low density lipoproteins* (VLDL), which carry endogenous triacylglycerols, cholesterol, and cholesteryl esters from the liver (their primary site of synthesis) to the tissues; the *low density lipoproteins* (LDL), which are formed during the breakdown of VLDL and are enriched in cholesterol and cholesteryl esters; and the *high density lipoproteins* (HDL), which transport endogenous cholesterol and cholesteryl esters back to the liver.

9·9 Phospholipids and Glycolipids Can Spontaneously Form Bilayers

In Chapter 2, we saw that soaps, being amphipathic, can form monolayers at water-air interfaces or micelles in suspension (Figure 2·7). Like soaps, phospholipids and glycolipids can form monolayers under certain conditions. In vivo, however, phospholipids and glycolipids tend to assemble into a structure vastly more interesting and biologically significant—a *lipid bilayer.*

Because they have two hydrocarbon tails, phospholipids and glycolipids do not pack well into micelles but fit nicely into lipid bilayers (Figure 9·21). Lipid bilayers form the structural basis of all membranes, including plasma membranes (membranes that surround cells) and intracellular membranes (membranes that surround or support the organelles within a eukaryotic cell). Note the orientation of the amphipathic lipid molecules in the assembly of such bilayers: the hydrophobic tails point toward the interior of the bilayer, whereas the hydrophilic heads are in contact with the aqueous solution on each surface. The positive and negative charges of the bilayer constituents, contributed by polar head groups such as those of phosphatidylcholines and sphingomyelins, provide both *leaflets* (or layers) of the bilayer with an ionic surface. An increase in solvent entropy provides the major driving force for the formation of lipid bilayers as it does for protein folding (Section 4·11A).

Synthetic vesicles consisting of phospholipid bilayers that enclose an aqueous compartment can be formed in high yield in the laboratory. Such structures, called *liposomes* (Figure 9·22), are generally quite stable. Because of their stability and impermeability to many substances, liposomes are often employed in biochemical research. Many of their properties resemble those of biological membranes, although it should be borne in mind that liposomes by definition are not naturally occurring.

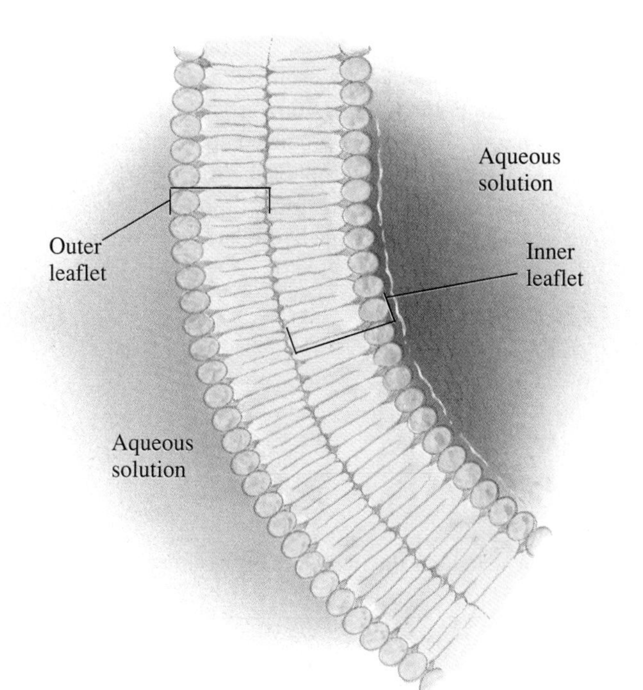

Figure 9·21
Schematic cross-section of a lipid bilayer. Such bilayers may be formed by amphipathic lipids, such as glycerophospholipids or sphingolipids. The bilayer is divided into two leaflets. In each leaflet, the polar head groups extend into the aqueous medium, and the nonpolar hydrocarbon tails point inward and are in van der Waals contact with each other. (Illustrator: Lisa Shoemaker.)

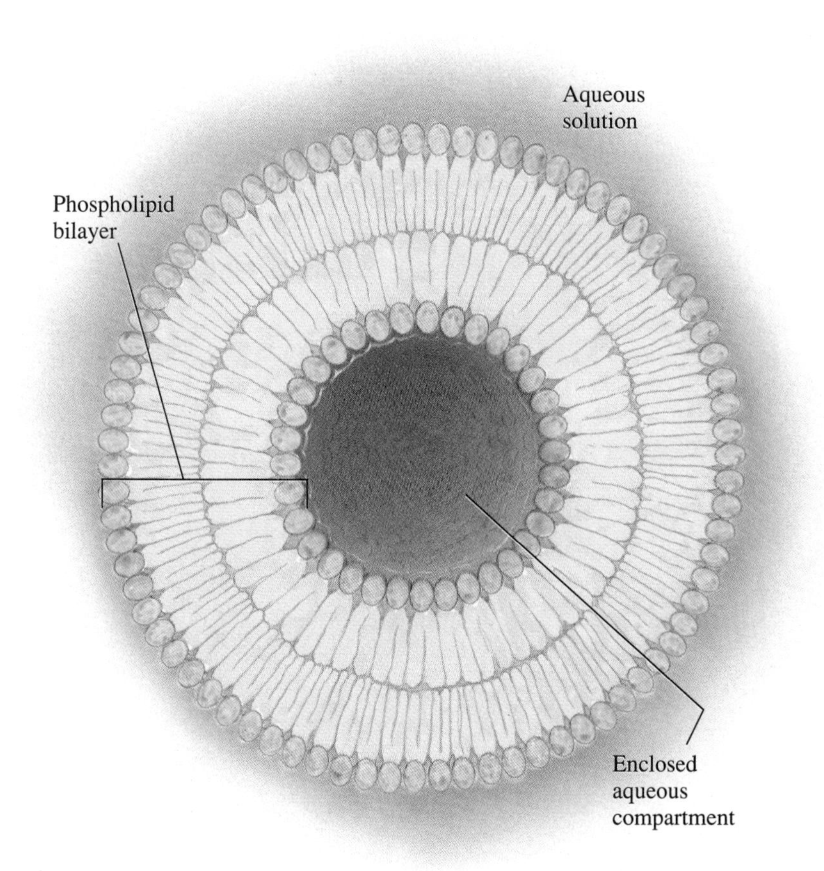

Figure 9·22
Schematic cross-section of a liposome. (Illustrator: Lisa Shoemaker.)

Not all amphipathic lipids form bilayers. Cholesterol, for example, is slightly amphipathic but cannot form a bilayer by itself since the polar —OH group is too small relative to the hydrophobic, fused hydrocarbon ring system. *Lysophosphoglycerides* are also amphipathic; they are produced upon hydrolytic removal of one of the two fatty acid moieties from glycerophospholipids (Box 9·1). Lysophospholipids can form micelles, but they do not form lipid bilayers because the polar heads are too large in relation to their single hydrocarbon tails to allow packing into such structures.

Box 9·1 Phospholipase A_2 Catalyzes the Hydrolysis of Glycerophospholipids to Lysophosphoglycerides

Lysophosphoglycerides, which are normally present in cells at low concentrations, serve as metabolic intermediates. High concentrations of lysophosphoglycerides disrupt membranes, thereby causing cells to lyse (hence their name). A number of snake, bee, and wasp venoms have high levels of phospholipase A_2, an enzyme that catalyzes the removal of the acyl moiety from the C-2 position of glycerophospholipids and thereby generates lysophosphoglycerides (Figure 1). Injection of such venoms into animals through bites or stings can lead to extensive hemolysis (lysis of blood cell membranes), which may be life-threatening.

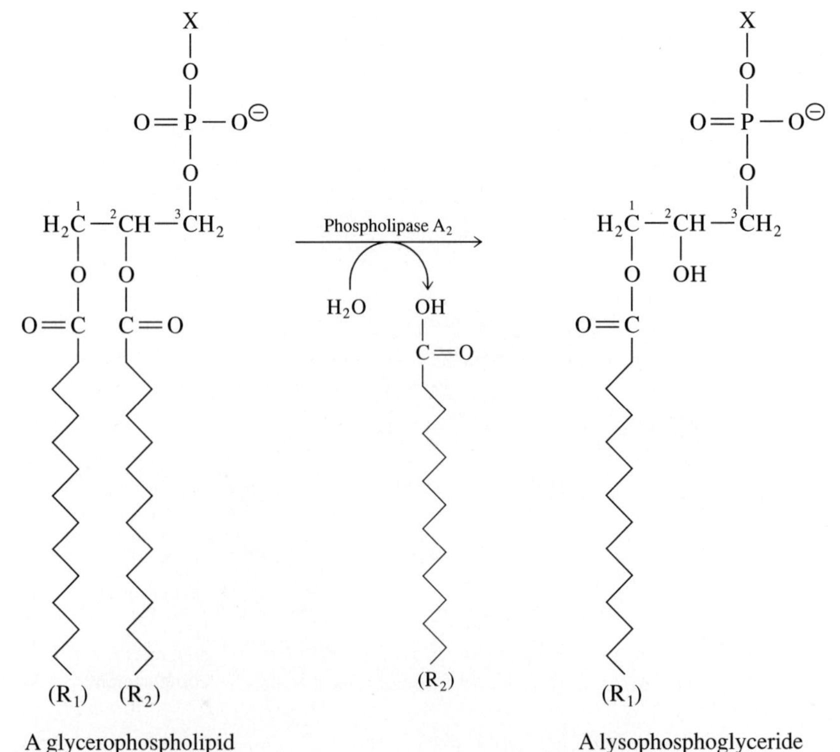

Box 9·1
Figure 1
Formation of a lysophosphoglyceride catalyzed by phospholipase A_2. Lysophosphoglycerides contain only one fatty acyl group and are unable to form bilayers.

Although phospholipase A_2 is an active ingredient of certain venoms, it is also required during normal cellular processes. Animals rely on pancreatic phospholipase A_2, which is secreted as a zymogen, for the intestinal digestion of membrane phospholipids from dietary sources. Moreover, intracellular phospholipase A_2, whose activity is highly regulated, catalyzes the release of arachidonate from membrane phospholipids, thereby initiating the synthesis of physiologically potent eicosanoids such as prostaglandins (Section 9·7).

9·10 Lipid Bilayers Are Fluid but Are Also Selectively Permeable Barriers

A lipid bilayer that contains no proteins is about 5 to 6 nm thick. Despite being thin, lipid bilayers are formidable barriers to the passage of molecules and serve as an integument that separates the living cell from its inanimate environment.

Lipid bilayers have many of the properties of fluids. For example, the lipid and protein components of a bilayer are not frozen but are in constant motion. In fact, diffusion of phospholipids within one leaflet of the bilayer, called *lateral diffusion,* has been shown to be very rapid. In a bacterial membrane about 2 μm long, a lipid molecule can diffuse from one end to the other in about one second. Given such freedom for lateral diffusion, a lipid bilayer can essentially be regarded as a two-dimensional solution.

In contrast with lateral diffusion, however, lipids do not readily undergo *transverse diffusion,* that is, the passage of the lipids from one leaflet of the bilayer to the other. Such diffusion can occur, but it usually requires the presence of enzymes that lower the activation energy of the reaction. Because the polar head of a phospholipid molecule is highly solvated, it would have to shed its solvation sphere and penetrate the hydrocarbon interior of the bilayer to move from one leaflet to the other. The potential energy barrier associated with such diffusion is so high that nonenzymatic transverse diffusion of membrane lipids occurs at about one billionth the rate of lateral diffusion (Figure 9·23).

Lipid bilayers also behave like fluids in that they undergo phase transitions. At low temperatures (those below the phase-transition temperature of the particular bilayer being studied), the bilayer exists in a well-ordered gel state. In such a state, the fatty acyl side chains assume an extended conformation and nestle together with maximal van der Waals contacts. When a lipid bilayer is heated, however, a phase transition analogous to the melting of a crystalline solid occurs. Synthetic membranes composed of a single type of lipid undergo a phase transition at a fairly distinct temperature (the melting point of the particular lipid of which the membrane is composed), whereas naturally occurring biological membranes, which contain a more heterogeneous population of lipids, exhibit more gradual transitions, often at temperatures between 30°C and 40°C. With greater rotational flexibility at higher temperatures, the hydrocarbon tails of fatty acyl groups become less extended in the liquid crystalline phase, and the thickness of the bilayer decreases by about 15%.

The lipid composition of a bilayer affects its fluidity and phase-transition temperature (Box 9·2). The presence of the *cis* double bonds of unsaturated fatty acid residues, for example, increases the fluidity of a membrane and lowers its phase-transition temperature, whereas a greater proportion of saturated fatty acid residues decreases fluidity and raises the phase-transition temperature. The intercalation of

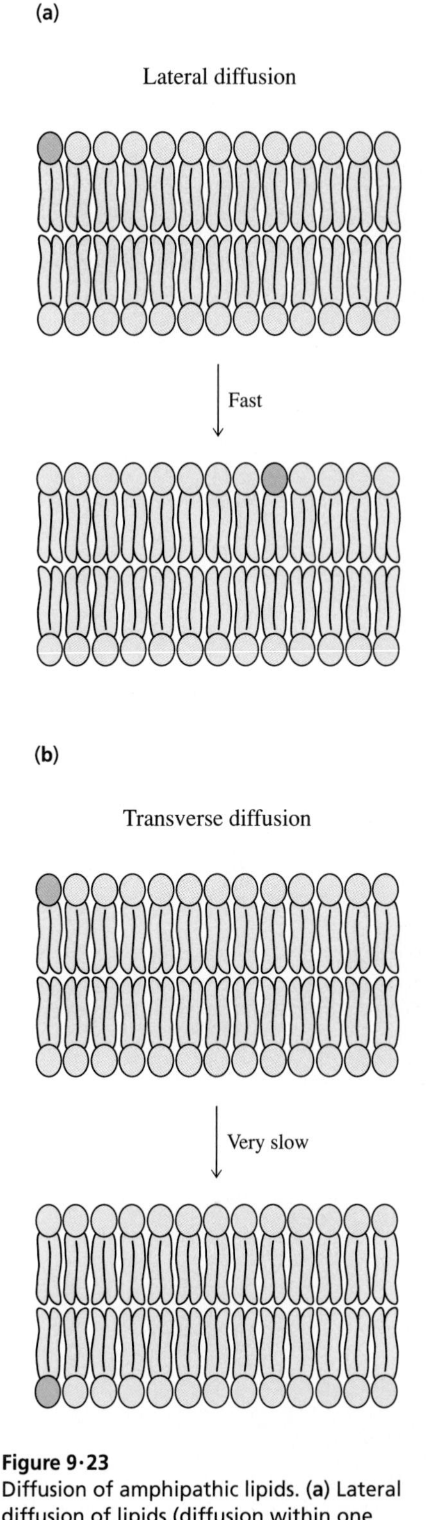

Figure 9·23
Diffusion of amphipathic lipids. **(a)** Lateral diffusion of lipids (diffusion within one leaflet of the bilayer) is relatively rapid. **(b)** Transverse diffusion of lipids (diffusion from one leaflet of the membrane to the other) is relatively slow, occurring at about one billionth the rate of lateral diffusion.

cholesterol molecules among the hydrocarbon tails of the phospholipids in a bilayer broadens the transition-temperature range, thereby modulating the fluidity of membranes; the rigid ring structure of cholesterol both restricts the mobility of unsaturated fatty acyl groups and thus decreases fluidity and, at the same time, disrupts the ordered packing of saturated fatty acyl groups and thereby increases fluidity. The presence of cholesterol in animal cell membranes thus helps to maintain fairly constant fluidity despite fluctuations in temperature or degree of fatty acid saturation.

The thin layer of hydrocarbon chains within the lipid bilayer of a sealed vesicle acts as a selectively permeable barrier that separates the inner aqueous compartment from its surroundings. In essence, this barrier allows small, uncharged or hydrophobic molecules to diffuse through the membrane while serving as an almost impenetrable barrier to most charged species. Thus, although water molecules are small enough to pass through lipid bilayers and to diffuse into or out of liposomes rather rapidly, the permeability of lipid bilayers to solvated ions such as potassium ($K^{\oplus}$) or sodium ($Na^{\oplus}$) is approximately one billionfold less.

Box 9·2 Fluctuation of Lipid Composition Keeps Bilayers Fluid Under Different Conditions

The low melting points of *cis* unsaturated fatty acids are reflected in the physical properties of the biological membranes from which they are isolated. Membranes with a high proportion of unsaturated fatty acyl groups, for example, are much more fluid than those having few unsaturated fatty acyl groups.

In living systems, relatively constant membrane fluidity is very important. The fluidity of membranes can be regulated through changes in the ratio of unsaturated to saturated fatty acyl groups within membranes. Bacterial membranes, for example, have different ratios of unsaturated to saturated fatty acyl groups depending on the temperature at which the bacteria are grown. More unsaturated fatty acyl groups are present if the bacteria are grown at a low temperature. In this way, the fluidity of bacterial membranes is maintained despite fluctuations in temperature.

In most warm-blooded organisms, the ratio of unsaturated to saturated fatty acyl groups within membranes varies less with temperature. An interesting exception is found in the membrane lipids of the reindeer leg. There, the number of unsaturated fatty acyl groups in membrane lipids increases as you move closer to the hoof. The lower melting point and greater fluidity of unsaturated fatty acyl groups in these membrane lipids permit the membranes to remain fluid and functional even at the low temperatures to which the hoof is exposed.

9·11 The Fluid Mosaic Model Postulates that Biological Membranes Are Assemblies of Proteins in a Fluid Lipid-Bilayer Matrix

Lipid bilayers provide biological membranes with their basic structural character. However, proteins are necessary components of membranes since they allow transport of molecules across the lipid bilayer, the transduction of signals across the membrane, and interactions between the plasma membrane and the cytoskeleton.

In 1972, S. Jonathan Singer and Garth L. Nicolson proposed a model of the structure of biological membranes, which is referred to as the *fluid mosaic model.* The model postulates that membrane proteins diffuse freely in the plane of the lipid bilayer, somewhat like protein icebergs in a lipid sea. Although some aspects of the original fluid mosaic model have been modified with subsequent research findings, it continues to provide valuable insights into the nature of biological membranes.

We now know that many, but not all, proteins diffuse freely within the lipid bilayer. There are, in fact, three broad classes of membrane proteins: *integral membrane proteins, peripheral membrane proteins,* and *lipid-linked proteins* (Figure 9·24). Integral membrane proteins, also referred to as intrinsic membrane or transmembrane proteins, contain α-helical regions composed of hydrophobic amino acid residues that are embedded in the lipid bilayer. Integral membrane proteins appear to span the bilayer completely, although at least one protein, cytochrome b_5, is anchored in the membrane by a hydrophobic α helix that does not traverse the entire bilayer.

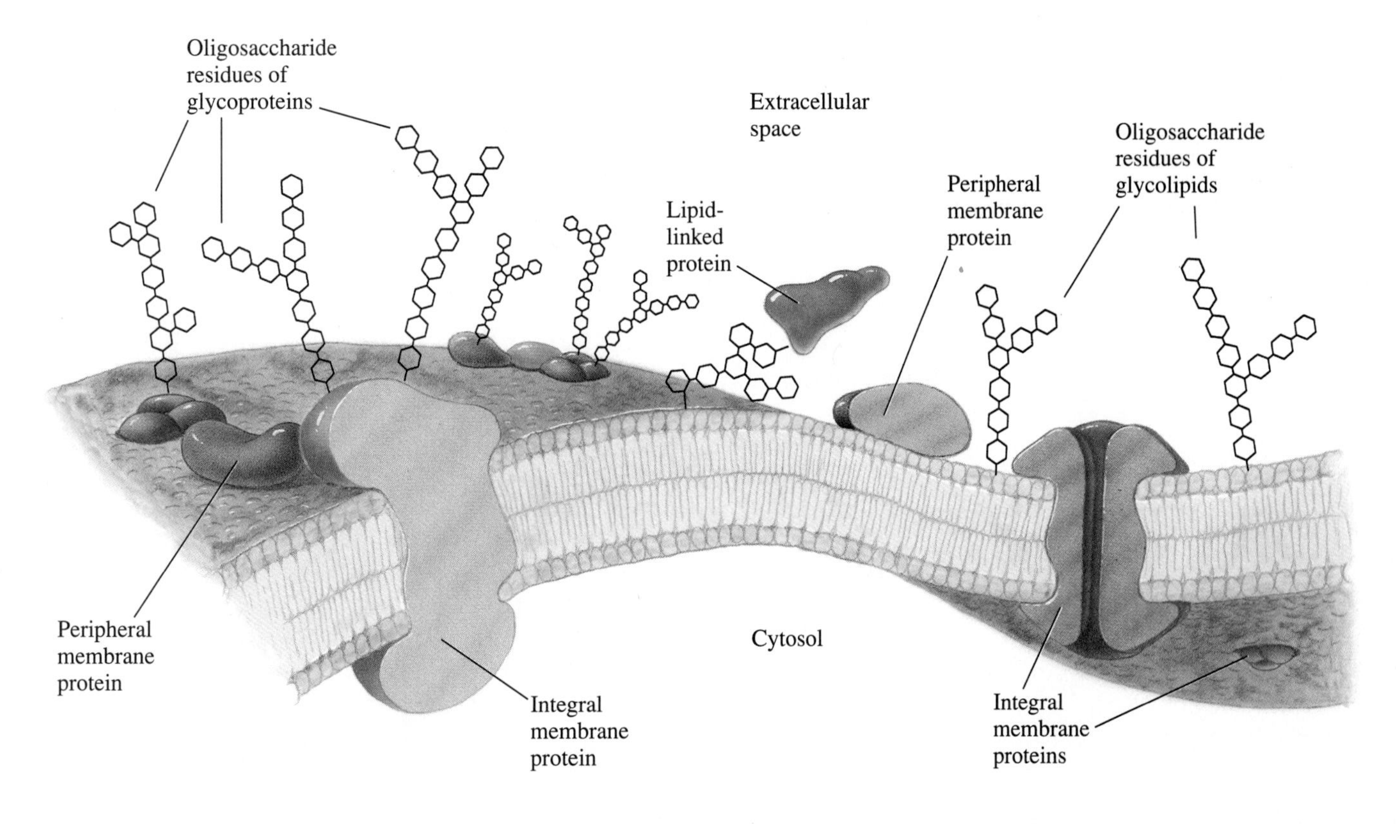

Figure 9·24
Structure of eukaryotic biological membranes. Biological membranes consist of a lipid bilayer and proteins associated with it in various ways. Integral membrane proteins are embedded in the lipid bilayer and appear to span both leaflets. Hydrophobic amino acid residues form van der Waals contacts with the hydrophobic tails of lipid molecules. Peripheral membrane proteins are weakly associated with one or the other leaflet of the membrane through ionic interactions and hydrogen bonds with the polar head groups of the lipid molecules or integral membrane proteins. Lipid-linked proteins are covalently attached to the membrane via a glycosyl-phosphatidylinositol anchor. (Illustrator: Lisa Shoemaker.)

Peripheral membrane proteins, also referred to as extrinsic membrane proteins, are more weakly bound to either an interior or an exterior membrane surface through ionic interactions and hydrogen bonding with the polar heads of the membrane lipids. Peripheral membrane proteins may also interact with integral membrane proteins. Because they are neither covalently attached to the lipid bilayer nor embedded within the lipid matrix of the bilayer, peripheral membrane proteins are more readily dissociated from membranes by procedures that do not require cleavage of covalent bonds or disruption of the membrane itself.

Finally, lipid-linked proteins are tethered to the outer face of certain plasma membranes through covalent bonding to glycosyl-phosphatidylinositol anchors. As shown in Figure 9·25, these anchors are grounded in the outer leaflet of the lipid bilayer by the 1,2-diacylglycerol portion of the phosphatidylinositol. A glycan of somewhat varied composition is attached to the inositol via a glucosamine residue, a phosphomannose residue links the glycan to an ethanolamine residue, and the C-terminal α-carboxyl group of the protein is linked to the ethanolamine via an amide bond. Both lipid-linked and integral membrane proteins are stably associated with membranes, whereas peripheral membrane proteins are much more easily disrupted from membranes.

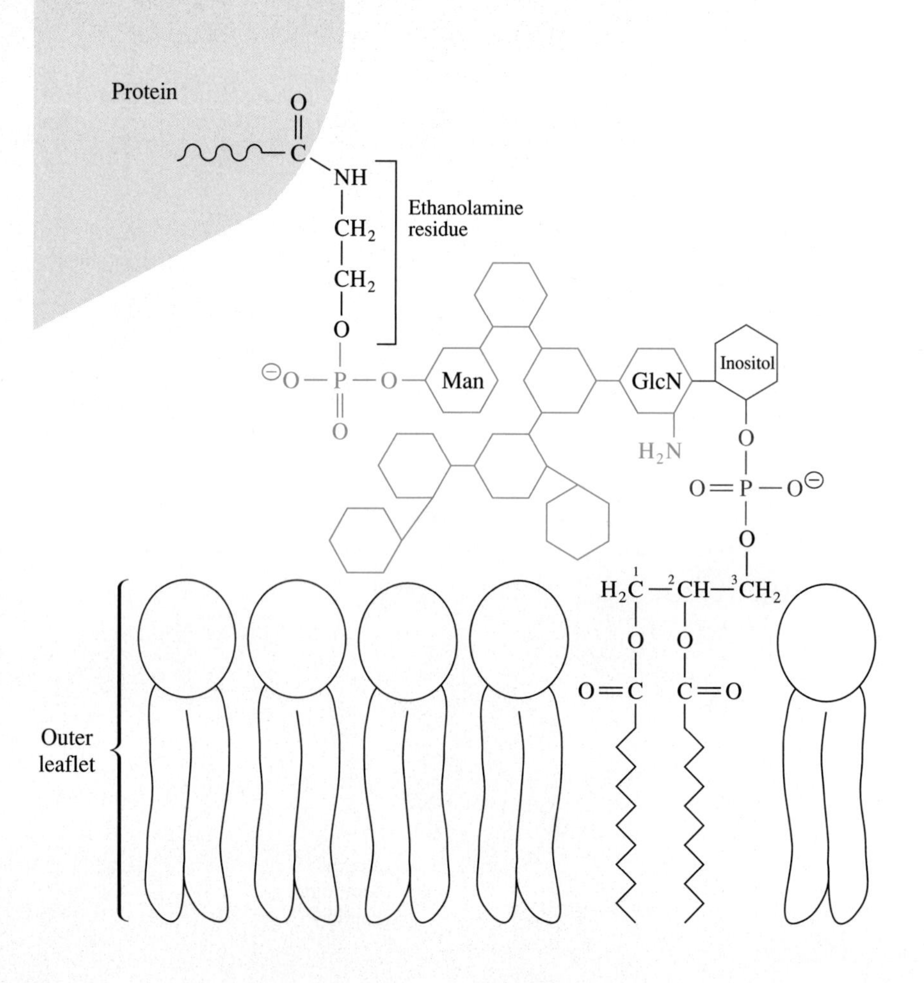

Figure 9·25
Composition of glycosyl-phosphatidylinositol anchors in *Trypanosoma brucei*. Lipid-linked proteins are covalently bound to an ethanolamine residue that is in turn bound to a glycan. The glycan can be of varied composition (blue hexoses), but usually features a phosphomannose residue to which the ethanolamine residue is attached and a glucosamine residue, which is attached to phosphatidylinositol, shown in red. The diacylglycerol portion of the phosphatidylinositol anchors the lipid-linked protein in the membrane. (Abbreviations: glucosamine, GlcN; mannose, Man.) [Based on data from Ferguson, M. A. J., and Williams, A. F. (1988). Cell-surface anchoring of proteins via glycosyl-phosphatidylinositol structures. *Annu. Rev. Biochem.* 57:285–320.]

L. D. Frye and Michael A. Edidin devised an elegant experiment to test the hypothesis that integral membrane proteins diffuse laterally within the lipid bilayer. They fused mouse cells with human cells to form heterokaryons (hybrid cells). Using red fluorescent–labelled antibodies that specifically bind to certain proteins in human plasma membranes and green fluorescent–labelled antibodies that specifically bind to certain proteins in mouse plasma membranes, they observed changes in the distribution of red and green fluorescent patches in the heterokaryon by fluorescence microscopy. The surface antigens were intermixed within 40 minutes after cell fusion (Figure 9·26).

Not all membrane proteins diffuse at random within the plane of the lipid bilayer; diffusion of certain proteins appears to be either restricted or directed in some cases. Many cell membranes, for example, are stabilized by a scaffolding of interior proteins called the *cytoskeleton*. Recent research has revealed that some membrane proteins are anchored to this cytoskeleton and are therefore relatively immobile. Moreover, certain membrane proteins, such as specific ligand receptors, are known to diffuse toward one another, forming protein-enriched patches. The mechanism of such directed diffusion is not yet understood.

In contrast to lateral diffusion, membrane proteins do not characteristically undergo transverse diffusion. As indicated in Chapter 8 and illustrated in Figure 9·24, the oligosaccharide chains of plasma membrane glycoproteins and glycolipids appear to be located exclusively on the outer (extracellular) surface. This suggests that glycoproteins, like glycolipids, are not able to cross to the inner leaflet.

The electron microscope is a powerful tool in the study of biological membranes. Many aspects of membrane structure have become known through *freeze-fracture electron microscopy*. In this technique, a membrane sample is rapidly frozen and then split along the interface between the leaflets of the phospholipid

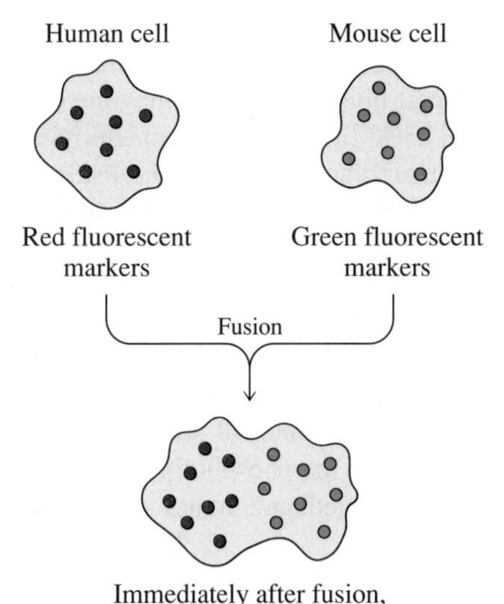

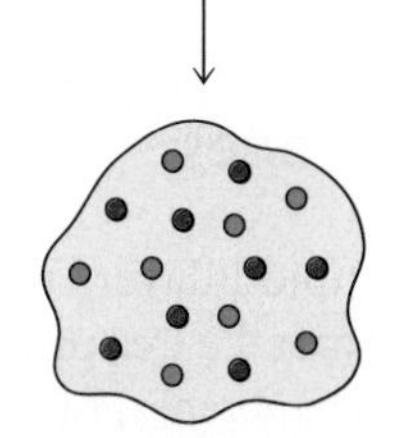

Figure 9·26
Frye and Edidin's experiment testing the free diffusion of integral membrane proteins. Human cells whose integral membrane proteins have been labelled with a red fluorescent marker are fused with mouse cells whose integral membrane proteins have been labelled with a green fluorescent marker. The initially localized markers become dispersed over the entire surface of the new cell within 40 minutes. This experiment demonstrates that at least some integral membrane proteins diffuse freely within biological membranes.

bilayer, where the intermolecular interactions are weakest (Figure 9·27). Water is evaporated from the exposed surface by a vacuum, and the exposed surface is then coated with a thin metal film, producing a replica of the leaflet surface. The leaflets of membranes rich in proteins, such as the plasma membrane of a red blood cell, shown in Figure 9·28, resemble a pitted landscape. In contrast, the leaflets of liposomes, which contain no proteins, are smooth.

Specialized integral and peripheral membrane proteins are characteristic of particular cellular and intracellular membranes. We shall consider the functions of some of these membrane proteins and protein aggregates in later chapters on oxidative phosphorylation, photosynthesis, and protein synthesis.

9·12 The Composition of Biological Membranes Is Highly Variable

We noted earlier that a lipid bilayer that contains no proteins is about 5 to 6 nm thick. In contrast, biological membranes are typically 6 to 10 nm thick due to the presence of proteins embedded in or associated with the bilayer. Generally, biological membranes average about 40% lipid and 60% protein (by mass), but specific membranes vary considerably in their compositions. The plasma membranes of red blood cells, for example, are exceptionally rich in proteins and have a lipid-to-protein ratio of 1:1.5, whereas myelin from nervous system tissue is exceptionally poor in proteins and has a lipid-to-protein ratio of about 4:1. The carbohydrate content of a membrane varies from 0% to a maximum of about 10%.

In addition to having a characteristic lipid-to-protein ratio, each biological membrane has a characteristic lipid composition. Membranes in brain tissue, for example, have a relatively high content of phosphatidylserines, whereas membranes in heart and lung tissue have high levels of phosphatidylglycerol and sphingomyelins, respectively. In the case of red blood cell membranes, phosphatidylcholines, phosphatidylethanolamines, and phosphatidylserines together account for nearly 50% of the total lipids, whereas sphingomyelins account for 18%, and cholesterol constitutes 23%. In contrast, phosphatidylethanolamines constitute nearly 70% of the inner membrane lipids of *Escherichia coli* cells, whereas diphosphatidylglycerols and phosphatidylglycerols compose some 12% and 18%, respectively.

In addition to being distributed differentially throughout different tissues, phospholipids are also distributed asymmetrically between the inner and outer leaflets of a single biological membrane. For example, sphingomyelins and phosphatidylcholines each account for almost half of the phospholipid molecules in the outer leaflet of the plasma membranes of human erythrocytes, but together they represent less than one-fifth of the phospholipids in the inner leaflet of the same membrane, where phosphatidylethanolamines and phosphatidylserines predominate.

9·13 Several Mechanisms Exist for the Transport of Matter Across Membranes

As discussed in Section 9·10, lipid bilayers are selectively permeable barriers whose hydrophobic interiors restrict the free passage of most molecules. This is an important feature of biological membranes, for without the physical separation of a living cell from its inanimate environment, there would be no life. However, it is equally important that water, oxygen, and all the other nutrients required for

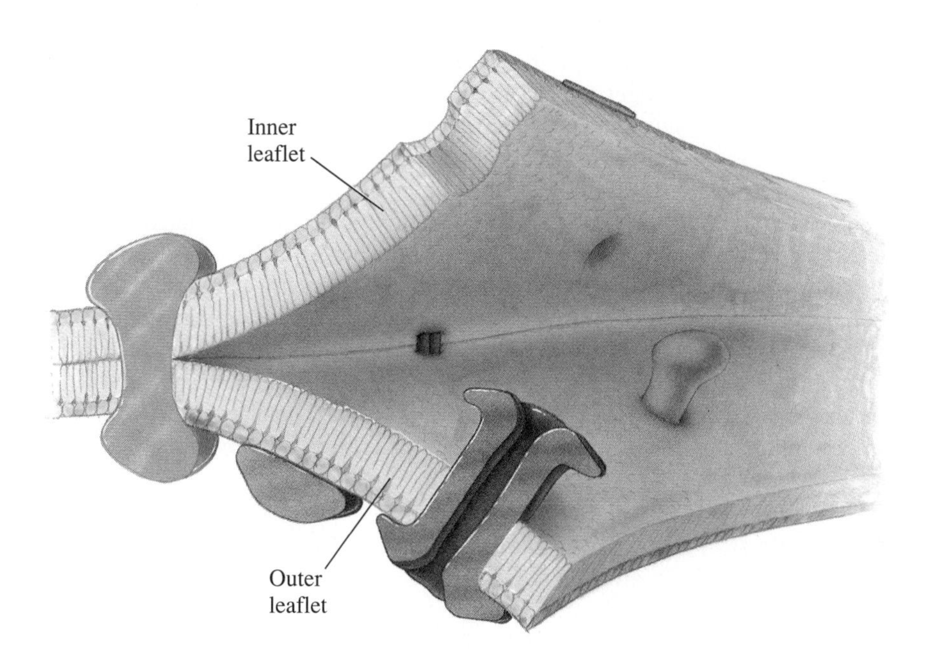

Figure 9·27
Separation of a lipid bilayer by freeze fracturing. A lipid bilayer can be split along the interface of the two leaflets, where intermolecular attractions are weakest. A replica of the exposed membrane surface is made by coating the separated leaflets with a thin metal film. Integral membrane proteins are detected as protrusions or cavities in the replica. (Illustrator: Lisa Shoemaker.)

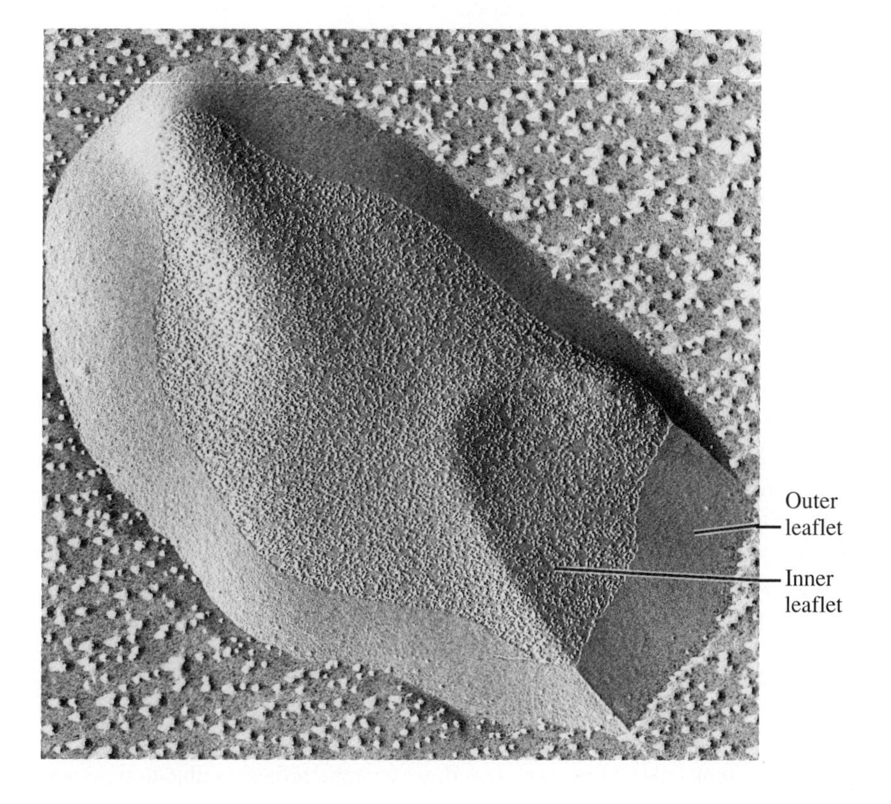

Figure 9·28
Electron micrograph of a freeze-fractured erythrocyte membrane. The bumps on the surface are integral membrane proteins. (Courtesy of Vincent T. Marchesi.)

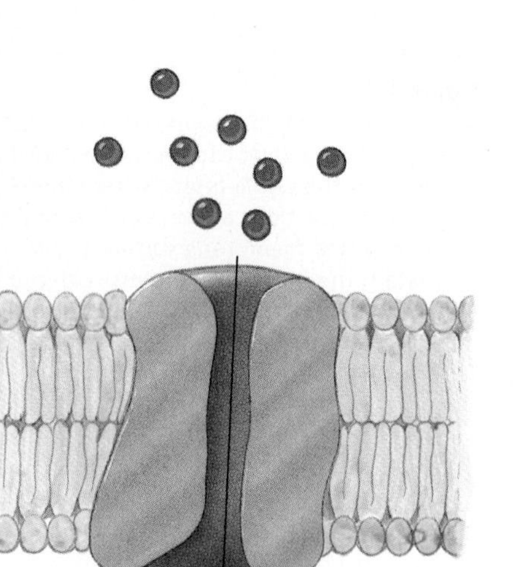

Figure 9·29
Membrane transport through pores. Pores form water-filled holes that allow appropriately sized molecules and ions to traverse the membrane in either direction. (Illustrator: Lisa Shoemaker.)

growth be able to enter a cell and that products generated by a cell for export (such as hormones, certain digestive enzymes, and toxins) as well as waste products (such as carbon dioxide and urea) be able to leave a cell. Thus, a wide variety of substances must be able to cross a membrane in order for a cell to survive.

Certain small, uncharged molecules can cross a membrane by simple diffusion. Water and oxygen, for example, are both able to diffuse relatively freely across a biological membrane. However, most molecules and ions, either because they are too large or because they are charged, cannot enter a cell by simple diffusion. For those molecules, there exist a variety of mechanisms that mediate the transport of matter across biological membranes. These mechanisms include facilitated diffusion and active transport (for the transport of small molecules and ions) and endocytosis and exocytosis (for the transport of macromolecules). Living cells use a combination of these mechanisms to move matter across membranes.

A. In Facilitated Diffusion, a Molecule Travels Down Its Concentration Gradient

If a drop of ink is placed on the surface of water in a beaker, the color soon spreads throughout the water by Brownian motion. The thermodynamic force that drives this diffusion is the increase in entropy that occurs as the initially concentrated ink particles become more randomly distributed. This same thermodynamic force moves solute molecules down their concentration gradient (from where they are most concentrated to where they are least concentrated). *Facilitated diffusion* is the protein-assisted diffusion of a molecule or ion across a biological membrane. The lipid-soluble proteins that mediate the transport of molecules across bilayers are called *membrane transport proteins.*

Membrane transport proteins are integral membrane proteins—they span both leaflets of the lipid bilayer. In some cases, these proteins are *pores,* meaning they form water-filled holes in the membrane through which appropriately sized molecules and ions can pass in either direction (Figure 9·29). In other cases, these proteins are *gates,* meaning they bind a specific molecule or ion on one face of the bilayer and then undergo a conformational change that allows them to release the molecule on the other face of the bilayer (Figure 9·30).

One example of a substance that crosses a membrane by facilitated diffusion is D-glucose, which is transported into human erythrocytes through a gate mechanism. The glucose transport protein is a glycoprotein (MW 55 000) that has two alternate conformations: one conformation exposes a glucose-binding site on the extracellular face of the membrane, and the other exposes a glucose-binding site on the cytoplasmic face of the membrane. Once a molecule of glucose has bound to the binding site on the extracellular face, a change in conformation of the transport protein moves the glucose molecule across the lipid bilayer. Glucose then dissociates from the protein, moving into the cytoplasm and exposing the second binding site. If no molecule of glucose binds the newly exposed site, the protein reverts to its original conformation.

A transport protein is usually specific for a certain molecule or group of structurally similar molecules. However, sometimes a single protein is able to transport two different molecular species across a membrane. An example of such a protein is band 3 protein in human erythrocytes. Band 3 protein transports bicarbonate anion ($HCO_3^{\ominus}$) out of the cell and chloride anion ($Cl^{\ominus}$) into the cell. In both cases, the molecules are being transported down their concentration gradients—$HCO_3^{\ominus}$ from a high interior concentration to a low exterior one and $Cl^{\ominus}$ from a high exterior concentration to a low interior one. The transport of two different molecular species in opposite directions across a membrane is called *antiport*. The transport of two different molecular species in the same direction across a membrane is called *symport* (Figure 9·31).

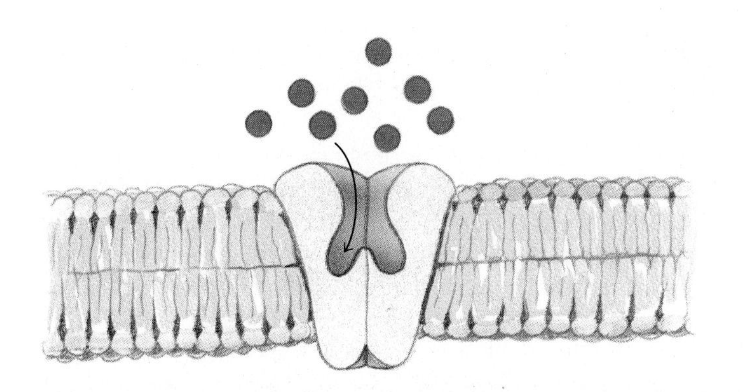

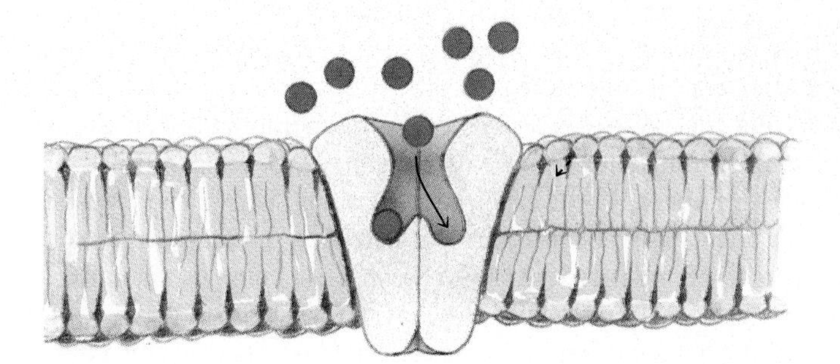

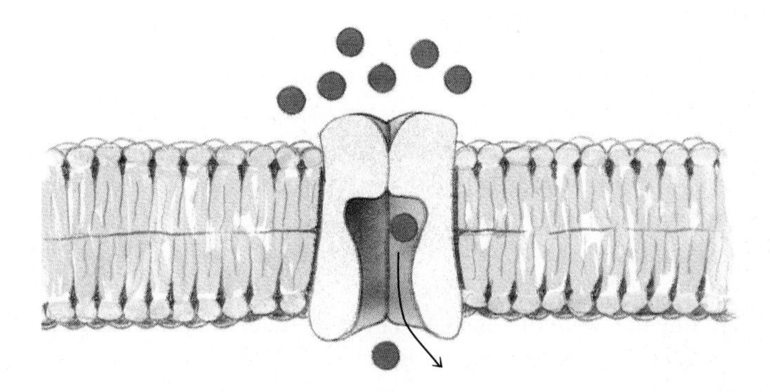

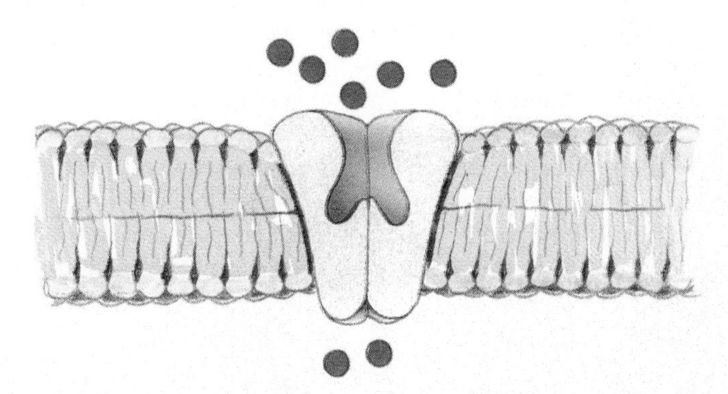

Figure 9·30
Membrane transport through gates. Gates bind to specific molecules or ions and then undergo a conformational change that allows the molecules or ions to be released on the other side of the membrane. The net movement of molecules or ions in either case is down their concentration gradients. (Illustrator: Lisa Shoemaker.)

(a)

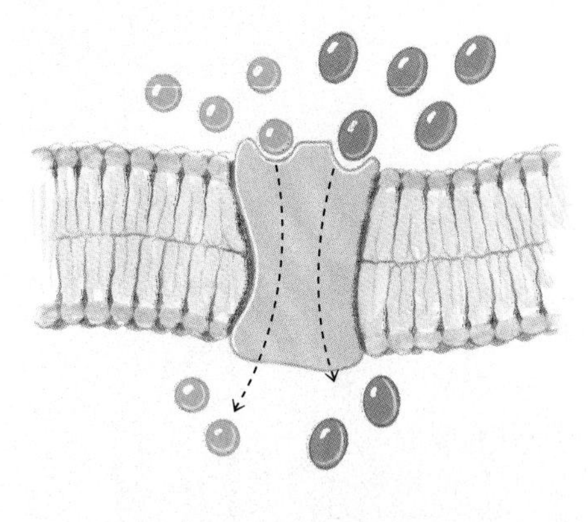

(b)

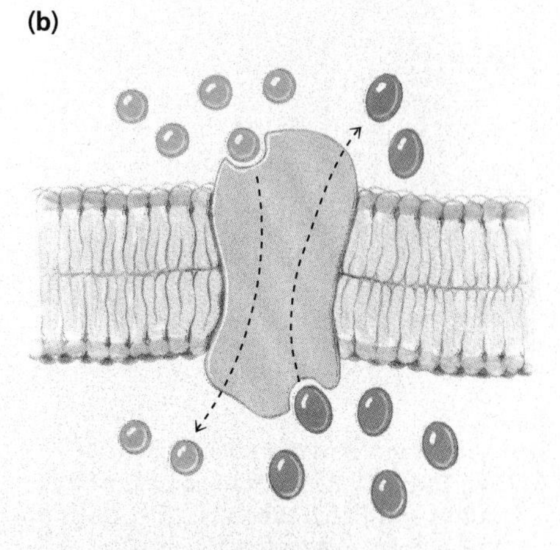

Figure 9·31
Two ways a transport protein can transport two different molecular species: **(a)** Symport. **(b)** Antiport. (Illustrator: Lisa Shoemaker.)

B. Active Transport Involves Transport of a Molecule Up Its Concentration Gradient

Not all molecules that move across a membrane are travelling down their concentration gradients. The net movement of a molecule *up* or *against* its concentration gradient is called *active transport.* Active transport, like facilitated diffusion, is always mediated by membrane transport proteins. Unlike facilitated diffusion, however, the proteins involved in active transport cannot be pores. Instead, they are always gates, often called *pumps*. Furthermore, entropy decreases during active transport (unlike facilitated diffusion), and a source of energy is needed to drive the transport process. There are two types of active transport, primary and secondary, categorized by the source of free energy used to drive the process.

Primary active transport is active transport fueled by primary sources of energy, such as light, hydrolysis of ATP, or electron transport (see Box 9·3 for a discussion of group translocation, a special type of primary active transport). One of the best-understood examples of primary active transport is the antiport of $Na^{\oplus}$ and $K^{\oplus}$ by the $Na^{\oplus}$-$K^{\oplus}$ ATPase. The plasma membranes of most mammalian cells maintain an intracellular concentration of $K^{\oplus}$ of about 140 mM in the presence of an extracellular $K^{\oplus}$ concentration of about 5 mM. The cytosolic concentration of $Na^{\oplus}$ is maintained at about 5 to 15 mM in the presence of an extracellular concentration of about 145 mM. The pump that maintains this imbalance is the $Na^{\oplus}$-$K^{\oplus}$ ATPase, an ATP-driven antiport system that pumps two $K^{\oplus}$ into the cell and ejects three $Na^{\oplus}$ for every molecule of ATP that is hydrolyzed. Each $Na^{\oplus}$-$K^{\oplus}$ ATPase hydrolyzes about 100 molecules of ATP per minute under optimal cellular conditions, which represents a significant portion of the total energy consumption of a typical cell.

The $Na^{\oplus}$-$K^{\oplus}$ ATPase is an integral membrane protein composed of two types of subunits (Figure 9·32). The stoichiometry of the protein is $\alpha_2\beta_2$. The α subunits (subunit MW 110 000) catalyze hydrolysis of ATP and have binding sites for $Na^{\oplus}$ on the cytosolic side and $K^{\oplus}$ on the extracellular side of the plasma membrane. The β subunits (subunit MW 55 000) are glycoproteins whose function in the complex is still unknown.

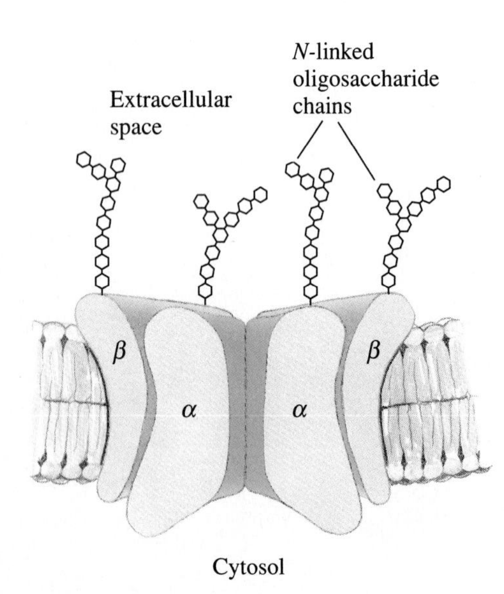

Figure 9·32
Structure and orientation of the $Na^{\oplus}$-$K^{\oplus}$ ATPase. The $Na^{\oplus}$-$K^{\oplus}$ ATPase consists of two α subunits and two β subunits. The catalytic α subunits carry out hydrolysis of ATP and transport $Na^{\oplus}$ and $K^{\oplus}$ across the cell membrane. The smaller β subunits are glycoproteins whose function is unknown. (Illustrator: Lisa Shoemaker.)

The mechanism of the $Na^{\oplus}$-$K^{\oplus}$ ATPase is outlined in Figure 9·33. Three $Na^{\oplus}$ bind to the $Na^{\oplus}$-$K^{\oplus}$ ATPase on the cytosolic side of the membrane. Phosphoryl-group transfer from the bound ATP molecule to an aspartate residue of the ATPase in the cytosol generates an energy-rich acyl-phosphate bond. A conformational change occurs, and the three $Na^{\oplus}$ are moved across the membrane and released outside the cell. Two $K^{\oplus}$ then bind to the $Na^{\oplus}$-$K^{\oplus}$ ATPase on the outside of the cell. Another conformational change occurs as the acyl-phosphate bond is hydrolyzed, and the two $K^{\oplus}$ are brought across the membrane and released inside the cell.

Secondary active transport is transport driven by ion gradients. The flow of ions down their concentration gradients provides a source of free energy that can be used to pull molecules into the cell up their concentration gradients. For example, the transport of lactose into *E. coli* cells is coupled to the flow of $H^{\oplus}$ into the cell. This transport is mediated by a protein called lactose permease, a transmembrane protein (MW 46 504) that carries out 1:1 symport of $H^{\oplus}$ and lactose. The influx of $H^{\oplus}$ down its concentration gradient pulls lactose into the cell up its concentration gradient.

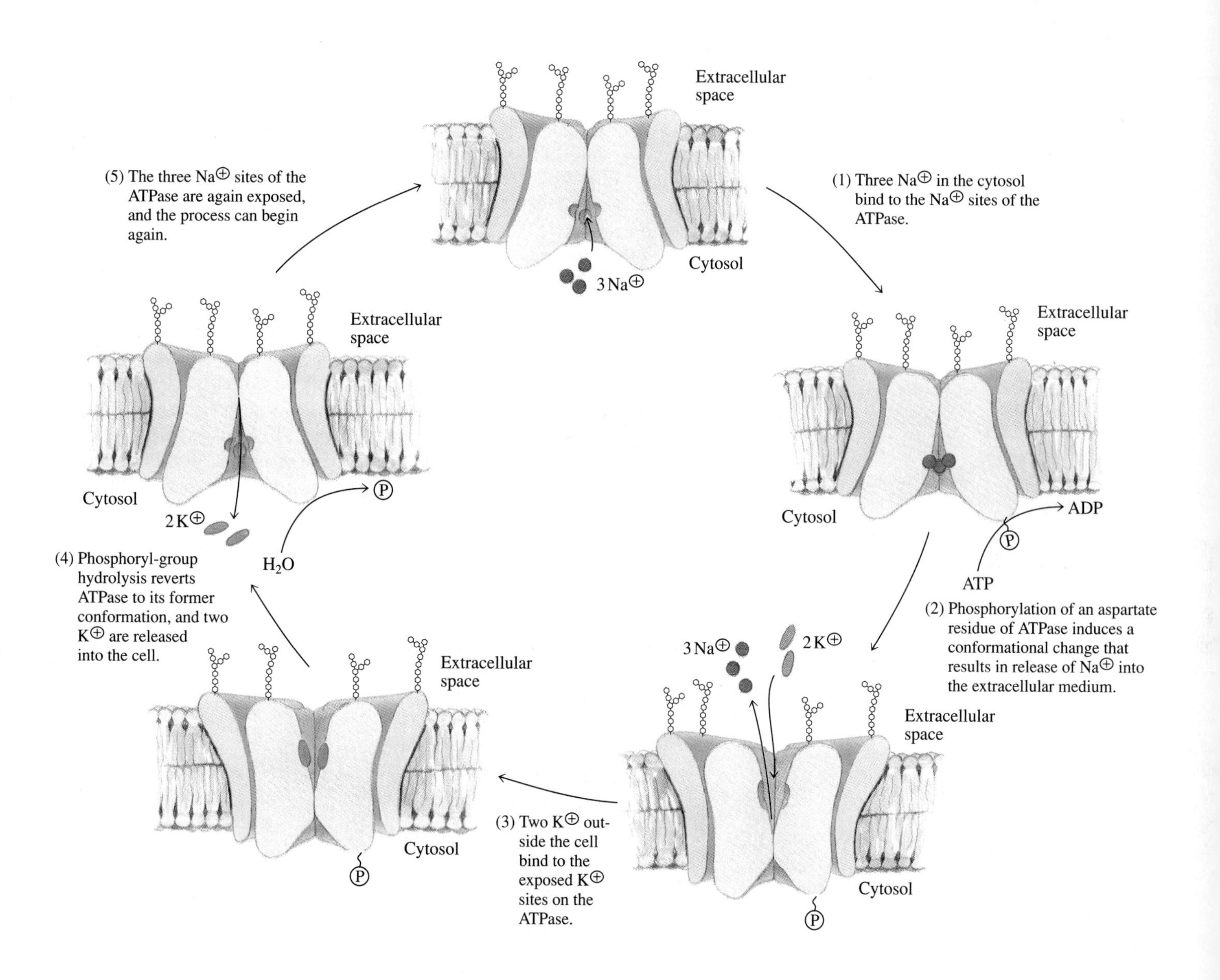

Figure 9·33
Mechanism of the Na⊕-K⊕ ATPase. The Na⊕-K⊕ ATPase expels three Na⊕ for every two K⊕ that it imports into the cell. The movement of both ions is up their concentration gradients. Energy to drive the process is supplied by the hydrolysis of ATP. (Illustrator: Lisa Shoemaker.)

Note that although we refer to the secondary active transport of lactose into *E. coli* cells as an example of proton (H⊕) symport, we could equally well call it an example of hydroxide ion (OH⊖) antiport. Because the *E. coli* cell and the surrounding medium are both aqueous phases, and because water rapidly equilibrates with its ion products, it is difficult to distinguish experimentally between symport and antiport when the ion products of water are involved. Throughout this book, we refer to transport of H⊕ in cases where the net effect is the movement of H⊕ in one direction. Keep in mind, however, that either H⊕ or OH⊖ could be the species being transported in these cases. Transport of H⊕ in one direction across a membrane is experimentally indistinguishable from the transport of OH⊖ in the other direction.

Box 9·3 Group Translocation Is a Special Type of Primary Active Transport

Group translocation is a process that couples transport across the membrane to the chemical modification of the transported species. An example can be found in the PEP-dependent sugar phosphotransferase system of *E. coli*. This system transports at least 10 different sugar species across the plasma membrane of *E. coli*.

The PEP-dependent sugar phosphotransferase system uses phosphoenolpyruvate (PEP) rather than ATP as the energy-rich phosphoryl-group donor. The translocation process involves three or four proteins, depending on the sugar being transported. These proteins are enzyme I, enzyme II, enzyme III, and HPr. Enzyme I and HPr are both soluble proteins found in the cytosol. Enzyme III interacts with enzyme II, a transmembrane protein.

The transport of glucose into an *E. coli* cell by the phosphotransferase system involves at least five steps (Figure 1). First, PEP donates its phosphoryl group to enzyme I. The phosphorylated enzyme in turn transfers this phosphoryl group to HPr. HPr then donates the phosphoryl group to enzyme III^{glc}, which is complexed to enzyme II^{glc}. Enzyme III^{glc} transfers the phosphoryl group to glucose as enzyme II^{glc} simultaneously translocates one glucose residue from the extracellular space to the cytosol. This transport represents a different type of active transport, one in which the translocated species is chemically modified during the transport process.

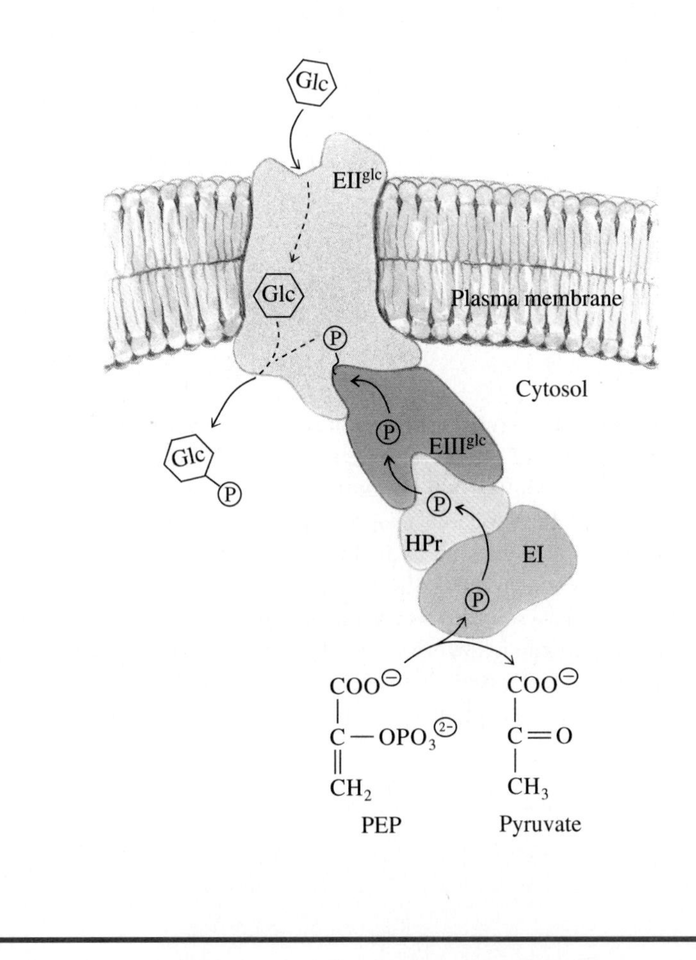

Box 9·3
Figure 1
The PEP:glucose phosphotransferase system of *E. coli*. During this type of transport, a sugar residue is covalently modified as it is translocated across the membrane. In this case, glucose is converted to glucose 6-phosphate. The phosphate group (PO_3^{2-}, designated Ⓟ) is ultimately derived from PEP. (Illustrator: Lisa Shoemaker.)

C. Endocytosis and Exocytosis Involve the Formation of Lipid Vesicles

The transport we have discussed so far involves the passage of molecules and/or ions across an intact membrane. But in some cases, material is too large or reactive to be transported directly across the membrane. In these cases, cells employ mechanisms such as *endocytosis* or *exocytosis* to move matter into or out of a cell, respectively. In both cases, transport involves creation of a lipid vesicle.

Endocytosis is the process by which matter is engulfed by a plasma membrane and brought into the cell within a lipid vesicle. As shown in Figure 9·34, the inside of the vesicle so formed is topologically equivalent to the outside of a cell. Thus, matter inside the lipid vesicle has not actually passed across a membrane. However, destruction of the lipid vesicle inside the cell allows release of the contents into the cytosol.

Figure 9·34
Endocytosis of coated vesicles. (**a**) Endocytosis begins with the binding of macromolecules to the plasma membrane of the cell. The membrane then invaginates, forming a coated vesicle that contains the molecules of interest. The inside of the vesicle is topologically equivalent to the outside of the cell. (**b**) Electron micrographs showing formation of a coated vesicle. (Courtesy of Margaret M. Perry.)

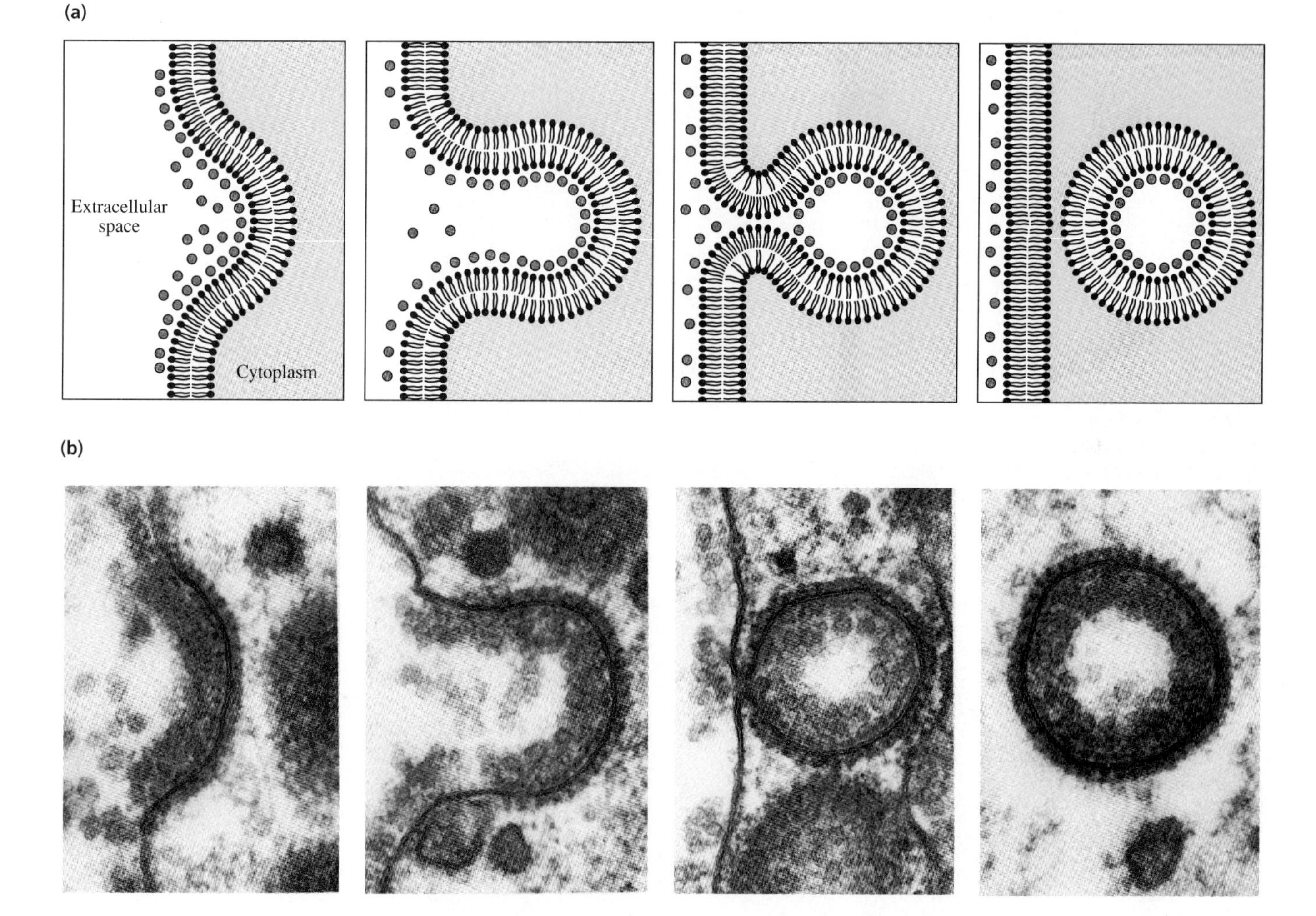

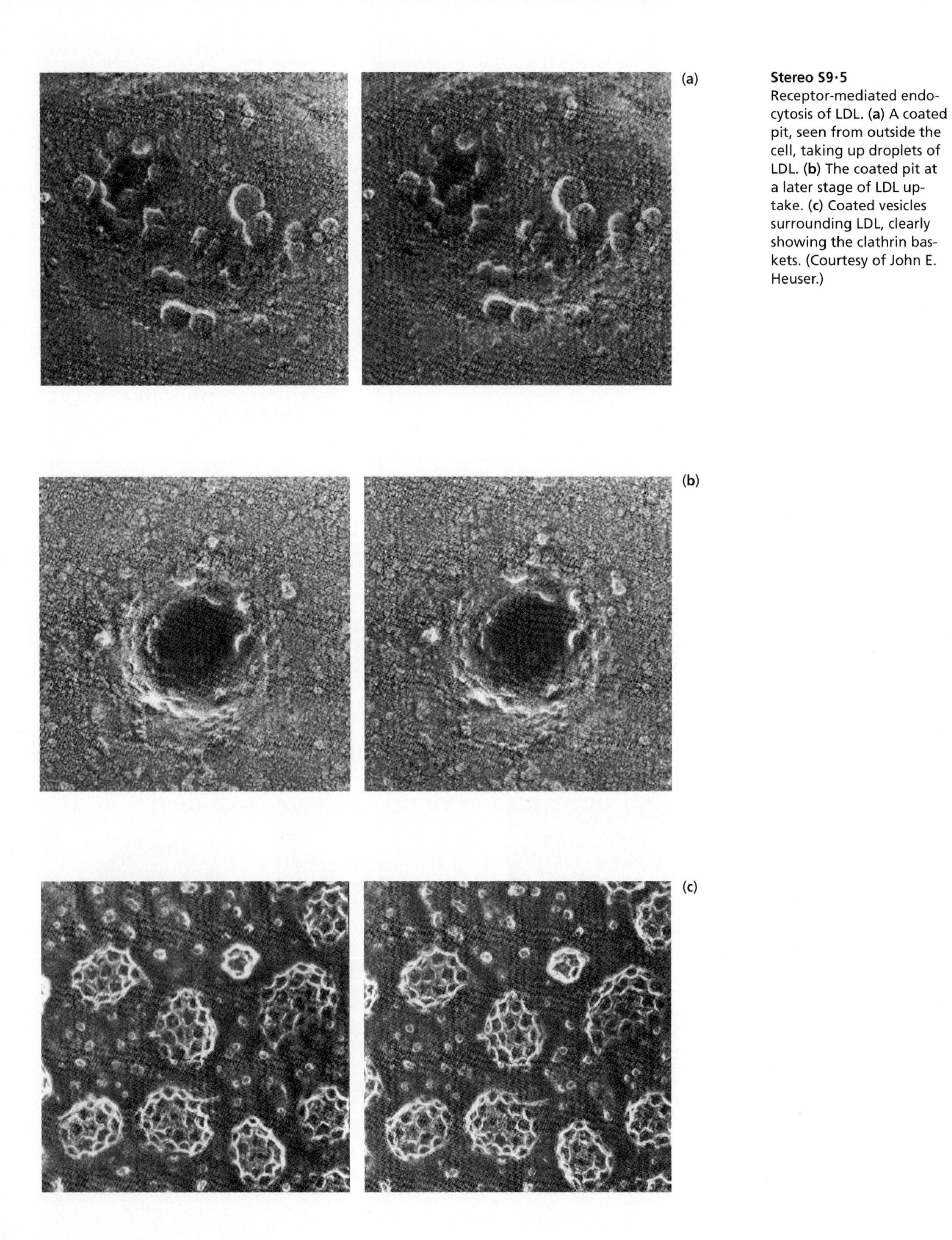

Stereo S9·5
Receptor-mediated endocytosis of LDL. (**a**) A coated pit, seen from outside the cell, taking up droplets of LDL. (**b**) The coated pit at a later stage of LDL uptake. (**c**) Coated vesicles surrounding LDL, clearly showing the clathrin baskets. (Courtesy of John E. Heuser.)

One special type of endocytosis involves the formation of coated vesicles. As noted in Section 9·8, lipids are transported to nonhepatic tissues packaged as soluble lipoproteins. Nonhepatic tissues obtain cholesterol for use in membrane synthesis from the plasma LDLs, which bind through their apoproteins to specific LDL receptors found in coated pits in the plasma membranes of such cells (Stereo S9·5). Once an LDL particle has bound to an LDL receptor, it undergoes endocytosis, becoming engulfed by the cell and entering it inside a coated vesicle.

The major protein of coated pits and coated vesicles is known as *clathrin,* a name reflecting its latticelike, or clathrate, structure. Clathrin-coated pits facilitate the endocytosis of receptor-bound LDL molecules. Once inside the cell, the LDL molecules travel from coated vesicles to endosomes (membrane-bound vesicles that lack clathrin) and finally to lysosomes. Once inside a lysosome, the apoprotein component of the LDL is degraded to its constituent amino acids while the cholesteryl esters are hydrolyzed, releasing free cholesterol that becomes incorporated into cellular membranes (Box 9·4).

Exocytosis is a process similar to endocytosis, except that the direction of transport is reversed. During exocytosis, material being secreted from the cell is enclosed in vesicles by the Golgi apparatus. The vesicles transport the material to the plasma membrane, with which they then fuse. The material inside the vesicles is thus released into the extracellular medium as free molecules. Zymogens released during digestion are exported from pancreatic cells in this manner.

Box 9·4 Atherosclerosis and LDL Receptors

Whereas some cholesterol is essential to the viability of animal cells, too high a concentration of LDL-cholesterol in the plasma can lead to formation of arterial plaques containing highly insoluble cholesteryl esters. The resulting condition is known as *atherosclerosis,* the degeneration and constriction of blood vessels by those cholesteryl ester deposits. Atherosclerosis is a major factor in human cardiovascular disease, especially heart attacks and strokes.

Starting in 1972, Michael S. Brown and Joseph L. Goldstein (jointly awarded a Nobel prize in 1985) undertook a study of *familial hypercholesterolemia,* an inherited disorder characterized by a greatly elevated plasma LDL-cholesterol level, which usually leads to heart disease and death before the age of twenty. In the course of their research, Brown and Goldstein discovered that the disease stems from the overproduction of endogenous cholesterol as a result of a deficiency in functional LDL receptors in the plasma membranes of the patients' cells. Such deficiency prevents normal delivery of cholesterol to the cells. This, in turn, precludes normal feedback regulation through inhibition of the intracellular enzyme that initiates cholesterol biosynthesis, called HMG-CoA reductase (Chapter 17). While the incidence of the homozygous form of familial hypercholesterolemia is rare (approximately 1 in 1 million), the more common heterozygous form of the disease (affecting about 1 in 500) involves partial deficiency of LDL receptors, leading to somewhat elevated plasma cholesterol and LDL levels and, hence, to increased risks of heart disease.

Summary

Lipids are water-insoluble organic compounds that can be extracted from biological cells or tissues using relatively nonpolar organic solvents. Lipids are quite diverse, both structurally and functionally. A number of amphipathic lipids are important components of biological membranes.

Fatty acids are relatively long-chained monocarboxylic acids. The majority of naturally occurring fatty acids contain an even number of carbon atoms, ranging from 12 to 20. Fatty acids that contain no carbon-carbon double bonds are classified as saturated fatty acids, those that contain one carbon-carbon double bond are classified as monounsaturated, and those that contain more than one carbon-carbon double bond are classified as polyunsaturated. Most of the double bonds found in unsaturated fatty acids have the *cis* configuration. As esters, saturated and unsaturated fatty acids are constituents of a wide variety of lipids.

Fatty acids are generally stored as complex lipids called triacylglycerols (fats and oils). Triacylglycerols are neutral and nonpolar. Waxes, also neutral, nonpolar lipids, are esters of long-chain aliphatic alcohols and fatty acids. Eicosanoids are physiologically important derivatives of the fatty acid arachidonate.

Glycerophospholipids are among the major amphipathic lipid components of biological membranes. They are esters of phosphatidates, such as phosphatidylethanolamine, phosphatidylserine, and phosphatidylcholine. Their polar heads include an anionic phosphodiester group as well as an ionic constituent, whereas their nonpolar tails are made up of fatty acyl residues esterified to C-1 and C-2 of the glycerol moiety. Plasmalogens are glycerophospholipids in which the C-1 oxygen of the glycerol 3-phosphate moiety is bound to a hydrocarbon chain as a vinyl ether.

Other major classes of lipids include the sphingolipids, steroids, and fat-soluble or lipid vitamins. The long-chain amino alcohol sphingosine provides the backbone for sphingolipids. Three major classes of sphingolipids are sphingomyelins, cerebrosides, and gangliosides. Steroids and the lipid vitamins are examples of polyprenyl compounds or isoprenoids, lipids synthesized from a five-carbon compound related to isoprene. Cholesterol, a steroid, is an important component of animal membranes and serves as the precursor to a variety of hormones. Other steroids are found in plants and other eukaryotes.

Nonpolar lipids do not form stable micelles and are not found in membranes. In contrast, amphipathic lipids, such as glycerophospholipids and sphingolipids, tend to assemble into bilayers spontaneously. At low temperatures, a lipid bilayer exists in a gel state, but at higher temperatures it undergoes a phase transition and becomes fluid. In general, saturated fatty acyl tails pack more densely than their unsaturated counterparts since *cis* double bonds create kinks in the hydrocarbon chains of unsaturated fatty acids; consequently, the phase-transition temperature for a lipid bilayer decreases as the concentration of unsaturated fatty acids in the membrane increases. The presence of cholesterol in animal cell membranes modulates their fluidity by increasing van der Waals interactions with unsaturated tails while decreasing van der Waals contacts among saturated chains. Above their phase-transition temperatures, lipid bilayers are essentially two-dimensional fluids. Amphipathic lipid molecules readily diffuse laterally within each leaflet but rarely cross from one leaflet to the other. The lipid composition of biological membranes varies considerably, and specific lipids are asymmetrically distributed within membranes.

Biological membranes are a fluid mosaic of proteins in a lipid bilayer. Integral membrane proteins traverse the lipid bilayer, whereas peripheral membrane proteins are less tightly bound to one of the membrane's surfaces. Lipid-linked proteins are anchored to the membrane through glycosyl-phosphatidylinositol linkages. Although some membrane proteins are anchored to the cytoskeleton, most diffuse freely within the plane of the bilayer. In contrast with their lateral diffusion, however, membrane proteins do not diffuse from one leaflet of the lipid bilayer to the other.

The thin layer of hydrocarbon within a lipid bilayer acts as a selectively permeable barrier, generally allowing only water and hydrophobic molecules to diffuse across it while serving as an almost impenetrable barrier to large or hydrophilic species. Such species require specific mechanisms for transport into or out of cells. These mechanisms of transport include facilitated diffusion, active transport, endocytosis, and exocytosis.

Facilitated diffusion involves the passage of molecules or ions down their concentration gradients and requires the presence of integral membrane proteins. These proteins may be pores or gates. Active transport involves the passage of molecules or ions up their concentration gradients and requires the presence of specific gated transport proteins. In primary active transport, the energy required to push a molecule up its concentration gradient comes from light, hydrolysis of ATP, or electron transport. In secondary active transport, the energy required to push a molecule up its concentration gradient comes from the passage of other molecules or ions down their concentration gradients. The passage of two molecules or ions in the same direction across a membrane is called symport. The passage of two molecules or ions in opposite directions across a membrane is called antiport. Matter that is too large or reactive to be transported directly across a membrane can be moved into or out of a cell by the mechanisms of endocytosis or exocytosis, respectively. In both cases, transport involves creation of a lipid vesicle.

Selected Readings

Aloia, R. C. (1983). *Membrane Fluidity in Biology* (New York: Academic Press).

Bretscher, M. S. (1985). The molecules of the cell membrane. *Sci. Am.* 253(4):100–108. A clear introduction to the structure and function of biological membranes.

Brown, M. S., and Goldstein, J. L. (1984). How LDL receptors influence cholesterol and atherosclerosis. *Sci. Am.* 251(5):58–66.

Cantor, C. R., and Schimmel, P. R. (1980). *Biophysical Chemistry* (San Francisco: W. H. Freeman and Company). Chapter 25 (in Part III) provides a physicochemical discussion of the structures and properties of lipid bilayers and membranes.

Cross, G. A. M. (1990). Glycolipid anchoring of plasma membrane proteins. *Annu. Rev. Cell Biol.* 6:1–39.

Frye, L. D., and Edidin, M. (1970). The rapid intermixing of cell surface antigens after formation of mouse-human heterokaryons. *J. Cell Sci.* 7:319–335.

Gennis, R. B. (1989). *Biomembranes* (New York: Springer-Verlag).

Harrison, R., and Lunt, G. G. (1980). *Biological Membranes: Their Structure and Function,* 2nd ed. (New York: John Wiley & Sons).

Jain, M. K. (1988). *Introduction to Biological Membranes* (New York: John Wiley & Sons).

Marchesi, V. T. (1985). Stabilizing infrastructure of cell membranes. *Annu. Rev. Cell Biol.* 1:531 561. A review of components of the cytoskeleton.

Rothman, J. E., and Lenard, J. (1977). Membrane asymmetry. *Science* 195:743–753. A description of the asymmetric distribution of lipids, carbohydrates, and proteins in biological membranes.

Singer, S. J. (1990). The structure and insertion of integral proteins in membranes. *Annu. Rev. Cell Biol.* 6:247–296.

Singer, S. J., and Nicolson, G. L. (1972). The fluid mosaic model of the structure of cell membranes. *Science* 175:720–731. This paper discusses the properties of membranes that led the authors to propose their model.

Starzak, M. E. (1984). *The Physical Chemistry of Membranes* (Orlando, Florida: Academic Press).

Storch, J., and Kleinfeld, A. M. (1985). The lipid structure of biological membranes. *Trends Biochem. Sci.* 10:418–421. A review of the structures of biological membranes and of areas of research interest and debate.

Tanford, C. (1980). *The Hydrophobic Effect: Formation of Micelles and Biological Membranes,* 2nd ed. (New York: John Wiley & Sons), pp. 181–218. A discussion of the role of the hydrophobic effect in the formation of micelles and lipid bilayers. Also included are serum lipoproteins and the interactions of membrane proteins and membrane lipids.

Vance, D. E., and Vance, J. E., eds. (1991). *Biochemistry of Lipids, Lipoproteins, and Membranes* (Amsterdam: Elsevier Science Publishing Company).

10

Nucleic Acids

Living organisms contain a set of instructions that specifies every step required for the organism to construct a replica of itself. The required information resides in the genetic material, or *genome,* of an organism. The genome is composed of deoxyribonucleic acid (DNA) in living organisms; however, some viral genomes are composed of RNA.

In biological systems, the information that specifies the primary structure of a protein is encoded in DNA. This information is enzymatically copied during the synthesis of ribonucleic acid (RNA), a process known as *transcription.* Some of the information contained in the transcribed RNA molecules is translated through the synthesis of polypeptide chains, which are subsequently folded and assembled to form protein molecules. Thus, we can generalize that the biological information stored in a cell's DNA flows from DNA to RNA to protein. We shall consider the biochemistry of this flow of information in Chapters 20–22; here, we examine the structure of nucleic acids.

The discovery of the substance that proved to be DNA was made in 1869 by Friedrich Miescher, a young Swiss physician working in the laboratory of the German physiological chemist Felix Hoppe-Seyler. Miescher treated white blood cells (which came from the pus on discarded surgical bandages) with hydrochloric acid to obtain nuclei for study. When the nuclei were subsequently treated with acid, a precipitate formed that proved to contain carbon, hydrogen, oxygen, nitrogen, and a high percentage of phosphorus. Because of its occurrence in nuclei, Miescher called the precipitate "nuclein"; later, when it was found to be strongly acidic, its name was changed to *nucleic acid*. Shortly after Miescher's discovery of DNA, Hoppe-Seyler isolated a similar substance from yeast cells; this substance is now known to be RNA.

Nucleic acids represent the fourth major class of biomolecules. Like proteins and polysaccharides, they are polymers composed of similar monomeric units that are covalently joined to produce large macromolecules.

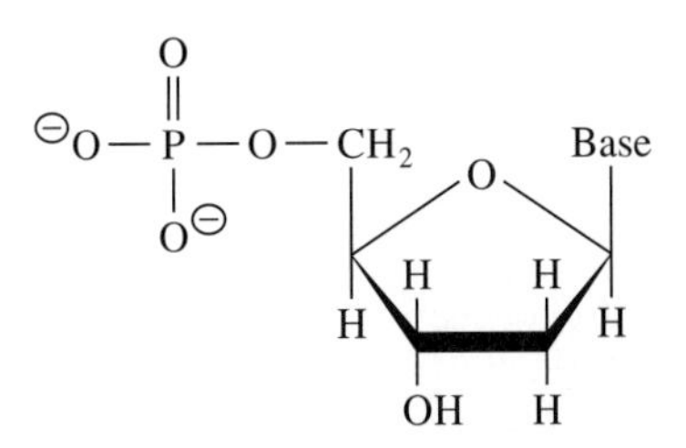

Figure 10·1
Structure of a nucleotide. The nucleotides found in nucleic acids are composed of a five-carbon sugar, a nitrogenous base, and a phosphoryl group. The sugar can be deoxyribose, as shown here, or ribose.

10·1 Nucleotides Are the Building Blocks of Nucleic Acids

Nucleic acids consist of polymerized nucleotide residues. *Nucleotides* are composed of a sugar, a weak base, and at least one phosphoryl group (Figure 10·1). Such a structure without a phosphoryl group is called a *nucleoside.* Thus ATP and

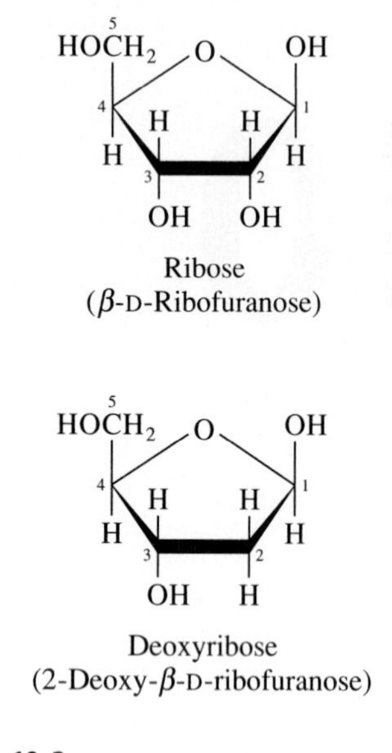

Figure 10·2
Structures of the two sugars found in nucleotides.

AMP are nucleotides, whereas the unphosphorylated form, adenosine, is a nucleoside. Like adenine in ATP, the other bases found in nucleotides are also composed of heterocyclic rings containing nitrogen. There are two general classes of nucleotides: *ribonucleotides,* in which the sugar moiety is ribose, and *deoxyribonucleotides,* in which the sugar moiety is 2-deoxyribose. The structures of these two sugars are shown in Figure 10·2.

A. There Are Two Classes of Bases in Nucleotides

The bases found in nucleotides are substituted *pyrimidines* and *purines*. The structures of pyrimidine and purine and systems for numbering the carbon and nitrogen atoms of each are shown in Figure 10·3. Pyrimidine is a heterocyclic compound that contains four carbons and two nitrogen atoms. Purine is a bicyclic structure consisting of pyrimidine fused to an imidazole ring. All of the carbon atoms of these cyclic molecules are sp^2 hybridized (that is, none are saturated); thus pyrimidine and purine are planar.

Unsubstituted purine and pyrimidine are not found in biological systems, but a number of substituted derivatives are. These derivatives are classified as pyrimidines or purines, depending on the parent molecule from which they are derived. The major pyrimidines found in nucleotides are uracil (U), thymine (T), and cytosine (C); the major purines are adenine (A) and guanine (G). The structures of these five bases are shown in Figure 10·4. Other, less common pyrimidines and purines also occur naturally (Section 21·6).

Adenine, guanine, and cytosine are found in both ribonucleotides and deoxyribonucleotides. In contrast, uracil is found mainly in ribonucleotides, and thymine in deoxyribonucleotides. Note that thymine is a substituted form of uracil; thus, it can also be called 5-methyluracil.

Each commonly occurring pyrimidine and purine can be drawn in two tautomeric forms. Adenine and cytosine can exist as either amino or imino forms, and

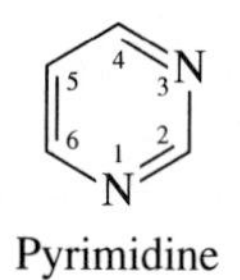

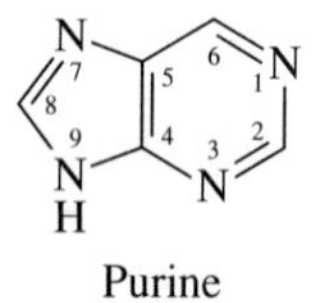

Figure 10·3
Structures of pyrimidine and purine. Although the structures of pyrimidines and purines are usually drawn with C-5 of the pyrimidines and C-8 of the purines to the right, the opposite orientation is shown here since it is the more common orientation in nucleotide structures.

PYRIMIDINES

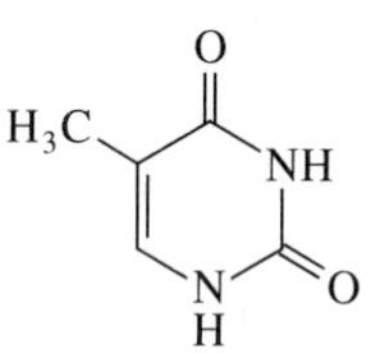

Uracil
(2,4-Dioxopyrimidine)

H_3C O NH N H O

Thymine
(5-Methyluracil)

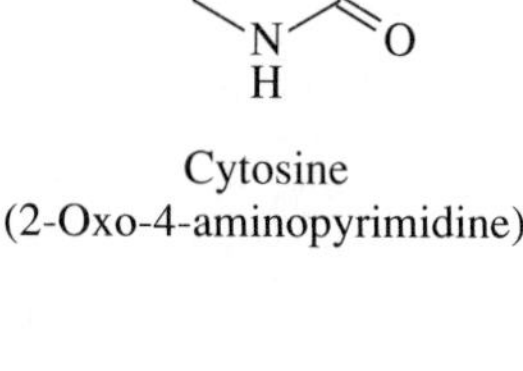

Cytosine
(2-Oxo-4-aminopyrimidine)

PURINES

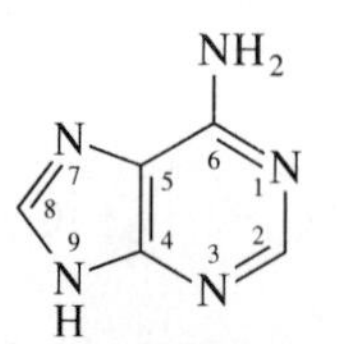

Adenine
(6-Aminopurine)

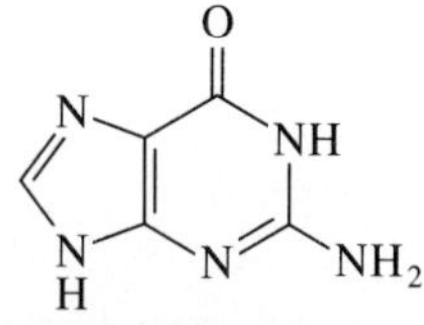

Guanine
(2-Amino-6-oxopurine)

Figure 10·4
Structures of the major pyrimidines, uracil, thymine, and cytosine, and of the major purines, adenine and guanine.

guanine, thymine, and uracil can exist as either lactam or lactim forms (Figure 10·5). The two forms for each base exist in equilibrium, but under the conditions found inside most cells, the amino and lactam tautomers are more stable and therefore predominate. Note that the ring carbon atoms retain their unsaturated sp^2 character in each tautomeric form.

The amino tautomers of adenine and cytosine can participate in the formation of hydrogen bonds; the amino groups serve as hydrogen donors, and the adjacent ring nitrogens (N-1 in adenine and N-3 in cytosine) as hydrogen acceptors. Cytosine also has a hydrogen-acceptor group at C-2. The lactam tautomers of guanine, thymine, and uracil can participate in the formation of three hydrogen bonds. In the case of guanine, the group at C-6 serves as a hydrogen acceptor; N-1 and the amino group at C-2 serve as hydrogen donors. In the cases of thymine and uracil, the groups at C-4 and C-2 serve as hydrogen acceptors, and N-3 serves as a hydrogen donor. As we shall see in the next section, these hydrogen-bonding patterns have important consequences for the three-dimensional structure of nucleic acids.

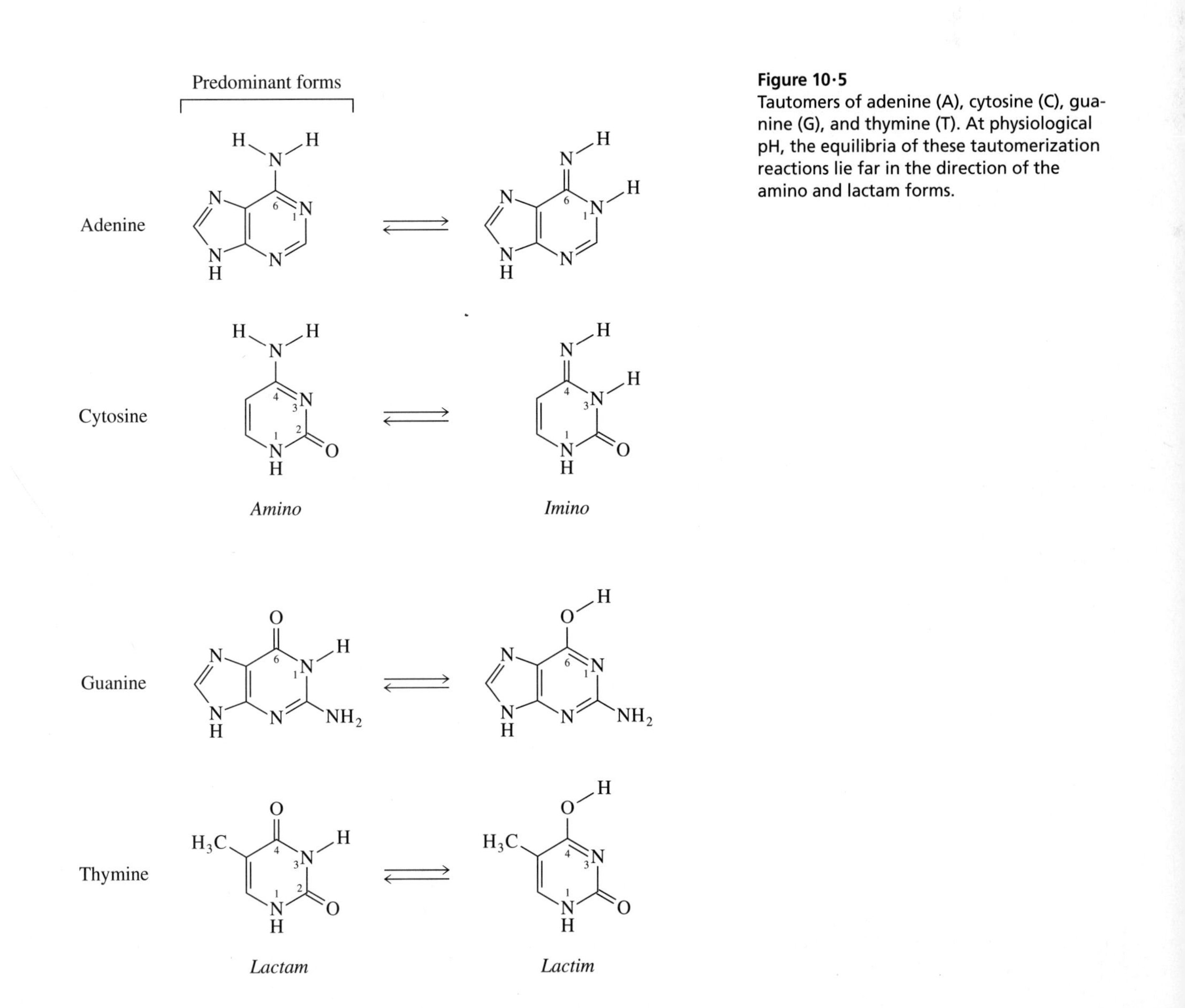

Figure 10·5
Tautomers of adenine (A), cytosine (C), guanine (G), and thymine (T). At physiological pH, the equilibria of these tautomerization reactions lie far in the direction of the amino and lactam forms.

B. Nucleoside Monophosphates Are the Monomers in Polynucleotides

The structures of the major ribonucleosides and deoxyribonucleosides are given in Figure 10·6. In each case, the base is connected to the sugar via a glycosidic linkage of the β configuration. Since this covalent bond involves one of the nitrogens of the heterocyclic base, the bond is referred to as a β-*N*-glycosidic bond. Note that the numbers of the carbon atoms of the ribosyl and deoxyribosyl residues are designated with primes in order to distinguish them from the numbered atoms in the purine and pyrimidine rings. Thus, adenosine 5′-monophosphate is the nucleotide in which a phosphate group is attached to the 5′ carbon atom of the sugar in the nucleoside adenosine.

The names of nucleosides are derived from the names of the bases they contain. For example, the ribonucleoside containing adenine is called adenosine. The other ribonucleosides are guanosine, cytidine, and uridine. The deoxyribonucleosides are usually called deoxyadenosine, deoxyguanosine, and deoxycytidine. Because thymine rarely occurs in ribonucleosides, however, deoxythymidine is also simply called thymidine. Single-letter abbreviations for bases are also commonly used to designate ribonucleosides: A, G, U, and C (for adenosine, guanosine, uridine, and cytidine, respectively). The deoxyribonucleosides are abbreviated dA, dG, dT, and dC.

Figure 10·6
Structures of some important nucleosides. (**a**) Ribonucleosides. The β-*N*-glycosidic bond of adenosine is shown in red. (**b**) Deoxyribonucleosides.

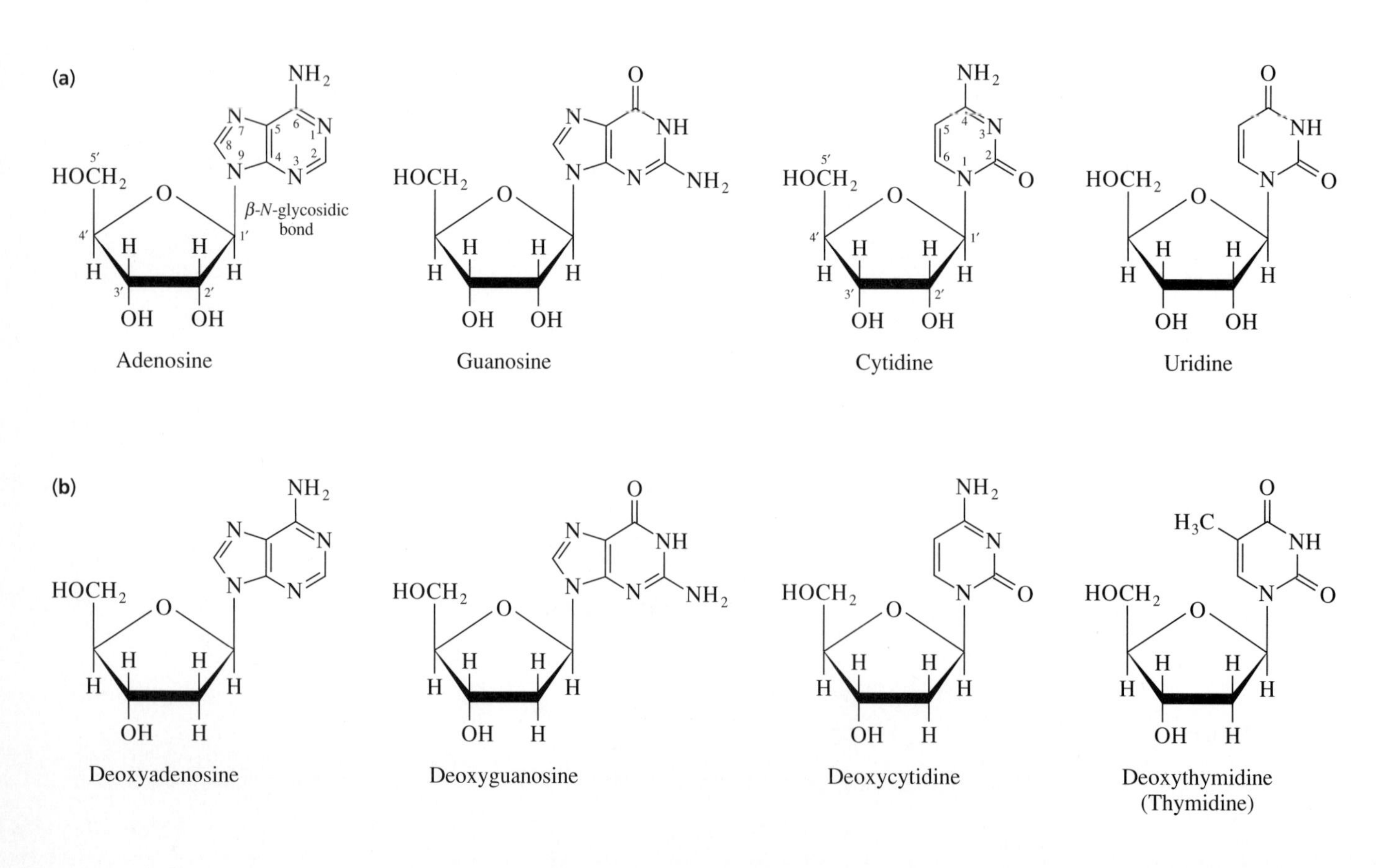

Ribonucleosides contain three hydroxyl groups to which phosphate can be esterified (2′, 3′, and 5′), whereas deoxyribonucleosides contain two such hydroxyl groups (3′ and 5′). In naturally occurring nucleotides, the phosphoryl groups are most commonly attached to the oxygen atom of the 5′-hydroxyl group; thus, a nucleotide is always assumed to be a 5′-phosphate ester unless otherwise designated. Nucleotides are acids and are anionic at physiological pH.

Nucleotides with one 5′-phosphoryl group are named to indicate the single phosphate. For example, the 5′-monophosphate ester of adenosine is called adenosine monophosphate (AMP). It is also simply called adenylate. Similarly, the 5′-monophosphate ester of deoxycytidine can be referred to as deoxycytidine monophosphate (dCMP) or deoxycytidylate. The 5′-monophosphate ester of deoxythymidine is sometimes called deoxythymidine monophosphate or deoxythymidylate instead of thymidine monophosphate or thymidylate since the former terms avoid ambiguity. Note that nucleotides with the phosphate esterified to the 5′ carbon are abbreviated as AMP, dCMP, and so on. Nucleotides with the phosphate esterified to a position other than 5′ are given similar abbreviations, but with position numbers designated (for example, 3′-AMP). Table 10·1 gives an overview of the nomenclature of bases, nucleosides, and nucleotides.

Table 10·1 Nomenclature of bases, nucleosides, and nucleotides

Base	Ribonucleoside	Ribonucleotide (5′-monophosphate)
Adenine (A)	Adenosine	Adenosine 5′-monophosphate (AMP); adenylate*
Guanine (G)	Guanosine	Guanosine 5′-monophosphate (GMP); guanylate*
Cytosine (C)	Cytidine	Cytidine 5′-monophosphate (CMP); cytidylate*
Uracil (U)	Uridine	Uridine 5′-monophosphate (UMP); uridylate*
Base	**Deoxyribonucleoside**	**Deoxyribonucleotide (5′-monophosphate)**
Adenine (A)	Deoxyadenosine	Deoxyadenosine 5′-monophosphate (dAMP); deoxyadenylate*
Guanine (G)	Deoxyguanosine	Deoxyguanosine 5′-monophosphate (dGMP); deoxyguanylate*
Cytosine (C)	Deoxycytidine	Deoxycytidine 5′-monophosphate (dCMP); deoxycytidylate*
Thymine (T)	Deoxythymidine (or thymidine)	Deoxythymidine 5′-monophosphate (dTMP); deoxythymidylate* or thymidylate*

*Anionic forms of phosphate esters predominant at pH 7.0

Mononucleotides can be further phosphorylated to form nucleoside diphosphates and nucleoside triphosphates. These additional phosphoryl groups form phosphoanhydride linkages. The structures of adenosine monophosphate (AMP), adenosine diphosphate (ADP), and adenosine triphosphate (ATP) are compared in Figure 10·7, and two three-dimensional views of ATP are shown in Stereo S10·1. It is the eight nucleoside triphosphates (ATP, GTP, CTP, UTP, dATP, dGTP, dCTP, and dTTP) that serve as substrates for RNA and DNA polymerases, the enzymes that catalyze the synthesis of cellular RNA and DNA molecules, respectively. During the polymerase reaction, inorganic pyrophosphate (PP_i) is cleaved from the triphosphate and a nucleoside monophosphate is added to the growing polynucleotide chain (Chapters 20 and 21). Thus, polynucleotides are composed of nucleoside monophosphate residues.

In addition to serving as building blocks of nucleic acids, nucleoside triphosphates have other important roles. Through the use of the potential energy of its phosphoanhydride linkages, ATP supplies many enzymes with the free energy needed to drive a host of reactions that would not otherwise occur. Other nucleoside triphosphates, such as GTP and CTP, are also important sources of energy in many coupled reactions. The metabolic role of ATP and other nucleoside triphosphates as energy-rich molecules is discussed in detail in the next chapter.

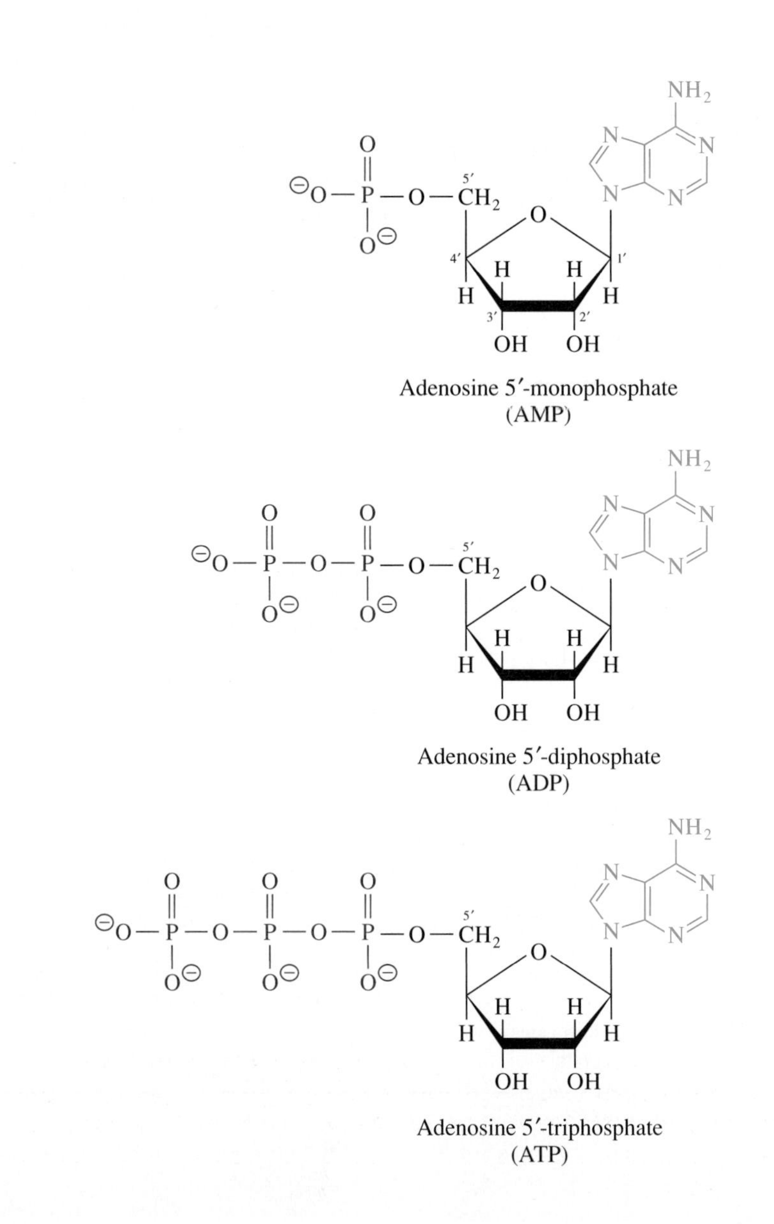

Figure 10·7
Structures of three common adenine ribonucleotides: adenosine 5′-monophosphate (AMP), adenosine 5′-diphosphate (ADP), and adenosine 5′-triphosphate (ATP). Each of the three nucleotides consists of an adenine (blue), a ribose (black), and one or more phosphoryl groups (red).

(a)

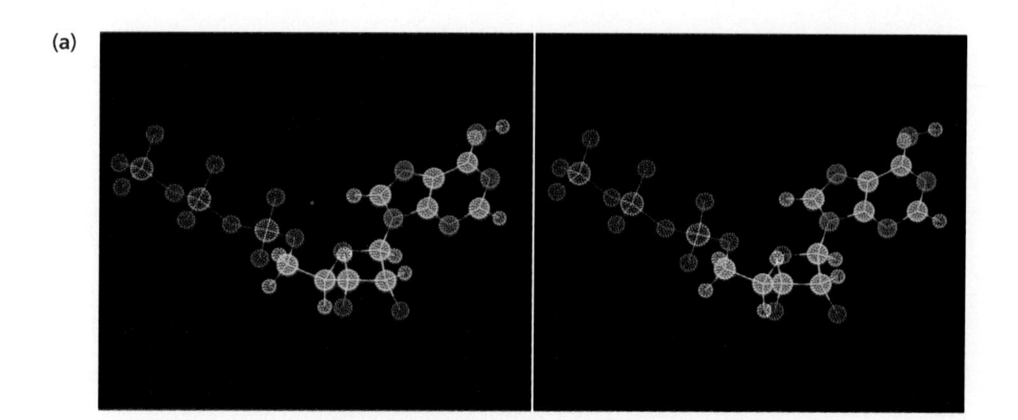

(b)

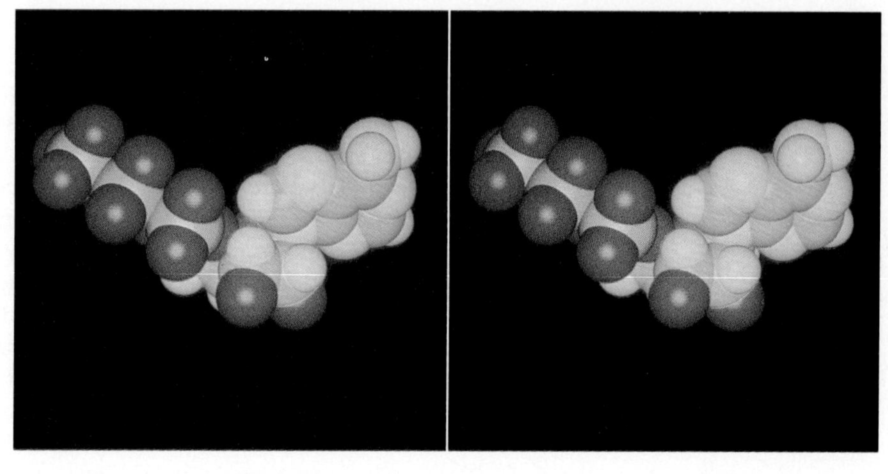

Stereo S10·1
Adenosine 5′-triphosphate (ATP). (**a**) Ball-and-stick model. Color key: carbon, green; hydrogen, light blue; nitrogen, dark blue; oxygen, red; phosphorus, orange. Some of the hydrogen atoms have been omitted for clarity.
(**b**) Space-filling model. Color key: carbon, green; hydrogen and nitrogen, blue; oxygen, red; phosphorus, orange. (Based on coordinates provided by P. Evans and P. Hudson; stereo images by Ben N. Banks and Kim M. Gernert.)

10·2 DNA Consists of Two Strands of Linear Polymers of Nucleotides

In 1944, Oswald T. Avery, Colin MacLeod, and Maclyn McCarty demonstrated that DNA was the molecule that carried genetic information. However, very little was known about the structure of this important molecule.

By 1950, it was clear that DNA is a linear polymer of 2′-deoxyribonucleotide residues linked by 3′–5′ phosphodiesters. Moreover, through analysis of hydrolysates, Erwin Chargaff had deduced certain regularities in the nucleotide compositions of DNA samples obtained from a wide variety of prokaryotes and eukaryotes. Chargaff noted that within the DNA of a given cell, dA and dT are present in equimolar amounts, as are dG and dC (Table 10·2, next page). Thus, the ratio of purines to pyrimidines is 1:1, even though the total mole percent of (G + C) may differ considerably from that of (A + T).

In 1953, nine years after the findings of Avery, MacLeod, and McCarty pointed to a genetic role for DNA, James D. Watson and Francis H. C. Crick proposed a model for the structure of DNA. This model was based on X-ray diffraction patterns that Rosalind Franklin and Maurice Wilkins obtained from DNA fibers, on the

Table 10·2 Base composition of DNA (mole %) and ratios of bases

Source	A	G	C	T	A/T*	G/C*	G + C	Purine/ Pyrimidine*
Escherichia coli	26.0	24.9	25.2	23.9	1.09	0.99	50.1	1.04
Mycobacterium tuberculosis	15.1	34.9	35.4	14.6	1.03	0.99	70.3	1.00
Yeast	31.7	18.3	17.4	32.6	0.97	1.05	35.7	1.00
Ox	29.0	21.2	21.2	28.7	1.01	1.00	42.4	1.01
Pig	29.8	20.7	20.7	29.1	1.02	1.00	41.4	1.01
Human	30.4	19.9	19.9	30.1	1.01	1.00	39.8	1.01

*Deviations from a 1:1 ratio are due to experimental variations.

chemical equivalencies noted by Chargaff, on model building, and on intuition. The model, which we now know to be essentially correct, accounted for the equal amounts of purine and pyrimidine by suggesting that DNA was double stranded and that A and G on one strand paired with T and C, respectively, on the other strand. Watson and Crick's proposed structure was that which is now referred to as the B conformation of DNA, or simply *B-DNA*.

DNA is the molecule that stores genetic information. Understanding its structure is critical to understanding the genetic code (Chapter 22) and the processes of DNA replication (Chapter 20) and transcription (Chapter 21). Because it is the storehouse of biological information, DNA serves most often as a template for reactions that require access to that information. Every cell contains dozens of enzymes and proteins that bind to DNA by recognizing certain structural features. In many cases, these proteins bind to a specific DNA sequence.

A. Nucleotides in DNA Are Linked by 3′–5′ Phosphodiesters

We have seen that the primary structure of a protein refers to the sequence of its amino acid residues linked by peptide bonds; in an analogous way, the primary structure of a nucleic acid is the sequence of nucleotide residues connected by *3′–5′ phosphodiester linkages*. A tetranucleotide representing a segment of a DNA chain illustrates such linkages (Figure 10·8). Notice that all the nucleotide residues within a polynucleotide chain have the same relative orientation; thus, like polypeptide chains, polynucleotide chains have directionality. One end of a linear polynucleotide chain is said to be 5′ (that is, there is no residue attached to its 5′-carbon) and the other is said to be 3′ (that is, there is no residue attached to its 3′-carbon). The forward direction (which is the direction of polymerization in the biosynthesis of both DNA and RNA) is 5′→3′; therefore, when not otherwise specified, structural abbreviations are assumed to read 5′→3′. Phosphates are often abbreviated "p." Thus, 5′-adenylate (AMP) can be abbreviated as pA, 3′-deoxyadenylate as dAp, and ATP as pppA. The tetranucleotide in Figure 10·8 can be abbreviated pdApdGpdTpdC or even AGTC when it is clear that the reference is to DNA.

Each phosphoryl group that participates in a phosphodiester linkage bears a negative charge at neutral pH (pK_a approximately 2). Consequently, nucleic acids are polyanions under physiological conditions. They are usually complexed to counterions such as Mg^{2+} or to cationic proteins (proteins that contain an abundance of the basic residues arginine and lysine).

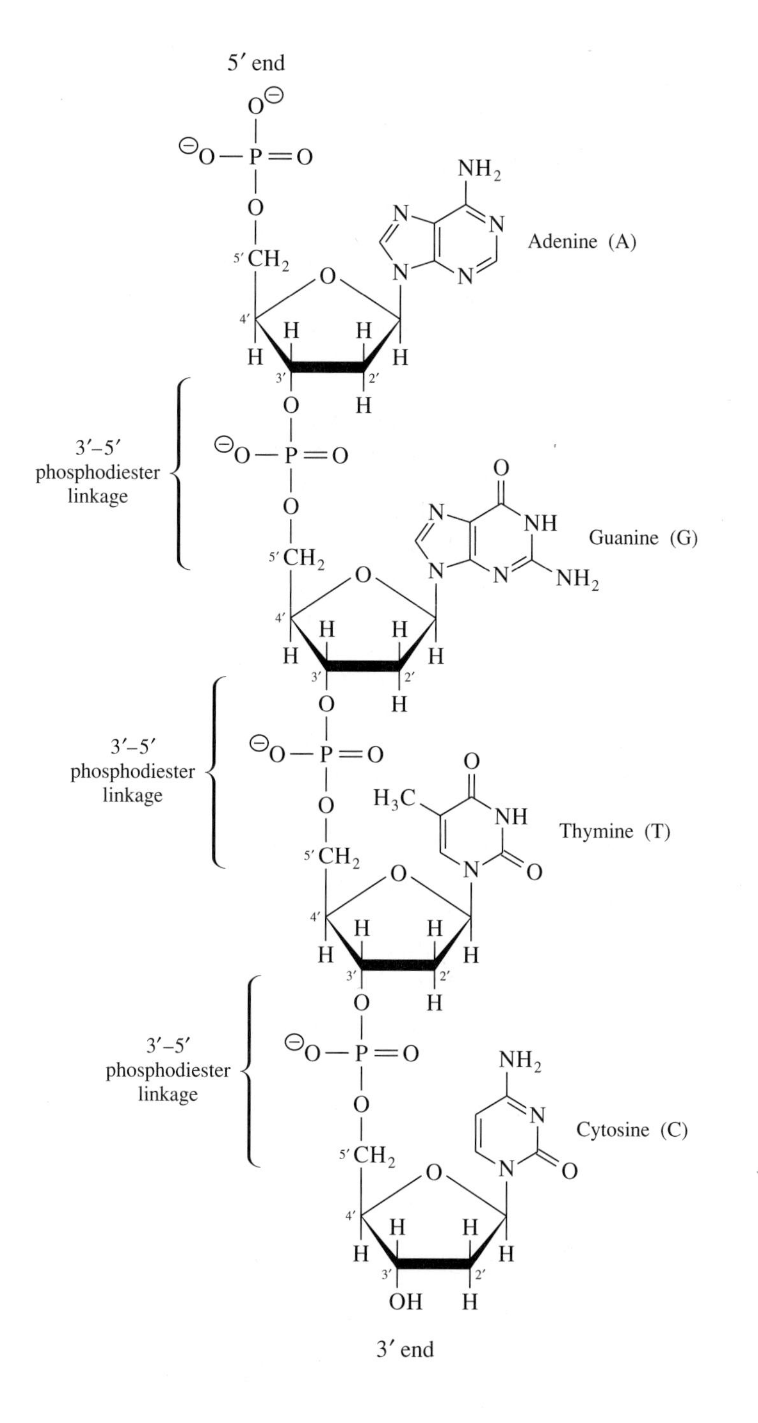

Figure 10·8
Structure of the tetranucleotide pdApdGpdTpdC. The nucleotide residues are linked by 3′–5′ phosphodiesters. This tetranucleotide has a terminal nucleotide with a 5′ phosphoryl group that is not bound to another nucleotide (the 5′ end) and a terminal nucleotide with a 3′ hydroxyl group not bound to another nucleotide (the 3′ end). The forward direction of a polynucleotide chain is defined as 5′→3′, that is, proceeding from the 5′ terminus to the 3′ terminus. Thus, this tetranucleotide can be represented by the abbreviation pdApdGpdTpdC.

B. Two Antiparallel Strands Come Together to Form a Double Helix

DNA is composed of two antiparallel strands of nucleic acid (Figure 10·9). Each base of one strand forms hydrogen bonds with a base of the opposite strand, forming a *base pair*. Only the lactam and amino tautomers of each base accommodate such hydrogen bonding. Guanine pairs with cytosine, and adenine with thymine; these base pairs maximize hydrogen bonding between potential sites. Accordingly, G/C base pairs have three hydrogen bonds and A/T base pairs have two. This feature of double-stranded DNA accounts for Chargaff's earlier discovery that the ratio of A to T and of G to C is 1:1 for a wide diversity of DNA molecules. Because A in one strand pairs with T in the other strand and G pairs with C, the strands are

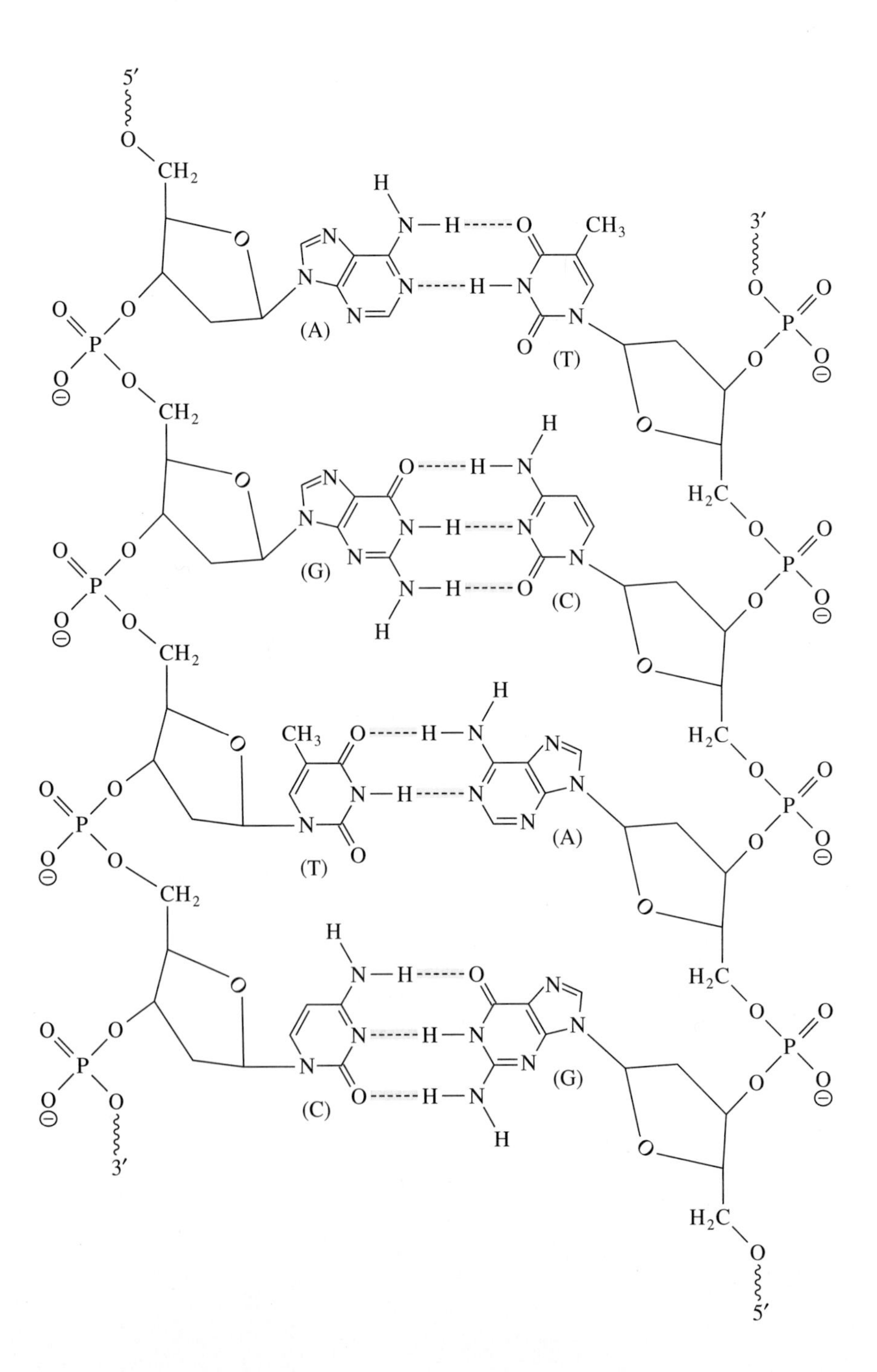

Figure 10·9
Structure of double-stranded DNA. The two strands run in opposite directions. Adenine in one strand pairs with thymine in the opposite strand, and guanine pairs with cytosine. In actuality, the base pairs are tilted so that the plane of the bases is perpendicular to the page.

complementary and can serve as a *template* for each other. The distance between the two sugar-phosphate backbones is the same for each base pair; consequently, the DNA molecule has a regular structure. Also, since the sugar-phosphate backbones of the two strands have opposite orientations, each end of double-stranded DNA is made up of the 5′ end of one strand and the 3′ end of another.

The actual structure of DNA differs from that shown in Figure 10·9 in two important aspects. In DNA, the base pairs would be perpendicular to the page so that we would see them edge on. Also, the two strands wrap around each other to form a two-stranded helical structure, or double helix. The DNA molecule may be visualized as a ladder that has been twisted into a helix. The "ladder" is formed by interaction of complementary bases on opposite strands. Each strand serves as a perfect template for the other; this interaction is responsible for the overall symmetry of double-stranded DNA. However, templating per se does not produce a helix. Interaction of adjacent base pairs brings the base pairs closer together and creates a hydrophobic interior that results in twisting of the sugar-phosphate backbone to form a helix. These stacking interactions stabilize the double helix. The templating and stacking features of DNA structure are shown in Figure 10·10.

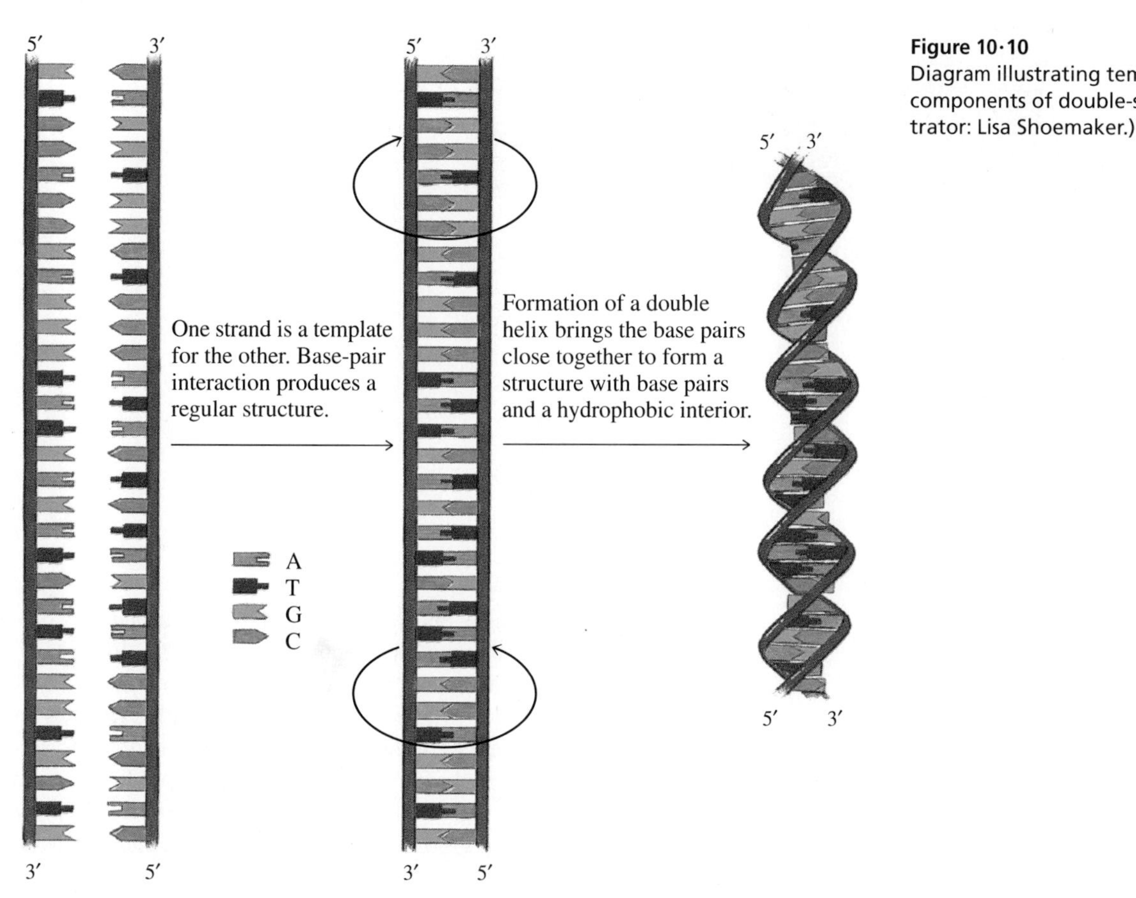

Figure 10·10
Diagram illustrating templating and stacking components of double-stranded DNA. (Illustrator: Lisa Shoemaker.)

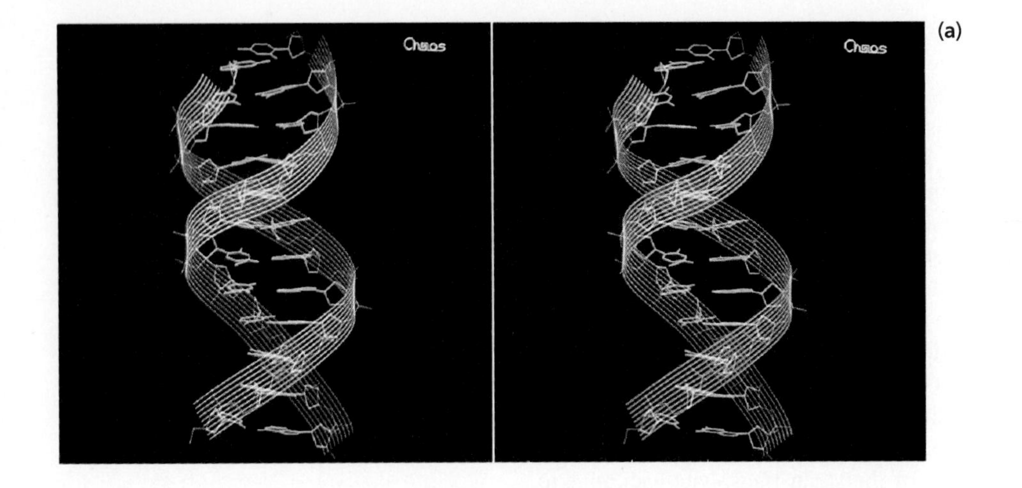

(a)

Stereo S10·2
B-DNA. **(a)** Stick model. Note that the base pairs (green) lie perpendicular to the sugar-phosphate backbone (orange), highlighted by purple ribbons. **(b)** Space-filling model. Color key: carbon, green; nitrogen, blue; oxygen, red; phosphorus, orange. (Based upon coordinates provided by H. Drew and Richard E. Dickerson; stereo images by Ben N. Banks and Kim M. Gernert.)

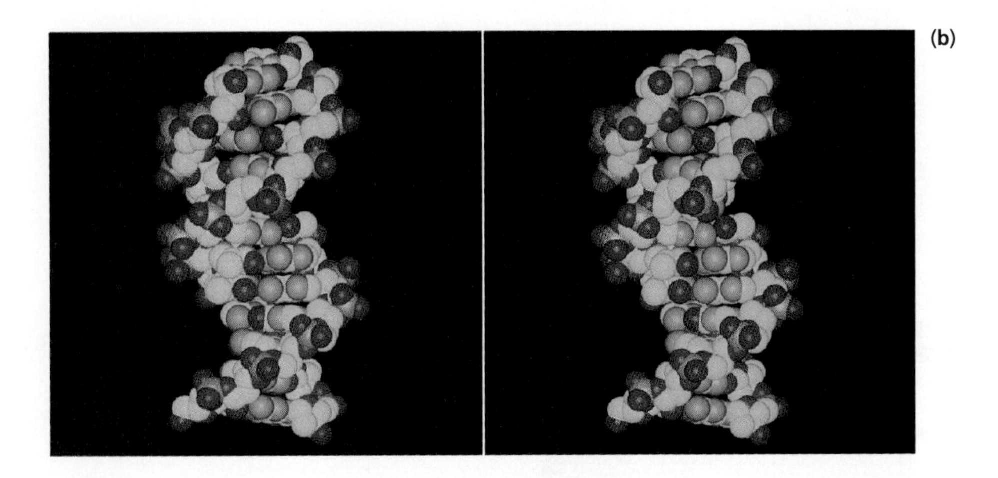

(b)

Because the two hydrophilic sugar-phosphate backbones wind around the outside of the helix, they are exposed to aqueous solvent (Stereo S10·2a). In contrast, the stacked, relatively hydrophobic bases are located in the interior of the helix almost perpendicular to the helical axis, where they are largely shielded from aqueous solvent (Stereo S10·2b). Some edges of the bases, however, are exposed to solvent. Because of the way in which the base pairs stack to form a double helix, the helix has two grooves of unequal width. The larger groove is called the *major groove;* the smaller groove, the *minor groove* (Figure 10·11). Within these grooves,

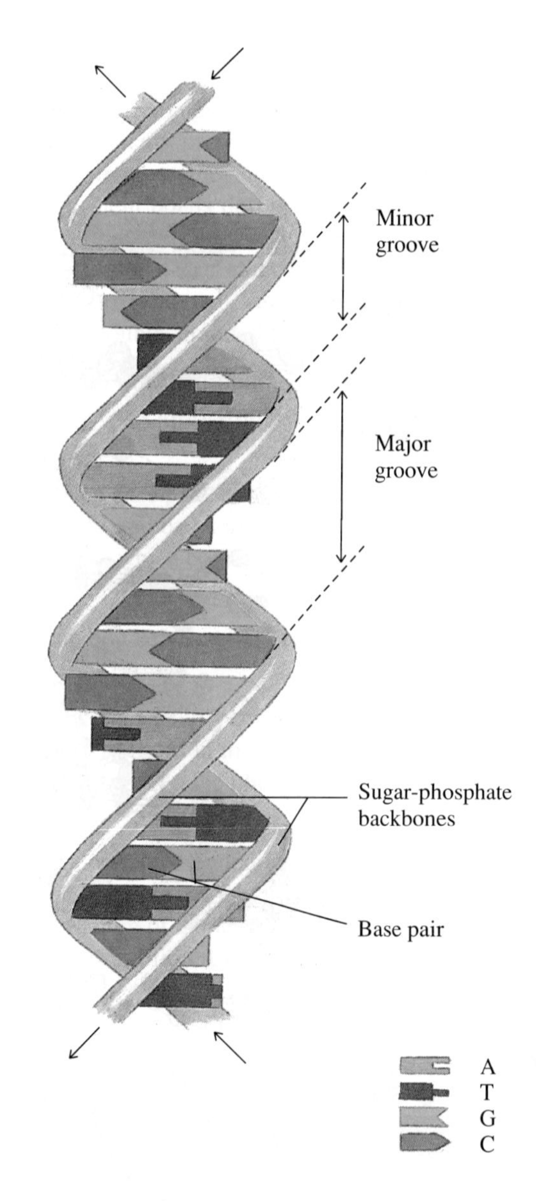

Figure 10·11
Three-dimensional structure of B-DNA. This stylized model of B-DNA depicts the orientation of the base pairs and the sugar-phosphate backbone and the relative sizes of the pyrimidine and purine bases. The sugar-phosphate backbone winds around the outside of the helix, and the bases protrude into the interior. Stacking of the base pairs creates two grooves of unequal width, the major and the minor groove. For ease of representation, a slight space has been drawn between the stacked base pairs, and the interaction between complementary bases has been schematized. (Illustrator: Lisa Shoemaker.)

the bases are exposed to solvent and are chemically distinguishable; thus, molecules that interact with particular base pairs through this groove can identify the base pairs without disrupting the helix. This is particularly important for proteins that must bind to DNA and "read" a specific sequence.

The double-helical structure proposed by Watson and Crick is an idealized model of the most common form of DNA, B-DNA. The Watson-Crick model is a right-handed helix with a diameter of 2.37 nm. The *rise,* or distance between one base pair and the next along the helical axis, is 0.33 nm, and the *pitch* of the helix, or distance it takes to complete one turn, is 3.40 nm. Since there are about 10.4 base pairs per turn of the helix, the angle of rotation between adjacent nucleotides within each strand is 34.6° (360°/10.4). These dimensions are shown in Figure 10·12 (next page), which also reveals that the base pairs can be tilted slightly and that the two bases in a base pair are not always perfectly coplanar.

Because most naturally occurring DNA is double stranded, the length of DNA molecules or segments is often expressed in terms of base pairs (bp). For convenience, longer structures are reported in thousands of base pairs, or *kilobase pairs,* commonly abbreviated as kb.

Table 10·5 Specificities of some common restriction endonucleases

Source	Enzyme[1]	Recognition sequence[2]
Bacillus amyloliquefaciens H	*Bam* HI	G↓GATCC
Bacillus globigii	*Bgl* II	A↓GATCT
Escherichia coli RY13	*Eco* RI	G↓A$\overset{*}{\text{A}}$TTC
Escherichia coli R245	*Eco* RII	↓C$\overset{*}{\text{C}}$TGG
Haemophilus aegyptius	*Hae* III	GG↓CC
Haemophilus influenzae R_d	*Hind* II	GTPy↓Pu$\overset{*}{\text{A}}$C
Haemophilus influenzae R_d	*Hind* III	$\overset{*}{\text{A}}$↓AGCTT
Haemophilus parainfluenzae	*Hpa* II	C↓CGG
Nocardia otitidis-caviarum	*Not* I	GC↓GGCCGC
Providencia stuartii 164	*Pst* I	CTGCA↓G
Serratia marcescens S_b	*Sma* I	CCC↓GGG

[1] The names of restriction endonucleases are abbreviations of the names of the organisms that produce them. Each abbreviated name is followed by a letter denoting the strain and a Roman numeral indicating the order of discovery of the enzyme in that strain.

[2] Recognition sequences are written 5′ to 3′. Only one strand is represented. The arrows indicate cleavage sites. Asterisks represent positions where bases can be methylated. Pu (purine) denotes that either A or G will be recognized. Py (pyrimidine) denotes that either C or T will be recognized.

H_3C

5′···NNGAATTCNN···3′
3′···NNCTTAAGNN···5′

CH_3

methylation ↑

5′···NNGAATTCNN···3′
3′···NNCTTAAGNN···5′

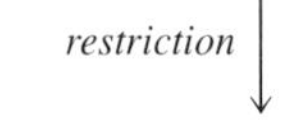

restriction ↓

5′···NNG 3′ 5′ AATTCNN···3′
3′···NNCTTAA 5′ 3′ GNN···5′

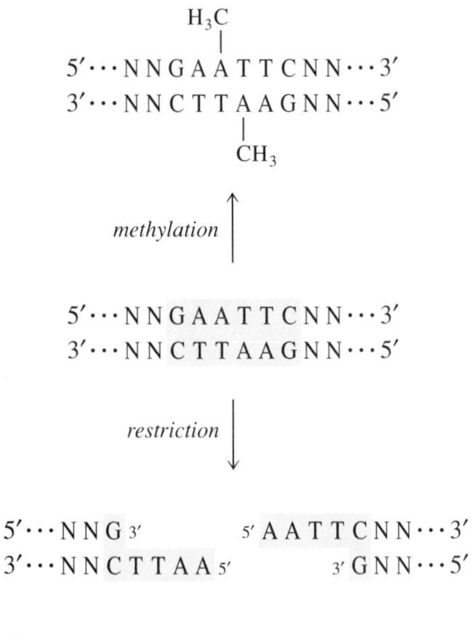

Figure 10·22
The *Eco* RI site. The recognition site for the restriction endonuclease *Eco* RI and for its corresponding methylase consists of the palindromic sequence GAATTC. The endonuclease cleaves both strands in the foreign DNA to produce fragments with staggered ends. The second adenine on both strands in the host DNA is methylated.

The specificities of a few representative restriction endonucleases are listed in Table 10·5. In nearly all cases, the nucleotide sequences that serve as recognition sites have a twofold axis of symmetry; that is, the 5′→3′ sequence of residues is the same in both strands within this four– to eight–base-pair segment of duplex DNA. Thus, the paired sequences "read" the same in either direction; such sequences are known as *palindromes*. (Palindromes in English include BIB, DEED, RADAR, and even MADAM I'M ADAM, provided one ignores punctuation and spacing.)

Consider, for example, *Eco* RI, one of the first restriction enzymes to be discovered. As shown in Table 10·5, *Eco* RI has a six–base-pair recognition sequence that is palindromic (each strand being d(GAATTC) in the 5′→3′ direction). The *Eco* RI enzyme is a homodimer and thus, like its substrate, possesses a twofold axis of symmetry. In the *E. coli* cell, the companion methylase to *Eco* RI converts the second adenine within this recognition sequence to *N*-6-methyladenine (Figure 10·22). Should any double-stranded DNA molecule that possesses an unmethylated d(GAATTC) sequence be exposed to *Eco* RI, the enzyme will catalyze hydrolysis of the phosphodiesters of both strands that link dG to dA, thus cleaving the DNA strand.

It is also evident from Table 10·5 and Figure 10·22 that some restriction endonucleases (including *Eco* RI, *Bam* HI, and *Hind* III) lead to *staggered cleavage*, which produces fragments that have a single-stranded extension at their ends. Such ends are called *sticky ends* because the single-stranded regions are complementary and can thus reform a double-stranded structure. Other enzymes, such as *Hae* III, *Hind* II, and *Sma* I, lead to *blunt-end cleavage* (cuts with no overlapping single-stranded segments).

C. *Eco* RI Binds Tightly to DNA

In order to recognize a specific sequence and cleave at a specific site, restriction endonucleases must bind tightly to DNA. The structure of *Eco* RI complexed to DNA has been determined by X-ray crystallography. The interaction between the enzyme and DNA seems to be similar to that of many other DNA-binding proteins. As shown in Stereo S10·3, each half of the *Eco* RI homodimer binds one side of the DNA molecule so that the DNA, which retains its helical structure, is almost surrounded. The enzyme recognizes the specific nucleotide sequence by contacting base pairs in the major groove. The minor groove (in the middle of the structure shown in Stereo S10·3a) is exposed to the aqueous environment.

Several basic amino acid residues line the cleft formed by the two *Eco* RI monomers. The side chains of these amino acids interact electrostatically with the sugar-phosphate backbone of DNA. In addition, two arginine residues (Arg-145 and Arg-200) and one glutamate residue (Glu-144) of each *Eco* RI monomer form hydrogen bonds with base pairs in the recognition sequence, thus ensuring specific binding. Other, nonspecific interactions with the backbone further stabilize the complex.

(a)

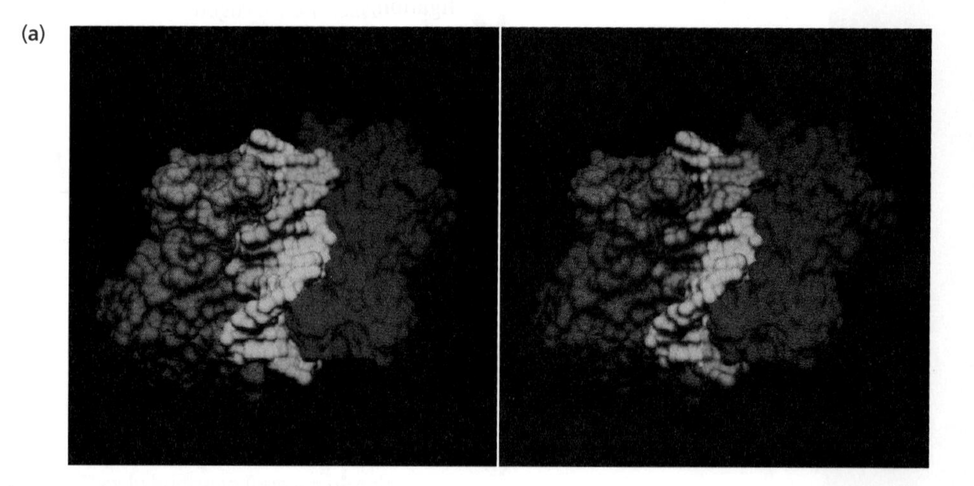

(b)

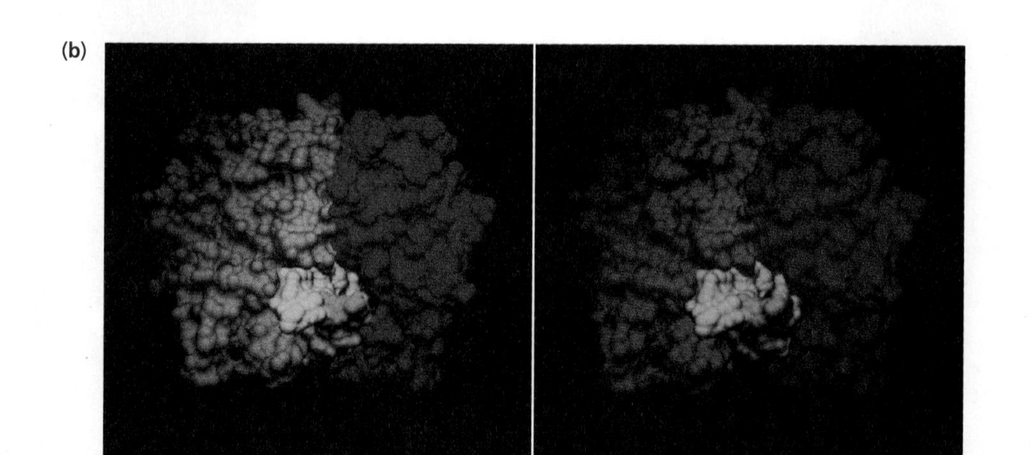

Stereo S10·3
Binding of restriction endonuclease *Eco* RI to DNA. *Eco* RI is composed of two identical monomers shown here in gold and red. The enzyme is bound to a fragment of DNA with the sequence CGCGAATTCGCG (recognition sequence underlined). The two strands of DNA are colored differently (blue and green) for emphasis. **(a)** Side view. **(b)** Top view. (Courtesy of John M. Rosenberg.)

Figure 10·2
Restriction
sites of clea
some restri
is a single s
ample. Dig
enzyme sh
of 10.0 kb a
shown in la

Figure 10·2
Digestion c
ous restrict
DNA is trea
electropho
separates f
The smalles
found at th
I digestion.
Kpn I diges
(1.5 kb) is n
the two fra
same size a
A mixture c
in lanes 1–4
England Bi
permission.

11

Bioenergetics: ATP and Other Energy-Rich Metabolites

In the preceding chapters, we examined the building blocks of the cell. All of the information we have covered will now be called upon as we examine cellular *metabolism*, the network of enzyme-catalyzed reactions in living cells. The area of biochemistry we will be examining is often termed *intermediary metabolism*, consisting of the central *catabolic*, or degradative, and *anabolic*, or biosynthetic, reaction sequences common to most cells.

Survival on the cellular level requires a constant exchange of matter and energy between the cell and its environment (Figure 11·1). Fuel molecules are constantly consumed by catabolic reactions, which harvest energy and biochemical building blocks. At the same time, the cell must continuously regenerate itself. Living cells harness the energy released by catabolism to fuel anabolic reaction sequences that

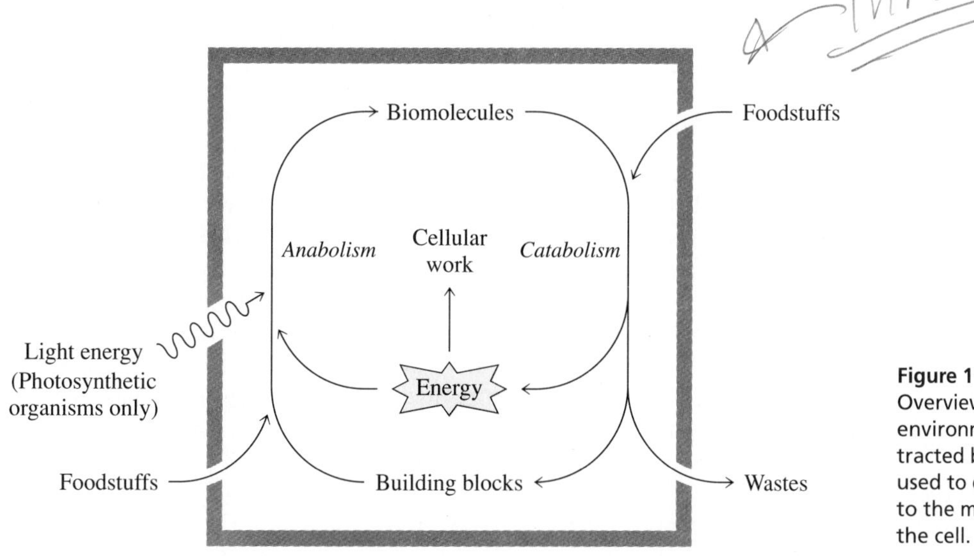

Figure 11·1
Overview of metabolism. Matter from the environment is absorbed and energy is extracted by catabolic processes. This energy is used to drive other reactions that contribute to the maintenance, growth, and division of the cell.

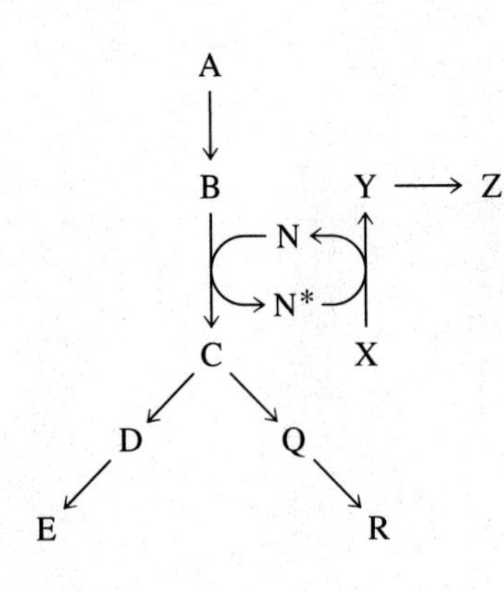

Figure 11·2
Schematic representation of pathway interconnections. C is a branchpoint from which two routes diverge. The coenzyme N, which can be converted to the form N*, links two separate pathways.

synthesize molecules needed to maintain the fitness of the cell and contribute to growth. Cells also use captured energy for the performance of cellular work, such as transport across cell membranes and cell movement.

The thousands of enzyme-catalyzed reactions that occur in living cells can be grouped into functional units called *metabolic pathways.* These are reaction sequences in which the product of one enzyme-catalyzed reaction is the substrate of the next (Figure 11·2). The study of metabolism examines the characteristics of the pathways, the biochemistry of pathway reactions, and the ways in which reactions and pathways are regulated.

An important generality distinguishes catabolic from anabolic pathways: catabolic routes usually funnel from a wide range of substances to a few end products, ultimately leading to the complete oxidation of biomolecules to CO_2 and H_2O. Anabolic pathways branch from a few starting points toward a large variety of end products. However, reaction sequences are not always easily categorized. In living cells, catabolic and anabolic reactions are active simultaneously, and catabolic pathways may yield molecules that are precursors or intermediates of anabolic pathways. In the cell, opposing pathways often operate simultaneously within the larger context of cell maintenance and growth.

Understanding the physical basis for the direction of biochemical reactions is the most fundamental component of an understanding of metabolism. Of key importance is the application of basic principles of thermodynamics. As a starting point for our examination of metabolism, we review these thermodynamic principles.

11·1 Thermodynamic Principles Underlie the Study of Metabolism

Each of the reactions of metabolism involves the transfer of both matter and energy. The transfer of matter is intuitive, represented by the movement of atoms among reacting species—movements that are captured in the familiar symbols of chemistry. Energy transfer in chemical reactions is less intuitive, and information about energy does not appear overtly in chemical symbols. A quantitative understanding of energy transfer is needed to determine why a reaction proceeds in a particular direction in a test tube and why it does so in a living cell.

Thermodynamics is the branch of physical science that deals with energy changes. Originally developed as a conceptual tool for understanding engines and engineering processes, the principles of thermodynamics supply a rigorous framework for analyzing all processes in which energy transformations occur. This includes the processing and consumption of energy within biological systems, a field of study called *bioenergetics.*

A. Enthalpy and the First Law of Thermodynamics

A chemical reaction, a cell, or an entire organism can be thought of as a *thermodynamic system.* Everything that is not part of the defined system constitutes the *surroundings* of the system. Once a thermodynamic system has been defined, it can be described by a set of *state functions,* quantities that mathematically and physically characterize the system. Pressure, temperature, and the concentrations of the system's components are examples of state functions. When dealing with thermodynamic transformations of systems, it is important to realize that state functions are *path independent.* The value of a state function does not depend on what occurred before or during the transformation of a system but only on the actual state of the system at a given moment.

The first law of thermodynamics states that the energy of the universe—that is, system plus surroundings—is constant. There can be no net energy change in the universe. However, net energy changes are possible within a given system. Consider a biochemical process in which energy from outside the system is used to convert two substrates (A + B) into two products (C + D).

$$\text{Energy} + A + B \rightleftharpoons C + D \tag{11·1}$$

The energy absorbed in the forward reaction is *exactly equal* to the energy released in the reverse reaction. The change in the internal energy of the system, ΔE, before and after the process can be expressed as

$$\Delta E = q - w \tag{11·2}$$

where q is defined as the heat absorbed by the system from its surroundings and w is defined as the work done by the system on its surroundings. The terms q and w are not state functions, as they are dependent on what occurs as the system moves from its initial to its final state. (We adopt the convention here in which uppercase roman letters are used for state functions and lowercase italic letters are used for path-dependent variables.) It should be clear that a change in the internal energy of the system, ΔE, is simply an accounting of the heat or work transactions that take place across the boundary between the system and its surroundings.

Work can be defined mathematically as

$$w = P\Delta V + w' \tag{11·3}$$

where P equals pressure, ΔV is the change in volume that occurs during the process, and w' represents work other than pressure-volume work. For most isolated biochemical reactions, $w' = 0$. Given this restriction, substituting Equation 11·3 in Equation 11·2 and rearranging gives

$$q_p = \Delta E + P\Delta V \tag{11·4}$$

where q_p is the heat absorbed at constant pressure. Since any combination of state functions must also be a state function, q_p is therefore a state function, called the *enthalpy change* (ΔH).

$$\Delta H = \Delta E + P\Delta V \tag{11·5}$$

Equation 11·5 relates the heat absorbed during a reaction at constant pressure to the change in energy of the system, the pressure, and the change in the volume of the system. In biological systems, volume remains essentially constant. Thus, ΔH and ΔE are effectively equivalent.

A process is *exothermic* when heat is evolved by the system and enters the surroundings. Such a process has a *negative* enthalpy change ($\Delta H < 0$). A process is *endothermic* when heat is absorbed by the system from the surroundings. Such a process has a *positive* enthalpy change ($\Delta H > 0$). It is important to recognize that changes in enthalpy are defined only for the system and not the surroundings.

Enthalpy changes do not determine the spontaneity of a reaction. The term *spontaneity* has a rigorous technical meaning in thermodynamics: a process is called spontaneous if it is thermodynamically favored—water flows downhill, not up. The downhill flow of water is spontaneous even if the water is trapped behind a dam. The term suggests nothing about the *rate* of the process. The lack of correlation between the enthalpy change and the spontaneity of a reaction is easily demonstrated. For example, mixing sulfuric acid (H_2SO_4) with water causes a great deal of heat to be evolved. Dissolving potassium chloride (KCl) in water causes heat to be absorbed from the environment, and the solution becomes cold. Thus, whether a process is spontaneous does *not* depend on whether heat is absorbed or evolved during the process.

B. Entropy and the Second Law of Thermodynamics

The criterion for spontaneity is found in the second law of thermodynamics. The second law states that in every spontaneous process, the *entropy* (S) of the universe (system plus surroundings) increases. Entropy can be considered a measure of disorder or randomness. Although the net entropy change for every process is positive, the entropy of a defined system, such as a cell, can decrease, provided that the decrease in order of the system is more than compensated for by an increase in the entropy of the surroundings. For any change of a system, the entropy of the universe (system plus surroundings) must increase. The tendency toward increasing entropy acts as a force that drives systems toward equilibrium.

In metabolic systems, the forces of entropy are countered by order-creating cellular processes. Metabolic pathways exist in a *steady state* that is far from the equilibrium state. The distinction between the steady-state condition and equilibrium is discussed in Box 11·1.

Box 11·1 Metabolic Pathways Are in a Steady State

Metabolic pathways are best described as being in a *steady state* rather than at equilibrium. It is therefore important that we distinguish equilibrium from steady state.

An equilibrium can be described by a minimum of two components in the reaction

$$A \rightleftharpoons B \quad (1)$$

Equilibrium occurs when the rate of conversion of A to B is equal to the rate of conversion of B to A.

By contrast, at least three components are needed to model a steady state:

$$A \longrightarrow B \longrightarrow C \quad (2)$$

Assuming there are no other reactants, only B in this system can be in a steady state. The steady state will occur if the rate of formation of B from A is equal to the rate of utilization of B to form C. Over time, the concentration of B in a steady state will not change, but A will be depleted and C will accumulate. In the steady-state condition, the concentration of B is maintained at the expense of continuous change in the concentrations of A and C.

A figurative description of the steady state is the "leaking bucket" model: a bucket has a small hole in the bottom, but a faucet adds water to it at a rate exactly equal to the rate of loss through the leak. Over time, the water level in the bucket remains constant, although the source is continually drained while the ground becomes flooded. In metabolic pathways consisting of several reactions, all of the intermediates are in a steady state, with the initial reaction in the pathway steadily replenished and the end product steadily removed.

11·2 The Gibbs Free-Energy Change Combines Enthalpy and Entropy Changes

The second law of thermodynamics establishes a direction for all processes—toward an increase in the entropy of the universe—and thus also provides a criterion for the spontaneity of processes. However, this is not an immediately practical function since the entropy of the universe cannot be measured. A useful criterion for spontaneity that combines the enthalpy change and entropy change of a system was developed by J. Willard Gibbs. The new state function, termed the (Gibbs) *free-energy change* (ΔG), is defined in terms of H and S for processes that occur at constant pressure:

$$\Delta G = \Delta H - T\Delta S \tag{11·6}$$

where T is the absolute temperature. The free-energy change, ΔG, is a measure of the energy available within a system to do work. The ability of a system to do work decreases as equilibrium is approached; at equilibrium, no energy is available to do work. The free-energy change is a thermodynamic state function that defines the equilibrium condition in terms of the enthalpy and entropy of the system at constant pressure. Since enthalpy reflects the first law of thermodynamics and entropy is a property of the second law, the concept of free energy unites the first two laws of thermodynamics. ΔG is a practical function, since it can be determined by measurements of the system alone.

If the sign of ΔG for a process is *negative,* the process is spontaneous; that is, it can proceed in the absence of energy provided from outside the system. Such processes are said to be *exergonic.* If the sign of ΔG is *positive,* the process cannot proceed spontaneously. For such a process to proceed, enough energy must be supplied from outside the system to make the free-energy change negative. Processes with a positive free-energy change are said to be *endergonic,* which is a way of saying the reverse process is thermodynamically favored. When the process is at equilibrium, the free-energy change is zero.

Whereas ΔG is composed of both enthalpy and entropy contributions, only the sum of these at a given temperature—as indicated in Equation 11·6—need form a negative value for a reaction to be spontaneous. Thus, even if ΔS for the system during a particular process is negative, a sufficiently negative ΔH can overcome the decrease in entropy, resulting in a ΔG that is less than zero. Similarly, even if ΔH is positive, a sufficiently positive ΔS can overcome the increase in enthalpy, resulting in a negative ΔG. Spontaneous processes that proceed due to a large positive ΔS are said to be "entropy-driven." Examples of entropy-driven processes include protein folding (Section 4·11A) and the formation of lipid bilayers (Section 9·9), both of which are associated with the hydrophobic effect (Section 2·4). The processes of protein folding and lipid-bilayer formation both result in states of decreased entropy for the protein molecule and bilayer components, respectively. However, the decrease in entropy is offset by a large increase in the entropy of surrounding water molecules.

The free-energy change of a reaction is dependent on the conditions under which the reaction occurs. Therefore, it is useful to have a set of reference reaction conditions established by convention. These reference conditions are referred to as the *standard state*. Chemists define the standard temperature as 298 K (25°C), standard pressure as 1 atmosphere, and standard solute concentration as 1.0 M. The notation ΔG° indicates the change in free energy under standard conditions as defined by chemists. The chemical standard state has been modified slightly for biological chemistry. Since most biochemical reactions occur at a pH near 7, the standard concentration of hydrogen ions in the biological standard state is 10^{-7} M (pH = 7.0) rather than 1.0 M (pH = 0.0). Free-energy change under the biological standard state is indicated by $\Delta G^{\circ\prime}$.

11·3 The Equilibrium Constant of a Reaction Is Related to the Standard Free-Energy Change

An important relationship exists between the actual free-energy change, or free-energy change under nonstandard conditions (ΔG), the standard free-energy change ($\Delta G^{\circ\prime}$), and the equilibrium constant (K_{eq}). The free energy of a solution of substance A is related to its standard free energy by

$$G_A = G_A^{\circ\prime} + 2.303\ RT \log[A] \qquad (11\cdot7)$$

where R is the universal gas constant (1.987 cal K^{-1} mol^{-1}) and 2.303 is a conversion factor (2.303 log = ln). Free energy is expressed in units of kcal mol^{-1}.

For the reaction

$$A + B \rightleftharpoons C + D \qquad (11\cdot8)$$

the actual free-energy change is the sum of the free energies of the products minus the free energies of the reactants.

$$\Delta G_{reaction} = (G_C + G_D) - (G_A + G_B) \qquad (11\cdot9)$$

By substituting the relationship of actual free energy to standard free energy (Equation 11·7), we can obtain

$$\Delta G_{reaction} = (G_C^{\circ\prime} + G_D^{\circ\prime} - G_A^{\circ\prime} - G_B^{\circ\prime}) + 2.303\ RT \log \frac{[C][D]}{[A][B]} \qquad (11\cdot10)$$

or

$$\Delta G_{reaction} = \Delta G^{\circ\prime}_{reaction} + 2.303\ RT \log \frac{[C][D]}{[A][B]} \qquad (11\cdot11)$$

If the reaction has reached equilibrium, then the ratio of concentrations in the last term of Equation 11·11 is, by definition, the equilibrium constant, K_{eq}. When the concentrations of the reactants and products are at equilibrium, the rates of the forward and reverse reactions are the same and $\Delta G_{reaction} = 0$. Thus,

$$\Delta G^{\circ\prime}_{reaction} = -2.303\ RT \log K_{eq} \qquad (11\cdot12)$$

If we know the value of $\Delta G^{\circ\prime}$ for a reaction, Equation 11·12 allows us to calculate K_{eq} and vice versa (Box 11·2).

Box 11·2 $\Delta G^{\circ\prime} = -2.303\ RT \log K_{eq}$

Consider the reaction S ⟶ P. If we know that the standard free-energy change for this reaction, $\Delta G^{\circ\prime}$, is −3.5 kcal mol^{-1}, then we calculate the equilibrium constant, K_{eq}, as follows:

$$\Delta G^{\circ\prime}_{reaction} = -2.303\ RT \log K_{eq} \qquad (1)$$

$$\log K_{eq} = \frac{-\Delta G^{\circ\prime}}{2.303\ RT} \qquad (2)$$

$$\log K_{eq} = \frac{-(-3.5\ \text{kcal mol}^{-1})}{2.303\ (1.987\ \text{cal K}^{-1}\text{mol}^{-1})\ (298\ \text{K})} \qquad (3)$$

$$\log K_{eq} = \frac{3500\ \text{cal mol}^{-1}}{1364\ \text{cal mol}^{-1}} = 2.6 \qquad (4)$$

$$K_{eq} = 4.0 \times 10^2 \qquad (5)$$

Thus, the ratio [P]/[S] at equilibrium is 400. The expression relating $\Delta G^{\circ\prime}$ and K_{eq} can be plotted as shown in Figure 1. Note that the vertical axis showing values for K_{eq} is logarithmic and covers several orders of magnitude. Small changes in $\Delta G^{\circ\prime}$ therefore represent large changes in K_{eq}.

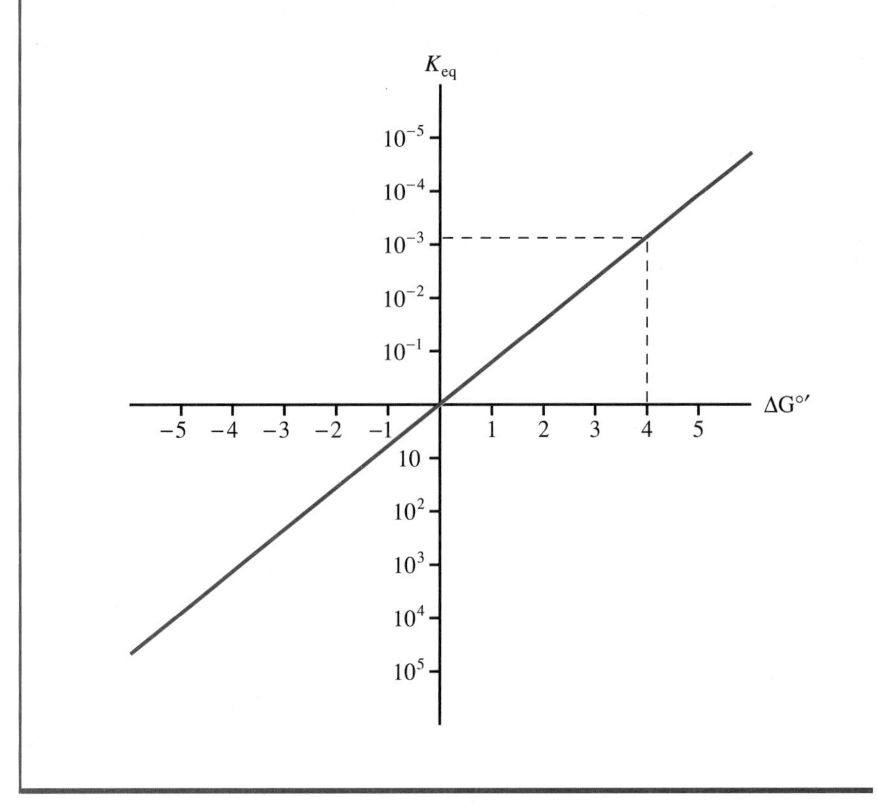

Box 11·2
Figure 1
Graph of $\Delta G^{\circ\prime}$ plotted against K_{eq}, derived from the formula $\Delta G^{\circ\prime} = -2.303\ RT \log K_{eq}$. Because the concentration term is logarithmic, small changes in $\Delta G^{\circ\prime}$ correspond to large changes in K_{eq}.

11·4 Actual Free-Energy Change, and Not Standard Free-Energy Change, Is the Criterion for Spontaneity in Cellular Reactions

For any enzymatic reaction within a living organism, the *actual* free-energy change (the free-energy change under cellular conditions) must be less than zero for the reaction to proceed. However, many metabolic reactions have *standard* free-energy changes that are positive. The difference between ΔG and $\Delta G^{\circ\prime}$ depends upon cellular conditions. The most important condition affecting free-energy change in cells is the concentrations of substrates and products of a reaction. For Equation 11·8 at equilibrium, the ratio of substrates and products is by definition the equilibrium constant, K_{eq}, and the free-energy change is zero.

$$K_{eq} = \frac{[C][D]}{[A][B]} \qquad \Delta G = 0 \qquad \textbf{(11·13)}$$

However, for a reaction not at equilibrium, a different ratio of products to substrates is observed, and the free-energy change is derived using Equation 11·11:

$$Q = \frac{[C]'[D]'}{[A]'[B]'} \qquad \Delta G = \Delta G^{\circ\prime} + 2.303\ RT \log Q \qquad \textbf{(11·14)}$$

where Q is the *mass action ratio*. It is this ratio relative to the ratio of products to substrates at equilibrium that determines the free-energy change for a reaction. In other words, the free-energy change is a measure of how far from equilibrium the reacting system is poised. Consequently, it is ΔG, and not $\Delta G^{\circ\prime}$, that is the criterion for assessing the spontaneity of a reaction and therefore its direction within a particular system.

In general, we may divide metabolic reactions into two types. Let us take Q to represent the steady-state ratio of reactant and product concentrations in a living cell. Reactions for which Q is close to K_{eq} are called *near-equilibrium reactions*. The free-energy changes associated with near-equilibrium reactions are quite small, and these reactions are readily reversible. Reactions for which Q is far from K_{eq} are called *metabolically irreversible reactions*. These reactions are greatly displaced from equilibrium, with Q usually two or more orders of magnitude from K_{eq}. Whereas ΔG must be at least slightly negative for all reactions in living cells, ΔG is a large negative number for metabolically irreversible reactions.

When flux through pathways changes, the intracellular concentration of metabolites varies but usually over a narrow range of no more than two- or threefold. Enzymes that catalyze near-equilibrium reactions are able to quickly restore levels of substrate and product to near-equilibrium status due to the large amount of enzymatic activity present.

By contrast, enzymes that catalyze metabolically irreversible reactions are present in limiting amounts in cells, insufficient to achieve near-equilibrium status for their reactions. All of the control points of pathways are metabolically irreversible reactions. The enzymes that catalyze these reactions are usually allosterically regulated. Metabolically irreversible reactions act as bottlenecks in metabolic traffic, controlling the flux through reactions further along the pathway.

Near-equilibrium reactions are not suitable control points. A near-equilibrium reaction cannot be accelerated, since it is already attaining near-equilibrium status, and inhibition of the reaction sufficient to regulate flux in only one direction (in other words, inhibiting the reaction to the point that buildup of substrate concentration makes the reaction irreversible) would entail a change in pathway conditions larger than that seen in cells. Flux through near-equilibrium reactions is therefore controlled only by changes in substrate and product concentrations. In contrast, flux through metabolically irreversible reactions is relatively unaffected by changes in metabolite concentration; flux through these reactions is primarily controlled by effectors that modulate the enzymatic rate.

11·5 ATP Is the Principal Carrier of Biological Energy

The energy produced by one biological process is frequently coupled to a second process that would not otherwise occur spontaneously, thereby allowing the second process to occur. In enzymatic reactions, coupling is achieved when the overall reaction includes a shared intermediate. We can view the two reactions taken together as a new reaction that has an overall ΔG of less than zero, meaning the reaction can proceed. Energy flow in metabolism depends on many coupled reactions in which ATP is the common intermediate.

ATP, a nucleoside triphosphate, contains one phosphate ester formed by linkage of the α-phosphate to the 5′ oxygen of ribose and two phosphoanhydrides formed by the α,β and β,γ linkages between phosphate groups. The free energy associated with ATP is transferred when the phosphoanhydrides are cleaved to produce either adenosine diphosphate (ADP) and inorganic phosphate (P_i) or adenosine monophosphate (AMP) and inorganic pyrophosphate (PP_i). Although the phosphoryl groups of ATP are usually transferred to acceptors other than water, hydrolysis reactions provide useful estimates of the free-energy changes involved.

Table 11·1 lists the standard free energies of hydrolysis ($\Delta G^{\circ\prime}_{hydrolysis}$) for ATP and AMP, and Figure 11·3 depicts the hydrolytic cleavage of each of the phosphoanhydrides of ATP. Note from Table 11·1 that whereas cleavage of the ester releases only 3.5 kcal mol^{-1}, cleavage of either of the phosphoanhydrides releases over 7 kcal mol^{-1} under standard conditions. For energy transfer within cells, only the phosphoanhydride bonds are commonly involved, with ADP or AMP formed as the

Table 11·1 Standard free energies of hydrolysis for ATP and AMP

Major ionic form of reactants and products	$\Delta G^{\circ\prime}_{hydrolysis}$ (kcal mol^{-1})
$ATP^{4-} + H_2O \longrightarrow ADP^{3-} + HPO_4^{2-} + H^{+}$	–7.3
$ATP^{4-} + H_2O \longrightarrow AMP^{2-} + HP_2O_7^{3-} + H^{+}$	–7.6
$AMP^{2-} + H_2O \longrightarrow$ Adenosine $+ HPO_4^{2-} + H^{+}$	–3.5

Figure 11·3
Hydrolysis of ATP to (1) ADP and inorganic phosphate (P_i) and (2) AMP and inorganic pyrophosphate (PP_i).

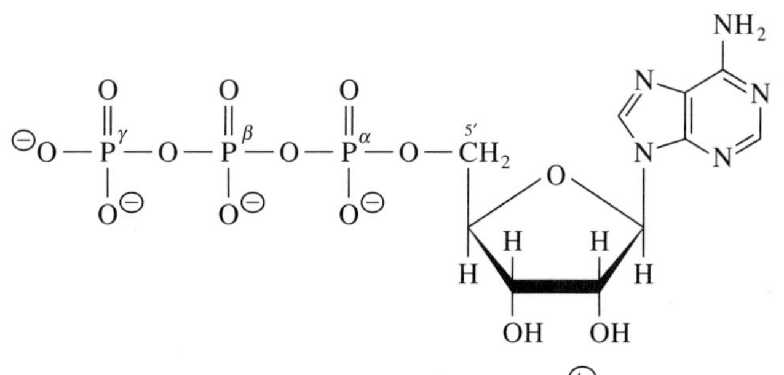

Adenosine 5′-triphosphate (ATP^{4-})

H_2O (1) (2) H_2O

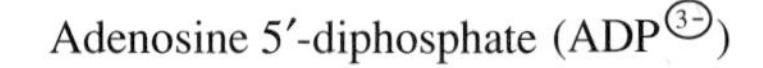

Adenosine 5′-diphosphate (ADP^{3-})

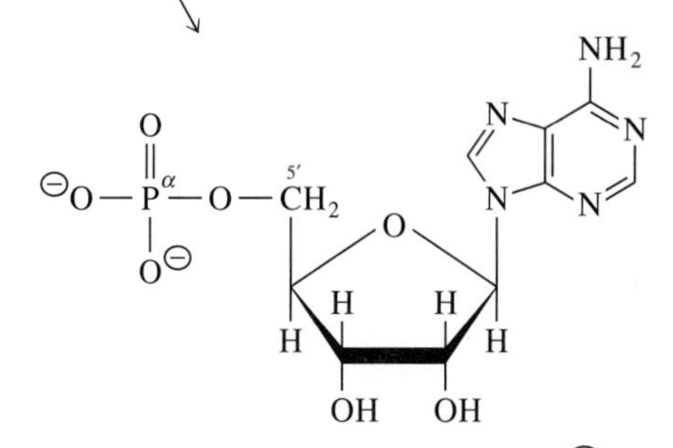

Adenosine 5′-monophosphate (AMP^{2-})

+

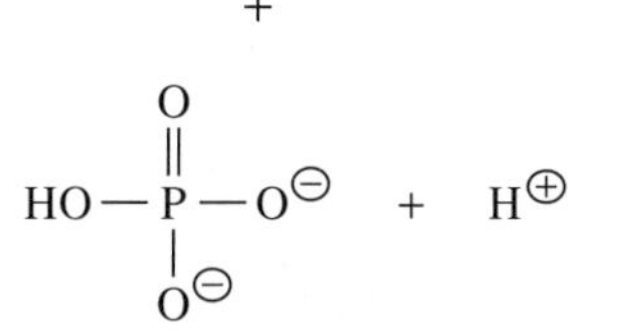

Inorganic phosphate (P_i)

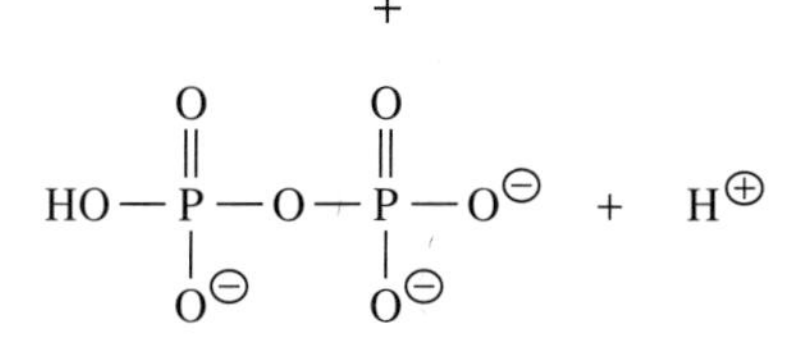

Inorganic pyrophosphate (PP_i)

product. In vivo, the ratio of ATP to its hydrolysis products is maintained extremely far from equilibrium, and the free-energy change associated with ATP hydrolysis in living cells is about -14 kcal mol^{-1}. Because of the free-energy change associated with cleavage of their phosphoanhydrides, ATP and the other nucleoside triphosphates—uridine triphosphate (UTP), guanosine triphosphate (GTP), and cytidine triphosphate (CTP)—are often referred to as *energy-rich metabolites*.

Hydrolysis of ATP is highly exergonic under both standard and cellular conditions partly because of the resulting decrease in electrostatic repulsion among the negatively charged oxygen atoms of the triphosphoanhydride group of ATP. The amount of electrostatic repulsion is in turn affected by pH. The reactants and products of ATP hydrolysis have ionizing functional groups with pK_a values between 6 and 7. Changes in pH alter the ionic states of all reactants and products in the reaction and strongly affect the free energy of hydrolysis. At high pH, when the ionizing groups are relatively dissociated, electrostatic repulsion is increased.

Electrostatic repulsion is also affected by the cellular concentration of magnesium ions (Mg^{2+}). Virtually all of the ATP and ADP molecules in a cell exist in complexes with Mg^{2+} (Figure 11·4), and hydrolysis of MgATP is less favorable than hydrolysis of uncomplexed ATP. Mg^{2+} partially neutralizes the negative charges on the oxygen atoms of ATP, thereby diminishing electrostatic repulsion. For simplicity, we usually refer to the nucleoside triphosphates as ATP, GTP, CTP, and UTP, but these molecules are actually in complexes with Mg^{2+} when they serve as substrates or coenzymes in cellular reactions.

Figure 11·4
Equilibria for formation of complexes between **(a)** ATP and Mg^{2+}, **(b)** ADP and Mg^{2+}, and **(c)** inorganic phosphate (HPO_4^{2-}) and Mg^{2+}. ("Ado" in the figure represents the adenosyl moieties of ATP and ADP.)

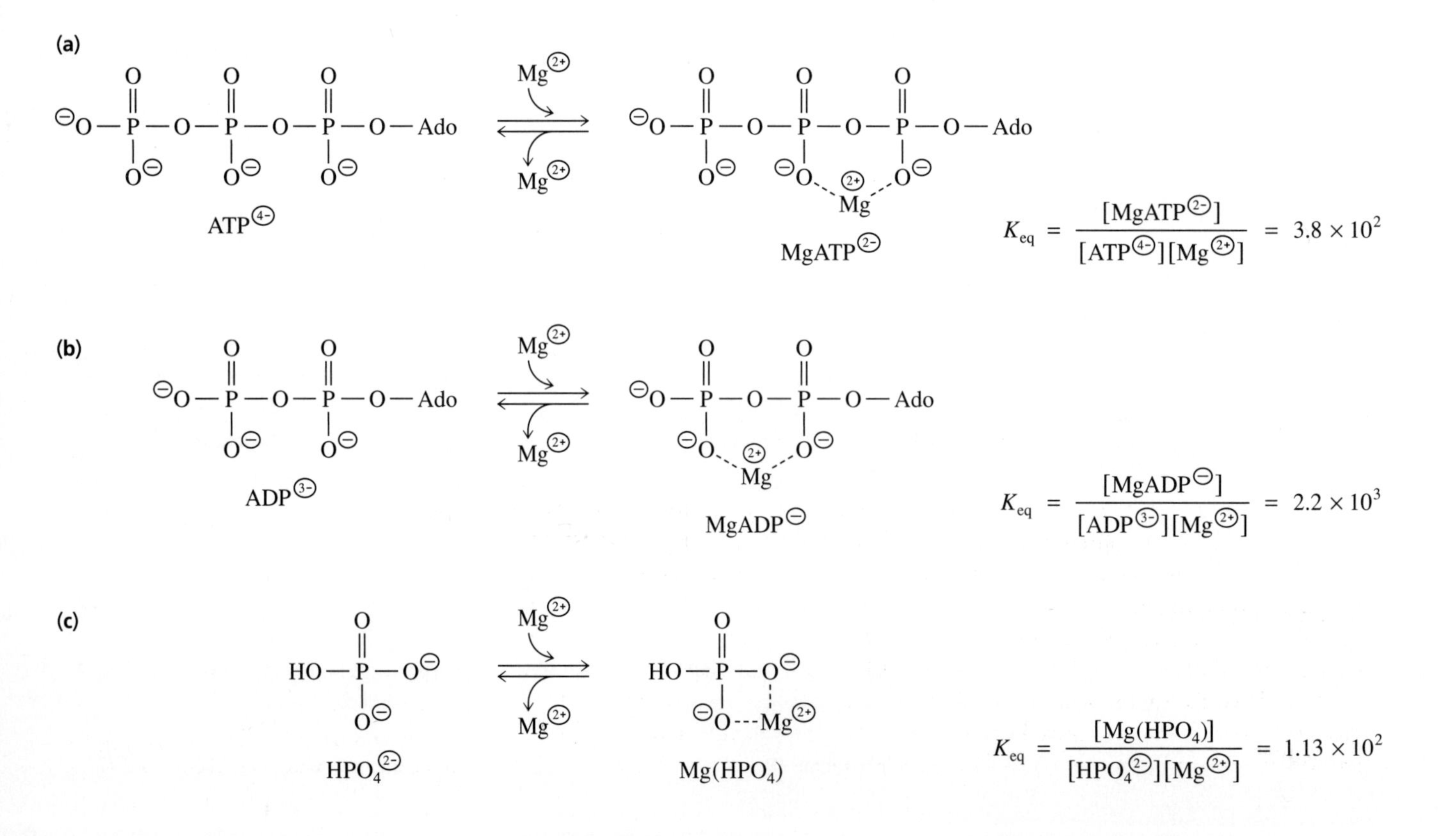

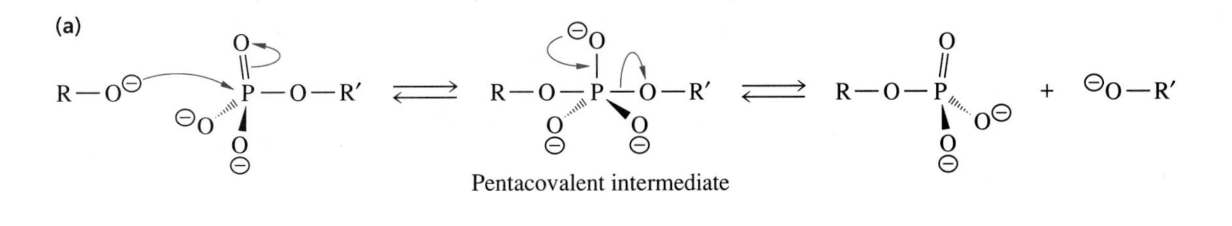

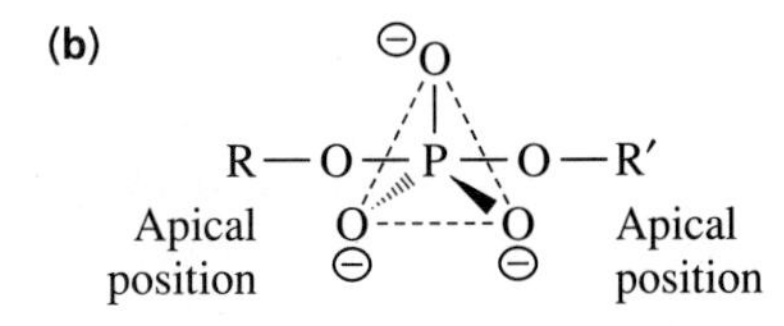

Figure 11·5
(a) In-line mechanism of phosphoryl-group transfer. **(b)** Pentacovalent intermediate of the in-line mechanism, in which the nucleophile and leaving groups occupy apical positions.

Solvation effects also make a significant contribution to the negative free energy of ATP hydrolysis. The products of hydrolysis—ADP and inorganic phosphate or AMP and pyrophosphate—are better solvated than is ATP itself. As ions are solvated (or in water, hydrated), they are separated from each other and electrically shielded. The decrease in the attraction of ions for each other is a thermodynamically favored result. Solvation of the products of ATP hydrolysis helps shift the equilibrium of the reaction toward hydrolysis.

Finally, the products of ATP hydrolysis are more resonance stabilized than ATP itself. The phosphoryl groups of ATP compete for the unpaired electrons on the bridging oxygen atoms of the phosphoanhydrides, whereas the products of hydrolysis assume a more stable electronic structure. Resonance stabilization of the products provides a large driving force for hydrolysis.

11·6 Hydrolysis of ATP Can Be Coupled to Biosynthesis of Other Molecules

Enzymes known as *kinases* (also called phosphotransferases) catalyze transfer of the γ-phosphoryl group from ATP (or, less frequently, from another nucleoside triphosphate) to another substrate. Kinase reactions are typically unidirectional (that is, kinases catalyze metabolically irreversible reactions). A few kinase reactions, however, such as that catalyzed by creatine kinase (Section 11·7), are near-equilibrium reactions.

Although the reactions they catalyze are sometimes discussed as phosphate-group transfer reactions, kinases actually transfer a phosphoryl group ($—PO_3^{2-}$) to their acceptors. This phosphoryl-group transfer occurs by a mechanism called *in-line nucleophilic displacement* (Figure 11·5). The energetics of various kinase reactions can be compared by assessing the standard free-energy changes of the reactions given in Table 11·2. (Under standard conditions, phosphoryl-group transfer potentials have the same values as the standard free energies of hydrolysis but are opposite in sign. Thus, the phosphoryl-group transfer potential is a measure of the free energy required for the formation of the phosphorylated compound.)

Table 11·2 Standard free energies of hydrolysis for common metabolites

Metabolite	$\Delta G^{\circ\prime}_{\text{hydrolysis}}$ (kcal mol^{-1})
Phosphoenolpyruvate	−14.8
1,3-*Bis*phosphoglycerate	−11.8
Acetyl phosphate	−11.2
Phosphocreatine	−10.3
Phosphoarginine	−7.7
ATP to AMP + PP_i	−7.6
ATP to ADP + P_i	−7.3
Pyrophosphate	−8.0
Glucose 1-phosphate	−5.0
Glucose 6-phosphate	−3.3
Glycerol 3-phosphate	−2.2

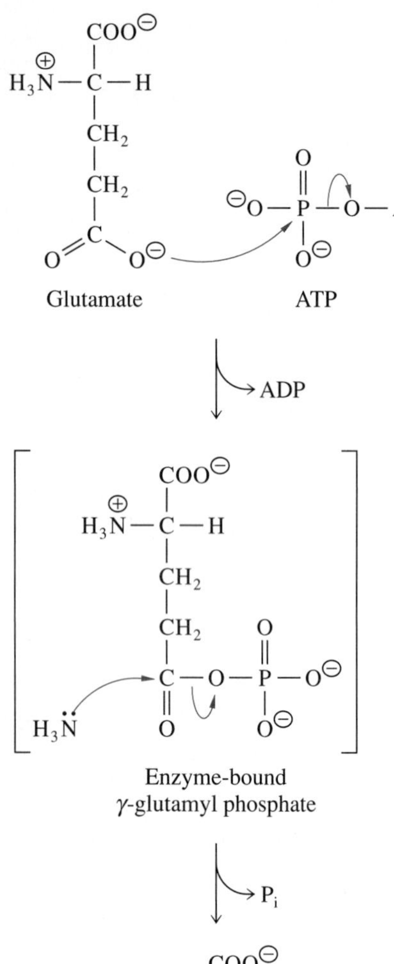

Figure 11·6
Conversion of glutamate to glutamine, catalyzed by ATP-dependent glutamine synthetase. Note that binding of the anhydride intermediate within the active site of the enzyme prevents reaction with water (hydrolysis).

The synthesis of glutamine from glutamate and ammonia illustrates how a cell can use the free energy of ATP to drive a biosynthetic reaction (Figure 11·6). This reaction is catalyzed by glutamine synthetase and provides an important means by which organisms can incorporate inorganic nitrogen into biomolecules as carbon-bound nitrogen. Just as a chemist must activate a carboxylic acid in order to bring about its reaction with ammonia, the enzymatic synthesis of amide bonds, including peptide bonds, depends on the activation of the carboxyl group of a substrate via synthesis of an anhydride intermediate. Glutamine synthetase catalyzes the nucleophilic displacement of the γ-phosphoryl group of ATP by the γ-carboxylate of glutamate, generating enzyme-bound γ-glutamyl phosphate, a high-energy intermediate, and releasing ADP. In the second step of the mechanism, ammonia acts as an attacking nucleophile, displacing the phosphate (a good leaving group) from the carbonyl carbon to generate the product, glutamine. Overall, a molecule of ATP is hydrolyzed to ADP + P_i for every molecule of glutamine formed from glutamate and ammonia.

11·7 The Energy of Other Metabolites Can Be Coupled to the Synthesis of ATP

Several energy-rich metabolites release more free energy upon hydrolysis than ATP under the same conditions. A variety of specific kinases catalyze the transfer of phosphoryl groups from such molecules to ADP, generating ATP.

Acetyl phosphate is a mixed anhydride generated by lactobacilli during anaerobic fermentation. Hydrolysis of acetyl phosphate yields acetate and inorganic phosphate (Figure 11·7). The standard free energy of hydrolysis for this reaction (–11.2 kcal mol^{-1}) is greater than the standard free energy of ATP hydrolysis (–7.3 kcal mol^{-1}). Accordingly, under *standard* conditions, the hydrolysis of acetyl phosphate can be coupled to the synthesis of ATP.

			$\Delta G^{\circ\prime}$ (kcal mol^{-1})
Acetyl phosphate + H_2O	$\longrightarrow$	Acetate + P_i + $H^{\oplus}$	−11.2
ADP + P_i + $H^{\oplus}$	$\longrightarrow$	ATP + H_2O	+ 7.3
Acetyl phosphate + ADP	$\longrightarrow$	ATP + Acetate	− 3.9

(11·15)

This reaction is strongly exergonic under standard conditions. The direction of the reaction is the same under cellular conditions, and the transfer of the activated phosphoryl group from acetyl phosphate to ADP provides lactobacilli with ATP needed for anaerobic metabolism.

The *phosphagens,* including phosphocreatine and phosphoarginine, are high-energy phosphate-storage molecules found in animal cells (Figure 11·8). Phosphagens are phosphoamides (rather than phosphoanhydrides). In the case of phosphocreatine, the standard free energy of hydrolysis is –10.3 kcal mol^{-1}. In the muscles of vertebrates, creatine kinase catalyzes replenishment of ATP through transfer of the activated phosphoryl group from phosphocreatine (Figure 11·9). In invertebrates, phosphoarginine is the source of the activated phosphoryl group.

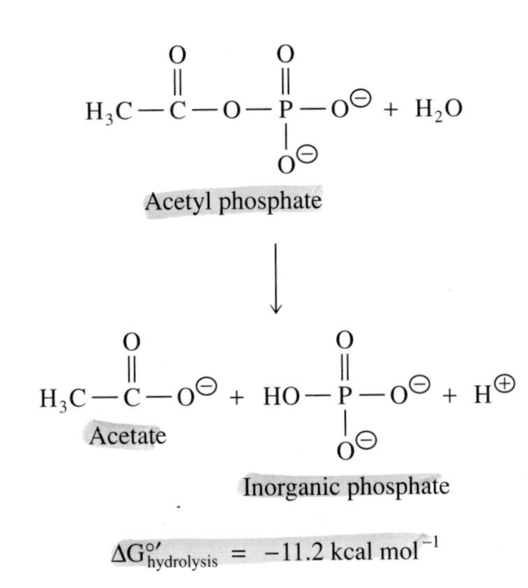

Figure 11·7
Hydrolysis of acetyl phosphate to acetate and inorganic phosphate.

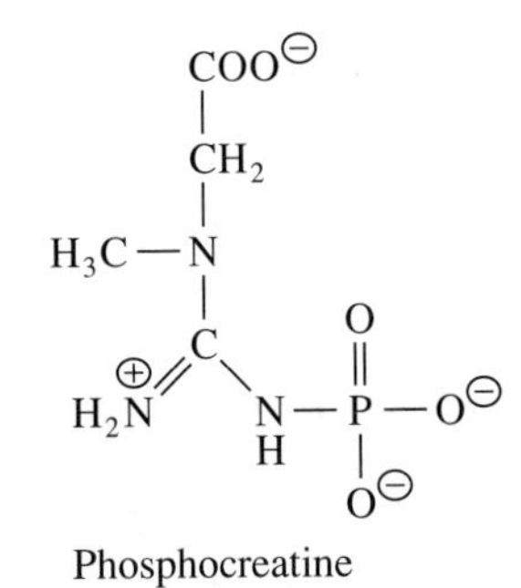

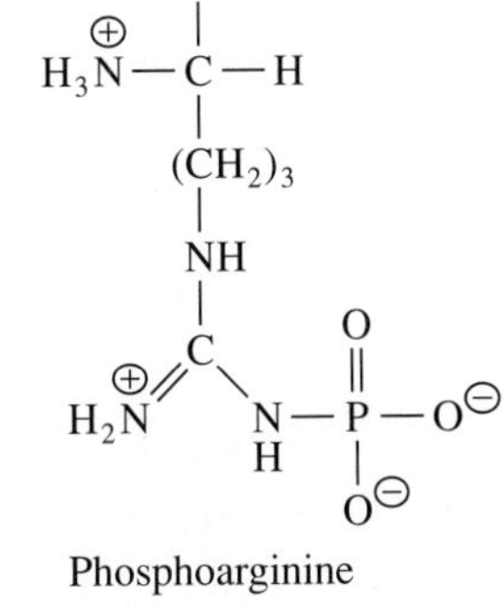

Figure 11·8
Structures of phosphocreatine and phosphoarginine.

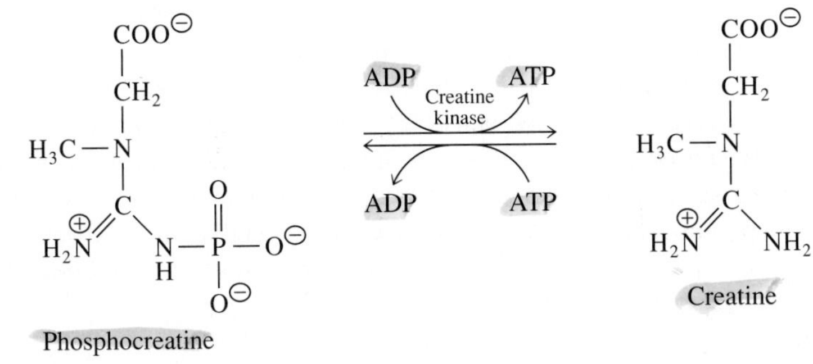

Figure 11·9
Phosphoryl-group transfer from phosphocreatine to ADP with formation of ATP and creatine. Under cellular conditions, the substrate and product concentrations are such that this is a near-equilibrium reaction.

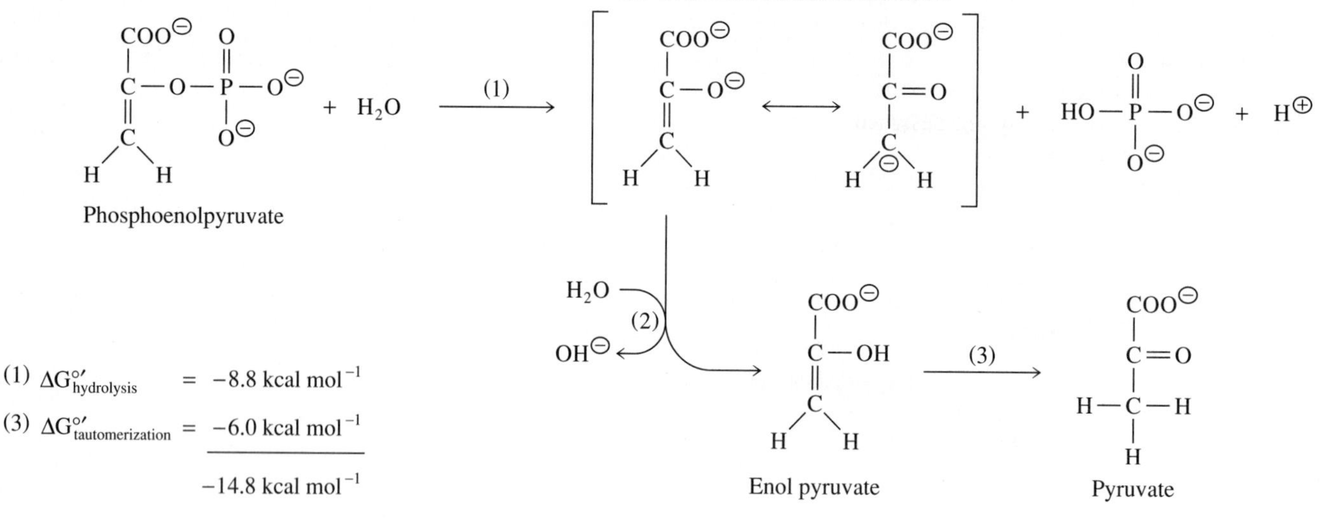

Figure 11·10
Hydrolysis of phosphoenolpyruvate. The reaction can be divided into three steps. (1) The hydrolysis step produces a resonance-stabilized enolate anion and inorganic phosphate. (2) The enolate anion is protonated. (3) Tautomerization converts the enol to a ketone.

Phosphoenolpyruvate, an intermediate of the glycolysis pathway (Chapter 12), has the most energy-rich phosphate bond known. The standard free energy of phosphoenolpyruvate hydrolysis is -14.8 kcal mol^{-1}. Phosphoenolpyruvate is neither a phosphoanhydride nor a phosphoamide, but an *enol ester.* The hydrolysis of phosphoenolpyruvate can be divided into three steps (Figure 11·10). First, cleavage of the P—O bond of the phosphate ester produces inorganic phosphate and a resonance-stabilized enolate anion. The unstable enolate anion is then protonated to yield enol pyruvate. Finally, tautomerization of enol pyruvate yields pyruvate. The free energy associated with phosphoenolpyruvate can be understood by considering the molecule as an enol that is locked into position by attachment of the phosphoryl group. When the phosphoryl group is removed, the molecule can assume the much more stable keto form.

Transfer of a phosphoryl group from phosphoenolpyruvate to ADP is catalyzed by the enzyme pyruvate kinase. $\Delta G^{\circ\prime}$ for the reaction is -7.5 kcal mol^{-1} ($-14.8 + 7.3$ kcal mol^{-1}). Thus, the equilibrium for this reaction under standard conditions lies far in the direction of transfer of the phosphoryl group from phosphoenolpyruvate to ADP, with formation of pyruvate and ATP. In cells, this reaction is an important source of ATP.

11·8 Nucleotide Kinases Catalyze the Formation of Nucleoside Triphosphates

All cells require eight nucleoside triphosphates for synthesis of nucleic acids. Ribonucleic acids are assembled from the four major ribonucleoside triphosphates, ATP, GTP, UTP, and CTP. DNA is synthesized from the four deoxyribonucleoside triphosphates, dATP, dGTP, dTTP, and dCTP. ATP serves as the source of phosphoryl groups transferred during the synthesis of nucleoside diphosphates from nucleoside monophosphates (for instance, ADP from AMP) and the synthesis of nucleoside triphosphates from nucleoside diphosphates (GTP from GDP, etc.). Such phosphoryl-group transfer reactions are catalyzed by several nucleotide kinases with various specificities.

Since all of the phosphoanhydrides involved have nearly equal standard free energies of hydrolysis (or formation), phosphoryl-group transfers between nucleoside phosphates have equilibrium constants close to 1.0. Of course, the intracellular concentrations of individual nucleoside mono-, di-, and triphosphates differ, depending on regulatory mechanisms designed to meet different metabolic needs. For example, the intracellular levels of ATP are far greater than dTTP levels. ATP is involved in many biochemical reactions and metabolic functions, whereas dTTP has only one known function, to serve as a substrate for DNA synthesis.

Ribonucleotides, in addition to their role in nucleic acid biosynthesis, also serve as building blocks for the biosynthesis of coenzymes. For example, ATP provides the adenylate moieties for $NAD^{\oplus}$, $NADP^{\oplus}$, FAD, and coenzyme A. ATP also provides the phosphoryl group for the 3′-phosphomonoester portion of coenzyme A and the 2′-phosphomonoester portion of $NADP^{\oplus}$, and we have already seen that ATP provides the pyrophosphate for the coenzyme thiamine pyrophosphate (TPP) (Figure 7·11).

11·9 The Free Energy of Biological Oxidation Reactions Can Be Captured in the Form of Reduced Coenzymes

The reactions of catabolism result in net oxidation of food molecules. The oxidation of one molecule is necessarily coupled with the reduction of another molecule. A molecule that accepts electrons and is reduced is an *oxidizing agent.* A molecule that loses electrons and is oxidized is a *reducing agent.* The net oxidation-reduction reaction, or *redox reaction,* is

$$A_{red} + B_{ox} \rightleftharpoons A_{ox} + B_{red} \quad (11 \cdot 16)$$

The electrons released in biological oxidation reactions are often transferred enzymatically to one of two oxidizing agents, either nicotinamide adenine dinucleotide ($NAD^{\oplus}$) or nicotinamide adenine dinucleotide phosphate ($NADP^{\oplus}$). The structures and functional mechanisms of these two coenzymes were discussed in Section 7·3. When $NAD^{\oplus}$ and $NADP^{\oplus}$ are reduced, their nicotinamide groups accept a hydride anion (Figure 7·5).

The reduced coenzymes NADH and NADPH are similar structurally, yet they have sharply different functions. NADH—produced during catabolic oxidation reactions—is converted to $NAD^{\oplus}$ during respiration with concomitant production of ATP. This process is called *oxidative phosphorylation* (Chapter 14). NADPH—generated by oxidation reactions in specialized pathways such as the pentose phosphate pathway (Section 15·7)—provides hydride anions for reductive biosynthetic processes such as fatty acid, amino acid, and nucleotide syntheses. The role of both of these coenzymes can be described as supplying *reducing power,* which is measured quantitatively as *reduction potential.*

A. Standard Reduction Potential Is Related to Standard Free Energy

Reduction potential can be measured quantitatively in *electrochemical cells.* A simple redox reaction involving the transfer of a pair of electrons from zinc (Zn) to copper (Cu^{2+}) illustrates the principle.

$$Zn + Cu^{2+} \rightleftharpoons Zn^{2+} + Cu \quad (11 \cdot 17)$$

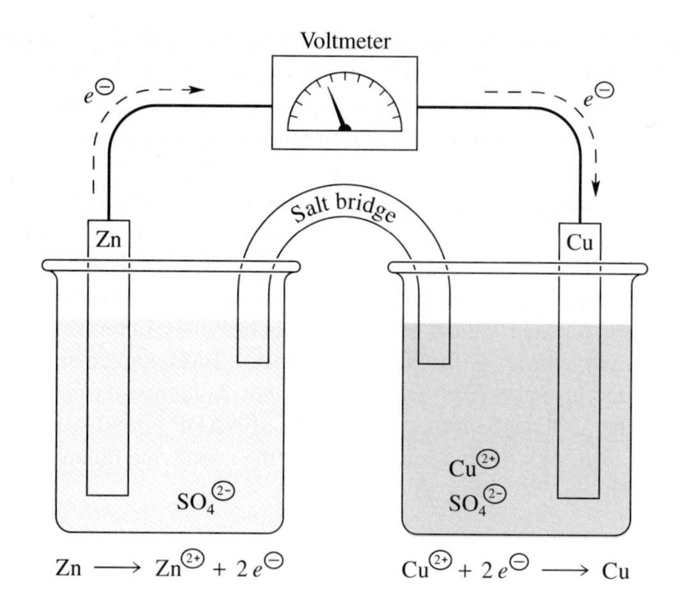

Figure 11·11
Schematic diagram of an electrochemical cell. Electrons flow through the external circuit from the zinc electrode to the copper electrode. The salt bridge permits the flow of counterions (sulfate ions in this example) without extensive mixing of the two solutions. The electromotive force is measured by the voltmeter connected across the two electrodes.

The reaction can be carried out in two separate solutions that divide the overall reaction into two half-reactions (Figure 11·11). The solutions are connected by a wire across two electrodes, allowing electrons to flow between the cells. A salt bridge preserves electroneutrality by providing a path for the flow of ions between the two solutions. This separation of ion flow from electron flow permits measurement of the latter, using a voltmeter.

The direction of the current through the circuit in Figure 11·11 indicates that Cu^{2+} has a greater affinity for electrons than Zn^{2+}. The reading on the voltmeter represents a potential difference—the difference in the reduction potential of the reaction on the left and the one on the right. The measured potential is the *electromotive force,* or emf.

As was the case with measurements of free energy, it is useful to have a reference standard for measurements of reduction potential. In the case of reduction potential, however, the reference is not simply a set of reaction conditions, but a reference half-reaction to which all other half-reactions can be compared. The chosen reference half-reaction is the oxidation of hydrogen gas (H_2), whose reduction potential under standard conditions (E°) is arbitrarily set equal to 0.0 V. The standard reduction potential of a given half-reaction is measured with a redox couple in which the reference half-cell contains a solution of 1 M H^+ and 1 atm $H_2(g)$, and the sample half-cell contains 1 M each of the oxidized and reduced species of the substance whose reduction potential is to be determined. Under standard conditions for biological measurements, the hydrogen-ion concentration in the sample half-cell is 10^{-7} M (pH 7.0). The voltmeter across the redox couple measures the difference in the standard reduction potential ($\Delta E^{\circ\prime}$) of the reference and sample half-reactions. Since the standard reduction potential of the reference half-reaction is 0.0 V, the measured potential is that of the sample half-reaction. The standard reduction potentials of some important biological reduction half-reactions are given in Table 11·3. Electrons flow spontaneously from more negative to more positive reduction potentials.

The standard reduction potential ($E^{\circ\prime}$) for the transfer of electrons from one molecular species to another is related to the standard free-energy change by the equation:

$$\Delta G^{\circ\prime} = -n\mathcal{F}\Delta E^{\circ\prime} \qquad \textbf{(11·18)}$$

where n is the number of electrons transferred, $\mathcal{F}$ is Faraday's constant (23.06 kcal V^{-1} mol^{-1}), and $\Delta E^{\circ\prime}$ is the difference between the standard reduction potentials of the oxidized and reduced species.

Recall from Equation 11·12 that $\Delta G^{\circ\prime} = -2.303$ RT log K_{eq}. Combining this equation with Equation 11·18, we get

$$\Delta E^{\circ\prime} = -2.303 \frac{RT}{n\mathcal{F}} \log K_{eq} \quad \textbf{(11·19)}$$

Under biological conditions, it is highly unlikely that all reactants in a system will be present in standard concentrations of 1 M. Just as the actual free-energy change for a reaction is related to the standard free-energy change by Equation 11·11, observed emf (ΔE) is related to the change in the standard reduction potential ($\Delta E^{\circ\prime}$) by the *Nernst equation:*

$$\Delta E = \Delta E^{\circ\prime} - 2.303 \frac{RT}{n\mathcal{F}} \log \frac{[A_{ox}][B_{red}]}{[A_{red}][B_{ox}]} \quad \textbf{(11·20)}$$

At 298 K, Equation 11·20 reduces to

$$\Delta E = \Delta E^{\circ\prime} - \frac{0.059}{n} \log Q \quad \textbf{(11·21)}$$

If we wish to calculate the emf of a reaction under nonstandard conditions, then we use the Nernst equation and substitute the actual concentrations of reactants and products.

Table 11·3 Standard reduction potentials for some important biological reduction half-reactions

Reduction half-reaction	$E^{\circ\prime}$ (V)
$\frac{1}{2} O_2 + 2 H^{\oplus} + 2 e^{\ominus} \longrightarrow H_2O$	0.82
$Fe^{3+} + e^{\ominus} \longrightarrow Fe^{2+}$	0.77
Photosystem P700	0.43
$NO_3^{\ominus} + e^{\ominus} \longrightarrow NO_2^{\ominus}$	0.42
Cytochrome *f*, $Fe^{3+} + e^{\ominus} \longrightarrow Fe^{2+}$	0.36
Cytochrome *a*, $Fe^{3+} + e^{\ominus} \longrightarrow Fe^{2+}$	0.29
Cytochrome *c*, $Fe^{3+} + e^{\ominus} \longrightarrow Fe^{2+}$	0.25
Cytochrome c_1, $Fe^{3+} + e^{\ominus} \longrightarrow Fe^{2+}$	0.22
Ubiquinone (Q) $+ 2 H^{\oplus} + 2 e^{\ominus} \longrightarrow QH_2$	0.10
Cytochrome *b* (mitochondrial), $Fe^{3+} + e^{\ominus} \longrightarrow Fe^{2+}$	0.08
Fumarate $+ 2 H^{\oplus} + 2 e^{\ominus} \longrightarrow$ Succinate	0.03
Cytochrome b_5 (microsomal), $Fe^{3+} + e^{\ominus} \longrightarrow Fe^{2+}$	0.02
Oxaloacetate $+ 2 H^{\oplus} + 2 e^{\ominus} \longrightarrow$ Malate	−0.17
Pyruvate $+ 2 H^{\oplus} + 2 e^{\ominus} \longrightarrow$ Lactate	−0.18
Acetaldehyde $+ 2 H^{\oplus} + 2 e^{\ominus} \longrightarrow$ Ethanol	−0.20
FMN $+ 2 H^{\oplus} + 2 e^{\ominus} \longrightarrow FMNH_2$	−0.22
FAD $+ 2 H^{\oplus} + 2 e^{\ominus} \longrightarrow FADH_2$	−0.22
Glutathione (oxidized) $+ 2 H^{\oplus} + 2 e^{\ominus} \longrightarrow$ 2-Reduced glutathione	−0.23
Lipoic acid $+ 2 H^{\oplus} + 2 e^{\ominus} \longrightarrow$ Dihydrolipoic acid	−0.29
$NAD^{\oplus} + 2 H^{\oplus} + 2 e^{\ominus} \longrightarrow NADH + H^{\oplus}$	−0.32
$NADP^{\oplus} + 2 H^{\oplus} + 2 e^{\ominus} \longrightarrow NADPH + H^{\oplus}$	−0.32
Lipoyl dehydrogenase (FAD) $+ 2 H^{\oplus} + 2 e^{\ominus} \longrightarrow$ Lipoyl dehydrogenase ($FADH_2$)	−0.34
$2 H^{\oplus} + 2 e^{\ominus} \longrightarrow H_2$	−0.42
Ferredoxin (spinach), $Fe^{3+} + e^{\ominus} \longrightarrow Fe^{2+}$	−0.43

[From Loach, P. A. (1968). Oxidation-reduction potentials, absorbance bands and molar absorbance of compounds used in biochemical studies. In *Handbook of Biochemistry: Selected Data for Molecular Biology,* H. A. Sober, ed. (Cleveland, Ohio: CRC Press).]

B. The Reduction Potential of NADH Is an Important Source of Free Energy

In living cells, most of the reduced coenzyme NADH formed in metabolic reactions is oxidized by the respiratory electron-transport chain, with concomitant production of ATP from ADP + P_i (Chapter 14). NADH is formed in many reactions and acts as a conduit delivering the energy of biological oxidation reactions to the site of oxidative phosphorylation (Figure 11·12). The ultimate acceptor of the electrons from NADH is oxygen. It is possible to calculate the free energy associated with this overall redox reaction under standard conditions by adding the standard reduction potentials of the two half-reactions and using Equation 11·18. The two half-reactions, from Table 11·3, are

$$NAD^{\oplus} + 2\,H^{\oplus} + 2\,e^{\ominus} \longrightarrow NADH + H^{\oplus} \qquad E^{\circ\prime} = -0.32\ V \tag{11·22}$$

and

$$\tfrac{1}{2}\,O_2 + 2\,H^{\oplus} + 2\,e^{\ominus} \longrightarrow H_2O \qquad E^{\circ\prime} = 0.82\ V \tag{11·23}$$

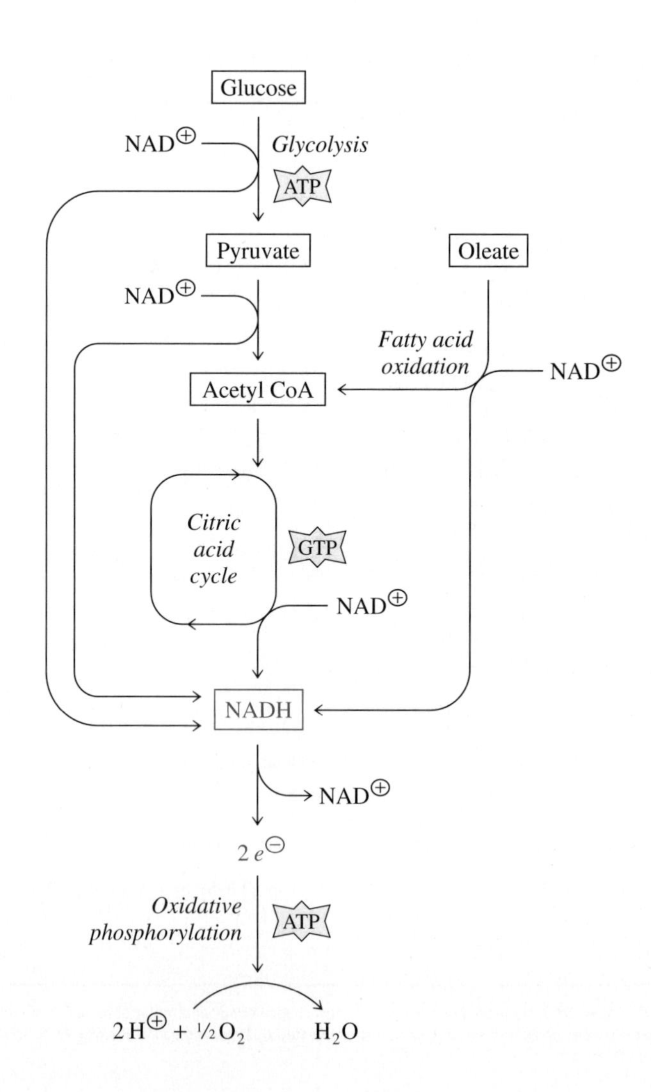

Figure 11·12
Metabolic pathways involved in energy production. Although ATP is produced in several pathways, the majority of ATP production is formed by oxidative phosphorylation, which is fueled mainly by the intermediary electron carrier NADH.

Since the NADH half-reaction has the more negative standard reduction potential, NADH will be oxidized whereas oxygen will be reduced. The net reaction is

$$\text{NADH} + \tfrac{1}{2}\text{O}_2 + \text{H}^{\oplus} \longrightarrow \text{NAD}^{\oplus} + \text{H}_2\text{O} \qquad \Delta E^{\circ\prime} = 1.14\ \text{V}$$

(11·24)

Using Equation 11·19,

$$\Delta G^{\circ\prime} = -(2)\,(23.06\ \text{kcal V}^{-1}\text{mol}^{-1})\,(1.14\ \text{V}) = -52.6\ \text{kcal mol}^{-1} \qquad \textbf{(11·25)}$$

The standard free-energy change for the formation of ATP from ADP + P_i is 7.3 kcal mol^{-1}. Under the conditions of the living cell, as noted earlier, it is approximately 14 kcal mol^{-1}. The energy released during the oxidation of NADH under cellular conditions is sufficient to drive the formation of several molecules of ATP. A more precise examination of the ratio of NADH oxidized to ATP formed will be given in Chapter 14, where other features of the process of oxidative phosphorylation that affect this ratio will be taken into account.

11·10 Acyl-Group Transfer Is Also Important in Metabolic Processes

Thus far we have considered reactions that involve phosphoryl-group transfer and electron transfer. Another important class of reactions involves transfer of acyl groups from an acyl coenzyme A molecule to an acceptor molecule.

The structure and functional mechanism of coenzyme A was examined in Section 7·9. Acyl groups are bound to coenzyme A as thioesters. Sulfur is in the same group of the periodic table as oxygen; however, thioesters are less stable than oxygen esters. This is because resonance structures that contribute to the stability of oxygen esters are not available to thioesters. The standard free energy of hydrolysis of acetyl CoA molecules is –7.5 kcal mol^{-1}, about that of ATP (Figure 11·13).

ATP, pyridine nucleotide coenzymes (NADH and NADPH), and coenzyme A are some of the primary carriers of activated groups in metabolic reactions. These carriers mediate the exchange of activated groups in a wide variety of biochemical reactions and are integral to the completion of many metabolic processes.

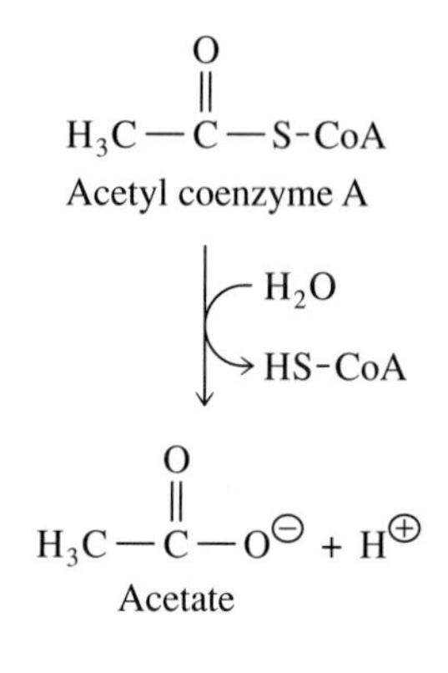

Figure 11·13
Standard free energy of hydrolysis of acetyl coenzyme A to coenzyme A and acetate.

Summary

Metabolic pathways consist of sequential, enzyme-catalyzed reactions. The direction that a chemical or enzyme-catalyzed reaction will take depends on the change in free energy. The standard free-energy change of a reaction is related to the equilibrium constant of the reaction by the formula $\Delta G^{\circ\prime} = -2.303\ RT \log K_{eq}$. Within cells, the change in free energy that occurs during a given reaction depends primarily on the concentrations of reactants and products. Free-energy change under nonstandard conditions for the reaction A + B $\longrightarrow$ C + D is derived by the formula

$$\Delta G_{\text{reaction}} = \Delta G^{\circ\prime}_{\text{reaction}} + 2.303\ RT \log \frac{[C][D]}{[A][B]}$$

In metabolic pathways, each of the individual reactions proceeds with a negative free-energy change. Since the free-energy change is dependent on concentration, the actual free-energy change, and not the standard free-energy change, is the criterion for the direction of a reaction in a cell. The concentrations of reactants and products in many metabolic reactions in cells approach the equilibrium state, i.e., the concentrations are near K_{eq}; such reactions are called near-equilibrium reactions. Reactions for which the steady-state concentrations of reactants are far from

K_{eq} (in other words, reactions that are far from equilibrium) are called metabolically irreversible reactions.

ATP plays a central role in bioenergetics. The energy released by one biological process is often conserved in the form of ATP, to be used by other, energy-requiring processes. The energy of ATP is released in the form of transfer of one or both of its terminal phosphoryl groups.

Many kinases catalyze transfer of the γ-phosphoryl group from ATP to another substrate, or, in some instances, catalyze phosphoryl-group transfer from another energy-rich substrate to ADP to form ATP. In addition to ribonucleoside triphosphates and deoxyribonucleoside triphosphates, there are several other metabolites with activated phosphoryl groups. Examples include the mixed anhydride acetyl phosphate, the phosphagens phosphocreatine and phosphoarginine, and the enol ester phosphoenolpyruvate.

The free energy of biological oxidation reactions can be captured in the form of reduced coenzymes. This form of energy is measured as reduction potential, the quantitative measure of a molecule's ability to donate or accept electrons. Standard reduction potential is related to standard free-energy change by the formula $\Delta G^{\circ\prime} = -n\mathcal{F}\Delta E^{\circ\prime}$. Under nonstandard conditions, reduction potential is given by the Nernst equation:

$$\Delta E = \Delta E^{\circ\prime} - 2.303\frac{RT}{n\mathcal{F}}\log\frac{[A_{ox}][B_{red}]}{[A_{red}][B_{ox}]}$$

Acyl-group transfer is another bioenergetically important process. A major reaction among this class is the highly exergonic transfer of acyl groups from coenzyme A to acceptor molecules.

Selected Readings

Bridger, W. A., and Henderson, J. F. (1983). *Cell ATP* (New York: John Wiley & Sons). A lucid treatment of the chemistry and metabolic function of ATP, including in-depth discussion of the reasons for constant cellular ATP concentrations and pathologies associated with elevated or depressed ATP concentrations.

Harold, F. M. (1986). *The Vital Force: A Study of Bioenergetics* (New York: W. H. Freeman and Company). Chapters 1 and 2.

Ingraham, L. L., and Pardee, A. B. (1967). Free energy and entropy in metabolism. In *Metabolic Pathways,* 3rd ed., Vol. 1, D. M. Greenberg, ed. (New York: Academic Press).

Klotz, I. M. (1967). *Energy Changes in Biochemical Reactions* (New York: Academic Press).

Newsholme, E. A., and Leech, A. R. (1983). *Biochemistry for the Medical Sciences* (New York: John Wiley & Sons). Chapters 1 and 2.

van Holde, K. E. (1985). *Physical Biochemistry,* 2nd ed. (Englewood Cliffs, New Jersey: Prentice Hall).

12

Glycolysis

In the summer of 1897, the doctrine of vitalism was discredited by an experiment that was to mark the birth of biochemistry. A German chemist, Martin Hahn, was attempting to extract proteins from yeast by grinding yeast cells in a mortar with fine sand and diatomaceous earth. The mixture was wrapped in cheesecloth and placed in a press. The resulting yeast extract was difficult to preserve. Remembering that fruit preserves are made by adding sugar, Hans Buchner, a colleague, suggested adding sucrose to the yeast extracts. The idea was tested by Hans's brother Eduard, who discovered that bubbles evolved from the mixture. Eduard Buchner concluded that *fermentation* was occurring, a process Louis Pasteur had characterized as "life without air." In a manuscript of the following year reporting his findings, Eduard Buchner concluded that "such a complicated apparatus as the yeast cell presents is not required. It is considered that the bearer of the fermenting action of the press juice is much more probably a dissolved substance, doubtless a protein; this will be called *zymase*." For his discovery of cell-free fermentation, Eduard Buchner was awarded the Nobel Prize in 1907.

The use of cell-free preparations led to new discoveries, including the isolation and purification of enzymes. Today we recognize that the "zymase" of yeast extracts is actually a mixture of enzymes that together catalyze the reactions of *glycolysis,* the first pathway to be examined in our study of metabolism.

12·1 Glycolysis Is a Ubiquitous Cellular Pathway

The pathway of glycolysis (Greek, *glykys*, sweet; *lysis*, dissolution) is a sequence of enzyme-catalyzed reactions that converts glucose to pyruvate. The conversion of one molecule of glucose to two molecules of pyruvate is accompanied by the net conversion of two molecules of ADP to ATP. The glycolytic pathway is found in virtually all cells; for some it is the only ATP-producing pathway.

The net reaction of glycolysis is shown in Equation 12·1. In addition to the production of ATP, two molecules of NAD⊕ are reduced to NADH.

$$\text{Glucose} + 2\,\text{ADP} + 2\,\text{NAD}^{\oplus} + 2\,\text{P}_i \longrightarrow 2\,\text{Pyruvate} + 2\,\text{ATP} + 2\,\text{NADH} + 2\,\text{H}^{\oplus} + 2\,\text{H}_2\text{O} \quad (12\cdot1)$$

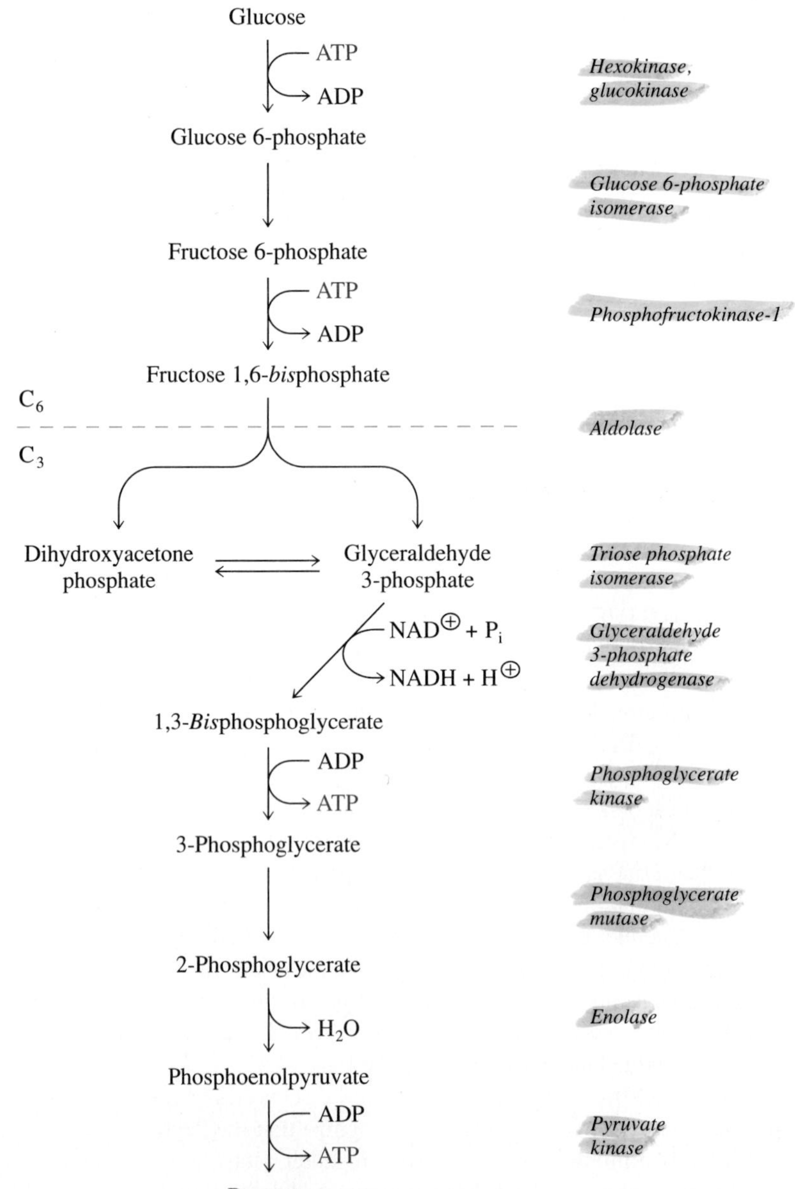

Figure 12·1
Conversion of glucose to pyruvate by glycolysis. The glycolytic pathway is present in virtually all cell types. Note the division between hexoses (C_6) and trioses (C_3). Following the interconversion reaction involving glyceraldehyde 3-phosphate and dihydroxyacetone phosphate, each reaction of glycolysis is traversed by two triose molecules for each hexose molecule metabolized. ATP is consumed in the hexose stage and generated in the triose stage.

The pyruvate produced by glycolysis has several possible fates. Under anaerobic conditions in muscle, pyruvate is converted to lactate, accompanied by the regeneration of NAD⊕. Under anaerobic conditions in certain microorganisms, formation of products other than lactate is possible. For instance, in yeast, pyruvate is converted to ethanol. In the presence of oxygen, pyruvate can be fully oxidized to CO_2 and H_2O by subsequent metabolic pathways.

The glycolytic pathway can be divided into two stages, as shown in Figure 12·1. Above the dotted line, all pathway intermediates are hexoses. At the dotted line, the hexose fructose 1,6-*bis*phosphate is split, and thereafter the intermediates of the pathway are trioses. Note that the two trioses produced by the splitting reaction undergo interconversion, with one of the trioses, glyceraldehyde 3-phosphate, continuing through the pathway. Thus, all subsequent steps of the triose stage of glycolysis are traversed by two molecules for each molecule of glucose metabolized.

In the hexose stage of glycolysis, two molecules of ATP are converted to ADP. In the triose stage, four molecules of ATP are formed from ADP for each molecule of glucose metabolized. Note that the pathway contains two ATP-forming steps, each of which is traversed by both molecules arising from the split hexose molecule. Thus, the net production of ATP by glycolysis is two molecules of ATP per molecule of glucose.

ATP consumed per glucose:	2 (hexose stage)	
ATP produced per glucose:	4 (triose stage)	**(12·2)**
Net ATP production per glucose:	2	

We shall now examine each of the reactions of glycolysis in turn. The individual reactions are listed in Table 12·1.

Table 12·1 The enzymatic reactions of glycolysis

No.	Chemical reaction	Enzyme
1	Glucose + ATP ⟶ Glucose 6-phosphate + ADP + H⊕	Hexokinase, glucokinase
2	Glucose 6-phosphate ⇌ Fructose 6-phosphate	Glucose 6-phosphate isomerase
3	Fructose 6-phosphate + ATP ⟶ Fructose 1,6-*bis*phosphate + ADP + H⊕	Phosphofructokinase-1
4	Fructose 1,6-*bis*phosphate ⇌ Dihydroxyacetone phosphate + Glyceraldehyde 3-phosphate	Aldolase
5	Dihydroxyacetone phosphate ⇌ Glyceraldehyde 3-phosphate	Triose phosphate isomerase
6	Glyceraldehyde 3-phosphate + NAD⊕ + P_i ⇌ 1,3-*Bis*phosphoglycerate + NADH + H⊕	Glyceraldehyde 3-phosphate dehydrogenase
7	1,3-*Bis*phosphoglycerate + ADP ⇌ 3-Phosphoglycerate + ATP	Phosphoglycerate kinase
8	3-Phosphoglycerate ⇌ 2-Phosphoglycerate	Phosphoglycerate mutase
9	2-Phosphoglycerate ⇌ Phosphoenolpyruvate + H_2O	Enolase
10	Phosphoenolpyruvate + ADP + H⊕ ⟶ Pyruvate + ATP	Pyruvate kinase

12·2 Hexokinase Catalyzes the Phosphorylation of Glucose to Form Glucose 6-Phosphate, Consuming One Molecule of ATP

In the first reaction of glycolysis, glucose is phosphorylated at the 6 position, producing glucose 6-phosphate. This phosphoryl-group transfer reaction is catalyzed by hexokinase and consumes a molecule of ATP (Figure 12·2).

A dramatic conformational change in hexokinase accompanies the binding of glucose. As can be seen in Stereo S12·1a, yeast hexokinase contains two domains joined by a hinge region. When the enzyme binds glucose, the two domains close and rotate by about 12°, positioning the C-6 hydroxyl group of glucose for catalysis (Stereo S12·1b). Conformational changes of this type have been termed *induced fit*. In the closed conformation, the active site is relatively hydrophobic. Thus, the C-6 hydroxyl group of glucose and the γ-phosphoryl group of ATP are unsolvated, precluding hydrolysis of ATP to ADP + P_i. The juxtaposition of substrates facilitates the phosphorylation of glucose by ATP.

The reaction catalyzed by hexokinase is metabolically irreversible and is one of the regulated steps of glycolysis. The key regulatory factor is the concentration of the product, glucose 6-phosphate, which allosterically inhibits hexokinase. Hexokinase is subject to regulation by glucose 6-phosphate concentration in most mammalian cells, with the exception of liver and pancreas cells, which contain an isozyme of hexokinase called glucokinase that is unaffected by the concentration of glucose 6-phosphate. (Isozymes are nonidentical proteins that catalyze the same reactions.) Glucokinase has characteristics suited to the physiological role of the liver

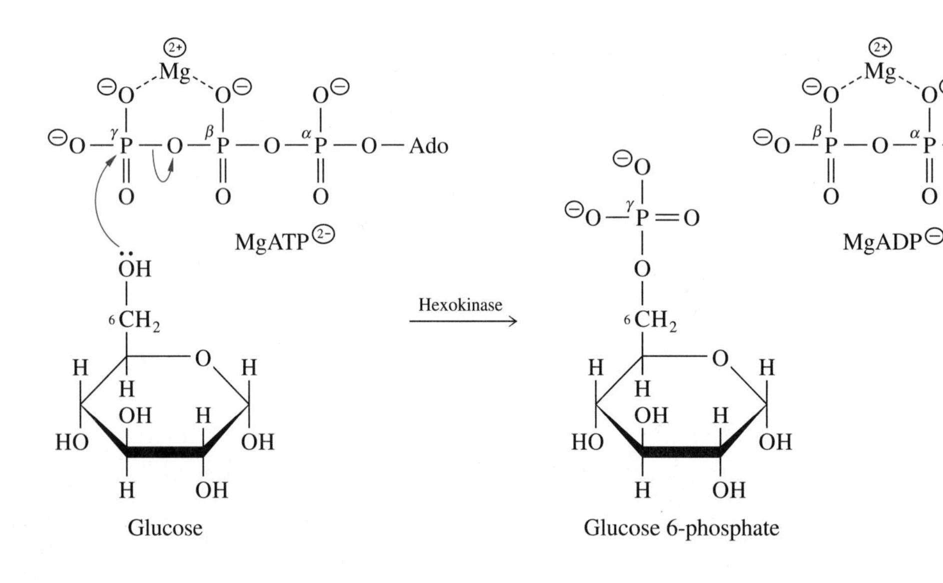

Figure 12·2
Phosphoryl-group transfer reaction catalyzed by hexokinase. This reaction occurs by attack of the C-6 hydroxyl group of glucose on the γ-phosphorus of $MgATP^{2-}$. $MgADP^{-}$ is displaced and glucose 6-phosphate is generated. Mg^{2+}, shown explicitly here, is also present in the other kinase reactions in this chapter, although it is not shown. ("Ado" represents the adenosyl moieties of ATP and ADP.)

and pancreas in managing the supply of glucose for the entire body. The K_m of glucokinase for glucose, that is, the concentration at which glucokinase is half-saturated, is about 10 mM, far greater than the K_m of other hexokinases for glucose (about 0.1 mM). In most mammalian cells, the intracellular concentration of glucose is controlled by regulated glucose transporters spanning the cell membrane and is far lower than the blood glucose concentration. The phosphorylation of glucose is regulated in these cells by internal signals, especially the concentration of glucose 6-phosphate, thereby responding, in tandem with regulated glucose transport, to the needs of the cell. In contrast, glucose freely enters liver and pancreas cells, and the concentration of glucose in these cells matches the concentration in the blood. Blood glucose concentration is typically 5 mM, and after a meal may rise to about 10 mM. Since the K_m of glucokinase for glucose is 10 mM, glucokinase in liver and pancreas is never saturated with glucose. Therefore, liver and pancreas tissue can respond to increases in blood glucose concentration by proportionate increases in the phosphorylation of glucose, serving the needs of the entire organism rather than the individual cell. Thus, the characteristics of the different isozymes suit distinct physiological purposes.

Stereo S12·1
Yeast hexokinase, space-filling models. Yeast hexokinase contains two structural domains connected by a hinge region. Upon binding of glucose (green), these domains close, shielding the active site from water. **(a)** Open conformation. **(b)** Closed conformation. (Coordinates for (a) provided by T. Steitz, C. Anderson, and R. Stenkamp; coordinates for (b) provided by W. Bennett, Jr. and T. Steitz; stereo images by Ben N. Banks and Kim M. Gernert.)

(a)

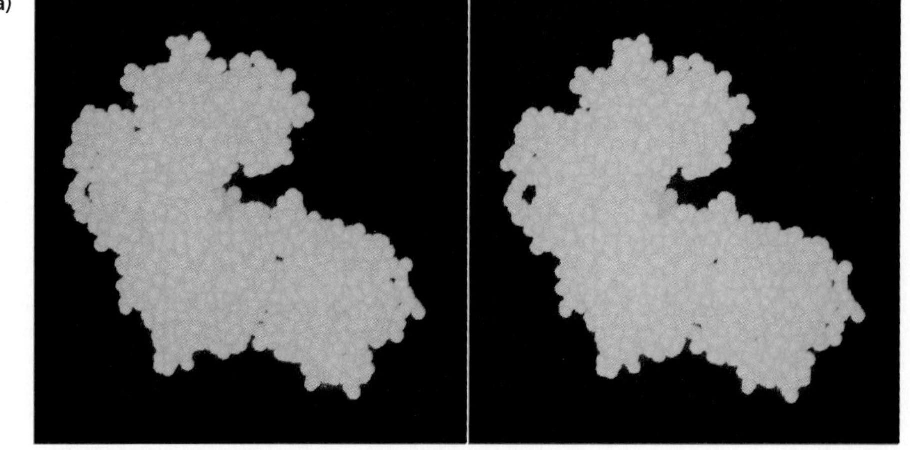

(b)

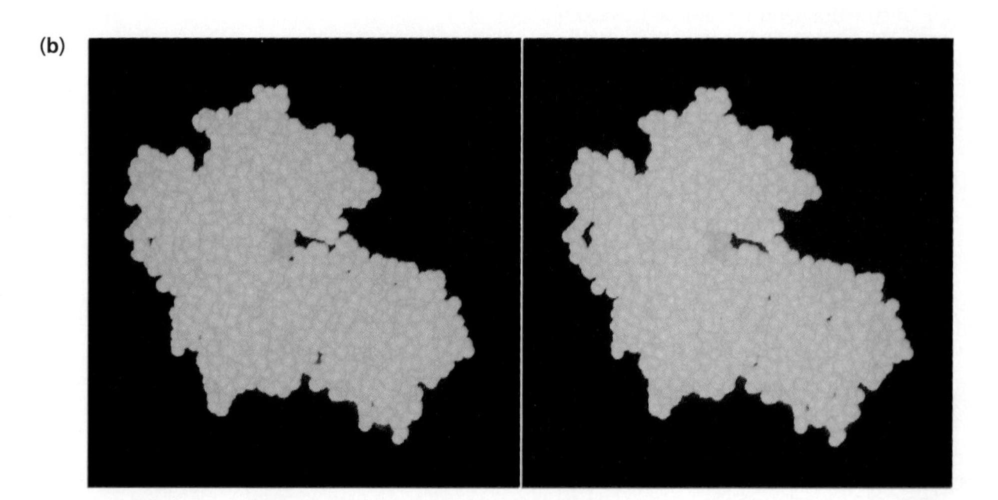

12·3 Glucose 6-Phosphate Isomerase Catalyzes the Conversion of Glucose 6-Phosphate to Fructose 6-Phosphate

In the second step of glycolysis, glucose 6-phosphate isomerase catalyzes the conversion of glucose 6-phosphate to fructose 6-phosphate. This reaction is an example of an aldose-ketose isomerization (Figure 12·3). The pyranose and furanose ring forms are the forms most stable in solution. The ring form of glucose 6-phosphate shown in Figure 12·3 (α-D-glucopyranose 6-phosphate) preferentially binds to glucose 6-phosphate isomerase. The open-chain form of glucose 6-phosphate is then generated in the active site of the enzyme, and an aldose-to-ketose conversion occurs through an enediolate intermediate (Figure 12·4). Glucose 6-phosphate isomerase catalyzes a near-equilibrium reaction in cells and therefore is not a control point for glycolysis. (The relationship between equilibrium status and regulation is discussed in Section 11·4.)

Figure 12·3
Conversion of glucose 6-phosphate to fructose 6-phosphate. The reaction is an aldose-ketose isomerization catalyzed by glucose 6-phosphate isomerase.

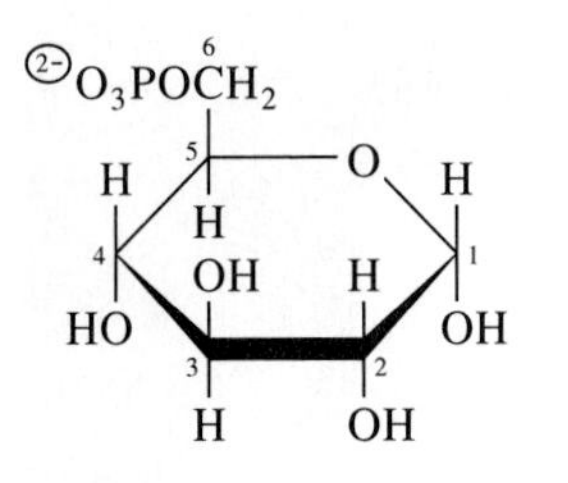

Glucose 6-phosphate
(α-D-glucopyranose form)

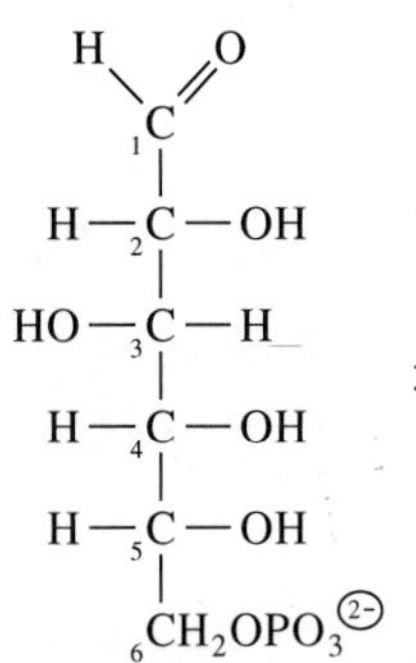

Glucose 6-phosphate
(open-chain form)

Glucose
6-phosphate
isomerase

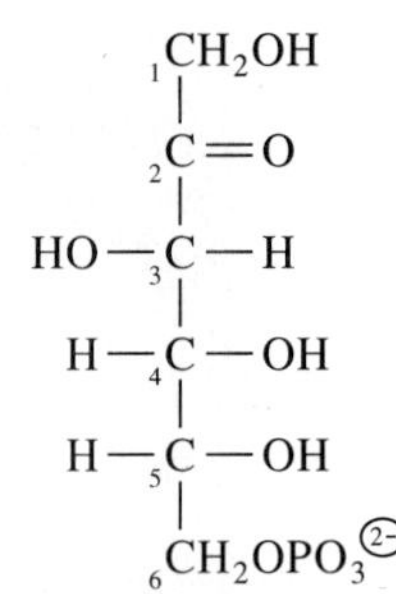

Fructose 6-phosphate
(open-chain form)

Fructose 6-phosphate
(α-D-fructofuranose form)

Figure 12·4
Mechanism of the reaction catalyzed by glucose 6-phosphate isomerase. A *cis*-enediolate intermediate is formed by proton abstraction, and the aldose-to-ketose conversion is completed by addition of a proton to C-1.

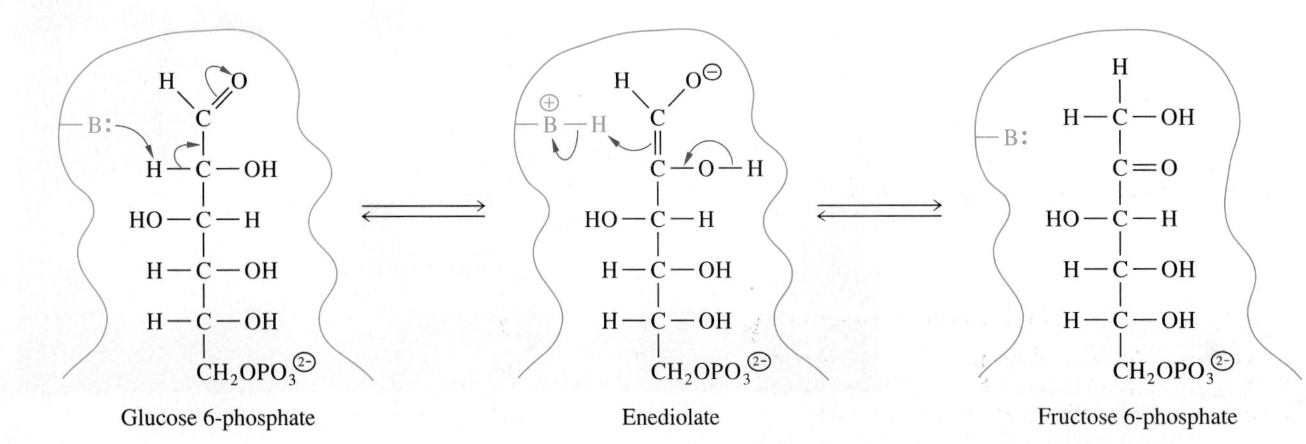

12·4 Phosphofructokinase-1 Catalyzes a Key Regulatory Step of Glycolysis and Consumes a Second Molecule of ATP

Phosphofructokinase-1 (PFK-1) is a critical regulatory enzyme for glycolysis in most cells. This enzyme catalyzes the conversion of fructose 6-phosphate to fructose 1,6-*bis*phosphate (Figure 12·5). Like hexokinase, PFK-1 catalyzes a metabolically irreversible reaction that consumes a molecule of ATP and results in the phosphorylation of a hydroxyl group of the sugar substrate.

ATP is both a substrate of PFK-1 and an allosteric inhibitor of the enzyme in vitro. ATP causes a decreased affinity of PFK-1 for fructose 6-phosphate, whereas ADP and AMP are allosteric activators that increase the affinity of PFK-1 for fructose 6-phosphate. In most cells, however, the concentration of ATP varies little, despite changes in the rate of its formation and utilization. Studies using the nondestructive technique of nuclear magnetic resonance spectroscopy have shown that ATP concentrations in vivo in muscles of several amphibians, birds, and mammals do not measurably change over an observation period of several hours. Significant changes in ADP and AMP do occur, principally because they are present in cells in much smaller concentrations than ATP, and very small changes in ATP cause proportionally larger changes in ADP and AMP. However, variations in ADP and AMP do not appear to be of major importance to the regulation of PFK-1 in living cells except under unusual conditions such as anoxia.

Citrate, an intermediate of the citric acid cycle (Chapter 13), is a physiologically important inhibitor of the reaction catalyzed by PFK-1. The citric acid cycle, which we shall examine in the next chapter, connects the partial oxidation of glucose to pyruvate with complete oxidation of pyruvate to CO_2 and H_2O. An elevated concentration of citrate serves as an indicator that ample substrate is entering the citric acid cycle. The regulatory effect of citrate on PFK-1 is an example of *feedback inhibition* that regulates the supply of pyruvate to the citric acid cycle.

Fructose 2,6-*bis*phosphate (Figure 12·6), discovered in 1980, is a potent *activator* of PFK-1, effective in the micromolar range. Fructose 2,6-*bis*phosphate is formed from fructose 6-phosphate through the action of the enzyme phosphofructokinase-2 (PFK-2). Surprisingly, in animals a different active site of the same protein catalyzes the dephosphorylation of fructose 2,6-*bis*phosphate, reforming fructose 6-phosphate. This activity of the enzyme is called fructose 2,6-*bis*phosphatase (Equation 12·3).

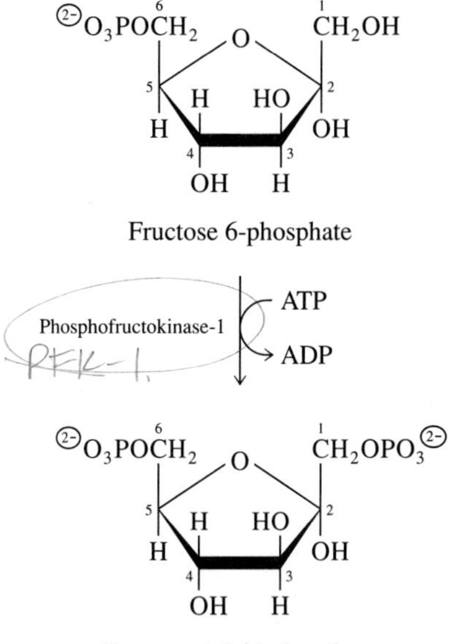

Figure 12·5
Conversion of fructose 6-phosphate to fructose 1,6-*bis*phosphate. The reaction is catalyzed by phosphofructokinase-1. In most cells, this reaction is a critical regulatory step of glycolysis.

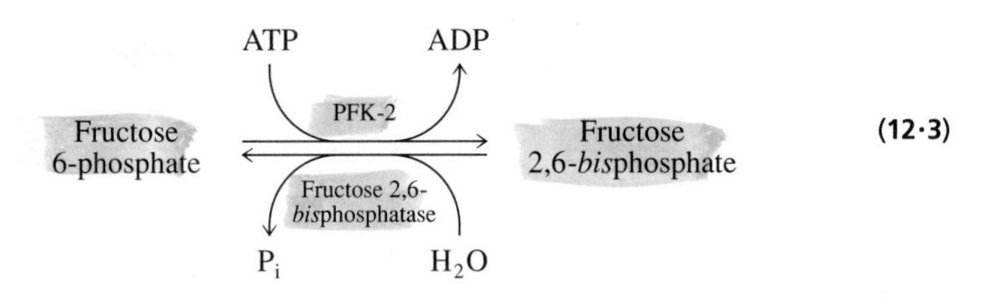

(12·3)

The bifunctionality of this enzyme presents a problem of nomenclature. The enzyme could be called phosphofructokinase-2/fructose 2,6-*bis*phosphatase. Here it is called simply phosphofructokinase-2, the "2" indicating that the enzyme catalyzes phosphorylation of fructose at the C-2 position. It is convenient that the numbering of PFK-1 and PFK-2 reflects both the order of discovery and the numbering of the carbon atoms acted upon by the enzymes. The dual activities of PFK-2 control the steady-state concentration of fructose 2,6-*bis*phosphate.

In liver, the activity of PFK-2 is linked to the action of glucagon, a hormone produced by the pancreas in response to low blood sugar. When the concentration of glucagon in the blood is elevated, a series of events is triggered in liver cells that culminates in the phosphorylation of a serine residue in PFK-2 catalyzed by

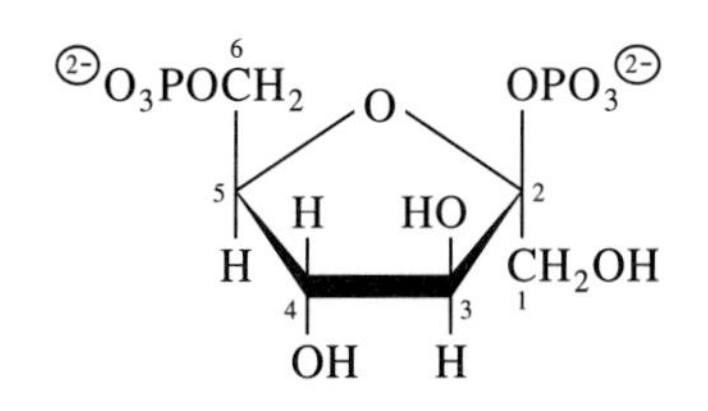

Figure 12·6
Structure of β-D-fructose 2,6-*bis*phosphate, the predominant form in cells.

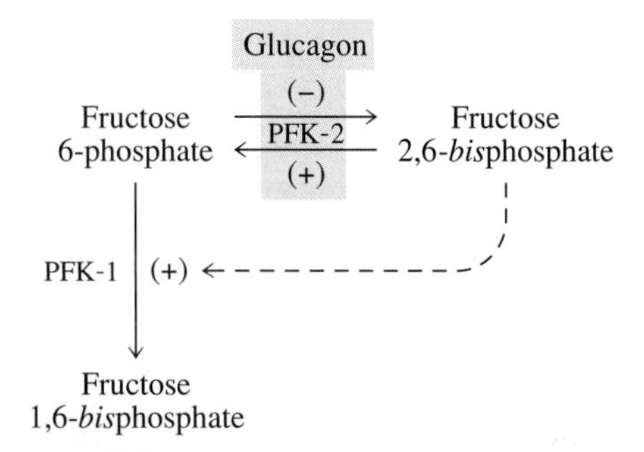

Figure 12·7
Control of PFK-1 by glucagon. When the concentration of glucagon is high, formation of the potent PFK-1 activator fructose 2,6-*bis*phosphate is decreased, and degradation of the activator is increased. As a result, the rate of glycolysis is slowed.

cAMP–dependent protein kinase (see Box 12·1). Phosphorylation decreases the kinase activity of the bifunctional enzyme and increases its phosphatase activity (Figure 12·7). Thus, the level of fructose 2,6-*bis*phosphate falls, PFK-1 becomes less active, and glycolysis is depressed. Under conditions in which glucose is rapidly metabolized, the concentration of fructose 6-phosphate increases, and more fructose 2,6-*bis*phosphate is formed, since fructose 6-phosphate is both a substrate of PFK-2 and a potent inhibitor of fructose 2,6-*bis*phosphatase. Thus, in liver cells, control of glycolysis by glucagon and glucose is accomplished through control of the bifunctional enzyme whose activity establishes the steady-state concentration of fructose 2,6-*bis*phosphate.

While other cells have regulatory systems different from that of liver, many contain fructose 2,6-*bis*phosphate, and most known isozymes of PFK-1 are stimulated by this compound. Furthermore, the isozymes of PFK-1 in most cell types are inhibited by citrate. There is a further connection between citrate and fructose 2,6-*bis*phosphate: citrate is an inhibitor of PFK-2. Thus, citrate attenuates the activity of PFK-1 in two ways: by direct inhibition and by blocking the formation of the PFK-1 activator, fructose 2,6-*bis*phosphate.

Box 12·1 Glucagon and the Liver: Hormone Action and Second Messengers

The action of hormones has been investigated by both physiologists and biochemists. Hormones regulate at the level of the whole organism, representing a control system of interest to physiologists, and they trigger a series of intracellular events that have been the focus of study by biochemists. Glucagon is a polypeptide hormone secreted into the blood by the pancreas. The effect of glucagon on liver cells is especially well understood. In Chapter 15, we shall examine in detail the regulatory role of several hormones associated with carbohydrate metabolism, including glucagon. Here, we give a brief overview of how glucagon affects the regulation of glycolysis.

The binding of glucagon to a receptor in the liver cell membrane is coupled to the formation inside the cell of a messenger molecule, cyclic adenosine 3′,5′-monophosphate, or cAMP (Figure 1). Formation of cAMP is catalyzed by adenylate cyclase, which is embedded in the plasma membrane of the cell with its active site facing the cytosol.

Since glucagon is considered the first messenger to the cell, cAMP is called a *second messenger*. Once formed, cAMP activates an enzyme in the cytosol known as cAMP-dependent protein kinase, also called protein kinase A. cAMP-dependent protein kinase is composed of a dimeric regulatory subunit and two catalytic subunits. The completely assembled enzyme is inactive. When the concentration of cAMP increases, four molecules of cAMP bind to the regulatory subunit of cAMP-dependent protein kinase, causing release of the catalytic subunits, which are then active (Figure 2). cAMP-dependent protein kinase can profoundly alter cell metabolism by catalyzing the phosphorylation of a number of interconvertible enzymes, either increasing or decreasing their activities. Phosphorylation of the enzymes is reversed by the action of protein phosphatases that catalyze removal of the covalently bound phosphoryl group, restoring the enzymes to their nonphosphorylated states. Since the enzymes subject to phosphorylation and dephosphorylation are typically regulatory enzymes, this type of covalent modification serves as a mechanism for the regulation of flux through entire pathways. The glycolytic enzymes PFK-2 and pyruvate kinase (Section 12·11) are both substrates of cAMP-dependent protein kinase. The action of cAMP-dependent protein kinase on these enzymes results in a decrease in the rate of glycolysis.

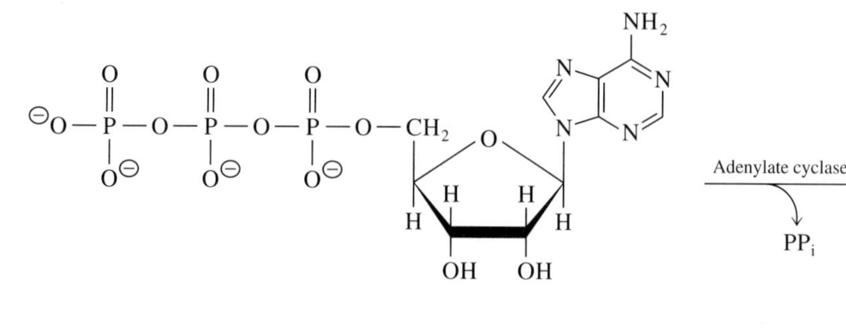

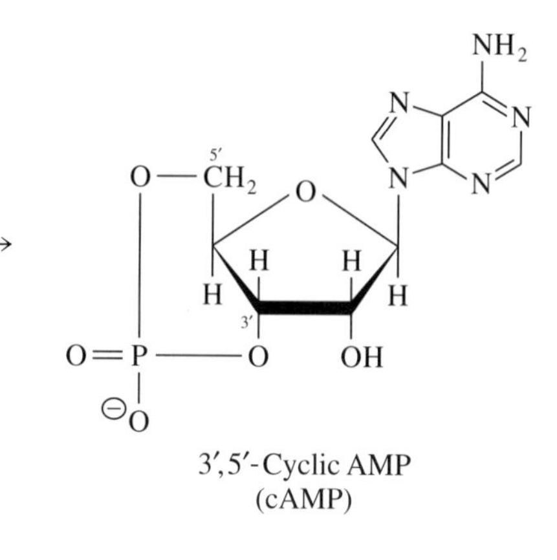

Box 12·1
Figure 1
Conversion of ATP to cyclic AMP (cAMP), catalyzed by adenylate cyclase.

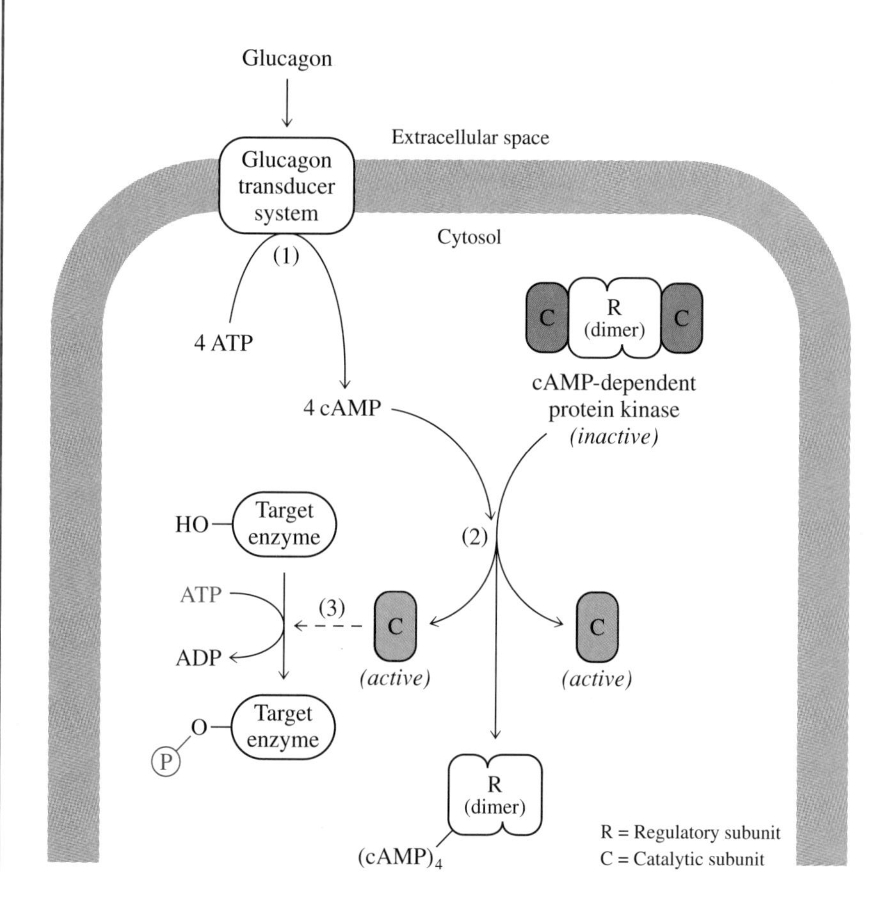

Box 12·1
Figure 2
First and second messengers in hormone action. (1) The glucagon transducer system includes the glucagon receptor, coupling proteins, and the enzyme adenylate cyclase. When glucagon binds to its receptor, adenylate cyclase is activated, and the second messenger cyclic AMP (cAMP) is formed inside the cell. (2) cAMP binds to the regulatory subunit (R) of cAMP-dependent protein kinase, which releases activated catalytic subunits (C). (3) The C subunits catalyze the phosphorylation of target enzymes. Phosphorylation can either increase or decrease the activities of the target enzymes.

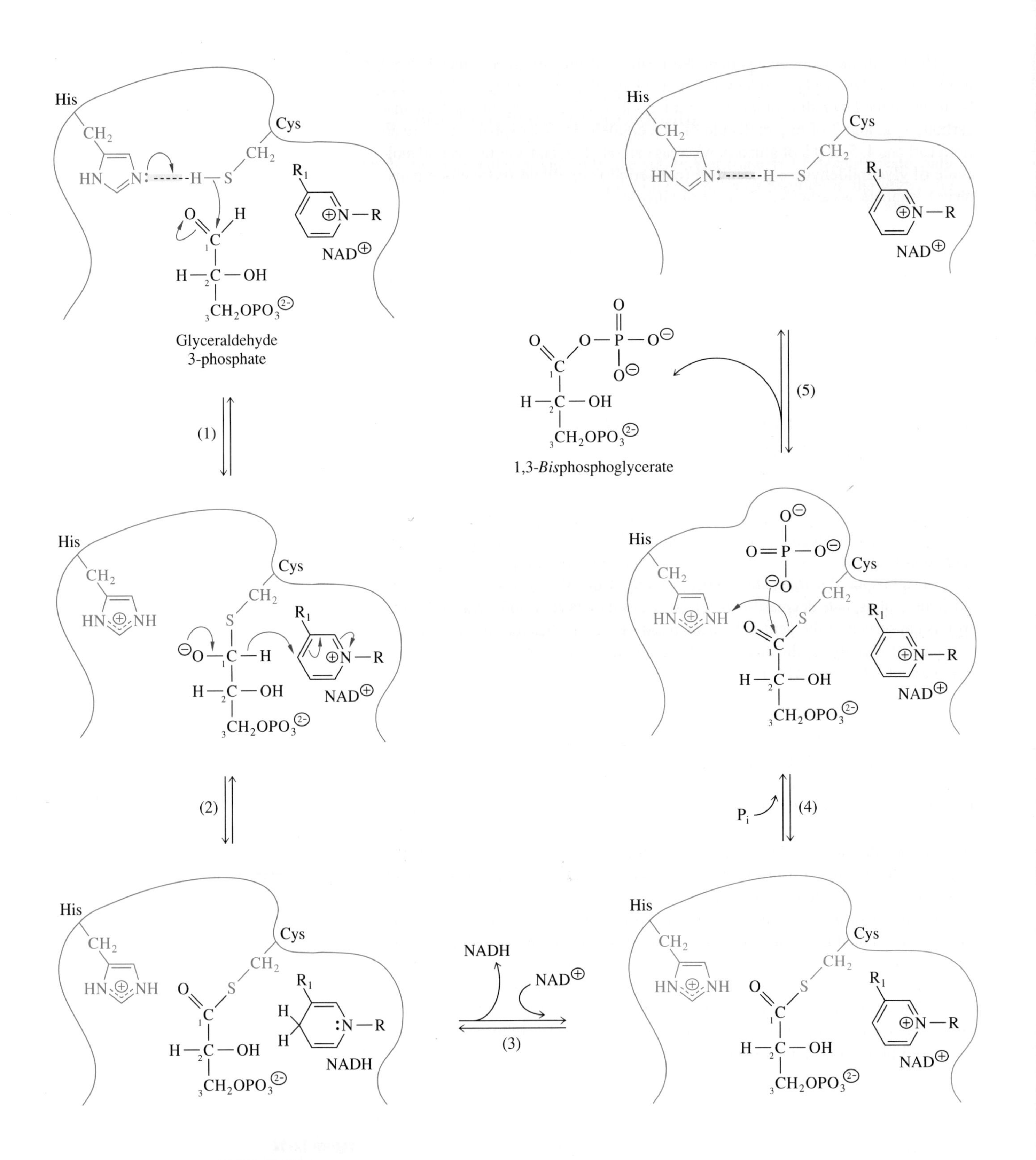

Figure 12·13
Reaction mechanism of glyceraldehyde 3-phosphate dehydrogenase. (1) An ionized cysteine sulfhydryl group attacks C-1 of glyceraldehyde 3-phosphate, forming a thiohemiacetal. (2) A hydride ion from the thiohemiacetal reduces NAD⊕ and forms a thioacyl-enzyme intermediate. (3) NADH dissociates and is replaced by NAD⊕. (4) Phosphate binds and displaces the thioacyl-enzyme intermediate, resulting in the formation of 1,3-*bis*phosphoglycerate, which dissociates from the enzyme (5).

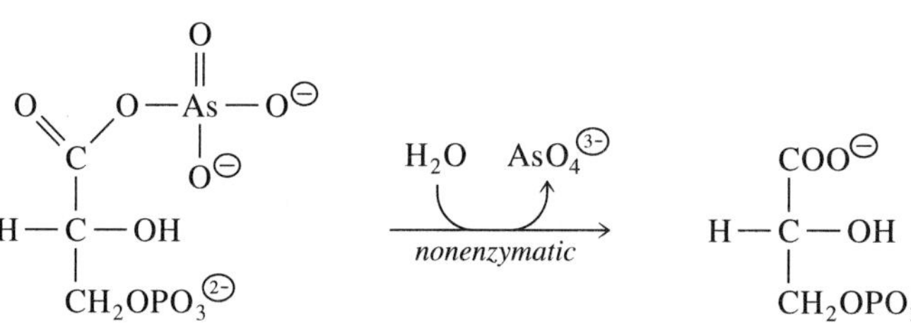

Figure 12·14
Spontaneous hydrolysis of 1-arseno-3-phosphoglycerate. Inorganic arsenate can replace inorganic phosphate as a substrate for glyceraldehyde 3-phosphate dehydrogenase, forming the unstable 1-arseno analog of 1,3-*bis*phosphoglycerate.

(Figure 12·14). This nonenzymatic hydrolysis produces 3-phosphoglycerate and regenerates inorganic arsenate, which can again react with a thioacyl-enzyme intermediate. Glycolysis can proceed from 3-phosphoglycerate, but the ATP-producing reaction involving 1,3-*bis*phosphoglycerate is bypassed. As a result, there is no net formation of ATP from glycolysis, with potentially lethal consequences.

12·8 ATP Is Generated by the Action of Phosphoglycerate Kinase

Phosphoglycerate kinase catalyzes phosphoryl-group transfer from the energy-rich mixed anhydride 1,3-*bis*phosphoglycerate to ADP, generating ATP and 3-phosphoglycerate (Figure 12·15). The formation of ATP by transfer of a phosphoryl group from a high-energy compound to ADP is termed *substrate-level phosphorylation.* This reaction is the first ATP-generating step of glycolysis, yet it is a near-equilibrium reaction in cells and therefore is not a regulatory step. This contrasts with the other glycolytic reactions that involve the production or consumption of ATP, each of which is a regulatory step of the pathway.

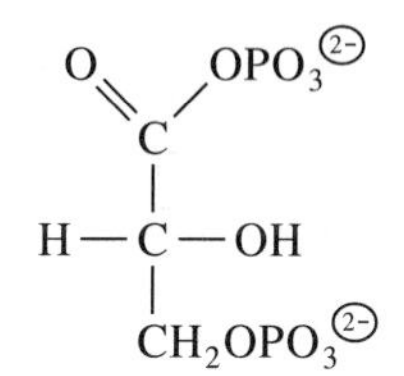

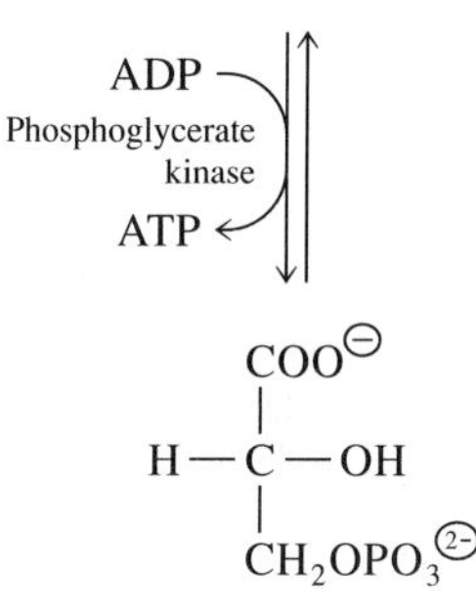

Figure 12·15
Phosphoryl-group transfer from 1,3-*bis*phosphoglycerate to ADP, catalyzed by phosphoglycerate kinase. This reaction is the first ATP-producing step of glycolysis.

12·9 Phosphoglycerate Mutase Catalyzes the Conversion of 3-Phosphoglycerate to 2-Phosphoglycerate

Phosphoglycerate mutase catalyzes a near-equilibrium reaction in which 3-phosphoglycerate and 2-phosphoglycerate are interconverted (Figure 12·16). Two distinct mechanisms are known for phosphoglycerate mutase isozymes. One is associated with animal and yeast phosphoglycerate mutases, the other with plant phosphoglycerate mutase.

The mechanism of the animal and yeast enzyme is characterized by formation of a 2,3-*bis*phosphoglycerate (2,3BPG) intermediate, as shown in Figure 12·17 (next page). The active enzyme is phosphorylated at a histidine residue prior to binding 3-phosphoglycerate. After substrate binding occurs, the enzyme transfers its phosphoryl group to the substrate to form 2,3BPG. In the next step, the phosphate from the 3 position of 2,3BPG is transferred back to the enzyme, leaving 2-phosphoglycerate.

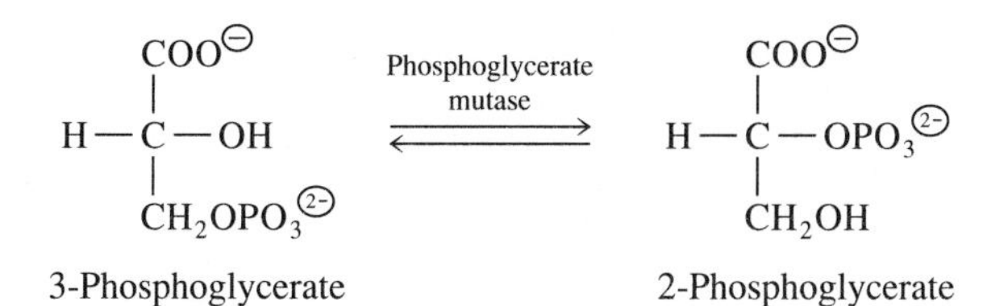

Figure 12·16
Conversion of 3-phosphoglycerate to 2-phosphoglycerate, catalyzed by phosphoglycerate mutase.

Figure 12·17
Mechanism of conversion of 3-phosphoglycerate to 2-phosphoglycerate in animals and yeast. A lysine residue at the active site of phosphoglycerate mutase binds the carboxylate anion of the substrate. A histidine residue in the enzyme, which must be initially phosphorylated in order for 3-phosphoglycerate to bind, donates its phosphoryl group to form the *bis*phosphate intermediate. Rephosphorylation of the enzyme yields 2-phosphoglycerate.

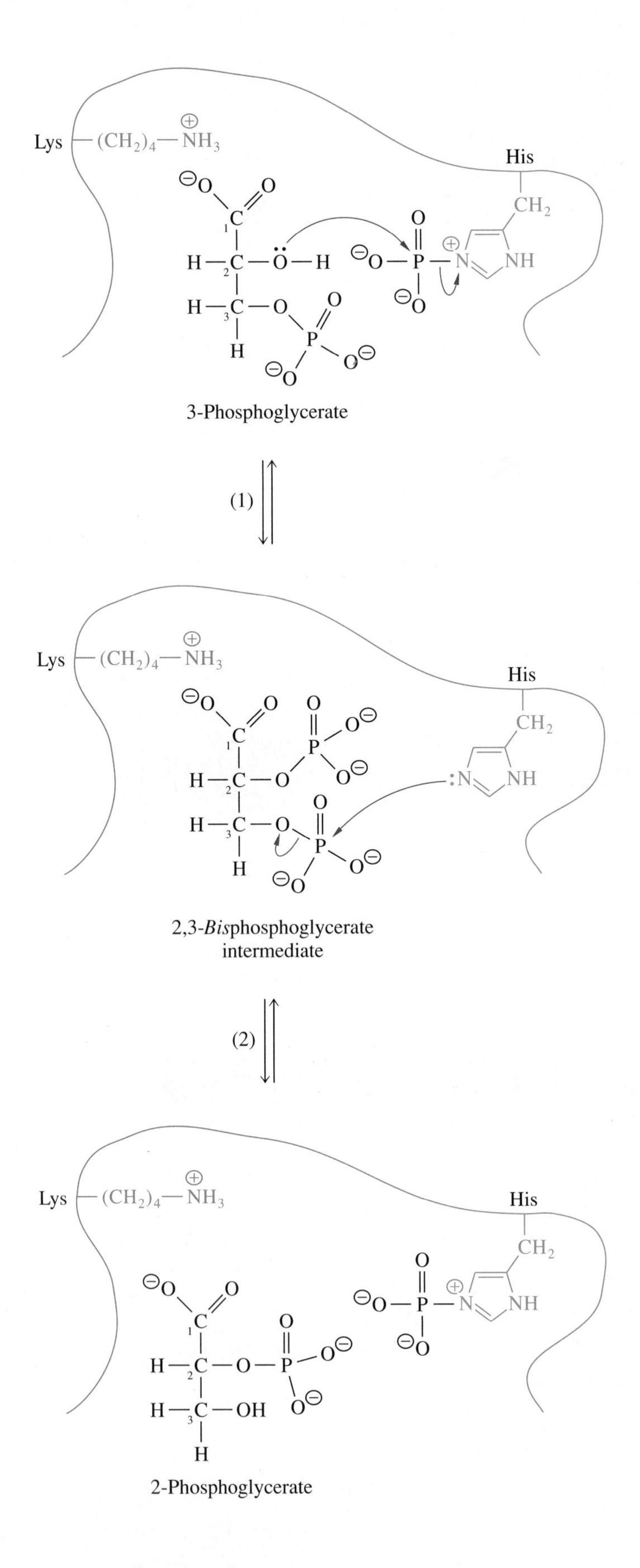

Plant phosphoglycerate mutase does not form a 2,3BPG intermediate. Instead, 3-phosphoglycerate binds to plant phosphoglycerate mutase and transfers its phosphoryl group to the enzyme. The phosphoryl group is transferred back to the substrate in the 2 position to form 2-phosphoglycerate. Note that despite their differences, both types of phosphoglycerate mutase have an intermediate that is poised to produce either 3-phosphoglycerate or 2-phosphoglycerate.

12·10 Enolase Catalyzes the Conversion of 2-Phosphoglycerate to Phosphoenolpyruvate

2-Phosphoglycerate is converted to phosphoenolpyruvate in a near-equilibrium reaction catalyzed by enolase. Enolase is a metalloenzyme that requires Mg^{2+} for activity. Magnesium ions have two roles in this protein: to stabilize the dimeric (active) form of the enzyme and to aid in binding the substrate to the enzyme. Enolase catalyzes the conversion of 2-phosphoglycerate to phosphoenolpyruvate by the reversible *trans* elimination of water (Figure 12·18).

Fluoride ions (F^{-}) inhibit enolase by forming a complex with Mg^{2+} in the active site. Fluoride was one of the first enzyme inhibitors used in laboratory investigations. However, it also inhibits a number of other metalloenzymes, and its usefulness as an investigative reagent is limited by its lack of specificity.

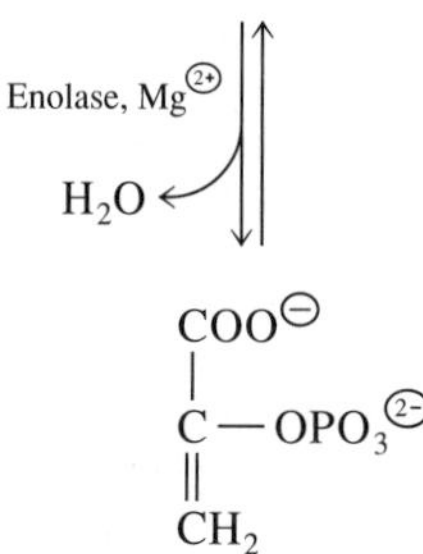

Figure 12·18
Conversion of 2-phosphoglycerate to phosphoenolpyruvate. This reaction, catalyzed by enolase, converts the phosphomonoester 2-phosphoglycerate to the energy-rich enol-phosphate ester phosphoenolpyruvate.

12·11 Pyruvate Kinase Catalyzes Phosphoryl-Group Transfer from Phosphoenolpyruvate to ADP, Forming Pyruvate and ATP

The pyruvate kinase reaction, illustrated in Figure 12·19, is the second substrate-level phosphorylation and the third metabolically irreversible reaction of glycolysis. Pyruvate kinase is subject to regulation by both allosteric effectors and covalent modification.

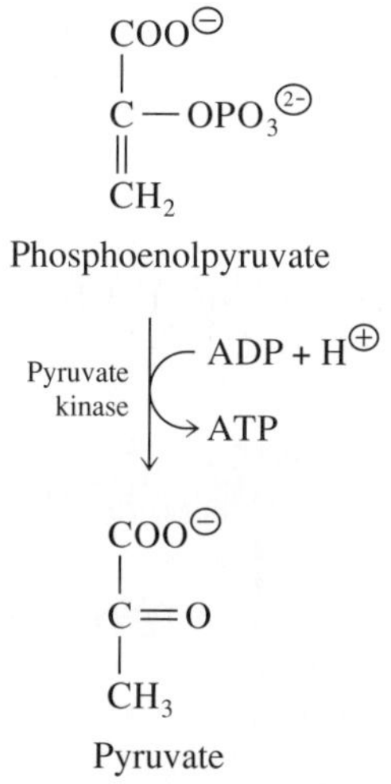

Figure 12·19
Formation of pyruvate from phosphoenolpyruvate, catalyzed by pyruvate kinase. Phosphoryl-group transfer from phosphoenolpyruvate to ADP generates ATP in this metabolically irreversible reaction.

COO^{-}
$C-OPO_3^{2-}$
CH_2
Phosphoenolpyruvate

Pyruvate kinase
$ADP + H^{+}$
ATP

COO^{-}
$C=O$
CH_3
Pyruvate

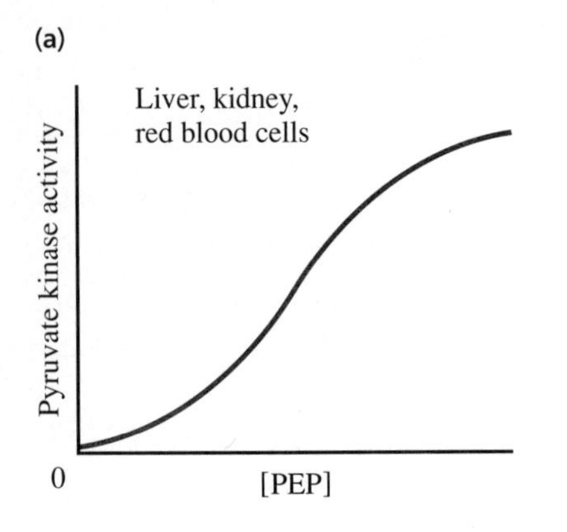

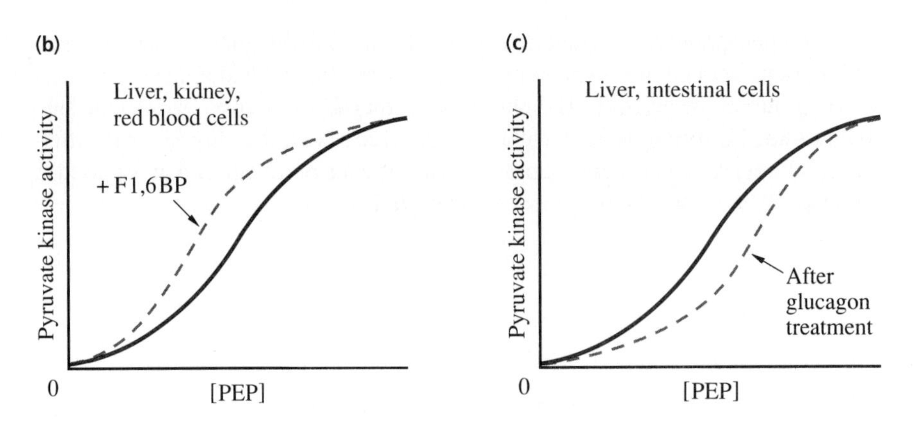

Figure 12·20
Regulation of pyruvate kinase. **(a)** A plot of phosphoenolpyruvate concentration versus enzyme activity for isozymes in some cells yields a sigmoidal curve, similar to the shape of the oxygen-binding curve of hemoglobin. **(b)** The presence of fructose 1,6-*bis*phosphate shifts the curve to the left, indicating that fructose 1,6-*bis*phosphate is an activator of the enzyme. **(c)** Incubation of cells with glucagon, which results in phosphorylation of pyruvate kinase, shifts the curve to the right, indicating a relatively less active pyruvate kinase.

There are several isozymes of pyruvate kinase in most organisms. For some of these, such as the isozymes found in liver, kidney, and red blood cells of animals, a sigmoidal curve is obtained when velocity is plotted against phosphoenolpyruvate concentration (Figure 12·20a). The presence of fructose 1,6-*bis*phosphate shifts the curve to the left (Figure 12·20b). This shift is evidence of allosteric activation. It may be appreciated from the figure that even if the concentration of substrate does not change, the enzyme activity is greater in the presence of the allosteric activator, up to the point at which the concentration of phosphoenolpyruvate is saturating. Recall that fructose 1,6-*bis*phosphate is the product of the reaction catalyzed by PFK-1. Its concentration can be expected to increase when the activity of PFK-1 is increased. Since fructose 1,6-*bis*phosphate is an activator of pyruvate kinase, activation of PFK-1 (which catalyzes an early step of the glycolytic pathway) will cause subsequent activation of pyruvate kinase (the last enzyme in the pathway). This type of regulation is called *feed-forward activation*.

The isozyme of pyruvate kinase found in mammalian liver and intestinal cells is subject to an additional type of regulation, covalent modification by phosphorylation of the enzyme. cAMP-dependent protein kinase (which also catalyzes the phosphorylation of PFK-2) catalyzes the phosphorylation of pyruvate kinase. Pyruvate kinase is less active in the phosphorylated state. The change in kinetic behavior when pyruvate kinase is phosphorylated is depicted in Figure 12·20c, showing the response in cells incubated with glucagon, which activates cAMP-dependent protein kinase. Dephosphorylation of pyruvate kinase is catalyzed by a protein phosphatase.

Pyruvate, the product of pyruvate kinase, has several possible fates. Under aerobic conditions, pyruvate is oxidized further by pathways leading ultimately to CO_2 and H_2O. Under anaerobic conditions, pyruvate forms lactate in many cell types and may have other fates, as in yeast, where pyruvate is converted to ethanol and CO_2.

12·12 Pyruvate Can Be Anaerobically Metabolized to Ethanol in Yeast

The conversion of glucose to pyruvate is accompanied not only by the synthesis of ATP, but also by the reduction of NAD⊕ to NADH at the glyceraldehyde 3-phosphate dehydrogenase step (Section 12·7). In order for glycolysis to operate continuously, the cell must have a means of regenerating NAD⊕. This is accomplished under aerobic conditions primarily by the process of *oxidative phosphorylation* (Chapter 14), which requires oxygen.

In the absence of oxygen (the anaerobic state), yeast cells convert pyruvate to ethanol and CO_2, in the process oxidizing NADH to NAD⊕. Two reactions are involved. First, pyruvate is decarboxylated to acetaldehyde in a reaction catalyzed by pyruvate decarboxylase (Section 7·5; the mechanism is shown in Figure 7·12). Next, alcohol dehydrogenase catalyzes the reduction of acetaldehyde to ethanol by transferring electrons from NADH. These reactions and the cycle of NAD⊕/NADH reduction and oxidation in alcoholic fermentation are shown in Figure 12·21.

Figure 12·21
Anaerobic conversion of pyruvate to ethanol in yeast. Pyruvate is first decarboxylated by the action of pyruvate decarboxylase. Subsequently, NADH produced by the glyceraldehyde 3-phosphate dehydrogenase reaction can be reoxidized to NAD⊕ by the action of alcohol dehydrogenase, which catalyzes reduction of acetaldehyde to ethanol. Regeneration of NAD⊕ allows fermentation to continue under anaerobic conditions.

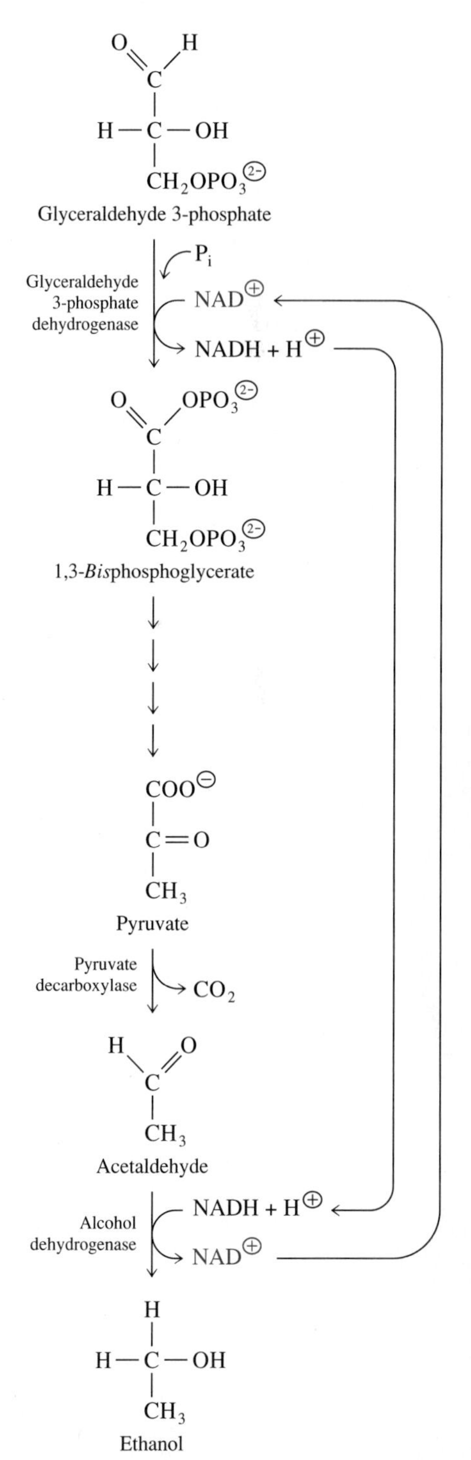

12·13 Pyruvate Can Be Converted to Lactate in Most Cells

Unlike yeast, most organisms lack pyruvate decarboxylase and therefore do not produce ethanol from pyruvate. Instead, pyruvate is reduced to form lactate in a reaction catalyzed by lactate dehydrogenase (Figure 12·22). Once formed, lactate has no other metabolic fate than reconversion to pyruvate. Hence, lactate is commonly considered a metabolic dead end. Since lactate formation catalyzed by lactate dehydrogenase regenerates NAD⊕ from NADH, the pathway of glycolysis is complete, with NAD⊕ becoming available for the glyceraldehyde 3-phosphate dehydrogenase reaction, as shown for fermentation in Figure 12·21. No net oxidation or reduction takes place during either alcoholic fermentation or anaerobic glycolysis to lactate.

Regardless of the final product—lactate or ethanol—glycolysis generates 2 mol ATP per mol of glucose consumed. Oxygen is not required in either case. This feature is not only essential for anaerobic organisms, but also for some specialized cells in multicellular organisms. In most cells, the majority of ATP is produced by oxidative phosphorylation (Chapter 14), which is a strictly oxygen-dependent process. Yet, in the cornea of the eye, for instance, oxygen availability is limited by poor blood circulation. Anaerobic glycolysis meets the need for ATP in the absence of sufficient oxygen for oxidative phosphorylation.

When skeletal muscle is active, lactate and pyruvate are transported out of muscle cells and carried via the circulatory system to the liver, where lactate is converted to pyruvate by the action of hepatic lactate dehydrogenase. Pyruvate can be further metabolized in various ways by the liver: a fraction of it is aerobically oxidized to CO_2 via the citric acid cycle (Chapter 13), and some of it is transaminated to alanine (Chapter 18), which is used in protein synthesis (Chapter 22). Pyruvate in liver cells can also be converted back to free glucose by gluconeogenesis (Section 15·4). Liver is the principal organ responsible for regulating the supply of blood glucose to other cells of the body, including skeletal muscle cells. Thus, a

Figure 12·22
Anaerobic conversion of pyruvate to lactate in muscle. The NADH produced by the glyceraldehyde 3-phosphate dehydrogenase reaction of glycolysis can be reoxidized to NAD⊕ under anaerobic conditions by this reaction, catalyzed by lactate dehydrogenase.

Pyruvate ($COO^\ominus$–C=O–CH_3) + NADH,H⊕ ⇌ (Lactate dehydrogenase) NAD⊕ + Lactate ($COO^\ominus$–HO–C–H–CH_3)

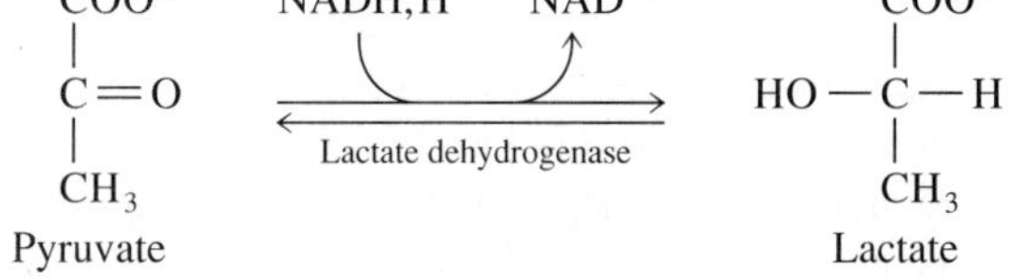

portion of the metabolic burden of muscular contraction is borne by the liver, which metabolizes the by-products of muscle activity (lactate and pyruvate) and synthesizes blood glucose. This interorgan metabolism, in which liver receives lactate and pyruvate and furnishes glucose to skeletal muscle, where the glucose is glycolytically metabolized to generate ATP, is referred to as the *Cori cycle*, in honor of the biochemists Gerti and Carl Cori, whose work contributed much toward its elucidation.

There are a number of isozymes of lactate dehydrogenase. The enzyme is tetrameric, with two types of subunits, designated M for muscle type and H for heart type. These form active enzyme in a variety of combinations: M_4, M_3H, M_2H_2, MH_3, and H_4. Thus, five separate isozymes are generated from only two distinct subunits.

Whereas some isozyme differences have obvious pertinence—for instance, pyruvate kinase of liver can be inactivated by phosphorylation but the muscle isozyme cannot—it is not always clear why isozymes have arisen. In particular, the differences between various isozymes of lactate dehydrogenase serve no apparent function. It is known that kinetic differences exist between skeletal muscle (M_4) and heart (H_4) lactate dehydrogenases in vitro, but these are of no consequence to the regulation of cellular metabolism in vivo: in both muscle and heart tissue, lactate dehydrogenase catalyzes near-equilibrium reactions. (For further discussion, see Box 12·2.)

Box 12·2 In Vitro Versus In Vivo

The relevance of kinetic observations made in vitro is never certain. The case of the lactate dehydrogenase isozymes illustrates the danger of assigning relevance based upon limited information.

In experiments conducted in vitro, the lactate dehydrogenase of skeletal muscle (M_4) catalyzes the conversion of pyruvate to lactate more efficiently than heart lactate dehydrogenase (H_4). Furthermore, the muscle isozyme in vitro has a greater affinity for pyruvate than does the heart isozyme, and pyruvate inhibits the heart isozyme. These properties would seem to fit the requirement for the heart to perform continuous aerobic work, since heart cells can utilize lactate from the blood, convert it to pyruvate, and completely oxidize the pyruvate molecules. It would also seem consistent with the ability of certain types of skeletal muscle to perform rapid spurts of activity for which high rates of glycolytic activity are needed. However, this picture is at odds with what is known about the regulation of metabolism. As mentioned earlier, lactate dehydrogenase catalyzes a near-equilibrium reaction in cells. Therefore, changes in substrate affinity cannot be a regulatory mechanism. Furthermore, not all isozymes of lactate dehydrogenase even remotely fit the expectations derived from observations in vitro. For example, liver has the M_4, or purely muscle-type isozyme, yet the lactate dehydrogenase reaction is catalyzed in both directions in liver, depending upon the metabolic situation: toward lactate when glycolysis is active and toward pyruvate when the pathway leads to the formation of glucose during gluconeogenesis (Section 15·4). Other mechanisms are now well known that explain how glycolysis and oxidative metabolism are regulated. These mechanisms involve phosphofructokinase-1, pyruvate kinase, and enzymes of oxidative metabolism to be considered in later chapters.

Kinetic data are obtained in vitro using purified enzymes under conditions in which *initial velocity* can be determined. This means that the product concentration is initially zero, and it remains essentially zero over the course of the determination. Thus, under the assay conditions, the reaction is far from equilibrium. In cells, the reaction may be near equilibrium, as is the case for lactate dehydrogenase. Thus, while initial velocity measurements allow us to determine kinetic constants, conclusions about physiological relevance require knowledge of intracellular conditions, including the concentrations of substrates and products.

12·14 The Overall Free-Energy Change of Glycolysis Is Highly Negative

As explained in the discussion of thermodynamics in Chapter 11, the free-energy change of a reaction must be negative for the reaction to occur. Obviously, all of the reactions of glycolysis in cells must have negative free-energy changes for the pathway to proceed toward product. However, the free-energy changes of the individual reactions in the glycolytic pathway are not equal: they are, by definition, large for the metabolically irreversible reactions and near zero for the near-equilibrium reactions.

In Figure 12·23, the standard free-energy changes for the glycolytic reactions and the actual free-energy changes for glycolysis in erythrocytes are plotted in a free-energy diagram. Values for the free-energy changes of the individual reactions

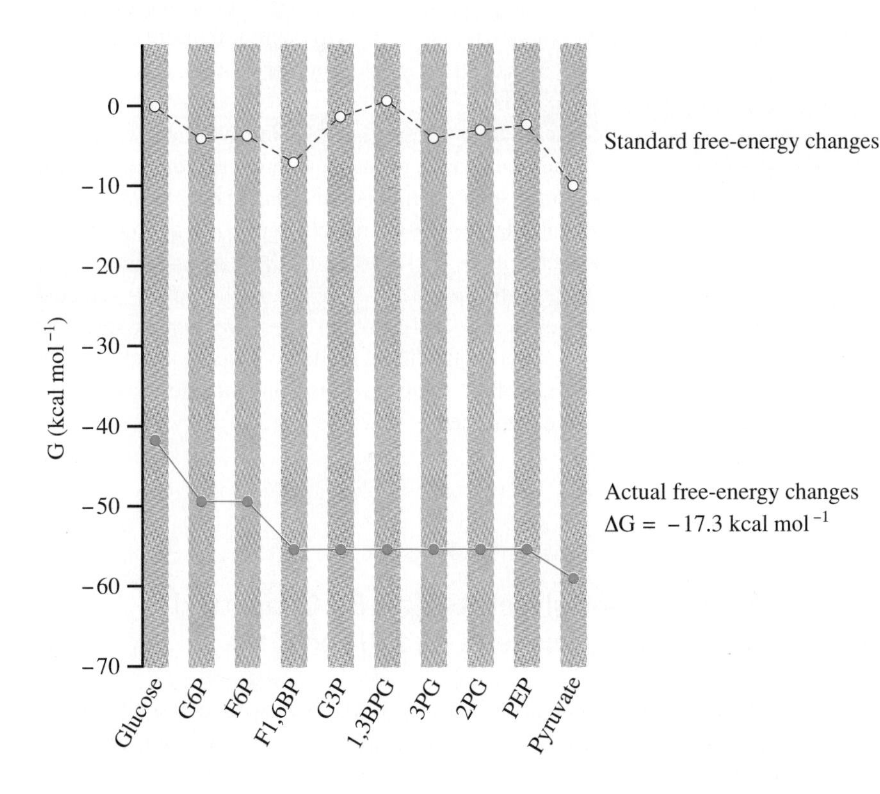

Figure 12·23
Standard and actual free-energy changes for the reactions of glycolysis. The vertical axis measures free energy in kcal mol^{-1}. Glucose in its standard state is arbitrarily assigned the value zero. The reactions of glycolysis are plotted in sequence horizontally. The upper plot (red) tracks the standard free-energy changes, and the bottom plot (blue) shows actual free-energy changes under cellular conditions. The data for actual free-energy changes are taken from erythrocytes. (Adapted from Hamori, E. (1975). Illustration of free energy changes in chemical reactions. *J. Chem. Ed.* 52:370–373.)

Table 12·2 Comparison of standard free-energy changes ($\Delta G^{\circ\prime}$) and actual free-energy changes (ΔG) in erythrocytes for the enzyme-catalyzed reactions of glycolysis

Enzyme	$\Delta G^{\circ\prime}$ (kcal mol^{-1})	ΔG (kcal mol^{-1})
Hexokinase	−4.0	−8.0
Glucose 6-phosphate isomerase	+0.4	near equilibrium
Phosphofructokinase-1	−3.4	−5.3
Aldolase	+5.7	near equilibrium
Triose phosphate isomerase	+1.8	near equilibrium
Glyceraldehyde 3-phosphate dehydrogenase	+1.5	near equilibrium
Phosphoglycerate kinase	−4.5	near equilibrium
Phosphoglycerate mutase	+1.1	near equilibrium
Enolase	+0.4	near equilibrium
Pyruvate kinase	−7.5	−4.0

are given in Table 12·2. The vertical axis of this figure is a free-energy scale. Glucose in the standard state is arbitrarily given a value of zero. The glycolytic intermediates are arranged sequentially along the horizontal axis. The figure rewards careful study by imparting a visual sense of the difference between free-energy changes under standard conditions ($\Delta G^{\circ\prime}$) and free-energy changes under cellular conditions (ΔG).

It can be readily seen from the blue plot, which tracks the actual free-energy changes for the reactions of glycolysis, that each reaction has a negative free-energy change, which must be so for the reaction to proceed toward product. It follows that the overall pathway, which is the sum of the individual reactions, must also be negative. This sum is depicted by brackets on the right side of the graph.

Notice that the actual free-energy changes are large only for the reactions catalyzed by hexokinase, phosphofructokinase-1, and pyruvate kinase—the metabolically irreversible steps of the pathway and also the steps with important regulatory roles. The ΔG values for the other steps are very close to zero. In other words, they are near-equilibrium reactions in cells.

In contrast, the standard free-energy changes for the same reactions exhibit no consistent pattern. Although there are large negative free-energy changes at the same steps that are highly exergonic under actual conditions, it is apparent that this is coincidental, as there are equally large free-energy changes for reactions known to be near-equilibrium in the cell. Furthermore, some of the standard free-energy changes for the reactions of glycolysis are *positive*, indicating that under standard conditions flux through these reactions would occur in the direction of substrate rather than product. Finally, it is apparent that there is a large difference between the free energy of the intermediates at their standard-state concentrations and at the much lower concentrations that exist in cells.

12·15 1,3-*Bis*phosphoglycerate Can Be Converted to 2,3-*Bis*phosphoglycerate in Red Blood Cells

An important function of glycolysis in red blood cells is the production of 2,3-*bis*phosphoglycerate (2,3BPG), an allosteric inhibitor of the oxygenation of hemoglobin (Section 4·18).

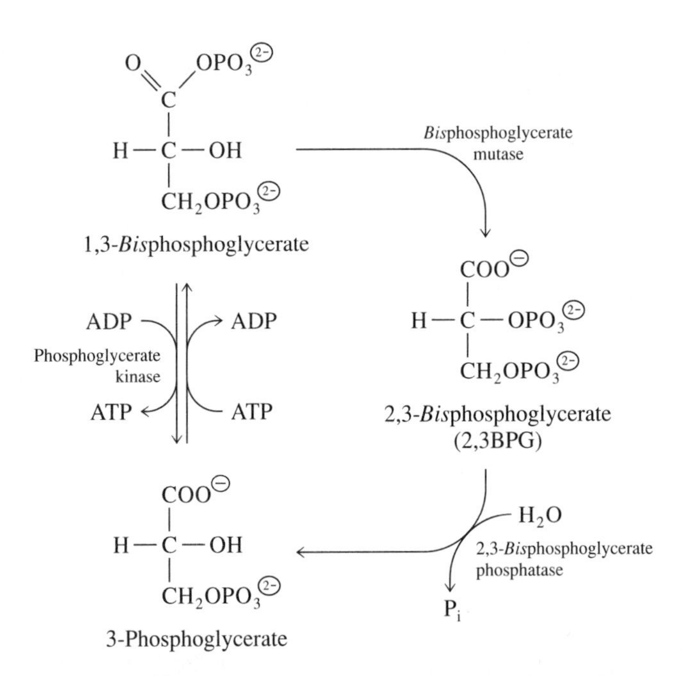

Figure 12·24
Formation of 2,3-*bis*phosphoglycerate (2,3BPG). In red blood cells, *bis*phosphoglycerate mutase catalyzes conversion of 1,3-*bis*phosphoglycerate to 2,3BPG, an allosteric inhibitor of the binding of oxygen to hemoglobin. 2,3-*Bis*phosphoglycerate phosphatase catalyzes the conversion of 2,3BPG to 3-phosphoglycerate, returning the intermediate to glycolysis. This shunt bypasses the generation of ATP by phosphoglycerate kinase.

Erythrocytes contain the enzyme *bis*phosphoglycerate mutase, which catalyzes the transfer of a phosphoryl group from C-1 to C-2 of 1,3-*bis*phosphoglycerate, forming 2,3-*bis*phosphoglycerate. As shown in Figure 12·24, 2,3-*bis*phosphoglycerate phosphatase catalyzes the hydrolysis of 2,3-*bis*phosphoglycerate to 3-phosphoglycerate, which can re-enter the glycolytic pathway and be catabolized to pyruvate.

The shunting of 1,3-*bis*phosphoglycerate through these two enzymes bypasses phosphoglycerate kinase, which catalyzes one of the two ATP-generating steps of glycolysis. However, only a small portion of glycolytic flux in blood cells—about 10%—is diverted through the mutase and phosphatase. In exchange for diminished ATP generation, this bypass provides a regulated supply of 2,3-*bis*phosphoglycerate, which is crucial to efficient transport of O_2.

12·16 Sucrose Can Be Catabolized via Glycolysis

The disaccharide sucrose (cane or beet sugar) and the monosaccharide fructose are used as sweetening agents in many foods and beverages, and they account for a significant fraction of the simple carbohydrate in human diets. The enzyme sucrase, anchored to the luminal intestinal surface by a stretch of about 30 amino acids, catalyzes the hydrolysis of ingested sucrose to glucose plus fructose. Both sugars are transported across intestinal epithelial cells and into the bloodstream. The bulk of the fructose is metabolized by the liver.

Fructose is a good substrate for hexokinase, but not for glucokinase, the isozyme of hexokinase found in liver cells. In liver, fructose is phosphorylated by a specific fructokinase that catalyzes the ATP-dependent formation of fructose 1-phosphate. A fructose 1-phosphate aldolase catalyzes cleavage of fructose 1-phosphate to dihydroxyacetone phosphate and free glyceraldehyde by a mechanism similar to that shown in Figure 12·9, which shows cleavage of fructose 1,6-*bis*phosphate to dihydroxyacetone phosphate and glyceraldehyde 3-phosphate. Glyceraldehyde in liver is then phosphorylated to glyceraldehyde 3-phosphate by

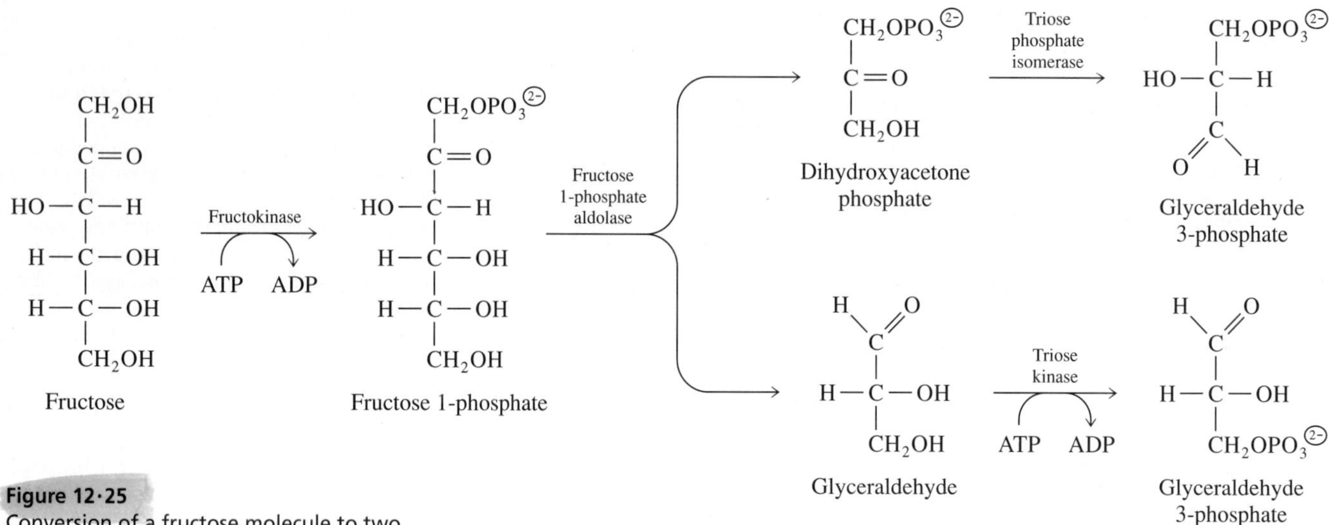

Figure 12·25
Conversion of a fructose molecule to two molecules of glyceraldehyde 3-phosphate by the actions of fructokinase, fructose 1-phosphate aldolase, triose phosphate isomerase, and triose kinase.

triose kinase, consuming a second molecule of ATP. These metabolic steps, together with the conversion of dihydroxyacetone phosphate to a second molecule of glyceraldehyde 3-phosphate by triose phosphate isomerase, are shown in Figure 12·25. Both molecules of glyceraldehyde 3-phosphate can then be metabolized to pyruvate by the remaining steps of glycolysis (triose stage).

The metabolism of 1 mol fructose to pyruvate produces 2 mol ATP and 2 mol NADH. This is the same yield as the conversion of glucose to pyruvate. However, fructose catabolism bypasses phosphofructokinase-1 and its associated regulation. Thus, diets rich in fructose or sucrose can lead to overproduction of pyruvate, which is a precursor for the synthesis of fats and cholesterol. This represents a major concern in human nutrition.

12·17 Lactose, Like Sucrose, Can Be Catabolized via Glycolysis

The disaccharide lactose (milk sugar) provides a major source of energy for nursing mammals, including human infants. Nearly all infants and young children are able to metabolize lactose by virtue of intestinal lactase, which catalyzes hydrolysis of lactose to glucose and galactose, both of which are absorbed from the intestine and transported in the circulatory system.

As shown in Figure 12·26, galactose, the C-4 epimer of glucose, can be converted to glucose-1-phosphate by a pathway that involves the recycling of uridine diphosphate-glucose (UDP-glucose, introduced in Section 7·1). In the liver, galactose is phosphorylated by the action of galactokinase, consuming a molecule of ATP. The product of this reaction is galactose 1-phosphate. Galactose 1-phosphate reacts with UDP-glucose in a reaction catalyzed by galactose 1-phosphate uridylyltransferase. The products of this reaction are glucose 1-phosphate and UDP-galactose. Glucose 1-phosphate can enter the glycolytic pathway after conversion to glucose 6-phosphate in a reaction catalyzed by phosphoglucomutase, an enzyme similar to the phosphoglycerate mutase discussed in Section 12·9. UDP-galactose, the other product of the reaction catalyzed by galactose 1-phosphate uridylyltransferase, is recycled by conversion to UDP-glucose, a reaction catalyzed by UDP-glucose 4′-epimerase.

Overall, four enzymes are required to convert galactose to glucose 6-phosphate for entry into glycolysis and further metabolism to pyruvate: galactokinase, galactose 1-phosphate uridylyltransferase, UDP-glucose 4′-epimerase, and phosphoglucomutase. Conversion of 1 mol galactose to 2 mol pyruvate produces 2 mol ATP and 2 mol NADH, the same yield as the conversions of glucose and fructose. Although there is a requirement for UDP-glucose, which is formed from the ATP equivalent UTP and glucose, only small (catalytic) amounts are needed since the molecule is recycled, and the additional amount of energy required to form UDP-glucose is therefore insignificant.

Infants fed a normal milk diet require the pathway of galactose metabolism to thrive. Those afflicted with the genetic disorder galactosemia are usually severely deficient in galactose 1-phosphate uridylyltransferase. In such cases, galactose 1-phosphate accumulates in the cells. This can lead to a compromise in liver function, recognized by the appearance of jaundice. The liver damage is potentially fatal. Screening for galactose 1-phosphate uridylyltransferase in the red blood cells of the umbilical cord can lead to detection of galactosemia at birth, and the severe effects of this genetic deficiency can be avoided by excluding lactose from the diet.

Lactose intolerance, resulting from lactase deficiency, is a very common condition in human adults. When dietary lactose is not hydrolyzed by the action of lactase, it remains in the intestinal lumen. The ingested disaccharide accumulates and upsets the normal osmotic balance between the lumen and the intestinal epithelial cells. Water then flows into the lumen from surrounding cells. Consumption of lactose by a lactose-intolerant individual results in abdominal distention and cramps. Thus, loss of a single enzyme of carbohydrate metabolism can have profound clinical consequences.

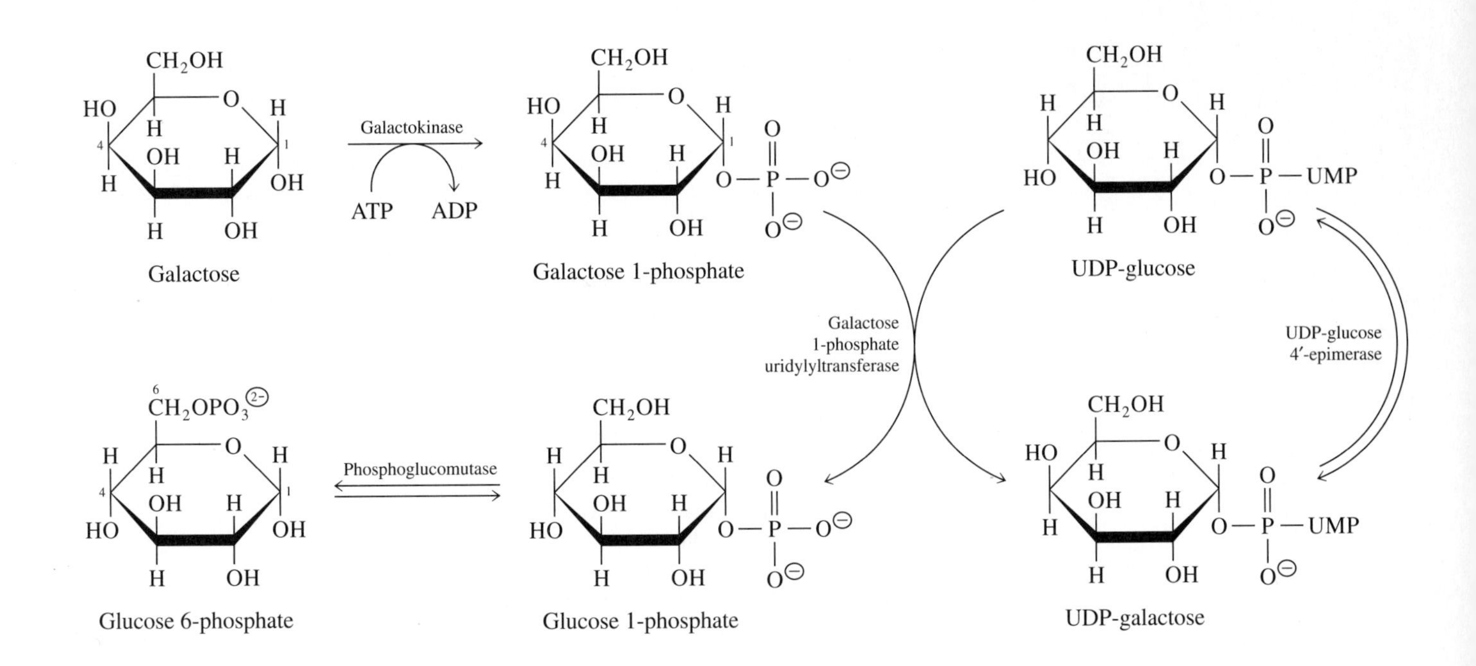

Figure 12·26
Conversion of galactose to glucose 6-phosphate. The metabolic intermediate UDP-glucose is recycled in the process. The overall stoichiometry for the pathway is galactose + ATP ⟶ glucose 6-phosphate + ADP.

Summary

Glycolysis is a ubiquitous pathway for the catabolism of monosaccharides. For each molecule of hexose that is converted to pyruvate, there is a net production of two molecules of ATP from ADP + P_i and two molecules of NAD⊕ are reduced to NADH. Glycolysis may be divided into two stages: a hexose stage, in which ATP is consumed, and a triose stage, in which a net gain of ATP is realized.

Three of the glycolytic reactions are metabolically irreversible in cells. These are the steps catalyzed by hexokinase, phosphofructokinase-1, and pyruvate kinase. Regulation of these enzymes, and thus of glycolysis, involves both allosteric interactions and covalent modifications. Regulation by covalent modification is mediated by enzymes that are not part of the glycolytic pathway but that act on enzymes of the pathway.

Anaerobic reoxidation of glycolytically produced NADH to NAD⊕ can occur through reductive metabolism of pyruvate. During alcoholic fermentation in yeast, pyruvate is cleaved to acetaldehyde in a reaction catalyzed by pyruvate decarboxylase, and acetaldehyde is reduced to ethanol, with concomitant oxidation of NADH to NAD⊕, in a reaction catalyzed by alcohol dehydrogenase. In anaerobic glycolysis to lactate, NADH is oxidized to NAD⊕ during the reduction of pyruvate to lactate, which is catalyzed by lactate dehydrogenase.

Specialized pathways allow fructose and galactose to enter glycolysis. Disease states are associated with missing or defective enzymes in the metabolism of these sugars.

Selected Readings

Cullis, P. M., and Knowles, J. R. (1987). Acyl group transfer–phosphoryl group transfer. In *Enzyme Mechanisms*, M. I. Page and A. Williams, eds. (London: Royal Society of Chemistry), pp. 178–220.

Hamori, E. (1975). Illustration of free energy changes in chemical reactions. *J. Chem. Ed.* 52:370–373.

Hers, H.-G., and Van Schaftingen, E. (1982). Fructose 2,6-*bis*phosphate 2 years after its discovery. *Biochem. J.* 206:1–12.

Hoffmann-Ostenhof, O., ed. (1987). *Intermediary Metabolism* (New York: Van Nostrand Reinhold Company). This collection of classic papers in metabolism contains the 1897 paper by Eduard Buchner from which the quotation in the introduction was taken.

Lienhard, G. E., Slot, J. W., James, D. E., and Mueckler, M. M. (1992). How cells absorb glucose. *Sci. Am.* 266(1):86–91. Accessible review of a topic relevant to this chapter. (The glucose transporter is described in this text in Section 9·13A.)

Pilkis, S. J., El-Maghrabi, M. R., and Claus, T. H. (1988). Hormonal regulation of hepatic gluconeogenesis and glycolysis. *Annu. Rev. Biochem.* 57:755–783.

Pilkis, S. J., and Granner, D. K. (1992) Molecular physiology of the regulation of hepatic gluconeogenesis and glycolysis. *Annu. Rev. Physiol.* 54:885–909.

Saier, M. H. (1987). *Enzymes in Metabolic Pathways* (New York: Harper & Row). Readily accessible accounts of enzyme mechanisms.

Schaub, J., van Hoof, F., and Vis, H. L. (1991). *Inborn Errors of Metabolism* (New York: Raven Press). Examines many genetic disorders, including the disorders of lactose and galactose metabolism discussed in this chapter.

13

The Citric Acid Cycle

In the laboratory, organic compounds composed of carbon, hydrogen, and oxygen can be completely oxidized to CO_2 and H_2O by combustion. During this process, energy is released in the form of heat. In the cells of aerobic organisms, pyruvate formed by glycolysis can be oxidized to CO_2 and H_2O through a series of enzymatic steps, during which much of the energy released is captured in the form of energy-rich compounds. The first enzymatic step in the conversion of pyruvate to CO_2 and H_2O is an oxidative decarboxylation reaction involving coenzyme A (CoASH). The product of this reaction, in addition to CO_2, is acetyl CoA, a molecule consisting of a two-carbon moiety attached to the group carrier coenzyme A (Section 7·9, Figure 7·25). Subsequent oxidation of the acetyl group of acetyl CoA is carried out by the *citric acid cycle*. The energy released in the oxidation reactions of the citric acid cycle is transferred as reducing power to the coenzymes $NAD^{\oplus}$ and ubiquinone (Q, Section 7·12), forming NADH and ubiquinol (QH_2).

The citric acid cycle is also known as the tricarboxylic acid cycle, because there are tricarboxylate intermediates, and as the Krebs cycle, after the biochemist Hans Krebs who discovered it. Proposed in 1937, the citric acid cycle was only the second metabolic cycle to be discovered. The first was the urea cycle (Chapter 18), outlined by Krebs in 1933. A third metabolic cycle, the glyoxylate cycle, was proposed by Krebs and Hans Kornberg in 1957. The glyoxylate cycle, a variation of the citric acid cycle, is described later in this chapter.

The citric acid cycle is the hub of aerobic metabolism. The aerobic catabolism of carbohydrates, fats, and amino acids merge at the reactions of the citric acid cycle, and intermediates of the citric acid cycle are the starting point for many biosynthetic pathways. The citric acid cycle is thus both catabolic and anabolic, or *amphibolic*.

The enzymes of the citric acid cycle are found in the cytosol of prokaryotes and in eukaryotic mitochondria (which are descended from bacteria). Before pyruvate produced in the cytosol by glycolysis can enter the citric acid cycle, it must be converted to acetyl CoA. In eukaryotes, enzymatic conversion to acetyl CoA occurs after transport of pyruvate into the mitochondrion.

13·1 Pyruvate Enters the Mitochondrion via a Transport Protein Embedded in the Inner Mitochondrial Membrane

The mitochondrion is delimited by a double membrane, but only the inner membrane presents a barrier to the passage of small molecules such as pyruvate. Small molecules pass through the outer membrane via an aqueous channel formed by a protein called *porin* that allows free diffusion of molecules of MW <10 000. Embedded in the inner membrane is a protein, pyruvate translocase, that allows transport of pyruvate from the intermembrane space to the interior space of the mitochondrion, known as the mitochondrial matrix. In Figure 13·1, pyruvate is shown

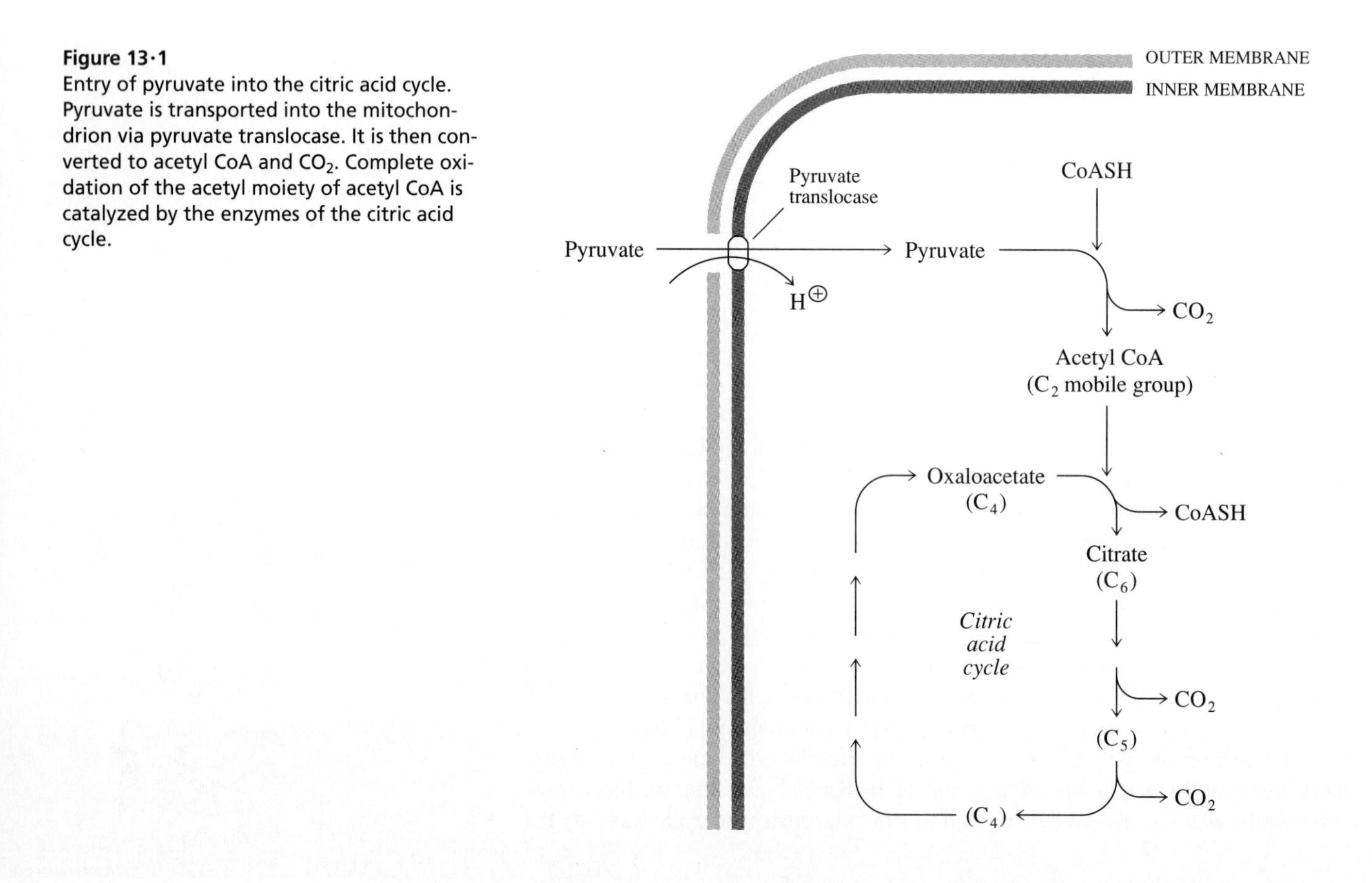

Figure 13·1
Entry of pyruvate into the citric acid cycle. Pyruvate is transported into the mitochondrion via pyruvate translocase. It is then converted to acetyl CoA and CO_2. Complete oxidation of the acetyl moiety of acetyl CoA is catalyzed by the enzymes of the citric acid cycle.

crossing the membrane in symport with H⊕. Once inside the mitochondrion, pyruvate is converted to CO_2 and acetyl CoA, which is further oxidized by the reactions of the citric acid cycle.

13·2 The Pyruvate Dehydrogenase Complex Converts Pyruvate to Acetyl CoA

In both prokaryotes and eukaryotes, the conversion of pyruvate to acetyl CoA and CO_2 is catalyzed by a complex of enzymes and cofactors known as the pyruvate dehydrogenase complex. The overall reaction is:

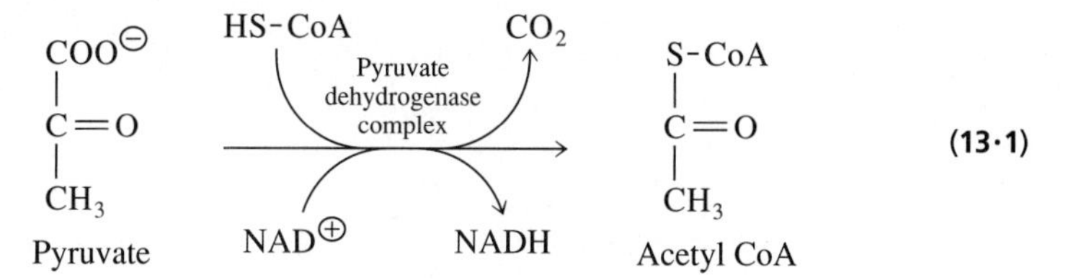

(13·1)

The pyruvate dehydrogenase complex is a *multienzyme complex,* that is, a noncovalently linked assembly of enzyme molecules that catalyze successive reactions. The product formed by one reaction of the enzyme complex does not diffuse into the medium but is immediately acted upon by the next component of the system. This is referred to as *channelling* of metabolites. Channelling greatly increases catalytic efficiency (Box 13·1).

The individual components and the reaction sequence of the pyruvate dehydrogenase complex are shown in Figure 13·2. The complex consists of multiple copies of three enzymes: pyruvate dehydrogenase (E_1), dihydrolipoamide acetyltransferase (E_2), and dihydrolipoamide dehydrogenase (E_3).

E_1 catalyzes the decarboxylation of pyruvate and the transfer of the remaining two-carbon fragment to E_2. Pyruvate first reacts with the prosthetic group of E_1,

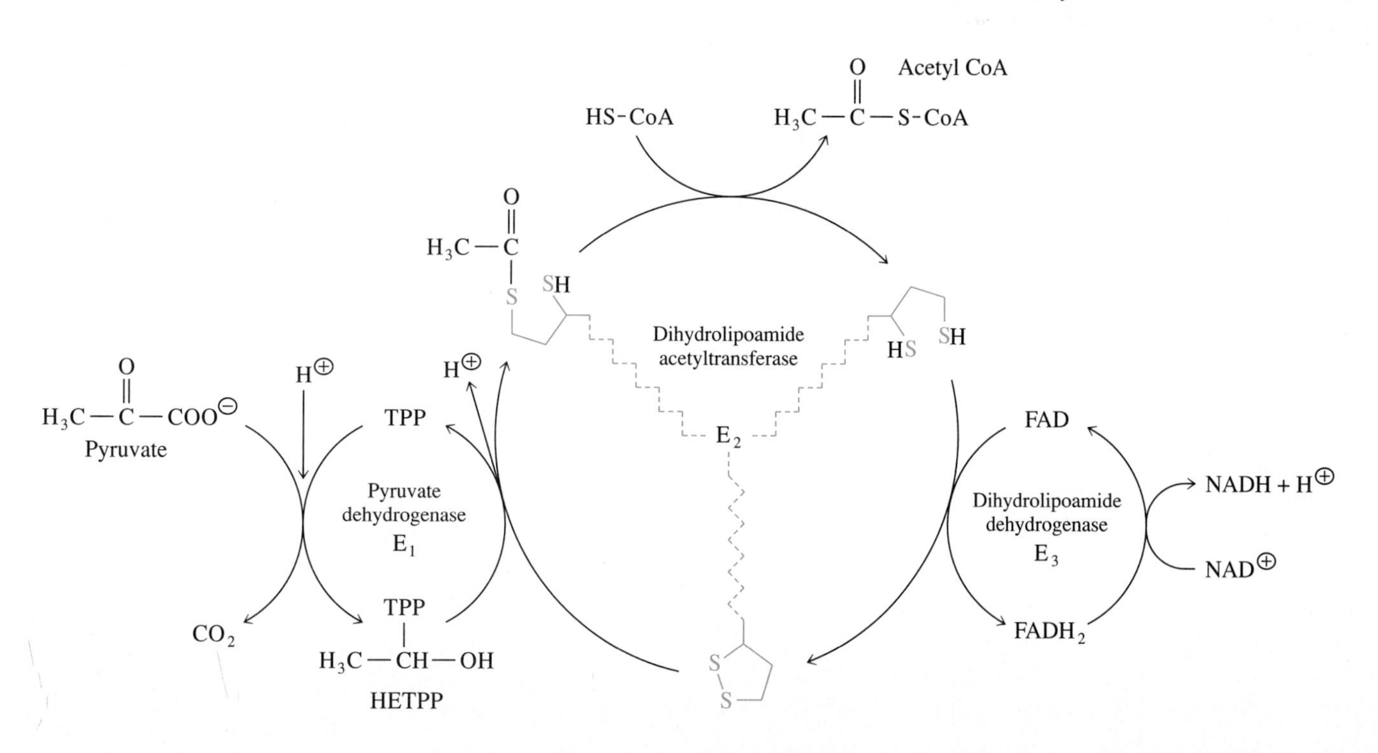

Figure 13·2
Reactions of the pyruvate dehydrogenase complex. A product is released at each step: CO_2 at pyruvate dehydrogenase (E_1), acetyl CoA at dihydrolipoamide acetyltransferase (E_2), and NADH at dihydrolipoamide dehydrogenase (E_3). The swinging arm of E_2 is a lipoamide prosthetic group formed by amide linkage of lipoic acid to a lysine residue of the enzyme.

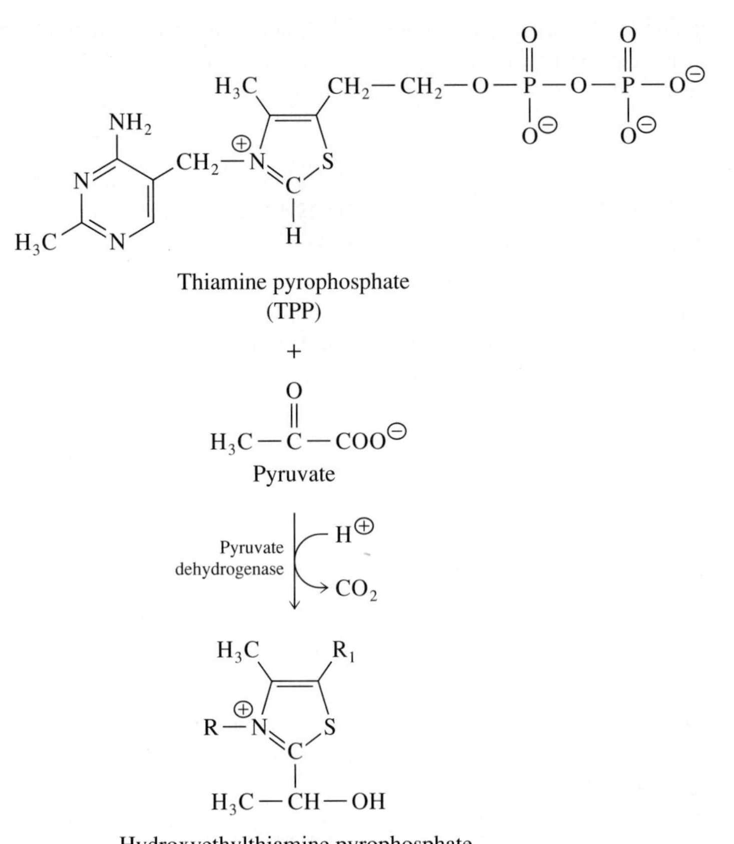

Figure 13·3
Formation of hydroxyethylthiamine pyrophosphate (HETPP) in the pyruvate dehydrogenase reaction.

thiamine pyrophosphate (TPP, Section 7·5); after release of CO_2, the resulting intermediate is hydroxyethylthiamine pyrophosphate (HETPP), as shown in Figure 13·3. The mechanism of this reaction is similar to that of pyruvate decarboxylase, diagrammed in Figure 7·12 as an example of the role of TPP. In that reaction, HETPP is cleaved, releasing acetaldehyde. In the reaction catalyzed by E_1, the two-carbon hydroxyethyl fragment is instead transferred to a lipoamide prosthetic group of E_2. This prosthetic group consists of lipoic acid covalently bound in amide linkage to a lysine residue of E_2 (Figure 13·4).

The lipoamide prosthetic group acts as a swinging arm that moves between the reactive sites of E_1 and E_3. The transfer of the two-carbon hydroxyethyl fragment from E_1 to the lipoamide prosthetic group of E_2 is thought to involve first the oxidation of HETPP by the disulfide form of the lipoamide group to form acetyl TPP, followed by transfer of the acetyl group to the dihydro form of lipoamide (Figure 13·5). Next, CoASH reacts with the acetyl group, forming acetyl CoA and leaving the lipoamide in the reduced, sulfhydryl form (top of Figure 13·2).

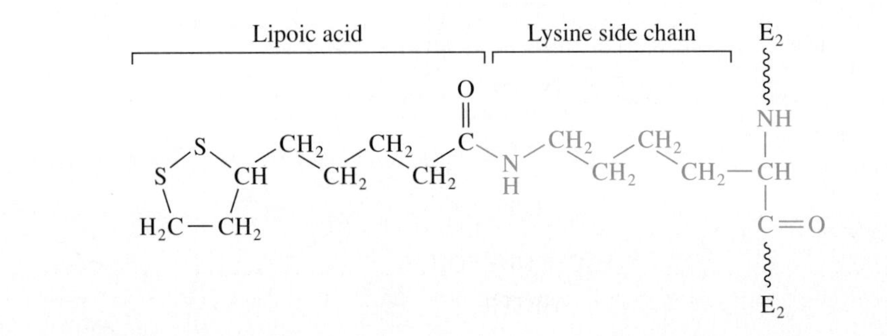

Figure 13·4
Lipoamide prosthetic group of E_2 of the pyruvate dehydrogenase complex.

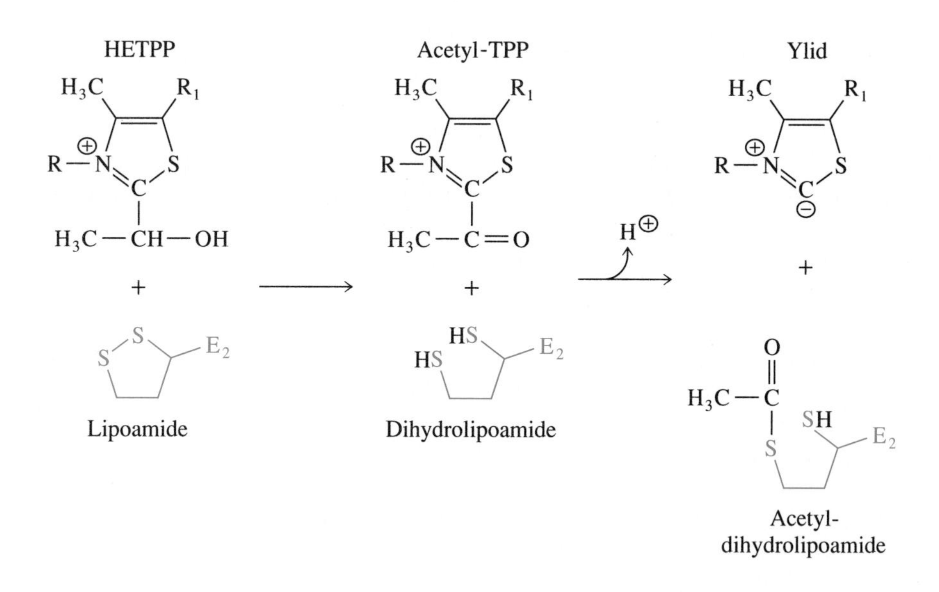

Figure 13·5
Possible mechanism for conversion of the hydroxyethyl group of HETPP to the acetyl group of acetyl-dihydrolipoamide. In this reaction, oxidation of HETPP to acetyl-TPP is coupled to reduction of the disulfide of the lipoamide group. The acetyl group is then transferred to a sulfhydryl group of dihydrolipoamide.

E_3 catalyzes the reoxidation of the reduced lipoamide of E_2, allowing E_2 to participate in another round of catalysis. The prosthetic group of E_3, flavin adenine dinucleotide (FAD, Figure 7·8), oxidizes the reduced lipoamide, resulting in formation of the reduced coenzyme, E_3-$FADH_2$. $NAD^{\oplus}$ is then reduced by E_3-$FADH_2$, so that E_3-FAD and NADH are formed and the catalytic cycle is completed.

An important distinction between FAD and $NAD^{\oplus}$ is that FAD remains tightly bound to the enzyme with which it is associated, whereas $NAD^{\oplus}$ is released to solution. Using the terminology introduced in Chapter 7, FAD is a prosthetic group, whereas $NAD^{\oplus}$ is a cosubstrate. In the pyruvate dehydrogenase complex, the E_3-$FADH_2$ complex must be reoxidized by reaction with $NAD^{\oplus}$ to regenerate the original holoenzyme and complete the catalytic cycle. Reduced NADH dissociates into the mitochondrial matrix where it serves as a mobile carrier of reducing power.

The pyruvate dehydrogenase complex is a large assembly with multiple copies of E_1 and E_3 surrounding a core consisting of many E_2 monomers. Table 13·1 lists the components and stoichiometry of the complexes in mammals and in *E. coli*. The molecular weight of the complex in *E. coli* is about five million; in mammals, it is about nine million. A protein species present only in mammalian pyruvate dehydrogenase complexes, called protein X, contains a bound lipoyl group. Befitting its name, the function of this protein is unknown. The eukaryotic complex also has associated with it several copies of a kinase and a phosphatase. These enzymes have regulatory roles that will be examined in Section 13·15.

Table 13·1 Components of the pyruvate dehydrogenase complex in mammals and *E. coli*

Enzyme	Coenzyme	Oligomeric form		Number of oligomers per complex	
		Mammals	*E. coli*	Mammals	*E. coli*
Pyruvate dehydrogenase (E_1)	TPP	$\alpha_2\beta_2$	α_2	20–30	12
Dihydrolipoamide acetyltransferase (E_2)	Lipoic acid, CoASH	α_{60}	α_{24}	1	1
Dihydrolipoamide dehydrogenase (E_3)	FAD, $NAD^{\oplus}$	α_2	α_2	6	6
X	Lipoic acid	Monomer	—	6	—
Kinase	None	$\alpha\beta$	—	2–3	—
Phosphatase	None	$\alpha\beta$	—	>3	—

[Adapted from Patel, M. S., and Roche, T. E. (1990). Molecular biology and biochemistry of pyruvate dehydrogenase complexes. *FASEB Journal* 4:3224–3233.]

Box 13·1 Channelling Versus Diffusion

Given the potential benefits of channelling, the question naturally arises: Why don't all pathways use this strategy? Channelling certainly improves catalytic efficiency, since the substrates and products pass directly from one active site to the next, never visiting the bulk solution. Indeed, some investigators are sympathetic to theories of extensive channelling of metabolic intermediates within pathways, interpreting data that cannot be easily explained otherwise as evidence of channelling.

However, channelling may not be a widespread phenomenon. Perhaps the most significant disadvantage of channelling is that intermediates do not diffuse into bulk solvent, which eliminates the possibility of metabolic branch points. Branch points are common in metabolic pathways. For example, virtually every intermediate of glycolysis is associated with other cellular pathways. Thus, although catalytic efficiency might be increased by channelling, metabolic flexibility would be lost.

Channelling has been proposed for a number of pathways, including sections of glycolysis and the citric acid cycle, less often because of compelling positive findings of complexed enzymes than because of anomalous data that resist more conventional explanation. In some cases, findings that have been ascribed to channelling have succumbed after a time to more pedestrian explanations. In other cases, puzzles remain, with channelling a plausible explanation. Finally, there are examples where the evidence for channelling is unambiguous, as is the case for the pyruvate dehydrogenase complex.

Channelling must be recognized as a possibility when good reason exists to suppose that a pathway segment might employ such a mechanism. Equally important is prudence when weighing the evidence for channelling. The extent of channelling is a question that engenders lively debate among biochemists, due in large part to our incomplete understanding of how pathways are integrated and regulated in cells.

13·3 The Citric Acid Cycle Generates Energy-Rich Molecules

Acetyl CoA formed from pyruvate or as the product of other pathways (such as the catabolism of fatty acids or amino acids) can be oxidized by the citric acid cycle. The reactions of the citric acid cycle are listed in Table 13·2 and shown in Figure 13·6. Before examining the reactions individually, we shall consider two general

Table 13·2 The enzymatic reactions of the citric acid cycle

Step	Reaction	Enzyme
1	Acetyl CoA + Oxaloacetate + H_2O $\longrightarrow$ Citrate + CoASH + $H^{\oplus}$	Citrate synthase
2	Citrate $\rightleftharpoons$ Isocitrate	Aconitase (Aconitate hydratase)
3	Isocitrate + $NAD^{\oplus}$ $\longrightarrow$ α-Ketoglutarate + NADH + CO_2	Isocitrate dehydrogenase
4	α-Ketoglutarate + CoASH + $NAD^{\oplus}$ $\longrightarrow$ Succinyl CoA + NADH + CO_2	α-Ketoglutarate dehydrogenase complex
5	Succinyl CoA + GDP (or ADP) + P_i $\rightleftharpoons$ Succinate + GTP (or ATP) + CoASH	Succinyl-CoA synthetase
6	Succinate + Q $\rightleftharpoons$ Fumarate + QH_2	Succinate dehydrogenase complex
7	Fumarate + H_2O $\rightleftharpoons$ L-Malate	Fumarase (Fumarate hydratase)
8	L-Malate + $NAD^{\oplus}$ $\rightleftharpoons$ Oxaloacetate + NADH + $H^{\oplus}$	Malate dehydrogenase

Net equation:
Acetyl CoA + 3 $NAD^{\oplus}$ + Q + GDP (or ADP) + P_i + 2 H_2O $\longrightarrow$ CoASH + 3 NADH + QH_2 + GTP (or ATP) + 2 CO_2 + 2 $H^{\oplus}$

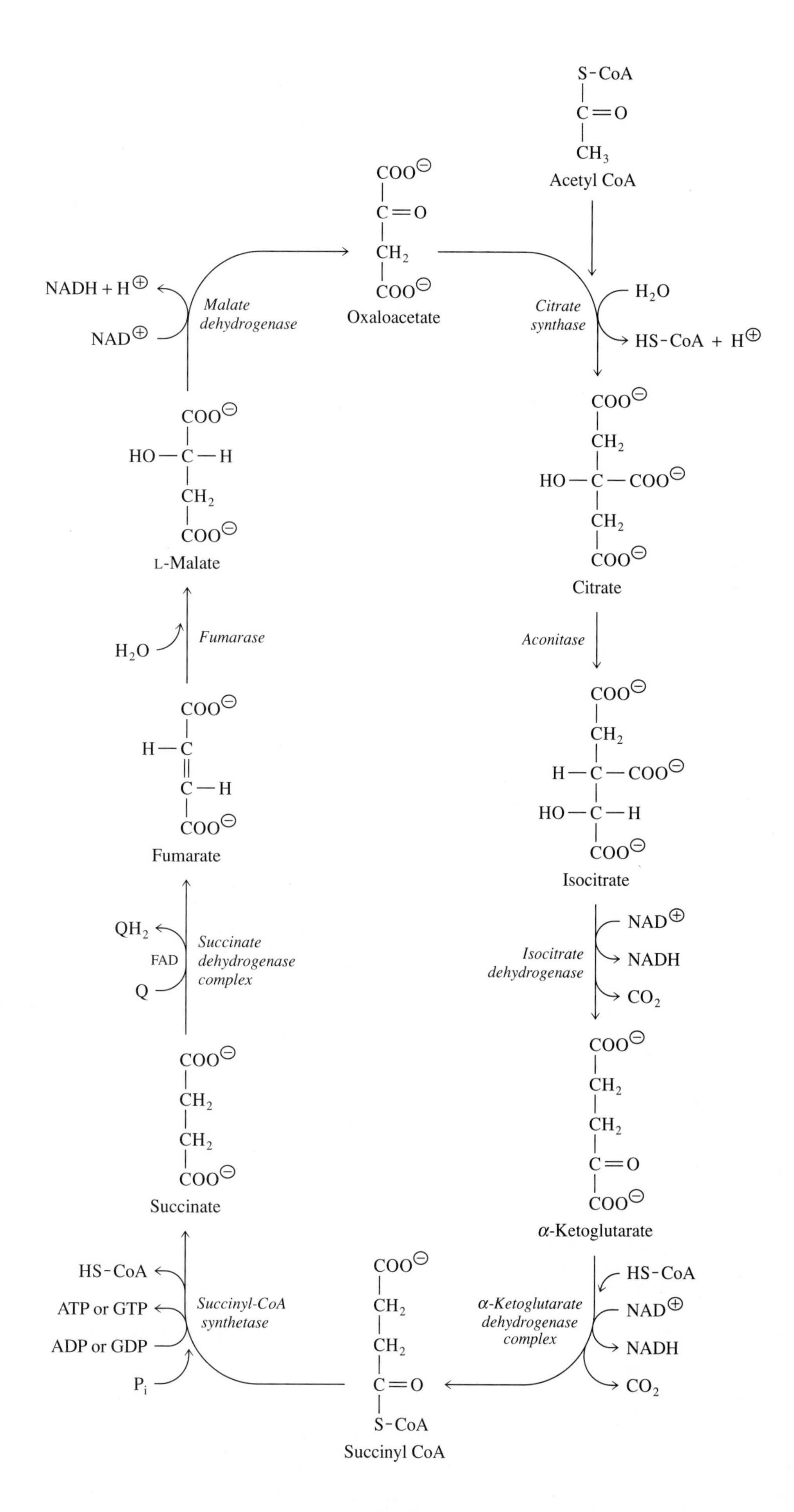

Figure 13·6
The citric acid cycle. In each turn of the cyclic pathway, the two-carbon acetyl group of an acetyl CoA molecule enters the cycle by condensing with oxaloacetate. Two molecules of CO_2 are subsequently released, the mobile coenzymes NAD⊕ and Q are reduced, a phosphoryl group is transferred to GDP (or ADP), and the original acceptor molecule, oxaloacetate, is reformed.

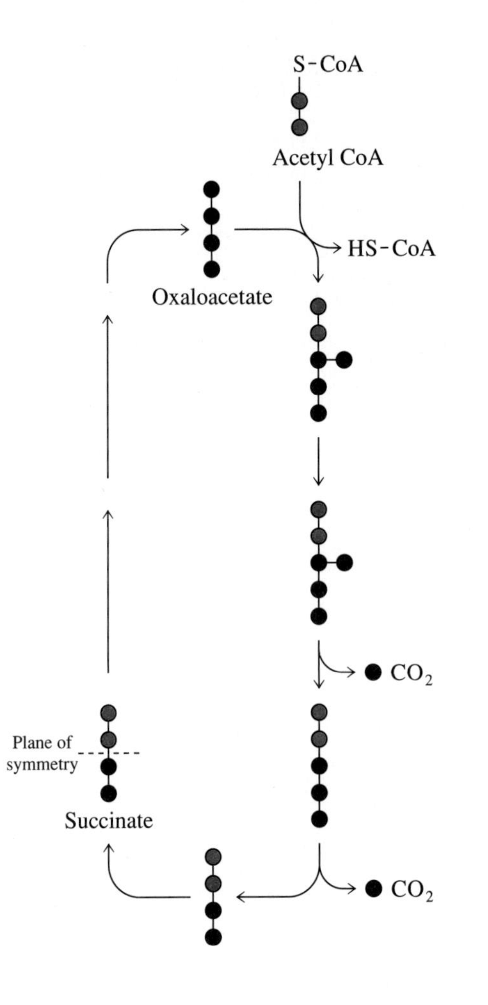

Figure 13·7
Fates of the carbon atoms from oxaloacetate and acetyl CoA during one turn of the citric acid cycle. The plane of symmetry noted for succinate means that the two halves of the molecule are chemically equivalent; thus, carbon from acetyl CoA (shown in red) is uniformly distributed in four-carbon intermediates leading to oxaloacetate. Carbon from acetyl CoA that enters in one turn of the cycle is thus lost as CO_2 only in the second and subsequent turns.

features of the pathway: the flow of carbon and the production of energy-rich molecules.

In the first reaction of the citric acid cycle, the two-carbon acetyl group of acetyl CoA condenses with the four-carbon molecule oxaloacetate to form the six-carbon intermediate citrate. Two molecules of CO_2 are released in subsequent reactions, followed by several steps that lead to regeneration of oxaloacetate, the original condensation partner for acetyl CoA. Note that the two carbon atoms that enter the cycle by condensation with oxaloacetate are not the same carbon atoms subsequently lost as CO_2 (Figure 13·7). However, the *balance* of carbon is such that for each two-carbon group from acetyl CoA that enters the cycle, two molecules of CO_2 are released during a complete turn of the cycle. The two carbons of acetyl CoA that entered the cycle become half of the symmetrical four-carbon molecule succinate in the fifth reaction. The two halves of this symmetrical molecule are chemically equivalent so that, in terms of carbon tracing, carbons arising from acetyl CoA are evenly distributed in molecules arising from succinate.

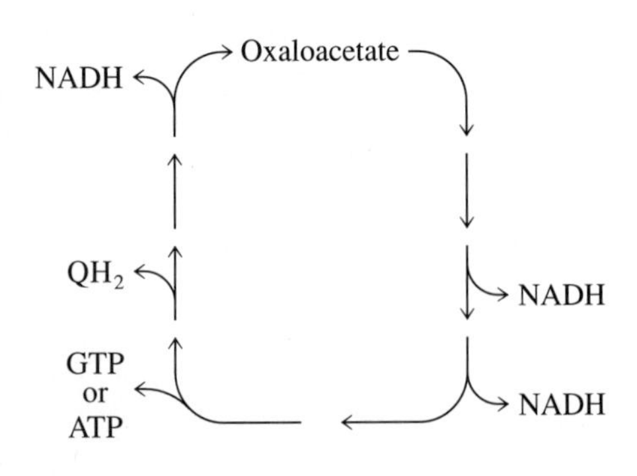

Figure 13·8
Production of energy-rich compounds by the citric acid cycle. The reduced coenzymes NADH and QH_2 are oxidized by the respiratory electron-transport chain, and some of the energy released from their oxidation is used for the formation of ATP from ADP and P_i via oxidative phosphorylation. One nucleoside triphosphate—either GTP or ATP, depending on the cell type—is produced by substrate-level phosphorylation.

Most of the energy released in the reactions of the citric acid cycle is conserved in the form of the reduced coenzymes NADH and QH_2 (Figure 13·8). Oxidation of these molecules by the electron-transport chain leads to the production of ATP in a process called *oxidative phosphorylation* (Chapter 14). Most of the ATP produced by aerobically metabolizing cells is generated by this process. In addition, one reaction of the citric acid cycle produces a nucleoside triphosphate via *substrate-level phosphorylation*. The triphosphate produced can be either GTP or ATP, depending on the cell type.

13·4 Acetyl CoA Enters the Citric Acid Cycle by Condensing with Oxaloacetate to Form Citrate

In the first reaction of the citric acid cycle, acetyl CoA reacts with oxaloacetate to form citrate and CoASH.

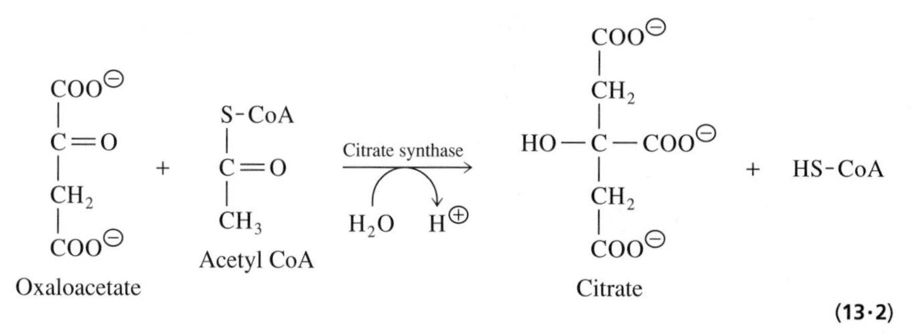

(13·2)

The reaction, catalyzed by citrate synthase, appears to be metabolically irreversible. The uncertainty about its steady-state condition arises from the technical difficulty of measuring precisely the concentrations of the reactants and products in the mitochondrial compartment, especially the concentration of oxaloacetate, which is present in very small quantities.

In the first step of the citrate synthase reaction, a proton is abstracted from the methyl group of acetyl CoA by a histidine residue of the enzyme. The resulting carbanion of the substrate is stabilized by resonance.

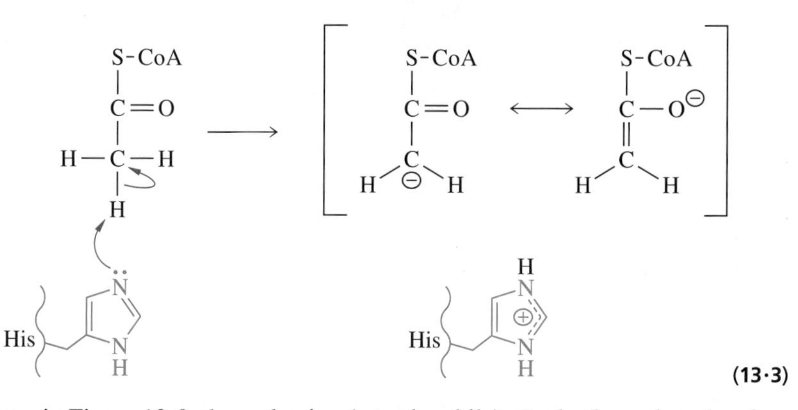

(13·3)

As shown in Figure 13·9, the carbanion (a nucleophile) attacks the carbonyl carbon of oxaloacetate to form an enzyme-bound intermediate, citryl CoA. This energy-rich thioester is hydrolyzed to release the products, citrate and CoASH.

Figure 13·9
Reaction catalyzed by citrate synthase.

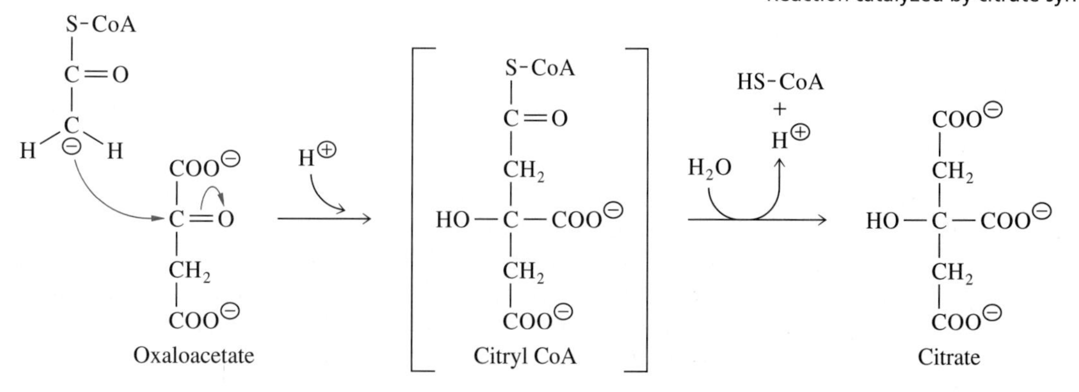

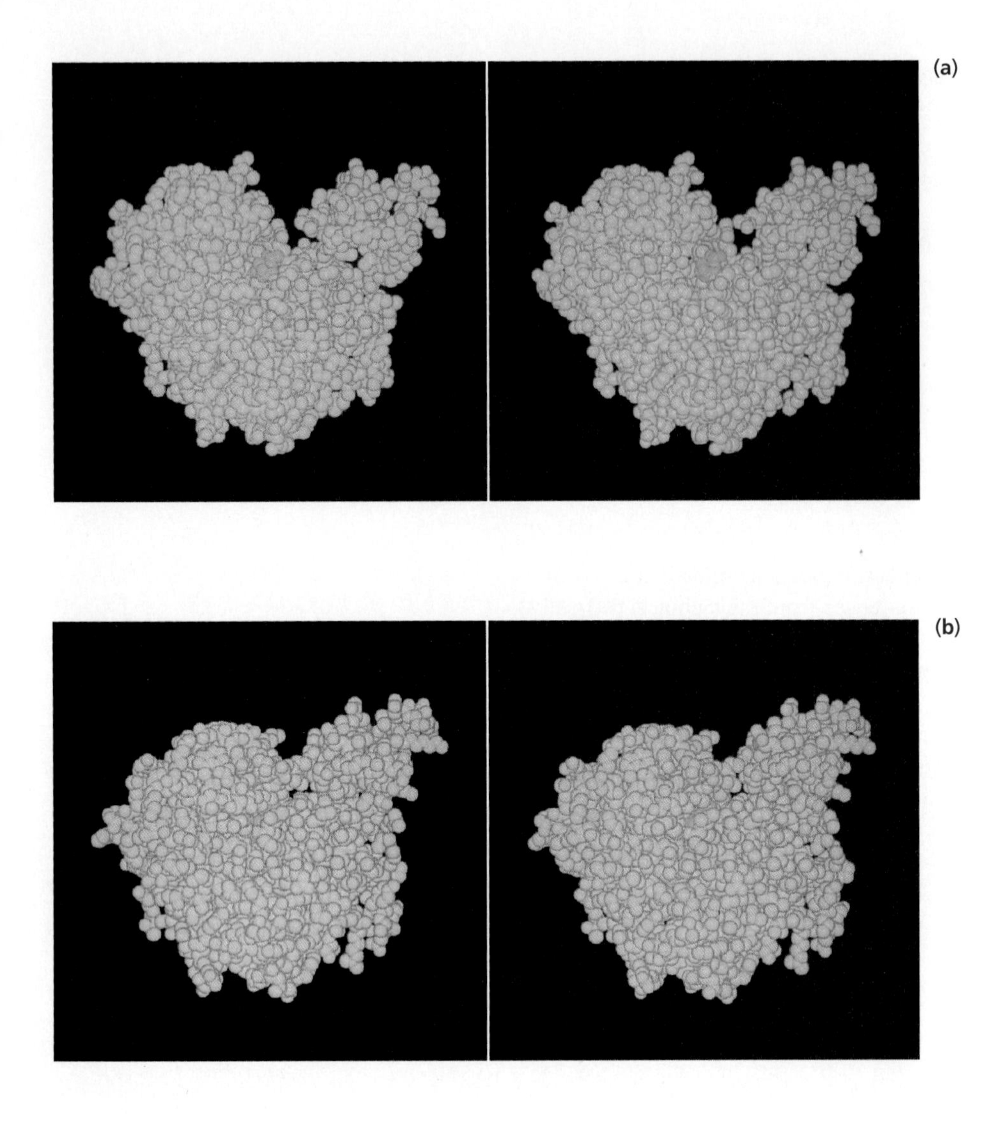

Stereo S13·1
Citrate synthase, space-filling model. **(a)** Open conformation. **(b)** Fully closed conformation. The product citrate (green) is positioned in the active site. (Based on coordinates provided by S. Remington, G. Weigand, and R. Huber; stereo images by Ben N. Banks and Kim M. Gernert.)

Citrate synthase (Stereo S13·1) undergoes several conformational changes upon binding its substrate and the intermediate citryl CoA. The closing of the enzyme in the presence of reactive ligands is an example of induced fit, a property that citrate synthase shares with hexokinase (Stereo S12·1). The conformational changes undergone by citrate synthase and hexokinase prevent side reactions by shielding the reactants from water.

13·5 Aconitase Catalyzes the Conversion of Prochiral Citrate to Chiral Isocitrate

Aconitase, the common name for aconitate hydratase, catalyzes the near-equilibrium conversion of citrate to isocitrate. Citrate is a tertiary alcohol and thus cannot be oxidized. The action of aconitase converts citrate to an oxidizable secondary alcohol. The name of the enzyme is derived from the intermediate of the reaction, *cis*-aconitate. The reaction proceeds by elimination of water to form a carbon-carbon double bond, followed by stereospecific addition of water to form isocitrate (Figure 13·10).

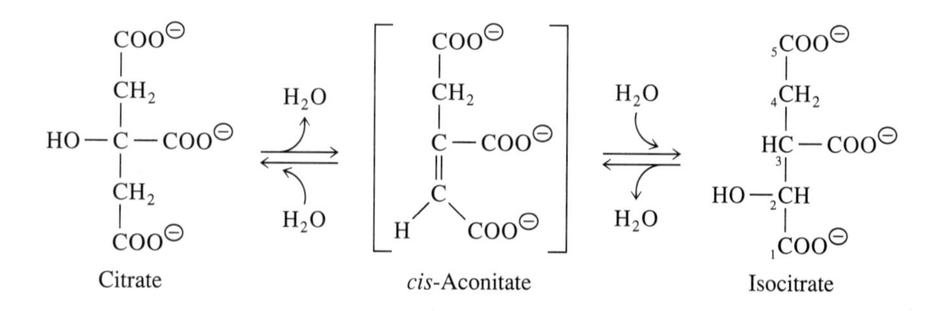

Figure 13·10
Reaction catalyzed by aconitase. The enzyme is named for its intermediate, *cis*-aconitate.

Aconitase contains a covalently bound iron-sulfur prosthetic group that assists in correct positioning of the substrate. In other iron-sulfur proteins, the iron is involved in electron transfer and undergoes redox changes, as discussed in the next chapter.

The proper positioning of citrate in the active site of aconitase is essential for the stereospecific reaction that follows. Citrate is a <u>*prochiral*</u> molecule, meaning that it has a carbon atom that has three types of substituents (C_{aacd}). A prochiral molecule can become chiral (C_{abcd}) by substituting a fourth type for one of the identical substituents. When the citric acid cycle was proposed by Krebs, the inclusion of the citrate-to-isocitrate reaction was a major barrier to its acceptance because labelling studies indicated that only a single isomer of isocitrate was produced in cells. At the time, conversion of a prochiral molecule to a single chiral isomer was unknown. In a conceptual breakthrough, Alexander Ogston suggested in 1948 how a chiral active site of an enzyme could distinguish between chemically equivalent groups on the citrate molecule. Ogston envisioned citrate binding in a manner he called three-point attachment, with nonidentical groups being the points of attachment (Figure 13·11). Once the substrate was bound to the enzyme, the two $—CH_2COO^{\ominus}$ groups of citrate would have specific orientations and thus would no longer be equivalent.

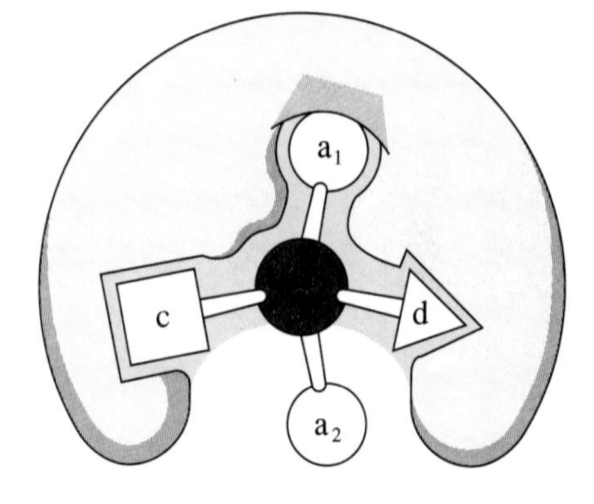

Figure 13·11
Nonequivalence of identical groups when a prochiral molecule is bound by three-point attachment. In a chiral active site, the chemically identical groups a_1 and a_2 of a prochiral molecule can be distinguished by the enzyme. The enzyme cannot bind a_2 when *c* and *d* are correctly bound.

The conversion of citrate to isocitrate creates not just one but two chiral centers in the isocitrate molecule. The stereochemistry of this molecule is complex and not clearly represented by DL nomenclature. Instead, the <u>*RS system of configurational nomenclature*</u> must be used. The *RS* system is summarized in Box 13·2. Using *RS* notation, the product of the aconitase reaction is 2*R*,3*S*-isocitrate.

Fluoroacetate, produced by a few plants, is highly poisonous. It is converted to fluoroacetyl CoA and then incorporated into fluorocitrate by the action of citrate synthase.

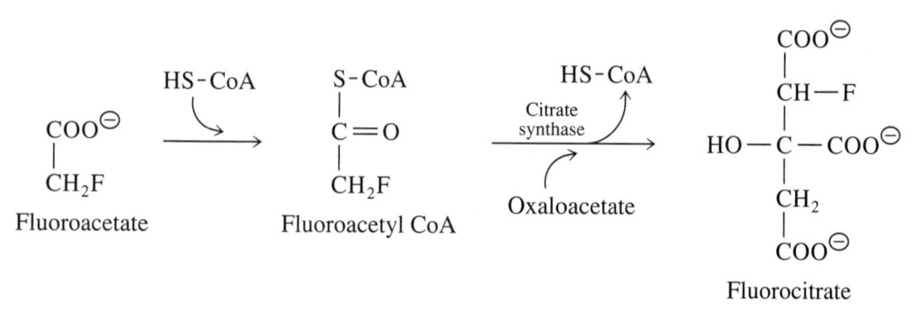

(13·4)

The citrate analog fluorocitrate is a potent inhibitor of animal aconitase, and aerobic metabolism via the citric acid cycle is blocked in its presence. The lethal properties of fluoroacetate have led to its use as a rat poison.

Within the citric acid cycle, three reactions are metabolically irreversible and are therefore potential sites of control. These are the reactions catalyzed by citrate synthase, isocitrate dehydrogenase, and the α-ketoglutarate dehydrogenase complex. Citrate synthase catalyzes the first reaction of the citric acid cycle, which would seem a suitable control point. However, mechanisms for the regulation of this enzyme are not well established. ATP inhibits the enzyme in vitro, but significant changes in intramitochondrial ATP concentration are unlikely in vivo, and therefore ATP is probably not a functional regulator.

Mammalian isocitrate dehydrogenase is allosterically activated by Ca^{2+} and ADP and inhibited by NADH. The enzyme in mammals is not subject to covalent modification. In *E. coli,* however, phosphorylation of a serine residue of isocitrate dehydrogenase catalyzed by a protein kinase virtually abolishes the activity of the enzyme (Section 5·15). The same protein molecule that contains the kinase activity also has a phosphatase activity located on a separate domain that catalyzes hydrolysis of the phosphoserine residue, reactivating isocitrate dehydrogenase. The kinase and phosphatase activities are reciprocally regulated: isocitrate, oxaloacetate, pyruvate, and the glycolytic intermediates 3-phosphoglycerate and phosphoenolpyruvate allosterically activate the phosphatase and inhibit the kinase (Figure 13·25). Thus, when the concentrations of glycolytic and citric acid cycle intermediates in *E. coli* are high, isocitrate dehydrogenase is active. When phosphorylation abolishes the activity of isocitrate dehydrogenase, isocitrate is diverted to the glyoxylate cycle, a specialized pathway present in some microorganisms and plants. The glyoxylate cycle is discussed in the next section.

The α-ketoglutarate dehydrogenase complex catalyzes a reaction that is analogous to that of the pyruvate dehydrogenase complex. The complexes are very similar, yet the α-ketoglutarate dehydrogenase complex has quite different regulatory features. No kinase or phosphatase is associated with the α-ketoglutarate dehydrogenase complex. Instead, calcium ions bind to E_1 of the complex and decrease the K_m of the enzyme for α-ketoglutarate, leading to an increase in the rate of formation of succinyl CoA. NADH and succinyl CoA have been shown to be inhibitors of the α-ketoglutarate complex in vitro, but it has not been established that they have a significant regulatory role in living cells.

The regulation of the pyruvate dehydrogenase complex and the citric acid cycle is summarized in Figure 13·26.

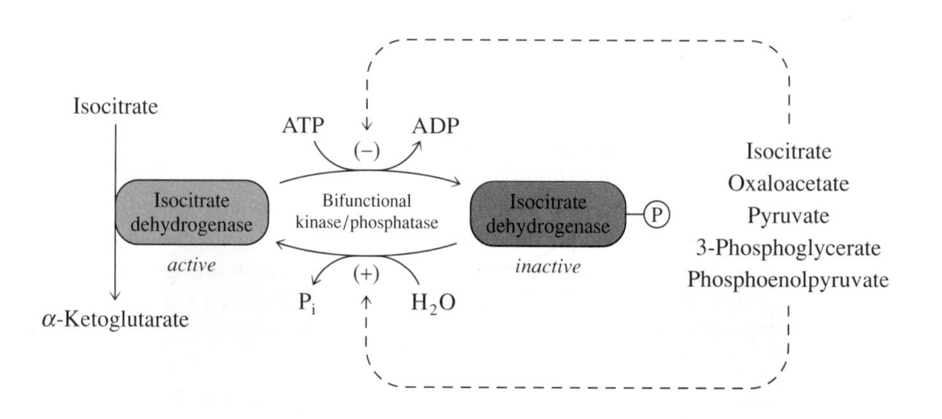

Figure 13·25
Regulation of isocitrate dehydrogenase in *E. coli* by covalent modification. A bifunctional enzyme catalyzes phosphorylation and dephosphorylation of isocitrate dehydrogenase. The two activities of the bifunctional enzyme are reciprocally regulated allosterically by intermediates of glycolysis and the citric acid cycle.

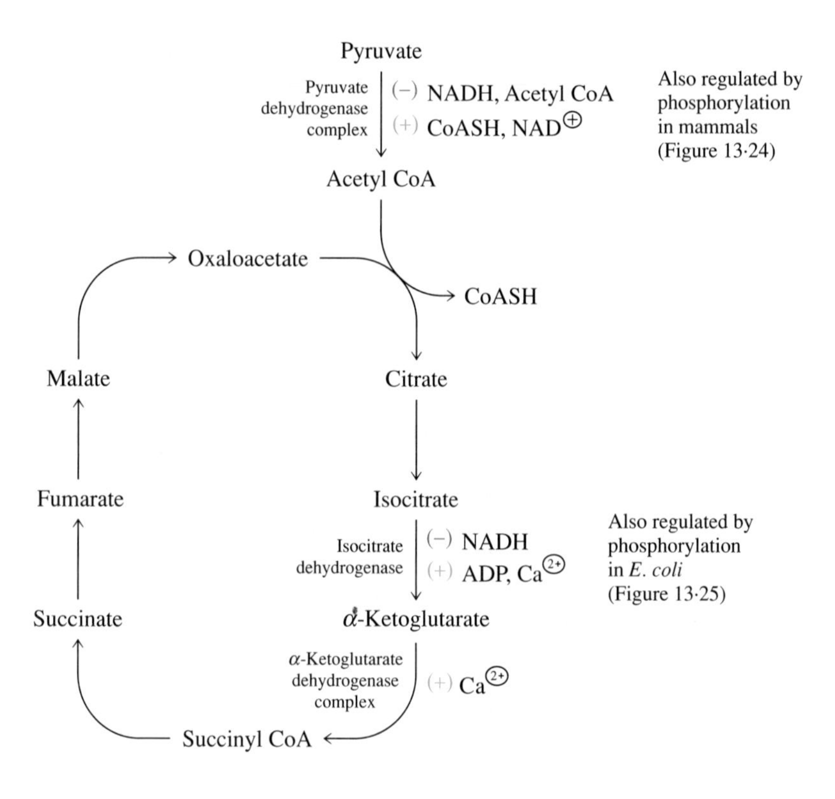

Figure 13·26
Regulation of the pyruvate dehydrogenase complex and the citric acid cycle.

13·16 The Glyoxylate Cycle Permits Acetyl CoA to Be Incorporated into Carbohydrates

The glyoxylate cycle, a modification of the citric acid cycle, is a biosynthetic route that leads to the formation of glucose from acetyl CoA. The pathway occurs in plants, bacteria, and yeasts, but not in animals. In oily seed plants, the glyoxylate cycle is especially active. Stored seed oil is converted to carbohydrates that sustain the plant during germination. Microorganisms such as yeasts and certain bacteria employ the glyoxylate cycle to sustain growth on two-carbon substrates such as acetate, which can be incorporated into acetyl CoA in a reaction catalyzed by acetate thiokinase.

$$\underset{\text{Acetate}}{H_3C-COO^{\ominus}} + HS\text{-}CoA \xrightarrow[\text{Acetate thiokinase}]{\text{ATP} \;\to\; \text{AMP, PP}_i} \underset{\text{Acetyl CoA}}{H_3C-\overset{\overset{\displaystyle O}{\|}}{C}-S\text{-}CoA} \tag{13·10}$$

The sequence of electron carriers in the respiratory chain was deduced through the use of inhibitors that bind to specific components of the electron-transport chain and block the transfer of electrons. The resulting changes in oxidation state are detected by changes in light absorption of the electron carriers. Early on, it was established that NADH was the first substrate of the respiratory chain and that O_2 was the final acceptor. Thus, it could be anticipated that any compound that caused a block in respiration (as evidenced by a diminished O_2 consumption by isolated mitochondria) should cause electron carriers closer to O_2 to be oxidized and those closer to NADH to be reduced. For example, rotenone binds to a site on Complex I and blocks the transfer of electrons to Q. As a result, the electron carriers of Complex I are reduced and those of complexes III and IV are oxidized. This suggests the order:

$$\text{NADH} \longrightarrow \text{I} \longrightarrow \text{III, IV} \longrightarrow O_2 \qquad \textbf{(14·4)}$$

The state in which some carriers are reduced and others oxidized is called a *crossover*, and experiments of this type are called *crossover analyses*. Other specific inhibitors are antimycin A, which binds to Complex III, and cyanide, which binds to Complex IV.

Box 14·1
Figure 1
Heme groups of cytochromes *a*, *b*, and *c*, shown in their reduced (ferrous) states. All share a similar, highly conjugated ring system, but the side groups differ.

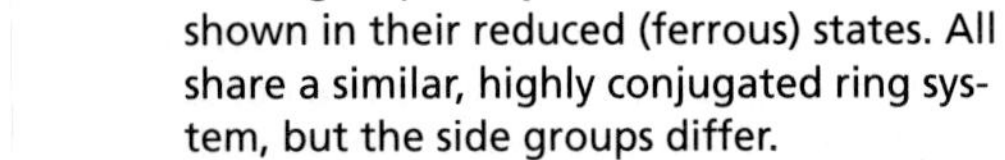

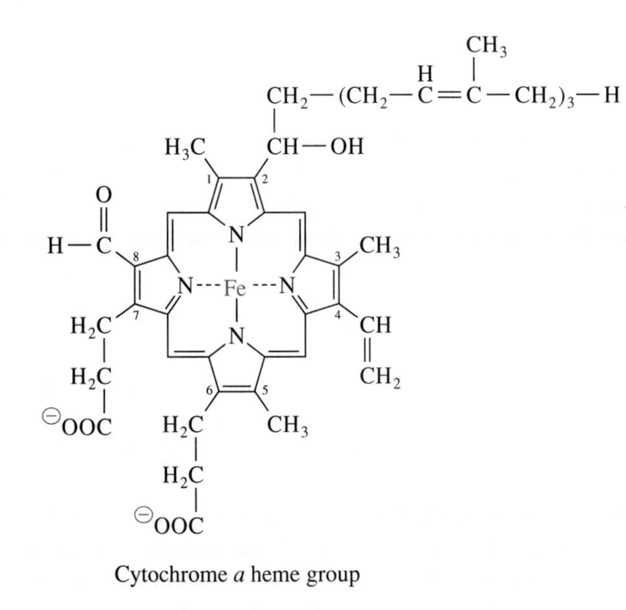

Cytochrome *a* heme group

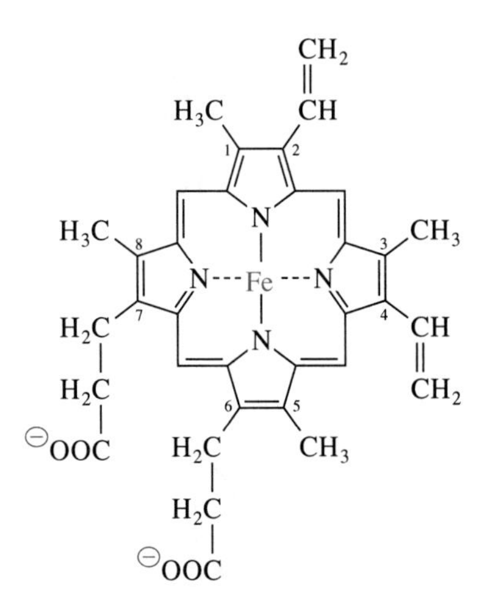

Cytochrome *b* heme group

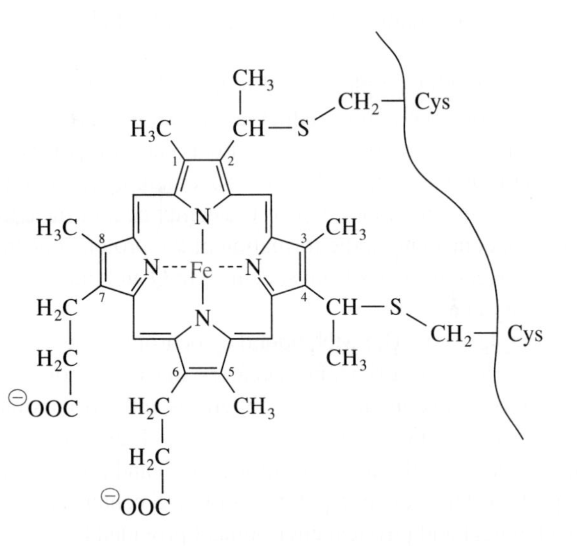

Cytochrome *c* heme group

Box 14·1 The abc's of Cytochromes

Cytochromes (*cyto*, cell and *chromo*, color) were first discovered by C. A. MacMunn in 1886 and were rediscovered by David Keilin in 1925. It was Keilin who recognized that these hemoproteins were involved in respiration. He classified them on the basis of their visible absorption spectra into the categories *a*, *b*, and *c*. The heme prosthetic groups of the three types of cytochromes have slightly different structures (Figure 1). Absorption spectra of reduced and oxidized cytochrome *c* are shown in Figure 2. Although the most strongly absorbing band is the Soret band near 400 nm, the band labelled α in the spectrum of reduced cytochrome *c* is used to characterize cytochromes. Thus, cytochrome *b* in Complex II is denoted cytochrome b_{560} to indicate the peak in the absorption band of the cytochrome in its reduced form. In some cases, nondescriptive subscript designations, such as cytochromes c_1 and a_3, are used to distinguish cytochromes of a given class. Peak wavelengths for reduced cytochromes are given in Table 1.

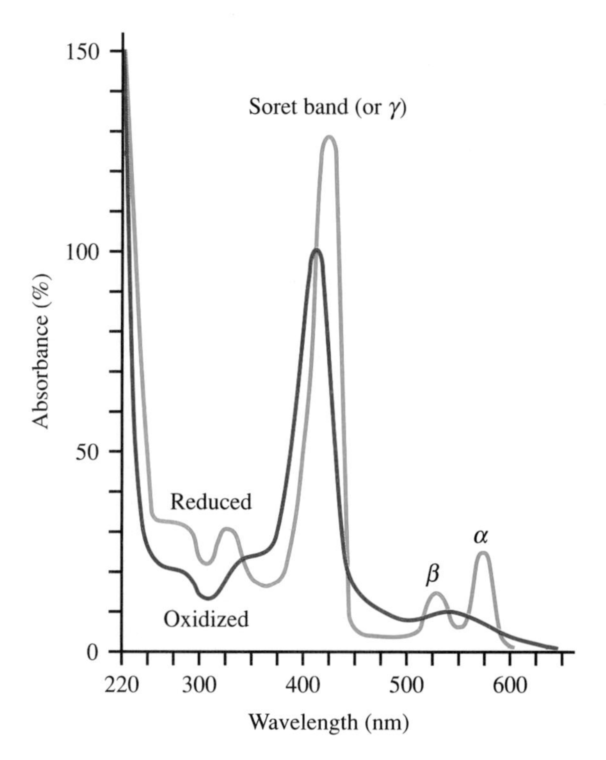

Box 14·1
Figure 2
Absorption spectra of oxidized and reduced horse cytochrome c. The reduced cytochrome shows three absorbance peaks in the visible portion of the spectrum; these peaks are designated α, β, and γ, starting with the peak of longest wavelength. Upon oxidation, the γ band, also called the Soret band, decreases in intensity and shifts to a slightly shorter wavelength, whereas the α and β peaks disappear, leaving a single broad band of absorbance. Units are absorbances in percent, normalized to the oxidized form of the Soret band as 100%.

Box 14·1
Table 1 Absorption maxima (in nm) of major spectral bands in the visible absorption spectra of the cytochromes

	Absorption band		
Heme protein	α	β	γ
Cytochrome *c*	550–558	521–527	415–423
Cytochrome *b*	555–567	526–546	408–449
Cytochrome *a*	592–604	Absent	439–443

14·3 The Energy Stored in a Proton Concentration Gradient Has Electrical and Chemical Components

Before considering the individual reactions of the electron-transport chain, including electron transport and proton translocation, we shall first examine in detail the nature of the energy stored in a proton concentration gradient. We shall see that the energy of the gradient is analogous to the electromotive force of electrochemistry.

We can depict a generalized reduction-oxidation, or redox, reaction with oxygen the acceptor of electrons as:

$$XH_2 + \tfrac{1}{2}O_2 \rightleftharpoons X + H_2O \qquad \textbf{(14·5)}$$

Recall that in Figure 11·11, we showed this type of reaction separated into two half-reactions that occurred in separate phases, half-cells linked by a salt bridge and connected by a wire across electrodes in each cell. The reaction of the electrochemical cell can be schematized as shown in Figure 14·5a. Movement of electrons through the wire is the result of a potential difference between cells; the measured potential is the electromotive force (emf), described in Section 11·9. The movement of electrons is vectorial; that is, it has magnitude and direction, as opposed to the scalar reactions involving $H^{\oplus}$ that occur in single phases, with $H^{\oplus}$ in each phase removed and supplied by an apparatus such as the salt bridge.

We can imagine an arrangement in which the conduction of electrons and ions shown in Figure 14·5a is switched. As shown in Figure 14·5b, the vectorial component is now proton transfer, with electrons removed and supplied at either half-reaction. The potential that develops between the two phases, by analogy to the electromotive force, is the *proton motive force*. Fitting this abstract picture to the mitochondrion, the path for electrons is the respiratory chain of the inner mitochondrial membrane. Vectorial movement of $H^{\oplus}$ produces a proton gradient across the inner membrane, as shown in the lower half of Figure 14·5b. The overall proton circuit, illustrated in Figure 14·1, is analogous to an electrical circuit. In fact, Mitchell has encouraged use of the term *proticity,* paralleling electricity, or *protonic potential* to describe the proton circuit, though these terms have gained limited circulation.

The chemiosmotic theory proposes that a concentration gradient of protons across a membrane, generated by the electron-transport chain, is the energy reservoir that drives ATP synthesis. As a result of the movement of protons from the matrix to the cytosol against a concentration gradient, chemical and electrical potential energies are generated. The free energy of the chemical contribution to the total potential energy generated is related to the pH difference, ΔpH, between the two compartments:

$$\Delta G_{chem} = 2.303\,nRT(pH_{in} - pH_{out}) \qquad \textbf{(14·6)}$$

The free energy of the electrical portion is proportional to the *membrane potential,* $\Delta\psi$, which is the term used for the potential difference that occurs across a membrane, and is represented by:

$$\Delta G_{elec} = n\mathcal{F}\Delta\psi \qquad \textbf{(14·7)}$$

By convention, the sign of $\Delta\psi$ is positive when an ion is transported from negative to positive. The total free-energy change is the sum of these expressions:

$$\Delta G = n\mathcal{F}\Delta\psi + 2.303\,nRT\Delta pH \qquad \textbf{(14·8)}$$

Dividing Equation 14·8 by $n\mathcal{F}$ gives an expression in units of volts:

$$\frac{\Delta G}{n\mathcal{F}} = \Delta\psi + \frac{2.303\,RT\Delta pH}{\mathcal{F}} \qquad \textbf{(14·9)}$$

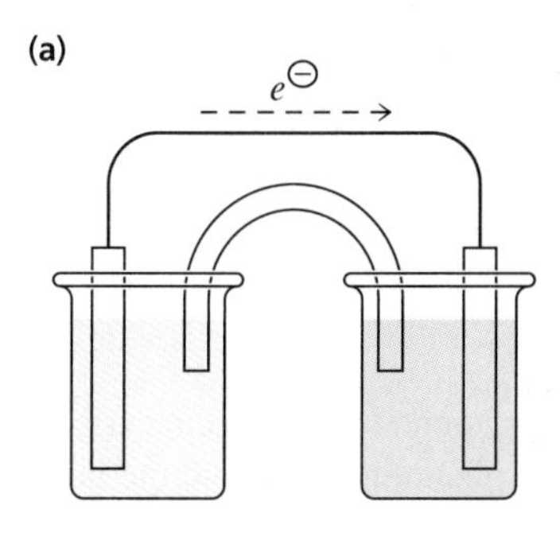

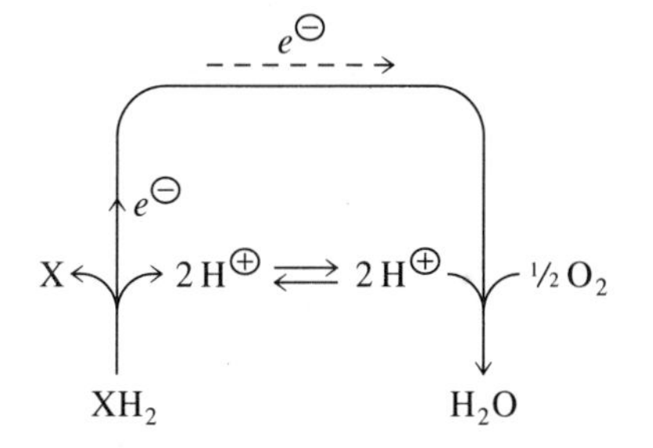

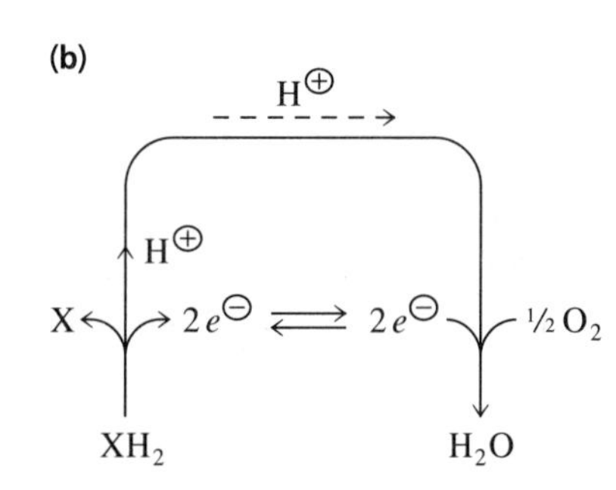

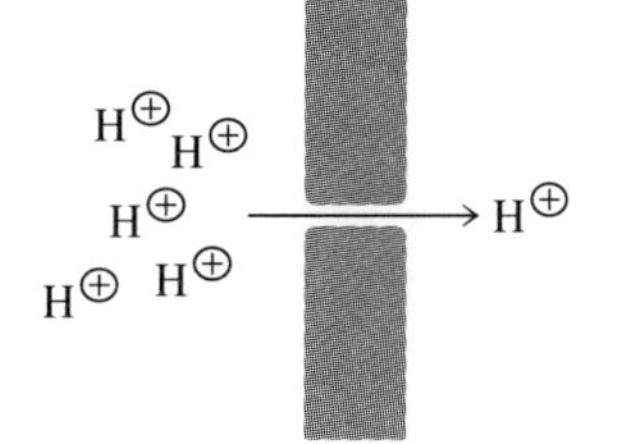

Figure 14·5
Analogy of electromotive and proton motive force. **(a)** The reaction of an electrochemical half-cell is schematized to show oxidation and reduction in separate phases and conduction of electrons between phases. The measured electrical potential between cells is the electromotive force. **(b)** If the configuration is reversed, i.e., the pathway of electrons is substituted by a pathway for protons, the potential is the proton motive force. Relating the schematic pathway in (b) to energy transduction in membranes, as electrons are transported within the membrane by the electron-transport chain, H⊕ is transported across the membrane, creating a proton concentration gradient that serves as a reservoir of electrical and chemical potential energy. The flow of H⊕ down this gradient, schematized in the lower part of (b), is coupled to the formation of ATP in mitochondria.

We can redefine the constants on the right side of Equation 14·9 as

$$Z = \frac{2.303\,RT}{\mathcal{F}} \qquad \textbf{(14·10)}$$

At 25°C, Z = 0.059 V. If we then define the terms on the left side of Equation 14·9 as Δp, we arrive at the expression

$$\Delta p = \Delta\psi + Z\Delta pH \qquad \textbf{(14·11)}$$

In Equation 14·11, Δp is the proton motive force, measured in volts. This expression compactly represents the electrical and chemical components of the proton motive force, which, according to chemiosmotic theory, supplies the driving force for oxidative phosphorylation.

When protons re-enter mitochondria coupled to the synthesis of ATP, some of the potential of the proton concentration gradient is consumed. A requirement of the chemiosmotic theory is that the membrane must be impermeable to H⊕. Otherwise, the proton concentration gradient would be dissipated as H⊕ leaked back into the matrix, a situation that would be analogous to a short in an electrical circuit. In fact, the inner mitochondrial membrane is quite impermeable to protons.

The proton concentration gradient is often referred to as a pH gradient. This is an oversimplification that omits the charge gradient, the other source of free energy contained in the proton concentration gradient. In a typical liver mitochondrion, the amount of free energy available from the charge gradient contribution is actually greater than that due to the chemical (pH) difference. Given typical values of $\Delta pH = 0.75$ and $\Delta\psi = 0.14$ V:

$$\begin{aligned}\Delta p &= \Delta\psi + Z\Delta pH \\ &= 0.14V + (0.059V)(0.75) = 0.18V\end{aligned} \qquad (14\cdot12)$$

Thus, nearly 80% of Δp is attributable to the charge gradient. It might seem curious, then, that the term *chemiosmosis* is applied to this phenomenon, since the label would appear to emphasize purely chemical (concentration) differences. This problem occurred to the originator of the theory. Peter Mitchell noted in 1967 that others had

> suggested that when chemiosmotic coupling is associated more with the development of an electrical potential across a membrane than with the development of a concentration difference of an ion across the membrane, the underlying feature is in essence "electrochemical" rather than "chemiosmotic".... Some readers may therefore prefer ... calling it electrochemiosmotic coupling; but I do not propose to make use of this rather cumbersome qualification.

Having examined the quantification of energy storage in proton concentration gradients, we shall now consider the individual reactions of the electron-transport chain and the coupling of the reactions of electron transport to the translocation of protons.

14·4 Complex I Transfers Electrons from NADH to Ubiquinone

Complex I, NADH-ubiquinone reductase (also called NADH dehydrogenase), catalyzes the transfer of two electrons from NADH to Q. Figure 14·6 illustrates the sequence of reactions within Complex I. NADH donates electrons to Complex I two at a time as a hydride ion ($H^{\ominus}$), whereas Q accepts electrons one at a time, passing through a semiquinone anion intermediate ($Q\cdot^{\ominus}$) before reaching its fully reduced state, ubiquinol (QH_2).

$$Q \xrightarrow{+e^{\ominus}} Q\cdot^{\ominus} \xrightarrow{+e^{\ominus},\,+2H^{\oplus}} QH_2 \qquad (14\cdot13)$$

In the first step of electron transfer through Complex I, NADH + $H^{\oplus}$ reduces FMN, forming $FMNH_2$. FMN can accept electrons two at a time as a hydride ion and donate them to the next oxidizing agent one at a time.

$$FMN \xrightarrow{+H^{\oplus},\,+H^{\ominus}} FMNH_2 \xrightarrow{-H^{\oplus},\,-e^{\ominus}} FMNH\cdot \xrightarrow{-H^{\oplus},\,-e^{\ominus}} FMN \qquad (14\cdot14)$$

Thus, FMN is a transducer that converts the flow of electrons to the one-electron steps that characterize the rest of the electron-transport chain. $FMNH_2$ transfers electrons to a cluster of Fe-S centers, obligatory one-electron acceptors, which pass the electrons to Q.

Coupled to the movement of electrons through Complex I, protons are translocated from the matrix to the intermembrane space. While it is clear that protons entering from NADH and the matrix are transferred to $FMNH_2$, it is not known by what path these protons are transported across the membrane. Many current studies are focussed on the stoichiometry of proton translocation—that is, the number of protons moved across the membrane for each electron pair passing from NADH to Q. The exact ratio will be important in confirming the mechanism of proton translocation.

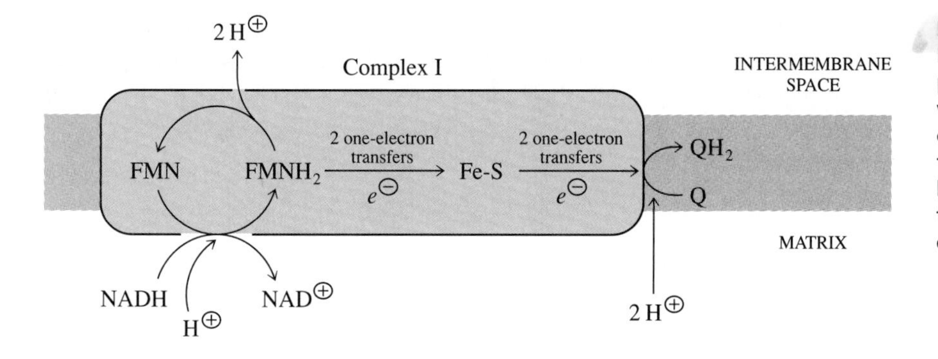

Figure 14·6
Electron transfer and proton flow in Complex I. Electrons are passed from NADH to Q via FMN and a cluster of Fe-S centers. The chemical mechanism for transfer of protons from $FMNH_2$ across the membrane is not known, and the stoichiometry of protons translocated per electron pair transferred is controversial.

The true measure of the role of Complex I in mitochondrial bioenergetics is how much it contributes to the formation of the proton concentration gradient and the resulting formation of ATP. In experiments with actively respiring mitochondria, it is possible to determine the amount of ATP formation that accompanies oxidation of an electron donor. The ratio of molecules of ADP phosphorylated to atoms of oxygen reduced, termed the *P/O ratio,* is about 3:1 for the transfer of electrons from NADH through the entire electron-transport chain to O_2. ATP is formed when protons re-enter the matrix via ATP synthase, and it appears that the re-entry of three protons is required to drive the formation of ATP. Most evidence suggests that three or four protons are transported by Complex I for each electron pair transferred through it. Thus, Complex I contributes to the formation of about one molecule of ATP.

14·5 Complex II Transfers Electrons from Succinate to Ubiquinone

Complex II, succinate-ubiquinone reductase (also called the succinate dehydrogenase complex), accepts electrons from succinate and, like Complex I, catalyzes the reduction of Q to QH_2. We previously considered the succinate dehydrogenase–catalyzed oxidation of succinate to fumarate as one of the reactions of the citric acid cycle (Section 13·9).

The transfer of two electrons from succinate to ubiquinone involves an FAD prosthetic group and Fe-S centers (Figure 14·7). A cytochrome b_{560} copurifies with the complex, but its role in electron transfer is not known. Little free energy is released in the reaction sequence catalyzed by Complex II, and the complex does not contribute to the proton concentration gradient across the inner mitochondrial membrane but serves to introduce electrons from the oxidation of succinate into the electron-transport sequence at the level of QH_2. Q is a lipid-soluble mobile carrier that can accept electrons from Complex I or II and donate them to Complex III and thence to the rest of the electron-transport chain. Reactions in several other pathways also donate electrons to Q. We shall consider one of them, the reaction catalyzed by glycerol 3-phosphate dehydrogenase, later in this chapter.

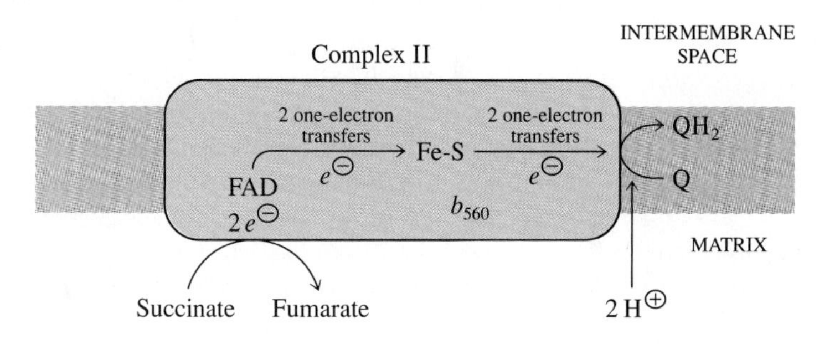

Figure 14·7
The flow of electrons through Complex II. The role of cytochrome b_{560}, which copurifies with the complex, is not clear. Complex II does not contribute to the proton concentration gradient, but serves as a tributary that introduces electrons into the electron-transport sequence at the level of QH_2.

The transfer of electrons from succinate to O_2 via the electron-transport chain results in an ATP yield of about two rather than the three ATP generated from the transfer of electrons from NADH to O_2. The reason for this should be clear: electrons introduced to the electron-transport chain at the level of QH_2 bypass Complex I, which accounts for the formation of about one molecule of ATP.

14·6 Complex III Transfers Electrons from QH_2 to Cytochrome *c*

Complex III, ubiquinol-cytochrome *c* reductase, contains several subunits including an Fe-S protein, cytochrome *b*, which includes on a single polypeptide chain two cytochrome hemes, b_{560} and b_{566}, and cytochrome c_1. Electrons are accepted from the complex by the one-electron carrier cytochrome *c*, a small peripheral membrane protein that moves along the cytosolic face of the inner membrane and transfers an electron to Complex IV.

The sequence of electron transfers and proton movements within Complex III had long proved elusive until a cyclic pathway termed the *Q cycle* was suggested in 1975 by Peter Mitchell, who proposed the chemiosmotic theory. Later investigators, including Bernard Trumpower, have contributed to our present understanding of the Q cycle.

When the electron carriers of Complex III were first identified, the linear sequence:

$$Q \longrightarrow b \longrightarrow c_1 \longrightarrow c \qquad \textbf{(14·15)}$$

was proposed, based upon the finding that antimycin A caused a crossover between cytochrome *b* and the *c* cytochromes. This sequence was placed in doubt, however, by the finding that mitochondria, slowly respiring in the presence of limited oxygen, exhibited a transiently *reduced* cytochrome *b* when pulsed with oxygen. Furthermore, this reduction of cytochrome *b* was magnified if antimycin was also present. Based on the linear sequence, one would expect that a sudden increase in the rate of cytochrome *c* oxidation resulting from the increase in oxygen availability would be followed sequentially by increases in the oxidation of cytochrome c_1, cytochrome *b,* and ubiquinol. The finding that cytochrome *b* was instead reduced was dubbed the "oxidant-induced reduction" and could not be rationalized by the linear sequence.

A recent formulation for the Q cycle, which not only explains this paradox but also describes how electrons are moved between the reduced molecule QH_2 (from either Complex I or II) and cytochrome *c,* is diagrammed in Figure 14·8. Note that three forms of Q are involved: QH_2, Q, and the semiquinone anion, $Q\cdot^{\ominus}$. The first two forms are soluble in the lipid bilayer and move between the cytosolic and matrix faces of the membrane; $Q\cdot^{\ominus}$, however, exists in two separate pools near the water phase of each side of the membrane. Tracing the sequence of events in the Q cycle, QH_2 diffuses to the cytosolic face of the membrane. The three arrows emerging from QH_2 at this point in Figure 14·8 represent a dismutation: $2\,H^{\oplus}$ are released to the cytosol, an electron is transferred to an Fe-S protein (and thence to cytochrome c_1) and the ubiquinone species remaining is $Q\cdot^{\ominus}$. Electron flow from $Q\cdot^{\ominus}$ is directed to b_{566} and then to b_{560} (which has a lower reduction potential than b_{566}). Reduced b_{560} then reduces Q on the matrix side to $Q\cdot^{\ominus}$. Matrix side Q for this reduction results from the diffusion of Q through the membrane. In order to complete the Q cycle, a second molecule of QH_2 has to repeat the process just described, with a second electron donated ultimately to cytochrome c_1, an additional $2\,H^{\oplus}$ released to the cytosol, and another electron arriving at b_{560}, which can be used, along with $2\,H^{\oplus}$ from the matrix, to convert matrix side $Q\cdot^{\ominus}$ to QH_2.

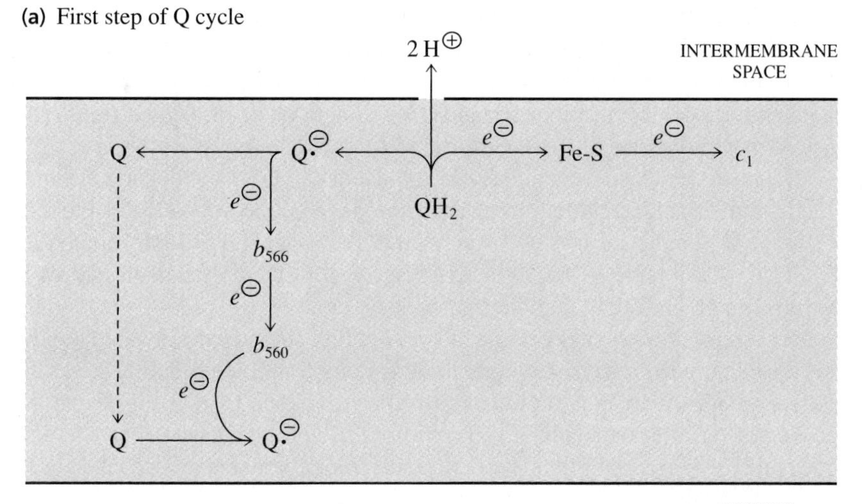

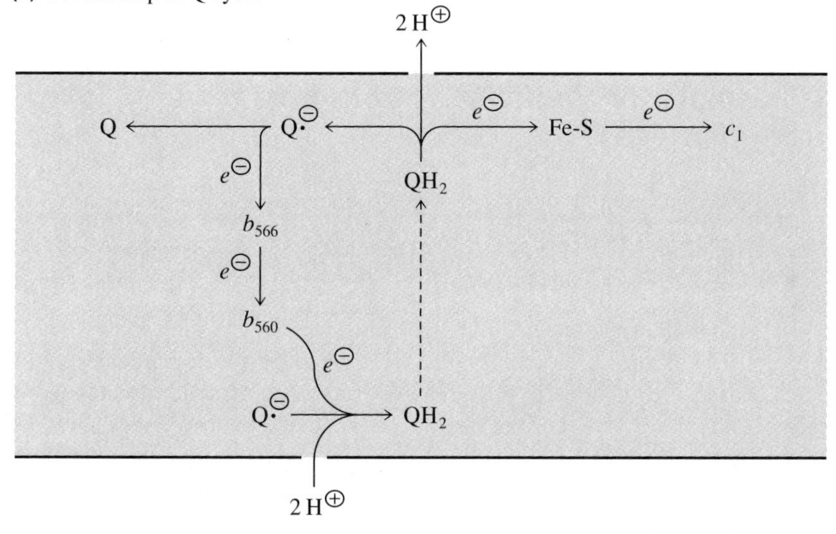

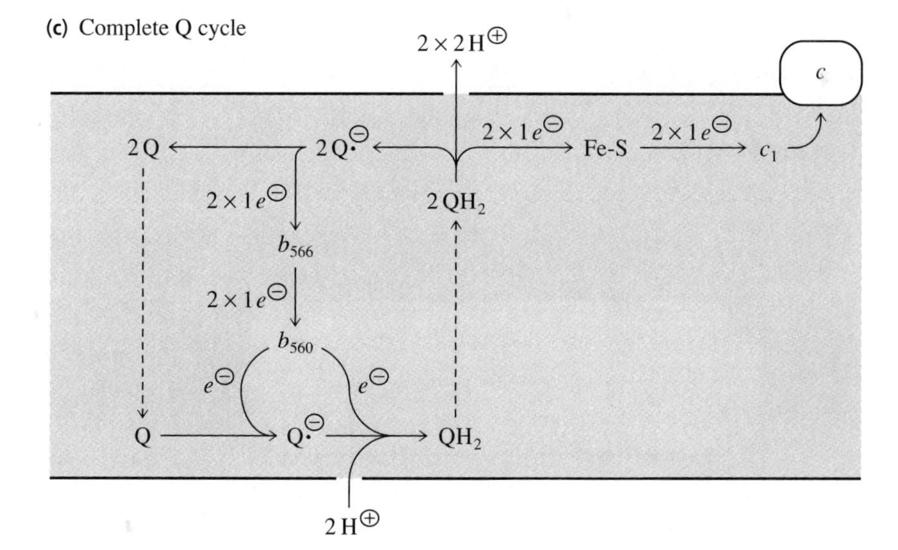

Figure 14·8
Proposed Q cycle. **(a)** QH_2 at the cytosolic face of the membrane donates an electron to the Fe-S protein of Complex III and $2\,H^{\oplus}$ to the cytosol, forming $Q\cdot^{\ominus}$. The Fe-S protein then reduces c_1, and the electron continues down the respiratory chain. $Q\cdot^{\ominus}$ donates an electron to b_{566}, with formation of Q, which diffuses to the matrix face of the membrane. Reduced b_{566} donates an electron via b_{560} to Q at the matrix face of the membrane to form $Q\cdot^{\ominus}$. **(b)** Completion of the cycle requires a second molecule of QH_2 to repeat the first part of the cycle, producing another reduced c_1, another $2\,H^{\oplus}$, another Q, and reduced b_{560}, which this time reduces $Q\cdot^{\ominus}$ to QH_2, consuming $2\,H^{\oplus}$ from the matrix. The result is the oxidation of two QH_2 molecules and the formation of one QH_2 molecule, as shown in the overall pathway **(c)**. Electrons from Complex III are donated to the mobile carrier cytochrome *c*. (Adapted from Trumpower, B. L. (1990). The protonmotive Q cycle: energy transduction by coupling of proton translocation to electron transfer by the cytochrome bc_1 complex. *J. Biol. Chem.* 265:11 409–11 412.)

The cyclic flows in Complex III are possible because Q and QH_2 diffuse between the two faces of the mitochondrial membrane and because the two hemes of cytochrome *b* together span the membrane to form a path for electrons. We can now understand the paradoxical experiments. Providing a pulse of oxygen to oxygen-deprived mitochondria will oxidize the *c* cytochromes and the Fe-S protein of Complex III, but as QH_2 becomes oxidized, it is converted to $Q\cdot^{\ominus}$ and donates electrons to cytochrome *b,* at least transiently. Thus, the pulse of oxygen must lead to a *reduction* of b_{566} and b_{560}. It is now known that the exact site of inhibition by antimycin A is between b_{560} and $Q\cdot^{\ominus}$ or QH_2. Thus, in the presence of antimycin, electrons can go no further than the hemes of *b* cytochromes.

Measurement of the Δp generated by Complex III indicates a value slightly lower than that generated by Complex I, with Complex III contributing somewhat less free energy than the amount required to drive the formation of one full molecule of ATP per molecule of QH_2 oxidized. Before the debut of the chemiosmotic theory, when many researchers were searching for an energy-rich metabolite capable of forming ATP by direct phosphoryl-group transfer, it was assumed that Complexes I, III, and IV each contributed to ATP formation with one-to-one stoichiometry. Energy transduction by formation of a proton concentration gradient permits Complex III to contribute to the overall formation of ATP even though the ratio of ATP formed per electron pair transferred is a value less than 1.0.

14·7 Complex IV Transfers Electrons from Cytochrome *c* to O_2

Complex IV, cytochrome *c* oxidase, is the last component of the respiratory electron-transport chain. This complex catalyzes the four-electron reduction of molecular oxygen (O_2) to water ($2\,H_2O$) and pumps protons into the intermembrane space.

The mammalian enzyme has 12 polypeptides with a total molecular weight of about 200 000. Some of the polypeptides are encoded in the nuclear genome, whereas others are encoded in the mitochondrial DNA. Among the polypeptide chains of Complex IV are those of cytochromes *a* and a_3, whose hemes have identical structures but different standard reduction potentials resulting from different environments within the oligomeric complex. Complex IV also contains two copper ions that alternate between the Cu^{2+} and $Cu^{\oplus}$ states as they participate in electron transfer.

Figure 14·9 shows that cytochrome *c* oxidase contributes to the proton concentration gradient in two ways. The first is a pump mechanism that transports $H^{\oplus}$ through cytochrome *c* oxidase via a protein subunit located close to cytochromes *a* and a_3. The second mechanism by which cytochrome *c* oxidase contributes to the proton concentration gradient is by consuming matrix $H^{\oplus}$ that reacts with oxygen to form water. While not involving the actual translocation of $H^{\oplus}$ across the membrane, this *annihilation mechanism* nevertheless contributes to the formation of Δp.

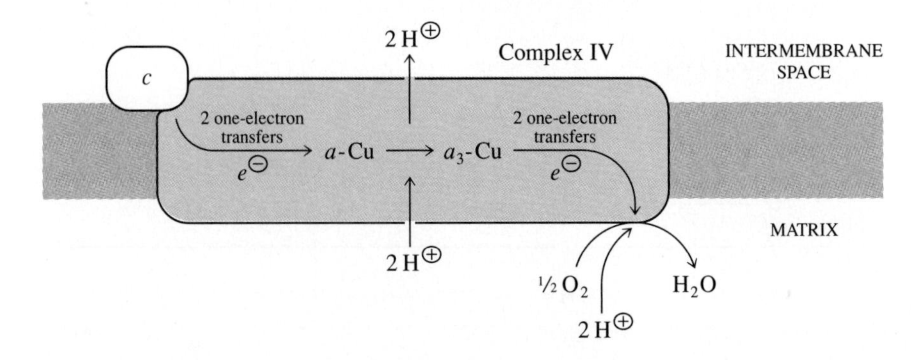

Figure 14·9
Proposed pathway for the flow of electrons and protons in Complex IV. In the *a* cytochromes, the iron atoms of the heme groups and the copper atoms are both oxidized and reduced as electrons flow. Complex IV contributes to the proton concentration gradient in two ways: proton pumping from matrix to intermembrane space occurs in close association with electron transfer between the *a* cytochromes, and the formation of water subtracts protons from the matrix.

14·8 Complex V, or F_OF_1 ATP Synthase, Couples the Re-entry of Protons into the Matrix with the Formation of ATP

Distinct from the other complexes that can be resolved from the mitochondrial inner membrane, Complex V, or the F_OF_1 ATP synthase, does not contribute to the proton concentration gradient but rather consumes it, using the energy for the synthesis of ATP from ADP and P_i. The "F" designation in the enzyme name signifies a coupling factor—ATP synthase couples the phosphorylation of ADP to the oxidation of substrates in the mitochondrion. The F_1 component contains the catalytic subunits; when isolated in solubilized form from membrane preparations, it catalyzes the hydrolysis of ATP. For this reason, it has traditionally been referred to as F_1 ATPase (ATP hydrolase). The F_O component is a proton channel that spans the membrane. The passage of protons through the channel into the matrix is coupled to the formation of ATP. The F_O component is named for its sensitivity to oligomycin, an antibiotic that binds in the channel, preventing the entry of protons and thereby inhibiting ATP synthesis.

Figure 14·10 shows the relation of the F_1 and F_O components with respect to the bacterial plasma membrane and depicts the apparent subunit composition of the *E. coli* ATP synthase. The molecular interactions among the subunits within each component have not been fully elucidated. However, the subunit composition of the bacterial F_1 component appears to be $\alpha_3\beta_3\gamma\delta\varepsilon$, and that of the F_O component is thought to be $a_1b_2c_{10-12}$. The c subunits of F_O interact to form the channel for passage of $H^{\oplus}$ through the membrane.

The F_1 component of the mitochondrial ATP synthase is similar to its bacterial counterpart. It is important to note, however, that in contrast to bacterial F_1, which extends into the cytosol, mitochondrial F_1 protrudes into the matrix. A multisubunit stalk joins the F_1 knob to the transmembrane F_O channel. The mitochondrial F_O component has a more complex structure and subunit composition than the bacterial form and is not as well characterized.

Protons entering the F_O channel bind to acidic amino acid residues. As they enter the matrix, these protons shift the equilibrium position of the ADP + $P_i \rightleftharpoons$ ATP + H_2O reaction in favor of ATP synthesis. How the passage of protons accomplishes this is the subject of active research. ATP, ADP, and P_i individually bind very tightly to the F_1 component, but the combination of ADP and P_i binds weakly. Proton flux appears to be required to force these two molecules toward each other in the active site. ATP formation at the active site is readily accomplished, but the ATP produced is in a thermodynamic pit (Section 6·10); that is,

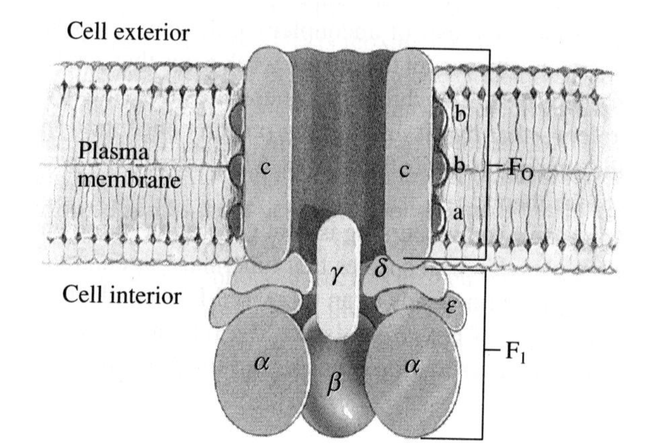

Figure 14·10
Schematic cross-section of F_OF_1 ATP synthase in *E. coli*. The F_1 component in *E. coli* protrudes into the cytosol, whereas in mitochondria, F_1 protrudes into the matrix. The F_O component spans the membrane, forming a proton channel. F_1 is believed to consist of three α and three β subunits associated to form a knoblike hexameric structure, and one γ, one δ, and one ε subunit, which connect F_1 to the F_O channel. The stoichiometry of F_O in *E. coli* is thought to be $a_1b_2c_{10-12}$. Mitochondrial F_O has a more complex structure and subunit composition and is not as well characterized. (Illustrator: Lisa Shoemaker.)

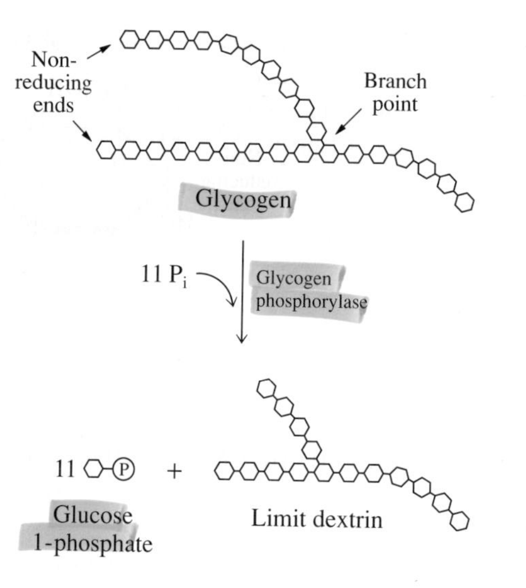

Figure 15·2
Degradation of glycogen. Glycogen phosphorylase catalyzes the sequential phosphorolytic cleavage of α-(1 → 4) glucosidic bonds from the nonreducing ends of glycogen chains, stopping four residues from an α-(1 → 6) branch point. One molecule of glucose 1-phosphate is formed for each glucose residue removed by glycogen phosphorylase.

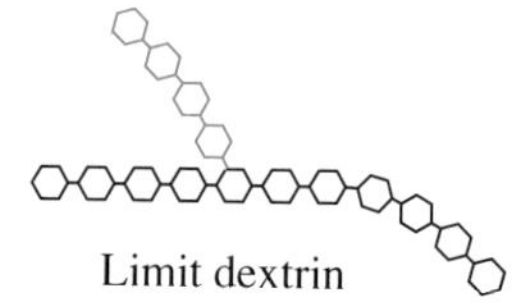

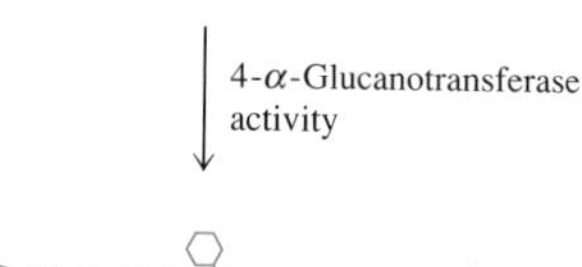

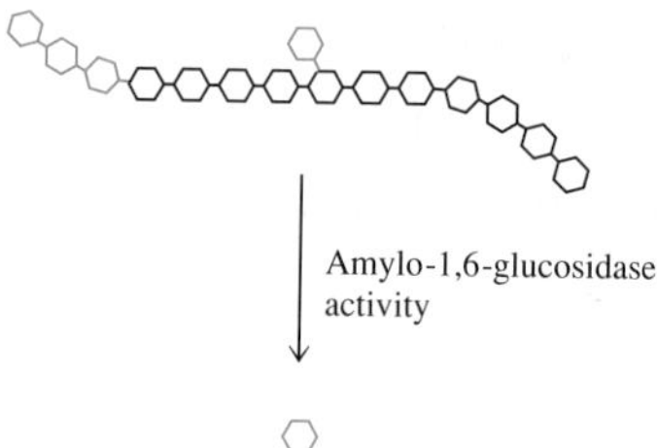

Figure 15·3
Enzymatic activities of the glycogen debranching enzyme. Both activities are catalyzed by a single enzyme. The 4-α-glucanotransferase activity catalyzes transfer of a trimer from a branch of the limit dextrin to a free 4′ end of the glycogen molecule. The amylo-1,6-glucosidase activity catalyzes release of the remaining α-(1 → 6)-linked glucose residue.

branch point (an α-(1 → 6) glucosidic bond), producing a limit dextrin (Figure 15·2); limit dextrins were described in Section 8·6A. The limit dextrin can be further degraded by the action of the glycogen debranching enzyme, which has two separate activities (Figure 15·3). The first, called 4-α-glucanotransferase, catalyzes the relocation of three glucose residues from a branch to a free 4′ end of the glycogen molecule. The original linkage and the new linkage are both α-(1 → 4). The second activity of glycogen debranching enzyme, called amylo-1,6-glucosidase, catalyzes hydrolytic removal of the remaining α-(1 → 6)-linked glucose residue. One glucose molecule is thereby produced for each branch in the original glycogen polymer, along with a considerably larger number of glucose 1-phosphate molecules generated by the action of glycogen phosphorylase.

In cells, glucose 1-phosphate is rapidly interconverted with glucose 6-phosphate in a near-equilibrium reaction catalyzed by phosphoglucomutase (Figure 15·4). Glucose 6-phosphate is an intermediate in a number of cellular pathways, including glycolysis, glycogen synthesis, and the pentose phosphate pathway. In liver, the principal end product of glycogenolysis is glucose, formed from glucose 6-phosphate by the glucose-6-phosphatase reaction.

$$\text{Glucose 6-phosphate} + H_2O \xrightarrow{\text{Glucose 6-phosphatase}} \text{Glucose} + P_i \qquad (15\cdot2)$$

This enzyme also catalyzes the final reaction of gluconeogenesis (Section 15·4).

Figure 15·4
Interconversion of α-D-glucose 1-phosphate and glucose 6-phosphate, a near-equilibrium reaction catalyzed by phosphoglucomutase. Glucose 1-phosphate is produced by glycogenolysis, and glucose 6-phosphate enters a number of cellular pathways.

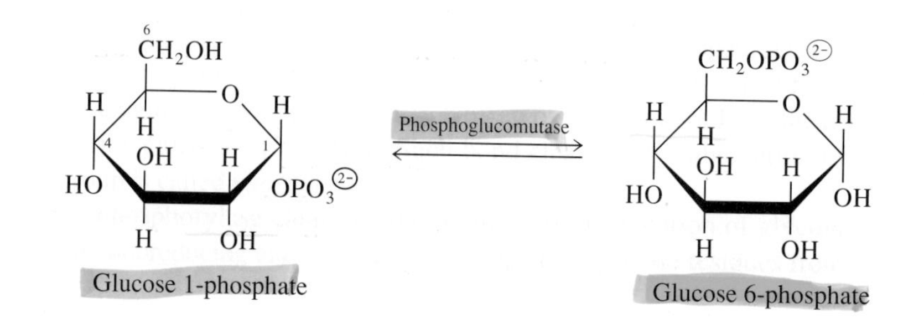

15·2 Glycogen Synthesis and Glycogen Degradation Require Separate Pathways

Based on the results of in vitro studies, it was widely believed until the late 1950s that glycogen phosphorylase was responsible for both synthesis and breakdown of glycogen. However, the concentrations of reactants and products in vivo do not favor glycogen synthesis via reversal of the glycogen phosphorylase reaction. The discovery of a uridine nucleotide derivative of glucose and an enzyme that used this compound as a substrate for glycogen synthesis changed the prevailing view. It is now known that synthesis and degradation of glycogen require separate enzymatic steps. In fact, it is a general rule of metabolism that different routes are required for opposing degradative and synthetic pathways.

Three separate enzyme-catalyzed reactions are required for the incorporation of a molecule of glucose 6-phosphate into glycogen (Figure 15·5). First, phosphoglucomutase catalyzes the near-equilibrium conversion of glucose 6-phosphate to glucose 1-phosphate. Glucose 1-phosphate is then activated by reaction with uridine triphosphate (UTP), forming UDP-glucose and inorganic pyrophosphate (PP_i). This reaction is catalyzed by UDP-glucose pyrophosphorylase (Figure 15·6). UDP-glucose is formed in the active site of UDP-glucose pyrophosphorylase when an oxygen atom of the phosphoryl group of glucose 1-phosphate attacks the phosphorus atom of the α-phosphoryl group of UTP. The PP_i released in this reaction is rapidly hydrolyzed to 2 P_i by the action of inorganic pyrophosphatase. Hydrolysis

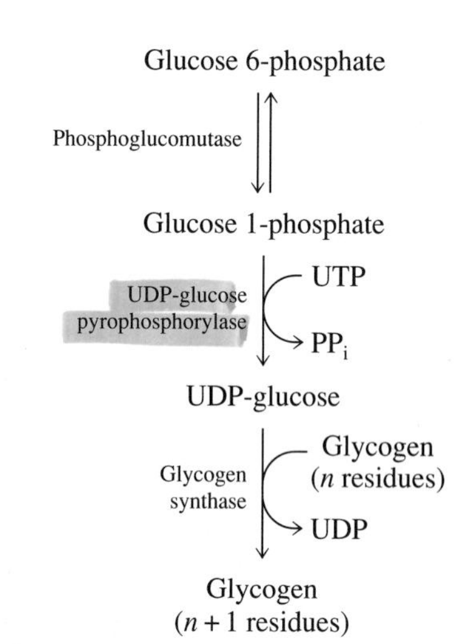

Figure 15·5
Synthesis of glycogen. The three-step pathway is composed of the sequential action of phosphoglucomutase, UDP-glucose pyrophosphorylase, and glycogen synthase.

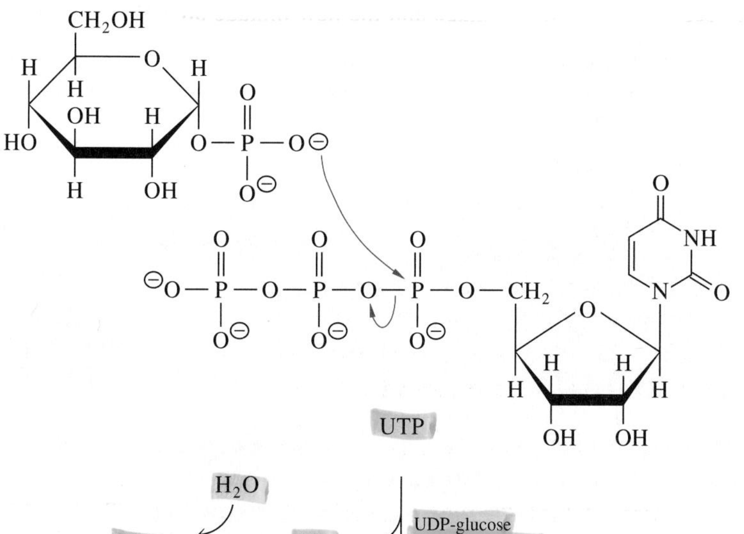

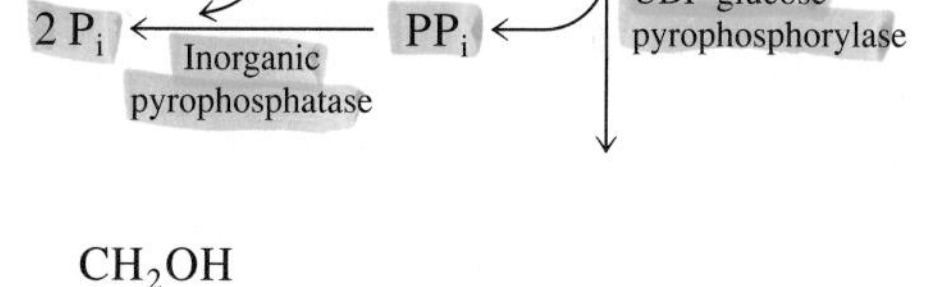

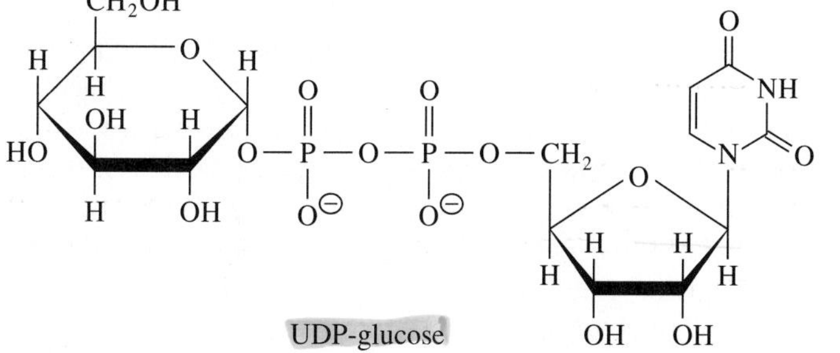

Figure 15·6
Formation of UDP-glucose, catalyzed by UDP-glucose pyrophosphorylase. The reaction mechanism involves nucleophilic attack by an oxygen atom of the phosphoryl group of glucose 1-phosphate on the phosphorus atom of the α-phosphoryl group of UTP. PP_i is released and rapidly hydrolyzed by the action of inorganic pyrophosphatase.

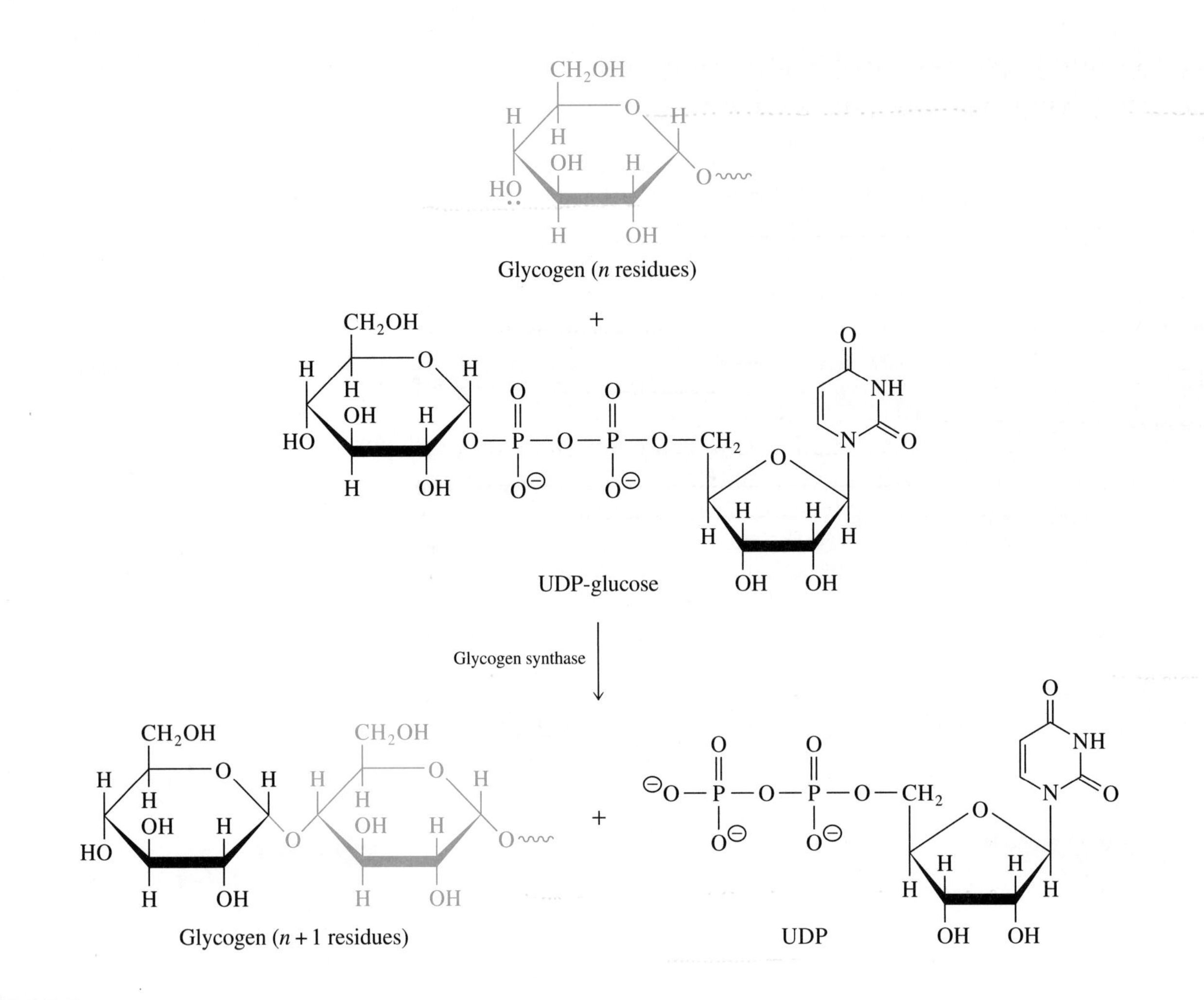

Figure 15·7
Addition of a glucose residue to the nonreducing end of a glycogen primer molecule, catalyzed by glycogen synthase.

of PP_i is highly exergonic and contributes to the metabolic irreversibility of the reaction catalyzed by UDP-glucose pyrophosphorylase. In the last step of glycogen synthesis, glycogen synthase catalyzes the addition of the glucose residue from UDP-glucose to the nonreducing end of glycogen (Figure 15·7). The glycogen synthase reaction is rate-limiting for the pathway of glycogen synthesis. Hormones that control the rate of glycogen synthesis do so by regulating the activity of glycogen synthase.

Glycogen synthase requires a preexisting glycogen primer of at least four glucose residues. The primer consists of up to eight α-(1 $\rightarrow$ 4)-linked glucose residues attached by the 1′-OH of the reducing end in glycosidic linkage to a single tyrosine residue of the protein glycogenin (MW 37 000). The primer is formed in two steps. Attachment of the first glucose residue appears to require a specific glucosyltransferase and UDP-glucose as glucosyl donor. Completion of the primer with up to seven additional UDP-glucose molecules is catalyzed by glycogenin itself. Thus, glycogenin is both a protein scaffold for the construction of glycogen as well as an enzyme. Each completed molecule of glycogen (which can contain several hundred thousand glucose residues) contains a single molecule of glycogenin. Linear extension of the glycogen primer is catalyzed by glycogen synthase.

Another enzyme, amylo-(1,4 $\rightarrow$ 1,6)-transglycosylase, forms branches in glycogen. This enzyme, also known as the branching enzyme, removes an oligosaccharide of at least six residues from the reducing end of an elongated chain and attaches it by an α-(1 $\rightarrow$ 6) linkage to a position at least four glucose residues from the nearest α-(1 $\rightarrow$ 6) branch point.

15·3 The Regulation of Glycogen Metabolism Is a Model for the Regulation of Biochemical Systems

The control of glycogen metabolism in mammals was the first metabolic regulatory system to be well understood at the intracellular level. It is now apparent that the regulatory systems of other pathways follow patterns similar to those of glycogen metabolism. In fact, the enzymes that control glycogen metabolism are also the controlling enzymes for several other pathways, including glycolysis, fatty acid synthesis, gluconeogenesis, and cholesterol synthesis.

We shall consider the regulation of glycogen metabolism at four levels: the whole organism, the cell-surface receptor, signal transduction between receptors and the interior of the cell, and events within the cell. At the intracellular level, we shall examine all of the the regulatory enzymes and proteins and their targets to give a comprehensive picture of how glycogen metabolism is regulated.

A. Hormones Are Messengers at the Level of the Whole Organism

The role of glycogen is to store glucose in times of plenty (after feeding) and to supply glucose in times of need (during fasting or "fight-or-flight" situations). The major sites involved in glycogen metabolism are the pancreas, adrenal gland, liver, and muscle. These sites are connected by the bloodstream, through which they communicate by hormones and shared pathway end products, such as glucose and lactate. The pancreas and adrenal gland supply hormones that affect glycogen metabolism, but only muscle and liver play quantitatively significant roles in the storage and utilization of glycogen in mammals. Other cell types, such as adipocytes, can also take up glucose and synthesize glycogen, but such cells account for a relatively minor portion of overall glycogen metabolism.

Glycogen metabolism is regulated differently within muscle and liver, befitting the different functions of these tissues. Muscle is of quantitative importance in large part because so much of the total mass of mammals consists of muscle. In muscle, glycogen is used as a store of readily available fuel for muscle contraction. In contrast, the glycogen stores in liver are largely converted to glucose that exits liver cells and enters the bloodstream to be transported to other tissues.

The principal hormones involved in the control of glycogen metabolism are insulin, glucagon, and epinephrine. *Insulin* is a small protein of 51 amino acid residues that is secreted by the β cells of the pancreas in response to elevations in blood glucose concentration. Thus, high levels of insulin are associated with the fed state of an organism. Insulin elicits increased uptake and intracellular utilization or storage of glucose in target cells such as muscle and adipose tissue, resulting in a decrease in blood glucose concentration. Diabetes mellitus is a disease marked by excessive and uncontrolled blood glucose concentrations. Type I diabetes mellitus ("juvenile-onset") is due to a lack of insulin, often the result of damage to the pancreas.

Glucagon, another protein hormone (containing 29 amino acid residues), is secreted by the α cells of the pancreas. Glucagon is released in response to low blood glucose concentrations. Glucagon increases blood glucose concentration by activating glycogen degradation. The effect of glucagon is opposite that of insulin, and an elevated glucagon concentration is associated with the fasted state. A characteristic of Type I diabetes in addition to a low insulin concentration is an inappropriately high glucagon concentration.

Unlike insulin and glucagon, which are peptides, *epinephrine* (also known as adrenaline) is a catecholamine derived from the amino acid tyrosine (Figure 15·8). Epinephrine is released from the adrenal glands in response to neural signals that

* function of insulin.

* glucagon function.

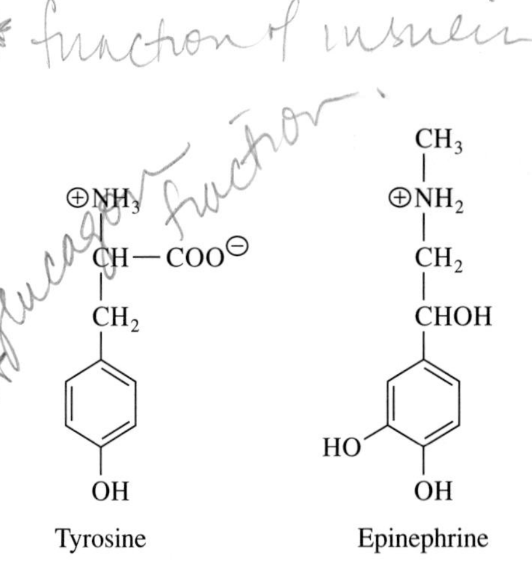

Figure 15·8
Tyrosine and epinephrine. Epinephrine is a catecholamine derived from tyrosine.

trigger the "fight-or-flight" response. Among its diverse physiological effects, epinephrine stimulates increased breakdown of glycogen, resulting in elevated intracellular levels of glucose 6-phosphate. This increase in glucose 6-phosphate leads to enhanced glycolysis in muscle and to an increase in glucose released to the bloodstream from liver.

B. Tissue-Specific Hormone Action Depends on the Presence of Cell-Surface Receptors

When hormones are released from their sites of synthesis (endocrine glands), they enter the bloodstream, where they could potentially communicate with all cells of the organism. Selectivity in hormone response is achieved by the presence in given cells of specific hormone receptors. Many cells, for example, possess insulin receptors; these cells are said to be "insulin responsive." By virtue of their *lack* of insulin receptors, red blood cells and brain cells are said to be "insulin insensitive." Insulin causes muscle cells to increase their uptake of glucose, which results in a decrease in the concentration of blood glucose. Type II or "adult-onset" diabetes is a disease in which insulin receptors of tissues such as muscle are poorly responsive to insulin. Blood glucose concentrations are higher than normal, but the increase is due to a depressed uptake of glucose rather than a deficit in insulin as in Type I diabetes.

Liver cells are the only cells rich in glucagon receptors, making glucagon extremely selective in its target. In contrast, a large number of tissues are responsive to epinephrine. Epinephrine binds to *adrenergic receptors,* of which there are a number of subtypes. Two of these—the α_1 and β receptors—are considered in detail below. They elicit distinct and well understood intracellular responses. In cells that contain both receptor subtypes, one is commonly predominant, so that both responses are not evident in the same cell. The presence of different adrenergic receptors also varies between species.

C. G Proteins Are Signal Transducers that Link Extracellular and Intracellular Events

A family of intracellular proteins called *G proteins* serves as transducers between hormone receptors in the plasma membrane and signalling systems within the cell. The G signifies "guanine-nucleotide binding," a universal characteristic of this class of proteins.

The cyclic activation and deactivation of G proteins is illustrated in Figure 15·9. When a hormone binds to its receptor on the outer face of the plasma membrane, the receptor undergoes a conformational change that allows it to bind an inactive G protein on the inner face of the membrane. Inactive G proteins are commonly complexed with GDP. Binding of the receptor to the inactive G protein causes replacement of GDP by GTP, thereby activating the G protein. The activated G protein then dissociates from the receptor and binds to and activates its target enzyme. All G proteins catalyze the hydrolysis of bound GTP, producing the inactive GDP-bound form of the protein. Thus, the intrinsic GTPase activity is a built-in attenuation mechanism.

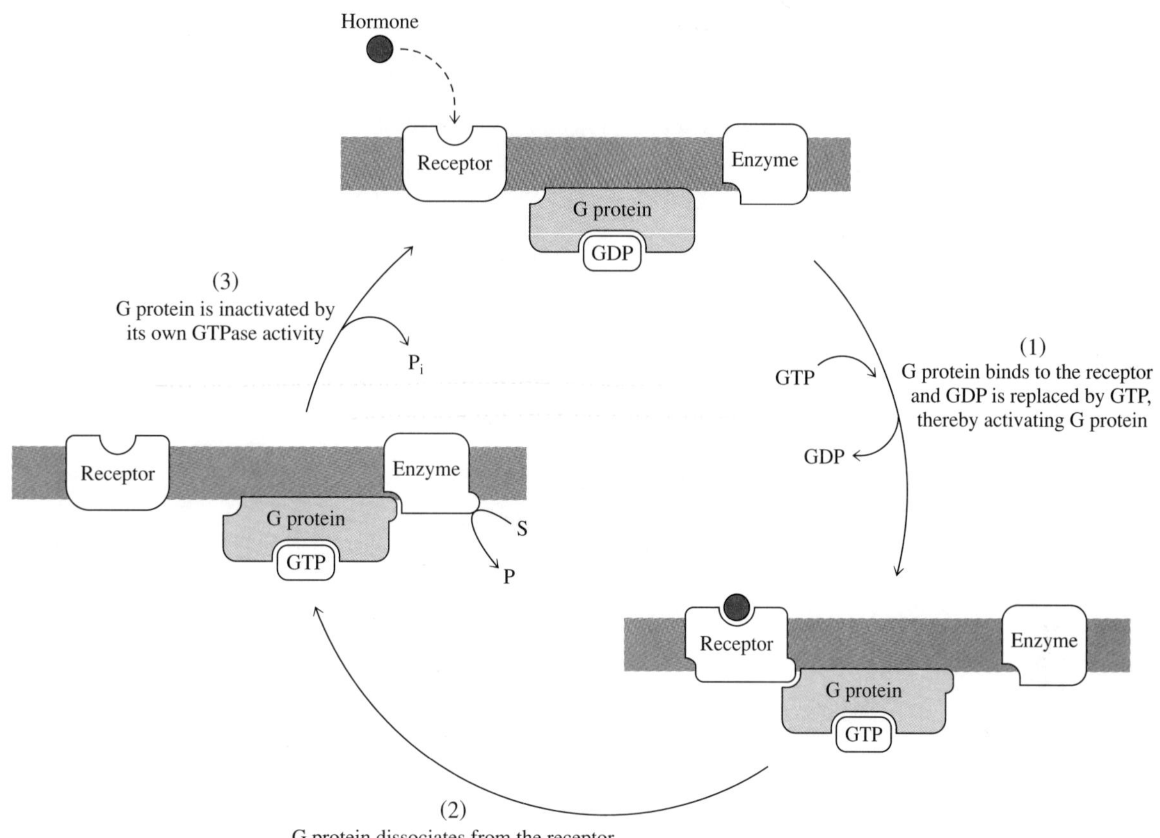

Figure 15·9
Model of G protein activity.

*See lecture notes for exact diagram

G proteins amplify hormone signals. Each receptor activated by a hormone activates many G protein molecules, and each G protein–target enzyme complex catalyzes many reactions before the G protein becomes inactivated. These series of amplifications are called *cascades* and are a common feature of biochemical regulatory systems.

Several hormones that have plasma-membrane receptors act through the intermediacy of a G protein. Glucagon, for example, binds to a glucagon receptor that activates a G protein called G_s. The β-adrenergic receptor of epinephrine also stimulates G_s. G_s in turn stimulates the activity of adenylate cyclase, which catalyzes the formation of cyclic AMP (Box 12·1, page 12·8). Cyclic AMP (cAMP) is an intracellular signal molecule that leads to glycogen degradation, as discussed in Section 15·3E. There are a number of other G proteins in addition to the one associated with the activation of adenylate cyclase. For example, the α_1-adrenergic receptor of epinephrine in liver and other tissues activates a G protein that triggers the series of intracellular events described in the next section.

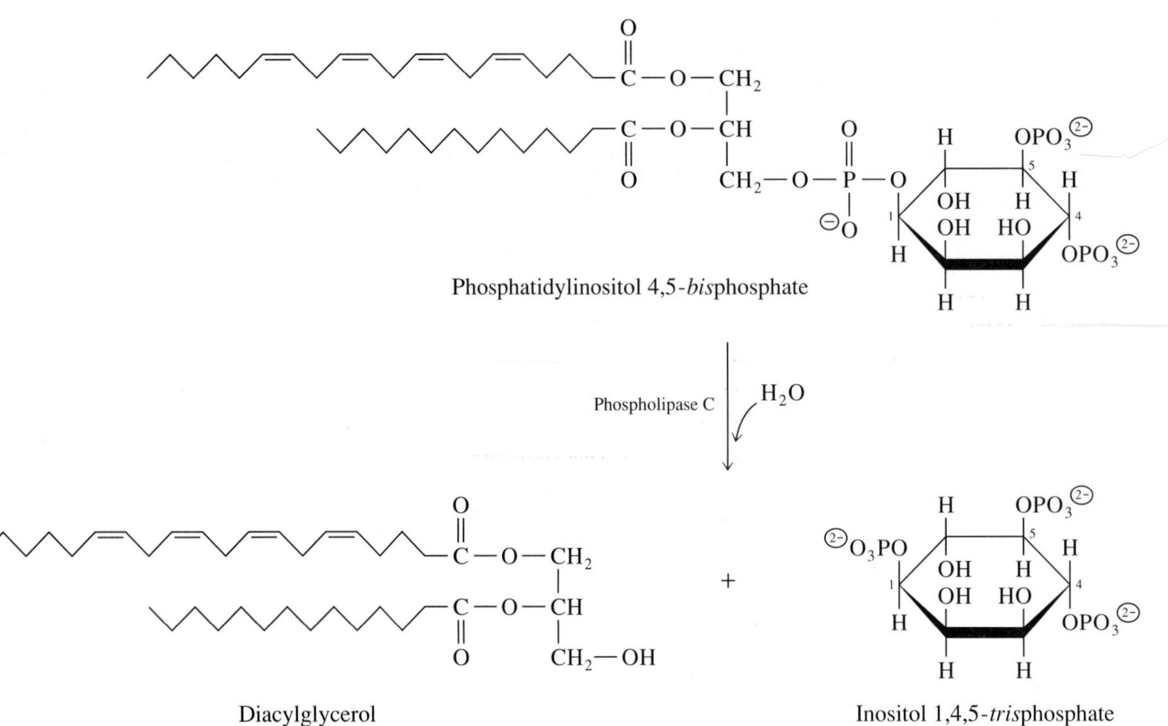

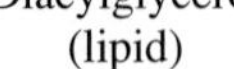

Figure 15·10
Reaction catalyzed by phospholipase C. The substrate, phosphatidylinositol 4,5-*bis*phosphate, is a phospholipid of the inner leaflet of the plasma membrane. The products, diacylglycerol and inositol 1,4,5-*tris*phosphate, are both regulatory molecules.

D. Intracellular Control of Glycogen Metabolism Is Mediated by Second Messengers

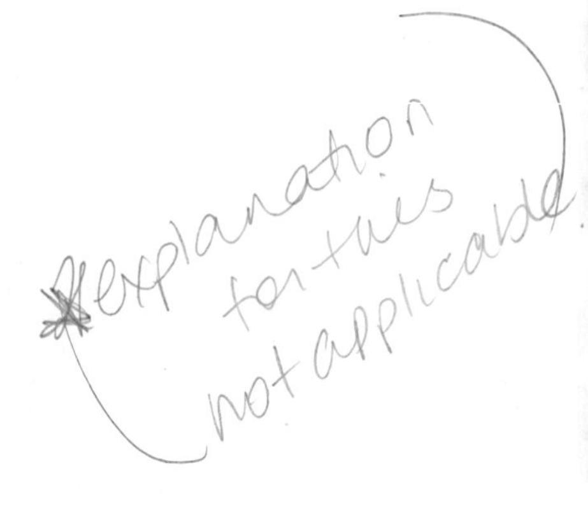

Similar events follow the binding of glucagon at the glucagon receptor and of epinephrine at the β-adrenergic receptor. Both cases result in an increase in the concentration of cAMP. The binding of epinephrine to the α_1-adrenergic receptor invokes an entirely different regulatory system, one that involves activation of the membrane-bound enzyme phospholipase C. The substrate of the reaction catalyzed by phospholipase C is a phospholipid of the plasma membrane, phosphatidylinositol 4,5-*bis*phosphate, and both products are regulatory molecules (Figure 15·10). One product is the water-soluble molecule inositol 1,4,5-*tris*phosphate, which causes release of calcium from the lumen of the endoplasmic reticulum into the cytosol of the cell. Calcium has diverse physiological effects, an example of which is seen in the next section. The other product, diacylglycerol, is a lipid molecule that activates protein kinase C, a membrane-bound regulatory enzyme. One of the known targets of protein kinase C is the insulin receptor. When phosphorylated, the insulin receptor binds insulin poorly. Thus, the action of protein kinase C, triggered by epinephrine, attenuates the action of insulin. A summary of the effects of epinephrine binding at the α_1-adrenergic receptor is shown in Figure 15·11.

Both inositol 1,4,5-*tris*phosphate and diacylglycerol, like cAMP, are *second messengers* in that they transmit information from the first messenger—the hormone. The signal molecule calcium could be considered a "third messenger," though this terminology is not in use. All of these molecules are signal molecules that control the activities of regulatory proteins.

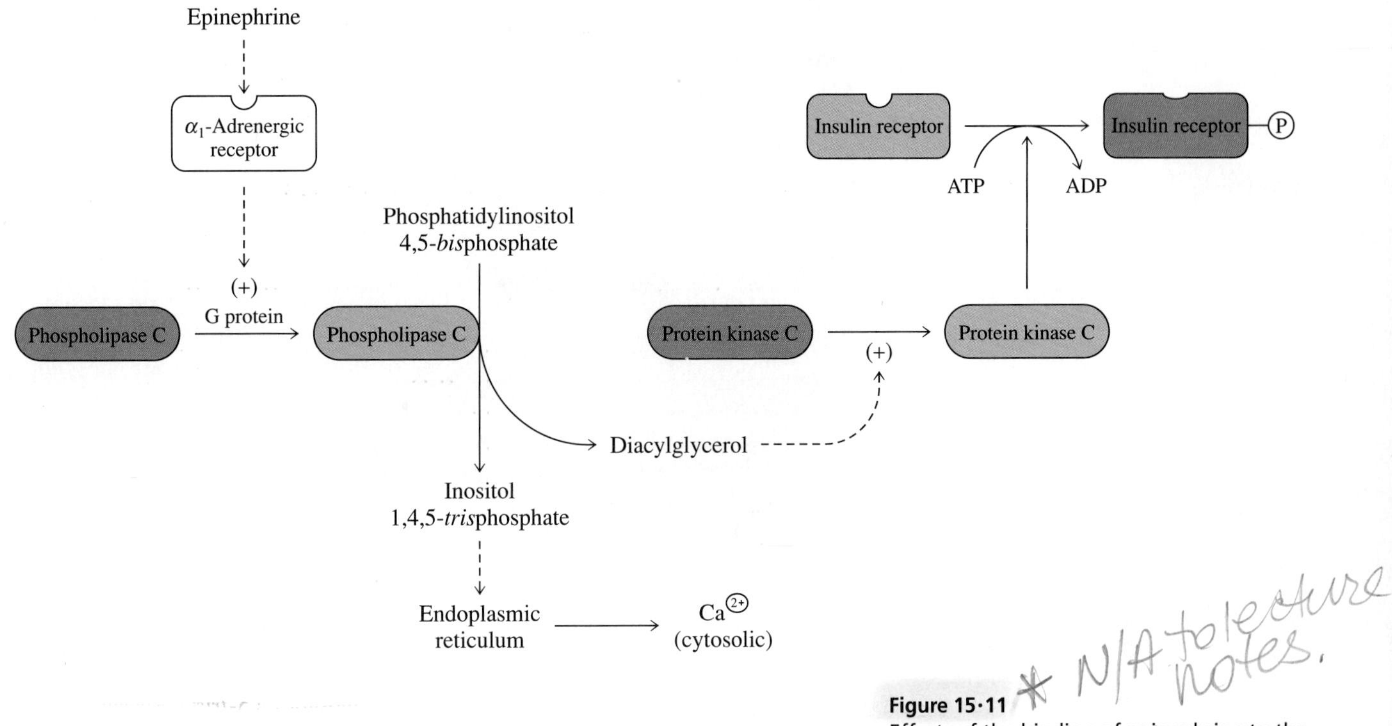

* N/A to lecture notes.

Figure 15·11
Effects of the binding of epinephrine to the α_1-adrenergic receptor. Epinephrine stimulates the activation of phospholipase C via the intermediacy of the α_1-adrenergic receptor and a G protein. Phospholipase C catalyzes the formation of the signal molecules inositol 1,4,5-*tris*phosphate and diacylglycerol. Diacylglycerol activates protein kinase C, which, among other activities, catalyzes the phosphorylation and inhibition of the insulin receptor. Inositol 1,4,5-*tris*phosphate stimulates release of calcium ions from the lumen of the endoplasmic reticulum to the cytosol.

E. Intracellular Regulation of Glycogen Metabolism Involves Interconvertible Enzymes

The enzymes and proteins involved in the intracellular regulation of glycogen metabolism are listed in Table 15·1. The core of this regulation system is the control of the two rate-limiting steps of glycogen metabolism, namely, the reactions catalyzed by glycogen phosphorylase (degradation) and glycogen synthase (synthesis). These enzymes are regulated reciprocally: when one is active, the other is inactive. The principal mechanism of intracellular regulation in glycogen metabolism is phosphorylation and dephosphorylation of specific serine residues of interconvertible enzymes and regulatory proteins. Glycogen phosphorylase and glycogen synthase are both interconvertible enzymes subject to the activity of protein

Table 15·1 Regulatory proteins of intracellular glycogen metabolism

Name	Description
Glycogen phosphorylase	Catalyzes rate-limiting reaction of glycogen breakdown
Glycogen synthase	Catalyzes rate-limiting reaction of glycogen synthesis
Cyclic AMP–dependent protein kinase	Catalyzes phosphorylation of phosphorylase kinase and glycogen synthase
Phosphorylase kinase	Catalyzes phosphorylation of glycogen phosphorylase
Calmodulin	Calcium-binding protein found free in cytosol and as a subunit of phosphorylase kinase
Protein phosphatase-1	Catalyzes dephosphorylation of the phosphoproteins of glycogen metabolism
Inhibitor-1	Inhibits protein phosphatase-1

	active	inactive
Glycogen synthase	—OH	—(P)
Glycogen phosphorylase	—(P)	—OH

Figure 15·12
The active and inactive states of glycogen synthase and glycogen phosphorylase. Both are interconvertible enzymes regulated by phosphorylation and dephosphorylation of specific serine residues.

kinases and protein phosphatases. The phosphorylated form of glycogen synthase is the *inactive* state, whereas the phosphorylated form of glycogen phosphorylase is the *active* state (Figure 15·12). The active and inactive forms of these two enzymes are frequently indicated by appending the letter *a* for active (i.e., glycogen phosphorylase *a*) and *b* for inactive.

The route to increased degradation of glycogen (that is, the route that culminates with glycogen phosphorylase in its phosphorylated, active state and glycogen synthase in its phosphorylated, inactive state) begins with an increase in the concentration of cyclic AMP, which is caused by cascades triggered by epinephrine (in liver and muscle) or by glucagon (in liver) in response to low blood glucose concentration. cAMP activates cAMP-dependent protein kinase, which catalyzes both the phosphorylation and activation of phosphorylase kinase and the phosphorylation and inactivation of glycogen synthase (Figure 15·13). Phosphorylase kinase in turn catalyzes the phosphorylation and activation of glycogen phosphorylase. Thus, by increasing the production of cAMP, either glucagon or epinephrine can cause both an increase in glycogenolysis and a simultaneous suppression of glycogen synthesis.

Phosphorylase kinase is a large enzyme (MW 1.3×10^6 in skeletal muscle) consisting of multiple copies of four distinct subunits. Two of the larger subunits contain serine residues that are phosphorylated when cAMP-dependent protein kinase is active. The smallest of the subunits is a regulatory protein called *calmodulin*, which binds calcium (Stereo S15·1). Calcium binding alone leads to partial activation of phosphorylase kinase. The enzyme is most active when it is both phosphorylated and in the presence of high concentrations of calcium. The fourth subunit of phosphorylase kinase contains the active site that catalyzes the phosphorylation and activation of glycogen phosphorylase, leading to the mobilization of glucose reserves.

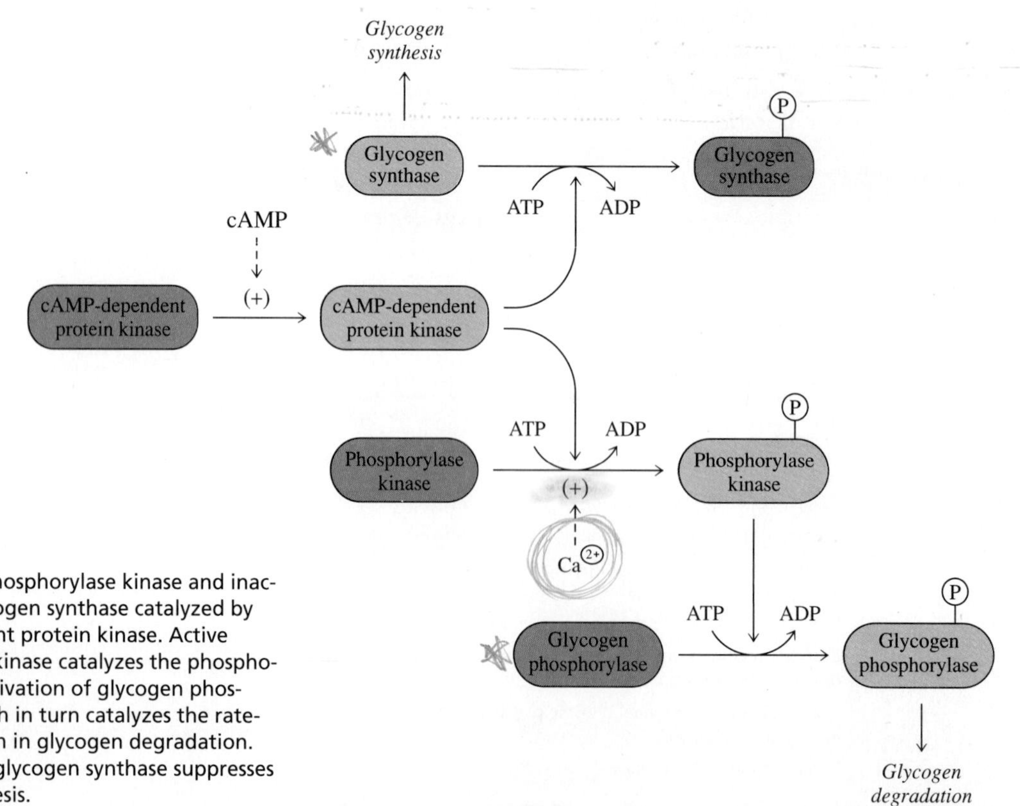

Figure 15·13
Activation of phosphorylase kinase and inactivation of glycogen synthase catalyzed by cAMP-dependent protein kinase. Active phosphorylase kinase catalyzes the phosphorylation and activation of glycogen phosphorylase, which in turn catalyzes the rate-limiting reaction in glycogen degradation. Inactivation of glycogen synthase suppresses glycogen synthesis.

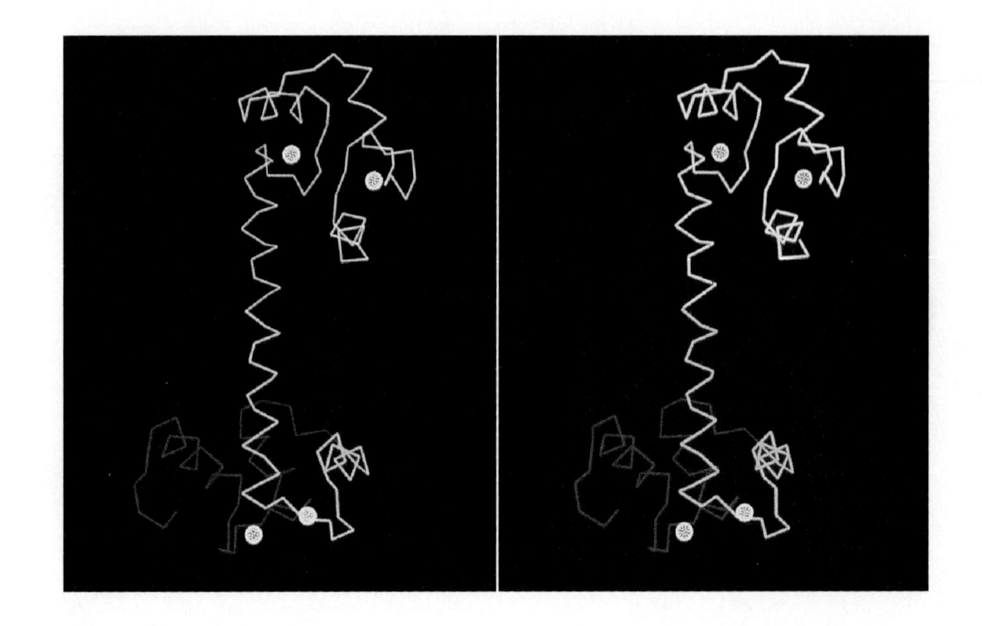

Stereo S15·1
Calmodulin, ribbon model. Four bound Ca^{2+} ions are shown in white. (Based upon coordinates provided by Y. Babu, C. Bugg, and W. Cook; stereo image by Richard J. Feldmann.)

The effects of cAMP are abolished by cAMP phosphodiesterase, which catalyzes the hydrolysis of cAMP to AMP (Figure 15·14). cAMP phosphodiesterase ensures that the effects of cAMP are short-lived. When cAMP phosphodiesterase is inhibited, the rate of conversion of cAMP to AMP is decreased. Coffee and tea contain caffeine and theophylline, respectively, which are methylated purine derivatives that inhibit cAMP phosphodiesterase. In the presence of these inhibitors, the effects of cAMP, and thus the stimulatory effects of the hormones that lead to its production, are prolonged and intensified.

Once the concentration of cAMP falls, the dephosphorylated forms of phosphorylase kinase, glycogen phosphorylase, and glycogen synthase are restored by the action of protein phosphatases. Four major protein phosphatases have been found in cells. One of these, protein phosphatase-1, catalyzes most of the protein

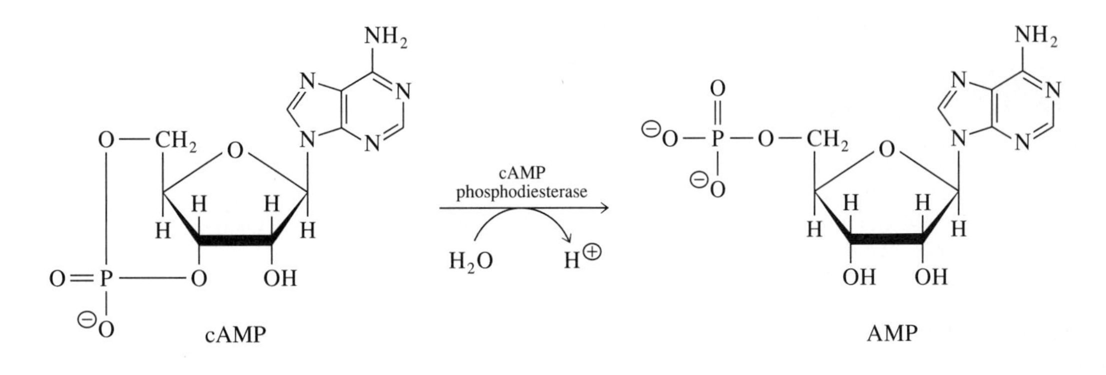

Figure 15·14
Hydrolysis of cAMP, catalyzed by cAMP phosphodiesterase. The effects of cAMP are abolished when cAMP phosphodiesterase catalyzes the hydrolysis of cAMP to AMP.

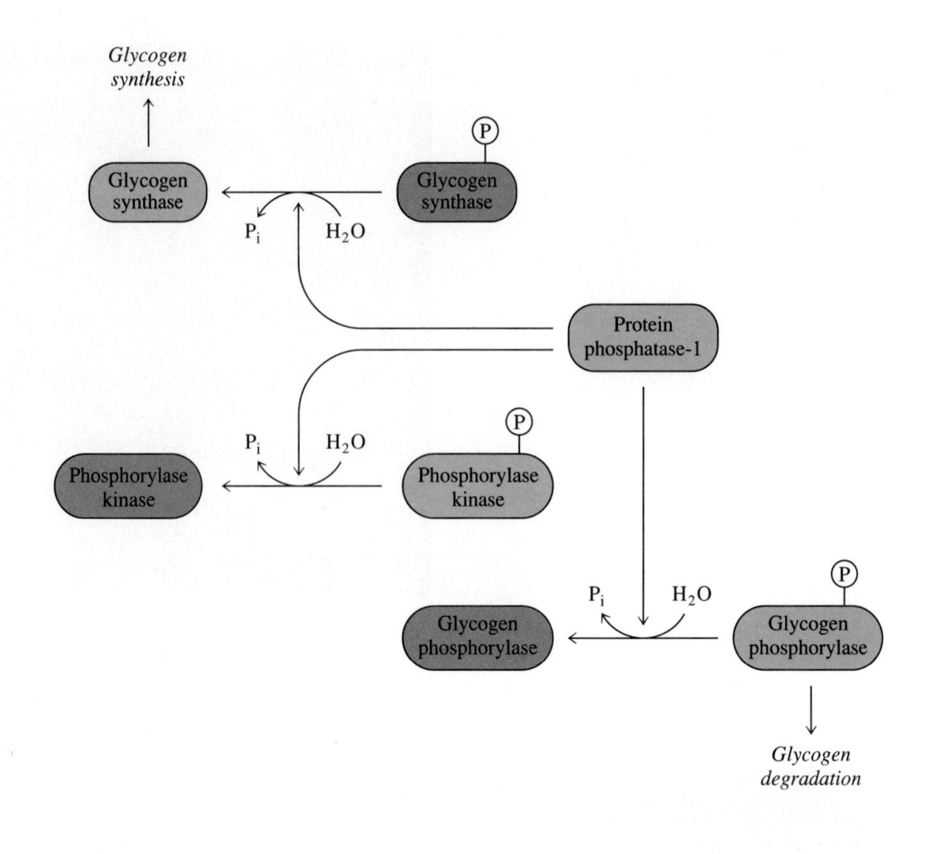

Figure 15·15
Dephosphorylation of target enzymes catalyzed by protein phosphatase-1. Protein phosphatase-1 catalyzes hydrolysis of phosphate ester bonds of the phosphorylated enzymes glycogen synthase, phosphorylase kinase, and glycogen phosphorylase. The result is an increase in glycogen synthesis and a decrease in glycogen degradation.

dephosphorylations of glycogen metabolism. This single enzyme largely accounts for the dephosphorylation of phosphorylase kinase, glycogen phosphorylase, and glycogen synthase (Figure 15·15).

The activity of protein phosphatase-1 is subject to an unusual form of allosteric inhibition by a small protein named inhibitor-1 (Figure 15·16). Inhibitor-1 is subject to phosphorylation catalyzed by cAMP-dependent protein kinase. When phosphorylated, inhibitor-1 is a potent allosteric inhibitor of protein phosphatase-1. The dephosphorylated form of inhibitor-1 has no effect on protein phosphatase-1. The regulatory action of inhibitor-1 is imposed on the previously described regulatory

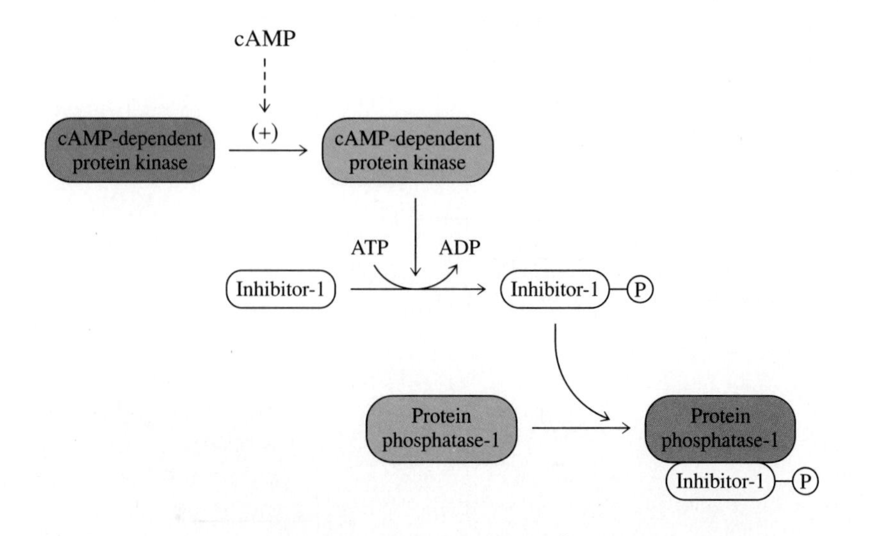

Figure 15·16
Inhibition of protein phosphatase-1 by inhibitor-1. Phosphorylated inhibitor-1 is a potent allosteric inhibitor of protein phosphatase-1. The effects of cAMP, ultimately leading to phosphorylation and activation of glycogen phosphorylase and phosphorylation and deactivation of glycogen synthase, are reinforced by inhibitor-1, which is activated by cAMP-dependent protein kinase.

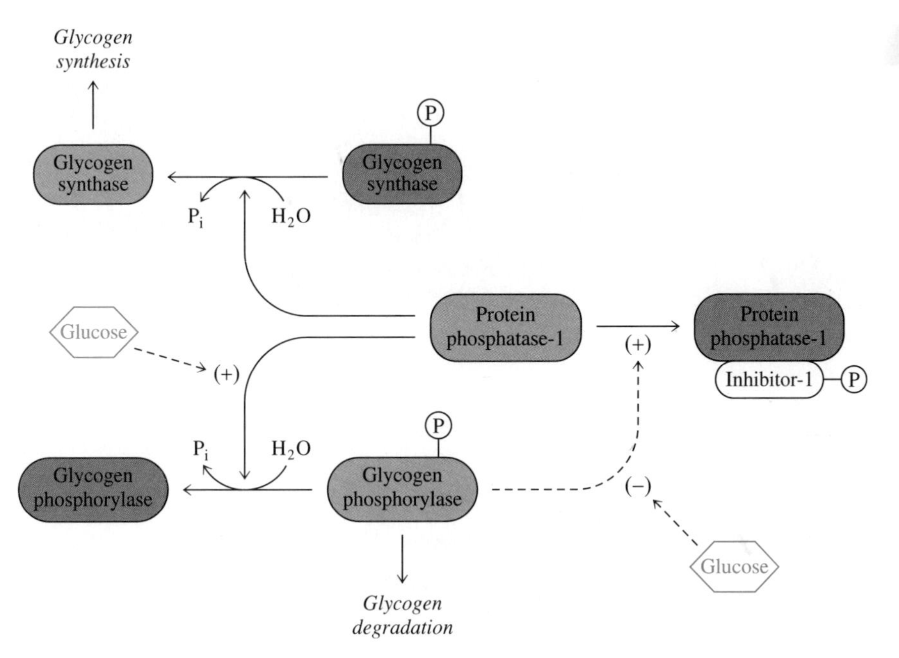

Figure 15·17
Regulation of glycogen metabolism by glucose in the liver. The presence of glucose in the liver has two effects, both of which result from its binding to the active form of glycogen phosphorylase. First, binding of glucose makes active glycogen phosphorylase a better substrate for protein phosphatase-1, the stimulating (+) effect noted in the figure. Second, since active glycogen phosphorylase is a potent inhibitor of protein phosphatase-1, glucose-driven depletion of active glycogen phosphorylase removes this inhibitor, the inhibitory (–) effect noted in the figure. The two effects lead to decreased glycogen breakdown and increased glycogen synthesis when glucose concentration is high.

cascade initiated by cAMP and works to reinforce the effects of cAMP by eliminating the counteracting effects of protein phosphatase-1. Interconvertible enzymes that would be dephosphorylated by the action of protein phosphatase-1—phosphorylase kinase, glycogen phosphorylase, and glycogen synthase—are fixed in their phosphorylated state when inhibitor-1 is active.

In liver, protein phosphatase-1 is also strongly inhibited by the active form of glycogen phosphorylase. The I_{50} for this inhibition (that is, the inhibitor concentration that depresses the activity of the enzyme by 50%) is just 2 nM. This regulatory relationship ensures that glycogen synthase cannot be activated by protein phosphatase-1 until glycogen phosphorylase is nearly all converted to its inactive form.

Finally, the presence of glucose in liver tilts the balance toward glycogen synthesis. Glucose binds to the active form of glycogen phosphorylase and makes it a better substrate for protein phosphatase-1 (Figure 15·17). Deactivation of glycogen phosphorylase removes a potent inhibitor of protein phosphatase-1, leading to activation of glycogen synthase. Just as the level of glucose in the blood mediates the initial release of insulin or glucagon, the level of intracellular glucose regulates glycogen metabolism in liver cells.

15·4 Glucose Can Be Synthesized from Noncarbohydrate Precursors by Gluconeogenesis

The liver has only a limited supply of glycogen for the provision of glucose to other tissues. After an overnight fast, most of liver glycogen is depleted, and the requirement for glucose is met by de novo synthesis from noncarbohydrate precursors, such as lactate and alanine. This process is called *gluconeogenesis.*

The pathway for gluconeogenesis from pyruvate is outlined in Figure 15·18. Many of the steps of gluconeogenesis are common to glycolysis—all of the near-equilibrium reactions of glycolysis are traversed in the reverse direction during gluconeogenesis. Enzymatic reactions unique to gluconeogenesis are required to bypass the highly exergonic reactions of glycolysis—the reactions catalyzed by pyruvate kinase, phosphofructokinase-1, and hexokinase.

Two enzymes are required to bypass the glycolytic reaction catalyzed by pyruvate kinase. First, pyruvate carboxylase catalyzes the conversion of pyruvate to oxaloacetate, coupled to hydrolysis of a molecule of ATP.

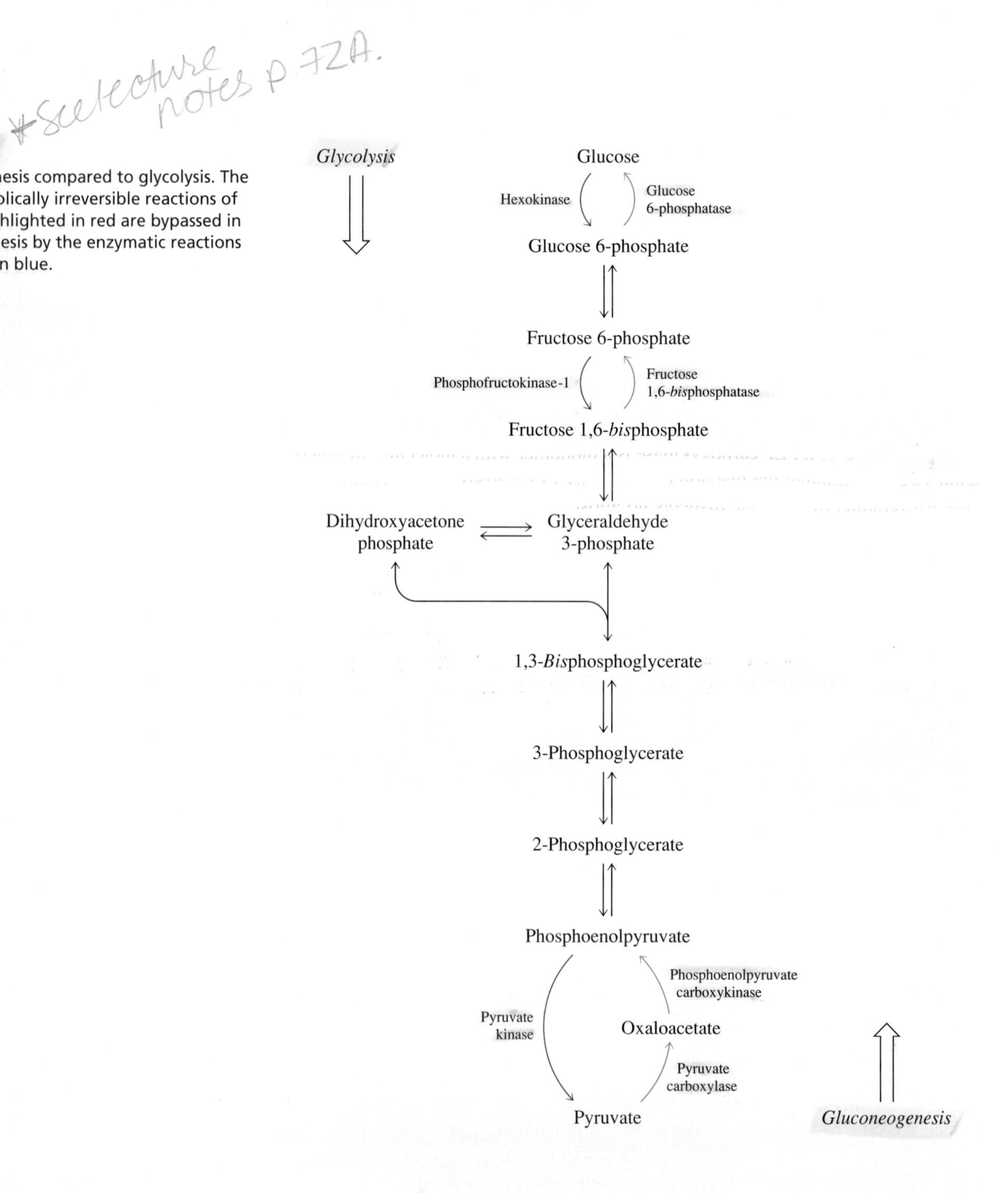

Figure 15·18
Gluconeogenesis compared to glycolysis. The three metabolically irreversible reactions of glycolysis highlighted in red are bypassed in gluconeogenesis by the enzymatic reactions highlighted in blue.

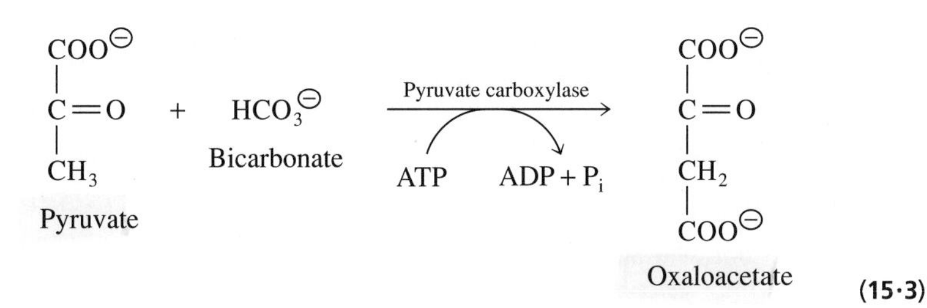

(15·3)

Pyruvate carboxylase has a molecular weight of 520 000 and is composed of four identical subunits. A molecule of biotin (Figure 7·18) is covalently linked to a lysine residue of each subunit. Biotin participates in the transfer of bicarbonate. The reaction mechanism for biotin-containing carboxylases was described in Section 7·7. Pyruvate carboxylase catalyzes a metabolically irreversible reaction and can be allosterically activated by acetyl CoA. This is the only regulatory mechanism known for this enzyme.

Following the pyruvate carboxylase reaction, phosphoenolpyruvate (PEP) carboxykinase catalyzes the conversion of oxaloacetate to phosphoenolpyruvate.

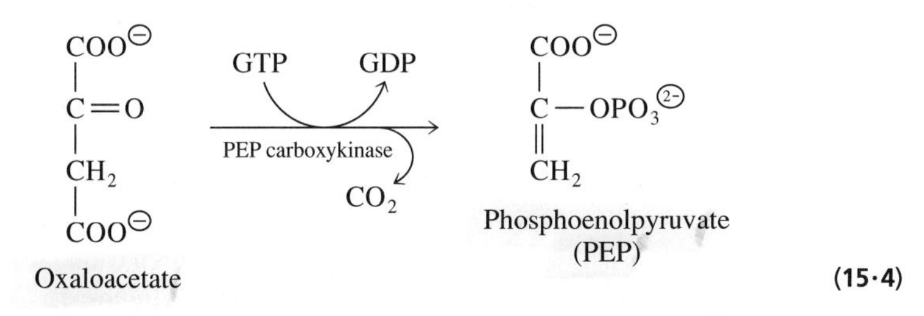

(15·4)

This decarboxylation reaction uses GTP as the donor of a high-energy phosphoryl group (Figure 15·19). PEP carboxykinase is a monomer with a molecular weight of about 70 000. Although the reaction catalyzed by PEP carboxykinase is metabolically irreversible, the enzyme displays no allosteric kinetic properties and has no

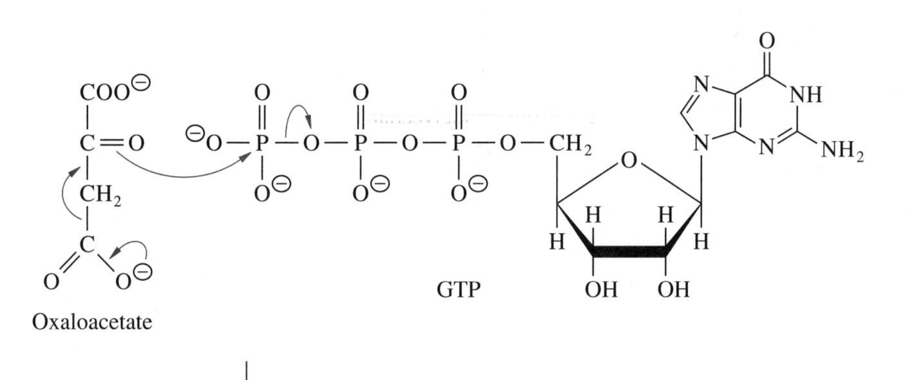

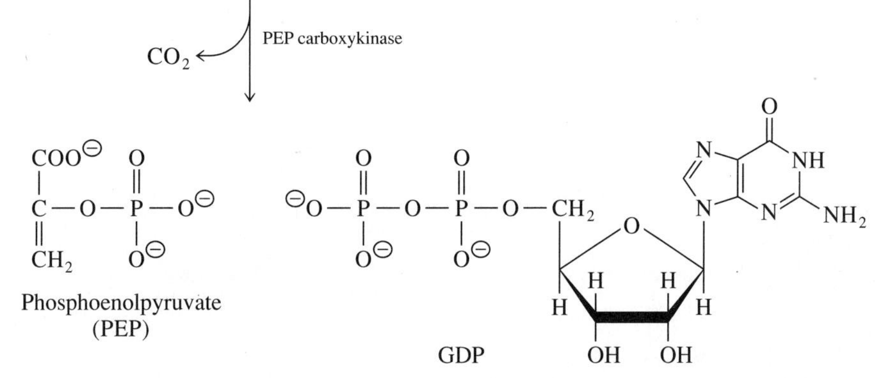

Figure 15·19
Mechanism of the reaction catalyzed by PEP carboxykinase. Decarboxylation of oxaloacetate is followed by nucleophilic attack on the phosphorus atom of the γ-phosphoryl group of GTP, yielding phosphoenolpyruvate, GDP, and CO_2.

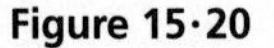

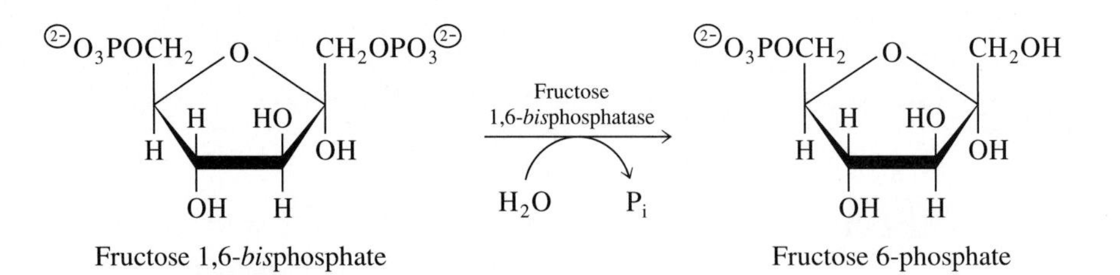

Figure 15·20
Formation of fructose 6-phosphate from fructose 1,6-*bis*phosphate, catalyzed by fructose 1,6-*bis*phosphatase.

known physiological effectors. However, the amount of PEP carboxykinase protein synthesized by cells sets an upper limit for the rate of gluconeogenesis. During progressive fasting in mammals, chronic production of glucagon by the pancreas leads to increased synthesis of PEP carboxykinase in liver, a process called *hormonal induction*. The increased synthesis of PEP carboxykinase is the result of a prolonged elevation of the intracellular concentration of cAMP, which triggers increased transcription of the PEP carboxykinase gene. After several hours of starvation, the amount of PEP carboxykinase rises, raising the rate of gluconeogenesis. Insulin is abundant in the fed state and acts in opposition to glucagon at the level of the gene, leading to a reduction in the synthesis of PEP carboxykinase.

The reactions of gluconeogenesis that convert phosphoenolpyruvate to fructose 1,6-*bis*phosphate are simply the reverse of the near-equilibrium reactions of glycolysis. However, the reaction catalyzed by phosphofructokinase-1 in glycolysis is metabolically irreversible. This reaction is bypassed by the third enzyme specific for gluconeogenesis, fructose 1,6-*bis*phosphatase, which catalyzes the conversion of fructose 1,6-*bis*phosphate to fructose 6-phosphate (Figure 15·20). Fructose 1,6-*bis*phosphatase is a tetrameric enzyme with a molecular weight of 150 000. The enzyme displays sigmoidal kinetics and is allosterically inhibited by the regulatory molecule fructose 2,6-*bis*phosphate. Recall that fructose 2,6-*bis*phosphate is a potent activator of phosphofructokinase-1, the enzyme that catalyzes the formation of fructose 1,6-*bis*phosphate in glycolysis (Section 12·4). Thus, the two enzymes that catalyze interconversion of fructose 6-phosphate and fructose 1,6-*bis*phosphate are reciprocally controlled by the concentration of fructose 2,6-*bis*phosphate.

Following the near-equilibrium conversion of fructose 6-phosphate to glucose 6-phosphate, the final enzyme required for gluconeogenesis is glucose 6-phosphatase. This enzyme catalyzes the hydrolysis of glucose 6-phosphate, producing glucose and inorganic phosphate (Figure 15·21). The glucose 6-phosphatase activity in mammals actually has two separate components. One is a transport protein embedded in the membrane of the endoplasmic reticulum that is specific for glucose 6-phosphate. The second is a relatively nonspecific hydrolase localized in the lumen of the endoplasmic reticulum that can catalyze dephosphorylation of a number of organic phosphates. Whereas all of the other enzymes required for gluconeogenesis are found in small amounts in many types of mammalian tissue, glucose 6-phosphatase is found only in the endoplasmic reticulum of liver and kidney cells.

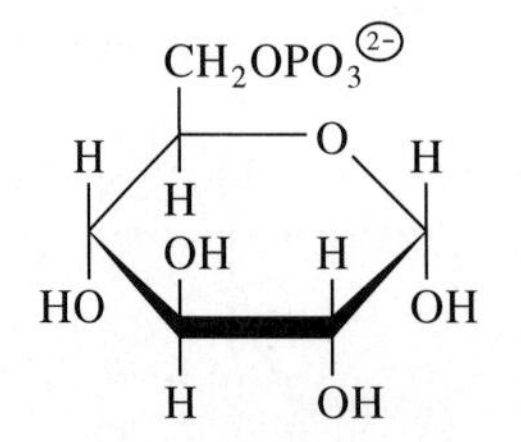

Glucose 6-phosphatase

H_2O P_i

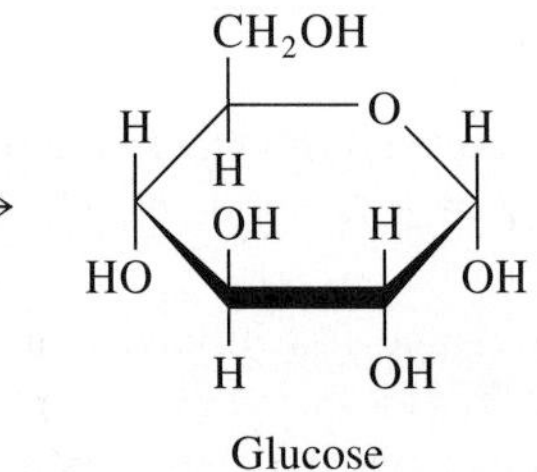

Figure 15·21
Formation of glucose from glucose 6-phosphate, catalyzed by glucose 6-phosphatase.

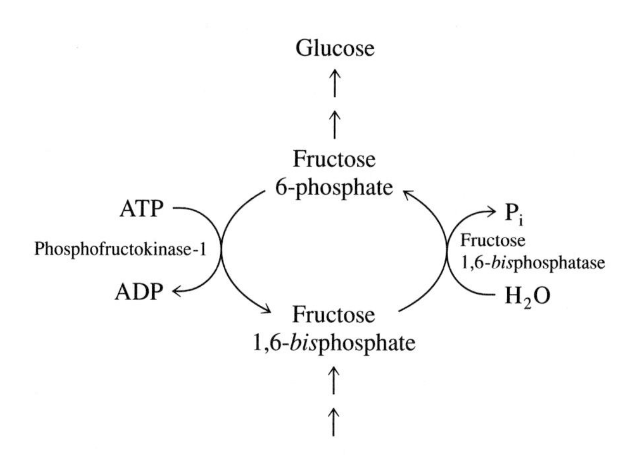

Figure 15·22
An example of a substrate cycle. The direction of flux shown is toward gluconeogenesis. However, phosphorylation of fructose 6-phosphate, catalyzed by phosphofructokinase-1, and dephosphorylation of fructose 1,6-*bis*phosphate, catalyzed by fructose 1,6-*bis*phosphatase, both occur simultaneously. The overall rate of gluconeogenesis can be controlled with great sensitivity by altering the rate of either or both of the enzymes of the substrate cycle.

Many mammalian tissues contain a partial set of gluconeogenic enzymes. For example, muscle contains PEP carboxykinase and fructose 1,6-*bis*phosphatase, but nonetheless muscle is not gluconeogenic. In tissues with at least partial gluconeogenic pathways, these enzymes catalyze so-called "futile cycles," reactions whose net formal balance is the hydrolysis of ATP. For example, the net reaction of phosphofructokinase-1 and fructose 1,6-*bis*phosphatase acting simultaneously is the hydrolysis of ATP to ADP plus P_i:

$$\begin{array}{lrcl} & \text{Fructose 6-phosphate} + \text{ATP} & \longrightarrow & \text{Fructose 1,6-}\textit{bis}\text{phosphate} + \text{ADP} \\ & H_2O + \text{Fructose 1,6-}\textit{bis}\text{phosphate} & \longrightarrow & \text{Fructose 6-phosphate} + P_i \\ \hline \text{Net} & H_2O + \text{ATP} & \longrightarrow & \text{ADP} + P_i \end{array} \qquad (15 \cdot 5)$$

When these reaction sequences were first considered, they were called futile cycles because their operation seemed only to expend ATP with no apparent gain. It was later realized that the presence of opposing, metabolically irreversible reactions that catalyze a cycle between two pathway intermediates provides a sensitive regulatory site. Control of one or both of these reactions can determine the direction and level of flux through the metabolic sequence. These reactions are now termed *substrate cycles*, and it is believed that they serve an important role in cells. In gluconeogenic tissues, where actual reversal of flux through the glycolytic/gluconeogenic pathways is possible, both the direction and the amount of flow through the pathway is controlled by substrate cycles (Figure 15·22).

15·5 Variations in the Gluconeogenic Pathway Arise from Cellular Localization of Enzymes and from the Identity of the Pathway Precursor

All of the near-equilibrium reactions of gluconeogenesis occur in the cytosol, as does the reaction catalyzed by fructose 1,6-*bis*phosphatase. The components of the glucose 6-phosphatase activity, as noted above, are localized in the endoplasmic reticulum.

Pyruvate carboxylase is found in mitochondria. Thus, cytosolic pyruvate must be transported into mitochondria in order to enter the gluconeogenic pathway. Recall from our discussion of the citric acid cycle that the outer mitochondrial membrane presents no barrier to molecules with molecular weights less than 10 000. A specific transporter allows entry of pyruvate across the inner mitochondrial membrane via symport with a proton.

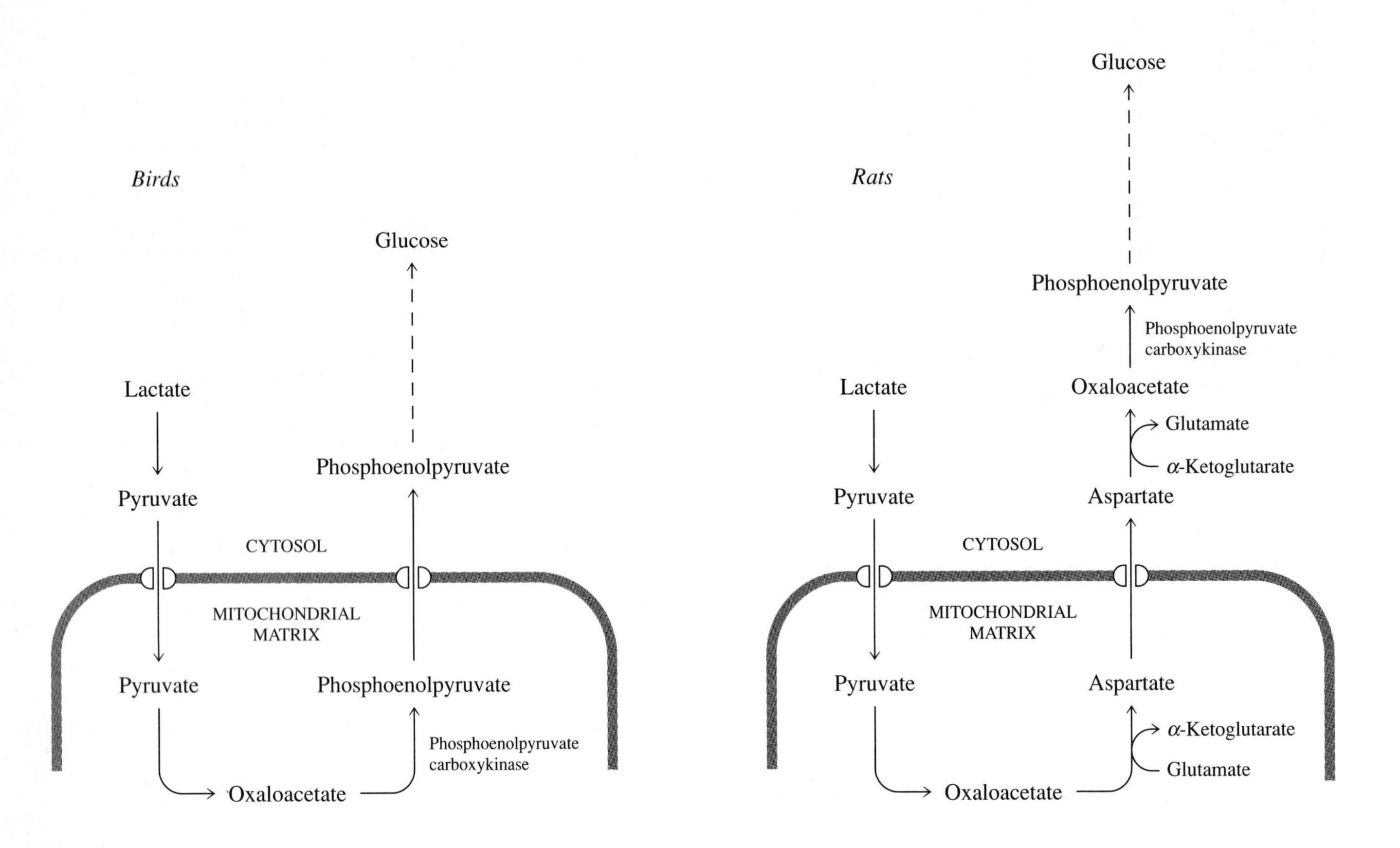

Figure 15·23
Gluconeogenesis in birds and rats. The early stages of gluconeogenesis from lactate are compared for bird liver tissue (left) and rat liver tissue (right), which differ in the localization of PEP carboxykinase.

The localization of the PEP carboxykinase reaction is species-specific. In bird liver, the enzyme is found in mitochondria; in rat liver, it is found in the cytosol; in human liver, it is found in mitochondria and in the cytosol. Cytosolic and mitochondrial PEP carboxykinase are distinct enzymes, and the hormonal induction discussed above occurs only with the cytosolic form.

The localization of PEP carboxykinase in mitochondria or in the cytosol results in different routes for gluconeogenesis. Initial steps in the pathway of gluconeogenesis from lactate in birds and rats are compared in Figure 15·23. The pathways diverge after the formation of mitochondrial oxaloacetate. In birds, mitochondrial PEP carboxykinase produces mitochondrial PEP, which is transported into the cytosol. In rats, oxaloacetate (which has no transport protein to carry it across the mitochondrial membrane) is first converted to aspartate, which has a transport protein in the inner mitochondrial membrane. Once in the cytosol, aspartate is converted back to oxaloacetate, which serves as the substrate for cytosolic PEP carboxykinase.

A second source of variation in gluconeogenesis stems from the necessity for replenishment of the NADH oxidized in the glyceraldehyde 3-phosphate dehydrogenase reaction. The reverse situation was seen in glycolysis. As described in Section 12·13, ongoing glycolysis requires that the $NAD^{\oplus}$ reduced to NADH in the glyceraldehyde 3-phosphate dehydrogenase step be continually renewed.

The mixture of substrates for gluconeogenesis includes those that may or may not provide cytosolic NADH. Two extreme situations for gluconeogenesis are illustrated in Figure 15·24. In one case, lactate is the pathway substrate; in the other, pyruvate is the substrate. In each case, the pathways are shown for species that have

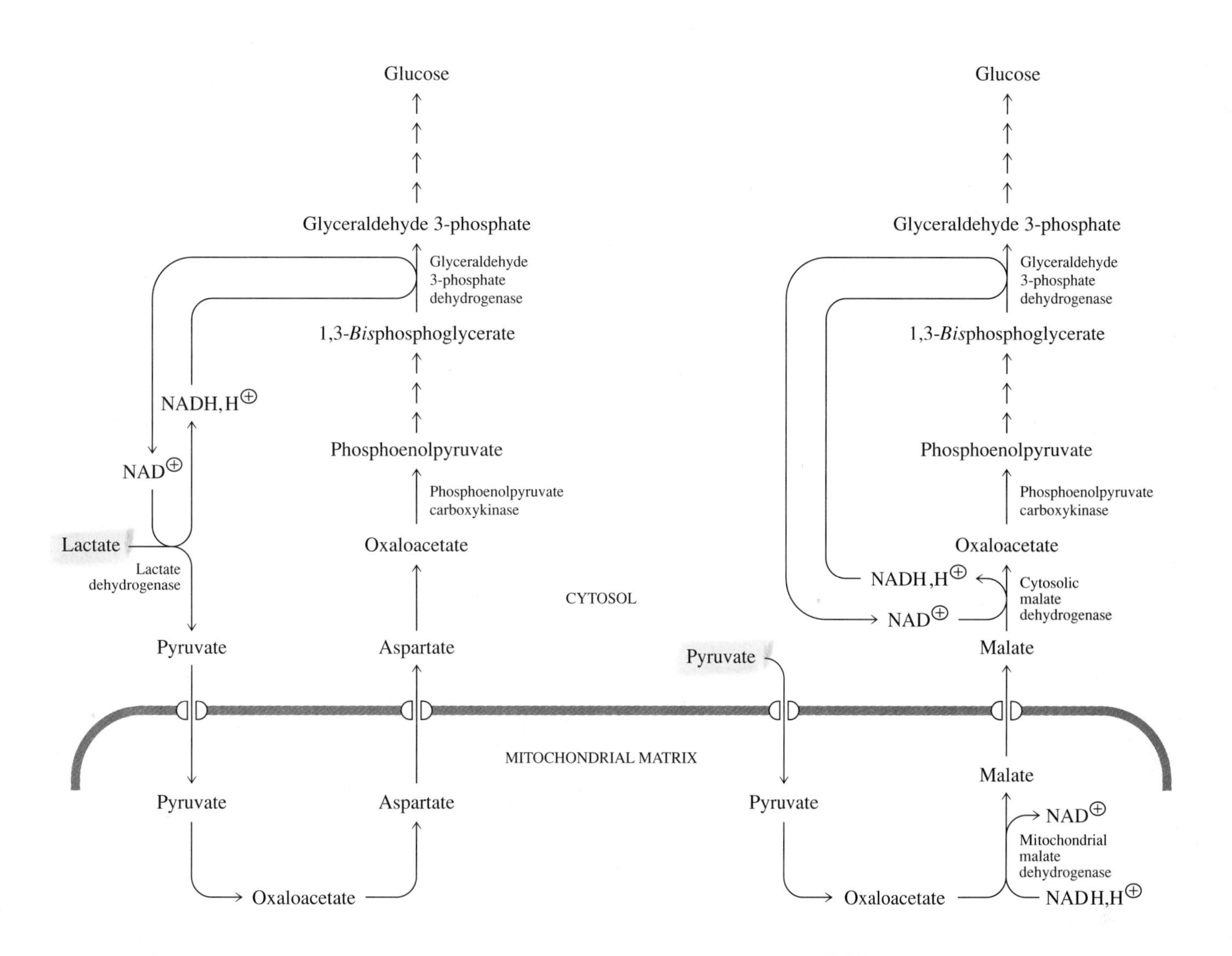

Figure 15·24
Gluconeogenesis from lactate and pyruvate. Because the supply of NADH must be renewed for the reaction catalyzed by glyceraldehyde 3-phosphate dehydrogenase, the pathway for gluconeogenesis varies with the redox state of the substrate.

a cytosolic PEP carboxykinase. When lactate is the substrate, the first step is catalyzed by lactate dehydrogenase, forming pyruvate and cytosolic NADH. Pyruvate is transported into the mitochondrion and cytosolic NADH is reoxidized further along in the gluconeogenic pathway in the reaction catalyzed by glyceraldehyde 3-phosphate dehydrogenase. Following carboxylation of pyruvate, the product, oxaloacetate, is converted to aspartate in a transamination reaction catalyzed by aspartate transaminase.

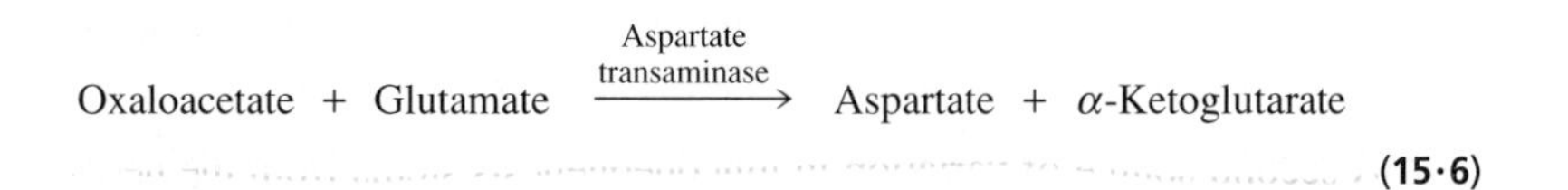
$$\text{Oxaloacetate} + \text{Glutamate} \xrightarrow{\text{Aspartate transaminase}} \text{Aspartate} + \alpha\text{-Ketoglutarate} \qquad (15\cdot6)$$

Aspartate can traverse the inner mitochondrial membrane via the glutamate-aspartate translocase (Section 14·12). Once in the cytosol, a separate aspartate transaminase catalyzes the reverse of the reaction in Equation 15·6. The oxaloacetate that is produced can then proceed through the remaining steps of the gluconeogenic pathway. As noted above, the NADH required for the glyceraldehyde 3-phosphate dehydrogenase reaction is generated in the initial reaction catalyzed by lactate dehydrogenase.

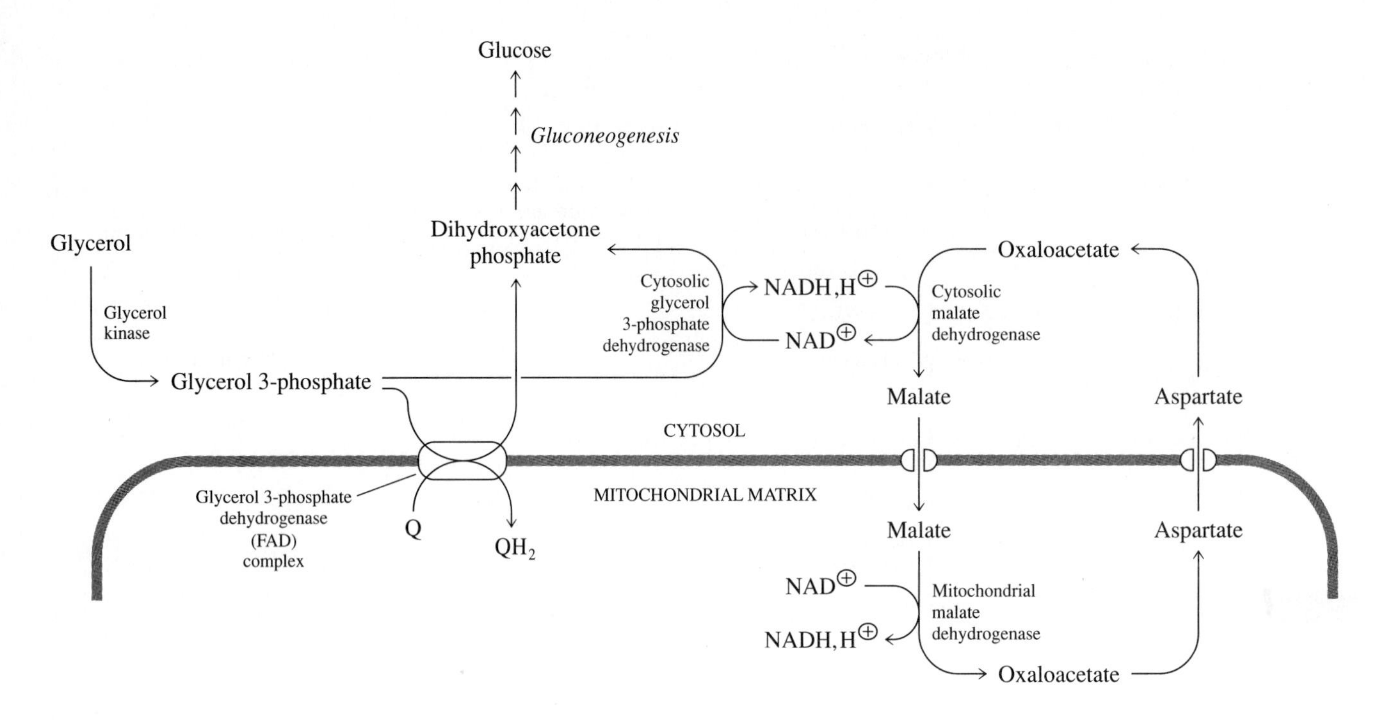

Figure 15·25
Gluconeogenesis from glycerol. Glycerol 3-phosphate can be oxidized in reactions catalyzed by either of two dehydrogenases, each of which results in the formation of reducing equivalents.

Gluconeogenesis from pyruvate must follow a different pathway, since the NADH-producing step catalyzed by lactate dehydrogenase is not involved. As shown on the right side of Figure 15·24, the pathway from pyruvate uses mitochondrial and cytosolic malate dehydrogenases, which effectively shuttle reducing equivalents from NADH in the mitochondrion to NADH in the cytosol.

Finally, glycerol, a product of the breakdown of triacylglycerols from adipose tissue, can be converted to glucose by a route that begins with phosphorylation to glycerol 3-phosphate, catalyzed by the enzyme glycerol kinase (Figure 15·25). Glycerol 3-phosphate enters the gluconeogenic scheme by conversion to dihydroxyacetone phosphate. This oxidation reaction can be catalyzed by a flavin-containing glycerol 3-phosphate dehydrogenase embedded in the inner mitochondrial membrane. The cytosolic face of this enzyme binds glycerol 3-phosphate, and after electrons are removed from the substrate, they are passed to ubiquinone (Q) and subsequently to the rest of the mitochondrial respiratory electron-transport chain. Oxidation of glycerol 3-phosphate can also be catalyzed by cytosolic glycerol 3-phosphate dehydrogenase, in which case NADH is a coproduct. In liver, where most gluconeogenesis in mammals occurs, both reactions are employed, as indicated in Figure 15·25. However, the reaction catalyzed by cytosolic glycerol 3-phosphate dehydrogenase produces NADH that is not needed for gluconeogenesis from glycerol. Thus, the malate-aspartate shuttle is required to carry the reducing equivalents from the cytosol to the mitochondrion, where mitochondrial NADH is oxidized by the electron-transport chain. This pathway is shown on the right of Figure 15·25.

15·6 Gluconeogenesis Is Regulated by Hormones and by Substrate Supply

Short-term regulation of gluconeogenesis, meaning regulation that can occur within minutes and does not involve synthesis of new protein, is exerted at two sites in the pathway—the substrate cycles between pyruvate and phosphoenolpyruvate and between fructose 1,6-*bis*phosphate and fructose 6-phosphate.

We have seen that a long-term regulatory effect of glucagon is the enhancement of gluconeogenesis due to a rise in the quantity of cytosolic PEP carboxykinase. Elevated levels of glucagon also lead to inactivation of pyruvate kinase, as noted in Section 12·11 during our discussion of the regulation of glycolysis. We can now appreciate an important feature of substrate cycles: modification of any enzyme in the cycle can alter the flux through two opposing pathways. For example, inhibition of pyruvate kinase leads to stimulation of gluconeogenesis. This can also be viewed as causing more phosphoenolpyruvate to enter the pathway leading to glucose rather than being converted to pyruvate by pyruvate kinase (Figure 15·26). Glucagon also regulates the fructose 6-phosphate/fructose 1,6-*bis*phosphate cycle. cAMP-dependent protein kinase catalyzes the phosphorylation of phosphofructokinase-2, as described in Section 12·4. The resulting decrease in the concentration of fructose 2,6-*bis*phosphate removes a potent activator of phosphofructokinase-1 and relieves the inhibition of fructose 1,6-*bis*phosphatase, activating gluconeogenesis.

A second form of regulation for gluconeogenesis is the concentration of substrate itself. The principal substrates for the gluconeogenic pathway in mammals are amino acids (mostly alanine and glutamine) and lactate. The amino acids arise from breakdown of muscle protein and are converted by pathways described in Chapter 18 to intermediates of the citric acid cycle. Under certain metabolic conditions, oxaloacetate produced by the citric acid cycle is diverted to the gluconeogenic pathway. The concentration of amino acids in blood is not saturating for the gluconeogenic pathway, and an increase in the concentration of free amino acids results in a greater conversion to glucose. In this way, products of pathways other than gluconeogenesis—and external to the liver itself—can control gluconeogenesis. Similarly, the concentration of lactate in the blood is below the saturation level for gluconeogenesis. Lactate is produced largely by muscle, reflecting the large mass and high rate of glycolytic activity in muscle tissue. Thus, gluconeogenesis is limited by the supply of all its major substrates.

It should be recognized that the pathways of gluconeogenesis and glycogen metabolism are not mutually exclusive. In liver, substrates may traverse the gluconeogenic pathway and go on to form glycogen rather than glucose as long as glycogen synthase is active. Glucose formation increases as glycogen is depleted. This corresponds both to changes in the activity of gluconeogenic enzymes and to the progressive inactivation of glycogen synthase. An organism is therefore able to maintain a constant blood glucose concentration. The pathways of glucose metabolism are not abruptly switched on or off, but rather are continuously adjusted according to the minute-to-minute metabolic needs of the organism.

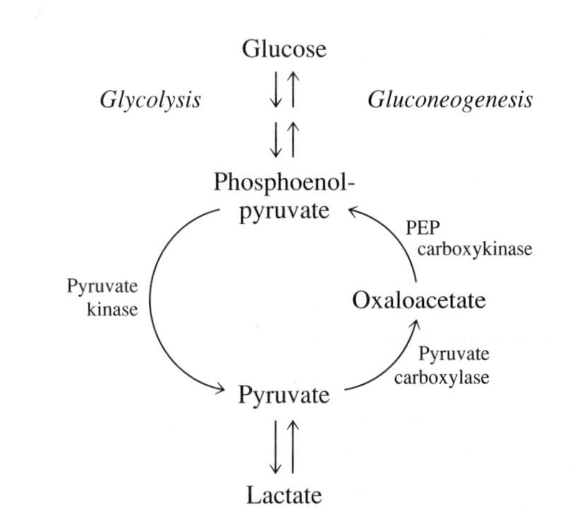

Figure 15·26
The substrate cycle between phosphoenolpyruvate and pyruvate. Any of the three enzymes involved in the substrate cycle can affect not only the rate of flux between phosphoenolpyruvate and pyruvate but also the direction of flux toward either glycolysis or gluconeogenesis.

Figure 15·27
The pentose phosphate pathway. The pathway can be divided into an oxidative and a nonoxidative stage. The oxidative stage produces the five-carbon sugar phosphate ribulose 5-phosphate, with concomitant production of NADPH. The nonoxidative stage produces the glycolytic intermediates glyceraldehyde 3-phosphate and fructose 6-phosphate.

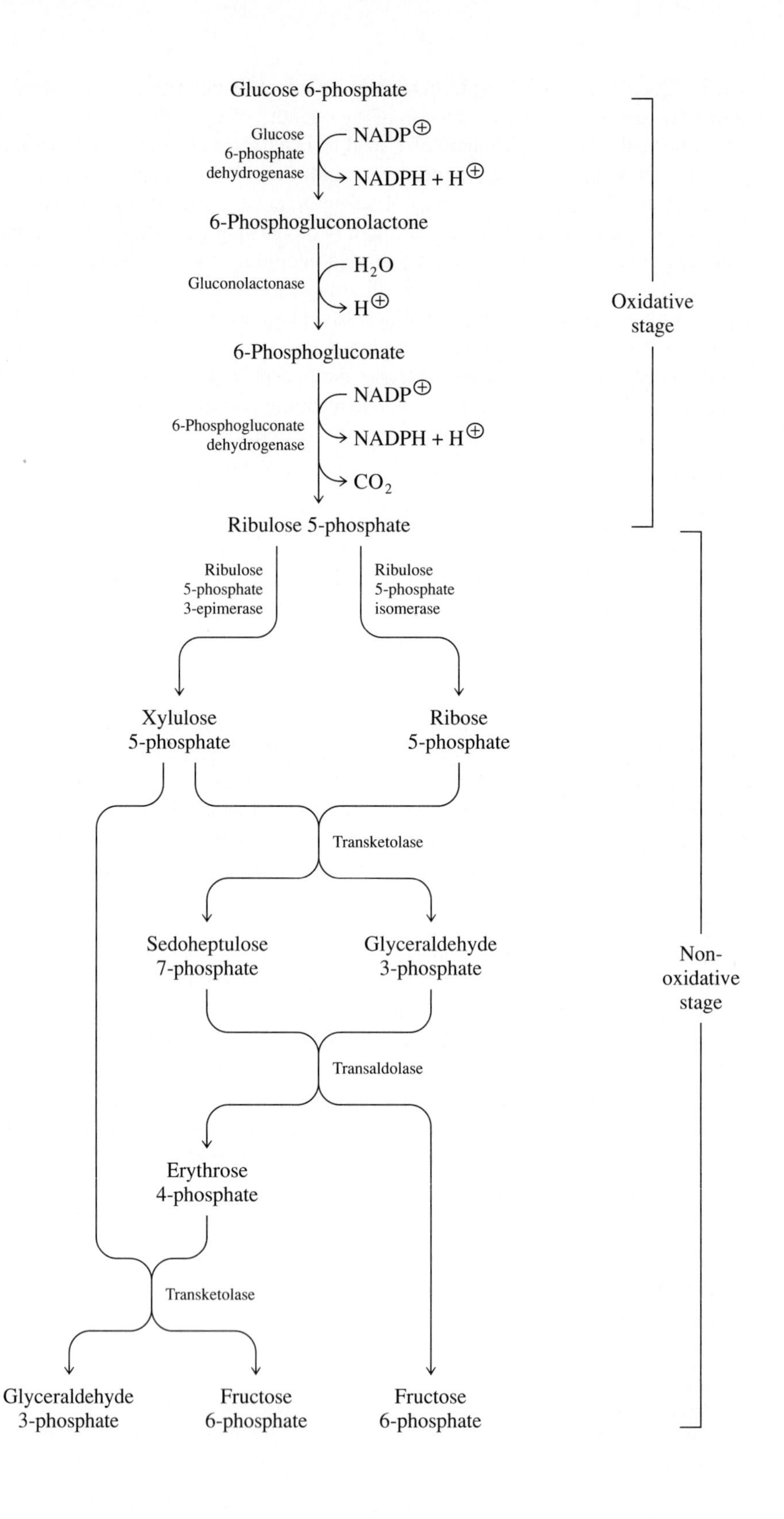

15·7 The Pentose Phosphate Pathway Produces NADPH and Ribose 5-Phosphate

The pentose phosphate pathway is the final pathway in our examination of glucose metabolism. Production of NADPH and the formation of ribose 5-phosphate for nucleotide biosynthesis are both functions of the pentose phosphate pathway. NADPH is a cellular reductant that is used in fatty acid and cholesterol synthesis. NADPH also serves as the reductant in a mechanism that protects against oxidative damage arising from the nonenzymatic formation of $O_2 \cdot^{\ominus}$ and H_2O_2 from O_2 (Box 15·1). In red blood cells, the concentration of oxygen is very high, and the potential for oxidative damage is great. Consequently, there is a large demand for NADPH, and the pentose phosphate pathway in red blood cells can account for about 10% of the total consumption of glucose. In other cells, such as muscle and brain, the pentose phosphate pathway accounts for little of the overall consumption of glucose.

The pentose phosphate pathway can be divided into an oxidative stage and a nonoxidative stage, as illustrated in Figure 15·27. In the oxidative stage, NADPH is produced as glucose 6-phosphate is converted to ribulose 5-phosphate. In the nonoxidative stage, ribulose 5-phosphate is converted to fructose 6-phosphate and glyceraldehyde 3-phosphate, intermediates of the glycolytic pathway. Ribose 5-phosphate is an intermediate of the nonoxidative stage.

The oxidative stage of the pentose phosphate pathway is shown in Figure 15·28. The first reaction, catalyzed by glucose 6-phosphate dehydrogenase, is the conversion of glucose 6-phosphate to 6-phosphogluconolactone. This step is rate-limiting for the entire pentose phosphate pathway. Glucose 6-phosphate dehydrogenase is allosterically inhibited by NADPH. By virtue of this simple regulatory feature, production of NADPH by the pentose phosphate pathway is self-limiting. The next enzyme of the oxidative phase is gluconolactonase, which catalyzes hydrolysis of 6-phosphogluconolactone to form the sugar acid 6-phosphogluconate. Finally, 6-phosphogluconate dehydrogenase catalyzes the oxidative decarboxylation of 6-phosphogluconate, producing a second molecule of NADPH, ribulose 5-phosphate, and CO_2.

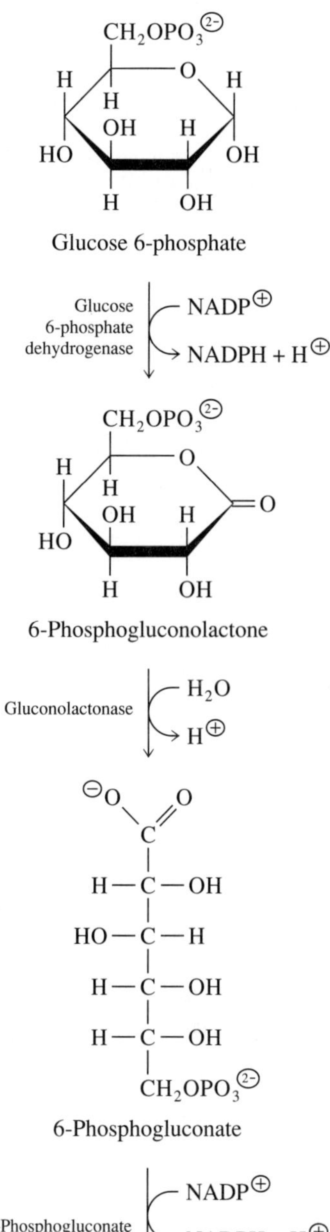

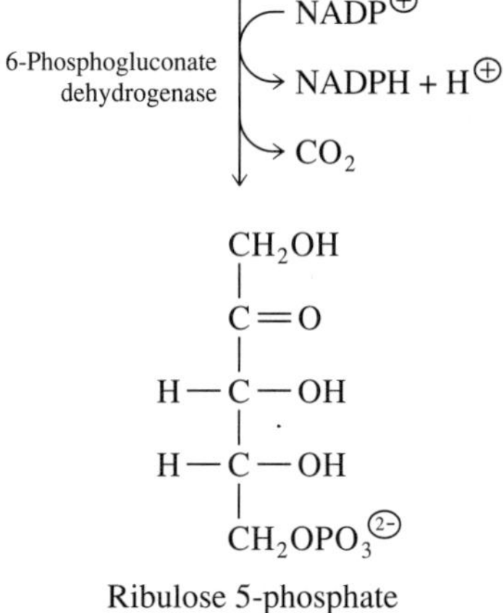

Figure 15·28
Oxidative stage of the pentose phosphate pathway. Two molecules of NADPH are generated for each molecule of glucose 6-phosphate that enters the pathway.

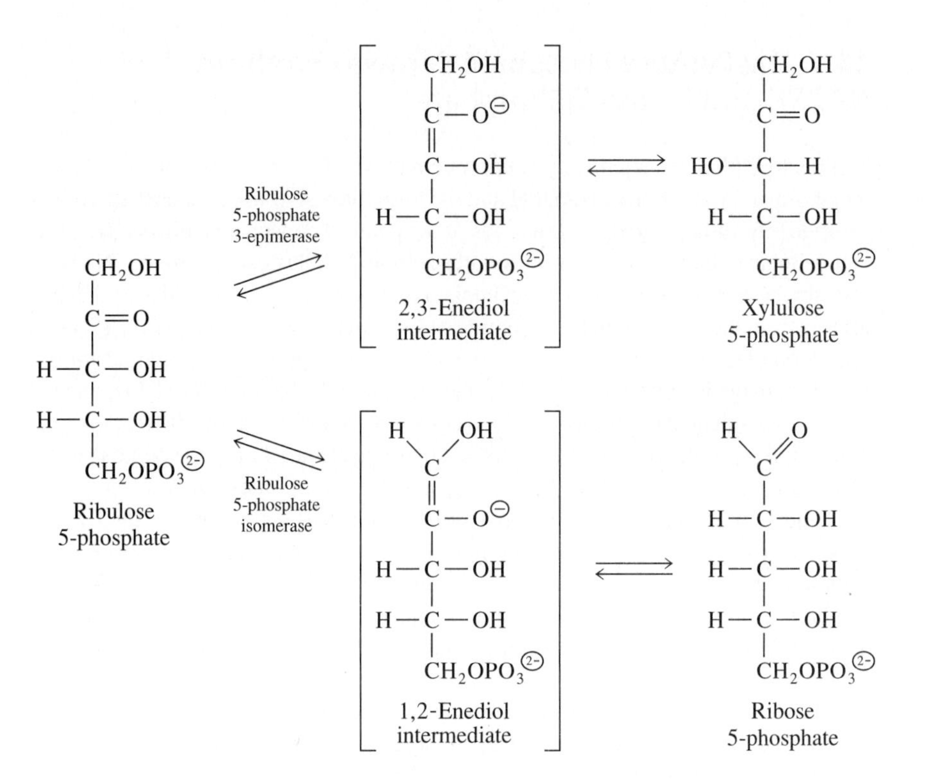

Figure 15·29
Conversion of ribulose 5-phosphate to xylulose 5-phosphate or ribose 5-phosphate. In either case, removal of a proton leads to formation of an enediol intermediate. Reprotonation forms either the ketose xylulose 5-phosphate or the aldose ribose 5-phosphate.

The nonoxidative stage of the pentose phosphate pathway in cells that have a high overall flux through the pathway consists entirely of near-equilibrium reactions. This stage of the pathway serves two functions: it provides five-carbon sugars for biosynthesis and introduces sugar phosphates into glycolysis or gluconeogenesis. There are two fates for ribulose 5-phosphate: an epimerase can catalyze the formation of xylulose 5-phosphate, or an isomerase can catalyze the conversion of ribulose 5-phosphate into ribose 5-phosphate (Figure 15·29). The latter is the precursor to the ribose portion of nucleotides and the polynucleotides RNA and DNA. However, only a very small amount of ribose 5-phosphate is siphoned from the pentose phosphate pathway to fill this need. The remaining steps convert the five-carbon sugars into glycolytic intermediates.

Box 15·1 NADPH and Oxygen Damage in Cells

O_2 is a reactive molecule that can form superoxide anion (the free radical $O_2\cdot^{\ominus}$) and hydrogen peroxide (H_2O_2) through enzymatic or nonenzymatic reactions in cells. H_2O_2 and $O_2\cdot^{\ominus}$ can interact to form the even more reactive and damaging hydroxyl radical, HO·. As part of the cellular protective mechanism against oxygen damage, a number of enzymes are responsible for metabolism of oxygen by-products. The superoxide anion is converted to O_2 and H_2O_2 by the action of superoxide dismutase (Section 6·7). H_2O_2 is removed by the action of either catalase (also introduced in Section 6·7) or glutathione peroxidase.

Glutathione (GSH), the tripeptide γ-glutamylcysteinylglycine, is the reduced cofactor in oxygen detoxification.

$$H_3\overset{\oplus}{N}-\underset{\displaystyle COO^{\ominus}}{CH}-CH_2-CH_2-\overset{O}{\overset{\|}{C}}-NH-\underset{\displaystyle \underset{\displaystyle SH}{CH_2}}{CH}-\overset{O}{\overset{\|}{C}}-NH-CH_2-COO^{\ominus}$$

Glutathione (GSH) **(1)**

Glutathione peroxidase is present in vertebrates and some invertebrates but not in bacteria or plants. In the presence of glutathione peroxidase and H_2O_2, the sulfhydryl groups of two molecules of glutathione react to form a molecule of glutathione disulfide (GSSG).

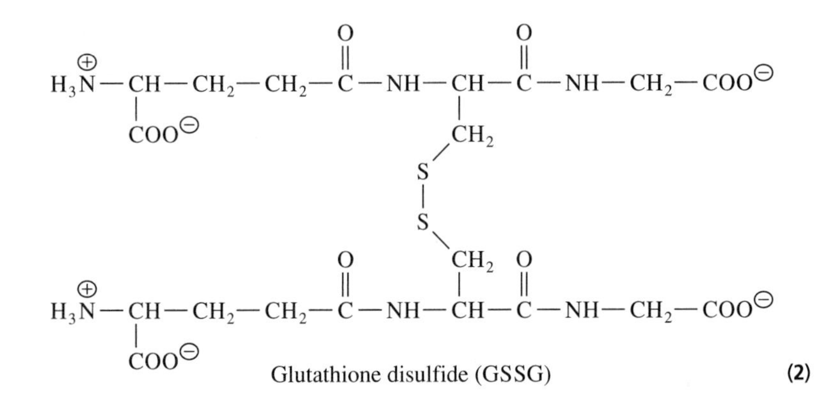

Glutathione disulfide (GSSG) **(2)**

NADPH is required for the restoration of reduced glutathione, catalyzed by glutathione reductase.

$$GSSG + NADPH + H^{\oplus} \longrightarrow 2\,GSH + NADP^{\oplus} \qquad \textbf{(3)}$$

In humans, a defect in the gene encoding either glucose 6-phosphate dehydrogenase or glutathione reductase can lead to hemolytic anemia, the destruction of red blood cells.

Specialized white blood cells called phagocytes use the reducing power of NADPH in a different way. Phagocytic cells remove foreign materials such as bacteria from tissues by engulfing and destroying them, using proteolytic enzymes as well as the toxic oxygen metabolites discussed above. The production of the toxic oxygen species occurs through the action of a membrane-bound NADPH oxidase that catalyzes the production of superoxide anion.

$$NADPH + 2\,O_2 \longrightarrow NADP^{\oplus} + 2\,O_2{\cdot}^{\ominus} + H^{\oplus} \qquad \textbf{(4)}$$

Thus, the reducing power of NADPH is used in different cells for opposite purposes: protection against the toxicity of oxygen metabolites and formation of these metabolites as a defensive device.

15·8 Transketolase and Transaldolase Catalyze Interconversions Among Sugar-Phosphate Metabolites

Transketolase and transaldolase, enzymes with broad substrate specificities, catalyze the exchange of two- and three-carbon fragments between sugar phosphates. The combined actions of these enzymes thus allow the five-carbon sugar phosphates to be converted to three- and six-carbon sugar phosphates.

Transketolase is a thiamine-pyrophosphate–dependent enzyme that catalyzes transfer of a two-carbon fragment from a ketose phosphate to an aldose phosphate. The enzyme requires that the source of the two-carbon fragment, the ketose, have an *S* configuration at C-3. An example of the action of transketolase is shown in Figure 15·30. In terms of carbon transfer, the ketose phosphate is shortened by two carbons, and the aldose phosphate is elongated by two carbons. The mechanism of transketolase, which illustrates the essential role of thiamine pyrophosphate (TPP), is shown in Figure 15·31. First, the carbanion of TPP attacks the carbonyl carbon of the substrate molecule. The resulting TPP adduct undergoes proton extraction by a basic group of the enzyme, and subsequent electron rearrangement leads to fragmentation, releasing the first product of the overall reaction, glyceraldehyde 3-phosphate. The remaining two-carbon fragment bound to TPP is resonance-stabilized; one resonance form is a carbanion. This carbanion attacks the carbonyl carbon of ribose 5-phosphate, forming another TPP adduct. A second fragmentation, similar to the one in the first part of the reaction, yields the original TPP carbanion and the second product, sedoheptulose 7-phosphate.

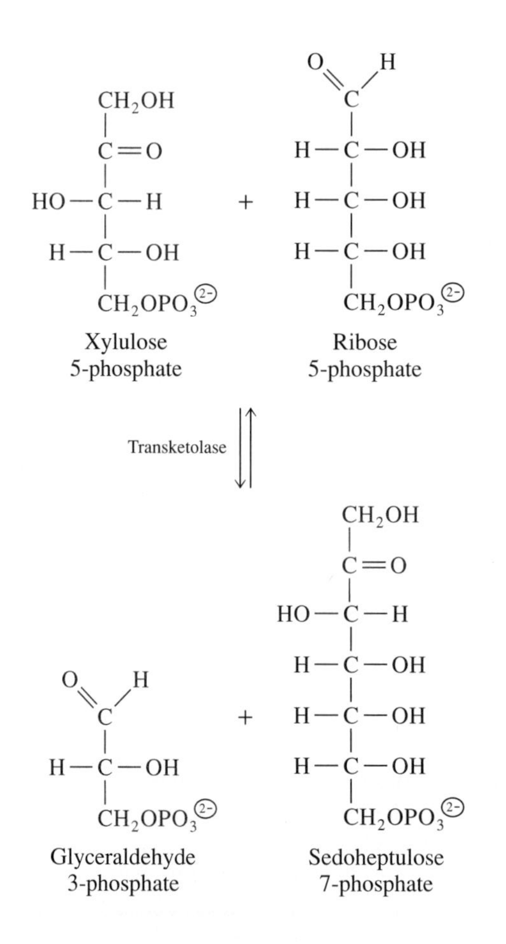

Figure 15·30
Reaction catalyzed by transketolase. Transketolase catalyzes the reversible transfer of a two-carbon fragment (C-1 and C-2) from xylulose 5-phosphate, shown in red, to ribose 5-phosphate, generating glyceraldehyde 3-phosphate and sedoheptulose 7-phosphate. Note that the number of carbon atoms balances and that the ketose-phosphate substrate (in either direction) is shortened by two carbon atoms while the aldose-phosphate substrate is lengthened by two carbon atoms. In this example, $5\,C + 5\,C \longrightarrow 3\,C + 7\,C$.

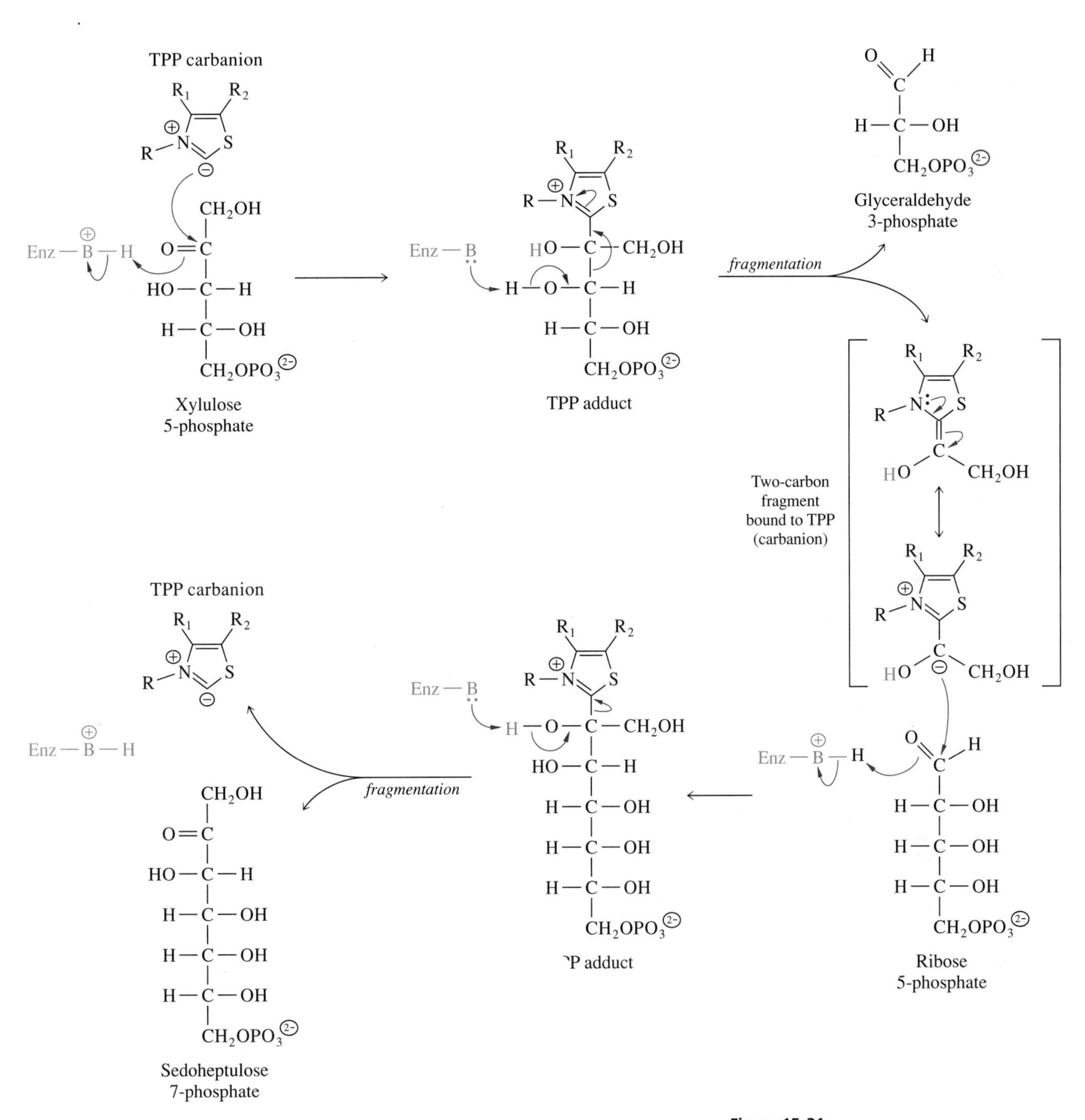

Figure 15·31
Mechanism of the reaction catalyzed by transketolase. Only the thiazolium ring of thiamine pyrophosphate (TPP) is shown.

Transaldolase catalyzes the transfer of a three-carbon fragment from a ketose phosphate to an aldose phosphate. The donor of the three-carbon fragment, the ketose phosphate, only binds to transaldolase if it has an *S* configuration at C-3. The transaldolase reaction of the pentose phosphate pathway converts sedoheptulose 7-phosphate and glyceraldehyde 3-phosphate to erythrose 4-phosphate and fructose 6-phosphate (Figure 15·32). The mechanism of the transaldolase reaction is shown in Figure 15·33. In the first step, a lysine residue of the enzyme condenses with the carbonyl group of sedoheptulose 7-phosphate to form a Schiff base. Deprotonation at C-4 by a base of the enzyme leads to cleavage of the bond between C-3 and C-4, releasing the first reaction product, erythrose 4-phosphate. A fragment containing the first three carbons of sedoheptulose phosphate remains bound to the enzyme. The carbanion of this fragment, a resonance form, attacks the carbonyl of glyceraldehyde 3-phosphate, and a new six-carbon sugar is formed, still bound to the enzyme as a Schiff base. Finally, hydrolysis of the Schiff base releases the second reaction product, fructose 6-phosphate.

A reexamination of the overall pentose phosphate pathway (Figure 15·27) shows that a six-carbon sugar (glucose 6-phosphate) is converted to a five-carbon sugar (ribulose 5-phosphate) and CO_2. Next, epimerase and isomerase reactions generate the proper stereochemical orientations for the sequential actions of transketolase and transaldolase. These reactions, in concert with another transketolase reaction, generate one three-carbon molecule (glyceraldehyde 3-phosphate) and two six-carbon molecules (fructose 6-phosphate). Thus, the products of the reactions described for the pentose phosphate pathway are glyceraldehyde 3-phosphate, fructose 6-phosphate, and CO_2. There are two possible fates for glyceraldehyde 3-phosphate and fructose 6-phosphate, since both are intermediates of glycolysis as well

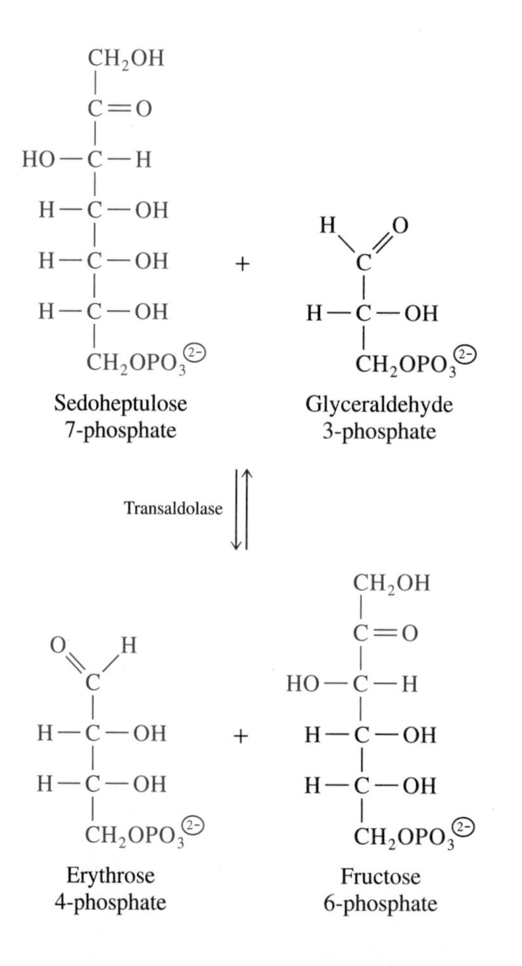

Figure 15·32
Reaction catalyzed by transaldolase. Transaldolase catalyzes the reversible transfer of a three-carbon unit (dihydroxyacetone) from sedoheptulose 7-phosphate, shown in red, to C-1 of glyceraldehyde 3-phosphate, generating a new ketose phosphate, fructose 6-phosphate, and releasing a new aldose phosphate, erythrose 4-phosphate. Note that the carbon atoms balance: $7\,C + 3\,C \longrightarrow 6\,C + 4\,C$.

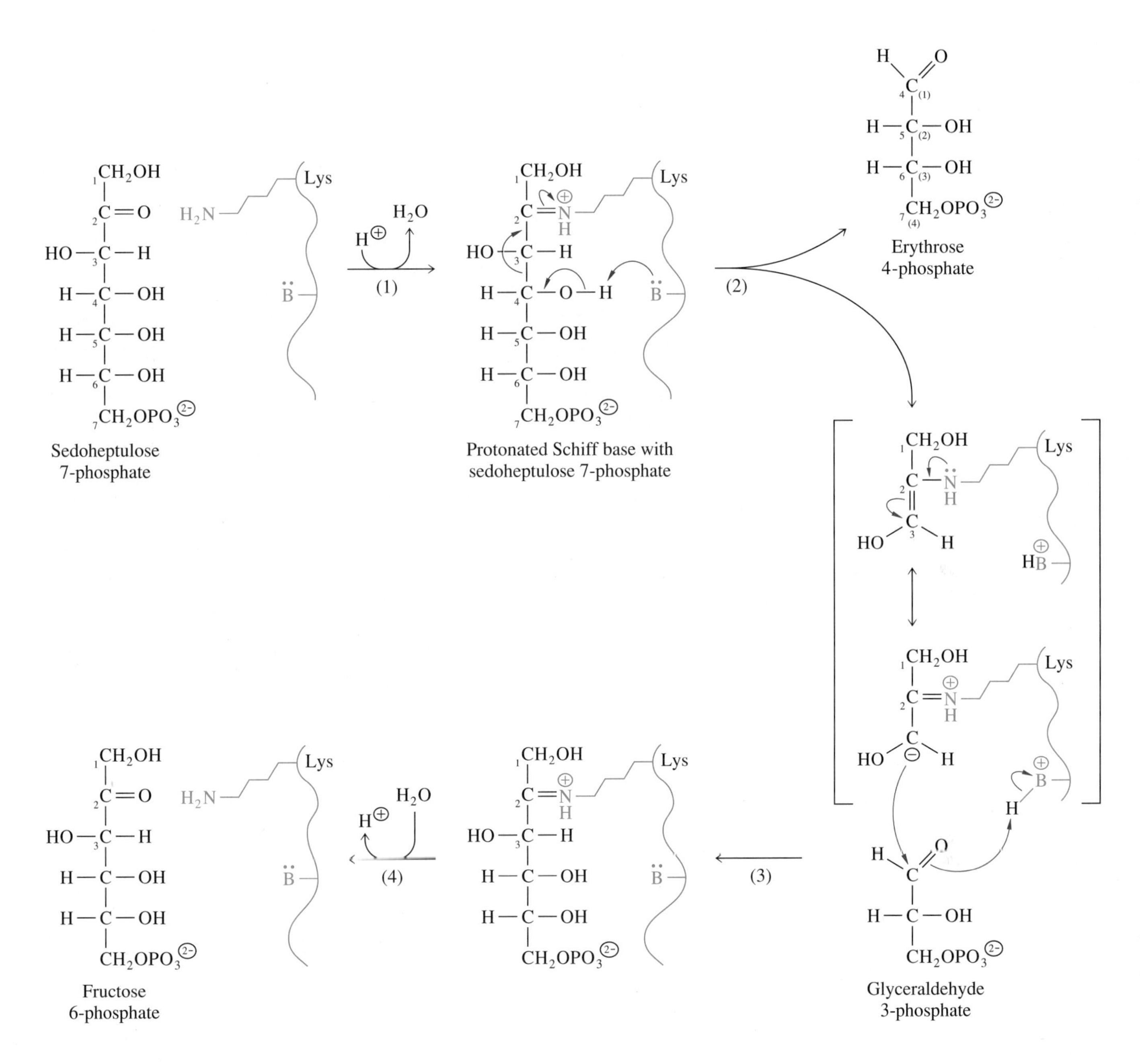

Figure 15·33
Mechanism of the reaction catalyzed by transaldolase, in which a three-carbon fragment (dihydroxyacetone) is transferred from sedoheptulose 7-phosphate to glyceraldehyde 3-phosphate.

as gluconeogenesis. If we combine the generation of glyceraldehyde 3-phosphate and fructose 6-phosphate by the pentose phosphate pathway with gluconeogenesis from these intermediates, the result is a cyclic pathway with the overall stoichiometry:

$$\text{6 Glucose 6-phosphate} \longrightarrow \text{5 Glucose 6-phosphate} + 6\,CO_2 + 12\,\text{NADPH} + P_i \quad (15\cdot7)$$

Indeed, an alternative name for the pathway is the *pentose phosphate cycle*. This formulation emphasizes that most of the glucose 6-phosphate that enters the pathway is recycled; one-sixth is converted to CO_2 and P_i, and there is net formation of NADPH. Quantitatively, formation of NADPH is the most important role of this pathway.

Summary

Glycogen is the glucose-storage polysaccharide of animals. In liver and muscle, glycogen degradation and glycogen synthesis are reciprocally regulated pathways that are controlled by the hormones epinephrine, glucagon, and insulin. Both epinephrine and glucagon stimulate the production of cAMP, which activates cAMP-dependent protein kinase. This leads to phosphorylation and inactivation of glycogen synthase. In addition, cAMP-dependent protein kinase catalyzes phosphorylation of glycogen phosphorylase kinase, which in turn catalyzes phosphorylation of glycogen phosphorylase, converting glycogen phosphorylase to its active form. Glycogen phosphorylase catalyzes degradation of intracellular glycogen to form glucose 1-phosphate, which can be converted to glucose 6-phosphate in a near-equilibrium reaction catalyzed by phosphoglucomutase. Epinephrine can also increase cytosolic calcium, which activates phosphorylase kinase, leading to activation of glycogen phosphorylase. In muscle, glucose 6-phosphate formed by glycogenolysis can be catabolized via glycolysis for the production of ATP. In liver, the glucose 6-phosphate can be hydrolyzed to glucose and P_i. The glucose leaves the cell and enters the bloodstream for use by other cells. The hormonally induced phosphorylations in both liver and muscle are reversed by a protein phosphatase that is itself regulated. Insulin counteracts the effects of epinephrine in skeletal muscle and glucagon in the liver.

Gluconeogenesis is the pathway for glucose synthesis from noncarbohydrate precursors, such as lactate and amino acids. Many of the gluconeogenic reactions are simply the reverse of the near-equilibrium reactions of glycolysis. Enzymes specific to gluconeogenesis catalyze the metabolically irreversible conversion of pyruvate to phosphoenolpyruvate (pyruvate carboxylase and phosphoenolpyruvate carboxykinase), fructose 1,6-*bis*phosphate to fructose 6-phosphate (fructose 1,6-*bis*phosphatase), and glucose 6-phosphate to glucose (glucose 6-phosphatase). The principal regulatory site of gluconeogenesis is the conversion of pyruvate to phosphoenolpyruvate; a secondary site is the reaction catalyzed by fructose 1,6-*bis*phosphatase. Different pathways for gluconeogenesis are required for different precursors, and there are some differences in the pathways between species.

The pentose phosphate pathway provides a secondary pathway for glucose 6-phosphate metabolism, initiated by the action of glucose 6-phosphate dehydrogenase. Glucose 6-phosphate dehydrogenase is also the major regulatory site of the pathway; the enzyme is allosterically inhibited by NADPH. The oxidative stage of the pentose phosphate pathway generates two molecules of NADPH per molecule of glucose 6-phosphate converted to ribulose 5-phosphate and CO_2. The nonoxidative stage of pentose phosphate metabolism includes isomerization of ribulose 5-phosphate to ribose 5-phosphate, a metabolite required for the biosynthesis of nucleotides and nucleic acids. Further metabolism of pentose phosphate molecules via the nonoxidative stage of the pentose phosphate pathway provides a mechanism for their conversion to triose-phosphate and hexose-phosphate intermediates of glycolysis and gluconeogenesis.

Selected Readings

Berridge, M. J. (1987). Inositol *tris*phosphate and diacylglycerol: two interacting second messengers. *Annu. Rev. Biochem.* 56:159–193.

Bourne, H. R., Sanders, D. A., and McCormick, F. R. (1990). The GTPase superfamily: a conserved switch for diverse cell functions. *Nature* 348:125–132. A discussion of the G protein activation cycle and the various functions of G proteins.

Bourne, H. R., Sanders, D. A., and McCormick, F. R. (1991). The GTPase superfamily: conserved structure and molecular mechanism. *Nature* 349:117–127. A review of the common structural design and molecular mechanism of G proteins.

Cohen, P. (1989). The structure and regulation of protein phosphatases. *Annu. Rev. Biochem.* 58:453–508.

Hers, H. G., and Hue, L. (1983). Gluconeogenesis and related aspects of glycolysis. *Annu. Rev. Biochem.* 52:617–653.

Kaziro, Y., Itoh, H., Kozasa, T., Nakafuku, M., and Satoh, T. (1991). Structure and function of signal-transducing GTP-binding proteins. *Annu. Rev. Biochem.* 60:349–400.

Pilkis, S. J., El-Maghrabi, M. R., and Claus, T. H. (1988). Hormonal regulation of hepatic gluconeogenesis and glycolysis. *Annu. Rev. Biochem.* 57:755–783.

Pilkis, S. J., and Granner, D. K. (1992). Molecular physiology of the regulation of hepatic gluconeogenesis and glycolysis. *Annu. Rev. Physiol.* 57:885–909.

Smythe, C., and Cohen, P. (1991). The discovery of glycogenin and the priming mechanism for glycogen biosynthesis. *Eur. J. Biochem.* 200:625–631.

Taylor, S. S., Buechler, J. A., and Yonemoto, W. (1990). cAMP-dependent protein kinase: framework for a diverse family of regulatory enzymes. *Annu. Rev. Biochem.* 59:971–1005.

Wood, T. (1985). *The pentose phosphate pathway* (Orlando, Florida: Academic Press).

16

Photosynthesis

As we have seen in earlier chapters, metabolism is fueled by two forms of chemical energy: the phosphoryl-group transfer potential of ATP and the reducing power of the nicotinamide cofactors NADH and NADPH. ATP and the reduced forms of the nicotinamide cofactors are produced by glycolysis and the citric acid cycle, and additional ATP is produced by oxidative phosphorylation. Together, these pathways result in the complete oxidation of carbohydrates to carbon dioxide and water.

If the formation of ATP from ADP and P_i and the reduction of $NAD^{\oplus}$ and $NADP^{\oplus}$ are dependent on a supply of carbohydrate, then what is the ultimate source of carbohydrate? The answer is *photosynthesis*. Through the process of photosynthesis, carbohydrates are synthesized from atmospheric CO_2 and water. Photosynthesis is an energy-requiring process that is fueled by light energy. It completes the global carbon cycle—nonphotosynthetic organisms give off CO_2 through the oxidation of carbohydrates, whereas photosynthetic organisms capture CO_2 and reduce it to the level of carbohydrates (CH_2O).

Organisms capable of photosynthesis are called *phototrophs,* a diverse group that includes certain bacteria, cyanobacteria (blue-green algae), algae, nonvascular plants, and vascular (higher) plants. Although our discussion of photosynthesis will focus on higher plants, much has been learned from research on the simplest kinds of photosynthetic organisms (photosynthetic bacteria and unicellular algae). With the exception of anaerobic bacteria, all phototrophs give off oxygen as a product of photosynthesis. It is believed that the atmosphere surrounding the earth was transformed from a reducing to an oxidizing environment more than two billion years ago following the evolution of oxygen-producing bacteria. This change in the atmosphere represents a pivotal event in the course of biological evolution.

16·1 Photosynthesis Consists of Two Major Processes

The net reaction of photosynthesis is

$$CO_2 + H_2O \xrightarrow{\text{Light}} (CH_2O) + O_2 \tag{16·1}$$

where (CH_2O) represents carbohydrate.

Photosynthesis encompasses two major processes that can be described by two partial reactions.

$$\begin{array}{lcl} H_2O + ADP + P_i + NADP^{\oplus} & \xrightarrow{\text{Light}} & O_2 + ATP + NADPH + H^{\oplus} \\ CO_2 + ATP + NADPH + H^{\oplus} & \longrightarrow & (CH_2O) + ADP + P_i + NADP^{\oplus} \\ \hline \text{Sum:} \quad CO_2 + H_2O & \xrightarrow{\text{Light}} & (CH_2O) + O_2 \end{array} \tag{16·2}$$

In the first process, the so-called "light reactions," protons derived from water are used in the chemiosmotic synthesis of ATP from ADP and P_i, while a hydrogen atom from water is used for the reduction of $NADP^{\oplus}$ to NADPH. The reactions are characterized by the light-dependent production of oxygen gas derived from the splitting of the water molecules. Such reactions are possible because photosynthetic organisms can harness light energy by several processes and use it to drive metabolic reactions.

The second process of photosynthesis involves the utilization of NADPH and ATP in a series of reactions leading to the reduction of gaseous carbon dioxide to carbohydrate. Because these reactions do not directly depend on light, but only on a supply of ATP and NADPH, they are referred to as the "dark reactions." Although the terminology of "light" and "dark" reactions is widely accepted, both processes normally occur simultaneously, with the products of the light-dependent process used to drive the reactions of the "dark" process.

16·2 Photosynthesis in Algae and Green Plants Occurs in Chloroplasts

In algae and green plants, photosynthesis occurs in specialized organelles called *chloroplasts* (Figure 16·1). The main structural feature of the chloroplast is a membranous internal network called the *thylakoid membrane,* which is the site of the light-dependent reactions that lead to the formation of NADPH and ATP. The thylakoid membrane, also called the thylakoid lamellae, is a highly folded, continuous membrane network suspended in the aqueous matrix of the chloroplast. This aqueous matrix, called the *stroma,* is the site of the second portion of the photosynthetic process, the reduction of carbon dioxide to carbohydrate. The chloroplast is enclosed by a double membrane that is highly permeable to CO_2 and selectively permeable to other metabolites.

The aqueous space within the thylakoid membrane is called the *lumen*. As we shall see later, the pumping of protons across the thylakoid membrane into the lumen creates the proton motive force that drives the synthesis of ATP. The thylakoid membrane is folded into a network of flattened vesicles that occur as appressed stacks called *grana* (singular, granum) or as single, unstacked vesicles that traverse the stroma and connect grana. Regions of the thylakoid membrane located within grana and not in contact with the stroma are called *granal lamellae,* whereas regions in contact with the stroma are called *stromal lamellae*. We shall discuss later how membrane-embedded components involved in photosynthesis are differentially distributed between the granal and stromal lamellae.

(a)

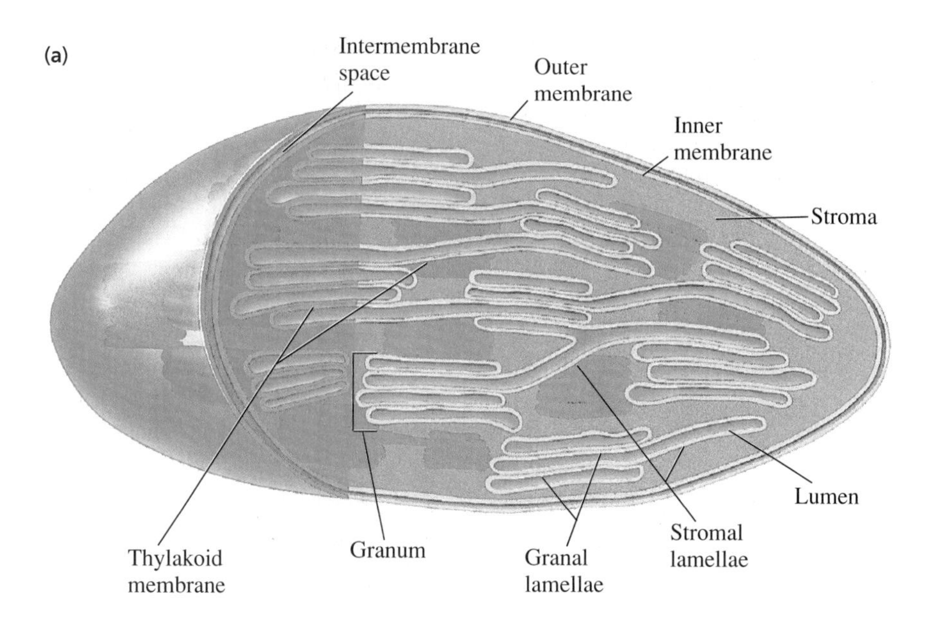

(b)

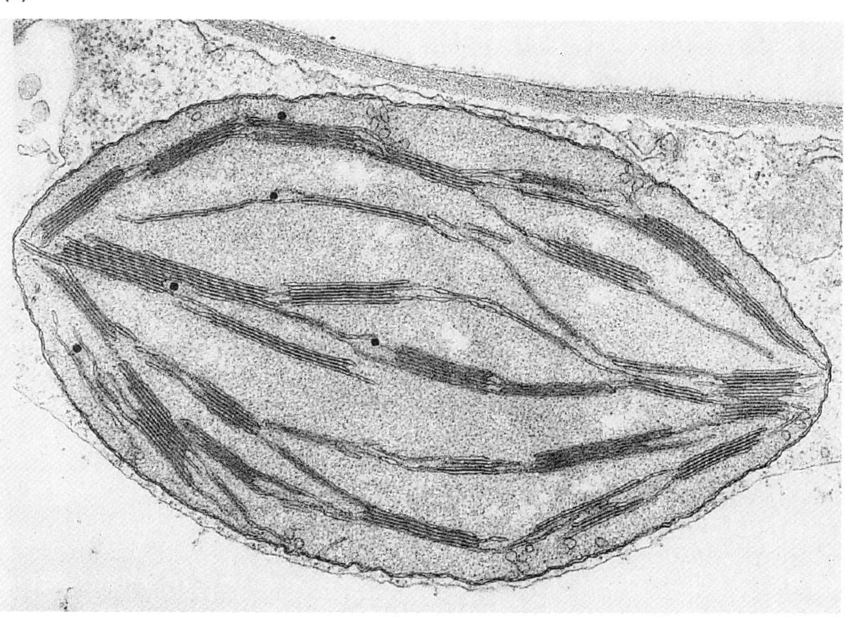

Figure 16·1
Structure of a chloroplast. **(a)** Illustration. A double membrane surrounds an aqueous region called the stroma. Located within the stroma is the thylakoid membrane, which is folded into flattened, saclike vesicles. These vesicles can occur in stacks called grana or as single vesicles that traverse the stroma and connect grana. Regions of the thylakoid membrane that are located within grana and are not in contact with the stroma are called granal lamellae. Regions exposed to the stroma are called stromal lamellae. The internal space enclosed by the thylakoid membrane is called the lumen. (Illustrator: Lisa Shoemaker.) **(b)** Electron micrograph of a spinach leaf chloroplast. (Courtesy of A. D. Greenwood.)

16·3 Chlorophyll and Other Pigments Are Components of Photosystems

Among the components embedded in the thylakoid membrane are various pigments that capture light energy for photosynthesis. *Chlorophyll* is usually the most abundant pigment and has the most significant role in the harvesting of light for photosynthesis. Four different types of chlorophyll occur among phototrophic organisms.

These are distinguished by the chemistry of the side chains attached to the main part of the molecule, as shown in Figure 16·2. Chlorophyll *a* (Chl *a*) and chlorophyll *b* (Chl *b*) are the two types of chlorophyll found in higher plants. Chl *a* is more abundant than Chl *b*. The major pigments in photosynthetic bacteria are bacteriochlorophyll *a* (BChl *a*) and bacteriochlorophyll *b* (BChl *b*). Two features common to the different chlorophylls are the hydrophobic phytol side chain, which anchors chlorophyll in the membrane, and the hydrophilic porphyrin ring. Together these features make chlorophyll molecules amphipathic. The porphyrin ring is similar to the heme prosthetic groups of hemoglobin, myoglobin, and the cytochromes. Unlike heme, which contains Fe^{2+}, chlorophyll contains Mg^{2+} chelated by the pyrrole nitrogen atoms of the ring. Chl *a* and Chl *b* both absorb light in the violet-to-blue region (absorption maximum 400–500 nm) and the orange-to-red region (absorption maximum 650–700 nm) of the electromagnetic spectrum. Because their structures are slightly different, however, the absorption maxima of Chl *a* and Chl *b* are not identical. Chlorophyll molecules are specifically oriented in the thylakoid membrane by noncovalent binding to proteins embedded in the membrane.

In addition to chlorophyll, several *accessory pigments* may be present in the photosynthetic membranes. These are carotenoids, which are present in all phototrophs, and the phycobilins, including phycoerythrin and phycocyanin, which are found in some algae and cyanobacteria. Like the chlorophylls, these molecules contain a series of conjugated double bonds that give them the light-absorbing properties characteristic of pigments. The absorption maxima of the accessory pigments

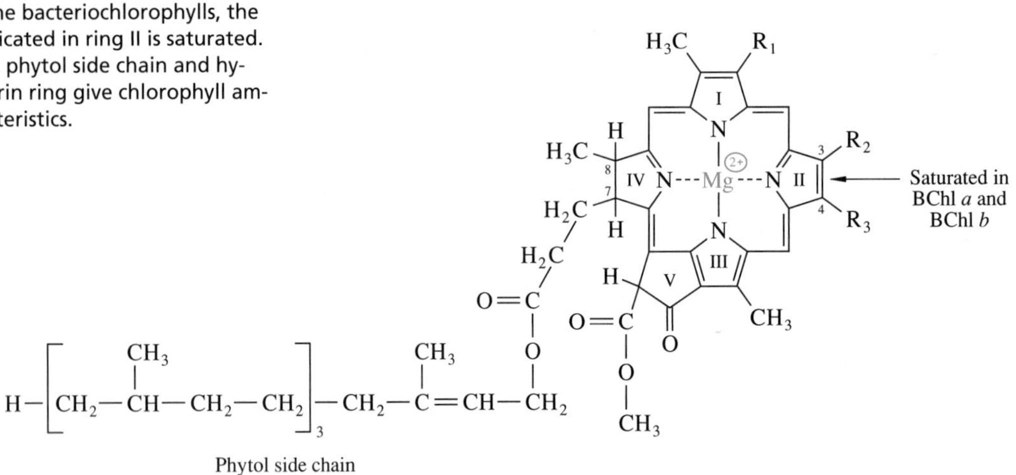

Figure 16·2
Structures of chlorophyll and bacteriochlorophyll pigments. Differences in substituent groups indicated as R_1, R_2, and R_3 are shown in the table. In the bacteriochlorophylls, the double bond indicated in ring II is saturated. The hydrophobic phytol side chain and hydrophilic porphyrin ring give chlorophyll amphipathic characteristics.

Chl species	R_1	R_2	R_3
Chl *a*	$-CH{=}CH_2$	$-CH_3$	$-CH_2-CH_3$
Chl *b*	$-CH{=}CH_2$	$-C(=O)-H$	$-CH_2-CH_3$
BChl *a*	$-C(=O)-CH_3$	$-CH_3$	$-CH_2-CH_3$
BChl *b*	$-C(=O)-CH_3$	$-CH_3$	$-CH{=}CH_2$

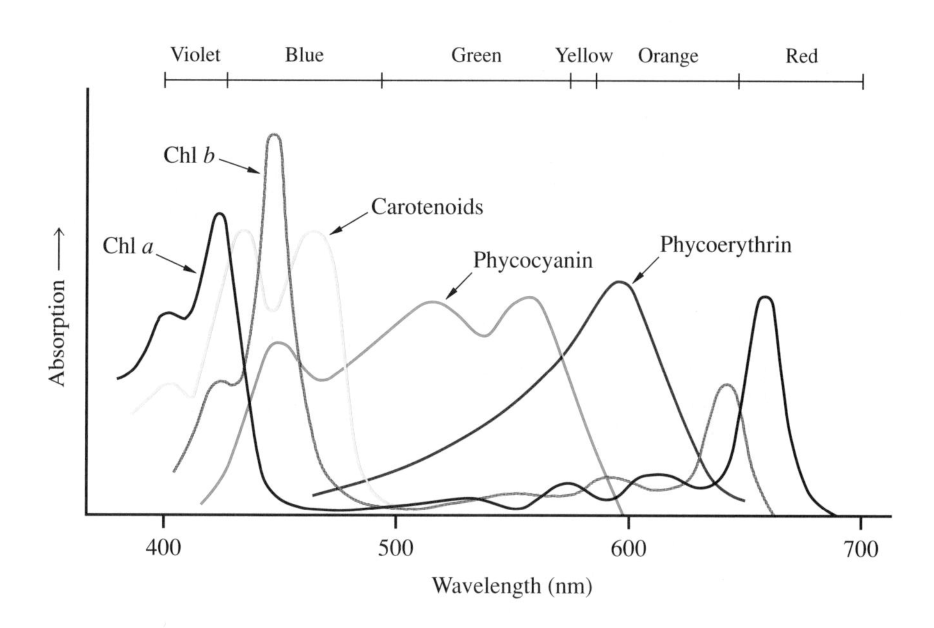

Figure 16·3
Absorption spectra of major photosynthetic pigments. Collectively, the pigments absorb radiant energy across the spectrum of visible light. [Adapted from Govindjee and Govindjee, R. (1974). The absorption of light in photosynthesis. *Sci. Am.* 231(6):68–82.]

differ from those of the chlorophylls. Figure 16·3 illustrates how the absorption spectra of accessory pigments complement those of the chlorophylls. Collectively, the pigments absorb a range of radiant energy that spans the spectrum of visible light. Since the absorption spectrum of a pigment determines its color, the profile of chlorophylls and accessory pigments determines the characteristic color of a photosynthetic tissue or organism.

Two distinct protein complexes called *photosystems* are the functional units of the light-dependent reactions of photosynthesis. Both photosystems, named photosystem I (PSI) and photosystem II (PSII), are localized in the thylakoid membrane; each contains a bed of pigments called an *antenna complex* that harvests light for the photosystem. Depending on the organism, each antenna complex may contain from 60 to 2000 pigment molecules. The actual site of photochemistry is a multisubunit protein complex within each of the photosystems called a *reaction center.* The reaction center contains two chlorophyll molecules called the *special pair,* which are distinct from other chlorophylls in that light energy from the antenna complex is funnelled to them. The reaction center of a photosystem is named according to the absorption maximum of its special pair. The PSI reaction center is called P700 (P for pigment and 700 for the absorption maximum of its special pair); the PSII reaction center is called P680.

The two membrane-spanning photosystems work in series, and although they are localized in different regions of the thylakoid membrane, they are linked by specific electron carriers. PSI is located in the stromal lamellae and therefore is exposed to the chloroplast stroma, whereas PSII is embedded in the granal lamellae, away from the stroma. A large antenna complex called the chlorophyll *a/b light-harvesting complex* (LHC) aids the photosystem in gathering light. The LHC, which is not considered part of the photosystem, contains about half the chlorophyll of the chloroplast. LHCs are distributed in the thylakoid membrane in three distinct fractions: one fraction is bound to PSI, another is bound to PSII, and the third is mobile and can serve in light harvesting for either PSI or PSII, as we shall see later in the chapter.

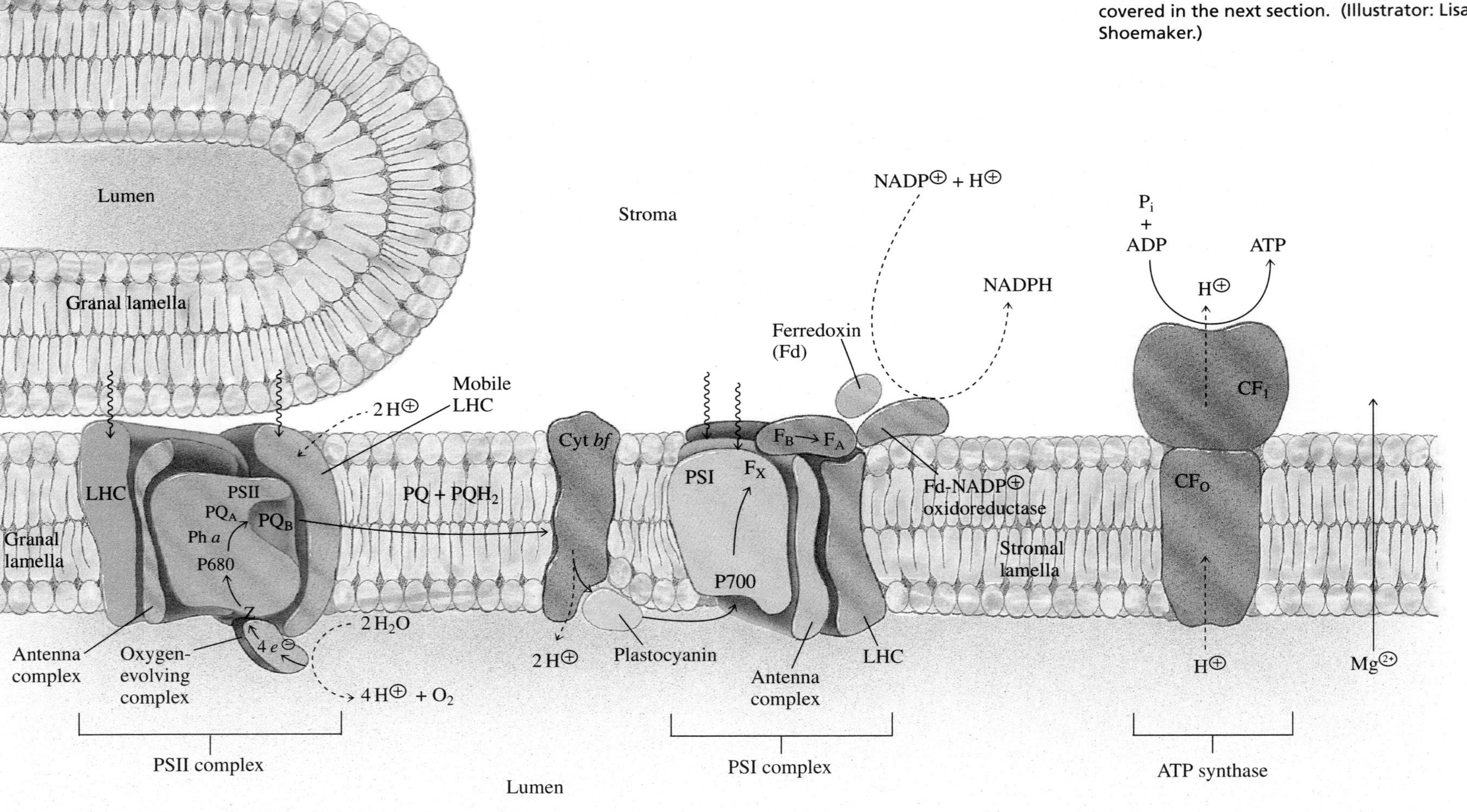

Figure 16·4
Components of the thylakoid membrane involved in photosynthesis. Also indicated are the processes of light capture (wavy arrows), electron transport (solid arrows), and proton displacement (dashed arrows), which will be covered in the next section. (Illustrator: Lisa Shoemaker.)

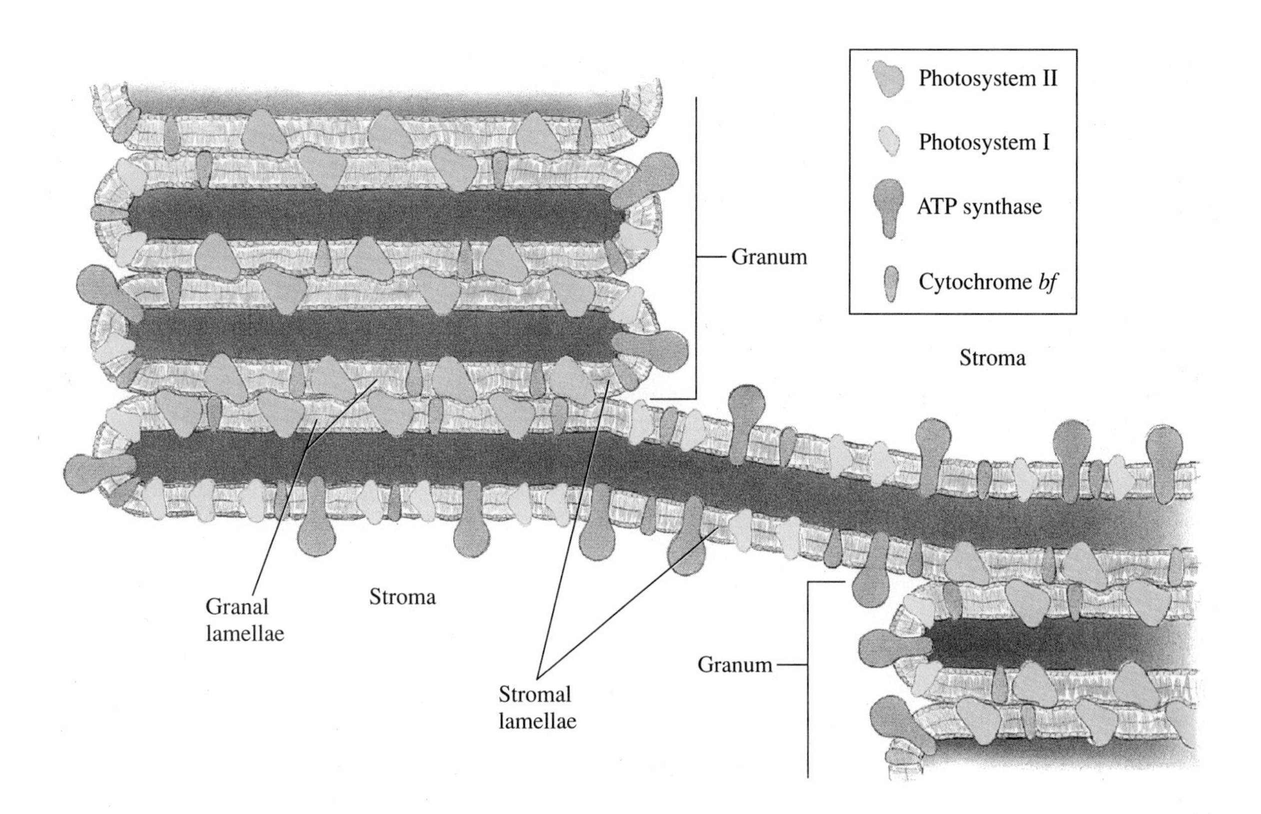

Figure 16·5
Distribution of photosynthetic components between stromal and granal lamellae. ATP synthase and PSI occur exclusively in the stromal lamellae, PSII is found in granal lamellae, and the cytochrome *bf* complex occurs in both the stromal and granal lamellae. (Illustrator: Lisa Shoemaker.)

In addition to the photosystems and light-harvesting complexes, other components participating in photosynthesis are embedded in, or associated with, the thylakoid membrane (Figure 16·4). These include the oxygen-evolving complex, the cytochrome *bf* complex, and the chloroplast ATP synthase. The oxygen-evolving complex is associated with PSII on the lumen side of the thylakoid membrane, whereas the cytochrome *bf* complex is in both the stromal and granal lamellae. ATP synthase is located exclusively in the stromal lamellae. Figure 16·5 illustrates the predominant locations within the thylakoid membrane of PSI, PSII, the cytochrome *bf* complex, and ATP synthase.

16·4 Noncyclic Electron Transport Results in the Reduction of NADP⊕ to NADPH

Photosynthetic electron transfer follows one of two routes. Electrons can be transferred linearly through a series of membrane-bound carriers, terminating with the transfer of electrons to NADP⊕, or they can be transferred through many of the same carriers, but in a cyclic manner, so that there is no terminal electron transfer to NADP⊕. During both noncyclic and cyclic electron transport, protons are pumped across the membrane from the stroma to the lumen, generating the proton motive force used to drive the formation of ATP. However, only noncyclic electron flow yields NADPH. We shall begin with a discussion of noncyclic electron transfer.

The absorption of a photon (quantum of light energy) by a pigment molecule in either the LHC or the antenna complex of PSII constitutes the initial event of the light reactions of photosynthesis. Note that the order of electron transfer is from PSII to PSI, and not from PSI to PSII, as might be expected. (The terminology reflects the order of discovery of the photosystems.) The energy of photon absorption excites the antenna pigment molecule to a high energy state. The energy of the excited pigment molecule is rapidly transferred (10^{-15} s) to other pigment molecules in the LHC and/or photosystem antenna complex until it is funnelled to the special pair in the P680 reaction center (Figure 16·6). The term *funnelling* appropriately describes this transfer of energy since, in the local environment of the reaction center, the special-pair chlorophylls have a higher reduction potential than the other chlorophylls. Because of this higher reduction potential, energy transfer between the pigments proceeds in an energetically downhill direction to the special-pair chlorophylls of P680. When the reduced special pair becomes energized, the reaction center is converted to an excited state, P680*, which is a powerful reductant.

Reduction potential is the basis of electron transfer between membrane-bound electron carriers in the chloroplast. Electron transfer proceeds from carriers of lower reduction potential (stronger reductants) to carriers of higher reduction potential (stronger oxidants). This same principle underlies mitochondrial electron transport. In photosynthesis, however, the absorption of light energy lowers the reduction potential of the reaction centers, P680 and P700, so that P680* and P700* have a far lower reduction potential (that is, they are stronger reductants) than their ground-state equivalents.

When the reduction potentials of the series of photosynthetic electron-transport components are plotted in sequence, the result is a zigzag figure called the *Z-scheme* (Figure 16·7). The Z-scheme shows the rise in reducing power created by the excitation of the photosystem reaction centers. Also shown are electron carriers between P680 and P700. The following discussion of the electron-transport sequence will refer to the Z-scheme in Figure 16·7 and to Figure 16·4, which illustrates the spatial relationships of the electron-transport components within the thylakoid membrane.

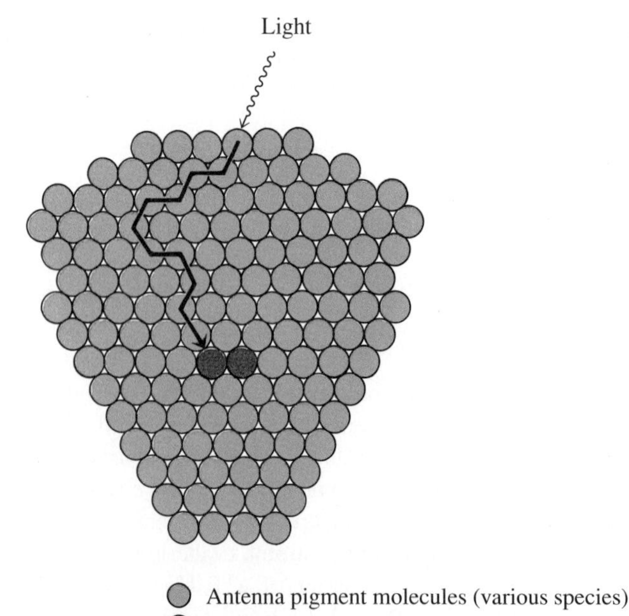

Figure 16·6
Schematic diagram illustrating the transfer of light energy between pigment molecules positioned in the antenna complex. The transfer process ends with the excitation of the special-pair chlorophyll molecules in the reaction center of a photosystem. (Illustrator: Lisa Shoemaker.)

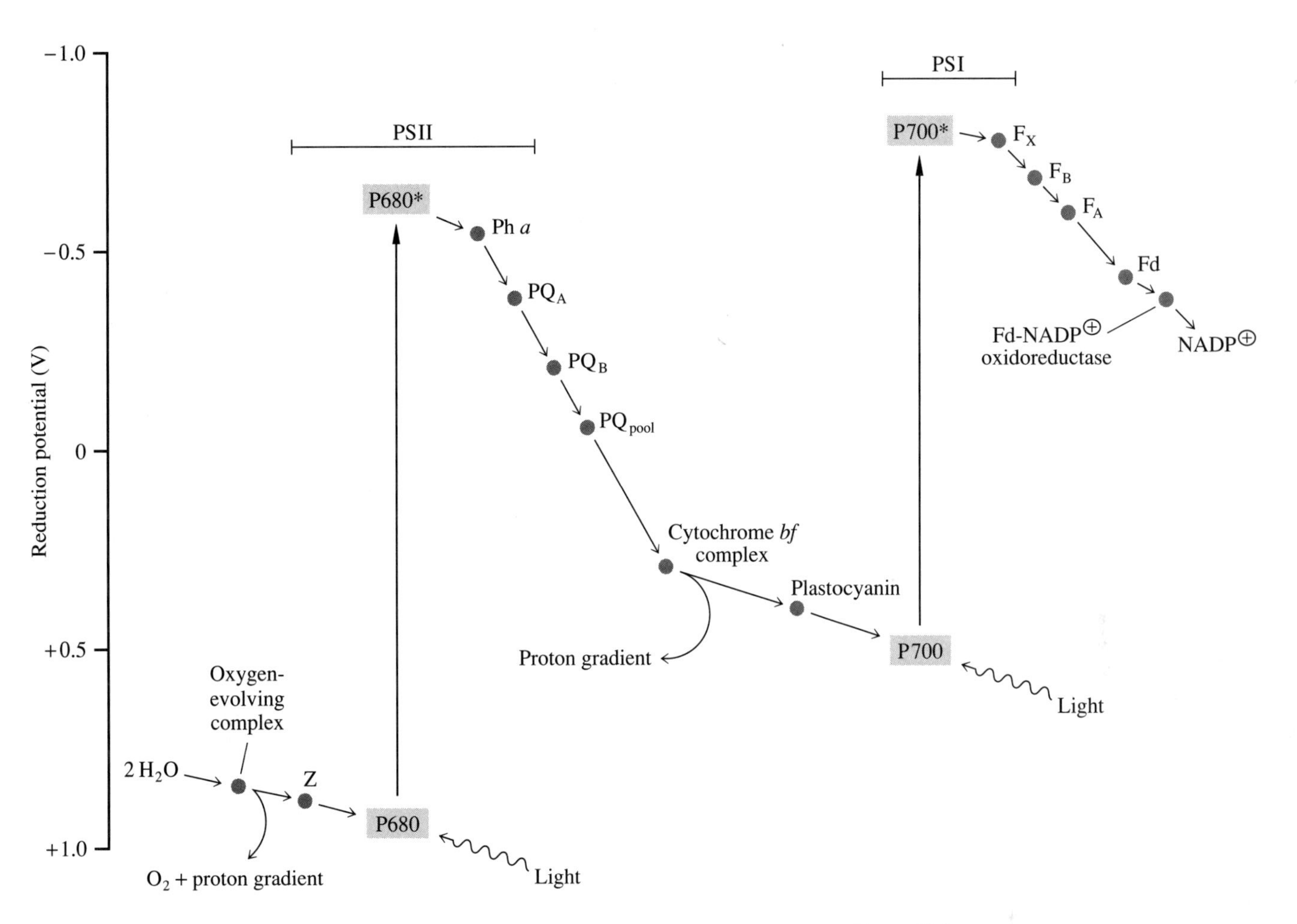

Figure 16·7
Z-scheme. The Z-scheme is a widely accepted model illustrating the reduction potentials associated with electron flow through photosynthetic electron carriers. The reduction potentials of the carriers vary with experimental conditions; therefore, the values shown are approximate. Abbreviations: Z, electron donor to P680; PQ_A, plastoquinone tightly bound to PSII; PQ_B, reversibly bound PQ undergoing reduction by PSII; PQ_{pool}, plastoquinone pool made up of PQ and PQH_2; F_X, F_B, and F_A, iron-sulfur centers; and Fd, ferredoxin.

The electrons needed for electron transport are obtained from water in a reaction catalyzed by the oxygen-evolving complex. Sometimes called the water-splitting enzyme, the oxygen-evolving complex is located on the lumen side of PSII and contains a cluster of four manganese ions. The manganese ions are specifically arrayed to accumulate and then sequentially transfer four electrons from two molecules of water to oxidized P680 (P680⊕). Once the Mn cluster has given up its four electrons, the following events related to water splitting occur: (1) two molecules of water are oxidized, (2) four protons are released into the thylakoid lumen, (3) one molecule of oxygen is evolved, and (4) four electrons reduce the Mn cluster. This stoichiometry can be observed experimentally by exposing chloroplasts in darkness to brief flashes of light that excite PSII. With each flash, P680 is reduced by an electron from the Mn cluster and is excited to P680*. After four flashes, when the Mn cluster is fully oxidized, oxygen is given off, and H⊕ movement across the membrane can be detected.

Electrons are transferred from the oxygen-evolving complex to P680⊕ via a carrier called Z, which poises the reaction center for excitation to P680*. Once excited, P680* donates its electron to pheophytin *a* (Ph *a*), which is identical to Chl *a* except that the magnesium ion of Chl *a* is replaced by two protons. Electron transfer proceeds from reduced pheophytin *a* to PQ_A, a plastoquinone tightly bound to a PSII polypeptide. Like mitochondrial ubiquinone (Figure 7·33), plastoquinone can be reduced by two sequential one-electron transfers. The bound PQ_A transfers two electrons, one at a time, to a second plastoquinone molecule, PQ_B, which is reversibly bound to a PSII polypeptide (Figure 16·8). PQ_B is thus reduced first to a

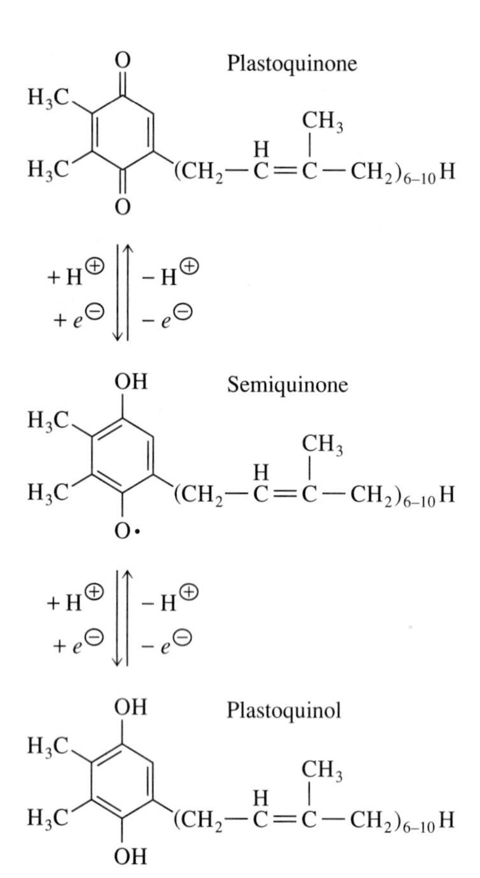

Figure 16·8
Reduction of plastoquinone to plastoquinol by two successive one-electron transfers.

semiquinone, PQH·, and then to a quinol, PQH_2. The fully reduced PQ_BH_2, which is soluble in the membrane due to its long hydrophobic tail, is released into the plastoquinone pool of PQH_2 and PQ.

The photosynthetic Q cycle, like the mitochondrial Q cycle (Section 14·7), carries out the oxidation of reduced quinone. The oxidation of PQH_2 occurs via the cytochrome *bf* complex. This complex is made up of cytochromes, which contain heme, and iron-sulfur proteins and is analogous to the mitochondrial ubiquinol-cytochrome *c* reductase. The two-step Q cycle (Figure 16·9) results in the oxidation of two molecules of PQH_2, the reduction of one molecule of PQ to PQH_2, the transfer of two electrons to plastocyanin and the net transfer of four protons into the lumen. In the first step of the Q cycle, PQH_2 is oxidized at a site adjacent to cytochrome *f*, and the two protons from the oxidized PQ are released into the lumen. One of the electrons from PQH_2 oxidation is funnelled through cytochrome *f* and reduces plastocyanin. The second electron, plus one stromal proton, converts a molecule of PQ to PQH· at a site away from the PQH_2 oxidation site. In the second step, a second molecule of PQH_2 is oxidized. Once again, one electron reduces plastocyanin; the other reduces PQH· to PQH_2. The fully reduced PQH_2 is then released into the plastoquinone pool. Since the cytochrome *bf* complex can accept only electrons and not protons from PQH_2, Q cycling by plastoquinone and the cytochrome *bf* complex serves as a proton pump, contributing to the proton motive force that drives ATP synthesis.

Plastocyanin is a small, copper-containing protein (MW 11 000) that is reduced by cytochrome *f* of the *bf* complex. Electron transport through plastocyanin occurs via redox changes in the copper atom, which is coordinated at the active site by two nitrogen and two sulfur atoms of four amino acid residues.

Once reduced by plastocyanin, P700 can become energized to P700* by pigment excitation that originates from the PSI antenna complex, the PSI-bound LHC, or under certain conditions, the mobile pool of LHC. Note that P700*, with a redox potential of approximately −0.80 V, is the strongest reductant in the chain of electron carriers. The electron from P700* is readily donated to a series of three iron-sulfur centers, designated F_X, F_B, and F_A. F_X is in the PSI complex, whereas F_A and F_B are in a polypeptide on the stromal side of the thylakoid membrane.

The electron delivered to the last iron-sulfur center, F_A, is transferred to a small iron-sulfur protein called ferredoxin (Fd, MW 10 700), which is dissolved in the chloroplast stroma. The low reduction potential ($E°' = -0.43$ V) of ferredoxin allows it to readily reduce $NADP^{\oplus}$ to NADPH ($E°' = -0.32$ V). The reduction of $NADP^{\oplus}$ by reduced ferredoxin is catalyzed by ferredoxin-$NADP^{\oplus}$ oxidoreductase, an enzyme loosely bound to the stromal side of the thylakoid membrane. The oxidoreductase has the prosthetic group FAD, which is reduced to $FADH_2$ by reduced ferredoxin in two one-electron transfers. $FADH_2$ in turn reduces $NADP^{\oplus}$ by donating two electrons and a proton in the form of a hydride ion. The pH difference across the thylakoid is increased by this step, since a proton from the stroma is consumed in the conversion of $NADP^{\oplus}$ to NADPH. Formation of NADPH completes the noncyclic electron-transport sequence.

Three separate steps in the photosynthetic electron-transport chain change the distribution of protons between the stroma and lumen. First, protons from water are released into the lumen by the oxygen-evolving complex. Second, stromal protons are transported to the lumen during the oxidation of PQH_2 by the cytochrome *bf* complex. Third, the uptake of a proton during reduction of $NADP^{\oplus}$ lowers the proton concentration in the stroma. As protons are moved from the stroma into the lumen, magnesium ions move from the lumen into the stroma so that charge balance is maintained between the two compartments. As we shall see later in the chapter, the increased level of Mg^{2+} in the stroma during active photosynthesis serves to regulate the synthesis of carbohydrates.

The membrane-bound ATP synthase in the presence of a transmembrane proton gradient catalyzes the formation of ATP from ADP and P_i. Since this process is

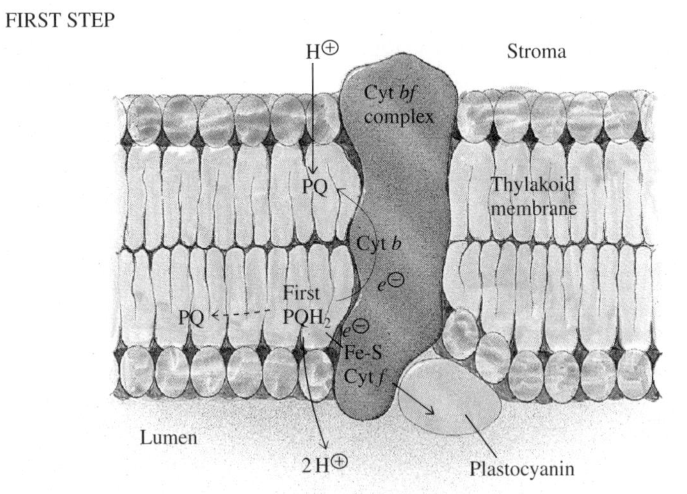

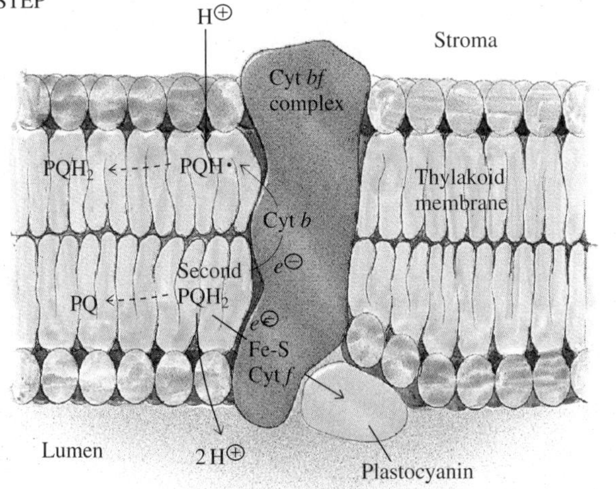

Figure 16·9
The photosynthetic Q cycle. The Q cycle is a two-step process. In the first step, PQH_2 is oxidized to PQ. In the process, two protons are released into the lumen; one electron is funnelled to plastocyanin via cytochrome *f;* and the other electron, along with a proton from the stroma, converts a molecule of PQ to PQH·. In the second step, a second molecule of PQH_2 is oxidized. Again, one electron is funnelled to plastocyanin; the other reduces PQH· to PQH_2, which is then released into the plastoquinone pool. Together, the two steps result in the oxidation of two molecules of PQH_2, the reduction of one PQ to PQH_2, the transfer of two electrons to plastocyanin, and the transfer of four protons into the lumen. (Illustrator: Lisa Shoemaker.)

light-dependent in plants, photosynthetic ATP formation is termed *photophosphorylation*. The ATP synthase of chloroplasts is a multiprotein complex similar to its mitochondrial counterpart in that it consists of two major particles, CF_O and CF_1. (The designations CF_O and CF_1 distinguish the chloroplast ATP synthase components from the mitochondrial F_O and F_1 components.) The CF_O particle is embedded in the thylakoid membrane and forms a channel for protons. CF_1 protrudes into the stroma and catalyzes the formation of ATP from ADP and P_i. The chloroplast ATP synthase is localized exclusively in the stromal lamellae.

16·5 Cyclic Electron Flow Contributes to the Proton Gradient Across the Thylakoid Membrane

It is estimated that for every two electrons transferred to reduce one molecule of NADP⊕ to NADPH, a proton motive force develops across the thylakoid membrane sufficient to synthesize one molecule of ATP. This one-to-one ATP/NADPH stoichiometry of the "light" reactions would create an imbalance with the stoichiometry of the carbohydrate-producing "dark" reactions, since, as we shall see, two molecules of NADPH and three molecules of ATP are consumed in the reduction of one molecule of CO_2 to (CH_2O). To compensate for this unbalanced coupling between the "light" and "dark" reactions, a modified sequence of electron-transport steps operates as a cycle to form ATP without the simultaneous formation of NADPH. This modified reaction sequence is called *cyclic electron transport*. As in noncyclic electron transport, electrons are transferred via the Q cycle from plastoquinone to the cytochrome *bf* complex, and then on to PSI and ferredoxin. In cyclic electron transport, however, the soluble ferredoxin donates its electron back to the PQ pool in a process mediated by the cytochrome *bf* complex via a cytochrome that is involved exclusively in cyclic electron flow. Through Q cycling, cyclic electron transfer supplements the proton motive force generated during noncyclic electron flow.

In general, the relative rates of cyclic and noncyclic electron flow are influenced by the relative amounts of NADPH/NADP⊕. When the stromal ratio of NADPH/NADP⊕ is high, the rate of noncyclic electron transfer is limited, due to the low availability of NADP⊕. Under these circumstances, cyclic electron flow is favored.

16·6 Capture of Light Energy by the Chloroplast and Energy Distribution Between the Photosystems Are Regulated

Fluctuations in the spectral quality and intensity of ambient light can create an imbalance between the amount of light energy being captured by PSII and PSI, respectively, since the two photosystems have different absorption maxima. Although this imbalance could limit the rate of electron transport, the problem is ameliorated by an elaborate short-term regulatory mechanism that affects the location of the mobile LHC, which preferentially surrounds PSII in the granal lamellae. Two tyrosine residues on a surface-exposed segment of the mobile LHC can be phosphorylated in a reaction catalyzed by a specific kinase. LHC phosphorylation causes electrostatic repulsion between adjacent mobile LHCs and leads to the migration of the phosphorylated LHC to the stromal lamellae, where PSI is located (Figure 16·10). As a result, light energy captured by the phosphorylated mobile LHC is funnelled to PSI.

The kinase that catalyzes LHC phosphorylation is activated when the PQH_2/PQ ratio is high, which results when plastoquinone is reduced by PSII more rapidly than it is oxidized by PSI. Phosphorylation of the mobile LHC alleviates this condition by increasing energy input into PSI. A phosphatase catalyzes the dephosphorylation of the LHC. Thus, under various light conditions the equilibrium between phosphorylated and dephosphorylated mobile LHC is determined by the redox state of the plastoquinone pool and results in a balanced energy distribution between PSII and PSI.

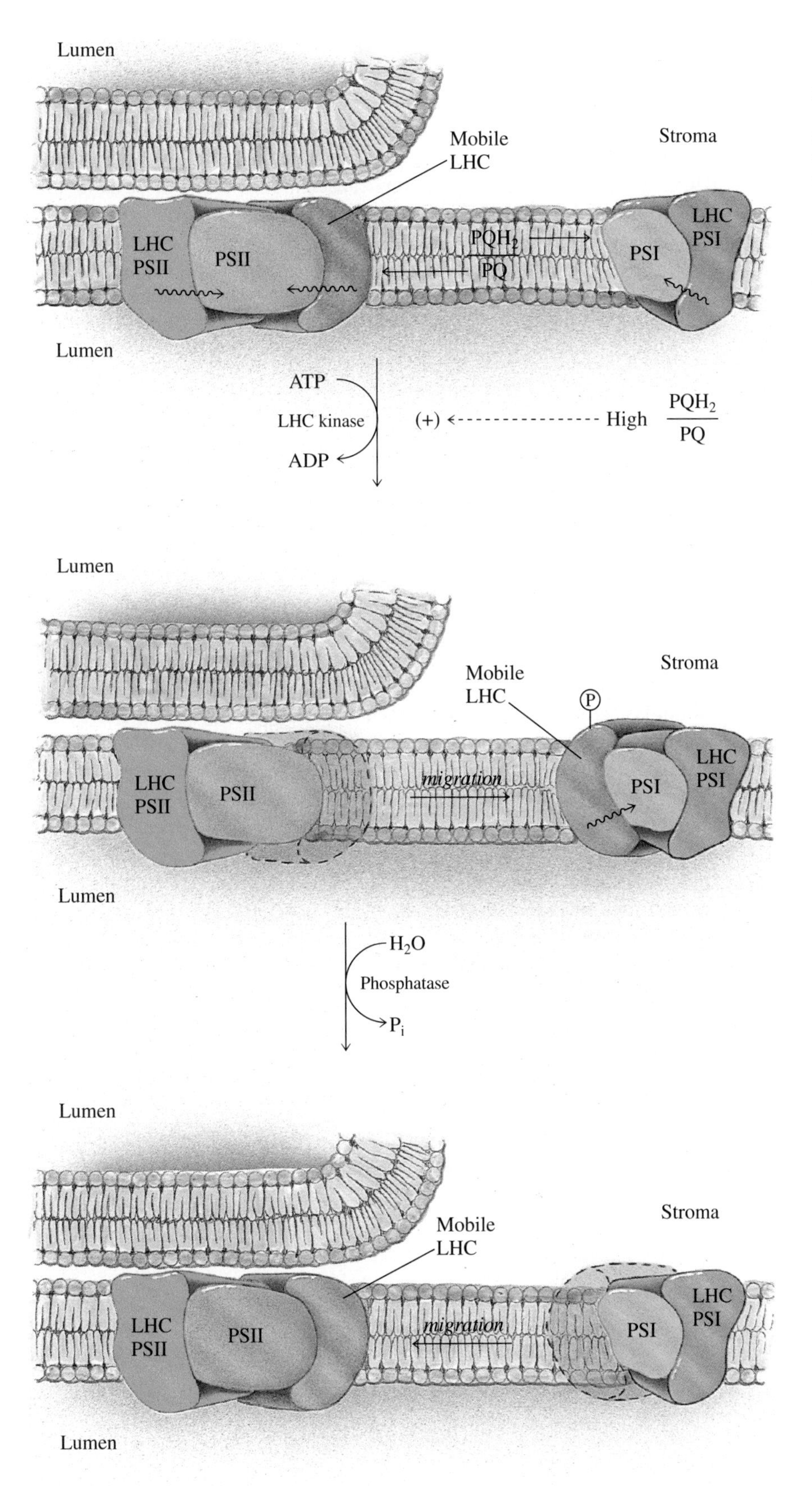

Figure 16·10
Regulation of light-energy input to photosystems by phosphorylation of the mobile LHC. Activated by a high PQH_2/PQ ratio, a specific LHC kinase catalyzes phosphorylation of the mobile LHC, causing it to migrate to the stromal lamellae. Preferential energy input into PSI increases the rate of PQH_2 oxidation and lowers the ratio of PQH_2/PQ. Dephosphorylation of the mobile LHC is catalyzed by a phosphatase. (Illustrator: Lisa Shoemaker.)

Long-term changes in photosynthetic capacity occur in response to longer-term changes in light intensities and spectral qualities. For example, the total number and the relative proportions of the various pigment complexes can change and the stoichiometries between PSII, PSI, and electron carriers can be modified. Under conditions of limited light, synthesis of chlorophyll and accessory pigments is increased in order to maximize light-gathering capacity. Conversely, in highly illuminated leaves, photosynthetic capacity is increased by higher levels of cytochrome *bf* complex, plastoquinone, ferredoxin, and ATP synthase per photosynthetic unit and a proportionately lower chlorophyll content of the light-harvesting components. It is not surprising that a process so fundamental to the survival of the plant can adapt over both the short and long term to optimize photosynthetic efficiency under different conditions.

16·7 Reactions of the Reductive Pentose Phosphate Cycle Assimilate CO_2 into Carbohydrates

The second major phase of photosynthesis is the reductive conversion of carbon dioxide into carbohydrates, a process powered by the ATP and NADPH formed during the light reactions of photosynthesis. The formation of carbohydrates occurs in the chloroplast stroma and is accomplished by a cycle of enzyme-catalyzed reactions that have three major consequences: (1) the fixation of atmospheric CO_2, (2) the reduction of CO_2 to carbohydrate, and (3) the regeneration of the molecule that accepts CO_2. The metabolic pathway leading to carbon assimilation has several names, including the reductive pentose phosphate cycle, the C_3 pathway (indicating that the first intermediate of the pathway is a three-carbon molecule), the photosynthetic carbon reduction cycle, and the Calvin, Calvin-Benson, or Calvin-Bassham cycle. We shall refer to the pathway as the *reductive pentose phosphate cycle*, or simply the RPP cycle. The name is appropriate because there are similarities between the RPP cycle and the oxidative pentose phosphate pathway (Section 15·7).

The substrate for the RPP cycle, CO_2, diffuses directly into photosynthetic cells or, in higher plants, enters photosynthetic cells via structures on the leaf surface called *stomata*. Stomata are composed of two adjacent cells on the surface of the leaf surrounding a cavity that is lined with photosynthetic cells. The aperture created by the stomatal cells changes in response to ion fluxes and the resulting osmotic uptake of water. The flux of ions across the stomatal cells is regulated by factors that reflect the suitability of conditions for carbon assimilation, such as availability of CO_2, temperature, and water status.

16·8 RuBisCO Catalyzes the Initial Step of the RPP Cycle

The fixing of gaseous CO_2 into an organic product is accomplished in the first step of the RPP cycle. This step is catalyzed by ribulose 1,5-*bis*phosphate carboxylase-oxygenase, abbreviated RuBisCO. RuBisCO makes up about 50% of the soluble protein in plant leaves; thus it is one of the most abundant enzymes in nature. Although the stromal concentration of RuBisCO active sites has been estimated to be

Stereo S16·1
Subunit organization of ribulose 1,5-*bis*phosphate carboxylase-oxygenase (RuBisCO), "top" and "side" views. Both views are shown with depth suppressed to emphasize structural organization. Large subunits are shown alternately yellow and blue; small subunits are purple. (Based on coordinates provided by David Eisenberg; stereo image by David Hyre and Kim M. Gernert.)

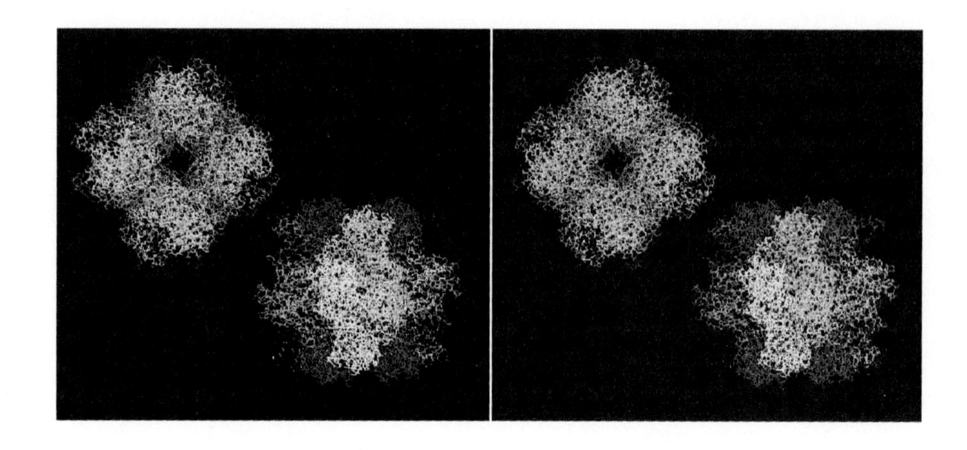

as high as 4 mM, the enzyme is so stringently regulated that the activity of RuBisCO can limit the rate of carbon assimilation. Physically, the RuBisCO of higher plants is composed of eight large subunits (MW 56 000 each) and eight small subunits (MW 14 000 each) for a total molecular weight of about 560 000. This composition is denoted L_8S_8 (Stereo S16·1). The eight large subunits associate to create the core (L_8) of the molecule, and the interfaces between these subunits form eight active sites. Four small subunits are localized at each end of the L_8 core. RuBisCO enzymes in phototrophs other than higher plants vary in their structure. For example, the RuBisCO of the purple bacterium *Rhodospirillum rubrum* consists of a dimer of large subunits that have about 30% sequence homology with the RuBisCO large subunit in higher plants.

To catalyze the fixation of CO_2, RuBisCO must be in an activated state. Activation of RuBisCO requires light, CO_2, Mg^{2+}, and correct stromal pH. In the light, RuBisCO activity increases in response to the higher pH and Mg^{2+} concentration in the stroma developed during proton pumping. The molecule of CO_2 required to activate RuBisCO reacts directly with the side chain of a lysine residue (Lys-202) that is located away from the active site of RuBisCO. The carbamate adduct formed between CO_2 and the lysine residue is similar to the reaction product of CO_2 with erythrocyte hemoglobin (Figure 4·35). During activation of RuBisCO, Mg^{2+} binds to the CO_2-lysine carbamate adduct and the β-COO^{-} of Asp-203. During catalysis, it is also coordinated at two sites formed by the bound substrate.

The reaction mechanism of RuBisCO is shown in Figure 16·11 (next page). The substrates for the reaction are a free molecule of CO_2 (not the CO_2 bound to Lys-202) and the five-carbon phosphorylated sugar ribulose 1,5-*bis*phosphate. The slow step in the reaction is the abstraction of a proton from ribulose 1,5-*bis*phosphate by a basic residue (—B:) of the enzyme to create the 2,3-enediol intermediate. Reaction of the enediol with CO_2 yields 2-carboxy-3-ketoarabinitol 1,5-*bis*phosphate, which is hydrated to an unstable 3-gem diol intermediate. The C-2—C-3 bond of the intermediate is immediately cleaved, generating a carbanion and one molecule of 3-phosphoglycerate. Stereospecific protonation of the carbanion yields a second molecule of 3-phosphoglycerate. This step completes the carbon-fixation stage of the pathway—two molecules of 3-phosphoglycerate are formed from CO_2 and the five-carbon sugar ribulose 1,5-*bis*phosphate.

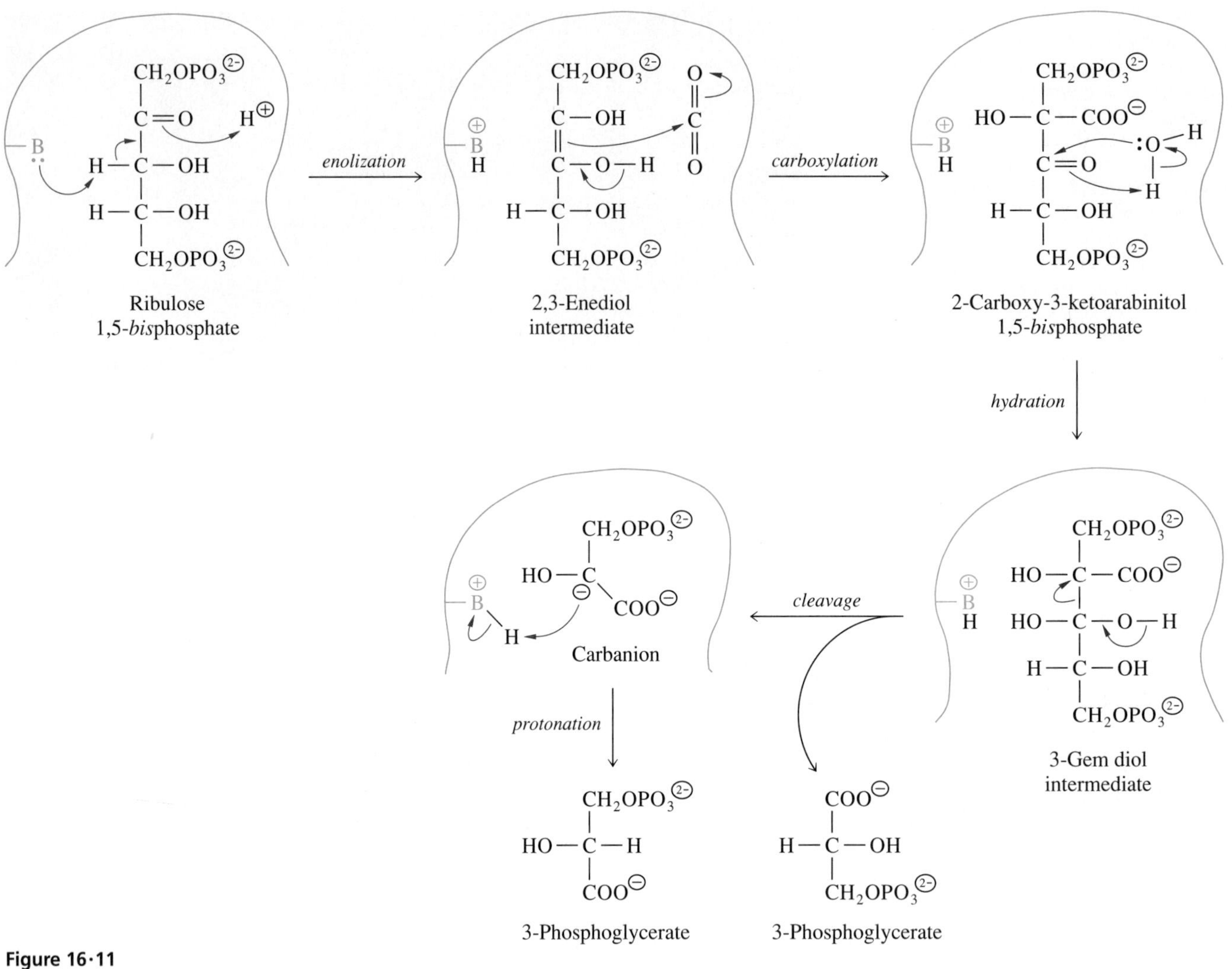

Figure 16·11
Mechanism of RuBisCO-catalyzed carboxylation of ribulose 1,5-*bis*phosphate to form two molecules of 3-phosphoglycerate.

16·9 After Fixation, Carbon Is Reduced, and the CO_2 Acceptor Molecule Is Regenerated

Figure 16·12 illustrates all of the steps of the RPP cycle. Note that the figure shows the steps for the assimilation of not one, but three molecules of carbon dioxide. This is because the smallest carbon intermediate in the RPP cycle is a C_3 molecule. Thus, three CO_2 molecules must be fixed before one C_3 unit can be removed from the cycle without decreasing the size of the pools of RPP cycle intermediates. The concentration of RPP cycle intermediates determines the concentration of the CO_2 acceptor molecule ribulose 1,5-*bis*phosphate and therefore has a large influence on the capacity of the pathway for carbon assimilation.

The reductive phase of the RPP cycle occurs in the two reactions following the RuBisCO reaction. First, 3-phosphoglycerate is converted to 1,3-*bis*phosphoglycerate in a reaction catalyzed by phosphoglycerate kinase and requiring a molecule of ATP. Next, 1,3-*bis*phosphoglycerate is reduced by NADPH in a reaction catalyzed by glyceraldehyde 3-phosphate dehydrogenase (named for the reverse reaction). An equilibrium is maintained between the reaction product, glyceraldehyde 3-phosphate, and its isomer, dihydroxyacetone phosphate, through the activity of triose phosphate isomerase.

Figure 16·12 (next page)
Reductive pentose phosphate (RPP) cycle. The cycle has three stages: CO_2 fixation, carbon reduction to (CH_2O), and regeneration of the CO_2 acceptor molecule. To avoid depleting the cycle of pathway intermediates, three molecules of CO_2 must be fixed before one triose phosphate (glyceraldehyde 3-phosphate or dihydroxyacetone phosphate) can exit the cycle.

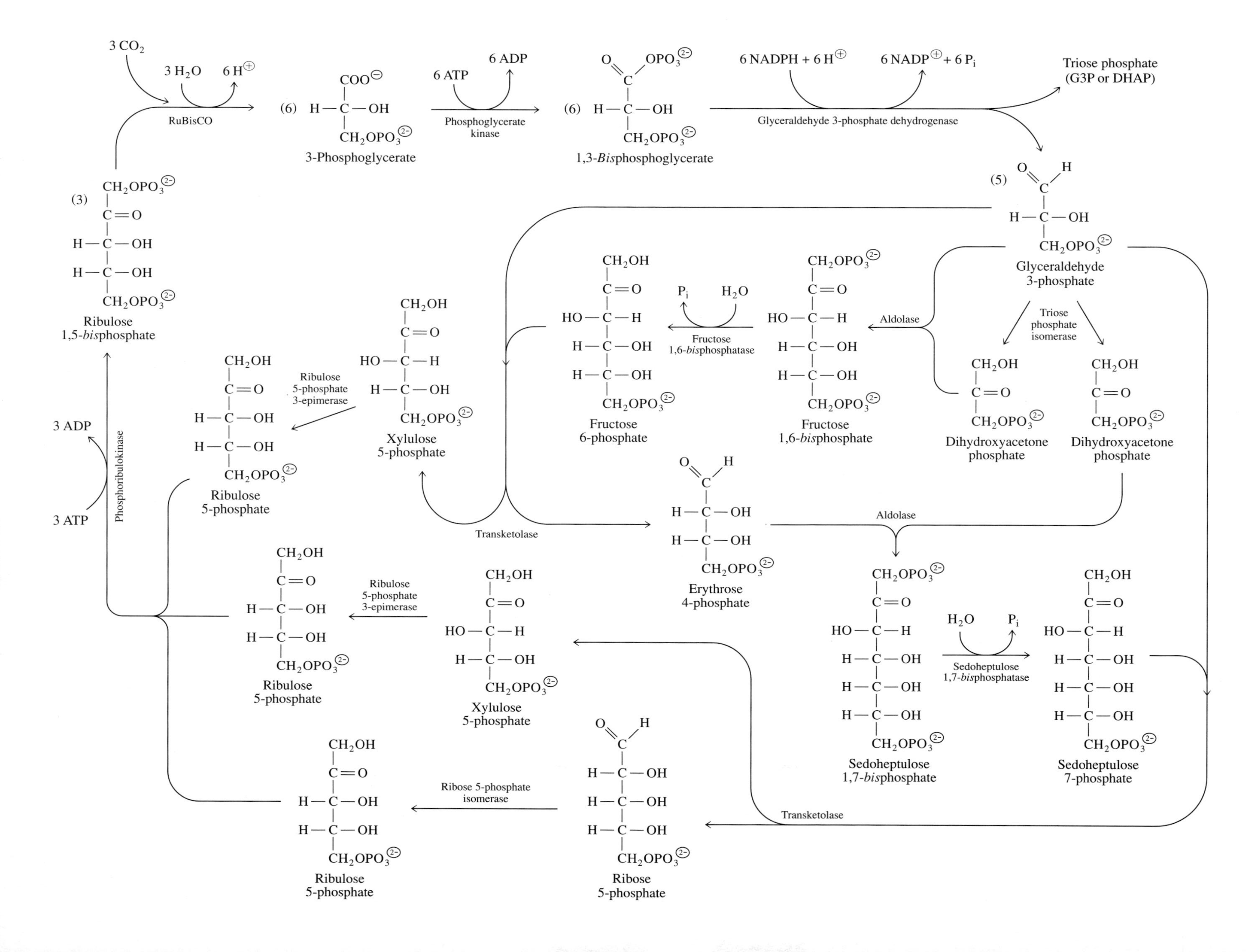

$3\ CO_2$
$3\ H_2O$
$6\ H^{\oplus}$
RuBisCO
(6) 3-Phosphoglycerate
6 ATP
6 ADP
Phosphoglycerate kinase
(6) 1,3-*Bis*phosphoglycerate
$6\ NADPH + 6\ H^{\oplus}$
$6\ NADP^{\oplus} + 6\ P_i$
Glyceraldehyde 3-phosphate dehydrogenase
Triose phosphate (G3P or DHAP)
(3) Ribulose 1,5-*bis*phosphate
(5) Glyceraldehyde 3-phosphate
Triose phosphate isomerase
Dihydroxyacetone phosphate
Dihydroxyacetone phosphate
Aldolase
Fructose 1,6-*bis*phosphate
P_i
H_2O
Fructose 1,6-*bis*phosphatase
Fructose 6-phosphate
Xylulose 5-phosphate
Ribulose 5-phosphate 3-epimerase
Ribulose 5-phosphate
Transketolase
Erythrose 4-phosphate
Aldolase
Sedoheptulose 1,7-*bis*phosphate
H_2O
P_i
Sedoheptulose 1,7-*bis*phosphatase
Sedoheptulose 7-phosphate
3 ADP
3 ATP
Phosphoribulokinase
Ribulose 5-phosphate
Ribulose 5-phosphate 3-epimerase
Xylulose 5-phosphate
Ribulose 5-phosphate
Ribose 5-phosphate isomerase
Ribose 5-phosphate
Transketolase

The sequence of reactions following the reductive phase of the cycle ends with the regeneration of the CO_2 acceptor molecule, ribulose 1,5-*bis*phosphate. Most of the reactions of the RPP cycle are part of the regeneration phase, which is stoichiometrically complex. Glyceraldehyde 3-phosphate is diverted into three different branches of the pathway and is interconverted between four-, five-, six-, and seven-carbon phosphorylated sugars. This regenerative phase, the so-called "sugar shuffle," is catalyzed by various isomerases, epimerases, aldolases, phosphatases, and transketolase. The last step is the phosphorylation of ribulose 5-phosphate by ATP to form ribulose 1,5-*bis*phosphate in a reaction catalyzed by phosphoribulokinase. All of the substrates, enzymes, and reactions of the RPP cycle are summarized in Table 16·1.

Table 16·1 Reactions of the RPP cycle of Figure 16·12, involving the fixation of three molecules of carbon dioxide for each molecule of triose phosphate to be used in carbohydrate synthesis

Reaction	Enzyme
3 Ribulose 1,5-*bis*phosphate + 3 CO_2 + 3 H_2O ⟶ 6 3-Phosphoglycerate + 6 H⊕	RuBisCO
6 3-Phosphoglycerate + 6 ATP ⟶ 6 1,3-*Bis*phosphoglycerate + 6 ADP	Phosphoglycerate kinase
6 1,3-*Bis*phosphoglycerate + 6 NADPH + 6 H⊕ ⟶ 6 Glyceraldehyde 3-phosphate + 6 NADP⊕ + 6 P_i	Glyceraldehyde 3-phosphate dehydrogenase
2 Glyceraldehyde 3-phosphate ⇌ 2 Dihydroxyacetone phosphate	Triose phosphate isomerase
Dihydroxyacetone phosphate + Glyceraldehyde 3-phosphate ⇌ Fructose 1,6-*bis*phosphate	Aldolase
Fructose 1,6-*bis*phosphate + H_2O ⟶ Fructose 6-phosphate + P_i	Fructose 1,6-*bis*phosphatase
Fructose 6-phosphate + Glyceraldehyde 3-phosphate ⇌ Erythrose 4-phosphate + Xylulose 5-phosphate	Transketolase
Erythrose 4-phosphate + Dihydroxyacetone phosphate ⇌ Sedoheptulose 1,7-*bis*phosphate	Aldolase
Sedoheptulose 1,7-*bis*phosphate + H_2O ⟶ Sedoheptulose 7-phosphate + P_i	Sedoheptulose 1,7-*bis*phosphatase
Sedoheptulose 7-phosphate + Glyceraldehyde 3-phosphate ⇌ Xylulose 5-phosphate + Ribose 5-phosphate	Transketolase
2 Xylulose 5-phosphate ⇌ 2 Ribulose 5-phosphate	Ribulose 5-phosphate 3-epimerase
Ribose 5-phosphate ⇌ Ribulose 5-phosphate	Ribulose 5-phosphate isomerase
3 Ribulose 5-phosphate + 3 ATP ⟶ 3 Ribulose 1,5-*bis*phosphate + 3 ADP	Phosphoribulokinase

Net reaction
3 CO_2 + 9 ATP + 6 NADPH + 5 H_2O ⟶ 9 ADP + 8 P_i + 6 NADP⊕ + Triose phosphate (glyceraldehyde 3-phosphate or dihydroxyacetone phosphate)

16·10 Light, pH, and Mg^{2+} Regulate the Activities of Some Enzymes of the RPP Cycle

Regulation of the RPP cycle occurs at the metabolically irreversible steps of the pathway, including the *bis*phosphatase reactions and the steps that consume ATP and NADPH. Specifically, these steps include the reactions catalyzed by fructose 1,6-*bis*phosphatase, sedoheptulose 1,7-*bis*phosphatase, phosphoglycerate kinase, phosphoribulokinase, and glyceraldehyde 3-phosphate dehydrogenase. All of these enzymes appear to be regulated by one or more of a few factors, including light, the concentration of stromal Mg^{2+}, and pH.

The activities of all the enzymes mentioned above except phosphoglycerate kinase are regulated by light. The mechanism of light activation involves the reduction of surface-exposed disulfides; this reduction increases the activity of the enzyme by causing a change in tertiary structure. Reducing equivalents for the formation of the —SH groups are provided by the photosynthetic electron-transport chain. Instead of reducing $NADP^{+}$, reduced ferredoxin donates its electrons to a small stromal protein, thioredoxin (Figure 16·13). Whereas redox changes in ferredoxin are between the Fe^{2+} and Fe^{3+} states of iron, in thioredoxin the redox center is a pair of sulfhydryl groups. Reduction of a cystine disulfide of thioredoxin is catalyzed by the enzyme ferredoxin-thioredoxin reductase. The thiol group of the soluble thioredoxin undergoes a spontaneous sulfhydryl-disulfide exchange with a surface disulfide of a light-modulated enzyme. Reduction activates the light-modulated enzyme.

The stromal concentration of Mg^{2+} and the pH, both of which increase during proton pumping into the lumen, are also regulatory signals for certain enzymes of the RPP cycle. Fructose 1,6-*bis*phosphatase and sedoheptulose 1,7-*bis*phosphatase both require an alkaline pH and a high concentration of Mg^{2+} for maximum activity. Phosphoribulokinase is subject to another type of pH regulation through its inhibition by 3-phosphoglycerate (3PG), the RuBisCO reaction product. Phosphoribulokinase is only inhibited by 3-phosphoglycerate in the $3PG^{2-}$ form. During active photosynthesis, when the stroma is alkaline, the predominant ionic form is $3PG^{3-}$. However, when photosynthetic activity decreases and the stromal pH drops, the level of $3PG^{2-}$ increases; consequently, there is increased inhibition of phosphoribulokinase. This inhibition prevents continued ATP consumption by phosphoribulokinase when the rate of photosynthesis declines. Since factors that affect the activities of the regulated RPP enzymes, such as pH and $[Mg^{2+}]$, are significantly different under photosynthetic versus nonphotosynthetic conditions, the rate of carbohydrate synthesis is coordinated with the rate of ATP and NADPH formation. Consider the importance of this coordination to a plant leaf that receives intermittent light. Rapid activation of the RPP cycle is important to achieve maximum carbon assimilation during the light, but in the dark efficient deactivation is necessary to prevent imbalances between the concentrations of RPP-cycle intermediates that would preclude rapid response to renewed photosynthetic activity.

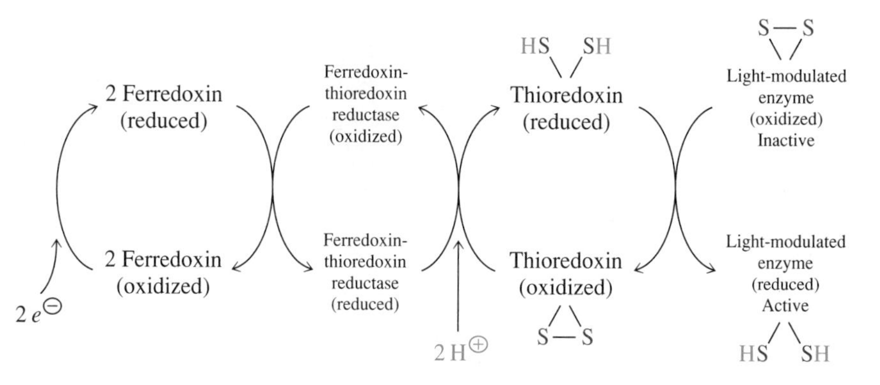

Figure 16·13
General scheme for the light activation of light-modulated enzymes of the RPP cycle. During light activation, electrons are diverted from electron transport and used to reduce disulfide groups of the inactive enzymes to sulfhydryl groups.

16·11 RuBisCO Catalyzes the Oxygenation of Ribulose 1,5-*Bis*phosphate

As its complete name indicates, ribulose 1,5-*bis*phosphate carboxylase-oxygenase catalyzes not only carboxylation but also oxygenation of ribulose 1,5-*bis*phosphate. The two reactions are competitive—CO_2 and O_2 compete for active sites on RuBisCO. Although the affinity of RuBisCO for CO_2 ($K_m = 12\,\mu M$) is much greater than for oxygen ($K_m = 250\,\mu M$) and CO_2 is more soluble in the stroma than O_2, the atmospheric concentration of oxygen (21%) is much higher than the level of carbon dioxide (0.03%). As a result, both reactions contribute significantly to the consumption of ribulose 1,5-*bis*phosphate in vivo. Under normal conditions, there is a carboxylation-to-oxygenation ratio of 3:1 to 4:1. In the preceding section, we discussed the outcome of the carboxylation reaction catalyzed by RuBisCO. We now turn to the metabolic sequence that follows the oxygenation of ribulose 1,5-*bis*phosphate by RuBisCO.

The products of the oxygenation reaction catalyzed by RuBisCO are one molecule of 3-phosphoglycerate and one molecule of phosphoglycolate. Figure 16·14 shows the net result of two oxygenation events. Two molecules of 3-phosphoglycerate enter the RPP cycle directly, and two molecules of 2-phosphoglycolate are metabolized to one molecule of CO_2 and one molecule of 3-phosphoglycerate, which enters the RPP cycle. *Photorespiration,* the light-dependent uptake of O_2 and release of CO_2, is the name given to the processes of oxygen uptake by RuBisCO and the metabolism of phosphoglycolate (Figure 16·14).

The metabolism of phosphoglycolate occurs in three different cellular compartments of mesophyll cells—the chloroplast, the peroxisome, and the mitochondrion. Phosphoglycolate is dephosphorylated to glycolate in the chloroplast; glycolate then enters the peroxisome, where it reacts with O_2 to form glyoxylate and hydrogen peroxide. Hydrogen peroxide is readily converted to O_2 and H_2O by the enzyme catalase, and glyoxylate is transaminated to glycine. Glycine exits the peroxisome and enters the mitochondrion, where in one pathway it undergoes oxidative decarboxylation in a cleavage reaction that yields 5,10-methylenetetrahydrofolate (Section 18·13), CO_2, and NH_3. CO_2 is either photosynthetically refixed or released from the leaf, and NH_3 is reassimilated to form glutamate. 5,10-Methylenetetrahydrofolate donates its methylene group to a second molecule of glycine (also from glyoxylate) to form serine. Serine exits the mitochondrion and enters the peroxisome, where it is deaminated to hydroxypyruvate; hydroxypyruvate is then reduced to glycerate. Glycerate enters the chloroplast, where it is phosphorylated by ATP to 3-phosphoglycerate. 3-Phosphoglycerate can be incorporated into the RPP cycle. In summary, two oxygenation events form two molecules of 3-phosphoglycerate (C_3) and two molecules of phosphoglycolate (C_2). The two phosphoglycolate molecules are metabolized to one molecule of 3-phosphoglycerate and one molecule of CO_2.

Since extensive searches of RuBisCO mutants have not uncovered a mutant enzyme that catalyzes *only* the carboxylation of ribulose 1,5-*bis*phosphate (a "RuBisC"), it is speculated that, although a seemingly wasteful process (one carbon atom is lost for every two O_2 fixed), photorespiration is physiologically essential. It may be that photorespiration regenerates ADP and $NADP^{\oplus}$ under conditions of low CO_2 concentration, for example, when stomata are closed and light intensity is high. During rapid photosynthesis, the capacity to form ATP and NADPH can be limited by the supply of stromal ADP and $NADP^{\oplus}$. Under these conditions, electron carriers can become over reduced, and the proton gradient across the thylakoid can become excessively high. With no means to "discharge," light-sensitive photosynthetic pigments of the LHC and photosystem antenna complexes can be oxidatively damaged.

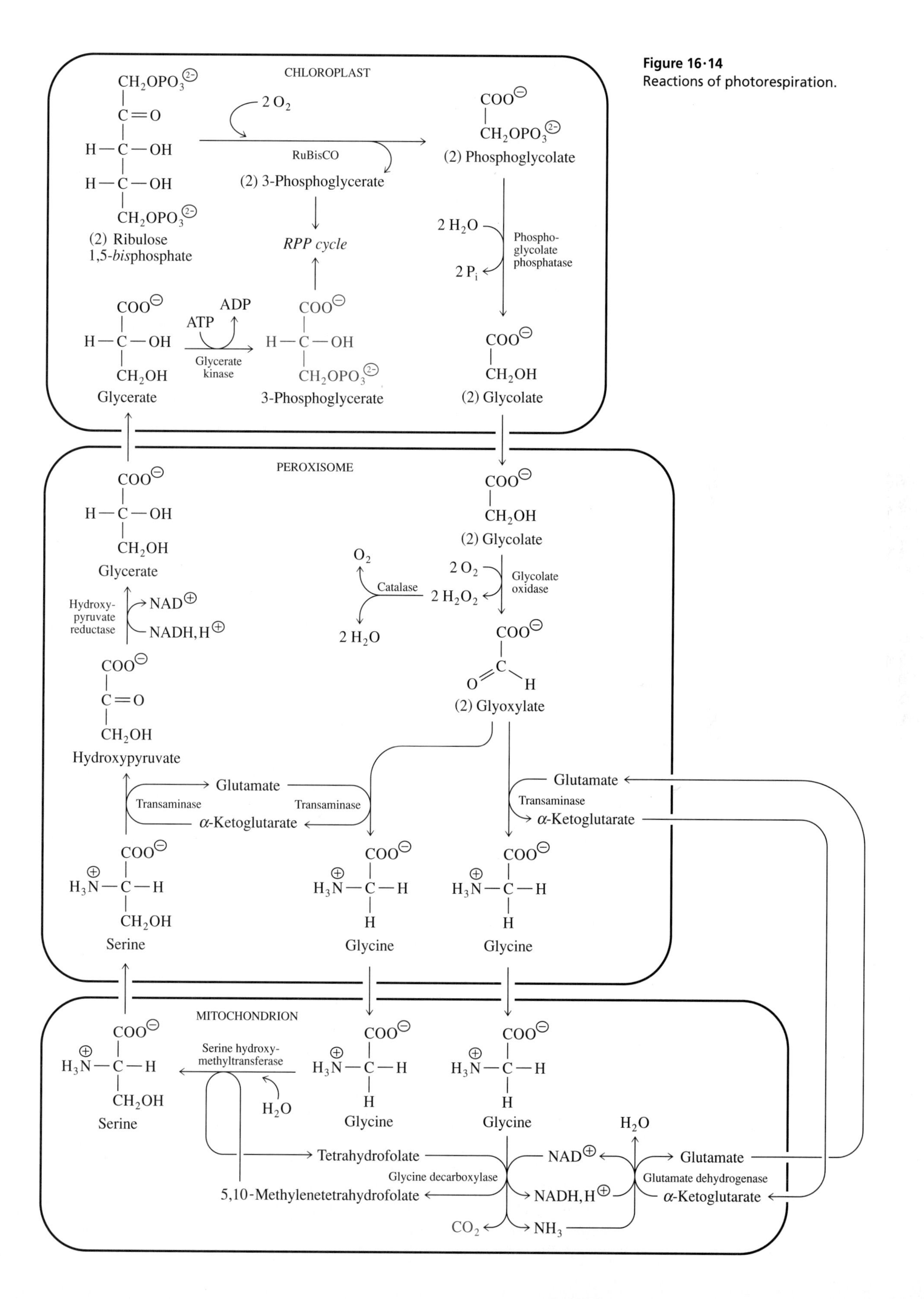

Figure 16·14
Reactions of photorespiration.

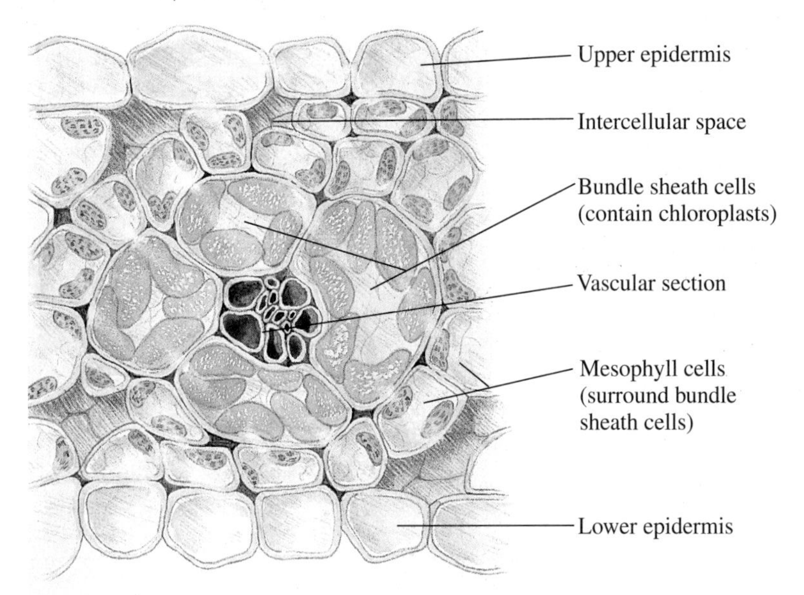

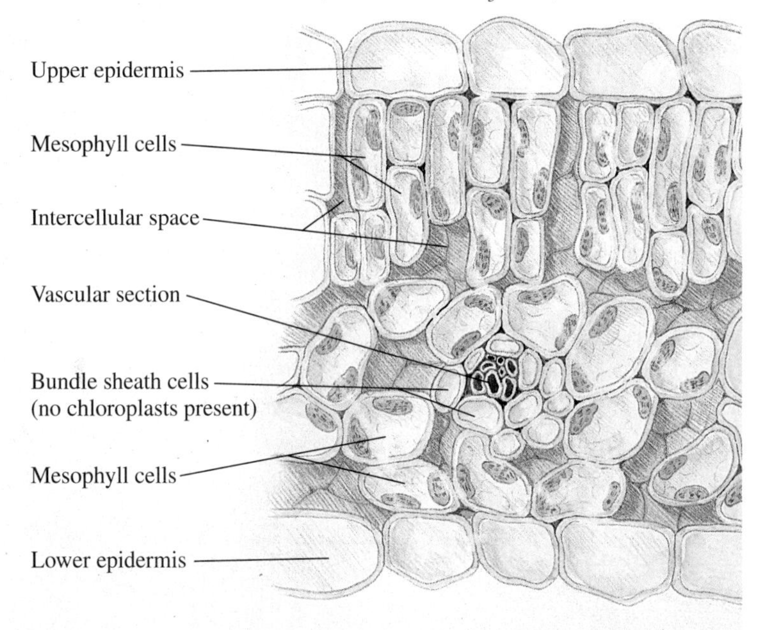

Figure 16·15
Illustrations of the structures of leaves of C_4 and C_3 plants. Notice that the large bundle sheath cells of C_4 plants contain chloroplasts and are completely surrounded by mesophyll cells. (Illustrator: Lisa Shoemaker.)

16·12 The C_4 Pathway Minimizes the Oxygenase Activity of RuBisCO by Concentrating CO_2

In several plant species, a second pathway for carbon fixation occurs in conjunction with the the RPP cycle. In this pathway, the initial product of carbon fixation is a four-carbon acid, rather than a three-carbon acid, and two distinct cell types are involved. Plants with this pathway have essentially no photorespiratory activity. Since the initial product of CO_2 fixation is a C_4 acid, the metabolic route is called the *C_4 pathway,* just as the RPP cycle is also called the C_3 pathway. C_4 plants include such economically important species as maize (corn), sorghum, and sugarcane, and many of the most troublesome weeds.

The C_4 pathway, which may be considered a prelude to the C_3 pathway, has two phases, carboxylation and decarboxylation, which occur in two different cell types within the leaf. In the initial phase, CO_2 is fixed to C_4 acids in the mesophyll cells (typically the site of the RPP cycle in C_3 plants). In the second phase, C_4 acids are decarboxylated in the bundle sheath cells, where CO_2 is released, refixed by RuBisCO, and incorporated into the RPP cycle. Figure 16·15 illustrates the cellular organization of C_4 and C_3 leaves in cross section. Two major structural features related to carbon assimilation distinguish C_4 from C_3 leaf types. First, in the C_4 leaf, mesophyll cells completely surround the bundle sheath cells, which surround the vascular tissue. Second, the C_4 bundle sheath cells are large and contain chloroplasts. In contrast, C_3 bundle sheath cells are nonphotosynthetic.

A schematic representation of the C_4 reaction sequence in the mesophyll and bundle sheath cells is shown in Figure 16·16. CO_2 is hydrated to bicarbonate ($HCO_3^{\ominus}$) in the mesophyll cytosol. Bicarbonate and phosphoenolpyruvate (PEP)

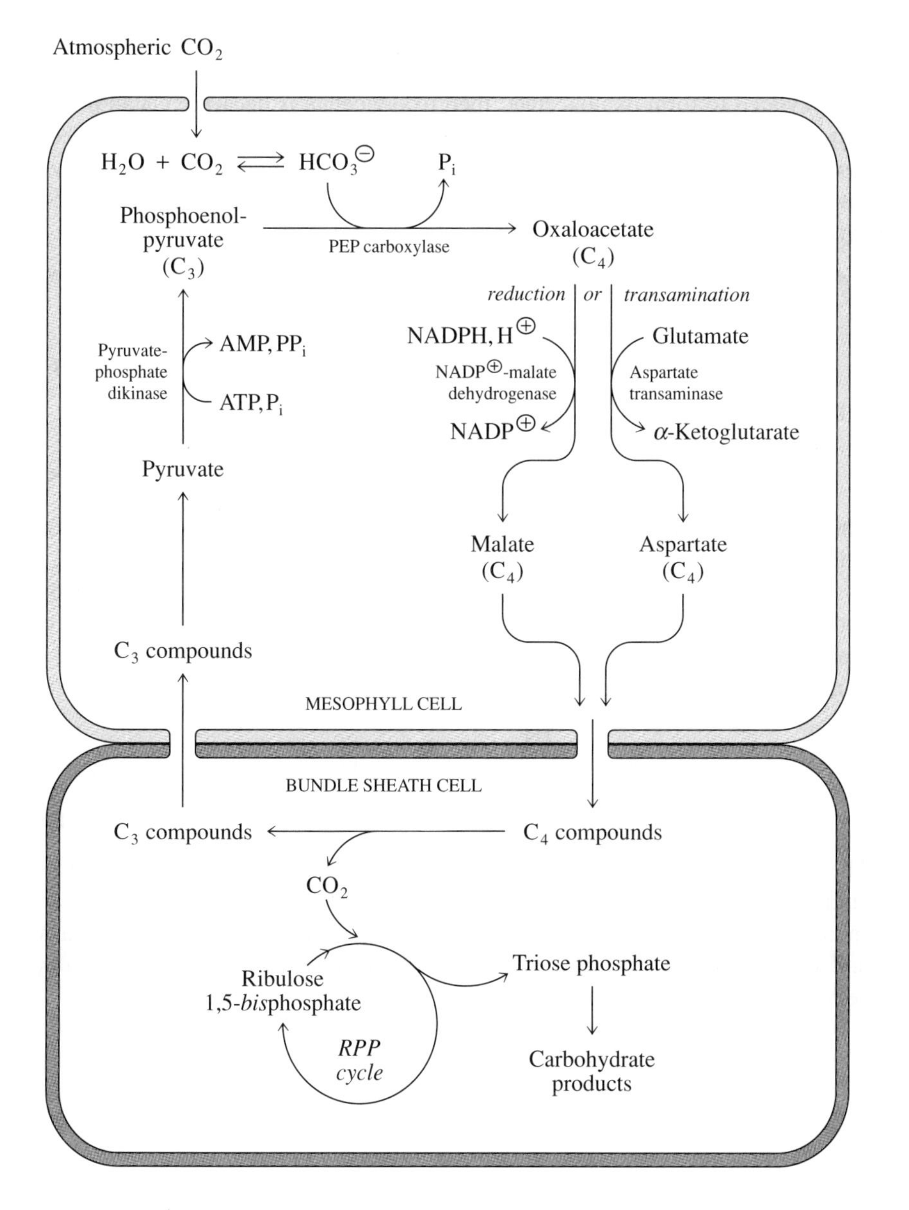

Figure 16·16
General scheme of the C_4 pathway. Four-carbon organic or amino acids are formed in the mesophyll cell, transported to the bundle sheath cells via plasmodesmata, and decarboxylated in the bundle sheath cells. The CO_2 released is refixed by the RuBisCO reaction and enters the RPP cycle. The remaining three-carbon unit is converted back to the CO_2 acceptor molecule, phosphoenolpyruvate.

Although there is an extra energy cost to form phosphoenolpyruvate for C_4 carbon assimilation, the absence of photorespiration gives C_4 plants a significant advantage over C_3 plants. C_4 plants have significantly higher quantum yields (molecules of CO_2 fixed per photon absorbed) than C_3 plants under their respective optimal photosynthetic conditions. For reasons related to the speed of intercellular metabolite movement, requirements for light activation, and the temperature optima for certain C_4 enzymes, C_4 plants are most photosynthetically efficient under conditions of high light intensity and high temperature.

16·13 Certain Plants Carry Out Carbon Fixation at Night to Conserve Water

A group of plants found primarily in arid environments and dry microclimates greatly reduce water loss during photosynthesis by carrying out a modified sequence of carbon-assimilation reactions that involves the accumulation of malate at night. Because the reaction sequence associated with the nocturnal accumulation of this acid was first discovered in the stonecrop family, *Crassulaceae,* this sequence has been called *Crassulacean acid metabolism,* or CAM (Figure 16·18). CAM is found in several other plant families as well, including the cacti, bromeliads, and orchids. CAM plants of economic importance include pineapple and many ornamental plants. CAM plants take up CO_2 into mesophyll cells through open stomata at night, since water loss through the stomata is much lower at cooler nighttime temperatures than during the day. CO_2 is fixed by the PEP carboxylase reaction, and the oxaloacetate formed is reduced to malate, which is then translocated into the vacuole. Transport of malate into the vacuole is necessary in order to maintain a near-neutral pH in the cytosol, since the cellular concentration of this acid can be as high as 0.2 M by the end of the night. The vacuoles of CAM plants generally occupy >90% of the total volume of the cell. During the next light period, when ATP and NADPH are formed by photosynthesis, malate is released from the vacuole and decarboxylated. Thus, the large pool of malate accumulated at night supplies CO_2 for carbon assimilation during the day. During decarboxylation of malate, leaf stomata are tightly closed so that neither water nor CO_2 can escape from the leaf. Thus, the level of cellular CO_2 can be much higher than the level of atmospheric CO_2. As in C_4 plants, the higher internal CO_2 concentration greatly reduces photorespiration.

In fact, CAM is analogous to C_4 metabolism in that the conventional C_3 pathway is preceded by a reaction sequence involving the formation of C_4 acids, catalyzed by PEP carboxylase. Also, the three alternative decarboxylation pathways (Figure 16·17) are the same in CAM as in C_4 metabolism. CAM and C_4 metabolism differ, however, in that the C_4 pathway involves the spatial separation of the carboxylation and decarboxylation phases of the cycle between mesophyll and bundle sheath cells, whereas CAM, which takes place in the mesophyll cells, involves the temporal separation of these phases into a day and night cycle. As in the C_4 pathway, PEP carboxylase catalyzes the reaction of bicarbonate and phosphoenolpyruvate to yield oxaloacetate, which is reduced to malate. In CAM plants, however, this reaction occurs only at night. The phosphoenolpyruvate required for malate formation is derived from starch, which is glycolytically converted to phosphoenolpyruvate. During the day, the phosphoenolpyruvate formed during malate decarboxylation (either directly by PEP carboxykinase or via malic enzyme and pyruvate phosphate dikinase) is converted to starch by gluconeogenesis and stored in the chloroplast. Thus, in CAM, not only are there large day/night changes in the pool of malate but also in the pool of starch (Figure 16·19).

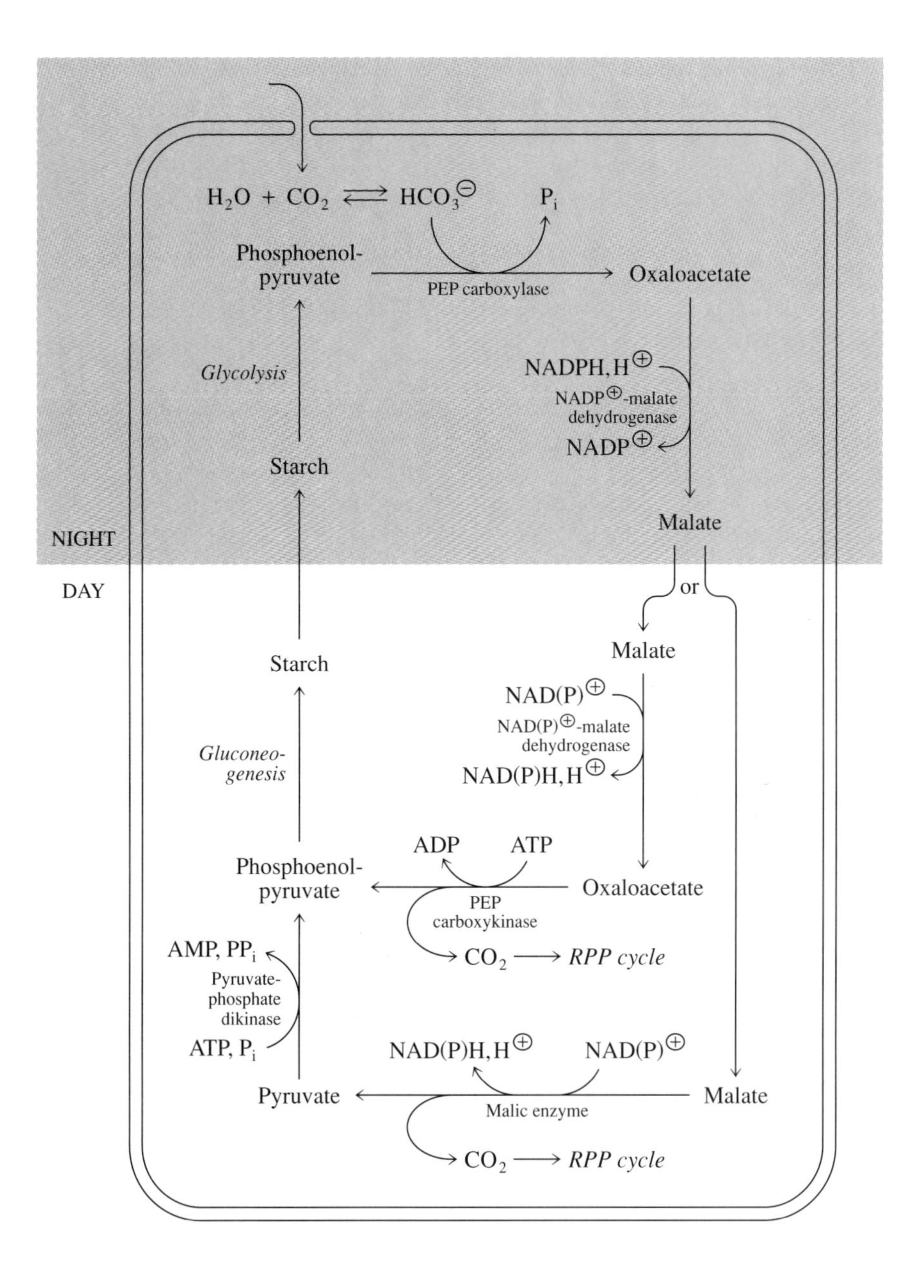

Figure 16·18
General scheme of Crassulacean acid metabolism (CAM). During the night, PEP carboxylase and NADP⊕-malate dehydrogenase catalyze the formation of malate. Phosphoenolpyruvate required for malate synthesis is derived from starch. The next day, decarboxylation of malate enriches the cell in CO_2 for the RPP cycle, while NADPH and ATP are formed. The decarboxylation of malate also yields phosphoenolpyruvate, which is subsequently converted to starch by gluconeogenesis.

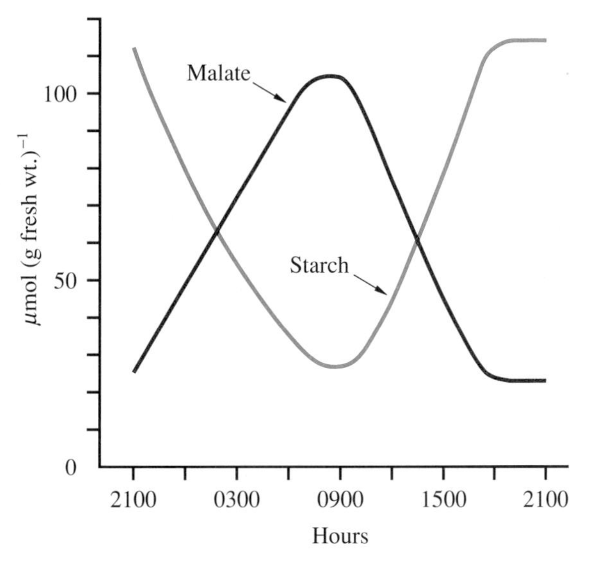

Figure 16·19
Reciprocal changes in starch and malate during the CAM cycle. Starch serves as a source of carbon for the formation of the three-carbon molecule phosphoenolpyruvate, which constitutes three-quarters of the carbon in the malate formed during nocturnal CO_2 fixation.

An important regulatory feature of the CAM pathway is the inhibition of PEP carboxylase by malate and low pH. During the day when the cytosolic concentration of malate is high and pH is low, PEP carboxylase is effectively inhibited. This inhibition is essential to prevent futile cycling of CO_2 and malate by PEP carboxylase and to avoid competition between PEP carboxylase and RuBisCO for CO_2.

16·14 Sucrose and Starch Are Synthesized from Metabolites of the RPP Cycle

Essentially all higher plants convert their recently fixed carbon into either starch or sucrose. Sucrose and some oligosaccharides based on sucrose are translocatable forms of carbohydrate that are moved through the phloem to plant organs that are not photosynthetically self-sustaining. These so-called "sink" organs include roots, seeds, and young, developing leaves. Thus, in the common terminology, "sink" organs obtain sucrose from "source" leaves. Some carbohydrate is retained in source leaves during the day, accumulating as starch. During the night, when there is no photosynthesis, starch is broken down, converted to sucrose, and used as the metabolic fuel for the source leaves themselves and other organs of the plant.

Sucrose is synthesized in the cytosol from triose phosphates (glyceraldehyde 3-phosphate and dihydroxyacetone phosphate) that originate either directly from the RPP cycle or less directly from starch. Chloroplast starch is broken down to glucose or glucose monophosphates and glycolytically converted to triose phosphates. The movement of triose phosphates from the chloroplast to the cytosol is mediated by the *phosphate translocator,* a proteinaceous carrier that spans the inner membrane of the chloroplast (Figure 16·20). The outer chloroplast membrane is readily permeable to triose phosphates and other small molecules. The translocation of chloroplast triose phosphate occurs in a strict exchange for inorganic phosphate from the cytosol. This stoichiometric requirement of the phosphate translocator regulates the rate of sucrose synthesis in relation to the ambient photosynthetic conditions by controlling the movement of triose phosphates, the substrate for sucrose formation, and inorganic phosphate, a product of sucrose formation.

Availability of chloroplast triose phosphate for export to the cytosol signals sustained photosynthesis, whereas availability of inorganic phosphate signals ongoing sucrose synthesis. When sucrose synthesis slows, cytosolic P_i is less available, and more triose phosphate is retained in the chloroplast, where it can be converted to starch. When photosynthesis slows, triose phosphates are less available for export to the cytosol for sucrose formation.

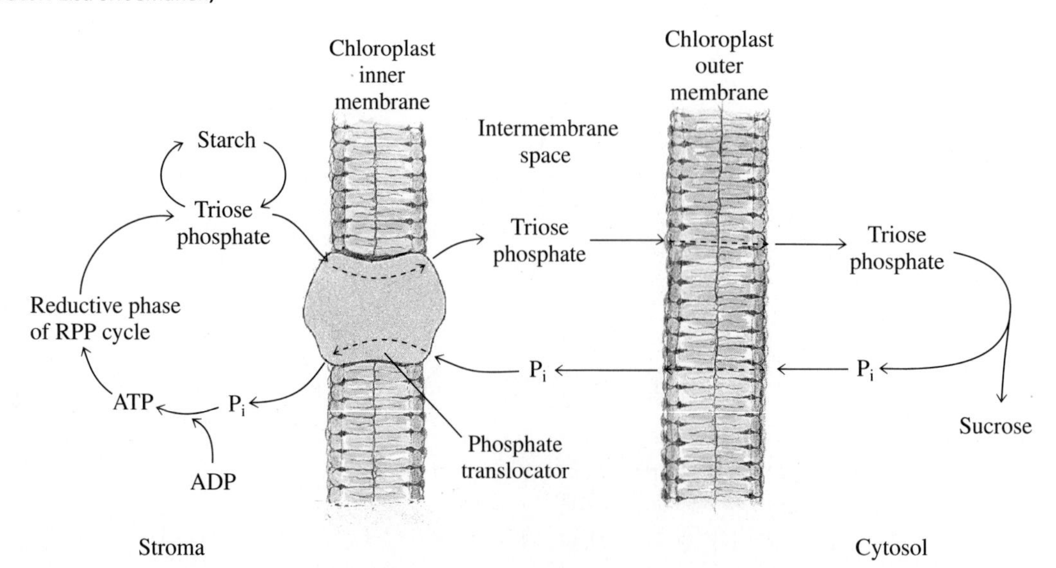

Figure 16·20
Export of stromal triose phosphates in exchange for cytosolic P_i via the phosphate translocator of the chloroplast inner membrane. (Illustrator: Lisa Shoemaker.)

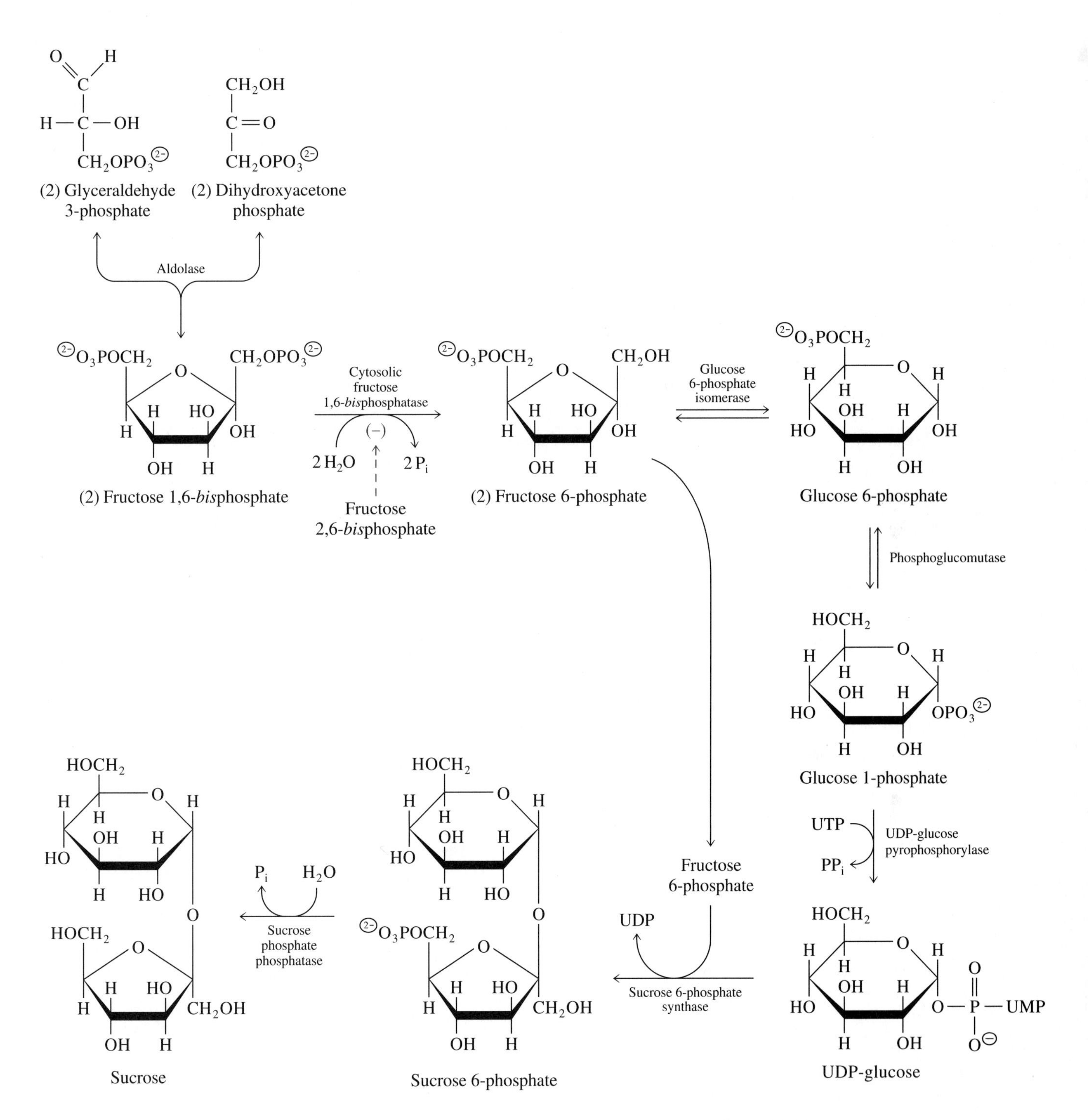

Figure 16·21
Pathway for the biosynthesis of sucrose from triose phosphates (glyceraldehyde 3-phosphate and dihydroxyacetone phosphate) in the cytosol of a photosynthetic plant cell. Four molecules of triose phosphate ($4\ C_3$) from the RPP cycle are consumed for each molecule of sucrose (C_{12}) synthesized. Two P_i and one PP_i are released. The pathway is regulated via the phosphate translocator and fructose 2,6-*bis*phosphate (F2,6BP).

The reaction sequence for sucrose synthesis is depicted in Figure 16·21. Note that inorganic phosphate and inorganic pyrophosphate are produced in the reactions catalyzed by fructose 1,6-*bis*phosphatase, UDP-glucose pyrophosphorylase, and sucrose phosphate phosphatase. The first metabolically irreversible step in the pathway is the hydrolysis of fructose 1,6-*bis*phosphate to yield fructose 6-phosphate and P_i. Not surprisingly, the activity of fructose 1,6-*bis*phosphatase is highly regulated, and interestingly, its regulation is via the allosteric effector molecule fructose 2,6-*bis*phosphate (Figure 12·6), which we encountered in our examinations of glycolysis and gluconeogenesis. Fructose 2,6-*bis*phosphate, which serves only a regulatory role in metabolism, is a potent inhibitor of the cytosolic fructose 1,6-*bis*phosphatase. The level of fructose 2,6-*bis*phosphate is determined by the relative rates

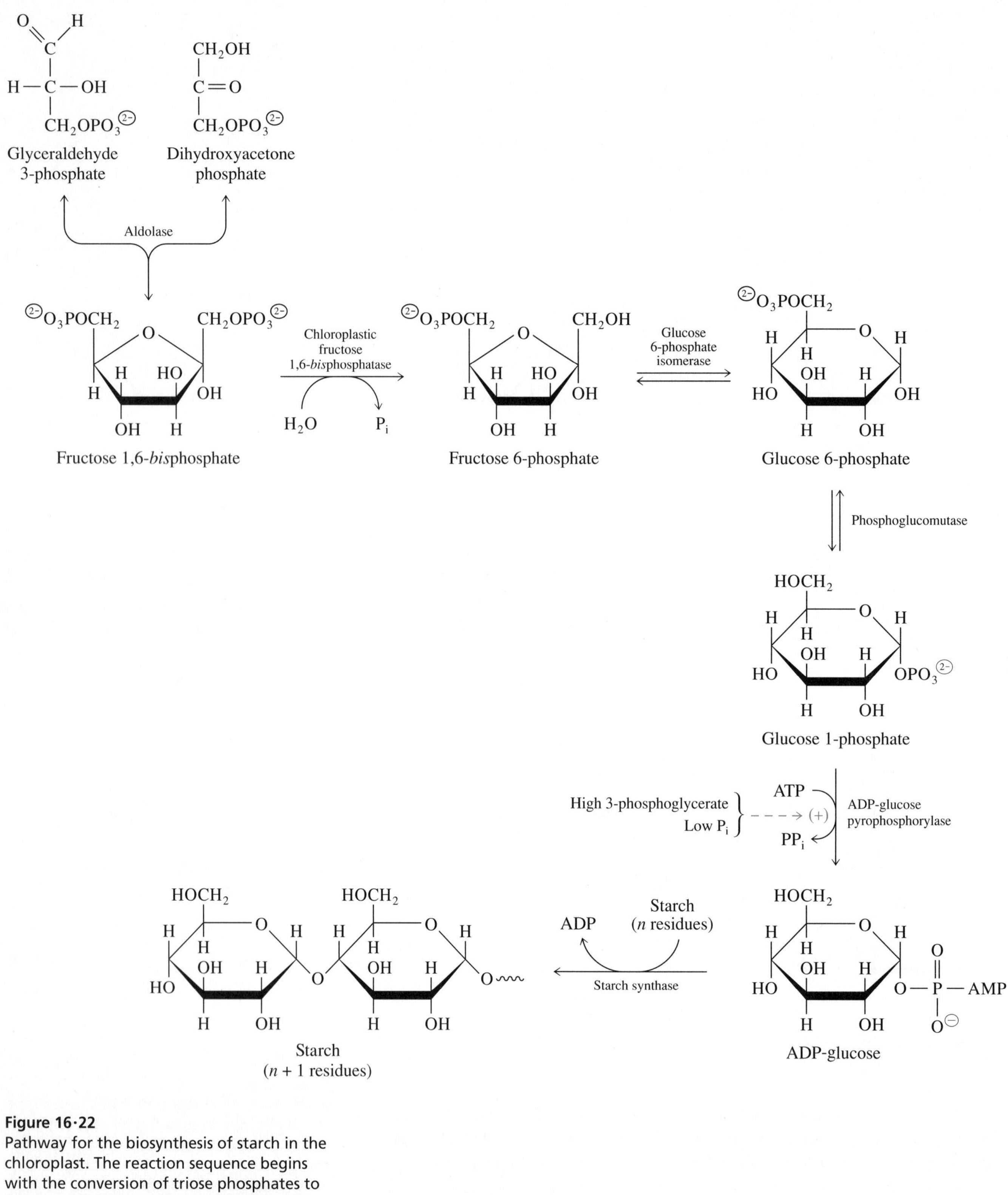

Figure 16·22
Pathway for the biosynthesis of starch in the chloroplast. The reaction sequence begins with the conversion of triose phosphates to fructose 1,6-*bis*phosphate and involves the conversion of fructose 6-phosphate to glucose 1-phosphate and then to ADP-glucose, which donates a glucose to the growing starch polymer.

of its synthesis and degradation, catalyzed in plants by the activities of fructose 2,6-*bis*phosphate kinase and fructose 2,6-*bis*phosphatase, respectively. Both of these enzyme activities are subject to complex regulation by a number of metabolites that reflect the suitability of conditions for sucrose synthesis.

Triose phosphate from the RPP cycle can also be converted to starch in the chloroplast (Figure 16·22). In starch synthesis as in sucrose synthesis, triose phosphate is condensed to fructose 1,6-*bis*phosphate and acted upon by fructose 1,6-*bis*phosphatase to yield fructose 6-phosphate, which is then isomerized to glucose 6-phosphate and converted to glucose 1-phosphate. In starch synthesis, glucose 1-phosphate is then converted to ADP-glucose (versus UDP-glucose in sucrose synthesis). In this reaction, catalyzed by ADP-glucose pyrophosphorylase, one molecule of ATP is consumed and pyrophosphate is released. Glucose from ADP-glucose is then donated to the growing starch polymer in a reaction catalyzed by starch synthase.

The synthesis of ADP-glucose catalyzed by ADP-glucose pyrophosphorylase is the regulated step in the pathway of starch biosynthesis. The activity of this enzyme is strongly influenced by the ratio of 3-phosphoglycerate to inorganic phosphate, which may change under different photosynthetic conditions. During rapid photosynthesis, ADP-glucose pyrophosphorylase will become activated, since the concentration of 3-phosphoglycerate, the product of the RuBisCO reaction, will be high, and the concentration of inorganic phosphate, used in ATP formation, will be low. Conversely, when the rates of photosynthesis and photophosphorylation decrease, the ratio of 3-phosphoglycerate to inorganic phosphate will decrease, and ADP-glucose pyrophosphorylase and therefore starch synthesis will be inhibited. This regulatory feature avoids excessive diversion of triose phosphates from the RPP cycle to starch synthesis when the rate of photosynthesis declines.

At night, starch serves as a source of carbon for growth and respiration. The starch molecule, a glucose polymer similar to glycogen, is phosphorolytically cleaved by the action of starch phosphorylase to generate glucose 1-phosphate. Glucose 1-phosphate is converted to glucose 6-phosphate and metabolized by glycolysis to triose phosphates, which are exported from the chloroplast for sucrose synthesis and subsequent respiration and/or transport from the cell. An alternate route for starch breakdown is hydrolysis by amylases to variously sized dextrins and eventually to maltose and then glucose. Glucose formed via this route is phosphorylated by the action of hexokinase and enters the glycolytic and sucrose biosynthetic pathways.

Summary

The photosynthetic processes encompass the light-dependent electron-transport reactions that form NADPH and ATP, and the subsequent utilization of NADPH and ATP in the conversion of atmospheric CO_2 to carbohydrate (CH_2O).

The chloroplast is the organelle in which photosynthesis occurs. In the chloroplast there are thylakoid membranes, which contain electron-transport components, including photosystem I and photosystem II (PSI and PSII), and the ATP synthase complex. The thylakoid network is suspended in the aqueous stroma. The stroma contains all of the enzymes and metabolites that participate in the conversion of CO_2 into (CH_2O) by the reductive pentose phosphate (RPP) cycle.

The absorption of light energy by chlorophyll and other accessory pigments culminates in the excitation of a special pair of reaction-center chlorophyll molecules. The excitation of these chlorophylls lowers their reduction potential. This event occurs in each of the two photosystems and drives the linear transfer of electrons from water, against an electrochemical potential, through the photosystems to

NADP⊕. The significant results of the light-dependent electron-transport process are: (1) protons are pumped across the photosynthetic membrane to generate a proton gradient that drives the chemiosmotic conversion of ADP and P_i to ATP, and (2) electrons are transferred to the terminal electron acceptor to yield NADPH.

NADPH and ATP are subsequently consumed during the reductive pentose phosphate cycle. There are three major stages in the RPP cycle: (1) the fixation of CO_2 by the enzyme RuBisCO, (2) the reduction of CO_2 to (CH_2O), and (3) the regeneration of the CO_2 acceptor molecule, ribulose 1,5-*bis*phosphate. The ATP and NADPH formed during the electron-transport reactions are consumed during the reduction and regeneration phases of the RPP cycle. Coordination of the RPP cycle with the rate of ATP and NADPH synthesis depends on the regulation of several RPP cycle enzymes by light, the concentration of Mg^{2+}, and pH. This coordination is essential to maintain balanced pools of RPP cycle intermediates during changing photosynthetic conditions.

There are three additional metabolic pathways that function together with the RPP cycle. The photorespiratory pathway catabolizes the reaction products that arise from the oxygenation of ribulose 1,5-*bis*phosphate by RuBisCO. The photorespiratory reaction sequence takes place in three different cellular organelles and recovers half of the carbon that is consumed by the oxygenation of RuBisCO. The C_4 and CAM pathways are found (individually) in certain plants in addition to the RPP cycle. Both pathways involve a preliminary carboxylation step catalyzed by PEP carboxylase, a decarboxylation, and refixation of the released CO_2 catalyzed by RuBisCO. In both pathways, the preliminary carboxylation has the effect of concentrating CO_2 at the site of RuBisCO, thereby essentially eliminating RuBisCO oxygenation. In the C_4 pathway, the processes of CO_2 fixation by PEP carboxylase and CO_2 fixation by RuBisCO occur in two different cell types. In the case of CAM, these two processes occur at different times during the day/night cycle.

The end products of carbon assimilation are starch and sucrose. Starch is synthesized in the chloroplast. Sucrose is formed in the cytosol from triose phosphates that are exported from the chloroplast to the cytosol via the phosphate translocator. The major regulatory control on the rate of synthesis of starch and sucrose is the ratio of phosphate esters to inorganic phosphate. This ratio is a general indicator of the rate of photophosphorylation and carbon assimilation.

Selected Readings

Arnon, D. I. (1984). The discovery of photosynthetic phosphorylation. *Trends Biochem. Sci.* 9:258–262.

Bassham, J. A., and Calvin, M. (1957). *The Path of Carbon in Photosynthesis* (Englewood Cliffs, New Jersey: Prentice Hall).

Bennett, J. (1991). Phosphorylation in green plant chloroplasts. *Annu. Rev. Plant Physiol. Plant Mol. Biol.* 42:281–311.

Edwards, G. E., and Walker, D. (1983). *C_3, C_4: Mechanisms and Cellular and Environmental Regulation of Photosynthesis* (Berkeley: University of California Press).

Goodwin, T. W., and Mercer, E. I. (1983). *Introduction to Plant Biochemistry,* 2nd ed. (New York: Pergamon Press).

Govindjee and Coleman, W. J. (1990). How plants make oxygen. *Sci. Am.* 262(2):50–58.

Hatch, M. D., and Boardman, N. K., eds. (1981). *Photosynthesis.* Vol. 8 of *The Biochemistry of Plants: A Comprehensive Treatise.* Stumpf, P. K., and Conn, E. E., eds. (New York: Academic Press).

Heber, U., and Krause, G. H. (1980). What is the physiological role of photorespiration? *Trends Biochem. Sci.* 5:32–34.

Staehlin, L. A., and Arntzen, C. J., eds. (1986). *Photosynthesis III: Photosynthetic Membranes and Light Harvesting Systems.* Vol. 19 of *Encyclopedia of Plant Physiology* (New York: Springer-Verlag).

Woodrow, I. E., and Berry, J. A. (1988). Enzymatic regulation of photosynthetic CO_2 fixation in C_3 plants. *Annu. Rev. Plant. Physiol. Mol. Biol.* 39:533–594

Youvan, D. C., and Marrs, B. L. (1987). Molecular mechanisms of photosynthesis. *Sci. Am.* 256(6):42–48.

17

Lipid Metabolism

Consider the flight of the migratory bird: some species can fly 1000 miles nonstop at 25 miles per hour. Sustained work such as this can be achieved only at the expense of lipid fuel; glycogen, the other important depot of stored energy, could supply ATP for muscle contraction for at best a fraction of an hour. Other instances of prolonged, intense output of work, such as the migrations of locusts and the record-chasing of marathon runners, are similarly fueled by the metabolism of triacylglycerols. In this chapter, we consider the pathways by which these molecules are oxidized for energy and synthesized for storage. We also consider the metabolism of membrane lipids, such as phospholipids, sphingolipids, and cholesterol, and the metabolism of eicosanoids, which are signal molecules. We begin our examination of lipid metabolism with dietary uptake as it occurs in mammals.

17·1 Absorption of Dietary Lipids Occurs in the Small Intestine in Mammals

Most of the lipids in the diet of mammals are triacylglycerols, with smaller amounts in the form of phospholipids and cholesterol. Digestion of dietary lipids begins when they mix with *bile salts* in the intestine to form an emulsion. Bile salts (Figure 17·1, next page) are cholesterol derivatives formed in the liver and secreted into the intestine from the gall bladder. An emulsion is a suspension of oil in water, like

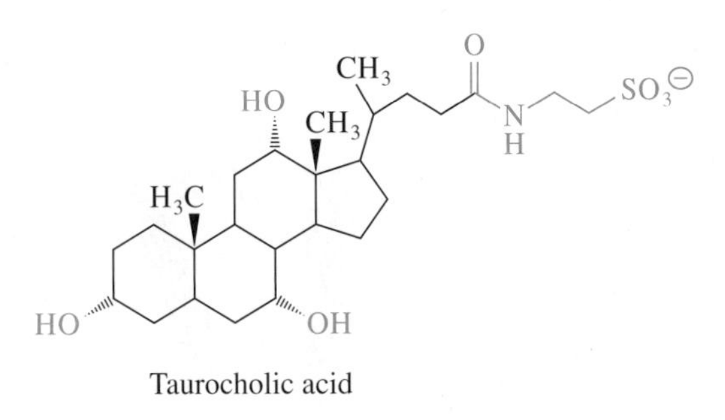

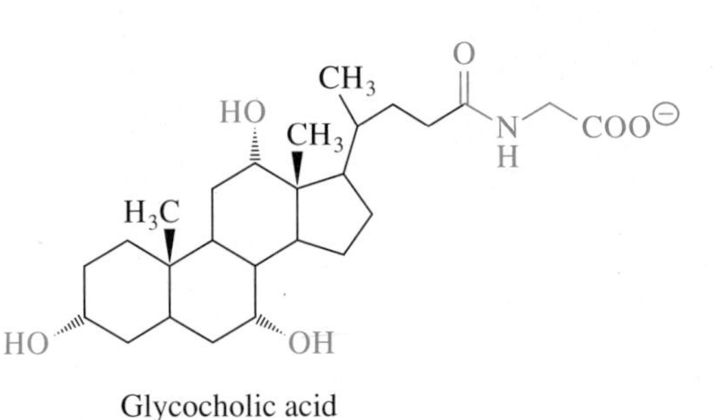

Figure 17·1
Bile salts. The cholesterol derivatives taurocholic acid and glycocholic acid are the most abundant bile salts in humans. Biles salts are amphipathic, providing an interphase between oil and water phases. The hydrophilic parts of the molecules, which in three dimensions would extend below the plane of the page, are shown in blue. The other parts of the molecules are hydrophobic.

the oil-in-vinegar suspension of salad dressing. Triacylglycerols present in the emulsion are subjected to enzymatic degradation by pancreatic lipase. A small protein named colipase assists in the binding of the water-soluble pancreatic lipase to the lipid-water interface. Pancreatic lipase catalyzes the hydrolysis of triacylglycerol molecules at the C-1 and C-3 positions, producing free fatty acids, the intermediates, 1,2 diacylglycerol and 2,3 diacylglycerol, and finally, 2-monoacylglycerol (Figure 17·2). Free fatty acids and monoacylglycerols mix with bile salts to form mixed micelles (micelles formed from more than one type of amphiphile). The fatty acids and monoacylglycerols are then taken up into the cells lining the inside of the intestine. The transfer of lipid molecules from extracellular micelles to the inside of intestinal cells is not well understood. There appears to be no carrier of fatty acids or monoacylglycerols for transport through the plasma membrane. After the uptake of fatty acids and monoacylglycerols, most of the bile salts (over 90%) are recirculated among intestine, blood, and liver in what is called enterohepatic circulation. This routing involves several membrane-bound transporters in both the

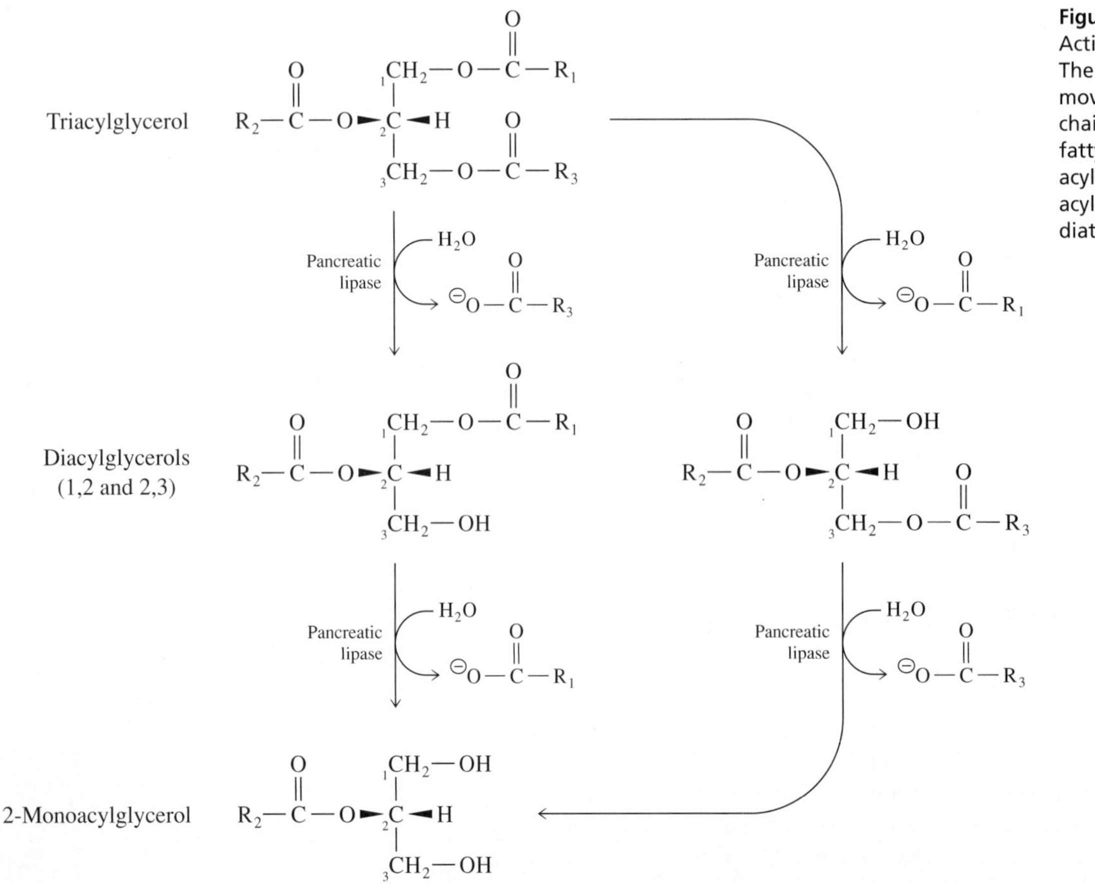

Figure 17·2
Action of pancreatic lipase. The enzyme catalyzes removal of C-1 and C-3 acyl chains, producing free fatty acids and 2-monoacylglycerol. 1,2 and 2,3 diacylglycerols are intermediates in the pathway.

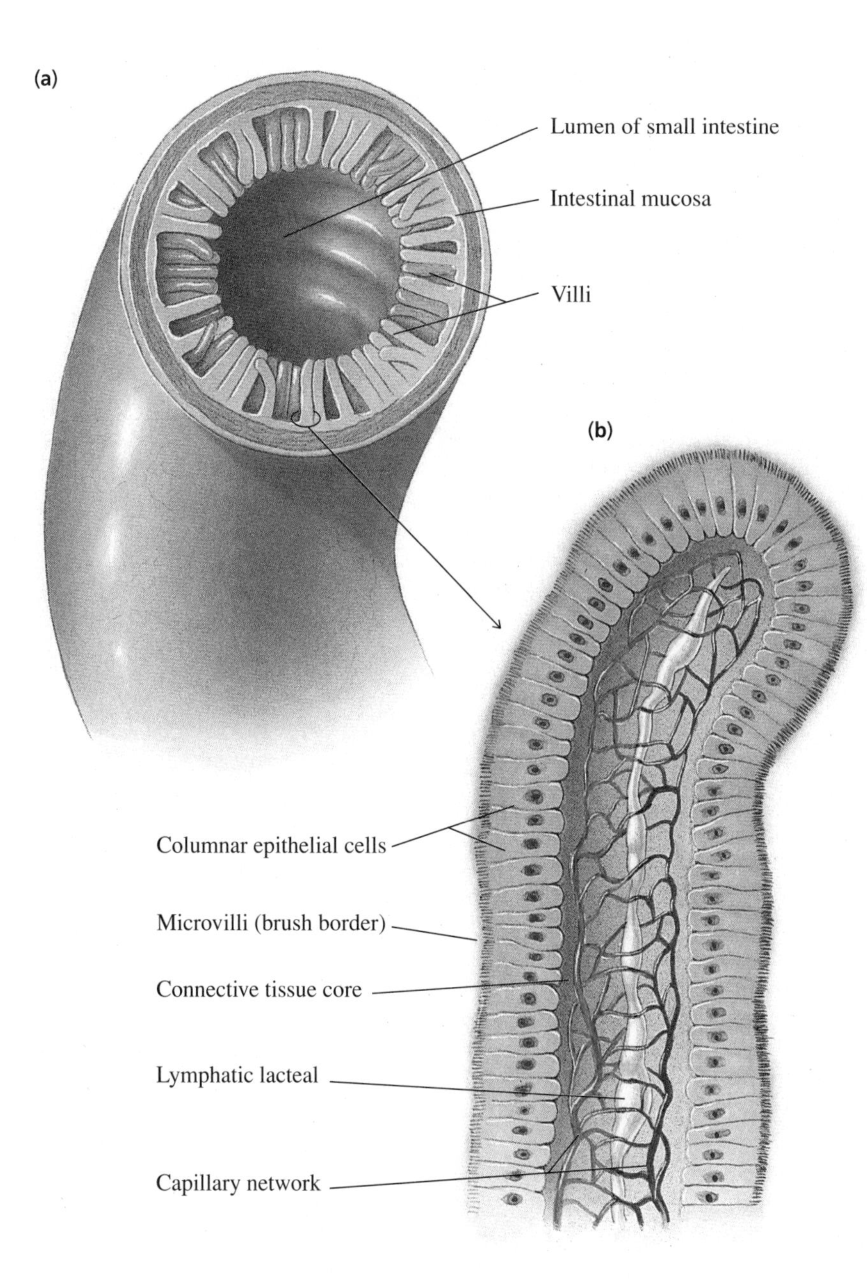

Figure 17·3
Villi of the small intestine. **(a)** Cross section of the small intestine. Villi of the intestinal mucosa project into the lumen of the small intestine. **(b)** Longitudinal section of a part of a villus, emphasizing the presence of columnar epithelial cells and microvilli. Each villus consists of a layer of columnar epithelial cells and a central core of connective tissue. Microvilli, consisting of infolded membranes that provide a large surface area, project from columnar epithelial cells into the lumen of the small intestine, collectively forming an absorptive brush border. Embedded in the connective tissue core are the lymphatic lacteal and the blood capillaries. Fatty acids and monoacylglycerols diffuse across the membranes of the microvilli and into the epithelial cells, where they are converted into triacylglycerols and packaged into chylomicrons for transport to the lymphatic lacteal. (Illustrator: Lisa Shoemaker.)

intestinal and liver cells. The small amounts of bile salts that are not extracted from the intestine are lost to the feces. This elimination of bile salts represents the only route for disposal of their precursor, cholesterol.

The villi of the intestine are specialized for the absorption of foods (Figure 17·3). The large number of villi increase the intestinal surface area, which is further increased by microvilli, highly infolded cell-surface membranes forming a brush border on the surface of villi. In the intestinal cells, fatty acids are converted to fatty acyl CoA molecules and then to triacylglycerols, which combine with absorbed cholesterol and specific proteins to form lipoprotein aggregates known as chylomicrons (Section 9·8). Chylomicrons transported out of intestinal cells pass through the lymphatic system and into the bloodstream.

The fate of phospholipids from the diet is similar to that of triacylglycerols. Pancreatic phospholipases secreted into the intestine catalyze removal of fatty acids from phospholipids present in food. These and other phospholipases are classified according to the ester subject to attack (Figure 17·4). The major phospholipase present in the pancreatic secretion is phospholipase A_2, discussed in Box 9·1, page 9·20.

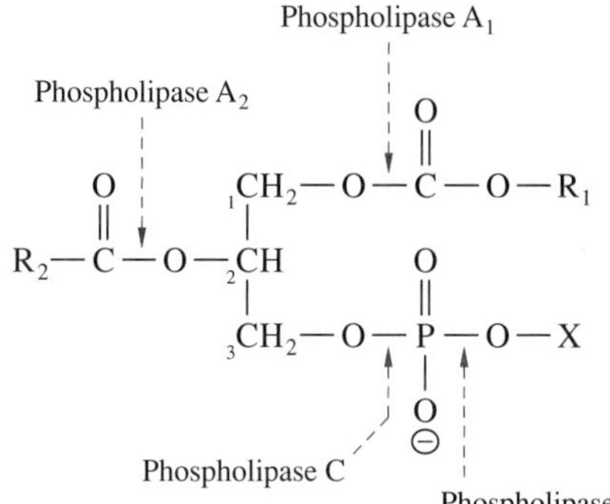

Figure 17·4
Specificities of four phospholipases. Cleavage by phospholipases A_1 or A_2 produces a lysophospholipid and a free fatty acid; cleavage by phospholipase C produces a diacylglycerol and a phosphorylated alcohol such as phosphocholine; cleavage by phospholipase D produces a phosphatidate and an alcohol.

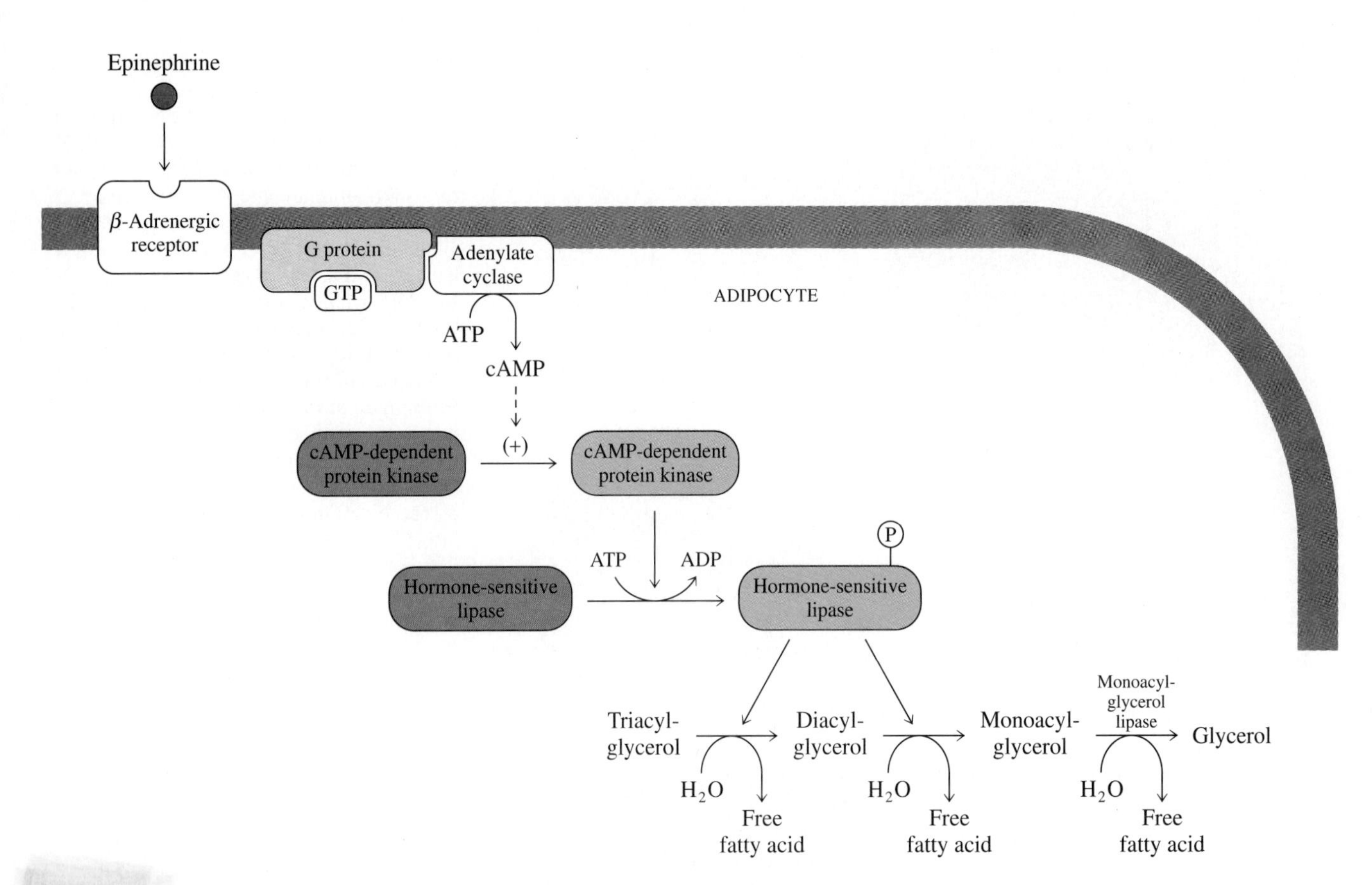

Figure 17·5
Hormonal regulation of triacylglycerol degradation. In adipocytes, cAMP-dependent protein kinase catalyzes the phosphorylation and resulting activation of hormone-sensitive lipase, which catalyzes hydrolysis of triacylglycerols to monoacylglycerols and free fatty acids. Hydrolysis of monoacylglycerols is catalyzed by monoacylglycerol lipase.

17·2 Fatty Acids Are Released from Adipocytes, the Main Storage Depots of Fat

Esters of fatty acids are stored in adipocytes. An adipocyte is essentially a fat droplet surrounded by a thin shell of cytosol in which the nucleus and other organelles are suspended. When the level of epinephrine in blood is elevated (as occurs during fasting or exercise), activation of the β-adrenergic receptors of adipocytes leads to the activation of adenylate cyclase, which catalyzes the production of cyclic AMP (cAMP). Elevated levels of cAMP in the cell lead to the activation of cAMP-dependent protein kinase. This activation sequence parallels the β-adrenergic regulation of glycogen metabolism, examined in Section 15·3. In adipocytes, cAMP-dependent protein kinase catalyzes the phosphorylation and resulting activation of hormone-sensitive lipase. This enzyme catalyzes the conversion of triacylglycerols to free fatty acids and monoacylglycerols. Although hormone-sensitive lipase can also catalyze conversion of monoacylglycerols to glycerol and free fatty acids, a more specific and more active monoacylglycerol lipase is present, which probably accounts for most of this catalytic activity. The sequence of triacylglycerol degradation is diagrammed in Figure 17·5. Glycerol and free fatty acids diffuse through the plasma membrane and enter the bloodstream. Glycerol is metabolized by the liver, where most of it converted to glucose via gluconeogenesis (Section 15·5). Fatty acids, which are poorly soluble in aqueous solution, travel through blood bound to serum albumin, one of the major serum proteins—serum albumin accounts for half of the total protein of serum. Fatty acids are carried to a number of tissues, such as heart, skeletal muscle, and liver, where they are oxidized.

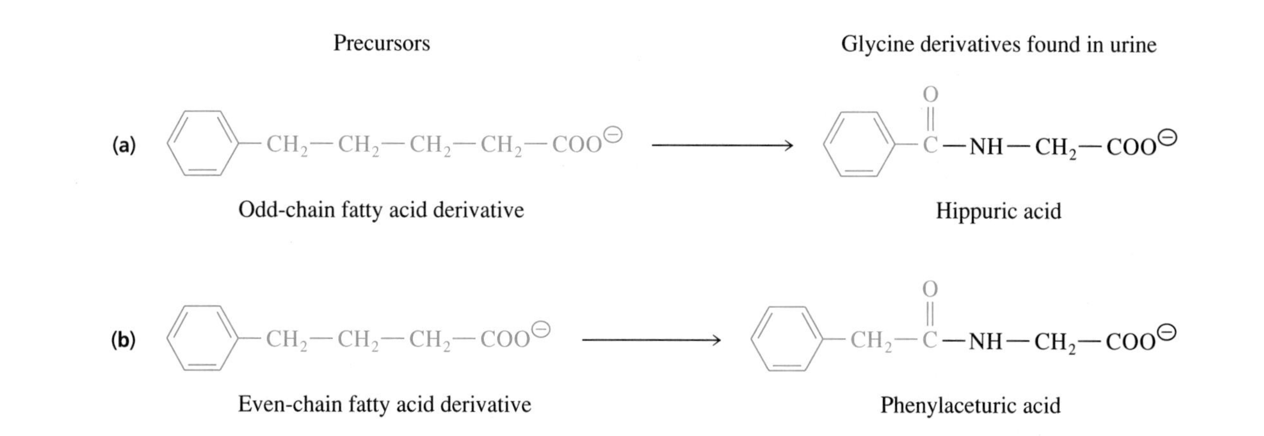

Figure 17·6
The Knoop experiment. **(a)** Dogs fed a phenylated odd-chain fatty acid derivative always produced a glycine derivative with a single carbon of the acyl chain remaining. **(b)** When an even-chain fatty acid derivative was fed to the dogs, two carbons of the acyl chain were left. Together, the results suggest the removal of two-carbon units during fatty acid oxidation.

17·3 Fatty Acids Are Oxidized by Removal of Two-Carbon Groups

In 1904, Franz Knoop conducted a classic biochemical experiment that revealed the pattern of fatty acid oxidation and led the way to the complete elucidation of the pathway of fatty acid degradation, commonly called the *β-oxidation pathway*. Knoop fed dogs fatty acid derivatives that contained phenyl groups attached to the terminal carbon and isolated phenyl compounds in the dogs' urine. Knoop's replacement of a hydrogen by a phenyl group, which allowed detection and isolation of the product, was the earliest metabolic labelling study. When derivatives of fatty acids containing an odd number of carbon atoms (called odd-chain fatty acids) were ingested, hippuric acid, the conjugate of benzoate and glycine, was detected (Figure 17·6a). When an even-chain fatty acid derivative was ingested, phenylaceturic acid, the conjugate of phenylacetate and glycine, was recovered (Figure 17·6b). Clearly, the acyl chain was not degraded one carbon at a time, or both types of fatty acids would have produced hippuric acid. If degradation proceeded by groups of more than two carbons, correspondingly larger phenylated derivatives would be produced. Knoop therefore proposed that fatty acids are shortened two carbons at a time by oxidation at the β-carbon.

The two-carbon fragments produced during β oxidation of fatty acids are now known to be transferred to coenzyme A to form acetyl CoA. The process by which fatty acids are degraded to produce acetyl CoA in cells can be divided into three stages: activation of fatty acids in the cytosol, transport into mitochondria, and degradation in two-carbon increments. We shall consider these stages in detail, focussing on the oxidation of an even-chain, saturated fatty acid. Specific reactions necessary for the oxidation of odd-chain and unsaturated fatty acids are considered in subsequent sections.

A. Fatty Acids Are Activated by Esterification to Coenzyme A

Upon entry to the cytosol, fatty acids are activated by conversion to thioesters of coenzyme A by the action of acyl-CoA synthetase:

$$R{-}COO^{\ominus} + HS\text{-}CoA \xrightarrow[\text{Acyl-CoA synthetase}]{ATP \;\to\; PP_i + AMP} R{-}\overset{O}{\overset{\|}{C}}{-}S\text{-}CoA \tag{17·1}$$

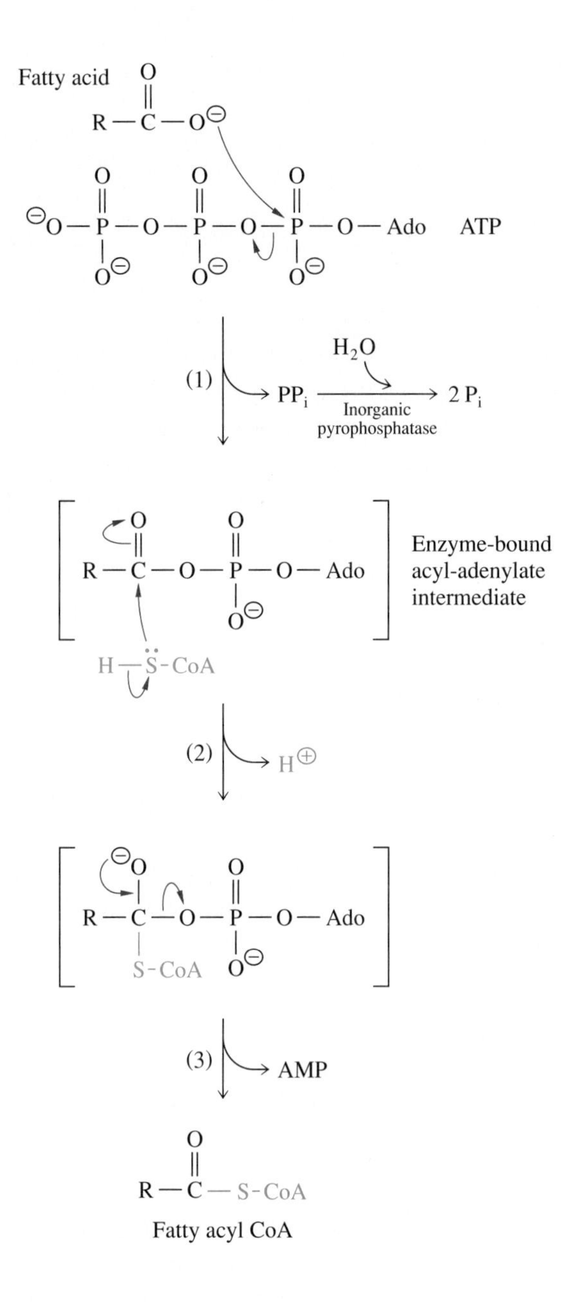

Figure 17·7
Mechanism of acyl-CoA synthetase. (1) An enzyme-bound acyl-adenylate intermediate is formed from a long-chain fatty acid and ATP. Pyrophosphate released in the process is hydrolyzed to two molecules of inorganic phosphate, with release of a large amount of free energy. (2) The thiol of CoASH reacts with the carbonyl carbon of the intermediate. (3) Cleavage releases AMP and fatty acyl CoA. ("Ado" represents adenosine.)

The pyrophosphate product is then hydrolyzed by the action of inorganic pyrophosphatase. Thus, two phosphoanhydride bonds, or two ATP equivalents, are consumed to form the CoA thioesters of fatty acids.

The mechanism of the activation reaction involves an acyl-adenylate intermediate formed by the reaction of a fatty acid with ATP (Figure 17·7). Nucleophilic attack by the sulfur atom of coenzyme A on the carbonyl carbon of the acyl group leads to release of AMP and the thioester fatty acyl CoA.

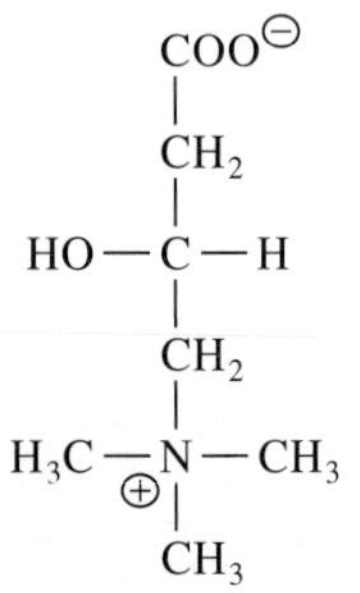

Figure 17·8
L-Carnitine.

B. Fatty Acyl CoA Is Transported into the Mitochondrial Matrix by a Shuttle System

Fatty acyl CoA formed in the cytosol cannot cross the mitochondrial inner membrane and enter the mitochondrial matrix, where the reactions of β oxidation occur. Transport is accomplished by an elaborate shuttle system initiated by esterification of the acyl group to L-carnitine (Figure 17·8). The shuttle process is carried out by

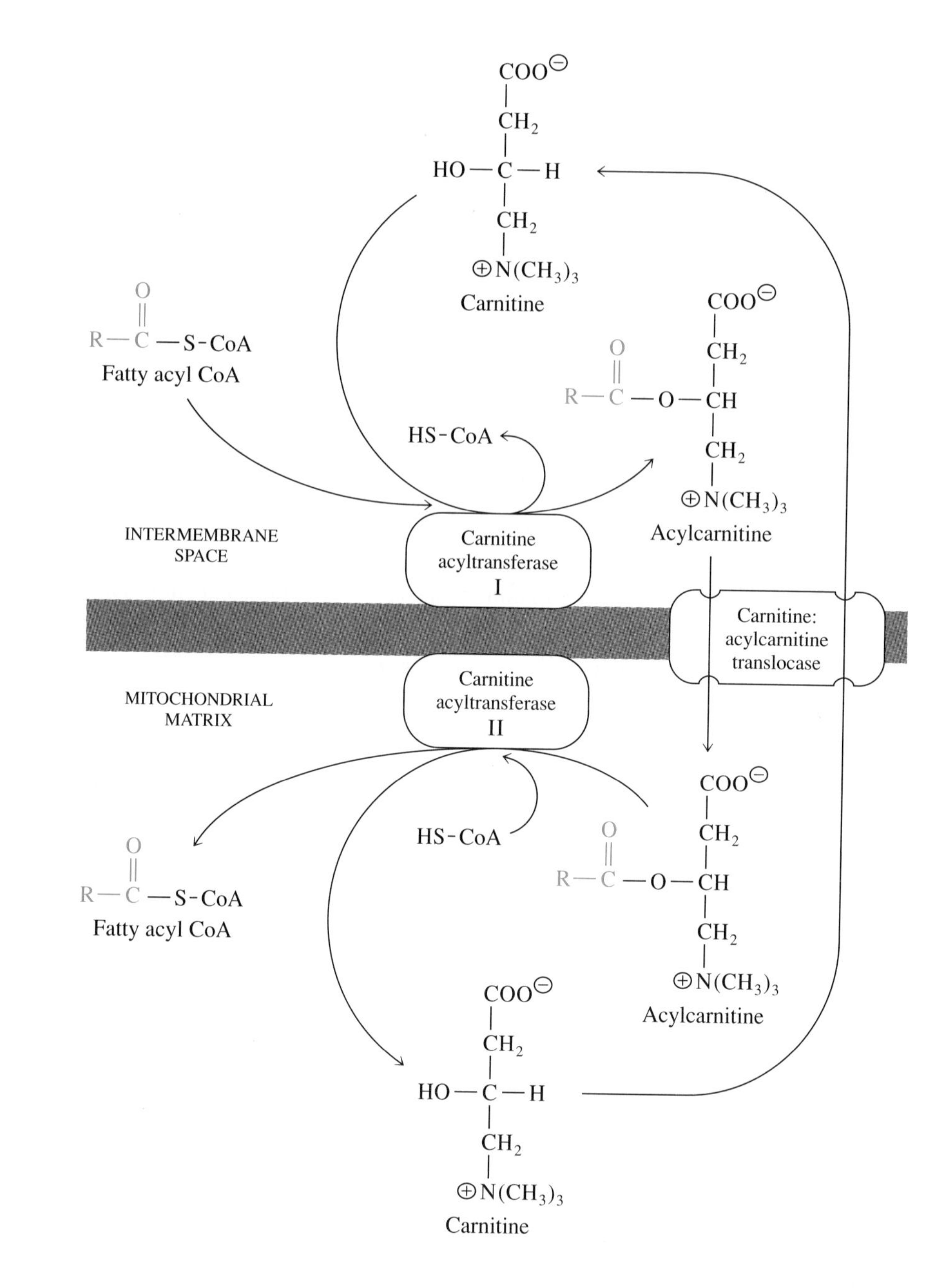

Figure 17·9
Carnitine shuttle system. In the intermembrane space, a fatty acyl CoA molecule reacts with carnitine to form acylcarnitine and CoASH in a reaction catalyzed by carnitine acyltransferase I. Acylcarnitine is transported across the inner mitochondrial membrane by carnitine:acylcarnitine translocase via antiport with carnitine from the matrix. In the matrix, membrane-associated carnitine acyltransferase II catalyzes the regeneration of fatty acyl CoA from acylcarnitine and CoASH.

two acyltransferases located on opposite sides of the mitochondrial inner membrane and a translocase protein embedded in the membrane (Figure 17·9). First, the fatty acyl CoA is converted to acylcarnitine in a reaction catalyzed by carnitine acyltransferase I (CAT I). This reaction is a key site of control of fatty acid oxidation, as we shall see in Section 17·6. Acylcarnitine then crosses into the mitochondrial matrix in exchange for free carnitine via the carnitine:acylcarnitine translocase. In the mitochondrial matrix, carnitine acyltransferase II (CAT II), an isozyme of CAT I, catalyzes the reverse of the reaction catalyzed by CAT I. Overall, the shuttle system results in removal of cytosolic fatty acyl CoA and generation of fatty acyl CoA in the mitochondrial matrix.

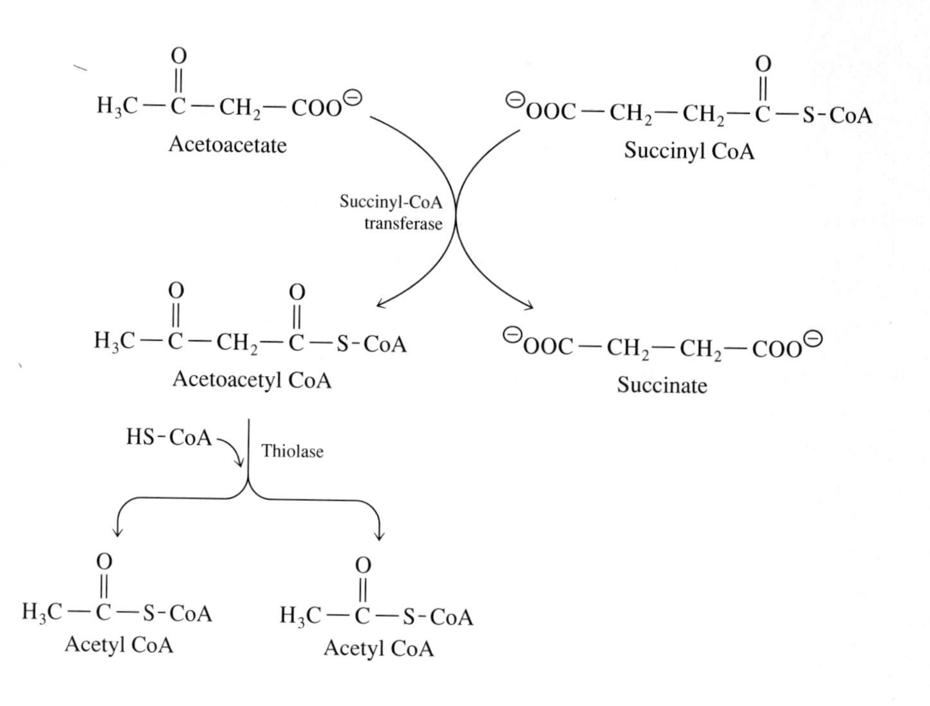

Figure 17·16
Conversion of acetoacetate to acetyl CoA. Succinyl-CoA transferase catalyzes the conversion of acetoacetate and succinyl CoA to acetoacetyl CoA and succinate. Acetoacetyl CoA is then converted to two molecules of acetyl CoA by the action of thiolase.

acetoacetyl CoA in a reaction catalyzed by succinyl-CoA transferase (Figure 17·16). This reaction siphons some of the succinyl CoA from the citric acid cycle. Energy that would normally be captured as GTP in the substrate-level phosphorylation catalyzed by succinyl-CoA synthetase (Section 13·8) is used instead to activate acetoacetate to its CoA ester. Acetoacetyl CoA is then converted to two molecules of acetyl CoA by the action of thiolase.

17·8 Fatty Acid Synthesis Occurs by a Pathway Different from Fatty Acid Oxidation

Fatty acid synthesis in mammalian cells occurs largely in liver and adipocytes and, to a lesser extent, in specialized cells under specific conditions, such as mammary gland cells during lactation. Fatty acids are synthesized and degraded by completely separate pathways. The key differences between the two processes in mammals are summarized in Table 17·1. In eukaryotes, fatty acid oxidation takes place principally in mitochondria, whereas synthesis takes place in the cytosol. The active thioesters in fatty acid oxidation are CoA derivatives, whereas in fatty acid synthesis the intermediates are bound as thioesters to acyl carrier protein (ACP). Separate

Table 17·1 Comparison of fatty acid oxidation and synthesis in mammals

	Oxidation	Synthesis
Localization	Mitochondria	Cytosol
Acyl carrier	CoA	Acyl carrier protein (ACP)
Carbon units	C_2	C_2
Acceptor/donor	CoA (C_2)	Malonyl CoA (C_3, donor reaction evolves CO_2)
Mobile redox cofactors	NAD⊕, ETF	NADPH
Organization of enzymes	Separate enzymes	A multifunctional enzyme

enzymes catalyze the reactions of oxidation, whereas a multifunctional protein of two identical polypeptide chains catalyzes most of the biosynthetic reactions in mammals. Synthesis and degradation both proceed in two-carbon steps. However, oxidation results in a two-carbon product, acetyl CoA, whereas synthesis requires a three-carbon substrate, malonyl CoA, which transfers a two-carbon unit to the growing chain. In the process, CO_2 is released. Finally, reducing power for synthesis is supplied by NADPH, whereas oxidation is dependent upon $NAD^{\oplus}$ and ETF.

Fatty acid synthesis in eukaryotes can be divided into three stages. In the first stage, mitochondrial acetyl CoA is transported into the cytosol. Next, carboxylation of acetyl CoA generates malonyl CoA, the substrate for the elongation reactions that extend the fatty acyl chain. The carboxylation of acetyl CoA is the regulated step of fatty acid synthesis. Finally, the actual assembly of the fatty acid chain is carried out by fatty acid synthase.

17·9 The Citrate Transport System Provides Acetyl CoA to the Cytosol

Fatty acid synthesis in the cytosol of eukaryotes requires that acetyl CoA be supplied from mitochondria, where it is produced. In the fed state, fatty acids are synthesized from acetyl CoA produced by carbohydrate metabolism. The export of acetyl CoA from mitochondria is accomplished by the *citrate transport system*. The transport is indirect, requiring a series of steps (Figure 17·17). In the first step, mitochondrial acetyl CoA condenses with oxaloacetate in a reaction catalyzed by citrate synthase. This is also the first step of the citric acid cycle. Next, citrate is transported out of the mitochondrion via the citric acid:dicarboxylic acid carrier; the dicarboxylate anion may be α-ketoglutarate or malate. In the cytosol, citrate is cleaved to form oxaloacetate and acetyl CoA in an ATP-requiring reaction catalyzed by citrate lyase.

Figure 17·17
The citrate transport system. The system achieves net transport of acetyl CoA from the mitochondrion to the cytosol and net conversion of cytosolic NADH into NADPH. Two molecules of ATP are expended for each round of the cyclic pathway.

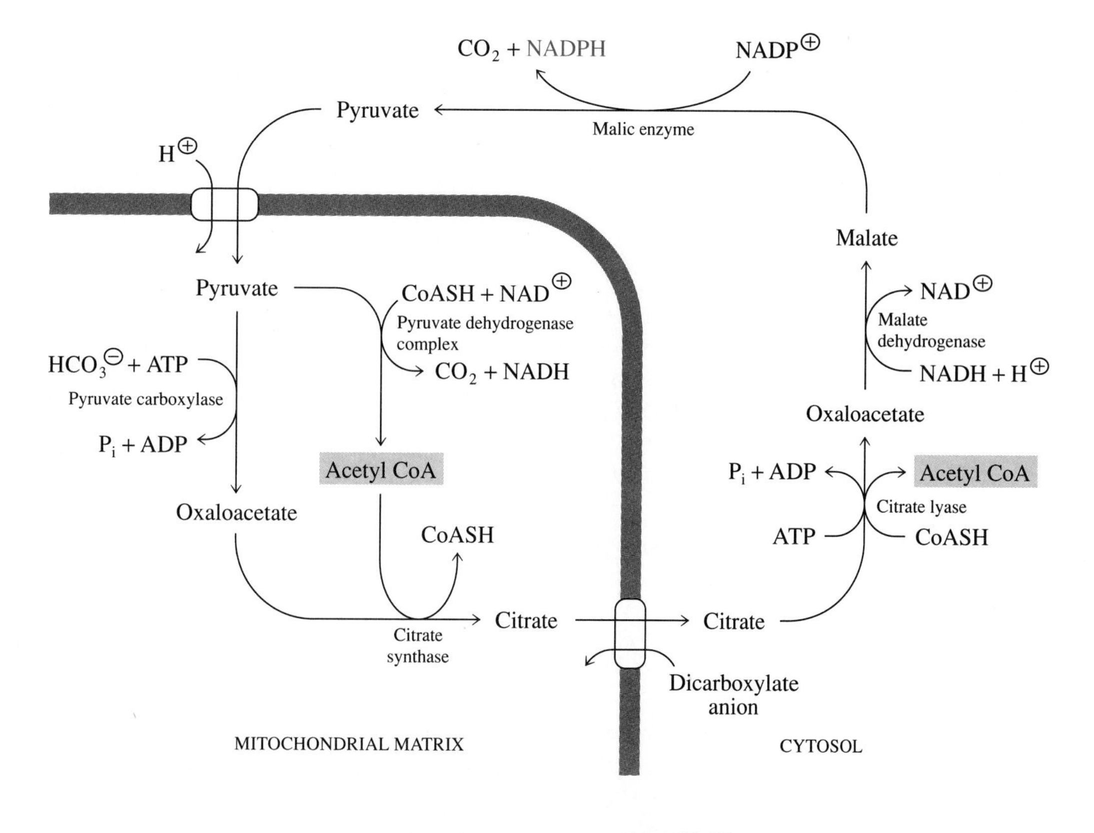

Once cytosolic acetyl CoA has been generated, further steps are required to return the remaining carbon of the original citrate to the mitochondrion. Cytosolic malate dehydrogenase, an isozyme of mitochondrial malate dehydrogenase, catalyzes the reduction of oxaloacetate to malate, with conversion of NADH to $NAD^{\oplus}$. Next, malate is decarboxylated to pyruvate in a reaction catalyzed by malic enzyme, with reduction of $NADP^{\oplus}$ to NADPH. With this pair of reactions, the citrate transport system serves not only to transport acetyl CoA from the mitochondrion to the cytosol but also to generate cytosolic NADPH. Reducing equivalents from NADPH are required for subsequent steps in fatty acid synthesis. The newly formed pyruvate enters the mitochondrion via pyruvate translocase. Once in the mitochondrion, pyruvate can be carboxylated to form oxaloacetate in an ATP-requiring step, or it can be converted to acetyl CoA by the action of the pyruvate dehydrogenase complex, thus completing the cycle of reactions. Cellular conditions determine the fate of pyruvate.

A second source of NADPH for fatty acid synthesis is the pentose phosphate pathway (Section 15·7), which contributes about half the NADPH required for the overall synthesis of fatty acids, the rest arising from the malic enzyme reaction of the citrate transport system.

17·10 Acetyl CoA Is Carboxylated to Form Malonyl CoA in a Regulated Step

The second stage of fatty acid synthesis consists of the carboxylation of acetyl CoA in the cytosol to form malonyl CoA in a reaction catalyzed by the biotin-dependent enzyme acetyl-CoA carboxylase. This reaction is the key regulatory step of fatty acid synthesis.

$$\underset{\text{Bicarbonate}}{HCO_3^{\ominus}} + \underset{\text{Acetyl CoA}}{H_3C-\overset{\overset{\large O}{\|}}{C}-S\text{-}CoA} \xrightarrow[\text{ATP} \;\; \text{ADP} + \text{P}_i]{\text{Acetyl-CoA carboxylase, Biotin}} \underset{\text{Malonyl CoA}}{{}^{\ominus}OOC-CH_2-\overset{\overset{\large O}{\|}}{C}-S\text{-}CoA} \quad (17\cdot5)$$

In bacteria, three separate protein subunits carry out this conversion: a carrier protein termed *biotin carboxyl carrier protein* (BCCP) and two enzymes, a biotin carboxylase and a transcarboxylase. In animals and yeasts, all of these activities are contained within a single polypeptide chain (Table 17·2).

Carboxylation of acetyl CoA proceeds in two steps. The ATP-dependent activation of $HCO_3^{\ominus}$ forms carboxybiotin:

$$HCO_3^{\ominus} + \text{Enz-Biotin} + \text{ATP} \rightleftharpoons \text{Enz-Biotin}-COO^{\ominus} + \text{ADP} + P_i \quad (17\cdot6)$$

Table 17·2 Components of acetyl-CoA carboxylase

E. coli	Animals and yeasts
Biotin carboxyl carrier protein 2 subunits, MW 22 500 each	Two copies of one polypeptide chain containing all catalytic sites 2 subunits, MW 230 000–290 000 each
Biotin carboxylase 2 subunits, MW 51 000 each	Total MW of complex: 460 000–580 000
Transcarboxylase 2 subunits, MW 30 000 each 2 subunits, MW 35 000 each	
Total MW of complex: 277 000	

This reaction is followed by the transfer of activated CO_2 to acetyl CoA:

$$\text{Enz-Biotin}-\text{COO}^{\ominus} + \underset{\text{Acetyl CoA}}{\text{H}_3\text{C}-\overset{\overset{\displaystyle\text{O}}{\|}}{\text{C}}-\text{S-CoA}} \longrightarrow \text{Enz-Biotin} + \underset{\text{Malonyl CoA}}{{}^{\ominus}\text{OOC}-\text{CH}_2-\overset{\overset{\displaystyle\text{O}}{\|}}{\text{C}}-\text{S-CoA}} \qquad (17\cdot7)$$

Acetyl-CoA carboxylase is the key regulatory enzyme of fatty acid synthesis. In vitro, citrate is an activator of the enzyme. However, it has not been established that citrate plays a role in the cellular regulation of this enzyme (Box 17·1). It is known that hormonal regulation is exerted through phosphorylation of acetyl-CoA carboxylase. Glucagon stimulates phosphorylation and resulting inactivation of the enzyme in liver, and epinephrine stimulates phosphorylation in adipocytes. Inactivation of acetyl-CoA carboxylase depresses the rate of fatty acid synthesis. It is also known that insulin stimulates phosphorylation of acetyl-CoA carboxylase, but apparently at a different site on the enzyme. The relationship between insulin-dependent phosphorylation and the activation state of the enzyme remains unclear.

Acetyl CoA-carboxylase is further regulated by fatty acyl CoA, which, in sufficient concentrations, allosterically inhibits the enzyme and, at lower concentrations, increases the activity of a protein kinase that catalyzes phosphorylation and the resulting inhibition of the enzyme. The ability of fatty acid derivatives to regulate acetyl-CoA carboxylase is physiologically appropriate—an increased concentration of fatty acids causes a decrease in the rate of the first committed step of fatty acid synthesis.

Box 17·1 In Vitro Versus In Vivo Revisited: Citrate and the Activation of Acetyl-CoA Carboxylase

In Box 12·2, we discussed how regulatory properties observed in vitro with purified lactate dehydrogenase isozymes led to plausible but probably incorrect conclusions about the regulation of the enzymes in vivo. The activation of acetyl-CoA carboxylase by citrate in vitro led early investigators to propose a feed-forward mechanism for the activation of fatty acid synthesis, since citrate in the cytosol is a precursor of acetyl CoA, the substrate for acetyl-CoA carboxylase. Another finding regarded as support for this mechanism was the observation that when purified acetyl-CoA carboxylase was exposed to citrate in vitro, polymerization of enzyme molecules into long chains occurred, which appeared to coincide with activation of the enzyme. Electron micrographs of polymerized enzyme were regarded as impressive visual evidence of an important control mechanism (Figure 1). The evidence was dramatic and unusual, and the conclusions about the regulation of acetyl-CoA carboxylase became quite well known.

Recently, experiments with liver cells have shown that changes in the concentration of citrate in vivo do not correspond to changes in the rate of fatty acid synthesis. In fact, since citrate lyase catalyzes a near-equilibrium reaction, whereas acetyl-CoA carboxylase catalyzes a metabolically irreversible reaction that is a control point for fatty acid synthesis, it may be predicted that stimulation of the activity of acetyl-CoA carboxylase should shift the equilibrium of the citrate lyase reaction toward product and decrease the concentration of citrate. Furthermore, recent studies have deflated the importance of the striking in vitro polymerization process. Activation of the enzyme has been shown to precede polymerization. Citrate clearly can activate the enzyme, and polymerization clearly occurs in vitro, but these phenomena are apparently not important factors in the cellular control of acetyl-CoA carboxylase.

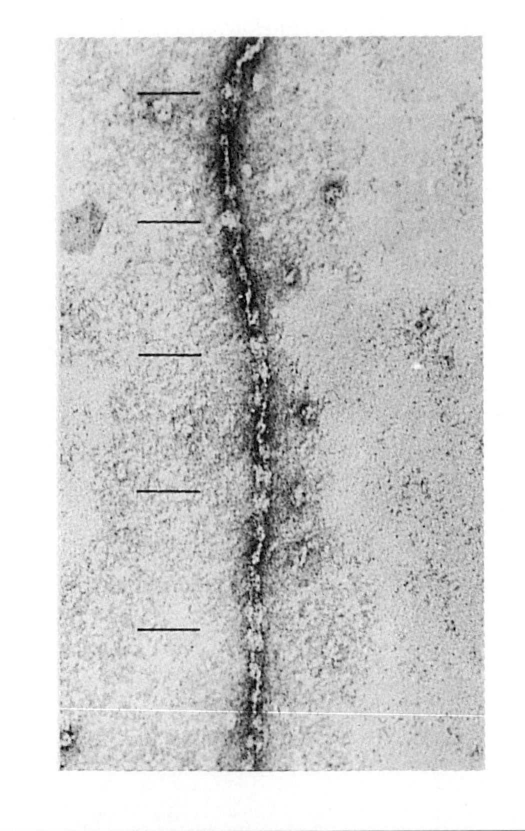

Box 17·1
Figure 1
Electron micrograph of a filamentous polymer of purified acetyl-CoA carboxylase. The physiological relevance of polymerization is in doubt. (Courtesy of M. Daniel Lane.)

17·11 The Final Steps of Fatty Acid Synthesis Are Catalyzed by a Multienzyme Pathway in *E. coli* and by a Multifunctional Enzyme in Mammals

Fatty acids are synthesized from acetyl CoA and malonyl CoA following transfer of these molecules to the prosthetic group phosphopantetheine, the same group found in coenzyme A. In *E. coli,* this prosthetic group is attached to ACP (Figure 17·18). Seven enzymes and ACP are required to synthesize fatty acids in *E. coli*. In mammals, a single polypeptide chain contains all of the catalytic activities, and the prosthetic group is incorporated within the multifunctional protein, whose active form is a dimer.

We shall first examine fatty acid formation in *E. coli* and then consider the differences imposed on the process when the catalytic activities are catalyzed by a multifunctional enzyme, as they are in the mammalian fatty acid synthase.

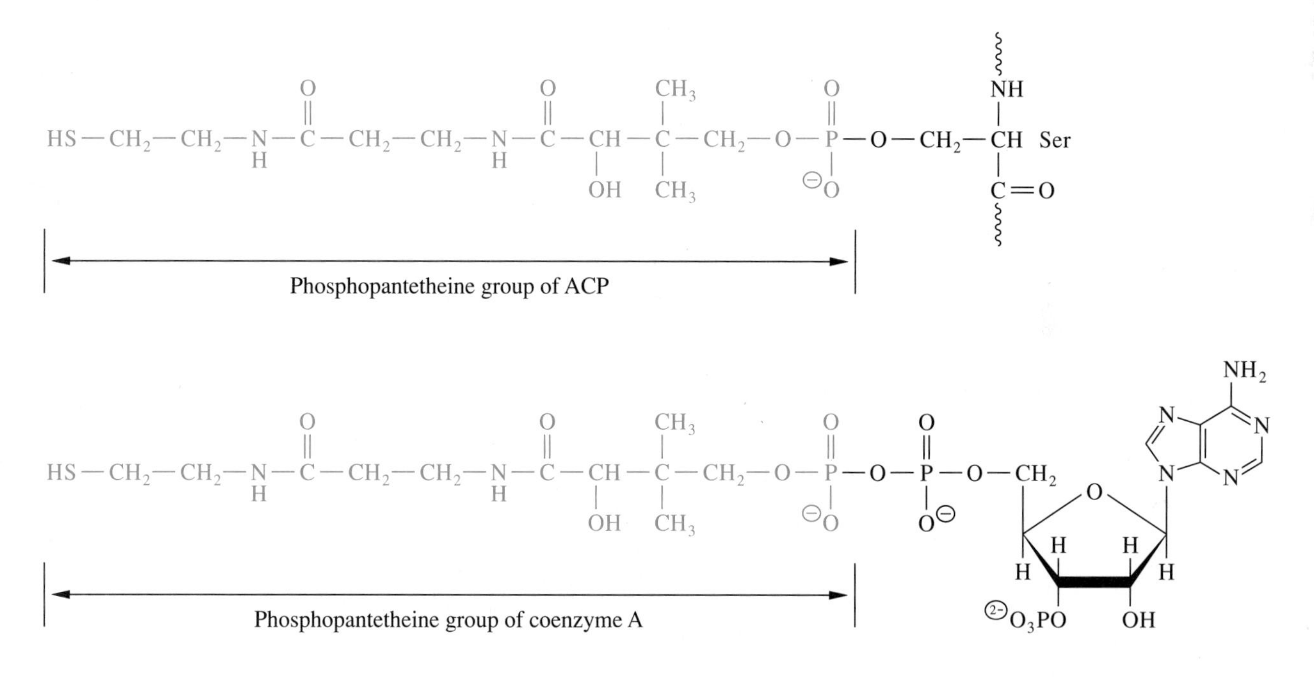

Figure 17·18
Phosphopantetheine in acyl carrier protein (ACP) and in coenzyme A.

A. Assembly of Fatty Acids Occurs in Five Stages

Fatty acid synthesis can be divided into five separate stages: *loading* of precursors by formation of two thioester derivatives, *condensation* of the precursors, *reduction, dehydration,* and a further *reduction.* The entire sequence is shown in Figure 17·19. The last four stages are repeated until a long-chain fatty acid is synthesized. Fatty acid synthesis is often called palmitate synthesis because palmitate is the preponderant product of these reactions. Other fatty acids are formed by ancillary reactions that are discussed in Section 17·12.

1. *Loading.* Two enzymes, acetyl CoA:ACP transacylase and malonyl CoA:ACP transacylase, are required for the loading steps, in which acetyl CoA and malonyl CoA are each transesterified to ACP.

2. *Condensation.* Ketoacyl-ACP synthase, also called the condensing enzyme, accepts an acetyl group from acetyl-ACP, and ACP-SH is released. Ketoacyl-ACP synthase then catalyzes transfer of the acetyl group to malonyl-ACP, evolving CO_2 from the latter substrate and forming acetoacetyl-ACP. Recall that the synthesis of malonyl CoA involves ATP-dependent carboxylation. This strategy of the cell in first carboxylating and then decarboxylating a compound used for a synthetic reaction results in a favorable free-energy change for the process at the expense of ATP consumed in the carboxylation step. A similar strategy is seen in mammalian gluconeogenesis: pyruvate (C_3) is carboxylated to form oxaloacetate (C_4), which is subsequently decarboxylated to form the C_3 molecule phosphoenolpyruvate (Section 15·4).

3. *Reduction.* The ketone of acetoacetyl-ACP is converted to an alcohol, thus forming D-β-hydroxybutyryl-ACP in an NADPH-dependent reaction catalyzed by ketoacyl-ACP reductase.

4. *Dehydration.* A dehydrase catalyzes the removal of water, with formation of a double bond.

5. *Reduction.* The product of dehydration, *trans*-butenoyl-ACP, undergoes reduction to form an acyl-ACP four carbons in length, butyryl-ACP, in a reaction catalyzed by NADPH-dependent enoyl-ACP reductase.

Synthesis continues by repeating the process from the condensation stage, with the growing acyl-ACP substituting for acetyl-ACP, and a new molecule of malonyl CoA entering with each round (Figure 17·20, page 17·24). Whereas in the condensation stage of the first round, acetyl CoA contributes the two-carbon unit that becomes the penultimate and terminal carbons of the growing molecule, in subsequent rounds malonyl CoA contributes the terminal two-carbon unit. Rounds of synthesis continue until a C_{16} palmitoyl group is formed. Palmitoyl-ACP is a substrate for thiolase, which catalyzes the formation of palmitate and ACP-SH:

$$\text{Palmitoyl-ACP} \xrightarrow[\text{Thiolase}]{H_2O} \text{Palmitate} + \text{HS-ACP} \qquad \textbf{(17·8)}$$

The overall stoichiometry of palmitate synthesis from acetyl CoA and malonyl CoA is:

$$\begin{aligned} &\text{Acetyl CoA} + 7\,\text{Malonyl CoA} + 14\,\text{NADPH} + 14\,H^{\oplus} \longrightarrow \\ &\text{Palmitate} + 7\,CO_2 + 14\,\text{NADP}^{\oplus} + 8\,\text{CoASH} + 6\,H_2O \end{aligned} \qquad \textbf{(17·9)}$$

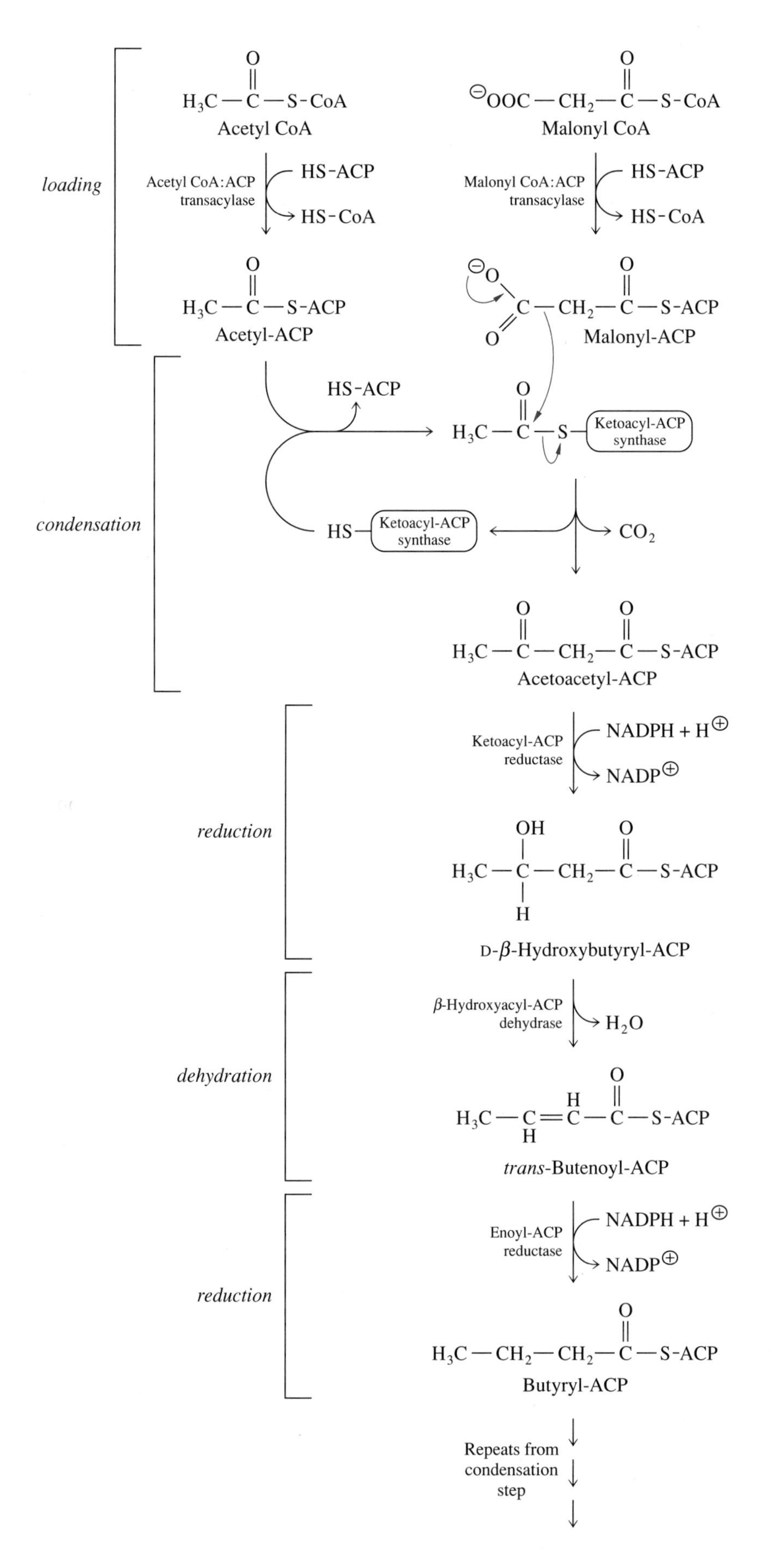

Figure 17·19
Stages in the biosynthesis of fatty acids from acetyl CoA and malonyl CoA in *E. coli*. In the loading stage, acetyl CoA and malonyl CoA are esterified to ACP. In the condensation stage, ketoacyl-ACP synthase (also called the condensing enzyme) accepts an acetyl group from acetyl-ACP, releasing ACP-SH. Ketoacyl-ACP synthase then catalyzes transfer of the acetyl group to malonyl-ACP to form acetoacetyl-ACP and CO_2. In the first reduction, acetoacetyl-ACP is converted to D-β-hydroxybutyryl-ACP in a reaction catalyzed by NADPH-dependent ketoacyl-ACP reductase. Dehydration of D-β-hydroxybutyryl-ACP results in the formation of a double bond, producing *trans*-butenoyl-ACP. Finally, reduction of *trans*-butenoyl-ACP produces butyryl-ACP. Synthesis continues by repeating the last four stages, with butyryl-ACP substituting for acetyl-ACP in the next condensation stage.

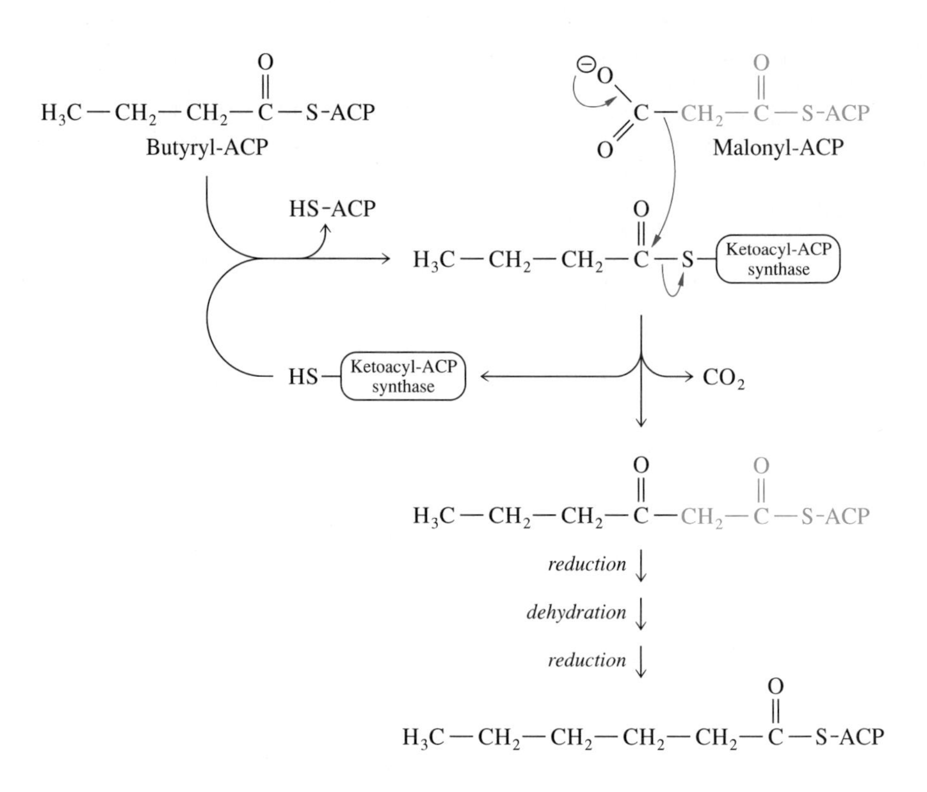

Figure 17·20
The condensation step for the second round of fatty acid biosynthesis. Butyryl-ACP formed by the reactions illustrated in Figure 17·19 reacts with ketoacyl-ACP synthase, which displaces ACP and binds the acyl group. The acyl group is then transferred to malonyl-ACP, with release of CO_2. This reaction is followed by the sequence of reduction, dehydration, and reduction outlined in Figure 17·19, with new molecules of malonyl CoA entering until a C_{16} chain is formed.

B. Mammalian Fatty Acid Synthase Is a Dimer of Identical, Multifunctional Polypeptides

The overall sequence of reactions for mammalian fatty acid synthesis is quite similar to the sequence in *E. coli,* yet the structural features of the two enzyme systems are strikingly different. Fatty acid synthase in mammals is a dimer of identical polypeptides, each containing all of the enzymatic activities of the separate enzymes that exist in *E. coli.* The two subunits of the dimer are arranged in head-to-tail fashion, as indicated schematically in Figure 17·21. Each half of the holoenzyme contains all of the activities for the synthesis of fatty acids, so that two fatty acids are formed simultaneously.

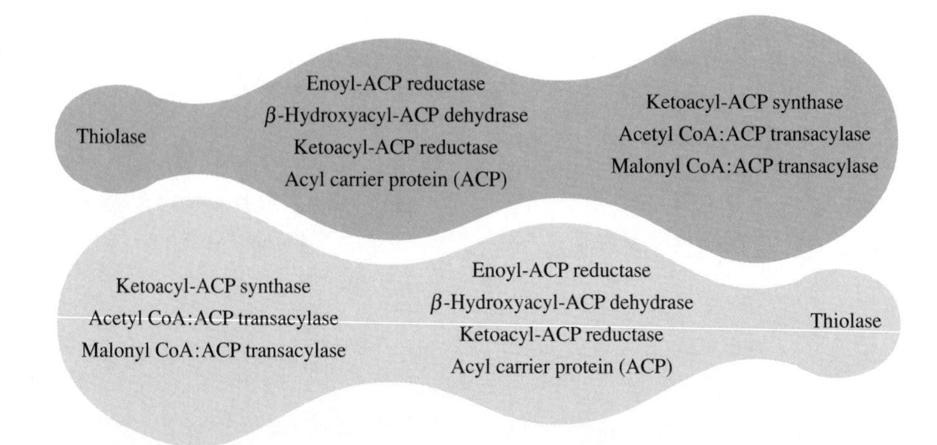

Figure 17·21
Mammalian fatty acid synthase. Each identical polypeptide chain contains seven enzymatic activities. The dimeric holoenzyme is composed of identical monomers arranged head-to-tail. Three domains can be identified in each monomer. Ketoacyl-ACP synthase, acetyl CoA:ACP transacylase, and malonyl CoA:ACP transacylase constitute a condensation domain. The remaining activities except for thiolase constitute a reduction and dehydration domain. Thiolase, in a separate domain, catalyzes release of fully formed C_{16} fatty acids.

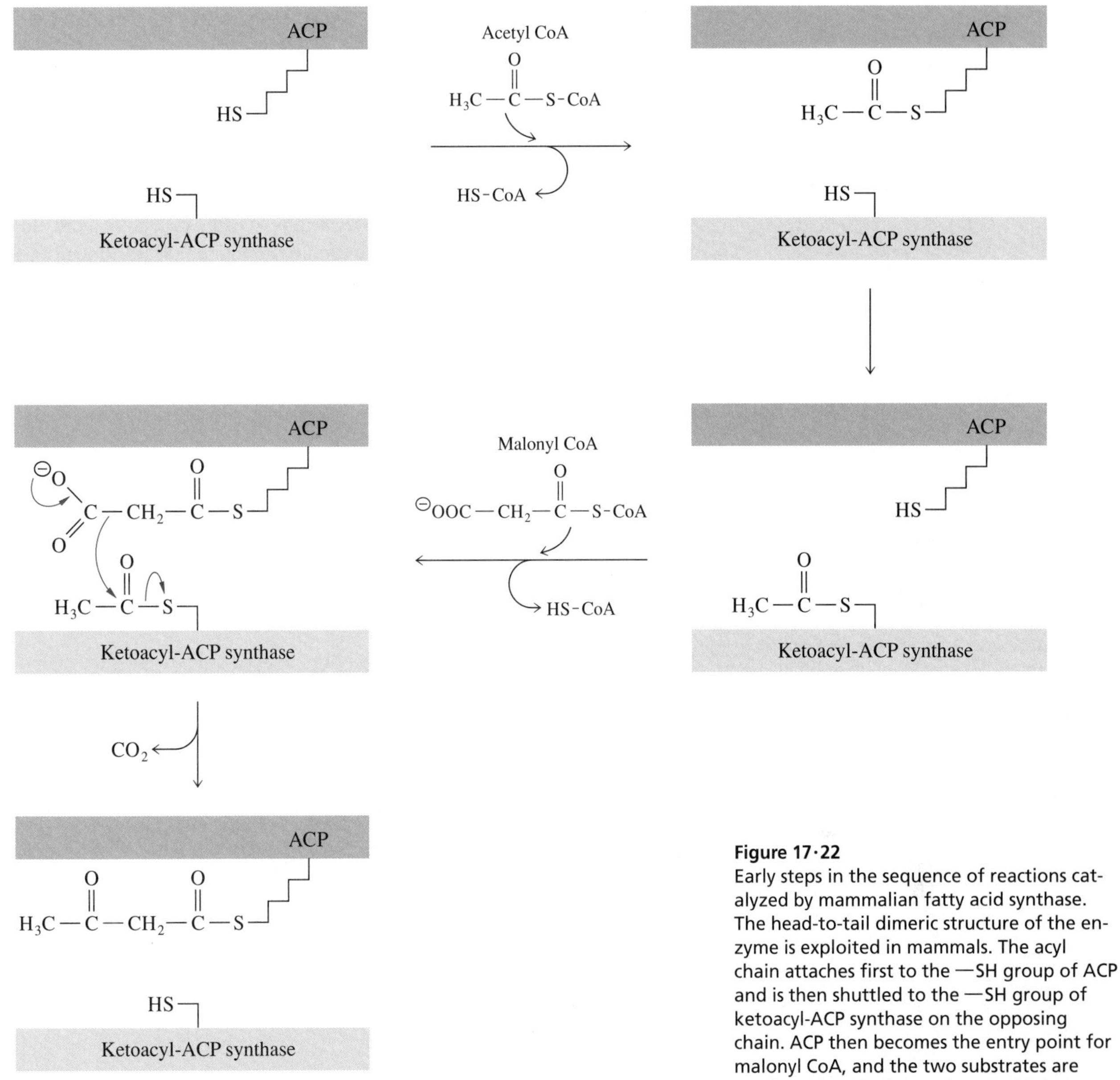

Figure 17·22
Early steps in the sequence of reactions catalyzed by mammalian fatty acid synthase. The head-to-tail dimeric structure of the enzyme is exploited in mammals. The acyl chain attaches first to the —SH group of ACP and is then shuttled to the —SH group of ketoacyl-ACP synthase on the opposing chain. ACP then becomes the entry point for malonyl CoA, and the two substrates are condensed, with the product remaining bound to ACP. Duplicate activities in the two monomers of the symmetrical dimer are active at the same time.

ACP, part of the synthase in mammals, is the site of entry for carbon destined for fatty acids. Acetyl CoA reacts with ACP and the acetyl group is then transferred to the sulfhydryl group of a cysteine residue of ketoacyl-ACP synthase, which resides on the facing subunit (Figure 17·22). ACP then binds malonyl CoA and the condensation step ensues, with the end product joined to ACP. During the remaining reduction, dehydration, and reduction steps, the carbon remains attached to ACP. For subsequent rounds of elongation, the reduced four-carbon chain attaches to the —SH group of ketoacyl-ACP synthase, ACP accepts an additional malonyl CoA, and condensation takes place as before.

17·12 Additional Enzymes Are Required for Further Chain Elongation and Desaturation

The most common product of fatty acid synthase of plants and animals is palmitate (16:0). Synthesis of a wide variety of other fatty acids requires a host of enzymes present in the endoplasmic reticulum as well as in mitochondria. For example, animal cells contain a number of desaturases that catalyze the formation of double bonds as much as nine carbons removed from the carboxyl end of a fatty acid. Only plant desaturases catalyze the desaturation of double bonds positioned farther than nine carbons from the carboxyl end. For this reason, a fatty acid such as linoleate 18:2 (9,12) is called an *essential fatty acid,* one that must be acquired in the diet of animals.

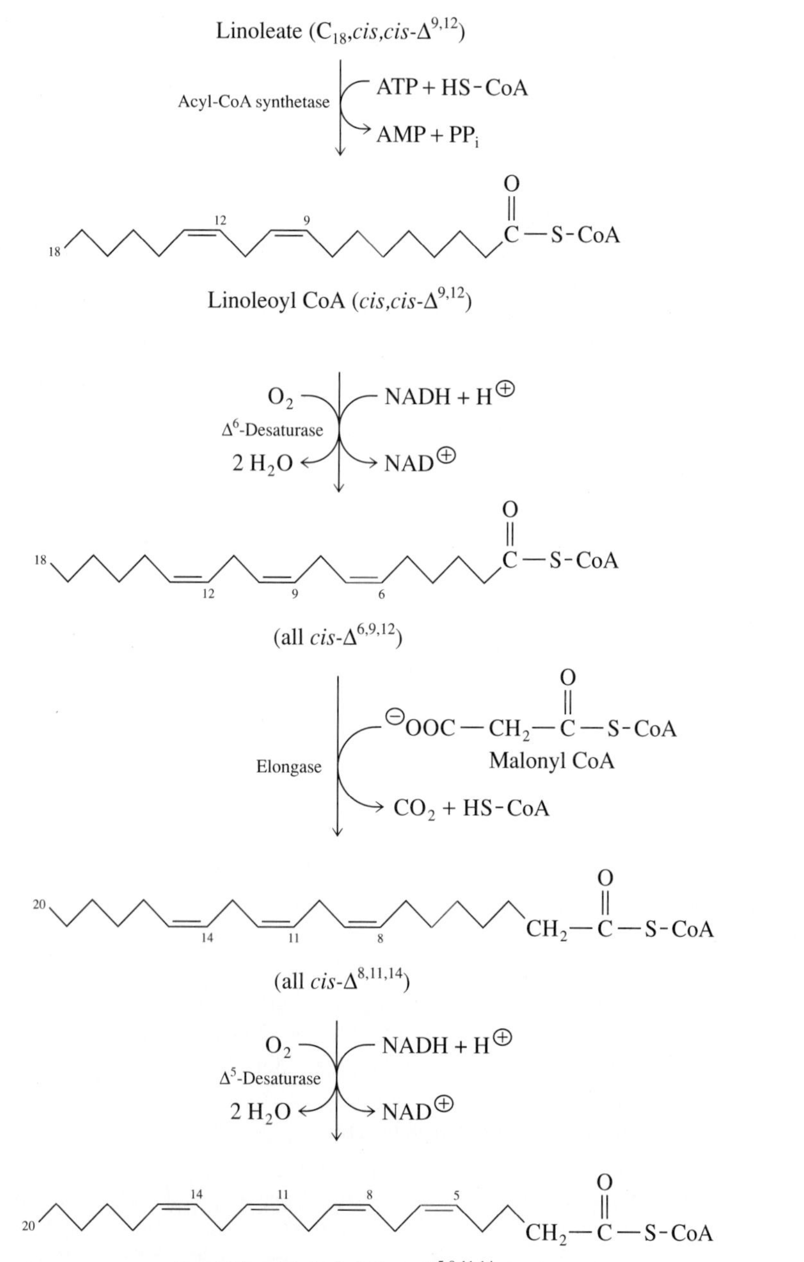

Figure 17·23
Elongation and desaturation reactions in the conversion of linoleate to arachidonoyl CoA. The activation step for linoleate is the same as that for fatty acid oxidation. Shown is an alternative to fatty acid oxidation—further desaturation and elongation of the molecule to form arachidonoyl CoA. The arachidonoyl moiety of arachidonoyl CoA is incorporated into triacylglycerols as well as phospholipids.

Linoleate is a precursor to arachidonoyl CoA, which is formed in a pathway involving a series of desaturation and elongation reactions (Figure 17·23). Elongation reactions employ malonyl CoA and thus depend on the activity of acetyl-CoA carboxylase (Section 17·10). However, the elongation steps are catalyzed by enzymes different from those of fatty acid synthase. The arachidonoyl moiety of arachidonoyl CoA can be incorporated into triacylglycerols or phospholipids. Arachidonate derived from phospholipid cleavage is a precursor of eicosanoids, which are shown in the figure below and discussed in the next section.

Figure 17·24
Pathways for the formation of eicosanoids. The two major pathways for eicosanoid formation begin with the products of the cyclooxygenase and lipoxygenase reactions. The cyclooxygenase pathway leads to a variety of prostaglandins, prostacyclin, and thromboxane A_2. The lipoxygenase pathway shown produces leukotriene A_4, a precursor to other leukotrienes, some of which trigger allergic responses.

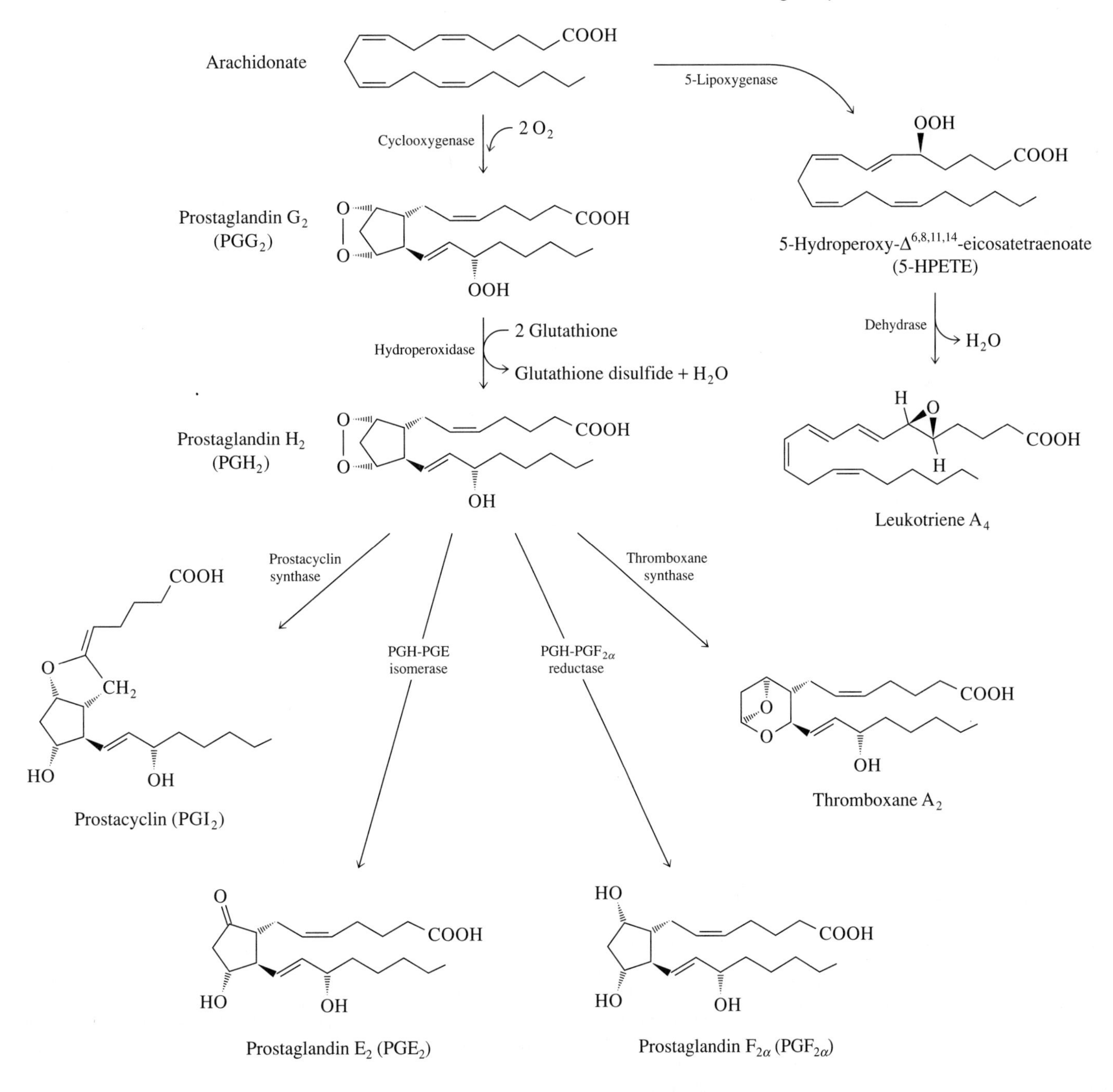

17·13 Eicosanoids, a Large Class of Signal Molecules, Are Derived from Arachidonate

Most arachidonate present in cells is found in the inner leaflet of the plasma membrane, esterified at C-2 of the glycerol backbone of phospholipids. Release of arachidonate to the interior of cells is catalyzed by phospholipase A_2 (Figure 17·4).

Once released, arachidonate can serve as the precursor for a large number of products called *eicosanoids,* a group of long-chain unsaturated fatty acid derivatives that serve as metabolic regulators. There are two general classes of eicosanoid regulator molecules. One class is derived from cyclization of arachidonate, catalyzed by cyclooxygenase (Figure 17·24). The product of this reaction is extremely unstable and is rapidly converted into an array of regulatory molecules with short lifetimes (on the order of seconds to minutes). These compounds, including the prostaglandins, prostacyclin, and thromboxane, have been termed *local regulators*. Unlike hormones, which are produced by glands and travel in the blood to their sites of action, eicosanoids typically act in the immediate neighborhood of the cell in which they are produced. For example, blood platelets produce thromboxane A_2, the only known naturally occurring thromboxane, which causes constriction of smooth muscles of arterial walls and leads to changes in local blood flow; the uterus produces contraction-triggering prostaglandins during labor. Eicosanoids also mediate pain sensitivity, inflammation, and swelling. Aspirin is effective in blocking these effects because its active ingredient, acetylsalicylic acid, irreversibly inhibits cyclooxygenase by transferring an acetyl group to an active-site serine residue of the enzyme. By blocking the activity of cyclooxygenase, aspirin prevents the formation of eicosanoids that are synthesized subsequent to the cyclooxygenase reaction.

The second class of eicosanoids are the products of reactions catalyzed by lipoxygenases. In Figure 17·24, 5-lipoxygenase is shown catalyzing the first step in the pathway leading to leukotriene A_4. Further reactions produce other leukotrienes, including a group of three that constitute what was once called the "slow-reacting substance of anaphylaxis" (allergic response) and are responsible for the occasionally fatal side effects of immunization shots.

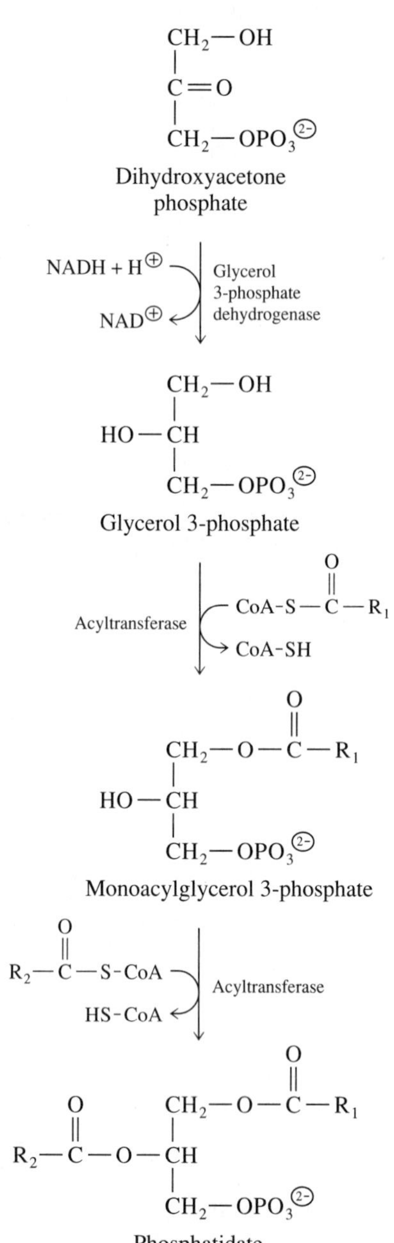

Figure 17·25
Pathway for the formation of a phosphatidate. The acyltransferase that catalyzes esterification at C-1 of glycerol 3-phosphate has a preference for saturated acyl chains; the acyltransferase that catalyzes esterification at C-2 has a greater affinity for unsaturated acyl chains.

17·14 Triacylglycerols and Neutral Phospholipids Are Synthesized from Diacylglycerol

Most fatty acids present in cells exist in esterified forms as triacylglycerols or glycerophospholipids. A useful division of phospholipids that reflects their metabolic origins classifies "neutral" (zwitterionic) phospholipids, such as phosphatidylethanolamine, separately from "acidic" (anionic) phospholipids, such as phosphatidylinositol.

Triacylglycerols and the neutral phospholipids phosphatidylcholine and phosphatidylethanolamine are synthesized by a common pathway. The first portion of this pathway is shown in Figure 17·25. First, dihydroxyacetone phosphate produced by glycolysis is reduced to glycerol 3-phosphate in a reaction catalyzed by glycerol 3-phosphate dehydrogenase. Glycerol 3-phosphate then serves as the backbone for acylation reactions catalyzed by two separate acyltransferases, with fatty acyl CoA

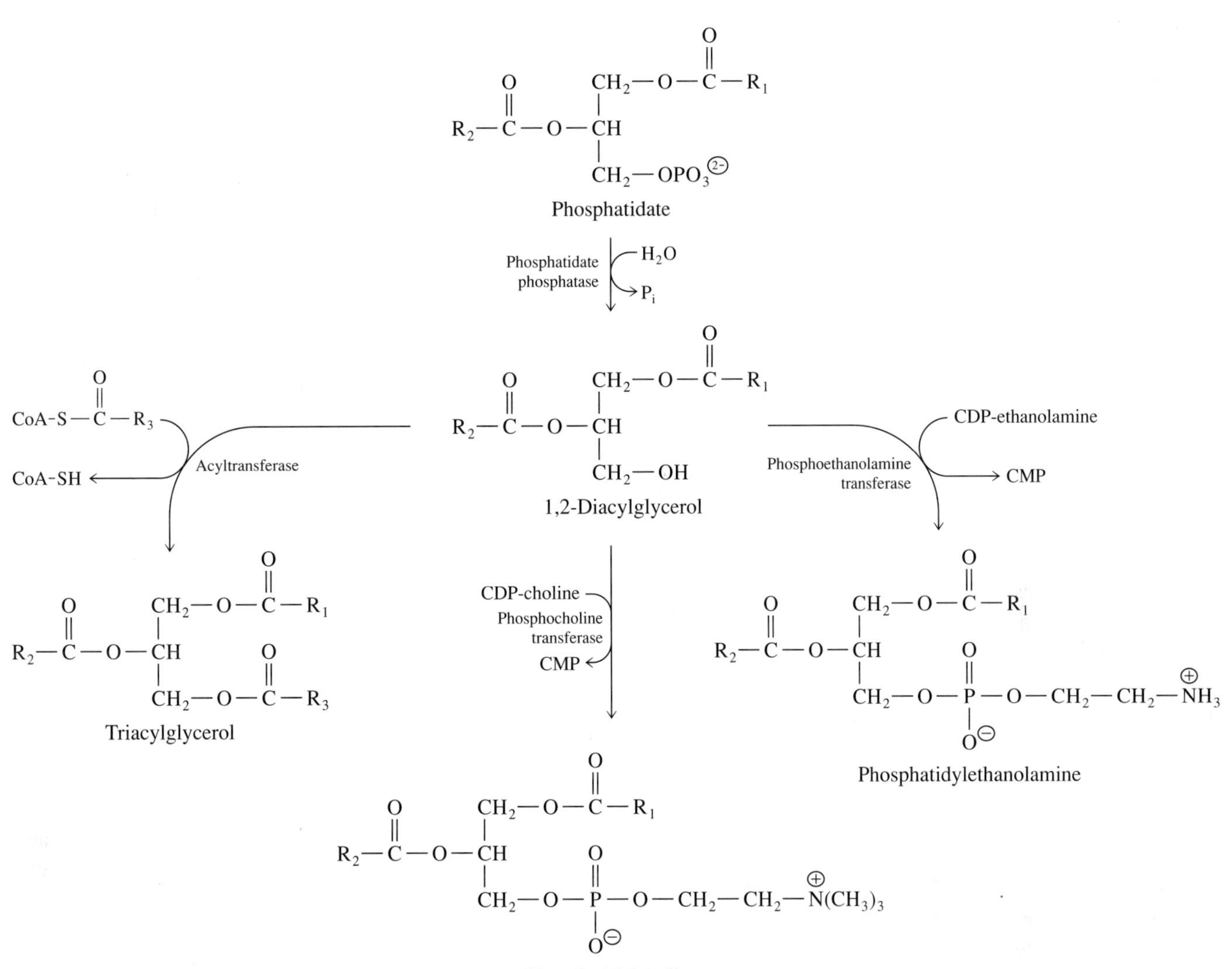

Figure 17·26
Neutral lipid synthesis. Formation of triacylglycerols, phosphatidylcholine, and phosphatidylethanolamine proceeds via a diacylglycerol intermediate. A cytidine nucleotide cosubstrate is the precursor of the polar head group for the phospholipids.

molecules serving as the sources of the acyl groups. The first acyltransferase, which has a preference for fatty acyl CoA molecules with saturated acyl chains, catalyzes esterification at C-1 of glycerol 3-phosphate; the second acyltransferase, which has more affinity for unsaturated species, catalyzes esterification at C-2 of monoacylglycerol 3-phosphate. The resulting molecule is phosphatidate. This term refers to a class of molecules whose properties depend upon the attached acyl groups.

The next step in the formation of neutral lipids is the dephosphorylation of phosphatidate, catalyzed by phosphatidate phosphatase. The product of this reaction, 1,2-diacylglycerol, can be either directly acylated to form a triacylglycerol or it can react with a CTP-derived cosubstrate, CDP-choline or CDP-ethanolamine, to form either of two phospholipids, phosphatidylcholine or phosphatidylethanolamine, respectively (Figure 17·26).

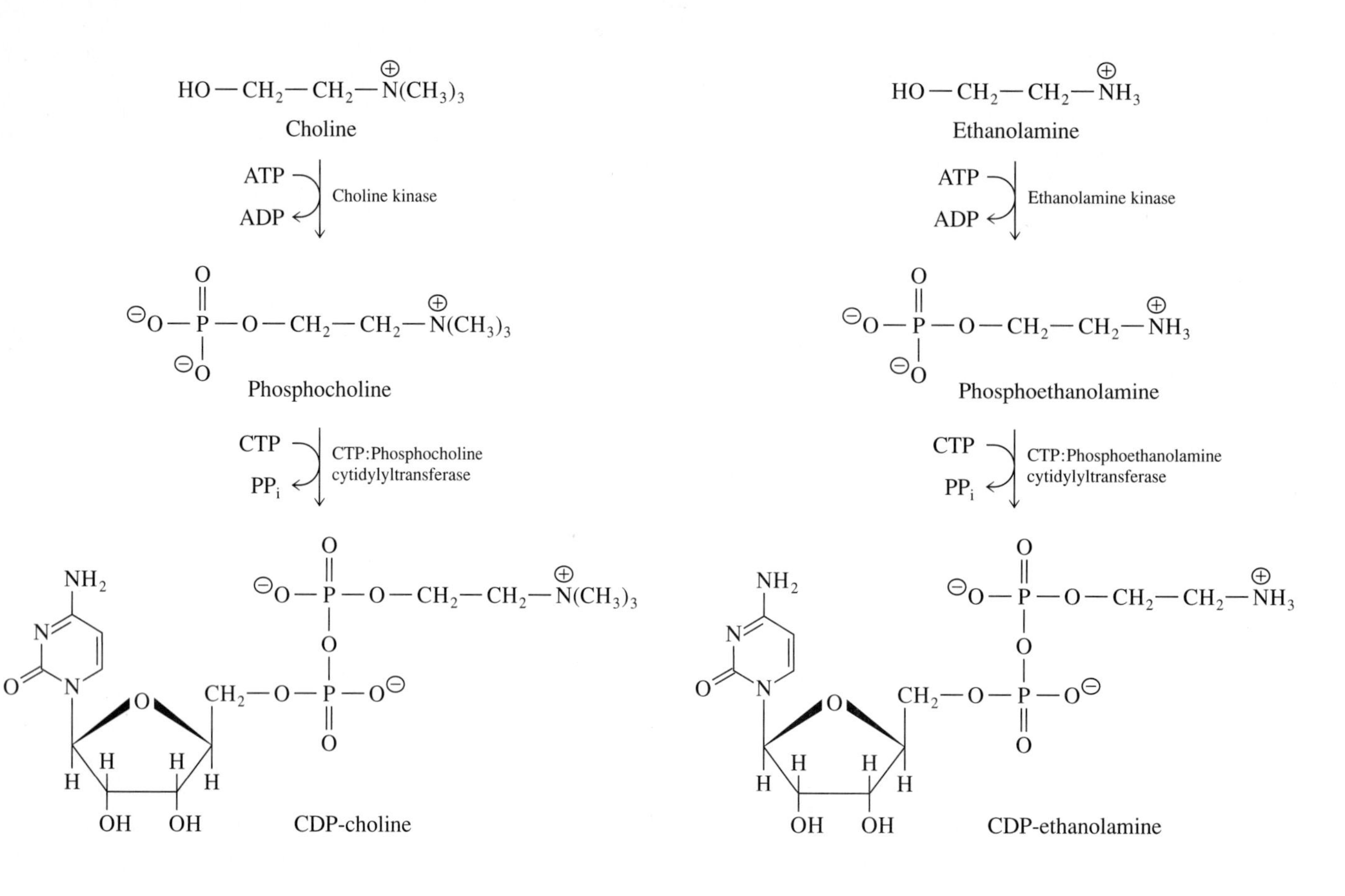

Figure 17·27
Formation of CDP-choline and CDP-ethanolamine.

Phosphatidylcholine synthesis requires CDP-choline formed through the phosphorylation of choline, catalyzed by choline kinase, followed by condensation with CTP to form CDP-choline, catalyzed by CTP:phosphocholine cytidylyltransferase (Figure 17·27). Reaction of diacylglycerol with CDP-choline completes the synthesis of phosphatidylcholine (Figure 17·26). A parallel series of reactions is employed to form phosphatidylethanolamine, with different kinase and transferase enzymes required for the analogous steps.

17·15 Biosynthesis of Acidic Phospholipids Proceeds from Phosphatidate

Acidic phospholipids have a net negative charge because their acid groups, usually phosphoric acid, are dissociated at physiological pH. The immediate precursor for acidic phospholipids is phosphatidate formed by the pathway shown in Figure 17·25. Condensation of CTP with phosphatidate to form CDP-diacylglycerol is the first step. Next, displacement of CMP by serine leads to phosphatidylserine in *E. coli*. In both prokaryotes and eukaryotes, displacement of CMP from CDP-diacylglycerol by inositol leads to phosphatidylinositol (Figure 17·28). Through successive phosphorylation, phosphatidylinositol may be converted to phosphatidylinositol 4-phosphate (PIP) and phosphatidylinositol 4,5-*bis*phosphate, or PIP_2 (Figure 17·29, page 17·32).

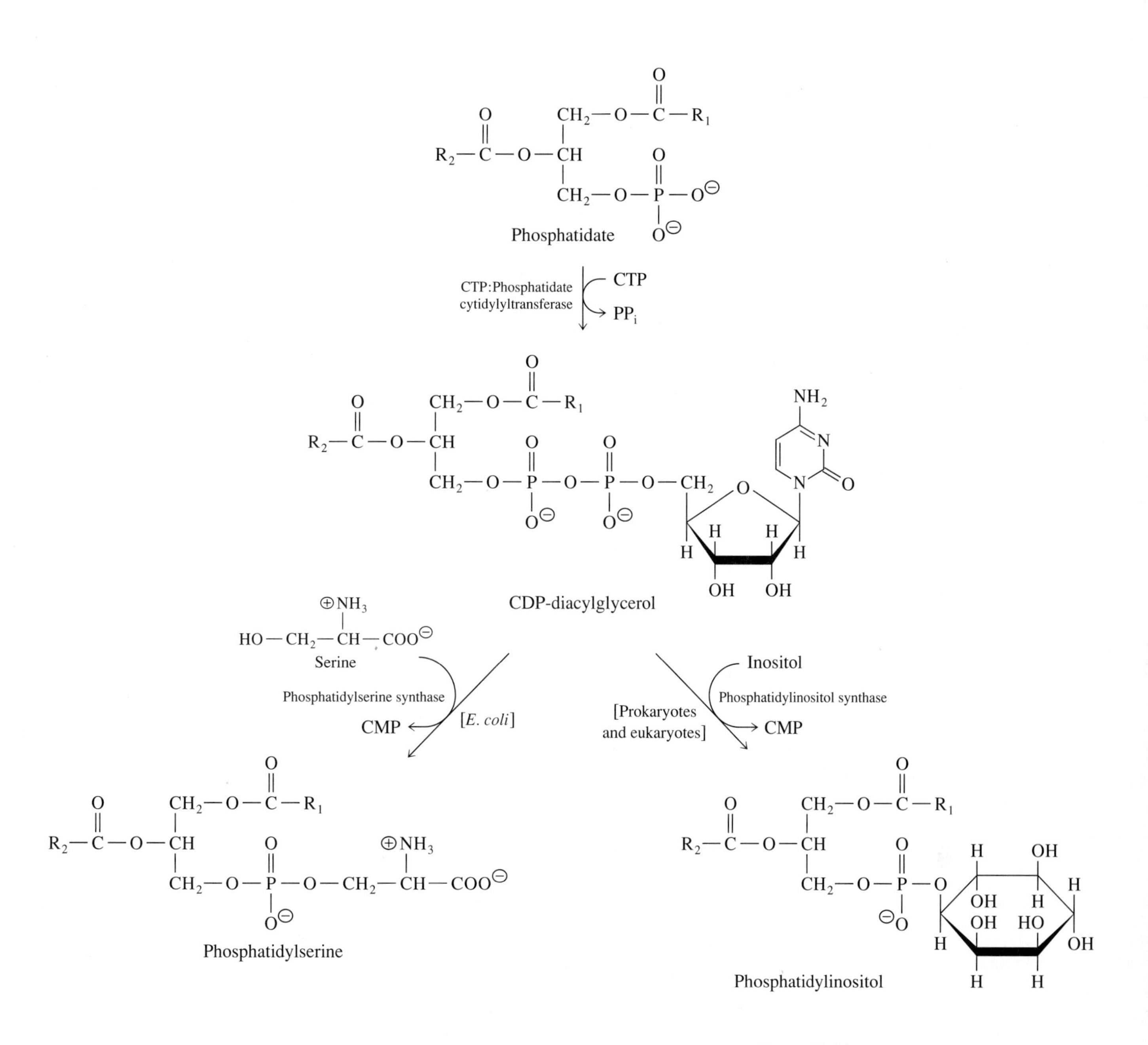

Figure 17·28
Synthesis of the acidic phospholipids phosphatidylserine and phosphatidylinositol. Phosphatidate accepts a cytidylyl group from CTP to form CDP-diacylglycerol. CMP is then displaced by an alcohol group of serine or inositol to form phosphatidylserine or phosphatidylinositol, respectively.

Mammalian phosphatidylserine formation is catalyzed by the base-exchange enzyme. This enzyme catalyzes the displacement of ethanolamine from phosphatidylethanolamine by serine and the reverse reaction (Figure 17·30). The conversion of phosphatidylserine to phosphatidylethanolamine also occurs via decarboxylation of phosphatidylserine in a reaction catalyzed by phosphatidylserine decarboxylase. The base-exchange reaction takes place in the endoplasmic reticulum, whereas the decarboxylation reaction takes place in the mitochondrion of eukaryotes and in *E. coli.*

In *E. coli* and in mitochondria, glycerol 3-phosphate itself can serve as a head group of phospholipids. Glycerol 3-phosphate displaces CMP from CDP-diacylglycerol to form phosphatidylglycerol phosphate, which is dephosphorylated to form phosphatidylglycerol (Figure 17·31). Here the pathways in *E. coli* and mitochondria diverge. In *E. coli,* phosphatidylglycerol condenses with CDP-diacylglycerol; in mitochondria, a second molecule of phosphatidylglycerol condenses with the first. The product is the same in both cases, diphosphatidylglycerol, or cardiolipin, as it is commonly known.

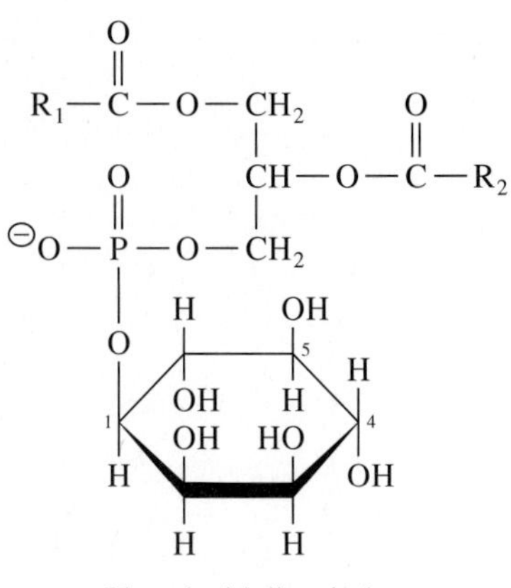

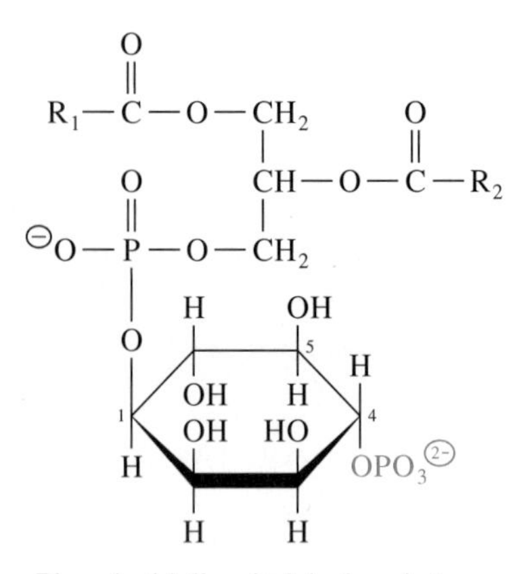

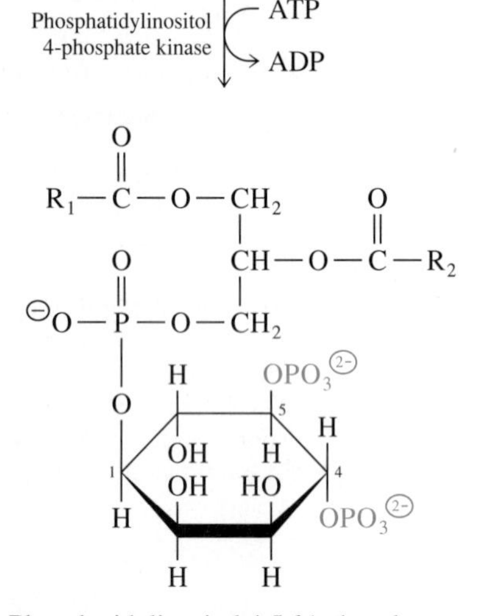

Figure 17·29
Pathways for the biosynthesis of phosphatidylinositol 4,5-*bis*phosphate from phosphatidylinositol.

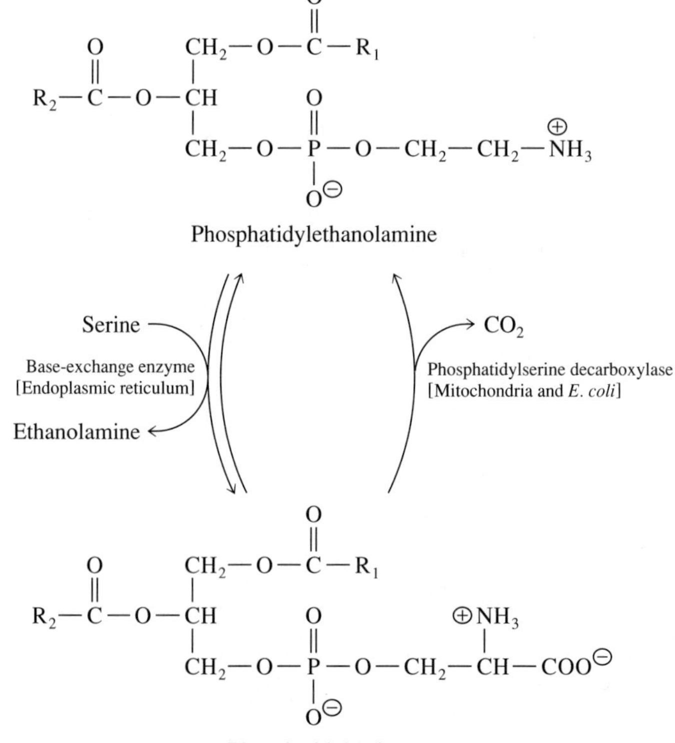

Figure 17·30
Interconversions between phosphatidylethanolamine and phosphatidylserine.

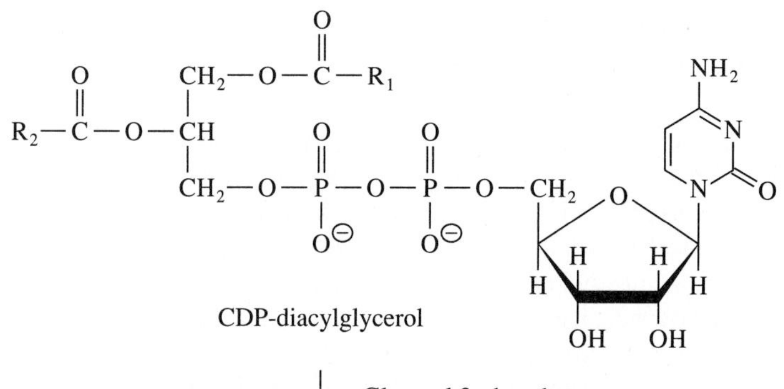

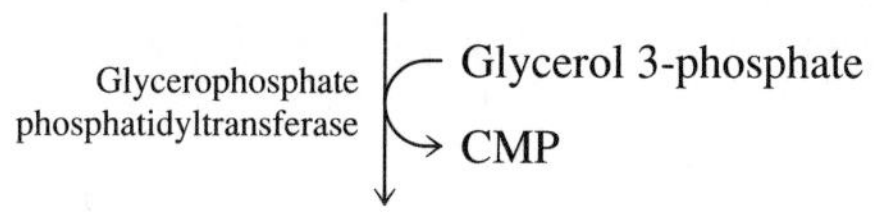

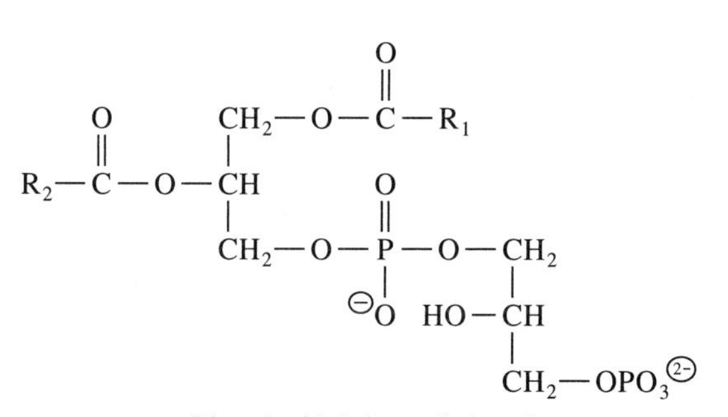

Phosphatidylglycerol phosphatase

H_2O

P_i

$$\begin{array}{l} R_2-\overset{O}{\overset{\|}{C}}-O-CH(CH_2-O-\overset{O}{\overset{\|}{C}}-R_1)-CH_2-O-\overset{O}{\overset{\|}{P}}(O^{\ominus})-O-CH_2-CH(OH)-CH_2-OH \end{array}$$

Phosphatidylglycerol

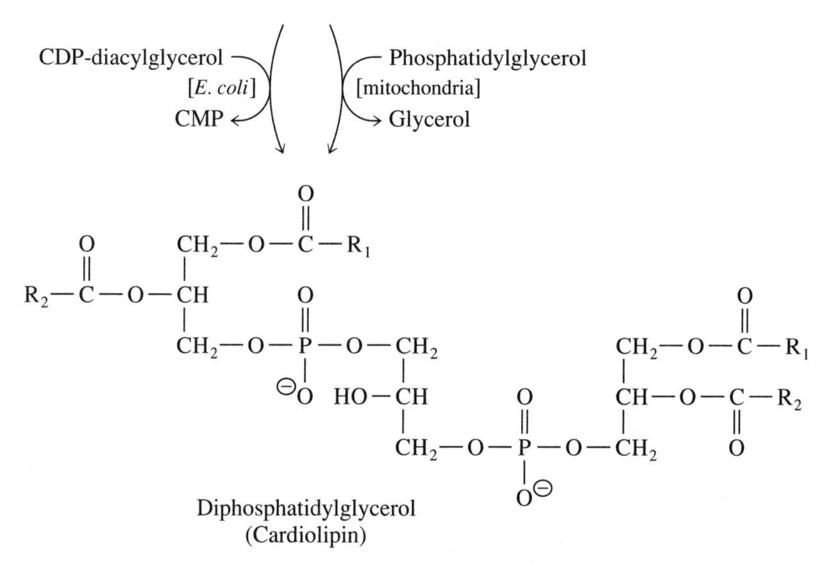

Figure 17·31
Formation of diphosphatidylglycerol. A molecule of glycerol 3-phosphate is incorporated into CDP-diacylglycerol by displacement of CMP to form phosphatidylglycerol phosphate. Next, a phosphatase catalyzes the formation of phosphatidylglycerol. Separate displacement reactions ensue in mitochondria (where phosphatidylglycerol displaces glycerol) and *E. coli* (where phosphatidylglycerol reacts with CDP-diacylglycerol, displacing CMP). Diphosphatidylglycerol, also known as cardiolipin, is found in the plasma membrane of prokaryotes and the mitochondrial inner membrane of eukaryotes.

17·16 Ether Lipids Are Synthesized from Dihydroxyacetone Phosphate

Ether lipids have an ether linkage in place of one of the usual ester linkages. Ether lipids are derived from dihydroxyacetone phosphate rather than from glycerol 3-phosphate, the precursor for most phospholipids. The overall pathway for the formation of ether lipids is shown in Figure 17·32. First, an acyl group from fatty acyl CoA is esterified to the alcohol moiety at C-1 of dihydroxyacetone phosphate, producing 1-acyldihydroxyacetone phosphate. Next, a fatty alcohol displaces the fatty acid to produce 1-alkyldihydroxyacetone phosphate, in an unusual reaction detailed in Figure 17·33. In this reaction, tautomerization (Step 1) and addition of a proton

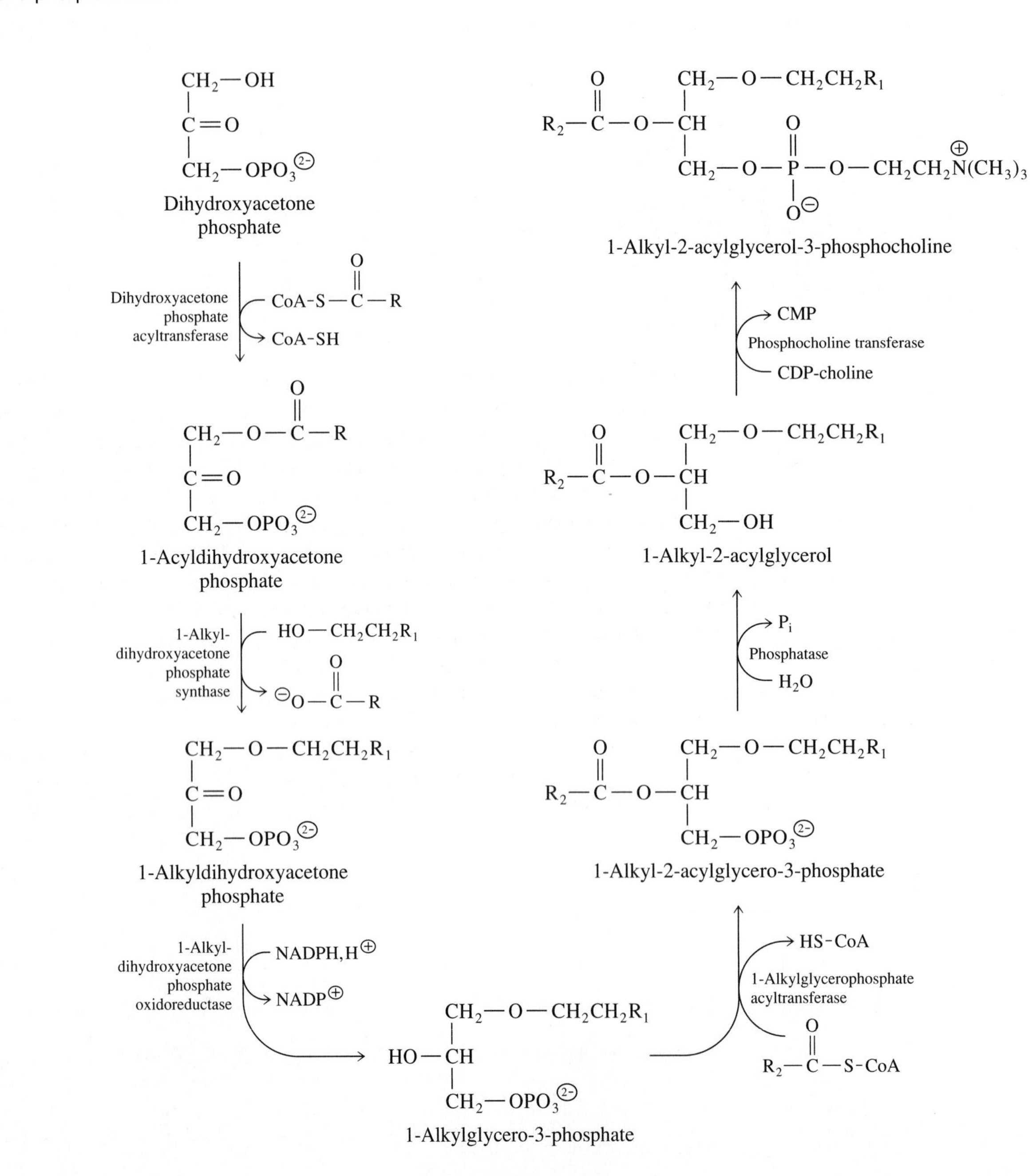

Figure 17·32
Formation of ether lipids. First, an acyl group from fatty acyl CoA is esterified to the alcohol moiety at C-1 of dihydroxyacetone phosphate, producing 1-acyldihydroxyacetone phosphate. A fatty alcohol displaces the fatty acid, producing 1-alkyldihydroxyacetone phosphate, which is reduced by NADPH to produce 1-alkylglycero-3-phosphate. 1-Alkylglycero-3-phosphate is then esterified at C-2 to produce 1-alkyl-2-acylglycero-3-phosphate, which undergoes dephosphorylation. Finally, 1-alkyl-2-acylglycerol receives a choline group from CDP-choline to produce 1-alkyl-2-acylglycerol-3-phosphocholine.

(Step 2) produces a carbonium ion. A fatty alcohol adds to the carbonium ion (Step 3), forming an ether and an ester bond at C-1. The fatty acid, a good leaving group, is then displaced in Step 4, and a second carbonium ion is formed. Proton extraction and tautomerization (Steps 4 and 5) forms the product, 1-alkyldihydroxyacetone phosphate.

The keto group of 1-alkyldihydroxyacetone phosphate is then reduced by NADPH to form 1-alkylglycero-3-phosphate (Figure 17·32). Reduction is followed by esterification at C-2 of the glycerol residue to produce 1-alkyl-2-acylglycero-3-phosphate. The subsequent reactions—dephosphorylation and transfer of the choline group—are the same as those shown earlier in Figure 17·26.

One class of ether lipids called *plasmalogens* has a vinyl ether linkage at C-1 of the glycerol backbone (see Figure 9·9). The physiological roles of plasmalogens and most other ether lipids are unknown. However, the role of one ether lipid called *platelet-activating factor* is known. An ether lipid with a palmitoyl group at C-1 of the glycerol backbone and an acetyl group at C-2, platelet-activating factor serves as a signal molecule that stimulates the aggregation of platelets during the process of blood clotting. The platelet-activating factor is extremely potent, effective at about 0.1 nM.

17·17 Sphingolipids Are Derived from Palmitoyl CoA and Serine

Sphingolipids are a class of membrane lipids that have sphingosine as their structural backbone. The backbone is derived from palmitoyl CoA and serine, which is the C_3 unit of sphingolipids analogous to glycerol in phospholipids. In the first step of the sphingosine biosynthetic pathway, serine condenses with palmitoyl CoA,

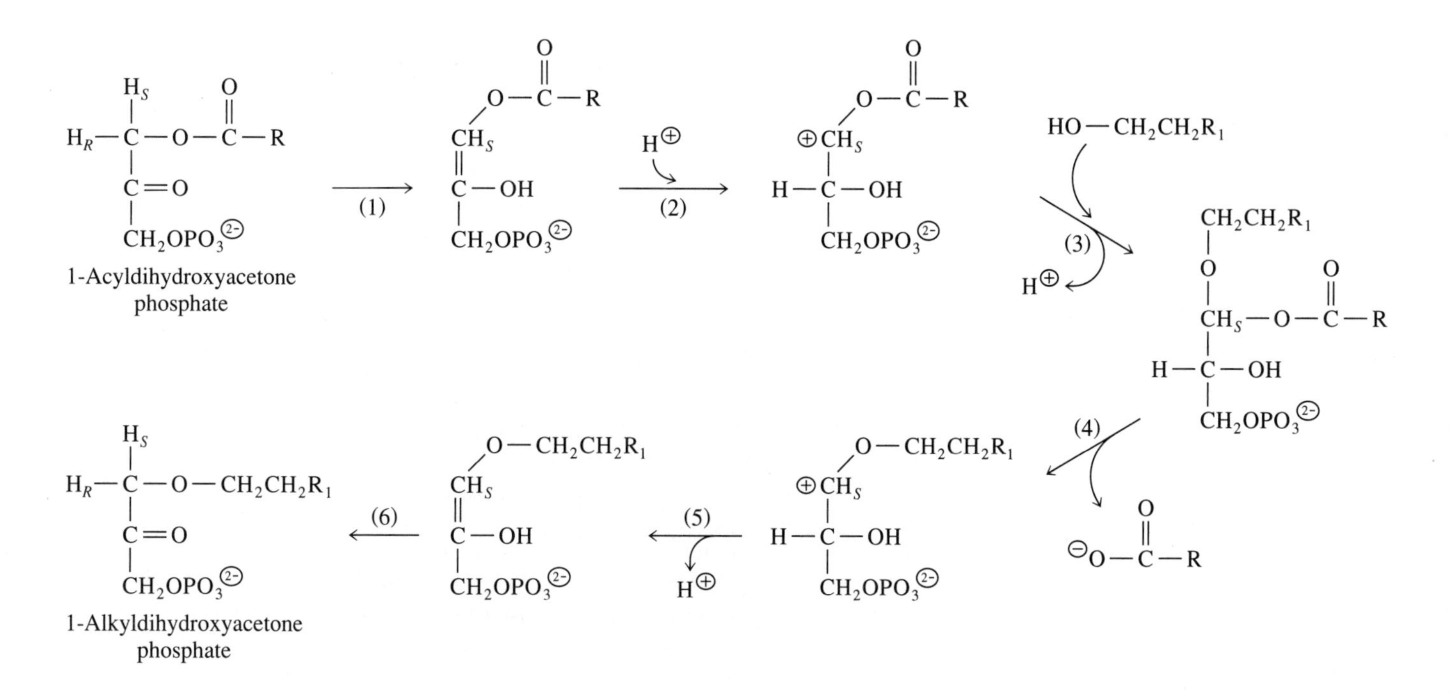

Figure 17·33
Proposed mechanism for ether bond formation. Tautomerization and addition of a proton forms a carbonium ion at C-1 of the dihydroxyacetone phosphate backbone. Addition of an alcohol (R—OH) to this moiety displaces the fatty acid at C-1, and deprotonation and tautomerization forms the 1-ether lipid, 1-alkyldihydroxyacetone phosphate.

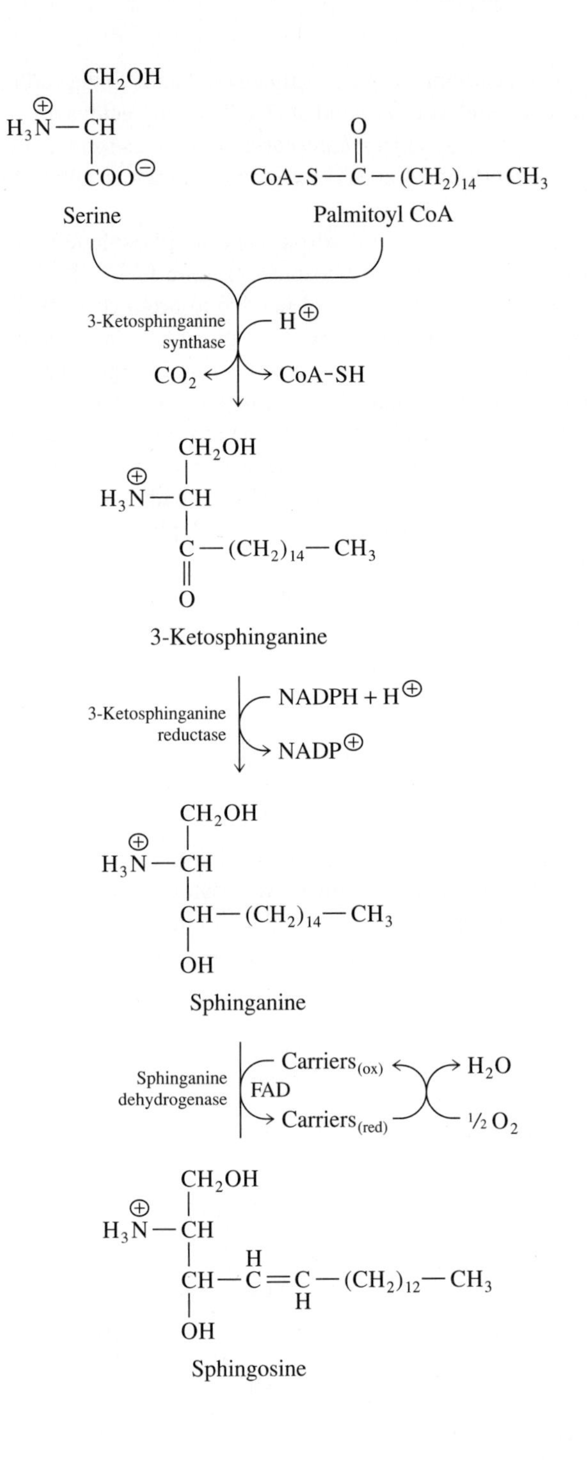

Figure 17·34
Biosynthesis of sphingosine.

producing 3-ketosphinganine (Figure 17·34). Reduction of 3-ketosphinganine, catalyzed by NADPH-dependent 3-ketosphinganine reductase, produces sphinganine. Finally, desaturation produces sphingosine in a reaction catalyzed by the flavin-containing sphinganine dehydrogenase.

Like most fatty acid–modifying enzymes, sphinganine dehydrogenase is embedded in the membrane of the endoplasmic reticulum, with its active site facing the cytosol. Although the details of the electron transfer needed to reoxidize the $FADH_2$ of the enzyme are not known, it is known that the endoplasmic reticulum contains a number of electron carriers (including hemoproteins and Fe-S proteins) that reduce O_2, but at a small fraction of the rate at which mitochondrial O_2 is reduced. In contrast to mitochondrial electron carriers, endoplasmic reticulum electron carriers are not involved in ATP production.

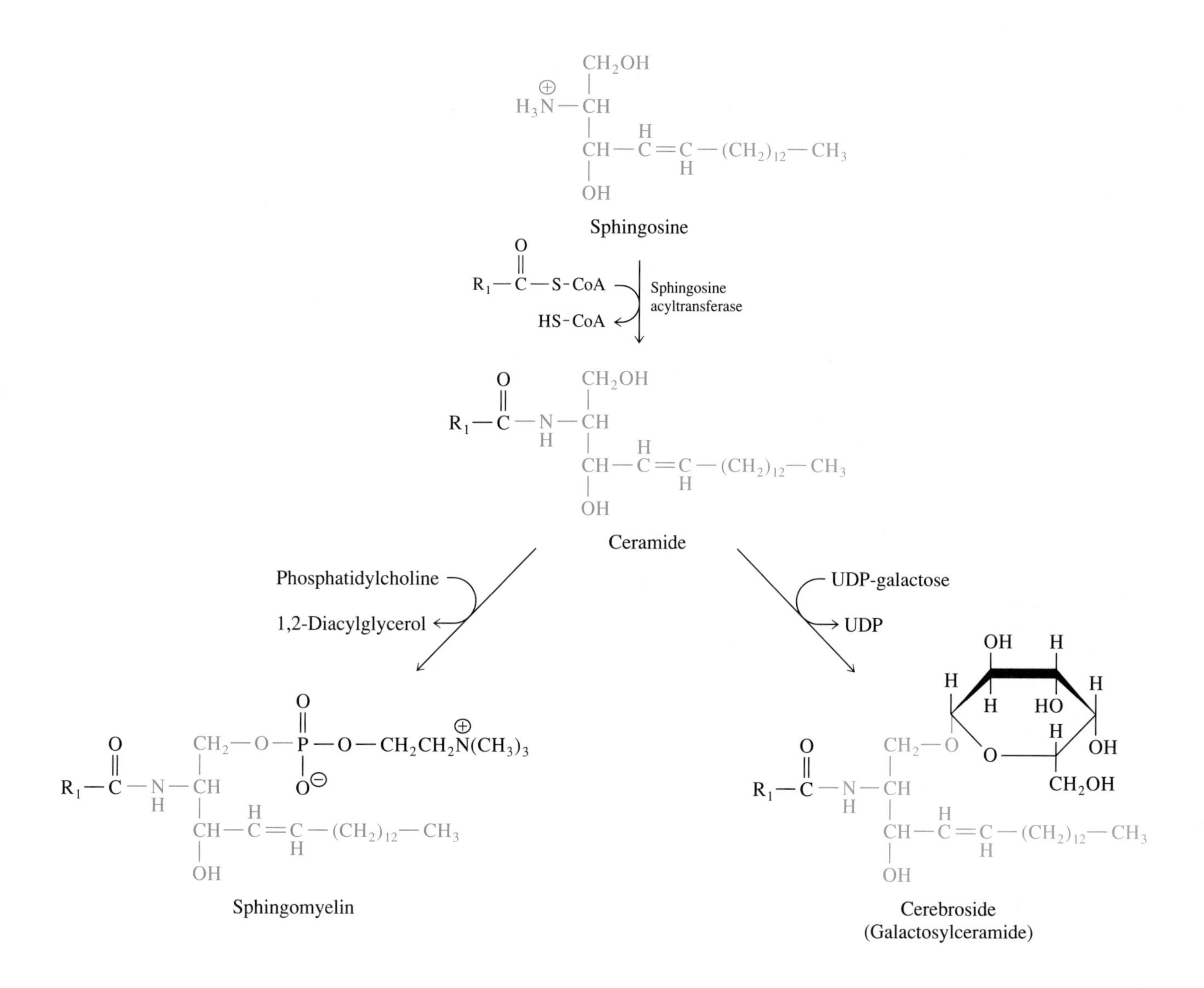

Figure 17·35
Formation of sphingomyelin and a cerebroside (galactosylceramide), starting from sphingosine.

Acylation of sphingosine produces a ceramide, which can then be modified by reaction with a phosphatidylcholine to form a sphingomyelin or with a UDP-sugar to form a cerebroside (Figure 17·35). More complex sugar-lipid conjugates can be formed by reaction with additional sugar moieties of UDP-sugars. The resulting molecules are gangliosides. One such ganglioside is shown in Figure 9·12b. The gangliosides are involved in antigen recognition in mammals and are part of the external leaflet of the plasma membrane, as are most sugar lipids.

17·18 Cholesterol Is Derived from Cytosolic Acetyl CoA

Although most animal cells are capable of synthesizing cholesterol, the pathway for cholesterol formation is substantially active only in liver cells. Part of the function of lipoproteins is the delivery of dietary and liver-derived cholesterol to the rest of the body's cells.

The first milestone in the elucidation of cholesterol synthesis was the discovery that all of the carbon arises from acetyl CoA, a fact that emerged from early radioisotopic-labelling experiments. Another discovery was the recognition that squalene, a C_{30} linear hydrocarbon, was an intermediate in the biosynthesis of the

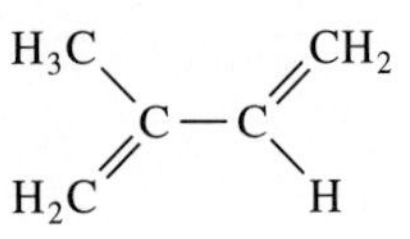

Figure 17·36
Structure of isoprene.

27-carbon cholesterol molecule and that squalene itself is formed from 5-carbon units related to isoprene, whose structure is shown in Figure 17·36. Thus, the stages of cholesterol biosynthesis were found to be:

$$\text{Acetate } (C_2) \longrightarrow \text{Isoprenoid } (C_5) \longrightarrow \text{Squalene } (C_{30}) \longrightarrow \text{Cholesterol } (C_{27}) \quad \textbf{(17·10)}$$

We shall divide our examination of cholesterol synthesis between the 5-carbon, 30-carbon, and 27-carbon stages.

A. Stage 1: Acetyl CoA to Isopentenyl Pyrophosphate

Cholesterol is synthesized from cytosolic acetyl CoA, which is transported from mitochondria via the citrate transport system. Since cytosolic acetyl CoA is also used for fatty acid synthesis, it is the branch-point metabolite of two major lipid biosynthetic pathways.

The first stage in the formation of cholesterol is the sequential condensation of three molecules of acetyl CoA, with the formation of hydroxymethylglutaryl CoA (HMG CoA) as an intermediate leading to mevalonate (Figure 17·37). The condensation steps are catalyzed by cytosolic isozymes of the mitochondrial enzymes involved in the formation of ketone bodies.

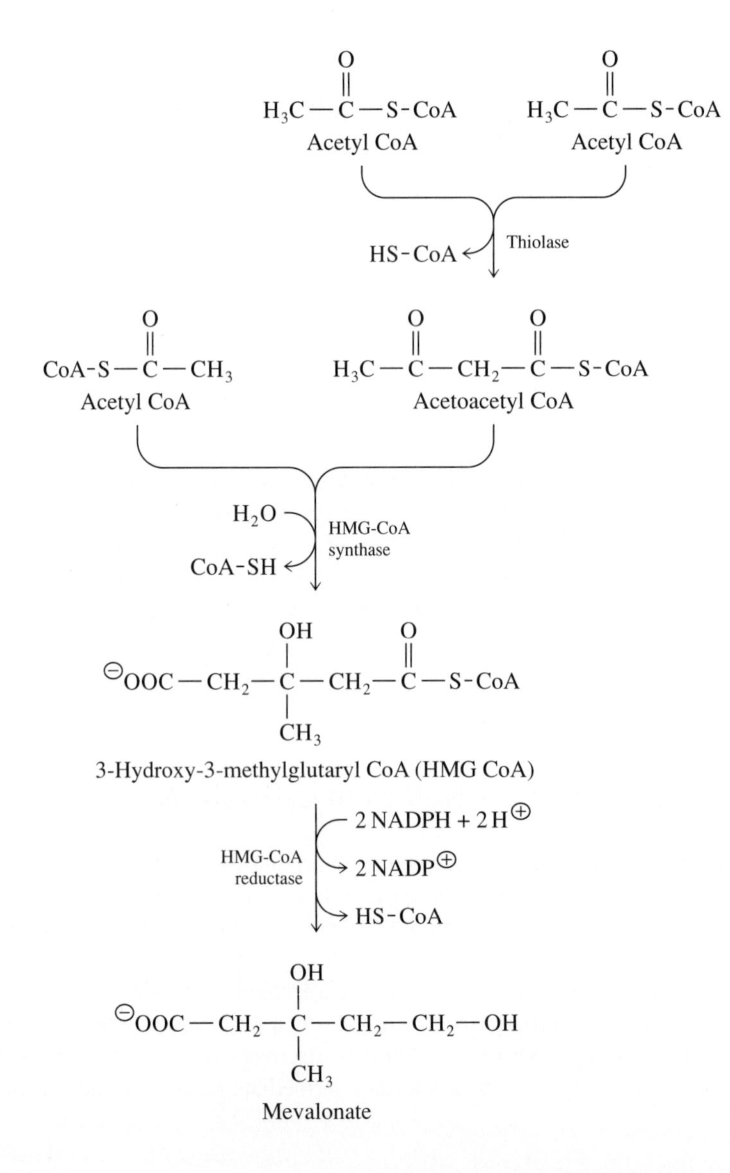

Figure 17·37
Initial reactions of cholesterol biosynthesis leading to the formation of mevalonate. HMG-CoA reductase catalyzes the rate-determining step for the synthesis of cholesterol.

HMG-CoA reductase, which catalyzes reduction of HMG CoA to mevalonate, is the rate-determining enzyme for the overall pathway of cholesterol synthesis. It is an interconvertible enzyme that is inactivated by phosphorylation. Additionally, the amount of enzyme present in cells is closely regulated. Insulin increases the activity of HMG-CoA reductase and glucagon has the reverse effect, although the steps between hormone reception and phosphorylation state remain unknown. Additionally, long-chain fatty acyl CoA molecules cause inhibition of the enzyme, which may result from both a direct allosteric effect on the enzyme and an effect on the kinase that catalyzes phosphorylation and resulting inhibition of HMG-CoA reductase. This is the same kinase that inactivates acetyl-CoA carboxylase. The activity of HMG-CoA reductase is also modulated by the concentration of cholesterol. Increased concentrations of cholesterol lead to formation of cholesterol derivatives, which allosterically inhibit the enzyme. In addition, high levels of cholesterol derivatives lead to increased degradation and decreased synthesis of the enzyme.

Mevalonate is converted to isopentenyl pyrophosphate in a series of enzyme-catalyzed steps that result in phosphorylation of mevalonate, a second phosphorylation, and decarboxylation (Figure 17·38).

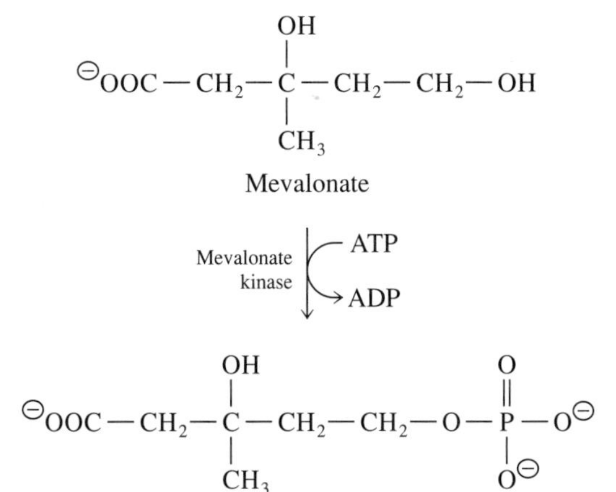

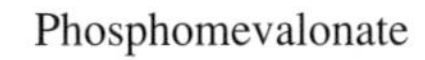

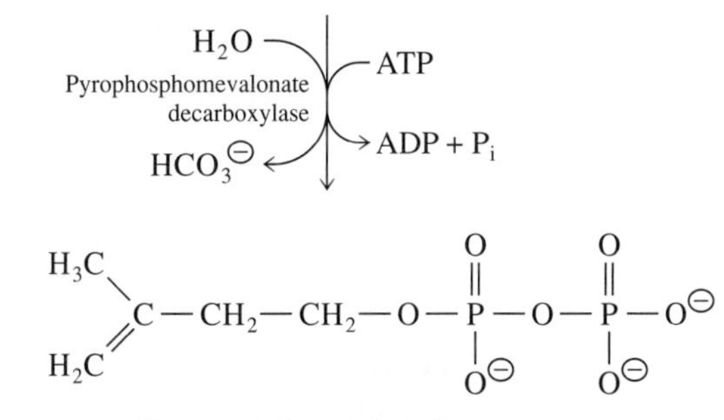

Figure 17·38
Formation of isopentenyl pyrophosphate. Two phosphorylations and a decarboxylation reaction convert mevalonate into the five-carbon molecule isopentenyl pyrophosphate.

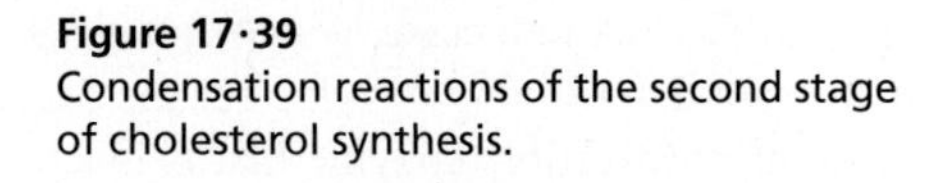

Figure 17·39
Condensation reactions of the second stage of cholesterol synthesis.

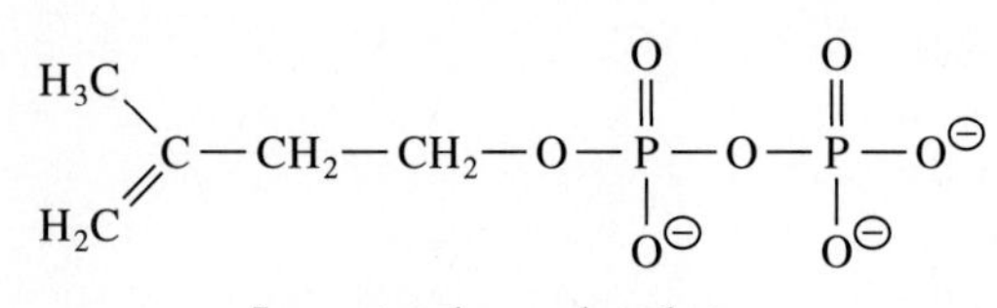

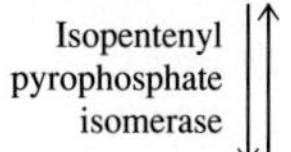

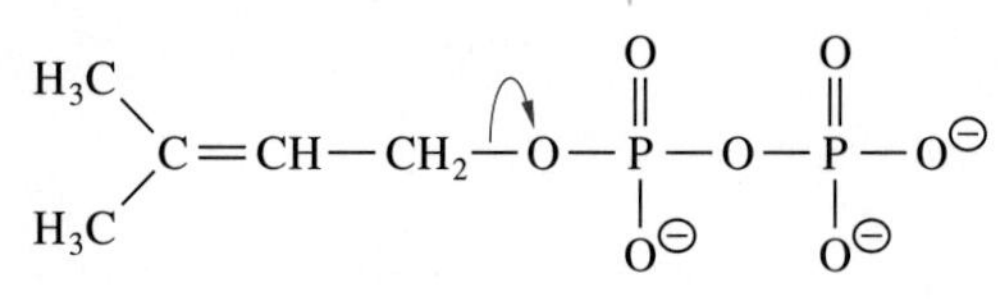

PP_i

Isopentenyl pyrophosphate

Prenyl transferase

$H^{\oplus}$

Geranyl pyrophosphate (C_{10})

Isopentenyl pyrophosphate (C_5)

Prenyl transferase

PP_i

Farnesyl pyrophosphate (C_{15})

Farnesyl pyrophosphate (C_{15})

Squalene synthase

$NADPH + H^{\oplus}$

PP_i

$NADP^{\oplus}$

Squalene (C_{30})

B. Stage 2: Isopentenyl Pyrophosphate to Squalene

An isomerase catalyzes the conversion of isopentenyl pyrophosphate to its isomer, dimethylallyl pyrophosphate, which then condenses in head-to-tail fashion with isopentenyl pyrophosphate to form the C_{10} molecule geranyl pyrophosphate (Figure 17·39). The condensation of the two C_5 molecules proceeds by formation of an intermediate carbonium ion of dimethylallyl pyrophosphate that is attacked by isopentenyl pyrophosphate. This basic reaction is repeated: the C_{10} molecule geranyl pyrophosphate condenses in head-to-tail fashion with isopentenyl pyrophosphate to form the C_{15} molecule farnesyl pyrophosphate. Two molecules of farnesyl pyrophosphate then condense in head-to-head fashion to form the C_{30} molecule squalene.

C. Stage 3: Squalene to Cholesterol

The steps between squalene and cholesterol are numerous and complex. One intermediate, lanosterol, accumulates in appreciable quantities in cells actively synthesizing cholesterol (Figure 17·40). The steps between squalene and lanosterol involve addition of an oxygen atom followed by cyclization of the chain to form the four-ring steroid nucleus. The cyclization occurs in a single step of concerted rearrangements of electrons from neighboring double bonds. The conversion of

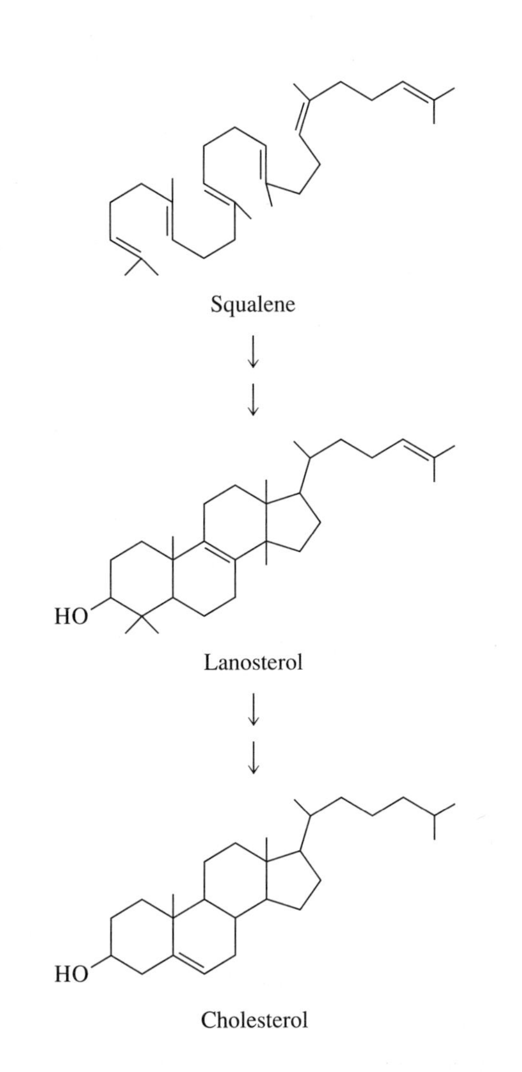

Figure 17·40
The final stage of cholesterol synthesis: squalene to cholesterol. The intermediates of this process are numerous, and the exact sequence of steps between lanosterol and cholesterol remains uncertain.

lanosterol to cholesterol occurs via a multistep pathway involving methyl-group shifts, oxidations, and decarboxylations. Many different enzyme activities have been implicated that provide several possible routes between lanosterol and cholesterol. The exact pathway is at present unknown.

17·19 Cholesterol Metabolism Is the Source of a Large Number of Other Cell Constituents

The pathway leading to cholesterol and pathways emanating from cholesterol contribute to a plethora of cellular constituents. Isopentenyl pyrophosphate, the C_5 precursor of squalene, and ultimately of cholesterol, is the precursor for a large number of other products, such as the fat-soluble vitamins A, E, and K, and ubiquinone in animal cells, and terpenes, plastoquinone, and the phytol side chain of chlorophyll in plants (Figure 17·41).

Cholesterol itself is a precursor of the bile salts, which facilitate fat digestion; vitamin D, which stimulates Ca^{2+} uptake from the intestines; steroid hormones that control sex characteristics, such as testosterone and β-estradiol; and steroids that control salt balance. The principal product of sterol synthesis is cholesterol itself, which modulates membrane fluidity and is an essential component of the plasma membrane of animal cells.

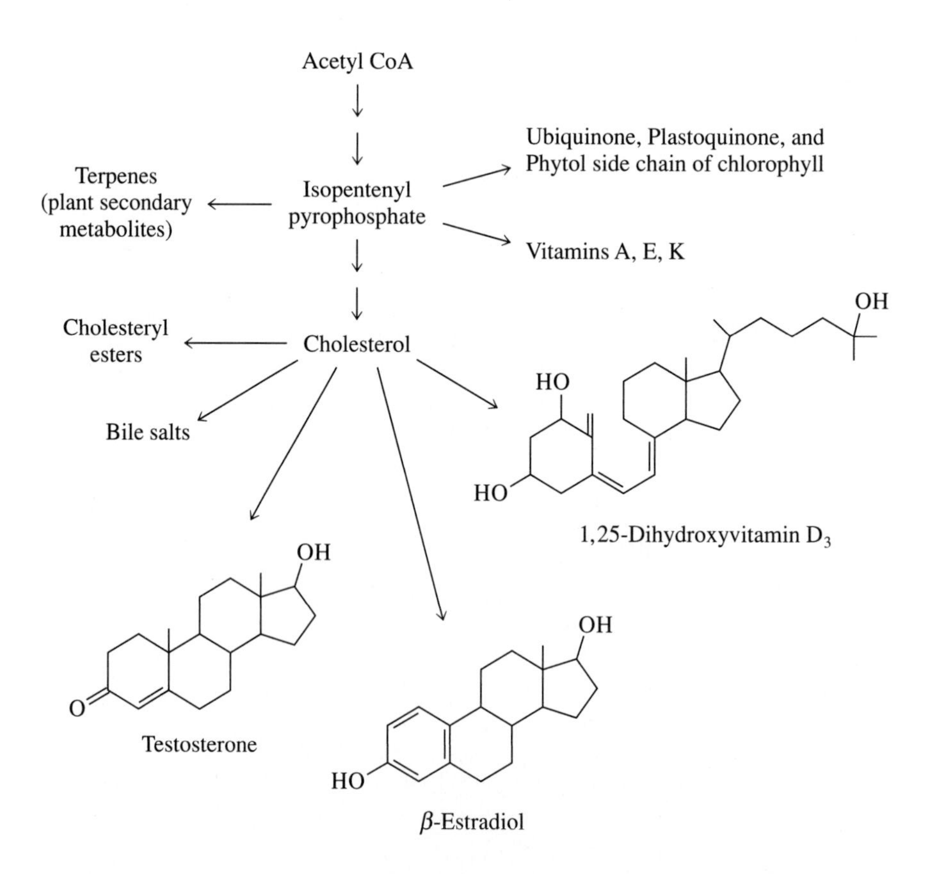

Figure 17·41
Diversity of products related to cholesterol and the pathway of cholesterol synthesis.

Summary

Most of the fat supplied in the diet of animals is in the form of triacylglycerols, which are digested in the small intestine. There they are mixed with bile salts and hydrolyzed to free fatty acids and 2-monoacylglycerols by pancreatic lipase. Fatty acids and monoacylglycerols are then taken up by intestinal cells.

Fatty acids are released from adipocytes by epinephrine-stimulated activation of hormone-sensitive lipase, which catalyzes the conversion of triacylglycerols stored in fat cells to free fatty acids and monoacylglycerols. Another more specific and active monoacylglycerol lipase catalyzes the hydrolysis of monoacylglycerol to free fatty acids and glycerol. The free fatty acids and glycerol then enter the bloodstream, bind transport proteins, and are delivered to tissues where they are oxidized for energy.

Fatty acids are degraded to acetyl CoA by the sequential removal of two-carbon fragments, a process called β oxidation. Before oxidation can take place, fatty acids are activated by esterification to CoA and then, in eukaryotic cells, transferred to L-carnitine for transport into mitochondria. In mitochondria, they are transferred to coenzyme A and oxidized in a series of reactions that produce large amounts of ATP.

The β-oxidation pathway for saturated fatty acids consists of four enzyme-catalyzed steps: oxidation, hydration, oxidation, and thiolysis. Oxidation of unsaturated fatty acids follows the same pathway until a double bond is reached. Then additional enzymes are required, an isomerase and a reductase.

β oxidation of odd-chain fatty acids produces propionyl CoA rather than acetyl CoA in the final reaction of the degradative pathway. Three enzyme-catalyzed reactions convert the resulting propionyl CoA to succinyl CoA, an intermediate of the citric acid cycle.

Fatty acid oxidation is regulated both at its sites of release and transport. Epinephrine stimulates fatty acid release from adipocytes. Glucagon and other cAMP-linked hormones such as epinephrine lead to stimulation of transport of fatty acids into mitochondria for oxidation.

The ketone bodies β-hydroxybutyrate and acetoacetate are water-soluble fuel molecules produced in the liver by the condensation of acetyl CoA molecules. A third ketone body, acetone, is produced in trace amounts in the blood by nonenzymatic decarboxylation of acetoacetate.

Fatty acid synthesis, which in animal cells takes place in the cytosol, occurs by a different pathway than fatty acid oxidation. Acetyl CoA needed for fatty acid synthesis is transported from the mitochondrion, where it is produced, to the cytosol via the citrate transport system. In addition, the citrate transport system, along with the pentose phosphate pathway, provides NADPH needed for the reactions of fatty acid biosynthesis. The conversion of acetyl CoA to malonyl CoA is the first committed step and the key regulatory step of fatty acid synthesis. The enzyme that catalyzes this reaction, acetyl-CoA carboxylase, is controlled by reversible phosphorylation in response to both hormonal signals and the presence of fatty acyl CoA.

The formation of long chain fatty acids from malonyl CoA and acetyl CoA occurs in five stages: loading, condensation, reduction, dehydration, and further reduction. This sequence is repeated from the condensation step until a long-chain fatty acid is released. In *E. coli,* the reactions of fatty acid synthesis are carried out by separate enzymes; in mammals, they are carried out by a multifunctional protein. The most common product of fatty acid synthesis is palmitate. Longer-chain and unsaturated fatty acids are produced by additional reactions.

Eicosanoids are metabolic regulators derived from arachidonate by two major pathways. The cyclooxygenase pathway leads to prostacyclin, prostaglandins, and thromboxane. Among the products of the lipoxygenase pathway are the leukotrienes.

Triacylglycerols and neutral phospholipids are synthesized by a common pathway, both being derived from phosphatidate via 1,2-diacylglycerol. Phosphatidate is also the precursor of the acidic phospholipids phosphatidylserine and phosphatidylinositol. Ether lipids have an ether linkage in place of one of the usual ester linkages. Sphingolipids have a backbone derived from palmitoyl CoA and serine. Acylation of sphingosine produces a ceramide, which can be modified by addition of phosphatidylcholine to form a sphingomyelin, or by addition of a sugar moiety to form a cerebroside. More complex sugar-lipid conjugates constitute the gangliosides.

Cholesterol is an essential component of animal cell membranes and a precursor to a large number of cell constituents. All the carbon atoms of cholesterol arise from acetyl CoA. The major regulatory step in cholesterol biosynthesis is the conversion of 3-hydroxy-3-methylglutaryl CoA (HMG-CoA) to mevalonate.

Selected Readings

Frerman, F. E. (1988). Acyl-CoA dehydrogenases, electron flavoprotein and electron transfer flavoprotein dehydrogenase. *Biochem. Soc. Trans.* 16:416–418.

Hardie, D. G. (1989). Regulation of fatty acid synthesis via phosphorylation of acetyl-CoA carboxylase. *Prog. Lipid Res.* 28:117–146.

Harwood, J. L. (1988). Fatty acid metabolism. *Annu. Rev. Plant Physiol. Plant Mol. Biol.* 39:101–138. Focusses on fatty acid metabolism in plants, with attention given to comparisons with bacteria and animals.

Mead, J. F., Alfin-Slater, R. B., Howton, D. R., and Popják, G. (1986). *Lipids: Chemistry, Biochemistry, and Nutrition.* (New York: Plenum Publishing Corporation). Authoritative treatment of the structure and metabolism of lipids.

Strålfors, P., and Belfrage, P. (1984) Reversible phosphorylation of hormone-sensitive lipase/cholesterol ester hydrolase in the hormone control of adipose tissue lipolysis and of adrenal steroidogenesis. In *Enzyme Regulation by Reversible Phosphorylation: Further Advances.* P. Cohen, ed. (Amsterdam: Elsevier Science Publishing Company), pp. 27–62.

Vance, D. E., and Vance, J. E., eds. (1991). *Biochemistry of Lipids, Lipoproteins, and Membranes.* (Amsterdam: Elsevier Science Publishing Company). Well-organized and highly readable.

Voelker, D. R., and Kennedy, E. P. (1982) Cellular and enzymatic synthesis of sphingomyelin. *Biochemistry* 21:2753–2759. E. P. Kennedy provides supporting evidence for phosphatidylcholine as the immediate donor of the phosphocholine moiety of sphingomyelin, recanting his earlier position that CDP-choline is the donor.

Wakil, S. J. (1989). Fatty acid synthase, a proficient multifunctional enzyme. *Biochemistry* 28:4523–4530.

18

Amino Acid Metabolism

The metabolism of amino acids involves a large number of enzyme-catalyzed interconversions of small molecules that contain the element nitrogen. The relatively inert gas N_2 is the ultimate source of biological nitrogen. Ammonia (NH_3) derived from N_2 is incorporated into intermediary metabolites, including the amino acids, via glutamate and glutamine.

The metabolism of amino acids is best considered from two points of view: the origins and fates of the amino acid nitrogen atoms and the origins and fates of their carbon skeletons. Nitrogen metabolism includes the major routes for the enzyme-catalyzed incorporation of nitrogen, beginning with N_2. This compound can only be metabolized or "fixed" by a few species. The product of nitrogen fixation is ammonia, which can be metabolized by all organisms. The metabolism of the precursors and degradation products of amino acids will be seen to intersect with many of the major pathways of metabolism discussed in previous chapters.

A few organisms can assimilate N_2 and simple carbon sources into amino acids. Others can synthesize the carbon chains of the amino acids but require nitrogen in the form of ammonia. Mammals have even less capacity for amino acid biosynthesis. They can synthesize only about half the amino acids they require; the rest—called *essential amino acids*—must be provided by dietary sources. The *nonessential amino acids* are those that mammals can produce in sufficient quantity metabolically. Whether amino acids are essential or nonessential is ascertained experimentally for individual organisms or species. The patterns of amino acid metabolism can be somewhat different across species, with variations in the pathways for synthesis of the same molecule. There also are a variety of routes for the degradation and disposal of the nitrogen-containing waste products of amino acid breakdown.

The role of amino acids in metabolism is distinct from that of the other major dietary components, fats and carbohydrates, because amino acids are not primarily used to supply cellular energy. Rather, amino acids are required principally to build proteins, which undergo constant turnover in cells. Each type of protein in the cell is degraded to its constituent amino acids at a different rate. In mammals, there are situations, such as fasting or a dietary excess of protein, during which metabolites from the catabolism of amino acids enter the pathway of gluconeogenesis.

Considering that there are 20 amino acids, many of which are converted to other metabolites, the number of enzymatic reactions in this chapter may seem overwhelming. However, the pathways of amino acid metabolism possess central features and common themes that reflect an underlying pattern of organization for this complex series of chemical conversions. We begin our examination of amino acid metabolism by considering the incorporation of nitrogen into cellular metabolites.

18·1 Nitrogen Is Cycled Through the Biosphere

Nitrogen in biological systems originates from gaseous N_2, which comprises about 80% of the atmosphere. The two atoms of molecular nitrogen are held together by an extremely strong triple bond (bond energy = 225 kcal mol^{-1}). This property makes nitrogen gas chemically unreactive, so it is not surprising that very specific and sophisticated enzyme systems are required to initiate the series of chemical reactions leading to amino acids.

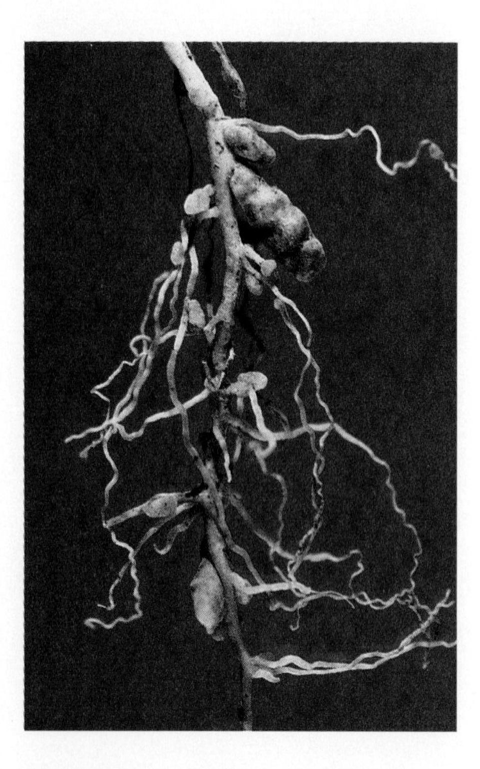

Figure 18·1
Root nodules of a clover plant. Symbiotic bacteria of the genus *Rhizobium* reside in these nodules, which provide a suitable environment for the reduction of atmospheric nitrogen to ammonia. (Photo by William R. West, supplied by Carolina Biological Supply Company.)

A. A Few Organisms Carry Out Nitrogen Fixation

The reduction of N_2 to ammonia is called *nitrogen fixation*. In the chemical industry, nitrogen is fixed for use in plant fertilizer by an energetically expensive process involving the use of high temperature and pressure in concert with special catalysts to drive the reduction of N_2 by H_2 to form ammonia. The availability of biologically useful nitrogen is often a limiting factor for plant growth, and the application of nitrogenous fertilizers is important in obtaining higher crop yields.

Most nitrogen fixation in the biosphere is carried out by a few species of bacteria or algae that have the ability to synthesize the complex enzyme nitrogenase. This multisubunit protein catalyzes the conversion of N_2 to two molecules of NH_3. Nitrogenase is present in members of the genus *Rhizobium* that live symbiotically in root nodules of many plants, including the legumes beans, peas, alfalfa, and clover (Figure 18·1). N_2 is also fixed by free-living soil bacteria such as *Azotobacter, Klebsiella,* and *Clostridium* and by cyanobacteria found in aqueous environments. Most plants require a supply of fixed nitrogen from the environment. Sources of this nutrient include excess fixed and oxidized nitrogen excreted by microorganisms, decayed animal and plant tissue, and fertilizer. Animals obtain fixed nitrogen exclusively through ingestion of plant and animal tissue.

The enzyme nitrogenase consists of two protein components, one containing iron and the other, iron and molybdenum. Both metalloproteins are highly susceptible to inactivation by O_2. Within the root nodules of leguminous plants, the protein leghemoglobin (Stereo S4·18, page 4·35) binds O_2 and thereby eliminates it from the immediate environment of the nitrogen-fixing enzymes of rhizobia.

A strong reducing agent—either reduced ferredoxin or reduced flavodoxin (a flavoprotein electron carrier from microorganisms)—is required for the enzymatic reduction of N_2 to NH_3. In the course of the six-electron reduction of a single molecule of N_2, 12 ATP molecules must be converted to ADP and P_i. The energy cost of biological nitrogen fixation is actually even higher, due to an apparently obligatory side reaction that detracts from the efficiency of the process. By present estimates, at least 16 ATP molecules are consumed per molecule of N_2 reduced.

The nitrogenase reaction is thought to proceed in three discrete steps, with two electrons transferred at each step. It has been postulated that diimine and hydrazine are reaction intermediates that remain bound to the enzyme.

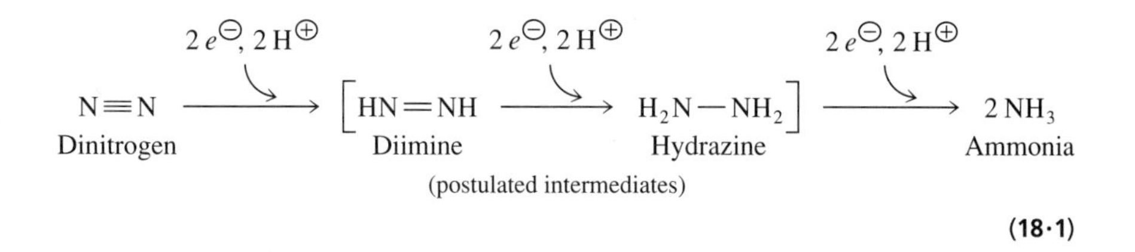

(18·1)

B. Plants and Microorganisms Can Convert Nitrate and Nitrite to Ammonia

During lightning storms, high-voltage discharges catalyze the oxidation of N_2. Upon reaction of N_2 with O_2 in the atmosphere, biologically useful nitrate ($NO_3^{\ominus}$) and nitrite ($NO_2^{\ominus}$) are formed and washed into the soil. Other sources of nitrate and nitrite are the oxidation of NH_3 by microorganisms such as *Nitrosomonas* and *Nitrobacter*. Most plants and microorganisms contain nitrate reductase and nitrite reductase, enzymes that together catalyze the reduction of the nitrogen oxides to ammonia. In higher plants, reduced ferredoxin, formed in the light reactions of photosynthesis (Section 16·4), passes its reducing power to either NAD$^{\oplus}$ or NADP$^{\oplus}$. The reduced pyridine nucleotide is a cosubstrate in the reaction, catalyzed by nitrate reductase, that converts nitrate to nitrite:

$$NO_3^{\ominus} \xrightarrow[\text{Nitrate reductase}]{NAD(P)H,\,H^{\oplus} \;\rightarrow\; NAD(P)^{\oplus}} NO_2^{\ominus} + H_2O \qquad \textbf{(18·2)}$$

Nitrite is reduced to NH_3 in a reaction catalyzed by nitrite reductase. This reduction utilizes reduced ferredoxin as a source of electrons. The intermediates of the reaction have not been isolated, but it is presumed that the sequence of steps is:

$$NO_2^{\ominus} \longrightarrow \underbrace{\left[NO^{\ominus} \longrightarrow NH_2OH\right]}_{\text{(postulated intermediates)}} \longrightarrow NH_3 \qquad \textbf{(18·3)}$$

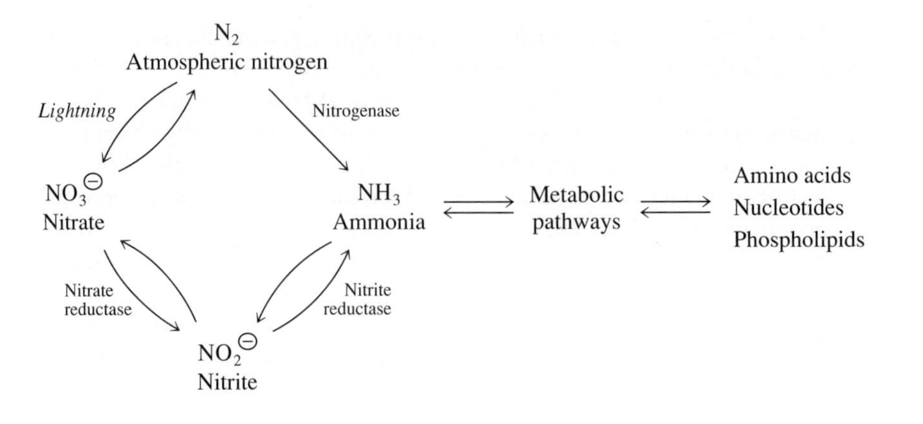

Figure 18·2
The nitrogen cycle. A few free-living or symbiotic microorganisms can convert N_2 to ammonia. Ammonia is incorporated into biomolecules such as amino acids and proteins, which later are degraded to form ammonia. Many soil bacteria and plants can carry out the reduction of nitrate to ammonia via nitrite. Several genera of bacteria can convert ammonia to nitrite. Others can oxidize nitrite to nitrate. Some other bacterial genera can reduce nitrate to N_2.

Some bacteria can convert ammonia to nitrite, whereas others can convert nitrite to nitrate. This formation of nitrate is called *nitrification.* Still other bacteria can reduce nitrate or nitrite to N_2 *(denitrification).* The overall scheme for interconversion of the major nitrogen-containing chemical species of the biosphere is depicted in Figure 18·2. The biological cycling of N_2 between nitrogen oxides, ammonia, nitrogenous biomolecules, and back to N_2 is termed the *nitrogen cycle*. By means of this cycle, nitrogen gas is converted to usable forms and assimilated into simple organic compounds, which then transfer nitrogen to a variety of low molecular weight metabolites and eventually to macromolecules. The cycle is completed when nitrogenous products arising from excretion or from the death of organisms are broken down to nitrogen oxides and N_2.

18·2 Glutamate Dehydrogenase Catalyzes the Incorporation of Ammonia into Glutamate

Ammonia generated biosynthetically from N_2 or nitrogen oxides is next assimilated into a large number of low molecular weight metabolites. Ammonia has a pK_a of 9.2 and therefore exists in neutral aqueous solution mainly as ammonium ions, $NH_4^{\oplus}$. In the catalytic centers of enzymes, however, the unprotonated nucleophile NH_3 is the reactive species. One highly efficient route for the incorporation of ammonia into the central pathways of amino acid metabolism is the reductive amination of α-ketoglutarate to glutamate, catalyzed in both plants and animals by glutamate dehydrogenase:

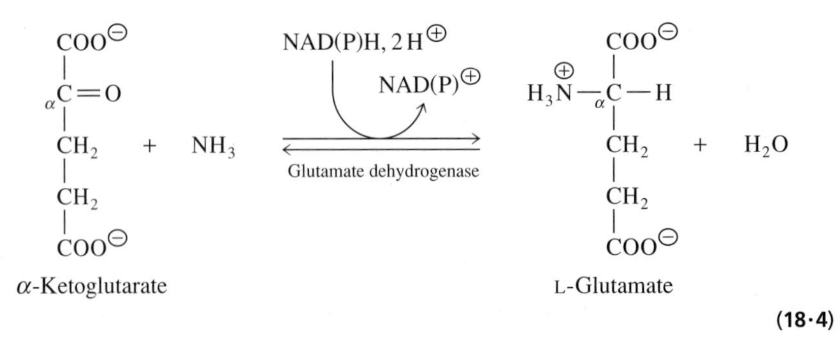

(18·4)

The glutamate dehydrogenases of some species or tissues are specific for NADH, whereas others are specific for NADPH; still others can use either cofactor. The reaction involves the condensation of ammonia with the carbonyl group of α-ketoglutarate, forming an enzyme-bound α-iminiumglutarate intermediate. This intermediate is reduced by the transfer of a hydride ion from NADH or NADPH (Figure 18·3).

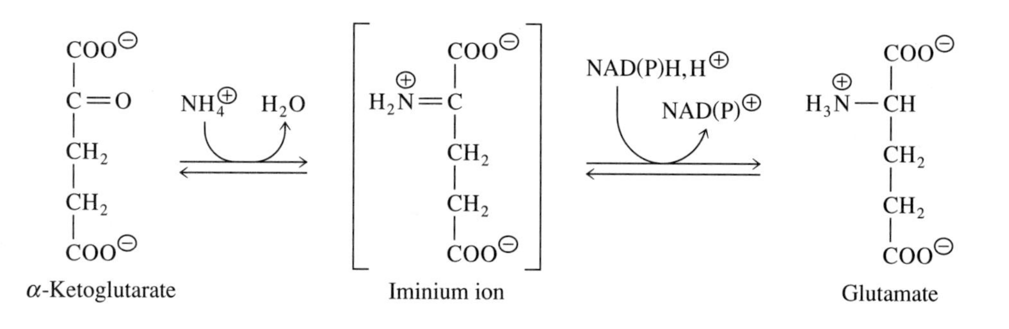

Figure 18·3
The glutamate dehydrogenase reaction.

The glutamate dehydrogenase reaction can have different roles in different organisms. In *E. coli,* its major role is glutamate formation when excess $NH_4^{\oplus}$ is present. In the mold *Neurospora crassa,* separate enzymes with distinct pyridine nucleotide specificities exist; an NADPH-dependent enzyme catalyzes primarily the reductive amination of α-ketoglutarate to glutamate, and an NAD$^{\oplus}$-dependent enzyme functions in oxidative catabolism of glutamate to α-ketoglutarate. In mammals, glutamate dehydrogenase usually plays a catabolic role, although the reverse reaction occurs in several tissues. The main direction of flow in this reaction seems to be toward glutamate (anabolic) in plants and microorganisms and away from glutamate (catabolic) in animals.

A number of compounds are allosteric effectors of liver glutamate dehydrogenase in vitro, yet we have seen that near-equilibrium reactions are not subject to regulation, and effectors of enzymes that catalyze reversible reactions in vivo are likely to have little physiological significance. Flux through these reactions is controlled simply by changes in the concentrations of substrates and products.

18·3 Glutamine Is an Important Carrier of Ammonia-Derived Nitrogen

Another reaction critical to the assimilation of ammonia in many organisms is the formation of glutamine from glutamate and ammonia, catalyzed by glutamine synthetase:

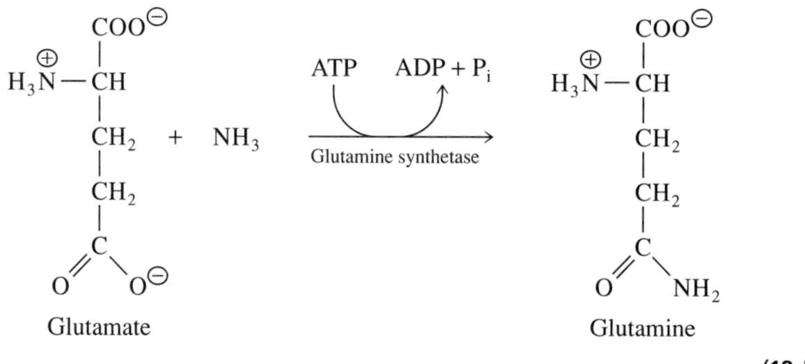

(18·5)

Glutamine is a nitrogen donor in many biosynthetic reactions. For example, the amide nitrogen of glutamine is the direct precursor of several of the nitrogen atoms of the purine and pyrimidine ring systems (Chapter 19).

Glutamine has a special role in mammals, where the formation of this compound reduces the circulating concentration of toxic $NH_4^{\oplus}$. In mammals, glutamine is mainly synthesized in muscle and transported via the circulatory system to other tissues, such as liver and kidney. The concentration of glutamine in blood (450–600 μM) is the highest of any amino acid.

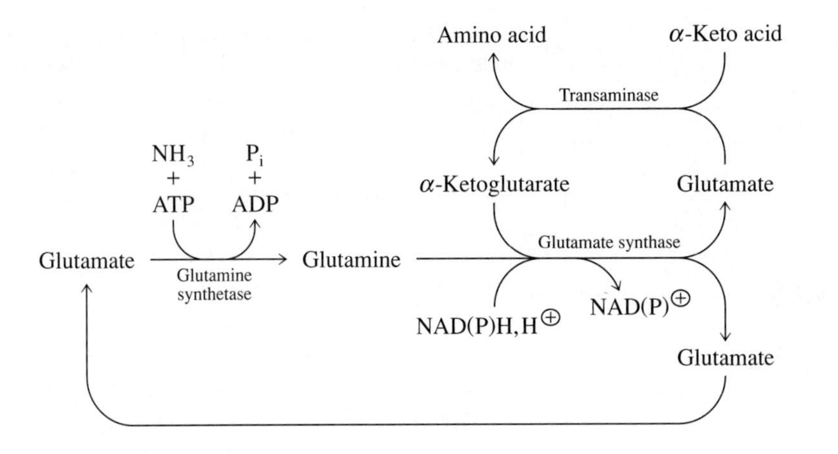

Figure 18·4
Combined action of glutamine synthetase and glutamate synthase in a pathway leading to the transamination of α-keto acids to form amino acids.

In prokaryotes, the amide nitrogen of glutamine can be transferred to α-ketoglutarate in a reductive amination reaction catalyzed by glutamate synthase. This enzyme requires a reduced pyridine nucleotide as a reducing agent.

$$\text{Glutamine} + \alpha\text{-Ketoglutarate} \xrightarrow[\text{Glutamate synthase}]{\text{NAD(P)H,H}^{\oplus} \;\rightarrow\; \text{NAD(P)}^{\oplus}} 2\ \text{Glutamate} \qquad \textbf{(18·6)}$$

The reaction has the same outcome as the glutamate dehydrogenase reaction—the assimilation of ammonia into glutamate. Coupled reactions catalyzed by glutamine synthetase and glutamate synthase are an important alternate to the glutamate dehydrogenase reaction in prokaryotes, especially when the concentration of available ammonia is low. Figure 18·4 shows how the combined actions of glutamine synthetase and glutamate synthase lead to the incorporation of ammonia into a variety of amino acids by formation of glutamate, followed by transamination with α-keto acids to form a number of amino acids. (Transamination reactions are examined in detail in Section 18·5). Note that the glutamate dehydrogenase reaction utilizes two molecules of reduced pyridine nucleotide per pair of glutamates formed, whereas the glutamate synthase route consumes one molecule of reduced pyridine nucleotide and one molecule of ATP.

18·4 Glutamine Synthetase in *E. coli* Is Regulated by a Sophisticated Mechanism

Glutamine synthetase plays a critical role in nitrogen metabolism since glutamine is a precursor of many other metabolites. The regulatory properties of glutamine synthetase of *E. coli* have been intensively investigated. It is regulated at three levels: by end-product inhibition (there are at least nine allosteric inhibitors), by covalent modification, and by regulation at the level of protein synthesis.

E. coli glutamine synthetase consists of 12 identical subunits (subunit MW 51 600). Each subunit has not only a catalytic site but also binding sites for the allosteric inhibitors. The nine recognized inhibitors are AMP, carbamoyl phosphate, CTP, histidine, tryptophan, glucosamine 6-phosphate, alanine, serine, and glycine (Figure 18·5). Six of the inhibitors contain a nitrogen atom obtained directly from the amide nitrogen of glutamine. The nitrogen atoms of alanine, serine, and glycine can be derived from glutamine indirectly. No single inhibitor blocks catalysis by glutamine synthetase. Instead, the degree of inhibition increases as more of the inhibitors bind, a process known as *cumulative feedback inhibition*. For example, when glutamine synthetase was tested with saturating concentrations of tryptophan, CTP, carbamoyl phosphate, or AMP, the residual activities were 84%, 86%, 87%,

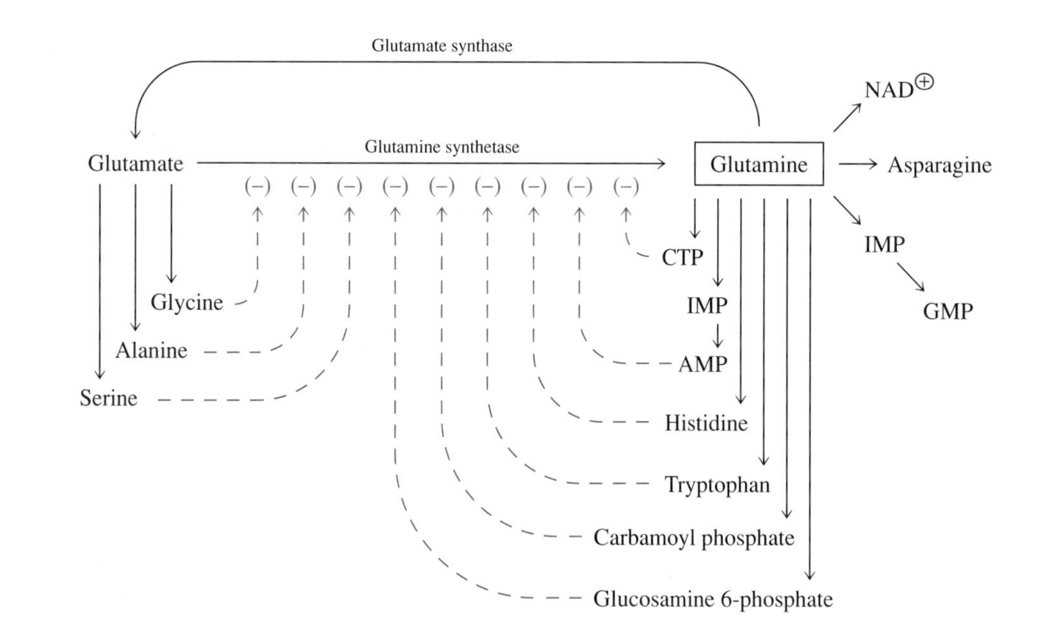

Figure 18·5
Allosteric feedback inhibition of glutamine synthetase in *E. coli*. Several compounds that contain a nitrogen atom ultimately derived from the amide group of glutamine are inhibitors of glutamine synthetase. The presence of a single inhibitor causes only partial inhibition. When all of the inhibitors are present, inhibition can be nearly total.

and 59%, respectively. When all four of these inhibitors were present simultaneously, the residual activity was 37%. The presence of all nine inhibitors blocks essentially all of the activity of the enzyme.

Covalent modification of glutamine synthetase in *E. coli* occurs as part of a regulatory cascade that includes the sequential addition of AMP moieties (adenylylation) to specific tyrosine residues, one present in each of the 12 subunits of this enzyme (Figure 18·6). Adenylylation, catalyzed by glutamine synthetase adenylyltransferase, decreases the activity of glutamine synthetase. The removal of the AMP group from glutamine synthetase, with concomitant reactivation of the enzyme, is also catalyzed by the adenylyltransferase protein. Which activity predominates is determined by another protein, P_{II}, that can form a complex with the

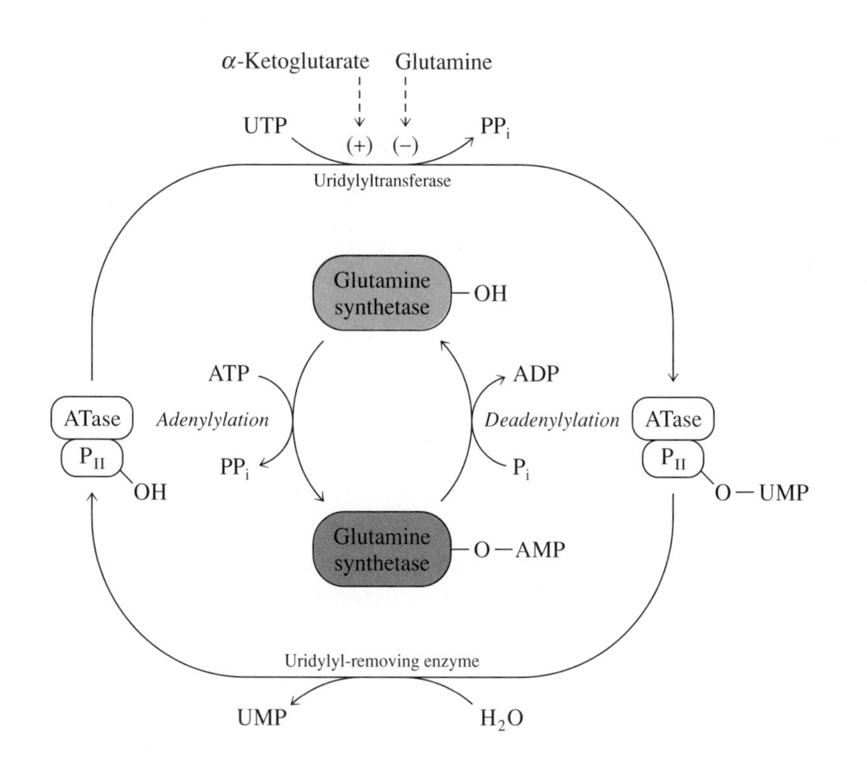

Figure 18·6
Regulation of glutamine synthetase by covalent modification. Glutamine synthetase is inactivated by adenylylation and activated by deadenylylation, both of which are catalyzed by the bifunctional enzyme glutamine synthetase adenylyltransferase (ATase). ATase is complexed to a regulatory protein called P_{II}. The reversible uridylylation of P_{II} controls ATase; UMP incorporation is controlled in turn by the allosteric effectors α-ketoglutarate and glutamine.

adenylyltransferase. P_{II} itself is subject to covalent modification by the reversible transfer of a UMP group to the phenolic oxygen of a specific tyrosine residue of the protein, in a reaction catalyzed by uridylyltransferase. The removal of UMP is catalyzed by uridylyl-removing enzyme. The uridylylation of P_{II} activates glutamine synthetase by stimulating deadenylylation of the enzyme. Uridylylation is stimulated by α-ketoglutarate and inhibited by glutamine. α-Ketoglutarate appears in the ammonia-assimilation pathway before the glutamine synthetase reaction and thus is a feed-forward activator. Glutamine, the product of the reaction, is a feedback inhibitor.

Allosteric regulation of the activity of an enzyme is extremely rapid, with the change of conformation occurring in roughly a millisecond. Changes in enzyme activity through covalent modification are significantly slower, taking closer to a second. Still longer-term adaptation to environmental changes is afforded by regulation at the level of gene expression. When ample quantities of usable nitrogen are available to *E. coli* from its environment, relatively low levels of glutamine synthetase are found within cells. Under these conditions, assimilation of ammonia can occur via the glutamate dehydrogenase reaction. In contrast, when *E. coli* grows in nitrogen-limited media, the alternate pathway utilizing glutamate synthase is followed, and higher levels of glutamine synthetase are produced through increased transcription and translation of its gene. The levels of glutamine synthetase in *E. coli* can vary over a 100-fold range.

Mammalian glutamine synthetases differ markedly from the bacterial enzyme in both their physical properties and regulatory behavior. The enzymes of liver and brain have been most thoroughly characterized. These proteins contain eight identical subunits, but they can also exist as tetramers. The mammalian glutamine synthetases are not regulated by covalent modification. Glycine, serine, alanine, and carbamoyl phosphate are inhibitors and α-ketoglutarate is an activator of mammalian liver glutamine synthetase. However, the range of modulation of the mammalian synthetase activity is not as extensive as that of the *E. coli* enzyme.

18·5 Transaminases Catalyze the Reversible Interconversion of α-Amino Acids and α-Keto Acids

Glutamate is a key intermediate in a large number of the reactions of amino acid metabolism, both synthetic and degradative. The amino group of glutamate can be transferred to many α-keto acids in reactions catalyzed by enzymes known as transaminases or aminotransferases. The basic reaction scheme for transaminases is shown in Figure 18·7.

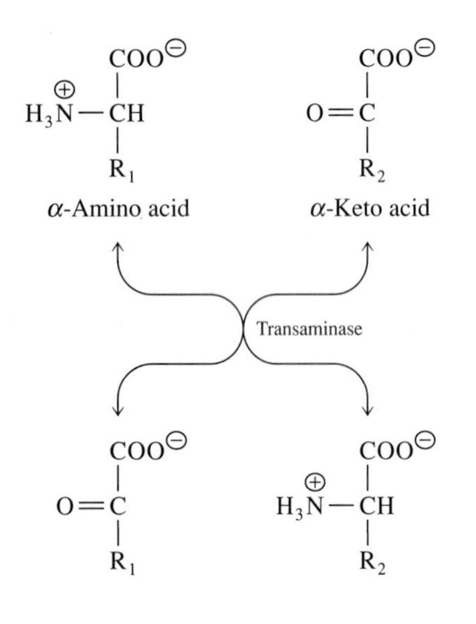

Figure 18·7
Transfer of an amino group from an α-amino acid to an α-keto acid, catalyzed by a transaminase.

In amino acid biosynthesis, the amino group of glutamate is transferred to various α-keto acids, generating the corresponding α-amino acids. Most of the common amino acids can be formed by transamination. However, the α-keto acid corresponding to lysine is unstable, so lysine is not formed by transamination. In addition, because the α-keto acid of threonine is a poor substrate for transamination, little threonine is formed by this type of reaction. In catabolism, one or more transamination reactions generates glutamate or aspartate. The amino groups of the glutamate and aspartate then enter pathways leading to nitrogen disposal, such as the urea cycle (Section 18·9) or the synthesis of uric acid (Chapter 19).

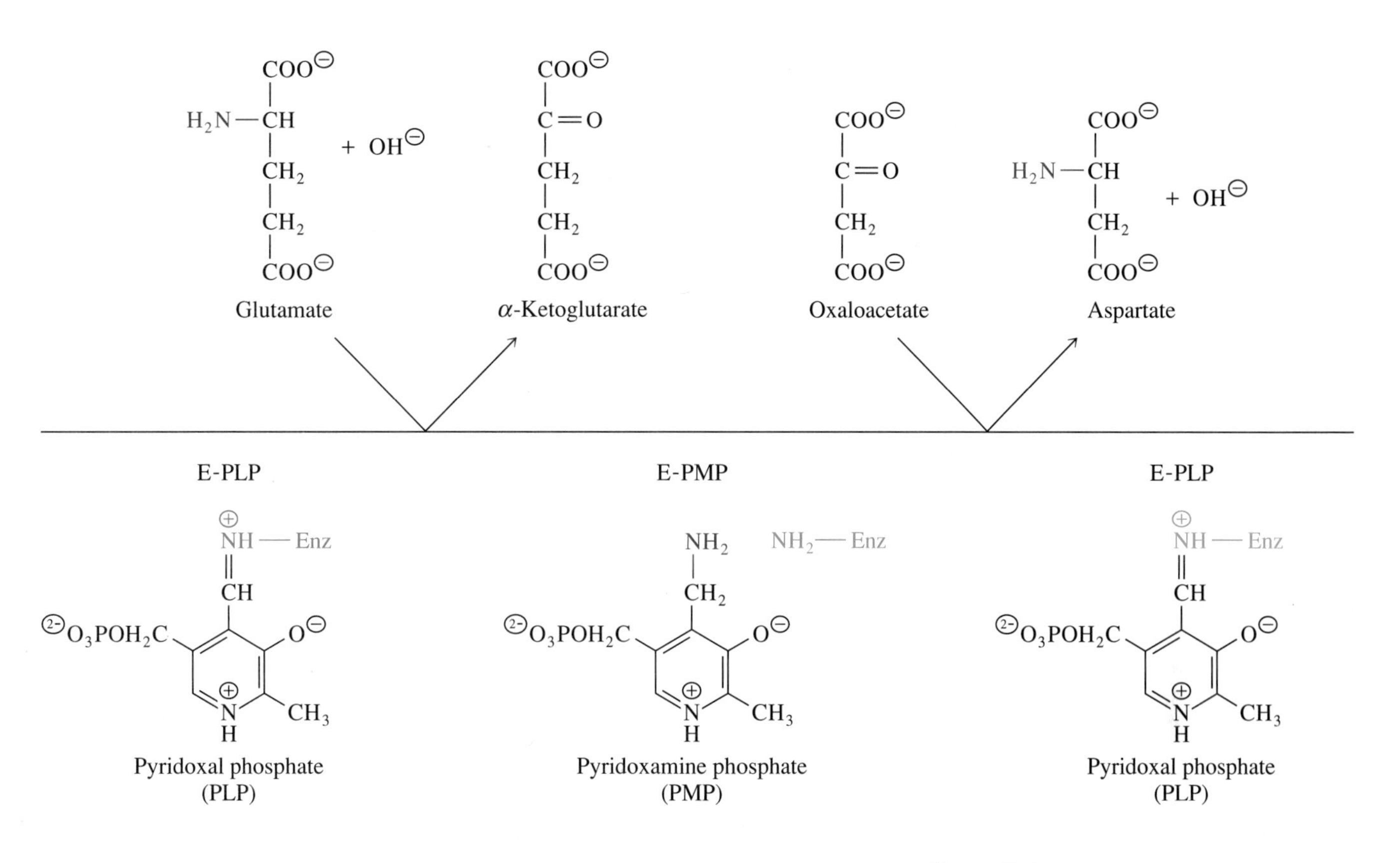

Figure 18·8
Ping-pong kinetic mechanism of aspartate transaminase. Part of the substrate is transferred to the enzyme and a stable enzyme--pyridoxamine phosphate intermediate (E-PMP) is generated. The second half of the mechanism mirrors the first half.

All known transaminases require the coenzyme pyridoxal phosphate (Section 7·6) and all have ping-pong kinetic mechanisms. The most thoroughly studied is aspartate transaminase.

$$\text{Glutamate} + \text{Oxaloacetate} \underset{}{\overset{\text{Aspartate transaminase}}{\rightleftharpoons}} \alpha\text{-Ketoglutarate} + \text{Aspartate} \qquad (18·7)$$

Figure 18·8 shows the kinetic mechanism of this enzyme. The amino group of glutamate is transferred to enzyme-bound pyridoxal phosphate, forming a stable enzyme–pyridoxamine phosphate (E-PMP) intermediate, with release of α-ketoglutarate. At this point, a second substrate, oxaloacetate, binds to the enzyme and reacts with E-PMP, forming aspartate and regenerating the original form of the enzyme. Each step in this reaction sequence is reversible.

The mechanism of the initial half-reaction of transamination is shown in Figure 18·9. The E-PLP holoenzyme exists in the form of an internal aldimine, with PLP covalently linked as a Schiff base to the side chain of an active-site lysine residue of the apoenzyme. PLP is further anchored to the enzyme by many noncovalent bonds. In Step 1, the nucleophilic —NH_2 form of the donor amino acid displaces the lysine residue by transimination (see Figure 7·16, page 7·13 for the

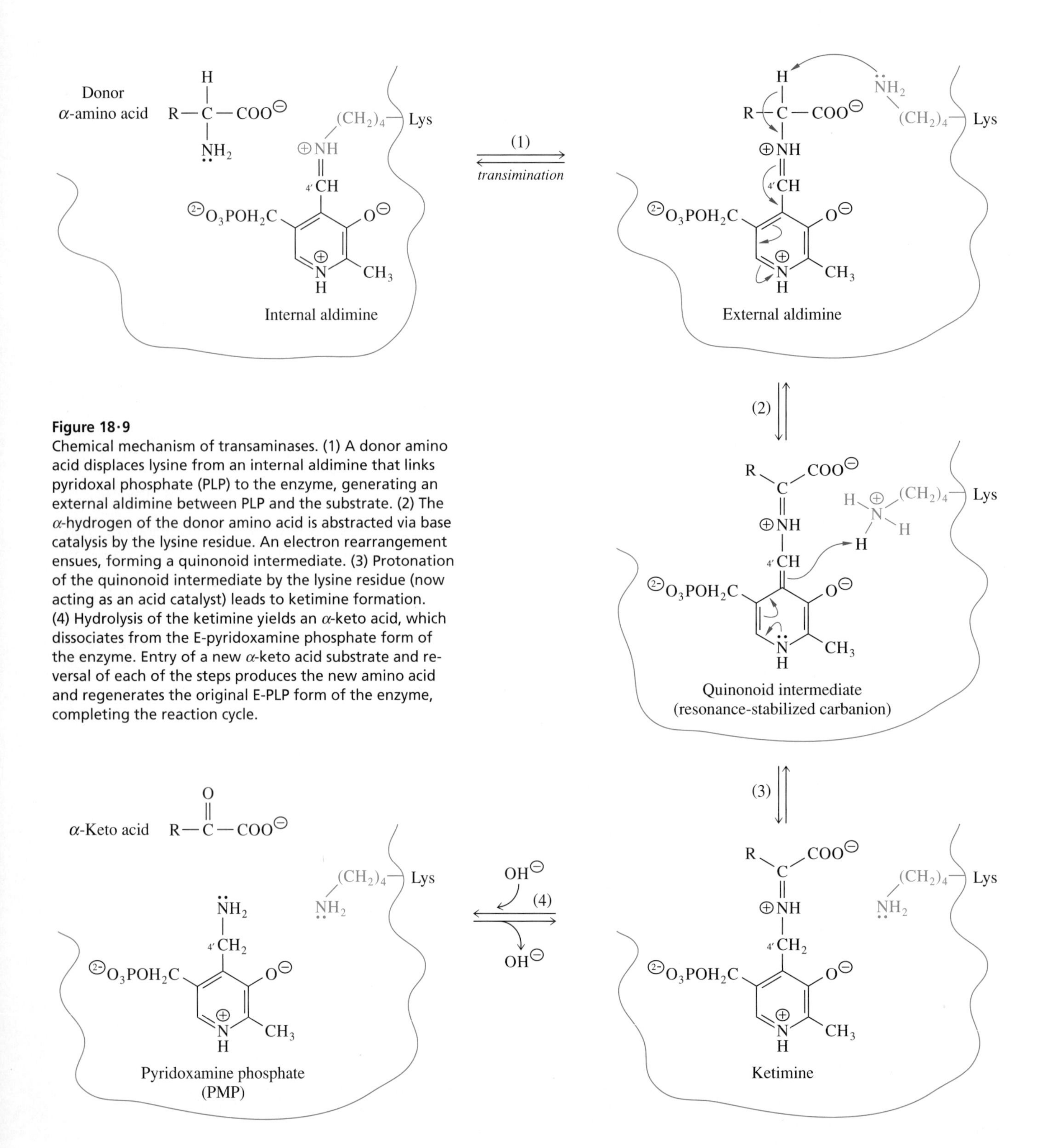

Figure 18·9
Chemical mechanism of transaminases. (1) A donor amino acid displaces lysine from an internal aldimine that links pyridoxal phosphate (PLP) to the enzyme, generating an external aldimine between PLP and the substrate. (2) The α-hydrogen of the donor amino acid is abstracted via base catalysis by the lysine residue. An electron rearrangement ensues, forming a quinonoid intermediate. (3) Protonation of the quinonoid intermediate by the lysine residue (now acting as an acid catalyst) leads to ketimine formation. (4) Hydrolysis of the ketimine yields an α-keto acid, which dissociates from the E-pyridoxamine phosphate form of the enzyme. Entry of a new α-keto acid substrate and reversal of each of the steps produces the new amino acid and regenerates the original E-PLP form of the enzyme, completing the reaction cycle.

mechanism of this reaction). Electron rearrangement, assisted by base catalysis involving the lysine residue, leads in Step 2 to a quinonoid intermediate. Further rearrangement in Step 3 leads to a ketimine, which is hydrolyzed to yield a free α-keto acid and the E-PMP form of the enzyme. In the second phase of the reversible and symmetrical reaction, a different α-keto acid becomes the substrate for the reverse sequence, starting with Step 4. The reaction sequence is completed when Step 1 is reached and E-PLP is formed, with release of an amino acid containing the carbon skeleton of the second α-keto acid.

The equilibrium constants for transaminase reactions are near 1, and the direction in which the reactions proceed in vivo depends on the supply of substrates and the removal of products. For example, in cells having excess α-amino nitrogen, the amino groups could be transferred via one or a series of transaminations to α-ketoglutarate to yield glutamate, which could undergo oxidative deamination. When amino acids are being actively formed, transamination reactions in the opposite direction will occur, with amino-group donation by glutamate.

18·6 Many Nonessential Amino Acids Are Synthesized Directly from Key Intermediates of Central Metabolism

Having examined the incorporation of nitrogen into amino acids, we shall now turn our attention to the origins of their carbon skeletons. Figure 18·10 depicts in schematic form the biosynthetic routes leading to the 20 common amino acids. In general, the nonessential amino acids are derived by short pathways from intermediates of glycolysis, the pentose phosphate pathway, or the citric acid cycle. The classification of amino acids as either nonessential or essential for mammals roughly parallels the number of steps in their synthetic pathways and, in related fashion, the energy required for their synthesis—amino acids with the largest energy requirements for synthesis are generally essential amino acids (Table 18·1).

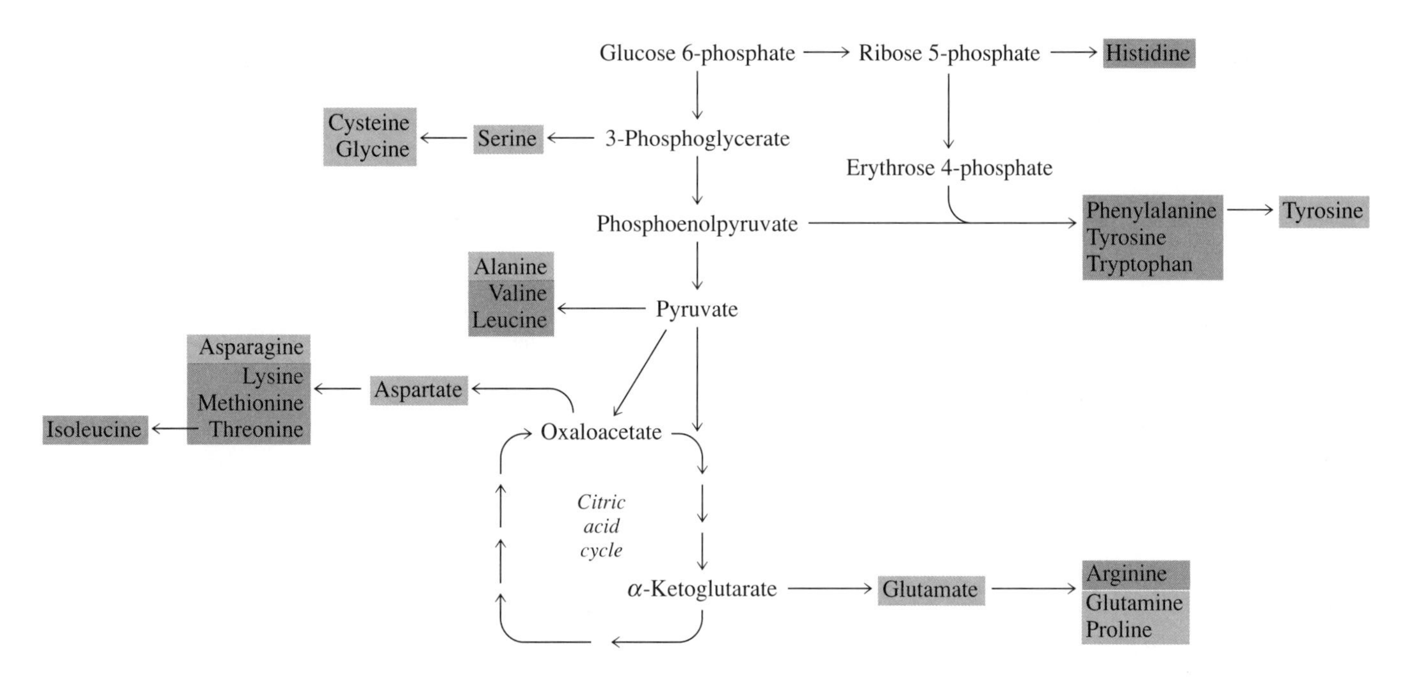

Figure 18·10
Biosynthesis of amino acids. Shown are the connections to glycolysis, the pentose phosphate pathway, and the citric acid cycle. Amino acids essential for humans are indicated in red; nonessential amino acids are indicated in blue.

Table 18·1 Essential and nonessential amino acids for humans, with energetic requirements for their biosynthesis

Amino acid	Moles of ATP required per mole of amino acid produced[1]	
	Nonessential	*Essential*
Glycine	12	
Serine	18	
Cysteine	19*	
Alanine	20	
Aspartate	21	
Asparagine	22	
Glutamate	30	
Glutamine	31	
Threonine		31
Proline	39	
Valine		39
Histidine		42
Arginine		44
Methionine		44
Leucine		47
Lysine		50 or 51
Isoleucine		55
Tyrosine	62*	
Phenylalanine		65
Tryptophan		78

*Formed from essential amino acids
[1]Moles of ATP required includes ATP used for synthesis of precursors and conversion of precursors to products.
[Adapted from Atkinson, D. E. (1977). *Cellular Energy Metabolism and Its Regulation* (New York: Academic Press).]

We have seen that glutamate can be formed by either the reductive amination or transamination of α-ketoglutarate. Other amino acids formed by simple transamination are alanine and aspartate. With the former, pyruvate is the amino-group acceptor:

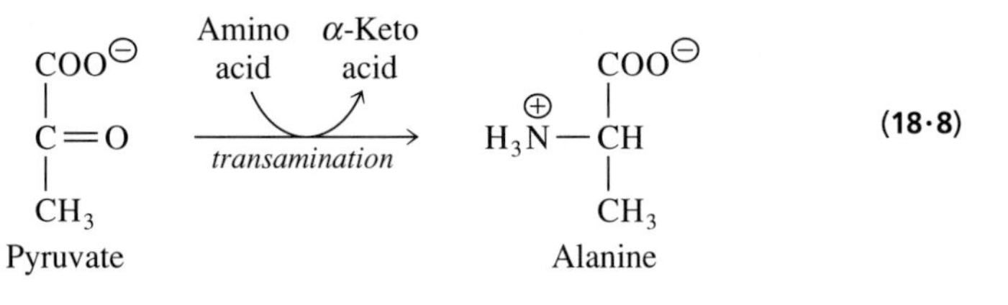

(18·8)

With the latter, oxaloacetate is the amino-group acceptor:

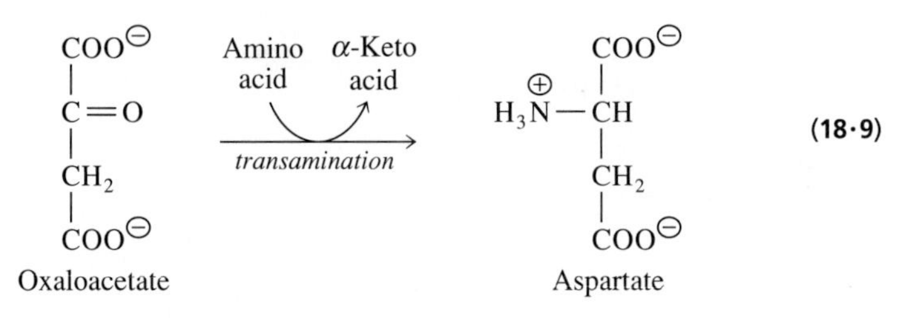

(18·9)

Glutamine, as we have seen, is formed from ammonia and glutamate in an ATP-dependent reaction catalyzed by glutamine synthetase. Asparagine, the other amide-containing amino acid, is synthesized in many animals in a reaction involving the transfer of the amide nitrogen of glutamine to aspartate, catalyzed by asparagine synthetase.

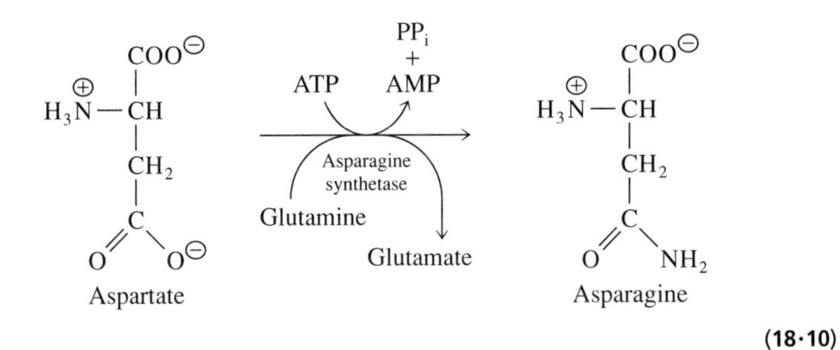

(18·10)

In some bacteria and plants, ammonia rather than glutamine is the source of the amide group of asparagine, via an ATP-dependent reaction that is mechanistically similar to that catalyzed by glutamine synthetase.

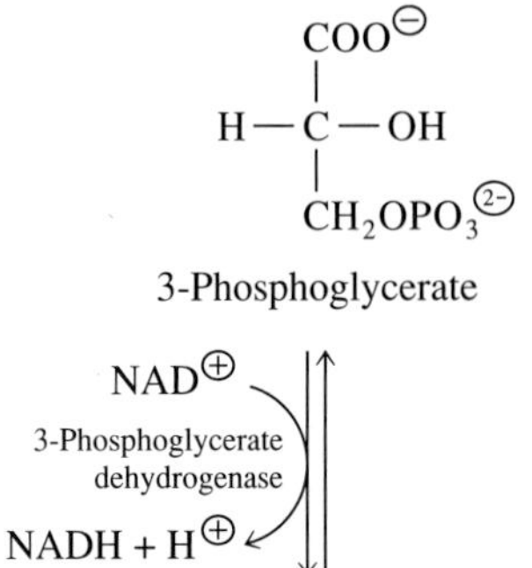

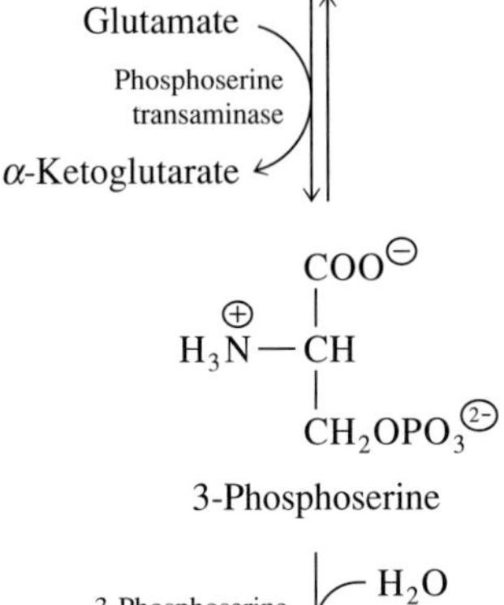

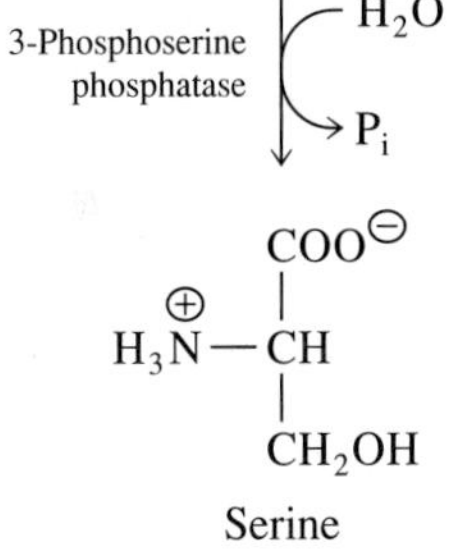

Figure 18·11
Serine biosynthesis.

A. Serine, Glycine, and Cysteine Are Derived from 3-Phosphoglycerate

Serine is the major biosynthetic precursor of both glycine and cysteine. Serine formation begins with 3-phosphoglycerate and involves three reactions (Figure 18·11). First, the secondary hydroxyl substituent of 3-phosphoglycerate is oxidized to a keto group by the action of 3-phosphoglycerate dehydrogenase, forming 3-phosphohydroxypyruvate. This compound undergoes transamination with glutamate to form 3-phosphoserine and α-ketoglutarate. Finally, 3-phosphoserine phosphatase catalyzes the formation of serine.

Serine is a major source of glycine via a reversible reaction catalyzed by serine hydroxymethyltransferase (Figure 18·12). The reverse reaction, which occurs in plant mitochondria, represents an alternate route to serine from that shown in Figure 18·11. The serine hydroxymethyltransferase reaction requires two cofactors: the prosthetic group pyridoxal phosphate and the cosubstrate tetrahydrofolate.

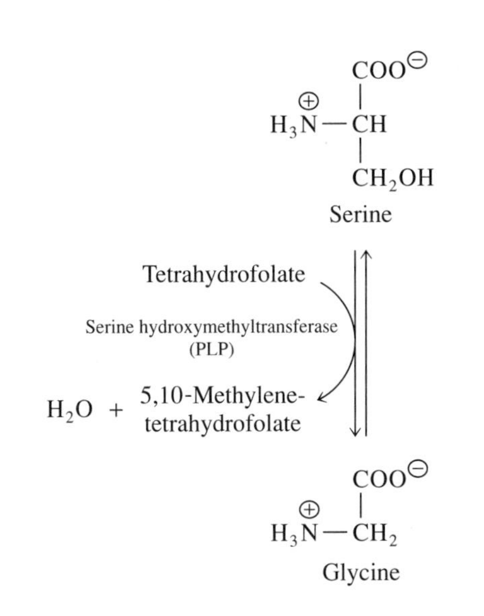

Figure 18·12
Biosynthesis of glycine. Pyridoxal phosphate (PLP) and tetrahydrofolate are both cofactors for this reaction, which is catalyzed by serine hydroxymethyltransferase.

Figure 18·13
Metabolic routes from serine and glycine.

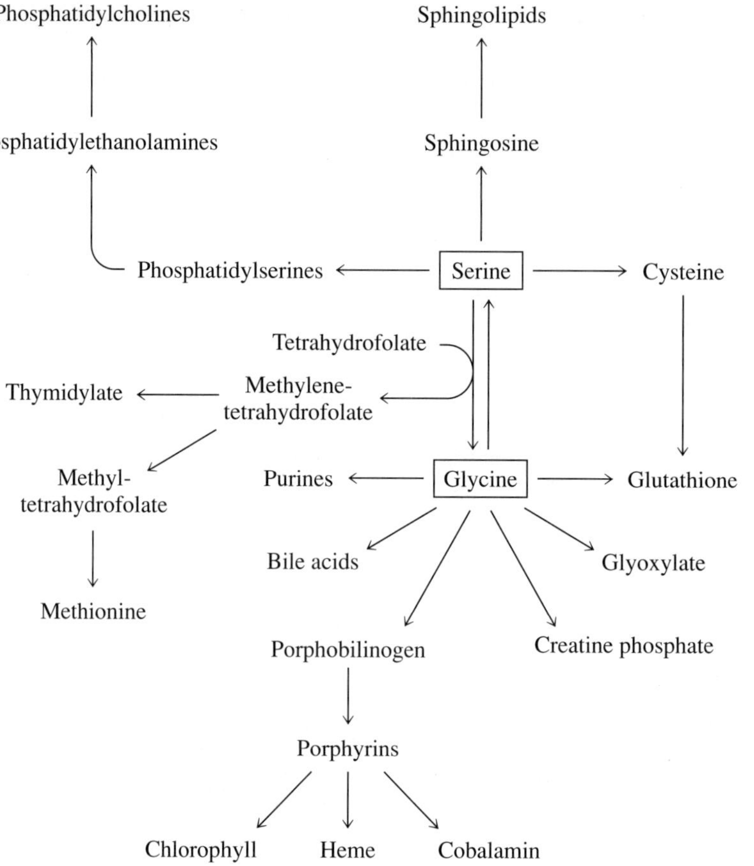

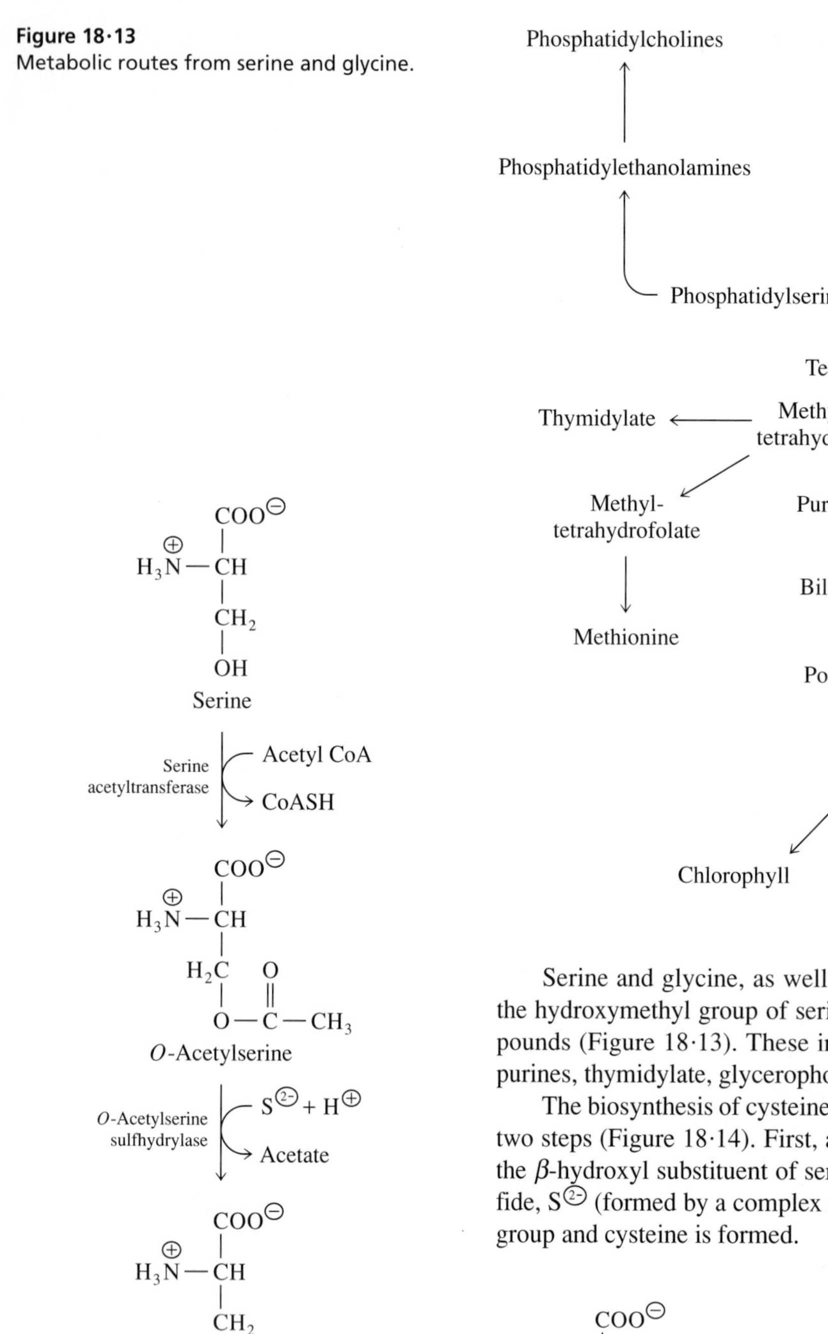

Figure 18·14
Biosynthesis of cysteine from serine in bacteria and plants.

Serine and glycine, as well as 5,10-methylenetetrahydrofolate derived from the hydroxymethyl group of serine, are metabolic precursors of many other compounds (Figure 18·13). These include the amino acids methionine and cysteine, purines, thymidylate, glycerophospholipids, and porphyrins (Box 18·1).

The biosynthesis of cysteine from serine in bacteria and plants is carried out in two steps (Figure 18·14). First, an acetyl group from acetyl CoA is transferred to the β-hydroxyl substituent of serine, forming *O*-acetylserine. Next, inorganic sulfide, S^{2-} (formed by a complex reduction of sulfate, SO_4^{2-}), displaces the acetate group and cysteine is formed.

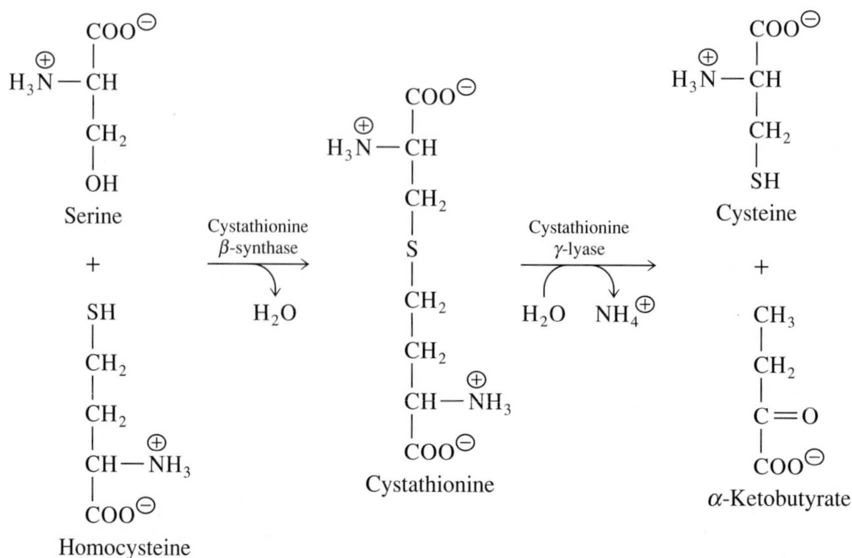

Figure 18·15
Biosynthesis of cysteine in mammals.

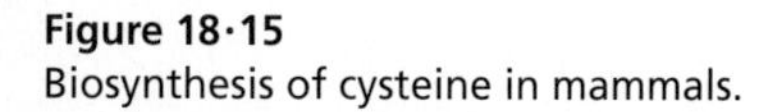

Cysteine biosynthesis in mammals also begins with serine. In the first step, serine condenses with homocysteine, a metabolite that occurs in the biosynthesis and catabolism of methionine. Since methionine is an essential amino acid for mammals, the availability of homocysteine for this reaction depends upon an ample supply of methionine in the diet of mammals. The product of the condensation reaction, cystathionine, is cleaved to α-ketobutyrate and cysteine (Figure 18·15).

Box 18·1 Glycine Is a Precursor of Heme

Glycine has several prominent synthetic roles, including its incorporation into purine nucleotides (Section 19·3) and porphyrins. The first step in the synthesis of the porphyrins of heme is condensation of the carbonyl group of succinyl CoA with the α-carbon of glycine to form δ-aminolevulinic acid (Figure 1). Two molecules of δ-aminolevulinic acid condense to form the substituted pyrrole porphobilinogen. Four molecules of porphobilinogen are converted in two steps to uroporphyrinogen, a porphyrin that is modified by alteration of its side chains and insertion of the iron atom to produce heme.

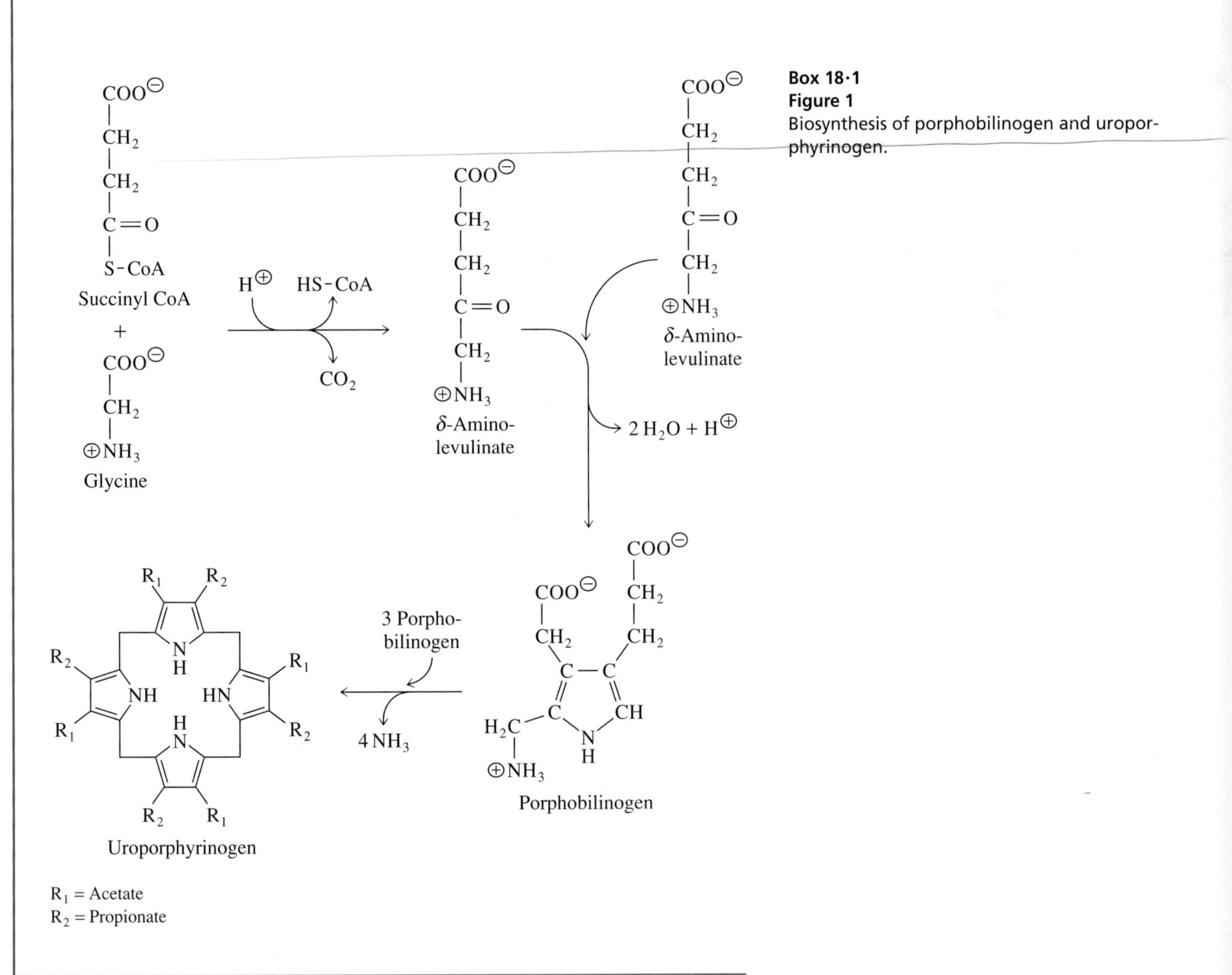

Box 18·1 Figure 1
Biosynthesis of porphobilinogen and uroporphyrinogen.

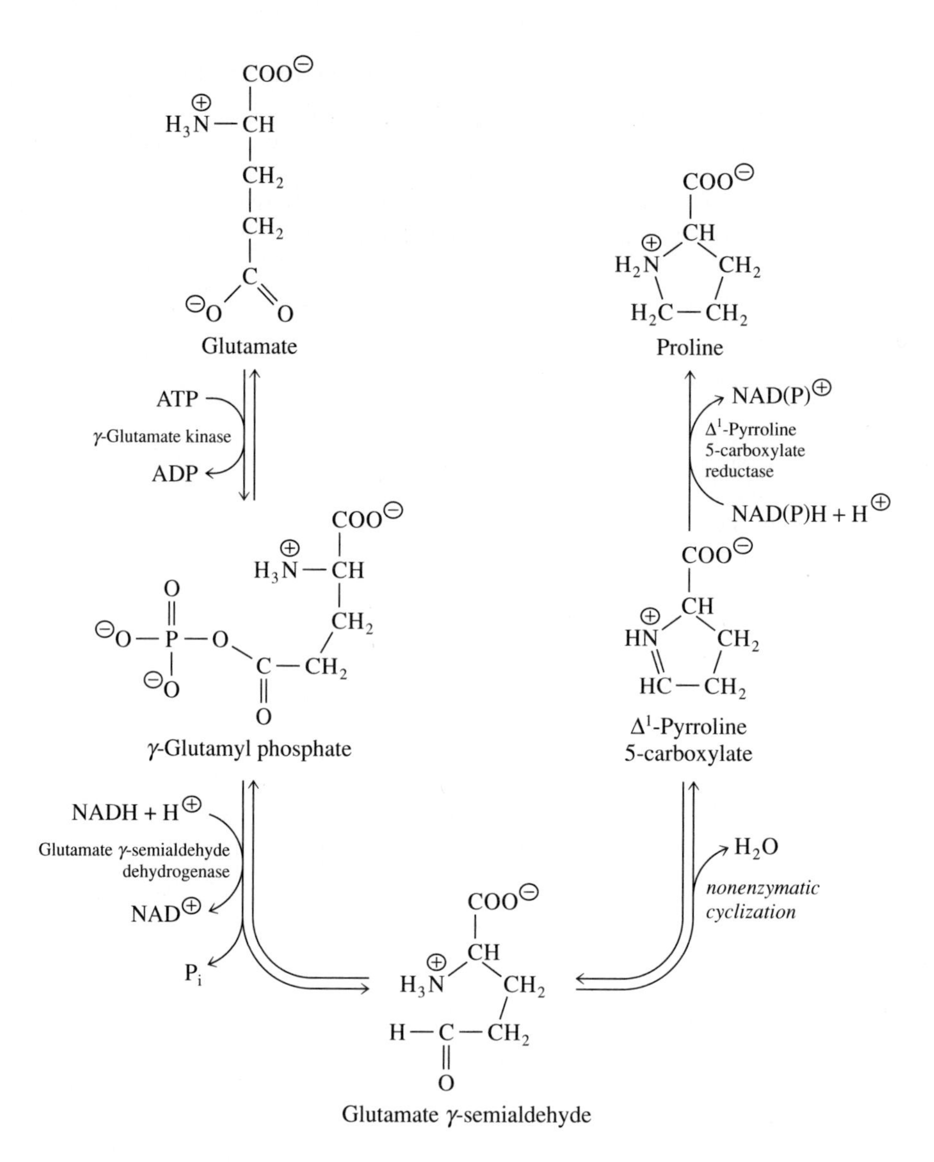

Figure 18·16
Biosynthesis of proline.

B. Proline Is Formed from Glutamate

The pathway for biosynthesis of proline is shown in Figure 18·16. First, γ-glutamate kinase catalyzes the phosphorylation of glutamate to form γ-glutamyl phosphate. This product is converted to glutamate γ-semialdehyde by addition of a hydride from NADH and release of P_i, catalyzed by glutamate γ-semialdehyde dehydrogenase. The mechanism of this enzyme is analogous to that of the glycolytic enzyme glyceraldehyde 3-phosphate dehydrogenase (Section 12·7). Glutamate γ-semialdehyde undergoes nonenzymatic cyclization by formation of an internal Schiff base to form Δ^1-pyrroline 5-carboxylate. Finally, a reductase catalyzes the formation of proline, with NADH or NADPH serving as cofactor. The specificity for the pyridine nucleotide varies with different organisms.

C. In Mammals, Tyrosine Can Be Formed from Phenylalanine

De novo synthesis of tyrosine follows a multistep, energy-dependent pathway in some organisms, such as bacteria, but in mammals tyrosine is formed from the essential amino acid phenylalanine in a single step catalyzed by phenylalanine hydroxylase (Figure 18·17). This reaction requires molecular oxygen and the reductant tetrahydrobiopterin (Section 7·8). One oxygen atom from O_2 is incorporated into tyrosine and the other is converted to water. Regeneration of the reduced form of biopterin, 5,6,7,8-tetrahydrobiopterin, from quinonoid dihydrobiopterin is catalyzed by dihydropteridine reductase in a reaction that requires NADH.

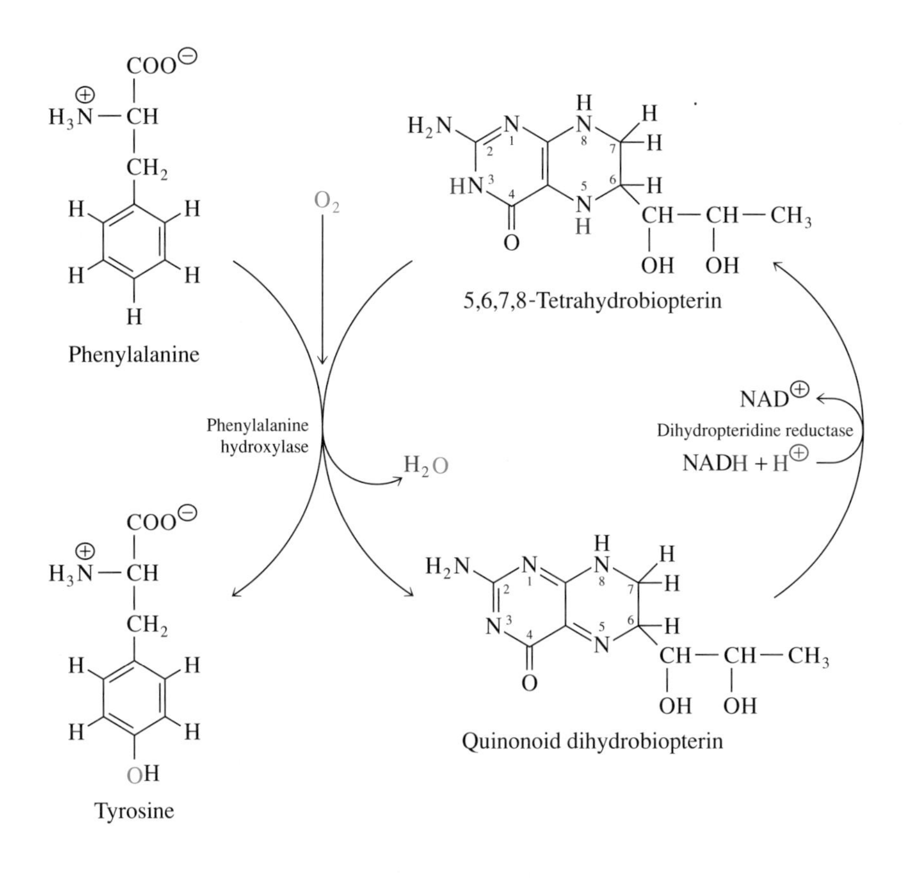

Figure 18·17
Conversion of phenylalanine to tyrosine. Tyrosine formation catalyzed by phenylalanine hydroxylase involves molecular oxygen and the cofactor tetrahydrobiopterin. The cofactor is regenerated in an NADH-dependent reaction catalyzed by dihydropteridine reductase.

Box 18·2 Phenylketonuria Is an Inherited Metabolic Disease

Early in this century, Sir Alexander Garrod observed abnormally high levels of amino acid catabolites in the urine of certain patients. After examining relatives of these patients, he related biochemistry to genetics by concluding that the metabolic defects in the individuals he studied were inherited as a recessive trait and that each disorder was due to the absence or reduced activity of an enzyme. This in turn led to the abnormal accumulation of a normal metabolite. Garrod termed diseases of this type inborn errors of metabolism; they are now commonly called *inherited metabolic diseases*. Studies of individuals afflicted with these diseases have aided the elucidation of metabolic pathways.

In North America and Europe, one of the most common disorders of amino acid metabolism is phenylketonuria. This condition is caused by a mutation within the gene that encodes phenylalanine hydroxylase. The result is an impairment in the ability of the affected individual to convert phenylalanine to tyrosine. The blood of children with the disease contains very high levels of phenylalanine and low levels of tyrosine. The phenylalanine, rather than being converted to tyrosine, is metabolized to phenylpyruvate by transamination. Elevated levels of phenylpyruvate and its derivatives cause severe and irreversible mental retardation. Phenylketonuria can be detected by testing for elevated levels of phenylalanine in the blood during the first days after birth. If the dietary intake of phenylalanine is strictly limited during the first decade of life, phenylalanine hydroxylase–deficient individuals can escape mental retardation. Elevated amounts of phenylalanine are also observed in individuals with deficiencies in dihydropteridine reductase or defects in the biosynthesis of the coenzyme tetrahydrobiopterin because each of these disorders results in impairment of the hydroxylation of phenylalanine.

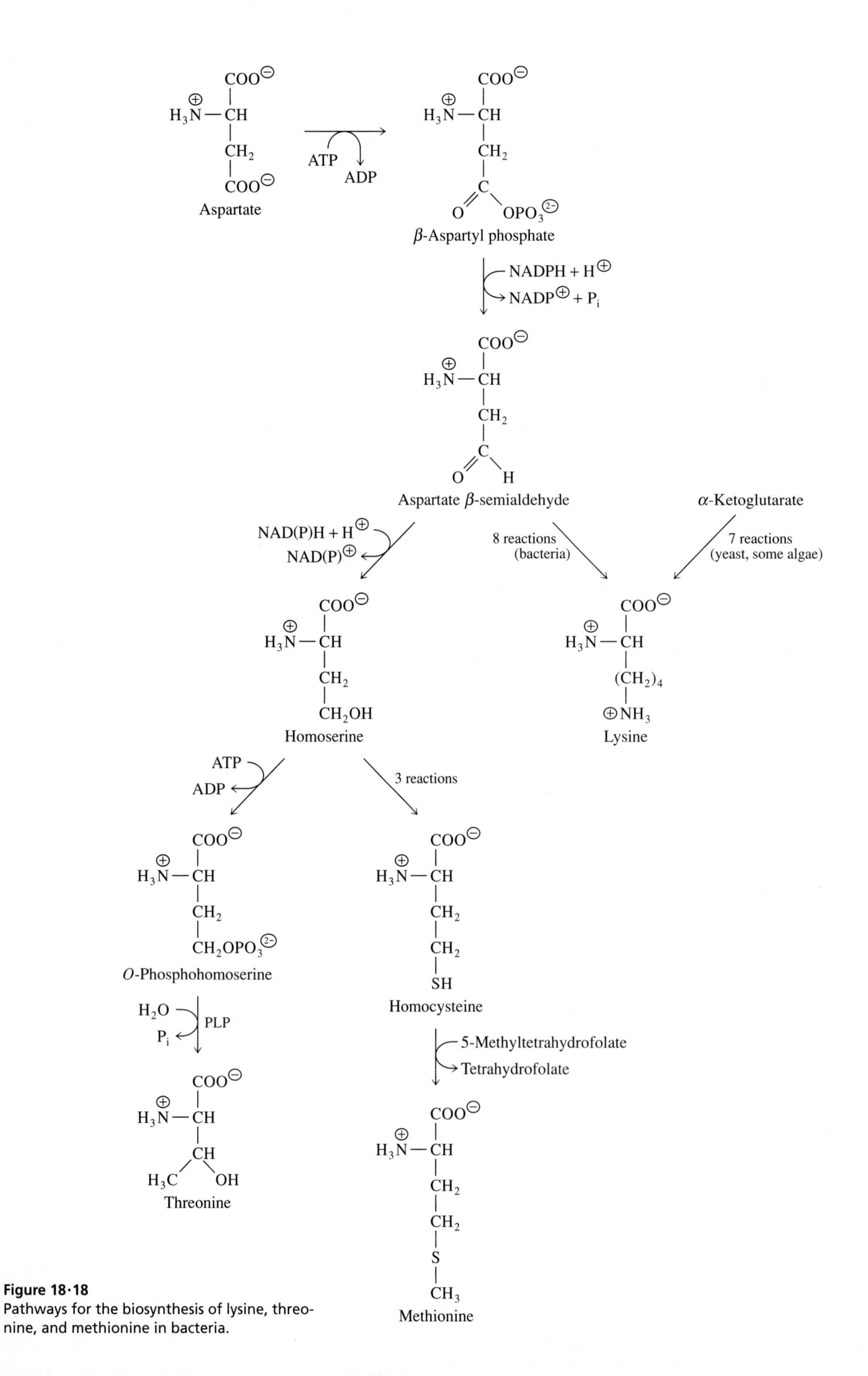

Figure 18·18
Pathways for the biosynthesis of lysine, threonine, and methionine in bacteria.

18·7 Bacteria and Plants Synthesize the Amino Acids that Are Essential for Mammals

Most of the research on the biosynthesis of amino acids that are essential for mammals has been performed with bacteria. Therefore, much of the following discussion relies on information obtained from experiments with bacterial enzymes. It is believed that plants synthesize these amino acids by similar pathways.

A. Aspartate Is the Precursor of Lysine, Threonine, and Methionine

In most microorganisms, aspartate is the precursor of lysine, methionine, and threonine. Recall that aspartate is formed in a transamination reaction involving oxaloacetate as the amino-group acceptor. The intermediate steps in the conversion of aspartate to lysine, methionine, and threonine are summarized in Figure 18·18. The first two reactions, leading to aspartate β-semialdehyde, are common to the formation of all three amino acids. In the branch leading to lysine, pyruvate is the source of carbon atoms added to the skeleton of aspartate β-semialdehyde, and glutamate is the source of the ε-amino group. An entirely different route to lysine, beginning with α-ketoglutarate, is operative in some eukaryotic microorganisms such as yeasts and some algae.

Homoserine, formed from aspartate β-semialdehyde, is a branch point for the formation of threonine and methionine. Threonine is derived from homoserine after phosphorylation to *O*-phosphohomoserine. The phosphate group of *O*-phosphohomoserine is removed and a hydroxyl group is added in a pyridoxal phosphate—dependent reaction to yield threonine. Methionine formation requires the synthesis of homocysteine in three steps from homoserine. Homocysteine is then methylated at its sulfur atom by 5-methyltetrahydrofolate (Section 7·8), forming methionine. An enzyme that catalyzes this reaction is also found in mammals, but its activity is too low to supply enough methionine for the needs of the cell. Therefore, methionine is an essential amino acid for mammals.

B. The Pathways for Synthesis of the Branched-Chain Amino Acids Isoleucine, Valine, and Leucine Share Enzymatic Steps

The biosynthesis of the branched-chain amino acids is depicted in Figure 18·19 (next page). Threonine is converted to isoleucine in five steps. In a very similar series of reactions, pyruvate is converted to valine. In fact, the same four enzymes catalyze the final four steps in isoleucine biosynthesis and the four steps of valine biosynthesis from pyruvate. The pathway for biosynthesis of leucine branches from that leading to valine.

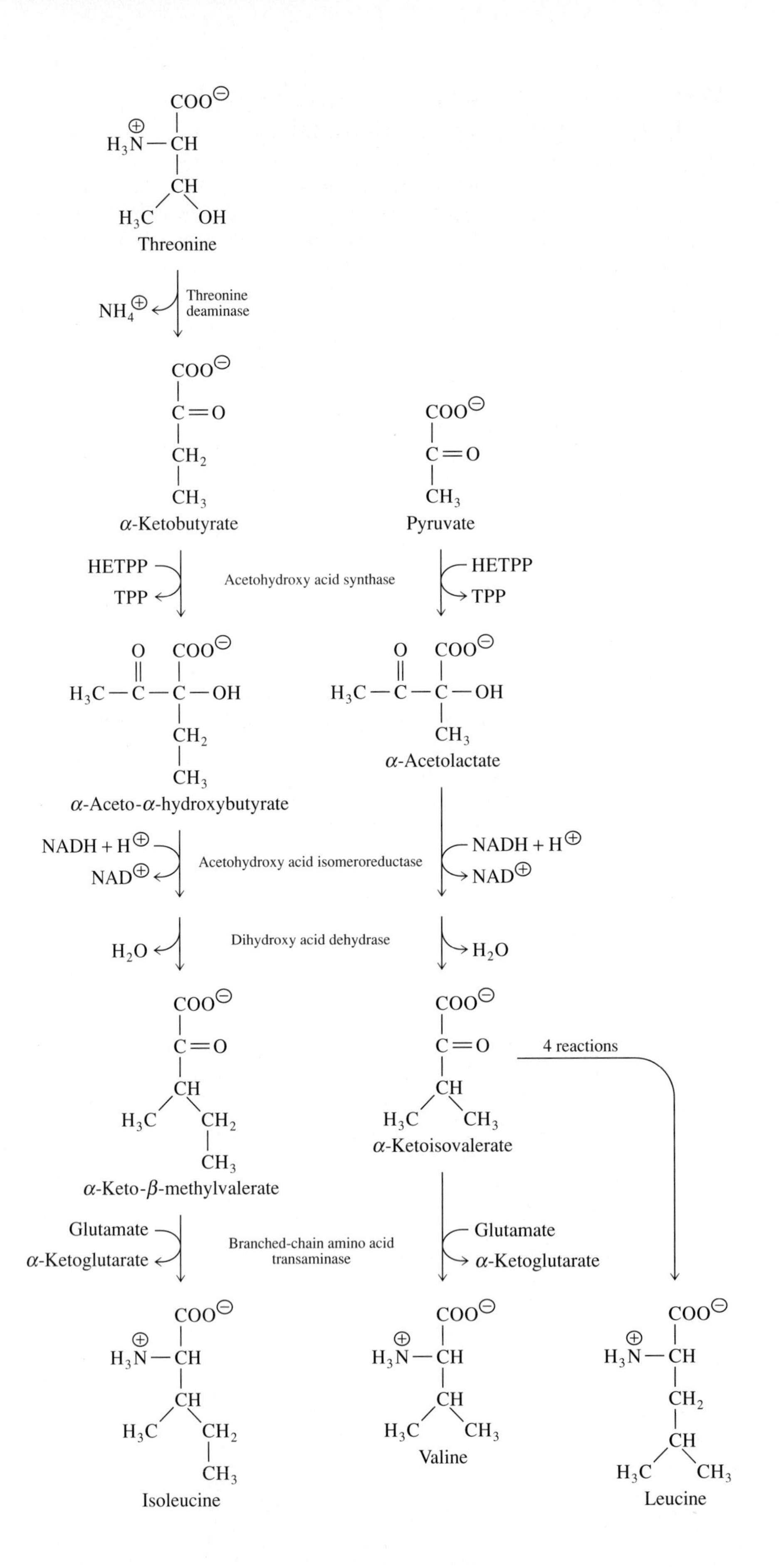

Figure 18·19
Pathways for the biosynthesis of the branched-chain amino acids in microorganisms and plants. The final four enzymes in the pathway from threonine to isoleucine also catalyze the four reactions in the pathway from pyruvate to valine. The carbon chain of α-ketoisovalerate, an intermediate in the formation of valine, is lengthened by one methylene group during the formation of leucine.

C. A Branched Pathway Leads to the Aromatic Amino Acids

During studies of bacterial mutants, Bernard Davis noted that some single-gene mutants required provision of as many as five compounds for growth—phenylalanine, tyrosine, tryptophan, *p*-hydroxybenzoate, and *p*-aminobenzoate (a component of folate). These compounds all contain an aromatic ring system. The inability of some of Davis's mutants to grow without these compounds was reversed if the compound shikimate was provided, showing that shikimate is an intermediate in the biosynthesis of these aromatic compounds. Further experiments by Frank Gibson showed that chorismate is formed in three steps from shikimate and that it is a key intermediate in aromatic amino acid synthesis. Chorismate is synthesized from phosphoenolpyruvate and erythrose 4-phosphate in seven steps (Figure 18·20). First, phosphoenolpyruvate and erythrose 4-phosphate condense to form a seven-carbon sugar derivative and P_i. In the enteric bacteria, three separately regulated isozymes each catalyze this reaction. The activities of these enzymes are allosterically inhibited by phenylalanine, tyrosine, or tryptophan. Three steps are required for transformation of the seven-carbon sugar phosphate to shikimate. First, cyclization of the open-chain form of the sugar produces 3-dehydroquinate. Dehydration introduces the first double bond of the ring, and an NADPH-dependent reduction forms shikimate. The C-3 hydroxyl group of shikimate undergoes phosphorylation, followed by addition to the C-5 hydroxyl group of a three-carbon moiety from phosphoenolpyruvate. Subsequent elimination of P_i introduces a second double bond to the ring to form chorismate. This compound is a branch-point metabolite that is converted via different pathways to phenylalanine, tyrosine, or tryptophan.

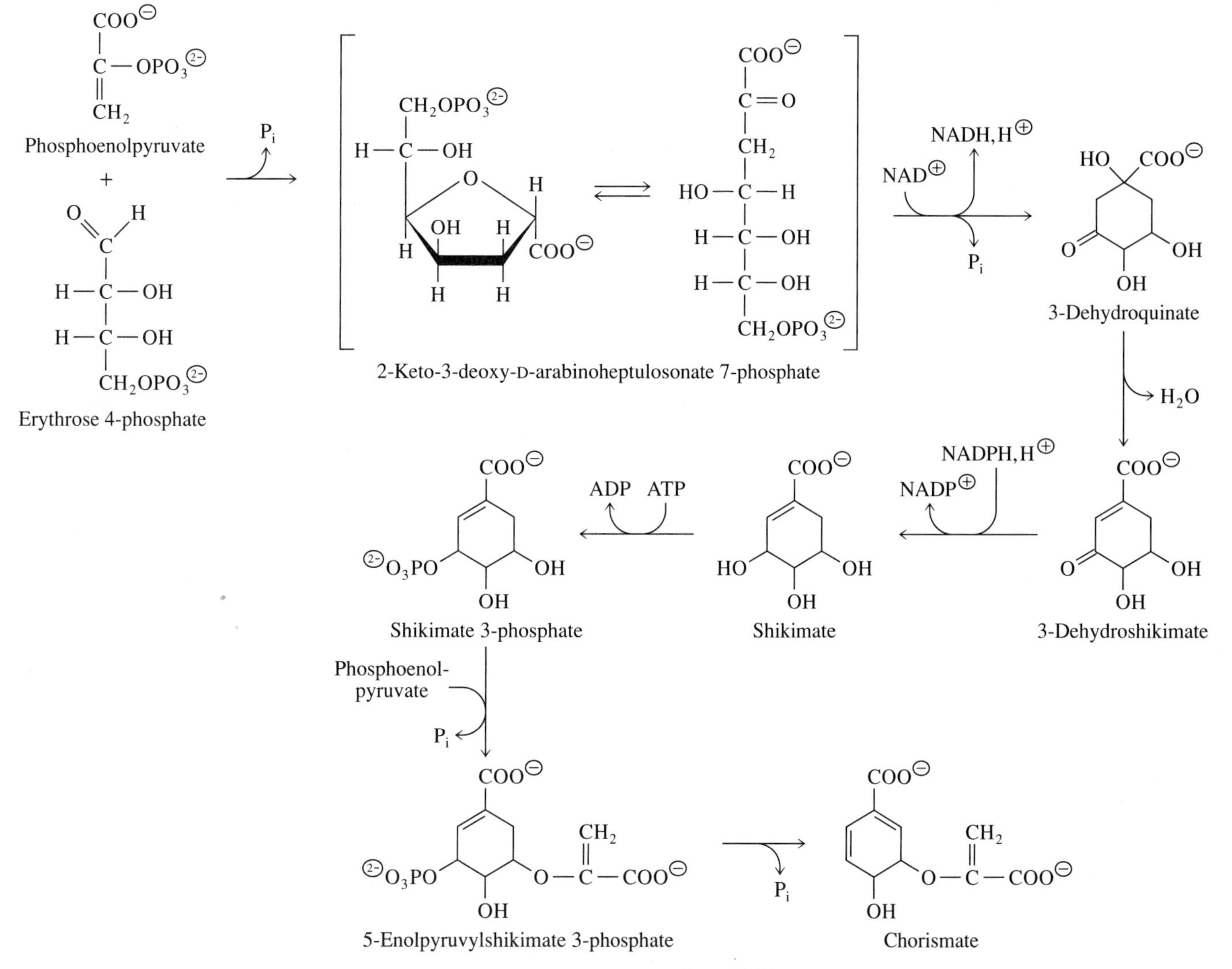

Figure 18·20
Biosynthesis of chorismate from phosphoenolpyruvate and erythrose 4-phosphate. Chorismate is a branch-point metabolite that leads via different pathways to phenylalanine, tyrosine, or tryptophan.

There are two pathways for the biosynthesis of phenylalanine from chorismate, with different organisms using one or the other of these routes (Figure 18·21). First, chorismate mutase catalyzes the rearrangement of chorismate to produce prephenate, a very reactive compound. In *E. coli*, several other bacteria, and some plants, water and CO_2 are eliminated from prephenate to form the fully aromatic product phenylpyruvate, which is then transaminated to phenylalanine. In the alternate pathway, utilized in some bacteria, prephenate is transaminated to form arogenate, from which water and CO_2 are removed to form phenylalanine.

Two pathways lead from chorismate to tyrosine, as shown in Figure 18·21. Following the enzymatic conversion of chorismate to prephenate, some organisms transaminate prephenate to produce arogenate, which is oxidatively decarboxylated to yield tyrosine. In other organisms, oxidative decarboxylation precedes transamination. The product of the oxidation reaction, 4-hydroxyphenylpyruvate, undergoes transamination to tyrosine.

The biosynthesis of tryptophan from chorismate in *E. coli* involves five enzymes, one of which catalyzes two consecutive reactions (Figure 18·22). It appears that tryptophan is synthesized in all organisms for which it is nonessential by the same pathway. In the first step, the amide nitrogen of glutamine is transferred to chorismate. Elimination of the hydroxyl group and the adjacent pyruvate moiety of chorismate produces the aromatic ring. The resulting compound, anthranilate, accepts a ribose 5-phosphate moiety from phosphoribosyl pyrophosphate, PRPP (Structure 18·11, next page).

Figure 18·21
Biosynthesis of phenylalanine and tyrosine from chorismate. Different organisms use one or the other of the two routes shown.

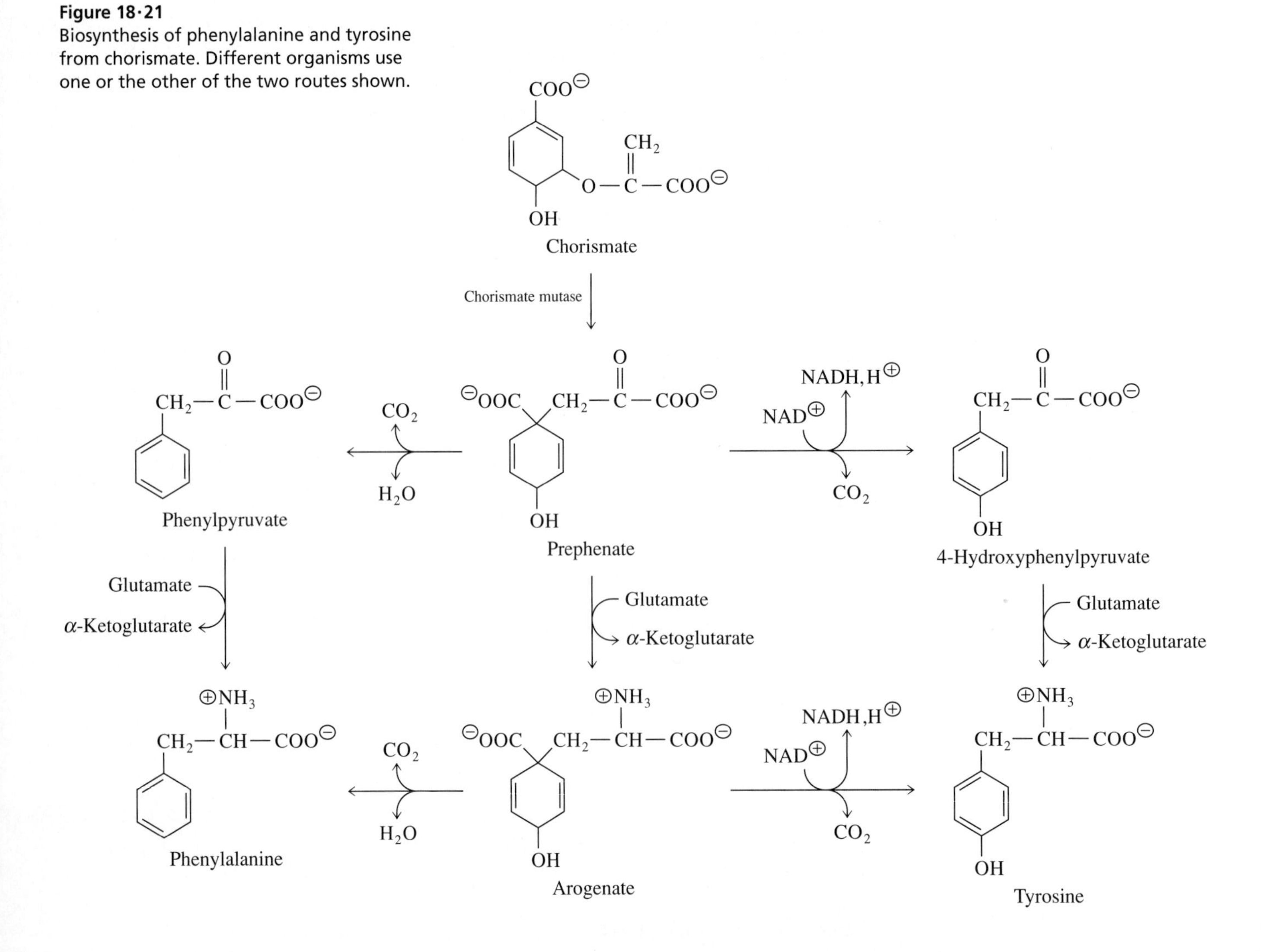

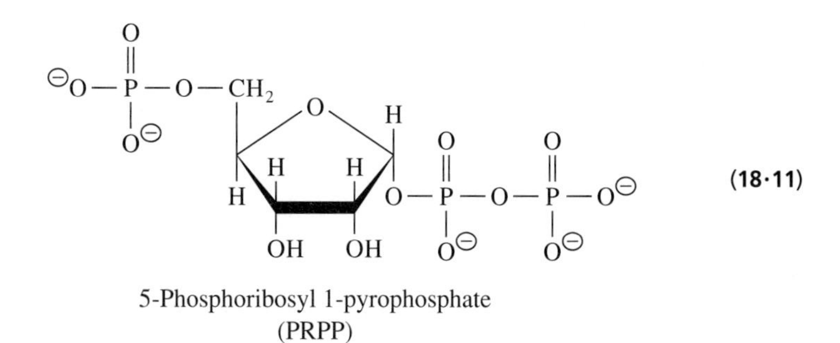

5-Phosphoribosyl 1-pyrophosphate
(PRPP)

(18·11)

Rearrangement of the ribose, decarboxylation, and ring closure form the indole component of indole glycerol phosphate.

The final stage in tryptophan biosynthesis is catalyzed by the enzyme tryptophan synthase. In some organisms, this enzyme contains two types of subunits, α and β. In other organisms, two catalytic domains are fused in a single polypeptide chain. Tryptophan synthase catalyzes two partial reactions. The α subunit or domain catalyzes the cleavage of indole glycerol phosphate to glyceraldehyde 3-phosphate and indole. The β subunit catalyzes the condensation of indole and serine in a reaction that requires pyridoxal phosphate as a cofactor. The indole produced in the reaction catalyzed by the α subunit does not dissociate to solvent but

Figure 18·22
Biosynthesis of tryptophan from chorismate in *E. coli.*

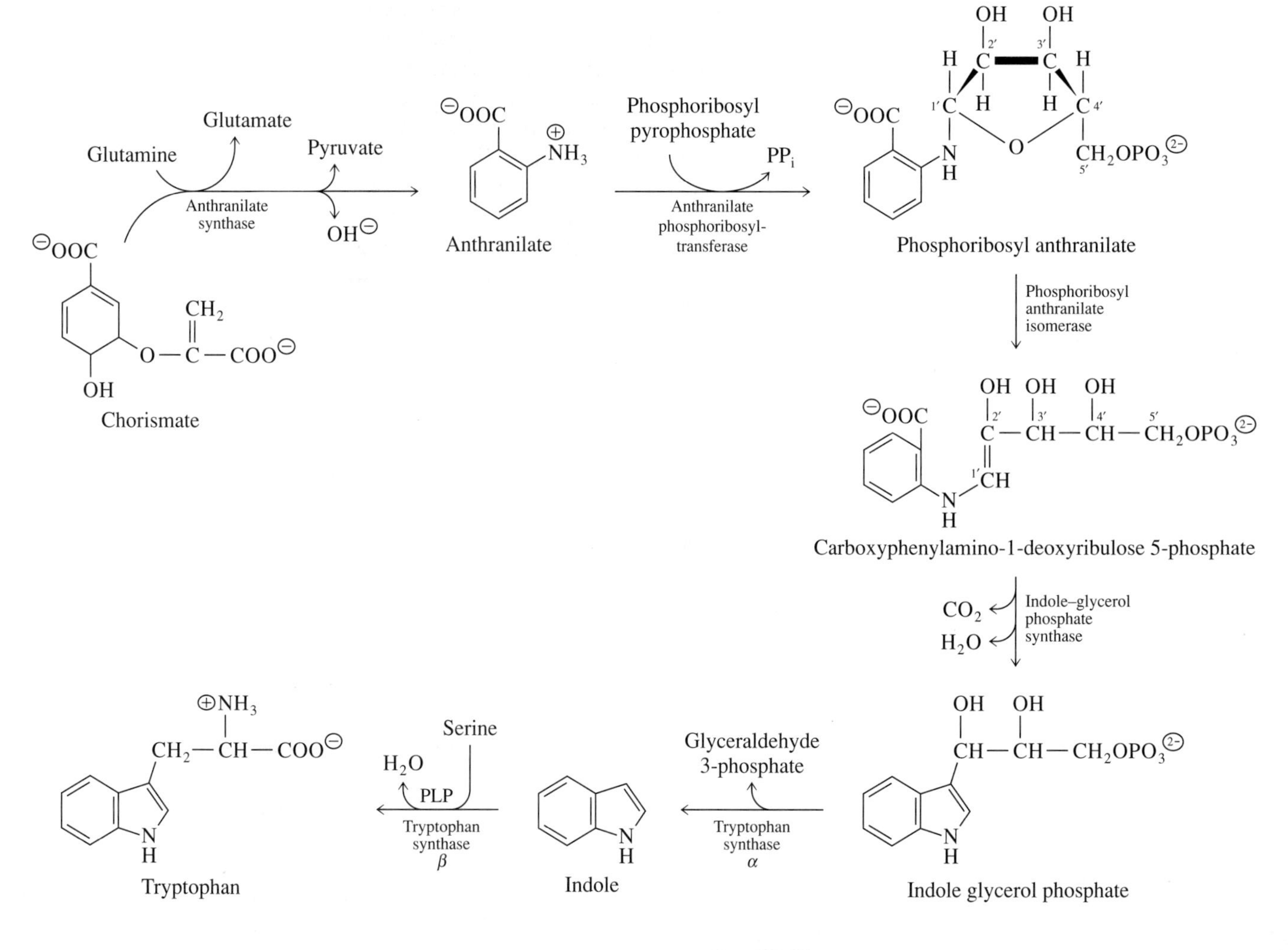

is instead transferred directly to the active site of the β subunit. When the three-dimensional structure of tryptophan synthase from *Salmonella typhimurium* was determined by X-ray crystallography, a tunnel joining the α and β active sites was discovered. The presence of a channel through the interior of a globular protein is most unusual. Because the diameter of the tunnel matches the molecular dimensions of indole, passage of indole through the tunnel has been invoked to explain the fact that indole does not enter free solution.

Short-term control of tryptophan biosynthesis is exerted by feedback inhibition of the first step in the pathway (conversion of chorismate to anthranilate) by tryptophan. High concentrations of tryptophan also inhibit the synthesis of the five enzymes of the pathway via a transcriptional control mechanism.

D. Phosphoribosyl Pyrophosphate, ATP, and Glutamine Are the Precursors of Histidine

Although mammals can synthesize small amounts of histidine, the quantity is often inadequate for the organism's needs. For example, human infants synthesize insufficient histidine for protein synthesis, and there is no strong evidence that adults synthesize any histidine. Therefore, histidine is classified as an essential amino acid.

In plants, simpler eukaryotes, and bacteria, histidine is synthesized from PRPP, which is also a precursor of tryptophan and of purine and pyrimidine nucleotides. The pathway for the biosynthesis of histidine begins with a condensation between the pyrimidine ring of ATP and PRPP (Figure 18·23). In subsequent reactions, the six-membered ring of the adenine moiety is cleaved, and a Schiff base intermediate is formed. Glutamine donates a nitrogen atom to the intermediate, releasing the amine portion of the Schiff base and initiating cyclization. The nitrogen atom from glutamine is incorporated into the imidazole ring of the product, imidazole glycerol phosphate, which undergoes dehydration, transamination by glutamate, hydrolytic removal of its phosphate, and oxidation from the level of a primary alcohol to that of a carboxylic acid in two sequential $NAD^{\oplus}$-dependent steps, forming histidine.

We have now examined the biosynthesis of all of the amino acids except arginine. The formation of arginine involves reactions that are also important in the operation of the urea cycle (Section 18·9), a series of reactions important in mammals for the disposal of surplus ammonia produced in the catabolism of amino acids.

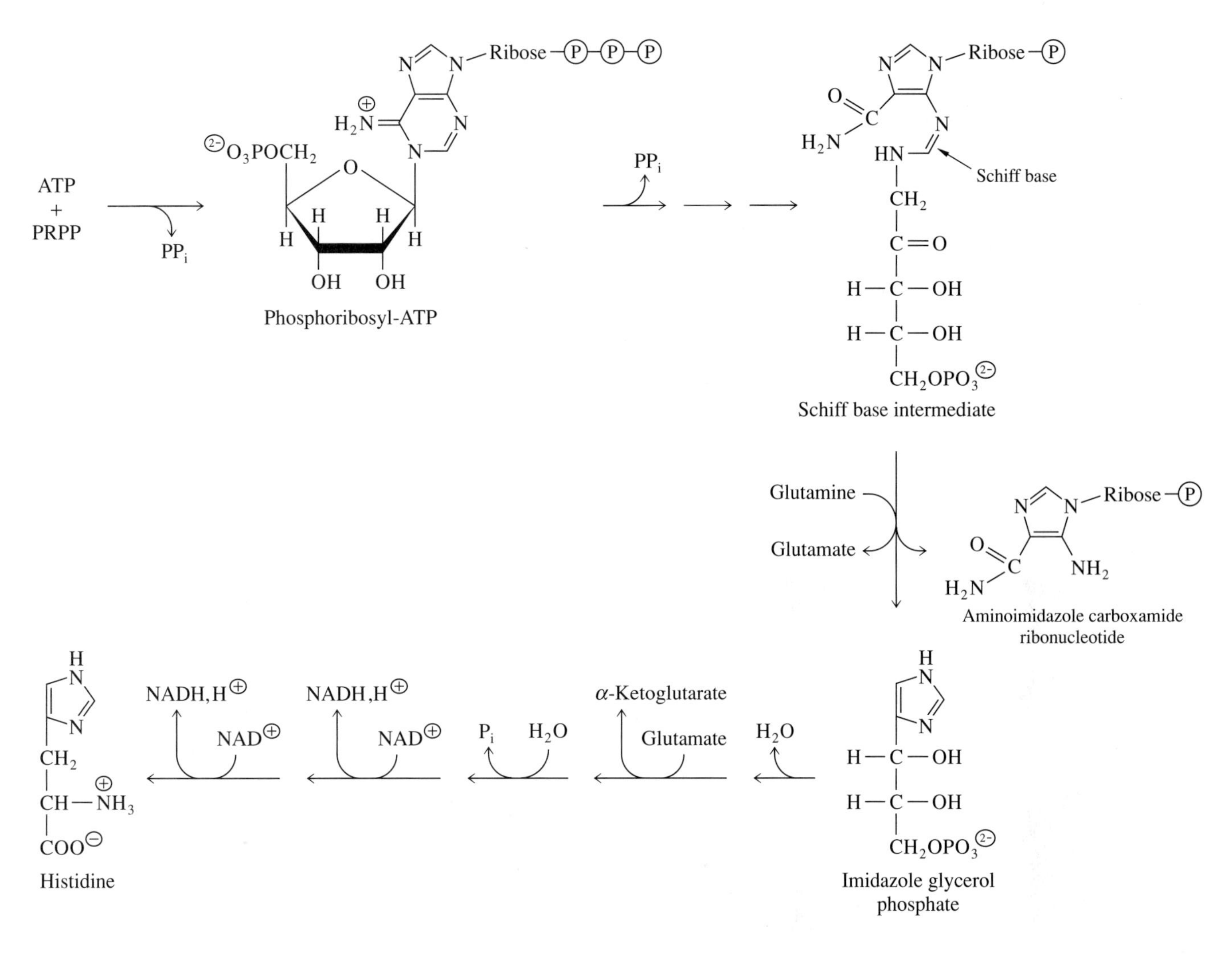

Figure 18·23
Outline of the biosynthesis of histidine.

18·8 Catabolism of Amino Acids Often Begins with Deamination, Followed by Degradation of the Remaining Carbon Chains

Excess amino acids, whether dietary in origin or derived from the normal turnover of body proteins, are, in many cases, first deaminated. Next, the carbon chains are altered in specific ways for entry into the central pathways of carbon metabolism. We shall first consider the metabolism of the ammonia arising from amino acid degradation, and then we shall examine the metabolic fates of the various carbon skeletons.

The deamination of amino acids occurs in several ways. In mammalian liver, α-ketoglutarate acts as an acceptor in transamination reactions with a number of amino acids to form glutamate, which is then oxidized in an NAD⊕-dependent reaction catalyzed by glutamate dehydrogenase. This reaction was shown (in the synthetic direction) on page 18·4.

The amide groups of glutamine and asparagine are hydrolyzed by specific enzymes—glutaminase and asparaginase, respectively—to produce ammonia and the corresponding dicarboxylic amino acids glutamate and aspartate.

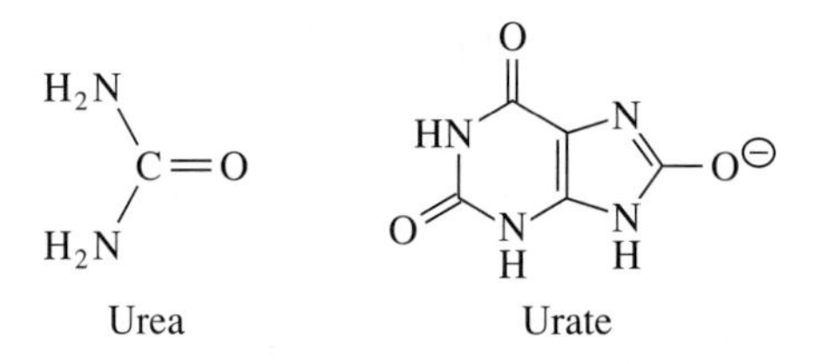

Figure 18·24 Urea and urate.

18·9 The Urea Cycle Converts Ammonia into Urea

Because ammonia is highly toxic to cells, its concentration is usually maintained at low levels. Different organisms have different pathways for the elimination of waste nitrogen. Many aquatic organisms are able to excrete ammonia directly across the cell membranes of gill tissue, to be flushed away by the vast amount of water in their surroundings. Most terrestrial vertebrates convert waste nitrogen to the less toxic product urea, an uncharged and highly water-soluble compound that can be carried in the blood to the kidney, where it is excreted as the major solute of urine (Figure 18·24). Birds and many terrestrial reptiles convert surplus ammonia to urate, a relatively insoluble compound that precipitates from aqueous solution to form a semi-solid slurry. While in their eggs, the embryos of these organisms excrete urate into membranous sacs. They continue to excrete urate throughout the course of their adult lives. Urate is also a product of the degradation of purine nucleotides by birds, reptiles, and primates. The further degradation of urate is considered in Chapter 19.

In mammals, the synthesis of urea, or ureogenesis, occurs exclusively in the liver. Urea is the product of a set of reactions called the *urea cycle*. This pathway was elucidated by Hans Krebs and Kurt Henseleit in 1932, several years before Krebs discovered the citric acid cycle. Several observations led to the proposal of the urea cycle. High levels of the enzyme arginase occur in the livers of all organisms that synthesize urea. Slices of rat liver can bring about the net conversion of ammonia to urea. Urea synthesis by these preparations is greatly stimulated when the amino acid ornithine is added; the amount of urea synthesized greatly exceeds the amount of added ornithine, suggesting that ornithine acts catalytically (Section 13·13).

The incorporation of ammonia into urea requires five steps. Carbamoyl phosphate, a direct precursor molecule, is formed first, and its nitrogen atom is incorporated into urea in the four reactions of the cycle (Figure 18·25). Two reactions occur in the mitochondria of liver cells, and the other three occur in the cytosol. The precursors of the two nitrogen atoms of urea are ammonia and aspartate. The carbon atom of urea comes from bicarbonate. The overall reaction for urea synthesis is:

$$\begin{aligned} NH_3 + HCO_3^{\ominus} + \text{Aspartate} + 3\,\text{ATP} &\longrightarrow \\ \text{Urea} + \text{Fumarate} + 2\,\text{ADP} + 2\,P_i + \text{AMP} + PP_i& \end{aligned} \qquad \textbf{(18·12)}$$

A total of four equivalents of ATP are consumed in urea synthesis. Three molecules of ATP are converted to two ADP and one AMP during formation of one molecule of urea, and hydrolysis of the molecule of inorganic pyrophosphate that is formed consumes a fourth ATP equivalent. Urea passes from the liver through the bloodstream to the kidney, where it is excreted in urine.

Leading into the urea cycle, carbamoyl phosphate is synthesized from ammonia, bicarbonate, and ATP in a mitochondrial reaction catalyzed by carbamoyl phosphate synthetase I. This enzyme is one of the most abundant in liver mitochondria, accounting for as much as 20% of the protein of the mitochondrial matrix.

Figure 18·25
The urea cycle. Two transporters connecting the mitochondrial matrix and the cytosol are required for the operation of the urea cycle: the citrulline-ornithine exchanger and the glutamate-aspartate translocase.

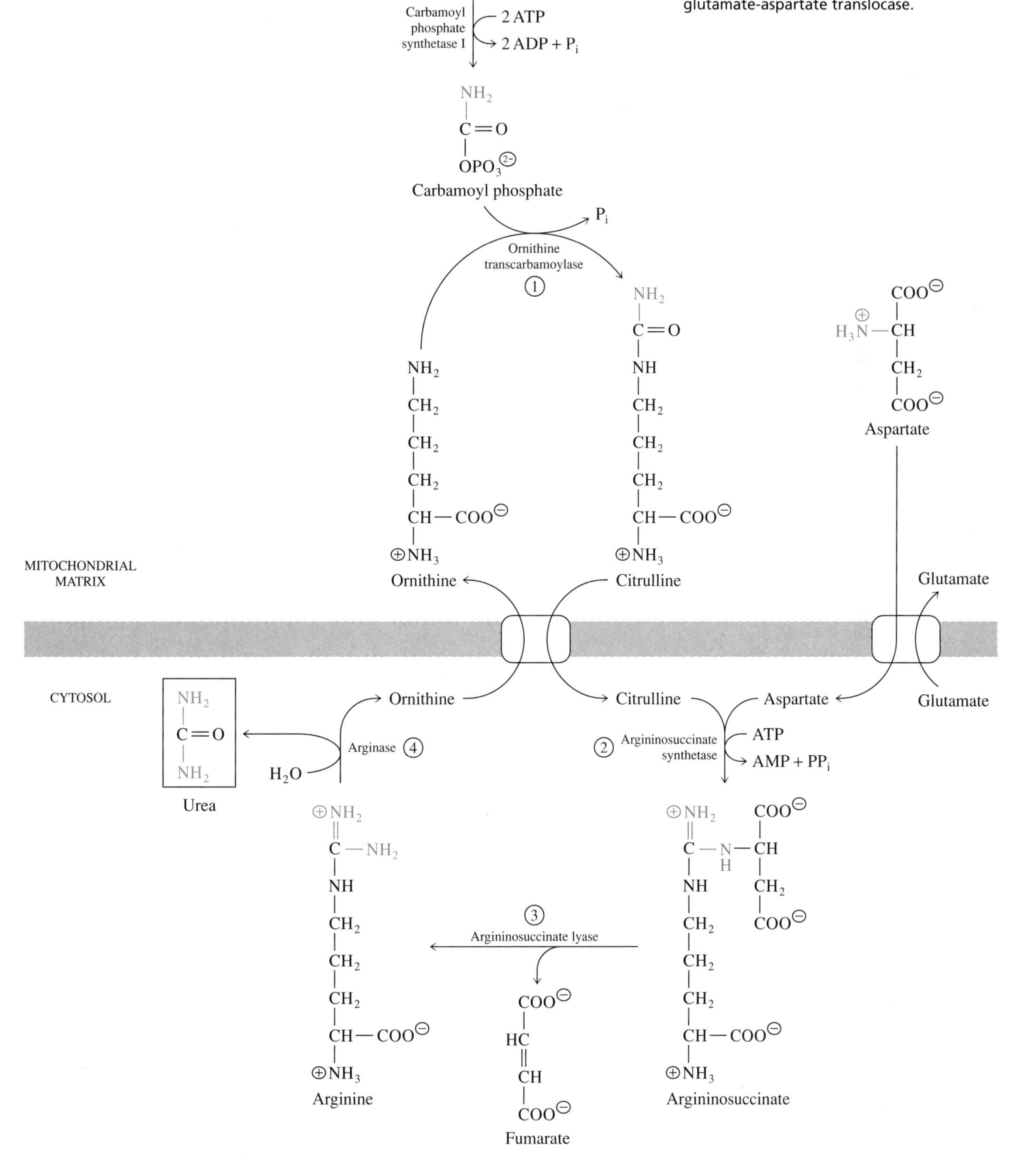

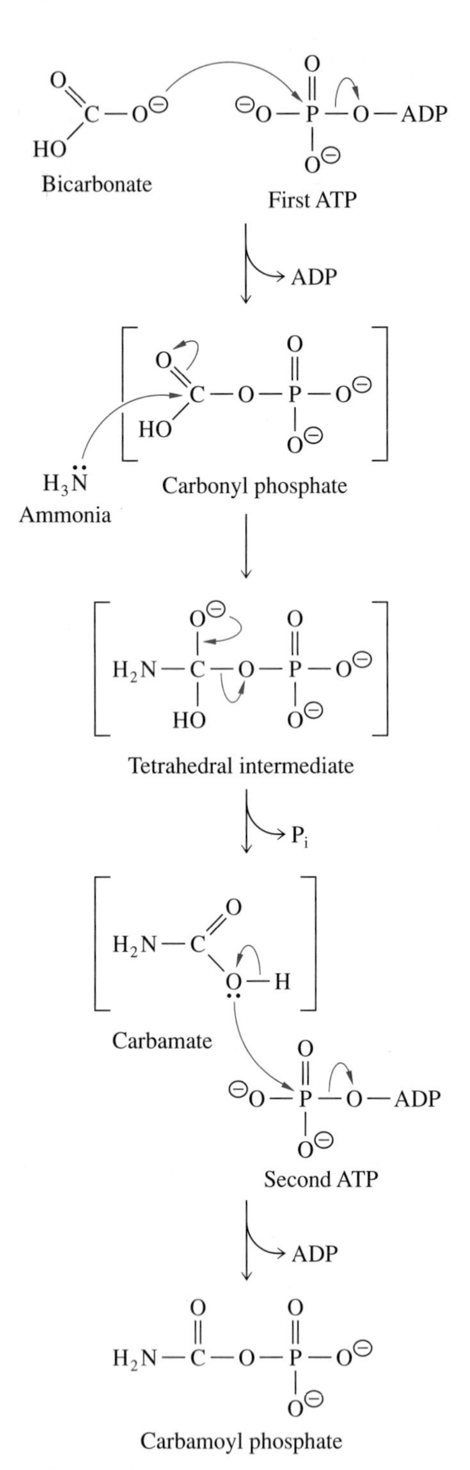

Figure 18·26
Synthesis of carbamoyl phosphate catalyzed by carbamoyl phosphate synthetase I. The reaction involves two phosphoryl-group transfers. In the first, nucleophilic attack by bicarbonate on ATP produces carbonyl phosphate and ADP. Next, ammonia reacts with carbonyl phosphate, forming a tetrahedral intermediate. Elimination of a phosphate group produces carbamate. A second phosphoryl-group transfer from another ATP forms carbamoyl phosphate and ADP. Structures in brackets remain enzyme bound during the reaction.

The mechanism of the reaction catalyzed by carbamoyl phosphate synthetase I involves three steps (Figure 18·26). First, carbonyl phosphate is formed by reaction of bicarbonate with the γ-phosphate of ATP; ammonia then displaces the phosphate to form carbamate, and finally, carbamoyl phosphate is formed by transfer of the γ-phosphate of a second molecule of ATP. A similar enzyme called carbamoyl phosphate synthetase II uses glutamine rather than ammonia as the nitrogen donor. This enzyme, located in the cytosol of liver and most other cells, catalyzes the formation of carbamoyl phosphate destined for the synthesis of pyrimidine nucleotides (Section 19·6).

The carbamoyl phosphate synthetase of mitochondria is allosterically activated by *N*-acetylglutamate formed by the reaction of glutamate with acetyl CoA, catalyzed by *N*-acetylglutamate synthase:

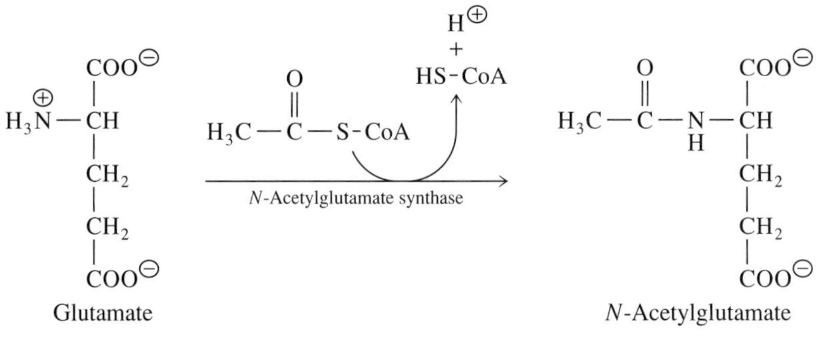

(18·13)

High rates of transamination during amino acid catabolism lead to elevated glutamate concentrations and concomitant increases in the concentration of *N*-acetylglutamate. Activation of carbamoyl phosphate synthetase I by *N*-acetylglutamate increases the supply of substrate for the urea cycle.

1. In the first reaction of the urea cycle, carbamoyl phosphate reacts in the mitochondrion with the urea cycle intermediate ornithine to form citrulline in a reaction catalyzed by ornithine transcarbamoylase. This step incorporates the nitrogen atom originating from ammonia into citrulline; citrulline thus contains half the nitrogen destined for urea. The mechanism of ornithine transcarbamoylase is shown in Figure 18·27. The nucleophilic δ-amino group of ornithine reacts with the carbonyl carbon of carbamoyl phosphate, releasing inorganic phosphate and yielding citrulline. Citrulline is then transported out of the mitochondrion in an electroneutral exchange with cytosolic ornithine, a compound that is produced in a subsequent reaction of the urea cycle.

2. The second nitrogen atom destined for urea is incorporated in the cytosol when citrulline condenses with aspartate to form argininosuccinate. This ATP-dependent reaction is catalyzed by argininosuccinate synthetase. Most of the aspartate in cells originates in the mitochondria, although in some circumstances aspartate is generated in the cytosol. For example, cytosolic asparaginase catalyzes the formation of aspartate from asparagine:

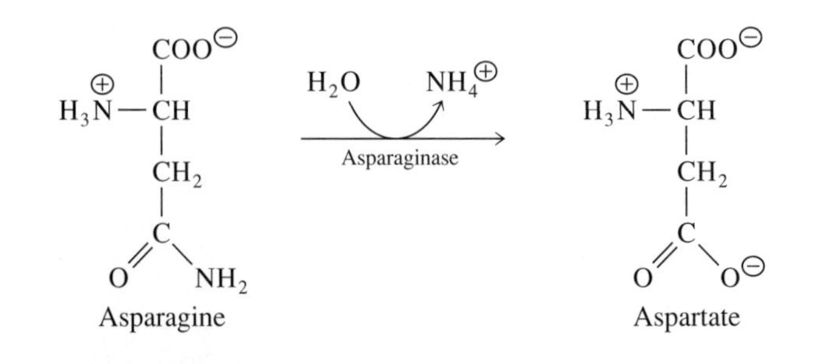

(18·14)

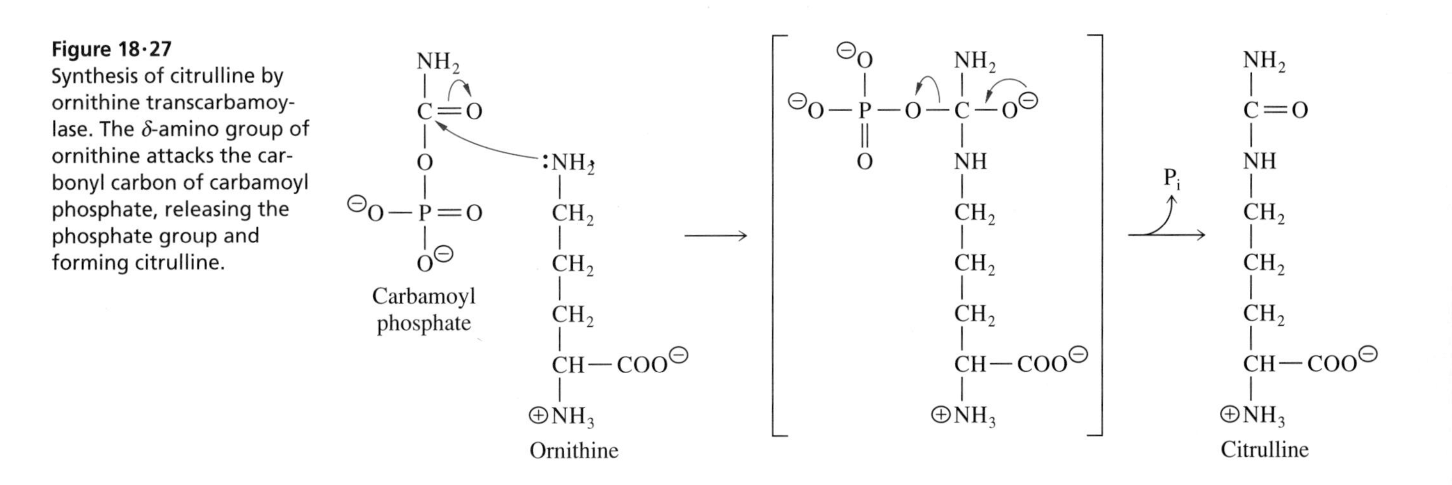

Figure 18·27
Synthesis of citrulline by ornithine transcarbamoylase. The δ-amino group of ornithine attacks the carbonyl carbon of carbamoyl phosphate, releasing the phosphate group and forming citrulline.

The mechanism of the reaction catalyzed by argininosuccinate synthetase is shown in Figure 18·28. First, a citrullyl-AMP intermediate is formed when the ureido oxygen of citrulline displaces PP_i from ATP. AMP becomes the leaving group when this oxygen is displaced by the amino group of aspartate.

3. Argininosuccinate is cleaved nonhydrolytically to form arginine plus fumarate in an elimination reaction catalyzed by argininosuccinate lyase (Figure 18·29, next page). Together, the second and third steps of the urea cycle exemplify a strategy for donation of the amino group of aspartate that will be encountered twice in the next chapter as part of purine biosynthesis. The key processes are an ATP-dependent condensation, followed by the elimination of fumarate.

Arginine is the immediate precursor of urea. The metabolic fate of the other product of this reaction, fumarate, requires explanation. It is known from experiments with isolated liver cells that the carbon skeleton of this cytosolic fumarate is converted to glucose and CO_2. Fumarate is hydrated to malate by the action of a cytosolic fumarase. Malate is oxidized to oxaloacetate by the action of malate dehydrogenase. Oxaloacetate then enters the pathway of gluconeogenesis (Figure 15·18), with three carbon atoms being incorporated into glucose and one carbon atom becoming CO_2 via the decarboxylation reaction catalyzed by phosphoenolpyruvate carboxykinase. Since each round of the urea cycle consumes a molecule of aspartate and releases a molecule of fumarate, one might guess that a plausible pathway for the

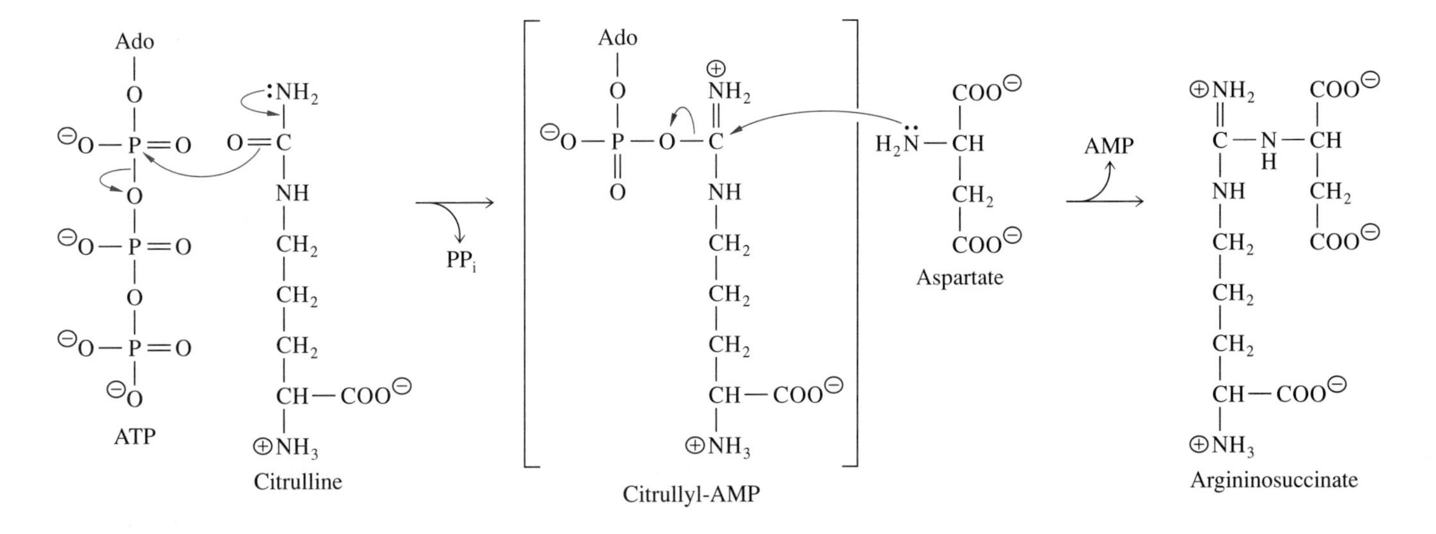

Figure 18·28
Synthesis of argininosuccinate. This reaction, catalyzed by argininosuccinate synthetase, proceeds by the formation of citrullyl-AMP. The amino group of aspartate displaces AMP. "Ado" in the figure represents the adenosine moiety.

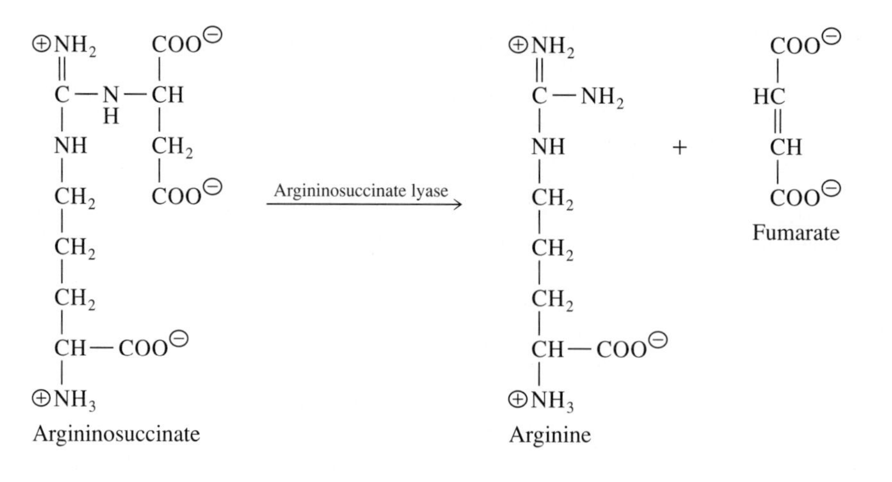

Figure 18·29
Conversion of argininosuccinate to arginine and fumarate, catalyzed by argininosuccinate lyase.

continued supply of aspartate would be fumarate to malate to oxaloacetate to aspartate, a route that is sometimes invoked in descriptions of the urea cycle. Yet experiments show that fumarate is metabolized to glucose. What then is the source of the carbon atoms in the aspartate molecule required for each round of the urea cycle? This aspartate is derived primarily from alanine formed in muscle by transamination of pyruvate arising from the catabolism of glucose. The glucose originates in the liver as a result of gluconeogenesis that is linked to the urea cycle. Glucose and alanine are transported between liver and muscle through the bloodstream. This interorgan transfer of glucose and alanine is called the *glucose-alanine cycle*. We shall consider it again in Section 18·12, where a summary is given of amino acid metabolism in humans.

4. In the final reaction of the urea cycle, the guanidinium group of arginine is hydrolytically cleaved to form ornithine and urea in a reaction catalyzed by arginase (Figure 18·30). Many mammalian tissues have some enzymes of the urea cycle, but only liver has all of them. In particular, liver is the only site with appreciable arginase activity, and liver is the only site of significant ureogenesis.

   Ornithine generated by the action of arginase is transported into the mitochondrion by the same exchange reaction that transports citrulline into the cytosol. Mitochondrial ornithine reacts with carbamoyl phosphate to support the continued operation of the urea cycle.

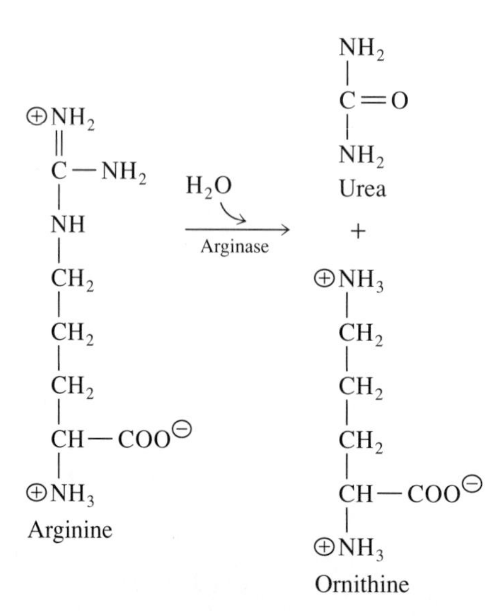

Figure 18·30
Hydrolysis of arginine to form ornithine and urea, catalyzed by arginase.

18·10 Ancillary Reactions Balance the Supply of Substrate into the Urea Cycle

The reactions of the urea cycle convert stoichiometric amounts of nitrogen from ammonia and from aspartate into urea. Many amino acids can participate as amino-group donors via transamination reactions with α-ketoglutarate, forming glutamate. Glutamate can either undergo transamination with oxaloacetate to form aspartate or deamination to form ammonia. Both glutamate dehydrogenase and aspartate transaminase are abundant in liver mitochondria and catalyze near-equilibrium reactions. These reactions can convert any proportion of amino acids and ammonia into the equal amounts of aspartate and ammonia that are consumed during operation of the urea cycle. Consider the case of a relative surplus of ammonia (Figure 18·31a). In this circumstance, the equilibrium of the reaction catalyzed by glutamate dehydrogenase will be shifted in the direction of glutamate formation; elevated concentrations of glutamate will shift the equilibrium of the reaction catalyzed by aspartate transaminase so that there is a proportionate increase in aspartate. In contrast, if an excess of aspartate exists, the reactions catalyzed by glutamate dehydrogenase and aspartate transaminase run in the opposite direction to provide ammonia for urea formation (Figure 18·31b).

Humans do not synthesize sufficient arginine to meet the needs of both protein synthesis and urea formation and so require an exogenous supply. In plants and microorganisms, the conversion of ornithine to arginine is an integral step in arginine biosynthesis. The synthesis of ornithine starts with acetylation of the α-amino

Figure 18·31
Balancing the supply of nitrogen for the urea cycle. Two situations are described: **(a)** NH_3 in extreme excess and **(b)** aspartate in extreme excess. Note that the directions of the reactions catalyzed by glutamate dehydrogenase and aspartate transaminase are reversed in the two cases. Adjustments to the flux through these reactions meet the requirement for equal amounts of nitrogen from ammonia and aspartate. More significantly, this mechanism allows the liver to synthesize urea when presented with any mixture of ammonia and amino acids, since near-equilibrium transamination reactions leading to either aspartate or glutamate correct imbalances.

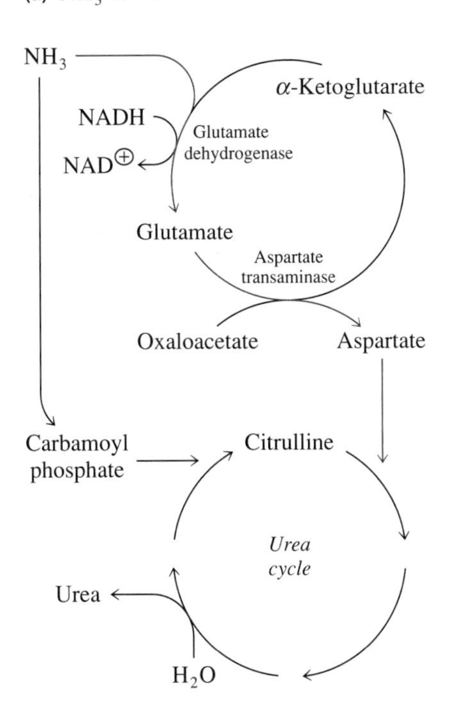

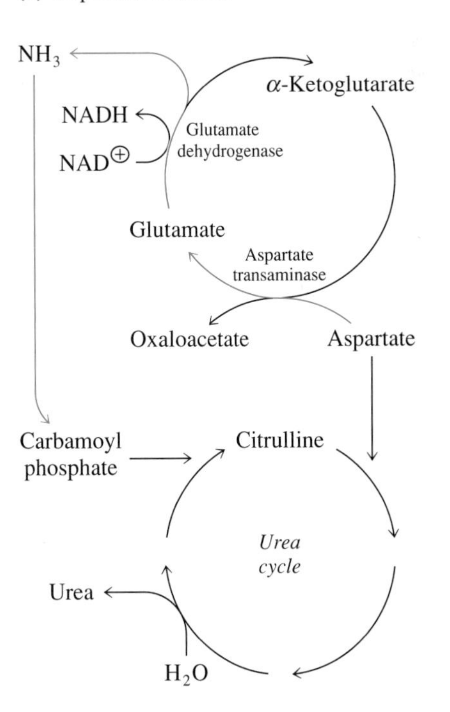

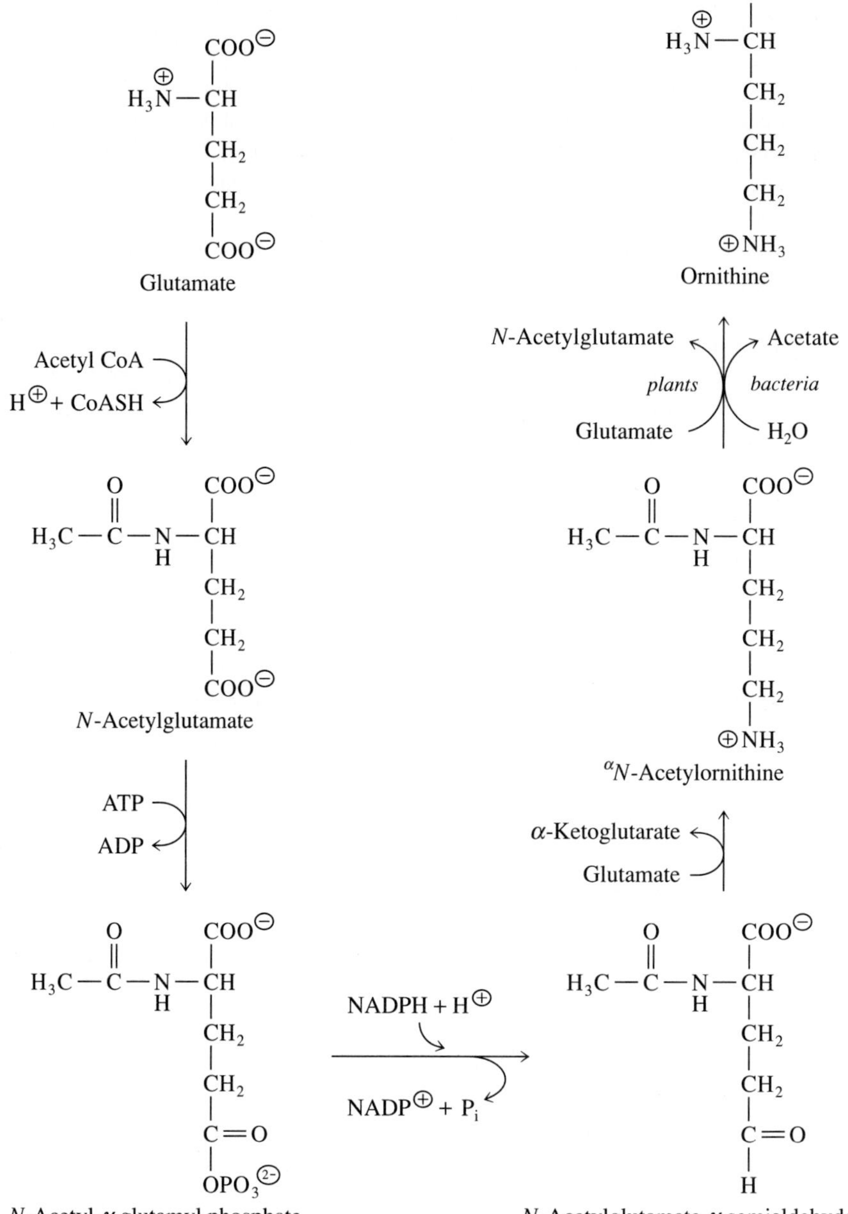

Figure 18·32
Biosynthesis of ornithine in plants and microorganisms.

group of glutamate by an acetyl group from acetyl CoA (Figure 18·32). Next, the γ-carboxylate is converted to an aldehyde by phosphorylation and reduction, reactions analogous to the first two steps of proline biosynthesis (Figure 18·16). The resulting *N*-acetylglutamate γ-semialdehyde undergoes transamination and deacetylation, forming ornithine. Ornithine is then converted to arginine by the same reactions as steps 1, 2, and 3 of the urea cycle (Figure 18·25). Acetylation of the amino group of glutamate in the first step of ornithine synthesis and its eventual deacetylation might seem wasteful. However, substitution of the amino group keeps it from reacting with the γ-aldehyde group to form an internal Schiff base.

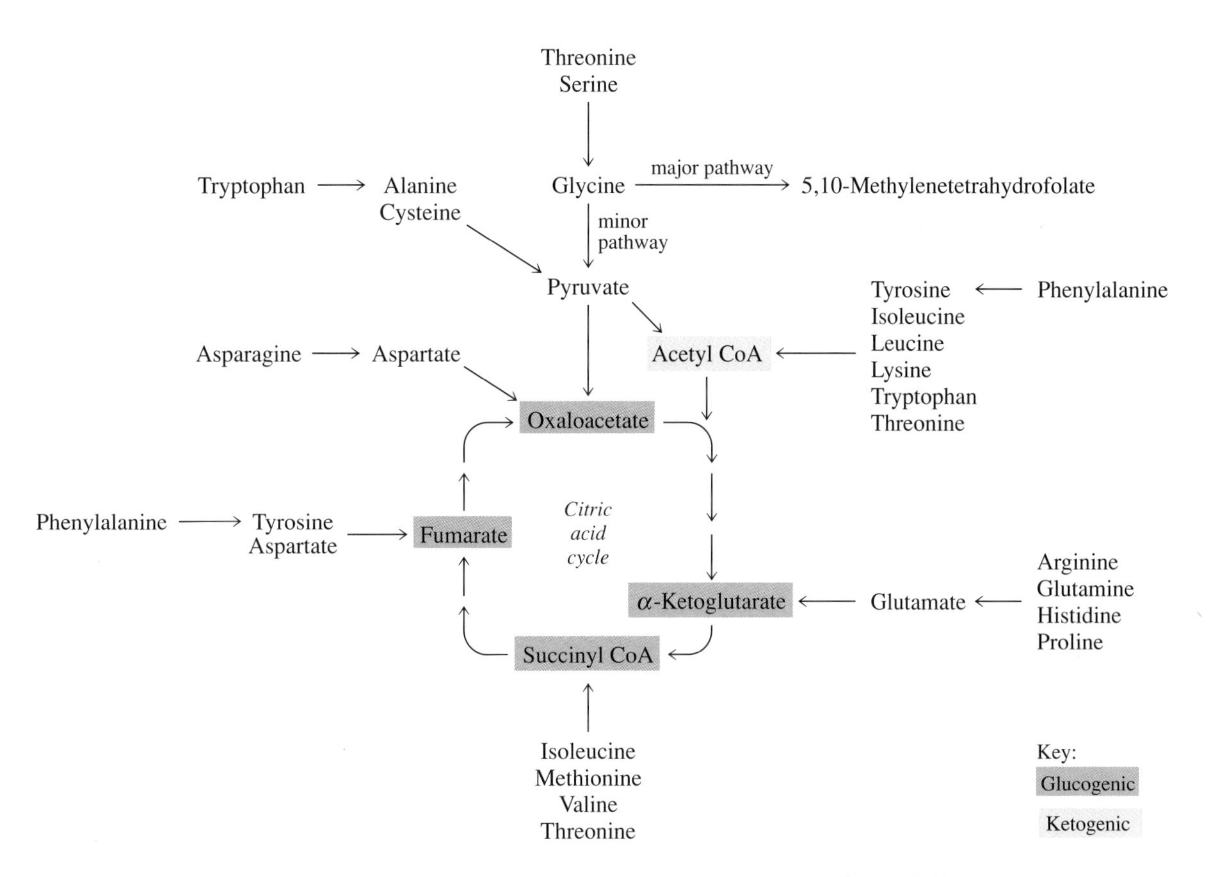

Figure 18·33
Conversion of the carbon skeletons of amino acids to pyruvate, acetyl CoA, or citric acid cycle intermediates for further catabolism.

18·11 The Pathways for Catabolism of the Carbon Chains of Amino Acids Converge with the Major Pathways of Metabolism

As shown in Figure 18·33, some amino acids are degraded to citric acid cycle intermediates, others to pyruvate, and still others to acetyl CoA. Amino acids that are degraded to citric acid cycle intermediates can supply the pathway of gluconeogenesis and are called *glucogenic*. Those that form acetyl CoA can contribute to the formation of fatty acids or ketone bodies and are called *ketogenic*. Those that form pyruvate can be metabolized to either oxaloacetate or acetyl CoA and so can be either glucogenic or ketogenic.

We shall examine the pathways of amino acid degradation in vertebrates, beginning with the simplest routes. Particular attention is given to common end products and the intersections of pathways at common intermediates.

A. Alanine, Aspartate, Glutamate, Asparagine, and Glutamine Are Degraded by Simple Transformations

Alanine, aspartate, and glutamate are synthesized by reversible transamination reactions, as we have seen. The breakdown of these three amino acids involves their re-entry into the pathways from which their carbon skeletons arose. By reversal of the original transamination reactions, alanine gives rise to pyruvate, aspartate to oxaloacetate, and glutamate to α-ketoglutarate. Since aspartate and glutamate are converted to citric acid cycle intermediates, both are glucogenic. Pyruvate formed from alanine can be either glucogenic or ketogenic, as noted above, depending upon whether it is converted to oxaloacetate or acetyl CoA.

The degradation of both glutamine and asparagine begins with their hydrolysis to glutamate and aspartate, respectively. Thus, glutamine and asparagine are both glucogenic.

Figure 18·34
Catabolism of proline and arginine.

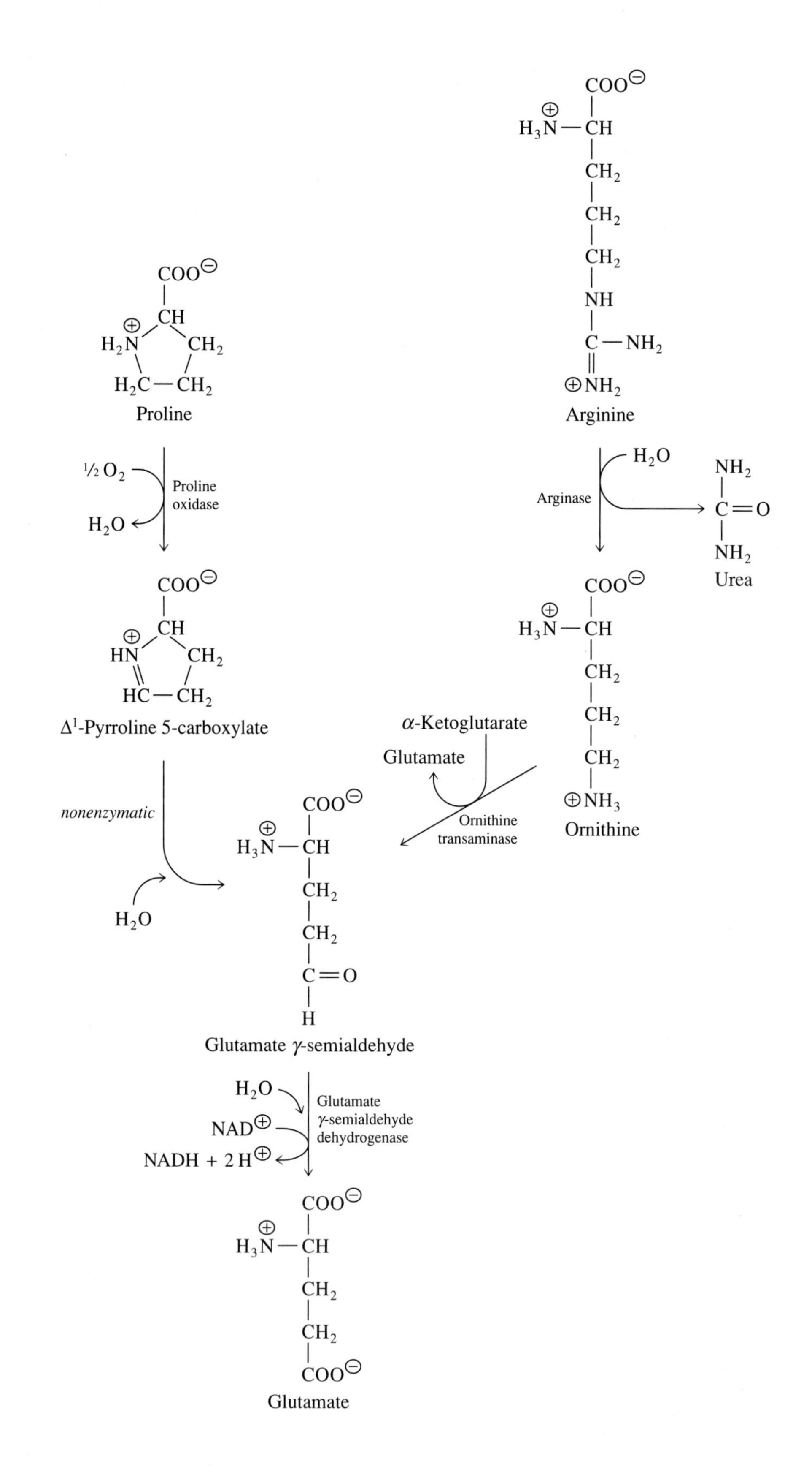

B. The Pathways for the Degradation of Proline, Arginine, and Histidine Lead to Glutamate

Figure 18·34 illustrates the convergence of the degradative pathways of proline and arginine. Proline forms glutamate in three steps. Following oxidation by molecular oxygen, the ring of the Schiff base pyrroline 5-carboxylate is opened by nonenzymatic hydrolysis. The product, glutamate γ-semialdehyde, is oxidized to form glutamate.

Arginine degradation commences with the reaction catalyzed by arginase, an enzyme of the urea cycle. The ornithine produced is transaminated to glutamate γ-semialdehyde, which, as in the pathway for proline degradation, is oxidized to form glutamate.

The major pathway for histidine degradation in mammals is shown in Figure 18·35. In the first step, histidine undergoes nonoxidative deamination to form urocanate and ammonia, catalyzed by histidine ammonia lyase. Next, urocanase catalyzes the addition of water to urocanate. The ensuing hydrolysis reaction opens the ring, forming *N*-formiminoglutamate. The formimino moiety is then transferred to tetrahydrofolate, forming 5-formiminotetrahydrofolate and glutamate.

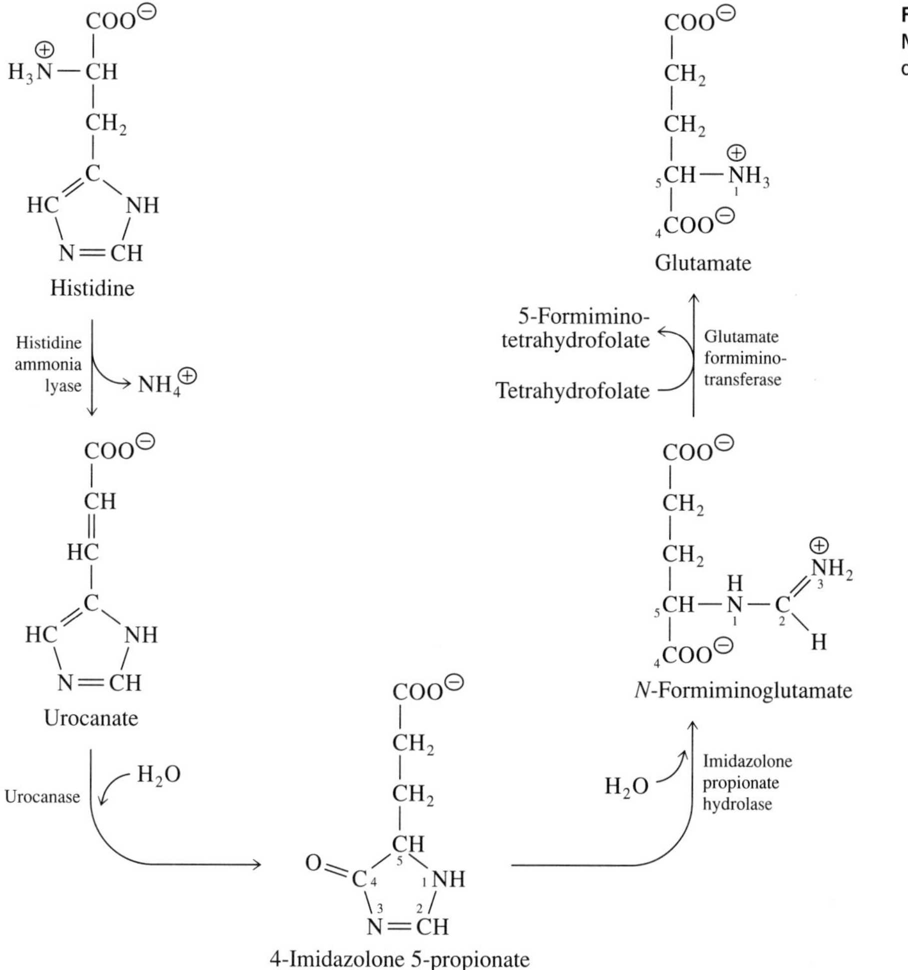

Figure 18·35
Major pathway for the catabolism of histidine to glutamate in mammals.

Box 18·3 The Messenger Compound Nitric Oxide Is Synthesized from Arginine

Nitric oxide (NO) is an unstable gaseous derivative of nitrogen gas with an odd number of electrons (·N=O). Although it is less reactive than other free-radical compounds, it can be toxic. Because nitric oxide in aqueous solution reacts rapidly with oxygen and water to form nitrates and nitrites, it exists in cells for only a few seconds.

A cytosolic enzyme found in mammals, nitric oxide synthase, catalyzes the formation of NO and citrulline from arginine (Figure 1). This oxidation of arginine requires NADPH; tetrahydrobiopterin, FMN, FAD, and a cytochrome are additional cofactors. Nitric oxide formed by this reaction has been implicated as a messenger molecule with involvement in several quite different physiological functions, including resistance to bacteria, dilation of blood vessels, and a role in the action of glutamate as a neurotransmitter.

When macrophages (a type of white blood cell) are treated with certain stimulants such as a bacterial cell wall lipopolysaccharide, they synthesize nitric oxide. Presence of the short-lived nitric oxide free radical allows macrophages to kill bacteria and tumor cells. The action of NO may involve interaction with superoxide anions ($\cdot O_2^{\ominus}$) to form more toxic reactants that account for the cell-killing activity.

Blood vessels are tubes of smooth muscle with a thin inner layer of endothelial cells. The endothelial cells contain nitric oxide synthase and receptors for binding circulating molecules such as acetylcholine that stimulate dilation of blood vessels. Binding of acetylcholine to the receptors triggers synthesis of nitric oxide, which diffuses to the muscle layer and causes it to relax. Nitric oxide activates the enzyme guanylate cyclase, which catalyzes the conversion of GTP to cyclic GMP. cGMP stimulates a protein kinase of muscle that causes the muscle to relax by an unknown mechanism. Nitroglycerin, used to dilate coronary arteries in the treatment of angina pectoris, exerts its effect by virtue of metabolic conversion to NO.

Nitric oxide also functions in brain tissue. The amino acid glutamate is a neurotransmitter, exciting neurons by binding to specific receptors and causing the opening of channels that allow Ca^{2+} to enter the neuronal cells. The nitric oxide synthase of brain is rapidly activated by the entering calcium ions, which bind to molecules of the regulatory protein calmodulin (Stereo S15·1, page 15·13) associated with the synthase. Although the exact role of nitric oxide in

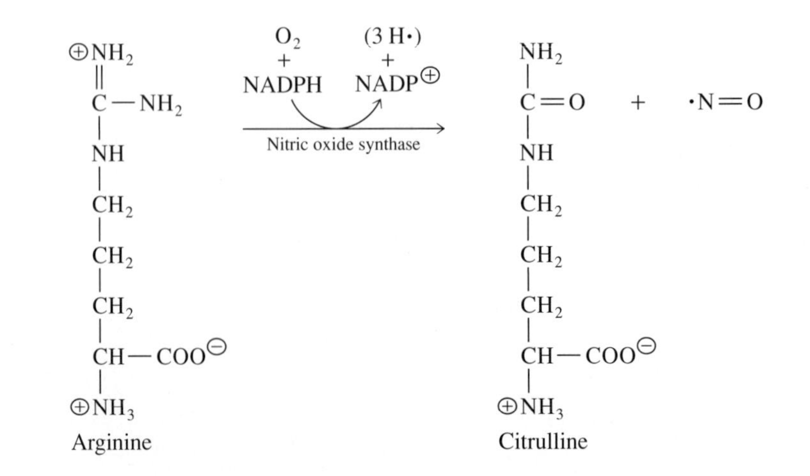

Box 18·3
Figure 1
Tentative stoichiometry of the reaction catalyzed by nitric oxide synthase.

brain is not known, it has been shown that it elicits an increase in the concentration of the second messenger cGMP. A sidelight to this research is the finding that nitric oxide formed in the brain of laboratory animals is involved in neural damage during strokes. Release of excess glutamate causes formation of abnormally high amounts of nitric oxide and appears to kill some neurons in the same way macrophages kill bacterial cells. Administration of an inhibitor of nitric oxide synthase to an animal produces some protection from the stroke. The roles of NO and cGMP in brain are being thoroughly examined and an active search for other simple messenger molecules that may function in ways similar to nitric oxide is underway.

C. Cysteine Is Converted to Pyruvate

In mammals, the major route of cysteine catabolism is a three-step pathway leading to pyruvate (Figure 18·36). Cysteine is first oxidized to cysteinesulfinate, which loses its amino group by transamination to form β-sulfinylpyruvate. Nonenzymatic desulfurylation produces pyruvate.

D. Serine Is Converted to Glycine, Which Is Degraded by the Glycine Cleavage System

Enzymes exist that catalyze the breakdown of serine and glycine to pyruvate, but this is not the major degradative pathway. In some organisms or tissues, a small amount of serine is converted directly to pyruvate by the action of serine dehydratase, a pyridoxal phosphate–requiring enzyme:

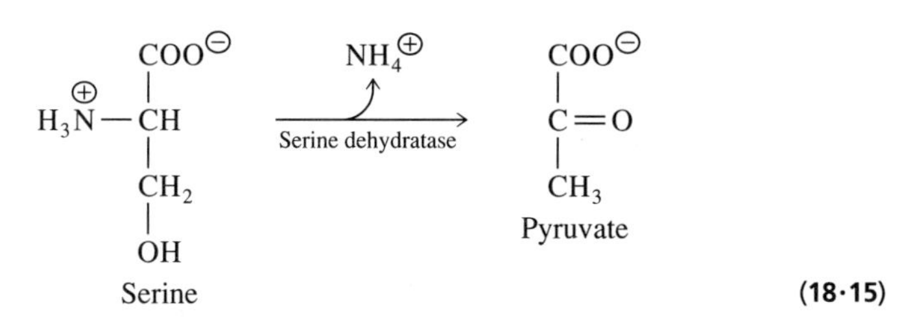

(18·15)

Similarly, a small amount of glycine may be converted to serine in a reaction catalyzed by serine hydroxymethyltransferase (Section 18·6A). Thus, glycine can be catabolized via serine to pyruvate.

The principal pathway of serine catabolism involves the formation of glycine, which is then broken down by the *glycine cleavage system,* the principal pathway for degradation of glycine in mammals.

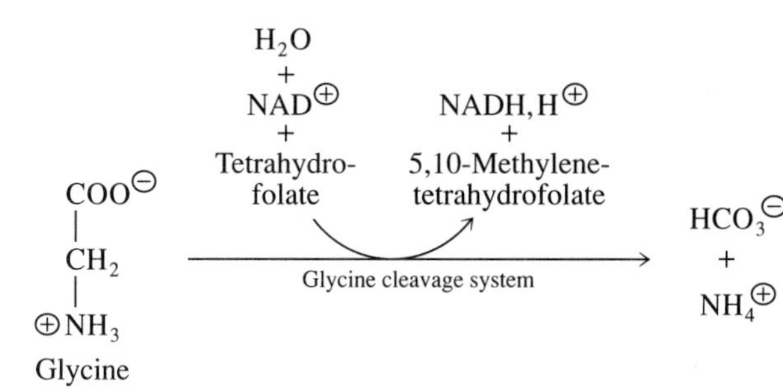

(18·16)

Cysteine

O_2, $H^{\oplus}$

Cysteinesulfinate

α-Ketoglutarate → Glutamate

β-Sulfinylpyruvate

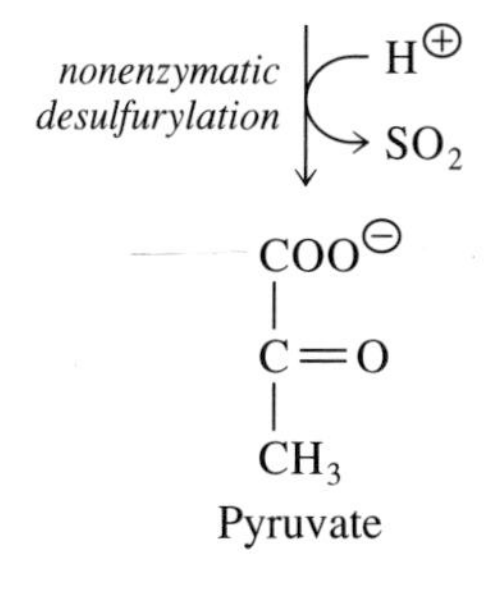

Figure 18·36
Conversion of cysteine to pyruvate in mammals.

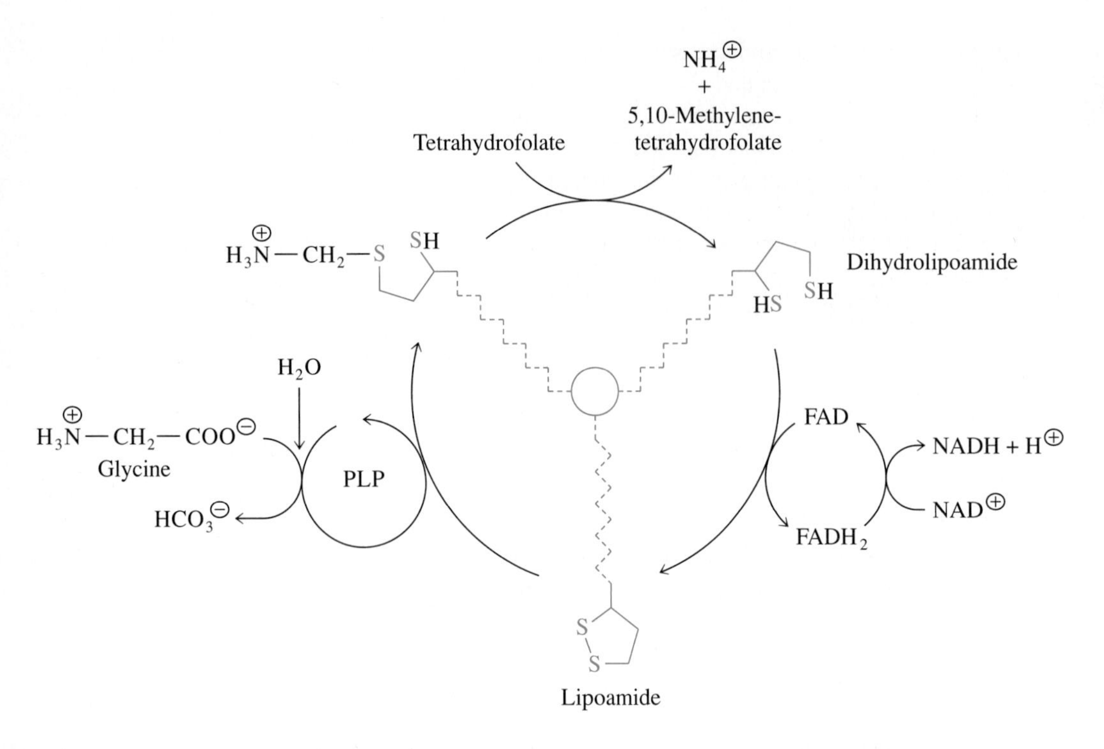

Figure 18·37
Components and mechanism of the glycine cleavage system.

Catalysis by this system, also present in plants and bacteria, requires a four-protein enzyme complex containing four nonidentical subunits. Pyridoxal phosphate, lipoamide, and FAD are present as prosthetic groups, and tetrahydrofolate is a cosubstrate. Initially, glycine is decarboxylated. Then $NH_4^{\oplus}$ is released, and the remaining one-carbon group is transferred to tetrahydrofolate to form 5,10-methylenetetrahydrofolate (Figure 18·37). The reduced lipoamide arm is oxidized by FAD, which in turn reduces the mobile carrier NAD$^{\oplus}$. Though reversible in vitro, the glycine cleavage system catalyzes an irreversible reaction in cells. The irreversibility of the reaction sequence is due in part to the K_m values for the products ammonia and methylenetetrahydrofolate, which are far above the concentrations of these compounds in vivo.

Glycine is also oxidized by the action of a mitochondrial flavoenzyme to form glyoxylate. Further oxidation of glyoxylate forms oxalate, which can precipitate in the kidney as calcium oxalate.

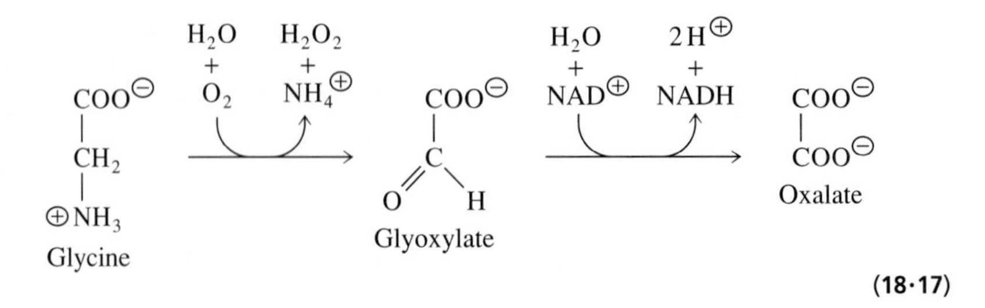

(18·17)

Alternatively, glyoxylate can undergo transamination to reform glycine.

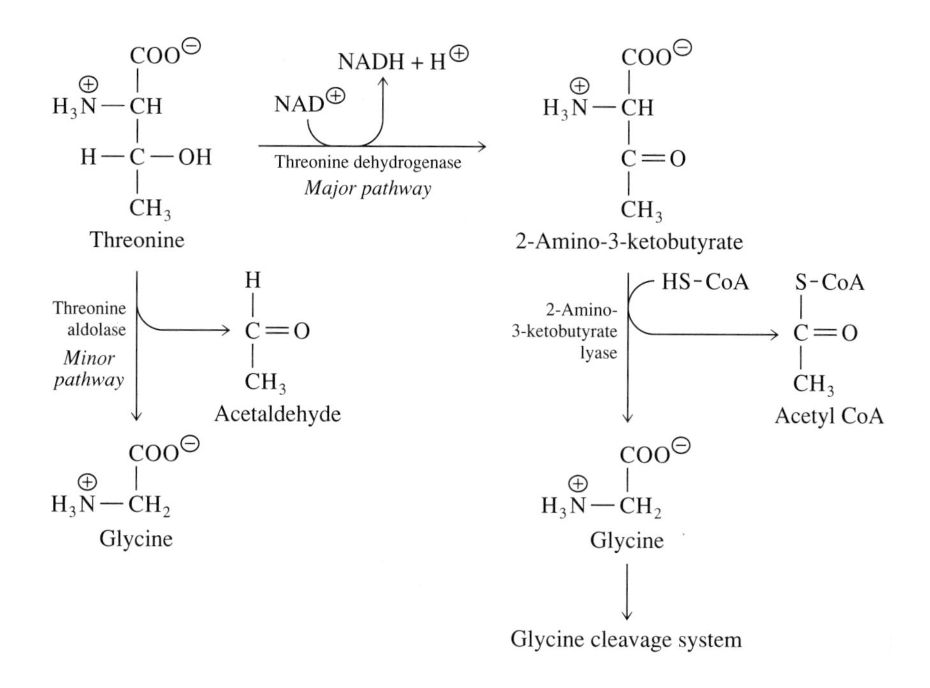

Figure 18·38
Alternative routes for the degradation of threonine.

E. Two Pathways for Threonine Degradation Lead to Glycine

There are several routes for the degradation of threonine. In the major pathway, threonine is oxidized to 2-amino-3-ketobutyrate in a reaction catalyzed by threonine dehydrogenase (Figure 18·38). 2-Amino-3-ketobutyrate can undergo thiolysis to form acetyl CoA and glycine. Another route for the catabolism of threonine consists of cleavage to acetaldehyde and glycine by the action of threonine aldolase, which in many tissues is actually a minor activity of serine hydroxymethyltransferase. Acetaldehyde can be oxidized to acetate by the action of acetaldehyde dehydrogenase, and acetate can be converted to acetyl CoA by acetyl-CoA synthetase.

A third route for the catabolism of threonine involves deamination to α-ketobutyrate, catalyzed by serine dehydratase (Figure 18·39), the same enzyme that catalyzes the conversion of serine to pyruvate (shown on page 18·37). α-Ketobutyrate can be converted to propionyl CoA, a precursor of the citric acid cycle intermediate succinyl CoA. The pathway, shown in detail in Figure 17·12, is as follows:

$$\text{Propionyl CoA} \longrightarrow \text{D-Methylmalonyl CoA} \longrightarrow \text{L-Methylmalonyl CoA} \longrightarrow \text{Succinyl CoA} \quad \textbf{(18·18)}$$

Threonine may thus produce either succinyl CoA or glycine and acetyl CoA, depending on the pathway by which it is catabolized.

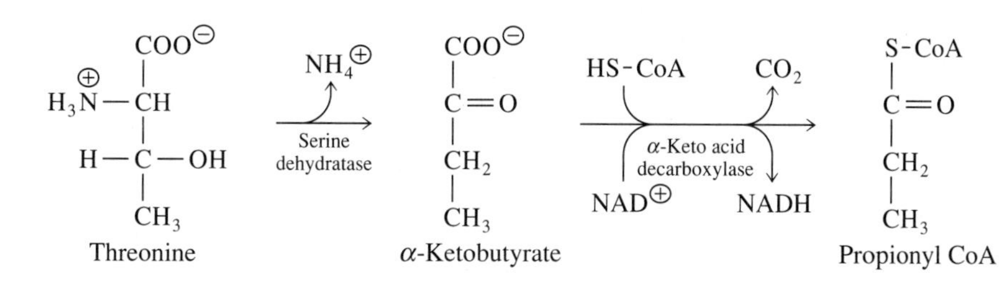

Figure 18·39
Conversion of threonine to propionyl CoA.

F. The Branched-Chain Amino Acids—Leucine, Valine, and Isoleucine—Are Degraded by Pathways that Share Common Steps

Leucine, valine, and isoleucine, the branched-chain amino acids, are degraded by related pathways, shown in Figure 18·40. The first step, transamination, is catalyzed in all three pathways by the same enzyme, branched-chain amino acid transaminase. In mammals, transamination of branched-chain amino acids occurs primarily in muscle, whereas the ensuing oxidative steps occur primarily in liver.

The second step in the catabolism of branched-chain amino acids is also catalyzed by the same enzyme in all three pathways, branched-chain α-keto acid dehydrogenase. In this reaction, the branched-chain α-keto acids undergo oxidative decarboxylation to form branched-chain acyl CoA molecules one carbon atom shorter than the precursor α-keto acids. The branched-chain α-keto acid dehydrogenase is a multienzyme complex that contains lipoamide and thiamine pyrophosphate and requires $NAD^{\oplus}$ and CoASH. Its catalytic mechanism is similar to those of the pyruvate dehydrogenase and α-ketoglutarate dehydrogenase complexes. The branched-chain α-keto acid dehydrogenase complex is inhibited by branched-chain acyl CoA molecules and by phosphorylation of one of its subunits. These effects regulate the overall degradation of the branched-chain amino acids.

The branched-chain acyl CoA molecules are oxidized by a third enzyme common to all three pathways, an FAD-containing acyl-CoA dehydrogenase. The electrons removed in this oxidation step are transferred via the electron-transferring flavoprotein (ETF, Section 17·3C) to ubiquinone (Q).

At this point, the steps in the catabolism of the branched-chain amino acids diverge. In the degradation of leucine, the intermediate formed after the initial three steps is carboxylated and hydrated to form hydroxymethylglutaryl-CoA (HMG-CoA), an intermediate in ketone-body synthesis in liver mitochondria (Section 17·7A). Each carbon of leucine is ultimately converted to acetyl CoA. Leucine is the only solely ketogenic amino acid.

In the pathway for degradation of valine in mammals, the product of the first three steps is converted to propionyl CoA. As in the degradation of threonine, propionyl CoA is converted via methylmalonyl CoA to succinyl CoA, which enters the citric acid cycle.

The isoleucine degradation pathway leads to both propionyl CoA and acetyl CoA. Isoleucine is therefore both glucogenic (succinyl CoA formed from propionyl CoA) and ketogenic (acetyl CoA). In contrast, degradation of valine produces only succinyl CoA, and leucine produces only acetyl CoA. Thus, although the initial steps in the degradation of the three branched-chain amino acids are similar, their carbon skeletons have different fates.

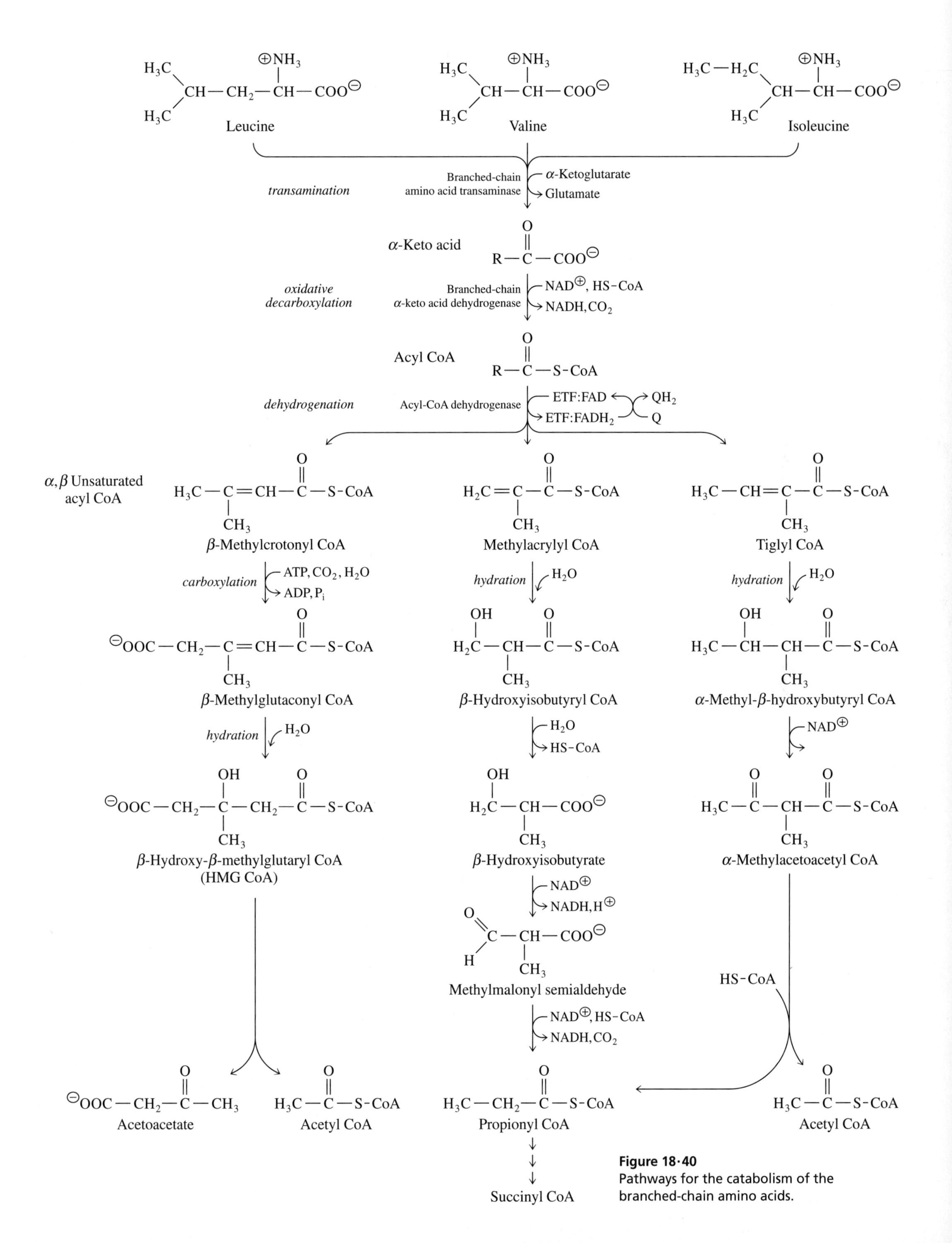

Figure 18·40
Pathways for the catabolism of the branched-chain amino acids.

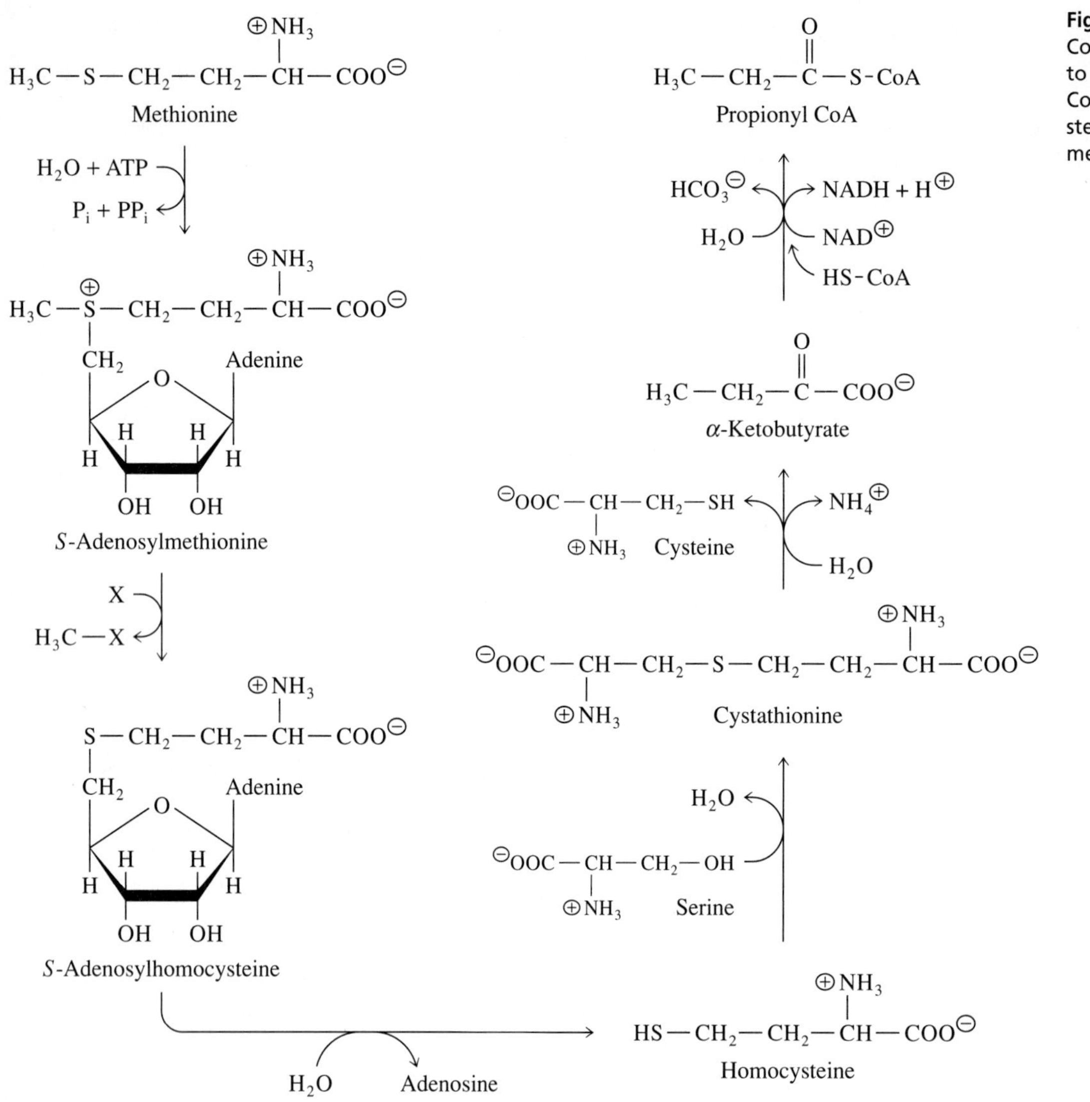

Figure 18·41
Conversion of methionine to cysteine and propionyl CoA. "X" in the second step is any of a number of methyl-group acceptors.

G. Methionine Degradation Involves Cysteine Synthesis

One major role of methionine is conversion to the activated methyl donor *S*-adenosylmethionine, SAM (Section 7·1). Transfer of the methyl group from SAM to a methyl acceptor leaves *S*-adenosylhomocysteine, which is degraded by hydrolysis to homocysteine and adenosine (Figure 18·41). Homocysteine can either be methylated by 5-methyltetrahydrofolate to form methionine, or it can react with serine to form cystathionine, which can be cleaved to cysteine and α-ketobutyrate. We encountered this series of reactions earlier as part of a pathway for the formation of cysteine (Figure 18·15). By this pathway, animals can form cysteine using a sulfur atom from the essential amino acid methionine. α-Ketobutyrate is converted to propionyl CoA by the action of an α-keto acid dehydrogenase. Propionyl CoA can be further metabolized to succinyl CoA, as explained earlier.

H. Phenylalanine, Tyrosine, and Tryptophan Are Catabolized via Oxidation, Deamination, and Ring-Opening Hydrolysis

The aromatic amino acids share a common pattern of catabolism. There is generally an oxidation at the earliest stage of the pathway, followed by the removal of nitrogen by transamination or hydrolysis, and then ring opening coupled with an oxidation.

The conversion of phenylalanine to tyrosine, catalyzed by phenylalanine hydroxylase (Section 18·6C), is important not only in the biosynthesis of tyrosine but also in the catabolism of phenylalanine. Catabolism of tyrosine begins with the removal of its α-amino group in a transamination reaction with α-ketoglutarate. Subsequent oxidation steps lead to ring opening and eventually to the final products, fumarate and acetoacetate (Figure 18·42). Like the fumarate formed in the urea cycle, this fumarate is cytosolic and is converted to glucose. Acetoacetate is a ketone body. Thus, the degradation of tyrosine is both glucogenic and ketogenic.

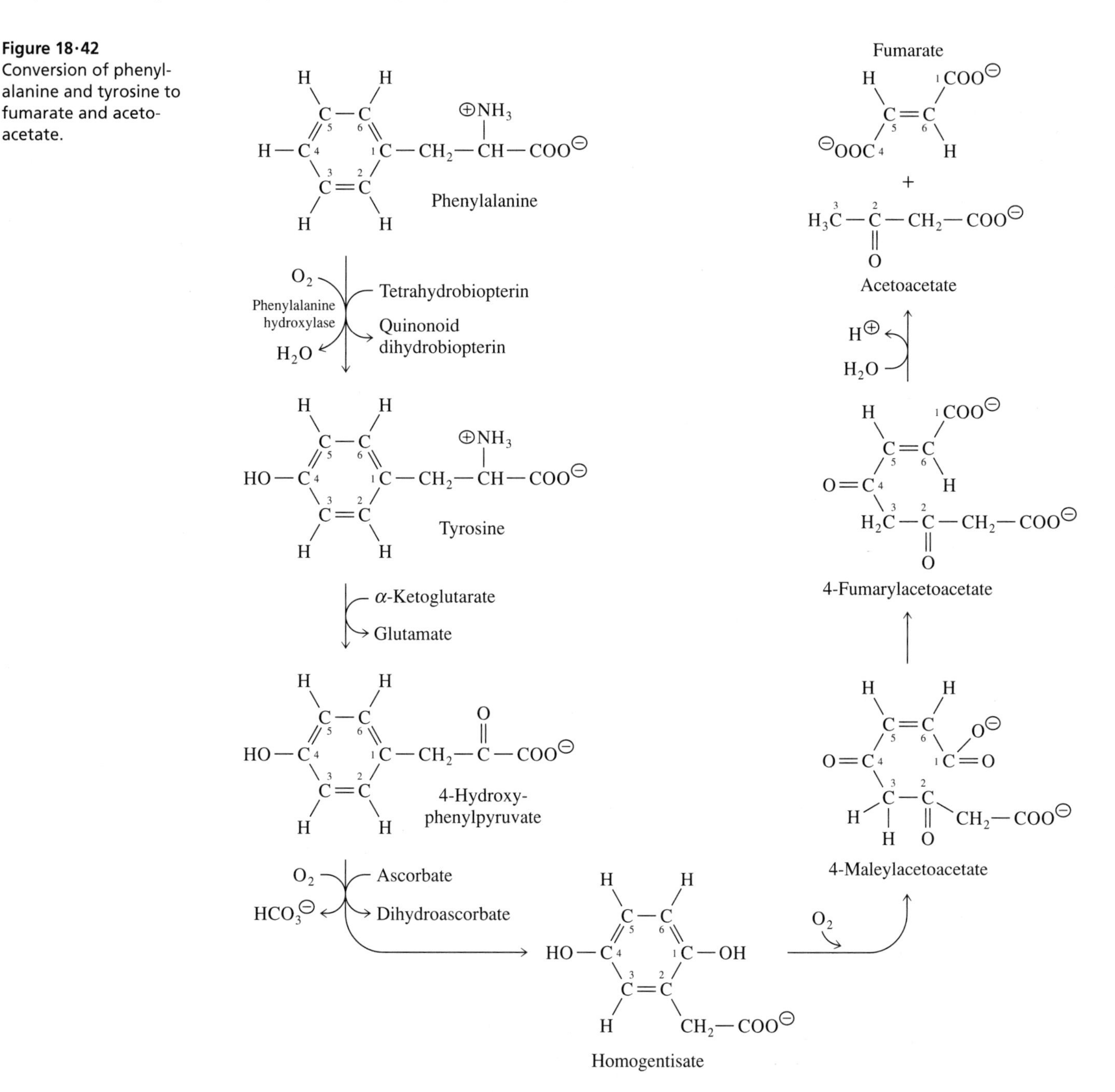

Figure 18·42
Conversion of phenylalanine and tyrosine to fumarate and acetoacetate.

Tryptophan, which has an indole ring system, has a more complex pathway of catabolism than phenylalanine or tyrosine, in that two ring-opening reactions are required. The major route of tryptophan oxidation in liver and many microorganisms is the kynurenine pathway (Figure 18·43). The first step is catalyzed by tryptophan 2,3-dioxygenase. O_2 is a substrate of the reaction, and both atoms of O_2 are incorporated into the product, *N*-formylkynurenine. The next three steps are hydrolysis, oxidation, and a second hydrolysis, in which alanine is released. Then another oxidation occurs, in which a second dioxygenase catalyzes the opening of the benzene ring, forming 2-amino-3-carboxymuconate ε-semialdehyde. As shown in Figure 18·43, this intermediate is a branch-point metabolite. The major degradative branch consists of a number of steps leading to α-ketoadipate and ultimately to acetyl CoA. The six steps from α-ketoadipate also occur in lysine catabolism, as we shall see in the next section. Alanine, formed early in tryptophan catabolism, is transaminated to pyruvate, which in turn is converted to either acetyl CoA or oxaloacetate. Thus, the catabolism of tryptophan is both ketogenic and glucogenic.

Proceeding down the other branch from 2-amino-3-carboxymuconate ε-semialdehyde, nonenzymatic ring closure forms quinolinate, a precursor of the pyridine nucleotide coenzymes NAD$^{\oplus}$ and NADP$^{\oplus}$. Small but significant amounts of the pyridine nucleotide coenzymes can be formed by this route in humans, sparing some of the dietary requirement for niacin. There are pathological conditions associated with impaired catabolism of tryptophan in which clinical pellagra occurs, strongly suggesting that the metabolic route from tryptophan to NAD$^{\oplus}$ is important under normal conditions.

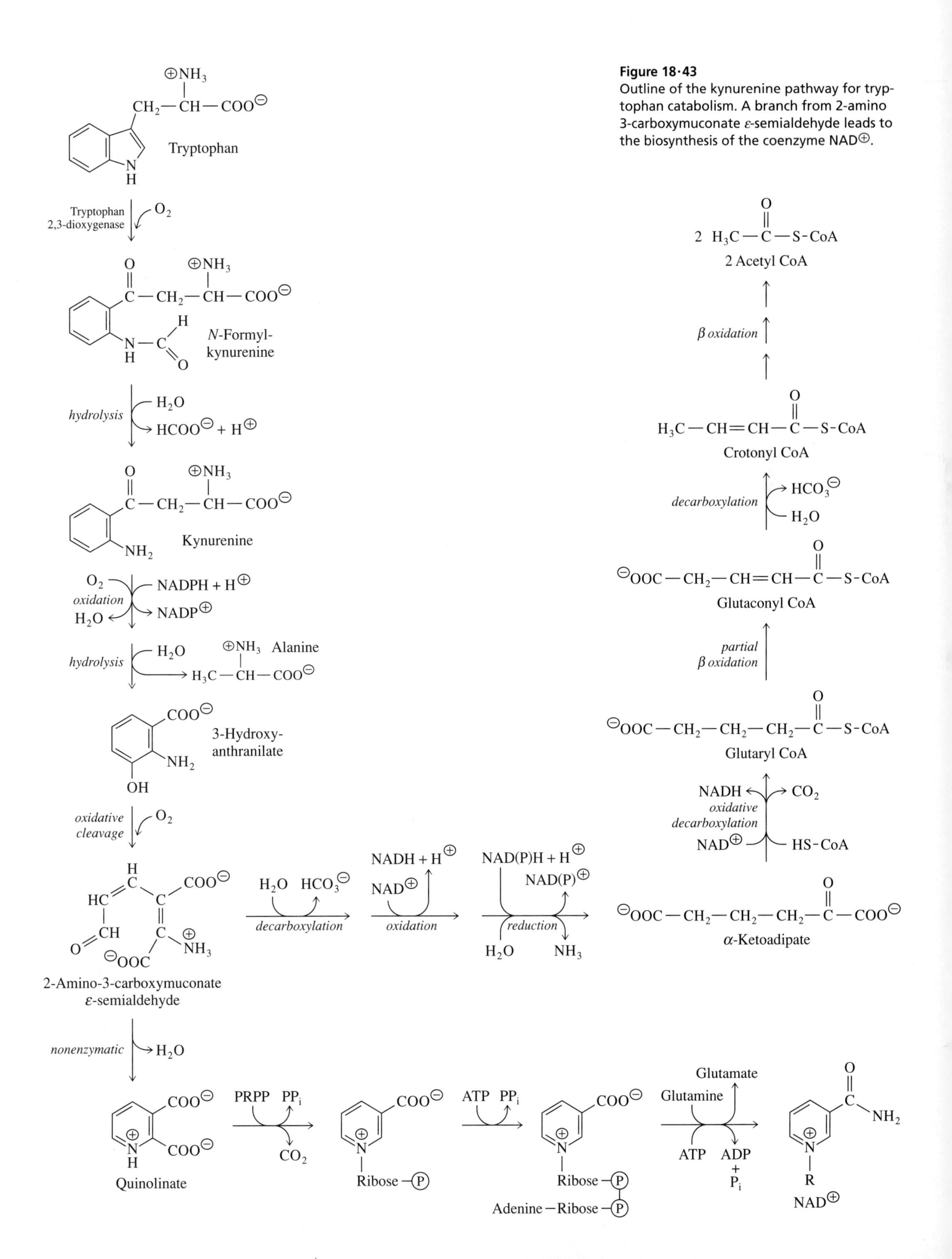

Figure 18·43
Outline of the kynurenine pathway for tryptophan catabolism. A branch from 2-amino 3-carboxymuconate ε-semialdehyde leads to the biosynthesis of the coenzyme NAD⊕.

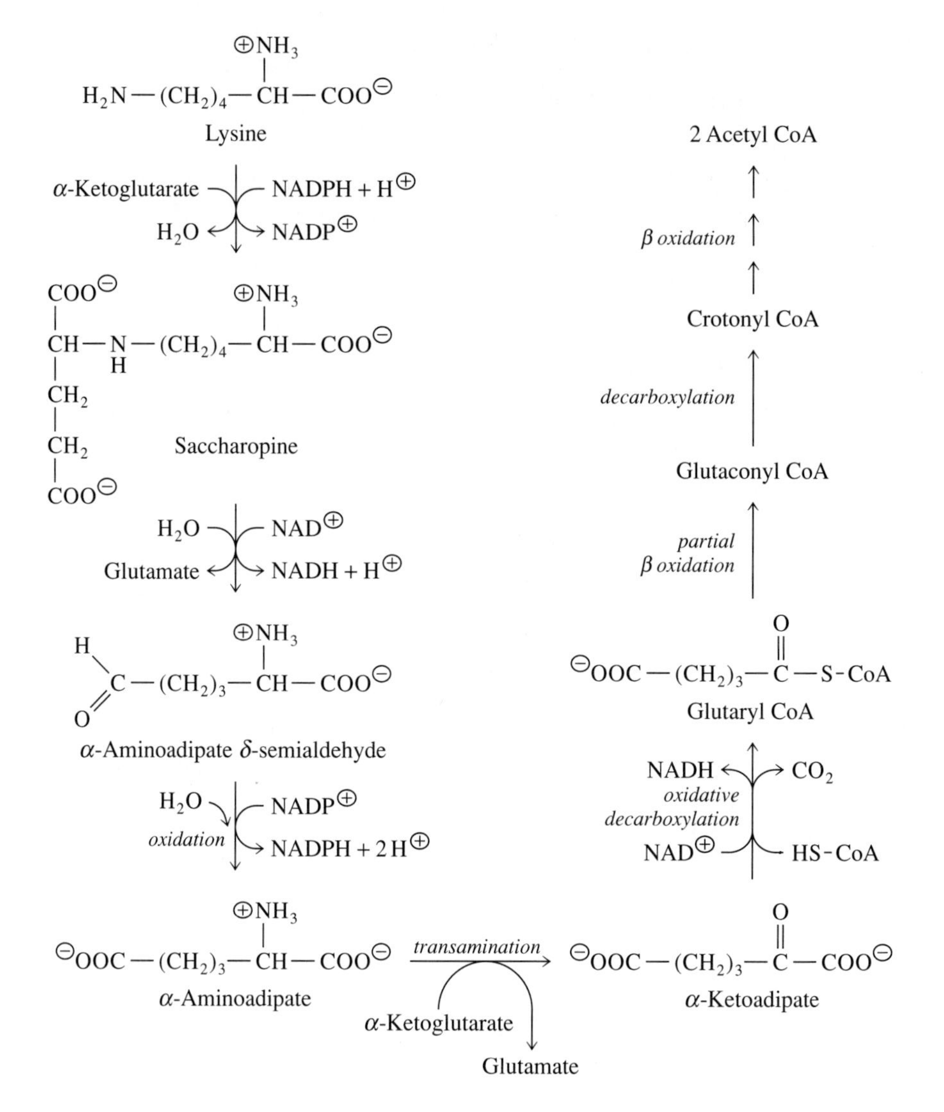

Figure 18·44
The saccharopine pathway for lysine degradation in mammalian liver.

I. The Pathway for Lysine Catabolism Merges with Tryptophan Catabolism at α-Ketoadipate

The main pathway for the degradation of lysine in both mammals and bacteria is via the intermediate saccharopine (Figure 18·44). Saccharopine is formed by the reductive condensation of lysine with α-ketoglutarate. Cleavage of saccharopine releases glutamate, accomplishing the transfer of the amino group of lysine to α-ketoglutarate and generating α-aminoadipate δ-semialdehyde. The latter molecule is oxidized to α-aminoadipate, which loses its amino group by transamination to become α-ketoadipate. α-Ketoadipate is converted to acetyl CoA by the same steps shown in Figure 18·43 for the oxidation of tryptophan.

18·12 The Metabolism of Amino Acids in Mammals Involves Exchange Between Organs

In mammals, the major organs involved in amino acid metabolism have distinct metabolic roles. In addition, there is extensive traffic of amino acids—particularly alanine and glutamine—among organs. We shall now examine three examples of interorgan relationships in mammals. All three involve a link between amino acid metabolism and gluconeogenesis.

Large amounts of alanine are synthesized de novo in muscle. The carbon skeleton of alanine arises from pyruvate formed by glycolysis. The amino group of alanine is obtained in transamination reactions with other amino acids. The alanine produced in muscle travels through the bloodstream to the liver, where its α-amino group is used for urea synthesis. Pyruvate formed in liver from the alanine is converted to glucose, which can return to the muscle tissue. This exchange of metabolites between muscle and liver, as noted earlier, is called the glucose-alanine cycle.

Gluconeogenesis from amino acids becomes important during the early stages of starvation (the hours after both ingested glucose and liver glycogen have been consumed). The body requires that the concentration of glucose in blood be relatively constant. During starvation, the source of most blood glucose is amino acids derived from proteolysis of muscle proteins. The α-amino groups from these amino acids are shuttled to the liver by the glucose-alanine cycle. The carbon skeletons of most of the amino acids are converted to pyruvate, which is transaminated to form alanine, and to α-ketoglutarate, which is converted to glutamine. As described above, alanine is converted to glucose in the liver. The carbon skeleton of glutamine is used for gluconeogenesis in the kidney.

Kidney and liver are the only organs in mammals in which gluconeogenesis occurs. In both organs, gluconeogenesis is linked to nitrogen excretion. The kidney does not synthesize urea but can produce large amounts of $NH_4^{\oplus}$ during conditions of acidosis. This state arises most commonly when metabolic acids such as lactate or ketone bodies are produced in excess, as in long-term starvation or in uncontrolled diabetes. Acidosis leads to the induction of two enzymes in kidney: glutaminase and phosphoenolpyruvate (PEP) carboxykinase. Hydrolysis of glutamine catalyzed by glutaminase, followed by oxidative deamination of the resulting glutamate catalyzed by glutamate dehydrogenase, leads to production of $NH_4^{\oplus}$ and α-ketoglutarate. The $NH_4^{\oplus}$ exits the kidney cells into the urine simultaneously with $Cl^{\ominus}$. The overall effect is removal of NH_4Cl from the body. NH_4Cl is acidic, and its removal provides a means to counteract acidosis. The α-ketoglutarate is converted to malate by reactions of the citric acid cycle and thence to glucose by the reactions of gluconeogenesis. The induction of kidney PEP carboxykinase increases the rate of gluconeogenesis, in coordination with the induction of kidney glutaminase, which increases the rate of ammoniagenesis. Thus, just as urea synthesis is linked to gluconeogenesis in liver, ammoniagenesis is linked to gluconeogenesis in kidney.

Summary

Amino acid metabolism can be approached from two points of view: the origins and fates of the amino acid nitrogen atoms, and the origins and fates of their carbon skeletons. Nitrogen is introduced into biological systems by reduction of chemically unreactive N_2 from the atmosphere to ammonia. This reaction, catalyzed by nitrogenase, is carried out by a few species of bacteria and algae. An alternate route for the assimilation of nitrogen, carried out by plants and microorganisms, is the reduction of nitrate to nitrite, catalyzed by nitrate reductase. Nitrite is reduced by the action of nitrite reductase to form ammonia, which can be utilized by all organisms.

Ammonia is assimilated into biological metabolites by several routes. Glutamate dehydrogenase catalyzes the reductive amination of α-ketoglutarate in a reversible reaction forming glutamate, which can then enter the central pathways of metabolism. Another important carrier of nitrogen is glutamine, which can be formed from glutamate and ammonia by the action of glutamine synthetase. Glutamate and glutamine are nitrogen donors in many reactions.

The amino group of glutamate can be transferred to many α-keto acids in reversible transamination reactions that form α-ketoglutarate and the corresponding α-amino acids. In the reverse direction, transamination reactions lead from a number of amino acids to glutamate or aspartate, which contribute amino groups to pathways for nitrogen disposal.

Nonessential amino acids are those that an organism can produce in sufficient quantity metabolically. Essential amino acids are those that must be supplied in the diet. Essential and nonessential amino acids are defined for individual species. The nonessential amino acids are generally those formed by short, energetically inexpensive pathways. Glutamate, alanine, and aspartate are formed by simple transamination, and glutamine and asparagine are formed by transfer of amide groups to the side chains of glutamate and aspartate. Serine, glycine, and cysteine are derived from 3-phosphoglycerate. Proline is formed from glutamate. Tyrosine is formed in a single reaction from the essential amino acid phenylalanine.

Of the amino acids that are essential in the diets of humans, lysine, threonine, and methionine are derived from aspartate. Pathways sharing common enzymatic steps lead to the branched-chain amino acids isoleucine, valine, and leucine. The aromatic amino acids arise from a pathway in which chorismate is formed in seven steps from phosphoenolpyruvate and erythrose 4-phosphate, followed by conversion of chorismate to phenylalanine, tyrosine, or tryptophan. Mammals can synthesize histidine but not in sufficient amounts, and thus they depend on histidine formed by plants, simpler eukaryotes, and bacteria from PRPP, ATP, and glutamine. There is no evidence that adult humans can synthesize arginine. Plants and microorganisms synthesize ornithine, which is converted to arginine by reactions that occur in the urea cycle.

Catabolism of amino acids often begins with deamination, followed by modification of the remaining carbon chains for entry into the central pathways of carbon metabolism. Glutamate or aspartate is the amino-group acceptor at the end of one or a series of transamination reactions. In mammals, nitrogen is disposed of via the formation of urea by the urea cycle. Urea is then excreted. The carbon atom of urea is derived from bicarbonate. One amino group is derived from ammonia, and the other, from the α-amino group of aspartate. Fumarate is released in the cytosol as a product of the urea cycle and enters the pathway of gluconeogenesis. Aspartate needed for continued operation of the urea cycle arises primarily from alanine, an intermediate of the glucose-alanine cycle between muscle, where glucose is consumed, and liver, where organic metabolites are converted to glucose for use in muscle and other organs.

The pathways for degradation of amino acids lead to intermediates of the citric acid cycle, pyruvate, or acetyl CoA. Amino acids that are degraded to citric acid cycle intermediates can supply the pathway of gluconeogenesis and are called glucogenic. Those that form acetyl CoA can contribute to the formation of fatty acids or ketone bodies and are called ketogenic. Amino acids degraded to pyruvate can be metabolized to either oxaloacetate or acetyl CoA and so can be either glucogenic or ketogenic.

Alanine, aspartate, and glutamate are degraded by reversal of the transamination reactions by which they were formed. Degradation of glutamine and asparagine begins with their hydrolysis to glutamate and aspartate, respectively. Proline, arginine, and histidine are degraded to glutamate. Cysteine is degraded to pyruvate. Serine is converted to glycine, which is catabolized by the multienzyme glycine cleavage system. Threonine may be converted by alternate routes to succinyl CoA or glycine and acetyl CoA. The branched-chain amino acids are degraded by pathways that share common steps leading to different fates; leucine is degraded to acetyl CoA, valine to succinyl CoA, and isoleucine to succinyl CoA and acetyl CoA. Methionine is degraded to succinyl CoA, with formation of cysteine as a by-product. Of the aromatic amino acids, phenylalanine is converted to tyrosine, which is degraded to fumarate and acetoacetate. Tryptophan is degraded to acetyl CoA, with formation of alanine as a by-product; alanine is converted to pyruvate and then to either oxaloacetate or acetyl CoA. Lysine catabolism merges with the pathway of tryptophan catabolism, leading to acetyl CoA.

Selected Readings

Adams, E., and Frank, L. (1980). Metabolism of proline and the hydroxyprolines. *Annu. Rev. Biochem.* 49:1005–1061.

Bender, D. A. (1985). *Amino Acid Metabolism,* 2nd ed. (Chichester: John Wiley & Sons). A monograph devoted to the pathways of amino acid metabolism, with emphasis on mammalian systems.

Cooper, A. J. L. (1983). Biochemistry of sulfur-containing amino acids. *Annu Rev. Biochem.* 52:187–222.

Hayashi, H., Wada, H., Yoshimura, T., Esaki, N., and Soda, K. (1990). Recent topics in pyridoxal 5′-phosphate enzyme studies. *Annu. Rev. Biochem.* 59:87–110.

Herrmann, K. M., and Somerville, R. L., eds. (1983). *Amino Acids: Biosynthesis and Genetic Regulation.* (Reading, Pennsylvania: Addison-Wesley Publishing Company). A collection of review articles on amino acid biosynthesis, mainly in prokaryotes.

Hyde, C. C., Ahmed, S. A., Padlan, E. A., Miles, E. W., and Davies, D. R. (1988). Three-dimensional structure of the tryptophan synthase $\alpha_2\beta_2$ multienzyme complex from *Salmonella typhimurium. J. Biol. Chem.* 263:17 857–17 871. Evidence from X-ray crystallography of a tunnel from one active site to another.

Kaufman, S. (1987). Aromatic amino acid hydroxylases. In *The Enzymes: Control by Phosphorylation, Part B,* Vol. 18, 3rd ed., P. D. Boyer and E. G. Krebs, eds. (New York: Academic Press), pp. 217–282. A description of the hydroxylation of phenylalanine, tyrosine, and tryptophan, with emphasis on regulation.

Krebs, H. A., Lund, P., and Stubbs, M. (1976). Interrelations between gluconeogenesis and urea synthesis. In *Gluconeogenesis: Its Regulation in Mammalian Species*, R. W. Hanson and M. A. Mehlman, eds. (New York: John Wiley & Sons), pp. 269–291.

Kvamme, E., ed. (1988). *Glutamine and Glutamate in Mammals.* Vol. 1. (Boca Raton, Florida: CRC Press, Inc.)

Scriver, C. R., Beaudet, A. L., Sly, W. S., and Valle, D., eds. (1989). *The Metabolic Basis of Inherited Disease,* Vols. 1 and 2. (New York: McGraw-Hill). Up-to-date descriptions of over 100 metabolic diseases, including many involving enzymes associated with amino acids, explained through molecular genetics.

Snyder, S. H. (1992). Nitric oxide: first in a new class of neurotransmitters? *Science* 257:494–496.

Snyder, S. H., and Bredt, D. S. (1992). Biological roles of nitric oxide. *Sci. Am.* 266(5):68–77.

Stacey, G., Burris, R. H., and Evans, H. J., eds. (1992). *Biological Nitrogen Fixation* (New York: Chapman and Hall). A large, thorough, advanced survey, including detailed reviews of nitrogenase structure and function.

Stadtman, E. R., and Ginsburg, A. (1974). The glutamine synthetase of *Escherichia coli*: structure and control. In *The Enzymes*, Vol. 10, 3rd ed., P. D. Boyer, ed. (New York: Academic Press), pp. 755–807.

Tyler, B. (1978). Regulation of the assimilation of nitrogen compounds. *Annu. Rev. Biochem.* 47:1127–1162.

Umbarger, H. E. (1978). Amino acid biosynthesis and its regulation. *Annu. Rev. Biochem.* 47:533–606.

19

Nucleotide Metabolism

Within cells, purine and pyrimidine bases are almost always found as constituents of nucleotides and polynucleotides. Some nucleotides, such as ATP, function as cosubstrates and others, such as cyclic AMP, are regulatory compounds. Nucleotides are also substrates for the biosynthesis of DNA and RNA (Chapters 20 and 21) and for the conversion of some B vitamins to their coenzyme forms (Chapter 7). Most organisms have the ability to synthesize purine and pyrimidine nucleotides, reflecting the importance of these molecules.

In this chapter, we shall first describe the biosynthesis of purine nucleotides from simpler precursors. We shall then discuss simple pathways by which preformed purines obtained from the breakdown of nucleic acids within cells or from the external environment (such as from the intestinal tract during the digestion of food) can be incorporated directly into nucleotides, a process called salvage. The ability to catalyze formation of purine nucleotides by the de novo pathway assures a measure of independence from dietary sources, and the salvage pathway provides for economy by recycling the products of nucleic acid breakdown.

We shall then describe the pathway for the de novo biosynthesis of pyrimidine nucleotides, a reaction sequence that differs in chemical strategy from that of the purine nucleotides, and then look at how pyrimidines are salvaged. Next, we shall present the conversion of the purine and pyrimidine ribonucleotides to their 2′-deoxy forms.

Finally, we shall examine the biological degradation of nucleotides. The chemical strategies for breakdown of the purine and the pyrimidine compounds also differ. Breakdown of purines leads to formation of potentially toxic compounds, whereas breakdown of pyrimidines leads to easily metabolized products.

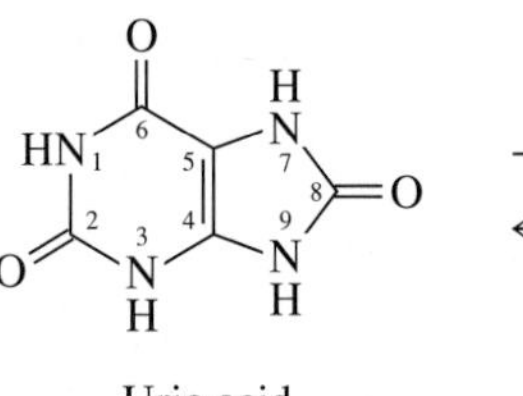

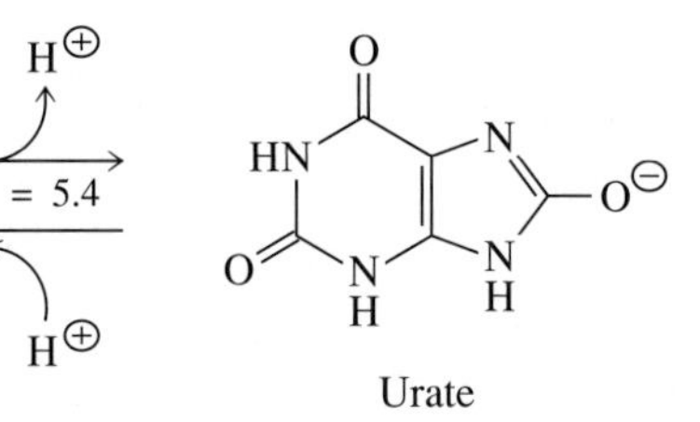

Figure 19·1
Uric acid (2,6,8-trioxopurine). With a pK_a of 5.4, uric acid at physiological pH exists primarily as the conjugate base, urate.

19·1 Early Investigations of Purine Biosynthesis Involved Isotopic Labelling

Uric acid (Figure 19·1) was the first purine to be discovered. In 1776, this compound was shown by both Karl W. Scheele and Torbern Bergmann to be present in human urine and bladder stones. A century later, it was discovered that most of the nitrogen ingested by adult hens was excreted as uric acid. Chemists of the nineteenth century established that, in birds and many reptiles, the major end product of the metabolism of nitrogen is uric acid, analogous to urea in mammalian metabolism.

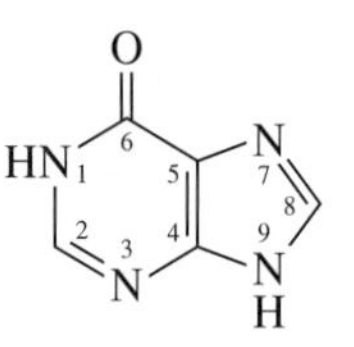

Figure 19·2
Hypoxanthine (6-oxopurine).

In 1936, H. A. Krebs and his colleagues performed experiments that paved the way for the complete elucidation of the pathway leading to purine nucleotides. They showed that in pigeons the incorporation of ammonia (NH_3) into uric acid occurs in two stages. Ammonia is first incorporated into hypoxanthine in the liver. Next, hypoxanthine is oxidized in the kidney to uric acid in a reaction catalyzed by xanthine oxidase. The structure of hypoxanthine is shown in Figure 19·2. Until the 1950s, many biochemists believed that the purines of nucleic acids arose by a pathway different from the pathway in birds that forms hypoxanthine as an intermediate in nitrogen excretion. However, experiments with isotopically labelled precursor molecules demonstrated that nucleic acid purines and uric acid arise from the same precursors and pathway. Homogenates of pigeon liver—a tissue in which purines are actively synthesized—were then used extensively as a convenient source of enzymes for studying the steps in purine biosynthesis. The pathway in avian liver has since been found in other organisms.

Isotopes of carbon and nitrogen became available as tracers for metabolic studies in the 1940s. The stable isotopes ^{13}C and ^{15}N were used at first, and the radioactive ^{14}C a few years later. Simple compounds such as $^{13}CO_2$, $H^{13}COO^{\ominus}$ (formate), and $^{\oplus}H_3N{-}CH_2{-}^{13}COO^{\ominus}$ (glycine) were administered to pigeons and rats by John M. Buchanan and his colleagues. They then isolated and chemically degraded excreted uric acid using a procedure that allowed labelled carbon and nitrogen atoms to be traced to their positions in uric acid. They observed that the carbon from carbon dioxide was incorporated into C-6 and carbon from formate into C-2 and C-8 of purines. Ultimately, the sources of the ring atoms were shown to be: N-1, aspartate; C-2 and C-8, formate via 10-formyltetrahydrofolate (Section 7·8); N-3 and N-9, amide groups from glutamine; C-4, C-5, and N-7, glycine; and C-6, carbon dioxide. These findings are summarized in Figure 19·3.

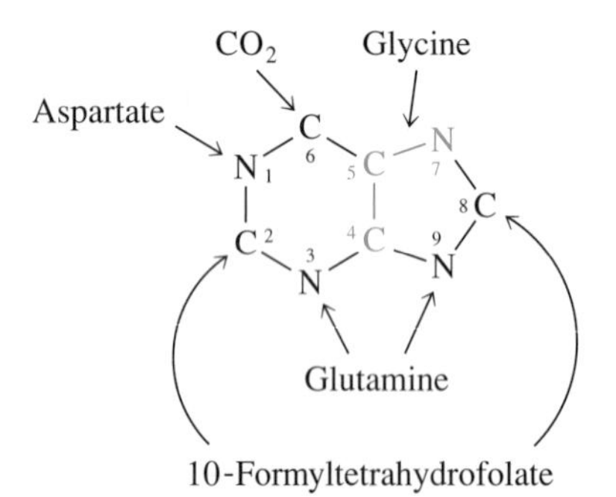

Figure 19·3
Sources of the ring atoms in purines synthesized de novo. Note that the N atom and both C atoms of a glycine molecule are incorporated into positions 4, 5, and 7 of the purine; the amide N atoms of two glutamine molecules provide N-3 and N-9; C-6 arises from CO_2; and aspartate contributes its amino N to position 1. C-2 and C-8 originate from formate via 10-formyltetrahydrofolate.

19·2 Phosphoribosyl Pyrophosphate (PRPP) Is Required for Nucleotide Biosynthesis

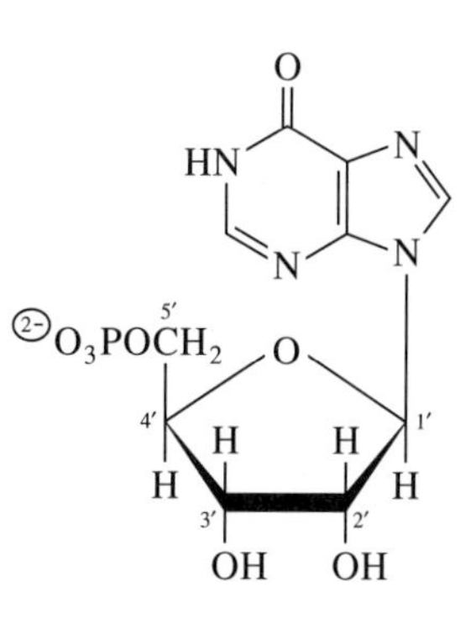

Figure 19·4
Inosine 5′-monophosphate (IMP or inosinate).

G. Robert Greenberg demonstrated that the purine ring structure of hypoxanthine is not synthesized as a free base but as a substituent of ribose phosphate. The end product of the biosynthetic pathway is not hypoxanthine, but its 5′-ribonucleotide, inosine 5′-monophosphate (IMP or inosinate; Figure 19·4). The hypoxanthine detected in avian liver by Krebs and his colleagues was formed by the action of degradative enzymes that catalyze removal of sugar and phosphate groups from ribonucleotides. The de novo pathway of purine synthesis involves a series of sugar phosphate intermediates with incomplete purine skeletons attached.

The source of ribose 5-phosphate for purine biosynthesis was identified by Buchanan and his colleagues as 5-phosphoribosyl 1-pyrophosphate (PRPP), a compound whose structure was elucidated in the laboratory of Arthur Kornberg in 1954. PRPP is synthesized from ribose 5-phosphate and ATP in a reaction catalyzed by PRPP synthetase (Figure 19·5); PRPP then donates its ribose 5-phosphate to serve as the foundation upon which the purine structure is built. PRPP is also a precursor for the biosynthesis of the pyrimidine nucleotides, although in that pathway it reacts with a preformed pyrimidine to form a nucleotide. In addition, PRPP is utilized in the salvage pathways and in the biosynthesis of histidine and tryptophan.

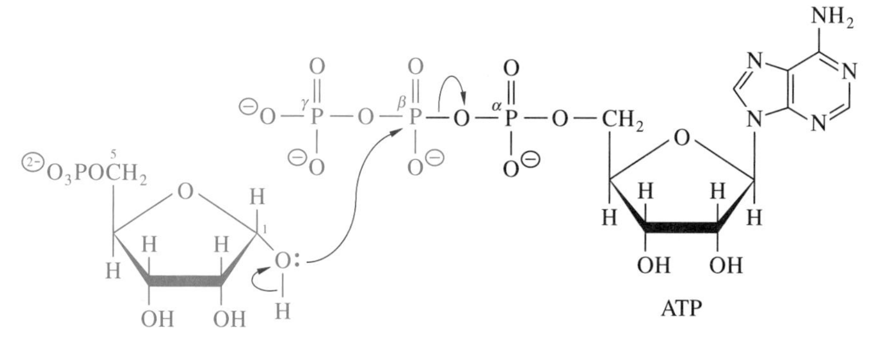

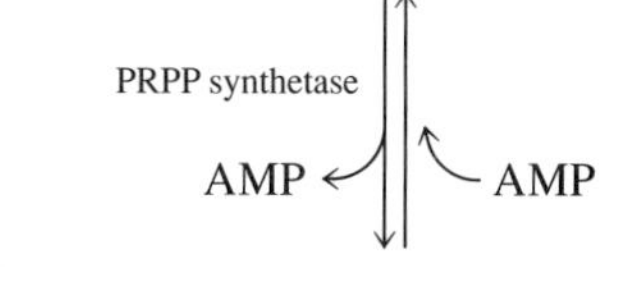

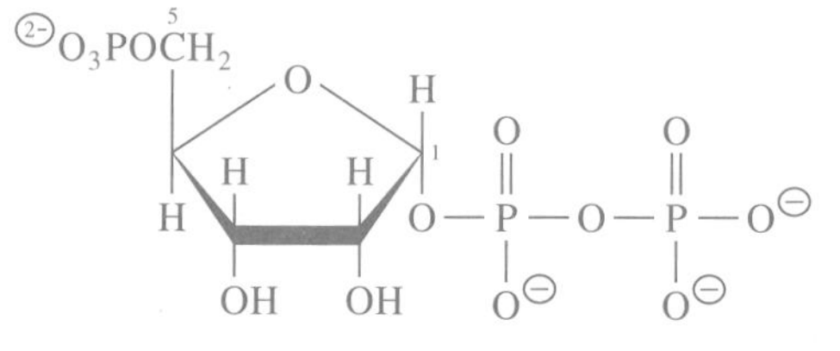

Figure 19·5
Synthesis of 5-phosphoribosyl 1-pyrophosphate (PRPP) from ribose 5-phosphate and ATP. PRPP synthetase catalyzes the transfer of a pyrophosphoryl group from ATP to the 1-hydroxyl group of ribose 5-phosphate.

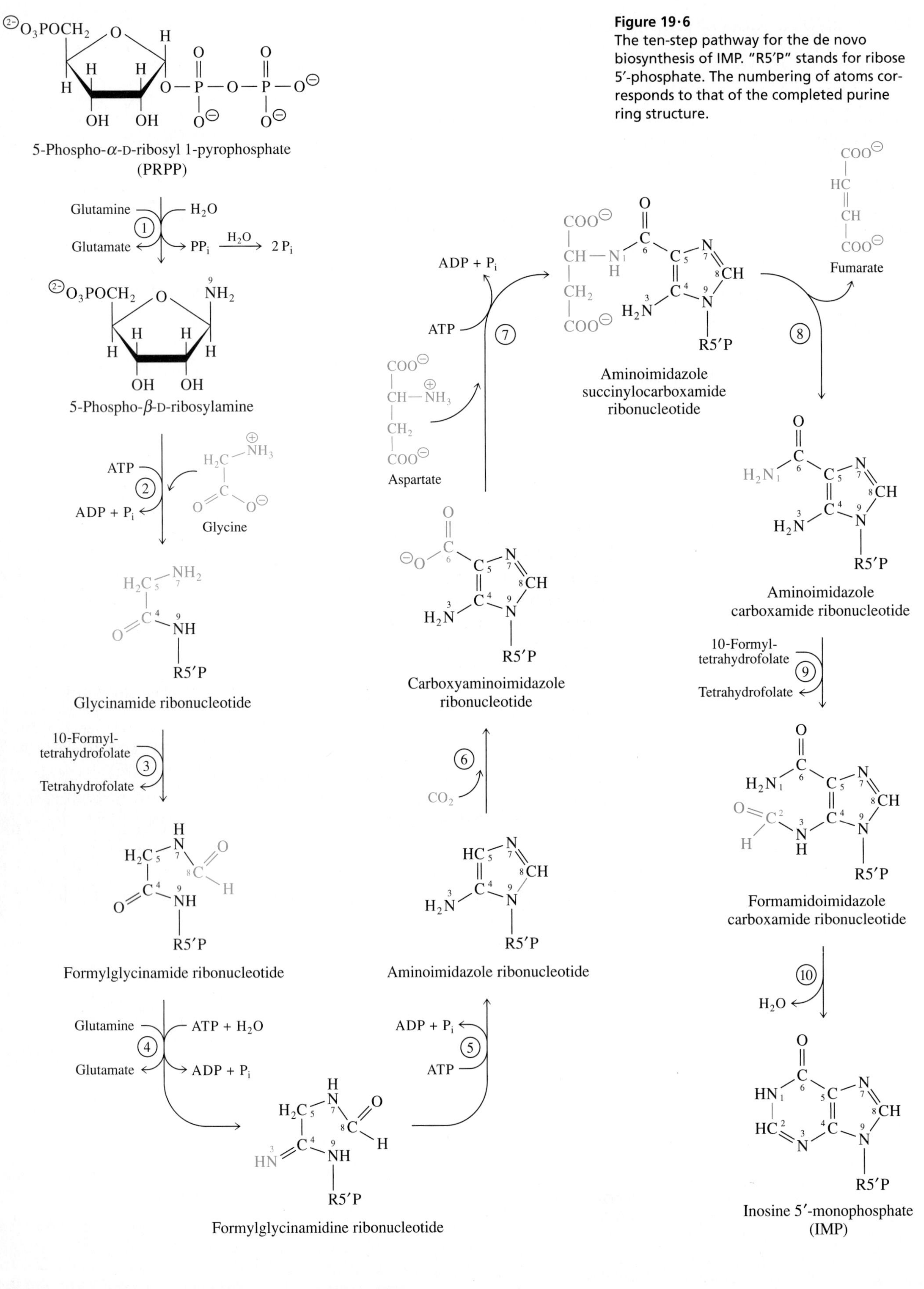

Figure 19·6
The ten-step pathway for the de novo biosynthesis of IMP. "R5′P" stands for ribose 5′-phosphate. The numbering of atoms corresponds to that of the completed purine ring structure.

19·3 The Purine Ring System of IMP Is Assembled in Ten Steps

The ten-step pathway for de novo synthesis of IMP and the structures of each intermediate were elucidated in the 1950s by the research groups of Buchanan and Greenberg. These workers used partially purified enzyme preparations from pigeon and chicken liver. The painstaking isolation and structural characterization of the intermediates and enzymes took about ten years. This work was greatly aided by knowledge of the precursors of all of the ring atoms, availability of an enzyme source with high levels of activity—homogenates of avian liver—and previous identification of the starting substrate (PRPP) and the ultimate product (IMP).

The de novo pathway to IMP is shown in Figure 19·6. The order in which atoms are added to the purine ring structure is: N-9 from glutamine; C-4, C-5, and N-7 from glycine; C-8 from 10-formyltetrahydrofolate; N-3 from glutamine; C-6 from CO_2; N-1 from aspartate; and C-2 from 10-formyltetrahydrofolate. Note that the five-membered imidazole ring is completed (Step 5) before the six-membered pyrimidine ring (Step 10). We shall discuss only a few of the steps in detail.

The pathway starts with displacement of the pyrophosphoryl group of PRPP by the amide nitrogen of glutamine, catalyzed by glutamine-PRPP amidotransferase. Notice that the anomeric configuration is inverted from α to β in this nucleophilic displacement; the β configuration persists in completed purine nucleotides. The amino group of the product, phosphoribosylamine, is then acylated by glycine to form glycinamide ribonucleotide. The mechanism of this reaction, in which an enzyme-bound glycyl phosphate is formed, resembles that of glutamine synthetase, which has glutamyl phosphate as an intermediate (Section 2·6).

Next, a formyl group is transferred from 10-formyltetrahydrofolate to the amino group destined to become N-7 of IMP. In Step 4, an amide is converted to an amidine, (R)HN—C=NH, in an ATP-dependent reaction that requires glutamine as the nitrogen donor. The enzyme that catalyzes this step, an amidotransferase, is irreversibly inhibited by antibiotics that are analogs of glutamine, such as azaserine and 6-diazo-5-oxo-norleucine (Figure 19·7, next page). These compounds, which are examples of affinity labels (Section 5·8), react with a sulfhydryl group of the enzyme.

Step 5 is a ring-closure reaction that requires ATP. In Step 6, CO_2 is incorporated by attachment to the carbon that becomes C-5 of the purine. This carboxylation is unusual in that neither biotin nor ATP is required. C-5 of the imidazole ring is an activated nucleophile because it is part of an enamine (H_2N—C=C), with NH_2 being analogous to the —OH group of an enol. The mechanism by which this nucleophile reacts with the electrophilic CO_2 is shown in Figure 19·8 (next page).

In the next two steps (7 and 8), the amino group of aspartate is incorporated into the growing purine ring system. First, the entire aspartate molecule condenses with the newly added carboxylate group to form an amide, specifically, a succinylocarboxamide. Then a lyase, adenylosuccinase, catalyzes a nonhydrolytic cleavage reaction that releases fumarate. This two-step process results in transfer of an amino group containing the nitrogen destined to become N-1 of IMP.

In Step 9, which resembles Step 3, 10-formyltetrahydrofolate donates a formyl group to the nucleophilic amino group of aminoimidazole carboxamide ribonucleotide. The amide nitrogen of the final intermediate condenses with the formyl group in a ring closure that completes the purine ring system of IMP.

The de novo synthesis of IMP consumes considerable energy. ATP is converted to AMP during synthesis of PRPP. Also, Steps 2, 4, 5, and 7 are driven by the conversion of ATP to ADP. Additional ATP energy is required for the synthesis of glutamine from glutamate and ammonia (Chapter 18).

Figure 19·7
Two analogs of glutamine. **(a)** Glutamine and the related antibiotics azaserine and 6-diazo-5-oxo-norleucine. **(b)** The diazo-containing antibiotics act as affinity-labelling reagents, initially binding to the glutamine-binding site of the amidotransferase that catalyzes Step 4 of purine biosynthesis (or the binding site of other glutamine-dependent amidotransferases). Inactivation involves the alkylation of a nucleophilic cysteine residue at the active site.

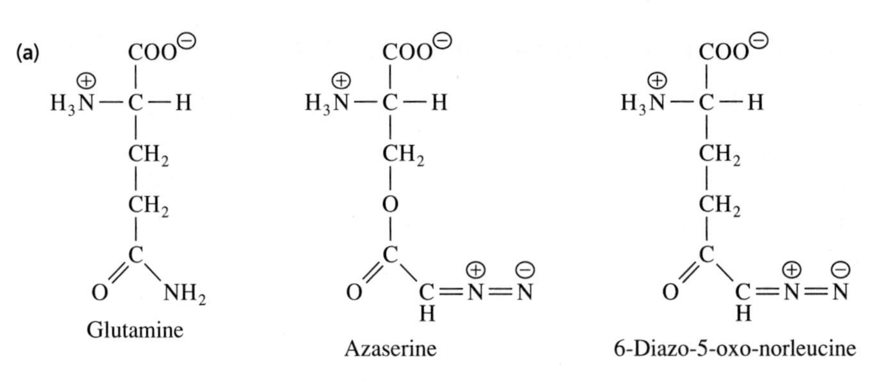

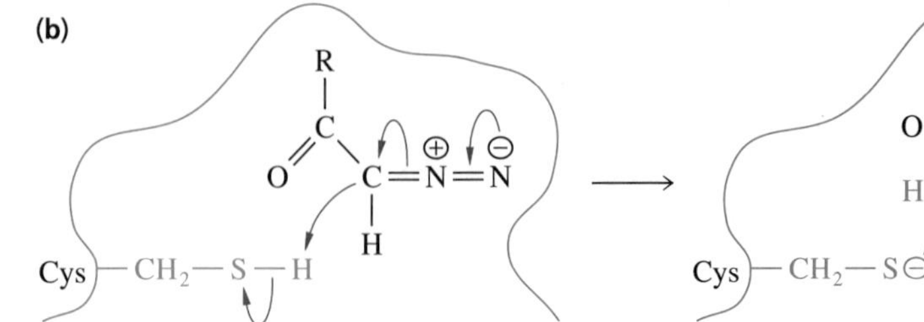

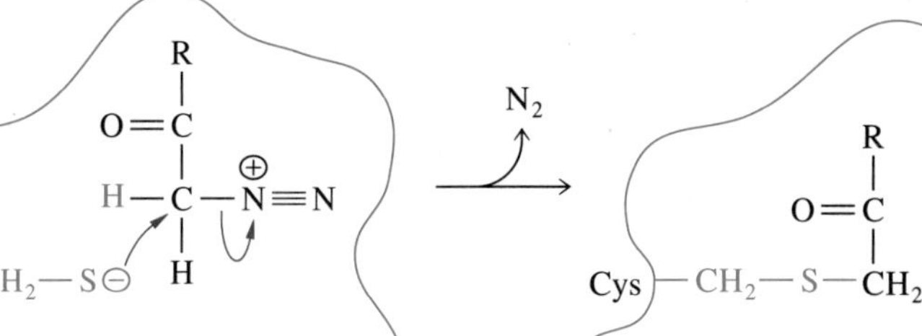

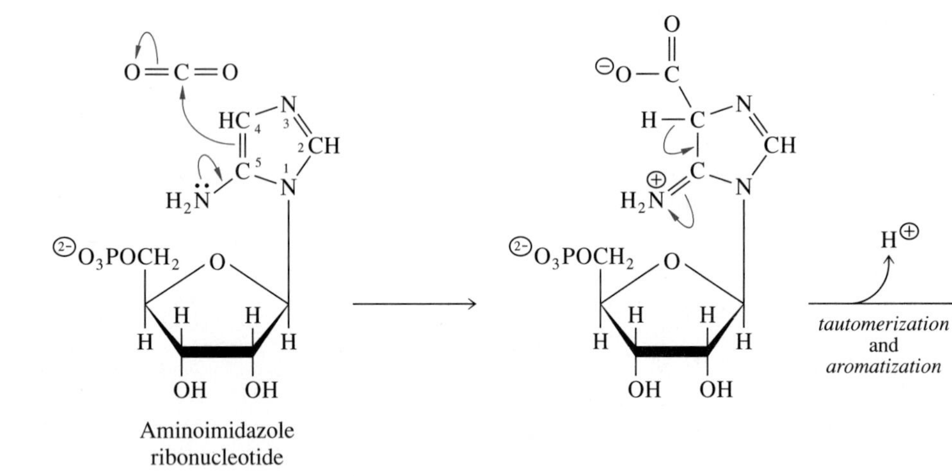

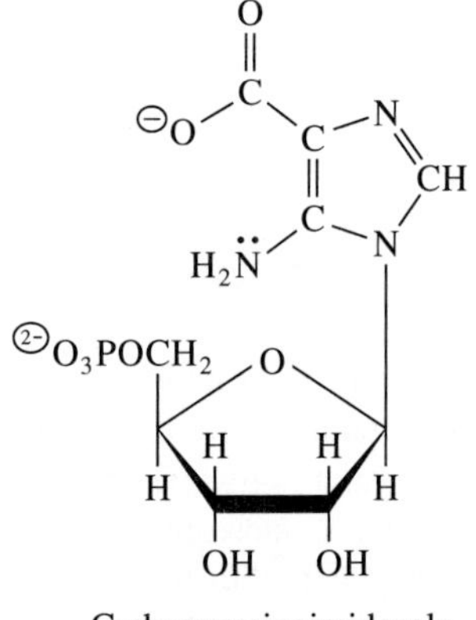

Figure 19·8
Mechanism of addition of carbon dioxide to aminoimidazole ribonucleotide. The enamine is an activated nucleophile that attacks the electrophilic carbon of CO_2. Subsequent tautomerization and aromatization produces carboxyaminoimidazole ribonucleotide.

19·4 AMP and GMP Are Synthesized from IMP

IMP can be converted to either of the major purine nucleotides, AMP or GMP, as shown in Figure 19·9. Two enzymatic reactions are required for each of these conversions.

The biosynthesis of AMP from IMP involves two steps that closely resemble Steps 7 and 8 in the biosynthesis of IMP. First, the amino group of aspartate condenses with the keto group of IMP in a reaction catalyzed by GTP-dependent adenylosuccinate synthetase. Then the elimination of fumarate from adenylosuccinate is

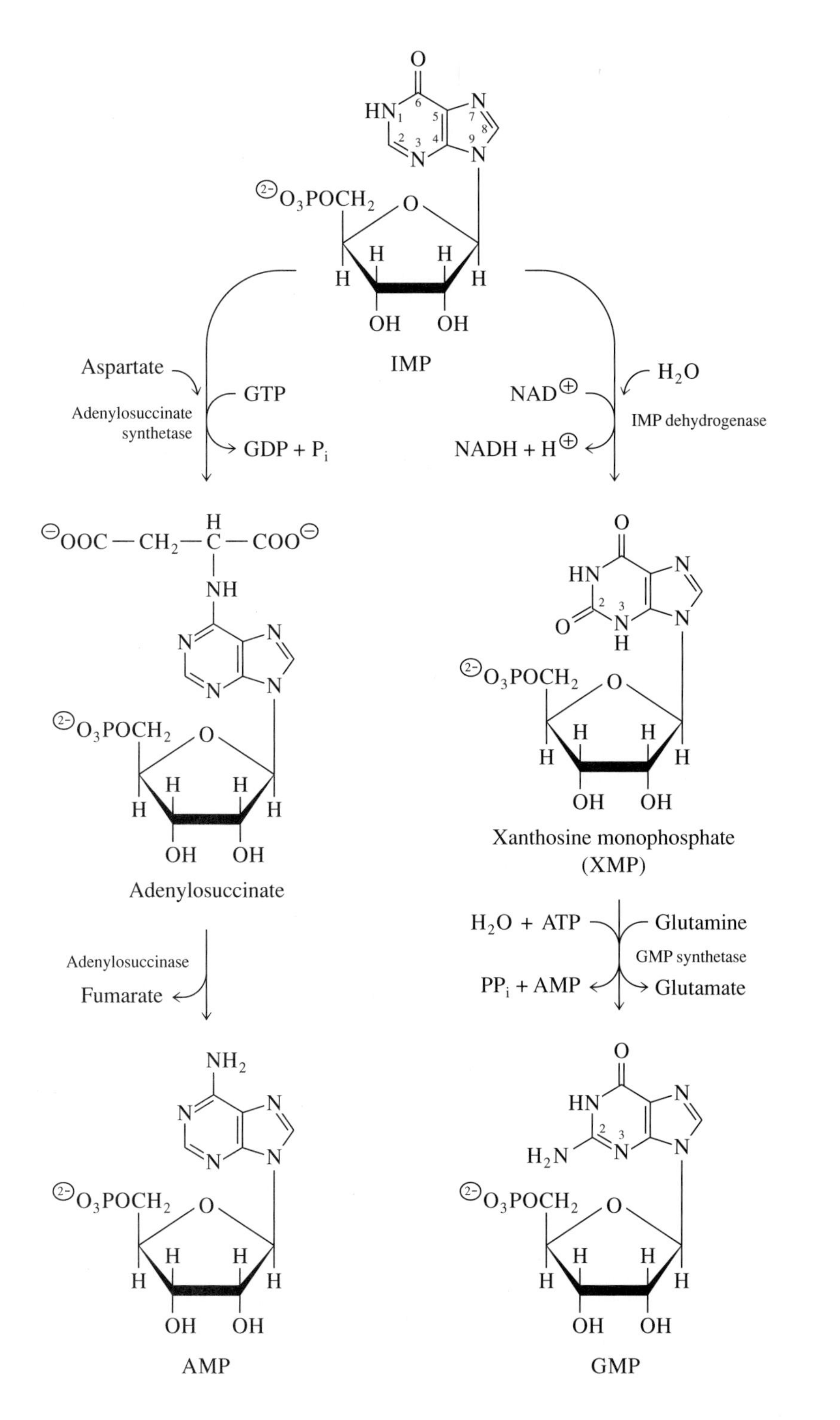

Figure 19·9
Pathways for the conversion of IMP to AMP or GMP.

catalyzed by adenylosuccinase, the same enzyme that catalyzes Step 8 of the de novo pathway. Steps similar to these amine-transfer reactions are involved in the biosynthesis of arginine in the urea cycle (Chapter 18).

In the conversion of IMP to GMP, C-2 is oxidized in a reaction catalyzed by IMP dehydrogenase. This reaction proceeds by addition of a molecule of water to the C-2$=$N-3 double bond and oxidation of the hydrate by NAD$^{\oplus}$. The product of the oxidation is xanthosine monophosphate (XMP). Next, in an ATP-dependent reaction catalyzed by GMP synthetase, the amide nitrogen of glutamine replaces the oxygen at C-2 of XMP. GMP synthetase, like other glutamine-dependent amidotransferases, is inactivated by azaserine (Figure 19·7).

Purine nucleotide synthesis is probably regulated in cells by feedback inhibition. Several enzymes that catalyze steps in the biosynthesis of purine nucleotides exhibit allosteric kinetic behavior in vitro. PRPP synthetase is inhibited by AMP, GMP, and IMP. The enzyme that catalyzes the first committed step in the pathway of purine nucleotide synthesis, glutamine-PRPP amidotransferase (Step 1 in Figure 19·6), is also allosterically inhibited by these and other purine nucleotides.

The paths leading from IMP to AMP and from IMP to GMP are also regulated by feedback inhibition. Adenylosuccinate synthetase is inhibited in vitro by AMP, the product of this two-step branch. IMP dehydrogenase is inhibited by both XMP and GMP. The feedback inhibition patterns of certain enzymes in this branched pathway to AMP and GMP are shown in Figure 19·10. Notice that the end products inhibit two of the initial common steps and the enzymes leading from the branch-point.

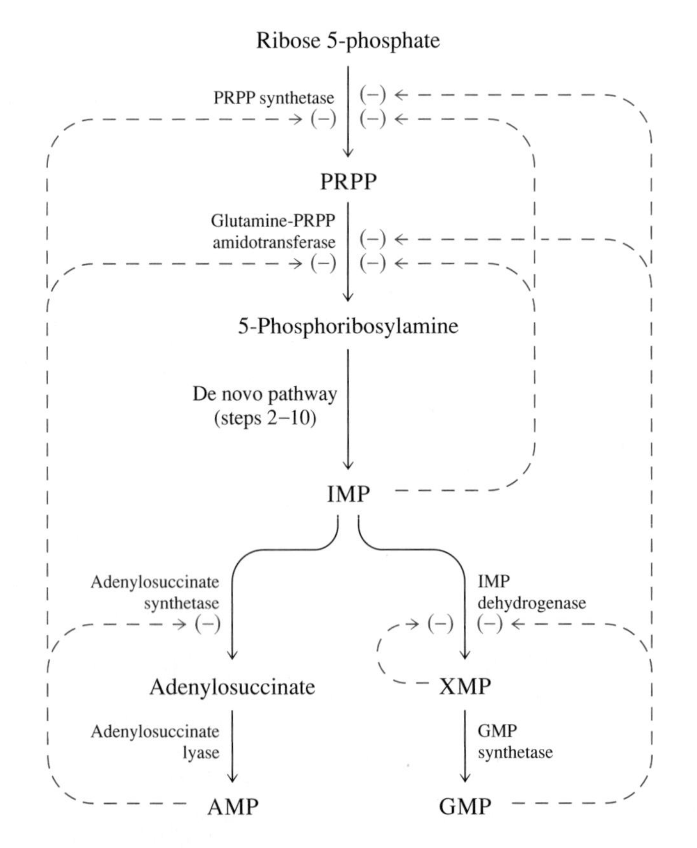

Figure 19·10
Regulation of purine nucleotide biosynthesis as indicated by results obtained in vitro.

19·5 Purine Molecules Arising from the Breakdown of Nucleotides Can Be Salvaged

During cellular metabolism and during the digestive process of animals, nucleic acids are degraded to mononucleotides, nucleosides, and eventually, heterocyclic bases. These catabolic reactions are catalyzed by enzymes that include ribonucleases, deoxyribonucleases, and a variety of nucleotidases and nucleosidases. Some of the purines formed in this way are further degraded to uric acid, but a considerable fraction is normally salvaged by direct conversion to purine ribonucleotides. Purines formed by intracellular reactions are salvaged to a greater extent than those formed during digestion. The recycling of intact purine molecules saves cellular energy.

PRPP is the donor of ribose 5-phosphate for the purine salvage reactions. Adenine phosphoribosyltransferase catalyzes the reaction of adenine with PRPP to form AMP and PP_i. Hydrolysis of PP_i catalyzed by inorganic pyrophosphatase renders the reaction metabolically irreversible. Hypoxanthine-guanine phosphoribosyltransferase catalyzes similar reactions—the conversion of hypoxanthine to IMP and guanine to GMP with concomitant formation of PP_i. The degradation of purine nucleotides to their respective purines and their salvage through reaction with PRPP are outlined in Figure 19·11.

In 1964, Michael Lesch and William Nyhan described a severe metabolic disease characterized by mental retardation, palsy-like spasticity, and a bizarre tendency toward self-mutilation. Individuals afflicted with this disease, called Lesch-Nyhan syndrome, rarely survive past childhood. Prominent biochemical features of the disease are the excretion of up to six times the normal amount of uric acid and a greatly increased rate of de novo purine biosynthesis. The disease is caused by a hereditary deficiency of the enzyme hypoxanthine-guanine phosphoribosyltransferase. Because the gene for this enzyme is on the X chromosome, the disease is usually restricted to males. Instead of being converted to IMP and GMP, respectively, hypoxanthine and guanine are degraded to uric acid. The PRPP normally used for salvage of hypoxanthine and guanine contributes to the de novo synthesis of excessive amounts of IMP, with the surplus IMP being degraded to uric acid. It is not known how this single enzyme defect causes the various behavioral symptoms. The catastrophic effects of the deficiency indicate that the purine salvage pathway is more than an energy-saving addendum to the central pathways of purine nucleotide metabolism.

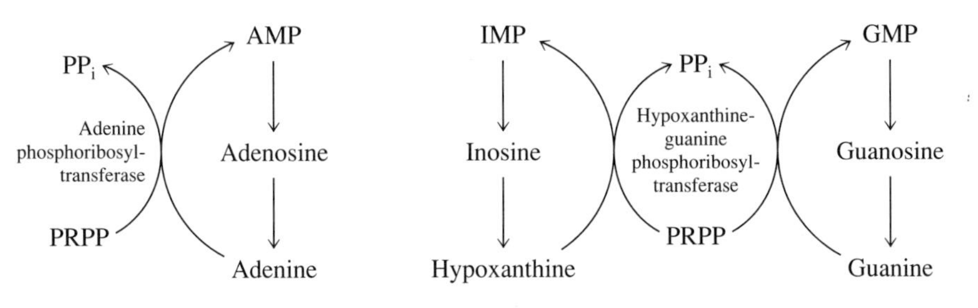

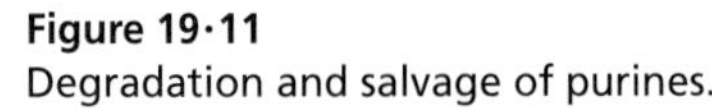
Figure 19·11
Degradation and salvage of purines.

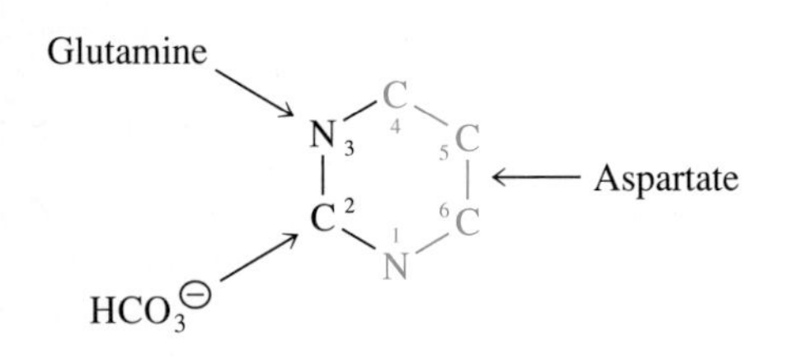

Figure 19·12
Sources of the ring atoms of pyrimidines synthesized de novo. Note that carbon atoms 4, 5, and 6, and N-1, are all contributed by aspartate.

19·6 The De Novo Pathway for Pyrimidine Synthesis Leads to UMP

The de novo biosynthetic pathway for pyrimidine nucleotide biosynthesis is simpler than the purine pathway and requires less consumption of ATP. There are three metabolic precursors of the pyrimidine ring: bicarbonate, which contributes C-2; the amide group of glutamine (N-3); and aspartate, which contributes the remaining atoms (Figure 19·12).

PRPP is required for biosynthesis of pyrimidine nucleotides, but the sugar-phosphate from PRPP is donated after the ring is formed rather than entering the pathway in the first step. A compound with a completed pyrimidine ring—orotate (6-carboxyuracil)—reacts with PRPP to form a pyrimidine ribonucleotide in the fifth step of the six-step pathway.

The six reactions of the pathway for de novo pyrimidine synthesis are shown in Figure 19·13. The first two steps generate a noncyclic intermediate that contains all of the atoms destined for the pyrimidine ring. This intermediate, carbamoyl aspartate, is enzymatically cyclized. The product, dihydroorotate, is subsequently oxidized to orotate. Orotate is then converted to the ribonucleotide orotidine 5′-monophosphate, which undergoes decarboxylation to form UMP. This pyrimidine nucleotide is the precursor of all other pyrimidine ribo- and deoxyribonucleotides. As we shall see, the enzymes required for de novo pyrimidine synthesis are organized differently in prokaryotes than in eukaryotes, and the pathways in these cell types are regulated differently.

The first step in the pathway of de novo pyrimidine biosynthesis is the formation of carbamoyl phosphate from bicarbonate, the amide nitrogen of glutamine, and ATP. This reaction, catalyzed by carbamoyl phosphate synthetase, requires two molecules of ATP, one to drive the formation of the C—N bond and the other to donate the phosphoryl group of the product.

Carbamoyl phosphate is also a metabolite in the pathway leading to the biosynthesis of arginine via citrulline (Chapter 18). In prokaryotes, the same carbamoyl phosphate synthetase is utilized in both pyrimidine and arginine biosynthetic pathways. This enzyme is allosterically inhibited by pyrimidine ribonucleotides such as UMP, the product of the pyrimidine biosynthetic pathway. It is activated by L-ornithine, a precursor of citrulline, and by purine nucleotides, the substrates, along with pyrimidine nucleotides, for the synthesis of nucleic acids.

In eukaryotic cells, there are two distinct carbamoyl phosphate synthetases (Chapter 18). A mitochondrial synthetase, called carbamoyl phosphate synthetase I because it was discovered first, catalyzes a reaction that utilizes an ammonium ion as the source of the amide of carbamoyl phosphate:

$$NH_4^{\oplus} + HCO_3^{\ominus} + 2\,ATP \xrightarrow{\text{Carbamoyl phosphate synthetase I}} \underset{\text{Carbamoyl phosphate}}{H_2N{-}\overset{\overset{\large O}{\|}}{C}{-}OPO_3^{2\ominus}} + 2\,ADP + P_i + H^{\oplus} \quad \textbf{(19·1)}$$

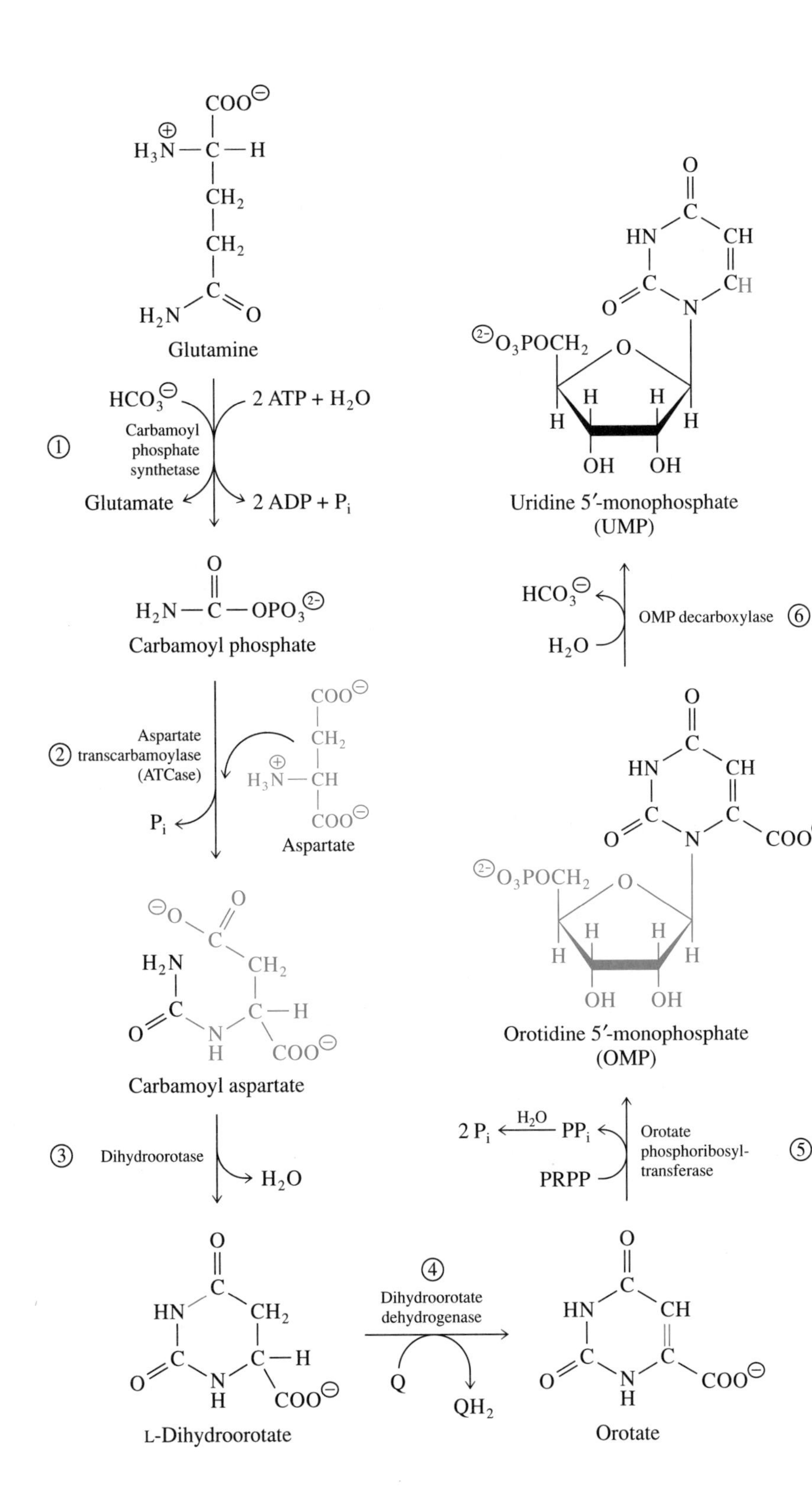

Figure 19·13
The six-step pathway for the de novo synthesis of uridylate (UMP). The enzyme names are those for prokaryotes. In at least some eukaryotes, Steps 1, 2, and 3 are catalyzed by a multifunctional protein called dihydroorotate synthase, and reactions 5 and 6 are catalyzed by a bifunctional enzyme, UMP synthase.

In mammalian liver, carbamoyl phosphate generated in mitochondria is utilized for synthesis of urea. The cytosolic synthetase, carbamoyl phosphate synthetase II, catalyzes the first committed step of pyrimidine synthesis. As noted earlier, the amide of carbamoyl phosphate in this reaction comes from glutamine:

$$\text{Glutamine} + HCO_3^{\ominus} + 2\,\text{ATP} + H_2O \xrightarrow{\text{Carbamoyl phosphate synthetase II}} \text{Carbamoyl phosphate} + \text{Glutamate} + 2\,\text{ADP} + P_i \quad \textbf{(19·2)}$$

Carbamoyl phosphate synthetase II is allosterically regulated. PRPP and IMP activate the enzyme, and several pyrimidine nucleotides inhibit it. The compartmentation of reactions between mitochondria and the cytosol allows separate control of each enzyme and the pathway it serves.

In the second step of UMP biosynthesis, the activated carbamoyl group of carbamoyl phosphate is transferred to aspartate to form carbamoyl aspartate. In this reaction, catalyzed by aspartate transcarbamoylase (ATCase), the nucleophilic nitrogen of aspartate attacks the carbonyl group of carbamoyl phosphate (Figure 5·21).

Prokaryotic ATCase (specifically ATCase from *E. coli*) was the first allosteric enzyme to be fully characterized. In *E. coli,* where carbamoyl phosphate synthetase generates an intermediate that can enter pathways leading either to pyrimidines or arginine, it is ATCase that catalyzes the first committed step of pyrimidine biosynthesis. This enzyme is inhibited by pyrimidine nucleotides and activated in vitro by ATP, as discussed in Section 5·14. Although ATCase in *E. coli* is only partially inhibited by CTP alone (50–70%), inhibition can be almost total when both CTP and UTP are present. UTP alone does not inhibit the enzyme. The allosteric controls, inhibition by pyrimidine nucleotides and activation by the purine nucleotide ATP, provide a means for carbamoyl phosphate synthetase and ATCase to balance the pyrimidine nucleotide and purine nucleotide pools in *E. coli.* The ratio of the concentrations of the two types of allosteric effectors determines the activity level of ATCase.

Eukaryotic ATCase is not feedback inhibited. Regulation by feedback inhibition is not necessary because the substrate of ATCase in eukaryotes is not a branch-point metabolite—the pathway leading to arginine and urea in the mitochondrion is separate from the pyrimidine biosynthetic pathway, in which five or six steps occur in the cytosol. Consequently, the pyrimidine pathway can be controlled by regulation of the enzyme preceding ATCase, carbamoyl phosphate synthetase II.

Dihydroorotase catalyzes the third step of UMP biosynthesis, the reversible closure of the pyrimidine ring. The product, dihydroorotate, is then oxidized by the action of dihydroorotate dehydrogenase to form orotate. In eukaryotes, dihydroorotate dehydrogenase is associated with the inner membrane of the mitochondrion. It is an iron-containing flavoprotein that catalyzes transfer of electrons to ubiquinone and thence to O_2 by the electron-transport chain.

Once formed, orotate displaces the pyrophosphate group of PRPP, producing orotidine 5′-monophosphate (OMP, or orotidylate) in a reaction catalyzed by orotate phosphoribosyltransferase. This reaction also has a salvage role—pyrimidines other than orotate can be acted upon by the same enzyme to form the corresponding pyrimidine nucleotides. As in the reactions catalyzed by adenine phosphoribosyltransferase and hypoxanthine-guanine phosphoribosyltransferase, the pyrophosphate produced is hydrolyzed, rendering Step 5 irreversible.

Finally, OMP is decarboxylated to form UMP. Product inhibition of OMP decarboxylase by UMP in vitro has been reported, but the main control point is earlier in the biosynthetic pathway.

19·7 Two Multifunctional Proteins Are Involved in Pyrimidine Biosynthesis in Mammals

Although the six enzymatic steps leading to UMP are the same in prokaryotes and eukaryotes, the structural organization of the enzymes (free versus associated) varies among organisms. For example, in *E. coli,* each of the six reactions is catalyzed by a separate enzyme. In mammals, a multifunctional protein in the cytosol known as dihydroorotate synthase contains separate catalytic sites (carbamoyl phosphate synthetase II, ATCase, and dihydroorotase) for the first three steps of the pathway. Dihydroorotate produced in the cytosol by Steps 1 through 3 passes through the outer mitochondrial membrane prior to being oxidized to orotate by dihydroorotate dehydrogenase. The substrate-binding site of this enzyme is located on the outer surface of the inner mitochondrial membrane. Orotate then moves to the cytosol where conversion to UMP takes place. A bifunctional enzyme known as UMP synthase catalyzes both the reaction with PRPP to form OMP and the rapid decarboxylation of OMP to UMP.

The intermediates formed in Step 1 and Step 2 (carbamoyl phosphate and carbamoyl aspartate) and OMP (from Step 5) are not normally released to solvent but remain enzyme bound and are channelled from one catalytic center to the next. Several multifunctional proteins, each catalyzing several steps, also occur in the pathway of purine nucleotide biosynthesis in some organisms. This sequestering of labile intermediates within multidomain enzymes prevents nonproductive degradation of intermediates, thereby conserving energy.

19·8 CTP Is Synthesized from UMP

UMP is converted to CTP in three steps. First, uridylate kinase (UMP kinase) catalyzes the transfer of the γ-phosphoryl group of ATP to UMP to generate UDP, and then nucleoside diphosphate kinase catalyzes the transfer of the γ-phosphoryl group of a second ATP molecule to UDP to form UTP. In these two reactions, two molecules of ATP are converted to two molecules of ADP.

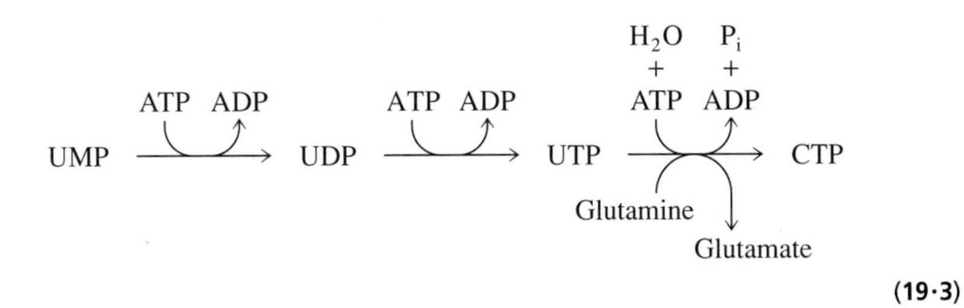

(19·3)

CTP synthetase then catalyzes the ATP-dependent transfer of the amide nitrogen from glutamine to C-4 of UTP, forming CTP. This reaction, shown in Figure 19·14, is analogous to Step 4 of purine biosynthesis (Figure 19·6).

CTP synthetase is allosterically inhibited by its product, CTP, and, in *E. coli,* allosterically activated by GTP. The activation involves both an increase in V_{max} and a decrease in the K_m of the enzyme for glutamine.

Thymidylate (dTMP) is also a derivative of UMP, but before thymidylate can be formed, UMP must undergo reduction to form dUMP. We shall examine the formation of deoxyribonucleotides next and then the formation of dTMP from dUMP.

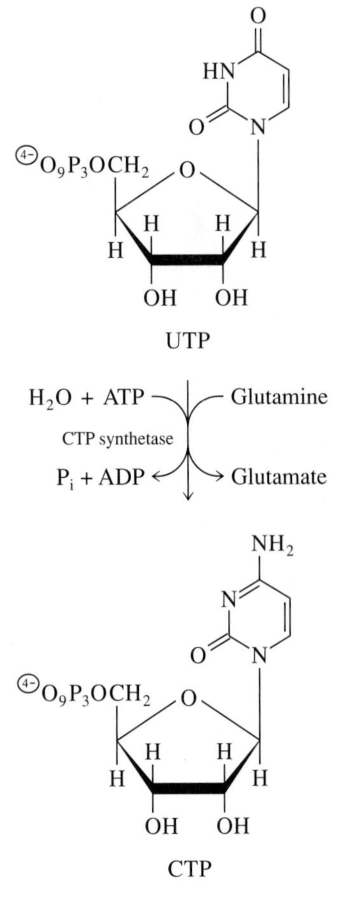

Figure 19·14
Conversion of UTP to CTP, catalyzed by CTP synthetase.

19·9 Deoxyribonucleotides Are Synthesized by Reduction of Ribonucleotides

The 2′-deoxyribonucleotides, whose principal role is to serve as triphosphate substrates for DNA polymerase, are synthesized through enzymatic reduction of ribonucleotides. In most organisms, this reduction occurs at the nucleoside diphosphate level. Peter Reichard and his colleagues, who overcame difficult technical barriers to elucidate this complicated reaction system, showed that all four ribonucleoside diphosphates—ADP, GDP, CDP, and UDP—are substrates of a single, closely regulated ribonucleoside diphosphate reductase. However, in some microorganisms, including species of *Lactobacillus, Clostridium,* and *Rhizobium,* ribonucleoside *tri*phosphates are the substrates for reduction by a cobalamin-dependent reductase. Both types of enzymes are referred to as ribonucleotide reductase, although more precise names are ribonucleoside diphosphate reductase and ribonucleoside triphosphate reductase, respectively.

NADPH provides the reducing power for the synthesis of deoxyribonucleoside diphosphates. A disulfide bond at the active site of ribonucleotide reductase is reduced to two thiol groups, which in turn reduce C-2′ of the ribose moiety of the nucleotide substrate by a complex free-radical mechanism. As shown in Figure 19·15, electrons are transferred from NADPH to ribonucleotide reductase via the flavoprotein thioredoxin reductase and the dithiol protein coenzyme thioredoxin. In the absence of thioredoxin (for example, in *E. coli* mutants that lack thioredoxin), another small dithiol protein, glutaredoxin, can replace reduced thioredoxin in deoxyribonucleotide formation. Glutaredoxin transfers electrons from reduced glutathione to ribonucleotide reductase. Once formed, dADP, dGDP, and dCDP are phosphorylated to the triphosphate level by the action of nucleoside diphosphate kinases. dUDP, as we shall see in the next section, is converted to dTMP via dUMP.

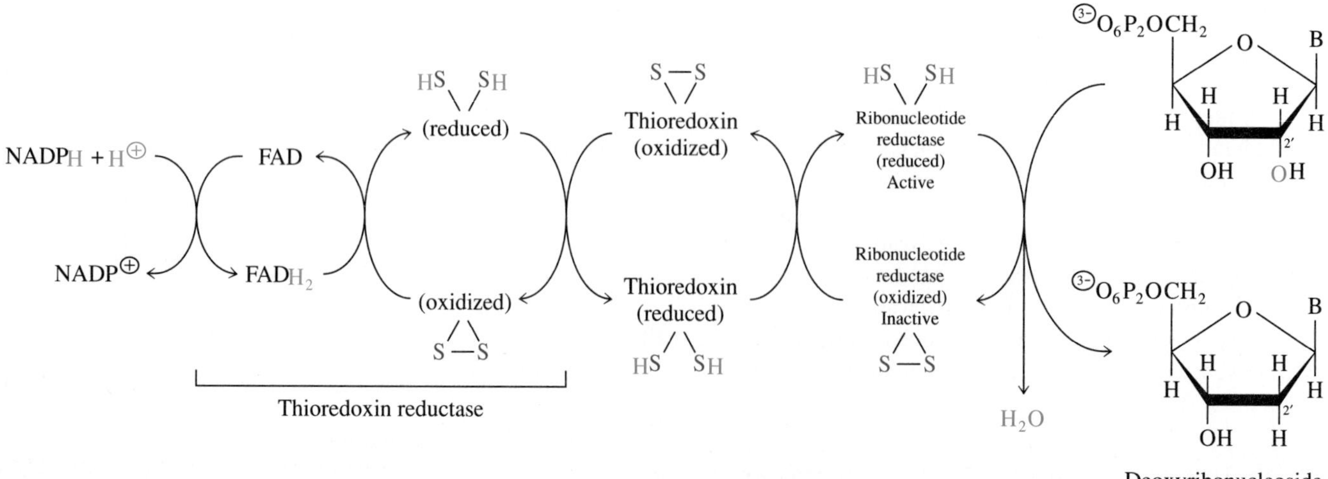

Figure 19·15
Reduction of ribonucleoside diphosphates with electrons derived from NADPH, transferred via thioredoxin. Three proteins are involved: the NADPH-dependent flavoprotein thioredoxin reductase, thioredoxin, and ribonucleotide reductase. (In this figure, B represents a purine (A, G) or pyrimidine (C, U) base.)

Under some circumstances, ribonucleotide reduction can be the rate-limiting step in DNA synthesis. Ribonucleotide reductases are strictly regulated by allosteric interactions. Both the substrate specificity and the catalytic rate of the enzyme are regulated in cells by the reversible binding of nucleotide metabolites. This regulation system has features that ensure that a balanced selection of deoxyribonucleotides is available for DNA synthesis. The allosteric modulators—ATP, dATP, dTTP, and dGTP—exert their effects by binding to ribonucleotide reductase at either of two regulatory sites. One allosteric site, termed the *activity site,* controls the activity of the catalytic site. A second allosteric site, called the *specificity site,* determines the substrate specificity of the catalytic site. The binding of ATP to the activity site activates the reductase; the binding of dATP to the activity site inhibits all enzymatic activity. When ATP is bound to the activity site and either ATP or dATP is bound to the specificity site, the reductase becomes pyrimidine-specific, catalyzing the reduction of CDP and UDP; binding of dTTP to the specificity site activates the reduction of GDP, and binding of dGTP activates the reduction of ADP. The main features of the allosteric regulation of ribonucleotide reductase are summarized in Table 19·1.

Table 19·1 Allosteric regulation of ribonucleotide reductase

Ligand bound to activity site	Ligand bound to specificity site	Activity of catalytic site
dATP	Enzyme inactive	
ATP	ATP or dATP	Specific for CDP or UDP
ATP	dTTP	Specific for GDP
ATP	dGTP	Specific for ADP

19·10 A Unique Methylation Reaction Produces dTMP from dUMP

Thymidylate (dTMP), required for DNA synthesis, is formed from UMP in four steps. UMP is phosphorylated to UDP, which is reduced to dUDP, and dUDP is dephosphorylated to dUMP, which is then methylated:

$$\text{UMP} \longrightarrow \text{UDP} \longrightarrow \text{dUDP} \longrightarrow \text{dUMP} \longrightarrow \text{dTMP} \qquad (19{\cdot}4)$$

The conversion of dUDP to dUMP may occur by two routes. dUDP can react with ADP in the presence of a nucleoside monophosphate kinase to form dUMP and ATP:

$$\text{dUDP} + \text{ADP} \rightleftharpoons \text{dUMP} + \text{ATP} \qquad (19{\cdot}5)$$

dUDP can also be phosphorylated to dUTP at the expense of ATP through the action of rather nonspecific nucleoside diphosphate kinases. dUTP is then rapidly hydrolyzed to dUMP + PP_i by the action of deoxyuridine triphosphate diphosphohydrolase (dUTPase):

$$\text{dUDP} + \text{ATP} \xrightarrow[\searrow \text{ADP}]{} \text{dUTP} \xrightarrow[\nearrow H_2O]{} \text{dUMP} + PP_i \qquad (19{\cdot}6)$$

The rapid hydrolysis of dUTP prevents it from being accidentally incorporated into DNA in place of dTTP.

dCMP may also serve as a source of dUMP via hydrolysis catalyzed by dCMP deaminase:

$$\text{dCMP} + H_2O \longrightarrow \text{dUMP} + NH_4^{\oplus} \qquad (19{\cdot}7)$$

dCMP deaminase in liver is subject to allosteric regulation: dTTP is an allosteric inhibitor of the enzyme, and dCTP is an activator.

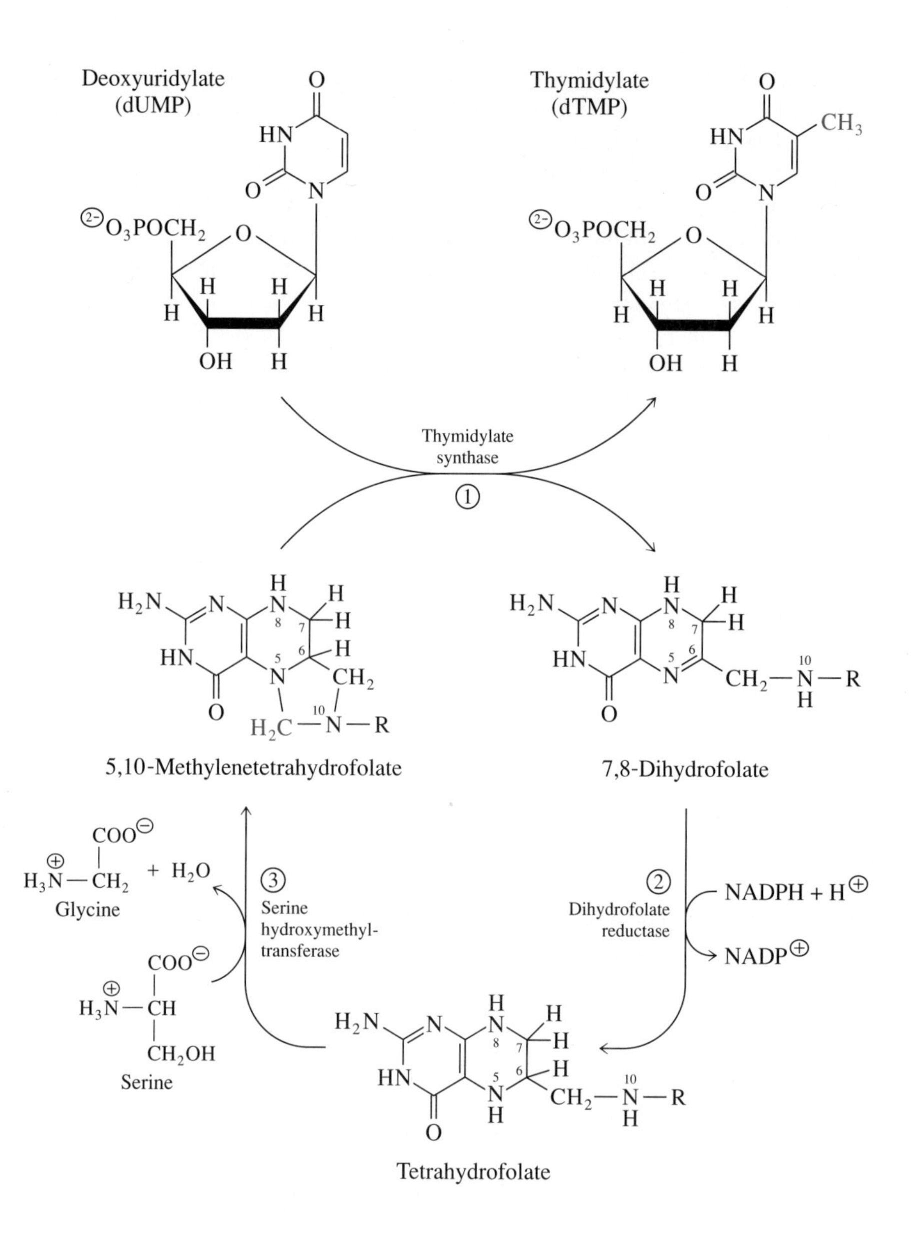

Figure 19·16
Cycle of reactions in the synthesis of thymidylate (dTMP) from dUMP. Thymidylate synthase catalyzes the first reaction of this cycle, producing dTMP. The other product of the reaction, dihydrofolate, must be reduced by NADPH in a reaction catalyzed by dihydrofolate reductase before a methylene group can be added to regenerate 5,10-methylenetetrahydrofolate. Methylenetetrahydrofolate is regenerated in a reaction catalyzed by serine hydroxymethyltransferase.

The conversion of dUMP to dTMP is catalyzed by thymidylate synthase. In this reaction, 5,10-methylenetetrahydrofolate (Section 7·8) is the donor of the one-carbon group (Figure 19·16). Note that the methyl ($—CH_3$) group in the product, dTMP, is more reduced than the methylene ($—CH_2—$) group in 5,10-methylenetetrahydrofolate, whose oxidation state is equivalent to that of a hydroxymethyl ($—CH_2OH$) group or formaldehyde. Thus, not only does methylenetetrahydrofolate provide the one-carbon unit, but it also serves as the reducing agent for the reaction, furnishing a hydride ion and being oxidized to 7,8-dihydrofolate in the process. Only *tetra*hydrofolate can accept another one-carbon unit for further reactions in one-carbon metabolism. Therefore, the 5,6 double bond of dihydrofolate must be reduced by NADPH in the reaction catalyzed by dihydrofolate reductase (Section 7·8). Serine hydroxymethyltransferase then catalyzes the transfer of the β-CH_2OH group of serine to tetrahydrofolate to regenerate 5,10-methylenetetrahydrofolate.

The conversion of dUMP to dTMP is the only reaction known in which the transfer of a one-carbon unit from tetrahydrofolate results in oxidation at N-5 and

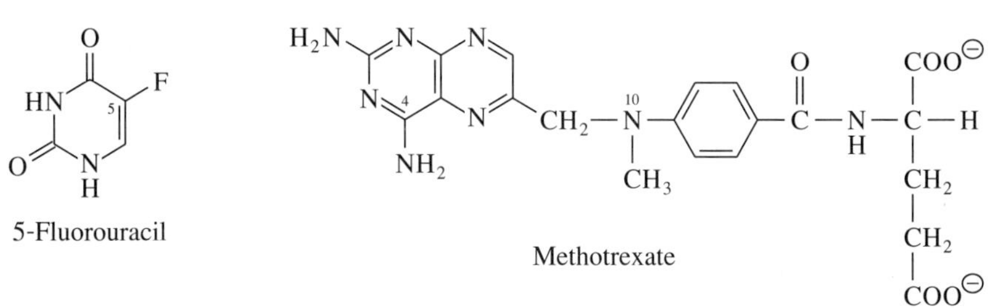

Figure 19·17
5-Fluorouracil and methotrexate, drugs designed to inhibit reproduction of cancer cells.

C-6 to produce dihydrofolate. The tetrahydrofolate of methylenetetrahydrofolate thus acts as a hydride-donating substrate in this reaction as well as a group-transfer coenzyme. This added role of the coenzyme contributes to a greater sensitivity of thymidylate synthase to depletion of tetrahydrofolate compared to most other tetrahydrofolate-requiring enzymes. Since dTMP serves as an essential precursor of DNA, any agent that lowers dTMP levels drastically affects cell division. Because rapidly dividing cells are particularly dependent on the activities of thymidylate synthase and dihydrofolate reductase, these enzymes have been major targets for anticancer drugs. Inhibition of either or both of these enzymes blocks the synthesis of dTMP and therefore the synthesis of DNA.

5-Fluorouracil and methotrexate, shown in Figure 19·17, have proven effective in combatting some types of cancer. 5-Fluorouracil is converted to its deoxyribonucleotide, 5-fluorodeoxyuridylate, by a pyrimidine salvage pathway. 5-Fluorodeoxyuridylate binds tightly to thymidylate synthase, inhibiting the enzyme and bringing the three-reaction cycle shown in Figure 19·16 to a halt.

Methotrexate, the most commonly used anticancer drug, is an analog of folate with an amino group in place of the oxygen atom at C-4 and a methyl substituent on N-10. Methotrexate is a potent and relatively specific inhibitor of dihydrofolate reductase, which catalyzes Step 2 of the cycle shown in Figure 19·16. The folate analog binds to the reductase extremely tightly by noncovalent interactions only. The resulting decrease in tetrahydrofolate levels greatly decreases the formation of dTMP, since dTMP synthesis is dependent on adequate concentrations of methylenetetrahydrofolate. Most normal cells undergo cell division more slowly than cancer cells and consequently are less sensitive to methotrexate. However, because methotrexate is toxic to all cells, it must be used with caution. Often, 5-formyltetrahydrofolate, which can be converted to methylenetetrahydrofolate (Section 7·8), is given to patients a brief period after they have been administered an otherwise lethal dose of methotrexate. This high dose–rescue approach enhances the therapeutic use of the drug.

dTMP can also be synthesized via the salvage of thymidine (deoxythymidine), which is catalyzed by ATP-dependent thymidine kinase.

$$\text{Thymidine} \xrightarrow[\text{Thymidine kinase}]{\text{ATP} \quad \text{ADP}} \text{dTMP} \tag{19·8}$$

Radioactive thymidine is often used as a highly specific tracer for monitoring intracellular synthesis of DNA because it enters cells easily and its principal metabolic fate is salvage leading to incorporation into DNA.

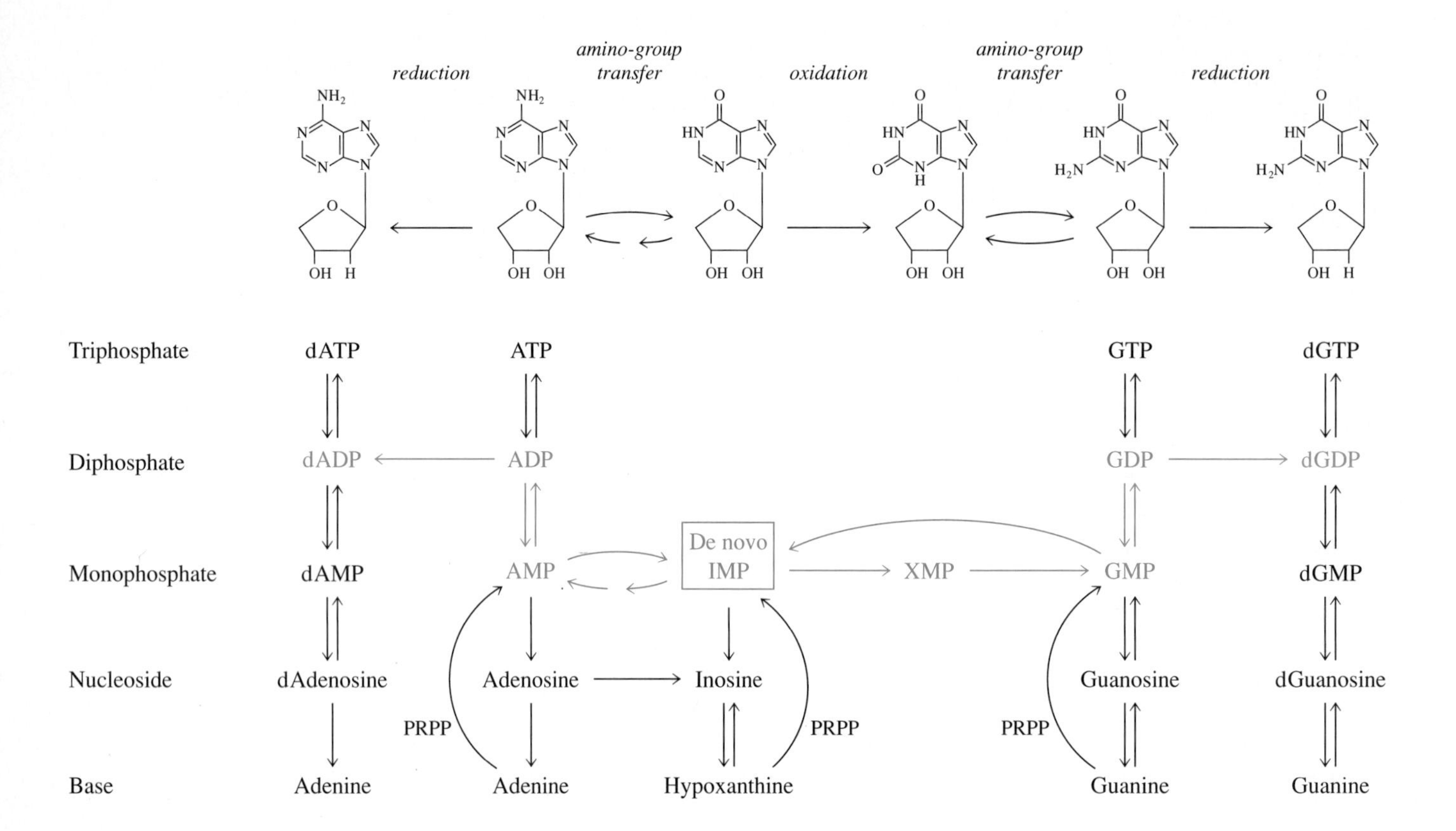

Figure 19·18
Interconversions of purine nucleotides and their constituents. [Adapted from T. W. Traut (1988). Enzymes of nucleotide metabolism: the significance of subunit size and polymer size for biological function and regulatory properties. *Crit. Rev. Biochem.* 23:121–169.]

19·11 Nucleotides and Their Constituents Are Interconverted: A Summary

A number of reactions exist for interconversions of nucleotides and their constituents. Many of the reactions we have seen already. The principal routes for purine nucleotide metabolism are shown in Figure 19·18. IMP, the first nucleotide product of the de novo biosynthetic pathway, is readily converted to AMP and GMP, their di- and triphosphates, and the deoxy counterparts of these nucleotides. As we shall see in more detail in the next section, nucleotides are also degraded into their constituents. Using purine bases from this degradation, salvage reactions can re-form nucleotides.

The major interconversions of pyrimidine nucleotide metabolism are illustrated in Figure 19·19. UMP formed by the de novo pathway can be converted to cytidine and thymidine phosphates. In addition, reactions catalyzed by phosphorylases and kinases can salvage free pyrimidines by converting them to the nucleotide level.

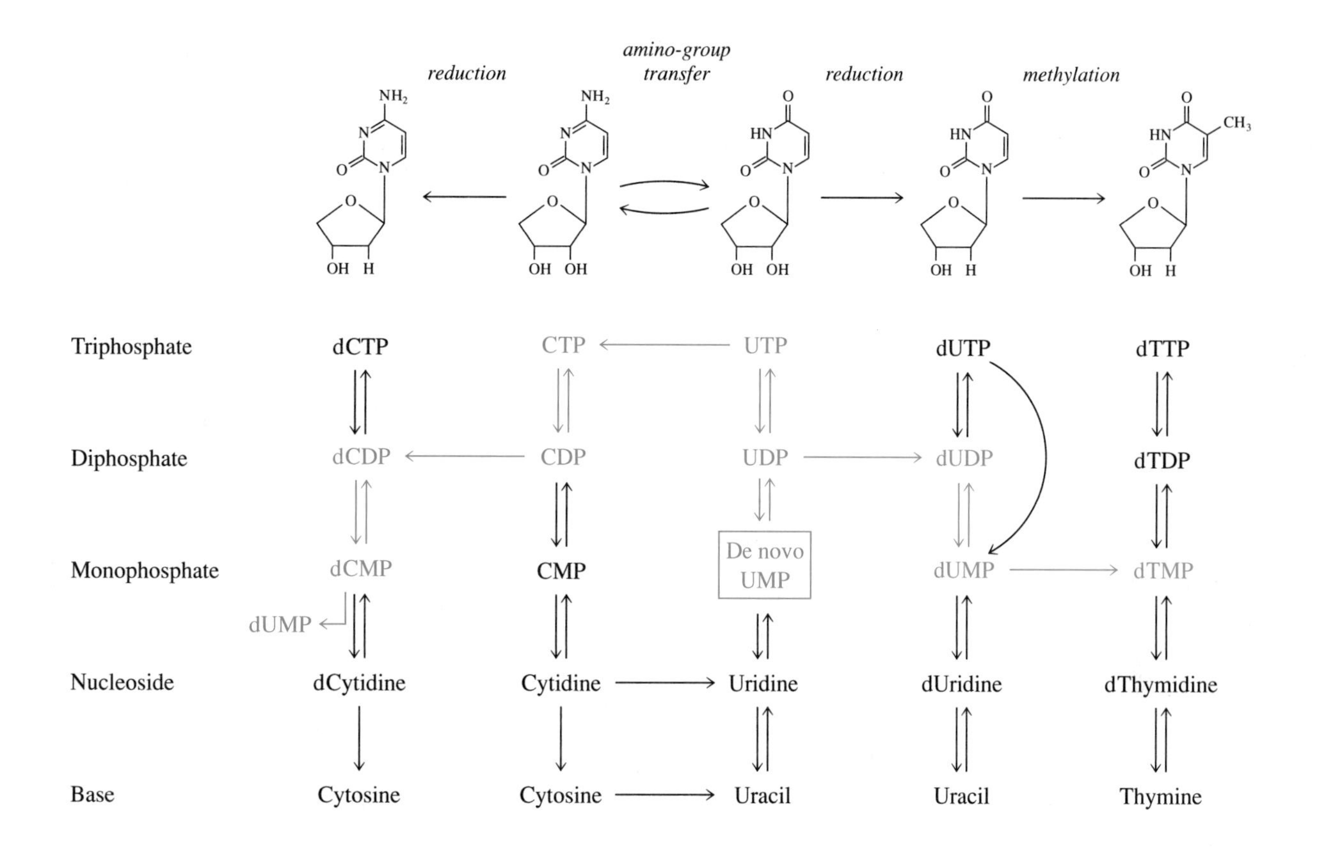

Figure 19·19
Interconversions of pyrimidine nucleotides and their constituents. [Adapted from T. W. Traut (1988). Enzymes of nucleotide metabolism: the significance of subunit size and polymer size for biological function and regulatory properties. *Crit. Rev. Biochem.* 23:121–169.]

19·12 Purine Breakdown by Primates, Birds, and Reptiles Leads to Uric Acid

Although most free purine and pyrimidine molecules are salvaged, some molecules are catabolized. These molecules originate from an excess of ingested nucleotides or from intracellular turnover (continual synthesis and degradation) of nucleic acids.

In birds, reptiles, and primates (including humans), purine nucleotides are converted to uric acid, which is excreted. For birds and reptiles, uric acid is also the excretion product for the disposal of surplus nitrogen from amino acid catabolism, a function served by urea in primates. Birds and reptiles lack enzymes that catalyze further breakdown of uric acid, but many organisms degrade uric acid to other products, as we shall see in the next section.

The pathways leading from AMP and GMP to uric acid are summarized in Figure 19·20. Hydrolytic removal of phosphate from AMP and GMP produces adenosine and guanosine, respectively. Adenosine is deaminated to inosine by the action of adenosine deaminase. Similarly, AMP is deaminated to IMP by the action of AMP deaminase. Hydrolysis of IMP leads to inosine, which can be converted to hypoxanthine by phosphorolysis. Phosphorolysis of guanosine produces guanine. Both of these phosphorolysis reactions (as well as phosphorolysis of several deoxynucleosides) are catalyzed by purine nucleoside phosphorylase and produce ribose 1-phosphate (or deoxyribose 1-phosphate) as well as a base. Adenosine is not a substrate of mammalian purine nucleoside phosphorylase.

Hypoxanthine formed from inosine is oxidized to xanthine, and xanthine is oxidized to uric acid, the final product in Figure 19·20. Either of the two enzymes xanthine oxidase or xanthine dehydrogenase catalyzes both reactions. In the reactions catalyzed by xanthine oxidase, electrons are transferred to O_2 to form hydrogen peroxide, H_2O_2. Xanthine oxidase, an extracellular enzyme in mammals, appears to be an altered form of the intracellular enzyme xanthine dehydrogenase, which generates the same products as xanthine oxidase but transfers electrons to $NAD^{\oplus}$ to form NADH. These two enzyme activities occur widely in nature and exhibit broad substrate specificity. Their active sites contain rather complex electron-transfer systems that include an iron-sulfur center, a molybdopterin, and FAD.

In most cells, guanine is deaminated to xanthine in a reaction catalyzed by guanase. Those animals that lack guanase excrete guanine. For example, pigs excrete guanine but metabolize adenine derivatives further to allantoin, the major end product of catabolism of purines in most mammals.

Gout is a disease of humans caused by the overproduction or inadequate excretion of uric acid. Uric acid is relatively insoluble, and when its concentration in blood is elevated, it can crystallize in cartilage and soft tissues, especially the kidney, and in toes and joints. Gout has several causes, including partial deficiency in hypoxanthine-guanine phosphoribosyltransferase activity, which results in less salvage of purines and more catabolic production of uric acid. It can also be caused by defective regulation of purine biosynthesis.

Gout can be treated by administration of allopurinol, a synthetic C-7, N-8 positional isomer of hypoxanthine.

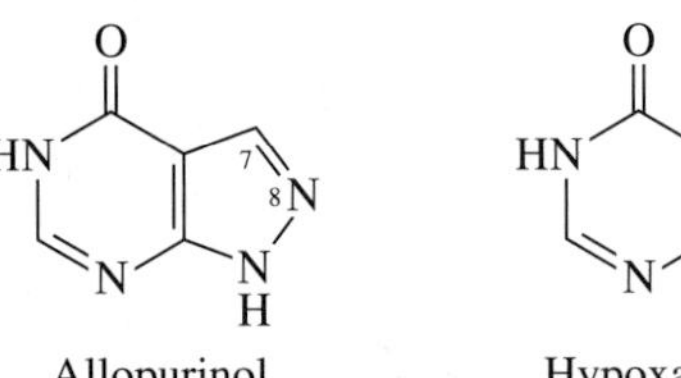

(19·9)

Because allopurinol is a powerful inhibitor of xanthine dehydrogenase, its presence prevents the formation of abnormally high levels of uric acid. Hypoxanthine and xanthine are more soluble than uric acid and can therefore be excreted.

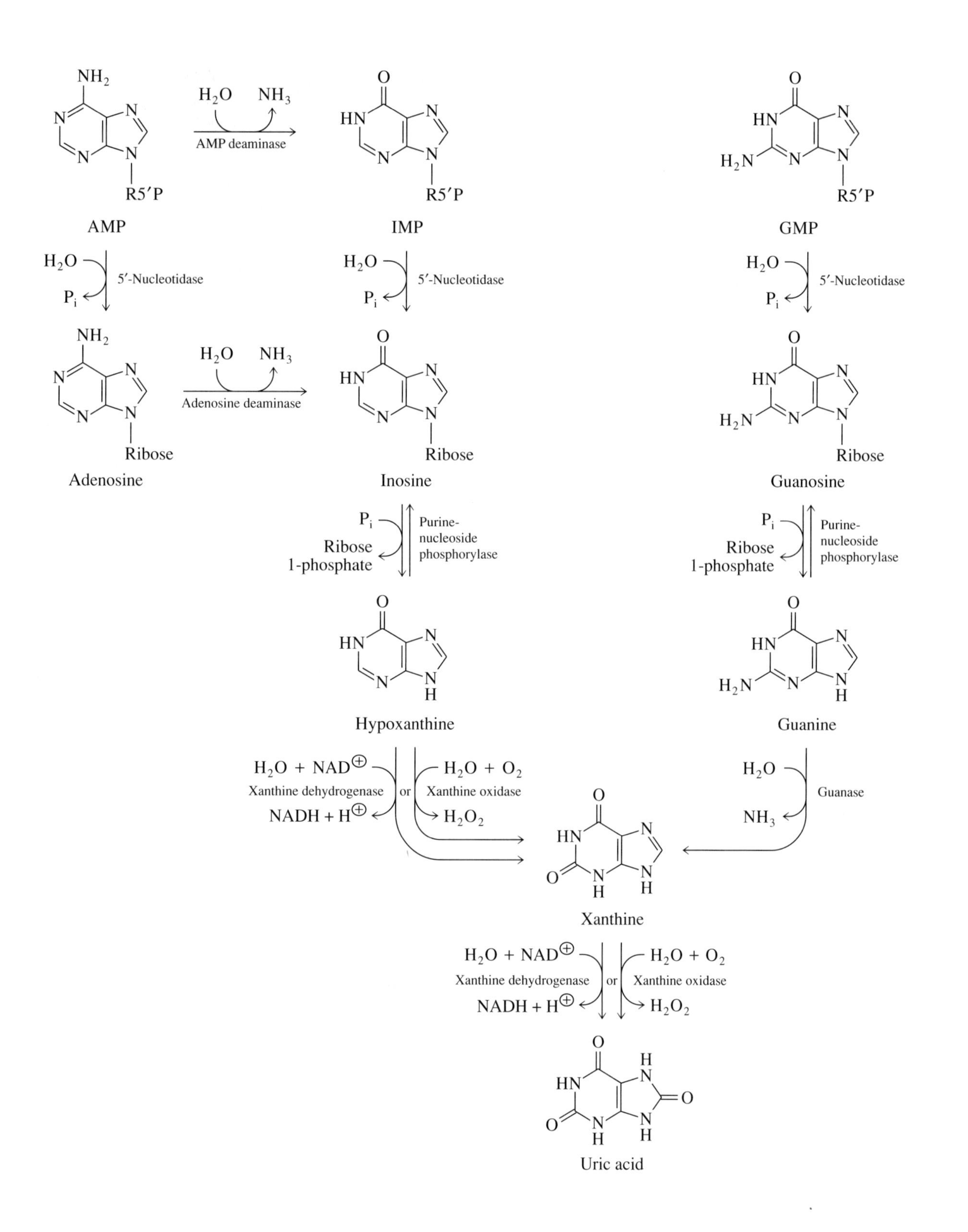

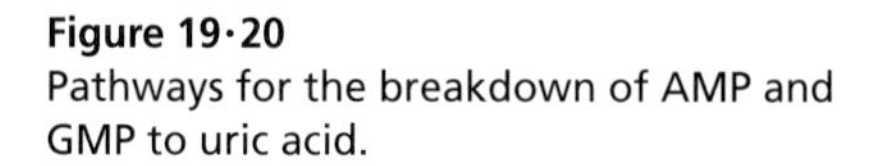
Figure 19·20
Pathways for the breakdown of AMP and GMP to uric acid.

Part Four

Biological Information Flow

20

DNA Replication and Repair

The transfer of genetic information from one generation to the next has puzzled biologists since the time of Aristotle. We now know that genetic information is carried in DNA, and that transferring information from a parental cell to two daughter cells requires exact duplication of the DNA. The structure of DNA, revealed in 1953 by James D. Watson and Francis H. Crick, immediately suggested a method of replication: since the two strands of double-helical DNA are complementary, the nucleotide sequence of one strand automatically specifies the sequence of the other. Watson and Crick proposed that the two strands of the helix unwind during DNA replication and each strand of DNA acts as a template for the synthesis of a complementary strand. In this way, DNA replication produces two daughter DNA duplexes, each containing one parental strand and one newly synthesized strand. This mode of replication is termed *semiconservative DNA replication* because one strand of the parental DNA is conserved in each newly synthesized molecule of double-stranded DNA (Figure 20·1).

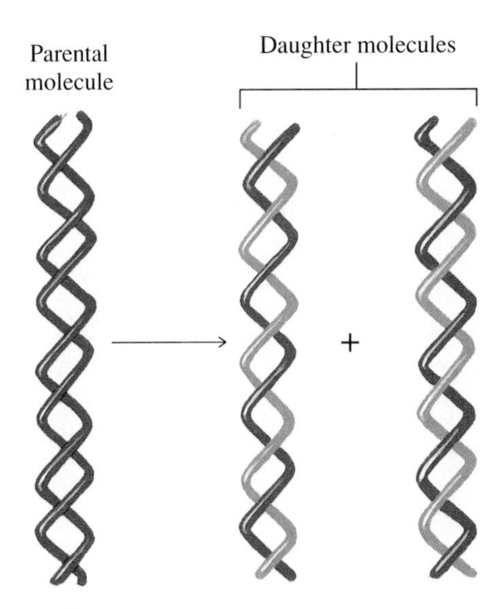

Figure 20·1
Semiconservative DNA replication. Each strand of DNA acts as a template for synthesis of a new strand. Each daughter molecule of DNA contains one parental strand and one newly synthesized strand. (Illustrators: Initial rendering—Lisa Shoemaker; Final rendering—Linda LeFevre Murray.)

Although in 1953 elucidation of the structure of DNA immediately suggested a mode of replication, the details of the mechanism remain an active topic of research. Biochemists have isolated and characterized enzymes that catalyze steps in DNA replication and, in many cases, have identified the genes that encode these proteins. Mutational analysis of these genes has been crucial in establishing the steps of the replication mechanism.

The results of these studies have produced an illuminating picture of how large numbers of polypeptides assemble into complexes that are capable of carrying out many functions simultaneously. The DNA replication complex is like a machine whose parts are made of protein. Some of the component polypeptides are partially active in isolation, but others are active only in association with the complete protein machine. The correct assembly of the replication complex at the site where DNA replication is to begin is an essential part of the initiation step of DNA replication. During the elongation stage, the complex replicates DNA semiconservatively, catalyzing the incorporation of nucleotides into the growing strands. At termination, the protein machine is disassembled, and the newly replicated strands of DNA separate.

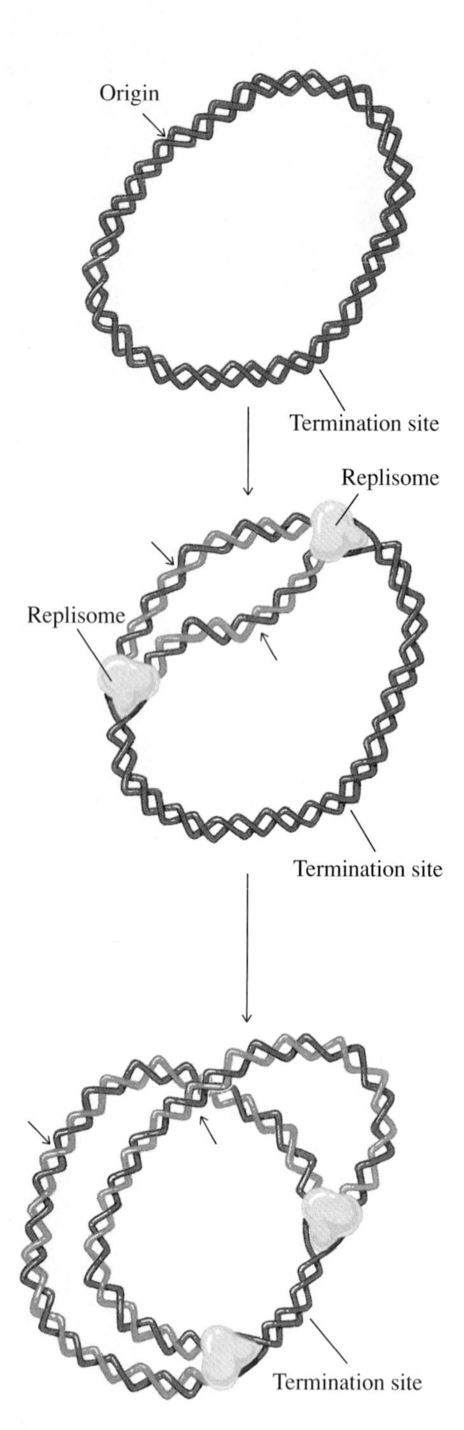

Figure 20·2
Bidirectional DNA replication in *E. coli*. Semiconservative DNA replication begins at the origin and proceeds both clockwise and counterclockwise. The synthesis of the new strands of DNA (light gray) occurs at the two replication forks, where the replisomes are located. When the replication forks meet at the termination site, the two chromosomes separate. (Illustrator: Lisa Shoemaker.)

The concept of a protein machine that carries out a series of biochemical reactions applies not only to DNA replication but also to transcription, RNA processing, and translation. Furthermore, there is increasing evidence that other processes of cellular metabolism are carried out by complexes of weakly associated enzymes and other macromolecules (Section 5·16).

The maintenance of genetic information from generation to generation requires that DNA replication be both rapid (because the entire complement of DNA must be replicated before each cell division) and accurate. In addition to having polymerization enzymes that are highly accurate, all cells have repair enzymes that fix damaged DNA and correct replication errors.

The overall strategy of DNA replication in eubacteria, archaebacteria, and eukaryotes appears to be conserved, although specific enzymes vary among organisms. Just as two different makes of automobile are similar even though individual parts cannot be substituted for one another, so too is the mechanism of chromosomal DNA replication similar in all organisms even though the individual enzymes may differ. We shall focus on chromosomal DNA replication and repair mechanisms in the bacterium *Escherichia coli* because many of its replication and repair enzymes are known and well characterized. These reactions are similar to those that occur in all organisms.

20·1 Chromosomal DNA Replication Is Bidirectional

The *E. coli* chromosome is a large, circular, double-stranded DNA molecule of 4.2 $\times 10^3$ kilobase pairs (kb). Replication of this chromosome begins at a unique site called the origin of replication and proceeds bidirectionally until the two replication complexes meet at the termination site, where replication stops (Figure 20·2). The protein machine that carries out the polymerization reaction is called a *replisome*. The replisome contains a number of different proteins that carry out the various reactions required for rapid and accurate DNA replication. One replisome is located at each of the two replication forks. An autoradiograph of a replicating *E. coli* chromosome is shown in Figure 20·3.

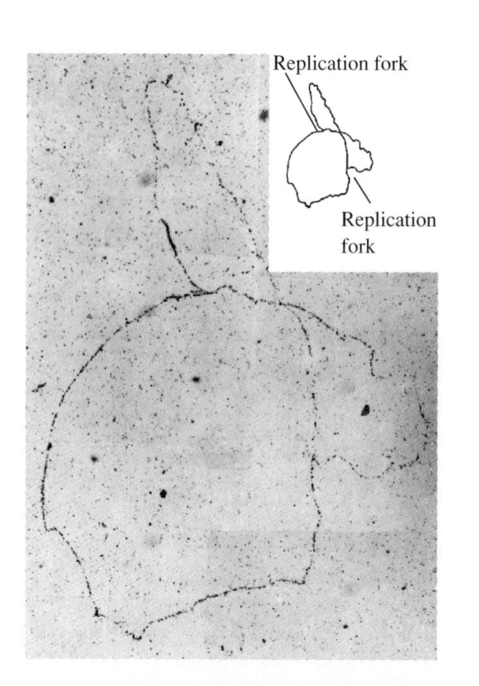

Figure 20·3
Autoradiograph of a replicating *E. coli* chromosome. The DNA has been labelled with [^{3}H]-deoxythymidine and has been detected by overlaying the replicating chromosome with photographic emulsion. The autoradiograph reveals that the *E. coli* chromosome has two replication forks. (Courtesy of John Cairns.)

In *E. coli,* the rate of movement of a replication fork is approximately 1000 base pairs per second. In other words, each of the two new strands is extended at the rate of 1000 nucleotides per second. Since there are two replication forks moving at this rate, the entire *E. coli* chromosome can be duplicated in 35 minutes.

Eukaryotic chromosomes are linear, double-stranded DNA molecules that are usually much larger than the chromosome of bacteria. For example, the size of the genome of the fruit fly *Drosophila melanogaster* is 1.65×10^5 kb (genome refers to the amount of DNA in one complete set of chromosomes), and the size of a mammalian genome is approximately 3×10^6 kb. In eukaryotes as in *E. coli,* replication is bidirectional, but whereas the *E. coli* chromosome has a unique origin, eukaryotic chromosomes have multiple sites where DNA synthesis is initiated (Figure 20·4). Although the rate of fork movement in eukaryotes is slower than in bacteria and eukaryotic genomes are much larger, the presence of a large number of independent origins of replication enables eukaryotic DNA to be copied in approximately the same time as prokaryotic DNA.

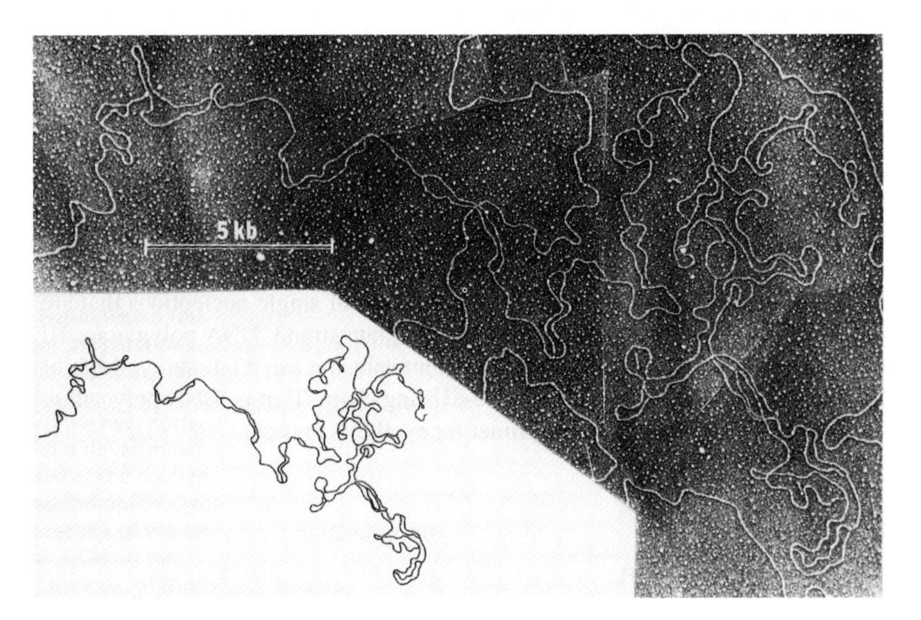

Figure 20·4
Electron micrograph of replicating DNA in the embryo of the fruit fly *Drosophila melanogaster.* Note the large number of replication forks. The drawing in the lower left corner highlights a single piece of DNA with replication forks. (Courtesy of David S. Hogness.)

20·2 DNA Polymerase Catalyzes the Polymerization Reaction at a Replication Fork

The synthesis of a new strand of DNA is achieved by adding nucleotides to the 3′ end of a growing chain. This polymerization activity is catalyzed by enzymes known as DNA-directed DNA polymerases or simply DNA polymerases. In *E. coli* cells, there are three different DNA polymerases, each designated by a roman numeral according to its order of discovery. DNA polymerase I repairs DNA and participates in the synthesis of one of the strands of DNA during replication; DNA polymerase II plays a role in DNA repair; DNA polymerase III, a key component of the replisome and the major DNA replication enzyme, is responsible for chain elongation during DNA replication.

DNA polymerase III is the most complex of the three DNA polymerases. The minimally active core of the enzyme consists of three subunits, α, ε, and θ. The holoenzyme, which is a larger, more active form of the enzyme, is isolated as a

B. DNA Polymerase III Holoenzyme Remains Bound to the Replication Fork During Chain Elongation

As noted earlier, the DNA replication complex catalyzes DNA synthesis inside the cell at a rate of approximately 1000 nucleotide residues per second. This is the fastest known rate of any in vivo polymerization reaction. The in vitro rate of polymerization by purified DNA polymerase III, however, is much slower, indicating that the enzyme, as isolated, lacks some components necessary for full activity. Only when the complete replisome is assembled does polymerization in vitro occur at approximately the rate found inside the cell.

Once DNA synthesis has been initiated, the enzyme remains bound to the replication fork until replication is complete. The 3′ end of the growing chain is associated with the active site of the enzyme while many nucleotides are added sequentially. Enzymes that remain bound to their nascent chains through many polymerization steps are said to be *processive.* (The opposite of processive is distributive; a distributive enzyme dissociates from the growing chain after the addition of each monomer.) Because DNA polymerase III as part of the replisome is highly processive during DNA replication, only a small number of DNA polymerase III molecules are needed to replicate the *E. coli* chromosome every generation. Processivity also accounts in part for the rapid rate of DNA replication.

The processivity of the DNA polymerase III holoenzyme is due to several things. One of these is the activity of the β subunit of the enzyme itself. This subunit has no activity on its own, but, when assembled into the holoenzyme, it acts as a clamp, which renders the polymerization activity of the enzyme extremely processive. The β subunit is probably assisted by several other subunits of the holoenzyme. The fact that DNA polymerase III is part of a larger protein machine at the replication fork further ensures that the enzyme remains associated with nascent chains during polymerization.

C. Polymerization Errors Can Be Corrected by Proofreading

DNA polymerase III holoenzyme is a 3′→5′ exonuclease as well as a polymerase. This exonuclease, whose active site lies primarily within the ε subunit, can catalyze hydrolysis of the phosphodiester that links the end residue to the rest of the growing polynucleotide chain. The simultaneous ability of the DNA polymerase III holoenzyme to catalyze chain elongation and to act as a phosphodiesterase allows the holoenzyme to proofread, or edit, newly synthesized DNA and correct mismatches. When DNA polymerase III recognizes a distortion produced by an incorrectly paired base, the exonuclease activity of the enzyme catalyzes the removal of the mispaired nucleotide before polymerization continues (Figure 20·7). Note that only incorrectly paired nucleotides are removed; the exonuclease is not active on correctly matched bases.

The wrong base is incorporated about once in every 10^5 elongation steps. The error rate is thus 10^{-5}. The 3′→5′ proofreading exonuclease activity has an error rate of about 10^{-3}. The combination of these two sequential reactions yields an overall error rate for polymerization of 10^{-8}, which is one of the lowest error rates of any enzyme. Nevertheless, replication errors are common when large genomes are duplicated (recall that even the *E. coli* chromosome consists of $>4 \times 10^3$ kb). Most of these errors can be repaired by DNA-repair enzymes (Section 20·7).

(a) If the last base has been correctly matched to its partner in the template strand, the holoenzyme advances.

DNA polymerase III holoenzyme

Correctly paired

Phosphodiester formation

DNA polymerase III holoenzyme advances

(b) If the next incoming residue is mismatched, the mispaired base is recognized by the holoenzyme at the proofreading site.

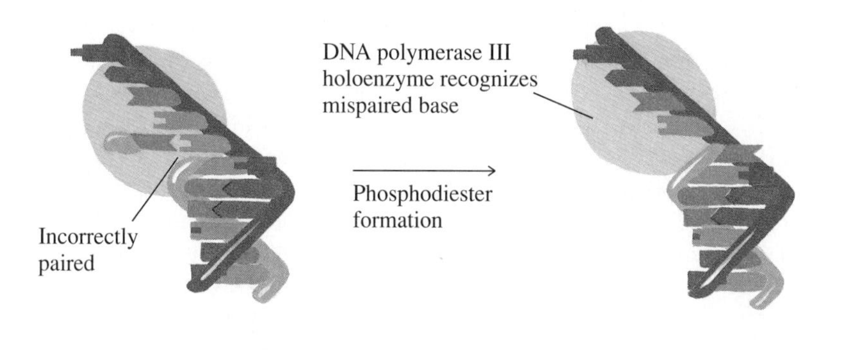

(c) The holoenzyme retreats and hydrolyzes the phosphodiester linkage between the mispaired nucleotide and the rest of the growing chain. The mispaired nucleotide then dissociates, and the holoenzyme inserts the correct nucleotide.

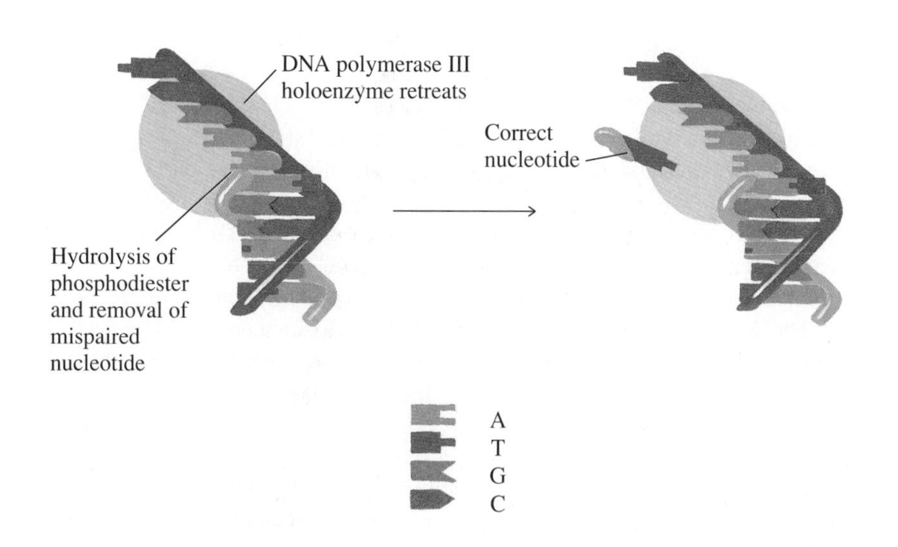

Figure 20·7
Proofreading mechanism of the *E. coli* DNA polymerase III holoenzyme. (Illustrators: Initial rendering—Lisa Shoemaker; Final rendering—Linda LeFevre Murray.)

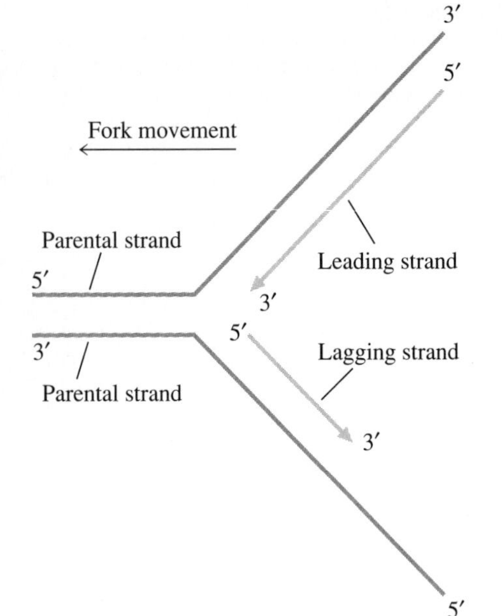

Figure 20·8
Schematic diagram of the replication fork. The two newly synthesized strands are of opposite polarity. On the leading strand, 5′→3′ synthesis moves toward the replication fork; on the lagging strand, 5′→3′ synthesis moves away from the replication fork.

20·3 The Two Strands at the Replication Fork Are Synthesized by Separate 10-Subunit Complexes of the DNA Polymerase III Holoenzyme Dimer

DNA polymerases catalyze chain elongation exclusively in the 5′→3′ direction. Since the two strands of DNA are antiparallel, 5′→3′ synthesis using one template strand occurs in the same direction as fork movement, whereas 5′→3′ synthesis on the other template strand occurs in the direction opposite fork movement (Figure 20·8). The newly synthesized strand formed by 5′→3′ polymerization in the direction of replication fork movement is called the *leading strand.* The newly synthesized strand formed by 5′→3′ polymerization in the direction opposite replication fork movement is called the *lagging strand.* One complex of the DNA polymerase III holoenzyme dimer is responsible for synthesis of the leading strand; the other complex of the DNA polymerase III holoenzyme dimer is responsible for synthesis of the lagging strand.

A. DNA Synthesis on the Lagging Strand Is Discontinuous

The leading strand is synthesized continuously from the origin of replication as the replication fork advances. Because the template for the lagging strand has a 3′→5′ orientation, however, the lagging strand is synthesized discontinuously in short 5′→3′ pieces in the direction opposite fork movement. These individual pieces of nascent lagging strand are then joined in a separate process. In Section 20·4, we shall explain the topology that makes this lagging-strand synthesis possible.

Discontinuous DNA synthesis was demonstrated by labelling newly synthesized DNA with [^{3}H]-deoxythymidine and examining the structures of replication intermediates. Growing *E. coli* cells were exposed briefly to a short pulse of [^{3}H]-deoxythymidine (which was readily taken up and converted to [^{3}H]-deoxythymidine triphosphate, or dTTP, a substrate for the polymerization reaction), synthesis of DNA was then interrupted, and the newly made DNA was isolated, denatured, and separated by size. Two types of labelled DNA molecules were found: very large DNA molecules that collectively contained about half the radioactivity of the partially replicated DNA and shorter DNA fragments consisting of about 1000 residues that collectively contained the other half of the radioactivity (Figure 20·9). The large DNA molecules arose from continuous synthesis of the leading strand; the shorter fragments arose from discontinuous synthesis of the lagging strand. The short pieces of DNA were named *Okazaki fragments* in honor of their discoverer, Reiji Okazaki.

B. Each Okazaki Fragment Begins with an RNA Primer

Discontinuous synthesis explains how the lagging strand is synthesized, but it does not explain how synthesis of each Okazaki fragment is initiated. No DNA polymerase can begin polymerization de novo; it can only add nucleotides to existing polymers. Initiation of the synthesis of Okazaki fragments can be explained by the presence of RNA primers, short pieces of RNA synthesized at the replication fork. Each primer, which is complementary to the lagging-strand template, is extended from its 3′ end by DNA polymerase to form an Okazaki fragment (Figure 20·10).

The synthesis of RNA primers is catalyzed by an enzyme called primase, the product of the dnaG gene in *E. coli.* Primase is a component of a larger complex called the *primosome,* which contains at least 16 polypeptides in addition to primase. The primosome, along with DNA polymerase III, is part of the replisome. The primosome catalyzes synthesis of a short RNA primer of approximately 10 nucleotides every second. Since the replication fork advances at a rate of approximately 1000 nucleotides per second, there is one primer synthesized for

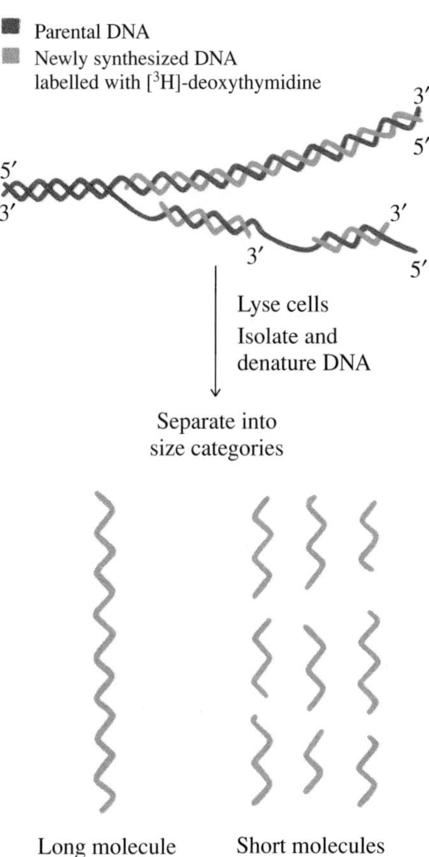

Figure 20·9
Discontinuous DNA synthesis demonstrated by analysis of newly synthesized DNA. Nascent DNA molecules are labelled in *E. coli* with [^{3}H]-deoxythymidine. The cells are then lysed, DNA is isolated and denatured, and single strands are separated on the basis of size. The newly synthesized molecules fall into two classes: molecules of great length and short fragments of approximately 1000 nucleotides. The long DNA molecules arise from continuous synthesis of the leading strand, and the short fragments arise from discontinuous synthesis of the lagging strand. (Illustrators: Initial rendering—Lisa Shoemaker; Final rendering—Linda LeFevre Murray.)

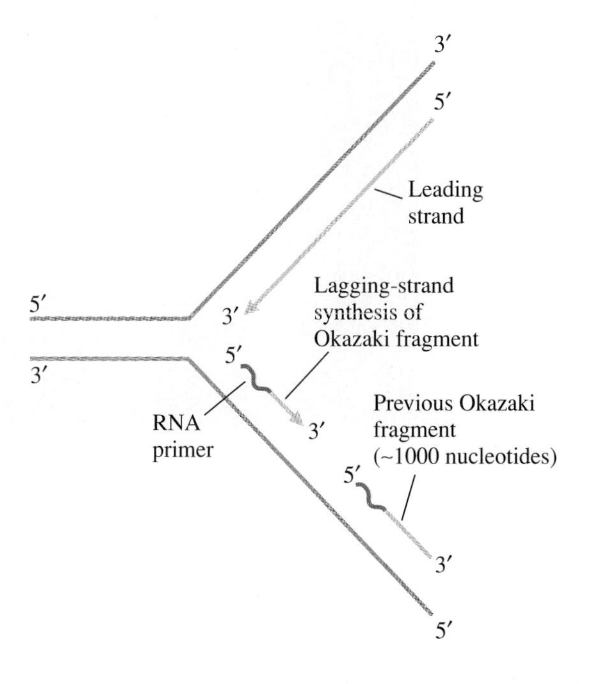

Figure 20·10
Schematic diagram of lagging-strand synthesis. A short piece of RNA serves as a primer for the synthesis of Okazaki fragments. The length of the Okazaki fragment is determined by the length of the gap between successive RNA primers.

approximately every thousand nucleotides. DNA polymerase III then catalyzes DNA synthesis in the $5'\rightarrow3'$ direction by extending each RNA primer. As more single-stranded DNA becomes exposed behind the holoenzyme as a result of continued unwinding at the replication fork, primase synthesizes additional primers for each new Okazaki fragment.

C. Okazaki Fragments Are Joined by the Combined Action of DNA Polymerase I and DNA Ligase

DNA polymerase I of *E. coli* was discovered by Arthur Kornberg in 1957. It was the first enzyme to be found that is capable of DNA synthesis using a template strand. In a single polypeptide, DNA polymerase I (Figure 20·11) contains a less efficient version of the activities found in the DNA polymerase III holoenzyme: a polymerase activity and a $3'\rightarrow5'$ proofreading exonuclease activity. DNA polymerase I also has a $5'\rightarrow3'$ exonuclease activity, an activity not found in DNA polymerase III.

The $5'\rightarrow3'$ exonuclease of DNA polymerase I catalyzes the removal of the RNA primer at the beginning of each Okazaki fragment, and the polymerase fills in the gap between Okazaki fragments as the RNA primer is degraded. This process is called *nick translation* (Figure 20·12). In nick translation, DNA polymerase I recognizes and binds to the nick (or gap) between the $3'$ end of the nascent DNA chain and the $5'$ end of the next primer. The $5'\rightarrow3'$ exonuclease then hydrolytically removes the first RNA nucleotide while the $5'\rightarrow3'$ polymerase adds a deoxynucleotide to the $3'$ end of the DNA chain. In this way, the enzyme moves the nick along the lagging strand. After completing 10 or 12 cycles of hydrolysis and polymerization, DNA polymerase I dissociates from the DNA, leaving behind two strands of newly synthesized DNA separated by a nick in the phosphodiester backbone. The removal of RNA primers by DNA polymerase I is an essential part of DNA replication because the final product must consist entirely of double-stranded DNA. DNA polymerase I has been widely used in the laboratory to sequence DNA (Box 20·1).

Figure 20·11
Space-filling model showing a fragment of DNA polymerase I bound to B-DNA. A portion of DNA polymerase I containing the polymerase activity and the proofreading activity is shown (in yellow) wrapped around the DNA template strand (blue) and the primer strand (red). This fragment, which is generated in the laboratory, is called the Klenow fragment. (Courtesy of Thomas A. Steitz and Lorena Beese.)

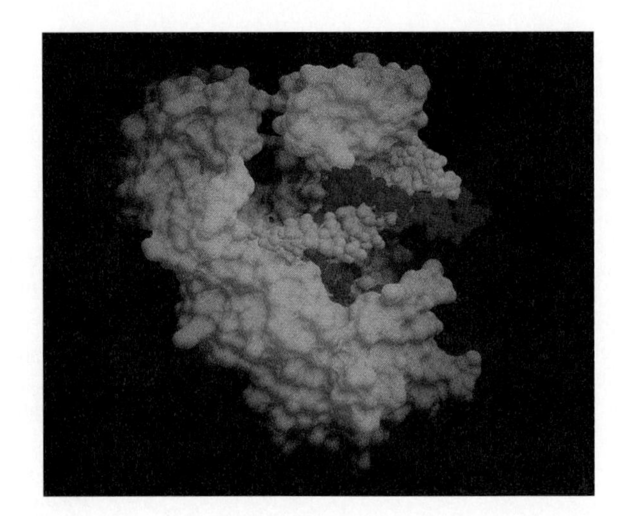

(a) The completion of lagging-strand synthesis leaves a gap between the Okazaki fragment and the preceding RNA primer.

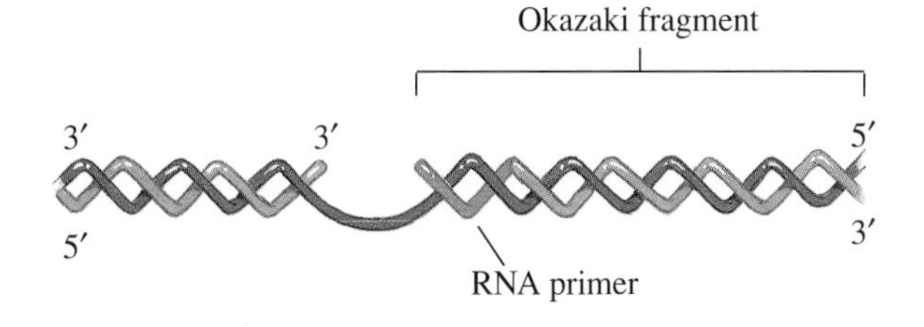

(b) DNA polymerase I extends the Okazaki fragment while its 5′ → 3′ exonuclease activity removes the RNA primer. This process, which results in movement of the gap along the lagging strand, is called nick translation.

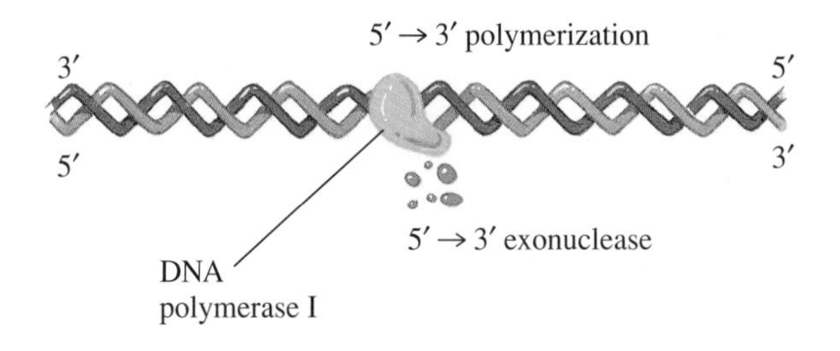

(c) DNA polymerase I dissociates after extending the Okazaki fragment 10–12 nucleotides. DNA ligase binds to the nick between Okazaki fragments.

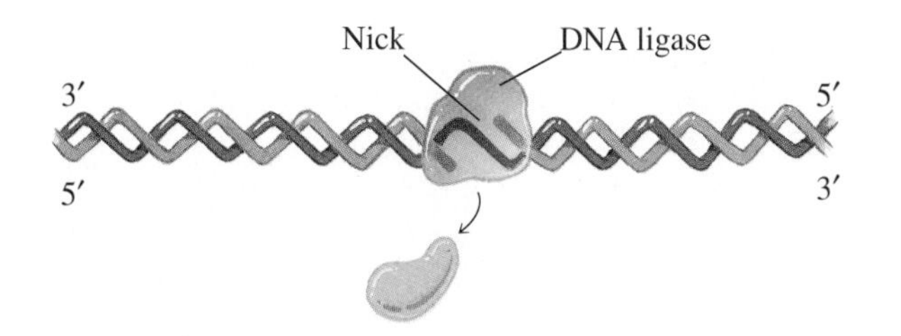

(d) DNA ligase catalyzes formation of a phosphodiester, which seals the nick, creating a continuous lagging strand. The enzyme then dissociates.

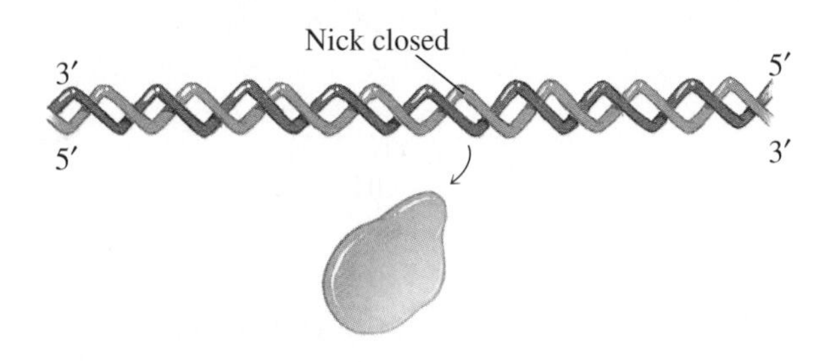

Figure 20·12
Joining of Okazaki fragments by the combined action of DNA polymerase I and DNA ligase. (Illustrator: Lisa Shoemaker.)

Figure 20·13
Mechanism of DNA ligase in *E. coli.* Using NAD⊕ as a cosubstrate, *E. coli* DNA ligase catalyzes the formation of a phosphodiester at nicks in DNA. The reaction can be divided into three steps. (1) The ε-amino group of a lysine residue of DNA ligase displaces nicotinamide mononucleotide (NMN) from NAD⊕. The result is an AMP–DNA ligase intermediate. (With DNA ligases that use ATP as the cosubstrate, pyrophosphate is displaced.) (2) The free 5′ phosphate group of the DNA displaces the enzyme, thereby forming an ADP-DNA intermediate. (3) The nucleophilic 3′ hydroxyl group on the adjacent DNA strand displaces AMP, generating a phosphodiester that seals the nick in the DNA strand.

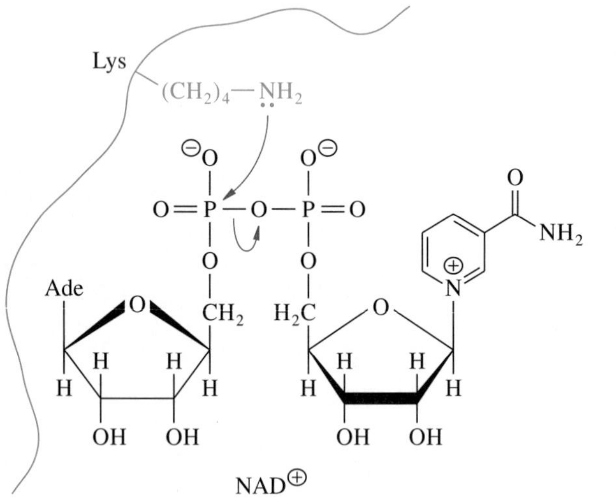

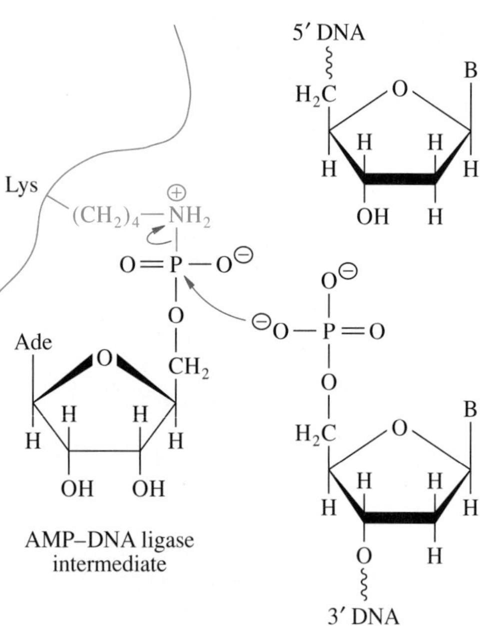

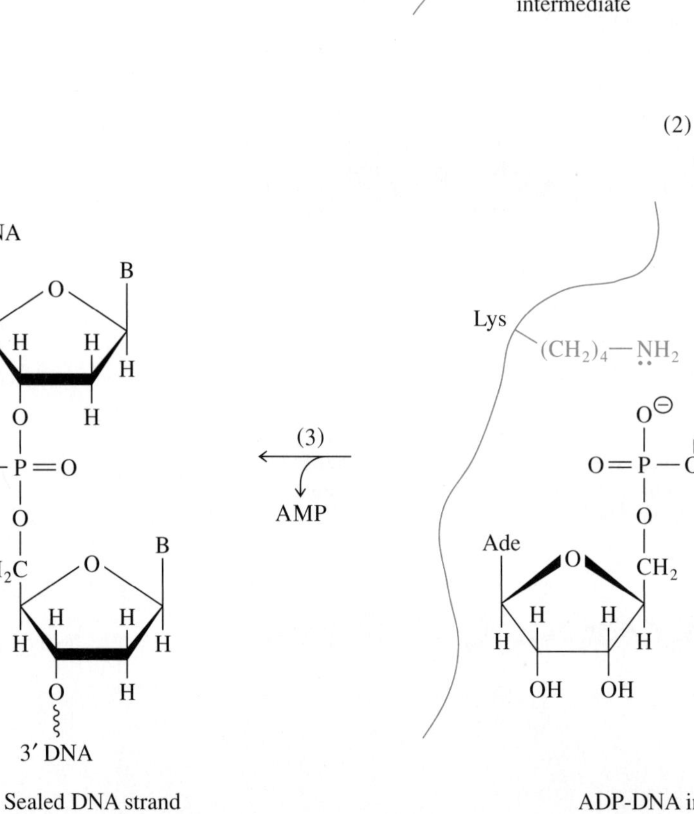

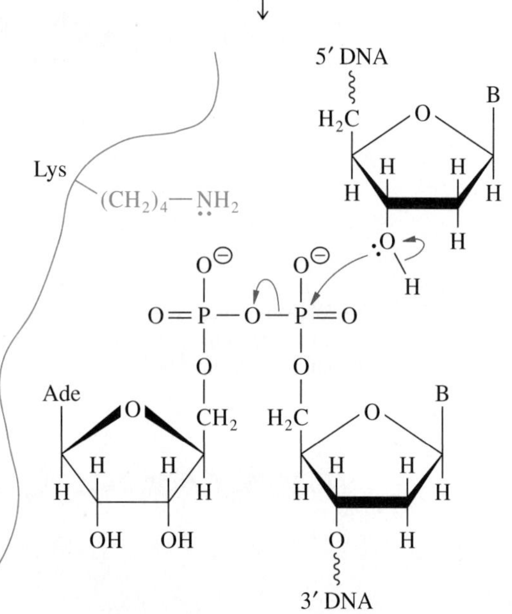

The last step in the synthesis of the lagging strand of DNA is the formation of a phosphodiester by bonding between the 3′-hydroxyl group at the end of one Okazaki fragment and the 5′-phosphate group of an adjacent Okazaki fragment. This step is catalyzed by DNA ligase. Like most other ligases, DNA ligase in eukaryotic cells and in phages that infect *E. coli* requires ATP as a cosubstrate. In contrast to other ligases, *E. coli* DNA ligase requires $NAD^{\oplus}$ as a cosubstrate.

The mechanism of *E. coli* DNA ligase is summarized in Figure 20·13. In the first step, the phosphorus atom bonded to the 5′-oxygen of the adenosine group of $NAD^{\oplus}$ undergoes nucleophilic attack by the ε-amino group of an active-site lysine residue. Nicotinamide mononucleotide ($NMN^{\oplus}$) is released, and an energy-rich AMP–DNA ligase intermediate is generated. The free 5′-phosphate group of the DNA then attacks the phosphoryl group of the AMP–enzyme complex, displacing the enzyme group and forming an energy-rich ADP-DNA intermediate at the 5′ position of the DNA. The activated 5′-phosphate of ADP-DNA then undergoes nucleophilic attack by the free 3′-hydroxyl group on the adjacent DNA chain, resulting in the displacement of AMP and sealing of the nick. The net reaction is:

$$\text{DNA}_{(\text{nicked})} + \text{NAD}^{\oplus} \longrightarrow \text{DNA}_{(\text{sealed})} + \text{NMN}^{\oplus} + \text{AMP} \qquad \textbf{(20·1)}$$

Box 20·1 Sequencing DNA Using Dideoxynucleotides

In 1976, Frederick Sanger developed a method of sequencing DNA enzymatically using *E. coli* DNA polymerase I. For this achievement, Sanger was awarded his second Nobel Prize. (His first Nobel Prize was awarded for developing a method of sequencing proteins.) The Sanger method of DNA sequencing is still widely used today, although other prokaryotic DNA polymerases are now used in place of the *E. coli* enzyme.

The Sanger sequencing method uses 2′,3′-dideoxynucleoside triphosphates (ddNTP), which differ from the deoxynucleotide substrates of DNA synthesis by the replacement of the 3′-hydroxyl group with a hydrogen (Figure 1). The dideoxynucleotides are recognized by DNA polymerase and are added to the 3′ end of the growing chain. But because these nucleotides lack a 3′-hydroxyl group, subsequent nucleotide additions cannot take place. Their incorporation therefore terminates the growth of the DNA chain. When a low level of a particular dideoxynucleotide is included in a DNA synthesis reaction, it will occasionally be incorporated in place of the corresponding dNTP and thus cause termination. The length of the resulting fragment of DNA indicates the position where the corresponding nucleotide would have been incorporated.

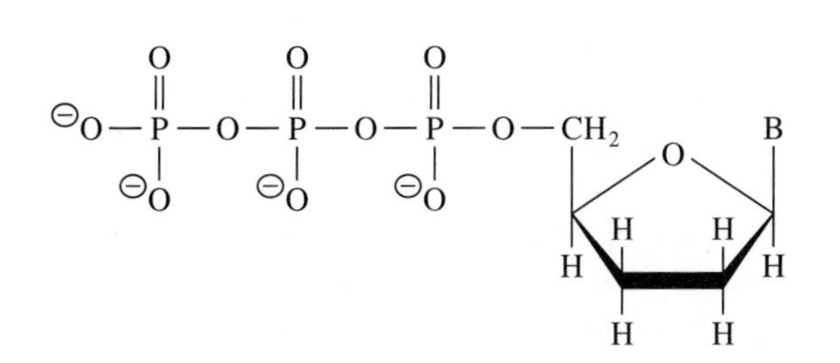

Box 20·1
Figure 1
Structure of 2′,3′-dideoxynucleoside triphosphate. B stands for any base.

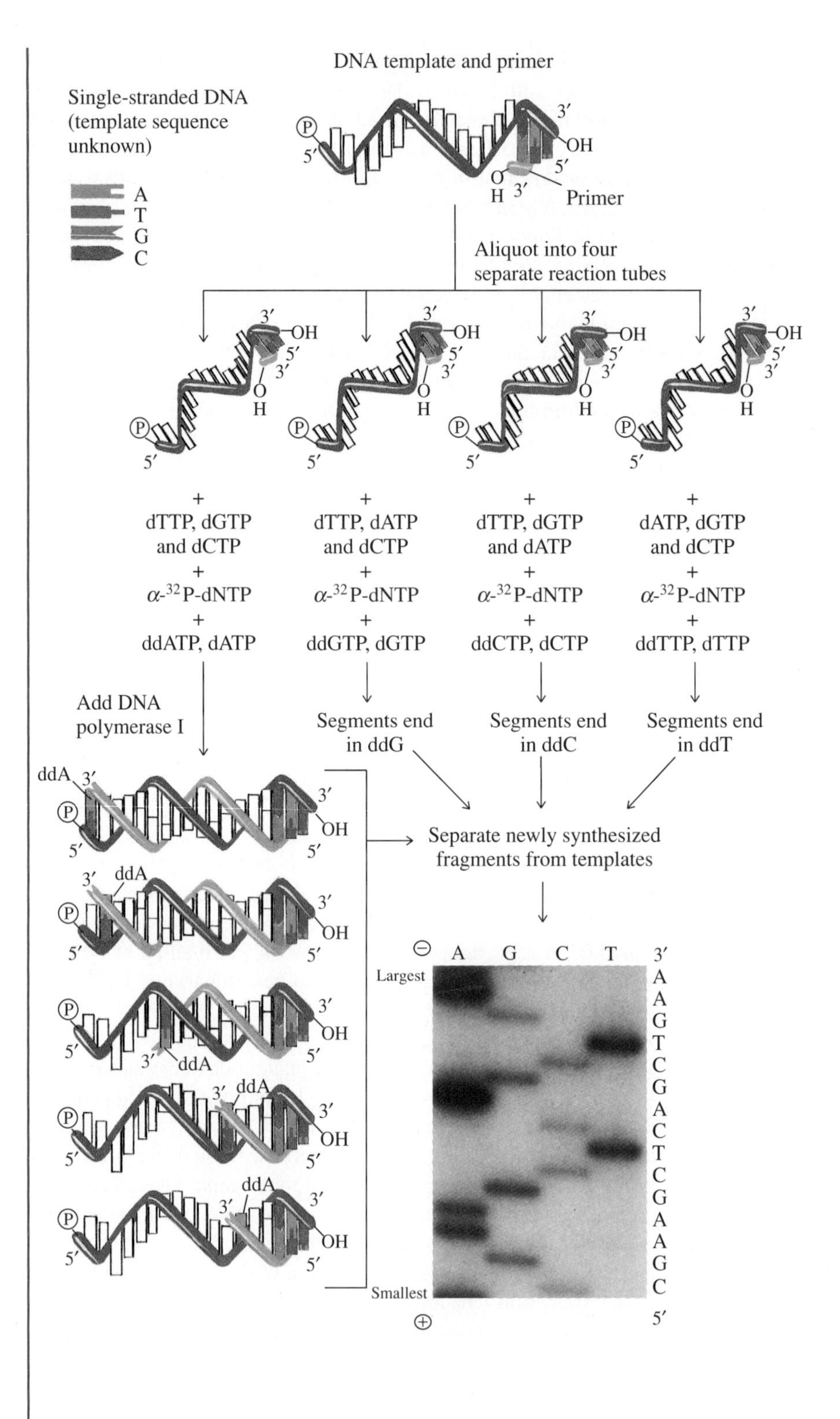

Box 20·1
Figure 2
Sanger method of sequencing DNA. In the Sanger sequencing method, addition of a low level of a particular dideoxynucleoside triphosphate (ddNTP) to the replicating mixture causes DNA synthesis to terminate when that ddNTP is incorporated randomly in place of the normal nucleotide. The positions of incorporated ddNTPs, as determined by the lengths of the fragments generated, indicate the positions of the corresponding nucleotide in the sequence. An electrophoresis sequencing gel allows the fragments generated during synthesis with each ddNTP to be visualized, and thus the sequence of the DNA can be read off the gel as shown by the vertical column of letters to the right of the gel. (Illustrators: Initial rendering—Lisa Shoemaker; Final rendering—Linda LeFevre Murray.)

As Figure 2 illustrates, DNA sequencing using ddNTP molecules involves several steps. The DNA to be sequenced is prepared as single-stranded molecules and mixed with a short oligonucleotide complementary to the DNA. This oligonucleotide acts as a primer for DNA synthesis catalyzed by DNA polymerase I. The oligonucleotide-primed material is then split into four separate reaction tubes. Each tube receives a small amount of one of the same α-^{32}P-labelled dNTPs, which will allow the newly synthesized DNA to be visualized by autoradiography. Then each tube receives an excess of the four nonradioactive dNTP molecules and a small amount of one of the four ddNTPs. For example, the A reaction tube receives an excess of nonradioactive dTTP, dGTP, dCTP, and dATP mixed with a small amount of ddATP. When DNA polymerase I is added to this reaction tube, a ddATP residue will occasionally be incorporated instead of dATP and thereby terminate the growing DNA chain. The result is production of newly synthesized DNA fragments of different lengths. Since each of the fragments ends with an A, the length of the fragment is a measure of the distance from the primer to one of the adenine residues in the sequence. Adding a different dideoxynucleotide to each reaction tube produces a different set of fragments: ddTTP produces fragments that terminate with a T, ddGTP with a G, and ddCTP with a C. The newly synthesized chains from each sequencing reaction are then separated from the template. Finally, the mixtures from the sequencing reaction are subjected to electrophoresis in adjacent lanes on a sequencing gel, where the fragments are resolved by size. Since each nucleotide of the DNA chain is the 3′ end of one and only one fragment, the sequence of the entire DNA molecule can be read from the gel, as shown in Figure 2.

20·4 Synthesis of Both Strands Occurs Simultaneously as the DNA Template Is Unwound at the Replication Fork

In the cell, polymerizations of the leading and lagging strands are tightly coupled. The proteins involved in the synthesis of both strands are assembled into the replisome, thereby physically associating the two replication processes. The replisome contains a primosome, DNA polymerase III, and additional proteins that are required for DNA replication.

Because the template for DNA polymerase III is an extended single strand of DNA, the two strands of the parental double helix must be unwound and separated before polymerization can begin. This unwinding is accomplished primarily by a class of proteins called helicases. In *E. coli,* the helicase dnaB is required for DNA replication. Since dnaB is one of the subunits of the primosome, which in turn is part of the larger replisome, the rate of DNA unwinding is directly coupled to the rate of polymerization as the replisome moves along the chromosome.

Another protein that is part of the replisome is *single-strand binding protein* (SSB), also known as helix-destabilizing protein. SSB binds to single-stranded DNA and prevents it from folding back on itself to form double-stranded regions such as hairpin loops. The replisome contains many molecules of SSB, which exist as tetramers (subunit MW 18 000). Each tetramer that binds to DNA covers about 32 nucleotides. The binding of SSB tetramers to DNA is cooperative; that is, binding of the first tetramer facilitates binding of the second, and so on. The presence of several adjacent SSB molecules on single-stranded DNA produces an extended, relatively inflexible DNA conformation that is free of secondary structure. Because it is free of inhibiting secondary structure, single-stranded DNA coated with SSB is an ideal template for synthesis of the complementary strand during DNA replication.

A model of DNA synthesis by the replisome is shown in Figure 20·14. The helicase and the primosome are located at the head of the replication fork, followed by a DNA polymerase III holoenzyme. As the helicase unwinds the DNA, primase synthesizes RNA primers on the lagging-strand template approximately once every second. One of the two 10-subunit complexes in the holoenzyme dimer synthesizes the leading strand continuously in the $5' \rightarrow 3'$ direction while the other half of the dimer extends the RNA primers on the lagging-strand template to form Okazaki fragments. The lagging-strand template is thought to be folded back into a large loop. This configuration not only allows both template strands to have the same apparent polarity, but also allows the proteins required for the synthesis of both leading and lagging strands to be contained in a single replisome at the replication fork.

(a) The lagging-strand template is looped back through the replisome so that the syntheses of both strands move in the same direction.

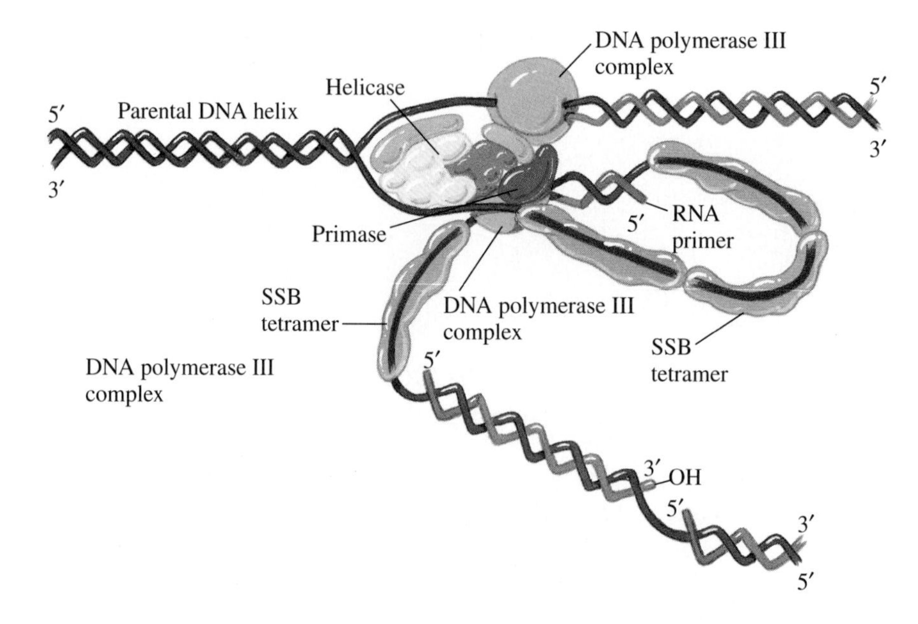

(b) As the helicase unwinds the helix, the primase synthesizes the next RNA primer of the lagging strand and the lagging-strand polymerase completes an Okazaki fragment.

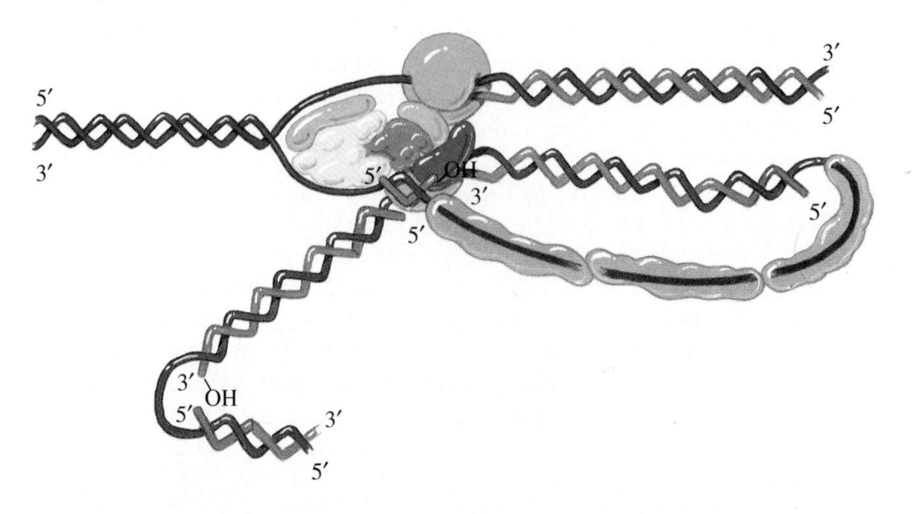

Figure 20·14
Simultaneous synthesis of leading and lagging strands at a replication fork by the replisome. The replisome contains several elements: the DNA polymerase III holoenzyme; a primosome containing primase, a helicase, and other subunits; and additional components including single-strand binding protein (SSB). One DNA polymerase III complex of the holoenzyme synthesizes the leading strand while the other polymerase complex synthesizes the lagging strand. The lagging-strand template is looped back through the replisome so that syntheses of the leading and lagging strands move in the same direction. (Illustrator: Lisa Shoemaker.)

When the replisome encounters an RNA primer of the previously synthesized Okazaki fragment, the lagging-strand template is released and rebound at the site of the newly synthesized primer. Although the two complexes of the DNA polymerase III holoenzyme are drawn in the model as equivalent, they function slightly differently: one remains firmly bound to the leading-strand template, whereas the other binds the lagging-strand template until it encounters an RNA primer, at which time it releases the template and continues synthesis at the next primer. This difference in function is probably due to slight differences in subunit composition between the two polymerase complexes. Hence, the holoenzyme is sometimes referred to as an asymmetric dimer. Note that the entire holoenzyme is extremely processive since half of it remains associated with the leading strand from the beginning of replication until termination while the other half processively synthesizes 1000 nucleotide stretches of the lagging strand.

(c) After encountering the RNA primer of an already synthesized Okazaki fragment, the lagging-strand polymerase releases the lagging-strand template while remaining bound to the replisome.

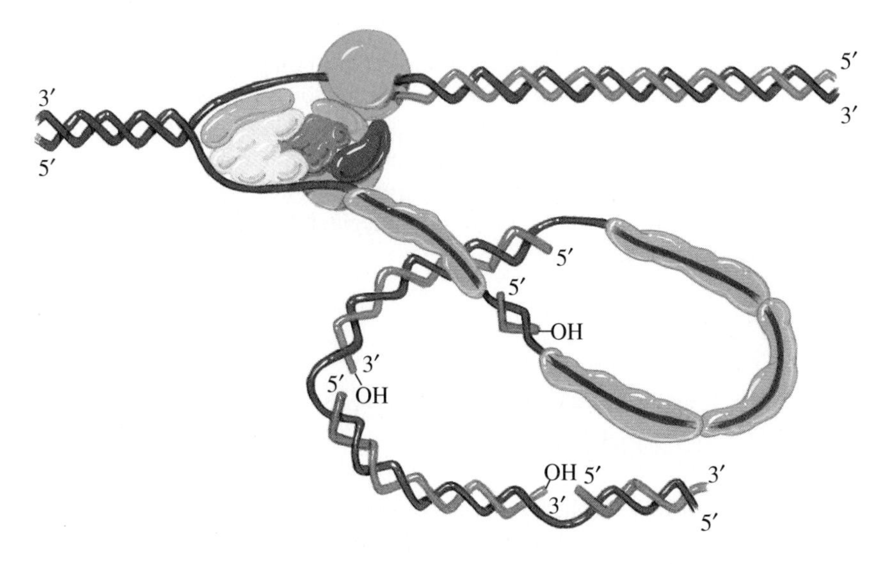

(d) The lagging-strand polymerase binds to a newly synthesized RNA primer and synthesizes another Okazaki fragment.

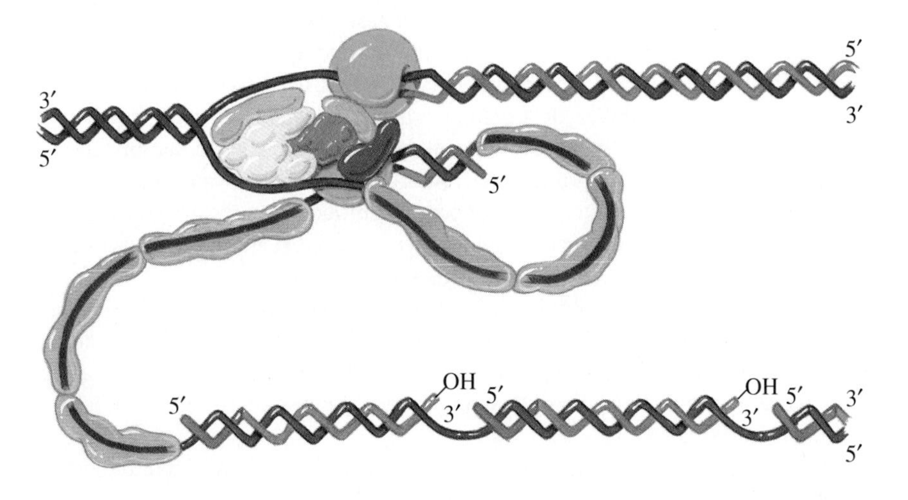

The replisome model explains how syntheses of the leading and lagging strands are coordinated. The structure of the replisome also ensures that all the components necessary for replication are available at the right time, in the right amount, and in the right place. This process of assembling a number of proteins into an interactive unit for carrying out multiple functions appears to be critical for a variety of cellular activities, including RNA synthesis, RNA processing, and protein synthesis.

20·5 Initiation and Termination of DNA Replication Require Additional Proteins

As noted earlier, DNA replication begins at a specific DNA sequence called the origin. The *E. coli* cell contains a protein (the product of the dnaA gene) that binds specifically to this region, causing local denaturation of DNA. Two replisomes assemble at this site and then proceed bidirectionally to replicate the chromosome (Figure 20·2). The initial RNA primers required for leading-strand synthesis are probably made by the primosomes at the origin.

Termination of replication in *E. coli* occurs at the termination site, a region opposite the origin on the circular chromosome. This region contains DNA sequences that are the sites of binding of a protein known as *terminator utilization substance* (tus). Tus prevents replication forks from passing through the region by inhibiting the helicase activity of the replisome. The termination site also contains DNA sequences that play a role in the separation of daughter chromosomes when DNA replication is completed.

20·6 DNA Replication in Eukaryotes Is Similar to DNA Replication in Prokaryotes

The mechanism of DNA replication in eubacteria, archaebacteria, and eukaryotes is fundamentally similar. In eukaryotes, as in *E. coli,* synthesis of the leading strand is continuous, and synthesis of the lagging strand is discontinuous. Furthermore, in both eukaryotes and bacteria, synthesis of the lagging strand is a stepwise process involving primer synthesis, Okazaki fragment synthesis, primer hydrolysis, and gap-filling polymerase activity. Eukaryotic primase, like prokaryotic primase, initiates an RNA primer approximately 10 ribonucleotide residues in length once every second on the lagging-strand template. However, because the replication fork moves more slowly in eukaryotes than in prokaryotes, each Okazaki fragment is only about 100–200 nucleotide residues in length, which is considerably shorter than the fragment length in prokaryotes. Since many of the enzymes that carry out DNA replication in eukaryotes have not been identified or characterized as fully as the enzymes in *E. coli,* the functions and sometimes the existence of eukaryotic enzymes are often inferred from those of prokaryotic enzymes.

Most eukaryotic cells contain at least four different DNA polymerases, designated α, β, γ, and δ. DNA polymerases α and δ are responsible for the chain-elongation reactions of DNA replication; DNA polymerase β is a DNA-repair enzyme found in the nucleus; and DNA polymerase γ plays a role in the replication of mitochondrial DNA.

DNA polymerase α is an tetrameric protein whose subunits have molecular weights of approximately 180 000, 75 000, 55 000, and 50 000, with some variations in size among species. The polymerase activity is contained in the subunit with a molecular weight of 180 000, and a primase activity is contained in the subunits with molecular weights of 55 000 and 50 000. In contrast to the separation of primase and polymerase activities in prokaryotes, in eukaryotes the primase and polymerase activities have not been purified as distinct proteins because the interactions between them are strong enough to withstand the relatively harsh treatments of purification. Like the DNA polymerase III holoenzyme of *E. coli,* DNA polymerase α requires an RNA primer in order to begin polymerization.

DNA replication in eukaryotic cells is extremely accurate; only one base in 10^9–10^{11} is incorrectly incorporated. This low error rate suggests that DNA replication involves an efficient proofreading step like that observed in *E. coli.*

Whereas there is only one origin in *E. coli* DNA replication, there are many origins in eukaryotic DNA replication. For example, the largest chromosome of the fruit fly *Drosophila melanogaster* contains some 6000 replication forks, which implies that there are at least 3000 origins (Figure 20·4). Eukaryotic origins are thought to be located at points where chromatin is attached to the proteins of the nuclear cytoskeleton. As eukaryotic replication forks merge with one another, replication bubbles form and increase in size until the daughter strands separate. Because of the large number of origins, the larger chromosomes of eukaryotes can still be replicated in less than one hour, even though the rate of fork movement is approximately one-twentieth the rate in prokaryotes.

The differences between eukaryotic and prokaryotic DNA replication arise not only from the larger size of the eukaryotic genome but also from the packaging of eukaryotic DNA into chromatin. The binding of DNA to histones and its packaging into nucleosomes, as described in Chapter 10, is thought to be responsible, at least in part, for the slower movement of the replication fork in eukaryotes. Eukaryotic DNA replication occurs with concomitant synthesis of histones. In each round of DNA replication, the number of histones is doubled. DNA replication and histone duplication involve different enzymes acting in different parts of the cell, yet both occur at about the same rate. It appears that existing histones remain bound to DNA during replication and that newly synthesized histones bind to DNA behind the replication fork shortly after synthesis of the new strands.

20·7 Damaged DNA Molecules Can Be Repaired

DNA is the only cellular macromolecule that can be repaired, probably because the cost to the organism of damaged DNA far outweighs the cost of energy expended in repairing the damage. Repair of other macromolecules is not profitable. For example, if a defective protein is formed as a result of an environmental insult, little is lost: the protein is simply replaced by a new, functional protein. If DNA is damaged, however, the entire organism may be in jeopardy because the instructions for the synthesis of a functional molecule may be altered. In single-celled organisms, if the damaged DNA is located in a gene encoding an essential protein, the damage may kill the organism. Even in multicellular organisms, the accumulation over time of defects in DNA can rapidly lead to a progressive loss of cellular functions or to deregulated growth such as that seen in cancer cells. DNA-repair mechanisms serve to protect not only individual cells but also subsequent generations from the effects of mutation. In single-celled organisms, whether prokaryotes or eukaryotes, DNA damage that is not repaired is passed on directly to the daughter cells following DNA replication and cell division. However, in the case of multicellular organisms, mutations can only be passed on to the next generation if they occur in the germ line. Germ-line mutations may have no noticeable effect on the organism that contains them but may have profound effects on the progeny, especially if the mutated genes are developmentally important.

Figure 20·15
Photodimerization of adjacent thymine residues. Ultraviolet light causes adjacent thymines in DNA to dimerize, thus distorting the structure of DNA. In double-stranded DNA, the thymines would be base paired with adenines on the opposite strand. For simplicity, only a single strand is shown here.

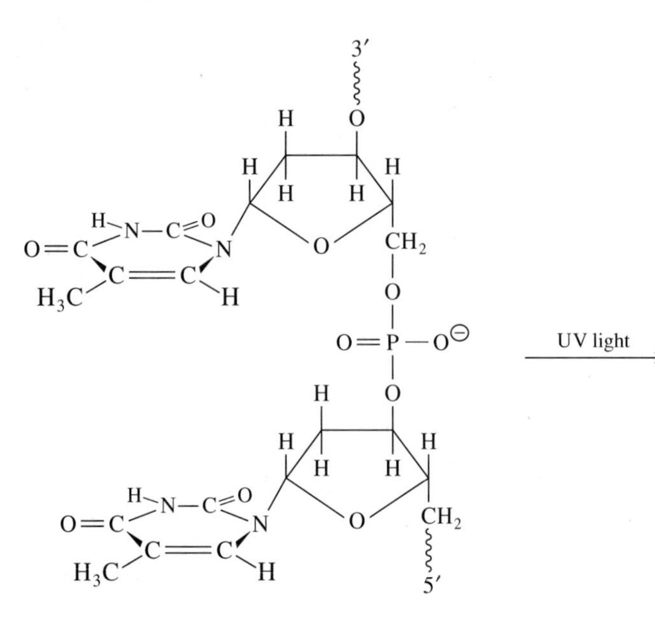

UV light

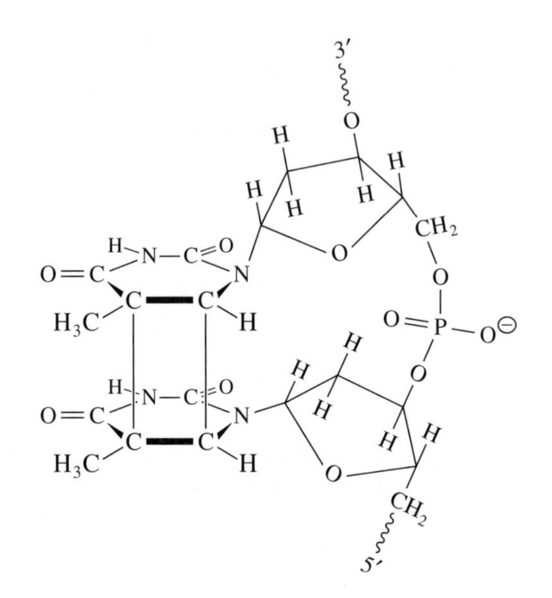

DNA is especially susceptible to damage by ultraviolet light, which causes dimerization of stacked thymines in double-helical DNA. This process is an example of photodimerization (Figure 20·15). DNA replication cannot occur in the presence of thymine dimers, probably because the dimers cause distortion of the template strand. Therefore, removal of thymine dimers is essential for survival.

Thymine dimers are repaired in all prokaryotic and eukaryotic organisms by a variety of similar processes. The simplest repair process is photoreactivation in which an enzyme known as the photoreactivating enzyme binds the distorted double helix at the site of the thymine dimer (Figure 20·16). As the DNA-enzyme complex absorbs visible light, the dimerization reaction reverses. The photoreactivating enzyme then dissociates from the repaired DNA, and normal A/T base pairs re-form.

DNA is damaged not only by ultraviolet light but also by other forms of ionizing radiation and naturally occurring chemicals, including acids and oxidizing reagents. These agents cause a variety of modifications, including methylation and deamination. DNA also undergoes spontaneous chemical modifications, such as depurination and depyrimidization (loss of heterocyclic bases).

Many of these damaged nucleotides, as well as mismatched bases that escape the proofreading mechanism of DNA polymerase, are recognized by specific repair enzymes that continually scan DNA in order to detect alterations. For example, damaged DNA can be repaired by a general excision-repair pathway whose overall features are similar in all organisms. In the first step of the pathway, an endonuclease recognizes distorted, damaged DNA and cleaves on both sides of the lesion, releasing an oligonucleotide containing 12 to 13 residues. In *E. coli,* this cleavage is performed by the UvrABC enzyme. Removal of the DNA oligonucleotide requires a helicase activity, which is often a part of the excision-repair enzyme. The result is a single-stranded gap that is analogous to the gap between Okazaki fragments after removal of the RNA primer. In the second step of the pathway, the gap is filled in by the actions of DNA polymerase I in prokaryotes and repair DNA polymerases in eukaryotes. The nick is then sealed by DNA ligase (Figure 20·17).

Another type of damage that DNA undergoes is hydrolytic deamination of adenine, cytosine, or guanine. (Because thymine does not have an amino group, it is not subject to deamination.) Spontaneous deamination of cytosine is one of the

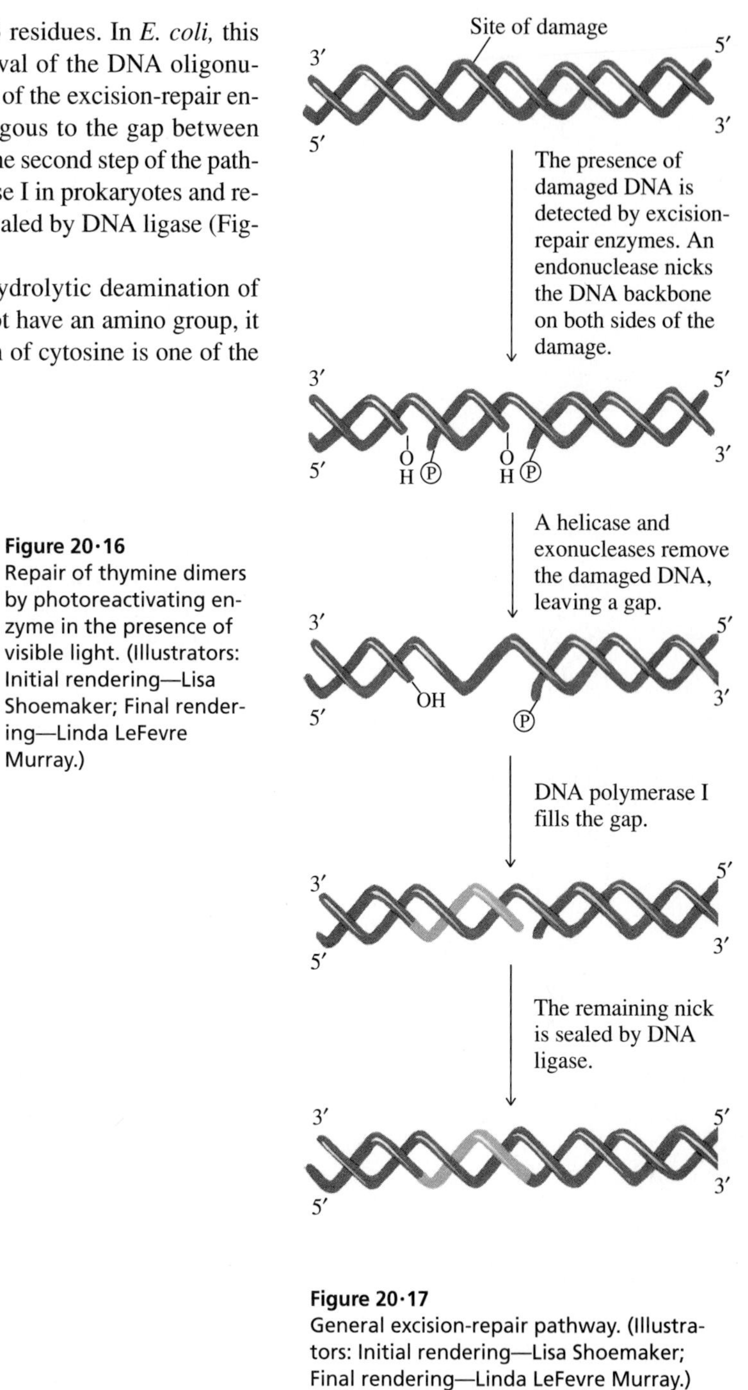

Figure 20·17
General excision-repair pathway. (Illustrators: Initial rendering—Lisa Shoemaker; Final rendering—Linda LeFevre Murray.)

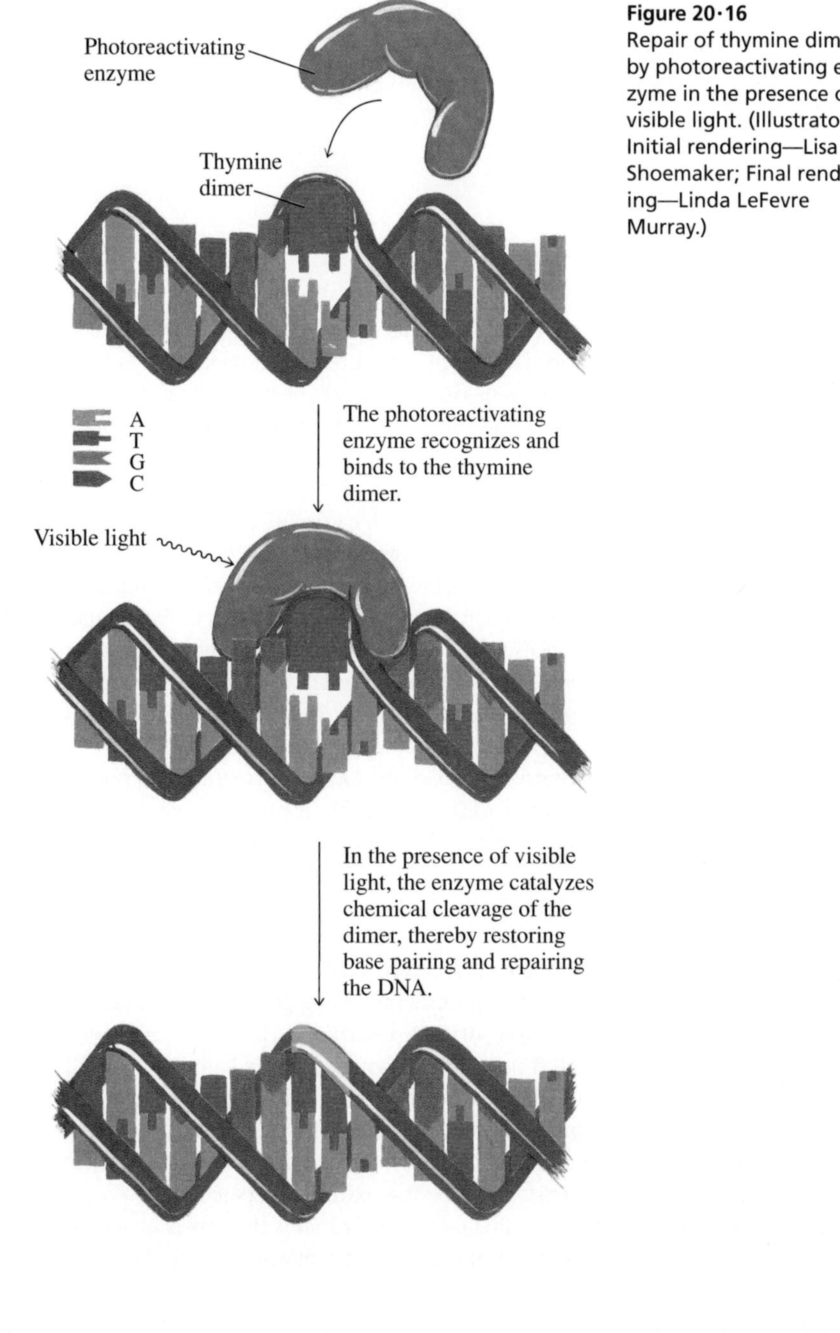

Figure 20·16
Repair of thymine dimers by photoreactivating enzyme in the presence of visible light. (Illustrators: Initial rendering—Lisa Shoemaker; Final rendering—Linda LeFevre Murray.)

most common DNA injuries (Figure 20·18). The deaminated bases form incorrect base pairs, and, if not removed, result in the addition of wrong bases during the next round of replication.

Enzymes called DNA glycosylases catalyze the removal of deaminated bases and other modified nucleotides by hydrolysis of the glycosidic bonds linking the modified bases to the sugars. For example, by hydrolyzing the *N*-glycosidic bond, uracil *N*-glycosylase removes a uracil produced by deamination (Figure 20·19). An endonuclease that recognizes the site at which the pyrimidine is absent then removes the deoxyribose phosphate, leaving a single-nucleotide gap in the duplex DNA. Excision-repair enzymes with exonuclease activity often extend this gap. In prokaryotes, DNA polymerase I then binds to the exposed 3′ end and fills in the gap. Lastly, the strand is sealed by DNA ligase.

Whereas deaminations of adenine and guanine are rare, deamination of cytosine is fairly common and would give rise to large numbers of mutations were it not for the replacement of uracil with thymine in DNA. If uracil were normally found in DNA, as it is in RNA, it would be impossible to distinguish between a correct uracil and uracil arising from deamination of cytosine. However, since uracil is not one of the bases in DNA, damage arising from deamination can be recognized and repaired. Thus, the presence of thymine in DNA increases the stability of genetic information.

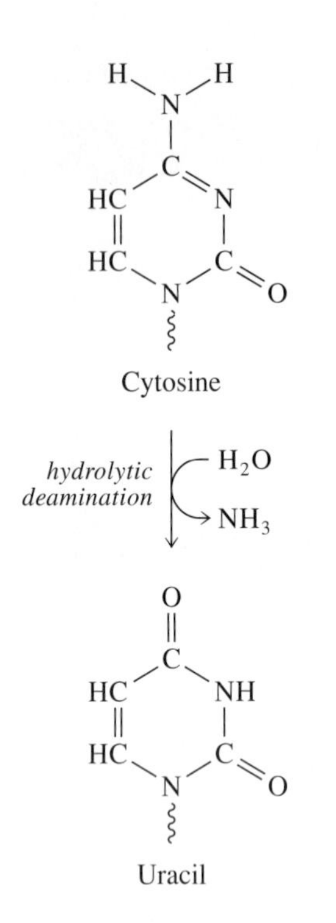

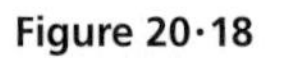

Figure 20·18
Hydrolytic deamination of cytosine. Deamination of cytosine produces uracil, which pairs with adenine rather than guanine.

Summary

In DNA replication, each strand of DNA serves as the template for synthesis of a complementary daughter strand. DNA replication begins at special sites termed origins and continues in both directions from the origin or origins to a termination site where replication stops. After one round of replication, each of the daughter molecules contains one strand from the parent and one newly synthesized strand. Thus, DNA replication is semiconservative.

Synthesis and repair of DNA are carried out by a family of enzymes known as DNA polymerases. Replication polymerases are located at replication forks and are part of a complex of proteins known as the replisome. In *E. coli,* the replication polymerase is DNA polymerase III, and in eukaryotic cells the main replication polymerase is DNA polymerase α or DNA polymerase δ. All DNA polymerases require a template and a primer in order to catalyze a reaction in which nucleoside triphosphates are the substrates for the addition of monomers to the 3′ end of a growing chain.

In addition to $5' \rightarrow 3'$ polymerization activity, DNA polymerase III contains a $3' \rightarrow 5'$ exonuclease that aids in proofreading the DNA. The enzyme recognizes the distortion caused by an improperly paired base and removes the mismatched nucleotide.

In DNA replication, one of the newly synthesized strands is called the leading strand; it is synthesized continuously in the $5' \rightarrow 3'$ direction. The other newly synthesized strand is called the lagging strand; it is made discontinuously in the $5' \rightarrow 3'$ direction. Lagging-strand synthesis requires small RNA primers, which are made by the primosome at the replication fork. Discontinuous pieces of DNA called Okazaki fragments are extended from the RNA primers by DNA polymerase III. In *E. coli,* each Okazaki fragment is approximately 1000 base pairs long.

Okazaki fragments are joined by the combined action of DNA polymerase I and DNA ligase in a repair reaction. DNA polymerase I contains a $5' \rightarrow 3'$ exonuclease that removes RNA primers from the Okazaki fragments and a $5' \rightarrow 3'$ polymerase that adds dNTPs to the gaps that remain after removal of the primers. DNA ligase then catalyzes the formation of phosphodiesters between the fragments.

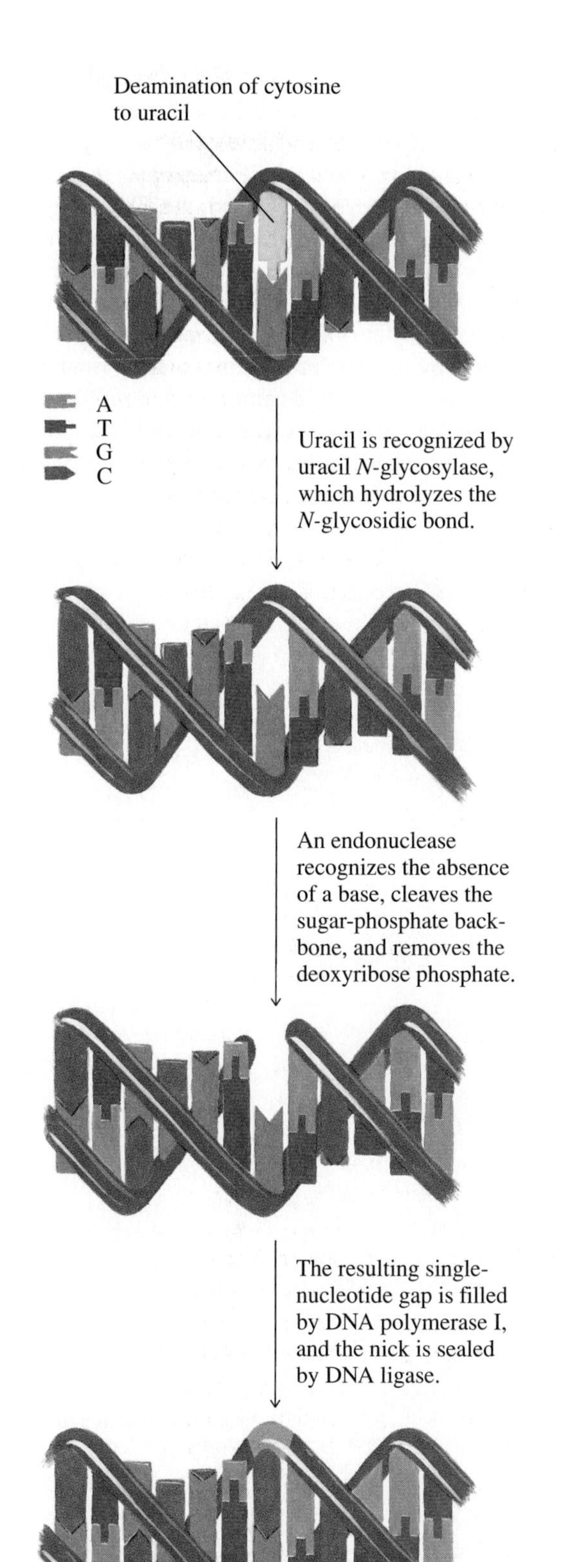

Figure 20·19
Repair of damage resulting from the deamination of cytosine. (Illustrators: Initial rendering—Lisa Shoemaker; Final rendering—Linda LeFevre Murray.)

The syntheses of the leading and lagging strands are tightly coupled in vivo, in part because the proteins required for the synthesis of both strands are contained in a single replisome. The DNA polymerase III holoenzyme, which is part of the replisome, is a dimer consisting of two polymerase complexes. One complex of the dimer is responsible for synthesis of the leading strand; the other complex of the dimer is responsible for synthesis of the lagging strand. The replisome also contains the primase, a helicase that is associated with the primase, and many molecules of single-strand binding protein.

Several enzymes are responsible for repairing damaged DNA. Formation of thymine dimers caused by ultraviolet light is repaired by photoreactivating enzymes or excision-repair enzymes. Deamination is repaired by base-specific *N*-glycosylases that hydrolyze the *N*-glycosidic bonds linking the deaminated bases to sugars. The sugar residue is then removed by an endonuclease, and the resulting gap in the DNA is repaired by the combined action of DNA polymerase I and DNA ligase.

Selected Readings

DNA Replication

Alberts, B. M. (1984). The DNA enzymology of protein machines. *Cold Spring Harbor Symp. Quant. Biol.* 49:1–12.

Alberts, B. M. (1985). Protein machines mediate the basic genetic processes. *Trends Genet.* 1:26–30.

Bramhill, D., and Kornberg, A. (1988). A model for initiation at origins of DNA replication. *Cell* 54:915–918.

Fersht, A. R. (1984). Basis of biological specificity. *Trends Biochem. Sci.* 9:145–147.

Harland, R. (1981). Initiation of DNA replication in eukaryotic chromosomes. *Trends Biochem. Sci.* 6:71–74.

Kornberg, A., and Baker, T. (1991). *DNA Replication.* 2nd ed. (New York: W. H. Freeman and Company).

Kornberg, A. (1984). DNA replication. *Trends Biochem. Sci.* 9:122–124.

Kuempel, P. L., Pelletier, A. J., and Hill, T. M. (1989). Tus and the terminators: the arrest of replication in prokaryotes. *Cell* 59:581–583.

Loeb, L. A., Liu, P. K., and Fry, M. (1986). DNA polymerase-α: enzymology, function, fidelity, and mutagenesis. *Prog. Nucleic Acid Res. Mol. Biol.* 33:57–110.

Lohman, T. M., Bujalowski, W., and Overman, L. B. (1988). *E. coli* single strand binding protein: a new look at helix-destabilizing proteins. *Trends Biochem. Sci.* 13:250–255.

Maki, S., and Kornberg, A. (1988). DNA polymerase III holoenzyme of *Escherichia coli:* III. distinctive processive polymerases reconstituted from purified subunits. *J. Biol. Chem.* 263:6561–6569.

Marians, K. J. (1984). Enzymology of DNA in replication in prokaryotes. *Crit. Rev. Biochem.* 17:153–215.

Marians, K. J., Minden, J. S., and Parada, C. (1986). Replication of superhelical DNAs *in Vitro. Prog. Nucleic Acid Res. Mol. Biol.* 33:111–140.

McHenry, C. S. (1988). DNA polymerase III holoenzyme of *Escherichia coli. Annu. Rev. Biochem* 57:519–550.

McMacken, R., Silver, L., and Georgopoulos, C. (1987). DNA replication. In Escherichia coli *and* Salmonella typhimurium: *Cellular and Molecular Biology,* F. C. Neidhardt, ed. (Washington, DC: American Society for Microbiology), pp. 564–612.

Sekimizu, K., Bramhill, D., and Kornberg, A. (1988). Sequential early stages in the *in Vitro* initiation of replication at the origin of the *Escherichia coli* chromosome. *J. Biol. Chem.* 263:7124–7130.

DNA Repair

Echols, H., and Goodman, M. F. (1991). Fidelity mechanisms in DNA replication. *Annu. Rev. Biochem.* 60:477–511.

Haseltine, W. A. (1983). Ultraviolet light repair and mutagenesis revisited. *Cell* 33:13–17.

Lindahl, T. (1982). DNA repair enzymes. *Annu. Rev. Biochem.* 51:61–87.

Walker, G. C. (1985). Inducible DNA repair systems. *Annu. Rev. Biochem.* 54:425–457.

21

Transcription and the Processing of RNA

Based on studies of the bread mold *Neurospora crassa,* George Beadle and Edward Tatum proposed in the 1940s that a single unit of heredity, or gene, in some way directed the production of a single enzyme. This relationship between genes and proteins was demonstrated in 1956 when hemoglobin from patients with the hereditary disease sickle cell anemia was shown to differ from normal hemoglobin by the replacement of a single amino acid (Section 4·19). Observations showed that genetic changes can manifest themselves as changes in the primary structure of a protein. We now know that information in the genome specifies the primary structure of every protein in an organism.

For the purposes of this chapter, we shall define a *gene* as a DNA sequence that is transcribed. In the genomes of most prokaryotes, there are approximately 2000 genes. Most of these are *housekeeping genes,* genes that encode proteins or RNA molecules that are essential for the normal activities of all living cells. For example, the enzymes involved in glycolysis and in the biosynthesis of macromolecules are encoded by such housekeeping genes, as are transfer RNA and ribosomal RNA. The number of housekeeping genes in unicellular eukaryotes, such as yeast and some algae, is similar to the number in prokaryotes.

In addition to housekeeping genes, all cells contain genes that are expressed only in special circumstances, such as during cell division. Multicellular organisms also contain genes that are expressed only in certain types of cells. For example, all cells in a maple tree contain the genes for the enzymes that make chlorophyll, but these genes are only expressed in cells that are exposed to light, such as cells on the surface of a leaf. Similarly, all cells in mammals contain insulin genes, but only certain pancreatic cells produce insulin. The total number of genes in multicellular eukaryotes is still not known with precision, but it is thought that insects contain approximately 7000 genes and mammals between 20 000 and 50 000 genes.

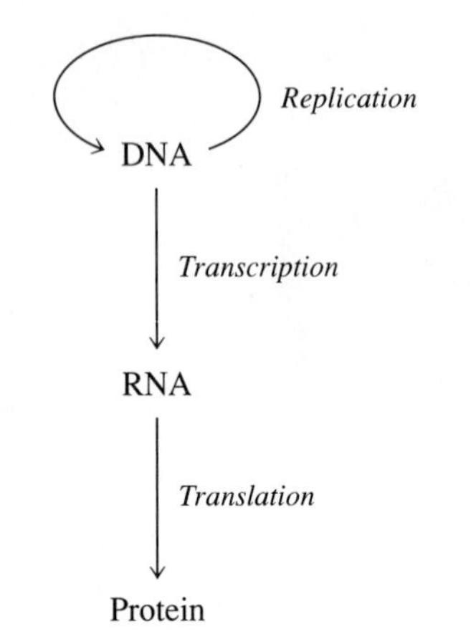

Figure 21·1
Central dogma of molecular biology. The flow of biological information is from DNA to RNA to protein. Genetic information is maintained from generation to generation by the replication of DNA.

In this chapter and the next, we shall examine how the information stored in DNA directs the synthesis of proteins. This process, which is summarized in Figure 21·1, has become known as the *central dogma* of molecular biology. In this chapter, we shall describe *transcription,* the process in which information stored in DNA is transferred to RNA and thus made available for protein synthesis, and *RNA processing,* the way in which RNA molecules are posttranscriptionally modified. In the next chapter, we shall examine *translation,* the process in which the information in mRNA molecules is interpreted and the appropriate proteins synthesized.

In the present chapter, we shall also briefly examine how gene expression is regulated. Although there are many steps at which the production of specific proteins can be regulated, one of the most important is initiation of transcription. We shall introduce the basic principles of transcriptional regulation, one of the most exciting and complex areas of biochemistry.

21·1 Several Classes of RNA Molecules Participate in the Transfer of Genetic Information from DNA to Protein

The transfer of information from DNA to protein requires the participation of several major classes of RNA molecules. *Transfer RNA* (tRNA) carries amino acids to the translation machinery. *Ribosomal RNA* (rRNA) makes up much of the ribosome. A third class of RNA molecules is *messenger RNA* (mRNA), whose discovery is due largely to the work of François Jacob, Jacques Monod, and their collaborators at the Pasteur Institute in Paris. In the early 1960s, these researchers showed that ribosomes participate in the synthesis of proteins by translating unstable RNA molecules (mRNA). They also made the important discovery that the sequence of an mRNA molecule is complementary to a segment of one of the strands of DNA. Finally, a fourth major class of RNA contains a variety of small RNA molecules that participate in various metabolic events, including RNA processing. Many of these small RNA molecules have catalytic activity.

A large percentage of the total RNA in the cell is rRNA, and only a small percentage is mRNA. But if we compare the rates at which the cell synthesizes RNA rather than the steady-state levels of RNA, we see a different picture (Table 21·1). Even though mRNA accounts for only 3% of total RNA in *Escherichia coli,* the *E. coli* cell devotes almost one-third of its capacity for RNA synthesis to the production of mRNA. In fact, this value may increase to about 60% if the bacterium is growing slowly and does not need to replace ribosomes and tRNA. The discrepancy between steady-state levels of various RNA molecules and the rates at which they are synthesized can be explained by the differing stabilities of the various RNA molecules: rRNA and tRNA molecules are extremely stable, whereas mRNA is rapidly degraded after translation. In bacterial cells, half of all newly synthesized mRNA is degraded by nucleases within three minutes. In eukaryotes, the average half-life of mRNA is about ten times longer. The relatively high stability of eukaryotic mRNA is a result of processing and modification events that prevent eukaryotic mRNA from being degraded during its transport from the nucleus, where transcription occurs, to the cytoplasm, where eukaryotic translation occurs.

Table 21·1 The RNA content of an *E. coli* cell

Type	Steady-state level	Synthetic capacity[3]
rRNA	83%	58%
tRNA	14%	10%
mRNA	3%	32%
RNA primers[1]	<1%	<1%
Other RNA molecules[2]	<1%	<1%

[1] RNA primers are those used in DNA replication; these are not synthesized by RNA polymerase.
[2] Other RNA molecules include several RNA enzymes, such as the RNA component of RNase P.
[3] Relative amount of each type of RNA being synthesized at any instant.
[Adapted from Bremer, H., Dennis, P. P. (1987). Modulation of chemical composition and other parameters of the cell by growth rate. In: Escherichia coli and Salmonella typhimurium: Cellular and Molecular Biology, Vol. 2, F. C. Neidhardt, ed. (Washington, DC: American Society for Microbiology), pp. 1527–1542.]

21·2 RNA Polymerase Catalyzes RNA Synthesis

About the time that mRNA was identified, researchers in several laboratories independently discovered an enzyme that catalyzed the synthesis of RNA if provided with ATP, UTP, GTP, CTP, and a template DNA molecule. The newly discovered enzyme was RNA polymerase. This enzyme catalyzes DNA-directed RNA synthesis, or transcription.

Although RNA polymerase was initially identified by its ability to catalyze polymerization of ribonucleoside triphosphates, further study of the enzyme revealed that it does much more. RNA polymerase is the core of a transcription complex that transcribes DNA (just as DNA polymerase is the core of a replication complex). During transcription, RNA polymerase partially unwinds the template DNA and synthesizes an RNA primer for further elongation. It also catalyzes processive elongation of the RNA chain while continuing to unwind and rewind the DNA. Finally, after synthesizing an RNA molecule, RNA polymerase terminates transcription.

Although the composition and assembly of the transcription complex varies considerably between different organisms, all transcription complexes catalyze essentially the same types of reactions. Therefore, we shall introduce the general process of transcription by discussing the reactions catalyzed by the well-characterized transcription complex in *E. coli*.

Table 21·2 Subunits of *E. coli* RNA polymerase holoenzyme

Subunit	MW
β'[1]	155 613
β[1]	150 618
σ	70 263[2]
α	36 512
ω	10 105

[1] The β and β' subunits are unrelated despite the similarity of their names.
[2] The molecular weight given is for the σ subunit found in the most common form of the holoenzyme.

A. RNA Polymerase Is a Multimeric Protein

Free RNA polymerase is isolated from *E. coli* cells as a multimeric protein composed of six subunits of five different types (Table 21·2). Five of these subunits combine in a stoichiometry of $\alpha_2\beta\beta'\sigma$ to form the holoenzyme; the ω subunit is usually present in less than stoichiometric amounts and may or may not be part of the functional holoenzyme. The subunits of the holoenzyme function together to carry out the reactions of transcription: the β' subunit contributes to DNA binding, the β subunit contains part of the active site of the enzyme, and the σ subunit plays an important role in the initiation of transcription. The role of the α subunit is unknown. Several different types of σ subunits exist within any one bacterium. The major form of the holoenzyme found in *E. coli* contains the subunit σ^{70} (MW 70 263). This holoenzyme is irregularly shaped, with a groove at one end (Figure 21·2). The groove is large enough to accommodate about 16 base pairs of double-stranded B-DNA.

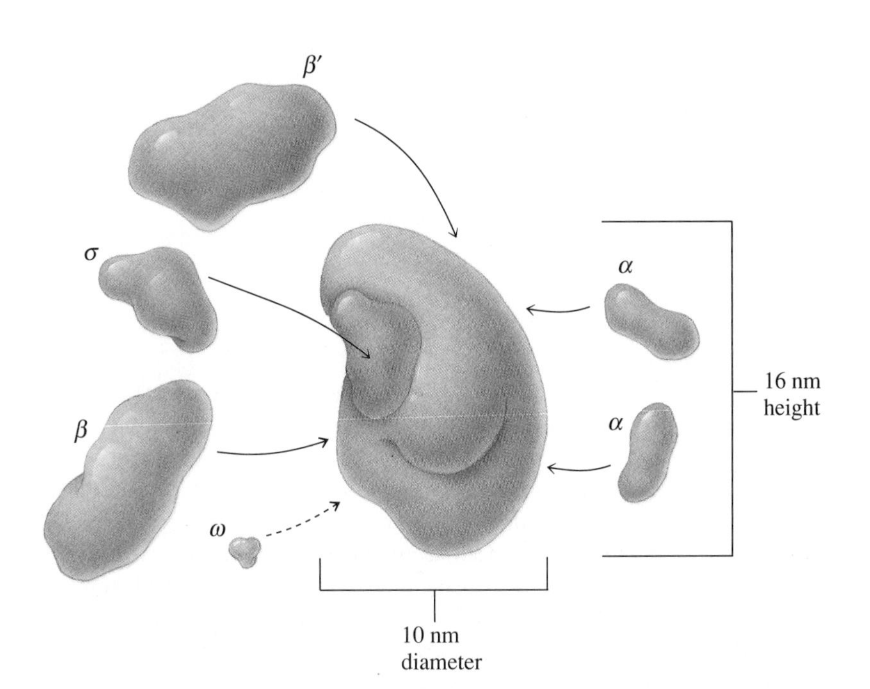

Figure 21·2
Subunit composition of an *E. coli* RNA polymerase holoenzyme. The holoenzyme is composed of four different subunits with the stoichiometry $\alpha_2\beta\beta'\sigma^{70}$; ω is present in less than stoichiometric amounts. The holoenzyme is depicted here and in subsequent diagrams as the core enzyme plus σ because σ dissociates from the holoenzyme after transcription is initiated. The overall shape of RNA polymerase is oblate, with a groove whose structure resembles the DNA-binding site of DNA polymerase I. (Illustrator: Lisa Shoemaker.) [Adapted from Darst, S. A., Kubalek, E. W., and Kornberg, R. D. (1989). Three-dimensional structure of *Escherichia coli* RNA polymerase holoenzyme determined by electron crystallography. *Nature* 340:730–732.]

The isolated holoenzyme can be further fractionated so that the σ subunit dissociates, leaving a core polymerase composed only of the α, β, and β' subunits ($\alpha_2\beta\beta'$). The weaker association of σ with the core polymerase reflects the role of σ in the cell; this subunit dissociates from the core polymerase after the initiation of transcription.

Archaebacterial and eukaryotic RNA polymerases, like eubacterial RNA polymerases, are multisubunit enzymes whose components function together to carry out the reactions of transcription. But whereas prokaryotic cells contain only one form of RNA polymerase, eukaryotic cells contain three different nuclear RNA polymerases. In eukaryotes, RNA polymerase I transcribes the genes for large rRNA molecules, RNA polymerase II transcribes the genes that encode proteins, and RNA polymerase III transcribes the genes that encode many small stable RNA molecules (including tRNA and small ribosomal RNA). There is considerable amino acid sequence homology between RNA polymerase subunits of distantly related eubacteria, archaebacteria, and eukaryotes, although the number and mass of the subunits may be quite different. This sequence homology suggests that all RNA polymerases evolved from a common ancestor.

B. Chain Elongation Is a Nucleotidyl-Group Transfer Reaction

RNA polymerase catalyzes chain elongation by a mechanism almost identical to that of DNA polymerase (Section 20·2A). Part of the growing RNA chain is base-paired to the DNA template strand, and incoming ribonucleoside triphosphates are tested in the active site of the polymerase for correct hydrogen bonding to the next unpaired nucleotide of the template strand. When the incoming nucleotide forms correct hydrogen bonds, RNA polymerase catalyzes nucleophilic attack by the 3′-hydroxyl group of the growing RNA chain on the α-phosphorus of the incoming ribonucleoside triphosphate. The result of this nucleotidyl-group transfer reaction is the formation of a new phosphodiester linkage and the release of pyrophosphate (Figure 21·3). Like DNA polymerase III, RNA polymerase catalyzes polymerization in the 5′→3′ direction and is highly processive when it is part of a complex that is bound to DNA. RNA polymerase differs from DNA polymerase in that RNA polymerase uses only ribonucleoside triphosphates (UTP, GTP, ATP, and CTP), whereas DNA polymerase uses only deoxyribonucleoside triphosphates (dTTP, dGTP, dATP, and dCTP).

The overall reaction of RNA synthesis can be summarized as follows:

$$\text{RNA}_n\text{–OH} + \text{NTP} \longrightarrow \text{RNA}_{n+1}\text{–OH} + \text{PP}_i \qquad \Delta G^{\circ\prime} = -2.2 \text{ kcal mol}^{-1} \tag{21·1}$$

Like the overall reaction of DNA synthesis, this reaction is thermodynamically assisted by the subsequent hydrolysis of pyrophosphate inside the cell. Thus, in vivo two phosphoanhydride linkages are ultimately disrupted for every nucleotide added to the growing chain.

The rate of transcription is rather slow compared to the rate of DNA replication. In *E. coli,* even though the flux of nucleotides into the active site of RNA polymerase is estimated to be greater than 10^4 molecules per second, the rate of transcription ranges from 30 to 85 nucleotides per second, which is less than one-tenth the rate of DNA replication. Apparently, most collisions between incoming ribonucleoside triphosphates and the active site are nonproductive, suggesting that RNA polymerase will form a new phosphodiester linkage only if the incoming ribonucleoside triphosphate fits the active site of the enzyme precisely. A precise fit into the active site requires base stacking and hydrogen-bond formation between the incoming ribonucleoside triphosphate and the template nucleotide.

Despite the requirement for a high-precision fit, RNA polymerase does make mistakes. The error rate of RNA synthesis is 10^{-6} (corresponding to one mistake for every one million nucleotides incorporated). This rate is higher than the overall error rate of DNA synthesis because RNA polymerase does not possess an exonuclease proofreading activity. Extreme precision in DNA replication is necessary to prevent mutations, which would be passed on to progeny. Accuracy in RNA synthesis is not as crucial to survival.

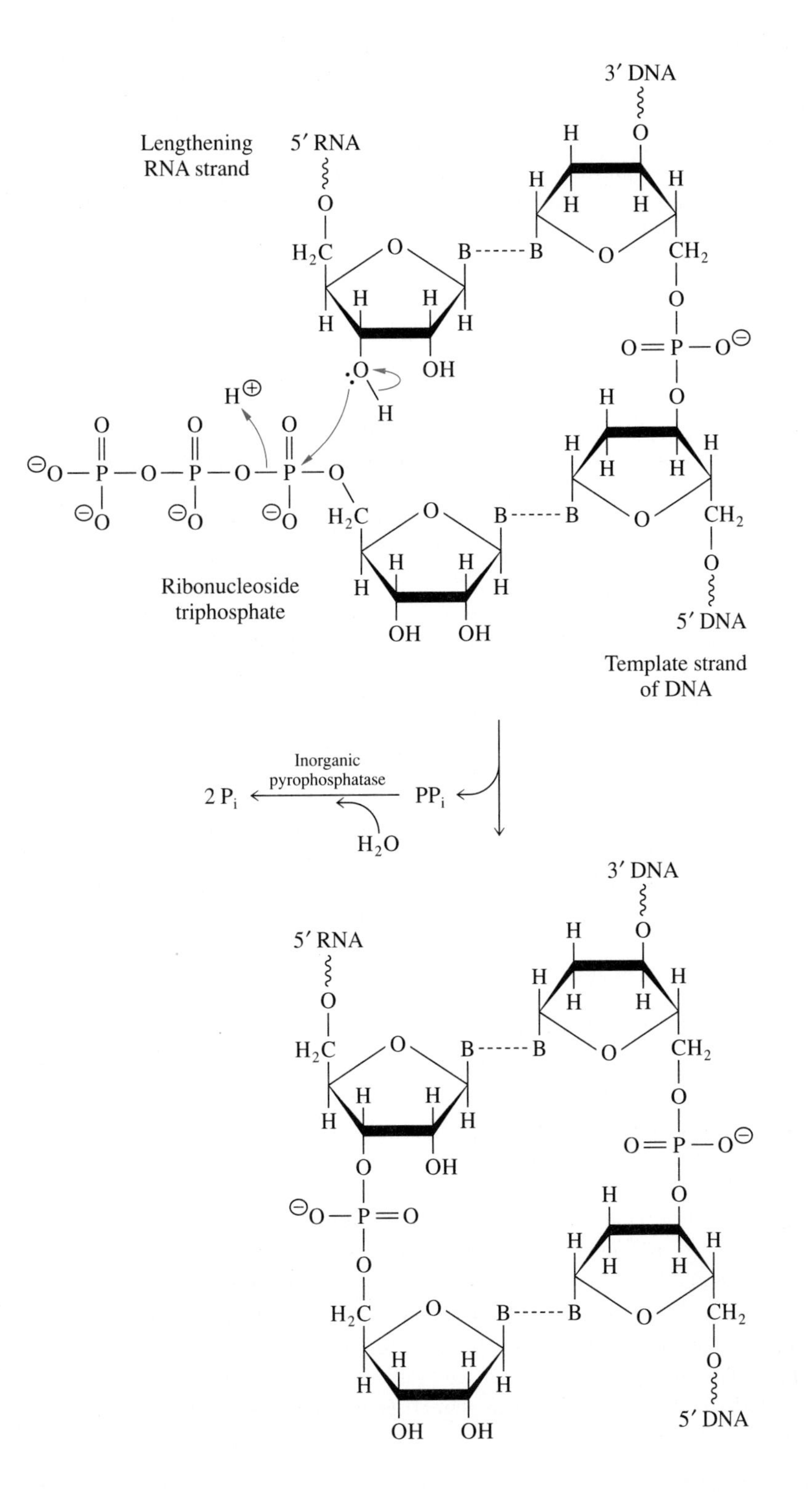

Figure 21·3
Reaction catalyzed by RNA polymerase. When an incoming ribonucleoside triphosphate correctly pairs with the next unpaired nucleotide on the DNA template strand, RNA polymerase catalyzes nucleophilic attack by the 3′-hydroxyl group of the lengthening RNA strand on the α-phosphorus of the incoming ribonucleoside triphosphate. As a result, a phosphodiester is formed and pyrophosphate is released. The subsequent hydrolysis of pyrophosphate catalyzed by inorganic pyrophosphatase provides additional thermodynamic driving force for the reaction. (For simplicity, in this and in subsequent figures, the bases are shown as B and hydrogen bonding between bases is designated by a single dashed line.)

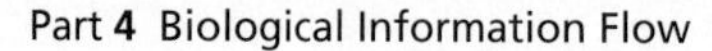

21·3 Transcription Is Initiated at Promoter Sequences

The elongation reactions of RNA synthesis are preceded by a separate and distinct initiation step in which a transcription complex is assembled at an initiation site and a ribonucleotide primer is synthesized. As in DNA replication, in transcription the assembly of an initiation complex requires initiation factors that can recognize and bind to the initiation site.

The sites of transcription initiation are called *promoters*. In bacteria, several genes are often cotranscribed from a single promoter; such a transcription unit is called an *operon*. In eukaryotic cells, each gene usually has its own promoter. There are hundreds of promoters in bacterial cells and thousands in eukaryotic cells.

The frequency of transcription initiation is usually related to the need for a gene product. For example, in cells that are dividing frequently, the genes for ribosomal RNA are usually transcribed frequently; every few seconds a new transcription complex begins transcription at the same promoter sequence of the rRNA gene. This process gives rise to structures such as those shown in Figure 21·4, where transcripts of increasing length are arrayed along the gene as many RNA polymerases transcribe the gene at the same time. In contrast, some bacterial genes may be transcribed only once every two generations. In these cases, initiation may only occur once every few hours.

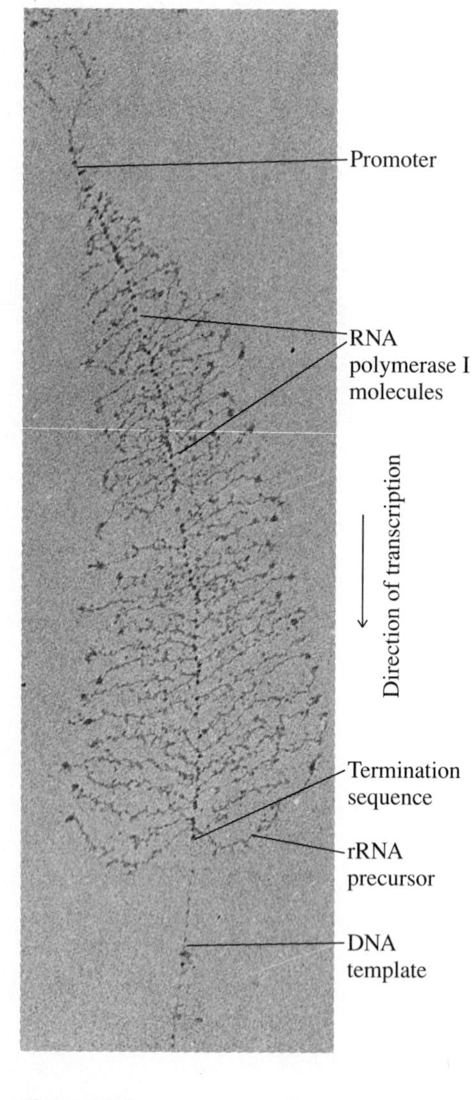

Figure 21·4
Electron micrograph of a single active ribosomal RNA gene isolated from an extrachromosomal nucleolus of an oocyte of the spotted newt, *Notophthalmus viridescens.* Because transcription initiation is so frequent, there are many RNA polymerases actively transcribing the gene at the same time. (Courtesy of Oscar L. Miller and Barbara R. Beatty.)

A. The Transcription Complex Is Assembled at Promoter Sequences

In both prokaryotes and eukaryotes, a competent transcription complex forms when the promoter sequence in DNA is recognized by one or more proteins that bind to the promoter and also to RNA polymerase. These DNA-binding proteins thus attract RNA polymerase to the promoter site. In bacteria, the σ subunit of RNA polymerase is required for promoter recognition and formation of the transcription complex. In eukaryotes, a variety of accessory proteins are needed for the assembly of the transcription complex. These eukaryotic proteins are collectively known as *transcription factors.*

The nucleotide sequence of a promoter is one of the most important factors affecting the rate of transcription of a gene. Soon after the development of DNA sequencing technology, many different promoters were examined. The start site, the point at which transcription actually begins, was identified and the region upstream of this site was sequenced to find out if the promoter sequences of different genes were similar. This analysis revealed a common pattern of the type which we now refer to as a *consensus sequence,* a sequence composed of the nucleotides found most often in each position. The consensus sequence of the most common type of promoter in *E. coli* is shown in Figure 21·5. This promoter is bipartite: there is a consensus sequence of TATAAT in the −10 region (relative to the transcription start site) and a consensus sequence of TTGACA in the −35 region. The promoter sequences in Figure 21·5 are all taken from the top strand of the gene since convention dictates that nucleotide sequences are written in the 5′→3′ direction (Box 21·1). However, since the actual binding site is double-stranded DNA, the complementary sequence is equally important.

The −10 region is known as a *TATA box,* and the −35 region is simply referred to as the *−35 region*. Together, the two regions define the promoter for the *E. coli* holoenzyme. These DNA sequences are recognized by σ^{70}, the most common σ subunit in *E. coli* cells. Not all *E. coli* promoters, however, have this particular consensus sequence. Some σ subunits in *E. coli* recognize and bind to promoters with quite different consensus sequences (Table 21·3). Furthermore, in other species of bacteria the promoter consensus sequences may be very different.

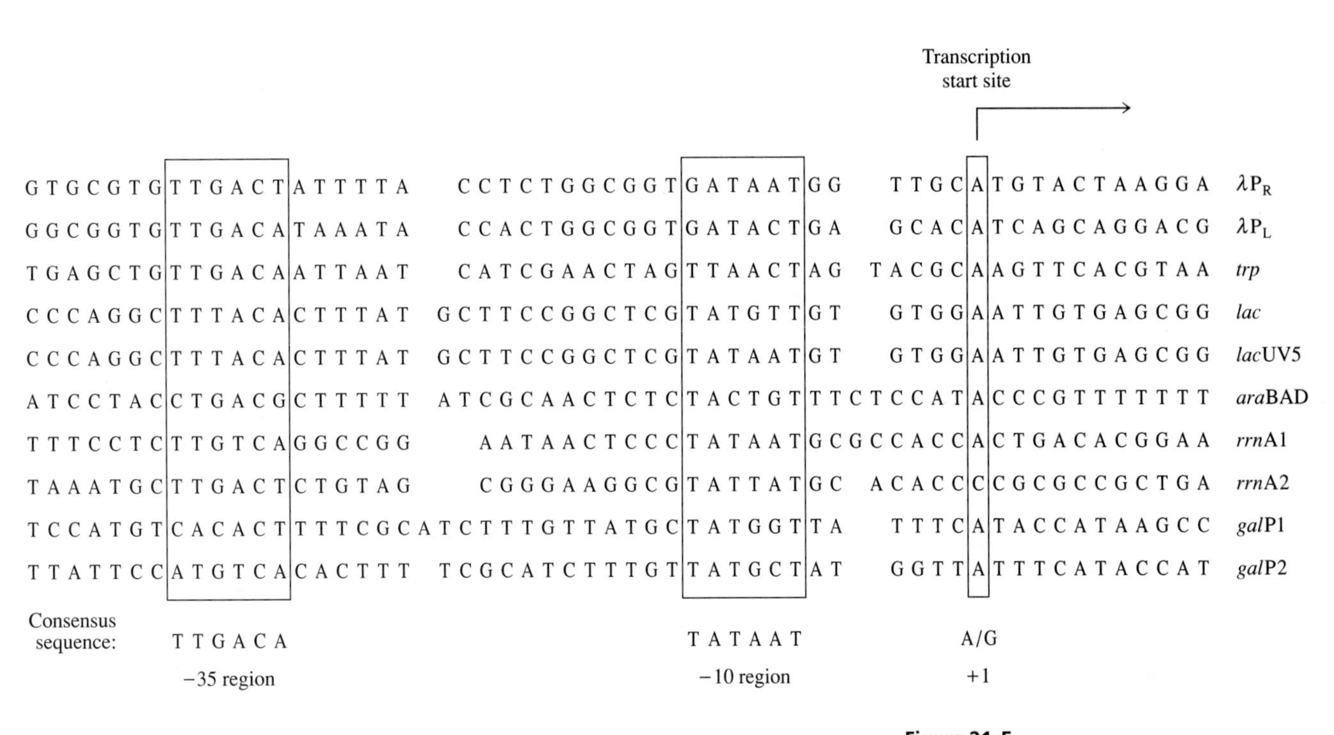

Figure 21·5
Promoter sequences from ten bacteriophage and bacterial genes. The nucleotide sequences are aligned so that their +1, −10, and −35 regions are in register. Notice the degree of sequence variation at each position. The consensus sequence has been derived from a much larger sample of promoters than is shown here. In the larger sample, for example, either purine can be found at the start site.

Even among those *E. coli* genes recognized by the σ^{70} subunit, very few have promoters that are exact matches to the consensus sequence shown in Figure 21·5. This sequence is called a consensus sequence precisely because it refers to the most common nucleotides at given positions and not to an exact sequence. In some cases, the match is quite poor, with G's and C's present at positions normally occupied by A's and T's. Such weak promoters are usually associated with genes that are transcribed infrequently. Other promoter sequences, such as the promoters for the rRNA operons, resemble the consensus sequence quite closely. These genes are generally transcribed very efficiently. Observations such as these suggest that the consensus sequence describes the most efficient promoter sequence for the RNA polymerase holoenzyme.

The promoter sequence of each gene has likely been optimized by natural selection to fit the requirements of the cell. An inefficient promoter is ideal for a gene whose product is not needed in large quantities, whereas an efficient promoter is necessary if large amounts of the gene product must be produced.

Table 21·3 *E. coli* σ factors and their consensus sequences

Factor	Genes transcribed	Consensus sequence −35	−10
σ^{70}	Housekeeping	TTGACA	TATAAT
σ^{32}	Heat shock	CTTGAA	CCCCAT-TA*
σ^{60}	Nitrogen metabolism	none	CTGGCAC-----TTGCA*
σ^{gp55}	Bacteriophage	none	TATAAATA

*Dashes indicate conserved spacing.

Box 21·1 By Convention, Genes Have a 5′→3′ Orientation

In Section 10·2, we introduced the convention that nucleic acid sequences are always written in the 5′→3′ direction. When a sequence of double-stranded DNA is displayed, the sequence of the top strand is written 5′→3′, and the sequence of the bottom, antiparallel strand is written 3′→5′.

Since our operational definition of a gene is a DNA sequence that is transcribed, we consider a gene to begin at the point where transcription starts (designated +1) and to end at the point where transcription terminates. Since RNA polymerization, like DNA polymerization, proceeds in the 5′→3′ direction, transcription of a gene begins at the 5′ end of RNA and terminates at the 3′ end. Consequently, in accordance with the convention established for writing nucleic acid sequences, the start site of a gene is presented on the left of a diagram and the termination site on the right. The top strand is the coding strand and the bottom strand is the template strand (Figure 1). Moving along a gene in the 5′→3′ direction is described as moving downstream; the 3′→5′ direction, upstream. Note that transcription proceeds in the 5′→3′ direction, but that the template strand is copied from the 3′ end to the 5′ end. Also note that the transcribed mRNA is identical in sequence to the coding strand except that uracils have replaced thymines.

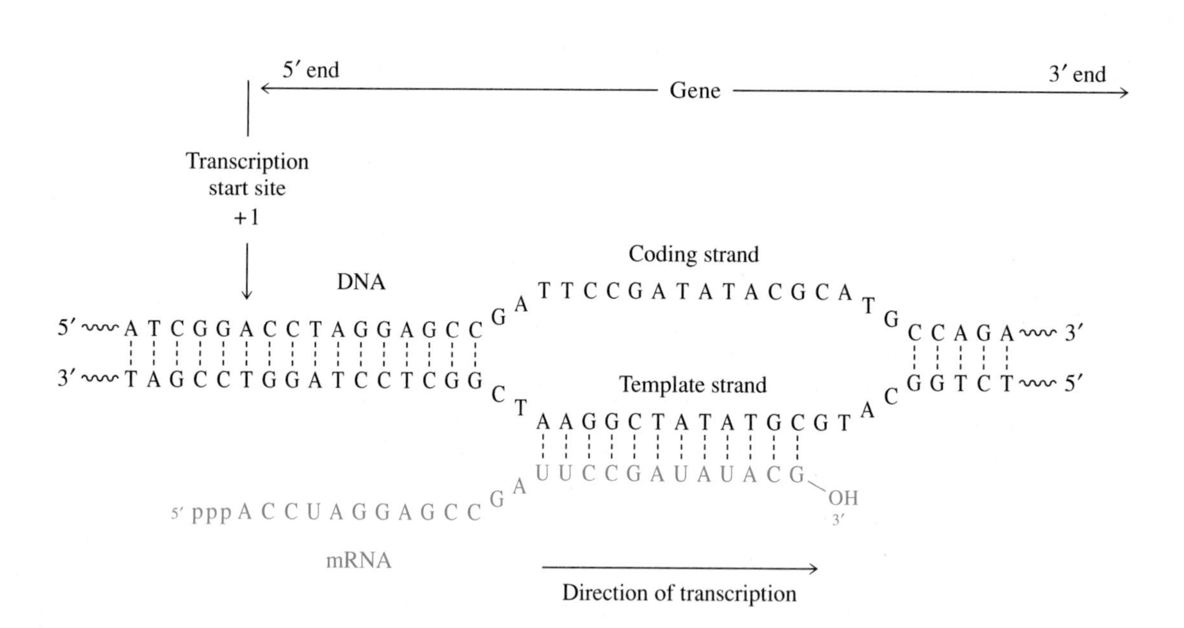

Box 21·1
Figure 1
The orientation of genes. The sequence of a hypothetical gene and the RNA transcribed from it are shown. Note that genes are transcribed from the 5′ end to the 3′ end but the actual template strand of DNA is copied from the 3′ end to the 5′ end. In RNA synthesis as in DNA synthesis, the growth of the nucleotide chain proceeds 5′→3′.

B. Promoter Recognition in *E. coli* Depends upon the σ Subunit

The effect of σ subunits, or σ factors, on promoter recognition can best be explained by comparing the binding properties of the core polymerase ($\alpha_2\beta\beta'$) and the holoenzyme ($\alpha_2\beta\beta'\sigma^{70}$). The core polymerase, which lacks a σ factor, binds to DNA nonspecifically; it has no greater affinity for promoters than for any other DNA sequence (its association constant, K_{assoc}, is approximately $10^{10}\,M^{-1}$). Once formed, this DNA:protein complex dissociates slowly ($t_{1/2} \cong 60\,\text{min}$). In contrast, the holoenzyme, which contains the σ subunit, binds more tightly to promoter sequences ($K_{assoc} \cong 2 \times 10^{11}\,M^{-1}$) than the core polymerase and forms more stable complexes ($t_{1/2} = 2$–$3\,\text{h}$). Although the holoenzyme binds preferentially to promoter sequences, it also has appreciable affinity for the rest of the DNA in a cell ($K_{assoc} \cong 5 \times 10^{6}\,M^{-1}$). The complex formed by nonspecific binding of holoenzyme to DNA dissociates very rapidly ($t_{1/2} \cong 3\,\text{s}$). These association constants reveal how the σ subunit affects the binding properties of the core polymerase: (1) σ decreases the affinity of the core polymerase for nonpromoter sequences, and (2) σ increases the affinity of the core polymerase for specific promoter sequences.

The association constants reveal nothing about the mechanism by which the RNA polymerase holoenzyme finds the promoter. We might expect the holoenzyme to search for the promoter by continuously binding and dissociating until it encounters a promoter sequence. Such binding would be a second-order reaction; its rate would be limited by the rate at which the holoenzyme diffuses in three dimensions. However, the rate constant for promoter binding has been measured experimentally as $k = 10^{10}\,M^{-1}\,s^{-1}$, which is two orders of magnitude greater than the maximum theoretical value for a diffusion-limited, second-order reaction.

The remarkable rate of the promoter-search reaction is achieved by one-dimensional diffusion of RNA polymerase along the length of DNA. During the short period of time (3 s) that the enzyme is bound nonspecifically, RNA polymerase can scan 2000 base pairs in its search for a promoter sequence. Several other sequence-specific DNA-binding proteins, such as restriction enzymes, locate their DNA-binding sites in a similar manner.

C. Formation of the Transcription Bubble and Synthesis of a Primer Are Rate-Limiting Steps in Initiation

Initiation of transcription is slow even though the holoenzyme searches for and binds to the promoter very quickly. In fact, initiation is often the rate-limiting step in transcription because, like initiation of DNA replication, it requires opening of the DNA helix and synthesis of an RNA primer for chain elongation. During DNA replication, these steps are carried out by a helicase and a primase; but during transcription, these steps are carried out by the RNA polymerase holoenzyme itself. Bacterial RNA polymerase is thus capable of initiating RNA synthesis de novo.

Unwinding of DNA at the initiation site is an isomerization reaction in which RNA polymerase (R) and the promoter (P) shift from a closed complex (RP_c) to an open complex (RP_o). In the closed complex, the DNA is double stranded; in the open complex, 18 base pairs of DNA are denatured, forming a *transcription bubble*. Formation of the open complex is usually the slowest step of the initiation events, probably because of the amount of energy required to denature 18 base pairs of DNA at the initiation site.

Once the open complex has formed, the template strand is positioned at the polymerization site of the enzyme. In the next step, a phosphodiester is formed between two ribonucleoside triphosphates that have diffused into the active site and formed hydrogen bonds with the +1 and +2 DNA nucleotides of the template strand. This reaction is slower than the analogous polymerization reaction during chain elongation in which one of the substrates (the growing RNA chain) is held in position by the formation of a short RNA:DNA helix.

Additional nucleotides are then added to the first dinucleotide to create a short RNA primer that is paired with the template strand. When the primer reaches approximately ten nucleotides in length, the RNA polymerase holoenzyme undergoes a transition from the initiation to the elongation mode, and the transcription complex moves away from the promoter along the DNA template. This last step is called promoter clearance. The initiation reactions are summarized in Equation 21·2:

$$R + P \underset{}{\overset{K_{assoc}}{\rightleftharpoons}} RP_c \xrightarrow[\textit{isomerization}]{} RP_o \xrightarrow[\textit{promoter clearance}]{} \qquad (21\cdot2)$$

As noted earlier, the holoenzyme has a much greater affinity for the promoter sequence than for any other DNA sequence. Because of this tight binding, it resists moving away from the initiation site. On the other hand, during elongation the core polymerase binds nonspecifically to all DNA sequences to form a highly processive complex. The transition from initiation of transcription to chain elongation is associated with a conformational change in the holoenzyme that causes the release of the σ factor. Without σ, the enzyme no longer binds specifically to the promoter and is able to leave the site of initiation. At this time, several accessory proteins bind to the core polymerase to create the complete protein machine required for

Figure 21·6
Initiation of transcription. (Illustrator: Lisa Shoemaker.)

(a) RNA polymerase holoenzyme has significant nonspecific affinity for DNA.

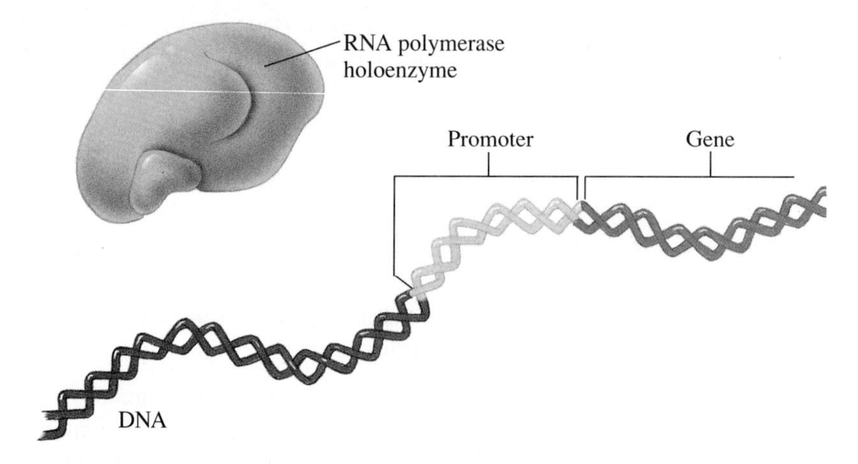

(b) After binding to DNA, the holoenzyme conducts a one-dimensional search along the DNA for a promoter sequence.

elongation. The binding of one of these accessory proteins, nusA (MW 65 000), may help convert RNA polymerase to the elongation form. NusA also interacts with other accessory proteins and plays a role in termination. Transcription initiation in *E. coli* is summarized in Figure 21·6.

(c) When a promoter sequence is found, the RNA polymerase holoenzyme and the promoter form a closed complex.

(d) Isomerization of the closed complex results in an open complex and the formation of a transcription bubble at the initiation site. A short RNA primer is then synthesized.

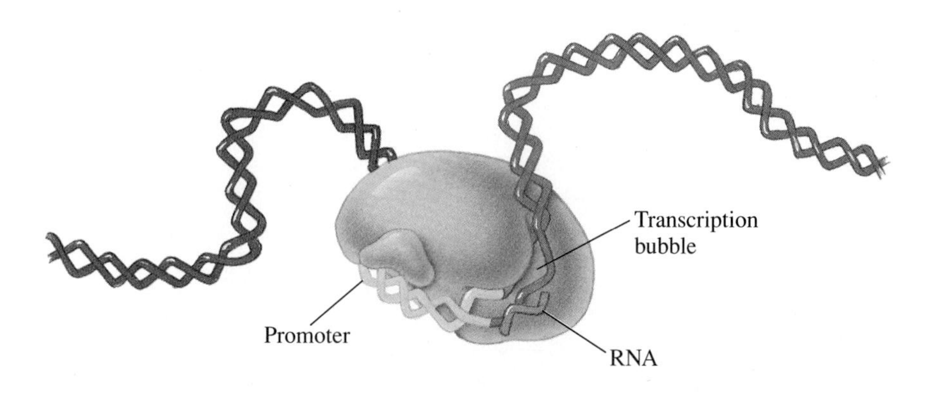

(e) The σ subunit dissociates from the core enzyme and RNA polymerase clears the promoter. After several accessory proteins, including nusA, join the transcription complex, the complex switches to the elongation mode.

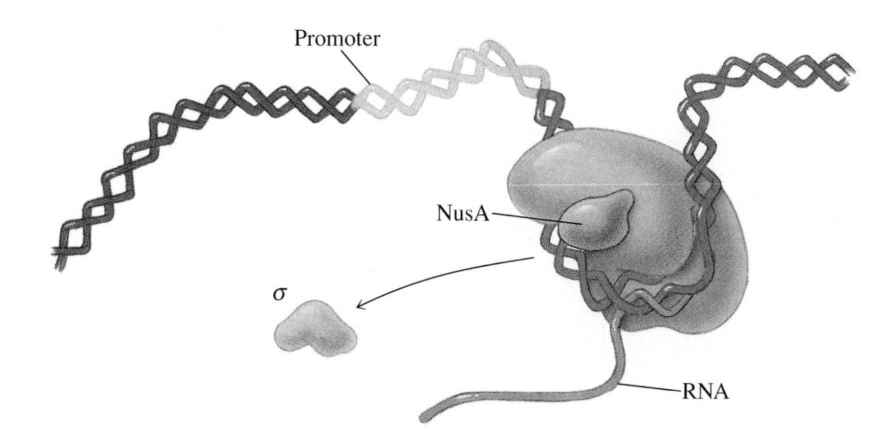

21·4 Termination of Transcription Requires Special Termination Sequences

Termination is a distinct event in transcription. Just as the transcription complex was actively assembled, it must be actively disassembled after elongation is complete. The simplest form of termination occurs at certain DNA sequences where the elongation complex is unstable. More complex forms of termination require a specific protein that facilitates disassembly.

The rate of RNA synthesis during elongation is sequence dependent. When the transcription complex is passing through a sequence rich in GC base pairs, for example, it is more difficult to denature the DNA template to form the transcription bubble. These sequences, called *pause sites*, can slow transcription by a factor of 10 to 100.

Pausing is exaggerated at sites on the DNA whose sequences exhibit dyad symmetry. These sequences are called palindromes (Section 10·6). When the palindromic sequence is transcribed, the newly synthesized RNA can form a hairpin structure (Figure 21·7). Formation of an RNA hairpin pulls the RNA strand through the transcription bubble and may destabilize the RNA:DNA hybrid in the elongation complex by prematurely stripping off part of the newly transcribed RNA. This partial disruption of the transcription bubble probably causes the transcription complex to cease elongation until the hybrid reforms. NusA increases pausing at such palindromic sites, perhaps by stabilizing the hairpin structure. Depending on the structure of the hairpin, the transcription complex may pause at such sites for 10 seconds to 30 minutes.

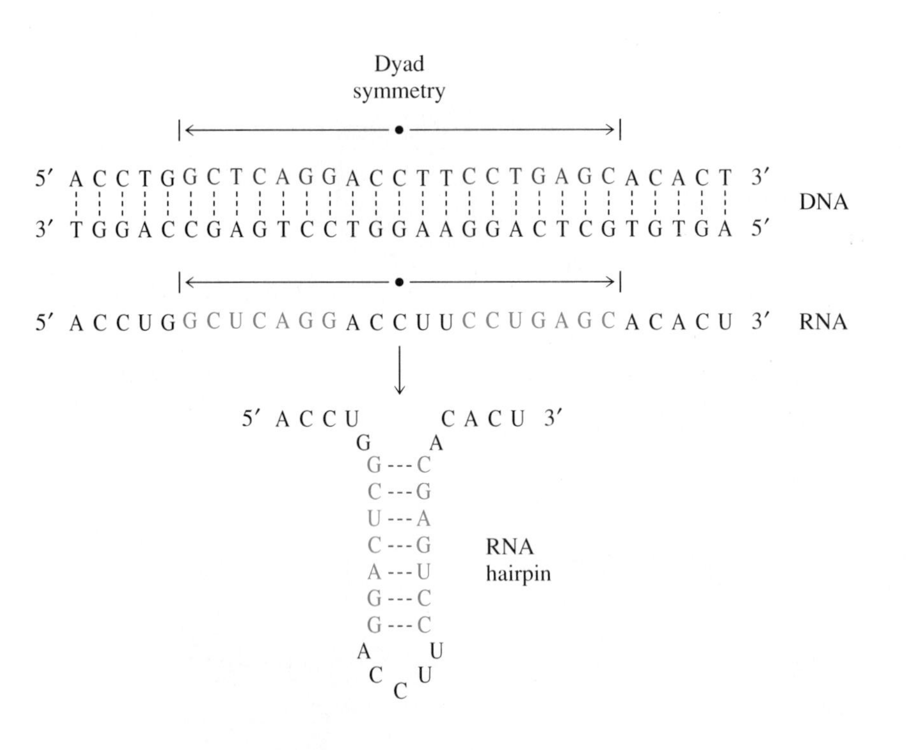

Figure 21·7
Formation of a hairpin structure in RNA. The DNA sequence contains a region of dyad symmetry. After transcription, the complementary sequences in RNA are able to form a region of secondary structure known as a hairpin. A three-dimensional representation of double-stranded RNA in such a structure is shown in Figure 10·16.

Transcription ends at DNA sequences called *termination sequences*. In *E. coli*, some strong pause sites are termination sequences, where the transcription complex spontaneously disassembles. In other cases, termination requires a specific regulatory protein named rho. Bacterial termination sequences that require rho are said to be rho-dependent. In the absence of rho, transcription at rho-dependent termination sequences does not stop, and RNA polymerase continues to copy the template strand of DNA.

Rho exists in the cytosol as a hexamer of identical subunits. It is a potent ATPase with an affinity for single-stranded RNA and may also act as an RNA-DNA helicase. Rho binds single-stranded RNA exposed behind a paused transcription complex in a reaction that is coupled to hydrolysis of ATP. Approximately 80 nucleotides of this RNA are wrapped around the protein, leading to the dissociation of the transcript from the transcription complex (Figure 21·8).

Termination at rho-dependent termination sequences appears to be due to both the destabilization of the RNA:DNA hybrid and the direct contact between the transcription complex and rho as rho binds RNA. Rho can also bind to nusA and other accessory proteins; this interaction may induce a conformational change that facilitates the dissociation of RNA polymerase from the template DNA.

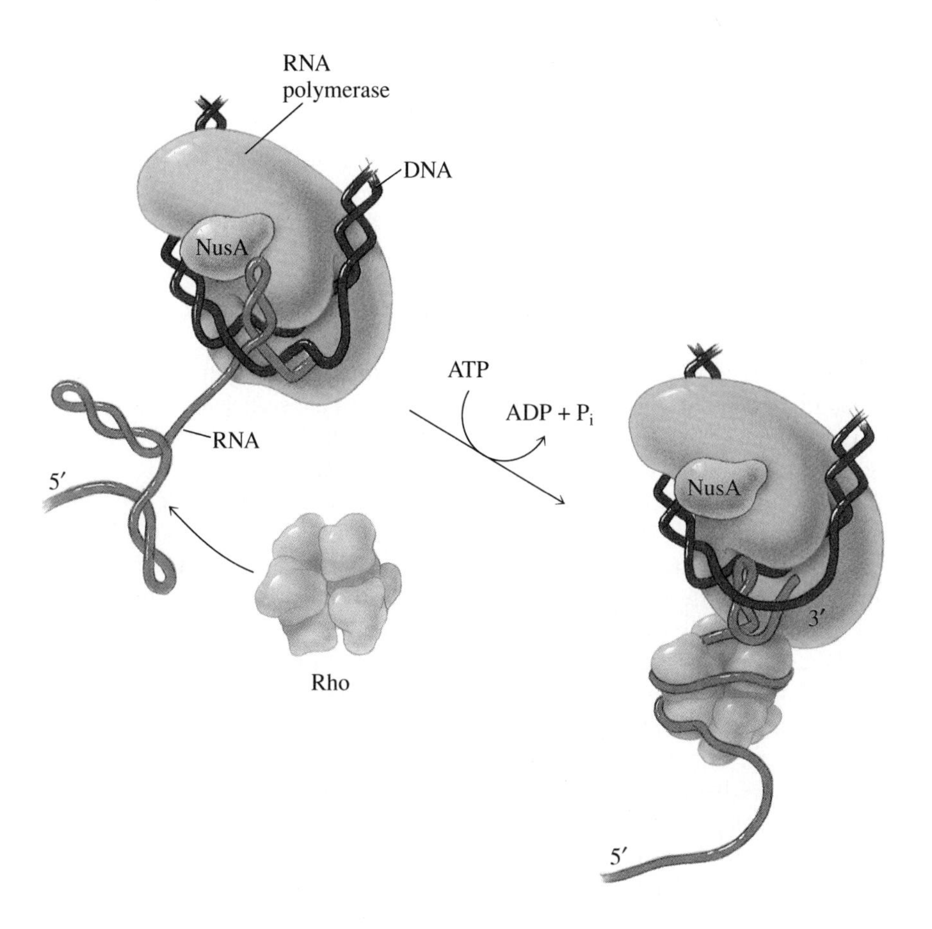

Figure 21·8
Rho-dependent termination of transcription in *E. coli*. RNA polymerase is stalled at a pause site, where rho binds to newly synthesized RNA. This binding is accompanied by ATP hydrolysis. Rho probably wraps the nascent RNA chain around itself, thereby destabilizing the RNA:DNA hybrid and terminating transcription. (Illustrator: Lisa Shoemaker.) [Adapted from Platt, T. (1986). Transcription termination and the regulation of gene expression. *Annu. Rev. Biochem.* 55:339–372.]

21·5 The Transcription of Genes Can Be Regulated

As noted in the beginning of this chapter, there are a large number of genes that are expressed in every cell. These housekeeping genes encode the proteins required for basic metabolic events such as glycolysis and for the biosynthesis of molecules such as amino acids and DNA. In general, housekeeping genes whose products are required continuously at high levels have strong promoters and are transcribed efficiently and continuously. Conversely, housekeeping genes whose products are required at low levels usually have weak promoters and are transcribed infrequently. Genes also exist that are expressed at high levels in some circumstances and not at all in others. Such genes are said to be regulated.

Regulation of gene expression can take place at any point in the flow of biological information but occurs most often at the level of transcription. A variety of elaborate mechanisms have evolved that allow cells to program development, to maintain basic functions, and to respond to environmental stimuli by altering the rate of production of certain RNA species. Since initiation of transcription requires both specific DNA sequences and soluble proteins, either of these elements can be manipulated to modulate transcriptional activity.

A. Activators Can Increase the Rate of Transcription from Poor Promoters

All cells contain activators, regulatory proteins that can increase the rate of transcription initiation at poor promoters. Generally, activators bind to specific DNA sequences near a promoter and interact directly with RNA polymerase, which is bound to the promoter. These protein:protein contacts are sufficient to convert the weak binding site to a stronger one. Activators can affect not only promoter binding but also the rate of open-complex formation or primer synthesis.

Activators are often allosteric proteins whose activity is controlled by small molecules, in much the same manner as allosterically regulated metabolic enzymes. The rate of transcription of regulated genes can thus be controlled by the concentration of effector molecules in the cell. If activators were always functional, then the rate of transcription would always be maximal.

One of the best-studied activators is the *E. coli* cAMP regulatory protein (CRP), a dimeric protein whose activity is controlled by cAMP. In the absence of cAMP, CRP has little affinity for DNA. When cAMP is present, however, it binds to CRP, converting it into a sequence-specific DNA-binding protein. The CRP-cAMP complex binds to specific DNA sequences near the promoters of more than 30 genes. While bound to DNA, this activator can contact RNA polymerase at the promoter site, leading to increased rates of transcription initiation (Figure 21·9).

Each subunit of the CRP dimer contains an α-helical domain, which in the presence of cAMP fits into adjacent sections of the major groove of DNA and contacts the nucleotides of the CRP-cAMP–binding site. In the absence of cAMP, the conformation of CRP changes such that the two α helices can no longer bind to the major groove (Figure 21·10). In this way, cAMP regulates the binding activity of CRP. The DNA-binding motif of CRP, in which two α helices fit into adjacent sections of the major groove of DNA, is common to many transcriptional regulatory proteins.

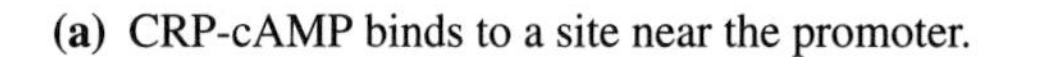

(a) CRP-cAMP binds to a site near the promoter.

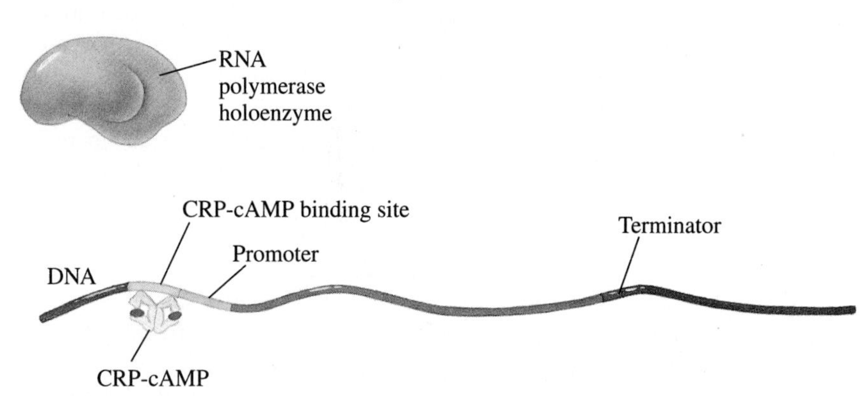

(b) RNA polymerase holoenzyme interacts with the promoter sequence and with the bound activator; consequently, the rate at which transcription is initiated is increased.

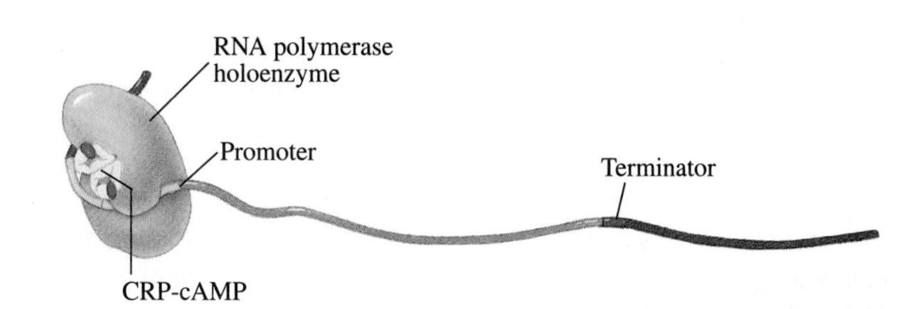

Figure 21·9
Activation of transcription initiation by CRP-cAMP. Some genes are transcribed inefficiently by RNA polymerase in the absence of additional activators. CRP-cAMP binds to a specific DNA sequence near the promoters of about 30 such genes and enables RNA polymerase to initiate transcription more efficiently. (Illustrator: Lisa Shoemaker.)

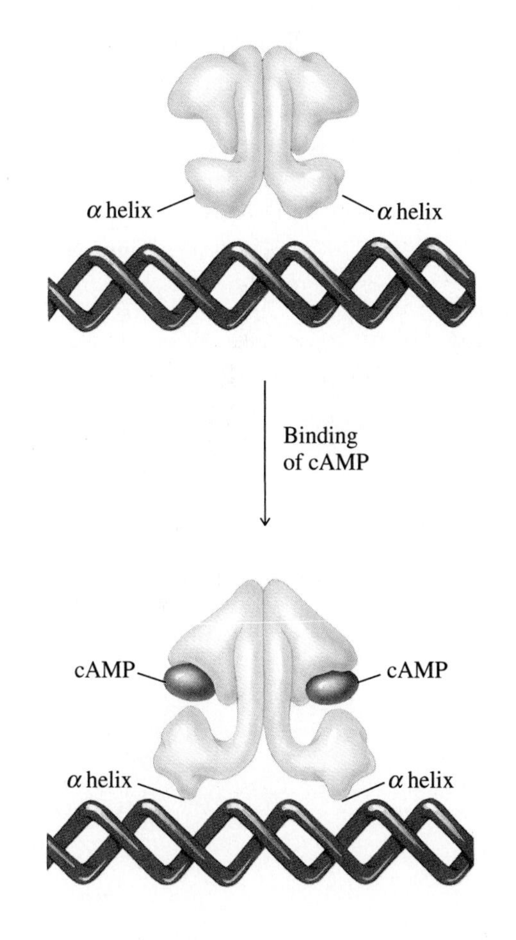

Figure 21·10
Allosteric changes in CRP caused by the binding of cAMP. Each monomer of the CRP dimer contains an α helix. In the absence of cAMP, the α helices cannot fit into adjacent sections of the major groove of DNA and hence cannot recognize the CRP-cAMP–binding site. When cAMP binds to CRP, the two α helices rearrange into the proper conformation for binding to DNA. (Illustrators: Initial rendering—Lisa Shoemaker; Final rendering—Sarah McQueen.)

In *E. coli,* the concentration of cAMP inside the cell is controlled by the presence of glucose outside of the cell. As explained in Box 9·3, when glucose is available, it is imported into the cell and phosphorylated by a complex of transport proteins. When glucose is not available, one of the glucose transport enzymes, Enzyme III^{glc}, phosphorylates adenylate cyclase, leading to its activation (Figure 21·11). Adenylate cyclase then catalyzes the conversion of ATP to cAMP, thereby increasing the levels of cAMP in the cell. As molecules of cAMP are produced, they bind to CRP, causing transcription to be activated.

Most of the genes that respond to CRP-cAMP encode products that enable the cell to overcome glucose deprivation. For example, enzymes that take up and metabolize lactose and other sugars are encoded by genes that contain a CRP-cAMP–binding site at their promoters. When cAMP levels rise, the rate of transcription of these genes is increased by the activator. This regulation of transcription by an allosteric activator allows the bacterium to use lactose and other sugars when glucose levels are low.

Transcriptional activators, such as CRP, are common in the cells of all species. Often, as in the case of *E. coli* CRP, the activity of the activator is itself regulated in response to environmental signals. In other cases, particularly in multicellular eukaryotes, specific activators are only present in some cell types. These activators control expression of genes that are tissue specific or developmentally regulated.

B. Repressors Can Inhibit Transcription

Cells also contain repressors, regulatory proteins that repress or inhibit transcription. Repressors work in many different ways; one of the simplest is by preventing RNA polymerase from binding to the promoter. In this case, the repressor-binding site is close to the promoter so that when the repressor is bound, there is not enough room for the RNA polymerase holoenzyme to bind to the promoter. In other cases, repressors permit the binding of RNA polymerase but inhibit initiation reactions, such as isomerization, primer synthesis, and promoter clearance.

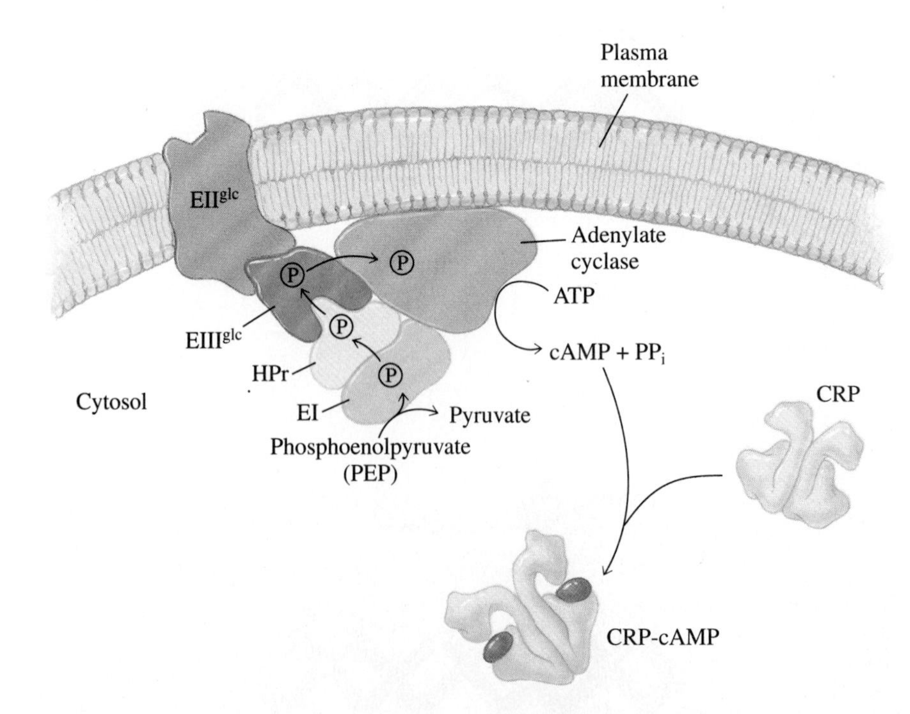

Figure 21·11
cAMP production. In the absence of glucose, Enzyme III^{glc} transfers its phosphoryl group to membrane-bound adenylate cyclase. Phosphorylated adenylate cyclase catalyzes the conversion of ATP to cAMP. As a result, levels of cAMP in the cell rise and cAMP binds to CRP. CRP-cAMP activates the transcription of a number of genes encoding enzymes that compensate for the lack of glucose as a carbon source. HPr and Enzymes I and II^{glc} are part of the PEP:glucose phosphotransferase system. (Illustrator: Lisa Shoemaker.)

An example of this latter type of repressor is the well-studied *lac* repressor in *E. coli*. This tetrameric regulatory protein controls transcription of an operon that encodes three proteins required for lactose metabolism. The *lac* repressor binds simultaneously to two sites near the promoter of the operon, thereby forming a loop of DNA (Figures 21·12 and 21·13); such loops are common near the promoters of both prokaryotic and eukaryotic genes. Although the binding of *lac* repressor does not prevent RNA polymerase from binding to the promoter, it does prevent the enzyme from initiating transcription. The mechanism of the binding of *lac* repressor to DNA is similar to that of CRP-cAMP. One α-helical region of each subunit lies in the major groove of DNA, where it directly contacts the nucleotides of the binding site.

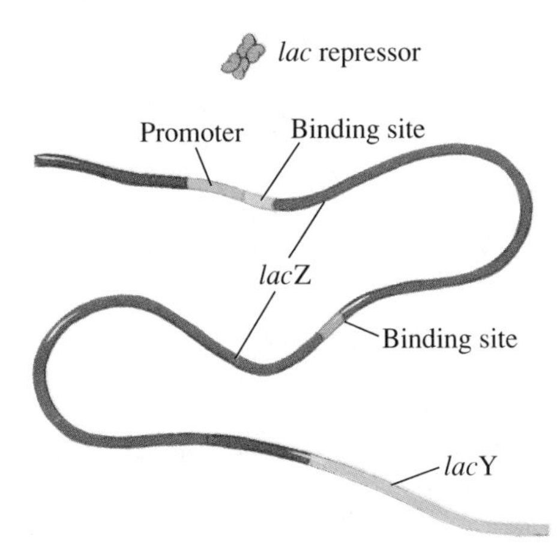

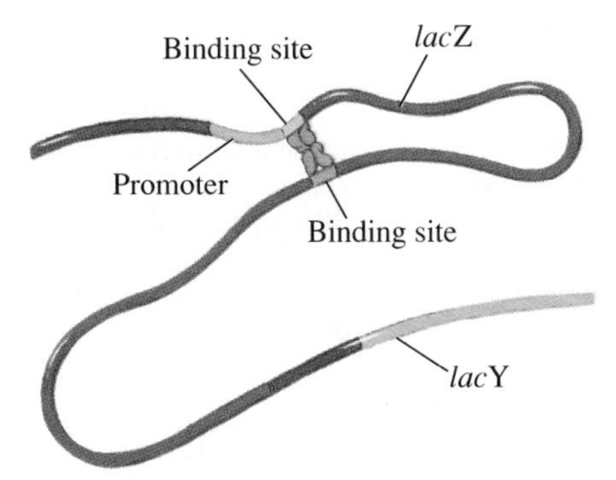

Figure 21·12
Binding of *lac* repressor to the *lac* operon. The tetrameric *lac* repressor interacts simultaneously with two sites near the promoter of the *lac* operon. As a result, a loop of DNA is formed. (The *lac* operon consists of three genes or coding regions. Only *lac*Z and part of *lac*Y are shown.) (Illustrators: Initial rendering—Lisa Shoemaker; Final rendering—Linda LeFevre Murray.)

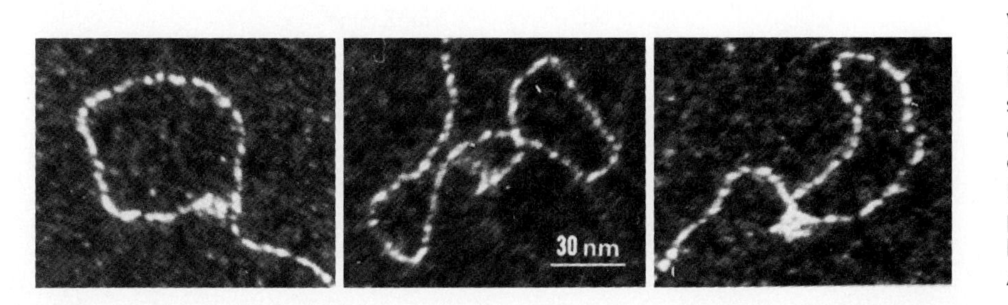

Figure 21·13
Electron micrographs showing DNA loops in vitro. These loops were formed by mixing *lac* repressor with a fragment of DNA bearing two synthetic *lac* repressor–binding sites. One binding site is located at one end of the DNA fragment and the other is located 535 base pairs away. DNA loops of 535 base pairs were formed when *lac* repressor bound simultaneously to the two binding sites. (Courtesy of Michèle Amouyal.)

The activity of *lac* repressor, and thus the expression of the *lac* genes, is regulated by *allo*lactose, the β-(1$\rightarrow$6) isomer of lactose. When lactose is available as a potential carbon source, some of the available lactose is converted to *allo*lactose; *allo*lactose then binds to *lac* repressor, causing a conformational change in the repressor that leads to its dissociation from the repressor binding site. The inactivation of the repressor allows RNA polymerase to initiate transcription of the operon (Figure 21·14).

Figure 21·14
Transcription of the *lac* operon regulated by *lac* repressor and lactose. (Illustrator: Lisa Shoemaker.)

(a) In the absence of lactose, *lac* repressor binds to DNA and thus prevents RNA polymerase from transcribing the *lac* operon even though the enzyme is bound to the promoter.

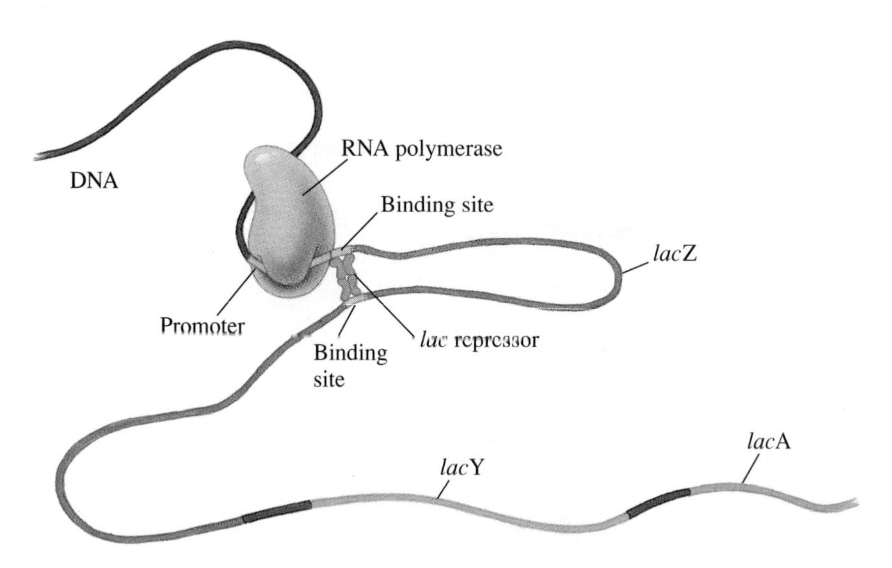

(b) In the presence of lactose, some of the lactose is converted to *allo*lactose, which binds to *lac* repressor. This binding results in a conformational change in the repressor that causes it to dissociate from the operon, allowing transcription to begin.

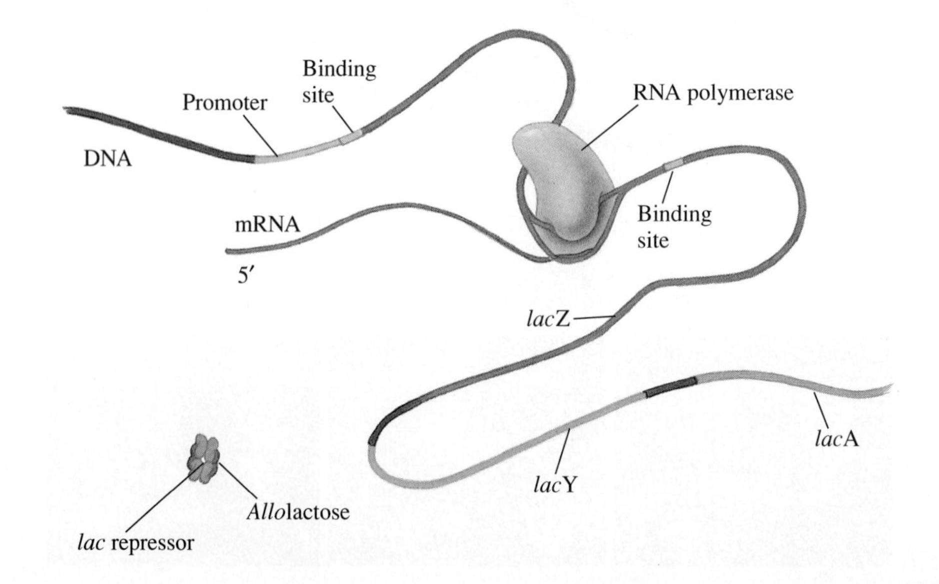

In the presence of lactose, the *lac* operon is transcribed because the repressor is inactivated. However, the rate of transcription and subsequent production of protein is low in the presence of the preferred carbon source, glucose. In the absence of glucose, transcription rates of the *lac* operon are increased by CRP-cAMP. Thus, transcription of the *lac* operon is both repressed and activated by regulatory proteins.

21·6 Many tRNA and rRNA Molecules Are Posttranscriptionally Modified

Many primary RNA transcripts are not released from the transcription complex in a fully active form; rather, they must be extensively altered before they can adopt their mature structures and functions. These alterations fall into three general categories: (1) the removal of nucleotides from primary RNA transcripts; (2) the addition of nucleotides with sequences not encoded by the corresponding genes; and (3) the covalent modification of certain bases. The reactions that transform a primary RNA transcript into a mature RNA molecule are referred to collectively as *RNA processing*. RNA processing is a crucial step in the production of most mature RNA and is thus an integral part of gene expression.

A. Many Different Processing Steps Are Required to Produce Mature tRNA

Mature tRNA molecules are generated in both eukaryotes and prokaryotes by the processing of primary transcripts. In prokaryotes, a tRNA primary transcript usually contains several precursors to mature tRNA molecules. These precursors are cleaved from the large primary transcripts and trimmed to their mature lengths by ribonucleases, or RNases (Section 10·6).

The endonuclease RNase P is responsible for the initial cleavage of most tRNA primary transcripts. RNase P recognizes the tertiary structure of the portions of the primary transcript that will become mature tRNA molecules. It then catalyzes cleavage of the primary transcript on the 5′ side of each tRNA precursor, releasing monomeric tRNA precursors with mature 5′ ends (Figure 21·15). Digestion with RNase P in vivo is rapid. Also, since tRNA molecules with immature ends cannot generally be isolated from living cells, digestion is thought to occur while the transcript is still being synthesized.

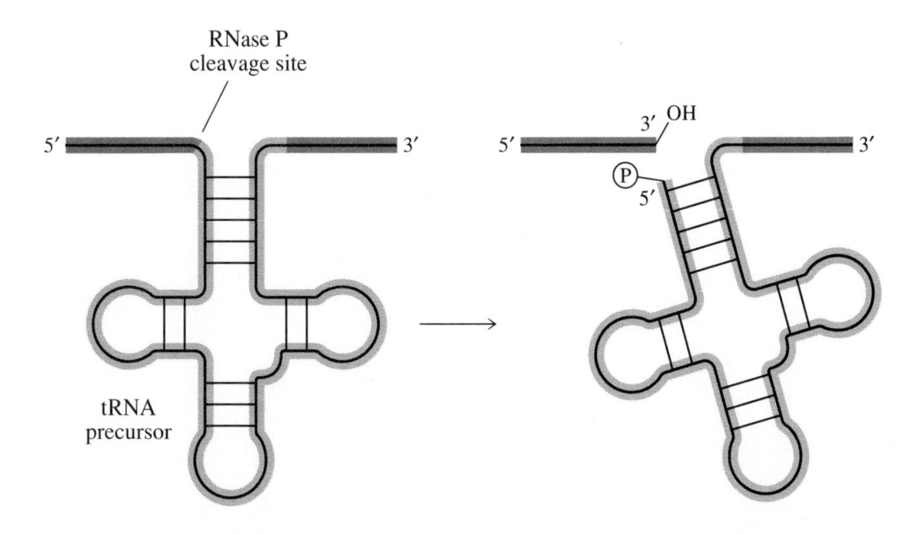

Figure 21·15
Cleavage of tRNA from a prokaryotic primary transcript. The mature 5′ terminus of each tRNA molecule is created by endonucleolytic cleavage catalyzed by RNase P. RNase P finds its cleavage site by recognizing the tertiary structure of the tRNA precursor.

RNase P was one of the first specific ribonucleases studied in detail, and much is known about its structure. The enzyme is actually a ribonucleoprotein. In *E. coli*, it is composed of a 377-nucleotide RNA molecule (130 kDa) and a 20 kDa protein. The RNA component (Figure 21·16) is catalytically active under certain conditions in vitro in the absence of protein. It was one of the first RNA molecules found to have enzymatic activity. In vivo, the protein component electrostatically shields the negatively charged catalytic RNA from its identically charged substrate and helps maintain the three-dimensional structure of the RNA.

RNase P activity has been found in a number of prokaryotes, and a counterpart has been identified in humans and yeast. Although the actual nucleotide sequence of the RNA component of RNase P differs between organisms, its tertiary structure is thought to be conserved, since protein subunits and RNA subunits isolated from different prokaryotes can be mixed to yield active RNase P.

Whereas RNase P cleaves tRNA precursors at the 5′ end, other endonucleases cleave tRNA precursors near the 3′ end. Subsequent processing of the 3′ ends of tRNA precursors requires the activity of an exonuclease, such as RNase D. RNase D catalyzes the sequential removal of nucleotides from the 3′ end of a monomeric tRNA precursor until the 3′ end of the tRNA structure is reached.

In order to function, all mature tRNA molecules, both prokaryotic and eukaryotic, require the sequence CCA at their 3′ end. Because the genes for tRNA molecules in all mammals and in some bacteriophages lack this sequence at the 3′ end and because this sequence is sometimes cleaved from the tRNA molecule by an exonuclease, these nucleotides must be added posttranscriptionally after all other types of processing at the 3′ end have been completed. The CCA sequence is added by the enzyme tRNA nucleotidyl transferase. A summary of the steps in the nucleolytic action of processing of prokaryotic tRNA precursors is shown in Figure 21·17.

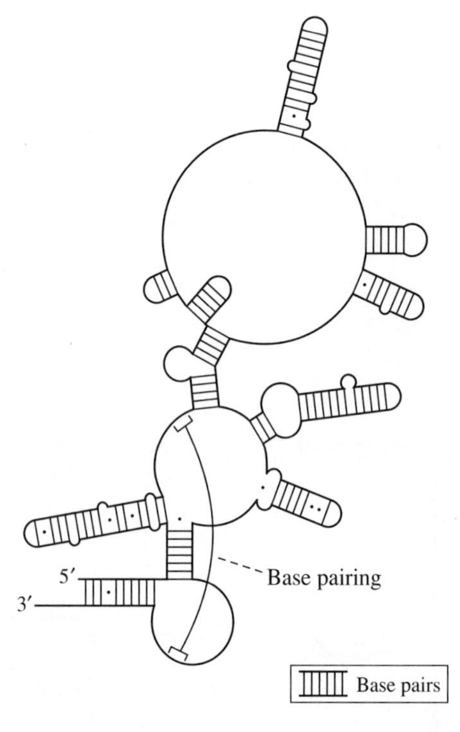

Figure 21·16
Proposed secondary structure of the RNA component of *E. coli* RNase P. This is one of several potential structures generated by aligning regions of possible secondary structure that have been conserved among molecules from different species. Complementary base pairing joins two distant regions of the molecule. Under certain conditions in vitro, this RNA molecule can catalyze the cleavage of tRNA precursors in the absence of protein. [Adapted from James, B. D., Olsen, G. J., Lui, J., and Pace, N. R. (1988). The secondary structure of ribonuclease P RNA, the catalytic element of a ribonucleoprotein enzyme. *Cell* 52:19–26.]

(a) After the endonuclease RNase P generates mature 5′ ends by cleaving the tRNA primary transcripts, other endonucleases release the monomers by cleaving them at their 3′ ends.

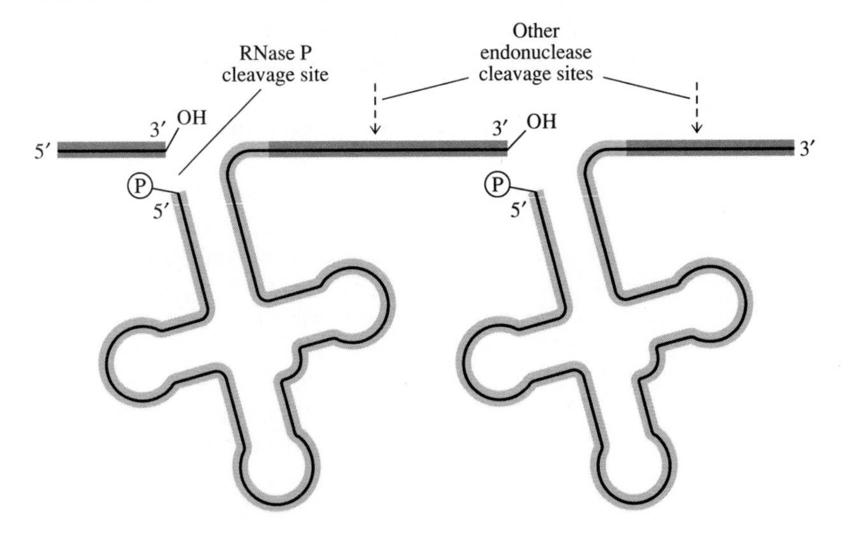

(b) The partially processed tRNA molecules are trimmed at their 3′ ends by the activity of the exonuclease RNase D.

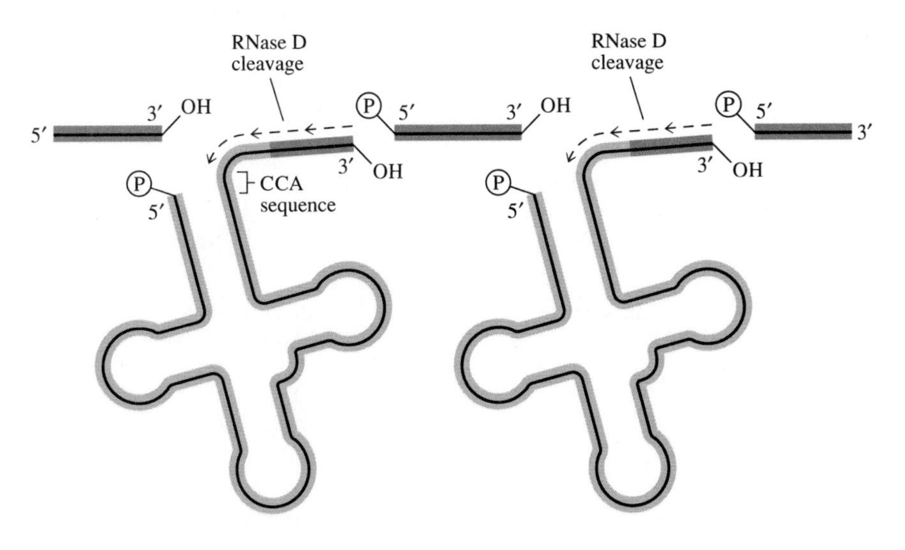

(c) If the 3′ terminal CCA sequence is present, cleavage by the exonuclease RNase D produces a mature tRNA molecule. If the 3′ terminal CCA sequence is missing, it is added by the activity of tRNA nucleotidyl transferase.

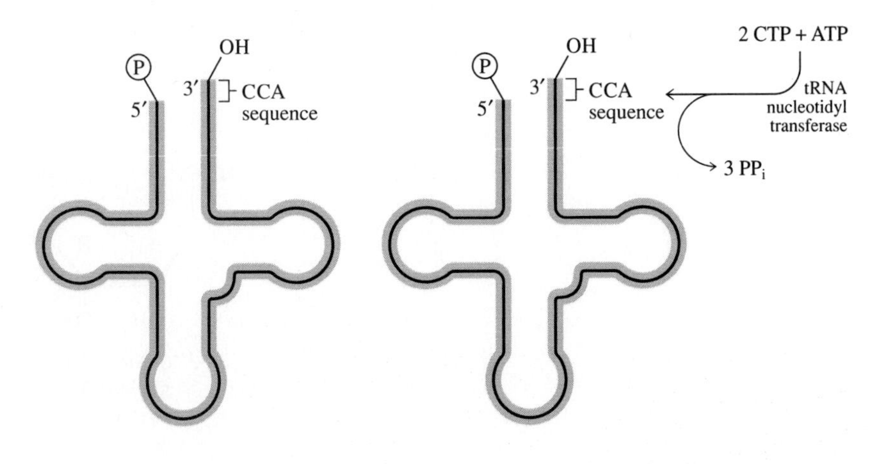

Figure 21·17
Summary of steps in the nucleolytic processing of prokaryotic tRNA precursors.

In both prokaryotes and eukaryotes, the nucleotides in mature tRNA molecules exhibit a greater diversity of covalent modifications than any other class of RNA molecule. Of the approximately 80 nucleotides in a tRNA molecule, 26 to 30 are covalently modified. Usually, each type of covalent modification occurs in only one location on each molecule. Some examples of modified bases and nucleosides are shown in Figure 21·18.

Figure 21·18
Examples of covalently modified nucleosides found in tRNA molecules. Modifications are shown in blue.

A

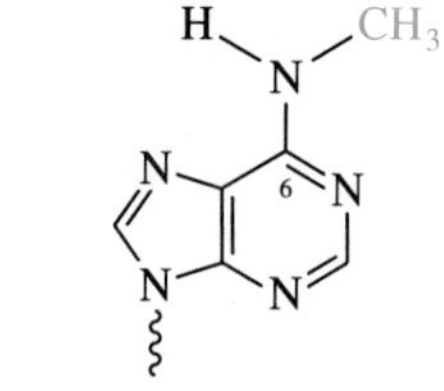

6-Methyladenosine
(m^6A)

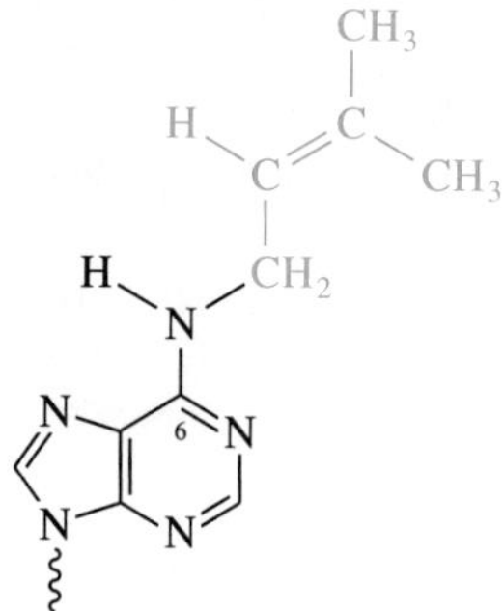

6-Isopentenyladenosine
(i^6A)

G

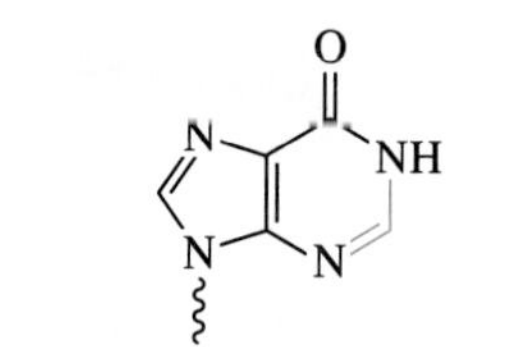

Inosine
(I)

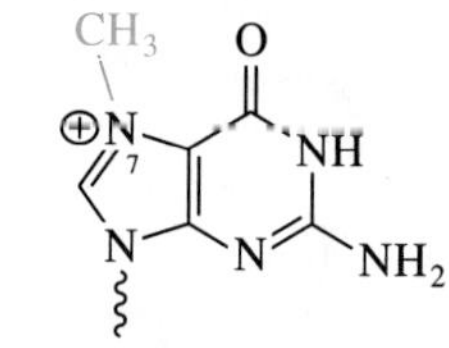

7-Methylguanosine
(m^7G)

U

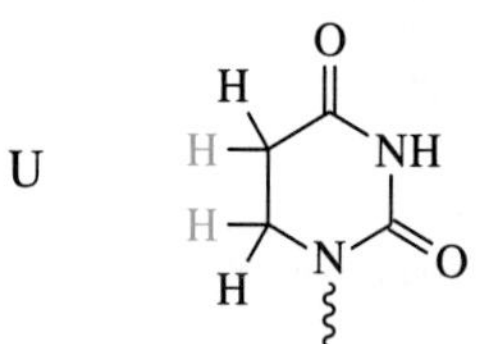

Dihydrouridine
(D)

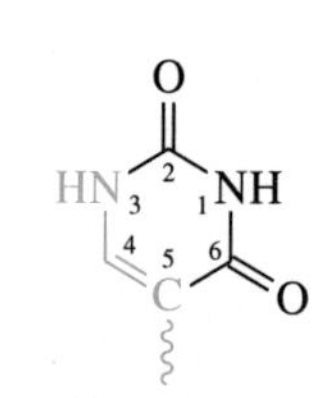

Pseudouridine
(Ψ)
(ribose at C-5)

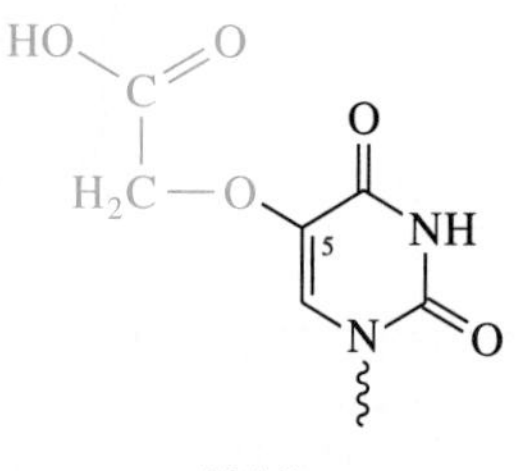

Uridine
5-oxyacetic acid
(cmo^5U)

C

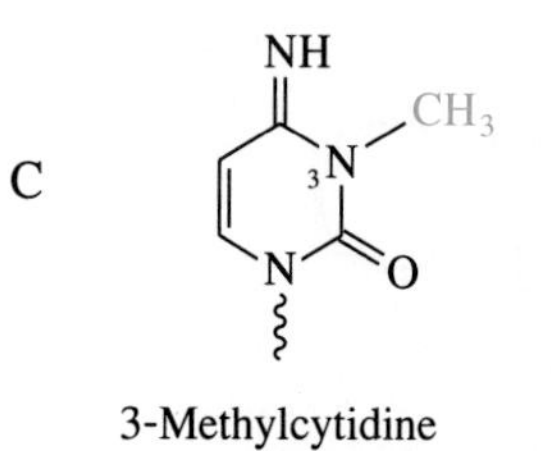

3-Methylcytidine
(m^3C)

5-Methylcytidine
(m^5C)

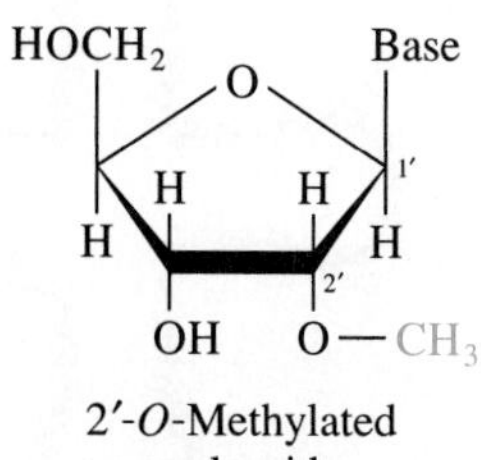

2′-*O*-Methylated
nucleoside
(Nm)

B. Ribosomal RNA Molecules Are Produced by Processing Large Primary Transcripts

Ribosomal RNA molecules in all organisms are produced as large primary transcripts that require subsequent processing, including methylation and cleavage by both endonucleases and exonucleases. In both prokaryotes and eukaryotes, rRNA processing is coupled to the assembly of ribosomes.

The primary transcripts of prokaryotic rRNA molecules are about 30S in size and contain one copy of each of the following species of rRNA: 16S, 28S, and 5S rRNA. (Note that S is the symbol for the Svedberg unit, a parameter that measures the rate at which particles move in the gravitational field established in an ultracentrifuge. Large S values are associated with large masses. The relationship between S and mass is not linear; therefore S values are not additive.) Since the genes for these three RNA molecules are transcribed from a single promoter, the same amount of each of these rRNA molecules is produced.

The 5′ and 3′ ends of mature rRNA molecules are usually part of a base-paired region in the primary transcript. In prokaryotes, a single endonuclease binds to this region and cleaves the precursor. In *E. coli,* this endonuclease is RNase III (Figure 21·19). Unlike the mature 5′ and 3′ ends of tRNA, the mature 5′ and 3′ ends of rRNA are produced in a single step. Many prokaryotic rRNA primary transcripts also contain tRNA precursors.

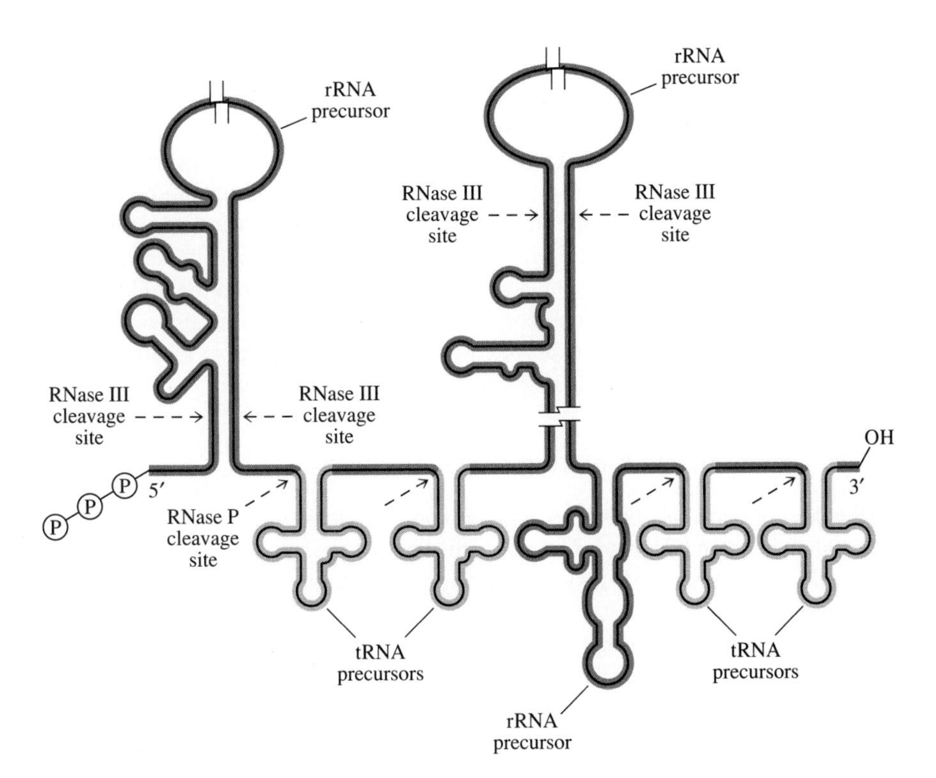

Figure 21·19
Endonucleolytic cleavage of ribosomal RNA precursors in *E. coli*. The rRNA primary transcript contains a copy of each of the three rRNA species and may also contain several tRNA precursors. Each rRNA precursor is cleaved from the large primary transcript by RNase III. This enzyme recognizes the tertiary structure of the primary transcript and, in cleaving the precursor from the primary transcript, generates the mature ends of the rRNA molecule. (Slash marks indicate portions of the rRNA primary transcript that have been deleted for clarity.)

Eukaryotic ribosomal RNA molecules are also produced by processing a larger precursor. The primary transcripts are between 35S and 47S in size and contain a copy of each of three RNA species: 18S, 5.8S, and 28S. (The fourth eukaryotic ribosomal RNA, 5S RNA, is transcribed as a monomer by RNA polymerase III and is processed separately.) The multimeric primary transcript is not separated into smaller subunits until it is completely synthesized and approximately 100 ribose groups within conserved sequences have been methylated. The transcript is made in the nucleolus, where initial processing occurs. Proper cleavage requires that each rRNA precursor bind to at least some of the proteins that will eventually be assembled with the mature rRNA in the ribosome.

21·7 Processing of Eukaryotic mRNA Molecules Involves Covalent Modification, Addition of Nucleotides, and Splicing

The processing of mRNA precursors is one of the biochemical processes that distinguishes prokaryotes from eukaryotes. In prokaryotes, the primary mRNA transcript is translated directly. In fact, as we shall see in the next chapter, translation often is initiated before transcription is completed. In eukaryotes, however, transcription and translation are carried out in separate compartments of the cell; transcription occurs in the nucleus and translation in the cytoplasm. This separation provides an opportunity for eukaryotic mRNA precursors to be processed in the nucleus without interfering with translation.

Mature eukaryotic mRNA molecules are produced by processing primary transcripts. This processing involves the same kinds of steps that are responsible for producing mature tRNA and rRNA molecules, namely, cleavage of a precursor, covalent modification of nucleotides, and addition and removal of terminal nucleotides. In some eukaryotes, nucleotides can be removed from the middle of an mRNA primary transcript in a process known as *splicing*. Splicing also occurs during the processing of some eukaryotic tRNA and rRNA precursors, although the mechanism is different from that of splicing of mRNA precursors.

A. Eukaryotic mRNA Molecules Have Modified Ends

All eukaryotic mRNA molecules undergo modifications that increase their stability and make them better substrates for translation. One way to increase stability is to modify the ends so that they are no longer susceptible to cellular exonucleases that degrade RNA.

The 5′ ends are modified before eukaryotic mRNA precursors are completely synthesized. The 5′ end of the primary transcript is a nucleoside triphosphate residue (usually a purine) that was the first nucleotide incorporated by RNA polymerase II. Modification of this end begins when the terminal phosphate group is removed by the activity of a phosphohydrolase (Figure 21·20). The resulting 5′-diphosphate group then reacts with the α-phosphorus of a GTP molecule to create a 5′–5′ triphosphate linkage. This reaction is catalyzed by the enzyme guanylyltransferase. The resulting structure is called a *cap*. The cap structure is further modified by methylation of the newly added guanine. In addition, the 2′-hydroxyl groups of the first two nucleosides in the original primary transcript may be methylated.

Figure 21·20
Formation of a cap structure at the 5′ end of an mRNA precursor in eukaryotes. (1) Formation of the cap begins when a phosphohydrolase catalyzes removal of the terminal phosphate at the 5′ end of the precursor. (2) The 5′ end then receives a GMP group from GTP in a reaction catalyzed by guanylyltransferase. (3) The added guanosine is then methylated at the N-7 position. (4) The 2′-hydroxyl group of the terminal ribose and of the penultimate ribose in the mRNA may also be methylated in consecutive steps. (The methyl group donor, *S*-adenosylmethionine, is not shown in reactions 3 and 4.)

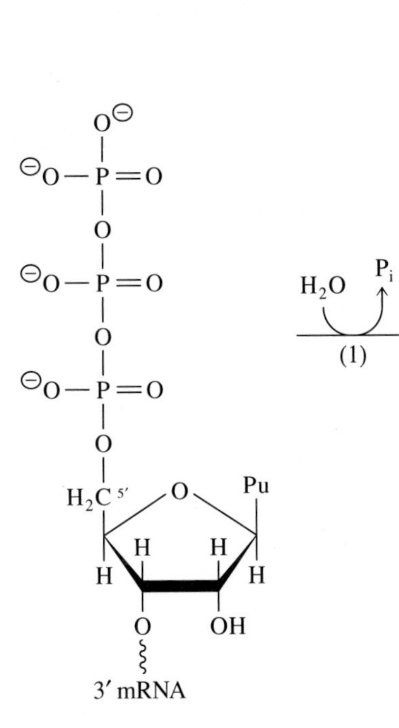

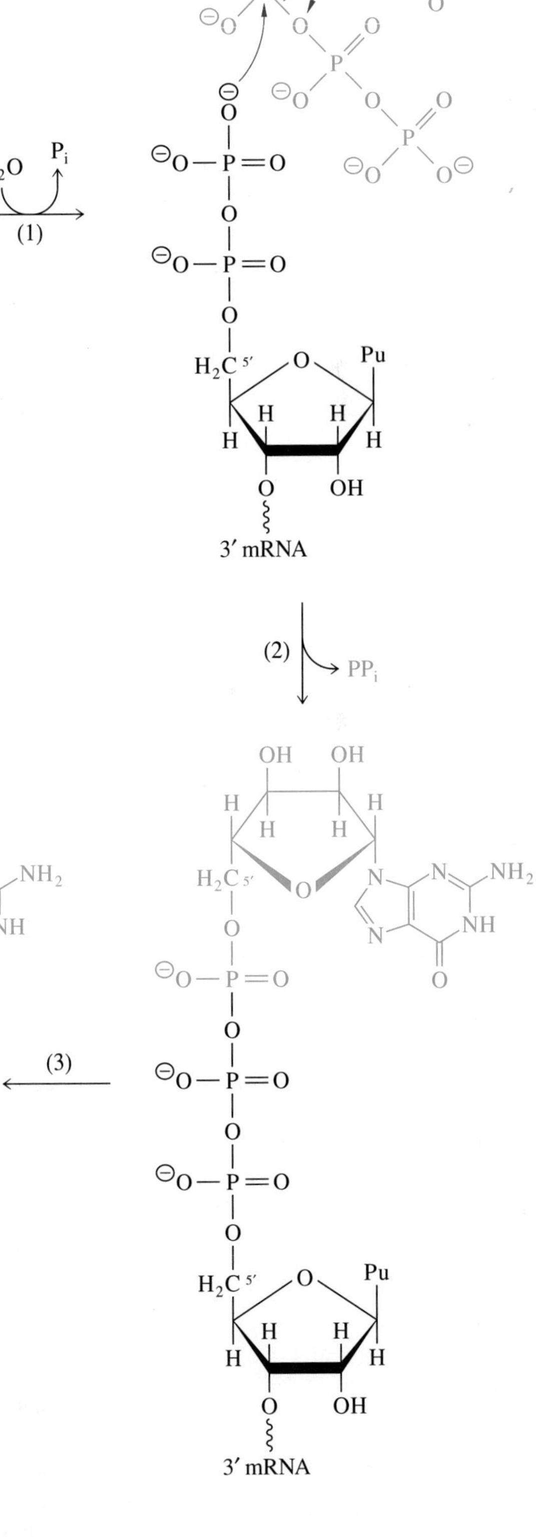

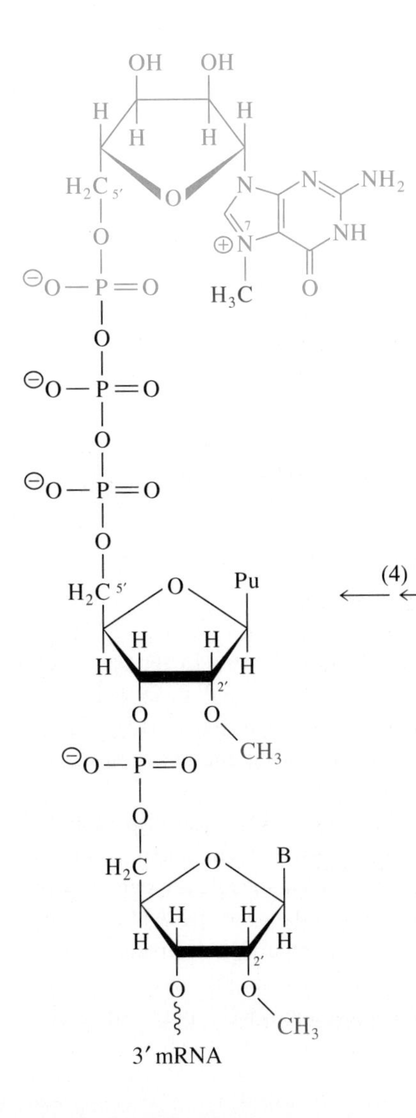

The 5′–5′ triphosphate linkage protects the molecule from 5′ exonucleases by blocking the 5′ end of the mRNA molecule. The cap also converts mRNA precursors into substrates for other processing enzymes in the nucleus, such as those that catalyze splicing. In mature mRNA, the cap serves as the attachment site for ribosomes during protein synthesis.

Eukaryotic mRNA precursors are also modified at their 3′ ends (Figure 21·21). Once RNA polymerase II has transcribed past the 3′ end of the coding region of DNA, the newly synthesized RNA is cleaved by an endonuclease near a specific site whose consensus sequence is AAUAAA. This site is called the *polyadenylation signal*. Cleavage generally occurs approximately 10 to 20 nucleotides downstream from this signal and probably depends upon additional sequences and perhaps upon the secondary structure of the mRNA precursor. The newly generated 3′ end of the molecule serves as the primer for the addition of multiple adenosine residues in a reaction catalyzed by poly A polymerase. In this reaction, which requires ATP, up to 250 nucleotides may be added to form a stretch of polyadenylate known as a *poly A tail*.

The poly A tails of mRNA precursors and mature mRNA molecules become tightly associated with a protein of molecular weight 78 000. This RNA:protein complex is presumed to stabilize mRNA by protecting it against degradation from the 3′ end. There are a few unusual examples of eukaryotic mRNA that do not have poly A tails.

B. Some Eukaryotic mRNA Precursors Are Spliced

Although splicing of mRNA is rare in prokaryotes, it is the rule in vertebrates and flowering plants. In some simpler eukaryotes, such as yeast and algae, only a few genes produce transcripts that are spliced, whereas in insects, such as *Drosophila melanogaster,* genes producing transcripts that are spliced and genes producing transcripts that are unspliced are both common.

Internal sequences that are spliced from the primary RNA transcript, and thus not part of the mature RNA molecule, are called *introns*. Sequences that are present in the primary RNA transcript *and* in the mature RNA molecule are called *exons*. The words *intron* and *exon* also refer to the regions of the gene (DNA) that encode corresponding RNA introns and exons. Note that since DNA introns are transcribed, they are considered part of the gene. The junctions of introns and exons are known as *splice sites* since these are the sites where the mRNA precursor will be cut and joined.

Figure 21·21
Polyadenylation of mRNA precursors. (Illustrators: Initial rendering—Lisa Shoemaker; Final rendering—Sarah McQueen.)

(a) Polyadenylation begins when RNA polymerase II synthesizes a polyadenylation signal sequence whose consensus sequence is AAUAAA.

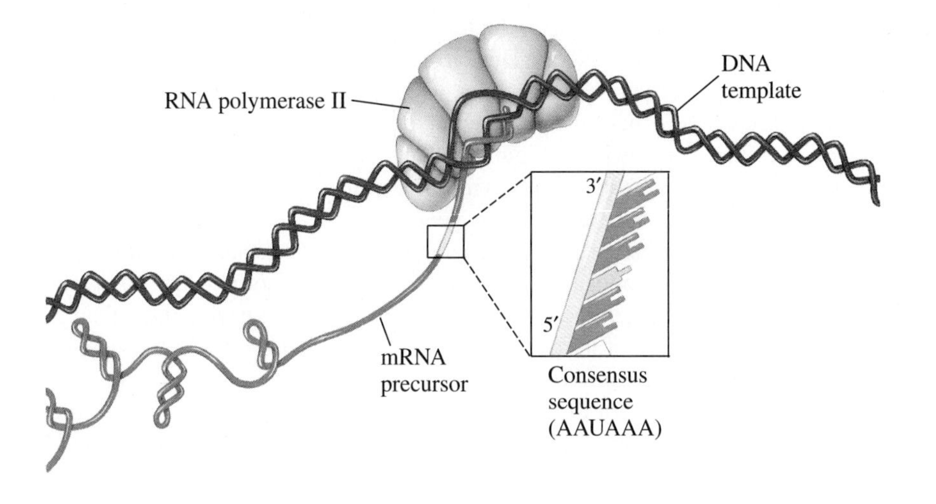

(b) The polyadenylation signal sequence is recognized by an endonuclease that cleaves the RNA at a point downstream from the recognition site. A new 3′ end is thus produced.

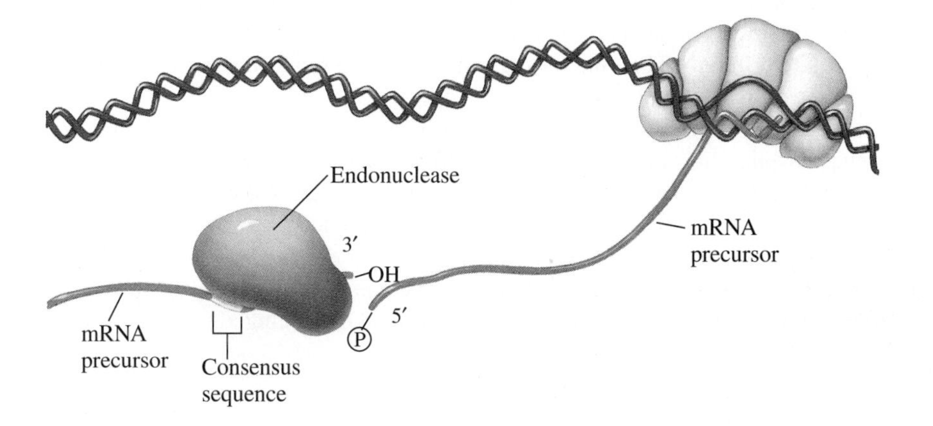

(c) The new 3′ end is polyadenylated by the activity of poly A polymerase.

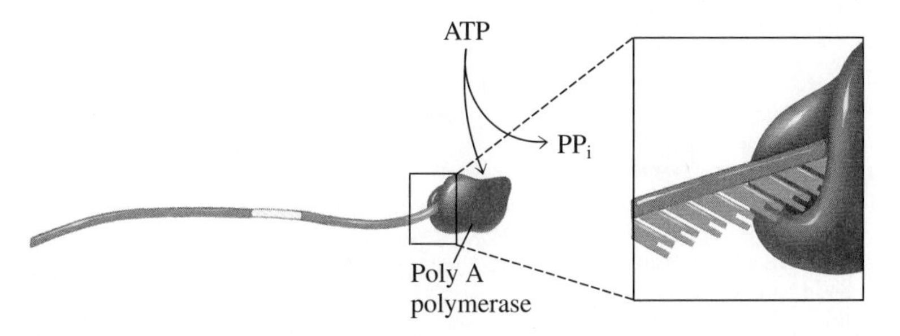

(d) Once the poly A tail has been added to the mRNA precursor, the tail becomes associated with a protein of molecular weight 78 000.

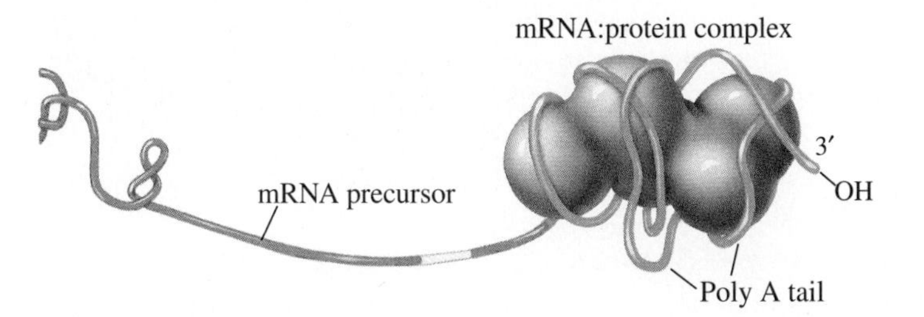

Because of the loss of introns, mature mRNA is often a fraction of the size of the primary transcript. For example, the gene for triose phosphate isomerase from maize contains nine exons and eight introns and spans over 3400 base pairs of DNA, whereas the mature mRNA, which includes a poly A tail, is only 1050 nucleotides in length (Figure 21·22a). The enzyme itself contains 253 amino acids.

The exons in eukaryotic genes are not positioned randomly; each exon usually encodes a discrete domain of the protein with defined secondary structure, such as a region of β sheets or α helices (Figure 21·22b). This organization, which is common in eukaryotes, contrasts sharply with the organization of most genes in prokaryotes. The gene for triose phosphate isomerase in *E. coli,* for example, has no introns, even though the enzymes from bacteria and from maize are very similar.

The gene for triose phosphate isomerase in maize is not unusual; many angiosperm and vertebrate genes contain multiple introns, some having more than 50 introns. The organization of introns in a gene can be visualized using *R-looping,* a technique in which mature mRNA is hybridized to DNA corresponding to the gene. In an electron micrograph of the hybrid, long stretches of DNA that are not represented in the mRNA sequence appear as single-stranded loops, or R-loops. The 16 introns of the gene for chicken egg white conalbumin, for example, are seen as 16 loops in an electron micrograph of the mRNA:DNA hybrid (Figure 21·23).

Introns in protein-coding genes vary in length from as few as 65 base pairs to as many as 10 000 base pairs. Except for short consensus sequences, the nucleotide sequence of an intron plays no known role in splicing or any other processing event. In fact, the intron sequences of identical genes from closely related species are often quite different. This finding suggests that these sequences are not under selective pressure.

Although entire intron sequences can be quite different, the nucleotide sequence at splice sites is similar in all mRNA precursors. The vertebrate consensus sequence at the two splice sites is shown in Figure 21·24. All of the sequences at splice sites closely approximate the consensus sequence, even though few match the consensus sequence exactly. An additional short consensus sequence is found within the intron near the 3′ end. This sequence, known as the *branch site* or the branch-point sequence, also plays an important role in splicing.

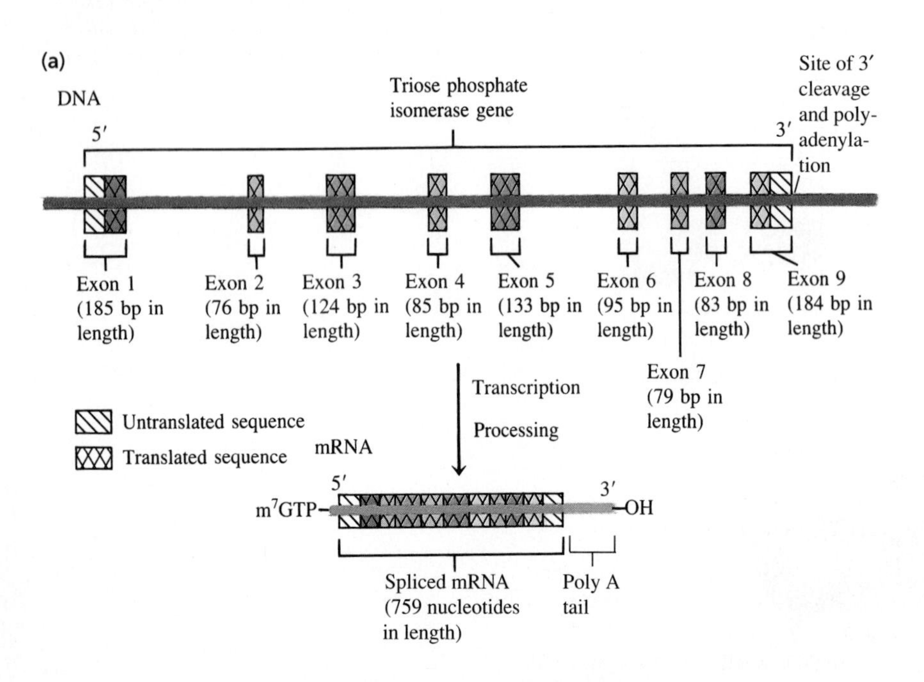

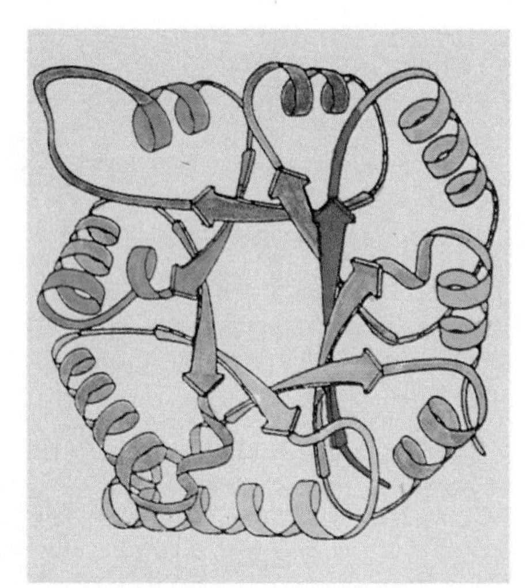

Figure 21·22
Triose phosphate isomerase gene from maize and the encoded enzyme. **(a)** The gene contains nine exons and eight introns. Exons consist of both translated and untranslated sequences. (Illustrators: Initial rendering—Lisa Shoemaker; Final rendering—Brian Eller.) **(b)** Three-dimensional structure of the protein showing the parts of the protein encoded by each exon. (Courtesy of Mark Marchionni.)

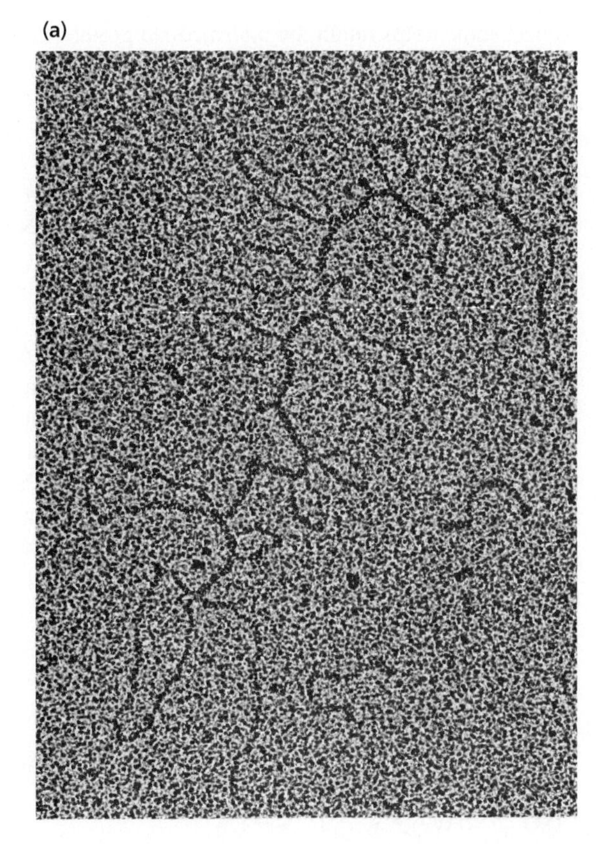

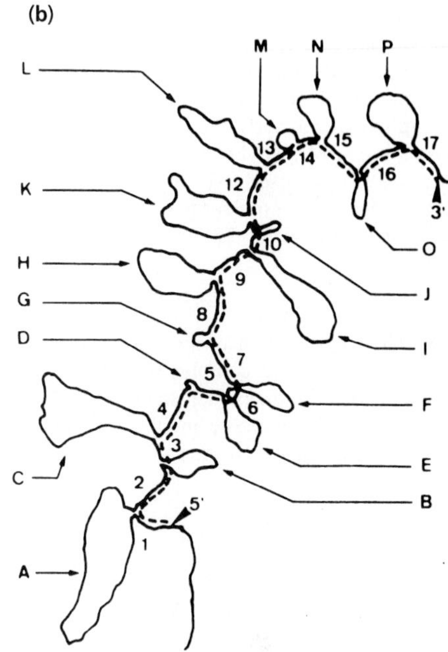

Figure 21·23
Organization of the chicken egg white conalbumin gene visualized by R-looping. **(a)** An electron micrograph of a hybrid mRNA:DNA molecule reveals that the gene contains 17 exons and 16 introns. Exons form double-stranded regions with the single-stranded mature mRNA. The introns form large, single-stranded loops because there is no corresponding sequence in the mature mRNA to which they can bind. **(b)** Tracing of the molecule in part (a), showing the positions of introns (A–P) and exons (1–17). (Courtesy of Pierre Chambon.)

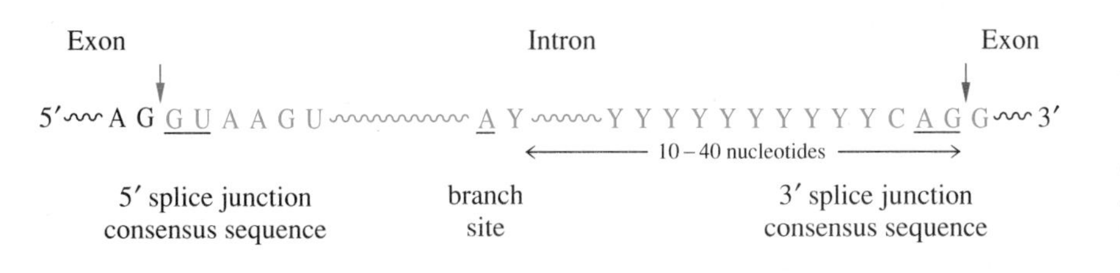

Figure 21·24
Consensus sequence at splice sites in vertebrates. The highly conserved nucleotides are underlined. Y represents any pyrimidine (U or C). The splice sites, where the RNA precursor will be cut and joined, are indicated by red arrows, and the intron is highlighted in blue.

C. Spliceosomes Catalyze Intron Removal

The splicing of mRNA precursors involves two transesterification reactions, one between the 5′ splice site and the branch-site adenosine and one between the 5′ exon and the 3′ splice site. The products of the two reactions are the joined exons and the excised intron in the form of a lariat-shaped molecule. The splicing reactions of mRNA are catalyzed by a large protein:RNA complex called the *spliceosome*. The spliceosome helps retain the intermediate products and position the splice sites so that the exons can be precisely joined (Figure 21·25).

The spliceosome is a large, multisubunit complex of 3×10^6 Da. It is composed of 45 proteins and 5 molecules of RNA whose total length is about 5000 nucleotides. The RNA molecules are called *small nuclear RNA* (snRNA) molecules and are associated with protein to form *small nuclear ribonucleoproteins,* or snRNPs. snRNPs are important not only in the splicing of mRNA precursors but also in other cellular processes.

There are more than 100 000 copies of snRNA molecules per cell nucleus in vertebrates. These are divided into five types called U1, U2, U4, U5, and U6. (U stands for uracil, a common base in these small RNA molecules.) All five snRNA molecules are extensively base-paired and contain modified nucleotides. Each snRNP is composed of one or two specific snRNA molecules plus a number of proteins that are bound to them. Some of the proteins are common to all snRNPs, whereas others are only found in one class of snRNPs.

Figure 21·25
Intron removal in mRNA precursors. Splicing of the precursor is catalyzed by the spliceosome, a multicomponent RNA:protein complex. (Illustrators: Initial rendering—Lisa Shoemaker; Final rendering—Sarah McQueen.) [Adapted from Sharp, P. A. (1987). Splicing of messenger RNA precursors. *Science* 235:766–771.]

(a) The intron sequences are organized by the spliceosome complex so that the adenosine at the branch site is positioned near the 5′ splice site and the 2′-hydroxyl group of the adenosine can attack the 5′ splice site.

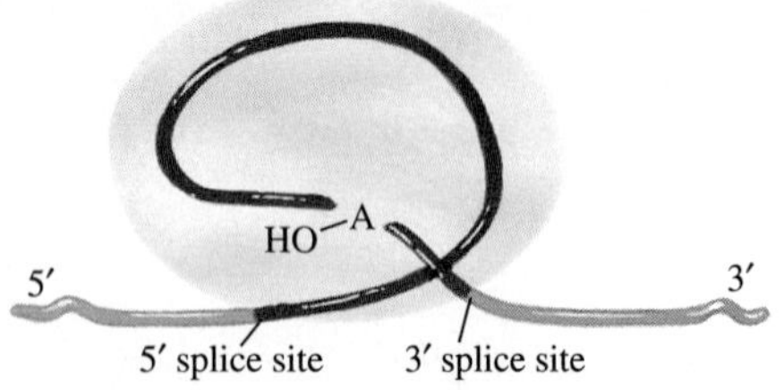

(b) As a result of this reaction, the 2′-hydroxyl group is attached to the 5′ end of the intron and the newly created 3′-hydroxyl group of the exon attacks the 3′ splice site.

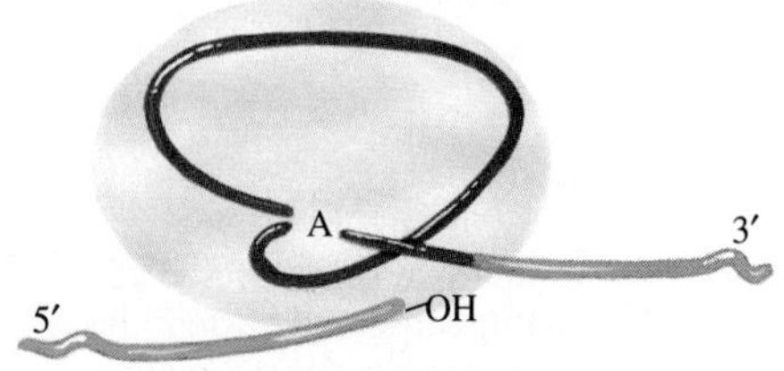

(c) As a result, the ends of the exons are joined, and the intron, in the form of a lariat-shaped molecule, is released.

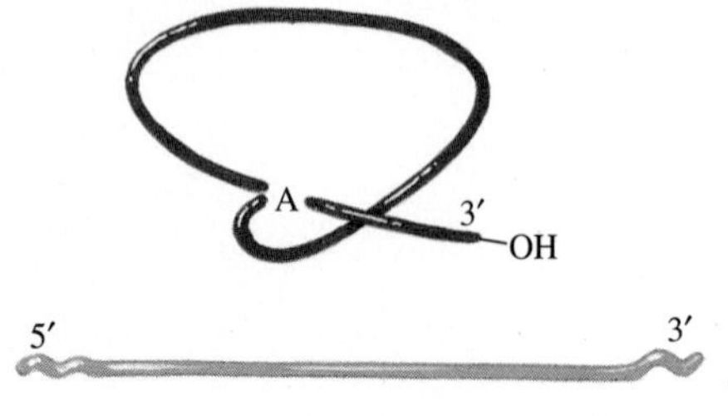

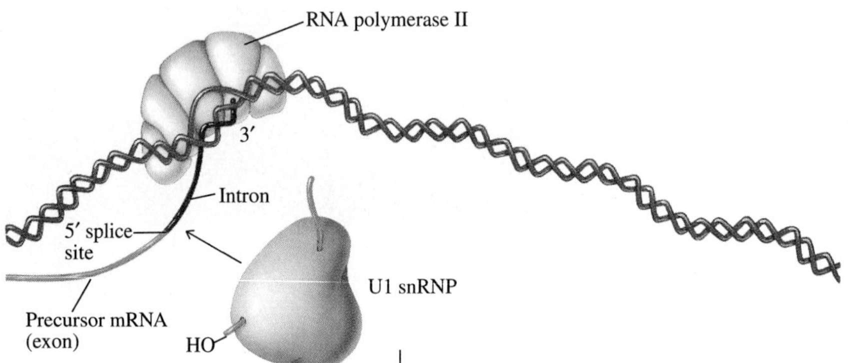

Figure 21·26
Formation of a spliceosome. (Illustrators: Initial rendering—Lisa Shoemaker; Final rendering—Sarah McQueen.)

(1) As soon as the 5′ splice site exits the transcription complex, a U1 snRNP binds to the 5′ splice site.

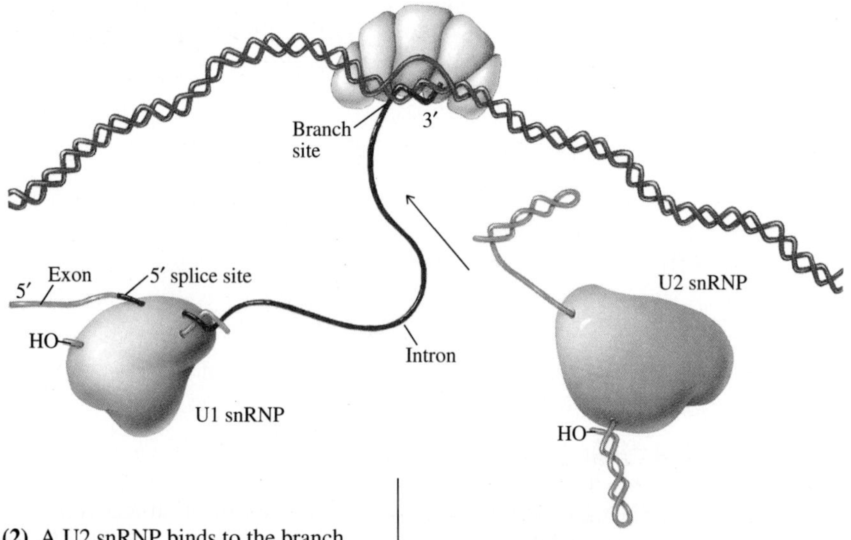

(2) A U2 snRNP binds to the branch site within the intron. *(See next page.)*

These various snRNPs assemble sequentially to form the spliceosome as shown in Figure 21·26. The formation of a spliceosome is initiated when a U1 snRNP binds to the newly synthesized 5′ splice site of the mRNA precursor. This interaction involves base pairing between the 5′ splice site and a complementary sequence near the 5′ end of U1 snRNA. A U2 snRNP then binds to the branch site of the intron, forming a stable complex that covers about 40 nucleotides. Next, a U5 snRNP and a protein called IBP associate with the 3′ splice site. Finally, U4/U6 snRNP joins the complex, and all snRNPs are drawn together to form the spliceosome. The assembly of the complete spliceosome also involves other protein factors in addition to the snRNPs and IBP.

The spliceosome does not form and splicing does not occur unless the 5′ and 3′ splice sites as well as the branch site are occupied by snRNPs. Since spliceosomes can be observed on nascent transcripts, it is thought that removal of the intron is the rate-limiting step in splicing. Although spliceosomes assemble during transcription, the intron is not usually removed until transcription is finished. Once the intron has been removed, the spliceosome disassembles.

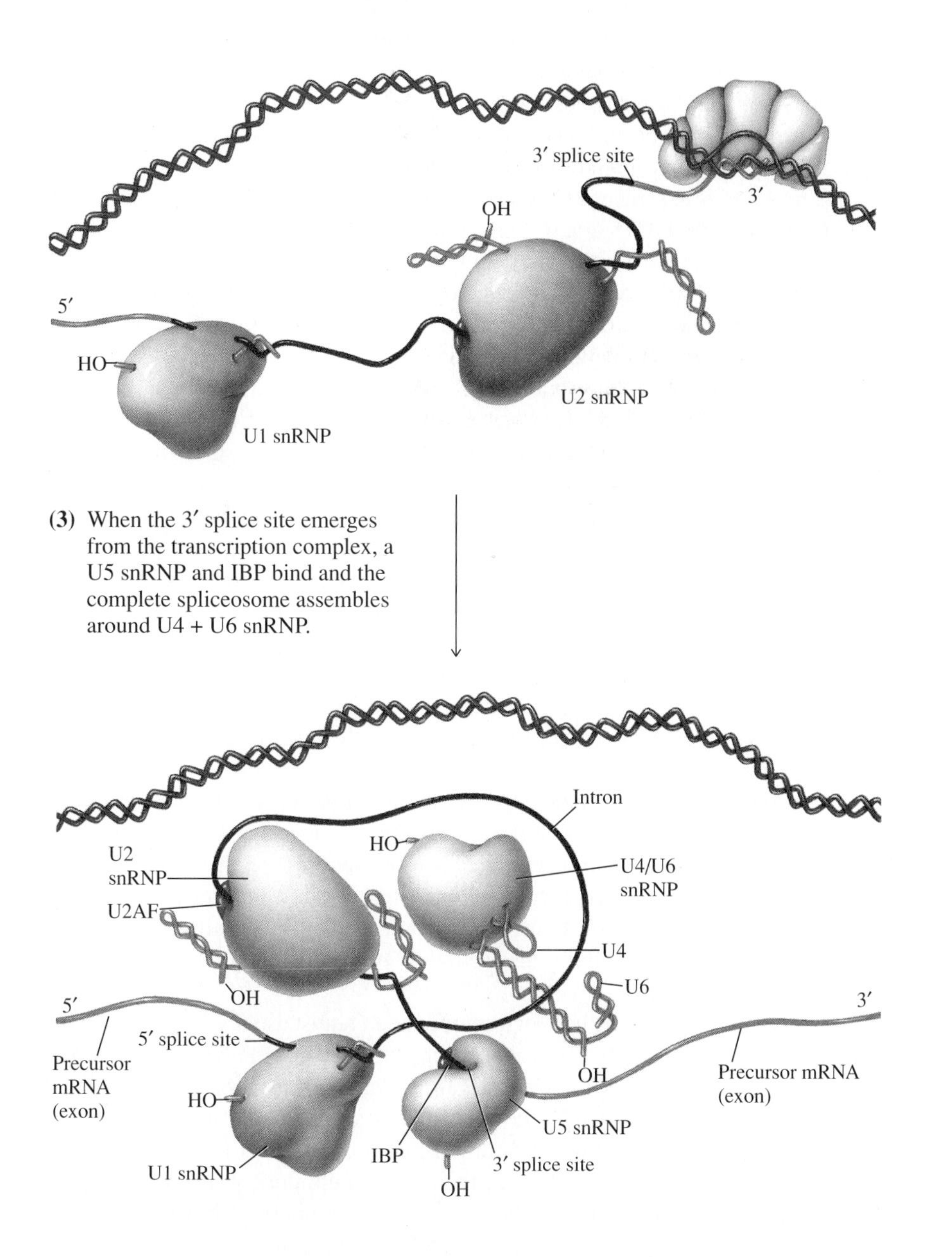

(3) When the 3′ splice site emerges from the transcription complex, a U5 snRNP and IBP bind and the complete spliceosome assembles around U4 + U6 snRNP.

The binding of U1, U2, and U5 snRNP to consensus sequences at the 5′ splice site, branch site, and 3′ splice site of the intron positions these three interactive sites properly for the splicing reaction. The spliceosome prevents the 5′ exon from diffusing away after cleavage and positions it to be joined to the 3′ exon.

Once a spliceosome has formed at an intron, it is quite stable and can be purified from cell extracts. Since the spliceosome, which is almost as large as a ribosome, is too large to fit through the nuclear pores, it prevents the mRNA precursor from leaving the nucleus before processing is complete.

Summary

The flow of biological information within a cell is from DNA to RNA to protein. The transfer of information from DNA to RNA is known as transcription. After transcription, RNA molecules can be modified or processed.

There are four classes of RNA molecules required for the transfer of genetic information from DNA to protein: transfer RNA (tRNA), ribosomal RNA (rRNA), messenger RNA (mRNA), and small RNA molecules such as small nuclear RNA (snRNA). Of all these classes, mRNA is the least stable.

RNA polymerase is the enzyme that catalyzes DNA-directed RNA synthesis, or transcription. This enzyme uses ribonucleoside triphosphates (ATP, UTP, CTP, and GTP) in a reaction similar to that catalyzed by DNA polymerase. Transcription begins near DNA sequences called promoters. RNA polymerase binds to the promoter and unwinds the DNA helix, exposing the template strand to the polymerase active site and creating an open complex. These steps in the initiation of transcription are followed by the synthesis of a short RNA primer. Once the primer has been synthesized, the enzyme catalyzes processive chain elongation. Finally, the enzyme catalyzes termination of transcription.

In all species, RNA polymerase is an multimeric protein. In *E. coli,* the holoenzyme of RNA polymerase has the composition $\alpha_2\beta\beta'\sigma$. The σ subunit readily dissociates from the core polymerase ($\alpha_2\beta\beta'$). This ability to dissociate reflects the nature of the σ subunit's role in transcription. In prokaryotes, σ is responsible for promoter recognition—it increases the affinity of RNA polymerase for specific promoter sequences and decreases its affinity for other DNA sequences. Once transcription starts, σ must dissociate in order for the polymerase to begin elongation. In eukaryotes, there are several accessory transcription factors required for formation of the transcription complex.

Not all genes are expressed in every cell. Genes that are expressed at varying levels in response to different circumstances are said to be regulated. Regulation of gene expression can occur at any point during the flow of biological information but occurs most often at the level of transcription.

Transcription can be regulated by the activity of proteins called activators and repressors. Activators and repressors are DNA-binding proteins that recognize specific sequences near the promoters of genes whose expression they regulate. Individual activators and repressors function in different ways, but in general, activators stimulate transcription from poor promoters by increasing polymerase binding, and repressors inhibit transcription by preventing polymerase binding or initiation of transcription.

After an RNA molecule has been synthesized, it must usually be processed before it becomes biologically functional. There are three general types of RNA processing: the removal of RNA nucleotides from primary transcripts, the addition of RNA nucleotides not encoded by the gene, and the covalent modification of bases. Only prokaryotic mRNA molecules do not require any RNA processing. tRNA molecules in prokaryotes are produced from long, multimeric primary transcripts. The release and subsequent processing of these monomeric tRNA precursors is carried out by several nucleases. One of these is the endonuclease RNase P. RNase

P is a ribonucleoprotein; the active site of this enzyme rests within the RNA component of the molecule. After RNase P releases the tRNA precursors from the primary transcript, a number of nucleotides within the tRNA precursors are covalently modified. rRNA molecules are also produced from long, primary transcripts. Release of rRNA monomers is catalyzed by the endonuclease RNase III.

The processing of mRNA precursors in eukaryotes involves capping, polyadenylation, and splicing. Splicing, the removal of introns and the joining of exons, is catalyzed by a complex called the spliceosome. The spliceosome is a large complex composed of a number of proteins and five species of small nuclear ribonucleoproteins, or snRNPs.

Selected Readings

General References

Alberts, B., Bray, D., Lewis, J., Raff, M., Roberts, K., and Watson, J. D. (1989). *Molecular Biology of the Cell,* 2nd ed. (New York: Garland Publishing), pp. 481–612.

Darnell, J., Lodish, H., and Baltimore, D. (1990). *Molecular Cell Biology,* 2nd ed. (New York: Scientific American Books), pp. 227–487.

Lewin, B. (1991). *Genes IV.* (New York: Oxford University Press).

Watson, J. D., Hopkins, N. H., Roberts, J. W., Steitz, J. A., and Weiner, A. M. (1987). *Molecular Biology of the Gene,* Vol. 1, 4th ed. (Menlo Park, California: Benjamin/Cummings Publishing Company), pp. 360–381.

RNA Polymerases and Transcription

Helmann, J. D., and Chamberlin, M. J. (1988). Structure and function of bacterial sigma factors. *Annu. Rev. Biochem.* 57:839–872.

Sentenac, A., and Kedinger, C. (1985). Eukaryotic RNA polymerases. *Crit. Rev. Biochem.* 18:31–90.

von Hippel, P. H., and Berg, O. G. (1989). Facilitated target location in biological systems. *J. Biol. Chem.* 264:675–678.

Regulation of Transcription

Harrison, S. C., and Aggarwal, A. K. (1990). DNA recognition by proteins with the helix-turn-helix motif. *Annu. Rev. Biochem.* 59:933–969.

Oehler, S., Eismann, E. R., Krämer, H., and Müller-Hill, B. (1990). The three operators of the *lac* operon cooperate in repression. *EMBO J.* 9:973–979.

Processing of RNA

Aebi, M., and Weissman, C. (1987). Precision and orderliness in splicing. *Trends Genet.* 3:102–107.

Björk, G. R. (1987). Modification of stable RNA. In Escherichia coli *and* Salmonella typhimurium: *Cellular and Molecular Biology,* Vol. 1, F. C. Neidhardt, ed. (Washington, DC: American Society for Microbiology), pp. 719–731.

Björk, G. R., Ericson, J. U., Gustafsson, C., Hagervall, T. G., Jönsson, Y. H., and Wikström, P. M. (1987). Transfer RNA modification. *Annu. Rev. Biochem.* 56:263–287.

Dreyfuss, G., Swanson, M. S., and Piñol-Roma, S. (1988). Heterogeneous nuclear ribonucleoprotein particles and the pathway of mRNA formation. *Trends Biochem. Sci.* 13:86–91.

James, B. D., Olsen, G. J., Liu, J., and Pace, N. R. (1988). The secondary structure of ribonuclease P RNA, the catalytic element of a ribonucleoprotein enzyme. *Cell* 52:19–26.

King, T. C., and Schlessinger, D. (1987). Processing of RNA transcripts. In Escherichia coli *and* Salmonella typhimurium: *Cellular and Molecular Biology,* Vol. 1, F. C. Neidhardt, ed. (Washington, DC: American Society for Microbiology), pp. 703–718.

Krämer, A. (1988). Presplicing complex formation requires two proteins and U2 snRNP. *Genes Dev.* 2:1155–1167.

Marchionni, M., and Gilbert, W. (1986). The triosephosphate isomerase gene from maize: introns antedate the plant-animal divergence. *Cell* 46:133–141.

Padgett, R. A., Grabowski, P. J., Konarska, M. M., Seiler, S., and Sharp, P. A. (1986). Splicing of messenger RNA precursors. *Annu. Rev. Biochem.* 55:1119–1150.

Reed, R., Griffith, J., and Maniatis, T. (1988). Purification and visualization of native spliceosomes. *Cell* 53:949–961.

Sharp, P. A. (1987). Splicing of messenger RNA precursors. *Science* 235:766–771.

22

Protein Synthesis

As we saw in the previous chapter, the transcription machinery is responsible for the synthesis of cellular RNA, including mRNA. The mechanism of RNA synthesis from a DNA template is conceptually quite simple: the base pairing of nucleotides allows DNA to serve as a direct template for RNA synthesis. Much less apparent, however, is the mechanism by which the sequence of nucleic acids in mRNA specifies the sequence of amino acids in a protein.

The sequence of nucleotides in a gene is encoded information that specifies the order of amino acids in a polypeptide. This encoded information is copied, or transcribed, into a message (mRNA) that is then translated by the protein synthesis machinery in the cell. With a few minor exceptions, all living organisms use the same genetic code to specify the amino acid sequence of a protein. Reading of this genetic code, and hence translation of a nucleic acid sequence into an amino acid sequence, is mediated by transfer RNA (tRNA) molecules. tRNA molecules are the adapters that form the interface between mRNA and proteins. One region of a tRNA molecule is covalently linked to a specific amino acid, while another region of the same tRNA molecule interacts directly with mRNA by complementary base pairing. It is the processive joining of the amino acids specified by an mRNA template that allows the precise synthesis of proteins.

The protein synthesis machinery of the cell is composed of a variety of proteins, RNA molecules, and protein:RNA complexes. In this chapter, we shall consider most aspects of protein synthesis, including the genetic code, the role of tRNA, the structure of ribosomes, and the reactions catalyzed by the protein synthesis machinery.

22·1 The Genetic Code Is Unambiguous, Degenerate, and Almost Universal

The specific code by which nucleic acids specify amino acids was not known until 1965. Ten years before that, George Gamow first proposed the basic structural units of the genetic code. He reasoned that, since the DNA "alphabet" consists of only four "letters" (A, T, C, and G) and since these four letters encode 20 amino acids, the genetic code might contain "words" made of a uniform length of three letters. Two-letter words constructed from any combination of the four letters produce a vocabulary of only 16 words (4^2), not enough for all 20 amino acids. In contrast, four-letter words produce a vocabulary of 256 words (4^4), far more than are needed. Three-letter words allow a possible vocabulary of 64 words (4^3), more than sufficient to specify each of the 20 amino acids but not excessive.

In a series of elegant experiments in the early 1960s, Francis Crick and Sydney Brenner demonstrated that the genetic code does in fact consist of three-letter words. Between 1962 and 1965, the actual code was cracked by a number of workers, chiefly Marshall Nirenberg and H. Gobind Khorana. It took ten years of hard work and experimentation to elucidate the way that mRNA encodes proteins. The subsequent development of methods for sequencing genes and proteins has allowed the direct comparison of the primary sequences of proteins with the nucleotide sequences of their corresponding genes. Each time a new protein and its gene are characterized, the genetic code is confirmed.

In principle, a code made up of three-letter words can be either overlapping or nonoverlapping. In an overlapping code, a letter is part of more than one word; changing a single letter will change several words simultaneously. For example, in the sequence shown in Figure 22·1, each letter is part of three different words in an overlapping code. In contrast, a particular letter is part of only one word in a nonoverlapping code. In this case, changing a particular letter will change one and only one word. The genetic code used by all living organisms is nonoverlapping.

In a nonoverlapping code of three-letter words, the message translated from a particular sequence of letters depends upon the point at which translation begins; shifting the starting point for the reading of the code by one letter to the left or to the right changes the entire message. Each potential starting point defines a unique sequence of three-letter words, or *reading frame*. Thus, correct translation of the genetic code depends upon establishing the correct reading frame for translation (Figure 22·2).

mRNA · · · A U G C A U G C A U G C · · ·

(a) Message read in overlapping triplet code
A U G
U G C
G C A
C A U
· · ·
· · ·

(b) Message read in nonoverlapping triplet code
A U G
C A U
G C A
U G C

Figure 22·1
Message read in **(a)** overlapping and **(b)** nonoverlapping three-letter codes. In an overlapping code, each letter is part of three different three-letter words (as indicated for the letter G in blue); in a nonoverlapping code, each letter is part of only one three-letter word.

mRNA	··· A U G C A U G C A U G C ···
Message read in reading frame 1	··· |A U G|C A U|G C A|U G C| ···
Message read in reading frame 2	··· A|U G C|A U G|C A U|G C ···
Message read in reading frame 3	··· A U|G C A|U G C|A U G|C ···

Figure 22·2
The three reading frames of mRNA. The same string of letters read in three different reading frames can be translated as three different messages. Thus, translation of the correct message requires selection of the correct reading frame.

The standard genetic code is shown in Figure 22·3. With a few minor exceptions, all living organisms use this genetic code, suggesting that all contemporary species are descended from a common ancestor that also utilized this code. This ancestral species probably lived about two billion years ago, or more than one billion years after life first arose on this planet.

The genetic code is made up of three-letter, nonoverlapping words, or *codons*. By convention, all nucleotide sequences are written $5' \rightarrow 3'$. Thus, UAC specifies tyrosine and CAU specifies histidine. The term *codon* usually refers to triplets of nucleotides in mRNA, but it can also apply to triplets of nucleotides in the DNA sequence of a gene. For example, one DNA codon for tyrosine would be TAC. The codons are always translated sequentially in the $5' \rightarrow 3'$ direction, beginning near the 5′ end of the message (the end first synthesized) and proceeding to the end of the coding region, which is usually near the 3′ end of the mRNA.

First position (5′ end)	Second position U	C	A	G	Third position (3′ end)
U	Phe	Ser	Tyr	Cys	U
	Phe	Ser	Tyr	Cys	C
	Leu	Ser	STOP	STOP	A
	Leu	Ser	STOP	Trp	G
C	Leu	Pro	His	Arg	U
	Leu	Pro	His	Arg	C
	Leu	Pro	Gln	Arg	A
	Leu	Pro	Gln	Arg	G
A	Ile	Thr	Asn	Ser	U
	Ile	Thr	Asn	Ser	C
	Ile	Thr	Lys	Arg	A
	Met	Thr	Lys	Arg	G
G	Val	Ala	Asp	Gly	U
	Val	Ala	Asp	Gly	C
	Val	Ala	Glu	Gly	A
	Val	Ala	Glu	Gly	G

Figure 22·3
The standard genetic code. The standard genetic code is composed of 64 triplet codons, whose mRNA sequences can be read from this chart. The left-hand column indicates the nucleotide found at the first (5′) position of the codon; the top row indicates the nucleotide found at the second (middle) position of the codon; and the right column indicates the nucleotide found at the third (3′) position of the codon. The codon AUG specifies methionine (Met) and is also used to initiate protein synthesis. STOP indicates a termination codon.

The standard genetic code has several prominent features.

1. It is unambiguous. In a particular organism or organelle, each codon corresponds to only one amino acid.

2. There are several codons for most amino acids. For example, there are six codons for serine, four for glycine, and two for lysine. Because of the existence of several codons for most amino acids, the genetic code is said to be *degenerate.* Different codons that specify the same amino acid are known as *synonymous codons*. The degeneracy of the genetic code minimizes the effects of mutations since the change of a single nucleotide often results in a codon that still specifies the same amino acid. The only amino acids with single codons are methionine and tryptophan.

3. The first two nucleotides of a codon are often enough to specify a given amino acid. For example, the four codons for glycine all begin with GG: GGU, GGC, GGA, and GGG.

4. Codons with similar sequences specify chemically similar amino acids. For example, the codons for the amino acid threonine differ from four of the codons for serine by only a single nucleotide at the 5′ position; the codons for the amino acids aspartate and glutamate begin with GA and differ only at the 3′ position. Codons that have pyrimidines at their second position usually encode hydrophobic amino acids. Therefore, mutations that alter either the 5′ or 3′ position of these codons will usually result in a chemically similar amino acid being incorporated into the protein.

5. Only 61 of the 64 codons specify amino acids. The three remaining codons (UAA, UGA, and UAG) are *termination codons,* or stop codons. Termination codons are not normally recognized by any tRNA molecule. Instead, they are recognized by specific proteins that cause newly synthesized peptides to be released from the translation machinery. The methionine codon AUG is also used to specify the initiation site for protein synthesis and is therefore frequently called the *initiation codon.*

22·2 Transfer RNA Molecules Are Required for Protein Synthesis

Transfer RNA molecules serve as interpreters of the genetic code. They are the crucial link between the sequence of nucleotides in mRNA and the sequence of amino acids in a polypeptide. In order for tRNA to fulfill this role, every cell must contain at least twenty tRNA species (one for every amino acid), and each of these tRNA molecules must also be able to recognize at least one mRNA codon.

A. All tRNA Molecules Have a Similar Three-Dimensional Structure

The nucleotide sequences of different tRNA molecules from many organisms have been determined. Despite diversity in their primary structures, the sequences of all analyzed tRNA molecules are compatible with a secondary structure that may be represented as a cloverleaf conformation (Figure 22·4). This cloverleaf structure is subdivided into several *arms,* regions composed of a loop or a loop and hydrogen-bonded stem. Double-stranded regions of these arms form a short, stacked right-handed helix similar to that of double-stranded DNA.

The 5′ end and the region near the 3′ end are base paired, forming a stem known as the *acceptor stem,* or amino acid stem. It is to this stem that the amino acid is covalently attached. The carboxyl group of the amino acid is linked to either the 2′- or 3′-hydroxyl group of the ribose of the 3′ adenylate. (Recall that mature tRNA molecules are produced by processing a larger primary transcript and that the nucleotides at the 3′ end of a mature tRNA molecule are invariably CCA.) The nucleotide at the 5′ end of all tRNA molecules is phosphorylated.

The single-stranded loop opposite the acceptor stem contains the anticodon, the three-base sequence that binds to a complementary codon in mRNA. The arm of the tRNA molecule that contains the anticodon is referred to as the *anticodon arm.* Two of the arms of the tRNA molecule are named for the covalently modified nucleosides found within them. (See Figure 21·18 for examples of modified sugars and bases in these nucleosides.) One of these arms contains thymidine (T) and pseudouridine (Ψ) followed by cytidine (C) and is referred to as the *TΨC arm.* Dihydrouridine (D) residues lend their name to the *D arm.* tRNA molecules also have a *variable arm* between the anticodon arm and the TΨC arm. The variable arm can range in length from about 3 to 21 nucleotides, and the length of the D arm is also somewhat variable. With a few rare exceptions, most tRNA molecules are between 73 and 95 nucleotides in length.

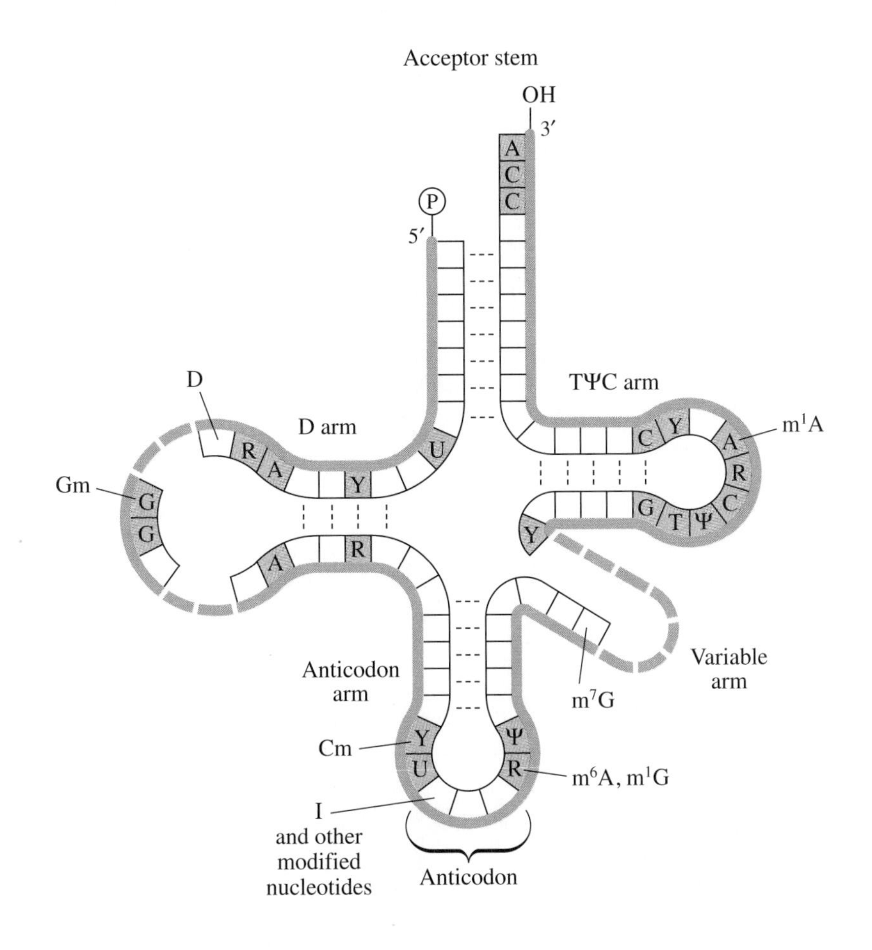

Figure 22·4
Cloverleaf secondary structure of tRNA. Hydrogen bonds of Watson-Crick base pairs are indicated by dashed lines between nucleotide residues. The molecule is divided into an acceptor stem and four arms. The acceptor stem is the site of amino acid attachment, and the anticodon arm is the region of the tRNA molecule that interacts with mRNA. The D and TΨC arms are named for modified nucleosides that are conserved within the arms. The number of nucleotide residues in each arm is more or less constant, except in the variable arm. Conserved bases (colored gray) and positions of common modified nucleosides are noted. Abbreviations other than standard nucleosides: R, purine; Y, pyrimidine; modified nucleosides: m¹A, 1-methyladenosine; Cm, 2′-*O*-methylcytidine; D, dihydrouridine; Gm, 2′-*O*-methylguanosine; m¹G, 1-methylguanosine; m⁷G, 7-methylguanosine; I, inosine; Ψ, pseudouridine; and T, thymidine.

The cloverleaf diagram of tRNA is a two-dimensional representation of a three-dimensional molecule. In three dimensions, the tRNA molecule is folded into an L shape as shown in Figure 22·5 and Stereo S22·1. The acceptor stem is at one end of the L-shaped molecule, and the anticodon is located in a loop at the opposite end. The resulting structure is compact and very stable, in part because of hydrogen bonds between the nucleotides in the D, TΨC, and variable arms. This base pairing may differ from the normal Watson-Crick base pairing that forms between nucleotides in double-stranded regions. Most of the nucleotides in tRNA are part of two perpendicular stacked helices (Figure 22·6); the stacking interactions between the bases are additive and make a major contribution to the stability of tRNA, as they do in double-stranded DNA (Figure 10·9).

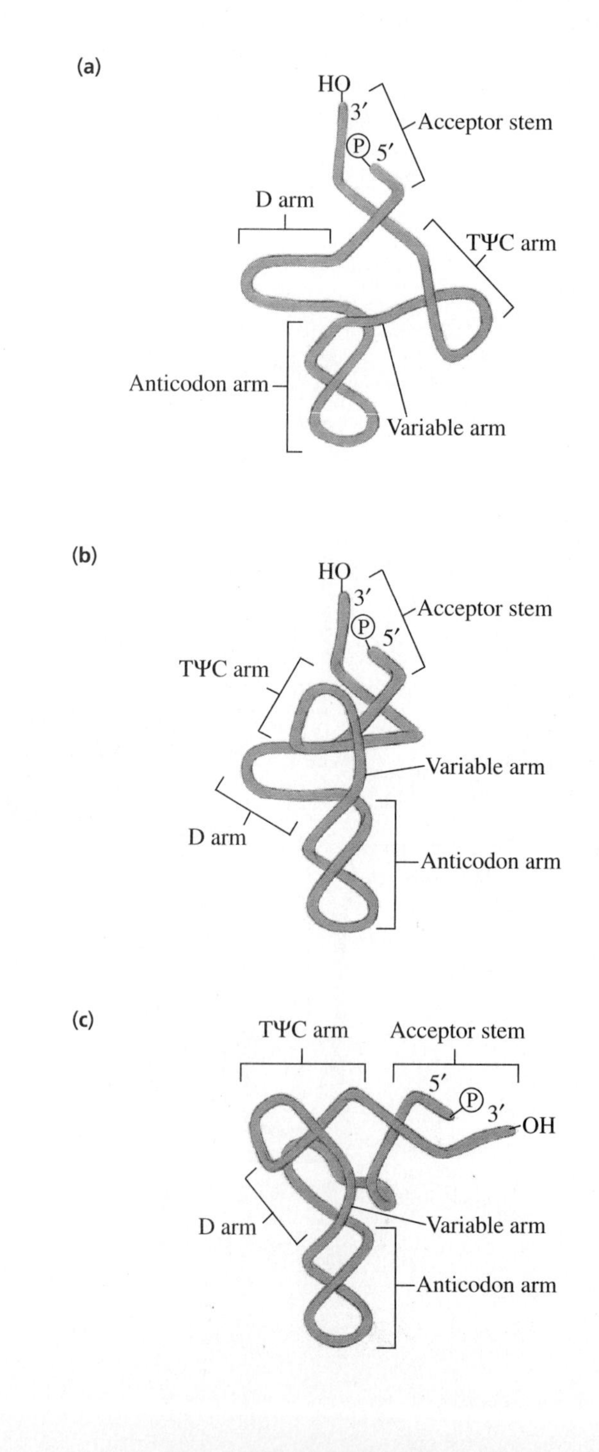

Figure 22·5
Relationship of the secondary and tertiary structures of tRNA. The ribose-phosphate backbone is indicated by a continuous ribbon. **(a)** Cloverleaf secondary structure. **(b)** The tertiary structure of tRNA involves base pairing and van der Waals interactions between the TΨC arm and the D arm, and **(c)** between the variable arm and the D arm. Mature tRNA molecules are L shaped. (Illustrators: Initial rendering—Lisa Shoemaker; Final rendering—Linda LeFevre Murray.)

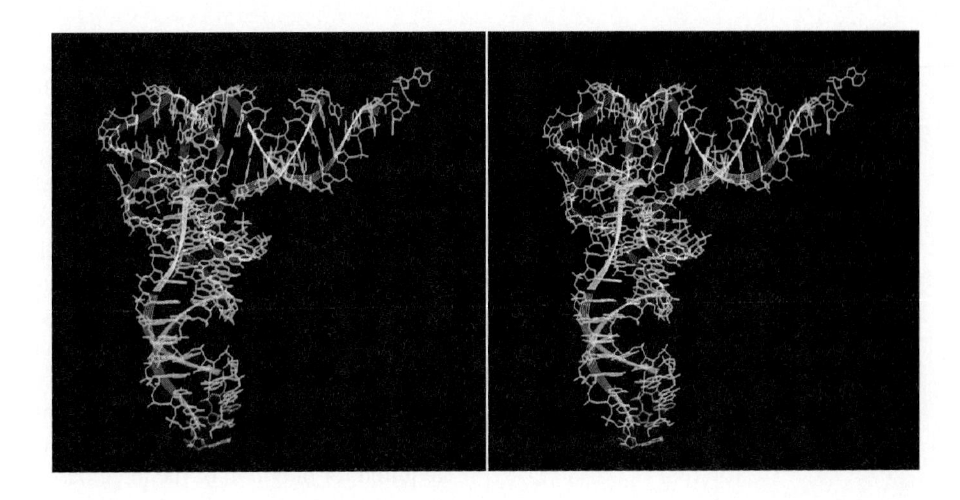

Stereo S22·1
Yeast $tRNA^{Phe}$. The residues are shown in green except for the anticodon, which is shown in blue. The ribbon highlights the sugar phosphate backbone and the different regions of secondary structure. Color key for ribbon: acceptor stem, orange; TΨC loop, purple; D loop, red; variable arm, yellow; anticodon loop, blue. (Based upon coordinates provided by E. Westhof and M. Sundaralingam; stereo image courtesy of Ben N. Banks and Kim M. Gernert.)

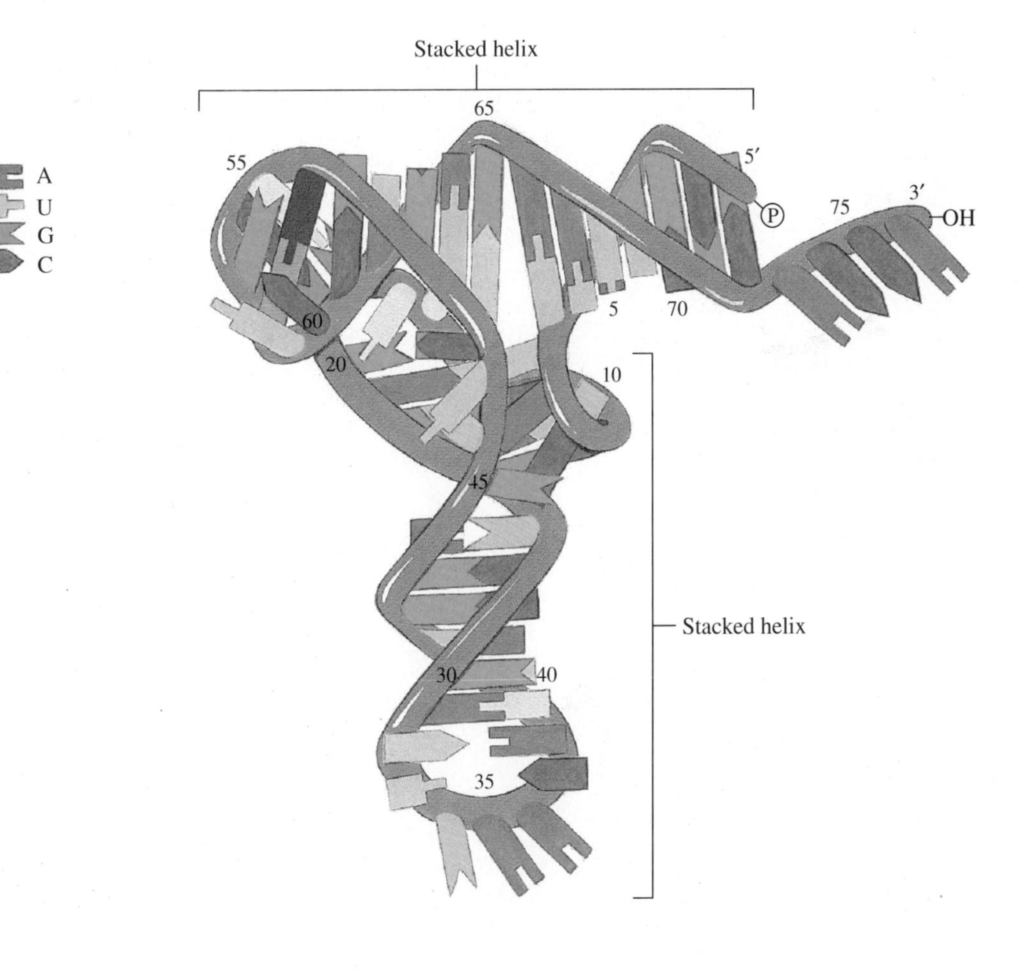

Figure 22·6
Tertiary structure of yeast $tRNA^{Phe}$. Most of the nucleotides lie within one of two perpendicular, stacked helices. This tertiary structure is stabilized by hydrogen bonds between nucleotides in the TΨC, D, and variable arm and by stacking interactions between the bases. In some cases, the hydrogen bonds are different from the Watson-Crick hydrogen bonds that form between nucleotides in double-stranded DNA. Red indicates thymine, and paler colors indicate modifications of standard bases. For example, pale yellow indicates modified uracil. (Illustrators: Initial rendering—Lisa Shoemaker; Final rendering—Linda LeFevre Murray.) [Adapted from Kim, S. H., Suddath, F. L., Quigley, G. J., McPherson, A., Sussman, J. L., Wang, A. H. J., Seeman, N. C., and Rich, A. (1974). Three-dimensional tertiary structure of yeast phenylalanine transfer RNA. *Science* 185:435–439.]

B. The tRNA Anticodons Base-Pair with mRNA Codons

tRNA and mRNA molecules interact through base pairing between anticodons and codons. The anticodon of a tRNA molecule therefore determines, in part, where the amino acid that is attached to its acceptor stem will be added to a polypeptide chain. A superscript is used to designate which amino acid the anticodon of the tRNA specifies. For instance, the tRNA molecule shown in Figure 22·6 has the anticodon GAA, which binds to the phenylalanine codon UUC. During protein synthesis, phenylalanine will be covalently attached to the acceptor stem of this tRNA. The molecule is therefore designated $tRNA^{Phe}$.

Much of the base pairing between the codon and the anticodon is governed by the rules of Watson-Crick base pairing: A pairs with U, G pairs with C, and the strands in the base-paired region are antiparallel. There is some variation, however. In 1966, Francis Crick proposed that complementary, Watson-Crick base pairing is required for only two of the three base pairs formed when the codon interacts with the anticodon. The codon must form Watson-Crick base pairs with the 3′ and middle bases of the anticodon, but other types of base pairing are permitted at the 5′ position of the anticodon. This alternate pairing suggests that the 5′ position is conformationally flexible. Crick dubbed this flexibility *wobble*. The 5′ position of the anticodon is thus sometimes called the *wobble position*.

Table 22·1 summarizes permitted base pairing between the wobble position of an anticodon and an mRNA codon. When guanine is at the wobble position, for example, it can pair with uracil, and vice versa. GU base pairing is also found in other parts of tRNA molecules. The base at the wobble position of many anticodons is often covalently modified, permitting additional flexibility in codon recognition. For example, guanosine at the 5′ anticodon position in several tRNA molecules is deaminated at C-2 to form inosine, which can form hydrogen bonds with adenosine, cytidine, or uridine (Figure 22·7). The presence of inosine at the 5′ position of the anticodon explains why $tRNA^{Ala}$ with the anticodon IGC can bind to three codons specifying alanine: GCA, GCC, and GCU (Figure 22·8).

Wobble allows some tRNA molecules to recognize more than one codon, but often several different tRNA molecules are required in order to recognize *all* synonymous codons. Different tRNA molecules that bind the same amino acid are called *isoacceptor tRNA molecules*. The term isoacceptor applies to tRNA molecules with different anticodons that bind the same amino acid, as well as to tRNA molecules with the same anticodon but different primary sequences.

Table 22·1 Rules for base pairing between the 5′ (wobble) position of the anticodon and the 3′ position of the codon

Base at 5′ (wobble) position of anticodon	Base recognized at 3′ position of codon
C	G
A	U
U	A or G
G	U or C
I*	U, A, or C

*I = Inosine

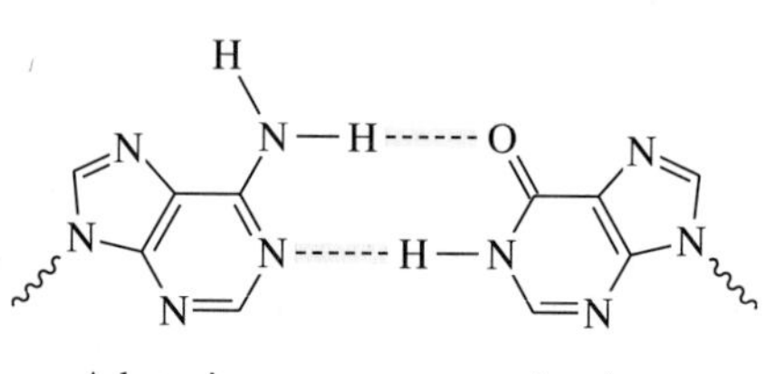

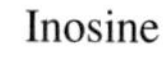

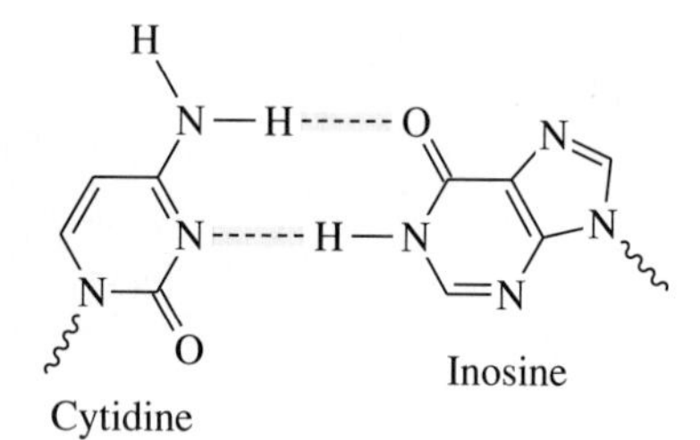

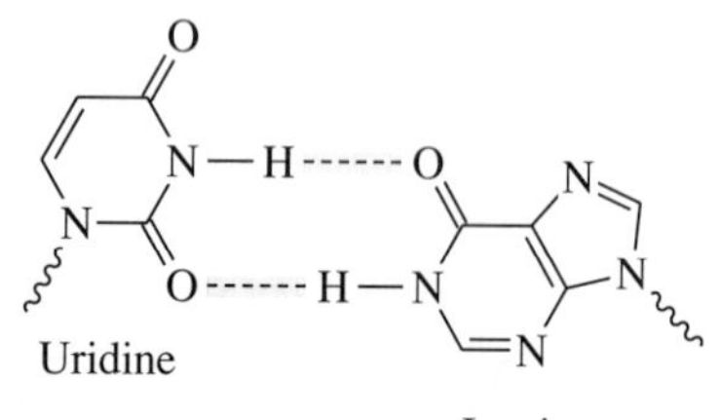

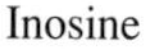

Figure 22·7
Inosine base pairs. Inosine is often found at the 5′ (wobble) position of a tRNA anticodon. Inosine can form hydrogen bonds with adenine, cytosine, or uracil. This versatility in hydrogen bonding allows the anticodon to recognize more than one synonymous codon.

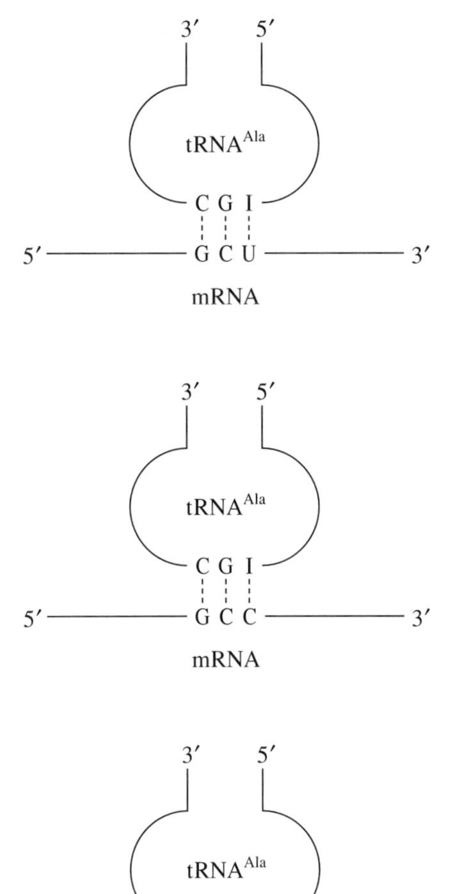

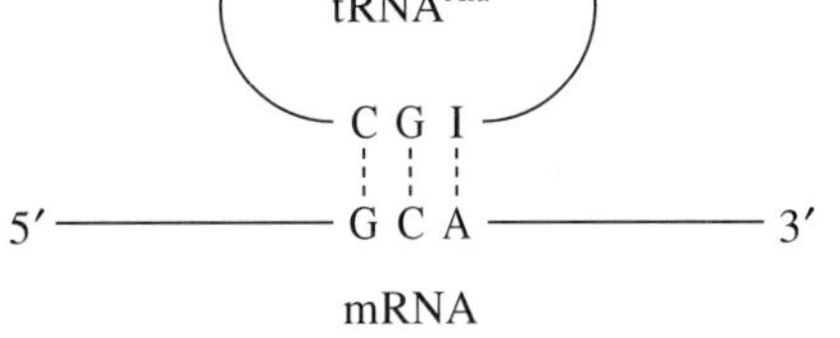

Figure 22·8
Base pairing at the wobble position. The tRNA^{Ala} molecule with the anticodon 5′-IGC-3′ can bind to any one of three codons specifying alanine (GCU, GCC, or GCA) because inosine can pair with U, C, or A. Note that the codon-anticodon interaction involves antiparallel RNA strands.

22·3 An Aminoacyl-tRNA Synthetase Catalyzes the Addition of an Amino Acid Residue to a tRNA Molecule

A particular amino acid is covalently attached to the 3′ end of each tRNA molecule. The product of this aminoacylation reaction is called an *aminoacyl-tRNA*. Because the aminoacyl-tRNA bond is a high-energy bond, the amino acid is said to be *activated* for subsequent transfer to a growing polypeptide chain. Thus aminoacyl-tRNA molecules are sometimes called *charged* or *activated* tRNA molecules. The specific type of aminoacylated tRNA molecule is indicated by a prefix; for instance, aminoacylated tRNA^{Ala} is called alanyl-tRNA^{Ala}. The enzymes that catalyze this aminoacylation reaction are called aminoacyl-tRNA synthetases.

Since 20 different amino acids are incorporated into proteins, at least 20 different aminoacyl-tRNA synthetases must be present in each cell. There are rarely more than 20 aminoacyl-tRNA synthetases, however, because a single synthetase is able to recognize many isoacceptor tRNA molecules. For example, there are six codons for serine and several different isoacceptor tRNA^{Ser} molecules. Each of these tRNA^{Ser} molecules is recognized by a single seryl-tRNA synthetase. The accuracy of protein synthesis depends on the ability of the aminoacyl-tRNA synthetases to catalyze the attachment of the correct amino acid to the appropriate tRNA species.

A. Aminoacyl-tRNA Molecules Are Synthesized in Two Steps

The activation of an amino acid by aminoacyl-tRNA synthetases is similar in principle to the activation of fatty acids (Section 17·9). The overall reaction is:

$$\text{Amino acid} + \text{tRNA} + \text{ATP} \longrightarrow \text{Aminoacyl-tRNA} + \text{AMP} + \text{PP}_i \quad (22\cdot 1)$$

The amino acid is attached to its tRNA molecule by formation of an ester linkage between the carboxylate group of the amino acid and the 2′- or 3′-hydroxyl group of the ribose at the 3′ end of the tRNA molecule.

This reaction occurs in two discrete steps (Figure 22·9). In the first step, the aminoacyl-tRNA synthetase catalyzes formation of a reactive intermediate. The carboxylate group of the amino acid attacks the α-phosphorus of ATP, displacing pyrophosphate and producing an aminoacyl-adenylate.

$$\text{Amino acid} + \text{ATP} \longrightarrow \text{Aminoacyl-adenylate} + \text{PP}_i \quad (22\cdot 2)$$

Formation of this high-energy intermediate activates the amino acid, and the aminoacyl-adenylate remains tightly but noncovalently bound to the aminoacyl-tRNA synthetase (see the description of the mechanism of tyrosyl-tRNA synthetase in Section 6·13). Synthesis of the aminoacyl-adenylate is thermodynamically aided by the subsequent hydrolysis of pyrophosphate inside the cell.

The second step of aminoacyl-tRNA synthesis is aminoacyl-group transfer from the intermediate to tRNA. The amino acid is attached to either the 2′- or 3′-hydroxyl group of the 3′-terminal adenylate residue of tRNA depending upon which aminoacyl-tRNA synthetase is catalyzing the reaction.

$$\text{Aminoacyl-adenylate} + \text{tRNA} \longrightarrow \text{Aminoacyl-tRNA} + \text{AMP} \quad (22\cdot 3)$$

The standard free-energy change for Reaction 22·3 is $-7\ \text{kcal mol}^{-1}$; the equilibrium lies overwhelmingly in the direction of aminoacyl-tRNA. Aminoacyl-tRNA is the substrate for protein synthesis, and its standard free energy of hydrolysis is approximately equivalent to that of a phosphoanhydride in ATP. Under cellular conditions, formation of aminoacyl-tRNA is favored and the intracellular concentration of free tRNA is very low. The energy that is stored in the aminoacyl-tRNA bond is ultimately used in the formation of a peptide bond during protein synthesis. Note that two phosphoanhydrides of ATP are cleaved for each aminoacylation reaction.

B. Each Aminoacyl-tRNA Synthetase Is Highly Specific for One Amino Acid and Its Corresponding tRNA Molecules

Aminoacyl-tRNA synthetases must distinguish not only between amino acids but also between tRNA molecules. The attachment of a specific amino acid to its corresponding tRNA molecule is a critical step in the translation of the genetic message. If it is not performed accurately, the wrong amino acid could be incorporated into a protein.

A given aminoacyl-tRNA synthetase binds ATP and also selects the proper amino acid substrate based on the charge, size, and free energy of binding of the amino acid. Amino acid selection initially involves exclusion of amino acids with inappropriate charges or hydrophobicity. For example, tyrosyl-tRNA synthetase readily distinguishes tyrosine from phenylalanine.

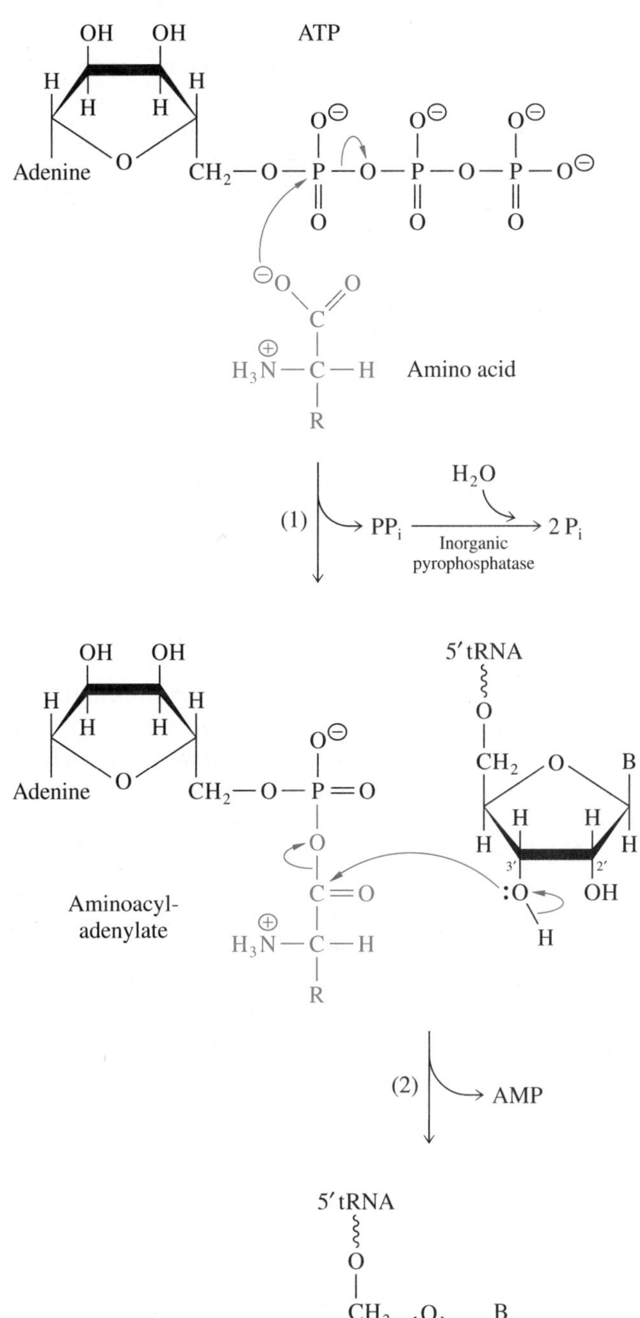

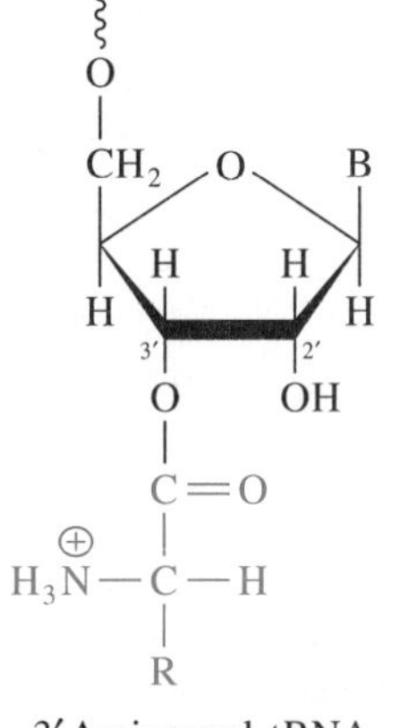

3′ Aminoacyl-tRNA

Figure 22·9
Synthesis of an aminoacyl-tRNA molecule.

The aminoacyl-tRNA synthetase then selectively binds a specific transfer RNA molecule. The proper tRNA molecule is distinguished by features unique to each tRNA structure. In particular, the acceptor stem, which lies on the inside surface of the L of the tRNA molecule, is implicated in the binding of tRNA to the corresponding aminoacyl-tRNA synthetase (Figure 22·10). In some cases, the synthetase recognizes the anticodon as well as the acceptor end of the tRNA. In other cases, the binding of tRNA molecules to the appropriate aminoacyl-tRNA synthetase requires only interactions between the enzyme and the acceptor end of the tRNA. In all cases, however, the net effect of the interaction is to position the 3′ end of the tRNA molecule in the active site of the enzyme.

C. Some Aminoacyl-tRNA Synthetases Have a Proofreading Activity

Isoleucine and valine are chemically similar amino acids, and both can be accommodated in the active site of isoleucyl-tRNA synthetase (Figure 22·11). Isoleucyl-tRNA synthetase mistakenly catalyzes the formation of the valyl-adenylate intermediate about one percent of the time. Based on this observation, we might expect valine to be attached to isoleucyl-tRNA and incorporated into protein in place of isoleucine about one time in a hundred. However, the rate of substitution of valine for isoleucine in polypeptide chains is only about one time in ten thousand. This lower rate of valine incorporation means that isoleucyl-tRNA synthetase also discriminates between the two amino acids *after* aminoacyl-adenylate formation. In fact, isoleucyl-tRNA synthetase catalyzes a proofreading step at the stage of the reaction in which the aminoacyl group is transferred from the aminoacyl-adenylate to the tRNA molecule. Although isoleucyl-tRNA synthetase may mistakenly catalyze the formation of valyl-adenylate, most of the time it catalyzes the hydrolysis of the incorrect valyl-adenylate to valine and AMP so that valyl-$tRNA^{Ile}$ does not form.

Not all aminoacyl-tRNA synthetases exhibit a proofreading activity. For example, no proofreading is required for tyrosyl-tRNA synthetase to distinguish tyrosine from phenylalanine because only tyrosine binds in the active site of tyrosyl-tRNA synthetase.

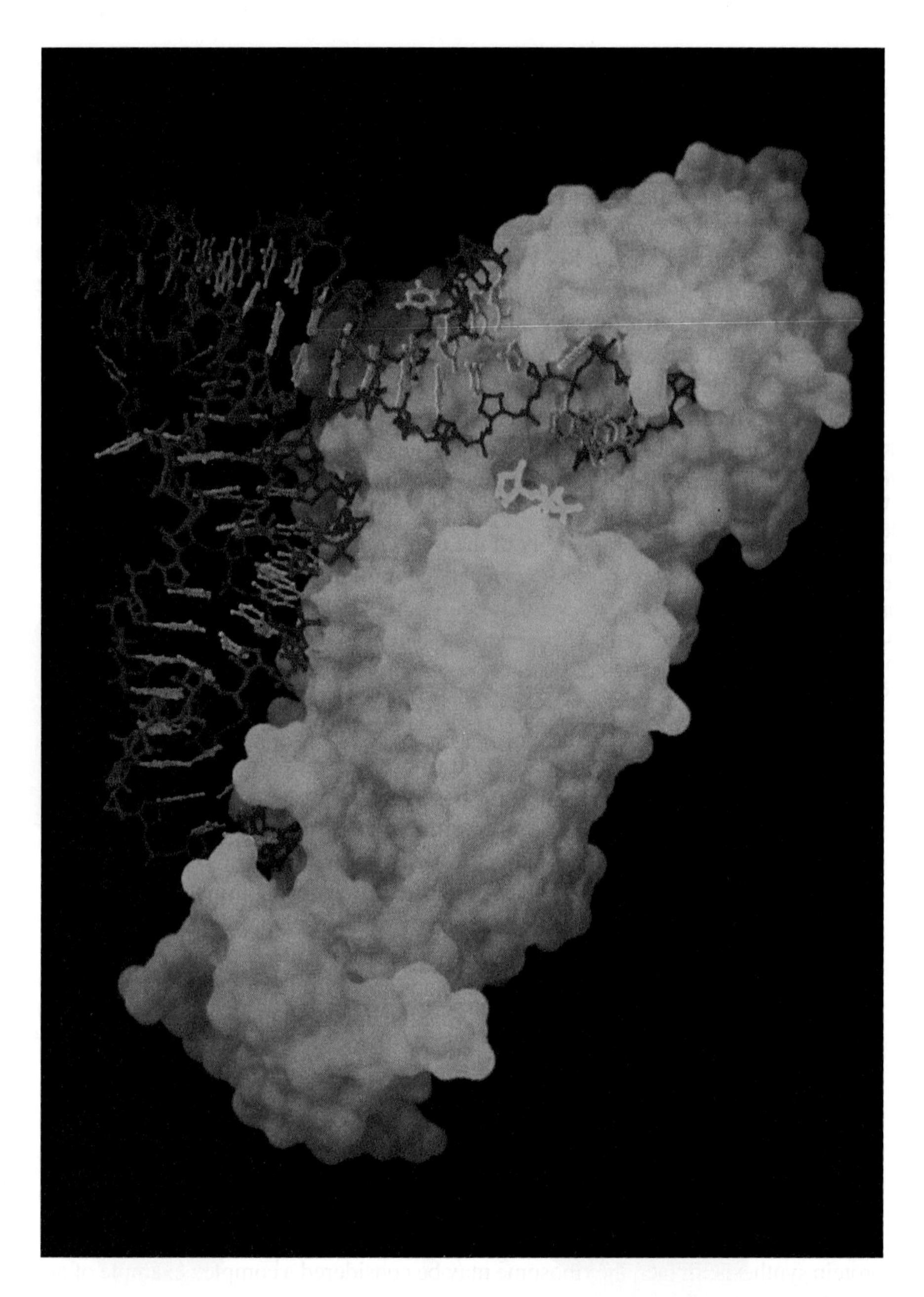

Figure 22·10
Structure of *E. coli* glutaminyl-tRNA synthetase bound to $tRNA^{Gln}$. The 3′ end of the tRNA is buried within a pocket on the surface of the enzyme. A molecule of ATP (green) is also bound at this site. The enzyme is interacting with the acceptor-stem region of the tRNA and also with the anticodon. (Courtesy of Thomas A. Steitz.)

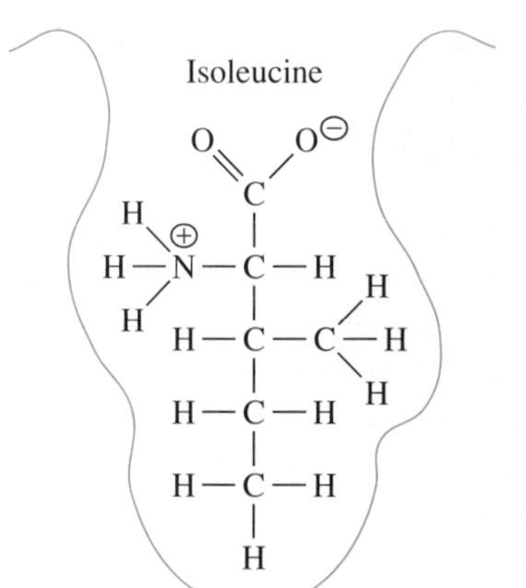

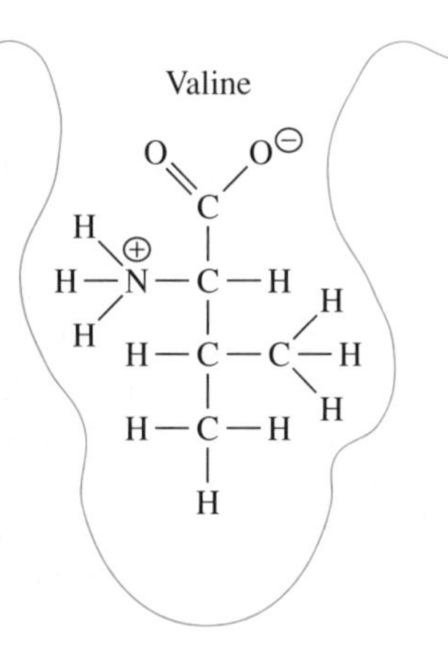

Figure 22·11
Model of the substrate binding site in isoleucyl-tRNA synthetase. Despite the similar size and charge of isoleucine and valine, van der Waals interactions allow isoleucyl-tRNA synthetase to bind to isoleucine about 100 times more readily than it binds to valine. However, even this degree of accuracy is not sufficient for protein synthesis.

Once the 30S complex has been formed at the initiation codon, the 50S ribosomal subunit binds to the 30S subunit. After the 50S subunit binds, the GTP bound to IF-2 is hydrolyzed, P_i is released, and the initiation factors dissociate from the complex. IF-2:GTP is subsequently regenerated when the bound GDP is exchanged for GTP. The steps in the formation of the 70S initiation complex are summarized in Figure 22·18.

D. Initiation in Eukaryotes Differs Slightly from Initiation in Prokaryotes

As in prokaryotes, initiation of translation in eukaryotes requires several initiation factors (Figure 22·19). One of these is known as the *cap binding protein* (CBP; also

Figure 22·18
Steps in the formation of the 70S initiation complex. (Illustrator: Lisa Shoemaker.)

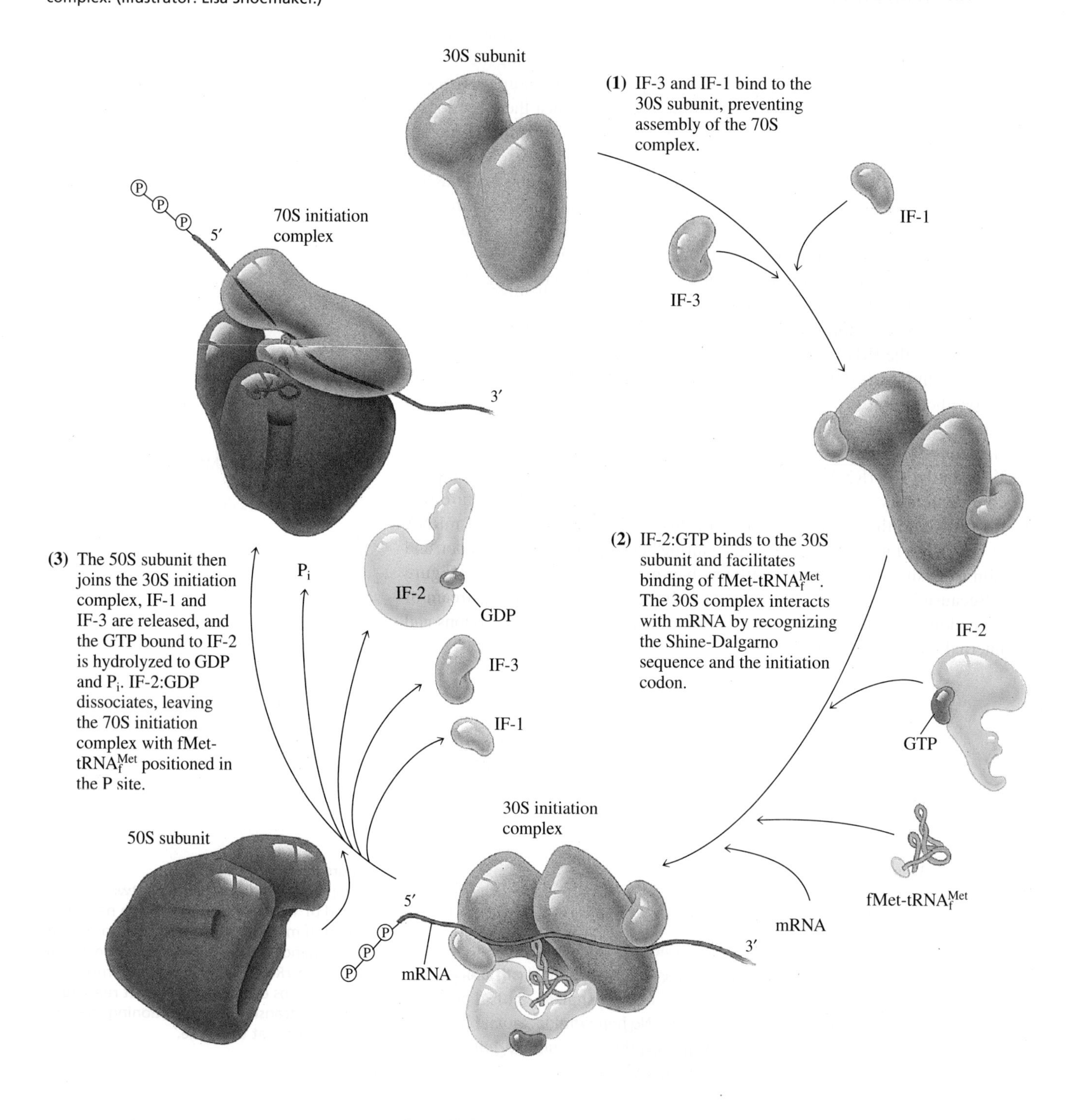

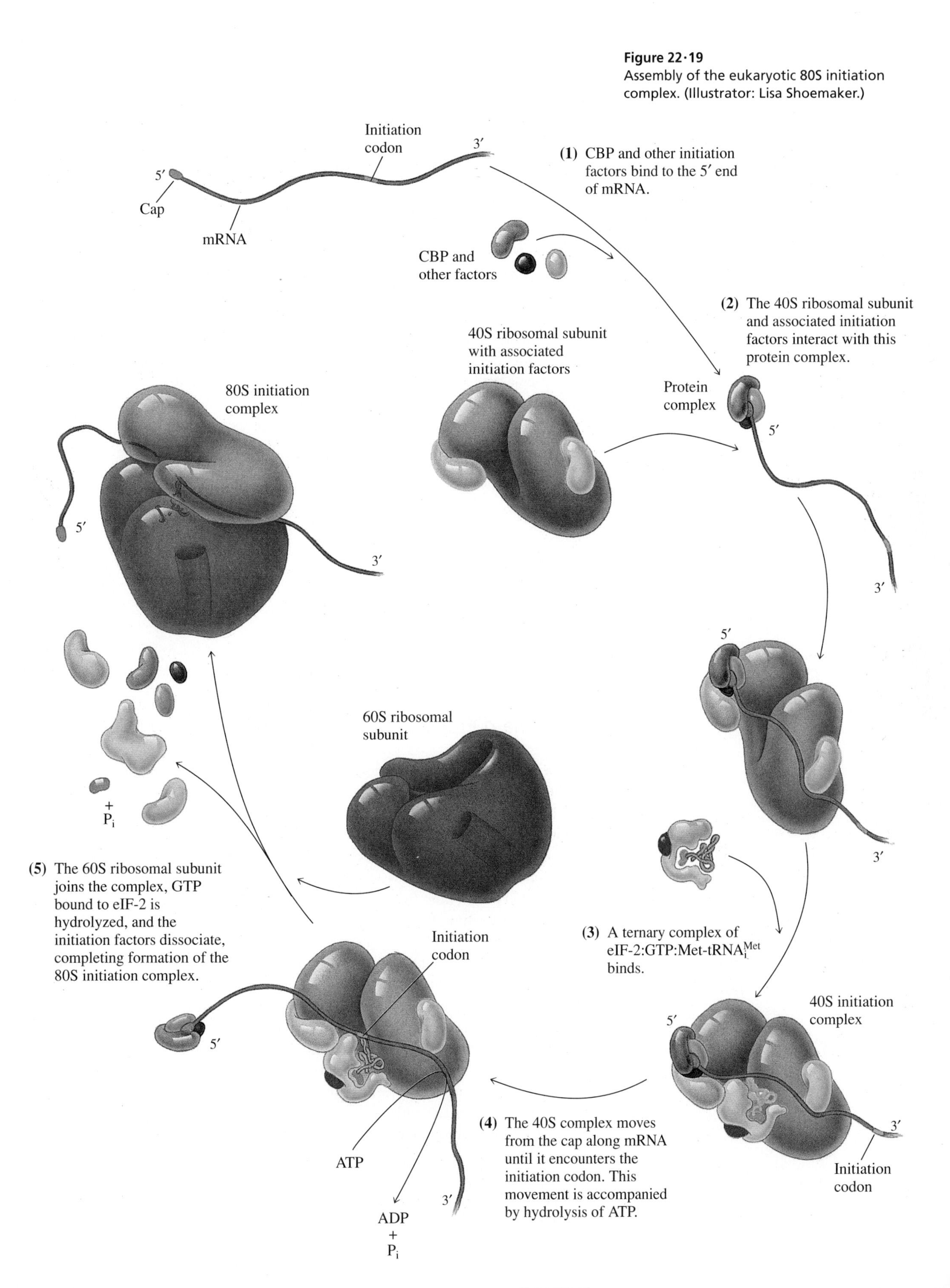

Figure 22·19
Assembly of the eukaryotic 80S initiation complex. (Illustrator: Lisa Shoemaker.)

called eIF-4 in mammals). CBP, along with several other protein factors, interacts specifically with the 5′ 7-methyl-guanine cap of eukaryotic mRNA. The mRNA: CBP complex is then bound by the small ribosomal subunit, which is itself bound to several initiation factors. Next, this complex is joined by a ternary complex of eIF-2:GTP:Met-$tRNA_i^{Met}$.

Once the ternary complex binds, the 40S initiation complex moves along the mRNA in the 5′ → 3′ direction until it encounters an initiation codon. This unidirectional search is coupled to the hydrolysis of ATP; when the search is complete, the small ribosomal subunit is positioned so that Met-$tRNA_i^{Met}$ interacts with the initiation codon in the P site. In the final step, the 60S ribosomal subunit binds to complete the 80S initiation complex, and all the initiation factors dissociate. Dissociation of eIF-2 is accompanied by GTP hydrolysis, as is the case with IF-2.

Eukaryotic mRNAs do not have Shine-Dalgarno sequences; instead, the 40S ribosomal subunit binds to the 5′ end of the message and scans the mRNA in the 5′ → 3′ direction for the initiation codon. Thus, in eukaryotes, the first AUG codon in the message usually serves as the initiation codon. Since this method of selecting the initiation codon permits only one initiation codon per mRNA, most mRNA molecules in eukaryotes encode only a single polypeptide and are said to be *monocistronic*. In contrast, prokaryotic mRNA molecules often have several coding regions, each beginning with an initiation codon that is associated with an upstream Shine-Dalgarno sequence. mRNA molecules that encode several polypeptides are said to be *polycistronic*.

22·6 Chain Elongation Occurs in a Three-Step Microcycle

Following initiation in all species, the mRNA is positioned so that the next codons can be translated during the chain-elongation stage of protein synthesis. The initiator tRNA occupies the P site in the ribosome, and the A site is ready to receive an aminoacylated tRNA in the first step of the chain-elongation reaction.

During chain elongation, each additional amino acid is added to the nascent polypeptide chain in a three-step, reiterative microcycle. The steps in this microcycle are (1) positioning the correct aminoacyl-tRNA in the A site of the ribosome complex, (2) forming the peptide bond, and (3) shifting the mRNA by one codon relative to the ribosome.

The translation machinery works relatively slowly compared to the enzyme systems that catalyze the other steps in the flow of biological information. Bacterial replisomes synthesize DNA at the rate of 1000 nucleotides per second, but proteins are synthesized at the rate of only 18 amino acid residues per second. This difference in rates reflects, in part, the difference between polymerizing four types of nucleotides to make nucleic acids and polymerizing 20 types of amino acids to make proteins. The testing and rejecting of incorrect aminoacyl-tRNA molecules takes time and slows protein synthesis.

The rate of transcription in prokaryotes is approximately 55 nucleotides per second, which corresponds closely to 18 codons per second, or the same rate that the mRNA is translated. In bacteria, translation initiation occurs as soon as the 5′ end of an mRNA is synthesized. This tight coupling between transcription and translation is not possible in eukaryotes because the two processes are carried out in separate compartments of the cell (the nucleus and the cytoplasm, respectively).

A. Elongation Factors Dock an Aminoacyl-tRNA in the A Site

At the start of the first chain-elongation microcycle, the A site is empty and the P site is occupied by the charged initiator tRNA. The tRNA molecule in the P site serves as the attachment point for the growing polypeptide chain; thus, at the end of each microcycle, the number of amino acid residues attached to the tRNA molecule in the P site has increased by one. The tRNA molecule to which the growing peptide chain is attached is called the *peptidyl-tRNA*.

The first step in chain elongation is the insertion of the correct aminoacyl-tRNA into the A site of the protein synthesis complex. In bacteria, this step is catalyzed by an *elongation factor* called EF-Tu.

EF-Tu is a monomeric protein that contains a binding site for GTP. There are approximately 135 000 molecules of EF-Tu per cell, making it one of the most abundant proteins in *E. coli*. EF-Tu:GTP associates with aminoacyl-tRNA molecules to form a ternary complex that fits into the A site of a ribosome. Almost all aminoacylated tRNA molecules in vivo are found in ternary complexes. The structure of EF-Tu is similar to that of IF-2 (which also binds GTP) and to that of G proteins described in Section 15·3C, suggesting that they all evolved from a common ancestral protein.

The EF-Tu:GTP complex recognizes common features of the tertiary structure of tRNA molecules and binds tightly to all aminoacyl-tRNA molecules except fMet-$tRNA_f^{Met}$. fMet-$tRNA_f^{Met}$ is distinguished from all other aminoacyl-tRNA molecules by the distinctive secondary structure of its acceptor stem.

Ternary complexes can diffuse freely into the A site in the ribosome. If the codon and anticodon do not match, the ternary complex is not stabilized in the A site and it leaves, soon to be replaced by another ternary complex. Given that the rate of chain elongation is 18 amino acids per second, one amino acid must be added to the polypeptide approximately every 50 ms. For every amino acid that is added, there are on average ten unsuccessful attempts at fitting a ternary complex into the A site, suggesting that ternary complexes enter and leave the A site about every 5 ms.

If correct base pairs are formed between the anticodon of an aminoacyl-tRNA in a ternary complex and the codon in the A site, then the complex is repositioned in a manner that allows EF-Tu:GTP to contact sites in the ribosome as well as the tRNA in the P site, as shown in Figure 22·20. These contacts trigger hydrolysis of GTP to GDP and P_i and an allosteric change in EF-Tu:GDP that causes it to release the bound aminoacyl-tRNA and dissociate from the chain-elongation complex. At the same time, aminoacyl-tRNA is fixed in the A site, where it is positioned for peptide-bond formation.

EF-Tu:GDP is unable to bind another aminoacyl-tRNA molecule until GDP dissociates. An additional elongation factor called EF-Ts catalyzes exchange of bound GDP for GTP (Figure 22·21). Note that one GTP molecule is hydrolyzed for every aminoacyl-tRNA that is successfully inserted into the A site.

Figure 22·20
The chain-elongation microcycle of protein synthesis in *E. coli.* (Illustrator: Lisa Shoemaker.)

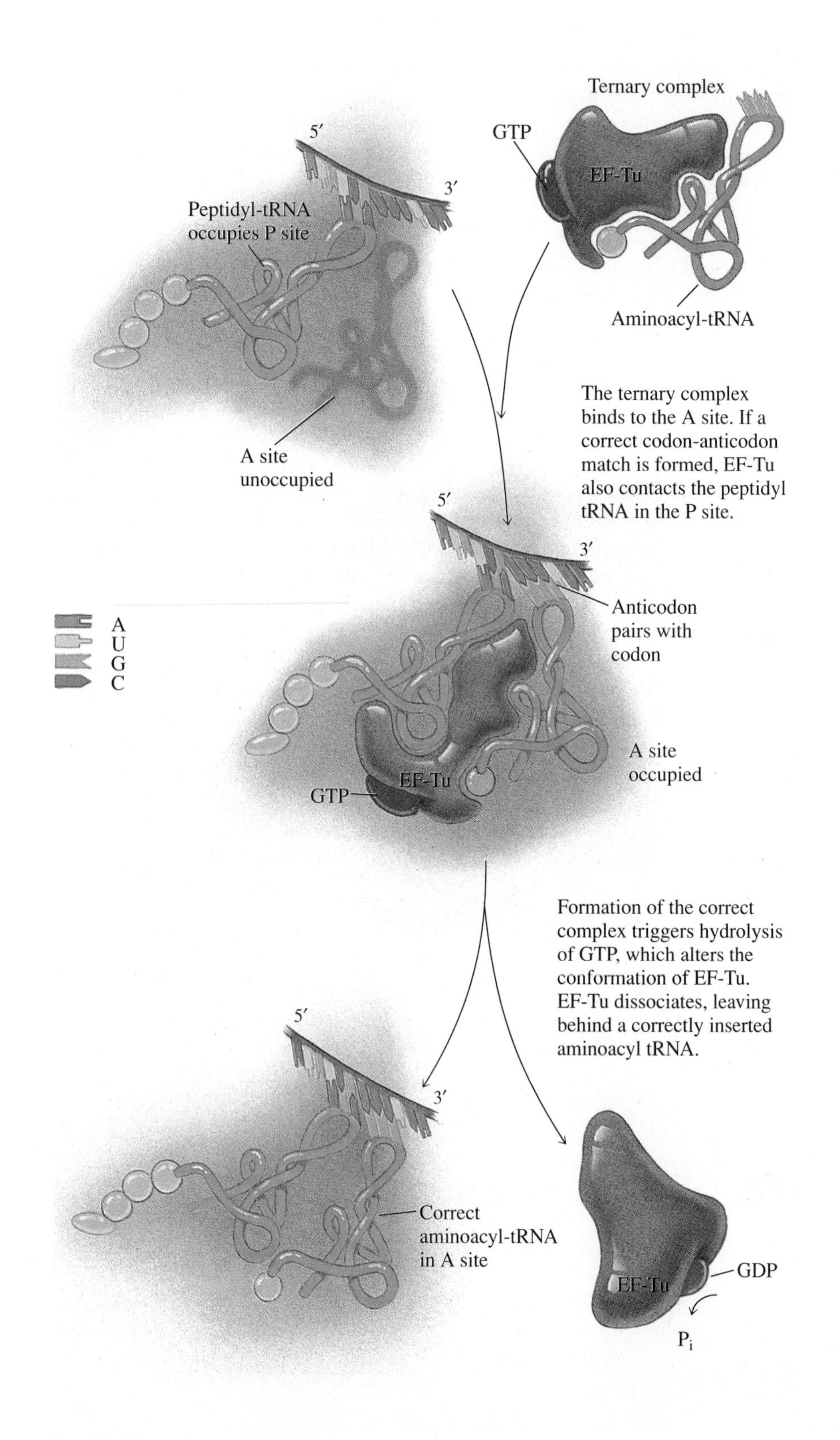

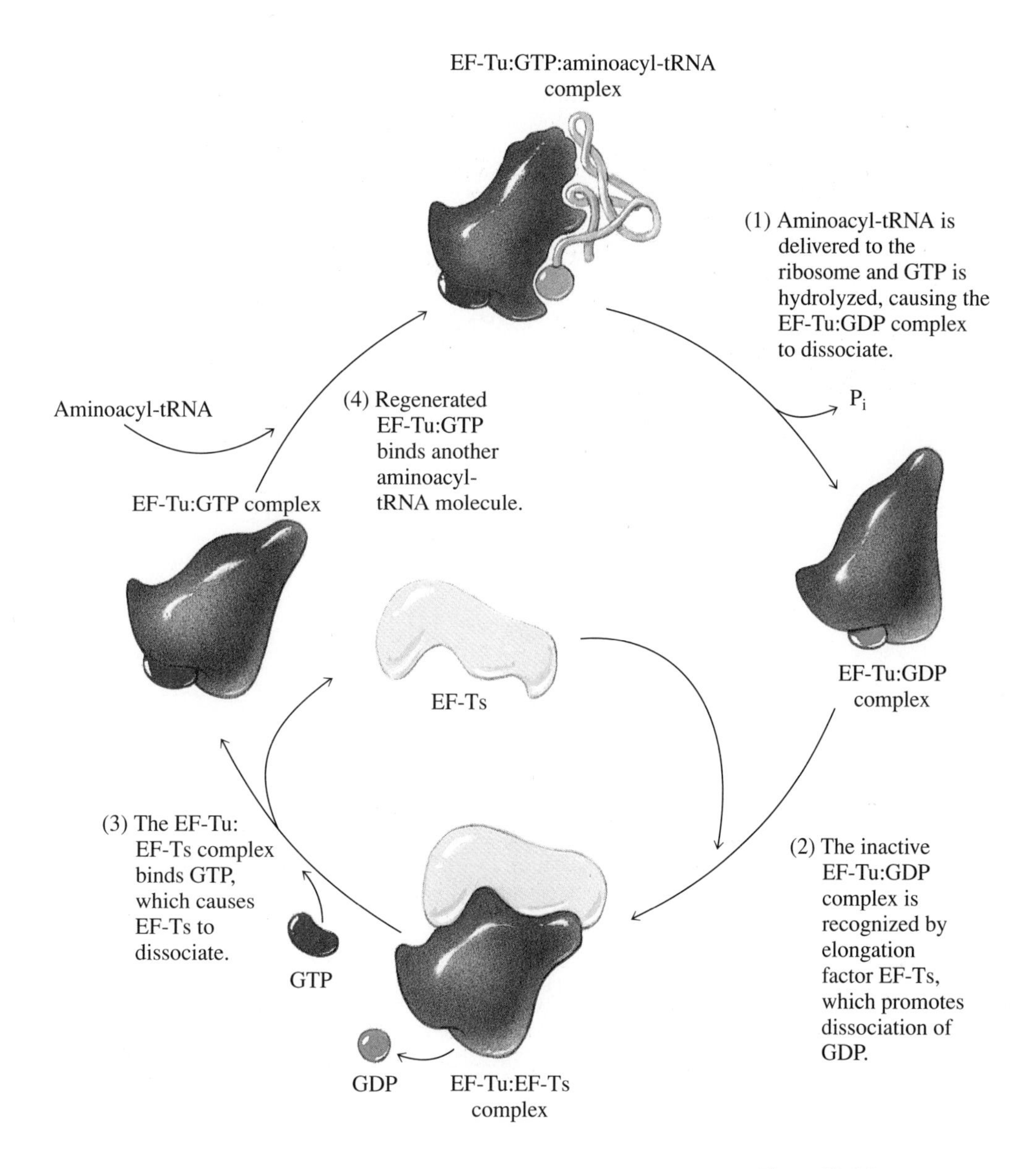

Figure 22·21
The cycling of EF-Tu:GTP. (Illustrators: Initial rendering—Lisa Shoemaker; Final rendering—Linda LeFevre Murray.)

B. Ribosomal Peptidyl Transferase Catalyzes the Formation of Peptide Bonds

Once it is bound in the A site, the aminoacyl-tRNA is positioned such that the amino group of the amino acid (a nucleophile) can attack the carbonyl carbon of the ester linking the nascent polypeptide chain to the peptidyl-tRNA. This nucleophilic displacement reaction, or aminoacyl-group transfer reaction, results in the formation of a peptide bond. The peptide chain, which has increased in length by one amino acid, is transferred from the tRNA in the P site to the tRNA in the A site (Figure 22·22). The enzymatic activity responsible for the formation of the peptide bond is referred to as *peptidyl transferase;* this activity is contained within the large ribosomal subunit. Both the 23S rRNA and the 50S ribosomal proteins contribute to the formation of the active site, and much of the catalytic activity appears to be due to the RNA component. The formation of the peptide bond is coupled to the hydrolysis of the energy-rich aminoacyl-tRNA bond. (Recall that formation of an activated amino acid is coupled to the hydrolysis of ATP.)

Several of the reactions of protein synthesis can be inhibited by common antibiotics (Box 22·1).

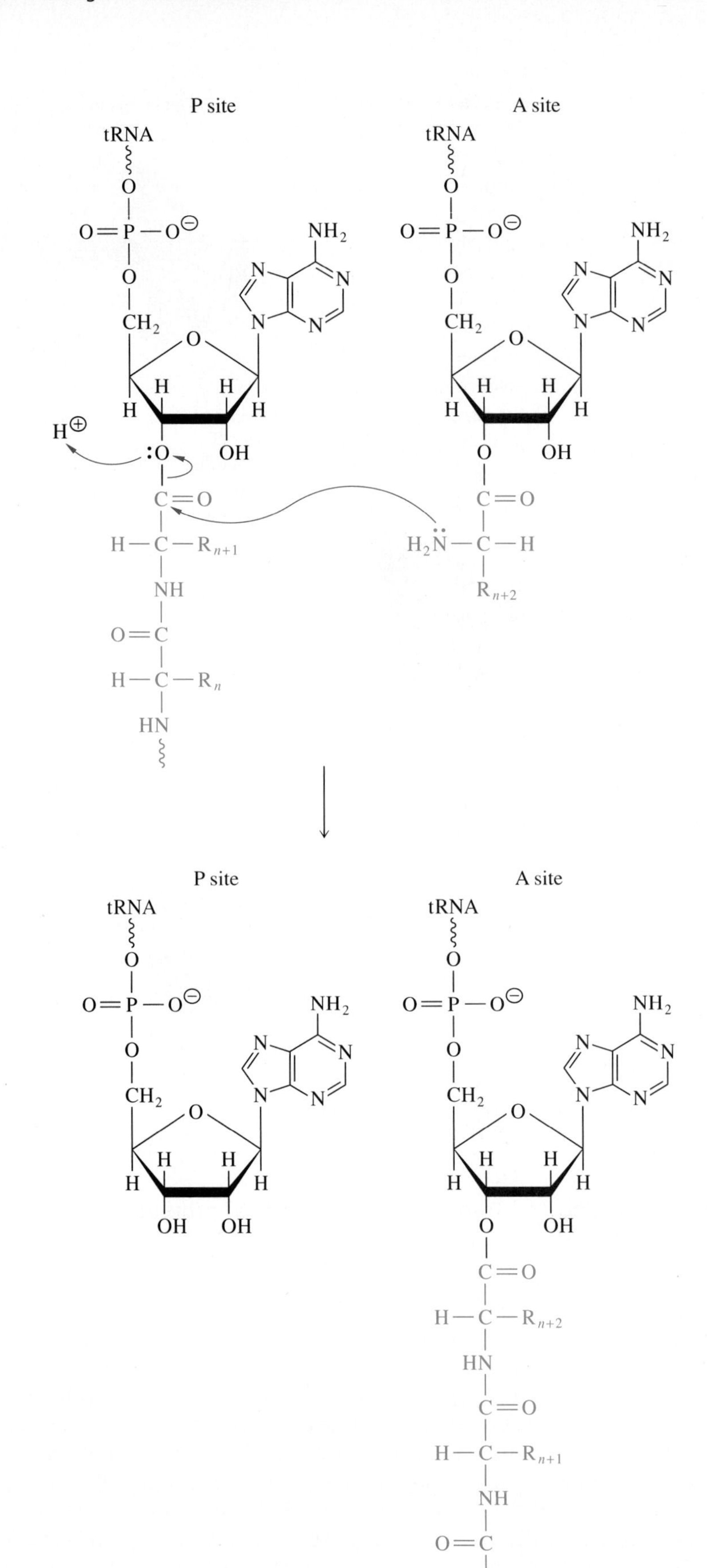

Figure 22·22
Formation of a peptide bond by a peptidyl-group transfer reaction. The carbonyl carbon of the peptidyl-tRNA undergoes nucleophilic attack by the amino group of the amino acid attached to the A site tRNA. This peptidyl-group transfer reaction results in the growth of the peptide chain by one residue and the transfer of the nascent peptide to the tRNA in the A site.

Box 22·1 Some Antibiotics Inhibit Protein Synthesis

Many microorganisms produce antibiotics, which they use as a chemical defense against competitors. Some antibiotics prevent bacterial growth by inhibiting the formation of peptide bonds. For example, the antibiotic puromycin has a structure that closely resembles the structure of the 3′ end of an aminoacyl-tRNA molecule. Because of this structural similarity, puromycin can enter the A site of a ribosome. The peptidyl transferase then catalyzes the transfer of the nascent polypeptide to the free amino group of puromycin (Figure 1). The peptidyl-puromycin is bound weakly in the A site and soon dissociates from the ribosome, thereby terminating protein synthesis.

Although puromycin effectively blocks protein synthesis in prokaryotes, it is not clinically useful since it also blocks protein synthesis in eukaryotes and is therefore poisonous to humans. Clinically important antibiotics, which include streptomycin, chloramphenicol, erythromycin, and tetracycline, are specific for bacteria and have little or no effect on eukaryotic protein synthesis. Streptomycin binds to one of the ribosomal proteins in the 30S subunit and inhibits initiation of translation. Chloramphenicol interacts with the 50S subunit and inhibits peptidyl transfer. Erythromycin binds to the 50S subunit, inhibiting the translocation step. Tetracycline binds to the 30S subunit, preventing the binding of aminoacyl-tRNA molecules to the A site.

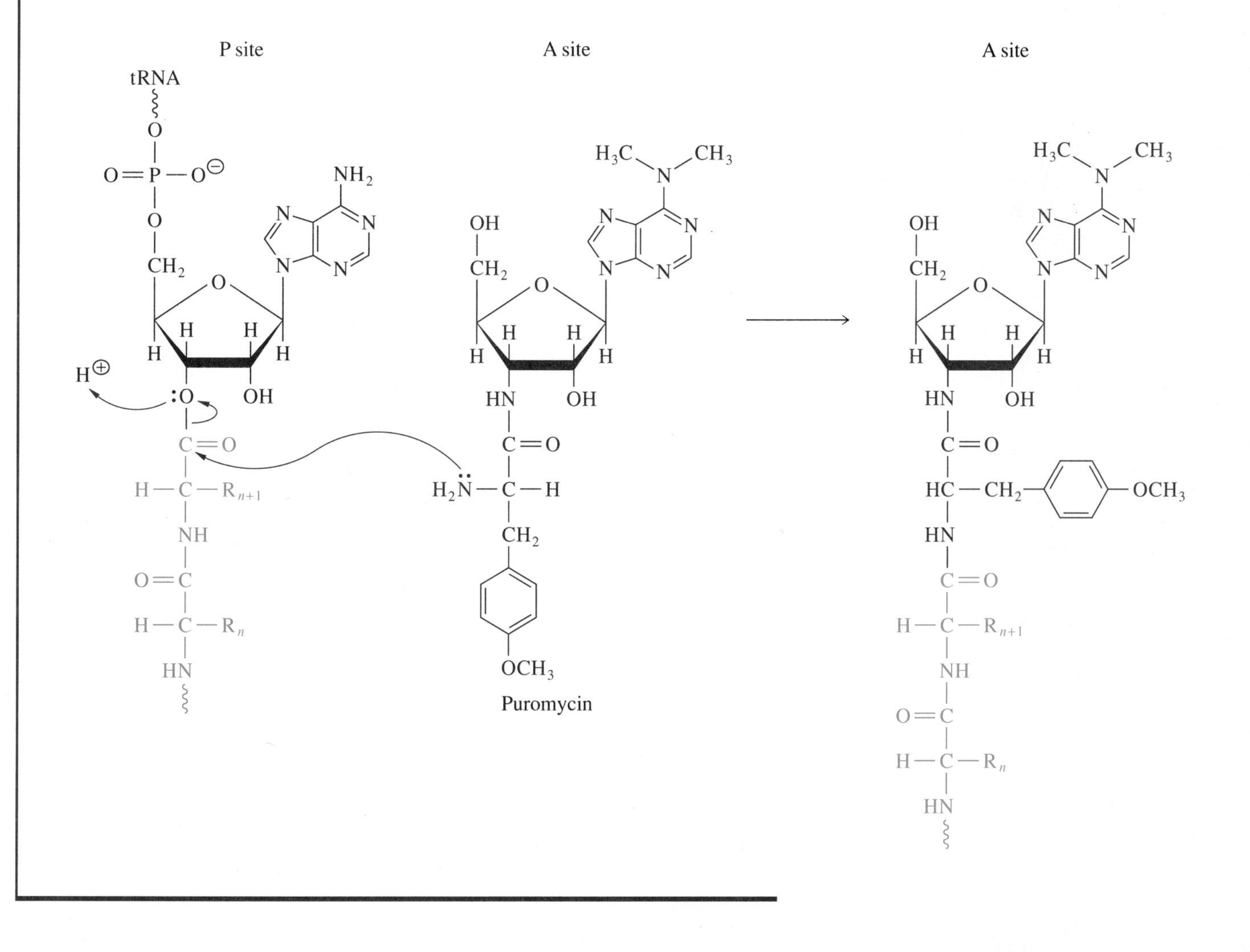

Box 22·1
Figure 1
Formation of a peptide bond between puromycin at the A site of a ribosome and the nascent peptide bound to the tRNA in the P site. The product of this reaction is bound only weakly in the A site and dissociates from the ribosome, thus terminating protein synthesis and producing incomplete, inactive peptides.

C. The Ribosome Moves by One Codon During Translocation

After the peptide bond has formed, the newly created peptidyl-tRNA is partially in the A site and partially in the P site, as shown in Figure 22·23. The deaminoacylated tRNA has been displaced somewhat from the P site; it occupies a position on the ribosome that is referred to as the *exit site,* or E site. Before the next codon can be translated, the deaminoacylated tRNA must be released, and the peptidyl-tRNA must be completely transferred from the A site to the P site, with concomitant movement of the mRNA relative to the ribosome by one codon. This *translocation* is the third step in the chain-elongation microcycle.

In prokaryotes, the translocation step requires the participation of a third elongation factor, EF-G. Like the other elongation factors, EF-G is an abundant protein: an *E. coli* cell contains approximately 20 000 molecules of EF-G, or roughly one for every ribosome. Like EF-Tu, EF-G has a binding site for GTP. Binding of EF-G:GTP to the ribosome releases the deaminoacylated tRNA from the E site and completes the translocation of the peptidyl-tRNA from the A site to the P site. EF-G itself is released from the ribosome only when its bound GTP is hydrolyzed to GDP, and P_i is released. The dissociation of EF-G:GDP leaves the ribosome free to begin another microcycle of chain elongation.

The polypeptide-chain–elongation reactions in eukaryotes are very similar to those described for *E. coli.* Three accessory protein factors participate in chain elongation in eukaryotes: EF-1α, EF-1β, and EF-2. EF-1α docks the aminoacyl-tRNA in the A site; its activity thus parallels that of *E. coli* EF-Tu. EF-1β acts analogously to EF-Ts, recycling EF-1α. EF-2 carries out translocation in eukaryotes.

The microcycle is repeated as each new codon in mRNA is translated. This results in the synthesis of a polypeptide chain that is usually several hundred residues long. Eventually, the protein synthesis complex reaches the end of the coding region, where translation is terminated.

22·7 Release Factors Help Terminate Protein Synthesis

In *E. coli,* three *release factors,* designated RF-1, RF-2, and RF-3, participate in the termination of protein synthesis. After formation of the final peptide bond in a polypeptide chain, the peptidyl-tRNA, which holds the nascent protein, is translocated from the A site to the P site, as usual. The translocation also moves the mRNA, positioning one of the three termination codons (UGA, UAG, or UAA) at the A site (Figure 22·24). The termination codons are not recognized by any tRNA molecules but, rather, are bound by one of the release factors. After the termination codon in the A site is tested by ternary complexes of EF-Tu:GTP:aminoacyl-tRNA without success, one of the much less abundant release factors will eventually diffuse into the A site. RF-1 recognizes UAA and UAG, and RF-2 recognizes UAA and UGA. RF-3 binds GTP and enhances the effects of RF-1 and RF-2.

The binding of the heterodimer RF-3:GTP and RF-1, or RF-3:GTP and RF-2, to mRNA at the A site alters the activity of the peptidyl transferase, causing it to hydrolyze the ester of the peptidyl-tRNA. Release of the final polypeptide product is probably accompanied by GTP hydrolysis and dissociation of the release factors from the ribosome. At this point, the ribosomal subunits dissociate from the mRNA, and initiation factors bind to the 30S subunit in preparation for the next round of protein synthesis. Whereas three release factors are required during termination in prokaryotes, only one release factor (RF), which is GTP dependent, is required in eukaryotes.

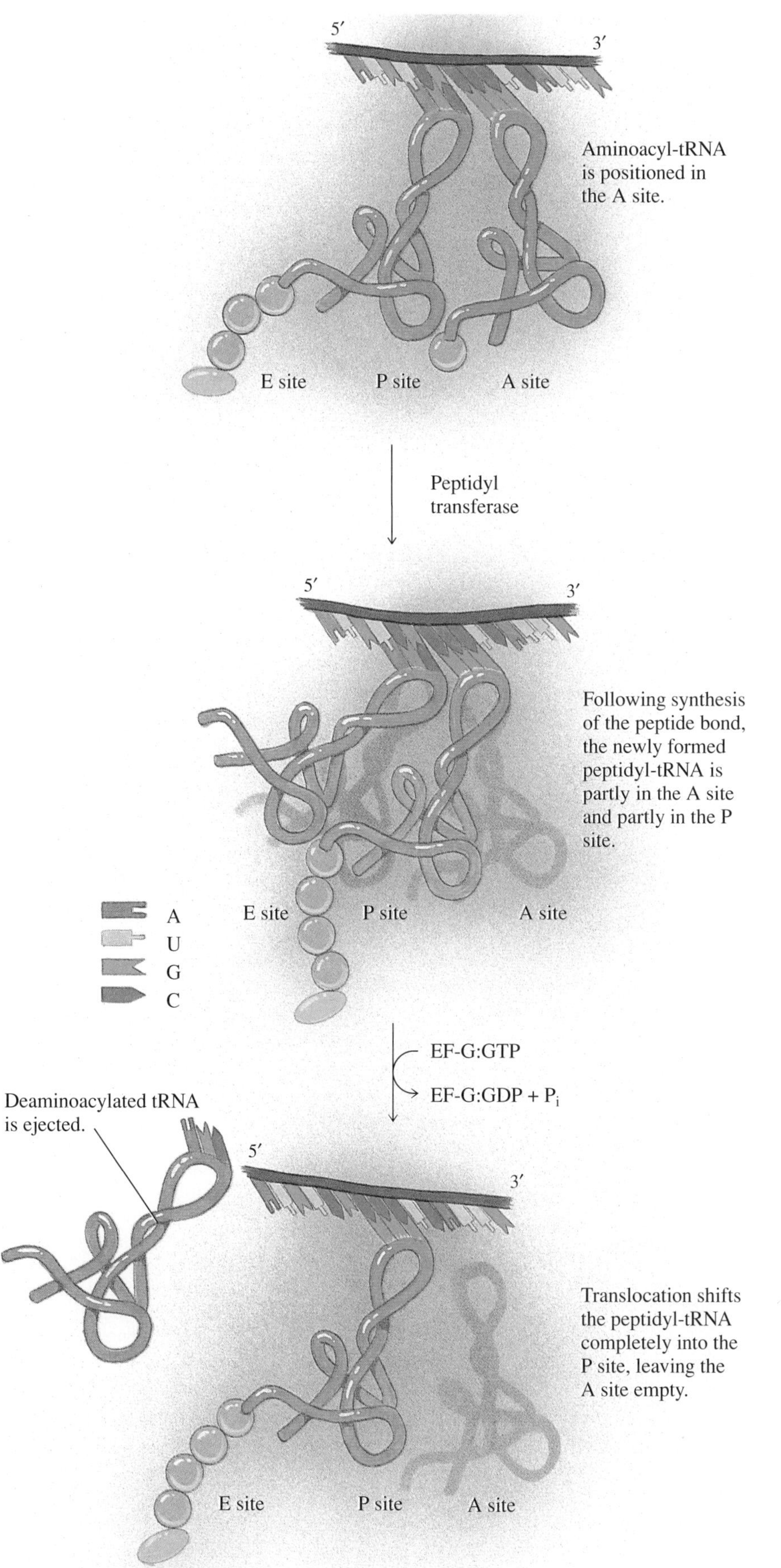

Figure 22·23
Translocation during protein synthesis in prokaryotes. (Illustrator: Lisa Shoemaker.)

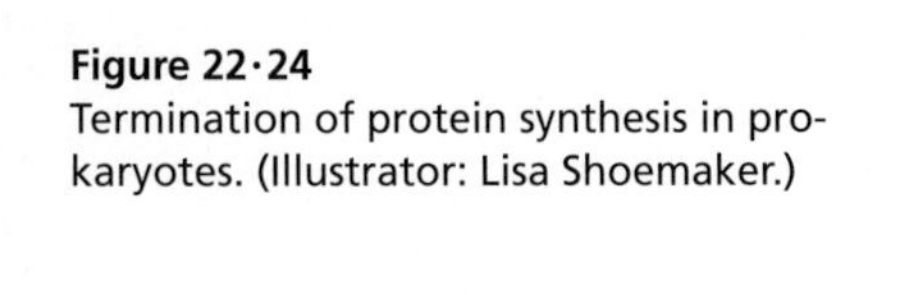

Figure 22·24
Termination of protein synthesis in prokaryotes. (Illustrator: Lisa Shoemaker.)

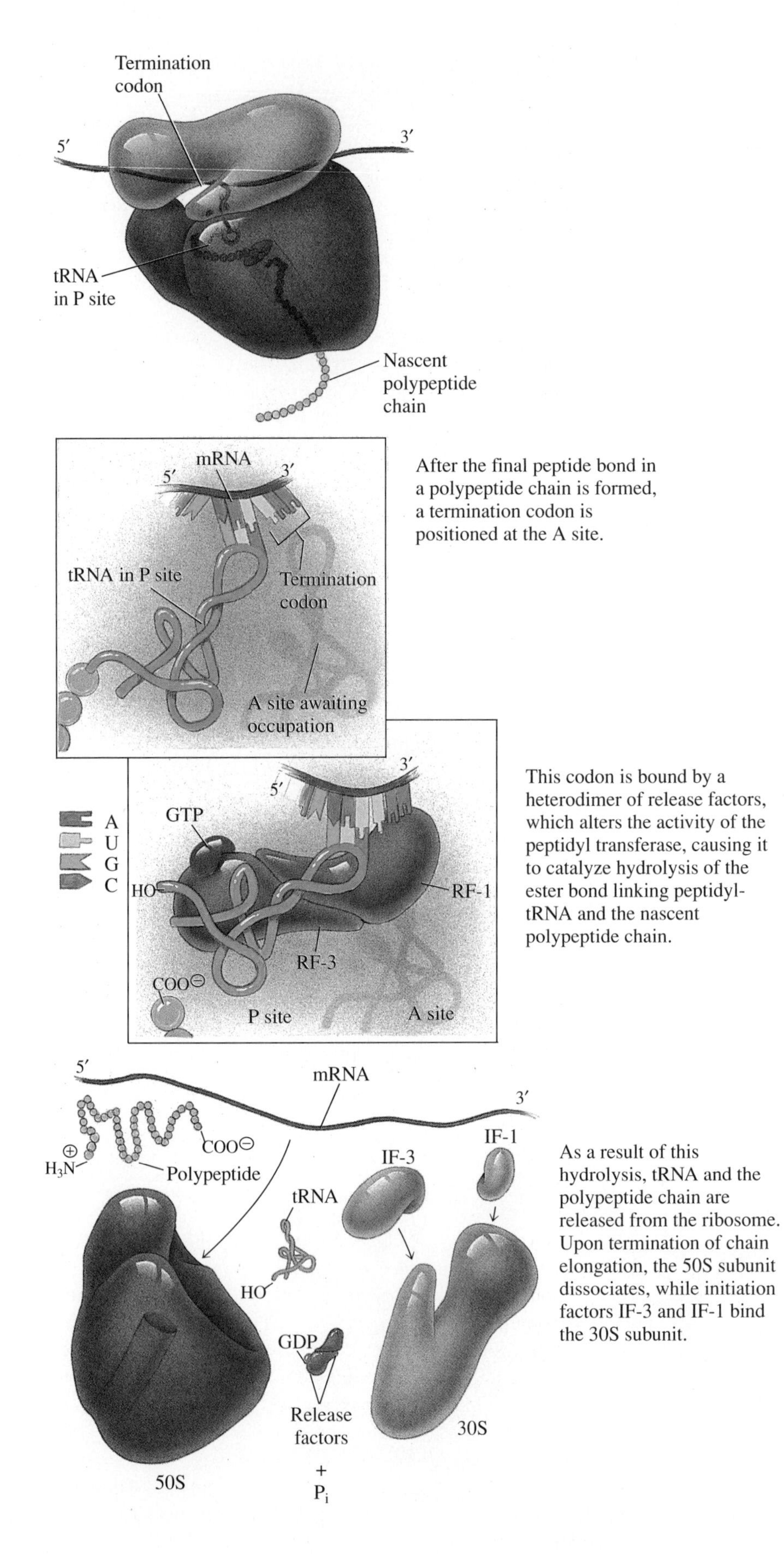

22·8 Protein Synthesis Is Energetically Expensive

Protein synthesis consumes much of the macromolecular biosynthetic capacity of the cell. Even when the cost of synthesizing small molecules is included, protein synthesis still seems to require a large fraction of all ATP and ATP equivalents in the cell. Where does all this energy go?

For each amino acid added to a polypeptide chain, four phosphoanhydrides are cleaved: ATP is hydrolyzed to AMP + 2 P_i during activation of the amino acid, and two GTP molecules are hydrolyzed to 2 GDP + 2 P_i during chain elongation. Whereas the hydrolysis of ATP is associated with the formation of new covalent bonds, the hydrolysis of GTP is coupled to conformational changes in the translation machinery. In this sense, GTP and GDP act like allosteric effectors. However, unlike most conformational changes induced by allosteric effectors, the conformational changes that occur during protein synthesis are associated with considerable releases of energy.

The hydrolysis of four phosphodiesters represents a large free-energy change. However, the synthesis of a peptide bond per se is much less expensive. The majority of the "extra" energy required for protein synthesis does not go into the bond energy but, rather, is used to compensate for the loss of entropy during protein synthesis. The decrease in entropy primarily results from the specific ordering of 20 different amino acids into a polypeptide chain. In addition, entropy losses accompany the linkage of an amino acid to a particular tRNA and the association of an aminoacyl-tRNA with a specific codon. These decreases in entropy are partially balanced by the release of energy stored in phosphodiester bonds.

22·9 Proteins May Be Posttranslationally Modified, Sorted, and Excreted

As the translation complex moves along the mRNA template in the $5' \rightarrow 3'$ direction, the nascent polypeptide chain grows in length. The 30 or so most recently polymerized amino acid residues remain buried in the ribosome, but amino acid residues closer to the N-terminus are extruded from the ribosome. These N-terminal residues start folding into the native protein structure even before the C-terminus of the protein is synthesized. As these residues fold, they are acted upon by enzymes that modify the nascent chain.

Modifications that occur before the polypeptide chain is complete are said to be *cotranslational,* whereas those that occur after the chain is complete are said to be *posttranslational.* Some examples from the multitude of cotranslational and posttranslational modifications include deformylation of the N-terminal residue in prokaryotic proteins, removal of the N-terminal methionine in prokaryotic and eukaryotic proteins, formation of disulfide bonds, cleavage by proteinases, phosphorylation, addition of carbohydrate residues, and acetylation.

One of the most important events that occurs co- and posttranslationally is the processing and transport of proteins through membranes. Protein synthesis occurs in the cytosol, but the mature forms of many proteins are located on the noncytosolic side of membranes. For example, many receptor proteins are embedded in the external membrane of the cell, with the bulk of the protein outside the cell. Other proteins are secreted from cells, and still others reside in lysosomes and other vesicles inside eukaryotic cells. In each of these cases, the protein synthesized in the cytosol must be transported across a membrane barrier. In fact, such proteins are synthesized by membrane-bound ribosomes that are attached to the plasma membrane in bacteria and the endoplasmic reticulum in eukaryotic cells.

The best-characterized transport system is the one that carries proteins from the cytosol to the plasma membrane for secretion. In eukaryotes, proteins destined to be secreted are transported across the membrane of the endoplasmic reticulum (ER) into the lumen of the ER, which is topologically equivalent to the outside of the plasma membrane. Once the protein has been transported across the ER, it can be transported by vesicles through the Golgi apparatus to the plasma membrane for release outside the cell (Figure 22·25).

Figure 22·25
Pathway by which proteins are secreted from eukaryotic cells. The proteins are actually transported *across* only one membrane—the membrane of the endoplasmic reticulum. (Illustrator: Lisa Shoemaker.)

Proteins are transported across the membrane of the endoplasmic reticulum cotranslationally.

In the lumen of the endoplasmic reticulum, proteins may be covalently modified.

Proteins are transported to the Golgi apparatus in vesicles that bud off of the endoplasmic reticulum.

Proteins to be secreted are carried from the Golgi apparatus to the plasma membrane in secretory vesicles. During transport, some of these proteins are activated by proteolytic cleavage.

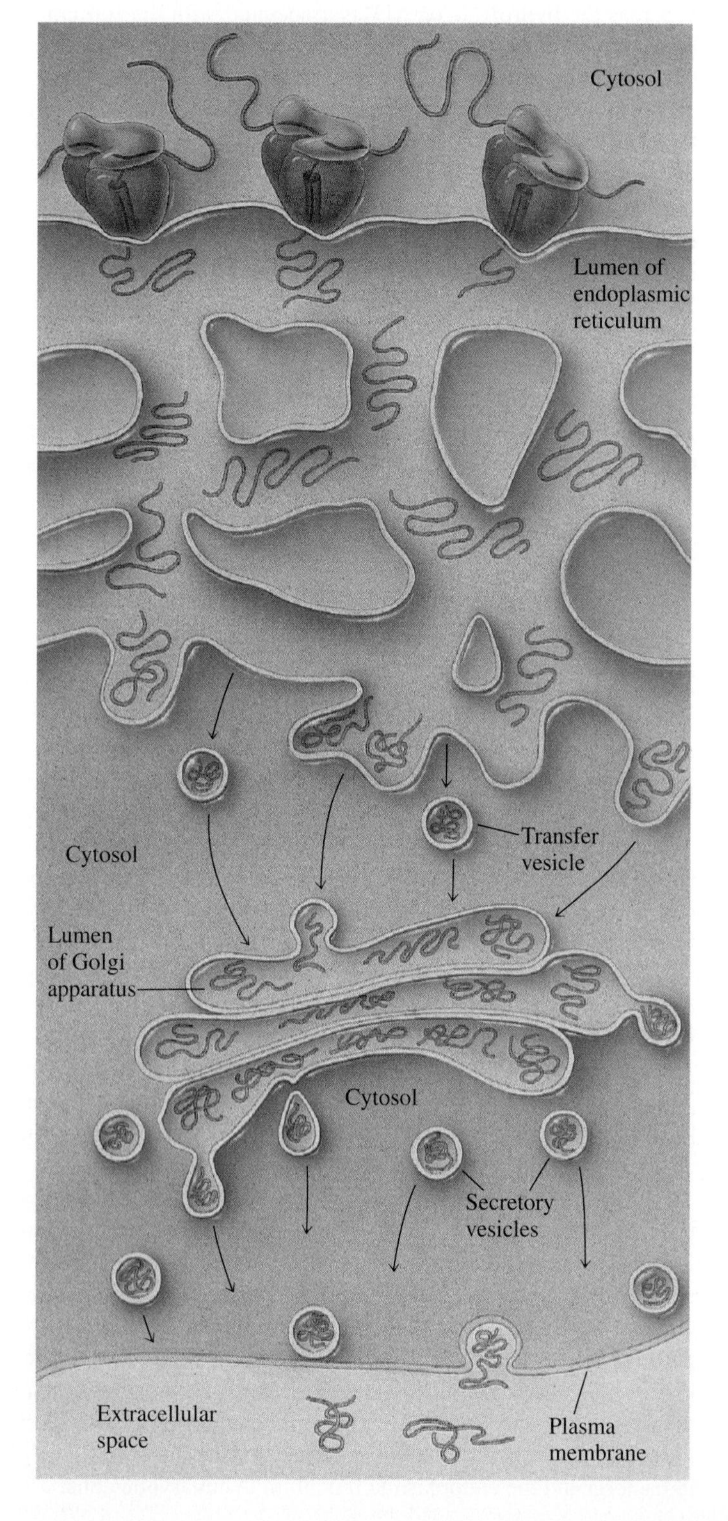

22·10 Protein Synthesis Can Be Regulated

During our discussion of RNA synthesis in Chapter 21, we learned that the flow of biological information in the cell is regulated. One level of regulation occurs during RNA synthesis—not all genes are transcribed at the same rate in all cells. Gene expression can also be regulated at the level of translation.

The translation of mRNA can be regulated at initiation, chain elongation, or termination, but it is most commonly regulated at initiation. Some proteins, for example, can bind to the 5′ end of their own mRNA and inhibit translation initiation. Thus, when the amount of protein accumulates to a certain level, no more will be synthesized. In other cases, the rate of initiation is regulated by controlling the activity of the initiation factors. For example, eIF-2 can be inactivated by phosphorylation, which occurs in response to conditions that are unfavorable for protein synthesis.

These sophisticated regulatory mechanisms have evolved to address the changing needs of the cell. As the availability of nutrients changes, the need for certain metabolic enzymes may change, or as a cell develops, different structural proteins may be required. Regulatory mechanisms allow the cell or cells of an organism to be dynamic—to adapt to changing requirements.

Summary

The translation of mRNA is the final stage in the flow of biological information. Accurate translation involves several protein and RNA species, including tRNA, mRNA, rRNA, ribosomal proteins, and a variety of initiation and elongation factors.

tRNA molecules are the adapters that form the interface between mRNA molecules and proteins. Each tRNA molecule interacts with a particular amino acid and is able to recognize a particular codon on mRNA. During translation, peptide bonds form between amino acids attached to tRNA molecules bound to adjacent mRNA codons.

The sequence of nucleotides in mRNA specifies the sequence of amino acids in a peptide by means of a genetic code. The genetic code, which is almost universal, has a number of prominent features: it is unambiguous and degenerate, the first two nucleotides of the three-letter code are often enough to specify a certain amino acid, codons with similar sequences specify chemically similar amino acids, and there are special codons for initiation and termination of peptide synthesis.

All tRNA molecules share a conserved secondary and tertiary structure. The secondary structure may be represented as a cloverleaf, with one hydrogen-bonded stem and three hydrogen-bonded arms. The tertiary structure resembles a letter "L," with the anticodon loop at one end of the structure and the acceptor stem at the other. The three-dimensional structure of the tRNA molecule is stabilized by a variety of intramolecular base pairs, including some that differ from normal Watson-Crick base pairs.

The tRNA interacts with the mRNA through the anticodon, whose 5′ position is conformationally flexible and can form certain non-Watson-Crick base pairs. This flexibility is dubbed wobble. The 5′ position of the anticodon can also be covalently modified to permit additional flexibility in codon recognition.

Aminoacyl-tRNA synthetases catalyze the addition of particular amino acids to the acceptor stem of particular tRNA molecules. Once attached to an amino acid, a tRNA molecule is called an aminoacyl-tRNA. The fidelity of protein synthesis depends, in part, on the accuracy of aminoacyl-tRNA synthetases in pairing a tRNA molecule with the appropriate amino acid.

Ribosomes are the RNA:protein complexes that catalyze the joining of amino acids bound to aminoacyl-tRNA molecules. All ribosomes are composed of two subunits: in prokaryotes, the 70S ribosome is composed of a 50S and a 30S subunit; in eukaryotes, the 80S ribosome is composed of a 60S and a 40S subunit. Prokaryotic ribosomes contain three rRNA molecules (5S, 16S, and 23S); eukaryotic ribosomes contain four (5S, 5.8S, 18S, and 28S).

Ribosomes have two aminoacyl-tRNA binding sites: the P site and the A site. While the polypeptide remains attached to the tRNA in the P site, the aminoacyl-tRNA molecule bearing the next amino acid to be added to the nascent polypeptide chain docks at the A site.

Translation can be divided into three stages: initiation, chain elongation, and termination. Translation begins with formation of an initiation complex. In all organisms, the initiation complex consists of an initiator tRNA, the mRNA template, the small ribosomal subunit, and several initiation factors. In prokaryotes, initiation takes place just downstream from Shine-Dalgarno sequences; in eukaryotes, initiation usually takes place at the initiation codon closest to the 5′ end of the mRNA.

Chain elongation, which requires the participation of accessory proteins called elongation factors, occurs in a three-step microcycle. The steps are (1) positioning of the correct aminoacyl-tRNA in the A site, (2) formation of the peptide bond, and (3) translocation of the ribosome relative to the mRNA by one codon. Termination of protein synthesis and disassembly of the translation complex require the activity of release factors, which recognize termination codons and cause hydrolysis of the ester bond of the peptidyl-tRNA.

During or after synthesis, proteins are often covalently modified or proteolytically processed. Some of these proteins are excreted by being transported across the membrane of the endoplasmic reticulum. In all organisms, protein synthesis, like other stages of biological information flow, can be regulated.

Selected Readings

Genetic Code

Crick, F. H. C. (1966). Codon-anticodon pairing: the wobble hypothesis. *J. Mol. Biol.* 19:548–555.

Crick, F. H. C. (1968). The origin of the genetic code. *J. Mol. Biol.* 38:367–379.

Crick, F. H. C., Barnett, L., Brenner, S., and Watts-Tobin, R. J. (1961). General nature of the genetic code for proteins. *Nature* 192:1227–1232.

Lengyel, P., Speyer, J. F., and Ochoa, S. (1961). Synthetic polynucleotides and the amino acid code. *Proc. Natl. Acad. Sci. USA* 47:1936–1942.

Nirenberg, M. W., and Matthaei, J. H. (1961). The dependence of cell-free protein synthesis in *E. coli* upon naturally occurring or synthetic polyribonucleotides. *Proc. Natl. Acad. Sci. USA* 47:1588–1602.

Structure of tRNA

Björk, G. R., Ericson, J. U., Gustafsson, C. E. D., Hagervall, T. G., Jönsson, Y. H., and Wikström, P. M. (1987). Transfer RNA modification. *Annu. Rev. Biochem.* 56:263–287.

Cigan, A. M., Feng, L., and Donahue, T. F. (1988). $tRNA_i^{Met}$ functions in directing the scanning ribosome to the start site of translation. *Science* 242:93–97.

Curran, J. F., and Yarus, M. (1987). Reading frame selection and transfer RNA anticodon loop stacking. *Science* 238:1545–1550.

Hou, Y-M., and Schimmel, P. (1988). A simple structural feature is a major determinant of the identity of a transfer RNA. *Nature* 333:140–145.

Kim, S-H. (1978). Three-dimensional structure of transfer RNA and its functional implications. *Adv. Enzymol.* 46:279–315.

McClain, W. H., and Foss, K. (1988). Changing the acceptor identity of a transfer RNA by altering nucleotides in a "variable pocket." *Science* 241:1804–1807.

Rich, A., and Kim., S. H. (1978). The three-dimensional structure of transfer RNA. *Sci. Am.* 238(1):52–62.

Robertus, J. D., Ladner, J. E., Finch, J. T., Rhodes, D., Brown, R. S., Clark, B. F. C., and Klug, A. (1974). Structure of yeast phenylalanine RNA at 3 Å resolution. *Nature* 250:546–551.

Schimmel, P. R., Söll, D., and Abelson, J. N., eds. (1979). *Transfer RNA: Structure, Properties, and Recognition.* (New York: Cold Spring Harbor Laboratory).

Aminoacyl-tRNA Synthetases

Freist, W. (1989). Mechanisms of aminoacyl-tRNA synthetases: a critical consideration of recent results. *Biochemistry* 28:6787–6795.

Schimmel, P. R. (1987). Aminoacyl tRNA synthetases: general scheme of structure-function relationships in the polypeptides and recognition of transfer RNAs. *Annu. Rev. Biochem.* 56:125–158.

Schimmel, P. (1989). Parameters for the molecular recognition of transfer RNAs. *Biochemistry* 28:2747–2759.

Ribosomes

Brimacombe, R. (1984). Conservation of structure in ribosomal RNA. *Trends Biochem. Sci.* 9:273–277.

Hill, W. E., Dahlberg, A., Garret, R. A., Moore, P. B., Schlessinger, D., and Warner, J. R., eds. (1990). *The Ribosome: Structure, Function, and Evolution* (Washington, DC: American Society for Microbiology).

Moore, P. B. (1988). The ribosome returns. *Nature* 331:223–227.

Noller, H. F. (1991). Ribosomal RNA and translation. *Annu. Rev. Biochem.* 60:191–227.

Nomura, M., Gourse, R., and Baughman, G. (1984). Regulation of the synthesis of ribosomes and ribosomal components. *Annu. Rev. Biochem.* 53:75–117.

Thompson, J. F., and Hearst, J. E. (1983). Structure-function relations in *E. coli* 16S RNA. *Cell* 33:19–24.

Walker, T. A., and Pace, N. R. (1983). 5.8S ribosomal RNA. *Cell* 33:320–322.

Yonath, A., Leonard, K. R., and Whittmann, H. G. (1987). A tunnel in the large ribosomal subunit revealed by three-dimensional image reconstruction. *Science* 236:813–816.

Yonath, A., and Wittmann, H. G. (1989). Challenging the three-dimensional structure of ribosomes. *Trends Biochem. Sci.* 14:329–335.

Protein Synthesis

Caskey, C. T. (1980). Peptide chain termination. *Trends Biochem. Sci.* 5:234–237.

Gualerzi, C. O., and Pon, C. L. (1990). Initiation of mRNA translation in prokaryotes. *Biochemistry* 29:5881–5889.

Hartz, D., McPheeters, D. S., and Gold, L. (1989). Selection of the initiator tRNA by *Escherichia coli* initiation factors. *Genes Dev.* 3:1899–1912.

Linder, P., and Pratt, A. (1990) Baker's yeast, the new work horse in protein synthesis studies: analyzing eukaryotic translation initiation. *BioEssays* 12:519–526.

McCarthy, J. E. G., and Gualerzi, C. (1990). Translational control of prokaryotic gene expression. *Trends Genet.* 6:78–85.

Salas, M., Smith, M. A., Stanley, W. M., Jr., Wahba, A. J., and Ochoa, S. (1965). Direction of reading of the genetic code. *J. Biol. Chem.* 240:3988–3995.

Index

An *F* following an entry indicates that a relevant *figure* is on that page, a *T* indicates a *table*, and *S* indicates a *stereo*.

continued

continued

continued

continued

continued

continued

continued

continued

Common Abbreviations in Biochemistry

ACP	acyl carrier protein
ADP	adenosine 5′-diphosphate
AMP	adenosine 5′-monophosphate (adenylate)
cAMP	cyclic adenosine 3′,5′-monophosphate
ATCase	aspartate transcarbamoylase
ATP	adenosine 5′-triphosphate
1,3BPG	1,3-*bis*phosphoglycerate
2,3BPG	2,3-*bis*phosphoglycerate
CDP	cytidine 5′-diphosphate
CMP	cytidine 5′-monophosphate (cytidylate)
CoA or CoASH	coenzyme A
CTP	cytidine 5′-triphosphate
d	2′-deoxy-
DHAP	dihydroxyacetone phosphate
DNA	deoxyribonucleic acid
DNase	deoxyribonuclease
E°	reduction potential
$E^{\circ\prime}$	standard reduction potential
EF-	elongation factor
emf	electromotive force
ETF	electron-transferring protein
$\mathcal{F}$	Faraday's constant
FAD	flavin adenine dinucleotide
$FADH_2$	flavin adenine dinucleotide (reduced form)
F1,6BP	fructose 1,6-*bis*phosphate
FMN	flavin mononucleotide
$FMNH_2$	flavin mononucleotide (reduced form)
F6P	fructose 6-*bis*phosphate
G	Gibbs free energy
ΔG	actual free-energy change
$\Delta G^{\circ\prime}$	standard free-energy change
GDP	guanosine 5′-diphosphate
GMP	guanosine 5′-monophosphate (guanylate)
cGMP	cyclic guanosine 5′-monophosphate
G3P	glyceraldehyde 3-phosphate
G6P	glucose 6-phosphate
GTP	guanosine 5′-triphosphate
H	enthalpy
ΔH	enthalpy change
ΔH°	standard enthalpy change
Hb	hemoglobin
HDL	high-density lipoprotein
HETPP	hydroxyethylthiamine pyrophosphate
IF-	initiation factor
eIF-	eukaryotic initiation factor
IMP	inosine 5′-monophosphate
IP_3	inositol 1,4,5-*tris*phosphate
K_{assoc}	association constant
K_{diss}	dissociation constant
K_{eq}	equilibrium constant
K_m	Michaelis constant
LDL	low-density lipoprotein
LHC	light-harvesting complex
Mb	myoglobin
$NAD^{\oplus}$	nicotinamide adenine dinucleotide
NADH	nicotinamide adenine dinucleotide (reduced form)
$NADP^{\oplus}$	nicotinamide adenine dinucleotide phosphate
NADPH	nicotinamide adenine dinucleotide phosphate (reduced form)
NMN	nicotinamide mononucleotide
NDP	nucleoside 5′-diphosphate
NMP	nucleoside 5′-monophosphate
NTP	nucleoside 5′-triphosphate
P_i	inorganic phosphate (or orthophosphate)
2PG	2-phosphoglycerate
3PG	3-phosphoglycerate
PEP	phosphoenolpyruvate
PFK	phosphofructokinase
PIP_2	phosphatidylinositol 4,5-*bis*phosphate
PLP	pyridoxal phosphate
PP_i	inorganic pyrophosphate
PQ	plastoquinone
PQH_2	plastoquinone (reduced form)
PRPP	5-phosphoribosyl 1-pyrophosphate
PSI	photosystem I
PSII	photosystem II
Q	ubiquinone
QH_2	ubiquinone (reduced form)
RF-	release factor
RNA	ribonucleic acid
hnRNA	heterogeneous nuclear ribonucleic acid
mRNA	messenger ribonucleic acid
rRNA	ribosomal ribonucleic acid
snRNA	small nuclear ribonucleic acid
tRNA	transfer ribonucleic acid
RNase	ribonuclease
snRNP	small nuclear ribonucleoprotein
RuBisCO	ribulose 1,5-*bis*phosphate carboxylase-oxygenase
S	entropy
ΔS	entropy change
ΔS°	standard entropy change
TDP or dTDP	deoxythymidine (or thymidine) 5′-diphosphate
TMP or dTMP	deoxythymidine (or thymidine) 5′-monophosphate (deoxythymidylate or thymidylate)
TPP	thiamine pyrophosphate
TTP or dTTP	deoxythymidine (or thymidine) 5′-triphosphate
UDP	uridine 5′-diphosphate
UMP	uridine 5′-monophosphate (uridylate)
UTP	uridine 5′-triphosphate
v	velocity
V_{max}	maximum velocity
v_0	initial velocity
VLDL	very-low-density lipoprotein
XMP	xanthosine 5′-monophosphate

Abbreviations for amino acids are given on page 3·5, and those for pyrimidine and purine bases are given on page 10·5.